Municipal Stormwater Management

Second Edition

Municipal Stormwater Management

Second Edition

Thomas N. Debo ▪ Andrew J. Reese

LEWIS PUBLISHERS

A CRC Press Company
Boca Raton   London   New York   Washington, D.C.

Library of Congress Cataloging-in-Publication Data

Debo, Thomas, N. (Thomas Neil), 1941–
　Municipal stormwater management / Thomas N. Debo, Andrew J. Reese. — 2nd ed.
　　p. cm.
　Updated ed. of: Municipal storm water management. c1995.
　Includes bibliographical refrences and index.
　ISBN 1-56670-584-3 (alk. paper)
　1. Urban runoff—Management. I. Reese, Andrew J. II Debo, Thomas N. (Thomas
Neil), 1941– Municipal storm water management. III. Title.
　TD657 .D43 2002

2002030199

This book contains information obtained from authentic and highly regarded sources. Reprinted material is quoted with permission, and sources are indicated. A wide variety of references are listed. Reasonable efforts have been made to publish reliable data and information, but the author and the publisher cannot assume responsibility for the validity of all materials or for the consequences of their use.

Neither this book nor any part may be reproduced or transmitted in any form or by any means, electronic or mechanical, including photocopying, microfilming, and recording, or by any information storage or retrieval system, without prior permission in writing from the publisher.

The consent of CRC Press LLC does not extend to copying for general distribution, for promotion, for creating new works, or for resale. Specific permission must be obtained in writing from CRC Press LLC for such copying.

Direct all inquiries to CRC Press LLC, 2000 N.W. Corporate Blvd., Boca Raton, Florida 33431.

Trademark Notice: Product or corporate names may be trademarks or registered trademarks, and are used only for identification and explanation, without intent to infringe.

Visit the CRC Press Web site at www.crcpress.com

© 2003 by CRC Press LLC
Lewis Publishers is an imprint of CRC Press LLC

No claim to original U.S. Government works
International Standard Book Number 1-56670-584-3
Printed in the United States of America　1　2　3　4　5　6　7　8　9　0
Printed on acid-free paper

Preface

The field of municipal stormwater management has expanded to the extent that it would take many volumes to exhaustively cover the subject. But, in the day-to-day design and management of the stormwater system, the engineer, planner, or administrator should not need to resort to a large number of documents to find needed information. It should be readily available in one spot, carefully explained where necessary, but not overly cluttered with detailed theory and derivation. That is the approach and major aim of this book.

From a technical design standpoint, the authors combed the literature to find and present clear treatments of each of the key design situations the practitioner will encounter. A wealth of knowledge was compiled from federal, state, and local design publications. In each case, the authors attempted to provide a stand-alone discussion, including all relevant charts, figures, tables, and example applications. In addition, much original information gathered from experience in many cities and counties is included.

To properly understand and manage stormwater requires more than a knowledge of design procedures and hydrologic or hydraulic practice. Actually making it happen in a real-world municipality, filled with the demands of regulations, financial pressures, stakeholder groups, and politics, is also an important part of stormwater management. While harder to quantify, the tools and understanding necessary for institutional, as well as technical, success are also included in this book. They are derived from literally hundreds of experiences and projects in cities and counties across the United States. A careful reading will provide a wealth of information and principles and a sense of what works and what does not.

Stormwater management is a fast-evolving field. Any treatment of it is only a snapshot of a changing picture. The authors included the most up-to-date information available in each of the key areas of the practice, including water quality best management practices and stormwater master and quality management plans. Also gathered and included are little known but valuable tables, charts, and procedures covering common designs and more specialized situations. English and metric versions of charts and nomographs are included where available and where deemed useful for the designer. The evolution of stormwater management, as any science or practice, builds on basic understandings and concepts. In each case, the authors attempted to provide, first, sufficient discussion to lay a foundation and framework upon and within which any new developments can be built.

The book is primarily designed for use by engineers, designers, and planners involved in municipal stormwater management. Information contained should cover most of the design applications within urban areas and most of the institutional aspects of stormwater management the planner or administrator faces daily. The book has also been designed to be applicable to civil and environmental undergraduate and graduate courses from planning or administration and engineering perspectives.

In addition to a general update of all information in the book, the major additions in this edition of the book are updated information on water quality best management practices, additional hydrology information, and new discussions related to the institutional aspects of urban stormwater management.

Thomas N. Debo
Andrew J. Reese

Authors

Dr. Thomas N. Debo is Professor Emeritus in the City Planning Program in the College of Architecture at the Georgia Institute of Technology and president of Debo & Associates, Inc., a consulting firm specializing in the urban stormwater management area.

He received his B.S. in Civil Engineering from Michigan Technological University in 1963 and his Master of City Planning and Ph.D. in Civil Engineering from the Georgia Institute of Technology in 1972 and 1975, respectively. He has over three decades of experience dealing with stormwater management, including development of hydrologic/hydraulic computer programs, engineering design handbooks, local ordinances and policies, and technical and engineering studies and designs. He has worked with numerous municipalities throughout the United States, state and federal agencies, and a large number of consulting and legal firms, and has extensive experience as an expert witness related to urban stormwater management issues. Dr. Debo has published several books and over 40 articles in professional journals and other publications. He has been very active in conducting stormwater-related workshops throughout the country and has been invited to give presentations and speeches extensively throughout the United States and several European countries. In recent years, Dr. Debo has been involved with the design and implementation of innovative techniques for the capture and reuse of urban stormwater and improvement of stormwater quality. He is a member of the American Society of Civil Engineers and a registered engineer in the State of Georgia. Dr. Debo is married and the father of four daughters and four grandchildren and presently resides in Atlanta, Georgia, and Morganton, North Carolina.

Andrew J. (Andy) Reese has over 27 years' experience in a wide variety of stormwater and water resource engineering and management roles.

In those years, he worked in a hydraulic and hydrologic research position with the Corps of Engineers, and then as a consultant. He has served in roles from site design, FEMA studies, master planning, stable channel design and sediment transport, hydraulic structures design, design criteria manual development, NPDES permitting, training, computer modeling and software development, public education programs, meeting facilitation, stormwater utility development, BMP programs, and numerous stormwater program assessments. He has managed many large and complex municipal stormwater projects. His experiences with well over 100 local governments in 48 states and for numerous state and federal agencies give him a unique view of stormwater management as it is practiced in the United States. He has been involved in the establishment of some of the most successful stormwater programs in the United States.

Reese has written and presented over 50 papers nationally and internationally on various stormwater-related subjects from complex modeling to program development, public involvement, and stormwater overviews. He is also a popular speaker and teacher, annually giving many presentations and short courses across the United States and internationally for national and regional conventions. He has master's degrees in business administration and hydraulic engineering. Reese is vice president for AMEC Earth & Environmental, Inc., a large international water resources, environmental, and geotechnical consulting firm. He is a registered professional engineer and hydrologist. He is married and the father of four teenagers residing near Nashville, Tennessee.

Contents

1 Introduction to Municipal Stormwater Management 1
1.1 Stormwater Management Paradigms 1
   Early Stormwater Paradigms 1
      Paradigm #1 — Run It In Ditches 1
      Paradigm #2 — Run It In Pipes 2
      Paradigm #3 — Run It In Stormwater Pipes 2
      Paradigm #4 — Keep It From Stormwater Pipes 3
      Paradigm #5 — Well, Just Do Not Cause Flooding 4
   The New Breed of Stormwater Paradigms 5
      Paradigm #6 — Do Not Pollute 6
      Paradigm #7 — It Is The Ecology 6
      Paradigm #8 — Water Is Water Is Watershed 7
      Paradigm #9 — Green And Bear It 9
      From Paradigm To Paradigm 9
1.2 Understanding Stormwater Management Problems and Solutions 11
1.3 The Organization of the Rest of the Book 15
References 16

2 Stormwater Management Programs 17
2.1 The Stormwater Management Program 17
2.2 Functional Perspective of Stormwater Management 18
   Long-Range Aspects 18
   Day-to-Day Stormwater Management 19
2.3 Overview of the Legal Aspects of Stormwater Management 20
2.4 Overview of the Technical Aspects of Stormwater Management 21
   Technical Manuals 21
   Computer Models 22
   Databases and Infrastructure Inventories 23
   Mapping, GIS, and Related 29
      GIS Systems 29
2.5 Overview of the Organizational Aspects of Stormwater Management 36
2.6 Overview of the Financial Aspects of Stormwater Management 38
2.7 Typical Stormwater Management Problem Areas 40
   Long-Term Problem Areas 40
   Legal/Financial/Organizational/Technical Underpinning Problem Areas 41
   Day-to-Day Problem Areas 42
2.8 Three Example Programs 42
   Charlotte, North Carolina 42
   Louisville MSD, Kentucky 45
   Tulsa, Oklahoma 47
2.9 Characteristics of Successful Programs 50
   Public Education and Involvement 50
   Key Champions 50
   Financing 50

	Basin-Wide Planning and Control	50
	Clear Procedures and Goals	51
	Focused Authority	51
	Strong Technical Guidance and Capabilities	51
	Guided Development	51
	Comprehensive Maintenance	51
	An Environmental Focus	52
2.10	The Stormwater Program Feasibility Study	52
2.11	Floodplain Management	53
	Flood Insurance Programs	55
Reference		57

3 Public Awareness and Involvement 59
3.1	Introduction	59
3.2	Defining the "Public"	59
3.3	Plan to Include the Public	61
3.4	Public Involvement and Education Techniques	63
	Public Participation	64
	Public Participatory Groups	64
3.5	The Stormwater Advisory Committee	66
	Defining the Group	67
	Media and the General Public	68
	Authority and Purpose	69
	Defining the Issues	69
	Informed Consent and Consensus Building	70
	Ground Rules and Objective Criteria	71
	Policy Papers	71
3.6	Volunteer Programs	73
3.7	Dealing with the Media	74
3.8	Risk Communications	76
3.9	Technical Communications	79
References		80

4 Ordinances, Regulations, and Documentation 81
4.1	Introduction	81
4.2	Legal Basis and Considerations	82
	Watercourse Law	82
	Riparian Doctrine	82
	Prior Appropriation Doctrine	83
	Diffused Surface Water Law	83
	Common Enemy Doctrine	83
	Civil Law Doctrine	84
	Reasonable Use Doctrine	84
	Liability and Damages	84
	Individual Liability	84
	Municipal Liability	84
	"Takings"	85
	Water Quality and Water Law	86
4.3	Municipal Ordinances	86
	Legal Authority and Context	86

	Technical Basis	87
	Administrative Apparatus	87
	Enforcement Provisions	87
	Special Water-Quality Considerations	87
	Detention	88
4.4	Drafting Local Ordinances and Regulations	89
	Identify Problems and Issues	89
	Formulation of Objectives	90
	Developing Policies	92
	Detention Analysis Example	93
	Drafting the Ordinance	95
	Stormwater Management Design Manual	97
4.5	Flexibility in a Stormwater Management Program	98
	Flexible Ordinance Provision	98
	Rigid Ordinance Provision	98
4.6	When to Adopt an Ordinance	100
4.7	The Complete Stormwater Management Program	102
	Land-Use Planning Aspects	102
	Municipal Role in Encouraging Innovative Solutions	102
	Administrative Problems	103
	Technical Requirements	103
	Staff and Financial Resources Needed	103
	Field Inspection	103
	Enforcement	104
	Legal Considerations	105
	Summary	105
4.8	Documentation	105
	Purpose of Documentation	106
	Types of Documentation	107
	Preconstruction	107
	Design	107
	Construction or Operation	108
	Documenting the Plan Development Process	108
	Documentation Procedures	109
	Storage Requirements	110
	Specific Content of Documentation Files	110
	Hydrology	110
	Bridges	111
	Culverts	111
	Open Channels	112
	Storm Drains	112
	Storage Facilities and BMPs	112
	Pump Stations	113
4.9	Documentation for Legal Proceedings	113
References		115

5   Financing Stormwater Management Programs 117
5.1	Financing Needs of Stormwater Management Programs	117
5.2	Major Stormwater Funding Methods	118
5.3	Stormwater Utility Overview	119
	Overview of the Concept	119

	Advantages of the Utility Concept	119
	Uniqueness Factor	122
5.4	Utility Funding Methods	123
5.5	Stormwater Utility Policy Issues	127
	General Policy Information	127
	Stormwater Credits	129
5.6	Steps in a Typical Financing Study	130
	Overview	130
	The Program Track	131
	The Public Track	132
	The Finance Track	136
	The Data Track	136
	Feasibility Studies	137
5.7	Typical Financing Feasibility Study Scope of Services	138
	Purpose	138
	Task 1 — Development of a Citizens Group	139
	Task 2 — Stormwater Program Description, Problems, and Needs	139
	Deliverables	139
	Task 3 — Program Objectives and Priorities	140
	Deliverables	140
	Task 4 — Projected Stormwater Program	140
	Deliverables	141
	Task 5 — Basic Funding Feasibility	141
	Deliverables	141
	Task 6 — Service Charge Billing, Collection, and Accounting Options	141
	Deliverable	142
	Task 7 — Public Communications Program	142
	Deliverables	142
	Task 8 — Action Plan Report	142
	Deliverables	143
References		143
Appendix A — Financing Stormwater Management Programs		146
6	**Data Availability and Collection**	**151**
6.1	Overview	151
	Introduction	151
	Data Collection Effort	151
6.2	Sources and Types of Data	152
	Data Categories	153
6.3	Major Data Topics	153
	Watershed Characteristics	153
	Contributing Size	153
	Slopes	155
	Watershed Land Use	156
	Streams, Rivers, Ponds, Lakes, and Wetlands	157
	Roughness Coefficients	157
	Stream Profile	157
	Stream Cross Sections	158
	Existing Structures	158
	Flood History	158

	Debris and Ice	158
	Scour Potential	159
	Controls Affecting Design Criteria	159
	Downstream Control	159
	Upstream Control	159
6.4	Data Acquisition, Survey Information, and Field Reviews	159
	Remote Data Acquisition	159
	Geographical Information Systems	160
	Telemetry	161
	Survey Information	162
	Field Reviews	162
6.5	Data Evaluation	163
	Data Accuracy	163
	Sensitivity Studies	164
	Statistical Analysis	165
	Time Series Analysis	165
	Extreme Event Analysis	166
	Geostatistical Analysis	166
	Other Statistical Methods	166
6.6	Precipitation Data Collection	167
	Overview	167
	Data Collection	168
	Factors Influencing Data Collection	168
	Storm Types	169
	Storm Movement	169
	Storm Decay	169
	Precipitation Data Acquisition	169
	Rain Gauges	170
	Radar Precipitation Data Acquisition	170
	Rain Gauging Networks	171
	Rainfall Average Depth Estimation	172
6.7	Flow Data Collection	174
	Introduction	174
	Basic Equipment and Techniques Overview	174
	Common Errors	178
6.8	Flow Data Collection — Natural Controls	178
	Selection of Gauging-Station Sites	178
	Velocity Measurements	179
	Velocity Meters	181
6.9	Flow Data Collection — Tracer Methods	182
	Salt-Velocity Method	182
	Salt or Color-Dilution Methods	182
6.10	Environmental Data Considerations and Collection	182
	Data Needs	183
	Environmental Sampling Objectives	183
	Typical Sampling Program Steps	185
	Key Constituents	185
	Metals	186
	Nutrients	186
	Human Pathogens	187
	Oxygen Demand	188

      Sediment 188
      Oil and Grease 189
  Pollution Sources...189
      Erosion and Sedimentation 189
      Atmospheric Fallout 189
      Vehicles 190
      Other Human Activities..190
      Impacts on Receiving Waters 191
  Land-Use Baseline Data..192
  Urban Runoff Quality Data Collection..194
      Site Selection 196
      Storm Event Sampling..196
      Sample Types 196
References...197
Appendix A — Sources of Data 201
Appendix B — Field Investigation Form and Checklist 204

7 Urban Hydrology 207
7.1   Introduction..207
7.2   Concept Definitions 208
7.3   Hydrologic Design Policies 209
       Drainage Basin Characteristics 209
       Stream Channel and Conveyance System Characteristics..........................210
       Floodplain Characteristics 210
       Meteorological Characteristics 210
7.4   Design Frequency and Risk...211
       Risk 211
       Risk-Based Analysis...212
       Design Frequency...213
          Cross Drainage 214
          Storm Drains 215
          Inlets 215
          Detention and Retention Storage Facilities..215
       Review Frequency..216
7.5   Hydrologic Procedure Selection 216
       Analysis of Stream Gauge Data..217
       Regression Equations...218
       Rational Method..219
       Unit Hydrograph Methods..219
          Synder's Unit Hydrograph..219
          SCS Synthetic Unit Hydrograph..220
       Continuous Simulation Models 220
       Summary 220
7.6   Calibration...221
7.7   Precipitation and Losses 222
       Intensity–Duration–Frequency Curves..223
       Rainfall Durations and Time Distributions 224
          Duration..224
          Distribution 225
          Example Information..225
          The "Balanced-Storm Approach" 225

	SCS 24-Hour Storm 226
	Huff Distributions 227
	Yen and Chow's Method 230
7.8	Rational Method 232
	Rational Formula 232
	Characteristics and Limits of the Rational Method 233
	Time of Concentration 235
	Rainfall Intensity 237
	Drainage Area 237
	Runoff Coefficient 238
	Example Problem — Rational Method 241
7.9	SCS Hydrologic Methods 242
	Basic Equations and Concepts 243
	Rainfall–Runoff Equation 243
	Curve Number 245
	Antecedent Moisture Conditions 247
	Connected Impervious Areas 247
	Unconnected Impervious Areas 247
	Composite Curve Numbers 249
	Rainfall 249
	Calculation of Lag Time and Time of Concentration 249
	Travel Time and Time of Concentration Empirical Equation 254
	Travel Time and Time of Concentration — Manning's Equation 256
	Drainage Area 259
	Procedures, Tables, and Figures 259
	SCS Peak Discharges 259
	Computations 260
	Limitations 260
	SCS Peak Discharge Example 263
	SCS Hydrograph Generation 265
	Rainfall Excess Using SCS Methods 265
	Rainfall Excess Example 266
	SCS Dimensionless Unit Hydrographs 268
	Dimensionless Unit Hydrograph Discussion 270
	Unit Hydrograph Applications 273
	Unit Hydrograph Example 273
7.10	Santa Barbara Urban Hydrograph Method 275
	Example: 10-Acre Site — Existing Conditions 276
	Input Data 276
	Output Table — 24-Hour Storm (Table 7-31) 276
	Example: 10-Acre Site — Developed Conditions 278
	Input Data 278
	Output Table — 24-Hour Storm 278
7.11	Water-Quality Volume and Peak Flow 280
	Water-Quality Volume Calculation 280
	Water-Quality Volume Peak Flow Calculation 280
7.12	Channel Protection Volume Estimation 282
	Basic Approach 282
7.13	Water Balance Calculations 286
	Basic Equations 286
	Rainfall (P) 286

	Runoff (Ro)	286
	Baseflow (Bf)	287
	Infiltration (I)	287
	Evaporation (E)	288
	Evapotranspiration (Et)	289
	Overflow (Of)	290
7.14	Downstream Hydrologic Assessment	291
	Reasons for Downstream Problems	292
	Flow Timing	292
	Increased Volume	292
	The Ten-Percent Rule	293
References		295

8 Storm Drainage Systems 299

8.1	Introduction	299
8.2	Concept Definitions	299
8.3	Pavement Drainage	300
	Design Steps	300
	Design Factors	301
	Return Period	302
	Spread	302
	Inlet Types and Spacing	302
	Longitudinal Slope	302
	Cross Slope	303
	Curb and Gutter Sections	303
	Roadside and Median Channels	303
	Bridge Decks	304
	Shoulder Gutters	304
	Median/Median Barriers	305
	Storm Drains	305
	Detention Storage	305
	Costs	305
8.4	Stormwater Inlet Overview	306
8.5	Design Frequency and Spread	307
8.6	Gutter Flow Calculations	308
	Uniform Cross Section	308
	Composite Gutter Sections	310
8.7	Grate Inlet Design	314
	Grate Inlets on Grade	315
	Grate Inlet in Sag	318
8.8	Curb Inlet Design	320
	Curb Inlets on Grade	320
	Curb Inlets in Sump	324
8.9	Combination Inlets	327
8.10	Energy Losses in a Pipe System	328
	Friction Losses	328
	Velocity Headlosses	328
	Entrance Losses	328
	Junction Losses — Incoming Opposing Flows	329
	Junction Losses — Changes in Direction of Flow	329
	Junction Losses — Several Entering Flows	329

8.11 Storm Drains..330
    Formulas for Gravity and Pressure Flow 333
    Hydraulic Grade Line 337
    Hydraulic Grade Line Design Procedure 338
8.12 Environmental Design Considerations..341
    Roadway Pollution 341
    Structural Roadway BMP Design Overview...344
References...345
Appendix A — Metric Design Figures 347

9   Design of Culverts 361
9.1 Introduction...361
9.2 Concept Definitions 362
9.3 Culvert Design Steps and Criteria 363
    Design Steps...363
        Determine and Analyze Site Characteristics 363
        Perform Hydrologic Analysis 364
        Perform Outlet Control Calculations and Select Culvert.............364
        Perform Inlet Control Calculations for Conventional
          and Beveled-Edge Culvert Inlets..364
        Perform Throat-Control Calculations for Side- and Slope-Tapered
          Inlets (Optional) 364
        Analyze the Effect of Falls on Inlet Control Section Performance (Optional)....364
        Design Side- and Slope-Tapered Inlets (Optional) 364
        Complete File Documentation 365
    Engineering and Technical Design Criteria 365
    Flood Frequency..366
    Velocity Limitations 366
    Buoyancy Protection 367
    Length and Slope 367
    Debris Control 367
    Ice Buildup 367
    Headwater Limitations 367
    Tailwater Conditions 368
    Storage — Temporary or Permanent 369
    Culvert Inlets 369
        Projecting Inlets or Outlets..369
        Headwalls with Bevels 369
        Improved Inlets 369
        Commercial End Sections 370
        Inlets with Headwalls 371
        Wingwalls and Aprons..371
        Improved Inlets 371
        Material Selection..371
        Culvert Skews..372
        Culvert Sizes and Shapes...372
        Weep Holes 374
        Outlet Protection 375
        Erosion and Sediment Control..375
        Environmental Considerations 375

	Safety Considerations	375
9.4	Culvert Flow Controls and Equations	376
	Design Equations	376
	Critical Depth — Outlet Control	376
	Tailwater Depth — Outlet Control	377
	Tailwater Depth > Barrel Depth — Outlet Control	377
	Tailwater < Barrel — Outlet Control	378
	Tailwater Insignificant — Inlet Control	378
	Inlet or Outlet Control	378
9.5	Design Procedures	378
	Tailwater Elevations	379
	Use of Inlet- and Outlet-Control Nomographs	379
	Storage Routing	384
9.6	Culvert Design Example	384
	Input Data	384
9.7	Long-Span Culvert	386
9.8	Design of Improved Inlets	387
	Bevel-Edged Inlet	388
	Side-Tapered Inlet	389
	Slope-Tapered Inlet	389
	Improved Inlet Performance	389
9.9	Design Procedures for Bevel-Edged Inlets	390
	Design Figure Limits	390
9.10	Flood Routing and Culvert Design	391
	Design Procedure	392
9.11	HY8 Culvert Analysis Microcomputer Program	393
	References	393
	Appendix A	395
	Appendix B	455

10	**Open Channel Design**	**515**
10.1	Introduction	515
10.2	Design Criteria	515
	Channel Types	515
	Vegetative Linings	516
	Flexible Linings	516
	Rigid Linings	516
	Rigid Low-Flow Channels	517
	General Design Criteria	517
	Channel Transitions	518
10.3	Hydraulic Terms and Equations	518
	Flow Classification	519
	Hydraulic Terms and Definitions	519
	Total Energy Head	519
	Specific Energy	519
	Kinetic Energy Coefficient	519
	Steady and Unsteady Flow	520
	Uniform Flow and Nonuniform Flow	520
	Gradually Varied and Rapidly Varied	520
	Froude Number	520

	Critical Flow	520
	Subcritical Flow	521
	Supercritical Flow	521
	Hydraulic Jump	521
	Equations	521
	Manning's Equation	521
	Continuity Equation	522
	Energy Equation	522
10.4	Manning's n Values	522
	Natural Channels	524
10.5	Manning's n Handbook	524
10.6	Best Hydraulic Section	527
10.7	Uniform Flow Calculations	528
	Direct Solution	529
	General Solution Nomograph	529
	Trapezoidal Solution Nomograph	532
	Trial-and-Error Solution	532
	Irregular Channels	533
	Average Roughness	533
	Irregular Channel Calculations	534
10.8	Critical-Flow Calculations	534
10.9	Vegetative Design	536
	Design Stability	536
	Design Capacity	539
	Erosion Control	539
10.10	Approximate Flood Limits	540
	100-Year Flood Elevation	540
	Setback Limits	541
10.11	Uniform Flow — Example Problems	541
	Direct Solution of Manning's Equation	542
	Irregular Channel (Example 1)	542
	Grassed Channel Design Stability (Example 2)	542
	Grassed Channel Design Capacity (Example 3)	543
10.12	Gradually Varied Flow	544
	Direct Step Method	545
	Standard Step Method	546
10.13	Gradually Varied Flow — Example Problems	548
	Direct Step Method	548
	Standard Step Method	550
10.14	Hydraulic Jump	552
10.16	Channel Bank Protection	554
	Riprap Design	554
	Riprap Design Example	558
	Grouted Riprap	558
	Gabions and Rock Mattresses	558
	Soil Cement	561
	Bioengineering	563
	Manufactured Bank-Protection Methods	564
10.17	Physical Stability of Channels	564
	Overview	564
	Channel Response to Change	565

- Proportionality Approaches 566
- Assessing Stream Stability 567
  - Stable Channel Design Approaches 568
  - Maximum Permissible Velocity Method 569
- Tractive Stress Method 572
- Regime Equations for Channel Proportions 574
  - Simons and Albertson Equations 575
  - Gravel Bed Equations 575
- Analytical Methods for Channel Proportions 576
  - Neill Method 576
- Stable Channel Design Examples 581
  - Permissible Velocity Example 581
  - Tractive Stress Design Example 581
  - Regime Equation Example 583
  - Neill Relations 583
- Stream Structures 583
  - Grade Control Structures 583
  - Sills 584
  - Drop Structures, Chutes, and Flumes 584
- 10.18 Environmental Considerations for Channels 585
  - Instream Mitigation Features 585
  - Riprap and the Environment 590
  - Environmental Consideration of Floodplains 592
    - Floodplain Environmental Activities 592
    - Greenway Planning 592
  - Stream Channel Preservation and Restoration 594
  - Restoration Guiding Principles 594
    - Preserve and Protect Aquatic Resources 594
    - Restore Ecological Integrity 595
    - Restore Natural Structure 595
    - Restore Natural Function 595
    - Work within the Watershed and Broader Landscape Context 595
    - Understand the Potential of the Watershed 595
    - Address Ongoing Causes of Degradation 595
    - Develop Clear, Achievable, and Measurable Goals 595
    - Use a Reference Site 595
    - Involve the Skills and Insights of a Multidisciplinary Team 596
    - Design for Self-Sustainability 596
    - Use Passive Restoration, when Appropriate 596
    - Use Natural Fixes and Bioengineering Techniques, where Possible 596
    - Monitor and Adapt where Changes are Necessary 596
  - Restoration Practices 596
  - Stream Health Assessment 598
- References 599

11 Storage and Detention Facilities 605
- 11.1 Introduction 605
- 11.2 Uses and Types of Storage Facilities 605
  - Uses of Storage Facilities 606
  - Types of Storage Facilities 607

11.3	Design Criteria	609
	General Design Criteria	610
	Specific Detention Design Criteria	611
	Specific Retention Design Criteria	611
	Outlet Works Design Criteria	611
	Location Design Criteria	611
11.4	Safe Dams Act	612
	Category 1	613
	Category 2	613
	Category 3	613
11.5	General Design Procedure for Storage Routing	613
	Stage–Storage Curve	614
	Stage–Discharge Curve	615
	General Procedure	615
11.6	Outlet Hydraulics	616
	Combination Outlets	617
	Perforated Risers	617
	Two-Way Drop Inlets	619
	Weirs	622
	Sharp-Crested Weirs	622
	Submerged Weir Correction	622
	Approach Velocity Correction	623
	Sharp-Crested Rectangular Weirs	623
	Trapezoidal and Cipolletti Weirs	626
	V-Notch Weirs	626
	Broad-Crested Weirs	627
	Submergence Effects	628
	Triangular Channel Broad-Crested Weirs	628
	Ogee Shapes	628
	Special Weirs	629
	Side-Channel Weirs	630
	Example of Side-Channel Weir	633
	Orifice Meters and Nozzles	634
	Drop Inlet Boxes	639
	Rooftop Detention	641
	Example Rooftop Design	642
	Modified Rational — Mass Balance Method	642
	Drywells for Roof Drains	643
	Detention Chambers and Pipes	644
	Parking Lot Detention	644
	Porous Pavement	644
	Pipes and Culverts	645
	Combination Outlets	645
	Multistage Outlet Design Procedure	646
11.7	Extended Detention Outlet Design	649
	Example Outlet Design	649
	Method 1: Maximum Hydraulic Head with Routing	649
	Method 2: Average Hydraulic Head and Average Discharge	650
	Extended-Detention Outlet Protection	651
11.8	Preliminary Detention Calculations	653
	Alternative Method	653

	Peak Flow Reduction...654
11.9	Routing Calculations 655
11.10	Example Problem 656
	Preliminary Volume Calculations...657
	Design and Routing Calculations..657
	Downstream Effects...659
11.11	"Chainsaw Routing" Technique for Spreadsheet Application......................................660
	Overview 660
	Stage–Storage..661
	Chainsaw Routing..662
	Numerical Instability...663
11.12	Modified Rational Method Detention Design...663
	Example Problem 667
11.13	Hand-Routing Method for Small Ponds 668
	Limitations..668
	Basic Approach Overview 669
	Emergency Spillway Approximation...670
	Details of Approach...671
	Design Example — Sizing of a Small Pond for Orifice Flow.......................................672
	Step 1 — Input Data.. 672
	Step 2 — Basic Calculations 673
	Step 3 — Preliminary Estimates 673
	Step 4 — Routing Number 674
	Step 5 — Orifice Size and Other Data 674
	Step 6 (Optional) — Recalculate Routing Number 674
	Step 7 (Optional) — Recalculate R and Other Data........................674
	Step 8 — Emergency Spillway Approximation 674
	Method Comparison..675
11.14	Land-Locked Retention..676
11.15	Retention Storage Facilities 677
11.16	Retention Facility Example Problem 678
11.17	Construction and Maintenance Considerations..679
11.18	Protective Treatment..680
	Trash Racks and Safety Grates..680
References..683

12	**Energy Dissipation**..**687**
12.1	Introduction..687
12.2	Recommended Energy Dissipators 687
12.3	Design Criteria..688
	Dissipator Type Selection..688
	Ice Buildup 689
	Debris Control 689
	Flood Frequency..689
	Maximum Culvert Exit Velocity 690
	Tailwater Relationship..690
	Material Selection..690
	Culvert Outlet Type...690
	Safety Considerations 691
	Weep Holes 691

12.4 Design Procedure 691
   Data Needs 691
   Procedure Outline 692
12.5 Local Scourhole Estimation 693
12.6 Riprap Aprons 695
   Design Procedure 696
   Design Considerations 698
   Example Problems 699
      Example 1 — Riprap Apron Design for Minimum Tailwater Conditions 699
      Example 2 — Riprap Apron Design for Maximum Tailwater Conditions 699
12.7 Riprap Basin Design 699
   Design Procedure 700
   Design Considerations 703
   Example Problems 704
      Example 1 704
      Example 2 705
      Example 3 706
12.8 Baffled Outlets 707
   Design Procedure 708
   Example Problem 709
12.9 Downstream Channel Transitions 711
12.10 Energy Dissipator Computer Model 714
References 714

13 Structural Best Management Practices 717
13.1 Introduction 717
13.2 Surface Waters, Pollution, and Mitigation Measures 718
   Surface Water Drivers for BMP Usage 718
   Pollution Dynamics 719
      Pollution Accumulation 719
      Urban Hot Spots 719
      Pollution Movement Forces 720
      Pollution Availability and Fate 720
      Stressors Associated with Urban Runoff 722
   Overall Approach: Structural and Nonstructural BMPs 723
   BMPs and Disease Vectors 725
13.3 Estimation of Pollutant Concentration and Loads 726
   Pollutant Loads 726
   Event Mean Concentrations 726
   Nationwide Regression Equations Method 728
   Example Application (Tasker and Driver, 1988) 730
   The Simple Method 733
   Acute or "Shock" Pollutant Loading Estimates 734
   Simulation Models 735
13.4 BMP Design Concepts 735
   Basis for Design Criteria for Structural BMPs 735
   Unified Sizing Criteria 739
      Stormwater Better Site Design 740
      Unified Stormwater Sizing Criteria 740
      Stormwater Credits for Better Site Design 740

13.5	Overview of Structural BMP Types	741
	Basic Structural BMP Types	741
13.6	Selection of Structural Management Measures	742
	Introduction	742
	General BMP Pollutant Removal Effectiveness	746
	Structural BMP Screening	749
	Step 1: Overall Applicability	750
	Step 2: Specific Criteria	753
	Step 3: Location and Permitting Considerations	755
13.7	Online Versus Offline Structural Controls	759
	Introduction	759
	Flow Regulators	759
13.8	Regional Versus On-Site Stormwater Management	761
	Introduction	761
	Advantages and Disadvantages of Regional Stormwater Controls	761
	Advantages of Regional Stormwater Controls	762
	Disadvantages of Regional Stormwater Controls	762
	Important Considerations for the Use of Regional Stormwater Controls	763
13.9	Using Structural Stormwater Controls in Series	763
	Combined Measures	763
	Stormwater Treatment Trains	764
	Calculation of Pollutant Removal for Structural Controls in Series	767
	Example Application	768
13.10	Routing with WQv Removed	768
13.11	Specific Structural BMP Design Guidance	769
	Overview	769
	Overall Design Approach	770
	Extended Dry Detention Basin	772
	Pond Sizing	774
	Outlets	775
	Pollutant Removal Information	777
	General Design Criteria	778
	Typical Required Specifications	779
	Typical Recommended Specifications	779
	Sizing Example	780
	Retention Pond	781
	Pollution Removal Efficiency	790
	Pond Sizing	790
	Typical Required Specifications	792
	Typical Recommended Specifications	794
	Typical Maintenance Standards for Extended Detention and Wet Ponds	794
	Alum Treatment System	795
	Design Example	796
	Constructed Stormwater Wetlands	801
	Stormwater Wetlands	804
	Pollution Removal Efficiency	812
	Wetland Design	812
	Permitting	814
	Typical Standard Specifications for Constructed Wetlands	815
	Typical Required Specifications	815
	Typical Recommended Specifications	817

- Operation and Maintenance Requirements 818
- Bioretention Areas 819
  - Pollution Removal Efficiency 819
  - Design 820
  - Typical Required Specifications 821
  - Typical Recommended Specifications 824
  - Maintenance 825
  - Design Example 825
- Sand Filtration Systems 829
  - Overview 829
  - Configurations 830
  - Maintenance 832
- Austin First-Flush Filtration Basin 833
  - Pollution Removal Efficiency 835
  - Design 835
  - Operation and Maintenance Requirements 839
- Surface Sand Filter 839
  - Design Information 840
  - Design Example 848
  - Linear (Perimeter) Sand Filter 854
  - Underground Sand Filter 856
  - Organic Sand Filters 860
- Infiltration Trenches 863
  - Infiltration Trench Design 865
  - Typical Operation and Maintenance Requirements 868
  - Typical Required Specifications 868
  - Typical Recommended Specifications 870
  - Design Example 871
- Porous Pavement 872
  - Pollutant Removal Efficiency 875
  - Porous Pavement Design 876
  - Operation and Maintenance Requirements 878
  - Typical Required Specifications 880
  - Typical Recommended Specifications 882
  - Design Example 882
- Modular Paving Blocks 883
  - Typical Design Specifications for Modular Blocks 887
- Grassed Swales 888
  - Pollutant-Removal Efficiency 891
  - Dry Swale Design 893
  - Wet Swale 896
  - Grass Channels 896
  - Typical Required Specifications: Dry and Wet Swales 896
  - Typical Recommended Specifications: Dry and Wet Swales 898
  - Typical Operation and Maintenance Procedures 898
- Filter Strips And Flow Spreaders 898
  - Pollutant-Removal Efficiency 900
  - Design of Filter Strips 901
  - Filter without Berm 904
  - Filter Strips with Berm 904
  - Filter Strips for Pretreatment 904

        Typical Required Specifications 905
        Typical Recommended Specifications 906
        Typical Operation and Maintenance Requirements 906
        Design Example 907
    Oil/Grit Separators (Water-Quality Inlet, Gravity Separator) 908
        Pollution-Removal Efficiency 910
        Oil/Water Separator Design 911
        Typical Required Specifications 915
        Recommended Specifications 916
        Operation and Maintenance Requirements 917
    Alum Treatment 917
        Pollutant-Removal Efficiency 919
        Design Criteria 919
    Proprietary Structural Controls 921
    Local Acceptance 921
    Multichambered Treatment Train 932
References 934

14 Stormwater Master Planning 945
14.1 The Role of Stormwater Master Planning 945
    Introduction 945
    Basic Master Plan Types 946
    Keys to Successful Master Planning 946
14.2 The Master Planning Process: The Scoping Study 947
    Step 1 — Make a Needs Analysis 948
    Step 2 — Determine Constraints to Possible Solutions 950
    Step 3 — Formulate the Technical Approach 951
    The End Product of the Scoping Study 951
14.3 Computer Model Choice for Master Planning 952
14.4 Five Basic Types of Master Plans: The Flood Study 953
    Major Creek Flooding 957
        Data Collection 959
        Preliminary Field Investigation 959
        Design Storm Development 960
        Hydrologic Model Development 962
        Hydraulic Model Development 966
        System Alternatives Development and Analysis 969
    Minor System Flooding 970
14.5 Five Basic Types of Master Plans: The Cost/Benefit Analysis Master Plan 971
14.6 Five Basic Types of Master Plans: Stormwater Quality Master Plan 977
14.7 Five Basic Types of Master Plans: The Ecological Study Master Plan 981
14.8 Five Basic Types of Master Plans: The Holistic Master Plan 982
    Introduction 982
    Basic Approach 984
    Example: Rapid Watershed Planning 985
        Step 1 — Identify Initial Goals and Establish a Baseline 985
        Step 2 — Set Up a Watershed Management Structure 986
        Step 3 — Determine Budgetary Resources Available for Planning 987
        Step 4 — Project Future Land-Use Change in the Watershed
        and Its Subwatersheds 987

  Step 5 — Fine-Tune Goals for the Watershed and Its Subwatersheds 987
  Step 6 — Develop Watershed and Subwatershed Plans 988
  Step 7 — Adopt and Implement the Plan 990
  Step 8 — Revisit and Update the Plan 990
References 991

15 Stormwater Quality Management Programs 995
15.1 Basic Urban Runoff Quality Understanding 995
 Introduction 995
 The Stormwater Quality Approach — History 995
15.2 Basic Findings 997
 Broad Pollution Categories 997
 Impacts of Urban Development 998
  Changes to Stream Flow 998
  Changes to Stream Geometry 999
  Water-Quality Impacts 1000
  Degradation of Aquatic Habitat 1000
  Constant EMC 1001
  Source Control 1002
  Hot Spots 1002
  First Flush and Treatment Volume 1004
  Treatment Train Concept 1005
 Water-Quality Standards 1007
15.3 Water Quality Act Overview 1009
 Phase I Basic Requirements 1010
 Municipal Reaction 1011
 Phase II Basic Requirements 1011
 Maximum Extent Practicable 1013
 Measurable Goals 1015
15.4 The Six Minimum Controls 1017
 General Permit Conditions 1017
 The Six Minimum Controls 1017
 Guidelines for Developing and Implementing this Measure 1019
  Forming Partnerships 1019
  Using Educational Materials and Strategies 1019
  Reaching Diverse Audiences 1019
  Guidelines for Developing and Implementing this Measure 1020
 Guidelines for Developing and Implementing this Measure 1022
  Guidelines for Developing and Implementing this Measure 1026
  Guidelines for Developing and Implementing this Measure 1030
  Guidelines for Developing and Implementing this Measure 1031
15.5 Costs of NPDES Phase II 1032
15.6 The Ten Commandments in Developing a SWQMP 1035
 Overall Approach 1035
 The Ten Commandments 1035
  #1 — Think Paradigm Shift 1036
  #2 — Fix Real Things 1036
  #3 — Bring Me in Early, I'm Your Partner; Bring Me in Late,
   I'm Your Judge 1037
  #4 — Do not Build New Problems 1038

#5 — Get the Foundations Right	1039
#6 — Integrate and Graduate	1040
#7 — Ride the Treatment Train	1041
#8 — Get a Tool Set	1041
#9 — Not All that Glitters is Gold	1042
#10 — Check the Pool	1044
15.7 The SWQMP Development Process	1045
The Nine-Step Planning Approach	1045
Step 1	1046
Step 2	1050
Step 3	1053
Step 4	1053
Step 5	1053
Step 6	1053
Evaluation of Best Management Practices	1054
Step 7	1057
Step 8	1057
Step 9	1058
EPA's Guidance Approach	1058
Self-Analysis	1058
Action Plan	1060
References	1061

16	**Site Design and Construction**	**1065**
16.1	Introduction	1065
16.2	Site Design Concepts — Overview	1065
	Conservation of Natural Features and Resources	1066
	Lower-Impact Site Design Techniques	1067
	Reduction of Impervious Cover	1067
	Using Natural Features for Stormwater Management	1067
	Site-Planning Goals	1067
	Watershed Basis	1070
16.3	Site Design Concepts — Implementation	1071
16.4	Site Design Concepts — Design Steps	1073
	Feasibility Study	1075
	Overview	1075
	Program Components	1078
	Site Characteristics	1078
	Planning and Regulatory Controls	1079
	Site Analysis	1080
	Overview	1081
	The Concept Plan	1086
	Overview	1086
	Development of the Stormwater Concept	1087
	Preliminary/Final Plan	1089
	Overview	1089
	Preliminary Plan	1089
	Calculation of Final Stormwater Control Volumes	1090
	Final Design	1091
	Construction	1091

  Preconstruction Meeting 1093
   Inspection/Observation during Construction.....................................1094
 Final Inspections...1094
  Final Inspection and Submission of Record Drawings 1094
  Maintenance Inspections...1094
16.5 Stormwater Aspects of Site Construction 1095
 Construction Costs..1096
 Construction Plans..1097
 Construction Considerations...1099
 Hydrology Considerations 1099
 Erosion- and Sediment-Control Considerations 1100
 Culvert Considerations 1101
 Bridge Considerations..1102
 Open-Channel Considerations..1104
 Storm Drain Considerations..1104
 Temporary Stormwater Management Facilities...1105
 Selection Factors..1105
 Example Format 1106
 As-Built Plans 1106
References.. 1107

17 Maintenance.. 1109
17.1 Introduction..1109
 Maintenance Categories...1110
 Inspection 1110
17.2 Municipal Maintenance Concepts.. 1111
 Introduction.. 1111
 Goals.. 1111
 Objectives..1112
 Policies...1113
 Policy Resolution...1113
 Extent and Level of Service..1114
  Extent of Service...1114
  Level of Service 1114
 Conditions Standards and Performance Standards 1115
17.3 Developing Municipal Maintenance Programs 1116
 Maintenance Approaches..1116
 Organizing the Drainage System 1119
 Infrastructure and Asset Management..1120
 Privatization of Services...1121
17.4 Detention Facilities 1122
 Maintenance Tasks..1123
 Storage Volume Control...1124
17.5 Storm Drains..1126
17.6 Culverts...1126
17.7 Bridges..1127
17.8 Ditches...1129
17.9 Stormwater Inlets..1130
17.10 Slope Drains..1131
17.11 Wash Checks and Energy Dissipators 1131

17.12 Bank Protection 1132
17.13 Underdrains 1132
17.14 Pavement Edge Drains 1132
References 1133

Index **1135**

1

Introduction to Municipal Stormwater Management

1.1 Stormwater Management Paradigms

We each have a stormwater management paradigm. A paradigm is what we think is true and right about a certain subject. It is the grid through which we put all information and input about a subject. In fact, it is everything we think is true about something. If we thought there were more to know, we would fit it into our paradigm somewhere, even if only to dismiss it as a silly idea. Whether our paradigm is, in fact, true and effective, is not the point. We believe it is. And, we only reluctantly change our ways and agree to agree with someone else's paradigm. Knowing your own paradigm is the first step in understanding stormwater management. The second is seeing how far you have to go, and fearlessly setting out.

Knowing what paradigm, or paradigms, your municipality is currently facing is key to making wise decisions about many aspects of change in the stormwater program. Some municipalities are just now facing the need for detention controls to reduce flooding impacts from new development. Others are facing Phase II stormwater regulations or the potential threat of a lawsuit or Total Maximum Daily Load (TMDL) study for a specific stream segment not meeting water quality requirements and standards.

Early Stormwater Paradigms

The evolution of stormwater practice in the United States is set against the backdrop of social change. Since the 1800s, the basic thrust in the United States has shifted from exploration, to cultivation, to industrialization, to urbanization, to gentrification. Gentrification is when we not only want a safe and efficient neighborhood, but a "green" one with walking paths and natural areas (Reese, 2001).

Paradigm #1 — Run It In Ditches

In the beginning, the rugged pioneers got tired of the horse commute and began to move to town. And, what was good enough for the country was good enough for the city. This led to the first stormwater paradigm: everything liquid, and things carried by liquids, should run in ditches, just like on the farm.

This worked for a while and was better than having no drainage system. When ditches and culverts were sized, there were some interesting rules. Stories say some early railroad openings were sized according to the time it took a horse rider to traverse the watershed perimeter — sort of a human planimeter. The drawbacks of this grand solution were soon apparent (Figure 1-1).

FIGURE 1-1
"Run it in ditches." (Photo Courtesy of Tom Reese.)

So, engineers came up with a solution to this dilemma, and unknowingly stepped into the second paradigm: "Let's put it all in pipes."

Paradigm #2 — Run It In Pipes

The urban infrastructure underwent a rapid change around the turn of the century (the turn before this last one, I mean). Downtown areas became clean and dry. All liquid waste went straight from the amazing flush toilets and sinks to the nearest river or stream.

Before there was a World War II, World War I was just called "The Great War." And, before separate sewers, combined sewers were just called "sewers." Today, we call this a "combined sewer" system, and you can find planned combined sewers in the center of most older large cities and unplanned ones everywhere (Figure 1-2).

Sewers solved the mire-in-the-street problem by putting it safely and efficiently in the streams. But soon, plagues convinced cities to shift from well water to surface water. We soon realized that all this mire-in-the-streams was being ingested by downstream dwellers. So, it was determined that we needed treatment plants for sewage and that treating all that "clean" stormwater runoff did not make economic sense. So, a new paradigm was born, and a solution to this dilemma: Run it in stormwater pipes.

Paradigm #3 — Run It In Stormwater Pipes

This urban stormwater design paradigm came into vogue about the time World War II ended, and reigned until the 1970s or so. Rapid postwar expansion and flight to the suburbs conspired to spur this paradigm on.

Technology has always come along to support changing paradigms. About this time, the Rational Method came into prominence, beating out several competitors, as the design method of choice. It was easy to do on a slide rule and made sense as a mass transfer equation. Intensity–duration–frequency curves became available in many places in the United States in the late 1950s, and everywhere in the early 1960s, with the publication of the Weather Bureau's TP40 and Western mountain rainfall maps.

The modern urban drainage infrastructure was born, consisting of an efficient drainage system with catch basins and pipes leading to the nearest stream. And that solved the problem.

FIGURE 1-2
Avenue "J" Chicago. (Photo Courtesy of the City of Chicago, IL.)

Or did it? Anecdotal evidence began to mount. Papers began to appear questioning this paradigm. It soon became apparent to standard engineers that the fruit of an efficient stormwater system is downstream flooding and channel erosion. The literature from the 1970s is replete with papers on the causes of flooding and a new idea to solve flooding forever — on-site detention. Modern stormwater *quantity* management was born.

Paradigm #4 — Keep It From Stormwater Pipes

The first stormwater detention ordinances appeared in the early 1970s and quickly spread across the country. Detention ponds are a promise with conditions. Rarely have those conditions been adequately met in a local on-site detention program. Those conditions include: comprehensive criteria, consideration of volume as well as peak, downstream assessment, realistic hydrograph routing, detailed plans review, as-built certification, long-term maintenance agreements, and strict enforcement. When these conditions are met, real peak flow reduction results have been obtained.

Take, for example, the impacts of volume using traffic flow as an analogy. After a football game lets out in a major city, there is a traffic jam for hours in the vicinity of the stadium. It is not just a peak flow problem. Each parking lot lets out only so many cars at once, like a detention pond. But, hundreds of lots are letting out cars at a controlled rate at once. The problem is a car volume problem, not a peak flow problem only.

This happens every time it rains. Each detention pond is like those smaller parking lots, but for runoff. The rainfall event is generating three to five times more water all at once, everywhere. So, directly below the detention pond, the peak of runoff may be controlled.

FIGURE 1-3
Keep it from stormwater pipes. (Photo Courtesy of French Wetmore.)

But downstream, maybe beyond the point where the total drainage area is ten times the area detained (see Chapter 4), the stream cannot even tell that there is detention. It is a runoff volume problem not a peak problem.

Here is another example. The Rational Method is exact for peak flow estimation of a constant flow of water off of a pane of glass, and it is okay for parking lots and rooftops with some grass and erratic rainfall. It is not a good detention designer. Thousands of ponds were (and are) designed using the Modified Rational and Bowstring methods. Most are undersized, some grossly. Allowable target flows are set too high, because the predevelopment C factors, conservative for peak flow, are sometimes nonconservative for setting low-flow target values. Volumes are set too low, because the actual runoff hydrograph is not truncated like a simple triangle. Ten-year control does nothing for two-year downstream flooding problems.

People living along streams downstream from new development were still getting flooded and saying it was not like this in the 1960s. And they were right. So, we needed a new and better paradigm to solve this problem. It was thought that simple on-site controls were the problem and that coordinated regional approaches made much more sense. It has now become too hard for the normal civil engineer to do higher-level stormwater design so … bring in the stormwater master planning experts (Figure 1-3).

Paradigm #5 — Well, Just Do Not Cause Flooding

The first mainframe hydraulics and hydrology models were converted for PC use in the mid- to late 1970s and became commonly available in the mid-1980s. MITCAT led the way. Mainframe programs, ported to PCs, became easier to use. Today, most engineers have more computing power under the palm of their hand than entire universities had in the 1960s.

With all of this new computing power, the literature in the 1970s began to reflect a new concept: stormwater master planning. The idea was that one could construct a hydrology model (how much water? how often?) and hydraulics model (how fast and high does the water from the hydrology model go?) of a watershed and then do "what if" analysis until a perfect solution to flooding problems (current problems and those only imagined) is found. The "regional treatment" mantra was everywhere mouthed by knowing experts.

By 1985, hundreds of master plans had been developed, and some were actually built the way they were planned. The problem was that the institutional side of master

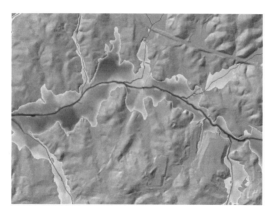

FIGURE 1-4
Technology making master plans effective. (Photo Courtesy of AMEC Earth & Environmental, Inc.)

planning and plan implementation was beyond the authority of most engineers doing the planning. For some, getting involved in the political and budgetary side of things was a new and not pleasant experience. The "hydro-illogical cycle" was born. In this cycle, local governments proceed from flooding to panic, to planning, and to procrastination until the next flood.

A lot of things conspire to cause chronic flooding. Only five of them are physical: more water than before, a clogged or broken system, a system designed too small in the first place, homes located in the wrong place, and an "act of God." All of the rest of the reasons are institutional in nature. So it takes a lot more than a model and a plan to make a stormwater master plan a success. Included for success must be consensus building, stormwater finance (stormwater utilities sprang up about now) and public relations.

With effective master plans, more and more flooding problems were solved (or better yet, avoided) on a neighborhood, if not watershed, basis (Figure 1-4). Many thought that was the end of real change and innovation in stormwater management.

We couldn't have been more wrong.

The New Breed of Stormwater Paradigms

A storwater paradigm pattern was emerging. First the literature begins to discuss how to solve current problems. New paradigms emerge. Each stormwater paradigm solves the immediate problem of the past paradigm and creates a more insidious problem of its own. Or, knowledge and technology create a real or perceived need for higher, more demanding, levels of stormwater management and regulation. It is not that the past paradigms disappear — they are thriving in various parts of the country.

About this time, the first Section 208 area-wide studies were just being completed. In them, there was learned discussion about CSO problems, detailed evaluation of wastewater point source controls and technology, expert industrial pollution source and assessment strategies, and vague concern about stormwater runoff.

No one important was sure what to do about this mysterious source of pollution. Actually, no one was even sure just how much of a real problem stormwater runoff was. Data were scarce, and reliable data were nonexistent. But, everyone suspected that urban streams were trashed, and runoff was an unindicted co-conspirator. Congress and the EPA focused their attention on stormwater pollution, and the age of stormwater *quality* was born.

FIGURE 1-5
Micro-pool extended detention. (Photo Courtesy of Center for Watershed Protection.)

Paradigm #6 — Do Not Pollute

The Phase I stormwater regulations grew out of the 1987 Water Quality Act, the 1986 publication of the National Urban Runoff Program (NURP) final report, and 305(b) reports.

Stormwater quality regulations hit some local governments from the blind side. Many cities and counties spent large sums of money on untested structural Best Management Practices and unworkable regulatory controls to meet ill-defined standards of Maximum Extent Practicable, enforced by understaffed regulators with no budget increase despite greatly increased workloads.

After a lot of thrashing about, some insightful articles about BMPs being pollutant deathtraps, horror stories about 80% failure rates for infiltration basins, and untargeted million-dollar stormwater quality models, some fairly effective programs emerged and are still emerging. The stage was set for the drafting of the Phase II regulations.

The urban infrastructure also went through a metamorphosis. Remember the first paradigm? Ditches started looking good again. Only now, they were designed and called "grassy swales" or "wetland swales." Riparian corridors had a makeover too, as engineered "buffers with flow spreaders" or "riparian filter strips." Complex detention ponds began to appear, along with a plethora of ultra-urban and commercial devices looking like mini wastewater treatment plants (Figure 1-5).

Designs became much more complex, as a raft of rules-of-thumb emerged to capture and treat a certain stormwater quality volume. We were not sure what the right volume was, or even if a volume approach was right, but we hoped to change the flow regime sufficiently to positively impact the stream water quality and overall stream health.

Few programs had anything remotely resembling a comprehensive monitoring program to measure success. The "Maximum Extent Practicable" standard for the Phase I regulations left a lot of room for innovation and mistakes. Many municipalities forged ahead with programs targeting the minimum controls and hoped that the resources they were expending were making a difference, even if it was not easily measured. Maybe there was a better way of measurement? Engineers would have been better off not asking those questions, because they led to the next paradigm.

Paradigm #7 — It Is The Ecology

During the 1990s, a new concept of measuring stream health and targeting stormwater program efforts began to emerge: biocriteria. If the macroinvertebrate community is thriving (Figure 1-6), and fish are diverse and healthy, maybe the streamwater quality is fine.

FIGURE 1-6
Mayfly Larva. (Photo Courtesy of Kane County, Illinois.)

So, under this paradigm, our stream restoration or conservation target is some measure of biological health, and our stormwater program is focused on how to attain and maintain this health.

This concept led to the development of various types of biocriteria. Rapid Bioassessment Protocols emerged and were modified for various states or regions. Groups one, two, and three taxa were popularly defined, scoring sheets were developed, and willing volunteers were trained.

The problem comes in when we try to come to agreement on what healthy is, how we measure it, and what the cause of ill health is. Is it natural seasonal variation? Has there been a die-off due to disease, not pollution? Is the sampling repeatable? Is this sluggish stream type naturally low in Group One Taxa, or is it impaired? Is the reference stream really applicable to this situation? Will that land use change really impact this rating, and, if so, by how much? There may be twenty-five reasons for a certain stream rating, with only half of them pertinent to stormwater runoff.

So, the understanding grew: If we can solve the biology problem, we have solved the problem, right? And, to solve that problem, we need to be able to change the way the riparian corridor is managed, the way the floodplain is regulated, the way the watershed land is used, the way wastewater is permitted, and the way runoff is developed in the headwaters. And, we come to an interesting conclusion.

Paradigm #8 — Water Is Water Is Watershed

There seems to be a not-so-secret conspiracy of understanding that emerged during the 1990s concerning the importance of watersheds and working at a watershed level. It began as the mantra of the environmental left and is now being adopted by everyone.

There was an organizational convergence. Whole federal agencies have reorganized to be watershed organizations. Thousands of nonprofit watershed organizations have grown up across the country. Local governments are thinking in terms of holistic watershed planning. (See Table 1-1 for more information on the informal understanding among various entities.)

There is also a great convergence occurring on the regulatory side. EPA is organizing many of its regulatory programs on a watershed basis. NPDES permitting is being synchronized within a watershed. Pollutant trading within a watershed emerged. If NPDES is a large wave, TMDL is a tsunami. This regulatory program, along with the Endangered Species Act, has the potential to trump all other water efforts within the next ten years. And, it will be done on a watershed basis.

TABLE 1-1

Responsibilities under the Watershed Approach

Federal Government	Local Government
Leadership	Land use control
Funding	Implementation
Technology	Local funding
Legislation	Partnering
	Participating
State Government	*Private Role*
Coordination	Initiating
Implementation	Influencing
Funding	Private funding
Regulating	Reporting problems
Organizing	Selling and doing

We are coming to the realization, even organizationally, that water is water. Wastewater, groundwater, drinking water, stormwater, rainfall, seawater, lakes, and atmospheric water are all part of one "circle of life." Our artificial organizational and political boundaries make no sense from this perspective. This has led to the formation of a large number of regional water planning and regulatory bodies. Of course, Florida and several other states have been organized this way for years (lands on other continents even longer). But even they are feeling a fresh wind of change coming.

Technology, ever the enabler of paradigm shifts, has taken quantum leaps in its ability to collect, deliver, and display massive amounts of data. GPS and GIS have shrunk in cost and burgeoned in capability. Airborne and satellite imagery, laser capabilities, and automated mapping have made it possible to collect and manage large blocks of data with relative ease. Computing power is rarely a limitation. Graphical capabilities have brought the ability to make a compelling visual argument in a public meeting within the reach of even modestly resourced groups. Models have slick input and output modules, allowing users to develop attractive tables and charts and real-time automated demonstrations. The Internet has become the communication medium of choice for many municipalities, groups, and individuals, serving as an immediate forum for ideas and thoughts to move rapidly across interest groups.

We are also coming to the understanding that everything that happens in the watershed impacts the stream and riparian corridor. So, under this paradigm, there are attempts to change the way things happen in watersheds: development standards, transportation, car washes, gardening, oil changing, trash disposal, landscaping, use of private lands, open space access, floodplain uses, zoning potential, subdivision ordinances, school curricula, and on and on.

Our thought has been, "Surely, when done right, this will solve all the stormwater problems." Well, the first public meeting you hold should be enough to change that thought forever. The world does not revolve around stormwater. It does not even revolve around water. We remember a carefully prepared-for public meeting, where the only person who showed up was someone who came to the room by mistake. When we enter the watershed, we enter a competition for the attention and resources of normal people, and we enter a world where our priorities may never float to the top.

Also, when you walk around a watershed, you inadvertently step on many kinds of toes. Developer toes, landowner toes, transportation planner toes, zoning board toes, sportsmen toes, council toes, etc. That is the bad news. The good news is that all share some things in common with you. Usually, they share a desire to live in a nice, attractive place, to make wise financial decisions about their resources, to leave a legacy for their children, and a need to meet regulatory requirements in the most cost-effective way (some needs are more voluntary than others).

Some of the drawbacks of the watershed approach that we have observed are that: it often bites off too big a land chunk for citizens to relate to; it involves watersheds, but most citizens do not fully understand what constitutes a watershed; it does not address the problems at the design and development process stage, where many occur in the first place; and it involves groups that do not necessarily have authority to accomplish the things needed to be done. So, local governments took a look at the watershed approach, at how local governments operate, and the urban reality of tight development sites, and at nature's way of dealing with rainfall, and developed a new paradigm — the last one we will talk about.

Paradigm #9 — Green And Bear It

There is a multifaceted green revolution going on in stormwater. The basic issues are these. Urban sprawl creates an environment that is ultimately not good for man or beast. This type of development is *not* inevitable. Through a combination of structural, nonstructural, and institutional practices, functional, environmentally friendly, sustainable, and beautiful living environments can be created. Surface and stormwater management play a large role in this movement.

There have been many popularly named approaches that address some or all of this, including: Low Impact Development (LID), Green Infrastructure, Better Site Design, Conservation Development, Zero Discharge, Sustainable Development, Multiobjective Floodplain Management, and so on. Green And Bear It creates a network of interconnected linear and patch areas that seeks to sustain life and enrich the quality of life by:

- Mimicking acceptable hydrology
- Enhancing natural diversity and beauty
- Balancing ecological preservation and conservation with economic growth and development
- Building systems that are sustainable and maintainable
- Working at a small, integrated scale with accumulated results
- Dealing with stormwater as a valuable resource

Some impressive successes have been trumpeted in conferences and government-funded publications. There has also been some strong grumbling about the fact that it is not as easy to move to, and maintain, this paradigm, as some would have it. Who will inspect hundreds, if not thousands, of small-dispersed devices? Who will educate all the homeowners? What happens if they fail (Figure 1-7)?

An interesting array of followers supports many of these concepts. Some in the development community are strong supporters because they reduce construction cost through reduction in infrastructure requirements, while retaining the per-lot value and selling price.

In any case, this paradigm returns to the small-scale, distributed, and accumulative approach that can accomplish lots of real things on the ground, because it is typically enforced by local governments (sometimes through state mandate). It can also be a failure if it is not strongly supported by local governments (politicians included).

From Paradigm To Paradigm

You may be thinking that you will never get past Paradigm Six (or Two). Regulatory requirements will prove you wrong. Most of the early paradigms were driven by the

FIGURE 1-7
A LID "rain garden." (Photo Courtesy of the Center for Watershed Protection.)

demands of the citizens to solve pressing problems. Many of the later ones are driven by regulations and local response to them.

It is much easier to know what the next paradigm is than to move into the next paradigm. As you progress through the paradigms, you will find that the goals and objectives that are the focus of the development begin to change and become more complex. What it takes to produce a design changes. There are more drawings, more studies, and more hearings. There are more layers of government to satisfy, more regulations to meet, and more stakeholders to convince. The time it takes can also grow greatly, especially if the local government is strong on regulations but weak on accelerated plans review.

Who it takes to produce a design also changes. Planners will someday rule the world. Until then, design teams will grow into multidisciplined teams, including biologists, landscape architects, ecologists, botanists, streambank restoration specialists, risk assessors, soil scientists, anthropologists, archeologists, public relations experts, graphics and database technicians, GIS experts, meeting facilitators, and even a stormwater engineer or two.

How much the public is involved has changed radically. We used to try to do "stealth projects." Now, we all have become experts in making complex things simple and at understanding differing views and saying, "I hear you say...." Bringing nontechnical people to technical consensus is not easy.

It is easy to "fail flawlessly" in moving to the newer paradigms. The roadway is littered with the roadkill of those who have demonstrated failure. They are characterized by one or more failings:

- Ignore funding needs
- Make regulations and programs contradictory and complex
- Make it impractical to apply in the real world
- Have insufficient state technical staff
- Make most provisions voluntary
- Skimp on public education and involvement
- Ignore grassroots groups and developers
- Drive compliance with lawsuits

- Review and enforce superficially
- Be adversarial

Somewhere in the United States, someone has already faced the things you are facing and has not only lived to tell about it, but to tell a story of success. There are many Web sites and list servers that give great insight, example ordinances, online BMP manuals, quick replies to e-mailed questions, design examples, and about anything you could imagine. This book also contains much helpful information. So charge on fearlessly.

1.2 Understanding Stormwater Management Problems and Solutions

Understanding what paradigm you and your municipality are going through is important in knowing how to proceed and what hurdles others have crossed ahead of you. It is also vital to understand the linkage between physical problems and deeper institutional root causes of those problems in your municipality. Many municipalities have not understood this linkage and, as a result, wrestle continuously with the same problems, never arriving at permanent solutions. Figure 1-8 illustrates the dynamics of this technical–institutional relationship using a "Five Whys" methodology that comes out of a Total Quality Management (TQM) consideration applied here to stormwater management (Reese, 1991, 1993; Senge, 1990).

Typically, a stormwater administrator, public works engineer, or political leader gets a drainage complaint call: "I have a flooding problem and I want you to fix it." This is Level 1: the complaint. The complaint could just as easily have been an erosion or pollution complaint. Following the same methodology would eventually lead to the same conclusions.

When the question is asked, "Why is there a flooding problem?" (Level 2), there is usually one of (or a combination of) five reasons: (1) obstructed or damaged structures, (2) high-risk residence location, (3) undersized structures, (4) increased flow due to the impacts of urban development, or (5) a flood in excess of design. The typical solution is to go and fix the problem or, more commonly, to tell the resident that the problem is not the municipality's responsibility.

However, if the Level 3 question is asked, "Why did this flooding problem (and ones like it) occur in the first place?" a matching set of four, more foundational, reasons is uncovered:

1. Structures are obstructed or damaged, because they are not inspected and maintained. Municipalities typically maintain very little of the drainage system. Most maintenance that is done is in response to complaints and performed within the street right-of-way only. Much of it is done only to protect streets or public property. The 50 to 80% of the drainage system closest to private houses and other structures is rarely or never inspected or maintained. This is true despite the fact that much of the water carried in these drainage systems is derived from public streets and is, in some sense, therefore, "public water." Over years of neglect, pipes and channels inevitably begin to fill with debris and sediment, structures begin to weather or are damaged, and erosion eats away at culvert headwalls and tail sections. Drainage systems work flawlessly when it is not raining. Finally, the system is tested and overwhelmed by a storm with an

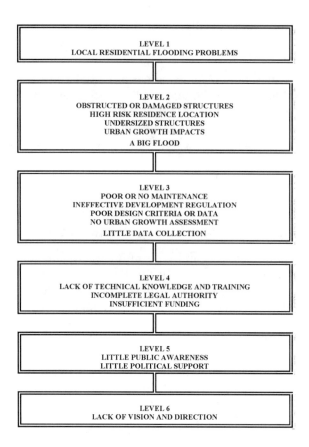

FIGURE 1-8
The Five Whys.

intensity often less than the system was designed to carry. Homes are flooded, roads are overtopped, damage is incurred, and complaints flood in.

2. Homes are located too close to streams, because nobody properly regulates the location. Because of the demands of the Federal Flood Insurance Program, most municipalities control the location and elevation of new construction within regulated floodplains. However, the vast majority of complaints are received from residents remote from regulated floodplains, in locations where there is no such control. Municipalities, which would not consider allowing development within the 100-year floodplain below the mandatory flood protection elevation, daily approve plans for developments in which a number of homes, unwittingly located within unregulated floodplains, would be inundated by smaller, more localized 100-year floods.

3. Structures are undersized because of poor or inappropriate methodology and incomplete data. Most municipalities allow drainage structures to be sized using the Rational Method. While this is not wrong, per se, designers and plans reviewers rarely understand the limits of this method. In cases where backwater effects predominate or under other special circumstances, such methods may give nonconservative results. Additionally, many municipalities have little data on actual rainfall values, inlet capacities, tidal influences, or actual expected future maintenance-related condition of structures. Without this information, designers may unknowingly produce inferior designs.

4. Upstream development floods downstream development, because it is not accounted for in the design of the downstream structures, or it is not accounted for far enough downstream in the drainage system. Few municipalities require designers to account for their own flow-related impacts or for the flow increases from expected development located upstream. Higher and faster peaks, greater flow volumes, increased velocities, warmer and dirtier water, and lower base flow are all the result of urban development. Those municipalities that account for impacts with a detention ordinance or policy rarely assess the impacts beyond the site boundaries. Therefore, the mitigative effect of the detention basin is not analyzed very far downstream; and the accumulative consequences of development, even with detention, result in growing systemic flooding problems.

The matter of "a big flood" is obviously caused by a big rain. The issue here is not that big rains come along but how the local government plans for, measures, and reacts to a big rain.

In urban hydrology, the rainfall depths contained in a typical intensity–duration–frequency (IDF) curve or table are *point* values. We would anticipate a typical ten-year design pipe or culvert to be overwhelmed by a storm event in excess of design once every ten years, on average. In most of the country, this is a thunderstorm with a radius of intense rainfall that may be less than a city block. For a local government area of many square miles, we would expect many storms in excess of the ten-year storm every year, so that every culvert in the area would face the same statistical chance of flooding. Thus, the "big flood" will happen several times each year to different parts of the area, so that the whole area experiences flooding according to the statistics reflected in the IDF curve.

First, the municipality should realize that every design contains an element of risk. How should a local government plan for and react to this natural situation? When the inevitable big storm hits, is the result minor nuisance flooding or is it homes and arterials under water? How does a municipality handle flows that overtop the system? What level of risk should the municipality tolerate in its design criteria? Should it be the same for all segments of the system?

Second, does the municipality have a plan for flood warning and evacuation? It is one thing for water to flow into a home or business. But, it is another for a street to be flooded with dangerous flows and the municipality not to be in a position to know which roads are impassable and to have taken steps to prevent motorists from inadvertently driving into the floodwaters.

And, finally, how can the municipality prove that the rainfall was in excess of design? Many of the flooding complaints received by local governments are the result of excess flows and not poor design, lack of maintenance, or inadequate planning. Pressure is then put on the leadership to fix a system that is not in need of fixing. With an adequate set of rain gauges, a municipality may be able to predict flooding and to plot rainfall hyetographs in near-real time. With this information, it can be shown that some flooding has nothing to do with a system problem but with the fact that, for example, a 50-year storm was experienced by a ten-year system.

If again the "why" question (Level 4) is asked, three basic causative factors emerge:

1. Municipalities do not require appropriate levels of technical analysis, because they are not sure what to require and how to implement these requirements. In spite of the wealth of computer software for drainage system analysis and the ability to remotely collect rainfall and runoff information, most municipalities have not had the time or the knowledge to investigate and invest in such solu-

tions. Junior or mid-level engineers without the authority or experience to make such changes often staff programs on a day-to-day basis. Their superiors have multiple other pressing duties and responsibilities and, without prompting and education, do not see stormwater as having the same importance or the same clear solutions as roads or solid waste. Thus, overall master planning for stormwater management is seldom done.

2. Municipalities do not impose certain flood mitigation measures or development controls or maintain off the public right-of-way, because there is no legal authority to do so, and there is little impetus to establish such authority. To extend control of development beyond federal mandates or to extend maintenance beyond the bare minimums requires gaining the support of political leaders, key staff members, and stakeholders. It is often difficult to stimulate such desires when so few of these individuals have anything clear to gain by doing so. Environmental Protection Agency (EPA) mandates, local citizen groups, and a big flood event are often the necessary catalysts to action.

3. Even if these last two factors were solved, the bottom line is that there is no stable, adequate, and equitable funding source for stormwater management. Stormwater usually cannot compete effectively with such things as solid waste and street repair for general tax-based funds. Therefore, a shift toward dedicated stormwater funding is occurring throughout the country. This can take the form of such things as sales taxes, earmarked tax revenues, and user fee systems (stormwater utilities).

Each of these three foundational issues (master planning, legal authority, and financing) will be covered in later chapters.

The more basic factors emerge with the next "why" question (Level 5). Even when key stormwater staff understands the problems, they must ask: Who else is aware that flooding, erosion, or pollution problems exist? Who supports a growth in stormwater management? Who must support it for a successful program to be established?

The public is usually little aware of flooding problems, and municipal staff has little long-term political support to solve such problems. If actions are not taken and decisions made within a month or two after a flood event, support quickly dries up and memories fade. Other pressing demands thrust aside flooding, erosion, and surface water pollution problems. However, the problem remains, largely invisible, until the next time a large storm moves through the area. Building and maintaining consensus and support for the stormwater program is necessary for its establishment and survival. The subject of public and political awareness is addressed in Chapter 3.

The basic reason for lack of success in stormwater management is the same as for lack of success in anything else: lack of vision or direction (Level 6). In almost every case where a successful program has been developed, one or several individuals had or developed a vision for what their streams and creeks should and could be. A Southwestern city wanted its citizens to be safe from major floods and developed a vision for attaining that goal. A Northwestern suburb wanted its streams to remain pristine spawning grounds for salmon and places for recreation. A Midwestern city wanted its miles of underground pipes to function as they were designed and to remove the thousands of traffic barricades, often set in concrete, throughout the city. An Eastern city wanted its streams to become urban greenway trails, attractive to its citizens. In each case, a dedicated few charted the course and set the pace for the many. They built consensus, informed the public, developed effective technology, established policies and procedures, and implemented well thought-out plans.

Had the original complaint been an erosion or sedimentation problem, an analogous set of four factors causing erosion and sedimentation can be given: the banks and bed have been attacked by the flow so as to reduce their ability to resist erosive forces or are not maintained; more sediment from upstream reaches overwhelms the stream's ability to carry it away, and sediment deposition occurs; or less sediment is present in the inflowing stream than the natural situation, which may cause the water to attempt to pick up additional sediment (up to its carrying capacity) from the bed and banks; the basic slope or channel configuration has been constructed or modified in such a way as to place the channel in an unnatural configuration and, without appropriate bank protection (termed "out of regime"), natural forces tend to bring the channel back in regime; or more water creates more and longer shear stresses on the sides and bed of the channel, causing more erosion.

Had the original complaint been pollution, a set of four other factors can be derived: pollution control devices are no longer operating properly, illegal dumping or illicit connections are present, pollution control devices were not placed or designed properly, or more uncontrolled development-related pollution is entering the system.

In each of these other two cases, the Five Whys could be asked. In each case, the questioning will merge at Level 3 in a manner similar to the flooding complaint.

Notice that the first levels of assessment contain primarily physical and technical problems for which structural technical solutions are appropriate. Water is impacted by some physical means. However, when the later levels are considered, the solutions are institutional, programmatic, and nonstructural in nature. People are impacted by administrative means. These foundational problems allow or generate the more visible physical problems. If the root institutional problems are not eventually solved, there will be a continual need to respond to an overwhelming number of flooding, erosion, and pollution complaints. Successful municipal stormwater management programs account for and deal with the technical and institutional aspects in a comprehensive and coordinated manner.

The Five Whys become a convenient stormwater program assessment tool that can be used in a variety of ways in identifying surface and root problems. For example, in Figure 1-9, the Five Whys are modified to assess causes of flooding problems for a certain community. In this case, only three levels of the tool are shown, indicating the major immediate causes and secondary causes of flooding in this community. This analysis eventually led to the creation of a comprehensive stormwater program designed to target the problems and deal with them, over time, in a comprehensive "building block" fashion.

1.3 The Organization of the Rest of the Book

The remainder of the book is organized into three sections based on the use of the Five Whys tool above. The first section deals with the institutional aspects of stormwater management: the foundations. The second section of the book deals with technical and design aspects. The third section deals with practical implementation aspects of stormwater management facilities and programs.

It is difficult to combine these three sections into one text without doing injustice to one or the other or without developing a book that is too lengthy for publication. Every effort has been made to ensure that this is a stand-alone text. Where detailed treatment is beyond

Physical Problem	Immediate Causes	Secondary Causes
Flooding in minor system and along drainage canals	System inadequately designed, poor criteria and construction	Poor or no design criteria in the past and incomplete criteria now
		Inadequate current local design criteria and/or knowledge
		Inadequate inspection and enforcement during construction
	System failed or blocked	Lack of maintenance of the major canals
		Beaver dams block ditches or canals
		Agricultural practices block ditches
		Lack of easements or easements too narrow or confined
		Lack of effective or fully enforced erosion control practices
	More water into system	Little thought given to downstream flooding impacts or "fit" of new construction
		Past lack of detention requirement
		No consideration of ultimate drainage sizes required for future systems
		Potentially inadequate current detection criteria due to lack of volume considerations
		Failing detention systems
		No knowledge of infrastructure location, condition or size
		Lack of enforcement of regulations or lack of taking advantage of regulatory opportunities to require consideration of impacts by OCRM
		No watershed master plans or models
	Poor home and structure location	Little knowledge of floodplains outside FEMA locations
		Lack of consistent enforcement of regulatory controls
		Lack of regulatory controls
		Lack of regulatory controls outside FEMA floodplains
		Outdated mapping

FIGURE 1-9
The "Five Whys" tool.

the scope of this text, appropriate references are given to lead the interested reader deeper and wider. The authors would appreciate any suggestions for improvements to this text, which can be sent to us through our publisher.

References

Reese, A.J., Successful Municipal Storm Water Management: Key Elements, *Proceedings of the 15th Annual Conference of the Association of Floodplain Managers*, June 10–14, Denver, Colorado, 1991, pp. 202–205.
Reese, A.J., Understanding stormwater problems, *APWA Reporter*, Jan. 1993, p. 7.
Reese, A.J., Stormwater Paradigms, *Stormwater Magazine*, July–August 2001.
Senge, P., *The Fifth Discipline: The Art and Practice of the Learning Organization*, Currency Doubleday, 1990.

2
Stormwater Management Programs

2.1 The Stormwater Management Program

The term "stormwater management" can take on as many different meanings as there are stormwater management programs. Some are narrow, single-purpose programs, while others are broad, basin-, or political-entity-wide. As we saw in the last chapter, all are in transition. There are also a number of general and specific ways to group the various functions and components of stormwater management for consideration. A stormwater management program may have many or few of these components, and each of these components may be well or poorly developed. We will look at several ways to understand and specify stormwater management prior to providing a method for beginning the assessment process for stormwater management. When combined with the Five Whys approach of the previous chapter, these simple tools become powerful ways to assess stormwater programs.

The basic underlying purpose of stormwater management is to *keep people from the water, to keep the water from people, and to protect or enhance the environment while doing so.* Keeping people from the water involves a wide-ranging number of nonstructural damage mitigation measures, including zoning, floodplain management, home acquisition, flood proofing, etc. Keeping the water from people involves direct interference with the water through the use of flow amount or flow elevation reduction measures, such as retention or detention storage facilities or channel enlargement.

Maintenance or enhancement of water quality is taking on added importance through the mandates of the 1987 Water Quality Act and its implementing regulations. Aesthetics and multiobjective planning of water resources are also growing rapidly, as expressed through stream restoration, greenway construction, and integration of flood control with parks or multiuse spaces. The water quality component is a relatively recent addition to the stormwater manager's job description and involves a wide array of often poorly understood structural and nonstructural approaches. By "poorly understood," we do not mean to imply that the practitioner does not understand how the programs or controls work, but that the ultimate, specific, and quantifiable goals and objectives of the program and the resultant impact on those goals and objectives of each program element are often not well understood or even measurable.

2.2 Functional Perspective of Stormwater Management

One way of looking at stormwater management is to divide it into specific general categories, which are either defined duties or physical products. These components are long-range planning guidelines and goals; day-to-day management tools, responsibilities, and procedures; and the legal, technical, organizational, and financial underpinnings required to implement the stormwater management program. Day-to-day management carries out the long-range goals and objectives using the tools and capabilities afforded by the underpinnings.

Long-Range Aspects

Long-range stormwater management aspects can be thought of as divided into two areas: program-wide and basin-specific. Program-wide stormwater management defines just what it is that a municipality wants to accomplish through its stormwater management program; what are its program priorities and objectives. A municipality must seek to establish answers to questions about its desire and authority to promote public safety and well being, protect the environment, encourage commerce, and control development. These basic objectives find definition in general and specific policy statements that may address such things as drainage levels-of-service, service areas, standards, construction requirements, erosion control, water quality goals, floodplain control, and zoning restrictions.

Typical basic policy statements are designed to accomplish the following seven foundational goals:

- Protect life and health
- Minimize property losses
- Enhance floodplain use
- Ensure a functional drainage system
- Protect and enhance the environment
- Encourage aesthetics
- Guide development

A stormwater ordinance, a comprehensive plan, and policy documents are the usual ways used to communicate these long-range goals and policies and to translate them into specific activities and programs.

The primary vehicle by which long-range, drainage-basin-specific, stormwater management is accomplished is the stormwater master plan. Master planning refers to a wide range of types of comprehensive plans that mitigate known flooding, erosion, and pollution problems; plan to avoid future problems and seek to take advantage of opportunities to enhance the environment; improve flood control capabilities; and seek multiobjective use goals.

The stormwater master planning process moves from identification of problems, needs, and opportunities through preliminary solutions and constraints, to final recommendations and actions. Much master planning has been driven almost solely by existing acute flooding problems. However, communities have realized that it is ten times cheaper to avoid problems than to attempt to solve them, and have taken steps to provide preventive measures (such as zoning or preplanned regional detention facilities) and remedial measures (such as channel improvement). Master planning is discussed in Chapter 14.

Day-to-Day Stormwater Management

The day-to-day aspects of stormwater management, supported by the legal, technical, and financial underpinnings, carry the stormwater management program to accomplishment of long-term goals, defined by the stormwater master plans and goal and policy statements. Day-to-day management tools and procedures can include a number of items, and the following are some examples:

- Calibrated hydrologic and hydraulic models are used to assess the impact of planned development or capital improvement structures on flood flows and water surface elevations for existing and ultimate development conditions. They are kept up to date as development occurs, or periodically. These models may be tied to GIS or other graphic systems.
- Engineering software is used to allow the reviewer to quickly check proposed development plans, to check the design of stormwater management facilities, and to determine if the design is optimal or simply adequate. Often, checklists for designs are filled in by the designer to facilitate such automated review. This software can be part of a hierarchical modeling structure.
- Stormwater-related databases can be anything from simple inventories of existing structures to a full-blown general application GIS system using overlays for topography, land use, soils, roads, utilities, etc. Procedures for maintaining the databases and decision making are also important. A computerized infrastructure management system (IMS) may be used to serve this role.
- Data collection systems are invaluable for the development and refinement of future hydrologic and hydraulic modeling efforts. A system of recording rain and crest-stage gauges is a necessity, at least in key basins, to improve the accuracy of results. Some municipalities have gone to remote collection of data and even remote control of floodgates, weirs, and other structures.
- Flood studies have normally been done as part of the National Flood Insurance Program. When kept up to date, they serve as a valuable reference to control development. Often, flood studies are extended beyond the one-square-mile limit set by the federal program to afford consistent control of development wherever flooding is likely to occur.
- A flood warning system or other hazard mitigation system has been found to be useful in areas of flash flooding and in areas where structural solutions are impractical (such as railroad underpasses).
- The implementation of a system to collect funds is a day-to-day activity, which is an integral part of most stormwater management programs. This may include databases, mailings, fee determinations, etc.
- Procedures for review, inspection, and enforcement of stormwater management requirements should be clearly defined and streamlined. A single authority is desirable where feasible.
- A computerized complaints system has proved valuable, not only for satisfying the needs of the public, but also for identifying nuisance problems, which may be leading-edge indicators of a growing problem within a basin. An equitable

system of prioritizing the complaints and dealing with them in a timely manner is also necessary. Customer service is also important, as is the "customer" mentality it fosters.

- A coordinated and automated maintenance management system ensures that the existing infrastructure is capable of functioning to its full capacity and meets the requirements of Government Accounting Standards Board (GASB) Rule 34 for local infrastructure depreciation or maintenance management. Limits of maintenance (both physical and financial) should be clearly defined to avoid making decisions under pressure. Maintenance should be preventative as well as remedial.
- Appropriate supporting literature can assist the manager in answering common questions or explaining often-used procedures simply and easily.

2.3 Overview of the Legal Aspects of Stormwater Management

For a stormwater program to be effective, it must be enforceable. Several stormwater or floodplain management programs have failed legal tests for a variety of reasons. However, most programs will stand these tests if they are not in violation of state legislation or municipal charters, equitable, fairly enforced, can be shown to be in the interest of the general public, and are well documented. State legislation may be necessary to establish local regulatory authority, to levy taxes or fees to finance such a program, or to allow creation of a special utility or control district.

There are specific programs, policies, and laws that establish the authority of a municipality to implement a stormwater management program. Some of these are as follows:

- Federal Flood Insurance program
- Federal wetlands regulation
- NPDES permitting
- State authority and legislation
- Local municipal charter
- Local ordinances
- Local regulations
- Local policies (written and unwritten)
- Checklists and informal guidance

Local control may take the form of detailed ordinances requiring the vote of municipal governing bodies for amendment or, preferably, a general ordinance, which incorporates an easily updated regulation or guidance document by reference. Municipal stormwater management ordinances are sometimes developed as a series of special ordinances, such as detention ordinance, floodplain ordinance, erosion control ordinance, floodway ordinance, etc., though a comprehensive single ordinance reduces inconsistencies and confusion and helps ensure that gaps do not exist. On the other hand, a single comprehensive ordinance can isolate the stormwater program from other related or parallel programs, limiting to some extent the ability to think and operate holistically.

Care must be taken to ensure there is a clear differentiation between actual legal requirements and mere statements of preferred policy. Many municipalities manage stormwater on the basis of "understandings" of how things will be done. It is important to write down and define these understandings. Lines of authority and division of authority and responsibility should be clearly understood and well documented. When taken to court, being consistent is just as important as being totally accurate from an engineering perspective.

Measurable, performance-oriented design criteria must be given, and goals should be stressed above technical methodology. For example, a requirement may be that a detention pond must reduce both the two-year and ten-year peak flows to preexisting conditions at the site exit and downstream at a testing point, where the newly controlled drainage area makes up 10% of the total drainage area above that testing point. Or, a pollution reduction best management practice (BMP) should reduce total suspended solids by 85%. However, within the confines of the performance criteria, innovation in design is allowed and encouraged.

Consistently applied, well-documented, and technically sound policies and procedures are vital to cities and counties administering development. This will become even more important as stormwater quality issues grow. Chapter 4 gives details on stormwater ordinances.

2.4 Overview of the Technical Aspects of Stormwater Management

The technical support component usually consists of the following:

- A stormwater management design manual or technical guide
- Computer models and some sort of master plan (master planning considered above)
- Databases or mapping of such things as an infrastructure, land use, and soils
- Mapping, GIS, CADD, maintenance management, and other automation applications
- Data collection for such things as rainfall, flow, water quality, etc.

Technical Manuals

Most municipalities have some form of a drainage or stormwater management manual, or, at least, drainage information contained in subdivision regulations, standards and specifications, or similar documents. The sections in the manual should, as a minimum, make clear reference to all technical aspects of regulatory requirements. The ordinance tells what must be done. The manual shows how.

The technical manual may be comprehensive or may make reference to commonly available design documents or methods. It should be designed to make use of local data and environment, local procedures where appropriate, and local capabilities. The actual design procedures should be clear and not cluttered by superfluous discussion. Special sections on theory, etc., may be separate from the step-by-step instruction, and design examples should abound.

No technical manual can cover all eventualities. There must always be room left for engineering innovation and the use of methods not found in the manual. However, if methods not found in the manual are used, it should be made clear in the manual and ordinance that there may be additional review-time and submittal requirements. Often, manuals are made available on the World Wide Web, as are specification drawings and ordinance requirements. Some cities are turning to the Web for management of many aspects of the development process.

Paired Software refers to the development of a hand calculation method in the manual that mimics the methodology in a user-friendly microcomputer environment. Thus, a technical manual not only stands on its own but also serves as a software user's guide helping engineers make the transition from hand methods to computers. Software should be well documented, user friendly, menu driven, interactive, and in the public domain, where available.

This concept can be extended beyond the manual to streamline master planning. "Hierarchical modeling" is a term used to extend the paired software concept into a three-level structure, where hand methods contained in the manual and supported by simple software at the review level are able to be input into complex basin models for assessment of overall impacts to a particular basin. Methods such as the U.S. Soil Conservation Service methodology for unit hydrograph generation or the Federal Highway Administration culvert procedure are conducive for this purpose. To be most efficient, this type of approach requires data submittal in a formatted form, often digitally.

Computer Models

There is a great deal of computer software that has been developed recently to cover most aspects of urban hydrology, hydraulics and stormwater quality, and design. Computer models use the computational power of computers to automate the tedious and time-consuming manual calculations. Most models also include extensive routines for data management, including input and output procedures, and possibly including graphics and statistical capabilities.

Computer modeling became an integral part of storm drainage planning and design in the mid-1970s. Several agencies undertook major software developments, and these were soon supplemented by a plethora of proprietary models, many of which were simply variants on the originals. The proliferation of personal computers in the 1990s has made it possible for virtually every engineer to use state-of-the-art analytical technology for purposes ranging from analysis of individual pipes to comprehensive stormwater management plans for entire cities.

Integration of GIS and modeling capabilities is becoming commonplace with both off-the-shelf and customized packages developed for individual municipalities. Figure 2-1 shows one such package.

In addition to the simulation of hydrologic and hydraulic processes, computer models can have other uses. They can provide a quantitative means to test alternatives and controls before implementation of expensive measures in the field. If a model has been calibrated and verified at a minimum of one site, it may be used to simulate nonmonitored conditions and to extrapolate results to similar ungauged sites. Models may be used to extend time series of flows, stages, and quality parameters beyond the duration of measurements, from which statistical performance measures may be derived. They may also be used for design optimization and real-time control.

A local staff or design engineer will typically use one or more of these pieces of software in stormwater facility design and review, according to the design objectives and available resources. However, it should be kept in mind that proper use of computer modeling packages requires a good knowledge of the operations of the software model and any

Stormwater Management Programs

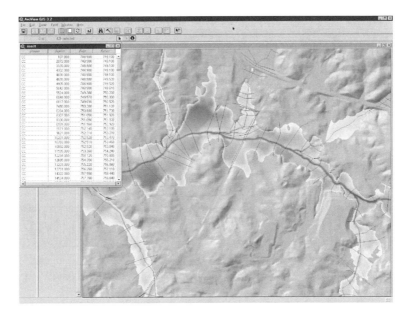

FIGURE 2-1
GIS-based computer modeling systems. (Courtesy of AMEC Earth & Environmental, Inc.)

assumptions that the model makes. The engineer should have knowledge of the hydrological, hydraulic, and water quality processes simulated, and knowledge of the algorithms employed by the model to perform the simulation.

In urban stormwater management, there are typically three types of computer models that are commonly used: hydrologic, hydraulic, and water quality models (for more information, see Chapter 14). Models can be simple, representing only a few measured or estimated input parameters, or can be very complex, involving 20 times the number of input parameters. The "right" model is the one that the user thoroughly understands, gives adequately accurate and clearly displayed answers to the key questions, minimizes time and cost, and uses readily available or collected information. Complex models used to answer simple questions are not an advantage. However, simple models that do not model key necessary physical processes are useless.

Databases and Infrastructure Inventories

There may be a large number of databases employed by the municipality in its management of stormwater, including all kinds of tracking (complaints, maintenance, easements, BMP locations, capital projects, etc.). Keeping them in an integrated, GIS-based system tied to infrastructure management capability and an open-architecture-type database structure will ensure current efficiency and future compatibility with new software and hardware. Perhaps the most important database is an inventory of stormwater structures.

The development of an inventory of the stormwater system is often the first step in developing a comprehensive stormwater management program. As with any other public infrastructure (water, wastewater, streets, etc.), having knowledge of the stormwater infrastructure is important for its proper and efficient management.

Relevant information includes location and classification of storm drains; drainage networks; structural stormwater control facilities; streams, ponds, and wetlands; industrial discharges and combined sewer outfalls; watershed boundaries; floodplains; existing and

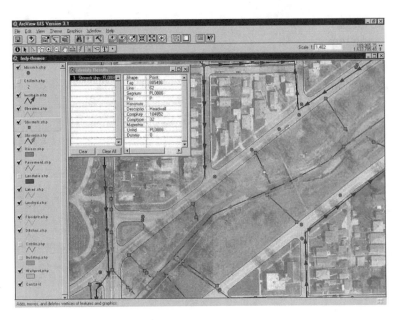

FIGURE 2-2
GIS-based inventory systems. (Courtesy of AMEC Earth & Environmental, Inc.)

proposed land use and zoning; and known water quality problem areas. This information can be collected and stored on paper maps or, ideally, in an integrated municipal GIS system. Figure 2-2 shows a typical GIS-based inventory system.

Perhaps it is easiest to understand the advantages of an inventory by stating what can be accomplished when a local city or county has effective inventory information. The uses of stormwater infrastructure inventory information include:

- *Complaint response* — The ability to quickly and effectively respond to a customer complaint by having online current information linked to addresses, past history of the address, a site map when arriving on the scene, and other information.
- *Maintenance management* — This includes a wide array of functions.
- *Remedial construction* — Quick turnaround construction on minor systems or minor repairs to larger systems that require little design time and would be handled with unit-cost, open-ended contracts.
- *Capital construction* — Programmed construction of larger items handled with pay-as-you-go or bonded capital funds.
- *Inventory control* — Handling of drainage structures in storage locations and warehouses.
- *Master planning* — System-wide analysis and planning of capital construction for problem areas and for areas facing new development.
- *Financial tracking* — Tracking of costs, efficiencies, crediting information, assessments, depreciation, system status for GASB 34 reporting, etc.
- *Materials testing* — Tracking of age and relative life-cycle costs of different materials.
- *Legal support information* — Tracking of easements, ownership, complaints, and other legal information.

- *Regulatory control* — Including NPDES monitoring reports, easements, permitting, negotiating requirements for new developments, flood insurance program, floodplain management, erosion control, permits issued, and other regulatory issues.

For each of these applications, information from an inventory serves as the basis for the program function. For example, knowing pipe sizes and general conditions allows for long-term budgeting and capital planning. Without this information, the city or county is left to simply respond on a reactive, case-by-case basis to needs and complaints as they occur.

Organizing an inventory can become a complex undertaking. Considerations should include types of structures inventoried, type of information needed, and program and purpose for collecting the information. For example, if the inventory information is to be used for detailed master planning of the whole system, then an accurate "Z" component is important, and the resources should be allocated to collect it. But if this type of analysis is not anticipated or will occur in only a few locations, the additional collection effort is probably not worth the effort.

The following are stormwater infrastructure components that can be included in a system-wide inventory:

- Streams and rivers
- Ditches
- Pipes
- Culverts
- Manholes
- Outfalls
- Inlets
- Bank or stream protection
- Stream enhancement
- Greenways, corridors
- Junctions
- Stormwater controls
- Detention ponds
- Dams
- Other structures
- Easements
- Floodplains
- Floodways
- Adjacent structures
- Waters of the state
- Regulatory outfalls
- Other

The type of information collected about stormwater infrastructure falls into several categories including:

- Structure type
- Size

TABLE 2-1

Stormwater Inventory Infrastructure Information and Uses

	Basic Information					Geometry				Condition						Administration/Management Information				
	Element Number	Element Type	Location	Elevation	Connectivity	Size	Geometry Information	Digital Picture	Material Type	Age	Structural Condition	Maintenance Condition	Environmental Condition	Upstream Land Use	Adjacent Flooding Potential	Ownership	Complaints	Utility Credits	Easement	Past Activity
Initial complaint response	X	X	X						X		X	X			X	X	X		X	X
Maintenance management	X	X	X		X	X	X	X	X	X	X	X	X			X	X		X	X
Remedial construction	X	X	X	X		X	X	X	X	X	X	X			X	X	X	X		X	X
Capital construction	X	X	X	X		X	X	X	X	X	X	X		X					X	X
Inventory control	X	X				X			X	X										
Master planning	X	X	X	X	X	X	X	X	X		X	X	X	X	X	X			X	X
Financial tracking	X		X													X	X	X	X	X
Material testing	X	X	X			X		X			X	X	X	X		X	X			X
Legal support information	X	X	X											X		X	X	X	X	X
Regulatory control	X	X	X			X		X				X	X	X	X	X			X	X

- Material type
- Maintenance condition
- Structural condition
- Elevation
- Connectivity
- Age
- Complaints
- Utility credits
- Geometry information
- Location

Integrating the information collected for each component with the uses of the information is the next step in an infrastructure inventory. Table 2-1 depicts an example chart that lists functional uses of inventory information and the types of information necessary to support those uses.

The following steps should be considered before undertaking a stormwater system inventory:

1. *Determine who will use the inventory information and what specific types of information and accuracy are required.* For example, if the information on street location simply needs a general location of streets, then a street centerline file would suffice. But if the actual edges of pavement are required, a much greater level of complexity is involved, including aerial photography and ground surveys. It is critically important to bring all stakeholders into the discussion of data needs, both for

FIGURE 2-3
Handheld data collector. (Courtesy of City of Charlotte, North Carolina.)

    cost sharing and to make sure the data are sufficiently accurate for the most stringent use.

2. *Determine the types of data and the methods of data collection.* It is important to insure accuracy in the collection, storage, and maintenance of the information. In many instances, new technology is available for some data and information. Data may already be available from a government agency (such as land use information from other aerial flights or satellite imagery). There is always a trade-off between the use of new technology and the opportunity for it to fail or for unknown errors to be introduced due to lack of familiarity with the technology. Inefficient use of GPS (Global Positioning Systems) has been an example of this, where some inventories have ended up being more expensive than conventional surveying when the extra time needed for training and to restore lost satellite lock is factored in. Figures 2-3, 2-4, and 2-5 show several data collection methods using a handheld data collector, GPS, and a PDA (Personal Digital Assistant) approach.

3. *Determine the technology and organizational responsibilities that will be used to store and maintain the data.* What types of hardware and software are necessary to

FIGURE 2-4
GPS data acquisition. (Courtesy of AMEC Earth & Environmental, Inc.)

FIGURE 2-5
PDA data collector. (Courtesy of Hansen Systems, Inc.)

collect and manipulate the information? Who will have access to the information and for what purposes? How will it be maintained and by whom?

4. *Collect the data and information.* Quality assurance is paramount. A single simple error repeated consistently can render the whole data effort fruitless. Checks should be made for random errors, systematic errors, blunders, and system assumption errors.

   a. *Random errors* occur due to human error and the inability of equipment to accurately read information. They tend to vary around the correct value both high and low. These can be taken into account in planning for the use of the data and can be minimized through repeated readings.

   b. *Systematic errors* are consistent ways of doing something, but doing it wrong. Systematic errors always render a reading high or low and can be dangerous. Close supervision at the beginning of the inventory should smoke out most of these types of errors.

   c. *Blunders* occur when a reading is simply wrong, and wrong to the point that the information is way off the mark. Blunders can usually be caught by error-checking software during the input process (i.e., automated out of bounds checking) or during graphical plotting of information to check for outliers. *System assumption errors* are those errors introduced during the planning of the inventory, when the planner simply makes a mistake on specifying how equipment is to be used or data read. A simple example of this might be the assumption that a certain datum is correct, when it is in error.

5. *Store and test the data.* Make sure it is accessible. Develop the data access software and programming. Concentrate on the procedures for data handling and access. Collect and manipulate trial data first, when possible, to make sure the system works as planned.

6. *Maintain the data.* Make sure that responsibilities for data updates have been assigned and budgeted. Procedures triggered by changes in the specific data (such as subdivision approval or use and occupancy permit issuance) should be set up and worked out.

7. *Develop and foster applications of the data.* Data is only as good as the use that is made of it. Therefore, it is vital that applications be developed for easy use of the inventory data and that a pool of potential users is trained in their use. Those applications envisioned early in the process should be quickly brought online. The greater amount of time that passes after the inventory, the less chance there is that full use of the information is ever made.

Mapping, GIS, and Related

There is an explosion of the ability to acquire, organize, access, and utilize data and information in support of stormwater management. Data acquisition systems include high-resolution satellite imagery, Light Detection and Ranging (LIDAR), aerial imagery, ground and aerial data acquisition using Global Positioning Systems (GPS), vehicle-mounted automated positioning and distance measuring equipment, and more conventional automated survey methods. Vast amounts of data can be collected and transposed into usable formats with a relative ease unforeseen even ten years ago.

Geographic Information System (GIS), CADD (Computer-Assisted Design and Drafting), and database applications software and hardware platforms can process, display, and analyze these data sets using the power available on a desktop or minicomputer. This information can be accessed and used remotely over the Internet, through local, and wide-area networks and through individual sites. Kiosks can be set up or front-end, Web-based systems developed for citizen access and use of even complex data sets. Satellite capabilities for delivery of remote and Web-based data are fast growing.

Yet, there are dangers as well. This type of technology has often outstripped the knowledge and ability of local staffs to use it. Many communities have spent large sums and invested resources for little practical return, a large but ill-planned database, little training, lack of computer availability for the practitioner, few applications, and a rapidly aging database. Trained personnel are often beyond the salary reach of local communities. Many are turning to privatized services. Others have determined to start slow and focused to try to immediately bring applicable results to the automation program without foreclosing on future opportunities or flexibility.

GIS Systems

The basic driver for most of these applications is a Geographic Information System (GIS). A GIS is a computer-based database system designed to spatially analyze and display data. A GIS stores information about a given area as a collection of thematic layers that can be linked together by geography or geo-referencing. This simple but extremely powerful and versatile concept has proven invaluable for solving many real-world stormwater problems from tracking complaints to master planning applications and infrastructure management.

A functional GIS integrates four key components: hardware, software, data, and trained users:

- *Hardware* — Desktop computers and digitizing equipment are the primary hardware components of a typical local GIS system.
- *Software* — GIS software provides the functionality and tools needed to capture or input, store, analyze, display, and output geographic information.
- *Data* — Generally, the most costly part of a GIS is data development. Some geographic data and related tabular data can be collected in-house or purchased

from commercial data providers. A GIS can also integrate tabular data or electronic drafting (CAD) data to build information into the GIS database.
- *Users* — GIS technology is of limited value without trained operators who understand the data, system, and organization, and how to apply the resources to achieve the desired results.

General-purpose geographic information systems essentially perform six processes or tasks, as follows:

Data Input

Before geographic data can be used in a GIS, the data must be converted into a suitable digital format. The process of converting data from paper maps into computer files is called digitizing. Modern GIS technology can sometimes automate this process fully for projects using scanning technology; some jobs may require some manual digitizing using a digitizing table. Today, many types of geographic data already exist in GIS-compatible formats. These data can be obtained from a number of different, often free, sources.

Data Conversion

It is likely that some needed data may not be in the correct format or proper map projection to use with your system. Most GIS software has the ability to do this conversion, but in some cases, this is better done by a contractor who specializes in data conversion. Be careful with third-party data; it is imperative that you understand the source, quality, age, accuracy, and limitations of a dataset. This and other information about a dataset is often provided in metadata that accompanies the dataset.

Query and Analysis

Once there is a functioning GIS containing geographic information, it can be used to answer questions such as:

- Who owns the land parcel being flooded?
- What is the distance between two stream locations?
- Which homes are located in the updated floodplain?
- How will the new development impact downstream properties?
- What types of infrastructure give us the most complaints, and where are they located?

GIS provides simple point-and-click query capabilities and sophisticated spatial analysis tools to provide timely information to stormwater managers and analysts. GIS technology can also be used to analyze geographic data to look for patterns and trends, and to undertake "what if" scenarios. Most modern GISs have many powerful analytical tools including the following:

- *Size analysis* — Provides specific information about a feature (e.g., What are the area and the perimeter of a parcel?)
- *Proximity analysis* — Determines relationships between objects and areas (e.g., Who is located within 100 feet of the streambank?)
- *Overlay analysis* — Performs integration of different data layers (e.g., What is the SCS curve number for this subwatershed, considering soils and land use?) (Figure 2-6 illustrates the overlay concept.)

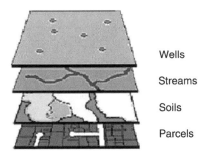

FIGURE 2-6
Example of overlay analysis. (Source: ESRI Web site)

- *Network analysis* — Analyzes the connectivity of linear features and establishes routes or direction of flow (e.g., Which pipes feed into this junction box?)
- *Raster analysis* — Utilizes a raster model to address a number of hydrologic issues (e.g., What does the three-dimensional model of this watershed look like? Where does the water flow?)

Data Display, Output, and Visualization

Geographic information systems excel at being able to create rich and detailed maps, graphs, and other types of output, which allow local staff, elected officials, and the general public to be able to visualize and understand complex problems and large amounts of information. These maps and charts can be integrated with reports, three-dimensional views, photographic images, and multimedia presentations.

GIS can be useful to a community in a wide variety of stormwater-related applications:

- GIS can be used for the mapping of surface features, land uses, soils, rainfall amounts, watershed boundaries, slopes, land cover, etc.
- A GIS can manage a stormwater system inventory and information about facility conditions, storm sewer networks, maintenance scheduling, and problem areas.
- GIS can be used to automate certain tasks such as measuring the areas of sub-watersheds, plotting floodplain boundaries, or assessing stormwater utility fees. Figure 2-7 shows an example of automated hydrologic mapping.
- A GIS can be used to evaluate water quality impacts and answer cause-and-effect questions, such as the relationship between various land uses and in-stream pollution monitoring results.
- "What-if" analyses can be undertaken with GIS. For example, various land use scenarios and their impacts on pollution or flooding can be tried in various combinations to determine the best management solutions or to determine the outcome of current decisions. When tied to hydrology, hydraulics, and water quality models, this type of analysis becomes a powerful tool to assess the impacts of new development on downstream properties. For example, Figure 2-8 shows the flooding impacts on a small tributary for a proposed new development approved during a rezoning.
- GIS databases can provide staff, elected officials, and citizens with immediate answers and ready information. For example, inventory, complaints, and other information about stormwater infrastructure (including pictures) can be placed in a database tied to geographic location.

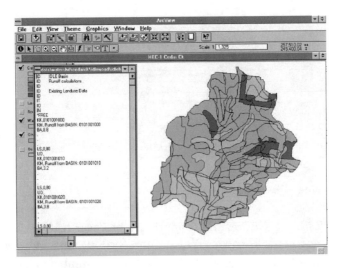

FIGURE 2-7
Automated hydrologic modeling. (Courtesy of AMEC Earth & Environmental, Inc.)

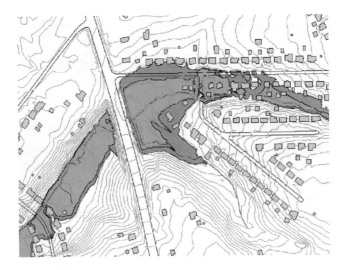

FIGURE 2-8
Impacts of new development. (Courtesy of AMEC Earth & Environmental, Inc.)

- Complex problems or changes over time, such as water quality improvements, can be easily visualized in maps and graphs generated by GIS systems.
- GIS maps can be used to educate or convince citizens and political leadership concerning a course of action or a project's viability. Figure 2-9 shows an advanced, three-dimensional, automated floodplain plotting and flyover capability used in public meetings.

GIS is closely related to several other types of mapping, database, and information systems, and can be used with these other information technologies.

CADD — Computer-Assisted Design and Drafting (CADD) systems evolved to create designs and plans of buildings and infrastructure. The systems are designed to do detailed drafting and drawing but have only limited capability to attach data fields to the electronic

FIGURE 2-9
Three-dimensional floodplain plot. (Courtesy of AMEC Earth & Environmental, Inc.)

drawing. As a result, these systems do not have the capability to perform spatial analysis. Fortunately, these drawing can be input to a GIS, saving significant digitizing efforts. Once in a GIS, attribute data can then be added to the graphic features.

DBMS — Database management systems (DBMS) specialize in the storage and management of all types of data, including geographic data. DBMSs are optimized to store and retrieve data, and many GISs rely on them for this purpose. They do not typically have the analytic and visualization tools common to GIS.

SCADA — SCADA stands for supervisory control and data acquisition system. These systems combine the ability to monitor information (e.g., rainfall, stream flow, flood level, etc.) remotely through telemetry. SCADA systems can also execute commands to do such things as open gates or close valves from a distance. Examples of the use of SCADA include automating stormwater pump station operation, automated alarms for flood warning, and automated lowering of traffic control barrier arms during high-water periods. SCADA systems can be combined with GIS to create comprehensive tracking systems.

Global Positioning Systems — The Global Positioning System (GPS) is a space-satellite-based-radio positioning system for obtaining accurate positional information for mapping or navigational purposes. GPS is made up of three distinct parts:

- *Satellites* — A constellation of 24 satellites orbiting the earth continuously emit a timing signal, provided by an on-board atomic clock, which is used to calculate the distance from each satellite to the receiver.
- *Receiver* — A GPS receiver located on the ground converts satellite signals into position, velocity, and time estimates.
- *Ground control* — The U.S. Department of Defense (DOD) developed and currently manages the maintenance of the satellite system. The DOD uses tracking antennas to constantly monitor the precise position of the NAVSTAR satellites. These positions can be used to correct for errors in the calculated positions of the roving receivers.

The GPS was built and is maintained by the U.S. government. The satellites orbit at an altitude of approximately 12,000 miles in a 12-hour pattern that provides coverage to the entire earth. The system is capable of serving an unlimited number of users free of charge.

A GPS receiver uses information from at least four of the 24 satellites to precisely triangulate its position on the Earth with about one-meter accuracy. If a receiver cannot "see" four satellites, it can calculate a less accurate estimate based on three satellites. Virtually all GPS receivers display basic positional information including latitude, longitude, elevation, and speed (if moving). Most receivers also display time, heading, the number of satellites in view, where those satellites are positioned in the sky, and signal quality. GPS receivers for data collection can collect the location (coordinates) and the attribute data of a given geographical feature.

Remote Sensing

Remote sensing is a technique for collecting observations of the earth using airborne platforms (airplanes and satellites), which have on-board instruments, or sensors. These sensors record physical images based on light, temperature, or other reflected electromagnetic energy. This sensor data may be recorded as analog data, such as photos, or digital image data. Figure 2-10 gives an example of low- (25 m), medium- (5 m), and high-resolution (1 m) satellite imagery.

Ground reference data are then applied to aid in the analysis and interpretation of the sensor data, to calibrate the sensor, and to verify the information extracted from the sensor data. Remotely sensed images have a number of advantages over on-the-ground observation, including:

- Remote sensing can provide a regional view.
- Remote sensing can provide repetitive looks at the same area over time.
- Remote sensors "see" over a broader portion of the electromagnetic spectrum than the human eye.
- Sensors can focus on specific bandwidths in an image and can also look at a number of bandwidths simultaneously.
- Remote sensors often record signals electronically and provide geographically referenced data in digital format.
- Remote sensors operate in all seasons, at night, and in bad weather.

The airborne platforms that carry remote sensing instruments can be any kind of aircraft or satellite observing the Earth at altitudes anywhere from a few thousand feet to orbits of hundreds of kilometers. Satellites may employ a variety of sensors for numerous applications. Currently, no single sensor is sensitive to all wavelengths. All sensors have

FIGURE 2-10
Low-, medium-, and high-resolution satellite imagery. (Source: Space Imaging, Inc.)

fixed limits of spectral sensitivity and spatial sensitivity, the limit on how small an object on the Earth's surface can be seen. The common types of sensors aboard satellites include:

- *Multispectral scanner (MSS) sensors* — Data are sensed in four spectral bands simultaneously: green, red, and two in near-infrared (these can sense as many as six bands including UV, visible, near-IR, mid-IR, and thermal).
- *Thematic mapper (TM) sensors* — Data are sensed in seven spectral bands simultaneously: blue, green, red, near-infrared, and two in mid-infrared. The seventh band detects only the thermal portion of the spectrum.
- *Radio detection and ranging (RADAR)* — Examples include Doppler radar systems used in weather and cloud cover predictions.

The appropriate band or combination of MSS bands should be selected for each interpretive use. For example, bands 4 (green) and 5 (red) are usually best for detecting cultural features such as urban areas, roads, new subdivisions, gravel pits, and quarries. The TM bands are more finely tuned for vegetation discrimination than those of the MSS, due in part to the narrower width of the green and red bands.

Examples of the growing number of remote sensing satellites include the U.S. Landsat satellites, the Indian Remote Sensing (IRS) satellites, Canada's RADARSAT, and the European Space Agency's Radar Satellite. Images from these satellites have spatial resolutions ranging from approximately 100 m to 15 m or better. The first commercial satellite capable of resolving objects on the ground as small as 1 m in diameter was recently launched. Several competing companies have similar offerings. For example, the IKONOS-2 features high spatial resolutions of 1 m panchromatic (black and white) and 4 m multispectral (color). Panchromatic data have a higher resolution, while multispectral data provide better interpretation. Additionally, the 1 m panchromatic spatial content can be combined with the spectral content of the 4 m multispectral data. This 1 m accuracy allows for a wide range of applications in stormwater management at a price typically less than $500 per square mile (with some minimum order restrictions).

Satellite imagery offers a diverse set of mapping products for projects ranging from land use and land cover evaluation to urban and regional planning, tax assessment and collection, and growth monitoring. In the case of stormwater runoff, multispectral imagery can be used to measure impervious surfaces, such as rooftops, streets, and parking lots. Pervious surfaces, such as tree- and grass-covered areas, can also be measured or delineated. Applying runoff coefficients to the area of each surface type can provide the best available estimates for nonpoint sourcewater pollution. By adding parcel boundaries, it is possible to provide estimates of runoff per parcel in order to calculate stormwater user fees. Similarly, designated land-use categories can be applied to the area of each surface type, and, in combination with the known soil coverage, can be used to calculate hydrologic curve numbers. Flood boundaries can be measured within a few meters accuracy in areas without tree cover using submeter multispectral fused imagery. Individual buildings and parcel boundaries can also be identified in order to assess flood vulnerability.

Light Detection and Ranging (LIDAR)

LIDAR is a technology used for collecting position and elevation data from a pulse scanning laser system flying in an aircraft (Figure 2-11). By using laser pulses aimed down toward the ground from a plane flying in traversing patterns at a known altitude, an accurate depiction of the topography of the land can be determined. These pulse

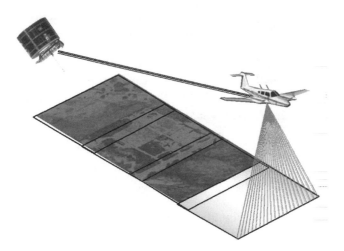

FIGURE 2-11
LIDAR imagery collection.

measurements can then be converted to three-dimensional points for inclusion in a GIS application. A LIDAR system is three technologies: a laser system used for ranging, combined with an Inertial Navigation System (INS) and a Global Position System (GPS) for an accurate definition of the aircraft position. A brief survey of published accuracy specifications reveals that LIDAR results in accuracies of 0.5' vertical and 2' to 4' horizontal, though errors can creep in in several ways.

2.5 Overview of the Organizational Aspects of Stormwater Management

In many municipalities, perhaps most, stormwater management grew as an adjunct duty of an office, which had primary responsibility in another area. Street maintenance forces often maintain culverts, pipes, catch basins, and tail ditches in conjunction with street maintenance and repair. However, their primary concern is not the drainage system but the street. Planners often administer the floodplain program as an ancillary duty to zoning. Because the stormwater function lacks focused leadership and is dispersed throughout the municipal organization, it is often neglected or deferred. Badly needed stormwater funds are diverted to more important needs. Those municipalities that have successful stormwater programs have given the program focus through organizational change. In most, the drainage system is seen as a public system in much the same way as the water supply and wastewater systems are viewed. This leads naturally into also seeing the stormwater function as a utility.

There is no right way to organize stormwater management. However, all stormwater management programs must perform certain functions. Table 2-2 lists these functions under nine "cost causation" categories. A typical municipality will have many of these functions dispersed throughout the municipality's governing structure.

Effective management of stormwater usually requires that a single staff position be identified as the central figure in stormwater control. This person should have primary control of all stormwater management functions (e.g., drainage, floodplains, water quality, erosion, and sediment control).

TABLE 2-2

Major Stormwater Management Functional Cost Centers

Administration	*Regulation and enforcement*
General administration	Code development and enforcement
General program planning and development	General permit administration
Interlocal coordination	Drainage system inspection and regulation
	Zoning and land use regulation
Public involvement and education	Special inspection programs
Public awareness and education	Flood insurance program
Public involvement	Multiobjective floodplain management
Standing citizen's group	Erosion control program
Billing and finance	*Capital improvements*
Billing operations	Major capital improvements
Customer service	Minor capital improvements
Financial management	Land, easement, and right-of-way
Capital outlay	Stormwater quality management
Indirect cost allocation	*Quality master planning*
Overhead costs	Retrofitting program
Cost control	Monitoring program
Support services	Structural BMP programs
	Nonstructural BMP programs
Engineering and planning	Pesticide, herbicide, and fertilizer
Design criteria, standards, and guidance	Used oil and toxic materials
Field data collection	Street maintenance program
Quantity master planning	Spill response and cleanup
Design, field, and operations engineering	Program for public education and reporting
Hazard mitigation	Leakage and cross connections
Zoning support	Industrial program
Multiobjective planning support	General community and residential program
GIS and database management	Illicit con and illegal dumping
Mapping	Landfills and other waste facilities
	CSO program
Operations	Groundwater protection
General maintenance management	Wellhead protection
General routine maintenance	Drinking water protection
General remedial maintenance	Watershed and TMDL Support
Emergency response maintenance	Septic and I&I program
Infrastructure management	
Public assistance	

Both large centralized organizations having almost all the functions in-house and small decentralized organizational structures hiring many of the functions both within municipal staff and outside have proven to be successful. Stormwater quality has been identified separately, though, in practice, it should be integrated with all the quantity functions in each functional area.

There are many ways to organize a stormwater management program, but most organizations fall into one of three categories: a separate entity, a sub-department of another organization (public works, engineering, water and sewer, etc.), and scattered responsibilities. Table 2-3 gives a summary of the pros and cons for each type of organization.

The separate entity maximizes command and control of the organization, but at a potentially higher cost and higher risk of visible failure. Also, other organizations necessary for the ultimate success of a comprehensive stormwater program may assume all aspects are well covered by the stormwater entity and may then ignore their responsibilities.

TABLE 2-3

Organizational Pros and Cons

Separate Entity	Subdepartment	Scattered
Pros	Pros	Pros
✓ Clear identity	✓ Support from others	✓ Low profile
✓ Independent	✓ Protection	✓ Little disruption
✓ Control budget	Cons	Cons
✓ High focus	✓ Loss of some control	✓ Coordination hassle
Cons	✓ Loss of some focus	✓ Loss of focus
✓ Costs	✓ Tendency for little change	✓ Loss of control
✓ Public acceptance		✓ Low priority
✓ Higher risk		
✓ Loss of cooperation		

The scattered approach is probably where most municipalities started and has the advantage of a low profile for a failing or ineffective program. The obvious cons can be overcome to some extent by elevating ultimate control of all resources to the level where all stormwater-related entities report and by mandating cooperation among entities through the use of an ad hoc stormwater committee facilitated by the high-level official or his or her representative.

The subdepartment approach is often the method of choice for programs that have just initiated a stormwater utility or some other public action. It tends to allow for control of resources without the potential for a large bureaucratic entity or the potential isolation of the separate entity. Colocation has its drawbacks in that the stormwater program may be "lost" in another department — especially water or sewer, where the stormwater budget is dwarfed and the stormwater priorities are often seen as less compelling and urgent.

2.6 Overview of the Financial Aspects of Stormwater Management

What happens in many municipalities is that a large flood event will trigger a public outcry for something to be done. This then results in a study of the problem. The time required to make the study is long enough for the memory of the flood to have faded in the minds of the political decision-makers and their constituents. Procrastination sets in. Funds are not allocated to make necessary improvements, the study is shelved, and the municipality awaits the next flood event. Someone has called this succession from flood to panic to planning to procrastination the "hydroillogical cycle," because it seems to parallel the "hydrologic cycle" of rainfall to runoff to evaporation and back again. Experience has shown that the key issue to be solved in breaking out of this cycle is primarily one of financing. When adequate financing is available, many of the other functions take care of themselves.

There are a large number of stormwater programs in the United States in various stages of development. One measure of the maturity of a stormwater program is the amount of money spent on it. Levels of funding for the basic programs (less any extraordinary items like large capital programs, large flood control maintenance programs, costs shared with counties, etc.) tend to be the same across the country. Table 2-4 indicates a typical range of program costs expressed in cost per developed acre (assuming a normal mix of urban development). Improvement or addition of the indicated capabilities is reflected in the cost ranges.

TABLE 2-4

Typical Stormwater Program Costs

Program Type	Cost/ Ac/Yr	Typical Program Features
Incidental	15–40	Reactive incidental maintenance and regulation as part of other programs, only emergency capital construction
Minimum	40–75	ADD: moderate responsive right-of-way maintenance, better regulation and inspection, minor capital construction program, more staff, erosion control, some planning
Moderate	75–120	ADD: additional maintenance program extent and levels of service, better regulation and inspection, master planning, better technology and GIS, minor capital program, general upgrade of capabilities, utility funding
Advanced	120–200	ADD: maintenance of the whole system, GIS-based master planning, regional treatment, basic water quality program, data collection, multiobjective planning, strong control of development and other programs, utility funding, advanced thinking about paradigms in stormwater
Exceptional	Over 200	ADD: advanced stormwater quality structural and nonstructural programs, advanced flood control and levels of service for maintenance, aesthetics become more important, many holistic and public involvement and education programs, catching up with all major and minor capital needs throughout the municipality, high level of technology, proactive stormwater infrastructure management, floodplain recovery program, greenways and riparian corridors, low-impact development concepts

Most programs were originally drainage programs appended to a street maintenance, water and sewer, or general public works function. Other public services take precedence over the stormwater infrastructure. In most municipalities, knowledge of the location and condition of the drainage infrastructure, even that on public rights-of-way, is not known. As services or requirements were added (such as FEMA administration), they were added to various departments or divisions of government. Therefore, operations may be part of a street's division or water and sewer. FEMA, and other planning may be part of a planning department. Engineering lodges in public works. Regulation and enforcement may be found in many or all of the other departments, depending on whether it is development-related and requires a subdivision or rezoning process for development. In few cases is there an individual who has the authority, knowledge, and desire to manage all aspects of the stormwater program. These would be termed "incidental" stormwater programs. All operations are complaint driven and normally limited to the right-of-way and a clear public interest. Planning is nil; regulation is limited to the minimum required for development. Technical guidance is poor or copied from other nearby municipalities or other government publications.

Better maintenance, perhaps with dedicated personnel, a more defined regulatory process, technical support at a higher level, and erosion control would be added to the incidental program to make it a minimum stormwater management program. Further additions would increase the quality of the program, as shown in Table 2-4.

Figure 2-12 shows a typical cost distribution among the nine cost causation functions for a fully mature stormwater program. Note that the three largest expenditures tend to be maintenance, water quality, and capital construction, and that water quality costs make up fully one-fourth of the total program costs.

The use of a user-fee-based system (often termed a "stormwater utility") of funding is growing in popularity. A true stormwater utility is simply a funding entity or umbrella under which exists any number of funding and operational activities related to stormwater management.

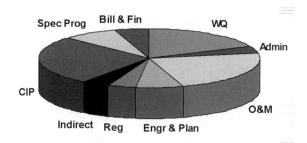

FIGURE 2-12
Typical stormwater program cost distribution.

Just as most municipalities now consider water and wastewater public systems, the drainage system should be considered a public system and a public responsibility. After all, public water makes up a large portion of the flow through the system, and systems quickly grow to such a size that a single property owner has not the technical or financial wherewithal to manage the system. Even if property owners had the resources, a piecemeal management of the system would eventually be as bad as none at all. And, just as the fee charged to water and wastewater customers is based on their use of those systems, stormwater charges are based on how much water an individual property owner puts through the system or how much of the system he uses. Total impervious area or some combination of total impervious area and total land area is often used as a surrogate for this use. When a person paves over a field or woods, he is putting a demand on the drainage system it was not designed to carry. Cost is incurred because of this demand, and the user should bear that cost. More information is given on stormwater utilities in Chapter 5.

2.7 Typical Stormwater Management Problem Areas

A series of studies and inventories was performed during 1986, 1987, 1992, and 2000 in some leading cities and counties. These inventories (combined with experience in a number of other municipalities across the country) uncovered common problem areas and isolated a set of characteristics of successful municipal stormwater programs. Certain characteristic problems are, or had been, the experience of most surveyed. Those municipalities that developed successful programs have exhibited fewer of these problem traits. No municipality exhibited all of them, and all experienced different levels of success and failure in each area. These typical problems and success characteristics are given below under the different categories of stormwater management: long-term, legal/financial/organizational/technical support, and day-to-day management.

Long-Term Problem Areas

- Specific goals or policies had never been articulated in writing. Therefore, such a thing as enforcement of design standards, stormwater quality programs, development plans approval, detention requirements, or longer-term planning was without concrete purpose or ability to assess achievement and was based on a vague sense of doing some good.

- Planners had an inability to get a grasp on what the extent of potential flooding, erosion, or pollution problems was or would be in the future. Capital improvement planning was incomplete and poorly coordinated with other planning functions.
- If stormwater master plans or other types of stormwater assessment were available, they were often difficult or cumbersome to interpret, out of date, not directly usable as analysis tools, or had used advanced engineering techniques or computer analyses incompatible with everyday design application or beyond the capabilities of engineering staff to modify easily and reuse. They tended to be static snapshots of pieces of an ever-emerging larger picture.

Legal/Financial/Organizational/Technical Underpinning Problem Areas

- Stormwater management was often seen as a low-priority item when compared to such things as roads and sanitation. This, in itself, is not a problem because it usually is of lower priority. However, the result was that stormwater was often managed piecemeal by various persons as an additional duty, poorly financed, and unplanned.
- There was little knowledge of the state of repair of the drainage infrastructure and what was actually out there. Therefore, future maintenance could not be scheduled or budgeted. The impact of this was that failed parts of the system remained defective for long periods.
- Many local flooding problems existed that were beyond the financial and technical ability of homeowners to repair but were not under the jurisdiction of the municipality. Causes of this were numerous, including poor design, poor maintenance, unanticipated upstream development overtaxing the system, and the unavoidable deterioration or natural adjustment of any man-made drainage way.
- Comprehensive and usable planning, policy, or design documents backed by effective ordinances and enforcement powers rarely existed or were enforced.
- Many runoff problems were caused by flow from areas outside the jurisdiction of the municipality. Interjurisdictional cooperation was lacking.
- Local, on-site detention was in a poor state of repair. This was particularly true of residential sites. The municipality did not have the funds, manpower, or legal ability to make the necessary repairs.
- Funding was inadequate and poorly targeted to meet actual needs. Impact or other fees could only be used in designated areas, and general funds were lacking for needed infrastructure maintenance. No analysis of needed funding had ever been done except on a year-to-year basis. Legal challenges to fees existed.
- Engineering methods used by designers were not uniform with regard to the type of method used, degree of accuracy, or presentation.
- Little data existed in directly usable form. Thus, many design engineers used inferior methods rather than develop the necessary information. Few rain or flow gauges existed even in basins of known future development.
- There was little knowledge of or concern for the environmental aspects of urban runoff in spite of the fact that Congress had mandated certain requirements in the future for virtually all municipalities.

Day-to-Day Problem Areas

- Planners had an inability to predict the impacts of potential developments or the impacts of required stormwater controls such as detention. Therefore, detention was required in a haphazard fashion or by using general rules of thumb.
- Day-to-day maintenance was driven by complaints and political pressures. Regular preventive maintenance was not done. Prioritization of maintenance was haphazard.
- Use and enforcement of proper erosion control measures was seen as unimportant or not possible.
- Those responsible for stormwater management spent a perceived disproportionate amount of time in plan review, often laboriously recalculating various drainage calculations by hand. There was little or no time or ability to assess whether designs were adequate or optimal.
- The development process was not defined well enough to ensure compliance with technical requirements, timely inspection, and as-built certifications.
- Many development control policies were simply understood by local engineers but never written, thought through in detail, or coordinated.

2.8 Three Example Programs

As part of the national surveys conducted, attempts were made to distill key characteristics of successful stormwater programs. Prior to listing these characteristics, it is instructive to look at several example stormwater programs. Three have been chosen for their apparent success, and because they each faced a different major problem with differing approaches and organizations. Programs from the "heartland" have been chosen, because they may better represent the normal situations faced by the majority of cities and counties throughout the United States: Charlotte, North Carolina; Louisville-MSD, Kentucky; and Tulsa, Oklahoma. Charlotte faced numerous off right-of-way flooding complaints, Louisville faced a 5-year backlog of drainage projects and major flooding, and Tulsa faced major damaging floods. Each eventually obtained a Phase I NPDES permit for stormwater management.

Charlotte, North Carolina

Charlotte is a city with a population of about one million. In 1993, the city began operating a stormwater utility to catch up with the large amount of drainage problems on private property caused by decades of inaction. This was done through a combination of maintenance projects and capital improvement projects. Up to that point its stormwater program was chronically underfunded. Charlotte suffered from several thousand unmet off right-of-way drainage needs, many of them long standing, and a frustrated council. It changed its extent of service to take responsibility for all storm drainage segments that carry public (i.e., street) water and defined its types and levels of service to transition into a more mature program over time.

The utility was formed as "Stormwater Services" (SWS) (Figures 2-13 and 2-14), an enterprise fund under the Engineering Department, responsible for all aspects of the stormwater program except for maintenance, which is done by the Charlotte Department

FIGURE 2-13
The Charlotte Stormwater Services logo (Charlotte, North Carolina).

FIGURE 2-14
A Charlotte press clipping (Charlotte, North Carolina).

of Transportation (CDOT). There is a separate utility district (CMUD) for water and sewerage. Later, Mecklenburg County joined the Stormwater Services in a transparent joint organization — the county handling the larger streams primarily through a floodplain recovery and greenway program.

In Charlotte, the drainage infrastructure has not traditionally been maintained as a complete system. In the past, the city has maintained pipes and culverts in the street rights-of-way. However, the majorities of pipes, channels, and streams are located on private property and have been the responsibility of the individual property owners for maintenance and improvement.

In the first year of operation, more than 4000 citizens called with qualifying service requests. After field inspection, staff realized that many required only localized infrastructure repairs spanning one to four properties. The highest-priority repairs addressed flooding that endangered public safety or damaged private property. Staff also identified more than 100 large capital improvement projects (Figure 2-15).

The price tag for these repairs was high. Estimates totaled in excess of $300 million. Annual revenue from the newly implemented stormwater utility fees was less than $15 million. At this rate, it would have been decades before some customers were served. As a result, they implemented a balanced portfolio strategy, dedicating the highest percentage of funds to the highest priorities and also dedicating funds to lower priorities to ensure they were not ignored. The balanced portfolio included watershed-wide flood control master planning, a quick-turnaround maintenance strategy, a channel erosion and stability program, support to areas targeted for neighborhood improvement, and an economic development fund to mitigate the challenges of infill development. It also initiated an agressive rain gauge program (Figure 2-16).

Since that time, Charlotte has become a nationally recognized leader in its ability to deal with remedial and minor construction needs, its customer service mentality, and its high-tech use of GIS in engineering and planning. More than 1600 of these small

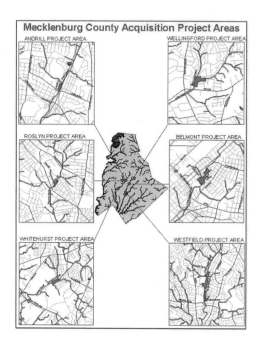

FIGURE 2-15
Charlotte, North Carolina, master planning.

repair projects have been completed through a quick-turnaround, privatized, unit-price contracting method.

Stormwater Services hires CDOT for maintenance, the county for water quality field operations, and private firms for some engineering, design, and operations support. Its operations and regulatory capabilities are mature but not notable, its water quality program is its current focus and is growing.

Customer service, community safety, and economic development remain its goals. Although the backlog is gone, experience shows stormwater services can expect 150–200 new repair-related service requests every year. In addition, it ahs a 15-year plan for addressing the large neighborhood-wide projects still on the waiting list and another 15-year plan for revisiting and addressing the 4000-plus lower-priority service requests char-

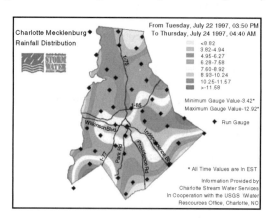

FIGURE 2-16
Charlotte, North Carolina, rain gauge system. (Courtesy of City of Charlotte Stormwater Services, North Carolina.)

acterized by erosion in channels or generally poor drainage without flooding. In this effort, there is emphasis with Mecklenburg County on floodplain recovery through property acquisition and multiobjective riparian corridor planning.

Charlotte has increased its investment in improving surface water quality by developing watershed-wide water-quality plans that include installation of best management practices, stream restoration, reestablishment of stream buffers, and public education and involvement.

The funding strategy is changing along with the program goals. In May 2000, the city issued stormwater revenue bonds ($36,355,000) with an AA+ rating. It anticipates additional bond sales as it continues to catch up with repairs and upgrades.

Louisville MSD, Kentucky

When the Louisville Metropolitan Sewer District (MSD) began (Figure 2-17), the district was collecting user fees to cover the costs of sanitary sewer construction, operation, and maintenance, but was not collecting any money to pay for storm sewers. During the first ten years of MSD's operation, nearly 70% of the sanitary sewer user fees were spent on storm sewer improvements rather than on sanitary sewer improvements.

In the mid-1980s, MSD formed a stormwater utility. Because the Kentucky legislature failed to pass legislation creating a drainage district, participation in the utility program was not mandatory, and several cities (Anchorage, Jeffersontown, St. Matthews, Shively, and Prospect) opted out of the program. The drainage utility's first bills were sent out in January of 1987 following an extensive public education program.

The stormwater utility of Louisville-Jefferson County is part of the Metropolitan Sewer District (MSD) and shares staff with other MSD programs. MSD is also responsible for the sanitary sewer system and the flood protection system of the Louisville metropolitan area. MSD is a subdivision of the Commonwealth of Kentucky, and the chief executives of the City of Louisville and Jefferson County appoint its governing board.

Today, MSD is responsible for a much larger wastewater collection and treatment network, which continues to expand; a comprehensive public stormwater drainage system for most of Jefferson County; the operation and maintenance of the community's Ohio River flood protection system; a computerized mapping and geographic information system; and several other programs — including stream monitoring and hazardous materials control — designed to protect and enhance the environment.

MSD is a nonprofit regional utility service. Its revenue comes from wastewater and stormwater service fees, plus charges for extending wastewater lines and connecting new customers. MSD does not receive supplementary income from taxes or from other local government agencies. All of the agency's revenue is used for operation, maintenance, and extension and improvement of services.

An eight-member board governs MSD. The mayor of Louisville appoints four members, with approval by the board of alderman; four are appointed by the Jefferson County Judge/Executive, with approval by Jefferson County Fiscal Court. Members serve three-

FIGURE 2-17
MSD logo (Louisville, Kentucky).

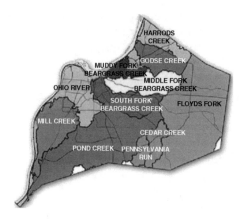

FIGURE 2-18
MSD watersheds (Louisville, Kentucky).

year overlapping terms and can be reappointed. The full board meets twice a month; committees meet as needed.

The drainage utility of MSD began with a backlog of drainage projects that dated back almost 50 years. Some drainage problems waiting to be resolved had been problems for the community since Louisville's settlement in the 1800s. MSD expected the number of drainage requests in the service area to double or triple after the first drainage bills were sent out, but the staff was overwhelmed by more than 18,000 new drainage service requests.

During the first three years of the program, the drainage division averaged around 38 new drainage service calls per day. The drainage utility staff quickly set up a system to prioritize the incoming service requests and schedule necessary construction and repair projects in the community. It began rebuilding the long neglected system.

Over the years, the utility assumed a leadership position in looking at stormwater on a watershed and holistic basis, integrating quantity and quality, and storm and sanitary sewers (Figure 2-18).

The utility initiated a greenways program to develop natural waterway banks and hired several biologists, a soils expert, and a landscape architect to address water quality issues. MSD recommended a floodplain ordinance to reduce potential flooding problems and extended the Ohio River floodwall system to provide more flood protection for the community.

The drainage utility is also responsible for several environmental education and service programs to teach the community how to protect its water resources, including river cleanups, teacher workshops, and hazardous materials collection programs (Figure 2-19).

In 1997, MSD began converting to a *watershed approach* for managing its wastewater and stormwater programs. MSD now groups many of its activities into six area teams: Beargrass, Floyds Fork, Mill Creek, North County, Pond Creek, and Morris Forman Wastewater Treatment Plant. The first five cover MSD's entire service area, while the Morris Forman Area Team is concerned with the operation and improvement of MSD's largest wastewater treatment plant. The Morris Forman Plant serves parts of several different watersheds. Some of the responsibilities of the six area teams include the planning, design, and construction of most of the projects in MSD's Capital Improvement Program and the planning of the Neighborhood Maintenance Program. MSD has divided each of the watersheds into more than 45 neighborhood areas that are basically defined by geographic boundaries and natural features. MSD uses these areas to group together service requests, so that its work can be planned and scheduled more efficiently. Approximately once a year, MSD sends newsletters to its customers who have contacted MSD regarding requests for service.

Stormwater Management Programs 47

FIGURE 2-19
Public education program (Louisville, Kentucky).

Tulsa, Oklahoma

Tulsa is a crossroads Midwestern community with a long history of flooding. The city has 375,000 people and extends across 200 square miles of gently rolling terrain in northeastern Oklahoma's Osage Hills. The area is one of transition between the Ozark Mountains to the east and the Great Plains to the west (Figure 2-20).

Every few years, Tulsa faced serious major creek flooding. In 1974, three major floods caused over $40 million in damage. Some homes were flooded three times that year. The 1976 Memorial Day was a 3-hour, 10-inch deluge. The resulting flood killed three and caused $75 million in damages to more than 3000 buildings. The 1984 Memorial Day Flood killed 14, injured 288, damaged or destroyed nearly 7000 buildings, and left $275 million in damages (Tulsa, 1994).

Angry, drenched victims waded out of the floods to demand help. They contended the city was not enforcing NFIP regulations. They tried to halt development, to avoid deeper flooding until existing problems could be solved. Developers objected strenuously (Figure 2-21).

FIGURE 2-20
Tulsa, Oklahoma, logo.

FIGURE 2-21
Angry, drenched citizens (Tulsa, Oklahoma).

Thus began a community debate over floodplain management, locally called "Tulsa's great drainage war," destined to last years. The city responded with a plan to widen part of Mingo Creek, including clearance of 33 houses in the right of way. The houses were removed just before the next flood.

By this time, the victims were becoming skilled lobbyists and were gathering sympathizers citywide. They stormed City Hall. Newly elected city commissioners responded with a wave of actions. They enacted a floodplain building moratorium; hired the city's first full-time hydrologist; developed comprehensive floodplain management policies, regulations, and drainage criteria; enacted stormwater detention regulations for new developments; instituted a fledgling alert and warning system; developed a stormwater utility; and began master drainage planning for major creeks.

The newly elected mayor and street commissioner had been in office for only 19 days, but both knew the issues well. In the darkest hours of the city's worst disaster, they pledged to make their response to reduce the likelihood that such a disaster would ever be repeated. Before daylight, they had assembled the city's first Flood Hazard Mitigation Team to develop the city's strategy. Within days, a new approach to Tulsa flood response and recovery was born (Figure 2-22).

As ultimately completed, the program included purchase and relocation of 300 flooded homes and a 228-pad mobile home park, $10.5 million in flood control works, and $2.1 million for master drainage plans. The total capital program topped $30 million, mostly from local capital sources, flood insurance claim checks, and federal funds. It was only the beginning.

FIGURE 2-22
Mingo Creek regional detention (Tulsa, Oklahoma).

FIGURE 2-23
Public acquisition of floodplains.

The 1984 flood also persuaded Tulsans that a coordinated, comprehensive stormwater management program was needed from the rooftop to the river. The Department of Stormwater Management in 1985 centralized responsibility for all city flood, drainage, and stormwater programs. A stormwater utility fee was established by ordinance in 1986 to operate the program. The utility fee ensures stable funds for maintenance and management, independent of politics. The ordinance allots the entire fee exclusively for floodplain and stormwater management activities.

Institutionalization and acceptance came in the 1990s, after Tulsans approved a change in city government from the mayor–commission to the mayor–council form. A new Department of Public Works consolidated all public works services. Stormwater management was reintegrated and finally institutionalized into the city structure.

Since comprehensive regulations were adopted in 1977, the city has no record of flood damages to any building that complies with those regulations. In the early 1990s, FEMA ranked Tulsa first in the nation for its floodplain management program, allowing Tulsans to enjoy the nation's lowest flood insurance rates. The program was also honored with FEMA's 1992 Outstanding Public Service Award; and the Association of State Floodplain Managers has twice given Tulsa its Local Award for Excellence (Figure 2-23).

Key program elements of the floodplain management program, considered a model for the nation, include:

- Public acquisition of floodplain lands
- Disclosure of flood-hazard information to purchasers and renters
- Flood alert, warning, and emergency management systems
- Public information, education, and awareness programs
- Development of a post-flood recovery plan
- Acquisition of frequently damaged properties
- Relocation of occupants from flood-prone areas
- Preservation of floodplain lands for park, recreation, and open-space purposes

2.9 Characteristics of Successful Programs

Those municipalities that appeared to have the most successful programs, including the three above, exhibited some common characteristics. Understanding these characteristics will help any municipality become more successful. First, they had found solutions for most of the problems stated above. But also, they had gone beyond simply overcoming most of the problems enumerated above to develop a coordinated program that took advantage of the unique political and physical environments in which they functioned. Several other specific comments, which do not fall under the three categories above, could be given.

Public Education and Involvement

Those successful programs executed public education and involvement programs to ensure the cooperation of the general public. One city conducted public meetings in every neighborhood, presenting problems that would be solved in that neighborhood. Many municipalities have established hotlines and logos and published attractive brochures describing the programs. To provide greater identity, some had special uniforms for the stormwater maintenance crews. Other municipalities created multimedia presentations, built information displays, held public forums, and provided informative news releases. Citizens had a sense of ownership of the surface water system. Many had citizen monitoring or adopt-a-stream-type programs. Student education and involvement were important. All had a mindset of customer service.

Key Champions

Most successful programs depended on key individuals or groups to champion the cause of stormwater management, and were adept at doing so. It is important to have key persons in politics, government staff, and citizen groups. The staff person must be low enough within the staff structure to care about stormwater management and high enough to make a difference. Many municipalities had citizen advisory committees. Most had developed open and effective communication and reporting channels with political leaders and other staff components.

Financing

Almost every successful program in a larger urban area had developed a stormwater utility user-fee form of funding. This funding mechanism has been found to be superior in being equitable to all citizens, generating the needed funding in a relatively painless way, and withstanding legal challenges. Many had secondary funding methods to enhance equity and shift costs to those extraordinary activities that caused the costs in the first place.

Basin-Wide Planning and Control

To help overcome multijurisdictional problems, several municipalities established a regional entity, which performed some stormwater work. The Denver, Colorado, metropolitan area established the Denver Urban Drainage and Flood Control District by special legislative act. The City of Albuquerque, New Mexico, established the Albuquerque Metropolitan Arroyo Flood Control Authority. The Louisville, Kentucky, metropolitan area administers stormwater through the Metropolitan Sewer District. Even when a multijurisdictional entity could not be established, all the successful programs managed stormwater in some way at a basin-wide level, through master plans, ordinances with master plan overlays, special funding methods, or watershed advisory groups.

Clear Procedures and Goals

Each had an efficient sense of purpose with well-defined goals and objectives in each key program area and the means to achieve these goals. The goals and policies provided for consistent treatment of similarly situated properties even when all needs could not be met.

Focused Authority

In many successful programs, the total stormwater function was consolidated or, at least, controlled and coordinated under one authority. This included maintenance (or at least maintenance management), capital budgeting, floodplain management, plans review, inspection, enforcement, emergency management, planning, and design. The most successful programs were not hidden as adjuncts into other utilities such as wastewater or water supply, but they shared equal importance. The stormwater administrator was senior enough to speak for the program at the highest levels.

Strong Technical Guidance and Capabilities

The most successful programs developed strong technical capabilities, resources, and approaches, in-house and through the use of outside consultants. Guidance manuals were clear and complete. Web sites were complete and informative. Master planning was done and used to guide development and capital improvements. Technical review was thorough, and state-of-the-art designs were strongly encouraged or required. Water quality was considered equally with flood control and efficient rainwater removal. Aesthetics and multiobjective best management practices were encouraged and sometimes mandated through greenway plans and landscaping requirements.

Guided Development

Each successful program had well-developed procedures to guide and control development, detailed submittal requirements, adequate inspection staffs, and streamlined enforcement capabilities. Development assumed responsibility in that it bore most or all of the cost of system improvement. Most administrators interviewed stated that there was some resistance and a need for reeducation at first, but the program ran smoothly after that. And most stated that there were no adverse repercussions from charging developments for drainage improvements. Often, master plans guided development.

Comprehensive Maintenance

Successful programs made a shift away from primarily reactive maintenance toward proactive and preventive maintenance. The problem of maintenance responsibility for residential subdivisions was overcome in many of the successful programs. Most provided defined levels of maintenance service to the whole system without regard to right-of-way location. Most private detention facilities were not maintained to function as designed. Government-operated maintenance of these facilities (including an abandonment program) proved to be most effective in several communities. Those municipalities with stormwater utilities often figured charges such that the local government could maintain all detention facilities if private maintenance failed. Strong crediting mechanisms often were used to encourage appropriate design, construction, and ongoing maintenance of drainage facilities.

An Environmental Focus

Successful programs embraced the various water-quality-related paradigms (from Chapter 1) at some level, and developed a sense of ownership of the surface water (and, in some cases, groundwater). In some cases, the original driver was regulatory, but, in most, the eventual driver was a desire to protect valuable riparian areas and water supplies, to increase the aesthetic value of these corridors, and to recognize that these steps made economic sense in the long run.

2.10 The Stormwater Program Feasibility Study

The stormwater practitioner needs a place to start, and that place is with a visionary plan of action. Figure 2-24 illustrates a "roadmap" for a stormwater feasibility study.

The feasibility study approach has several purposes. It allows the focus to be on the need for a program but also spawns an approach to funding of the program. It serves as a logical building block process that can take a citizens group through a rational consideration of key stormwater issues in a logical sequence. It allows for building of vision and provision of direction without making an a priori commitment to a direction or expenditure. Chapter 3 discusses the use of a citizens group in this way in detail.

The following steps are illustrated in Figure 2-24:

- *Program description* — Lays out the current program, responsibilities, resources, and interactions
- *Problems and needs* — Uses the Five Whys questioning methods and understanding of paradigms of Chapter 1 to gain a handle on the surface problems and their root causes, and on where the overall program will probably be going in the future

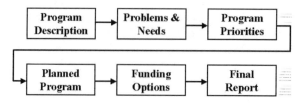

FIGURE 2-24
The stormwater feasibility study roadmap.

- *Program priorities* — Lays out the basic direction for the stormwater program, using functional categories described in Table 2-2
- *Planned program* — Provides the details of the stormwater program, including schedule, resources, costs, and interconnection of various aspects of the program
- *Funding options* — Analyzes various funding options to support various aspects of the program (More details are given in Chapter 5.)
- *Final report* — Gives the direction and implementation steps of the program changes

2.11 Floodplain Management

One of the main duties of the urban stormwater manager is to manage the floodplain as it interacts with the streams and open channels. This duty is assigned by virtue of need and federal, state, and local regulations. The official regulatory floodplain is defined as depicted in Figure 2-25. The Floodplain Management Act of 1978 requires the floodplain manager to perform a number of duties. These, combined with other duties, spell out the purposes and programs of floodplain management. Table 2-5 lists typical and atypical duties of a local floodplain manager.

There are literally hundreds (maybe thousands) of publications, brochures, and papers on how to manage various aspects of the floodplain. FEMA (1986) provides an overview of all the pertinent details of floodplain management and provides a conceptual framework for floodplain management that stands on four important legs:

- Reduce flood losses and threats to health and safety
- Preserve and restore the natural and beneficial uses of floodplains

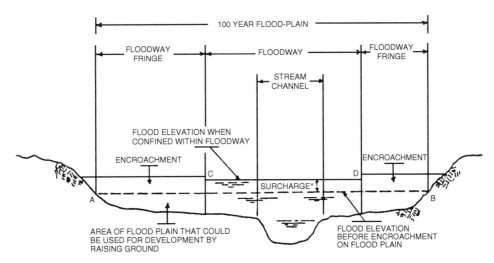

FIGURE 2-25
The floodplain and related definitions.

TABLE 2-5

Duties and Tools of a Floodplain Manager

1. Administer the National Flood Insurance Program
   Maintain flood maps
   Maintains records
   Administer flood map changes
   Provide information on home locations
   Maintain elevation and anchoring certificates
   Update maps periodically
   Coordinate with federal and state agencies
2. Administer the local Community Rating System
   Apply for rating annually
   Administrate measures for which points can be given
3. Adopt and administer local floodplain ordinances and regulations
   Develop design standards and criteria
   Review plans
   Monitor compliance
   Perform inspections
   Provide information
   Issue permits
   Hear appeals
   Administer related regulations such as zoning, subdivision, or building codes
4. Provide technical assistance to the community in floodplain and flooding matters
   Assistance on flood insurance
   Assistance on development design
   Assistance on flood reduction and avoidance measures
5. Perform floodplain planning
   Perform technical master planning
   Identify and seek to mitigate local flooding problems
   Eliminate repeated loss situations
   Seek to preserve and enhance natural values of floodplain
   Look into integrated uses and parks planning
   Work with federal and state agencies on flood reduction and stream restoration measures
6. Plan and manage flood fighting and emergency preparedness
   Develop preparedness plans
   Maintain flood warning and mitigation systems
   Flood response
   Stockpile equipment and supplies
   Manage or assist in flood response and recovery operations
   Maintain flood protection measures
   Maintain flooding records and statistics
7. Educate self and the public
   Keep up with new developments in the program
   Keep up with new technical approaches
   Coordinate with federal and state agencies
   Coordinate with other local agencies
   Conduct public education program
   Keep staff and political leaders educated and informed
   Maintain information
   Develop self-help programs
   Assist in disaster assistance programs
   Provide individual technical and regulatory assistance
   Coordinate with realtors, developers, engineers, and bankers

- Take a balanced view that minimizes exposure to loss rather than promotes floodplain abandonment on the one hand, or intense floodplain development on the other
- Develop and use the tools available to provide careful and technically sound consideration of all information and alternative uses of floodplains

In terms of flooding, floodplain management seeks to reduce human and property susceptibility to flooding, reduce the impacts of flooding, and reduce the actual flood levels and peaks. Reducing human susceptibility includes putting restrictions on the following: mode and time of day/season of occupancy of floodplains; in the ways and means of access; in the pattern, density, and elevation of structures; in the materials out of which structures are constructed, and their designs; and in the infrastructure and landscaping in flood-prone areas. This is done through regulations, disaster preparedness and assistance, floodproofing, and flood warning and mitigation measures.

Reducing the impacts of floods on individuals includes the actions of flood insurance, education, tax adjustments as incentives for wise development, flood emergency response measures, and postflood recovery operations. Nonstructural alternatives should be pursued aggressively, especially when taking into account nonpoint source control measures.

Floodplain management, when integrated with the overall stormwater management program, provides a regulatory means to improve the surface water system throughout the municipality. Flooding does not stop at the edge of the regulatory floodplain; neither does proper floodplain management. Flood-prone areas should be identified throughout the municipality, upstream from and adjacent to regulatory floodplains. Life and property losses due to persons entering floodplain areas should be dealt with just as strongly as those from persons living within floodplain areas. With proper, and patient, application of sound principles of floodplain management, great and positive changes have been demonstrated in various municipalities around the country.

Flood Insurance Programs

The Federal Emergency Management Agency (FEMA), through the National Flood Insurance Program (NFIP), offers a wide range of services and requirements. NFIP was created in 1968 by the National Flood Insurance Act and further defined and modified by several subsequent acts. The program has two major thrusts: to shift the burden of costs for flood losses from taxpayers at large to floodplain occupants, and to reduce losses due to flooding through floodplain regulation. The two major vehicles to do this are flood insurance availability to all homeowners and the required locally passed floodplain regulations and programs necessary for a municipality to retain eligibility for flood insurance.

Based on studies of the flooding in the Midwest in 1993, there is expected to be many basic changes on how the federal government looks at managing floodplains in the future, and the types of programs and support it will offer. It can be expected that the NFIP will also change. The Report of the Interagency Floodplain Management Review Committee (1994) recommends fundamental changes in philosophy and implementation including: a stronger and more coordinated floodplain management strategy with the three thrusts of avoiding, minimizing, and mitigating flood damages at all levels of government and the private sector; establishment of a National Floodplain Management Program by legislative mandate; and revitalization of the Federal Water Resources Council.

The vehicle to accomplish flood-prone designation is through the use of the 100-year floodplain delineation (termed the base flood) as conveyed in a series of maps and reports for each municipality in the program. Originally, municipalities entered the "emergency phase" of the program after the development of Flood Hazard Boundary Maps (FHBM) was complete. These maps do not show flood elevations and thus are not as accurate as recently completed studies. Insurance rates are higher for municipalities still in the emergency phase. They can convert to the regular program through a restudy of the municipality's streams and the adoption of stricter floodplain regulations after a model regulation published by FEMA.

As shown in Figure 2-25, FEMA has designated an area immediately adjacent to the stream as the regulatory floodway. This is an area of higher velocity and deeper flooding. It is determined in engineering hydraulic backwater profile studies by mathematically encroaching in the floodplain from both sides until the water surface rises a maximum of one foot. The area between the edge of the floodplain and the floodway line is termed the floodway fringe and is depicted in Figure 2-25. Development outside the regulatory floodplain is unconstrained by flooding considerations from a federal regulatory standpoint. Development within the floodway fringe must be elevated or flood-proofed to a minimum of 1 ft above the flood elevation (this elevation is termed the flood protection elevation). Development or fill within the floodway is prohibited unless a hydraulic study is done and submitted to FEMA showing that the fill or development does not raise upstream flood elevations. This is normally difficult to do without redefining channel dimensions or roughness or proving a flaw in the original study.

Prior to 1988, FEMA used three vehicles to convey the floodplain information:

- The Flood Insurance Rate Map (FIRM) shows 100-year and 500-year floodplain boundaries and flood elevations and an insurance designation.
- The Flood Boundary and Floodway Map (FBFM) shows the same 100-year and 500-year boundaries but without elevations. Instead, it shows the key cross sections used in developing the detailed study of the stream. These cross sections are keyed to the Federal Insurance Study (FIS).
- The FIS contains complete information on the municipality, the study methods and results, and other floodplain management information. Included are Floodway Data Tables and, generally, tables of elevations and river mileage of each cross section.

FEMA also used a wide range of designations to show various flooding depths. Due primarily to resistance by insurance companies unwilling to write many different policies for each of the designations and the apparent redundancy in the FIRM and FBFM, FEMA changed the designations and the way results are reported. Since 1988, one map depicts all necessary flood information found on the FIRM and the FBFM. And, since 1988, the flood designations are as follows:

A Areas within the 100-year floodplain boundary determined by approximate methods. Development within these areas requires a detailed hydraulic and hydrologic study submitted to the municipality.

AE Areas within the 100-year floodplain boundary determined by detailed methods.

AH Areas of 100-year shallow flooding with constant water surface elevation and average depths from 1 to 3 ft.

AO Areas of shallow flooding (usually sheet flow) where average depths are 1 to 3 ft.

A99 Areas currently within the 100-year floodplain to be protected by a federally authorized flood protection system to be constructed.

X   Areas outside the regulatory 100-year floodplain that are flood prone (normally areas upstream from the 1-mile cutoff for most detailed studies), protected from the 100-year flood by a levee; areas of sheet flow are less than 1 ft in depth.

D   Unstudied areas where flooding is possible but undetermined.

The program has had several drawbacks that have been minimized by some municipalities. The FIS program allows for encroachment sufficient to raise the flood elevation a maximum of 1 ft. While this rarely happens in practice (theoretically, a levee located at the floodway boundary would be necessary to accomplish a 1-ft rise), some municipalities have a stricter definition of the flood protection elevation requiring as much as 5 ft freeboard (though most are under 3 ft). Others allow encroachments at some lesser level. For example, Indiana allows only a 0.1 ft rise due to encroachment.

Another drawback is that the floodplain boundary is determined based on hydrology derived from current land use conditions. As development occurs, greater flows result, rendering the floodplain boundary inadequate. Some municipalities established full build-out boundaries or a combination of assumed partially reduced peaks due to detention requirements and full-build-out. Other municipalities seek to keep their flood maps current through frequent and mandatory use of FEMA's method for flood map updating: Letters of Map Revision (LOMRs), Conditional Letters of Map Revision (CLOMARs), and Physical Map Revisions (PMRs).

Reference

City of Tulsa, Oklahoma, Rooftop to River, 1994.

3
Public Awareness and Involvement

3.1 Introduction

Public awareness and education are the foundations of a successful urban stormwater management program. From before the program's inception and throughout its growth and service life, interaction with, and education of, the public is not an adjunct to the program's purpose; they are (or should be) a main purpose of the program. This is becoming especially important as the demands of the NPDES permit and its source control philosophy comes more into play in major and minor municipalities across the country.

Public awareness and education are carried out in stormwater management programs in two main ways: specific public awareness campaigns and ongoing "baseline" public information programs and activities. These two aspects differ in that a campaign has a beginning and an end, while the ongoing program goes through transformations but does not envision an ending. But, for either case, there are specific factors that should be kept in mind and methodologies that will make a stormwater management program successful.

3.2 Defining the "Public"

The first step in designing a public program is to ask the question: Who is the "public"? Who am I trying to reach and with what message? Who will have which specific concerns? Whom does this or that aspect of the program influence? Different sectors or segments of the public will participate or be interested in different issues and in different stages of the same issue:

- The development community will be vitally interested in regulatory and financial aspects.
- The environmental community will be vitally interested in water quality issues.
- Specific neighborhoods will be interested in specific provisions for drainage controls, regional stormwater treatment, safety, park integration, greenways, etc.
- Clubs or social organizations may be interested in participating in some programs.
- Schoolchildren can be interested in the environment or clean creeks.

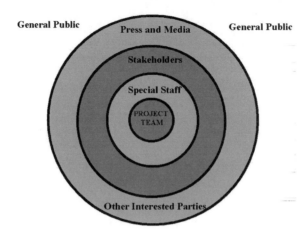

FIGURE 3-1
Public awareness strategy.

- Social classes may be interested in utility fees or charges.
- Tax-exempt and governmental properties will be interested in new user fees.
- Commercial and industrial concerns may have similar interests in fees and charges (and credits).
- Design professionals will have interests in the technical criteria and regulatory requirements.

Figure 3-1 illustrates the basic public awareness and involvement understanding. There are different levels of involvement for different people based on such things as job duties, technical expertise, level of concern, and willingness to invest time and effort. Different types of public education and involvement will be necessary to reach different groups.

The center core of the project is key staff: the decision-makers. They are the ones tasked with implementing or carrying on a successful stormwater management program. They are often assisted by outside expertise in public awareness program planning, development, and execution. They have full-time involvement. The next circle out contains the specialized staff and consultants necessary for part-time support (such as legal, GIS, or financial staff).

The next two levels of involvement come from stakeholders, political leaders, or a specialized citizens group. People organize around common interests: political, financial, moral, environmental, cultural, social, locational, etc. Because some aspects of the stormwater management program impact them or something they value, they will be stakeholders in the program's processes.

The difference in the two circles is the sense of "officialness" of the involvement or the intensity of the concern. Those most involved or those most affected by the stormwater management programs of the community should and do have the greatest influence. From experience, a small vocal group of dedicated stakeholders often controls the direction of stormwater management program growth and activities. Such a group can fill many roles, from policy setting to simple advisory duties and from technical to strictly policy-oriented activities. Stakeholders normally include the development community, representatives from any large institutions or industries, members representing socioeconomic groups, and environmental groups. The press is also involved at the outer of these two levels.

The final level is the general public. Stormwater is not normally vitally important to most of the local residents, unless there is flooding. It can be of interest to the general public if there is an effective public education and awareness program. For example, based on the results of a statistically derived phone survey conducted in an Eastern city, clean water is important to the average citizen; so much so that citizens were willing to pay over $5 per month to help clean it up.

The closer to the center of the target one moves, the greater the level of effort required to maintain involvement and influence. Also, as one tries to influence those in the more outer rings, it takes more effort to maintain that awareness.

3.3 Plan to Include the Public

For stormwater success, especially in campaigns such as initiation of a stormwater utility fee, it is important to have a "champion" in three areas: political leader, staff leader, and respected community leader. These champions must be well educated on the issues and articulate in explaining the goals of the program. The other factor for success (or at least a factor to avert disaster) is to identify early the needs, interests, and levels of concern of the various stakeholder groups, and plan a way to address those concerns. The greater the controversy, the greater the public interest.

The stepwise and thoughtful development of a public awareness plan is the most vital tool in the success of a public awareness program. The public should be involved at every stage of the process, but the stages must be carefully defined. The plan describes in detail the objectives, activities, sequencing, timing, costs, and responsibilities for every aspect of the public program. Development of the plan is important, because it (EPA, 1990):

- forces a careful analysis of how the public plan fits the stormwater management program
- brings together in planning and (hopefully) agreement all entities or departments that will be involved in the plan and program
- communicates to the public that they are of vital importance and that the public entity is "contracted" to involve them (Table 3-1 gives the elements of a typical public awareness plan)

TABLE 3-1

Elements of a Public Awareness Plan

1. Describe the history leading up to this point and major past players or causative events.
2. Describe the objectives and goals of the program and potential major issues likely to emerge in a campaign or in a general public awareness program.
3. Estimate the level of public interest likely to be generated or hoped to be generated, and develop ways to gauge public reaction or program effectiveness.
4. Identify the major potential stakeholders and their level and type of concern.
5. List and describe stages in the campaign, activity, or program the public awareness program must support, and the public involvement support stages.
6. Outline a sequential and interrelated plan of activities and products to support each stage of the public program and the stormwater management program.
7. List key milestones when the public program will be reassessed and midcourse corrections made, if appropriate.
8. Estimate costs and level of effort required for each phase of the program.

It must be remembered that the public program is to support and follow the stormwater management program, not lead and shape it. The program drives the public campaign, not vice versa. There is often a tendency for the public program to take on a life of its own, losing sight of the real-world objectives of the stormwater management program.

For example, assume a municipality has a project in mind that has the following three goals:

1. To obtain a federally mandated permit to discharge stormwater from the municipal separate storm sewer system to creeks and streams
2. To plan, organize, and take steps to establish the foundation for a comprehensive stormwater quality and quantity management program
3. To develop and carry out a stormwater utility to fund the stormwater management program

Based on these goals, the following general objectives could be developed (Ogden, 1992):

1. To *inform and educate* the stakeholders and the public about the following:
    a. Stormwater management programs and needs
    b. The federal stormwater quality regulations and permit requirements
    c. Stormwater management and program growth and direction
    d. Stormwater management program costs and financing alternatives, with emphasis on the stormwater utility
    e. The advisory committee and council actions and discussions
2. To *seek input* from the stakeholders and public on the following:
    a. The existing stormwater problems and future needs
    b. Desired stormwater management program direction, activities, and structure
    c. Willingness to pay for the program
3. To *involve* the stakeholders and public in program development the following ways:
    a. Through the use of a citizen task force
    b. By meeting with stakeholders and other groups
    c. Through the use of special events
    d. Through the use of hotlines
    e. Through the use of public hearings
4. To *gain consensus* for the stormwater quality management plan, stormwater management program, and the creation of the utility
5. To *monitor* the public education program in the following ways:
    a. By measuring public acceptance of the total program
    b. By monitoring all media, i.e., TV, newspaper, and radio, to keep track of public opinion
    c. To obtain feedback at all public events
    d. To obtain feedback through a hotline

The following target audiences could then be defined:

1. Area neighborhoods
   a. Business (owners)
   b. Residential
   c. Associations
   d. Churches
   e. Party precinct leaders
   f. Council wards
2. Local media
   a. Print
      i. News departments
      ii. Editorial writers
      iii. Feature writers
   b. Broadcast
      i. News departments
      ii. Public service directors
      iii. Talk show hosts
3. Elected officials
   a. Federal and state
   b. City council
   c. County commission
4. Government staff
   a. City or county managers
   b. City or county public service and information director
   c. Other staff
5. Developers
   a. Residential
   b. Commercial
6. Environmental groups
   a. Sierra Club
   b. Other groups (local, regional, national)
7. General public

With this basic information, the public awareness plan could be formulated following the guidance in Table 3-1 and assessing the techniques in the next section.

3.4 Public Involvement and Education Techniques

There are almost an infinite number of techniques, combinations of techniques, and variations to inform and educate the public. Table 3-2 is an example list of some of

TABLE 3-2

Example Techniques Used in a Public Involvement Plan

Media kit and white paper
Media tour
News releases and op-ed pieces
Feature stories on surface water problems
B-roll footage for local news stations
Paid advertisement
Developed pamphlets
Free government informational and educational pamphlets
Slide-based or other video
Bill stuffers
Talk shows
Public access programming
Speaker's bureau availability
Political leader meetings
One-on-one key person meetings
Booths
Special neighborhood or stakeholder meetings (such as chamber of commerce, etc.)
Information workshops
Customer service representatives
Hotline
Identity creation — logo, letterhead, vehicle markings
Phone survey research

the most commonly used in urban stormwater management public programs. Each has different advantages and effectiveness, and none of the techniques can be used by itself. Each must be used with others to create a synergy. Some are most effective to introduce general information to masses of people, while others are more "rifle shots" targeted toward even a single individual. Some are participatory, and some are quite passive.

The point is that each activity or product must be clearly understood within the context of the target audience, the objectives, the phase of the campaign or program, the nature of what is being advertised, the message communicated, etc.

Written public information pieces must be factual and unbiased. There will certainly be a backlash if a piece about a subject where controversy can be expected is seen to be a sales job for the community officials. When this happens, the staff and the program lose credibility — their greatest asset.

Public Participation

Some of the techniques are specially conceived to invite participation of various target audiences. Table 3-3 lists the most common participatory techniques, together with features, advantages, and disadvantages.

Public Participatory Groups

There is much to be wary of in public participatory groups. Generally speaking, for example, public hearings, at best, rarely provide useful information and are largely ignored by the general public. At worst, they turn into a shouting match, where misinformation rules, with little or no chance for logical and reasoned rebuttal.

Advisory groups can be very helpful under controlled and well-defined circumstances and will be discussed in detail in the next section.

TABLE 3-3

Participation Techniques Used in a Public Involvement Plan

Technique	Features	Advantages	Disadvantages
Advisory groups/ task forces	Stakeholders are invited to participate; may be technical, advisory, or directive in nature; should be balanced and representative	Provides oversight to the process or campaign; promotes communication between parties and gets all sides exposed; can provide a forum for reaching consensus; obtains ownership of solutions, policies, or programs	Consensus can fall apart if members have no authority to make or enforce decisions; if advice is not taken, there is potential for controversy; requires substantial commitment of staff time
Focus groups	Small discussion groups to give typical reactions of the general public; normally conducted by a professional facilitator; may be several parallel groups or sessions	Provides in-depth reaction and detailed input; good for predicting emotional reactions	May not be representative of the general public or a specific group; might be perceived as manipulative
Hotline	Widely advertised phone number to handle questions and provide information	Gives a sense of people knowing who to call; gives a customer service feel; provides one-stop service when done right; can handle two-way communication with the general public	Requires effective and trained hotline operator on standby
Interviews	Face-to-face interviews with key persons or stakeholders	Can be used to anticipate reactions or gain key individual support and provide targeted education	Requires extensive staff time and an effective interviewer
Hearings	Formal meetings in which people present formal speeches and presentations	May be used for introductory or wrap-up meetings; useful for legal purposes or to handle general emotional public input safely	Can exaggerate differences without opportunity for feedback or rebuttal; does not permit dialogue; requires time to organize and conduct
Meetings	Less formal meetings of persons to present information, ask questions, etc.	Highly legitimate form for public to be heard on issues; may be structured to allow public to be heard on issues and to allow small group interaction	May permit only limited dialogue; may get exaggerated positions or grandstanding; may be dominated by forceful individuals
Workshops	Smaller meeting designed to complete a task or communicate detailed or technical information	Very useful to handle specific tasks or to communicate, in a hands-on way, technical information; permits maximum use of dialogue and consensus building	Inappropriate for large audiences; may require several different workshops due to size limitations; requires much staff time in detailed preparations and many meetings
Plebiscite	Citywide election to decide a particular issue	Provides a definite and usually binding decision to end a stalemate or drawn-out process	Campaign can be expensive and time consuming; general public may be susceptible to uninformed or emotional arguments

TABLE 3-3

Participation Techniques Used in a Public Involvement Plan   *(Continued)*

Technique	Features	Advantages	Disadvantages
Volunteer programs	Programs where citizens volunteer to perform some necessary function in support of the stormwater management program	Builds shared ownership and a sense of personal pride; allows stakeholders to experience some of the realities of the program	May not be high-quality work; may be liability issues; takes a good deal of staff supervision
Polls	Carefully designed questions are asked of a statistically selected portion of the public	Provides a quantitative estimate of general public opinion	Susceptible to specific wording of questions; provides only a static snapshot of a changing public opinion; costly

Source: After Environmental Protection Agency, Sites for Our Solid Waste, EPA/530-SW-90-019, 1990.

Workshops can serve a valuable role in providing information of a factual or technical nature, while attempting to skirt the more emotionally charged side of things. This allows a mutual understanding to be reached and a common baseline of information to be gained before tackling controversial issues.

If the public is invited to participate, the stormwater manager should make sure its input is recognized and considered fully, and that the public knows it. This is accomplished through summary feedback of what was heard.

Elected officials, in particular, need to be kept informed of all aspects of the program or campaign. Elected officials should never be taken by surprise, should never find something out by reading it in the paper, and should never give a speech or talk on the subject without adequate preparation by the staff.

Remember that a key for any successful campaign is to ensure that the municipality has taken all appropriate steps to meet any expected concerns of the citizens. For example, before implementing a user fee or rate increase for a stormwater utility, it is prudent to define and document problems the fee will solve and to take a hard look at the program costs and activities, and ensure the following (EPA, 1989):

- Inefficiencies have been eliminated.
- Labor forces have been trimmed and are lean.
- The program is perceived as the minimum necessary to meet customer expectations or regulatory demands.

3.5   The Stormwater Advisory Committee

Recent emphases on citizen group or stakeholder involvement in a variety of urban stormwater-related, policy-making situations have led to the need for the technical professional to become proficient in facilitating such groups. Rarely does a professional facilitator have the technical skills and background necessary to lead such a group; therefore, it is incumbent on the urban stormwater professional to develop the techniques, approaches, and skills necessary to successfully bring a dissimilar group to agreement on a diverse suite of issues. Such skills are commonly needed in areas such as stormwater program development, funding and rate development, master planning, and watershed planning.

The role of the technical facilitator is to guide a group to consensus on a set of related issues in which most of the group has a stake and an opinion. The tools the facilitator uses include a roadmap, ground rules, objective criteria, policy papers and policy statements, objective criteria, and common sense.

There are many ways to gain public input, to involve the public, and to educate. Figure 3-1 indicates a way to consider levels of involvement. Information flows outward, and input and commentary flow inward (Reese, 1999, 2000). Level of involvement and access to information decrease the further you move from the center and the project team.

The project team is at the center of all actions and decisions. Special staff, such as legal counsel, is involved on an as-needed basis. The advisory group is at the next level of involvement. It has input on most key decisions, though access to project information is controlled and limited to a need-to-know basis.

Why this layer? From our experience, there are key individuals and key public sectors that wield disproportionate influence on local decision-making. If these key stakeholders can be convinced of a certain course of action or can be brought into agreement on a recommendation on a specific issue, the political leadership will most likely support it. This is especially the case if the decision involves spending funds on a stormwater project not seen as immediately beneficial to all citizens, or the decision results in specific fees or taxes. The insertion of a citizen-group between a political body and a potentially unpopular decision is a common method to "get more fingerprints on the knife."

Outside the advisory group layer is a layer consisting of other interested parties and the media. There may be other parties that have an interest but cannot, or should not, be on a committee. They can be corresponding members and receive meeting minutes or other information. The media has a special interest in, and needs particular access to, information, though less than an advisory group. Outside this group are the general public, mildly interested, and more easily influenced by the information obtained from other levels.

A second reason for the use of a citizens group can be summed in the old adage, "Bring me in early, I'm your partner; bring me in late, I'm your judge." Gaining the support of key stakeholders by including them in the process early is the best way to ensure success during the approval and implementation stages of a stormwater endeavor.

The inclusion of competent and successful citizens representing diverse viewpoints can also bring the following:

- Political influence to gain approval
- Influence on key stakeholder groups to gain their support
- Perceived legitimacy and potential ability to educate the public
- "Guinea pigs" with which to try out concepts and proposals
- Ownership of solutions
- Potential sources of financial partnership for solutions
- Technical or financial know-how
- Potential ability to handle or influence the media
- A source of good ideas

Defining the Group

The first step in designing a citizen group is to ask the question that was posed in Section 3.2 and to ensure broad participation. It is important that the citizens group have representation for each legitimate stakeholder group plus a few well-known or influential (and reasonable) citizens who can help keep the group on track and steady. A group of about 10 to 15 is ideal. This size group allows for some of the people to be absent without reducing the group below a quorum or seeming to be too small to be influential. This size also allows for lively discussion and a diversity of opinion, and it mitigates against one individual dominating the group.

Media and the General Public

It is usually important to plan and run a campaign to inform the general public about the group and its process and results. This can go a long way in making the eventual recommendations more acceptable to political leadership and the public. For controversial issues, it may be prudent for group members to agree not to speak to the media except through a chosen spokesperson.

Often, a frequently asked question (FAQ) sheet is developed to give to various team members, political leaders, and group members. The FAQ provides the party line on key questions as well as gives vital statistical information on the problems the group is attempting to address. It is important to have a logically thought-through presentation of the group's approach and findings. In one case, in selling the concept of a stormwater utility user fee system as a primary funding mechanism, the following logic was provided to all involved:

- Stormwater problems are real and unresolved.
- The problems can be solved.
- Government must lead.
- Benefits will result.
- Adequate, stable, equitable funding is created.
- A utility is the most practical vehicle for solutions.

Group members, if chosen carefully, can perform a service in selling the results of the process in general and to individuals and stakeholder groups (Figure 3-2). Often, a scripted slide show is developed, and group members use it to advertise what they are doing. In some cases, brochures can be developed with head shots of group members complete with quotes and sound bites about how they feel about the fairness of the process and its

FIGURE 3-2
A citizens group at work.

eventual results. Group members are seen as unbiased regular citizens by the media and can often convince the media and citizens of the benefits of a certain decision when staff or political leaders cannot.

Authority and Purpose

The next issue is to define the specific purpose of the group and its authority. There are several types of groups, and it is important to specifically determine which type you desire. There is much to be wary of in public participatory groups.

Advisory groups can be helpful under controlled and well-defined circumstances. There is a built-in dichotomy with such groups in that the members selected to represent stakeholders on such advisory groups are normally selected because they are decision-makers who carry weight with their constituency. Such people are rarely satisfied with only providing advice or being involved in situations in which their advice is not perceived as being heard. Alternately, the persons sent to the advisory committee may have little or no authority within their constituency but are sent to be eyes and ears. If decisions are reached that the stakeholders do not like, the participation of such low-level individuals will be somewhat meaningless, as issues are revisited politically.

It is helpful to ask and answer some key questions: What are the group's goals and objectives? Is the group advisory, or does it actually make policy? Who should appoint the group and on what basis? How will its recommendations be formulated and presented? What is the expected outcome and impact of decisions the group will make? How influential will it be?

Defining the Issues

Decisions in local government are made at different levels. Rezoning and other land use decisions are often reserved for elected officials only, as are certain legislative decisions. Much of the regulatory function is left to staff decision-making and authority. Plans reviewers, city engineers, maintenance foremen, etc., may make smaller decisions on a daily basis.

In the same way, the spectrum of decisions and issues with which the citizens group should deal should be presented in such a way that the committee can actually make decisions without needing to become technical experts or deal with all the minutiae surrounding its decisions or recommendations. Therefore, it is important to:

- Define issues in simple (not simplistic) terms where clear decisions or recommendations are needed and can be made. Do not leave the group with vague guidance or an unclear structure.
- Limit the number of decisions to an amount that can be handled in an appropriate number of meetings. Do not overwork the group.
- Structure the decisions such that the group follows a logical, building-block process, with a beginning and an end.
- State the decisions to be made in the form of a question to be answered.
- Provide well-defined limits to the types of decisions that can be made and the scope of the group's consideration.
- Provide the necessary background information to allow the group to make informed decisions.

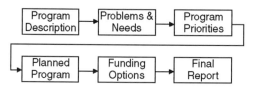

FIGURE 3-3
SWAC roadmap.

To provide vision and definition for the decision-making process, a roadmap is often used. Figure 3-3 is an example of a roadmap (similar to Figure 2-24) used in a group with a goal of determining the direction for a local stormwater program and the feasibility of funding stormwater through a stormwater utility. Often several of the blocks are shaded, representing those parts of the meeting process that are of current or past concern. This helps to keep the group on track and focused on current issues. It also tends to limit the ability of the group to backtrack through old issues. The whole group needs to agree that there is a need to revisit a completed issue to go back.

This roadmap depicts the objectives of the group in a logical order. The questions the group was to address were as follows:

- What are we spending now on stormwater, and what are we doing?
- What are the current stormwater program problems, needs, and issues?
- What should the program priorities be in stormwater management in the next five to ten years?
- How do these priorities translate into a program growth schedule and into a demand for financial resources and manpower?
- What options do we have for funding the program?
- How should we report our findings and recommendations to the appointing authority?

Informed Consent and Consensus Building

Finally, the process the group follows in its deliberations and decision-making is important. It is important to have a controlled but not overly rigid process. We favor a modified consensus-building model of decision-making, often termed "informed consent."

Informed consent is a decision-making process that works creatively to include all persons making the decision. It is a powerful decision process, as all members agree to the final decision. Informed consent takes into account and validates each participant. Everyone gets the opportunity to voice his opinion or block a proposal, if he feels strongly enough about a decision. It equalizes power over a group of people. Instead of simply voting for an item and having the majority win, the group has to sit down and find a solution to a problem that *everyone* is okay with. The solution that the group thinks is the most positive gets chosen, unless a member of the group finds the solution totally unacceptable. Informed consent is based on compromise and the ability to find common ground. It promotes participation, because each person has the power to make changes in the system and to prevent changes that he finds unacceptable. There are different levels of informed consent (full agree, agree with reservations, disagree but stand aside, full disagreement) and different ways to test for informed consent in the midst of a discussion (straw poll, thumbs-up, green-yellow-red, etc.). A more complete discussion of informed consent and consensus-building is beyond the scope of this book but can be found in the relevant literature.

In general, informed-consent-building has several disadvantages. Because it is a lengthier process to hash out ideas until all objections are resolved, group meetings might be longer, and some proposals might take more than one meeting to decide. Also, because some proposals may be dismissed if there is no hope of compromise, informed consent sometimes favors the status quo. It is more expensive, but the results of having all parties own the solutions are usually worth it. Remember, the citizen-group is not used for run-of-the-mill decisions but for those where controversy may result from decisions made, and where broad support is seen as necessary for progress and success.

Ground Rules and Objective Criteria

It is important, to keep order, that ground rules and, if necessary, objective criteria are agreed to at the first meeting. When the group agrees on these things, members tend to become more self–policing. This is especially true if the ground rules are posted prominently at every meeting, and ideas or suggestions are subjected to the objective criteria.

Ground rules are simple statements on how we intend to conduct ourselves while we are conducting a meeting. They might include the following:

- Wait to be recognized
- Stay on topic
- One question at a time
- Share time with others
- No sidebar conversations
- Arrive on time and stay until the end

Objective criteria are a set of standards by which we evaluate suggestions and compare competing suggestions or ideas. In one major Eastern city, the list developed by the group included that a new stormwater utility rate methodology:

- Be clear and understandable
- Promote revenue stability
- Reflect experience elsewhere
- Promote good behavior
- Not harm the city
- Be efficient
- Be cost-effective
- Be revenue neutral
- Be equitable and fair
- Promote good stewardship of financial and environmental resources

Policy Papers

It is convenient to work through the roadmap using a policy paper process. That process is illustrated in Figure 3-4. Initially, the staff and consultant team draft a policy paper.

FIGURE 3-4
The policy paper process.

The paper may be three to five pages long giving background information in nontechnical terms, policy options based on what other cities or counties have done and what might work in this situation and pros and cons, and making recommendations. The paper is mailed to all group members a week ahead of time. The paper is discussed at the next meeting. A brief presentation of the paper is made, and key specific questions to be answered are framed by the meeting facilitator. An attempt is made to arrive at consensus on the issue, or issues, in the paper. Once this is achieved, the group moves on to the next policy paper.

Between meetings, a policy statement is developed by the staff and consultant team and mailed. At the next meeting, a policy statement is discussed. This statement is a concise description of the issue and a brief statement that describes precisely the feelings and consensus of the group. This statement is modified as necessary and agreed to. There is resistance to going back to rehash the previous meeting unless there is a consensus that the disagreement or new information should be considered. If a member needs to miss a meeting, an alternate should be chosen to attend. Unless the group agrees, the ability of the member to change the policy paper after the fact is forfeited.

All policy papers and policy statements are collected into a report at the end of the process. They make a convenient record of the group's activities and can often simply be bound together with a preface. They serve as the final report to the appointing body, and, along with meeting minutes, can serve as a record of the proceedings sufficient for use in legal defense, should that become necessary.

In summary, of the many tools a stormwater manager has to educate and involve the public, the use of a citizens group can be rewarding and can greatly assist a local government in making key decisions about a stormwater program. Citizens groups are often used in such areas as master planning decisions, policy development, funding implementation, stormwater program development, etc.

Key points can be summarized in a checklist as follows:

- Have a controlled process.
- Have limited and defined goals.
- Have ground rules and objective criteria.
- Have defined group authority.
- Keep records and publicize your group.
- Report to the appointing authority.

- Do not have a "stacked" committee.
- Do not overcontrol discussion or decisions.
- Do not allow the group to get off track and wander aimlessly.
- Do not have a committee without influence.
- Do not ignore committee members' input.

3.6 Volunteer Programs

One highly successful method of public involvement and education is through volunteer programs. This has been especially successful in the areas of recycling and citizen monitoring of stream quality. Once skeptical, states and local agencies now see citizen monitoring as a key in the total sampling program (Nichols, 1992). These programs have more than tripled since 1988. Other examples of the growth and importance of volunteer programs include the following:

- The EPA has sponsored several conferences on the use of volunteers.
- A complete magazine, *The Volunteer Monitor*, is published in San Francisco, California.
- The EPA publishes a *National Directory of Citizen Volunteer Environmental Monitoring Programs* that lists programs by state.
- The Izaak Walton League has used citizen monitors since 1927.

The Izaak Walton League offers a training course in simplified monitoring techniques, including a simplified macroinvertebrate monitoring methodology, through its "Save Our Streams" program.

Urban stormwater management programs such as the Bellevue, Washington, "Stream Teams" (Bellevue, 1989); the Charlotte, North Carolina, "Streamwalk Program;" or the more recent Seattle, Washington, "Water You Doing" program (Seattle, 1992) have proven that citizens want to be involved in their streams. Those municipalities that have fostered a sense of community ownership of the streams have seen their surface water quality and aesthetics improve measurably over those who still see the streams as abandoned and unsightly urban fields. Volunteers can reach people and locations that staff cannot, and they can build community support in an effective way when staff personnel cannot.

There has been much learned about how to conduct a successful citizen participation program over the years. Generally, there are six basic principles that form the foundation for volunteer success (University of Wisconsin, 1993). These are given in Table 3-4.

The EPA compiled a sourcebook for citizen involvement and education materials for water quality. The publication contains information on hundreds of pamphlets, books, bumper stickers, catalogs, citizen action guides, fact sheets, software, handbooks and manuals, newsletters and magazines, booklets, posters, slide shows, student activities, videos, and other ideas (EPA, 1993a, 1993b). Since that time, and with the inception of the NFDES Phase II program, literally hundreds of sources, example programs, and Web sites have appeared.

TABLE 3-4

Keys for Successful Volunteer Programs

Program and Staff Commitment
The program and staff must, at all levels, continually acknowledge that the program counts on volunteers to be successful; that they have something important to contribute. Staffs must be trained and encouraged to cultivate this attitude. They should be trained on how to make best use of and train volunteers.

Recruitment Strategy
Includes development of volunteer job descriptions and targeting ways and locations to look for volunteers.

Communicating Effectiveness
Formal or informal assessments of achievement are important to let volunteers know they are important and to make corrections to problems.

Enjoyment
This is achieved through social interaction and a pleasant working environment. It is important to set a congenial working environment and find tactful ways to deal with problems.

Personal Development
Gaining knowledge and skills is important for many volunteers. Project staff should provide opportunities for personal development in skills necessary for the volunteer program.

Shared Ownership
The commitment of the volunteers rests on a sense that they own part of the program. Staff members need to acknowledge this sense of shared ownership of the program and be willing to share management and coordination duties. Citizen advisory committees have more of a sense of ownership if they are given responsibilities for setting agendas, schedules, and facilitating meetings. Ownership is also communicated when volunteers are given opportunities to speak publicly about the program and represent it to other potential volunteers or citizen groups.

3.7 Dealing with the Media

Tables 3-5 through 3-7 are lists of tips for developing a trusted and effective relationship with the media and for conducting interviews, television appearances, and news conferences effectively (Harris, 1986; AWWA, 1989; Newsom and Scott, 1981; Bernstein, 1988; Wilcox and Nolte, 1990).

TABLE 3-5

Media Tips — Television Interviews

Listen to questions and remember them in detail.
Smile when talking.
Speak in 30-second clips.
Be cordial.
Give education and information in an entertaining way.
Deliver a positive message about the subject.
Give the information asked.
Dress conservatively.
You know more than anyone on this subject, but brush up.
Talk at a tenth-grade educational level — no technical jargon.
All microphones are always "on."

TABLE 3-6

Media Tips — News Conferences

Remember that there is competition for the news media's attention, and they often have several conferences to choose from.
Make sure a news conference is more appropriate than an interview or press release.
Pick a good central location familiar and accessible to the media.
Choose a day early in the week and one without competing conferences.
Avoid the rush by holding it early in the afternoon, say 1:30.
Give notification in the form of a media advisory six working days prior to the conference.
Address information to the news-planning desk and include event, time, place, speakers, and background.
List your event in the local news calendar (termed the "daybook") one day prior.
Provide footage for TV media (have visuals, in any case).
Hire a photographer to photograph the event on film (black and white is fine).
Prepare the room, and remember to allow for sound hookups and camera platforms where necessary. Find the pay phones. Put up directional signs.
Brief a spokesperson and prepare (see Table 3-7).
Make up simple press kits: news release summarizing the conference; lists of names and titles of speakers; bio and photo of primary spokesperson; fact sheet on municipal program or campaign and the municipal department in charge; additional articles, brochures, etc.
Offer phone interviews to no-shows and send along press kits.
Follow-up the day before.

TABLE 3-7

Media Tips — Interviews

Be interviewed in a familiar setting.
Anticipate touchy questions and have answers ready.
If you cannot answer, say so. State that the information is not for public disclosure, and politely decline to answer.
Stay cordial. Do not argue.
Talk from the viewpoint of the public's interest.
Talk in personal terms whenever possible.
Allow time for the interviewer to meet the deadline.
Know your topic thoroughly.
Have printed support material on hand to give the interviewer.
Set ground rules for the interview, if possible.
Build personal rapport prior to the interview.
Avoid off-the-record remarks.
Help the reporter make only one stop for all necessary information and follow-up interviews.
Give the story the reporter is after.
Offer to get answers to questions unanswered and then get them to the reporter quickly.
Do not ask when the story will run or how big it will be.
State the most important fact at the beginning.
Do not repeat a question or words in a question if it is offensive, even to deny them. They could be printed.
Direct questions deserve direct answers.
Tell the truth, and do not exaggerate.
Keep in touch on a continuing basis.
Allow for the fact that the reporter knows little about stormwater.
Offer to help in the follow-up, if necessary.

The public, through the reporting of the print and broadcast media, knows the performance and objectives of any stormwater management program or campaign. The media can be the best friend and support or it can become the worst nightmare of the stormwater staff person or political leader.

Staff and political leaders often complain that reporters publish stories that are inaccurate, sloppy, misleading, incomplete, and biased against them. Editors and reporters state that municipal staff and leadership are often less than candid and defensive, provide partial information in an unwieldy or untimely fashion, hide key facts, and limit access to key information sources. They also state that there is a difference between bias and simple unfavorable news, which is not understood by the subjects of the reports.

There are some things that can be done to improve the relationship and make it a win-win situation. The reporter has a job with a tight deadline. He or she is looking for news, not information: controversy, provocative stories, interesting stories, human interest, etc. If a subject is dry, highly technical, without pictures or news footage or a "hook" for the reader or listener, it is not news.

Deadlines for general news in the daily morning paper are generally 4 p.m. the day before publication. For the evening paper, it is anywhere before 3 p.m. the day of publication. Feature stories normally should be sent or developed several weeks prior to the expected publication time to ensure they are given ample opportunity for inclusion. Television news can be included up to about 2 hours prior to airtime if there is no film involved. If film is involved, there is normally a longer lead-time required of about two additional hours. General news releases for radio should be done the day prior to broadcast, but there is great flexibility in radio broadcast media. Newsmagazines should be contacted individually for interest in stories and submittal deadlines and formats. Color photos are always a must to ensure consideration.

News releases should be short, use short words, and deal with the key elements in the first two paragraphs: who, what, when, where, why, and how. Head it with the municipal name and department, address, phone number, date, contact person, and "NEWS RELEASE" (AWWA, 1989).

3.8 Risk Communications

Stormwater management facilities and conveyance structures are dangerous when there is flooding. People are injured or killed, and property is damaged by flooding, erosion, and pollution. One important community education and awareness role for urban stormwater and floodplain managers is that of risk communication. Risk communication is the exchange of information between the stormwater manager and the general or particular (such as floodplain residents) public concerning a particular hazard and what can be done and is being done to manage the hazard and its consequences (EPA, 1990).

Examples of risk-based communication needs in a community pertaining to urban stormwater management include such things as communicating effectively to citizens: floodplain hazards, water quality hazards, spill hazards, stream crossing hazards, risks of flooding from a levee or flood wall project, dam break hazards, hazards from the construction of regional structural controls, etc.

The Federal Emergency Management Agency's (FEMA) Community Rating System (CRS) is a methodology by which communities can reduce the flood insurance rates its citizens pay through proper stormwater and floodplain risk management (FEMA, 1993). One of the

TABLE 3-8

Risk Communication Steps

1. Identify risk communication objectives for each step of a public awareness campaign or for each phase or aspect of a particular program.
2. Determine the information exchange needed to take place for each part decided in step one. Use a risk message checklist like the one shown in Table 3-9.
3. Identify the interests with whom information must be exchanged.
4. Develop appropriate risk messages for each targeted group or audience.
5. Identify the appropriate channels, techniques, or means for communicating risks to various segments of the public.

Source: After Environmental Protection Agency, Sites for Our Solid Waste, EPA/530-SW-90-019, 1990.

categories under which points can be gained is in the area of public information. Under this category, the community can gain credit if it promotes programs and activities such as:

- Issuance and maintenance of elevation certificates that indicate flood hazard elevations for new construction
- Risk communication of flood hazards and availability of federal insurance
- Real estate agent disclosure of flood hazards
- Maintenance of flood information at a public library
- Technical advice on flood hazards and flood protection measures

The Corps of Engineers (as well as other agencies) has turned to risk-based analysis for the sizing and evaluation of flood damage reduction projects. In this way, they have an opportunity to effectively communicate to the public the actual risks and uncertainties involved in the project and the likelihood of project failure or floods that overwhelm project capabilities (U.S. COE, 1993).

Risk communication is successful if it raises the level of understanding of the relevant issues or actions and satisfies those involved that they are adequately informed within the limits of available knowledge. Risk communication is difficult to do correctly due to the technical nature of the risks being communicated and the potential for either over- or under-reaction to the risks involved.

There are typical steps in the development of a generic risk communication program as developed by the EPA (EPA, 1990). These are given in Table 3-8. The risk message checklist (example given in Table 3-9) is a convenient list of questions to be asked whenever a community is considering construction or initiation of a program in which there are hazards and benefits.

People perceive risk in different ways. Some risks are seen as more acceptable than others. There are certain characteristics of risk perception that the stormwater manager should keep in mind when deciding how to communicate to the public concerning risk-related topics (Hance, Chess, and Sandman, 1987). Table 3-10 gives some basic principles when considering risk, as citizens perceive it.

If a crisis happens, be accessible, be accurate, and take responsibility for what is yours (AWWA, 1989). Tell the bad news right away. Explain what is confidential and why and tell all the facts you know but no more. Update frequently and regularly and monitor news coverage and correct any mistakes.

People who are affected by the impacts of decisions made by stormwater managers expect some kind of mitigation. Most impacts affect neighborhoods, and local residents demand some sort of equivalent benefit for any perceived damages. Mitigation is most effective if it can be in some way tied to the project. For example, siting a detention pond in a neighbor-

TABLE 3-9

Risk Message Checklist

Information about the Nature of Risks
What are the hazards of concern and their characteristics?
What is the probability of exposure to hazards?
What is the hazard's distribution or location?
What is the probability of each type of harm from the hazard?
What are the sensitivities of the population groups to hazards?
What is the total population risk?

Information about the Nature of the Benefits of the Program or Proposed Project
What are the benefits associated with the program or proposed project and their characteristics?
Will the benefits actually occur?
Who benefits and in what way?
How many people benefit and for how long?
Does any group get disproportionate benefits?
What is the total benefit?

Information on Alternatives
Are there any alternatives to the hazard?
What is the effectiveness of each alternative?
What are the risks and hazards of alternatives and of doing nothing?
What are the costs, and how are they distributed?

Uncertainties about Risks
What are the weaknesses in the available data?
What are the inherent assumptions?
How sensitive are the estimates to changes or inaccuracies in assumptions?
How sensitive are decisions made to changes in assumptions?

Information on Management
Who is responsible for decisions?
What issues have legal importance?
What constraints and resources are on the decisions?

Source: After Environmental Protection Agency, Sites for Our Solid Waste, EPA/530-SW-90-019, 1990.

TABLE 3-10

Basic Risk Principles

Voluntary risks are more acceptable than imposed risks.
Risks under individual control are more acceptable than those under government control.
Risks that seem fair are more acceptable than those that seem unfair.
Risk information that comes from a reputable source is better than information from a less trustworthy source.
Risks that are dreaded are less acceptable than those that carry less dread.
Risks that are unseen or come without warning are less acceptable than those that are detectable or come with warning.
Physical distance from a site influences acceptability of risk.
Rumor, misinformation, dispute, and sheer volume of information may interact to give a misperception of risk.

Source: After Environmental Protection Agency, Sites for Our Solid Waste, EPA/530-SW-90-019, 1990.

hood can be mitigated through the creation of a small neighborhood park in the area, especially if the neighborhood has young children. However, safety issues normally cannot be mitigated by anything less than reducing the unsafe aspect of a project (EPA, 1990). It is also true that most citizens see the present situation as risk free, while any changes made will be closely questioned. For example, a channel-widening project may alert citizens to the safety issues of a neighborhood creek, while there were few concerns prior to the project. Many mitigation issues concern procedure as much as result. The questions of who makes the decisions, how they are made, and who has input are vital.

3.9 Technical Communications

The problem of communicating technical information to a nontechnical public is a real one encountered every day by stormwater managers. Table 3-11 gives a checklist of steps for the development of a technical communication campaign.

In one city, a woman called the stormwater manager and, upon learning she had to pay a portion of an assessment to all homes within a watershed for a flood control facility, stated, "I want to know how I got in this watershed and I want out." In another instance,

TABLE 3-11

Technical Communication Checklist

Anticipate the issues that will emerge.
Use interviews and polls.
Identify studies that will be necessary to resolve these issues.
Develop an interim strategy for dealing with questions about these issues prior to study completion.
Identify and evaluate alternative strategies to mitigate potential impacts.

Get participation in developing the study plan.
Use an outside, unbiased consultant.
Use a scoping process to identify issues from every stakeholder.
Use workshops to identify issues.
Use technical citizens for technical input and nontechnical citizens to get a "sanity check."

Validate methodological assumptions.
Use Table 3-8 or rely on workshops and roundtable discussions.

Invite public involvement in consultant selection.
Reduce the sense of staff advocacy for a project.

Provide technical assistance to the public.
Use citizen education workshops prior to study evaluation.
Hire outside consultants to serve as advocates for the citizens.

Use an outside body to review a technical study.
Choose a technical citizen body or university staff.
Hire an outside, objective consultant.
Remember that an objective review might find the study inadequate and require parts of it to be redone.

Present technical information in understandable language.
Have nontechnical personnel such as public relations experts review language in handouts, etc., and translate technical information into common language.
Use advisory groups to pretest news releases, reports, etc.

Source: After Environmental Protection Agency, Sites for Our Solid Waste, EPA/530-SW-90-019, 1990.

an individual builder called the local municipality to inquire about the cost of obtaining a "hydrograph," which was required by local ordinance.

Managers often become experts in dealing with complex issues in simple ways. There are many pitfalls in this. Oversimplification can lead to false assumptions of knowledge. There is a predisposition for the public to mistrust technical information. Also, there is a history of misuse of such information for many subjects, leaving all parties assumed guilty. Unsupported rhetoric is often believed more than state-of-the-art studies. Who speaks is often more important than what is said.

Two goals of any stormwater manager are building the credibility of technical information in the eyes of the public and improving the relevance of technical studies to public concerns (EPA, 1990). The fundamental way to do this is to create visibility and participation on the front end of a study (such as a capital improvement plan or flood study) to reduce resistance to the study results. How a study is done is as important as what the findings are. Key stakeholders can oversee the plans for the study, oversee the methodology and assumptions, participate in hiring technical expertise, and review the results.

References

American Water Works Association, Public Information, Denver, CO, 1989.
Bellevue, Washington, Stream Team Guidebook, Storm and Surface Water Utility, 1989.
Bernstein, G., Meet the press, *Public Relations Journal*, March 1988.
Environmental Protection Agency, Building Support for Increasing User Fees, EPA/430-09-89-006, 1989.
Environmental Protection Agency, Sites for Our Solid Waste, EPA/530-SW-90-019, 1990.
Environmental Protection Agency, Office of Wastewater Management and Region V, Urban Runoff Management Information/Education Products, OWEC (EN-336), Washington, DC, 1993a.
Environmental Protection Agency, Guide to Federal Water Quality Programs and Information, EPA-230-B-93-001, 1993b.
Federal Emergency Management Agency, National Flood Insurance Program Community Rating System Coordinator's Manual, with Supplement 1, FIA 15, July 1993.
Hance, B.J., Chess, C., and Sandman, P.M., Improving Dialogue with Communities: Risk Communication Manual for Government. prepared for the New Jersey Department of Environmental Protection, Rutgers University, New Brunswick, NJ, 1987.
Harris, J., Get the most out of news conferences, *Public Relations Journal*, September 1986.
Newsom, D. and Scott, A., *This is PR*, Wadsworth Publishing, 1981.
Nichols, A., Citizens monitor water quality, *Water Environ. & Technol.*, March 1992.
Ogden Environmental and Energy Services, Inc., Public Education Plan for the City of Charlotte, NC — Draft, 1992.
Reese, A.J., Developing Technical Policy with Citizen Groups, IAHR — 8th International Conference Urban Storm Drainage, Sydney, Australia, August 30–September 4, 1999, pp. 1706–1713.
Reese, A.J., Developing Technical Policy with Citizen Groups, *Proceedings of the Association for Rainwater Storage and Infiltration Technology*, Vol. 36, Tokyo, Japan (in Japanese), 2000, pp. 51–56.
Seattle Engineering Department, Water You Doing, Seattle Municipal Building, Seattle, WA, 1992.
U.S. Army Corps of Engineers, Risk-Based Analysis for Sizing and Performance Evaluation of Flood Damage Reduction Projects, EC 1105-2 (DRAFT), November 1993.
University of Wisconsin, *Keeping Current*, Environmental Resources Center, Madison, WI (S.M. Steele and C Finley), March/April 1993.
Wilcox, D.L. and Nolte, L.W., *Public Relations Writing and Media Techniques*, Harper & Row, New York, 1990.

4

Ordinances, Regulations, and Documentation

4.1 Introduction

Stormwater ordinances and regulations are a basic tool or method available to local governments for the control and management of urban drainage and flood control problems. Specifically, ordinances or regulations can be used for floodplain management, on-site detention and retention, erosion and sediment control, development regulation, and water quality control and enhancement. There is a difference between ordinances, resolutions, regulations, policies, etc. However, they are considered together in this chapter because the implementation and application of each is often similar. There are a number of regulatory drivers and resulting ordinance types used in municipalities throughout the country.

Passage of the National Flood Insurance Act of 1968 and subsequent amendments provided incentive for local governments to adopt floodplain regulations and ordinances. If a community wishes its citizens to be eligible for subsidized flood insurance and certain postflood disaster assistance, it must meet minimum requirements of the National Flood Insurance Program (NFIP) which includes the adoption of floodplain regulations for designated flood hazard areas. Most local governments with designated flood hazards are part of the NFIP.

With regard to erosion and sediment control, some states passed legislation requiring local governments to enact erosion and sediment control regulations. Maryland was one of the first states to require such measures and continues to be a leader in this area. In 2001, it passed a comprehensive state-wide stormwater quality requirement. Another example is the State of Georgia, which passed an Erosion and Sedimentation Act of 1975 that requires the governing authority of each county and municipality to adopt a comprehensive erosion and sedimentation ordinance establishing procedures governing land-disturbing activities. The State of South Carolina passed a comprehensive Stormwater Management and Sediment Regulation Act. This act mandates that stormwater management and all municipalities within the state enact water quality controls. This is one of the most comprehensive stormwater management state laws.

Federally funded area-wide water quality planning activities in the late 1970s and early 1980s performed pursuant to Section 208, pl 92-500, The Federal Water Quality Control Act, resulted in recommending erosion and sediment control ordinances. Also, federal Environmental Protection Agency (EPA) grants for sewage treatment and interceptor construction require municipalities to adopt erosion and sediment control ordinances as a grant condition. Recently, EPA required municipalities to comply with water quality regulations related to the Clean Water Act.

The 1987 Water Quality Act and implementing regulations led many municipalities into regulatory control of illicit discharges, illegal connections, industrial discharges, and a variety of other nonpoint-discharge-related controls and ordinances.

All of these and other activities resulted in the development and adoption of numerous municipal ordinances, regulations, and policies related to stormwater management.

4.2 Legal Basis and Considerations

Ordinances, policies, regulations, and even city codes and charters, contain within their language applications of basic principles of law. These principles are based on common law, legislation, and case law. Therefore, understanding the legal basis of ordinances and regulations is important to understanding how to apply and amend them. Should a municipality be challenged legally, a clear understanding of the legal basis for decisions will lend credibility and viability to staff policy development and implementation.

Legal consideration of stormwater and floodwater (part of "water law") has developed across the United States based on property rights law in slightly different ways and usually in response to a narrowly defined crisis. Thus, it is not to be considered a seamless whole but a patchwork of cases and judgments, which together can comprise individual contradictory findings and generally consistent applications.

In the course of defining individual property rights, the right of society as a whole and of various governmental bodies is also defined. This definition has followed along separate lines depending on the various locations and states of water: groundwater law, watercourse law, and diffused surface water law (Cox, 1982). While the distinction is sometimes blurred, different decisions for and against municipalities have been based on these separate considerations.

Watercourse Law

Watercourse law developed differently in different groups of states, owing to their different concerns and history of cases. This grouping led to the development of two basic doctrines within watercourse law: the riparian doctrine, generally in the more humid East, and the doctrine of prior appropriation, generally in the arid West.

Riparian Doctrine

The basic concept of the riparian doctrine is that private water rights exist as an incidence of the ownership of land bordered or crossed by watercourses. All owners of such riparian land (defined as a contiguous tract of land in contact with a stream and within the same watershed as the stream) have a constitutionally protected right to certain use of water in the stream, while nonriparian landowners generally do not. That right cannot be taken without due process. A riparian right is reciprocal to other riparian rights on the same watercourse, and all rights along a watercourse must be respected equally.

In terms of water supply, the riparian doctrine does not fix amounts of water rights but allows use to vary with conditions. Unexercised rights cannot be lost by prescription and continue to be attached to property, developed or not.

Historically, the riparian rights doctrine was exercised through the theory of natural flow. According to this theory, each downstream landowner was entitled to the natural flow of a stream, except as diminished by domestic use. As water supplies changed and were made public, this theory of the application of the riparian doctrine gave way to the reasonable use theory of riparian rights.

In the reasonable use theory, each landowner has the right to make use of any water, provided that use is reasonable in relation to the use of other riparian landowners. The application of this reasonable use theory required considerable interpretation for individual cases and, therefore, much ambiguity and seeming (and real) contradiction in similar court cases (APWA, 1981). In application for flood control, a landowner has the right to protect his land from flooding, but only if this protection does not harm another riparian landowner. This then has implications in constructing flood walls, diking land that floods upstream property, undersizing conveyance structures at crossings, filling in floodplains to remove property from flooding, draining wetlands, etc.

While engineers have looked at development in terms of protecting upstream landowners from on-site flooding, judges have increasingly taken an approach that considers downstream property owners' rights not to have the watercourse altered to their detriment.

Prior Appropriation Doctrine

The prior appropriation doctrine evolves out of a scarcity of water resources related to common law application in mining claims and irrigation. Like the riparian doctrine, the appropriative right is a property right. But, it differs considerably in application. The basic premise of the appropriation doctrine is "first in time — first in right." Rights are not equal or attached inseparably to the land but are ranked in a hierarchy established by the date beneficial uses were first initiated and tied to place and type of use, not location. A long period of nonuse presumes abandonment and the allowance of transfer of that water right.

In relation to stormwater management for flood control and water quality, the prior appropriation doctrine has more indirect applications. Flood control activities should not conflict with established water rights. The scarcity of water has implications on water use and reuse in the West (both of which could be covered under NPDES stormwater and other permitting programs). As true water resources grow scarcer in the East, there may be instances of application of the prior appropriation doctrine beyond its present agriculture association of today.

Diffused Surface Water Law

Diffused surface water is the term applied to runoff after it has hit the ground but before it enters a defined watercourse. With few exceptions, the right to use diffused surface water belongs to the property on which it is found. Three doctrines have been developed that apply to diffused surface water. They have been in predominance in about a third of the states each. They are termed the common enemy doctrine, the civil law doctrine, and the reasonable use doctrine. The reasonable use doctrine is gaining more popularity in recent years, though it was the last to arise historically (APWA, 1981).

Common Enemy Doctrine

Under this property-law-based rule, landowners can do anything they want with the water on their land including diversion, concentration, diking, etc. It is a strict interpretation of property rights. Because water is a common enemy, property owners are required to protect themselves as best they can. A more recent modernization of this rule limited the modifications of water flow to those that are incidental to ordinary use, protection, or development of the land. In any case, the lower property is required to accept the flow from the upper property even with increased volume, velocity, and peak.

Civil Law Doctrine

Under this property-law-based rule, landowners should be liable for downstream injury caused by interference with the natural flow of water from their property. However, the upper property has an implied easement for the natural flow across the lower property and is termed the "dominant estate." The key word is "natural," and interpretation of that word has led to many court cases.

Reasonable Use Doctrine

This rule is a compromise rule that falls somewhere between the other two with regard to whether alteration of the natural drainage amount and pattern is lawful. In its application, diversion and concentration of runoff are expected as part of development but are limited to some reasonable amount, and proper engineering care and design are expected. Thus, there is a case-by-case application balancing the utility with the harm of the development.

Liability and Damages

The application of the reasonable use rule is different from the other two in that it is based on tort law rather than on property law. In tort law, liability is based on negligence. There is no need to show negligence, just simple injury or damage under property law.

Individual Liability

Defenses under property law include acts of God, third-party activities that contributed to the injury, or fault of the plaintiff. Under the reasonable use rule, factors considered in judgments include such things as the social value of the activity, its suitability for the location in question, the difficulty in preventing damage, the foreseeability of the harm, the motive with which the developer acted, and the impact on the activity if compensation is required for resulting injury (Cox, 1982; APWA, 1981). The extent of damages is based on the extent of the injury, the character of the injury, the social value of the injured activity, its suitability to the location in question, and the difficulty for the injured party to avoid harm.

Municipal Liability

Municipalities are not ordinarily liable for failure to provide drainage systems or poor system planning. Providing drainage systems is a legislative power with which courts do not ordinarily interfere. Liability for construction as opposed to design is a much more common cause of municipal liability. This is also normally true of poor or negligent maintenance of constructed facilities. Citizens should have a reasonable expectation and reliance on municipal constructed systems to operate as designed and planned. Municipalities have been found liable for concentrating and funneling drainage onto private property and for trapping water by taking away natural outlets. They have also been challenged in court in cases where inadequate openings were supplied for stream crossings. Also, theories of nuisance are being applied to cities to avoid the need to prove negligence. Repeated flooding, sewer backups, and urban development that render existing systems inadequate have been used in nuisance cases, with municipalities found liable.

It is clear that the "jury is not in" on a municipality's or individual's liability for stormwater decisions. However, it is clear that the reasonable use doctrine is being widely interpreted in recent cases. Jury awards have been given for concentration, increase, or diversion of flows, which may have passed the reasonable use test years ago. Cities and counties have been named as codefendants in a number of areas. A changing expectation of stormwater service levels is growing and spreading. Even developing a detailed estimate of the risk in a design may not protect from liability (Lewis, 1992; Kenworth, 1972). Some recent examples include the following:

- Assuming even one-time maintenance of a certain site may imply a continuing responsibility.
- Not maintaining sites where public water makes up a significant portion of the total flow may also open a municipality to some level of liability.
- Allowing accumulated flow increases, which damage downstream properties, may be seen as negligent, even if a municipality follows a standard practice of development controls.
- The courts may see drainage structures, which may back water into upstream property, even if properly designed according to common practice, and even for storms in excess of design, as inappropriate.
- Municipalities, which provide certain types of design assistance to developers (such as establishing flow elevations along creeks), may incur design liability.

"Takings"

The definition of a "taking" of private land by a municipality for stormwater control or flood control purposes is often a murky issue. Cases have been decided that seem to contradict each other. Generally:

- Ordinances and regulations adopted for a valid public purpose may substantially reduce land values without a taking.
- The impact of regulations on the entire land must be reviewed before a judgment of a taking is rendered.
- No property owner has a right to create a nuisance or threaten public safety.
- Regulations are, in general, a taking if they deny all use or all economic use of an entire property, including investment-backed expectations.

Municipalities can avoid takings in a number of ways (ASFPM, 1990):

- By providing variances or special permit abilities in ordinances to deal with difficult cases
- By emphasizing health and safety, avoidance of pollution, and prevention of nuisances in written findings of investigations or hearings in permit denials
- By tying regulations to state or federal programs where possible
- By applying large-lot zoning to areas where intense development would be seen as detrimental (such as floodplains or sensitive areas)
- By documenting with great care the need for regulations and the denial of permits in areas where land values are high

- By encouraging preapplication meetings, where mutually agreeable designs and developments can be envisioned prior to design investment
- By applying regulations consistently and fairly maximizing the opportunities for public hearings and minimizing decision-making based on politics or non-technical issues
- By adopting a development moratorium only for a clearly stated and legitimate reason and for a fixed period
- By coordinating regulatory, tax, and public works policies to ensure that the fiscal burden on landowners for community services is consistent with permitted uses
- By applying, in extreme cases, transferrable development rights to help relieve burdens on landowners
- By using acquisition rather than regulation to protect public interests and to avoid single landowners or groups of landowners bearing the cost for a general public good [Past examples of the use of acquisition include Tulsa, Oklahoma (municipal purchase of homes when they come up for sale), and Birmingham, Alabama (federal home purchase and relocation). In other cases, attempted residential condemnation proceedings were so politically unpopular that viable projects were eventually abandoned.]

Water Quality and Water Law

Many states and localities enacted water-quality control legislation, which is only partly based on the various doctrines and rules for watercourse law and diffuse surface water law. Florida's and Virginia's approaches to stormwater control are good examples of such legislation. For example, the reasonable use concept of the riparian doctrine can serve as a limit on quality changes as well as quantity changes. However, the application of the various water law theories has limited use for water pollution. Also, most legal actions involving water pollution involve violations of state or federal laws and are not handled in court by individuals. Most individual suits in the water quality area are not based on property law but on tort law. It is not uncommon for such suits to oppose pollution on the basis that it constitutes a nuisance or trespass or that it results from negligent conduct.

4.3 Municipal Ordinances

Stormwater management ordinances for municipalities normally have four major components, and today, they normally address stormwater quantity and quality (EPA Region V, 1990).

Legal Authority and Context

Depending on the type of state government, municipalities may or may not have the derived authority to control their stormwater management quantity and quality. Chapter 1 briefly describes basic derived legal authority for stormwater management. Often, stormwater provisions are scattered throughout other codes. In these cases, municipalities sometimes develop a compendium of all such related codes and publish, informally, such a collection for the aid of developers and engineers (not to mention regulators) in the municipality.

The context for whom and for what the ordinances are written is quite important. Grandfather and severability clauses are important in case any portion of the ordinance is invalidated.

Technical Basis

The ordinance should provide clear, technically based, and measurable performance standards rather than design criteria standards. For example, "control the 2- and 10-year storm such that the postdevelopment peak is equal to or less than the predevelopment 2- and 10-year peaks" is preferable to a set of plans and specifications on how detention ponds must look.

Administrative Apparatus

A clearly identified administrator of each ordinance should be identified in the ordinance. If procedures under the ordinance are required, checklists, forms, and steps should be clearly laid out and provision made within the local government for efficient implementation of the requirements. These should be included in implementing regulations or policy documents to ease the revision process.

Enforcement Provisions

Enforcement must be fair, consistent, stepped, and swift. Several steps of enforcement beginning with a warning and ending with a penalty large enough to stop and punish any would-be offender should be included. The mechanism and will power to carry out such enforcement actions must be provided with the ordinance. It is better not to have such an ordinance than to have it and not be able to carry out its provisions. The ability to obtain and execute a stop-work order is important in situations in which development is harming the environment or causing a health or safety hazard. Rigorously training and then empowering field inspectors as the first line of defense is the best method for effective enforcement.

Special Water-Quality Considerations

The drafting of an ordinance for stormwater quality involves considerations beyond those for stormwater quantity, such as the following (EPA Region V, 1990; Ogden, 1992):

- Many municipalities have, but have not taken advantage of, provisions in charters or codes for public health protection, which can include stormwater quality provisions. The municipality should spell out these or enabling legislation developed to clearly indicate the authority for stormwater quality. The legal authority for stormwater must be sufficient to prohibit illicit discharges and illegal dumping, allow for inspections, and require compliance with any program or policy provisions.
- Technical criteria for stormwater quality should be carefully thought out and spelled out. Capture of the first one or one-half inch of runoff can be shown to

make sense in most climates and for most urban-sized basins. However, other provisions (for example, prohibition of curb and gutter) must have appropriate technical backing, program, and political support before extension of pilot or literature results to the whole municipality is implemented.

- Technical analysis required, even tacitly, within an ordinance for general development should not be beyond the means of the general civil engineering public to perform. This can be accomplished through education and sound design criteria manuals.
- Administration of a stormwater quality ordinance must be based on easily measurable performance criteria and not be arbitrary or allow too much room for judgment, when such judgment is nearly impossible to provide by the available staff (or by any staff).
- Violations of stormwater quality criteria are more difficult to prove without exhaustive monitoring programs. The provision of self-monitoring for industry may be applicable in some cases. In most cases, the fact that a given BMP is being maintained to physical standards should be sufficient. In other words, it is easier and fairer to hold landowners and developers to physical design criteria than to criteria that must be stated and enforced in terms of measured pollution reduction. Therefore, it is incumbent on ordinance writers to attempt to link pollution reduction amounts to physical designs. These designs and their maintenance then become the measure of compliance, not the chemical condition of the receiving water. Adjustments can always be made in the criteria in the future, if acceptable results are not obtained.
- Stormwater quality ordinances must allow room for innovation and regionalization or physical treatment. This may be difficult, but sufficient safeguards can be built-in-terms of extra monitoring to ensure performance of unknown and untried BMPs or designs.

Detention

With regard to drainage and flood control, many local governments are requiring on-site detention facilities to reduce downstream runoff peaks, because they feel it is good stormwater management practice to do so. Residents downstream of new developments are increasingly questioning the wisdom of allowing upstream developments to cast more water downstream for them to deal with. Such concern puts pressure on local governments to require on-site detention in new developments. In addition, some states passed laws related to use of detention facilities. The State of Pennsylvania passed a Stormwater Management Act in 1978. This act requires the development of stormwater management plans for all watersheds within the state. Section 13 of this law, titled, Duty of Persons Engaged in the Development of Land, states the following:

> *Any landowner and any person engaged in the alteration or development of land, which may affect stormwater runoff characteristics, shall implement such measures consistent with the provisions of the applicable watershed stormwater plan as are reasonably necessary to prevent injury to health, safety, or other property. Such measures shall include such actions as are required:*
>
> *(1) to assure that the maximum rate of stormwater runoff is no greater after development than prior to development activities; or*
>
> *(2) to manage the quantity, velocity, and direction of resulting stormwater runoff in a manner which otherwise adequately protects health and property from possible injury.*

Implementation of item (1) almost always involves the use of detention storage facilities. Such a requirement is the major focus of many municipal stormwater ordinances that have been adopted. This has resulted in the proliferation of many small detention storage facilities in many urban areas. As a result of the increased use of ordinances and regulations to control quantity and quality problems related to storm runoff, municipalities have gained experience with implementing these regulations. Some areas are experiencing the positive and negative effects resulting from their implementation. From this experience, professionals involved with urban stormwater management are posing several interesting questions: Is the adoption of ordinances, which generally require the use of detention facilities on all developments, the answer to flooding and drainage problems? Do sediment control regulations adequately protect water resources? What engineering and technical methods should be included in a community's stormwater management program? What controls should be included in stormwater quantity and quality control?

If a detention ordinance is implemented, three additional factors should be present. First, the municipality should have a mechanism to encourage and implement regional storage facilities. Second, the detention pond should be designed to control downstream impacts, not just peak flows at the outlet. Third, a comprehensive inspection and maintenance program should be implemented.

4.4 Drafting Local Ordinances and Regulations

As stated earlier, stormwater management ordinances can be used to address a variety of subjects. Separate ordinances can be written for each of the areas identified, or a single comprehensive ordinance can be adopted to address all stormwater-related problems. For example, Austin, Texas, adopted a comprehensive-type watersheds ordinance that covers site development and subdivision regulations, watershed protection measures, maintenance, and enforcement (Austin, 1986). For the purposes of this section, a stormwater management ordinance will refer to a comprehensive ordinance that sets forth a local municipality's policies related to drainage, flood control, and water quality aspects as well as the legal framework for permitting implementation of the controls. A stormwater management program will include the ordinance and supporting program elements (i.e., technical tools needed, financing strategy, and engineering design manuals).

Before any municipality drafts and adopts a stormwater management ordinance, some preliminary study and evaluation should be made. The municipality should identify the problem areas that the ordinance must consider, and basic decisions should be made concerning the municipality's philosophy in dealing with stormwater management problems. The general format of the stormwater management ordinance should be developed to effectively reflect the goals and objectives of the municipality.

Table 4-1 and the discussion that follows describe a five-part process involved in drafting a stormwater management ordinance.

Identify Problems and Issues

Before any aspect of a stormwater management program is considered, the problems and issues related to stormwater quantity and quality must be determined. Identification of the problems will define in general terms the scope and extent of the stormwater management program. Although each municipality should identify its own problems and

TABLE 4-1

Process for Drafting Stormwater Management Ordinances

Part 1 — Identify Problems and Issues
Part 2 — Formulate Objectives
Part 3 — Develop Policies
Part 4 — Draft the Stormwater Management Ordinance
Part 5 — Develop Stormwater Management Design Manual

issues, listed below are some examples that have been addressed in several different municipalities:

- Control of local stormwater management problems
- Flood control
- Enhancement of water quality
- Control of erosion and sediment
- Economic efficiency in stormwater system design
- Improvement of engineering design and analysis
- Protection for present and future development
- Equity in stormwater design and implementation
- Control of development within floodplain areas
- Control of runoff from proposed developments
- Maintenance of stormwater management facilities
- Obtainment of funding sources for stormwater management programs
- Control of nuisance problems within the drainage system
- Enhancement of the local environment
- Addressing of citizen complaints and problems
- Provision of training and education related to stormwater management
- Efficiency in review and inspection procedures
- Encouragement of public participation in program development

The local municipality must determine which problems it needs to include in its stormwater management program and then proceed with the formulation of objectives.

Formulation of Objectives

Many times in developing stormwater management ordinances and programs, little consideration is given to stating objectives to be accomplished by the program, keeping the objectives up-to-date as the program is developed, or evaluating the adequacy and making appropriate changes to the objectives as the program is developed and implemented. The results of an extensive study of urban floodplain programs in five metropolitan areas indicated the important role objectives could play (Debo, 1976). Many of the persons interviewed as part of this study gave what they considered to be the objectives of their stormwater management program and also stated that this represented their opinion and not an official position of the local government. In contrast to what was done, when the subject of objectives was brought up during interviews, technical and nontechnical personnel felt that the establishment of a definite set of objectives is not only important but also should be one of the first steps in formulating a stormwater management program. Furthermore, many of those interviewed felt that objectives should be used to give direction and feedback to the overall program. Experience in several municipalities since this study was completed confirms the results and further indicates the importance of establishing definite objectives for the stormwater management program.

For example, Charlotte, North Carolina, developed a complete set of policy statements through a process of citizen, consultant, and staff interaction over a period of many months. Louisville MSD, Kentucky, established policy goals in each key area of stormwater management in the development of its program. These policy goals serve to guide future development of ordinances and program elements. Prince George's County, Maryland, is a good example of overall goals for a stormwater program (Prince George's County, 1984):

> ...to protect, maintain, and enhance the public health, safety, and general welfare by establishing minimum requirements and procedures to control the adverse results of increased stormwater runoff associated with land development. Proper management of stormwater runoff will minimize damage to public and private property, reduce the effects of development on land and stream channel erosion, assist in the attainment and maintenance of water quality standards, reduce local flooding, and maintain, nearly as possible, the pre-development runoff characteristics of the area.

It is also important that all personnel be informed of the stormwater management objectives and when changes are made. Such information should be published and promulgated by the municipality. In the municipalities studied, personnel within the local governments as well as outside consultants complained that they did not know about several aspects of the total program and were not kept informed of program changes (Debo, 1976). One means of keeping everyone up-to-date would be for the local government to publish a newsletter documenting changes and other aspects of its stormwater management program.

The opinions of numerous elected officials were obtained as part of this study concerning different aspects of their local stormwater management program. All interviewed said that objectives were essential. They felt objectives help evaluate long-term planning, help interpret regulations and ordinances, and give direction to the program. Some elected officials said objectives should be separately stated but contained in the ordinance, while others felt they should be contained in another document, such as a general plan or resolution. Others said objectives were inherent in any document and need not be written separately. The politicians did not offer any specific objectives, but several stated that they should be specific and simply stated in contrast to some of the more general and elaborate ones contained in many planning documents. Most of the politicians saw no particular problems in arriving at stated objectives for their stormwater management programs, although they could not give specific reasons why their present program existed without some stated objectives.

Without properly defined objectives, known to the people involved in the stormwater management program and actually representing the local municipality's view, the major effort of the personnel involved with the program is with the day-to-day problems and their immediate solutions. Little, if any, time is given to any of the following important aspects of the program, which would have an impact on the implementation of the program:

1. Long-range implications of present actions
2. The direction and scope of the stormwater management program
3. The effectiveness of implementing the different program elements
4. Changes that should be made in the program to achieve desired objectives

5. Determination of what is being accomplished by the present program and what the local community wants to accomplish

It should be pointed out that stating objectives will not guarantee that the above five aspects will receive proper consideration, but it should at least bring some of these aspects to the attention of the policy makers and those implementing the stormwater management program. Thus, it is important to have stated objectives for the stormwater management program rather than inherent objectives. Stated objectives should enhance evaluations of different aspects of the program and lessen the chance of misinterpretations and forgetting what the objectives are.

It is often the case that "specialty" ordinances are necessary. In this case, there will be narrowly defined objectives to meet specific needs in support of a more comprehensive program. These special ordinances must fit into the framework of an overall stormwater program. Some municipalities, for example, instituted detention ordinances without commensurate programs for inspection, enforcement, and maintenance. The result has been a general failure of detention due to clogging, structural failure, and vandalism. There have been municipalities that have actually begun abandonment proceedings for detention to eliminate redundant or ineffective detention ponds. Examples of such specialty ordinances include such topics as lawn and fertilizer control (Shoreview, 1985, 1988), a wetlands overlay district (Wayzata, 1988), a hazardous materials ordinance for spill control (Louisville MSD, 1988), or the plethora of floodplain management regulations.

Developing Policies

A governing body must officially adopt the stormwater management ordinance before it can be legally applied. As a result, the political process becomes the controlling force behind such a document. Therefore, it should be designed so that it can be understood by elected officials and aid them in the execution of their responsibility.

Elected officials should not be dealing with engineering criteria or design standards. Also, most elected officials are not administrators and should not be concerned with the detailed administrative aspects of a stormwater management program except as they relate to budget control. Elected officials should be more concerned with determining the overall policies within their jurisdiction. Thus, to make the stormwater management ordinance a viable part of the political process, it should serve as a policy statement for the municipality. Technical and engineering details related to the program should be included in other documents.

Some administrators prefer ordinances that are detailed and attempt to cover all problems that may arise. In addition, they tend to include engineering criteria and design standards in order to encourage uniformity. The administrators who must deal with the public feel they can be firmer in enforcing the provisions of the stormwater management program if the provisions are written into law. They contend that, if many members of the public, or a few influential ones, complain to elected officials, these officials may find it easier to use their political pressure to allow exceptions if a specific point of contention is not written into law. In contrast, other administrators feel that stormwater management ordinances should be limited to stating specific policies. Engineering criteria and technical details should be contained in other documents in order to fully discuss these criteria and details and keep them up-to-date. Elected officials could then refer technical questions to the engineering or public works departments and spend more time and effort dealing with policy matters. In order to prevent the engineering or public works departments from becoming completely autonomous from the public and political influence, engineering documents should be subject to the approval of elected officials, without formally being enacted into law. Or, they may be adopted, by reference, in the ordinance. This would encourage some uniformity and give administrators a ready reference for dealing with problems and combating political pressures to make exceptions. Bellevue, Washington, used the by-reference method to adopt a stormwater master plan and system map (Bellevue, 1974).

Generally, the ordinance and supporting documents should state what should be accomplished and why. Technical manuals and other guidance should provide the "how" information. Ordinances should give performance-based criteria but leave the means and methods of accomplishing these criteria up to other, more technical sources. For example, a detention ordinance may give the design criteria (such as "control the 2-year and 10-year storms to the predeveloped peak flow values at the outlet and to a designated point downstream"), while the design manual gives analysis and design specifics.

Some examples of policies that should be considered in a stormwater management ordinance are the following:

- Which aspects of the stormwater management ordinance should address frequent flood events (i.e., 10-year flood), and which should address rare events (i.e., 100-year flood), and which aspects should address both types of flood events?
- Should the ordinance provisions be applied only in large drainage areas or also in small ones, and if so, how small?
- Should the intent of the ordinance be to pass the increased runoff from new developments downstream or to control runoff on-site?
- Should the regulations and standards of the stormwater management program vary from one part of the municipality to another or be uniform throughout the municipality?
- Is the local government going to maintain the entire drainage system, certain portions of the system, or require private maintenance?
- Should flood storage facilities be centralized or decentralized, publicly owned or privately owned?
- What restrictions should be placed on development related to flood control and control of drainage problems?
- What erosion control and water quality enhancement practices should be required by the ordinance?

These and other questions must be answered in order to develop policies that can be included in the stormwater management ordinance. Sometimes, obtaining a consensus from the local politicians, personnel from the planning, engineering, or public works department can do this. Other times, technical studies may be needed to determine appropriate policies to accomplish specific tasks. Following is an example of the results of a study, which were used to develop a policy concerning downstream limits for hydraulic analysis of detention storage facilities.

Detention Analysis Example

After years of requiring engineers to analyze peak flows from proposed developments at the exit from these developments, municipalities are starting to realize that limiting the

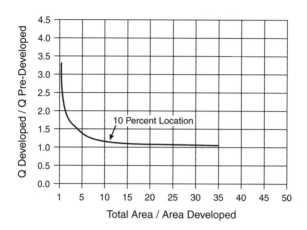

FIGURE 4-1
Determining downstream analysis limits.

hydraulic analysis of stormwater management facilities to the exit from the proposed development or detention site is not sufficient to prevent increased downstream flooding and drainage problems. Thus, there is a need for a policy to address this problem that could be incorporated into local ordinances and regulations. What is agreed is that hydraulic analysis should extend downstream from proposed developments, but how far downstream the limits of this analysis should extend is difficult to determine. Requiring that all analysis extend a standard distance downstream (i.e., one mile) may not be a sufficient distance for a very large development or may be much too far for very small developments. It must be remembered that any required analysis will cost money that must eventually be paid in higher development costs. Thus, some procedure needs to be developed to account for developments of different size, location, and land-use density. The policy should also not require expensive engineering and technical studies that will be a burden to the development process.

A study completed for the city and county of Greenville, South Carolina, and Raleigh, North Carolina, demonstrates how such a policy could be developed (Debo and Reese, 1992). This study used a hydrologic-hydraulic computer model to analyze the downstream effects of storm runoff from developments of different size, shape, physical characteristics, and location within larger drainage basins. This study also examined different size flood events and different types of downstream drainage systems. The results of numerous computer runs of watersheds located throughout the United States were analyzed, and Figure 4-1 shows a typical summary of these results. It can be seen from this figure that the effects of the development process stabilize at the point where the proposed development represents from 5 to 10% of the drainage area, depending on the size of the development and the amount of increased impervious area. This analysis was then used as the basis for developing the following policy statement that has been incorporated in the Greenville ordinance and several other municipal ordinances:

> In determining downstream effects from stormwater management structures and the development, hydrologic-hydraulic engineering studies shall extend downstream to a point where the proposed development represents less than ten (10) percent of the total watershed draining to that point.

Similar computer model analyses can be effectively used to develop other reasonable and adequate policies that can provide stormwater management facilities designed to protect local residents without undue cost to the development process. Modeling the effects of different policies can provide valuable information in the policy development process.

Drafting the Ordinance

To be an effective part of the municipality's stormwater management program, the initial draft of the stormwater management ordinance should be a team effort. The team should consist of representatives from the engineering, public works, planning, environmental management, legal, and any other affected departments. All of these departments will play a significant role in solving local flooding, drainage, and water quality problems, and excluding them from active participation in the drafting of the ordinance would be a major mistake. It should be remembered that the ordinance, if properly drafted and implemented, would play a major role in guiding the entire stormwater management program for the municipality. If the final ordinance is not acceptable to any of these departments, its prospects for implementation are usually quite low. The team effort in drafting the ordinance seems quite logical, but, in many municipalities, one or two departments have drafted ordinances without consulting the other departments.

Stormwater management ordinances will eventually affect, either directly or indirectly, many different interest groups in the local community, including citizens, developers, builders, realtors, investors, politicians, environmentalists, consulting engineers, landscape architects, surveyors, and local government employees. Some groups will be involved in the implementation and enforcement of the ordinance, while others will have to comply with specific ordinance provisions. In order to get as much local support as possible, it will be necessary to involve representatives from all of these groups in the formulation and drafting of the ordinance. Public participation should be integrated into the initial phases of the process (i.e., when problems and issues are established) and not used just to endorse the final draft of the ordinance.

The involvement of local stakeholders is usually important to obtain a buy-in to the process and to help avoid regulatory language that may prove to be unworkable or unreasonable. Grassroots consensus is normally preferable to top-down imposition of regulations. A balance of development and business interests and environmental or homeowner groups allows an even consideration of all aspects of policies. The use of "Blue Ribbon Committees" has worked quite well in many municipalities. Under this arrangement, representatives from local interest groups are placed on a committee whose function is to review and comment on proposed provisions of the ordinance. The committee also provides a means of public input into the process of drafting the ordinance. The final ordinance then goes to the local governing body with the support of this committee.

These stormwater ad hoc task forces or advisory committees can have varying levels of authority and responsibility, depending on how decisions are made within any given community. In one Midwestern city, the advisory board retained great authority in the development of policies and presented an independent report to council recommending the final policy statements. In another Eastern city, a group provided only advice to the staff and had little say in the final ordinance.

Many times, municipalities will hastily adopt a stormwater management ordinance in order to conform to federal or state mandates. To comply with these mandates, municipalities may be tempted to obtain ordinances from nearby municipalities or even municipalities thousands of miles away. The local municipality then changes the names in the ordinance to fit local names and adopts the ordinance. It seems quite obvious that flood, drainage, and water quality problems in Los Angeles, California, would be quite different from those in Tampa, Florida, or Atlanta, Georgia. Thus, an ordinance drafted for one area may not be appropriate to deal with the problems in another area. This distinction might not be quite so obvious with two municipalities that are geographically close. However, when using an ordinance as a policy statement, one has to consider more than just the physical differences between municipalities, as each may have different goals and objectives, local laws, or past flooding and water quality practices that might affect the specific provisions of an acceptable ordinance.

There are also numerous ordinances available in the stormwater management area, such as those developed by Maryland (1983) or northern Illinois (IDOT, undated) to assist its local counties and municipalities by providing minimums for their own program's development. Care and judgment should be exercised when using any model ordinance. Because model ordinances are written to cover many aspects of stormwater management, they are general in nature and are not aimed at specific local problems. The Maryland case is a good example in that it required the use of control measures, including infiltration practices. Many of those communities that required infiltration practices did not have the technical or organizational wherewithal to carry out the provisions of their own ordinances, resulting in a high failure rate of the infiltration practices (Galli, 1992; Schueler, Kumble, and Heraty, 1992).

While model ordinances and ordinances from other areas should be evaluated, each community should draft its own ordinance tailored to its goals, objectives, and problems. Municipalities would be wise to spend adequate time in formulating and drafting the ordinance before it is adopted. Otherwise, they may find that the ordinance must be continually revised or amended — a process that would eventually erode the public's confidence in the municipality's ability to deal with stormwater management problems.

If a municipality does not have an engineering or a planning department, or both, that can draft the ordinance, it should seek professional assistance from another level of government (regional, state, or federal) or employ the services of an experienced consulting engineering or planning firm. Obtaining assistance from a professional outside of the municipal government is usually recommended in order to provide some unbiased input to the ordinance development process. It is important to obtain the proper professional assistance during the drafting of the ordinance in order to prevent as many problems as possible.

Logically, development of a local ordinance using a model ordinance should follow these steps:

- Develop the means to provide input from politicians, local citizens, technical groups, and others interested in being involved.
- Identify and characterize the stormwater management problems in the local area.
- Establish goals and objectives for the ordinance.
- Examine each provision of the model ordinance to determine its applicability to the local area.
- Eliminate those provisions that do not apply, and make revisions where necessary to adapt the model provisions to local needs (this may also include adding some new provisions).
- Draft the local ordinance using the revised provisions.

- Have a legal review made of the ordinance to ensure its compatibility with local, state, and federal laws. In addition, the legal staff should be involved during development of the ordinance to avoid any legal problems.

Stormwater Management Design Manual

If the ordinance is designed to function as a policy statement, then the municipality should publish a manual that would contain the engineering criteria and design standards needed for implementation of the stormwater management program. This manual would contain the technical information needed for the flood and drainage aspects and the water quality control aspects of the program.

If it is desired that the manual have the force of law, then it must be incorporated by reference in the original enactment (ordinance) or adopted at some later time by the governing body. This will necessitate that the municipality formally adopt any changes to the manual after it is adopted. In most cases, the stormwater management design manual would not be officially enacted into law by the municipality but could be subject to its approval and kept in a form such that it could be continually updated and changed to conform to the latest criteria and procedures. One of the earlier comprehensive manuals, developed by the Denver Urban Drainage District, uses this procedure, and publishes the manual in loose-leaf notebooks so that it can be easily updated (Storm Drainage Criteria Manual, 1969). Following is a quotation from this manual, which further emphasizes the need for updating such a manual.

> *A compilation of engineering criteria such as the Urban Drainage Criteria Manual is a dynamic rather than static volume and needs to be reviewed and updated to keep it abreast of developments in the important and rapidly expanding field of urban storm drainage. It is the intent of those responsible for the conception and development of the Manual to periodically issue revisions to the Manual which incorporate new data, methods or criteria, and such other information as may be deemed appropriate.*

This manual has been updated through the years with several revisions.

The major thrust of a stormwater management design manual is to present criteria and examples such that a design engineer can determine generally what the municipality in different flood, drainage, and water quality designs expects. The manual should not be a limiting document forcing the engineer to use standard designs and procedures, but instead should present performance standards. Numerous design examples should be given. The manual should have the connotation of a document that suggests and informs rather than limits or prescribes. The manual should not replace good engineering judgment.

The municipality's engineering department should be assigned the responsibility of keeping the stormwater management design manual up-to-date. Periodic reviews of this document should be made by an elected official or a special committee selected for this purpose (i.e., every five years) or when major changes in the manual occur. This would enable the elected officials to be kept informed of engineering and technical changes, which might affect stated policies and objectives, and also provide a check to ensure a sound basis for a given update. If needed, elected officials or committee members could obtain outside opinions on engineering and technical matters to augment those provided by the local engineering department.

Stormwater management design manuals have recently been developed for several municipalities in North and South Carolina. As an example of the general content of these

TABLE 4-2

Stormwater Management Design Manual — Greenville, South Carolina

Table of Contents
Chapter 1 — Introduction
Chapter 2 — Stormwater Management Ordinance
Chapter 3 — Data Availability
Chapter 4 — Documentation
Chapter 5 — Hydrology
Chapter 6 — Storm Drainage Systems
Chapter 7 — Design of Culverts
Chapter 8 — Open Channel Hydraulics
Chapter 9 — Storage Facilities
Chapter 10 — Energy Dissipation
Chapter 11 — Infiltration Facilities Design
Chapter 12 — Best Management Practices

manuals, Table 4-2 gives the table of contents for the Greenville, South Carolina, Stormwater Management Design Manual.

4.5 Flexibility in a Stormwater Management Program

One of the initial steps in formulating a stormwater management program is to thoroughly evaluate the general policies and objectives to be sought by the program, and whether a flexible or rigid ordinance should be used to implement the program. A flexible ordinance would be limited to stating the general policies of the stormwater management program and allow judgment to be used to determine the best solution for given stormwater management problems. Rigid ordinances are specific, technically and administratively, and require standard solutions for stormwater management problems. As an example, the following ordinance provisions might be included in flexible and rigid ordinances.

Flexible Ordinance Provision

If the required hydrologic and hydraulic studies reveal that the proposed development would cause increased flood stages for the design storm, so as to increase the flood damages to existing developments or property, or increase flood elevations beyond the vertical limits set for the floodplain district, then the development permit shall be denied unless one or more of the following flood mitigation measures are used to solve anticipated flooding and drainage problems:

- On-site storage
- Off-site storage
- Existing drainage system improvement
- Floodproofing

Rigid Ordinance Provision

All developments over five acres must use on-site storage to limit the peak runoff rate from the site to that which would have been expected from predevelopment conditions.

The design of the storage facility should be based on the Rational Formula Method with a C = 0.3 for predeveloped conditions using a design storm of 10-year frequency.

A flexible stormwater management program encourages the design engineer to evaluate each stormwater management problem and try to devise an optimum solution for that problem. Also, the local government retains flexibility in administering and implementing its program. Thus, it is possible that different restrictions and designs could be applied to different areas of the community. However, a flexible stormwater management program will introduce several possible problems in administering the program:

1. The municipality will have to hire sufficient qualified personnel to evaluate each development on a case-by-case basis.
2. In addition to studying the drainage within a development, the drainage interrelationships between all the developments within a watershed and among watersheds will need evaluation.
3. The municipality will need to obtain the technical aids necessary to do the above evaluations.
4. The municipality should consider taking an active role in the design, operation, and maintenance of the drainage system.

In contrast, many municipalities have written rigid ordinances that legislate standard evaluations, designs, and solutions to stormwater management problems. Personnel from these municipalities feel it is easier to administer a rigid program, because a standard set of rules can be developed and applied to all developments. They also contend that the administration of a rigid program requires fewer and less-qualified personnel than would be required for a flexible program.

The major disadvantage of a rigid program is that it may force too much conformity in engineering designs. Each stormwater management problem involves certain aspects that are unique to a particular situation. These unique aspects are difficult to take into account in formulating general engineering rules and criteria. As an example, allowing an increase of 5 cfs or 10 cfs in a drainage design may be appropriate for a large drainage system but may be totally inappropriate for a small design, where the capacity of a grate inlet would be exceeded, and runoff would flow into an adjacent building. In addition, design methods and procedures change with time, while many times, provisions of ordinances and programs do not. Thus, a rigid program may not allow for the use of the most economical or cost-effective solutions.

In many instances, the basis for including in the stormwater management ordinance certain design and engineering criteria was that these criteria were adopted many years ago and had been accepted because of their continual use. Thus, the use and acceptance of engineering criteria can result from their inclusion in an ordinance or because of their continued use. Adopting a rigid program for administrative ease has advantages and disadvantages that should be thoroughly evaluated. It should be remembered that rigid stormwater management ordinances and programs could stifle much of the ingenuity and imagination that could be used to solve stormwater management problems.

In practice, most stormwater management programs incorporate ordinances that are somewhere between a flexible and rigid approach. A program with specific guidelines that also allows for deviation from the standard if it can be shown that such deviation is in the best interest of the municipality is usually the best approach. It is usually an advantage to have some uniformity in engineering designs, but standard solutions should not preclude the use of good engineering judgment in the evaluation of stormwater management problems. Each municipality will have to determine how much flexibility it will allow in its program, depending on local conditions (e.g., available staff and resources, severity of local problems, political climate). The decision is not a flexible or rigid ordinance, but how much flexibility should be incorporated into the ordinance.

4.6 When to Adopt an Ordinance

When to adopt a stormwater management ordinance can be almost as important as what is included in the ordinance. Stormwater management problems have some unique characteristics that separate them from the usual problems involved in other land use controls. As an example, if some land use controls (or zoning ordinances) are not used to prohibit certain combinations of development from taking place, you may end up living next to a large factory or other noncompatible use. Once the factory is constructed, local residents must coexist with it 24 hours a day, seven days a week. Thus, the conflicting use is always present. Urban stormwater management problems, especially flooding problems, are quite different. On a warm sunny day, the stream that meanders by a residential lot can be a pleasant neighbor and can offer hours of enjoyment for adjacent residents. But, on some occasions, which may be rare, this neighbor may expand, overtop its banks, and create a ravaging flood that damages property and destroys land areas. Days later, when the sun shines, this same neighbor is safely back in its banks.

Thus, the types of problems created by urban flooding have several unique characteristics:

- Urban flooding and water quality problems normally affect only a small portion of the total citizens in a community, those who happen to live near the ditches, streams, and rivers (though all citizens are affected if streets are flooded and unsafe).
- In most cases, those citizens damaged by flooding are the same ones who enjoy the benefits of the stream or river during nonflooding periods.
- During or shortly after flooding events, interest and public sympathy are high for those damaged, but this quickly diminishes with time.
- Flood-prone areas offer many advantages for development in spite of the potential dangers (e.g., aesthetic qualities of streams and rivers, relatively flat and easily developable land).
- Correction of stormwater management problems can range from simple, inexpensive floodproofing to complex, expensive flood control projects.
- Major flooding events may occur rarely with long periods where flood-prone areas are unaffected by floodwaters.
- Surface drainage problems cause many problems, particularly of an inconvenience nature that are difficult to assign monetary values to.

Figure 4-2 graphically depicts the interest cycle associated with flood-related problems. The interest cycle is referred to as a citizen and government interest and involvement hydrograph. The basic point depicted in the figure is that too many urban stormwater management ordinances are being adopted as a result of federal or state mandates or as an emotional result of a major flooding event. These ordinances are adopted without a community commitment to their implementation and the costs and resource needs this

FIGURE 4-2
Citizen and government interest and involvement hydrograph.

commitment would entail. An ordinance is only one element of a complete drainage program, is relatively inexpensive to write and adopt, and can represent only an outward commitment to correct the associated problems.

Adopting an ordinance early in the development of a stormwater management program can have several disadvantages:

1. The ordinance should represent the policies of the municipality relative to stormwater management problems. These policies should be thoughtfully developed after careful study.

2. Adopting an ordinance can be an end in itself. Once an ordinance is on the books, the body politics of a community can feel their job is done and the problems are solved. A typical response to future citizen complaints could be, "We have adopted an ordinance that should solve your problems." Obtaining further commitments such as money and staff can be difficult.

3. The ordinance should reflect the municipality's financial and technical abilities to implement its provisions. Thus, the financial and technical requirements associated with a stormwater management program should be known before the ordinance is adopted. The optimum time to adopt an ordinance will vary from one municipality to another, but an ordinance should not be adopted until the community has a definite commitment to its implementation and has developed a stormwater management program to a point where the ordinance reflects a realistic approach to deal with the local problems. Thus, municipalities should stop adopting ordinances that are only intended to comply with federal and state programs or result from emotional consequences, and, instead, should adopt well-thought-out and locally supported ordinances. It might be better not to adopt any ordinance rather than to add one more meaningless regulation to what might be a long list of existing ones.

4.7 The Complete Stormwater Management Program

Drafting and adopting a stormwater management ordinance, although an important part of any stormwater management program, cannot be viewed as encompassing a complete program. Several other important aspects of the program must be considered and are discussed in the following sections.

Land-Use Planning Aspects

A stormwater management program has several land-use planning aspects that deserve mention. One question that should be discussed is, "Where should the ordinance and other aspects of the program be applied in the planning process?" Should the ordinance be applied during the initial planning stages or not until specific development proposals are reviewed? The answer is, "The ordinance and other aspects of the drainage program should be applied at several stages in the planning process."

A stormwater management ordinance usually states specific objectives related to the stormwater management program. These objectives should be incorporated into an overall-planning program. The ordinance also states specific land-use policies (areas included in ordinance restrictions, uses allowed in specific areas, special permits, etc.) that should be included in all land-use planning efforts. Provisions of the ordinance have specific consequences with regard to the zoning process (improvements required, nonconforming uses, hydrologic and hydraulic studies required, and provisions related to land uses). Those provisions, which affect different land-use classifications, should be taken into account when preparing zoning districts and when deciding on zoning changes and variances.

Municipal Role in Encouraging Innovative Solutions

One of the major purposes of adopting a flexible stormwater management program is to encourage innovative solutions to stormwater management problems. In many municipalities, the private engineering sector has not been particularly innovative in finding new solutions to stormwater management problems. The primary interest of the engineer, most likely spurred by pressure from developers, is to get development plans approved by the local government so proposed construction can proceed. Also, many engineers feel that rigid rules and regulations, which must be followed in order to get plans approved, discourage innovation. Engineers and developers know it is much easier to get approval for standard designs than for new or untried ones. As a result, if innovative and imaginative designs are going to be applied to flooding problems, it will be up to the local government to take the lead and provide the climate necessary to encourage such designs. To accomplish this, the local government will have to hire qualified personnel who can use their knowledge and expertise to propose, encourage, and efficiently review new and different approaches to stormwater management problems. However, many times, those people who appropriate funds disagree with the need to hire sufficient qualified personnel and thus limit the effectiveness of a stormwater management program.

In defense of the "not-so-innovative," it should be noted that people in public works often need to be conservative. Mistakes or poor and ineffective designs are often lived with for a long time. Also, if a public works facility or concept fails, the consequences can be serious and sometimes life-threatening. Innovative solutions are often unproven, and care must be exercised in their application. Thus, the benefits of an innovative versus a tried-and-true approach must be weighed against consequences and the results of possible failure.

Administrative Problems

It is one thing to adopt regulations requiring detailed engineering evaluations and another to administer and enforce such regulations. Strict compliance with this ordinance could require costly engineering studies for many developments in which common sense would indicate that such studies may not be necessary to make needed decisions. Evaluation of upstream and downstream effects of proposed development and associated stormwater management facilities, evaluation of the effects of a proposed development on flood elevations, design and evaluation of storage facilities, and the interrelationships between several storage facilities and drainage structures all present complex hydrologic and hydraulic problems. Solutions are difficult, if not impossible, with traditional engineering methods, and most municipal engineering staffs and local consultants do not have the resources necessary to effectively implement such provisions. Thus, the administrative, planning, and engineering aspects of a stormwater management program should be coordinated so that one aspect does not produce problems for another.

Technical Requirements

Adequate administration of a stormwater management program requires that the municipality have the necessary technical aids to properly assess the effects of proposed developments. In some communities (e.g., Cobb County, Georgia; Greenville, South Carolina; Chapel Hill, North Carolina), hydrologic computer simulation models have been developed and used for this purpose. Such computer simulation models are extremely helpful in administering a comprehensive stormwater management program. If municipalities adopt a program dealing with such areas as the combined effects of several developments, drainage interrelationships between different watersheds and subwatersheds, floodplain map updates, and detailed evaluations of the drainage effects of proposed developments, then they should accept the responsibility of providing the technical tools and financial resources needed to administer the program. Development of these technical tools can be quite expensive and time-consuming; thus, a commitment to their development and use should be obtained before adopting the stormwater management ordinance.

Staff and Financial Resources Needed

Developing and implementing a comprehensive stormwater management program will necessitate employing qualified professionals with expertise in hydrology and water resources planning. Adequate funding of the program will also be critical to effectively implement all of the elements involved. The staff and financial resources needed should be assessed early in the development stage of the program so appropriate budgets can be established and political commitments of support can be obtained.

Field Inspection

Although it is relatively easy and inexpensive for a local government to adopt regulations, it becomes much more difficult and expensive to implement them. Not only should the municipality anticipate the need to hire qualified office personnel to administer the stormwater management program, but, also, sufficient qualified field personnel will be needed.

Adequate field inspection is a major problem associated with implementing stormwater management programs. Both the quantity and quality of inspectors are often below what would be adequate to do a good job of implementing the programs. Field inspection is especially essential for implementing the erosion, sediment, and water quality aspects of the program. Sediment ponds fill up, hay bale barriers get destroyed or moved, undisturbed buffer zones are often encroached by development, and berms and other erosion and sediment control facilities get damaged or for some reason do not operate as designed. Unless construction sites are adequately inspected, needed repairs and maintenance to erosion and sediment control facilities are usually not done, and these facilities quickly become ineffective. Although a final inspection can usually determine if the proper drainage facilities have been installed, there are also problems with not having more frequent inspections of these facilities. It is often impossible to tell what is beyond the physical features of facilities; for example, was structural backfill placed correctly, was properly graded riprap used, are the foundations placed as designed, etc.

Inspection is conducted and paid for in a number of ways:

- In order to decrease the field inspection work, some municipalities have been successful in using a system of spot-checking construction sites with large fines for violating stormwater management regulations. Random sampling with fines could encourage self-maintenance by owners, require fewer inspections, and raise funds for the inspection program. Thus, in lieu of hiring a large staff of inspectors, some form of random sampling combined with fines, as provided by law, could prove effective in enforcing ordinance provisions.
- In one Southern city, developers hire inspectors from an approved list. While the developer pays for the inspector, the inspector reports to the city.
- Many municipalities have conducted studies to require the development community to pay partial or complete cost of the plans review and inspection. The fee is levied on a variety of bases, including actual hourly accounting, number of lots, acreage or length of water, wastewater, and storm drain systems. For example, Hampton, Virginia, uses $0.25 per linear foot of these conveyance systems (Whitley, 1993).

Personnel and resources available plus local political and economic conditions will determine, to a large extent, what administrative procedures are used in implementing the program. Each municipality should adopt procedures that prove to be most effective for its particular conditions and circumstances.

A relatively new type of ordinance, which has stormwater management implications, is the "infrastructure permit" or the "private development permit." It allows for ongoing inspection of infrastructure as it is being built, with the inherent ability for stop-work orders to be enacted prior to the infrastructure adoption into the public system proceedings (Whitley, 1993).

Before any stormwater management program is adopted, the financial responsibility of providing adequate field inspection should be assessed and provided. Getting stormwater management facilities properly designed and included in the construction plans is an important part of the program, but getting them constructed, operating, and maintained has proved to be much more difficult.

Enforcement

Enforcement of ordinance provisions is important, if the desired program objectives are to be reached. Basically, enforcement can be divided into two areas. First, all the required paperwork such as development plans, hydrologic and hydraulic studies, drainage plans, and erosion and sediment control plans must be completed and approved by the municipality. Second, adequate inspection must be provided during and after the development phase to ensure that the proposed measures are installed and function as designed. In addition, adequate penalties must be provided to ensure compliance with the provisions of the ordinance.

Enforcement must be phased, swift, and effective. Phased enforcement allows for warnings and advice prior to fines or work stoppages. Swift enforcement reduces reconstruction costs and irreversible environmental damage. Effective enforcement is consistent and fair. Each development is handled in the same way without undue reliance on an individual inspector's subjective judgment. In many municipalities, the enforcement is uneven, unsupported by the political leadership, and cumbersome to execute. Inspector time lost in paperwork and court appearances significantly reduces the effectiveness of any program. Environmental courts have expedited such legal proceedings in some municipalities. However, there is no substitute for having adequate authority, well-documented and consistently and professionally executed and supported.

Legal Considerations

Before any municipality adopts a stormwater management ordinance, the municipality's legal staff should complete a detailed review of the ordinance to ensure that provisions do not conflict with federal, state, or local laws, or go beyond the bounds established by state law and municipal charter.

Summary

Stormwater management programs are becoming more and more complex in their administrative and technical aspects. Drafting and adopting a stormwater management ordinance is an important part of such a program, but it is not the complete program. Determining community objectives and policies related to drainage and flood control, determining water quality standards, acquiring the needed technical personnel, providing adequate field inspection, providing the technical tools, and giving overall direction to the program are also important parts of any stormwater management program. As a result of federal and state mandates, many municipalities are drafting and adopting stormwater management ordinances. It is now time to integrate these ordinances into a complete program in order to achieve the desired results. Without a complete program, these ordinances will prove ineffective and will only add to the frustration of those trying to deal with our complex urban stormwater management problems.

4.8 Documentation

An important part of the design or analysis of any stormwater management facility is the documentation. Appropriate documentation of the design of any stormwater management facility is essential because of the following:

- The importance of public safety
- Justification of expenditure of public funds

- Future reference by engineers (when improvements, changes, or rehabilitations are made to the facilities)
- Information leading to the development of defense in matters of litigation
- Public information

Frequently, it is necessary to refer to plans, specifications, and analysis long after the actual construction has been completed. Documentation permits evaluation of the performance of structures after flood events to determine if the structures performed as anticipated, or to establish the cause of unexpected behavior, if such is the case. In the event of a failure, it is essential that contributing factors be identified in order that recurring damage can be avoided.

The definition of hydrologic and hydraulic documentation as used in this chapter is the compilation and preservation of the design and related details as well as all pertinent information on which the design and decisions were based. This may include drainage area and other maps, field survey information, source references, photographs, engineering calculations and analyses, measured and other data, and flood history, including narratives from newspapers and individuals, such as highway maintenance personnel and local residents who witnessed or had knowledge of an unusual event.

Purpose of Documentation

Although the amount and detail of documentation should be commensurate with the size, scope, and importance of the project, this section presents the documentation that should be included in the design files and on the construction plans. While the documentation requirements for existing and proposed stormwater management facilities are similar, the data retained for existing facilities are often slightly different than that for proposed facilities, and these differences are discussed. This section focuses on the documentation of the findings obtained in using the other chapters of this book, and thus, readers should be familiar with all the hydrologic and hydraulic design procedures associated with this chapter. This section identifies the system for organizing the documentation of stormwater management facility designs and reviews so as to provide as complete a history of the design process as is practical.

The major purpose of providing good documentation is to define the design procedure that was used and to show how the final design and decisions were arrived at. Often, there is expressed the myth that avoiding documentation will prevent or limit litigation losses, as it supposedly precludes providing the plaintiff with incriminating evidence. This is seldom, if ever, the case, and documentation should be viewed as the record of reasonable and prudent design analysis based on the best available technology. Thus, good documentation can provide the following:

- Protection for the design engineer and the local agency by proving that reasonable and prudent actions were, in fact, taken (such proof should certainly not increase the potential court award and may decrease it by disproving any claims of negligence by the plaintiff)
- Identification of the situation at the time of design that might be important if legal action occurs in the future
- Documentation that shows that rationally accepted procedures and analysis were used at the time of the design that were commensurate with the perceived site importance and the flood hazard (this should further disprove any negligence claims)

- Provision of a continuous site history to facilitate future reconstruction
- Provision of the data necessary to quickly evaluate any future site problems that might occur during the facility's service life
- Expedition of plan development by clearly providing the reasons and rationale for specific design decisions

Types of Documentation

There are three basic types of documentation that should be considered. The types are preconstruction, design, and construction or operation.

Preconstruction

Preconstruction documentation should include the following, if available or within the budgetary restraints of the project:

- Aerial photographs
- Contour mapping
- Watershed map or plan including:
  - Flow directions
  - Watershed boundaries
  - Watershed areas
  - Natural storage areas
- Surveyed data reduced to include:
  - Existing hydraulic facilities
  - Existing controls
  - Profiles — roadway, channel, driveways
  - Cross sections — roadway, channels, faces of structures
- Flood insurance studies and maps by Federal Emergency Management Agency (FEMA),
- Soil Conservation Service soil maps
- Field trip report(s) that may include:
  - Videocassette recordings
  - Audiotape recordings
  - Still camera photographs
  - Movie camera films
  - Written analysis of findings with sketches
- Reports from agencies (local, state, or federal), newspapers, and abutting property owners

Design

Design documentation should include all the information used to justify the design, including the following:

- Reports from other agencies
- Hydrological report
- Hydraulic report
- Computer analysis output
- Any approvals received

Construction or Operation

Construction or operation documentation should include:

- Plans
- Revisions
- As-built plans and subsurface borings
- Photographs
- Record of operation during flooding events, complaints, and resolutions

It is important to prepare and maintain in a permanent file the as-built plans for every stormwater management structure to document subsurface foundation elements such as footing types and elevations, pile types and (driven) tip elevations, etc. There may be other information that should be included or may become evident as the design or investigation develops. This additional information should be incorporated at the discretion of the designer.

Documenting the Plan Development Process

Documentation should not be considered as occurring at specific times during the design or as the final step in the process, which could be long after the final design is completed. Documentation should rather be an ongoing process and part of each step in the hydrologic and hydraulic analysis and design process. This will increase the accuracy of the documentation, provide data for future steps in the plan development process, and provide consistency in the design, even when different designers are involved at different times of the plan development process. Accurate documentation should be provided during the following steps or phases of the plan development process:

1. Surface water environmental (environmental impact study) phase
2. Reconnaissance phase
3. Location phase
4. Survey phase (drainage surveys)
5. Design phase
6. Revised design phase
7. Construction phase to include as-built plans
8. Operational phase — documentation should be continuous over the structure's life cycle

The designer should be responsible for determining what hydrologic analyses, hydraulic design, and related information should be documented during the plan development process. The designer should make a determination that complete documentation has been achieved during the plan development process, which will include the final design.

Documentation Procedures

The designer should maintain a complete hydrologic and hydraulic design and analysis documentation file for each project or design. Where practicable, this file should include such items as:

- Identification and location of the facility
- Photographs (ground and aerial)
- Engineering cost estimates and actual construction costs
- Hydrology investigations
- Drainage area maps
- Vicinity maps, topographic and contour maps
- Interviews (local residents, adjacent property owners, and maintenance forces)
- Newspaper clippings
- Design notes and correspondence relating to design decisions
- History of performance of existing structure(s)
- Assumptions

The documentation file should contain design and analysis data and information that influenced the facility design and that may not appear in other project documentation.

Following are some recommended practices that the designer should follow related to documentation of hydrologic and hydraulic designs and analyses:

1. Hydrologic and hydraulic data, preliminary calculations and analyses, and all related information used in developing conclusions and recommendations related to stormwater management facility requirements, including estimates of structure size and location, should be compiled in a documentation file.
2. The designer should document all design assumptions and selected criteria, including the decisions related thereto.
3. The amount of detail of documentation for each design or analysis should be commensurate with the risk and the importance of the facility.
4. Documentation should be organized to be as concise and complete as practicable, so that knowledgeable designers can understand years hence what predecessors did.
5. Circumvent incriminating statements wherever possible by stating uncertainties in less-than-specific terms (e.g., the culvert *may* back water rather than the culvert *will* back water).
6. Provide all related references in the documentation file to include such things as published data and reports, memos and letters, and interviews. Include dates and signatures where appropriate.

7. Documentation should include data and information from the conceptual stage of project development through service life, so as to provide successors with all information.
8. Documentation should be organized to logically lead the reader from past history through the problem background, into the findings, and through the performance.
9. An executive summary at the beginning of the documentation should provide an outline of the documentation file to assist users in finding detailed information.
10. Include all completed forms used by local agencies, copies or references of standards used, and any items required for submittal, related to facility design.

Storage Requirements

Where and how to store and preserve records is an important consideration. Ease of access, durability, legibility, storage room required, and cost are the prime factors to consider when evaluating alternative methods of storage and preservation. For instance, microfilm and microfiche systems require relatively small storage spaces. However, the support systems are expensive, subject to down-time, and may produce poor copies, and both microfilm and microfiche systems require controlled humidity and temperature. Conversely, the storage of actual plans and documents requires much more space, and the records are not as durable. However, better reproductions are usually obtained, and the temperature and humidity requirements are not as critical. Combinations of the two are sometimes used.

The designer should maintain the documentation files, including microfilm, microfiche, magnetic media, etc., in a location where they will be readily available for use during construction, for defense of litigation, and for future replacement or extension. Only that documentation which is not retained elsewhere need be retained. Original plans, project correspondence files, construction modifications, and inspection reports are the types of documentation that usually do not need to be duplicated in the documentation file.

Hydrologic and hydraulic documentation should be retained with the project plans or in some other permanent location, at least until the stormwater management facility is totally replaced or modified as a result of a new stormwater management facility study. Procedures should be established to determine when documentation could be destroyed.

Specific Content of Documentation Files

The following items should be included in the documentation file. The intent is not to limit the data to only those items listed, but rather to establish a minimum requirement consistent with the stormwater management design procedures as outlined in the different chapters of this book. If circumstances are such that the stormwater management facility is sized by other than normal procedures or if the size of the facility is governed by factors other than hydrologic or hydraulic factors, a narrative summary detailing the design basis should appear in the documentation file. Additionally, the designer should include in the documentation file items not listed below but that are useful in understanding the analysis, design, findings, and final recommendations.

Hydrology

The following hydrologic-related items used in the design or analysis should be included in the documentation file:

- Contributing watershed area size and identification of source (map name, etc.)
- Design frequency and decision for selection
- Hydrologic discharge and hydrograph estimating method and findings
- Copies of all computer analyses
- Flood frequency curves to include design, 100-year flood, any other flood frequencies used for design or analysis, discharge hydrograph, and any historical floods
- Expected level of development in upstream watershed areas over the anticipated life of the facility (include sources of and basis for these development projections)

Bridges

The following items should be included in the documentation file related to the design and analysis of bridges:

- Design and 100-year highwater for undisturbed, existing, and proposed conditions
- Stage–discharge curve for undisturbed, existing, and proposed conditions
- Cross section(s) used in the design highwater determination
- Roughness coefficient (*n* value) assignments
- Information on the method used for design highwater determination
- Observed highwater, dates, and discharges
- Velocity measurements or estimates and locations (include both the through-bridge and channel velocity) for design and 100-year floods
- Performance curve to include calculated backwater, velocity, and scour for design, 100-year floods, and 500-year flood for scour evaluation
- Magnitude and frequency of overtopping flood
- Copies of all computer analyses
- Complete hydraulic study report
- Economic analysis of design and alternatives
- Risk assessment
- Bridge scour results
- Roadway geometry (plan and profile)
- Potential flood hazards to adjacent properties

Culverts

The following items should be included in the documentation file related to the design and analysis of culverts:

- Culvert performance curves
- Allowable headwater elevation and basis for its selection
- Cross section(s) used in the design highwater determinations
- Roughness coefficient assignments (*n* values)
- Observed highwater, dates, and discharges

- Stage discharge curve for undisturbed, existing, and proposed conditions to include the depth and velocity measurements or estimates and locations for the design, 100-year and other check floods
- Performance curves showing the calculated backwater elevations, outlet velocities and scour for the design, 100-year, and any historical floods
- Type of culvert entrance condition
- Culvert outlet appurtenances and energy dissipation calculations and designs
- Copies of all computer analyses and standard computation sheets given in Chapter 9 of this book
- Roadway geometry (plan and profile)
- Potential flood hazard to adjacent properties

Open Channels

The following items should be included in the documentation file related to the design and analysis of open channels:

- Stage discharge curves for the design, 100-year, and any historical water surface elevation(s)
- Cross section(s) used in the design water surface determinations and their locations
- Roughness coefficient assignments (n values)
- Information on the method used for design water surface determinations
- Observed highwater, dates, and discharges
- Channel velocity measurements or estimates and locations
- Water surface profiles through the reach for the design, 100-year, and any historical floods
- Design or analysis of materials proposed for the channel bed and banks
- Energy dissipation calculations and designs
- Copies of all computer analyses

Storm Drains

The following items should be included in the documentation file related to the design and analysis of storm drain systems:

- Computations for inlets and pipes, including hydraulic grade lines
- Copies of the standard computation sheets given in Chapter 8 of this book
- Complete drainage area map
- Design frequency
- Information concerning outfalls, existing storm drains, and other design considerations
- A schematic indicating storm drain system layout

Storage Facilities and BMPs

The following items should be included in the documentation file related to the design and analysis of storage facilities and structural BMPs that utilize storage facilities in their design:

- Routing calculations including inflow and outflow hydrographs and volumes for the design and 100-year floods
- Calculations of any volumes used in the facility design
- Design calculations of any riser pipes and outflow devices
- Computations for dam design including typical cross section
- Storage volume estimates
- Design or analysis of materials proposed for the banks and bottom of the storage facility
- Emergency spillway design
- Outlet velocity estimates
- Energy dissipation calculations and designs
- Maintenance plan
- Copies of all computer analyses

Pump Stations

The following items should be included in the documentation file related to the design and analysis of pump stations:

- Inflow design hydrograph from drainage area to pump
- Flood frequency curve for the attenuated peak discharge
- Maximum allowable headwater elevations and related probable damage
- Starting sequence and elevations
- Sump dimensions
- Available storage amounts
- Pump sizes and operations
- Pump calculations and design report
- Line storage and pit storage capacity

In addition, whenever computer programs are used for design or analysis, input data listings and output results of selected alternatives should be included in the documentation file. The file should also document what computer model was used, the version, and any other information that would enable future designers to duplicate the computer analysis that was used. If computer models are changed or new versions purchased, copies of the model or versions used for analysis or design should be kept.

4.9 Documentation for Legal Proceedings

Municipal ordinances and regulations outline the basic policies and procedures that must be following when designing and analyzing local stormwater management facilities and developing development plans. However, it should be remembered that these ordinances and regulations outline the minimum requirements, and it is up to the designer to determine if plans and design should go beyond the minimum requirements to adequately protect local citizens and property and comply with the intent of the regulations. In our litigious society, the designer should always be aware that development plans, stormwater management designs, and analysis may undergo the scrutiny of a legal proceeding. Thus, extreme care should be taken to establish documentation that will provide the information needed to document all factors that were involved in the design process. Following are some recommendations for developing the documentation considering possible legal proceedings:

- Fully document all assumptions that were used in the design and analysis, including variables that were used in any computer models.
- Any assumptions that are used should not be characterized as guesses, arbitrary estimates, or manipulated values needed to make some computer model run.
- Document from where data were obtained (e.g., field studies, publications, databases).
- Do not assume that available information is correct without doing some accuracy checking. Thus, one would not use a phrase like, "assumed values in previous model runs were correct," in documentation materials.
- Document any field studies or other observations that were used to calculate data for analysis and design or investigation of site conditions.
- In most cases, it is more appropriate to characterize the value used in design or analysis as a calculated value rather than an estimated value.
- If previous studies, analysis, computer models, etc., are used and changes are made as the result of changing conditions or errors that are discovered, be careful about the documentation of these changes. Avoid the use of words like "errors," "incorrect assumptions," "faulty data," etc. It would be better to document by stating that additional information or data necessitated changes, field studies indicated that appropriate changes should be made, or recent drainage and flooding problems indicated that a more conservative approach should be used.
- If previous studies are used, be sure to verify all variable values and assumptions to be sure they are reasonable and appropriate for future use.
- Although hydrology and hydraulics are not exact sciences, a lot of work has been done to make hydrologic and hydraulic studies more consistent and less dependent on engineering judgment. The narrative documentation that is included should give sufficient information so that another engineer or designer can duplicate the design that is approved and implemented.
- It is important to document who worked on the project, their duties (designer, reviewer, surveyor, etc.), qualifications, and who was responsible for checking the accuracy of study results.
- In most cases, it is best to provide excess documentation rather than inadequate documentation, because it will be difficult to augment the available documentation in future years.
- Finally, it should be remembered that a legal action might take place many years after the design is completed, facilities are constructed, municipal reviewers are not longer available, and individuals that worked on the project are retired or leave a firm to accept another position. Thus, complete and accurate documentation provides the best protection as the project is examined under the microscope of the legal system.

References

American Association of State Highway and Transportation Officials, Highway Drainage Guidelines, 1982.
American Association of State Highway and Transportation Officials, Model Drainage Manual, 1991.
American Public Works Association, Urban Storm Water Management, Chicago, IL, Special Report No. 49, 1981.
Association of State Flood Plain Managers (ASFPM), The Insider, Jan Kusler, pp. 5–6, 1990.
Austin, TX, Comprehensive Watershed Ordinance, Ordinance # 860508-V, 1986.
Bellevue, Washington, Ordinance 2003, An Ordinance Relating to Storm and Surface Water, February 1974.
Cox, W.E., Water Law Primer, *Proceedings of the Water Research Planning and Management Division ASCE*, Vol. 108, No. WR1, March 1982.
Debo, T.N., Survey and Analysis of Urban Drainage Ordinances and a Recommended Model Ordinance, thesis presented to the Georgia Institute of Technology, February 1976.
Debo, T. and Reese, A.J., Downstream Impacts of Detention, *Proceedings of NOVATECH 92*, Lyon, France, Nov 3–5, 1992.
Environmental Protection Agency (EPA), Region V, Stormwater Management Ordinances for Local Governments, Chicago, IL, December 1990.
Galli, J., Analysis of Urban BMP Performance and Longevity in Prince George's County, Maryland, Final Report, Metro Washington Council of Governments, August 1992.
Illinois Department of Transportation (IDOT), Model Flood Plain Ordinance.
Kenworth, W.E., *Urban Drainage: Aspects of Public and Private Liability*, Dicta, Denver, CO, July–August 1972.
Lewis, G.L., Jury verdict: frequency versus risk-based culvert design, *ASCE J. Water Res. Plann. & Manage.*, 118, 2, 166–185, March/April 1992.
Louisville MSD, Hazardous Materials Ordinance, Louisville, KY, November 1988.
Maryland, Model Stormwater Management Ordinance, Department of the Environment, 1983.
Ogden Environmental and Energy Services, Inc., Storm Water Quality Program Development — Charlotte, NC, Ogden-Nashville, TN, 1992.
Prince George's County, MD, Storm Water Ordinance, Bill No. CB-52-1984, 1984.
Schueler, T.R., Kumble, P.A., and Heraty, M.A., A Current Assessment of Urban Best Management Practices, Metro Washington Council of Governments, March 1992.
Shoreview, MN, An Ordinance Relating to Lawn Fertilizer Application, Ordinance No. 477, 1985.
Shoreview, MN, An Ordinance Relating to Lawn Fertilizer Application and Pesticide Control, Ordinance No. 477, 1988.
Wayzata, MN, An Ordinance ... Adding Wetlands District Definitions, Ordinance No. 515, 1988.
Whitley, F., Permitting Public Infrastructure Installation of Private Developments, *Public Works*, August 1993.

5
Financing Stormwater Management Programs

5.1 Financing Needs of Stormwater Management Programs

Stormwater management is rarely and then only temporarily the highest priority program within most municipalities. Unless a home or property is affected by flooding or drainage problems, other services provided by the municipality are much more important and will obtain more support from local politicians. If flood awareness is high, psychological factors could increase a temporary willingness to pay (Thunberg and Shabman, 1991). But, this is not generally true for the whole of the municipal population.

Improving the drainage system, installing regional detention facilities, and upgrading culverts under local roadways are not seen as projects that will gain votes or support from a large segment of the voting population. Single-purpose drainage bond issues have had major problems in gaining enough support for passage (Debo and Williams, 1979). Building roads, providing recreation areas, upgrading schools, and providing funds for police and programs dealing with drugs are much more likely to obtain funding before stormwater management projects.

Until recently, stormwater management in many municipalities has been ignored or received attention only when major problems resulted during major storm events. Unless developers could be required to provide needed stormwater management facilities as part of new developments, these facilities were not provided. Also, except for facilities under public roadways, maintenance of the drainage system was delegated to private adjacent landowners. As a result, little if any maintenance was done, and drainage systems within urban areas across the country were inadequate to provide the protection needed for adjacent property and structures. In most municipalities, stormwater management is seen as an added function of the public works and engineering department, and sufficient funds to implement this added function are not provided.

With growing public expectation for service, the passage of the EPA water quality regulations, and some state mandates to provide water quantity and quality management, municipalities are being required to develop and implement comprehensive stormwater management programs. Many of these programs will require spending several times the current drainage budgets to provide the needed services. As a result, municipalities are now struggling with the problem of how to fund these programs without adversely impacting other municipal services. Different approaches are being taken by different municipalities, depending on local politics, funding sources available, and citizen support for different funding programs. A phone survey found that citizens of one city were willing to pay over $5 per month for clean water in the streams, while there was little willingness to pay for flood control (Charlotte, 1991).

5.2 Major Stormwater Funding Methods

Municipalities employ a variety of funding methods, including service charges, several types of taxes, franchises and other fees, fines, and penalties. The various funding methods have distinctive characteristics that separate them legally, technically, and in terms of public perceptions. Four major categories of municipal revenue generation methods are taxes, service charges, exactions, and assessments:

- Taxes are intended primarily as revenue generators, and, with some exceptions (such as special local option sales or earmarked taxes), without any particular association with the activities or improvements that they fund. They can be used for the general purposes of local government. These include property tax, income tax, sales tax, etc.
- Service charges are not established simply to generate revenue but must be tied to the objectives of a specific program to which they are associated. For example, water and sewer service charges are structured to cover the cost of those programs, not to simply generate revenue that is used for other purposes as well. Thus, the total revenue generated must be tied to the cost of providing services and facilities, and the amount each ratepayer is charged must be related to the impact or use of the system (rational nexus).
- Exactions are related to the extension of an approval or privilege to use. Franchise fees for the privilege of using the right-of-way for cable and phone companies limited to a certain percentage of revenue by federal or state laws are an exaction. Licenses, tap fees, impact fees, fees in lieu of detention, capital recovery charges of all kinds, and the mandatory dedication of infrastructure during development are also exactions.
- Assessments are geographically or otherwise limited fees levied for improvements or activities of direct and special benefit to those who are being charged. The benefit must be direct — tied to a specific and measurable or estimable property improvement. And, it must be special — a benefit that is not realized generally in the community or area.

The distinctions of the four revenue categories are important. One of the critical issues that typically must be resolved if a utility service charge of any type is legally challenged is whether the service charge is clearly related to and incidental to the activities and improvements of the utility, or is, in fact, merely a means of creating revenue for all governmental purposes generally (a tax), or is a special assessment (which is supposed to reflect a direct and special benefit). Thus, a stormwater utility must be based on a stormwater program and not simply a perceived financial need or willingness to pay.

Municipal stormwater management programs have been funded using a number of mechanisms as the primary generator of funds, including property taxes, sales taxes, state revolving funds, road funding, user fees, bonding, and surcharges on other utility fees. By far the most common current funding method is property-tax-based. Other major revenue generators include franchise fees, income taxes, gasoline tax (for roadway-related drainage), sales taxes, and stormwater user fees.

A major new source of funding for stormwater management is in the form of a user fee system under the auspices of a stormwater utility (Cyre and Reese, 1992; Davis, Hatoum, and Rose, 1999). This form of funding has several advantages over other competing forms of finance, including its equitability, stability, and adequacy. The user fee concept of a stormwater-utility-based funding method is fast growing. In the early 1970s, there were only one or two true stormwater utilities in existence. In the early 1990s, there were over 200. By 2000, the number had grown to 400. This number is expected to more than triple in the next decade as the financial impacts of stormwater quality legislation reach many small municipalities.

5.3 Stormwater Utility Overview

Overview of the Concept

A stormwater utility falls primarily under the second of these funding categories: a service charge. It is based on the premise that the urban drainage system is a public system, similar to a wastewater or water supply system. When a demand is placed on either of these two latter systems, the user pays. In the same way, when a forested or grassy area is paved, a greater flow of water is placed on the drainage system. This is the demand. The greater the demand (i.e., the more the parcel of land is paved), the greater the user fee should be. A stormwater utility differs from the other two water-related utilities in several key ways. First, there is no way to remove or discontinue services for nonpayment. Second, the service is provided to all citizens without choice (though mandatory water and sewer service makes this difference less distinctive). Third, the demand placed on the system can only roughly be measured or approximated. Also, the actual service rendered to a particular property is often difficult to quantify. Despite these drawbacks, the utility concept for stormwater financing is a viable and growing funding method.

A stormwater utility must be seen as an umbrella under which individual communities address their own specific needs in a manner consistent with local problems, priorities, and practices. It is understood in three ways: a means of generating revenue, a program concept, and an organizational entity. It is important when establishing a stormwater utility to determine which of these three the utility actually is. If the only reason for establishment of a utility is to generate revenue and to free additional tax revenues, there will be trouble. Citizens who thought they were getting stormwater for free and now see a line item on a bill will pick up the phone and expect better service than before. Therefore, it is important to offer a better level of service (program concept) if a utility is to be formed.

Any organizational concept might be appropriate, and many variations on the concepts presented in Chapter 2 have been successful. Development of adequate funding is the bottom-line solution that allows other, more technical, programmatic solutions to flourish and be successful. With the expected needs for organizational changes within most municipalities to manage stormwater and the demands placed on municipalities by the EPA, NPDES permit for stormwater discharges, the stability and adequacy of a utility is a great advantage.

Advantages of the Utility Concept

A stormwater utility provides a vehicle for the following:

- Consolidating or coordinating responsibilities that were previously dispersed among several departments and divisions
- Generating funding that is adequate, stable, equitable, and dedicated solely to the stormwater function
- Developing programs that are comprehensive, cohesive, and consistent year to year

The stormwater user fee (utility) form of funding has several advantages over other sources of funding, including revenue stability, adequacy, flexibility, and perceived equity (SAFE).

A stormwater utility is *stable*, because it is not as dependent on the vagaries of the annual budgetary process as taxes. As stated earlier, stormwater cannot compete with schools, police, and solid waste for tax-based funding, and should not. When the stormwater budget varies from year to year depending on competing needs, the opportunity to leverage federal funds, and how recent the last flooding episode was, managers cannot plan anything but a minimal caretaker program planning for the troughs of the ebb and flow of funding and trying to take advantage of the peaks when, and if, they occur. They cannot manage a capital improvement program or a proactive maintenance program, and they cannot fund new initiatives such as greenways, partnering, and riparian corridors. This is not satisfying for citizens who need to know when and if their flooding problem will be fixed. Many have given up calling. Often, it takes a disaster or loss of human life to pump funds into the program. A disaster is almost inevitable, because systems do not self-improve over time *m* they deteriorate. Only when a flood in excess of design tests a drainage system can the actual level of decay be known (Doll, Lindsey, and Albani, 1998).

A utility provides a stable revenue stream that actually allows stormwater managers to plan for and schedule maintenance, equipment purchases, and capital construction programs. It allows them to leverage revenues to obtain greater funding from outside sources. A utility in Georgia actually received almost more revenue in matching funds several years in a row than it did through utility revenue (Keller, 2001).

And, it is *adequate*, because a typical stormwater program can be financed with payments below the normal customer willingness to pay (EPA, 1989). The normal citizen willingness to pay varies across the country and according to economic status. In areas that have a thriving economy and a strong environmental sensitivity, citizens are willing to pay in excess of $7 per household per month (with correspondingly larger fees for nonresidential properties). In other areas, the willingness to pay is $3 to $4 per household per month.

Typical fees are hard to compare, because they are based on differently configured programs. Some municipalities have included a charge for their own streets in the rate to be paid by the municipality. This would tend to reduce the monthly rate by about 12 to 20% and shift this cost to tax-based funding. Other municipalities have elected to fund portions of the program through other methods, such as state street funds or bonding. The definition of what is included under the stormwater utility is also differently interpreted. This is especially true when stormwater-quality-related programs are funded fully or partially by a utility charge. Figure 5-1 provides an indication of the range of monthly charges for 206 cities across the country based on older surveys (APWA, 1991; Hartigan, 1989; HDR, 1990; LeClere, 2000; Black and Veatch, 1996, 2002; FASU, 1995) and recent updating. The average charge is about $3.80, and the median charge is $3.00 per month.

Table 2-4 indicated that stormwater programs, funded at a moderate level, cost in the $100 per developed acre per year range. Experience has shown that a stormwater utility charge of $1 per household per month (and uncapped comparable charges for nonresidential properties) will generate between $25 and $45 per acre per year for a normal

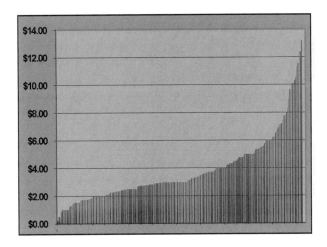

FIGURE 5-1
Range of monthly charges.

self-contained municipality (i.e., not a bedroom community). If the willingness to pay is, say, $3 per household per month, then the utility can generate between $75 and $135 dollars per acre per year — or a moderate stormwater program. Thus, a municipality can stay below the willingness of the normal citizen to pay and still generate enough revenue to operate a stormwater program. This does not even take into account other sources of revenue, such as fees or supplemental tax-based funding (Treadway and Reese, 2000).

A stormwater utility is *flexible*, because the rate structure can include any number of modifiers and secondary funding methods to meet a variety of objectives. For example:

- The user fee is rarely restricted in its use, except that it must be related to stormwater. Given the vast array of things termed "stormwater," this is almost no limitation.
- There is the possibility of using secondary funding methods to shift costs to accomplish almost any program objective and to enhance equity. For example, development-related fees could be added to the utility rate structure to shift costs toward (or away from) the development community.
- Credits can be developed to encourage good stormwater behavior in a number of ways (see below).
- Differing rates can be charged for, say, urban and rural services districts, or by watershed.
- Specific fees and rates can be developed to take into account environmental costs.

A stormwater utility is *equitable* because the cost is borne by the user on the basis of demand placed on the drainage system. Stormwater looks and is managed a lot more like water and sewer than like police, schools, and fire. Few people would think it equitable to pay for water or sewer on the basis of property value. So, it only makes sense to pay for stormwater on the same basis — the more you pave, the more you pay. All citizens can intuitively grasp this concept, and the vast majority feels it is fair.

Figure 5-2 shows a typical house and a typical small commercial establishment. If the house has 2500 square feet of impervious area (rooftops, driveway, sidewalk, etc.) and the commercial establishment has ten times that, then the commercial establishment would

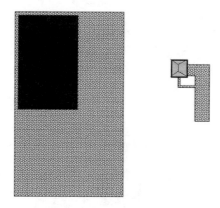

FIGURE 5-2
Impervious area methodology.

pay ten times the fee that a single-family residence would pay. If the 2500 number is taken as "typical" for the municipality, then it can be termed the Equivalent Residential Unit (ERU). Then, the commercial site is ten ERUs.

There are those entities that would not like the utility concept even if it can be seen as more fair than other funding methods. Those who would pay more under the utility concept include nonprofits (including schools), large paved areas with cheap or no buildings, those on fixed incomes, and developers, if the fees are high (though owners, not builders, pay the stormwater fee).

Uniqueness Factor

No two utilities are identical, just as no two communities are alike (Lindsey, 1988). Therefore, it is not prudent to follow a prefabricated "one size fits all" approach but to seek to carefully understand the makeup of the community, its problems, and its goals. There must be a clear understanding of the community's stormwater systems, capabilities, and issues. A questionnaire can be helpful in gaining understanding of a community. An example questionnaire is provided in Appendix A at the end of this chapter.

Some communities have simply attempted to clone a stormwater program concept or rate methodology adopted by another city or county. Any city or county should carefully guard against that temptation if wants to be successful. A stormwater program or rate structure cloned from somewhere else cannot sustain intense scrutiny if a staff, advisory committee, elected officials, or interest groups take a really hard look at it. The program then often fails politically. Conversely, a well-thought-out program and funding strategy can survive the scrutiny that would accompany controversy if questions should develop during or after the program development process.

The real danger of the cloning approach is that it inevitably falls short of efficiently and effectively solving the local stormwater problems, because it is not founded on addressing them. The local problems, needs, and circumstances must drive the form, priorities, and pace of transition of the program. The success of programs in Charlotte, North Carolina; Louisville, Kentucky; Tulsa, Oklahoma; Cincinnati, Ohio; Mecklenburg County, North Carolina; Bellevue, Washington; and elsewhere, are based not on offering the same solutions everywhere, but rather on tailoring the program and financing strategies to the local needs. Solving real short-term and long-term stormwater problems in a municipality must be the focus of the program and utility development.

5.4 Utility Funding Methods

There are a number of funding methods, under the utility's umbrella, available for financing stormwater capital improvement projects and stormwater infrastructure maintenance programs. In most cases, the method selected in any given municipality is based on a few factors:

- The authority to use different funding methods available through state statutes or home rule charters
- The scope of program or projects to be funded
- Related local funding policies and practices
- The general financial health of the local governments and their individual revenue funds
- The political atmosphere in which elected officials must make funding decisions

A base funding method (or methods) must include the following if the stormwater management program is to be successful:

- Sufficient revenue capacity to meet short-term program and capital project needs
- A relatively stable revenue stream
- An appropriateness for most, if not all, of the different types of projects and activities
- Flexibility to adjust as the program evolves during the next decade and beyond

The financing method package developed for a particular utility is divided into three modules: (1) the basic rate methodology; (2) modification factors that can be applied to any of the rate concepts to enhance equity, reduce costs, and meet other objectives; and (3) the secondary funding methods that can be adopted in concert with the service charges. Typical modification factors include flat rates for single-family residences, fixed costs per account, a crediting mechanism, location charges or credits, etc. Secondary funding methods are evaluated primarily to ensure that they would be consistent with the basic rate methodology.

The development of a preferred rate methodology involves the following steps:

1. Selection of a menu of appropriate funding methods for the program
2. Definition of a primary funding method
3. Selection of a basic rate concept from a menu of options
4. Selection of appropriate rate modifiers from a menu of options
5. Selection of appropriate secondary funding methods from the remaining funding methods

Normally, a rate structure analysis looks at basic rate methodologies. Each method has advantages and disadvantages, and there are newer concepts related to runoff coefficients being discussed (Minner et al., 1998). Some typical basic methods for calculating demand on the system consist of consideration of:

- Impervious area
- Impervious area and gross area
- Impervious area and impervious percentage
- Gross area and an intensity of development factor
- Gross area only, with extensive use of modifying factors
- Runoff coefficient

Over half of the utilities in existence use the impervious area method for rate development. The basic rate methodology is selected on the basis of (Rosholt and Pigott, 1991):

- Workability, ease of administration
- Fairness and equity
- Consistency with local policy
- Ability to meet revenue requirements
- Objectivity versus discretional determination of individual costs
- Applicability to entire service area
- Effectiveness of desired incentives

After the basic rate methodology is selected, modification factors are examined. The modification factors are used to enhance equity or improve ease of utility implementation and management without unduly sacrificing equity. Typical modification factors considered might include:

- A flat-rate, single-family residential charge
- A base rate for certain costs that are fixed per account
- Basin-specific surcharges for major capital improvements
- A surcharge for properties located in floodplains
- Credits against the monthly service charge for properties that have on-site detention or retention systems
- A water quality impact factor
- A development and land use factor
- A level of service factor

Evaluation criteria are employed in the funding study to assess each method of primary or secondary funding:

- Financial impact on citizens and businesses
- Equity and public acceptance
- Revenue sufficiency
- Timeliness, and the process required for implementation
- The cost of development, implementation, and upkeep
- Consistency with capital project and program needs

While stormwater is fundamentally different from wastewater or water supply, many of the funding methods available in these utilities have analogous methods for stormwater.

Bonding, short-term financing, and credit enhancers are available to the water and wastewater utility (Raftelis, 1989; AWWA, 1986) and the stormwater utility.

Capital planning for stormwater utilities has traditionally not been very sophisticated. Water and wastewater utilities have used various capital-budgeting techniques (such as return on investment, internal rate of return, or net present-value) for a number of years. There may be opportunities for capitalization grants under a state revolving fund (SRF) program, though this is not a certainty from state to state (Houmis, 1992; Davis, Major, and Dunlap, 1989; EPA, 2001). Because stormwater does not normally have treatment plants and does not own or operate much of its system, such methods are difficult to apply. However, as infrastructure rehabilitation and replacement needs increase and master planning for major flood control projects takes place, such accounting methods have taken on more importance.

Recipients may use loans for the planning, design, and construction of publicly owned wastewater treatment facilities or to build or rehabilitate sewer collection systems. Urban wet weather flow control activities, including stormwater and sanitary and combined sewer control measures, are also eligible for funding.

EPA encourages its state partners to use watershed planning and improved priority setting systems to choose projects that address the greatest remaining environmental challenges. For example, a state can make a 0% loan to a community for 20 years, saving the community 50% of the total project costs over a similar loan at 7.5%.

The more popular financing methods available to most municipalities include the following (Raftelis, 1989; EPA, 1988; Cyre, 1987; EPA, 1990):

1. *Stormwater drainage services charges* — This method of financing fits well with stormwater and flood control program needs in most municipalities. It is normally the primary financing mechanism for the utility and is often called "the utility." The functions and costs associated with stormwater management are similar to those for a water or sanitary sewer utility. Thus, this stormwater utility would be set up and function in a manner similar to other existing utilities. The principal difference between water and sanitary sewer service charges and stormwater service charges is the basis of measurement employed for setting rates. The typical basis for stormwater charges is the runoff production potential of a particular property. Usually, to avoid excessive utility administration and start-up costs, only industrial/commercial/multifamily sites are considered on an individual basis, while single-family residences are covered under blanket charges.

2. *Revenue bonding* — This method of municipal funding is most commonly used for facility construction, the purchase of equipment, or other major capital outlays. Unlike general obligation bonds, which are backed by the full faith and credit of the issuing governmental agency, revenue bonds are backed by revenues from specifically defined sources. Thus, it is intended to work in conjunction with some other ongoing funding system. "Anticipation notes" are somewhat similar, with the anticipated revenues from a future utility covering the notes issued to generate funds to set up the utility.

3. *System development charges* — System development charges are also best used in conjunction with other funding methods as a mechanism for balancing financial participation in the cost of services and facilities. They provide a funding mechanism through which owners of properties that develop in the future share in the cost of projects built-in anticipation of their needs. They can properly be described as a deferral mechanism through which a property owner's financial participation in a project is delayed until development occurs that uses the additional capacity originally built into the system.

4. *Special assessments and improvement districts* — A number of different methods of levying special assessments on benefited properties have been used throughout the United States for stormwater and flood control improvements. Projects funded through special assessments must have a special benefit to the properties included in the assessment area, and charges for each parcel must be consistent with the relative benefit to each property. Special assessment mechanisms are most often used for small local projects, because they localize the cost of projects, which serve a limited area and a limited constituency.

5. *Plan review and inspection fees* — This method of funding is typically used by public works departments and utilities that operate stormwater drainage systems and must maintain and regulate their use. The fee is used to cover the cost of plan review and inspection of project sites during the construction of the proposed development or to fund ongoing inspection programs of private structural facilities, such as detention ponds or water-quality best-management practices. Many municipalities charge only a token fee, while others try to determine the actual costs involved and set a fee to cover these costs.

6. *In-lieu-of-construction fees* — Many municipalities require the construction of a stormwater detention facility for all nonresidential properties unless it can be demonstrated that construction of such a facility would adversely affect stormwater peak discharge rates downstream from the site. The in-lieu-of-construction fee provides a funding mechanism, whereby construction of numerous individual on-site detention basins or other stormwater conveyance structures can be waived, with developers contributing a fee to a fund to build a regional facility or make improvements in stormwater conveyance at locations remote from the development site. One drawback of this funding method is that the facility must be built ahead of the development and, therefore, an alternate funding source must be available. Rarely do the funds for construction become available prior to the construction start.

7. *General facilities charges* — Utility service charge rate structures sometimes incorporate surcharges for facilities that are necessary to provide adequate service to all ratepayers. These charges are known by various names, but the term "general facilities charges" describes their purpose. General facilities charges for stormwater and flood control improvements would be most appropriate for regional detention facilities or drainage systems serving only roads, parks, or other public properties used by the citizens in a municipality. Such charges could also be used for the development of a database or mapping information that serves all properties generally rather than just a certain sector or area.

8. *Impact fees* — Impact fees are also designed to be used in conjunction with other funding mechanisms. They are most appropriate as a mechanism to balance financial participation with the cost of service, especially the short-term cost resulting from the development of private property. Impact fees must pass the "rational nexus" test and should only be used where impacts to be mitigated are specific rather than general, a consistent methodology is available for quantifying the impact, a separate accounting for each impact fee project can be maintained, and a mechanism exists for returning the unused portion of an impact fee after a mitigation period has passed. Impact fees have lately come under close legal scrutiny (Schlette, 1989). In some cases, the cost of providing enough information to develop an appropriate impact fee method might make this funding method unattractive to the local municipality.

9. *Developer extension/latecomer fees* — Stormwater management facilities built as part of private developments often should be oversized to accommodate service beyond the immediate confines of a single project. This results in additional construction costs to serve the needs of these external areas. A financing method can be employed that allows an original developer to be compensated for these front-end expenses by developers subsequently building in the same area and served by the oversized facilities. These fees are particularly useful where an overall development master plan is available that defines future stormwater requirements.

Other secondary funding methods include sales taxes, gaming taxes, surcharges, grants and loans, penalties, and many more.

5.5 Stormwater Utility Policy Issues

General Policy Information

The analysis of stormwater utility funding has numerous policy implications (Hardten, Bensen, and Thomson, 1990; Priede, 1990; Scholl, 1991a, 1991b; Zielke, 1990). Many issues involve deciding how service charges should be implemented and applied to specific properties in a consistent and fair manner. Timing is also important. Some issues will be resolved early in the process; some will wait until the utility is functioning.

Although policy-making in the highest sense is reserved to the Mayor and Council, day-to-day policy decisions are, in fact, often made at several levels. The Municipal Council formally adopts many of the major policy decisions that guide the municipality.

The Mayor also makes policy decisions often based on Municipal Council positions. Municipal Manager and staff administrators pursuant to the general directives spelled out by the Mayor and Council make other policy decisions. Recognizing this dispersed policy-making environment, a simple hierarchy is recommended for the level of review of the important issues:

- Key staff and consultants
- Other involved staff
- Advisory committee
- Manager's office
- Municipal council

An initial screening of possible issues must consider the following (Cyre, 1986):

- Impacts of policy decision alternatives on costs and manpower
- Appropriate level(s) of municipal government at which the issue should be addressed and resolved
- Relationship of each specific issue to other policy issues
- Priority and timing associated with the issue given the municipality's objective of implementing alternative funding for stormwater management

Policy issues in the development of a stormwater utility can be divided into those dealing primarily with program issues, those dealing primarily with funding issues, and those dealing primarily with billing technical issues. Following is a list of typical policy issues in the three categories:

1. Program-related policy issues
    a. Program mission
    b. Major program priorities
    c. Program service description
    d. Service area
    e. Extent of service
    f. Levels of service
    g. Stormwater quality strategy
    h. Organization and staffing
    i. Privatization
    j. Interlocal agreements and responsibilities
    k. Relationship with other programs
    l. Public relations
    m. Public input or advisory groups
2. Funding-related policy issues
    a. Types of stormwater services funded
    b. Basis for cost distribution
    c. Prior investment
    d. Future use of stormwater systems
    e. Accounting method
    f. Rate methodology
        i. Basic funding methodology
        ii. Secondary funding methods
        iii. Modification factors
    g. Overall funding strategy
    h. Credits
    i. Equivalent residential unit (ERU) base
    j. Public streets and property
    k. State and federal property
3. Billing-related policy issues
    a. Billing and collection methods
        i. New stand-alone system
        ii. Independent database system tie-in
        iii. Modification of existing billing system
    b. Appeals and adjustments
    c. Billing period
    d. Collections and delinquencies

i. Water bill tie-in
ii. Property liens
iii. Enforcement procedural issues
e. Management reporting
f. Master account file development process and accuracy
i. Use of other databases
ii. Resolution procedures for discrepancies
iii. Number and type of data fields required
g. Impervious area methodology
i. Rounding and ranges
ii. Use of street centerline data
iii. Impervious or total area measurement accuracy
h. Master account file database maintenance and updating processes
i. Billing cost allocations
j. Customer service procedures
k. Billing owners or tenants
l. Case exceptions including:
i. Multiple owners
ii. Undivided interest, common areas
iii. Multistory condominiums
iv. Stormwater-only accounts
v. Consolidated billing
m. Use of GIS, mapping, or CADD
n. Information to put on bill

Stormwater Credits

A rate modifier, which fits within the overall rate structure, is the use of a crediting mechanism to reduce the fee a property owner would pay. Credits typically do not have significant total utility revenue reduction potential (often less than 5%) but may have large potential in reducing the resistance to the utility concept from large-fee payers or others who would qualify for a credit. There is a difference between a one-time credit (often termed an offset) and an ongoing credit. Credits are one of only a few ways stormwater utilities have to encourage sound development using a "carrot" rather than a "stick." As such, they carry an importance far beyond their actual revenue significance (Reese, 1996; Doll, Lindsey, and Albani, 1998; Doll and Lindsey, 1999).

Urban stormwater utility credits are becoming more important as cities are looking for support for using utilities for covering federal and state stormwater quality program costs under the National Pollutant Discharge Elimination System (NPDES) and other programs, and for ways to encourage reducing pollution and flooding. Stormwater credits are often granted to improve equity and to provide incentives to implement or carry out an overall community stormwater management plan or to advance some other social or environmental objective. Credits in use by stormwater utilities throughout the United States derive their primary basis in different ways:

- A class of ratepayers (such as elderly or disadvantaged)
- The class of property (such as tax exempt, agricultural, or publicly-owned property)
- Location within the watershed or service area
- Activities that improve the system beyond normal expectations
- Ongoing activities on the property that reduce impact
- Ongoing activities on the property that reduce the city's cost of service

In configuring a credit, it is important that the credit only be offered on the same basis that cost is accrued and fees are charged — impact on the drainage system or imposition of cost on the municipality. Thus, it is typically not legal, even if it is moral, to offer a credit within the rate structure for elderly, disadvantaged, churches, or other nonprofits. A credit can be offered for activities that reduce the municipality's cost, reduce peak flow, reduce pollution, or do some of the municipality's work for it. If a local government wishes to offer a fee reduction based on economic need or property class, it should do it apart from the rate structure, and use general fund revenues to do so.

Some examples include the following:

- A credit for detention or retention ponds
- A credit for pollution controlling devices (BMPs)
- A credit for performing maintenance on the city's system
- A credit to all schools for performing education of schoolchildren in fulfillment of Phase II requirements
- A credit for obtaining and maintaining an industrial NPDES permit
- A credit for oversizing of detention ponds

5.6 Steps in a Typical Financing Study

Overview

Each municipality will need to develop a financing program that takes into account the local politics, existing municipal programs, available resources, state programs, etc. Thus, each program will be tailored for that specific municipality, but there are some common steps that should be followed by all municipalities. Rate-making for stormwater is similar to that for water and wastewater, and similar procedures and opportunities for automation and use of GIS apply (Raftelis, 1989; Hargett and Eggeman, 1992; Tomaselli, 1989).

While there are many ways to approach utility development (Hodges, 1991; Hartigan, 1989; Rosholt, 1990; Cyre, 1990), it generally follows along four parallel tracks, each playing off the other. The four tracks (not including administration and project management) are illustrated in Figure 5-3 and are as follows:

- *Program track* — What will the program accomplish, and when will it be accomplished? How will it be organized? What are its priorities?
- *Utility track* — Who makes up the rate base? What factors will be used? What secondary funding methods will be used? How will a large number of policy issues be answered?

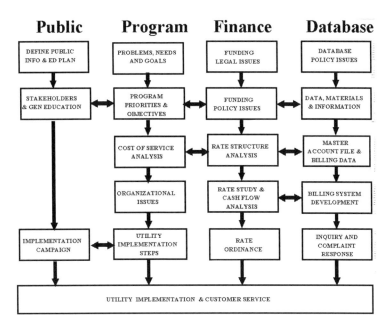

FIGURE 5-3
Setup of a stormwater utility.

- *Billing track* — How will the bill be sent out? How will the master account file be developed? What databases now exist?
- *Public education track* — How and when will the public be involved in the creation of the stormwater utility?

The efforts involved in the development of the utility can be costly and time-consuming. However, the total setup costs amount to only a few months' revenue. Often, these funds are obtained by establishing the legal entity of the stormwater enterprise fund and obtaining a loan from another fund, such as the water or wastewater reserve fund. Anticipation notes are another way to fund the establishment of the utility (Raftelis, 1989).

The Program Track

The program track drives the rest of the project in that it contains the reasons the rest of the project exists in the first place. It is important to hit the ground running with projects, not planning. The headlines from the day the bill goes out should be similar to those in Figure 5-4 below. Yet, it is equally important to control expectations and not oversell the program's ability to fix all complaints the first month. A building-block program makes the most sense — one that balances fixing immediate needs while laying foundations for a sound stormwater program.

The cost-of-service analysis is typically the final expression of the program track. It is a detailed, budget document that lays out a three- to seven-year stormwater program in detail. It serves as the basis for the rate study. Thus, it is important to think through the logical steps in program development and to estimate all costs as accurately as possible. The stormwater functional table given in Chapter 2 serves as a good way to get a handle on the functions in the local municipality, and the Five Whys analysis serves as a good way to zero in on problems and needs and program priorities. Examples of

FIGURE 5-4
Headlines from day one in Charlotte, North Carolina.

each of these and of a preliminary cost-of-service analysis are shown in Figures 5-5, 5-6, and 5-7, respectively.

The Public Track

The public track normally consists of three phases and begins with a public education plan (as outlined in Chapter 3). Dealing with the media in an effective way is critical during all three phases. Often, informational brochures, FAQ sheets, and other means are used to promote support for the utility. Answering the key questions of what, who, and how much are vital at the front end (Luken and Swensen, 2001). Meeting with and educating the editorial board of the local newspaper can yield dividends, as illustrated in Figure 5-8.

The key is selling the stormwater program and making a compelling case for stormwater. In every municipality, there are a number of good reasons to support the stormwater program, including deteriorating infrastructure, flooding complaints, environmental protection, aesthetics, beaches, stream access, greenways, etc. It is important to find and package these reasons in a way that they are understandable to the general public and compelling. Statistics, pictures, testimonials, horror stories, and clear vision and objectives are important.

The three phases are as follows:

- *General public awareness phase* — The goal is to raise the awareness of stormwater problems and objectives throughout the community.
- *Buildup phase* — This phase lasts from about three months until the day the first bill goes out. Its goal is to educate and build support for the utility.

Existing Program Costs by Department			FUNCTIONAL AREA	FTE
Engineering	Public Works	Codes Enf.		
			ADMINISTRATION	
$ 13,650	$ 9,100		Division Director of Public Works (15% & 10%)	0.25
$ 11,300			Deputy County Engineer (20%)	0.20
$ 20,230			Administrative Assistant (75%)	0.75
	$ 6,860		Executive Assistant (20%)	0.20
			SPECIAL PROGRAMS	
$ 12,160			Hotline Representative (50%)	0.50
$ 7,430			Civil Engineer II (15%) - GIS	0.15
$ 22,280			Civil Eng. Designer - CADD/GIS (60%)	0.60
			FINANCE, BILLING AND INDIRECT	
			WATER QUALITY	
$ 11,300			Deputy County Engineer (20%)	0.20
			ENGINEERING AND PLANNING	
$ 28,260			Deputy County Engineer (50%)	0.50
$ 17,050			Property Manager (40%)	0.40
$ 24,310			Right-of-Way Technician	1.0
$ 17,340			Civil Engineer II (35%)	0.35
$ 23,000			Support to HSWCD	0.30
		$ 19,500	Floodplain Management (50%)	0.50
			OPERATIONS AND MAINTENANCE	
	$ 52,000		Crew District 1 (20%)	2.6
	$ 52,000		Crew District 2 (20%)	2.6
	$ 52,000		Crew District 3 (20%)	3.0
	$ 65,000		Crew District 4 (20%)[1]	2.4
	$ 65,000		Crew District 5 (20%)[1]	2.4
	$ 58,500		Storm Drain Crew (100%)[1]	2.0
	$ 15,480		Trucking (10%)	2.2
	$ 200,000		Materials	
	$ 175,000		Equipment (30% miles, 50% fuel/oil)	
$ 40,000			Beaver Control	2.0
$ 32,500			Engineering Tech (2@50%)	1.0
			REGULATION AND ENFORCEMENT	
$ 5,650			Deputy County Engineer (10%)	0.1
		$ 19,500	Floodplain Management (50%)	0.5
$ 34,290			Engineering Tech (2@50%)	1.0
			CAPITAL IMPROVEMENTS	
$ 320,750	$ 750,940	$ 39,000	TOTALS	27.7
$ 1,110,690			Existing Program Total	

Note: 30% Fringe added to all labor values.
[1] Not yet hired

FIGURE 5-5
Example of functional table application.

- *Customer support phase* — The goal is to provide friendly and effective customer service in answering questions and responding to complaints.

In each phase, it is important to keep in mind the various "publics" and how the message should be crafted and delivered to each. Figure 5-9 illustrates a summary table of the public sectors in one city.

Physical Problem	Immediate Causes	Secondary Causes
Sediment-based pollution	Lack of effective erosion control	(see above under Erosion)
Beach closings	Pollution from stormwater outfalls	Lack of comprehensive stormwater quality program
	Pollution from leaking and flooded sanitary systems	Need for system upgrade (non-stormwater issue)
	Pollution from septic systems	Lack of comprehensive septic inspection and enforcement program
Intracoastal Waterway	Excess discharge of point source pollutants versus assimilative capacity	Lack of tertiary treatment
	Non-point source discharges	Lack of comprehensive non-point pollution program in the County
	Poor flushing characteristics of the Waterway	

FIGURE 5-6
Example of the Five Whys application.

Existing		Program Elements	FUNDING FOR STORMWATER PROGRAM							
			FY2001	FY2002	FY2003	FY2004	FY2005	FY2006	FY2007	FY2008
	1	**Administration & Finance**								
$33,250	2	Administrative Assistance	$33,250	$33,250	$33,250	$33,250	$33,250	$33,250	$33,250	$33,250
	3	Indirect costs and overhead	$150,000	$150,000	$150,000	$150,000	$150,000	$150,000	$150,000	$150,000
	4	Billing & customer service	$50,000	$50,000	$50,000	$50,000	$50,000	$50,000	$50,000	$50,000
	5	Public Awareness & Education	$20,000	$20,000	$20,000	$20,000	$20,000	$20,000	$20,000	$20,000
	6	Additional Administrative Assistance			$33,250	$33,250	$33,250	$33,250	$33,250	$33,250
	7									
	8	**Planning & Engineering**								
$33,250	9	Existing Engineering Assistance	$33,250	$33,250	$33,250	$33,250	$33,250	$33,250	$33,250	$33,250
	10	Inventory of System	$300,000	$200,000	$5,000	$5,000	$5,000	$5,000	$5,000	$5,000
	11	New Design Standards		$30,000						
	12	Engineer	$79,800	$79,800	$79,800	$79,800	$79,800	$79,800	$79,800	$79,800
	13	Engineering Tech II - Plans Review	$50,000	$50,000	$50,000	$50,000	$50,000	$50,000	$50,000	$50,000
	14	Master planning		$200,000	$200,000	$75,000	$10,000	$10,000	$10,000	$10,000
	15	GIS/Technology Upgrade		$30,000	$30,000	$5,000	$5,000	$5,000	$5,000	$5,000
	16									
	17	**Operations & Maintenance**								
$216,207	18	Original Maintenance Crew	$216,207	$216,207	$216,207	$216,207	$216,207	$216,207	$216,207	$216,207
$30,950	19	Equipment and Materials	$40,000	$40,000	$40,000	$40,000	$100,000	$100,000	$100,000	$100,000
	20	New Maintenance Crew		$216,207	$216,207	$216,207	$216,207	$216,207	$216,207	$216,207
	21	Equipment and Materials		$100,000	$100,000	$100,000	$100,000	$100,000	$100,000	$100,000
	22	Sweeper Operator		$33,250	$33,250	$33,250	$33,250	$33,250	$33,250	$33,250
$100,000	23	Remedial maintenance program	$100,000	$250,000	$500,000	$500,000	$300,000	$300,000	$300,000	$300,000
	24									
	25	**Regulation & Enforcement**								
	26	New Inspector	$50,000	$50,000	$50,000	$50,000	$50,000	$50,000	$50,000	$50,000
	27									
	28	**Water Quality**								
	29	TMDL support	$10,000	$10,000	$10,000	$10,000	$30,000	$30,000	$30,000	$30,000
	30	Phase II		$50,000	$70,000	$90,000	$100,000	$120,000	$120,000	$120,000
	31	Green space, development sppt.	$20,000	$20,000	$20,000	$20,000	$20,000	$20,000	$20,000	$20,000
	32									
	33	**Capital Construction**								
$500,000	34	Original Capital Construction	$500,000	$500,000	$500,000	$500,000	$500,000	$500,000	$500,000	$500,000
	35	Capital construction - new	$200,000	$200,000	$200,000	$200,000	$750,000	$750,000	$750,000	$750,000
	36									
	37	**Contingency Funds**	$30,000	$30,000	$30,000	$30,000	$30,000	$30,000	$30,000	$30,000
$913,657	38	**TOTALS**	$1,882,507	$2,561,964	$2,670,214	$2,565,214	$2,915,214	$2,935,214	$2,935,214	$2,935,214
	39	**TOTALS WITH 4% Inflation**	$1,957,807	$2,771,020	$3,003,628	$3,000,938	$3,546,804	$3,713,982	$3,862,541	$4,017,043
	40									
	41	rate @ $600,000 per dollar	$3.14	$4.27	$4.45	$4.28	$4.86	$4.89	$4.89	$4.89
	42	rate @ $700,000 per dollar	$2.69	$3.66	$3.81	$3.66	$4.16	$4.19	$4.19	$4.19
	43	rate @ $800,000 per dollar	$2.35	$3.20	$3.34	$3.21	$3.64	$3.67	$3.67	$3.67

FIGURE 5-7
Example cost-of-service spreadsheet.

8A THE CHARLOTTE OBSERVER Monday, February [illegible]

The Charlotte Observer

ROLFE NEILL, Chairman and Publisher
RICHARD OPPEL, Editor **JOHN LUBY,** General Manager
GENE WILLIAMS, Executive Vice President
ED WILLIAMS, Editor of the Editorial Pages **JANE SHOEMAKER,** Managing Editor
TOM BRADBURY, JERRY SHINN, Associate Editors

Editorials

Don't delay drainage plan

■ Task force outlines good policies for a storm water utility, but its suggestion that city wait deserves to go down the drain.

The citizens Storm Water Task Force has given the city council something it never asked for and probably doesn't want: a recommendation to delay establishment of a storm water utility beyond the council's January 1993 target. As task force Chairman Sam Smith and three other members suggested in a minority report, that's an idea that deserves to go down the sewer with the storm water.

Delay would only allow more time and opportunity for political maneuvering and second-guessing about rates and the way costs are assessed, which could go on for years if the council lets it.

It isn't as if the council were rushing into something. Storm water isn't a new problem. Some drainage systems were built more than 80 years ago, many more were built more than 30 years ago, when development and resulting runoff were minimal compared to what they are today. As a result, lawns that once flooded once every few years now flood several times a year; hundreds of catch basins, pipes and culverts are too small or broken and need repair or replacement, and thousands more are clogged. It has been clear for years that the costs of meeting those needs have risen far above anything that could be routinely covered by general revenues.

Since at least the mid-1980s the council has been discussing the idea of a utility to fund repairs and improvements and spread the cost fairly. It approved the concept more than three years ago. The federal Environmental Protection Agency's new water quality regulations are not driving the council's decision. They came later and served only to impose a deadline on what the council already had decided to do.

and has devoted many hours, working with city staff and consultants, to determining the magnitude of the problem, estimating the costs of capital improvements and maintenance, and developing recommendations for a rate structure. (Advising the council on timing wasn't part of the task force charge.)

Except for the proposal for delay, the task force recommendations are sound. They include a rate structure based roughly on how much particular properties contribute to the problem. Fees would reflect the amount of impervious surface — that is, rooftops and paving that don't absorb water.

Preliminary estimates suggest the cost would be between \$3 and \$4 a month for residences, although that could drop if streets and other public properties are added to the rate base, which is now considered likely. Large commercial and industrial tracts would pay a lot more, of course, but those costs, in the case of shopping centers, for example, would be divided among many tenants. There will be credits for drainage control structures that many industrial and commercial sites already have in place. And if it turns out that the proposed rate structure would produce some backbreaking bills here and there, there's plenty of time before next January for the council to find a balance between what's fair and what's reasonable.

One reason the task force cited in calling for delay was to allow time for the city and county to consolidate their plans, since county government is developing a storm water utility, too. But county government doesn't have to deal with the urban drainage problems the city faces, or with a federal deadline. The city should move ahead on schedule. Whether the county wants to hop on board is up to the

FIGURE 5-8
Stormwater editorial. (Reprinted with permission from The Charlotte Observer. Copyright owned by The Charlotte Observer.)

Summary Table

	Applicable Media Options										
	Presentations	Informational Brochure	Fact Sheets	White Papers	News Articles		Informational Meetings		Individual Meetings	Customer Service	
Developers	✓	G	✓		G	G	✓	✓	✓		G
Businesses	✓	G	✓		G	G	✓			✓	G
Industry	✓	G	✓		G	G	✓			✓	G
Minorities		G			G	G		✓	✓	✓	✓
Schools		G			G	G			✓		G
Churches	✓	G	✓		G	G	✓	✓			G
Homeowners		✓			✓	✓		✓		✓	✓
Landlords	✓	G	✓		G	G	✓				G
City Commissioners	✓	G	✓		G	G	✓	✓	✓		✓
Press		G	✓	✓	✓	G	✓	✓	✓		✓
Top 100		G	✓		G	G	✓		✓		G
City Staff		G	✓		G	G			✓		G

FIGURE 5-9
The public sectors.

The Finance Track

The finance track is paired to the program track in that it takes the program elements and determines the best ways to fund them. There are myriad policy decisions to be made (see below). This track moves through a basic feasibility and policy analysis to a rate structure analysis and finally to a rate study and ordinance. The feasibility analysis quickly culls the methods and approaches that will not work well to allow for focus on those that might. The rate structure analysis builds the rate structure consisting of a basic rate methodology, secondary funding methods, and rate modifiers (discussed below).

The rate study is just one element of funding policy. There may be other funding methods blended with the stormwater user fee. In fact, blended stormwater funding is becoming the rule. The rate structure and final rate, and period for which the rate is expected to remain constant, are based on the program. The actual rate methodology depends on policy decisions and also on the ability to efficiently apply data and technology as determined in the data track.

It is important to be able to satisfy judicial standards for the rate to withstand a potential court challenge. The key is that the charge be considered a fee and not a tax or special assessment (although a special assessment was attempted in Florida and was upheld at the Supreme Court level). Twenty-five percent of all utilities are challenged in court. Few, if any, stormwater utilities have failed court challenges if they are fair and reasonable, the costs are related to the services rendered, they are legal by charter or legislation, and the proper procedures were followed in setting up the utility. It is also important, when challenged in court, that the utility and program concept be consistent with community expectations, allow for change and anomalies, and provide for appeals and adjustments.

A Natural Resources Defense Council survey of laws in all 50 states found that, in almost all cases, municipalities can legally create stormwater utilities. If a particular state has no statute specifically delegating that authority to municipalities, precedent or the state attorney general's office can help determine what is necessary to set up a utility. Typically, two ordinances are needed: the first to establish the utility and the second to set the rate structure. Depending on state and local law, a general referendum may be needed for the first ordinance, or the city council or county board of supervisors may vote on it (Kaspersen, 2000; Cyre, 2000).

The Data Track

The data track seeks to answer two questions: (1) How do I calculate a bill for each applicable property?, and (2) How do I physically deliver the bill to the ratepayer?

If possible, it is always cheaper to make the calculation a database exercise rather than a hand-digitizing exercise. Thus, if there is a database from which one can derive a surrogate for impervious area that would be preferable. Sometimes, the tax assessor's data are sufficient to develop an algorithm to derive a sufficiently accurate estimate of impervious area or intensity of development. The courts have not indicated that it must be an engineering calculation but simply be fair and reasonable.

When hand-digitizing is necessary, having a flat rate for residential property is most cost-effective. Graphical data were not collected with measurement of impervious area by parcel in mind. Thus, polygons are often not closed, parcel lines are inaccurate, and building outlines are only approximate. Figures 5-10 and 5-11 illustrate inaccurate and accurate graphical data.

When delivering the bill, there are several options: the tax bill, another utility bill, or a stand-alone system. The final solution depends on the coverage of the different billing systems, costs, legality, and preference. Cities often use another utility bill to handle the

Financing Stormwater Management Programs

FIGURE 5-10
Inaccurate graphical data.

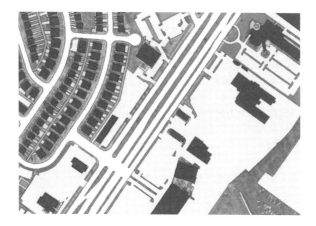

FIGURE 5-11
Good graphical data.

stormwater bill, while counties that do not have a single utility countywide use the tax bill. But, there is great variability.

Feasibility Studies

A feasibility study takes a group of staff and citizens on a walk through all the key aspects of the utility development without committing to utility development until all concerned agree it is the right way to go. This has several advantages when the "go/no-go" decision has not yet been made. It allows for a testing of the waters before committing to a utility. This gives political leaders a sense of safety and allows there to be more "fingerprints on the knife" for a "go" decision to impose a new fee. The feasibility study also finds pitfalls prior to falling into them, saves time and money, generates a good estimate of costs and schedule, develops a scope of services, and smokes out the problem sectors and stakeholders.

A typical feasibility study consists of a group as discussed in Chapter 3, appointed by the political leadership (with suggestions from staff), that is asked to assess the stormwater program and answer the following questions:

1. What are we spending and doing now for stormwater in the municipality?
2. What are the significant problems, needs, and issues now facing the municipality?
3. How have others solved them?
4. What should our priorities be over the next five years?
5. What will that cost be in terms of dollars and resources?
6. How should we pay for it?

You can see from this set of questions that the program drives the decisions, and the fact that a utility may be recommended as part of the study is not the main focus but a support to the main focus.

5.7 Typical Financing Feasibility Study Scope of Services

Purpose

The purpose of this project is to assess the city's stormwater management and water resources program, make recommendations for future directions and changes, and assess the feasibility of funding the water resources program with a stormwater utility (user fee) and other methods. In the following discussions, the consultant is assumed to be hired by the city to lead the development of the financing feasibility study and represent the city's interest during the study development.

Figure 5-12 illustrates the roadmap the consultant is proposing for this project. We will take a group of citizens and staff through a consideration of the following questions, the answers leading to program and funding directions:

1. What is the city currently doing in terms of stormwater management?
2. What are the stormwater-related problems, issues, needs, and opportunities currently faced by the city? How have other cities solved similar problems?
3. What stormwater program priorities should guide the city in the next three to five years?
4. What specific program improvements should the city make, and what will the costs be?
5. What is (are) the best way(s) to pay for these program improvements?
6. How should the funding method(s) be implemented?

The scope of work presented below describes the process of evaluating the city's long-term stormwater management options and developing the necessary analyses and reports to support initial decisions.

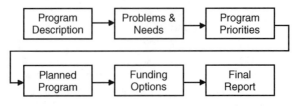

FIGURE 5-12
Proposed roadmap for a project.

This project primarily involves the use of senior consultants with long experience in stormwater programs and utility development.

Task 1 — Development of a Citizens Group

The consultant will work with city staff to identify candidates and appoint members to a citizens stormwater action committee (SWAC). The SWAC will attend a series of meetings, as described in later tasks, to learn more about the stormwater program and make informed recommendations for action by city council. The consultant will advise in SWAC member selection, prepare information and educational pieces for SWAC members, and prepare a press release for the news media. The consultant will also provide a draft SWAC introductory letter for distribution by the city. Two, one-hour conference calls with city staff are included in this task. No meetings are included under Task 1.

Task 2 — Stormwater Program Description, Problems, and Needs

The consultant will work with city staff to update its assessment of the current stormwater program. The program will be described in terms of stormwater-related resources, organization, and function, in such a way as to facilitate discussion with a citizens group. The consultant will prepare a memorandum that describes the current stormwater program in terms of resources, manpower, and functions. It will be prepared in a format suitable for discussion with the SWAC.

The consultant will also work with city staff to describe and quantify (where appropriate) the types of needs, problems, and issues facing the city. The goal will be to develop a compelling case for the inception of increased stormwater financing. Key areas of concern will be identified, quantified, and described, and pictures or graphics will be developed where available. The consultant will prepare a policy paper that identifies, describes, and assesses the city's stormwater program, problems, needs, and issues at the physical and institutional levels. Jurisdictional issues and a brief description of current technology, data, and information will also be included in the policy paper.

A meeting will then be held with the SWAC to initiate the action plan process and to discuss the existing program and issue policy papers. Also discussed will be ways that other local governments have used with success in dealing with similar problems, and ideas for solutions. The consultant will prepare a draft cover letter, meeting agenda, and draft policy paper for distribution by the city prior to the SWAC meeting. The city will make all necessary arrangements and schedule the meeting facility. Based on SWAC input, revisions to the draft policy paper will then be incorporated into a final policy paper by the consultant.

Deliverables

- Kickoff meeting
- Memorandum — Existing stormwater program
- Policy paper #1 — Stormwater program, problems, needs, and issues (draft and final)
- SWAC mail-out materials (draft letters, agenda, informational and educational pieces)

- SWAC meeting #1 — Introductions, education, existing program description
- SWAC meeting #2 — Overview and issues identification

Task 3 — Program Objectives and Priorities

The consultant will work with key city staff to identify a preliminary list of objectives and priorities to guide the stormwater program for the next five to ten years. Objectives will be related to functional areas and cost centers. Priorities will serve as the basis for development of a projected five- to ten-year stormwater program, and any other potential organizational or staff changes recommended for the program. One meeting and several conference calls with city staff are included to identify objectives and priorities.

Based on information obtained, the consultant will prepare a draft policy paper that identifies stormwater program objectives and priorities. Discussion will focus on basic program directions, levels of service, and other information required to address the problems and needs previously identified.

The consultant will prepare a draft cover letter, meeting agenda, and draft policy paper for distribution by the city prior to the SWAC meeting. The city will make all necessary arrangements and schedule the meeting facility. The initial list of stormwater program objectives and priorities will be discussed with the SWAC to gain the benefit of its input. Based on SWAC input, revisions to the draft policy paper will be incorporated into a final policy paper by the consultant.

Deliverables

- Policy paper #2 — Program objectives and priorities
- SWAC meeting #2 — Program objectives and priorities

Task 4 — Projected Stormwater Program

The consultant will work with key city staff to translate the identified stormwater program objectives and priorities into a five- to ten-year projected stormwater program. The projected program will be specific enough to make gross cost and manpower estimates. One meeting and several conference calls with city staff are included to develop the projected program.

Based on information obtained during the conference calls, the consultant will prepare a draft policy paper that describes the projected stormwater program. It is anticipated that the projected program will include the development of specific parts of the stormwater program along with major policy decisions, where necessary, to develop cost estimates. Program items might include the key functional areas of administration and management, billing and finance, indirect cost allocation, special services (PR and GIS), engineering and planning, operations and maintenance, regulation and enforcement, capital construction, and water quality.

The consultant will prepare a draft cover letter, meeting agenda, and draft policy paper for distribution by the city prior to the SWAC meeting. The city will make all necessary arrangements and schedule the meeting facility. The projected program, including a discussion regarding organizational options, will be presented to the SWAC to gain its input. Based on SWAC input, revisions to the draft policy paper will be incorporated into a final policy paper by the consultant.

Deliverables

- SWAC meeting # 3 — Projected stormwater program
- Policy paper #3 — Projected stormwater program

Task 5 — Basic Funding Feasibility

The consultant will prepare a draft policy paper that describes the basic funding feasibility of a stormwater program. Feasibility of major alternate funding methods will be assessed in the context of the city's problems, needs, issues, resources, goals, and functional requirements. The objective of this task is to eliminate unrealistic approaches. One meeting and several conference calls with city staff to discuss the draft policy paper are included in Task 5.

The consultant will prepare a draft cover letter, meeting agenda, and draft policy paper for distribution by the city prior to the SWAC meeting. The city will make all necessary arrangements and schedule the meeting facility. The consultant will meet with the SWAC to discuss the use of various funding options and to educate it on the stormwater user-fee concept. One objective of this task is to have the SWAC make recommendations on the best way(s) to fund the stormwater program. Based on SWAC input, revisions to the draft policy paper will be incorporated into a final policy paper by the consultant.

Deliverables

- SWAC meeting # 4 — Funding options
- Policy paper #4 — Funding options

Task 6 — Service Charge Billing, Collection, and Accounting Options

The demands of billing for, collecting, and properly accounting for stormwater service charges and other forms of funding will be investigated. The consultant will:

- Identify and evaluate materials and information needed during the development and implementation of stormwater service charges
- Identify gaps and deficiencies in the information, systems, and procedures, and suggest remedies to those data gaps
- Recommend how and when the materials and information should be assembled
- Propose data evaluation criteria consistent with the types of information to be used in various aspects of the funding implementation work
- Recommend scheduling of associated analytical, system development, and implementation work
- Develop a detailed methodology and schedule for the development of the master account file

The results of this investigation will be a recommendation to the city on how to assemble the base master account file. Recommendations will also be made on how to generate the data for billing. Preliminary findings influencing the feasibility of various funding and rate options will be summarized in the action plan report.

One meeting with city staff is envisioned to provide the basic information necessary for the development of this section of the action plan report. Key staff and documents necessary to be made available for this meeting will be identified ahead of time by the consultant and the city. The city is responsible for making these personnel and documents available at an agreed time and place.

Deliverable

- Memorandum on billing and collection options

Task 7 — Public Communications Program

The purpose of this task is to assist the city in communicating the city's stormwater needs and goals to the public; begin building key public support for stormwater funding; and develop a public involvement strategy for implementation, should a decision be made to move forward with the recommended program.

The consultant will work with the city to develop two fact sheets suitable for distribution to the news media; develop one PowerPoint or slide presentation for use by key city staff in public meetings or small group settings; provide a maximum of five "SWAC Updates" for city staff to distribute to city council and others on a periodic basis.

The consultant will also work with the city in the development of a brief public education and utility campaign support plan, should the second phase of the project be approved.

The city will be responsible for final approval and printing of all materials and for the provision of any and all original graphics for materials.

Deliverables

- Two fact sheets
- One PowerPoint presentation
- Five "SWAC Update" handouts (maximum)
- Public involvement strategy plan

Task 8 — Action Plan Report

The consultant will work with key city staff and the SWAC to prepare an action plan report for presentation to city council. The action plan report will be a compilation of the previous policy papers, including a summary paper detailing SWAC decisions and recommendations.

The consultant will develop the draft action plan report and submit up to five copies of the draft report to city staff for distribution and review by city staff and others. Initial revisions will be made to the report based on one complete set of review comments provided to the consultant prior to the SWAC meeting.

The consultant will prepare a draft cover letter, meeting agenda, and draft executive summary on SWAC decisions and recommendations for distribution by the city prior to the SWAC meeting. The city will make all necessary arrangements and schedule the meeting facility. The draft policy paper will be presented to the SWAC to gain its input and final concurrence. Based on final input from the SWAC, revisions to the draft policy paper will be incorporated into a final policy paper by the consultant.

The consultant will then publish the final action plan report, including a section on SWAC decisions and recommendations. Up to ten copies of the final report will be submitted to city staff for distribution to the SWAC and others. The final report will then be presented to key staff and elected officials, as determined by city staff. The consultant will make a presentation.

A complete scope of services schedule and cost report will be prepared by the consultant to identify and develop a proposal to further implement the city-wide stormwater program. The report will include recommendations for immediate subsequent steps, estimated costs, and schedules.

This task will lead to a go/no-go decision by city staff and political leaders.

Deliverables

- Action plan report, including SWAC recommendations (draft and final)
- SWAC meeting # 5 — Action plan report and SWAC recommendations
- Meetings with and presentations to city staff and city council
- Scoping and implementation report

References

American Water Works Association, Water Rates, AWWA Manual M1, AWWA, Denver, CO, 1986.
American Water Works Association, Water Rates and Related Charges, AWWA Manual M26, AWWA, Denver, CO, 1986.
American Water Works Association, Water Utility Capital Financing, AWWA Manual M29, AWWA, Denver, CO, 1988.
American Public Works Association, Financing Stormwater Facilities: A Utility Approach, APWA, 1991.
Black and Veatch, Stormwater Utility Survey, Kansas City, MO, 1996, 2002.
Charlotte, NC, Results of Citizen Survey for Stormwater Management, 1991.
Cyre, H.J., Key Feasibility Issues in Developing and Implementing a Stormwater Management Utility, Int. Public Works Congress and Eq. Show, New Orleans, LA, Sept. 23, 1986.
Cyre, H.J., Refinements in Stormwater Utility Rate Structures, ASCE 14th Annual Water Res. Plann. and Manage. Conf., Kansas City, MO, March 16–18, 1987.
Cyre, H.J., Financing Urban Stormwater Management Case Studies of Change in Progress, Proc. WPCF Annual Conf., Washington, DC, Oct. 7–11, 1990.
Cyre, H.J. and Reese, A.J., Stormwater Utilities in the United States of America, Proc. NOVATECH 92, Lyon, France, Nov 3–5, 1992.
Cyre, H.J., The Stormwater Utility Concept in the Next Decade (Forget the Millennium), EPA Natl. Conf. on Tools for Urban Water Resource Manage. and Prot., Conference Draft, Cincinnati, OH: U.S. EPA Office of Research and Development, 2000.
Davis, P.E., Major, A.E., and Dunlap, J.N., Making the State Revolving Fund Work: The Tennessee Experience, Water Environ. and Technol., November 1989.
Davis, K., Hatoum, W., and Rose, D., Prepared for a rainy day, *Water Environ. and Technol.*, June, 1999, pp. 36–41.
Debo, T.N. and Williams, J.T., Voter reaction to multiple-use drainage projects, *ASCE J. WR Plann. and Manage.*, 105, WR2, September 1979.
Doll, A. and Lindsey, G., Credits bring economic incentives for on-site stormwater management, Watershed and Wet Weather Tech. Bull., 4, 1, 12–15, 1999.

Doll, A., Lindsey, G., and Albani, R., Stormwater Utilities: Key Components and Issues. Advances in Urban Wet Weather Pollution Reduction, Conf., Water Environment Federation, Cleveland, OH, June 28–July 1, 1998, pp. 293–302.

Doll, A., Scodari, P., and Lindsey, G., Credits as Economic Incentives for On-Site Stormwater Management: Issues and Examples, EPA Natl. Conf. on Retrofit Opportunities for Water Resour. Prot. in Urban Environ. in Chicago, IL, February 9–12, 1998.

Environmental Protection Agency, Financing Marine and Estuarine Programs: A Guide to Resources, EPA Office of Water, Report Number 503/8-88/001, 1988.

Environmental Protection Agency, Building Support for Increasing User Fees, EPA, Office of Water, Report Number 430/09-89-006, 1989.

Environmental Protection Agency, Financing Mechanisms for BMPs, EPA Region V, Chicago, IL, December 1990.

Environmental Protection Agency, Office of Water, The Clean Water State Revolving Loan Program, http://www.epa.gov/owm/factsht2.htm, 2001.

Florida Association of Stormwater Utilities, Stormwater Utilities Survey, Tallahassee, FL, 1995.

Hardten, R.D., Bensen, R.B., and Thomson, K.D., How Much to Charge and How to Collect it: Stormwater Rate Setting and the Billing System, Proc. WPCF Annu. Conf., Washington, DC, October, 7–11, 1990.

Hargett, C.W. and Eggeman, G.L., Database Utility Rate Model, Public Works, August 1992.

Hartigan, J.P., Use of Stormwater Utility to Meet New Water Quality Requirements, Presented to Virginia Sec. ASCE, October 20, 1989.

HDR Engineering, Inc., Stormwater, News Bulletin, February 1990.

Hodges, R.H., How to Create a Stormwater Utility, Public Works, October 1991.

Houmis, N., Infrastructure financing, *Am. Consulting Engr.*, Winter, 2, 1, 1992.

Kasperson, J., The stormwater utility — will it work for your community ? *Stormwater*, [Online] Available: http://www.forester.net/sw_0011_utility.html, November/December, 2000.

Keller, B., Buddy, can you spare a dime? What's stormwater funding? *Stormwater* 2, 2, [Online] Available: http://www.forester.net/sw_0103_buddy.html, 2001.

LeClere, J., Trends in managing stormwater utilities. Watershed protection techniques, 2, 4, 500–502, [Online] Available: http://www.stormwatercenter.net, 2000.

Lindsey, G., Financing Stormwater Management: The Utility Approach, Sed. & Stormwater Adm., State of Maryland, 1988.

Lindsey, G., Update to Survey of Stormwater Utilities, Proc. WPCF Annu. Conf., Washington, DC, October 7–11, 1990.

Luken, K.M. and Swenson, S., A stormwater management plan your communities, businesses, and residents will support. *Stormwater*, 2, 2, [Online] Available: http://www.forester.net/sw_0103_plan.html, 2001.

Minner, M. et al., Cost apportionment for a storm-water management system: differential burdens on landowners from hydrologic and area-based approaches, *Appl. Geogr. Stud.*, 2, 4, 247–260, 1998.

Pardiwala, S.D. and Leserman, J.R., Gaining Acceptance for Utility Rate Increases, Public Works, June 1992.

Priede, N., Managing Stormwater Utilities, Proc. WPCF Annu. Conf., Washington, DC, October 7–11, 1990.

Raftelis, G., *Water and Wastewater Finance and Pricing*, Lewis Publishers, 1989.

Reese, A.J., Storm-water utility user see credits, *ASCE J. Water Resour. Plann. and Manage.*, 122, 1, 49–56, January/February 1996.

Rosholt, J.E., Program for Successful Implementation of a Stormwater Utility, Proc. WPCF Annu. Conf., Washington, DC, October 7–11, 1990.

Rosholt, J.E. and Pigott, S.P., Financing and Service Charge Alternatives for Storm and Surface Water Management, URS Consultants, Inc., personal communication, 1991.

Schlette, T.C., Funding wastewater projects with impact fees, *Water Environ. and Tech.*, November 1989.

Scholl, J.E., Rate Structure Development for Stormwater Management Utility Billing, CH2M Hill, Gainesville, FL, personal communication, 1991a.

Scholl, J.E., Stormwater management utility billing rate structure, *Water Environ. and Tech.*, January 1991b.

Thunberg, E. and Shabman, L., Determinants of landowner's willingness to pay for flood hazard reduction, *AWRA Water Res. Bull.*, 27, 4, August, 657–665, 1991.

Tomaselli, L.K., A Geographic Information Systems Approach to Fiscal Impact Analysis, Ph.D. dissertation, University of Minnesota, Minneapolis, MN, 1989.

Treadway, E. and Reese, A.L., Financial strategies for stormwater management, *APWA Reporter*, February 12–14, 2000.

Zielke, L.J., Legalizing Stormwater Service Fees and Avoiding Issues of Taxation, Proc. WPCF Annu. Conf., Washington, DC, October 7–11, 1990.

Appendix A — Financing Stormwater Management Programs

Questionnaire — Municipality of Anyplace Stormwater Management

Distribution of Questionnaire

The following functional or interested personnel should each receive a copy of the questionnaire:

- Planning design
- Engineering
- Maintenance
- Operations
- Enforcement
- Inspection
- Zoning
- Regulation
- Plans review
- Financial
- Floodplain management
- Codes and standards
- Emergency service
- Police
- Fire
- Legal
- Parks and recreation
- Health and safety
- Complaints
- Water and sewer
- Utility
- Council
- Mayor
- Other staff

and:

- Key community technical individuals such as state, federal, or university offices
- The development community
- Concerned citizens group representatives

Purpose of Questionnaire

The Municipality of Anyplace has contracted with the Consultant Team to conduct an analysis of stormwater management in the Municipality of Anyplace. One purpose of this study is to look at the policies regulating stormwater management in Anyplace and to make recommendations for improvements and modifications as well as to develop an ordinance to regulate stormwater management activities.

The study is not budgeted to allow for in-depth interviews with all individuals involved in some aspect of stormwater management. This questionnaire is designed to serve as a basis for and to supplement the interviews that will be conducted.

Directions to Respondents

Respondents should use a separate piece of paper to answer the questions and may wish to dictate answers for secretarial transcription. *There may be some areas where individuals feel they are not qualified to respond. They should not hesitate to leave these questions unanswered or simply insert the name and office of the individual whom they feel is best qualified to address the question.* A broad distribution of this questionnaire should ensure that all questions are answered. Respondents should answer those questions they can in sufficient detail so that their responses can stand alone and be complete. Do not hesitate to include reports, documents, etc., with the response.

The consultant wishes to thank respondents for their participation in this information-gathering process.

Please Provide Your Answers on A Separate Sheet Keyed to Question Numbers Here

A. Physical drainage problems

  A1. *Complaints* — About how many drainage complaints are on file? How are complaints handled? What is the response procedure? What type of flooding complaints predominate?

  A2. *Flooding* — How would you characterize the flooding problems in the Municipality of Anyplace? Are they primarily related to uncontrolled overflow of creeks, overland flow, undersized systems, major or minor systems, nuisance or life-threatening, etc.? Are they in open or closed conveyances? Are any of the problems related to aging of the system, clogged systems, damaged systems, etc.?

  A3. *Erosion* — How would you characterize erosion problems? Are they minor backyard problems, major creek erosion, only in spots along creeks, minor feeder streams, only downstream from urban development sites, etc.? What are the predominant causes?

  A4. *Water Quality* — Describe any Municipality of Anyplace activities in urban water quality. What type of information is available? Sampling program? Are there any known areas of water quality impairment?

  A5. *Locations* — What is the geographical extent of the various types of problems? Are they concentrated in a few areas along the major watercourses, widely scattered minor problems? Name the areas or reaches and streets. Do they vary from one part of the municipality to another? How? Are they seasonal, year-round, or limited to special-weather events?

  A6. *Impact* — How directly do the problems impact people? What are some specific examples? Is there significant exposure to personal hazard or risk or significant property damage due to flooding?

A7. *Safety concerns* — Have drainage conditions ever posed a problem for fire, police, or emergency aid vehicles, whether by flooding major transportation routes or contributing to poor driving conditions? Does potential exist even though no known problems have yet been encountered? Where? Have there been any other safety concerns related to flooding or conveyance structures?

A8. *Sanitary system* — Have there been any water-quality-related health or safety problems due to leaking sanitary systems, septic system malfunction, etc?

A9. *Toxics* — Are there any known areas of hazardous or toxic materials storage that may pose a problem during flooding conditions? Describe the emergency response capability of the municipality.

A10. *Responses* — Describe the municipal responses to flooding, erosion, water quality, and safety concerns. What types of solutions are typical? Are there special designs that seem to work best?

A11. *Root causes* — Can you characterize any deeper problems you feel might be deeper root causes of drainage problems?

A12. *Limits* — What are the major constraints on responding to complaints or correcting flooding, quality, and erosion problems?

B. Administration, finance, and program development

B1. *Organization and responsibility* — Use a functional table to identify who, if anyone, has primary and secondary responsibility in each area now performed in the Municipality of Anyplace. Provide an organization chart if available for one or more agencies. Describe principal stormwater-related duties, annual budgets related to stormwater management, and legal authority.

B2. *Financial overview* — What are the basic financing methods used to support stormwater management in all its various aspects within the municipality? What other financing methods are presently used for other municipal programs that are well accepted by the community? Which methods are least popular? What level of acceptance do the municipal utility service charges have generally? What is the level of acceptance of property, sales, and other taxes that generate local revenue?

B3. *Financial priorities* — Is there a perceived balance in funding various municipality programs or is there a perception that one or more programs receive strong support while others lack public or political support?

B4. *Utility charge structure* — Does the Municipality of Anyplace presently charge schools, public hospitals, municipal and other government agencies, churches, other nonprofit charitable organizations, state or federal buildings or properties for water and sanitary sewer services?

B5. *Utility condition* — What is the financial condition of the municipality's utilities? Has the municipality issued revenue bonds for capital improvements or other purposes? What was the bond rating?

B6. *Economic condition* — What is the condition of the local economy? What is the level of unemployment? Does the Municipality of Anyplace have large concentrations of low-income and elderly citizens? Are delinquencies on water and sewer utility charges increasing or decreasing recently? How many residential mortgage foreclosures have there been recently?

B7. *Bonding health* — What is the municipality's general-obligation bond rating? How much uncommitted general obligation bonding capacity exists?

B8. *Willingness to pay* — How much do you think residents would be willing to pay monthly for drainage control if it were funded through a service charge?

B9. *Billing area* — What is the present area within the municipal limits? Is annexation being aggressively pursued? Does the planning function go beyond municipal limits?

B10. *Public involvement* — What types of special-interest groups or other governmental authorities have some impact on municipal stormwater management? Is there a constituency supporting stormwater in the general public or neighborhood groups?

B11. *Political awareness and support* — Describe the political concern and involvement in stormwater management. Are politicians concerned and knowledgeable?

B12. *Current stormwater issues* — What issues or circumstances have led to the current interest in stormwater management? What are the key issues in the minds of staff, political leaders, and the public?

B13. *Current other issues* — Are there any other issues now ongoing that may impact the ability of the municipality to respond aggressively to stormwater needs?

B14. *Program planning* — What plans are now underway to change, add to, or in any way impact any existing stormwater programs or activities?

B15. *Limits* — In your opinion, what now limits the stormwater management program from an organizational, financial, and administrative point-of-view?

C. Planning, design, and engineering

C1. *Design criteria* — What types of design guidance or design criteria documents or software packages exist that are used by engineers in Anyplace? What policies (written or unwritten) exist in the Municipality of Anyplace for design criteria?

C2. *Technical studies* — What types of technical studies exist that describe the stormwater system and address quality, quantity, or erosion problems? Has there been any stormwater master planning? Include soil surveys and flood program documents.

C3. *Design methods* — What are the typical design methodologies used for common drainage designs, whether or not they are included in municipal documents or manuals?

C4. *Water supply* — What is the municipal water supply? If it is surface waters, how are they protected?

C5. *Hazard mitigation* — Is there any type of flood hazard mitigation program in place such as flood warning, emergency evacuation, floodplain zoning, flood-proofing, public education, etc.?

C6. *Limits* — In your opinion, what now limits the stormwater management program from a planning, engineering, and design point-of-view?

D. Operations and maintenance
   D1. *Basic scope and operation* — What portions of the stormwater conveyance system are maintained (e.g., catch basins, minor streams, crossings, etc.)? Who performs this maintenance? What is done typically (schedule, activities, etc.)?
   D2. *Priority* — Is the maintenance program for each part of the conveyance system periodic or driven mainly by complaints? How is maintenance prioritized? What is the emergency response capability?
   D3. *Budget and manpower* — What is the level of manpower, organization, equipment, and budget for each entity that performs maintenance? Is any maintenance contracted out (e.g., remedial, minor reconstruction)?
   D4. *Erosion* — How are erosion and sediment control addressed in the program?
   D5. *Limits* — In your opinion, what now are the major constraints on expanding the maintenance system and operation (legal, financial, etc.)?
E. Regulation and enforcement
   E1. *Development process* — Describe the development process as it relates to grading and drainage. What permits are required, and when? What inspections are performed, when, and by whom?
   E2. *Written guidance* — Describe the documents, policies, checklists, etc., that govern all aspects of stormwater management (for example, ordinances, regulations, written policies, etc.).
   E3. *Authority* — Who establishes and enforces stormwater-related regulations? What is the involvement of various agencies? What enforcement penalties are available?
   E4. *FEMA* — Describe the municipality's implementation of the FEMA requirements.
   E5. *Floodplain management* — Describe the Municipality of Anyplace's efforts in greenway planning and general floodplain management. Has there been any other park and recreation planning in conjunction with stormwater management?
   E6. *Water quality* — Which department within the municipality will have responsibility for any water quality regulations (including pretreatment, toxicity, and water quality)? How do other departments support this department? Who would run a stormwater quality program?
   E7. *Other government levels* — Describe the role of state and federal government in stormwater management in the Municipality of Anyplace in flood control, erosion, and environmental quality.
F. Capital improvements and expenditures
   F1. Describe the municipality's capital improvement program for stormwater management. Have there been any recent projects? How were they financed?
   F2. What is the expected level of need for major or minor capital improvements?
   F3. Describe the municipality's land-acquisition or easement program.

6

Data Availability and Collection

6.1 Overview

Introduction

The need for accurate and applicable data and information to assist the stormwater manager has never been greater. With the onset of complex water-quantity control and the need to assess and seek to control stormwater quality, the usefulness of pertinent data cannot be overstated. Knowing what data are needed, what may be available, where to find it, what to do if data are missing or partial, how to assess data, and how to plan and organize a data collection effort are part of the day-to-day responsibilities of the modern stormwater manager. Yet, it is an area in which many stormwater managers have little training or experience.

The purpose of this chapter is to outline the types of data that are normally required for stormwater management analysis and design, possible sources, and details of common types of data collection. Identification of stormwater management data needs should be a part of the early planning phase of any project or program. Several categories of data may be relevant to a particular stormwater management project, including published data on precipitation, soils, land use, topography, streamflow, instream water quality, point and nonpoint discharges and flood history. Field investigations are generally necessary to determine drainage areas, identify pertinent features, obtain high-water information, and survey channel sections and bridge and culvert crossings. Because of its complexity and the relative difference in character, environmental sampling is considered separately at the end of this chapter.

Data Collection Effort

The principal use of stormwater management data is to establish the hydrologic, hydraulic, and water-quality characteristics of a watershed or conveyance system in order to evaluate stormwater runoff, flooding conditions, and water-quality values. Depending on the type of study or design being done, existing and future watershed conditions may be considered. Data should be collected before calculations are initiated, using the following general guidelines:

1. Identify data needs, sources, accuracy required, and uses.
2. Collect published data based on sources identified in Step 1 and information presented in the following sections of this chapter.
3. Compile and document the results of Step 2, and compare data needs and uses with published data availability. Identify any additional field data needs.
4. Collect field data based on needs identified in Step 1 and Step 3.
5. Compile and document the results of Step 4.

The efforts necessary for data collection and compilation should be tailored to the importance of the project. Not all of the data discussed in this chapter will be needed for every project. Some data will be necessary for submission to a municipal engineering or public works department as part of engineering studies and analysis. Other data will be needed to make engineering and design decisions, apply hydrologic and hydraulic procedures, etc. Finally, data may be needed for administration or judicial review. Examples might include historic flood elevations, recent flood elevations and associated damages, recent land use and drainage facility changes, etc.

A well-planned data collection program leads to a more orderly and effective analysis and design that is commensurate with:

- Project or program scope
- Project or program cost
- Complexity of site hydraulics
- Regulatory requirements

Data collection for a specific project must be tailored to the following:

- Site conditions
- Scope of the engineering analysis
- Social, economic, and environmental requirements
- Unique project requirements
- Regulatory requirements

6.2 Sources and Types of Data

There are many potential sources of data typically required for stormwater management projects. Identifying these sources can be difficult, and making the subsequent necessary contacts can be time-consuming. In addition to the data sources discussed in this chapter, existing watershed master plans, flood studies, or other local studies provide the best sources of data for the watersheds studied and provide a starting point for watersheds not studied. Appendix A at the end of this chapter gives a list of data sources that will prove useful for stormwater design and analysis projects.

Often, the types and accuracy of existing data determine the types and detail of studies that need to be done. For example, historical rainfall–runoff data for a particular watershed will naturally lead to individual storm calibration of hydrologic models. If such data are not available, synthetic calibration or data transfer is necessary.

Data Categories

There are many ways to categorize the types of data needed for various studies. Some data are continuous, while others are discreet or binary (yes or no) data. Some are man-made, while others are natural. Table 6-1 provides a listing of commonly encountered primary urban stormwater management data acquisition categories. Development of estimates for these data needs can be done using field or office methods or a combination of both. For more specialized studies, many more categories and types of data could be listed. Specific parameters required for each category depend on the methods chosen for making calculated estimates.

6.3 Major Data Topics

Following is a brief discussion of some of the more common major data topics that relate to stormwater management facility analysis and design.

Watershed Characteristics

The three most important general physical watershed characteristics for municipal stormwater management are contributing size, slope, and land use.

Contributing Size

The size of the contributing drainage area expressed in acres or square miles is determined from some or all of the following:

- Direct field surveys with conventional surveying instruments
- Use of topographic maps* together with field checks to determine any changes in the contributing drainage area such as may be caused by the following:
  - Terraces
  - Urban storm drain systems
  - Lakes, sinks
  - Debris or mud-flow barriers
  - Reclamation and flood control structures
  - Irrigation diversions
- Use of state highway planning survey maps
- Aerial maps or aerial photographs that may be available from local municipalities, state agencies, and real estate companies

In determining the size of the contributing drainage area, any subterranean flow or any areas outside the physical boundaries of the drainage area that have runoff diverted into the drainage area being analyzed should be included in the total contributing drainage

---

* Note that U.S. Geological Survey (USGS) topographic maps are available for many areas of the United States. Topographic maps can also be obtained from municipal and county entities and local developers.

Table 6-1

Categories of Primary Data Needs for Urban Stormwater Management

Precipitation
- Rainfall intensity, duration, frequency
- Rainfall temporal distribution
- Rainfall areal extent
- Storm movement
- Annual and seasonal precipitation
- Past studies

Losses
- Interception
- Depression storage
- Evaporation parameters
- Infiltration and soil parameters
- Rainfall runoff gauging information
- Past studies

Open Channel Flow
- Channel system layout
- Channel shape and geometry parameters
- Channel slope
- Flow resistance
- Natural controls
- Channel structures' locations, types, and routing information
- Channel conveyance structures
- Historical data on peak flows
- Stage–discharge rating curves
- Velocity and discharge measurements
- Flow frequency information from stream gauging
- Debris and ice-flow information
- Adjacent or regional information or parameters
- Past studies

Basin Characteristics
- Size
- Basin slope
- Basin shape — length, width
- Drainage density
- Land use — agricultural and urban
- Surface and subsurface geology
- Karst or depression storage information
- Past studies

Flood Damages
- Stage–frequency curves
- Property appraisals and types
- Business types and susceptibility to flooding
- Generalized stage–damage curves
- Historical damage information
- Future development and population estimates
- Existing flood-control structural or nonstructural programs
- Transportation information

Pipe Systems
- Inlet size, type, location
- Outlet types and protection
- Backwater and submergence information
- Location sizes, slopes, and shapes of pipes
- Ground cover
- Flow bypass information
- Connectivity
- Maintenance and structural conditions and blockages
- Ages of system
- Maintenance records

Table 6-1 *(Continued)*

Categories of Primary Data Needs for Urban Stormwater Management

- Material types
- Special structures
- Past studies

Water Quality
- Buildup–wash-off parameters
- Land-use loadings
- Event Mean Concentration (EMC) values — regional, site-specific, national
- Receiving water-quality information — storm and base flow
- Designated uses of receiving waters
- Specific modeling parameters
- Best Management Practice (BMP) design information and efficiency values
- Rainfall and runoff volumes
- Chemical sampling information
- Biological or fish tissue testing information
- Known use impairment
- Atmospheric deposition
- Point sources and types
- Nonstructural programs and effectiveness
- Existing structural BMPs

Groundwater
- Surface and subsurface geology
- Soils information and parameters
- Observation well information
- Groundwater level measurements
- Well, geologic, or stratigraphic log information
- Aquifer test information
- Infiltration test information
- Electrical, seismic, or gravimetric information
- Laboratory analysis of soils
- Groundwater quality analysis
- Down-hole groundwater quality data

Sediment Transport and Geomorphic Studies
- Channel sinuosity and planform statistics
- Channel type: meandering, braided, straight
- Sediment transport measurements: suspended load, bed load, and wash load
- Flow duration information or curves
- Bed and bank shear stress
- Turbulence and dispersion
- Bed and bank material and sizes and roughness
- Vegetation
- Bedforms and bar locations
- Local or general scour information
- Historical information on channel size, form, or alignment changes
- Surface and subsurface geology
- Backwater information from receiving stream
- Existing bank or bed protection locations and types
- Dredging records, potomology studies, reservoir sedimentation records
- Past studies

area. In addition, it must be determined if floodwaters can be diverted out of the basin before reaching the site.

Slopes

The slope of the stream, the average slope of the watershed (basin slope), and other characteristics of the terrain important to the study or design should be determined.

Hydrologic and hydraulic procedures in other chapters of this book are dependent on watershed slopes and these other physical characteristics.

The average watershed slope together with the length and flow retardance are the major factors affecting the runoff rate through the watershed (SCS, 1968). The determination of average watershed slope can be standardized in such a way as to produce consistent and fairly representative results. The determination of average watershed slope consists of (SCS, 1977):

1. Selecting four to eight representative slope sections that typify the slopes found in the watershed and drawing lines from the top of these slopes to the bottom
2. Calculating the slope for each line as the elevation difference divided by the length of the line
3. Adjusting each slope by multiplying it by the estimated percentage of the watershed it represents; a weighted product is then computed to obtain the average slope for the entire watershed

Watershed Land Use

Watershed land-use information is particularly important, especially how it relates to the amount of impervious area for existing and future land use, and how this impervious area is distributed throughout the watershed. Directly connected impervious areas in smaller watersheds exhibit runoff characteristics different from areas where runoff flows over grassy areas first. If the demarcation line between pervious and impervious areas is clear, two separate hydrologic models are sometimes developed for the directly connected areas and the nondirectly connected areas. If the impervious areas are located in concentrated places in the watershed, the watershed is often broken into subareas to reflect consistent land use within each subarea. Defining land-use categories to break along zoning category lines is often convenient to permit wider application of results to land-use decisions.

It may not be necessary to collect specific land-use categories if the impervious areas are known from aerial photographs, digital mapping, multispectral photography and raster processing, or some other means. In terms of flow, the only concern is whether the area is directly connected and small, or simply impervious. However, for water-quality concerns, it may be necessary to determine the actual land use categories, specific industrial category, and age of the neighborhood, because different categories can exhibit different event mean concentrations of certain pollutants. Also, some land uses are known for higher incidences of illicit connections, point sources, or illegal dumping. For watershed land use, the following should be done:

- Present and expected future land use, particularly the location, degree of anticipated urbanization, and data source, should be defined and documented.
- Information on existing use and future trends may be obtained from:
  - Aerial photographs (conventional and infrared) and satellite images
  - Zoning maps and master plans
  - USGS and other maps
  - Municipal planning or public works agencies
- Specific information about particular tracts of land can often be obtained from the courthouse, filed plans in a public works or engineering office, owners, developers, realtors, and local residents. Care should be exercised in using data

from some of these sources because their reliability may be questionable, and these sources may not be aware of future development within the watershed that might affect specific land uses.

- Existing land-use data for small watersheds can best be determined or verified from a field survey. Field surveys should also be used to update information on maps and aerial photographs, especially in basins that have experienced changes in development since the maps or photos were prepared. Infrared aerial photographs may be particularly useful in identifying types of urbanization and vegetation at a point in time. Aerial photography to track crop use may be available from the Department of Agriculture or SCS state representatives.

Streams, Rivers, Ponds, Lakes, and Wetlands

At all streams, rivers, ponds, lakes, and wetlands that will affect or may be affected by the proposed structure or construction, there are data essential in determining the expected hydrology and that may be needed for regulatory permits. The following are the data that should be secured:

- Outline the boundary (perimeter) of the water body for the ordinary highwater.
- Determine the elevation of normal as well as high water for various frequencies.
- Obtain a detailed description of any natural or manmade controls, spillway, outlet works, or emergency spillway works, including cross sections, dimensions, elevations, and operational characteristics.
- Obtain a description of adjustable gates and soil- and water-control devices.
- Determine the use of the water resource (stock water, fish, recreation, power, irrigation, municipal or industrial water supply, etc.).
- Note the existing conditions of the stream, river, pond, lake, or wetlands as to turbidity and silt.
- Determine riparian ownership(s) as well as any water rights, if appropriate.

Roughness Coefficients

Roughness coefficients, ordinarily in the form of Manning's n values, should be estimated for the entire flood limits of the stream. A tabulation of Manning's n values with descriptions of their applications can be found in Chapter 10. Some municipalities have developed a Manning's n book for their area that shows pictures of local stream channels and floodplain areas with appropriate Manning's n values given (Greenville, South Carolina, 1992).

Stream Profile

Stream bed profile data should be obtained, and these data should extend sufficiently upstream and downstream to determine the average slope and to encompass any proposed construction or aberrations. Identification of headcuts that could migrate to the site under consideration is particularly important. Profile data on live streams should be obtained from the water surface, where there is a stream gauge relatively close; the discharge, date, and hour of the reading should be obtained.

Stream Cross Sections

Stream cross-section data should be obtained that will represent the typical conditions at a structure site as well as at other locations where stage–discharge and related calculations will be necessary. The actual locations may depend on the type of modeling being done. Cross sections should be typical for the reach in question and may, in fact, not be a particular section but a typical composite. Cross sections should be taken wherever channel curvature makes the parallel streamline assumption invalid, where roughness changes significantly, and where conveyance changes by an abnormal amount.

Existing Structures

The location, size, description, condition, observed flood stages, and channel section relative to existing structures on the stream reach and near the site should be secured in order to determine their capacity and effect on the stream flow. Any structures, downstream or upstream, that may cause backwater or retard stream flow should be investigated. Also, the manner in which existing structures have been functioning with regard to such things as scour, overtopping, debris and ice passage, fish passage, etc., should be noted. With bridges, these data should include span lengths, type of piers, and substructure orientation, which usually can be obtained from existing structure plans. The necessary culvert data include other things such as size, inlet and outlet geometry, slope, end treatment, culvert material, and flow line profile. Photographs and high-water profiles or marks of flood events at the structure and past flood scour data can be valuable in assessing the hydraulic performance of the existing facility.

Improvements, property use, and other developments adjacent to the proposed site upstream and downstream may determine acceptable flood levels. Incipient inundation elevations of these improvements or fixtures should be noted. In the absence of upstream development, acceptable flood levels may be based on freeboard requirements. In these instances, the presence of downstream development becomes particularly important as it relates to potential overflow points along the road grade or drainage system.

Flood History

The history of past floods and their effects on existing structures are of exceptional value in making flood hazard evaluation studies and are needed information for sizing structures. Information may be obtained from newspaper accounts, local residents, flood marks, or other positive evidence of the height of historical floods. Changes in channel and watershed conditions since the occurrence of the flood should be evaluated in relating historical floods to present conditions.

Debris and Ice

The quantity and size of debris and ice carried or available for transport by a stream during flood events should be investigated and such data obtained for use in the design of structures. In addition, the times of occurrence of debris and ice in relation to the occurrence of flood peaks should be determined; and the effect of backwater from debris and ice jams on recorded flood heights should be considered in using stream flow records. Data related to debris and ice considerations can be obtained from municipal records and state agencies.

Scour Potential

Scour potential is an important consideration relative to the stability of the structure over time. Scour potential will be determined by a combination of the stability of the natural materials at the facility site, tractive shear force exerted by the stream, and sediment transport characteristics of the stream. Data on natural materials can be obtained from agencies or by tests at the site.

Bed and bank material samples sufficient for classifying channel type, stability, and gradations, as well as a geotechnical study to determine the substrata if scour studies are needed, will be required. The various alluvial river computer-model data needs will help clarify what data are needed. Also, these data are needed to determine the presence of bed forms, so a reliable Manning's n as well as bed form scour can be estimated.

Controls Affecting Design Criteria

Many controls will affect the criteria applied to the final design of stormwater management structures including allowable headwater level, allowable flood level, allowable velocities, and resulting scour, and other site-specific considerations. Data and information related to such controls can be obtained from federal, state, and local regulatory agencies and site investigations to determine what natural or manmade controls should be considered in the design. In addition, there may be downstream and upstream controls that should be documented.

Downstream Control

Any ponds or reservoirs, along with their spillway elevations and design levels of operation, should be noted, as their effects on backwater and streambed aggradation may directly influence the proposed structure. Also, any downstream confluence of two or more streams should be studied to determine the effects of backwater or streambed change resulting from that confluence.

Upstream Control

Upstream control of runoff in the watershed should be noted. Conservation and flood-control reservoirs in the watershed may effectively reduce peak discharges at the site and may also retain some of the watershed runoff. Capacities and operation designs for these features should be obtained. The Soil Conservation Service, Corps of Engineers, Bureau of Reclamation, consulting engineers, and other reservoir sponsors often have complete reports concerning the operation and design of proposed or existing conservation and flood-control reservoirs.

The redirection of floodwaters can significantly affect the hydraulic performance of a site. Some actions that redirect flows are irrigation facilities, debris jams, mud flows, and highways or railroads.

6.4 Data Acquisition, Survey Information, and Field Reviews

Remote Data Acquisition

It is clear that efficient remote data acquisition, interpretation, analysis, and use is a fast-changing and emerging science and art. Aerial photography is a common method for obtaining physical representations of the terrain and digital base maps. Aerial photography involves decisions on scale, accuracy, important information, ground control, processing equipment, final format of information, coverages to be mapped, etc. There is the possibility of using multispectral photography to capture information not available in the visible light range, such as vegetation type differences, moisture, heat radiation, impervious areas, etc. The wide array of mapping products can be confusing. For example, there is a great deal of difference in terms of accuracy and potential applications between mapping and aerial photography, and, within photography, a great difference between aerial photography, scaled photographs, rectified photographs, and digital orthophotography. The type and accuracy of use drives the technological decisions. It is good advice to seek several opinions before investing in new mapping or photography.

Satellite data acquisition has evolved over the last several years to become a valuable tool, even for urban stormwater managers. The most common source of remotely sensed information and data in the United States is the USGS EROS Data Center located north of Sioux Falls, South Dakota (for customer service, phone: 605-594-6507). It was established in 1971 to support the NASA LANDSAT program. It is responsible for collecting, processing, archiving, and distributing a wide range of remotely sensed land data and has an inventory of over 3 million satellite and 7 million aerial photographs of the Earth's surface. It is accessible through CD-ROM disk readers or through an online system called the Global Land Information System (GLIS).

Global positioning systems (GPS) use satellites to accurately determine position in three-space through signals received from such satellites. Such use of satellite information allows for accurate and fast surveys and positioning, even in remote areas. These systems are often tied to Geographical Information Systems (GIS) for automated mapping and even real-time information processing.

A loose consortium of federal, state, and local agencies and industry is developing a protocol and format for the compatible exchange of spatial data and information, called the National Spatial Data Infrastructure (NSDI). NSDI will make possible the interchange of all sorts of spatial data by making their format compatible. Industry, for its part, is building standardized products that use or are compatible with these formats. The main coordinating entity for this is the USGS at its Reston, Virginia, National Mapping Division, Branch of Geographic Data Coordination (phone: 703-648-5725).

Geographical Information Systems

Geographical information systems (GIS) are an emerging tool somewhat akin to the emergence of CADD over ten years ago. GIS can provide the following types of information related to stormwater applications (after Berry, 1994a):

- Mapping of land and underground features, land uses, soils, rainfall amounts, watershed boundaries, slopes, land cover, etc., to any scale and in any combination is easy once the information is in the system.
- Database and map linking provide management-type answers to where things are and give ready information about them. For example, information about stormwater pipes can be placed in a database tied to the pipe.
- Changes over time can be easily portrayed in GIS format or in charts and graphs generated in GIS systems. For example, temporal water-quality changes can be graphically portrayed as color changes moving across the map in a demonstration or as a series of maps showing the color changes.

- Interrelationships between mapped objects can be determined by GIS. This works well for such things as determining stormwater runoff with the intersection of soils, land use, and watershed basin boundaries. Distances and slopes can be automatically determined for time-of-concentration calculations. Three-dimensional plots can be developed to get an idea of the appearance of a site for multiobjective use.
- Suitability type analysis is easy with GIS through the use of various rating schemes. For example, suitability for park reserves can be determined by intersecting soils, land use, vegetation, ownership, utilities, and species information. The best areas (those with the highest rating) can be colored to "jump off the map" for even the most obtuse citizen.
- Cause-and-effect questions about natural or man-made systems can be asked and answered, such as autocorrelation between rainfall gauge sites or the relationship between industrial SIC-coded locations and in-stream pollution monitoring results.
- What-if analysis can easily be accomplished with GIS. Land-use decisions and their impacts on pollution or flooding can be tried in endless combinations to determine the best solutions or to determine the outcome of current decisions.
- Convincing graphics is the final major category of GIS use. Overhead GIS projections or maps have been used to educate or convince citizens and political leadership concerning a course of action or a project's viability. Proper implementation of GIS applications for stormwater management involves planning for stormwater-only applications and integrating these applications with other potential users within the municipality. For example, the street centerline network serves as a good skeleton on which to hang other coverages and features. But, the ultimate accuracy with which it is established depends on its ultimate use and users.

Four basic issues must be fully considered before and during implementation (Berry, 1994b): hardware/software, databases, applications, and human impact. Hardware and software issues are relatively simple to solve once the basic uses for the system are identified. However, some municipalities have found that the "devil is in the details," and that some systems delivered at lower cost do not perform quite as expected when it comes to database development or transfer. Then, the cost can be enormous. The database development can cost many times the hardware and software development cost. Some databases are available and should be used if at all possible.

Telemetry

The first telemetered data was transmitted via phone lines in the 1930s. Since that time, there have been thousands of applications using hundreds of different types of telemetry systems, including phone lines, radio, satellite, and microwave. These systems are used in handling data such as water level, temperature, flow rate, gate movement, precipitation, humidity, and other data types. There are two general ways to transmit the data: analog and digital. The digital method is now by far the most common. Digital output is either serial or parallel (one bit of information at a time or all bits at a time).

The ability for two-way communication with remote sites is also valuable for the urban stormwater manager (termed Supervisory Control and Data Acquisition or SCADA). There is the possibility to remotely control flood- or pollution-control gates, dams, weirs, and all types of hydraulic devices from a central control room. For example, for flood control, this capability has been demonstrated in Bellevue, Washington. This is bringing about the rapid convergence of automated mapping and facilities management (AM/FM) and SCADA. For example, interior drainage pumping stations can be remotely controlled to catch first flush of pollutants and pump back to sanitary treatment sites, and can also be opened for necessary interior drainage flood control. A series of detention ponds can be controlled for maximum flood reduction through a series of rain gauges, a SCADA system, and a knowledge-rule-based hydrologic model. Combined sewer capacities can be maximized using the same principles.

Survey Information

Complete and accurate survey information is necessary to develop a design that will best serve the requirements of a site. The individual in charge of the drainage survey should have a general knowledge of stormwater management design, and, as such, should coordinate the data collection with the designer. The amount of survey data gathered should be commensurate with the importance and cost of the proposed structure and the expected flood hazard.

At many sites, photogrammetry is an excellent method of securing the topographical components of drainage surveys. Planimetric and topographic data covering a wide area are easily and cost effectively obtained in many geographic areas. A supplemental field survey is required to provide data in areas obscured on the aerial photos (underwater, under trees, etc.).

Data collection should be as complete as possible during the initial survey in order to avoid repeat visits. Thus, data needs must be identified and tailored to satisfy the requirements of the specific location and size of the project early in the project design phase. Coordination with all departments requiring drainage-related survey data before the initial fieldwork is begun will help ensure the acquisition of sufficient, but not excessive, survey data. An example form and checklist for collecting field data are provided in Appendix B at the end of this chapter.

Field Reviews

Watershed, stream reach, and site characteristic data, as well as data on other physical characteristics, can be obtained from a field reconnaissance of the site. Some level of field or aerial drainage survey of the site and its contributing watershed should always be undertaken as part of the analysis and design. Survey requirements for small stormwater management facilities such as 36-inch culverts are much less extensive than those for major facilities such as bridges. However, the purpose of each survey is to provide an accurate picture of the conditions within the zone of hydraulic influence of the facility. Data that can be obtained or verified include the following:

- Contributing drainage area characteristics
- Stream reach data — cross sections and thalweg profile
- Existing structures
- Location and survey for development, existing structures, etc., that may affect the determination of allowable flood levels, capacity of proposed stormwater management facilities, or acceptable outlet velocities
- Drift and debris characteristics

- General ecological information about the drainage area and adjacent lands
- High-water elevations, including the date of occurrence

Much of these data must be obtained from an on-site inspection. It is often much easier to interpret published sources of data after an on-site inspection. Only after a study of the area and a complete collection of required information should the designer proceed with the design of the stormwater management facility. All pertinent data and facts gathered through the survey should be documented as explained in Chapter 4.

There are several criteria that should be established before making the field visit. Does the magnitude of the project warrant an inspection, or can the same information be obtained from maps, aerial photos, or by telephone calls? What kind of equipment should be taken, and most important, what exactly are the critical items at this site? Photographs should be taken. As a minimum, photos should be taken looking upstream and downstream from the site as well as along any contemplated structure centerline in both directions. Details of the streambed and banks should also be photographed, along with structures in the vicinity both upstream and downstream. Close-up photographs complete with a scale or grid should be taken to facilitate estimates of the streambed gradation.

6.5 Data Evaluation

Once the needed data have been collected, the next step is to compile it into a usable format. The designer must ascertain whether the data contains inconsistencies or other unexplained anomalies that might lead to erroneous calculations or results. The main reason for analyzing the data is to draw all of the various pieces of collected information together, and fit them into a comprehensive and accurate representation of the hydrologic and hydraulic characteristics of a particular site.

Experience, knowledge, and judgment are important parts of data evaluation. It is in this phase that reliable data should be separated from that which is less reliable and historical data combined with that obtained from measurements. Historical data should be reviewed to determine whether significant changes have occurred in the watershed and whether these data can be used. Data acquired from the publications of established sources such as the USGS can usually be considered valid and accurate. Maps, aerial photographs, Landsat images, and land-use studies should be compared with one another and with the results of the field survey, and any inconsistencies should be resolved.

Data Accuracy

Errors can creep into any measuring or sampling program and may take the forms of bias, random measuring errors, and gross mistakes and errors. Bias is a consistent over- or underestimation of the true values of a measurement. Errors may enter measurements through such things as improperly zeroing an instrument or not taking into account tape sag or vertical angles in distance measurements. Bias errors can only be reduced by carefully planning and constantly checking the measurement activities.

Random errors are inherent in reading any instrument or measuring device or in performing any measurement activity. These errors can be reduced through repetition of the measurements and attention to quality-control procedures. Gross mistakes and errors can be caught through careful and deliberate measurement processes, repetition, quality assurance programs, and commonsense checks. While bias was previously defined, two other terms are important to data collection: precision and accuracy. Precision is a measure of the closeness of agreement among individual measurements; the scatter. Accuracy is a measure of how closely the measurements reflect the true value. Little or no bias and high precision lead to high accuracy.

It is important to establish data accuracy needs or availability before data are collected. For example, the publication "Accuracy of Computed Water Surface Profiles," U.S. Army Corps of Engineers, December 1986, focuses on determining relationships between:

- Survey technology and accuracy employed for determining stream cross-sectional geometry
- Degree of confidence in selecting Manning's roughness coefficients
- Resulting accuracy of hydraulic computations

The report also presents methods for determining the upstream and downstream limits of data collection for a hydraulic study requiring a specified degree of accuracy.

It is also important to realize that data collected for one purpose are often then used for a different purpose by the same or a different agency. The accuracy of the data should be sufficient to meet defined needs, and their limitations should be stated.

Sensitivity Studies

Often, sensitivity studies can be used to evaluate data and the importance of specific data items to the final design. Sensitivity studies consist of conducting a design or study with a range of values for specific data items. Following are the steps in such a study:

Step 1: Evaluate the model(s) that will be used, and determine which independent parameters are key for sensitivity testing. Determine which parameters are correlated with one another, as well as the dependent variables. Attempt to isolate only basic independent parameters.

Step 2: Determine, if possible, the relationship of the dependent variable with the independent input parameters mathematically. Differentiating the equation with respect to the parameter will then provide an exact measure of the sensitivity.

Step 3: For other parameters, determine a standard set of variables that will serve as the starting point for the sensitivity study. This set should center on the expected values over the range of the study conditions.

Step 4: Vary each key parameter in turn, holding all other parameters at the standard value. The range of variation should cover any expected range for the study site conditions. Be aware of parameters that may have secondary amplifying effects on one another at specific high or low ranges.

Step 5: Develop a dimensionless plot of the varying output variable (such as peak runoff, pollutant load, etc.) and varying parameter divided by the standard values. This allows for percentage comparisons. Check the plot for the reasonableness of the variation. Further investigate any apparently anomalous relationships.

Step 6: Note those parameters that cause greater variation in the dependent variable for the conditions expected to be tested. Establish standards for measurement accuracy attainable, and determine the effects of measurement or estimation error on the final results.

The effect on the final design can then be established. This is useful in determining what specific data items have major effects on the final design and the importance of possible data errors. Time and effort should then be spent on the more sensitive data items, making sure these data are as accurate as possible. This does not mean that inaccurate data are accepted for less sensitive data items, but it allows prioritization of the data collection process, given a limited budget and time allocation.

Statistical Analysis

Statistical analysis in urban stormwater management usually falls into one of four topics: time series analysis, extreme event analysis, geostatistical analysis, or general statistical analysis (containing hypothesis testing, probability function fitting, extrapolation, etc.).

For hydrologic data or other types of data that have inherent randomness, statistically developed sampling may be important. While there is a physical reason for all variability, much of it cannot be accounted for within the model we are using to attempt to mimic nature. Three reasons account for much of the difficulty in predicting hydrologic or environmental phenomena (Hirsch et al., 1993): inherent randomness, sampling error, and incorrect understanding of the processes involved. For example, in frequency analysis, there is rarely sufficient data to accurately estimate the risk of exceedance of a certain flow, pollution loading, rainfall intensity, etc. Assumptions are made about the probability distributions the maxima follow. However, these distributions are only approximations of the real distribution of maxima. When a short period of record is extended, large errors can result in predicting, for example, the flood event for spillway design.

Time Series Analysis

This analysis is often called trend sampling, systematic sampling, or sampling along a line. Many hydrologic and environmental variables lend themselves to this type of sampling, including rainfall, peak flows, volumes, pollution peaks or concentrations, low flows, etc. Urban runoff time series analysis is often used in continuous modeling applications or to establish statistical properties of a particular parameter for other analysis. Thus, a time series is nothing more than a sequence of values arrayed in chronological order that can be characterized by statistical properties (Chow, 1964). It is a subset of overall statistical applications to measured parameters termed stochastic hydrology (Yevjevich, 1972a).

To develop a time series, a continuous variable (such as discharge) is sampled over a discrete time interval. The series of data then subjected to statistical tests to determine means, variances, correlations, etc. A time series is said to be *autocorrelated* if the measurements of the variable from time-step to time-step depend somewhat on each other. This can happen, for example, when flows depend on slow storage release in a watershed and thus are related to each other (Salas, 1993). Series are said to be *cross correlated* if two different measurement locations show a relationship to each other. This can happen when two or more rain gauges are in the same region and thus influenced by the same storm. Trends may become apparent when a random and nonrandom element of a variable are separated through statistical analysis.

Using time series analysis, rain gauges can be corrected, data can be transferred, statistics can be derived and used to generate years of simulated data, seasonal variation can be quantified, relationships can be tested and quantified, etc. The reader is referred to the given references for further information.

Extreme Event Analysis

This analysis is really a subset of general statistical fitting of data to probability distributions. However, its use is so common in hydrology and environmental work that a separate brief discussion is warranted. It has been shown that distributions of the largest events over N periods of time with M events occurring in each period approach a limit as M increases. Depending on the distribution of the M variables, the final probability distribution (called an extremal distribution) will take one of three basic forms (Chow, 1964). A generalized extremal distribution has been developed that encompasses all three of the forms (Hosking, Wallis, and Wood, 1985). Different distributions are used to attempt to model different phenomena through the use of plotting positions and goodness-of-fit testing (Stedinger, Vogel and Foufoula-Georgiou, 1993):

- The normal distribution is used for average annual stream flow or average annual pollution loadings, where the central limit theorem dictates a normal distribution of the averages of many observations.
- Lognormal transformations are used for variables that are positively skewed, such as event mean concentrations for urban stormwater runoff (USEPA, 1983).
- A Gumbell distribution is used for maxima of certain events such as rainfall (Hershfield, 1961).
- A Weibull distribution characterizes distributions of minima bounded by zero, such as low flows.
- A Log-Pearson Type III distribution is recommended by the Water Resource Council for flood distributions (Interagency Committee, 1982).

Geostatistical Analysis

This is the analysis of spatially distributed information and the drawing of inferences from that data. For example, it is used in urban stormwater management to estimate rainfall distributions over an area from a number of rain gauges or to estimate groundwater height or transmissivity properties based on well data (Kitanidis, 1993). In stormwater quality, it is used to estimate the distribution of pollutants in a water body from samples or to estimate regional contamination based on a grid of measurements (Gilbert, 1987).

Geostatistical analysis can estimate unknown quantities or plot a contour of quantity estimates, evaluate the reliability and risk involved in the estimates, and predict the effect on the results of adding another sampling or measuring location. The two most commonly used methods in geostatistics are weighted moving average methods and trend surface analysis. Weighted moving averages is the most flexible of the methods and lends itself well to computer applications. Kriging is a method for applying a weighting to the values from contiguous locations based on location on a laid out grid (Clark, 1979).

Other Statistical Methods

Other methods include a wide variety of statistics used in various aspects of hydrology and environmental science, including (Yevjevich, 1972b, Chow, 1964, Bowker and Lieberman, 1972):

- Probability functions and frequency distributions to fit certain phenomena
- Combinatorial analysis
- Risk and uncertainty

- Statistical analysis of empirical distributions
- Descriptive statistics
- Estimation statistics
- Regression and correlation analysis
- Sampling theory
- Hypotheses testing
- Goodness of fit
- Multivariate analysis

A note of caution is in order with statistical analysis. Spurious correlation of parameters, false regression fitting, biased data and interpretation, unknown mathematical errors, etc., all can come into play (Gumbel, 1926; Benson, 1965). Correlations of dimensionless groupings of numbers will look better than they are if the same variable is used in each of the dimensionless groups. The use of logarithmic plotting can cause an otherwise poor fit to look better when large value differences are hidden in the log compression. Percentage comparisons of large numbers where small variations make a difference will always look good, while they will always look poor for smaller numbers. Means, modes, and medians are all termed "averages" at times. Even fitting a straight line to data can take on complications depending on the purpose for that fitting (Hirsch and Gilroy, 1984). Graphical scale selection or distortion can inadvertently magnify variations or hide them. Sampling or monitoring can miss the events below the lower detection limit, biasing the data. It is easy to lie with statistics, even when one is not trying to (Huff, 1954).

6.6 Precipitation Data Collection

Overview

Depending on the data needed and the physical conditions in the field, there are many different techniques and variations on these techniques for collecting data and information. New technology is making the acquisition of data and information easier. For example:

- It is possible through multispectral digital analysis and aerial photography or satellite imagery to estimate the percentage imperviousness of an area without direct measurements or mapping or to classify land uses.
- New survey equipment and techniques allow for the location of points in space through a *point-and-shoot* technique without recourse to a reflector or back sighting.
- Handheld data loggers allow for the logically complex acquisition of inventory information quickly and easily and the automated transfer of such information directly into CADD or GIS equipment.
- NEXRAD radar precipitation information is available for most of the United States, allowing for convenient GIS and spatially distributed model applications in hydrology, while software using NEXRAD is on the market (Smith, 1993).
- Topographic maps can be obtained for areas throughout the United States on CD-ROM.
- Large online databases of hydrologic data are available nationwide.

This acquisition methodology matches the need for more and better data by the newer computer models designed to take advantage of faster microcomputers and better graphical interfaces.

However, there will always be the need for the accurate measurement of such data as flow, rainfall, channel dimensions, etc. The next sections discuss such measurements for the benefit of the stormwater manager who needs to develop estimates of such values for design, plans review, and modeling of noncomplex systems. The collection of precipitation data and flow data are the two most important surface water quantity data collection needs.

Data Collection

The collection and proper interpretation of precipitation data for the development of design storms or for continuous hydrologic modeling is one of the more complex field operations in urban stormwater management. Even on relatively small urban catchments, lack of knowledge about the spatial distribution of rainfall is the greatest source of error in runoff simulation (Shilling, 1984). There are a number of assumptions inherent in precipitation estimation that may not be fully true in any field operation, such as (Arnell et al., 1984):

- Measurements reflect the true values of precipitation — there is no gauge, radar, or human error in measurement.
- Data are consistent — and, there are no internal changes to the system during the data collection period (double mass analysis is used to check this assumption).
- Data are homogeneous — there are no external changes to the system during data collection (double mass analysis is again used to check this assumption).
- Data are stationary — there is no cyclic climatic change, trends, or periodicity happening (statistical testing is used to check this).
- Data are independent — there is no correlation between events (statistical testing is used).
- Data series are sufficiently long to reflect the population.
- Precipitation extremes follow a prescribed probability distribution.
- Estimated parameters fully describe the presumed probability distribution.

For a complete discussion of the topic, see Arnell et al. (1984).

For example, gauge data collected at a distant airport or National Weather Service site are often used to try to calibrate hydrologic models to known storm runoff. The gauge data are then assumed to be uniform and are applied uniformly over the watershed. The correlation between measured runoff hydrograph and measured rainfall hyetograph is often poor. This poor correlation can lead to harmful calibrations when the modeler sets other parameters unrealistically to fit rainfall and runoff. The obvious solution is to increase the number of gauges and number of storms for calibration and to distribute the rainfall more realistically among gauges.

Factors Influencing Data Collection

There are a number of factors that make the measurement and application of precipitation data difficult. Depending on the catchment characteristics (size, slope, land use, orientation, etc.), the need for accuracy, and the available time and financial budget, there are different approaches and considerations that may be important. It should be remembered that, for small catchments, the peak flow estimate is dependent on the estimate of the peak intensity of the rain, while, for large catchments, the peak outflow is more dependent on accurate estimates of the volume of the storm.

Storm Types

Precipitation in the United States generally falls into one of two patterns. *Stratiform* systems are the frontal or precipitation line systems that move out of the Pacific and Gulf carrying large amounts of relatively steady rainfall over large areas. These are often called extratropical cyclones (Gilman, 1964; Smith, 1993). *Convective* systems are made up of local thunderstorms that form quickly, often repeatedly, releasing short intense cloud bursts over limited areas. These midlatitude thunderstorms are dependent on the land use over which they form. It has been shown on several continents that urban areas generate a greater frequency of high-intensity thunderstorms due to the heating effects of the paved surfaces (Huff, 1975; Vogel and Huff, 1975; Niemczynowicz, 1987). This shadow effect is felt in the urban area and downwind, and especially during hot summer afternoons.

In smaller urban catchments, the convective storms normally cause flooding of the minor feeder streams, ditches, neighborhoods, and streets. But, for the large creeks and streams, it is often the frontal systems with their long and intense storms that cause the major system flooding, though there are many exceptions.

Storm Movement

Studies in both Lund, Sweden (Niemczynowicz, 1987), and St. Louis, Missouri (Vogel, 1984), show that storm movement has an important effect on hydrograph generation, especially when the storm is moving downstream along the watershed axis. Many hydrologic models have the ability to switch rainfall on and off for calibration purposes. For long basins oriented along the axis of normal storm movement (such as might be the case when orographic effects predominate or where prevailing winds dominate), this may be appropriate for design purposes.

Storm movement information may be approximated through the use of high-altitude wind, long-term wind, and wind rose data available from the National Weather Service (Niemczynowicz, 1988). In fact, Niemczynowicz (1988) has shown that the use of storm movement in model calibration allows for the use of fewer rain gauges (3 instead of 12 over a 25-square kilometer area) to achieve the same level of accuracy. This is because modeling storm movement automatically takes into account an areal reduction and a peak duration increase. The peak intensity moves as a longer duration streak across the watershed rather than falling all at once simultaneously throughout the watershed.

Storm Decay

Storm cells build and decay over time. The scale of such a cycle varies considerably but can be neglected for most smaller urban catchments, perhaps on the order of 10 square miles for most of the United States. For example, average path length for cells was about 5 miles in St. Louis, Missouri (Huff, 1975).

Precipitation Data Acquisition

The National Oceanic and Atmospheric Administration (NOAA) National Climatic Data Center, (NCDC) Climate Services Branch, is responsible for the collection of precipitation data. Data on hourly, daily, and monthly precipitation at a large number of stations across the United States is available to the public on diskette, microfiche, or hard copy. Orders can be placed by calling (704) 259-0682 or by writing the center at Federal Building, Asheville, North Carolina, 28071-2733.

The National Weather Service (NWS) of NOAA can provide historic, present, and predicted weather to aid in sampling operations. Weather radio is available from 380 NWS stations via weather band radio and provides continuous broadcasts of the most current weather information (EPA, 1992).

There are basically two ways to physically obtain rainfall data: rain gauges and radar or satellite imagery (USGS, 1984).

Rain Gauges

Gauges are of two types: recording and nonrecording. All gauges have a collector that leads to a funnel and then to some sort of holding container with a measuring device. Nonrecording gauges are of two standard sizes: 4 or 8 inches with either a graduated container or a dipstick. Snow boards or snow stakes or a modified rain gauge are used for snowfall measurement. Recording gauges are of three types, all with a clock-driven drum and continuous pen trace:

- Weighing-type gauges continuously record the weight of the empty container and rainfall. This type of gauge has the disadvantage in that it cannot empty itself. But, it is effective for solid precipitation, because melting is not necessary prior to recording amounts and thus is good in colder climates.
- Tipping-bucket-type gauges work using a balanced pair of small buckets. When sufficient rainfall enters the collecting bucket, the gauge is unbalanced. It tips, placing the second bucket in position to collect rain, while the first empties into a holding container. Each tip of the bucket records a specific amount of rainfall: normally not greater than 2 mm.
- Float-type gauges drain rainfall into a container with a float-mounted pen trace. Several gauges have automatic emptying devices or siphoning mechanisms. Heaters are necessary in freezing weather, and water must be retained in the bottom of the float chamber at all times. This may require filling during long dry periods.

The accuracy of rain gauges is not a given. It has been shown that wind-induced turbulence can reduce the catch of the gauge by as much as 50% (Court, 1960). A wind speed of 10 mph causes about a 15% deficiency, while a speed of 25 mph causes a 40% deficiency (Wilson, 1954). Because turbulence varies with height above ground, the lower the gauge, the better. Shields are available that improve rain and snow catch (Gray, 1970). Also, the gauge should be in a nonexposed area but open to rainfall.

Radar Precipitation Data Acquisition

This data acquisition method has advanced greatly since its inception in the 1940s (Vogel, 1984). Radar can provide temporal resolutions to 5 min and spatial resolutions to 1 km. The Next Generation Weather Radar System (NEXRAD) is a network of 120 radar sites able to provide spatial resolution of about a 4 by 4 km square and one hour, though finer temporal and spatial information can be produced by these WSR-88D systems.

Radar operates by sending and receiving echoes of radar waves in the length range of from 1 to 10 cm. The backscatter is then interpreted based on the radar equipment and storm target characteristics, and reflectivity versus rainfall equations are developed (Smith, 1993). Automated methods are being developed that will provide incredible power to the precipitation modeler in terms of accuracy and spatial detail. This type of raster or grid-based data fits well with GIS hydrologic modeling applications for model calibration and real-time forecasting and flood warning. Radar estimates are subject to a number of types of error. It is normally prudent to use field gauge observations to calibrate the radar and provide "ground truth."

A system in Chicago pioneered this type of forecasting using radar information and rain gauge information (to calibrate the radar) (Vogel, 1984). By using a combination system such as this, many fewer rain gauges can be employed, while spatial and temporal accuracy is maintained. Satellite imagery is also used for rainfall measurement, though it is not as accurate, being based on infrared determination of relationships between cloud height and brightness and actual rainfall.

Rain Gauging Networks

The design of a gauging network depends on the analytic purpose, economics, climatic stress, and access. The shorter the duration, the smaller the area of a storm, or the more variability in geography, the more gauges are necessary to obtain absolute estimates. The effect of increasing gauge density is to increase the average rainfall value over the watershed. The reason is that, with more gauges, the chances of one of the gauges recording the peak rainfall track is higher.

To measure convective storms with some degree of accuracy, a dense rain gauge grid is required. For short-term rainfall with a 1 to 10 min storm increment (averaging time), with rain cells reported to vary between 1 to 5 km², a gauge spacing of about 1.8 km will explain about 90% of the variation (Huff, 1979). Simanton and Osborn (1980) have shown that the correlation between values of contiguous gauges falls dramatically when they are separated more than 5 to 10 km for air-mass thunderstorms. McGuiness (1963) developed an equation for watersheds in Ohio to estimate the error when a lesser gauge density is used for thunderstorm-type rainfall as follows:

$$E = 0.03\ P^{0.54}\ G^{0.24} \qquad (6.1)$$

where E = absolute error (in); P = "true" value of rainfall (in); and G = gauge network density (sq mi/gauge).

Alternately Hershfield (1965), using data collected from 15 watersheds across the United States, provides information by which the distance between gauges to obtain a correlation coefficient of 0.9 can be obtained from the 2-year 1-h and 24-h storms. Figure 6-1 gives the appropriate information.

It may not be cost-effective to attempt such a gauge density. For smaller catchments, a single rain gauge is often placed near the centroid of the watershed to monitor a few storms. As mentioned above, the use of storm movement can enhance even sparse data.

A uniform network is normally found to be best when developing gauge spacing, though the spacing should reflect orographic or other changes. Special networks in these areas work well and can be integrated into an overall grid. For larger areas, such as over a whole urban area, the concept of *unit source areas* can save time and money (Gray, 1970). A unit source area is a representative gauging network that represents a much larger area. These smaller areas are gauged intensively, and the results are then extrapolated to other areas. Care must be taken when performing modeling of the whole area that concentrations of rainfall from the unit source area are not extrapolated throughout the area as if a thunderstorm of limited extent was covering the whole of the city simultaneously.

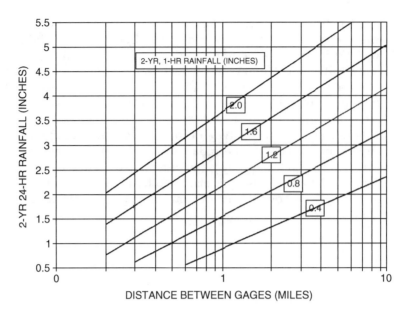

FIGURE 6-1
Estimating gauge spacing distance as a function of the 2-year, 1-hour, and 24-hour storms.

A more dense grid system can often be greatly reduced over time as certain gauges are found to be representative of greater areas (some gauges are redundant) and knowledge is gained of storm cell size and spacing.

Rainfall Average Depth Estimation

We are seldom interested in precipitation at a point but are in its areal distribution, so that runoff can be predicted. If only one gauge is available, the areal reduction equation provided in Chapter 7 can be used. The application of areal reduction factors to small areas has been proven in Swedish studies (Niemczynowicz, 1982). It has even theoretically been shown that the areal reduction factor could be greater than 1 (Nguyen Rouselle, and McPherson, 1981).

The average depth of rainfall over an area can be estimated from the totals from several gauges in the area using several methods. All of the methods applied mathematically or graphically have the same form:

$$P = x_1 p_1 + x_2 p_2 + \ldots x_n p_n \tag{6.2}$$

where P is the mean areal precipitation at a watershed or site of interest, and p_i and x_i are the n station rainfalls and weighting factors, respectively. The sum of all the xs is 1. The only difference in the methods is the method of weighting the point precipitation values. Kelway (1974) and Gottschalk and Jutman (1983) investigated the various methods without reaching strong conclusions on which is most accurate. Because precipitation is non-normally distributed, none of the methods reduces the uncertainty in estimate, because the true value between observation points is never known (Arnell et al., 1984). An arithmetic mean can be used for flat areas with similar climatic conditions, where each of the xs is equal to $1/n$.

The Thiessen polygon method, a graphical application (Thiessen, 1911), assumes that the amount of rainfall at any gauge site can be applied to the distance halfway to the

Data Availability and Collection

nearest station in any direction. The applicable areas are found by drawing lines between each station. The perpendicular bisector of each line is then joined where they cross, enclosing an area around each gauge site.

The isohyetal method consists of plotting rainfall totals for the storm or time increment on a map and drawing rainfall isohyets (rainfall contours) through the points. A modification of this method, the percentage-of-mean-annual method (Gilman, 1964), is used for storm averages in areas with orographic effects. The following steps are employed:

- The ratio of the storm amount to the mean or seasonal average amount is calculated for each gauge site and plotted on a map.
- Isolines of the ratios are estimated and plotted.
- The means of the ratios are determined for each incremental area between isolines.
- Each mean ratio is then multiplied by the appropriate mean or seasonal amount to obtain rainfall depth averages.
- Multiplying each incremental area by the depth for that area, adding and dividing by the total area, then gives the area weighted average for the total area of concern.

The reciprocal–distance method has been used to estimate point thunderstorm-induced rainfall amounts with accuracy in both the South (Dean and Snyder, 1977) and Southwest (Simanton and Osborn, 1980). In this method, the rainfall at any point is an average of the gauge readings from nearby gauges weighted by the distance to those gauges to some power. If the point rainfall calculated can be related to an incremental area, such as a watershed subbasin at its centroid, then average rainfall over an area can be calculated. The equation for n gauges is of the following form:

$$P_p = [P_1/d_1^b + P_2/d_2^b + \ldots P_n/d_n^b]/[1/d_1^b + 1/d_2^b + \ldots 1/d_n^b] \qquad (6.3)$$

where P_p = weighted average point rainfall (in); P_i = rainfall at gauge i (in); d_i = distance from point to gauge (mi); and b = distance exponent (1 in Southwest, 2 in South Piedmont area).

Missing gauge information can be estimated from three surrounding gauges by weighting it using the normal annual precipitation or some other seasonal or annual precipitation measure.

$$P_p = 1/3\,[(N_p/N_1)P_1 + (N_p/N_2)P_2 + (N_p/N_3)P_3] \qquad (6.4)$$

where N = normal annual or seasonal precipitation, and other variables are as indicated in the previous equation.

This method only works well when there is a high correlation between the gauges used. If there is not, a better approach would be to sketch an isohyetal map and estimate the missing data from that or correlate using fewer gauges.

As mentioned earlier, a kriging technique has been applied to spatial precipitation data as well as fitting of polynomials and Fourier series (Thorpe, Rose, and Simpson, 1979). This method does not provide better estimation of spatial distribution but allows a better estimate of the confidence limits on the estimate. The most promising future development is the application of better data acquisition through the use of radar and GIS applications.

6.7 Flow Data Collection

Introduction

The collection of flow data in urban applications is normally limited to discharge and velocity measurement in open systems (ditches, feeder streams, channels and streams, and in partially full conduits) and pressure flow in closed conduits. The measurement of flow includes the use of devices (structures, meters, etc.) and of techniques that may or may not involve complex devices. Basic references for further information and details on various techniques include Shelley and Kirkpatrick (1975), USGS (1983), Brater and King (1976), Grant (1989), Leupold & Stevens (1991), Rantz (1982) and Bureau of Reclamation (1974).

The measurement of urban runoff flow is complicated by several factors including:

- The poor quality of the flow and the amount of sediment and debris washed into the flow
- The likelihood of vandalism of gauging sites and devices
- The flashy and intermittent nature of the flows
- High velocities and presence of many sharp bends leading to nonuniform or abnormal velocity distributions throughout much of the flow area

Therefore, the selection of flow measuring devices and the locations of measurement are of prime importance. Flow measurement can be targeted toward continuous measurement, peak flow only, low flow, or miscellaneous measurements in a relatively continuous flow. One-time measurements can also be made to calibrate a continuous device.

Basic Equipment and Techniques Overview

Flow measurement belongs to two broad categories: quantity measurement and rate measurement. In quantity measurement, the primary measurement device weighs volumes of water or measures certain masses of water. For rate measurement, the primary device interacts with the fluid in a certain way dependent on one or more properties of the fluid. This interaction with a continuous stream of fluid is then counted or measured in such a way that dependence on fluid properties, physical laws, or empirical equations convert the interaction to a flow velocity and rate. Under these two basic categories, there are many basic ways to measure flow and a number of instruments or techniques under each of the methods. Table 6-2 is a summary of the most common ways to measure flow in open channel and closed systems.

Common types of measuring devices for pressure flow applications include venturi meters, ultrasonic meters (both Doppler and transit time), flow nozzles, orifice meters, magnetic flow meters, rotameter, elbow meters, and pitot tube meters. The most common methods for flow measurement in open channels and partially full pipes include flumes, weirs, pitot tubes, flow meters, orifices, nozzles, and slope and velocity measurements with application of theoretical or empirical formulae.

Shelley and Kirkpatrick (1975) give some basic criteria in assessing which type of equipment might be applicable for any situation, including:

1. Range — covering a 100 to 1 range is preferable.
2. Accuracy — an accuracy of ±10% of the reading is necessary; ±5% is highly desirable, and repeatability of ±2% is desired in all instances.

TABLE 6-2

Categories of Primary Flow Measurement Devices

Classification	Type	Subtype
Quantity Devices		
Gravimetric	Weigher, tilting trap, weigh dump	
Volumetric	Metered tank or bucket	
	Displacement piston	
	Nutating disk	
	Sliding or rotating vane	
Rate Devices		
Differential	Venturi	
Pressure	Flow nozzle	
	Round-edge orifice	
	Square-edge orifice	Concentric, eccentric
		Segmented, gate, or variable area
	Centrifugal	Elbow, long radius bend, guide vane speed ring
	Impact tube	Pitot-static, pitot venturi
	Linear resistance	Pipe section, capillary tube, porous plug
Variable area	Gate	
	Cone and float	
Head-area	Weir	Broad crested, sharp crested, and combination
	Flume	Venturi, Parshall, Palmer-Bowlus, cutthroat, san dimas, trapezoidal, SCS types HS, H, and HL
	Open flow nozzle	
Flow velocity	Float	Simple, integrating
	Tracer	
	Vortex	Vortex and eddy shedding
	Rotating element (meter)	Horizontal, vertical
Force-displacement	Vane	
	Jet deflection	
Thermal	Hot and cold tip	
Other	Electromagnetic, acoustic, Doppler, dilution, electrostatic, nuclear resonance	

Source: From Shelley, P.E. and Kirkpatrick, G.A., EPA 600/2-75-027, 1975, Washington, DC, 1975, and Leupold & Stevens, Inc., Steven Water Resources Data Book, 5th ed., Leupold & Stevens, Beaverton, OR, 1991.

3. Flow effects — the unit should be able to retain its accuracy over a range of conditions found in urban flow.

4. Gravity and pressure flow — it is desirable for the unit to have the ability to handle both types of flow in a closed conduit situation.

5. Submergence or backwater sensitivity — backwater or even reversed flow should be sensed by the unit, even if it cannot continue to function accurately under such conditions.

6. Solids — sediments and floatables should move through the unit without clogging or blockage.

7. Flow obstruction — the unit should function without undue obstruction of the flow.

8. Headloss — the unit should induce as little headloss as possible.

9. Manhole operation — the unit would be able to be installed and operate within a manhole.

10. Power requirements — the unit should require little or no power and be able to operate remotely.
11. Site requirements — unit should not require extraordinary site support facilities.
12. Installation restrictions — the unit should be flexible enough to operate in a number of installation types without restrictions or limitations.
13. Simplicity and reliability — the unit should be simple to operate and calibrate, with a minimum of moving parts, and be reliable over a long period and many different flow and site types.
14. Unattended operations — the unit should function reliably while unattended.
15. Maintenance requirements — routine maintenance should be minimal, and troubleshooting should be able to be handled in the field with ease.
16. Adverse ambient effects — the unit should be free from being affected by adverse climatic or chemical environments.
17. Ruggedness — the unit should be able to withstand rough handling and be as vandal- and theft-proof as possible.
18. Self-contained — the unit should be self-contained as much as possible.
19. Precalibration — the unit should not require field calibration at each site.
20. Ease of calibration — any necessary calibration should be accomplished with ease in the field.
21. Maintenance of calibration — the unit should operate accurately for extended periods of time.
22. Cost — the unit should be affordable in terms of installation, maintenance, and operation.
23. Portability — the unit should be able to be transported easily to one or many measurement points using standard equipment.

In addition, the units should also have the following characteristics:

- Adaptability — the unit should be able to measure a wide variety of flow types, in different modes, and interface with many types of secondary communication and recording equipment.
- Submersion proof — the unit should be able to be submerged without damage.

Not every type of equipment will need to have all these characteristics for every application. But, given the applications, different characteristics will be crucial. For example, gravimetric and volumetric methods work well for low-flow and illicit connection measurement when a simple bucket or tank can collect all the flow for a certain time. Flow meter sampling works well for specific one-time applications or measurements of flow in open channels but does not work to supply a continuous record of flow. For a continuous record, a fixed structure with a known stage–discharge or pressure–discharge relationship works well, and thus, orifice plates and venturi tubes work well for pressure flow, while weirs, flumes, and orifices work well for channel and partially full pipe flow.

With this set of criteria (numbers 1 through 23), Shelley and Kirkpatrick evaluated a number of types of flow measurement equipment. Table 6-3 gives a summary for the most common types of urban stormwater runoff measurement equipment. The numbers across the top refer to the numbered criteria on the previous pages. Much of the work was subjective.

TABLE 6-3
Evaluation of Primary Flow Measurement Devices

Equipment type	1	2	3	4	5	6	7	8	9	10	11	12	13	14	15	16	17	18	19	20	21	22	23
Gravimetric — all	G	G	H	Y	L	H	H	H	P	M	H	H	P	Y	H	M	F	Y	Y	G	F	H	N
Volumetric — all	P	G	H	Y	L	H	H	M	P	L	H	H	F	Y	H	M	F	Y	Y	G	F	H	N
Venturi tube	P	G	S	N	L	S	S	L	P	L	H	S	G	Y	M	M	G	Y	Y	G	G	H	Y
Flow nozzle	P	F	S	N	L	H	H	H	P	L	H	S	G	Y	H	M	F	Y	Y	G	P	L	N
Elbow meter	P	F	S	N	L	S	S	L	P	L	H	S	G	Y	L	M	G	Y	Y	F	G	L	Y
Sharp-crested weir	F	F	M	N	M	H	H	H	F	L	M	M	G	Y	H	M	G	Y	Y	G	P	L	N
Broad-crested weir	F	F	S	N	H	H	M	M	G	L	M	M	G	Y	L	M	G	Y	N	F	F	L	Y
Parshall flume	G	F	S	N	M	S	S	L	F	L	M	S	G	Y	L	M	G	Y	Y	G	G	M	Y
Palmer-Bowles[a]	F	F	S	N	M	S	S	L	G	L	S	S	G	Y	L	M	G	Y	Y	G	G	L	Y
Cutthroat flume	G	F	S	N	L	L	S	L	P	L	S	S	G	Y	L	M	G	Y	Y	G	G	L	N
Trapezoidal flume	G	F	S	N	L	M	S	L	F	L	M	M	G	Y	L	M	G	Y	Y	G	G	L	Y
SCS flumes	G	F	S	N	H	M	S	H	G	L	M	M	G	Y	M	M	G	Y	Y	G	F	L	Y
Open flow nozzle	G	F	S	N	H	M	S	H	G	L	M	M	G	Y	M	M	G	Y	Y	G	F	L	Y
Float velocity	G	P	H	N	L	S	S	L	G	L	S	S	G	N	L	H	G	N	-	G	-	L	Y
Tracer velocity	F	F	M	Y	L	S	S	L	F	M	S	S	F	Y	M	S	F	N	N	G	G	H	Y
Rotating meter	F	F	S	Y	L	H	M	L	F	L	S	S	G	N	H	H	G	N	Y	G	G	L	Y
Force-momentum	P	G	S	N	M	M	M	L	P	H	H	H	P	Y	H	S	P	Y	Y	G	G	H	N
Doppler meter	P	G	S	Y	L	H	S	L	F	M	M	M	F	Y	M	S	F	Y	Y	G	G	L	N
Optical meter	F	P	S	N	L	S	S	L	F	L	S	S	G	N	L	H	G	N	Y	G	G	L	Y
Dilution	G	G	M	Y	L	H	S	L	G	M	S	S	F	Y	M	S	F	Y	N	G	G	H	N
Acoustic meter	G	G	S	Y	L	M	S	L	F	M	M	M	F	Y	M	S	F	Y	Y	G	G	H	N
Electromagnetic	F	G	S	Y	L	S	S	L	P	H	M	M	F	Y	M	S	F	Y	Y	G	G	H	N
Hot-tip meter	F	P	S	Y	L	H	M	L	F	M	M	M	F	Y	H	M	F	Y	Y	G	F	H	N

[a] Includes the Leopold–Lagco Flume.

Note: F — Fair, G — Good, H — High, L — Low, M — Medium or Moderate, N — No, P — Poor, S — Slight, Y — Yes.

Source: From Shelley, P.E. and Kirkpatrick, G.A., EPA 600/2-75-027, 1975, Washington, DC, 1975.

Common Errors

There are avoidable and unavoidable errors that will be encountered in the installation and operation of any flow measurement site. Mort (1955), Thomas (1957), USBR (1974), Leupold and Stevens (1991) and Grant (1989) provide descriptions of these errors and ways to minimize or avoid them. The most commonly mentioned errors are as follows:

- Faulty fabrication or construction
- Improper gauge or head measuring location or head measurement
- Incorrect zero setting
- Use of device outside its range of accuracy
- Improper maintenance or checking against settling, shifting, etc.
- Turbulence in approach channels
- Unnoticed clogging and debris

6.8 Flow Data Collection — Natural Controls

The development of a gauging site for the installation of a permanent stage-discharge station is similar to the development of artificial controls such as a weir or flume, with the exception that the stage–discharge relationship is developed, not through known relationships between head and discharge, but through the measurement of point velocities in the flow area and the integration of these point velocities (and the small cross-sectional areas they represent) to produce an estimate of discharge for a given stage. When this is done for a number of stages, a permanent relationship between stage and discharge is developed.

Selection of Gauging-Station Sites

The selecting of a gauging site is guided by several considerations. First, the purpose for which the data are needed will often dictate the location or the reach of the stream. For example, if the purpose is to measure discharge in conjunction with a proposed project, the site will be near the project. The other criterion is that the site be an adequate control.

The physical element that governs the stage–discharge relationship is called the control. The control may be a point where the flow passes through critical depth, such as a waterfall or break in the grade. It may be a constriction in a channel that controls the flow depth upstream. Rarely will a single control dictate depth over the whole range of flows. Often, the control will shift from a slope break point for lower flows to channel constrictions for larger flows. There are three attributes of a satisfactory control: stability, permanence, and sensitivity.

Stability refers to the one-to-one correspondence between stage and discharge over all expected conditions. If the cross section is too close to a downstream tributary that may cause backwater or a dam with variable gate control, there may be times when a certain stage reflects a different flow than the one measured during stream gauging. Changing vegetation may also be reflected in different stages for given flows in different seasons. Vegetation that suddenly leans at a given velocity will cause gauge readings to be inaccurate, as will low and dense canopy cover levels that interfere with high flow levels during the summer season.

Permanence means the control will remain constant over a long period of time. Rock outcroppings have this characteristic, while sandbars and pools do not. Streams that are undergoing degradation or sedimentation due to urban development will have to be remeasured periodically. Sandbed streams scour during high flows and redeposit sediment on the backside of the flow hydrograph. The depth of this scour will need to be measured or estimated to gain accuracy in high-flow measurements.

Sensitivity means that, at low flow, a small change in flow will be reflected in a relatively large change in stage. V-shaped channels or those with small low-flow channels are often best for this. Channels with small pools at the site during low flow will ensure a measurement can be made for even the lowest flows.

The ideal gauge site has the following criteria (Rantz, 1982):

- The general course of the stream is straight up- and downstream for some distance (200 to 300 ft for medium-sized urban streams).
- The total flow is confined to one channel for all stages without bypass flow.
- The streambed is not subject to scour and fill and is free from excessive aquatic growth.
- Banks are permanent, high enough to contain the flow, and free from heavy brush.
- Unchanging natural controls are present in the form of a bedrock outcrop or riffle at low flow and channel constriction at higher flows, or a falls that remains unsubmerged.
- A pool is present upstream from the control at low stages.
- The site is far enough upstream from tidal influence or the influence from confluence with a tributary.
- A satisfactory reach for measuring discharge is available within reasonable proximity of the gauge site.
- The site is accessible for installation and maintenance of the gauging station.

Rarely will any site meet all of these criteria. The stormwater manager should weigh pluses and minuses to maximize accuracy and mitigate any factors that will cause error. See Rantz (1982) for ways to minimize poor locational aspects of a site.

Velocity Measurements

Velocity measurements are taken in a series of verticals across a section in the vicinity of the gauging station location. The site for taking velocity measurements should preferably have the following characteristics:

- Reach is straight, and streamlines are parallel to each other.
- Velocities are greater than 0.5 ft/s, and depths are greater than 0.3 ft.
- Streambed is relatively uniform and free from boulders, aquatic growth, excessive turbulence, and dead flow areas.
- The section is close enough to the gauging station site to avoid intervening tributary inflow or the need to calculate storage for rapidly changing flow situations.

After the site has been selected, the width perpendicular to the flow is measured, and vertical sections are laid out. No vertical strip subsection should contain more than 10% of the flow. For medium to large urban streams, this may equate to 25 to 30 subsections, though no sections should be less than 1 ft apart. A tagline stretched across the stream is used to mark the locations.

Velocity is measured in the vertical using one of two common methods: the 0.6 depth method and the 0.2–0.8 depth method. For depths less than 2.5 ft, use the method where the velocity is measured once per vertical subsection at a point 0.6 of the depth from the surface. This is the theoretical mean velocity in the vertical. If the depth is greater, use the 0.2 and 0.8 points, and take the arithmetic mean.

This measurement procedure assumes that the stream can be waded. For details on other means of performing measurements, see Rantz (1982). A rule of thumb for wading safety that has been proven in experiments on mountain streams in Colorado is if the sum of the velocity in feet per second and depth in feet is greater than your height in feet, it is unsafe to wade in the stream. For example, unless you are over 6 feet, do not wade a stream that is 2 ft deep flowing at 4 ft/sec. Use a float such as a ball, sticks, or leaves and a stopwatch, to estimate the velocity for safety purposes. When wading, the hydrographer should stand at least 1.5 ft downstream from the flow meter holding the rod to which the meter is attached in a vertical position.

Figure 6-2 shows the layout for flow calculations. Distances are measured from one bank and always referenced to that bank. Proper use of the meters can produce accuracies of ±5%.

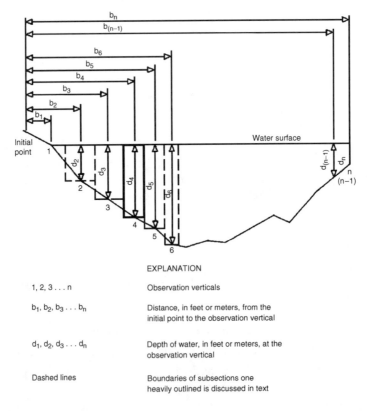

FIGURE 6-2
Flow measurement layout. (From Rantz, S.E., Measurement and Computation of Streamflow, 2 Vols., U.S. Geological Survey Water Supply Paper 2175, 1982.)

Equation 6.5 gives the method for calculation of discharge in a subsection. The nomenclature is given in Figure 6-2.

$$q_x = v_x \,[(b_{x+1} - b_{x-1})/2]\, d_x \tag{6.5}$$

Where q_x = discharge through subsection x (cfs); v_x = mean velocity at vertical x (ft/s); b_{x-1} = distance from initial point to preceding vertical (ft); b_{x+1} = distance from initial point to next vertical (ft); d_x = depth of water at vertical x (ft).

For example, the discharge at subsection 4 (heavily outlined in Figure 6-2) is:

$$q_4 = v_4 \,[(b_5 - b_3)/2]\, d_4$$

For the first and last sections, the equations are, respectively:

$$q_1 = v_1 \,[(b_2 - b_1)/2]\, d_1 \tag{6.6}$$

$$q_n = v_n \,[(b_n - b_{n-1})/2]\, d_n \tag{6.7}$$

Fulford and Sauer (1986) have shown that considerable time can be saved if the cross section is fairly uniform and well defined by simply estimating the velocities in many of the verticals rather than measuring them. This is especially true for velocities in sections that are nearly uniform. They found that a linear interpolation scheme was as accurate as any complex scheme, and that only five vertical measurements were necessary for fairly uniform channel sections. Mean absolute errors of under 5% were measured. This method is especially effective for determining flows for other stages after the initial cross-sectional survey and detailed measurements are done for one stage. In this case, the velocity is measured in five representative verticals. The mean velocity in a subsection between two measured verticals (verticals 2 and 3) is as follows:

$$v_i = (a/L)\, v_3 + [(L-a)/L]\, v_2 \tag{6.8}$$

Where v_i = interpolated velocity in an intermediate subsection (ft/s); a = distance between verticals, v_i is calculated and v_2 is measured (ft); L = distance between the two measured velocity verticals (ft); and v_2, v_3 = measured velocities at section 2 and 3, respectively (ft/s).

Bank velocity is approximated as zero, giving the estimated velocity between the bank and the first measured velocity vertical as:

$$v_i = (a/L)\, v_1 \tag{6.9}$$

Velocity Meters

Current meters come in two types: vertical-axis with cups or vanes and horizontal-axis (propeller) with vanes. The two give similar results for similar situations. The Price AA vertical-axis cup meter is a USGS standard for most situations in which depths exceed 1.5 ft. A Price pygmy meter is used for depths to 0.3 ft. For lesser depths, use a portable weir or flume. Other popular meters include the Ott, Neyrpic, Haskell, Hoff, and Braystoke. Details on these meters and their use can be found in Rantz (1982) and from the various manufacturers.

6.9 Flow Data Collection — Tracer Methods

Discharge can be estimated using any of a number of related tracer methods in which a known quantity, rate, or concentration of a substance is injected into the flow at some upstream location. It is then observed, measured, or weighed at some downstream location, and the total flow is determined.

Salt-Velocity Method

In this method, a slug of salt solution is forced into the stream and measured downstream at two separate locations through the use of electrodes. Because saltwater conducts electricity better than freshwater, the electrodes will register a hydrograph of electrical conductance with a recognizable center of mass. The time taken between the passage of the center of mass at the two electrode sites can be used along with the distance to calculate the mean velocity.

Salt or Color-Dilution Methods

In this method, salt solution of known concentration, C_1, is introduced at a constant rate q_s. After the solution has flowed far enough to mix completely, it produces a uniform concentration, C_2. The natural salt concentration of the flow is C_0. Then, the discharge can be calculated directly from the following:

$$Q = q_s \, [(C_1 - C_2) / (C_2 - C_0)] \qquad (6.10)$$

This method will also work with any number of dyes and measurements taken using fluorescence or color analysis. Dyes used include fluorescein or potassium permanganate. This method also works with radioisotopes and the use of a Geiger counter. See Rantz (1982) for complete details.

6.10 Environmental Data Considerations and Collection

The need for environmental data in engineering analysis and design stems from the need to investigate and mitigate possible impacts of urban runoff due to specific development or other ongoing activities in a watershed. Nowhere in urban stormwater management is the complexity of the situation mismatched by the dearth of good data, information, and analysis techniques. Neither the actual physiological effects of various stormwater pollution constituents nor the concentrations required to cause these effects are generally known, even with the most complete measurements. The great variation in flow volumes, concentrations, bioavailability, chemical interactions, durations of high pollution loading, etc., make it difficult to do more than roughly estimate actual impacts of pollution in many situations. For example, the toxicity of metals found in urban stormwater runoff depends on such things as pH, organic and inorganic ligands, hardness, other chemicals, and temperature (Davies, 1986). Herricks (1986) has gone so far as to say that there are no simple cause-and-effect relationships in the natural environment. A complete understanding of the typical stormwater quality situation is not possible under current scientific knowledge.

However, great improvements in urban stormwater quality can be made without a complete understanding of cause and effect. Rules of thumb about design criteria, chemical concentrations, and capture and treatment volumes can be useful in reducing pollution loadings and actual impacts, with sufficient accuracy. Measurements may be significant in fact without being statistically significant due to too few data points to form a reliable confidence interval (Loftis et al., 1991). This suggests that measuring should be balanced with doing. While it may never be known, for example, just how much reduction in lead loading will prevent damage to a certain organism in a local stream, it can usually be said with some certainty that abnormally high lead levels should be reduced, and logical engineered steps can be taken to do just that. Then, results should be monitored. This "do something" approach is recommended by Sonnen (1983, 1986) and tacitly adopted by USEPA in its "maximum extent practicable" standard set in the stormwater quality regulations (see Chapter 15).

Data Needs

Environmental data needs may be summarized as follows:

- Information necessary to define the environmental sensitivity of the facility's site relative to impacted surface waters is needed, e.g., water use, water quality and standards, aquatic and riparian wildlife biology, and wetlands information. Some of this information is available in the water-quality standards and criteria published by state water control boards, studies from various federal agencies (USEPA, USGS, etc.), and targeted literature searches.
- Physical, chemical, and biological data for many streams are also available from state and federal water pollution control agencies, the USGS, municipalities, water districts, and industries that use surface waters as a source of water supply. In unique instances, a data collection program possibly lasting several years and tailored to the site may be required.
- Information necessary to determine the most environmentally compatible design is needed, e.g., circulation patterns and sediment transport data. Data on circulation, tides, water velocity, water quality, and wetlands are available from the U.S. Coastal and Geodetic Survey, U.S. Army Corps of Engineers, universities, marine institutes, and state, federal, and local agencies and organizations. Information on sediment transport is vital in defining the suitability of a stream for most beneficial uses, including fish habitat, recreation, and water supply. Such information may be essential for projects in critical water-use areas, such as near municipal or industrial water supply intakes.
- Information necessary to define the need for and design of mitigation measures should be obtained, e.g., fish characteristics (type, size, migratory habits), fish habitat (depth, cover, pool–riffle relationship), sediment analysis, and water-use and quality standards. Fish and fish habitat information is available from state and federal fish and game agencies.

Environmental Sampling Objectives

The development of a monitoring program depends on the study objectives, physiography of the site, duration of the data collection effort, expertise of the staff, and funding for the study (Terstriep, 1986). It is therefore important to first establish clear objectives for the sampling program. Sonnen (1983) and Reinelt, Horner, and Mar (1988) suggest a list of typical types of, or objectives for, environmental sampling:

- Monitoring of a physical process for research purposes
- Problem identification or reconnaissance monitoring
- Alternative solution monitoring
- Design performance monitoring
- Regulatory compliance or enforcement monitoring
- Operational performance, trend, or ambient monitoring

While research monitoring to establish physical principles requires a great amount of data over a long time period and over many monitoring situations, problem identification monitoring may require only five or six storms at several sites and in receiving streams. Constituents in the latter type of study could include only indicator species such as BOD, TSS, fecal coliform, and nitrogen or phosphorus. An approach for problem identification monitoring that quickly isolates locations and approaches was developed by Reinelt, Horner, and Mar (1988). Their approach is to define objectives, choose levels of detail, analyze the watershed to find critical monitoring locations, develop the monitoring program and statistical support for the results, and prioritize the monitoring tasks based on program objectives.

Alternative solution monitoring must be sufficient for calibration of a model used to test different alternatives. In design monitoring, the only data collected are the minimum necessary to obtain input to the design criteria model. In environmental design, the parameters are often estimated based on typical values of such things as event mean concentrations.

Compliance monitoring normally consists of measuring one or several constituents over a period of time (or forever) and comparing results to a regulatory standard. If no standard exists, then the monitoring is expected to report on trends or the effectiveness of a BMP program. Statistical testing may be important in this type of monitoring, though most of the tests were derived from an assumption of a normal distribution of the variable. This is not always the case with environmental data, as asymmetrical and skewed data are the norm (Gilbert, 1987). In these cases, other methods or transformation of the data can be accomplished. Many urban stormwater variables (such as event mean concentration or flood flows) appear to follow a log-normal distribution (Driscoll, 1986). When monitoring operations, two different objectives can be met: checking the operation against the design expectation and building a record of long-term performance. One must always be aware that the samples taken to estimate long-term performance may be tainted by auto- or serial-correlation (lack of independence of events). This correlation is dependent on the temporal and spatial scale and procedures used in the measurements (Loftis, McBride, and Ellis, 1991). For example, measurement of dissolved oxygen in the same place at 8:00 a.m. each morning would not provide trend analysis for any other time of day or in another spot in the stream. There is no right way to do it except to use sound judgment and to fully document assumptions and procedures.

It is also important to consider which types of and concentrations of constituents are likely to be encountered or are of importance, and what tests are necessary for detection. For example, for a designated use of swimming, some measure of the presence of pathogenic organisms should be of importance. For irrigation, the presence of dissolved solids is important (USEPA, 1983). In stormwater quality considerations, thresholds often exist and should be identified if possible. Considering dissolved oxygen, for example, any value above the threshold for survival of a certain species may be acceptable. Any value below a level where the species cannot survive will be unacceptable. Making improvements in the dissolved oxygen value that do not raise it above the survival threshold will not achieve any beneficial result with regard to that species, even though funds and efforts are expended. In the same way, the mere presence of a toxic metal in the environment does not automatically mean it is available biologically for uptake. It may be physically bound to sediment particles or chemically bound in some other way and thus unavailable.

The impacts of stormwater on quality are physical and chemical. Therefore, a simple analysis of chemical constituents will not suffice for assessment of the quality of a particular stream. Erosion and sedimentation in many cases does far more harm to a benthic community than does chemical contamination.

Typical Sampling Program Steps

A typical sampling program uses the following steps (Gilbert, 1987; USEPA, 1992; Valiela and Whitfield, 1989; Loftis, McBride, and Ellis, 1991; USGS, 1984; Ogden, 1992):

Step 1: Clearly define the study and sampling objectives.

Step 2: Define the time-space population of interest developing as recognizable boundaries on the population as possible.

Step 3: Perform a literature search sufficient to determine best methods, ranges of variables to be encountered, possible problems, data trends, etc.

Step 4: Define the hypothesis to be tested, and estimate the sampling size required.

Step 5: Define the types of tests and sample analysis to be made on any samples or the types of statistical or quality assurance analyses to be made of measurements.

Step 6: Collect field information necessary to define a sampling plan.

Step 7: Define the types of samples to be collected (e.g., composite, grab, continuous) or measurements to be made.

Step 8: Define field procedures and field equipment, including risks and safety procedures.

Step 9: Define quality assurance procedures for all phases of the program from collection through analysis or measurement through quality checks.

Step 10: Conduct the study according to plans.

Step 11: Summarize, plot, and analyze results as appropriate.

Step 12: Assess possible error and uncertainty.

Step 13: Assess results compared to objectives, and report any changes in procedure for future studies.

Key Constituents

An understanding of the typical pollutants found in urban stormwater runoff and typical concentrations will help direct any sampling program. Stormwater runoff quality has been characterized through the results of numerous studies. The National Urban Runoff Program (NURP) was an EPA-sponsored program from 1978 to 1983. It included 28 projects across the United States (USEPA, 1983). The runoff-related findings of the NURP study are instructive when considering a stormwater quality sampling and monitoring program (USEPA, 1983). These findings are as follows:

- Heavy metals are the most prevalent priority pollutant in urban runoff. Freshwater acute criteria violations and chronic exceedances were common for copper (47 and 82%) and lead (23 and 94%). Zinc and cadmium had chronic exceedance frequencies of 77 and 48%, respectively.
- Organic pollutants were detected less frequently than heavy metals and at lower concentrations. Sixty-three of a possible 106 organics were detected. The most common were a plasticizer, four pesticides, three phenols, four polycyclic aromatics, and a single halogenated aliphatic. Organics tend to be site- or municipality-specific rather than a general national problem.
- Coliform bacteria are present at high levels and can be expected to exceed EPA water quality criteria during and immediately after storm events.
- Nutrients are present in moderate quantities (an order of magnitude less than most POTWs) in almost all samples.
- Oxygen-demanding substances are present in concentrations equivalent to secondarily treated effluent from POTWs, although no NURP site specifically attributed low DO to urban runoff.
- Total suspended solids in runoff is high compared to treated effluent (an order of magnitude). While Total Suspended Solids (TSS) has no EPA water quality standard, TSS is a carrier of adsorbed toxics and other pollutants and may cause sedimentation or other physical stream problems.

The NURP program sampled for 120 priority pollutants. Of these, 77 were detected in urban stormwater runoff from commercial, residential, and light industrial sites. Those detected in 10% of the samples are listed in Table 6-4. A significant number of these pollutants exceeded various freshwater criteria.

Different constituents have different impacts and sources. Some are sampled in different ways or have specialized handling and preservation considerations. Reference should be made to basic sources such as USEPA (1992 a,b), Standard Methods (1989), and USEPA (1983) and to the literature to determine the various forms of each constituent and in-stream chemical interaction.

Metals

Trace metals are often toxic to man and animals. Concentrations of zinc, cadmium, lead, and copper were found in frequencies and concentrations high enough to warrant concern, though actual impairment due to urban stormwater runoff is not easy to prove. Other metals (arsenic, beryllium, chromium, cyanide, mercury, nickel, selenium, and thallium) were also detected in a few samples and in minute quantities. Large fractions of the metals adsorb to sediments and are not immediately available for bioaccumulation (though they can be re-suspended during high flows). The dilution of urban runoff when it enters receiving streams is also a factor in the consideration of metals' impacts on receiving waters. Copper and zinc are the most soluble of the metals and therefore of the most concern.

Nutrients

Nitrogen (N) and phosphorous (P) (in all the various compounds) are found in urban runoff and contribute to undesirable algal blooms in lakes and ponds. Control of the flow of the limiting nutrient is essential to slow the eutrophication process of water bodies. Available nitrogen is generally obtained from the nitrates plus ammonia nitrogen compounds. The available phosphorous fraction is generally the soluble orthophosphorus plus

TABLE 6-4

Priority Pollutants Detected in at Least 10% of NURP Sample

Metals and Inorganic	Frequency of Detection
Antimony	13
Arsenic	52
Beryllium	12
Cadmium	48
Chromium	58
Copper	91
Cyanads	23
Lead	94
Nickel	43
Selenium	11
Zinc	94
Pesticides	—
Alpha-hexachlorocyclohexane	20
Alpha-endosulfan	19
Chlordane	17
Lindane	15
Halogenated Aliphatics	—
Methane, dichloro-	11
Phenols and Cresols	—
Phenol	14
Phenol, pentachloro-	19
Phenol, 4-nitro-	10
Phthalate Esters	—
Phthalate, bis(2-ethylhexyl)	22
Polycyclic Aromatic Hydrocarbons	—
Chysene	10
Fluoranthene	16
Phenanthrene	12
Pyrene	15

Source: From U.S. Environmental Protection Agency, EPA PB 84-18552, December 1983.

a fraction (20 to 30%) of the particulate. The theoretical uptake rate of these nutrients in algae is 7.5 N to 1 P. If the ratio of N:P is greater than 10, the limiting nutrient is most likely phosphorous. If the ratio is less than 5, it is nitrogen. In between, it may vary (Wanielista et al., 1981). Physicochemical and biological processes determine the eventual fate of the nutrients in that they may remain in solution or suspension, settle out, chemically interact and precipitate or adsorb to sediments, be taken up biologically, or be released back to solution through chemical changes and plant growth and death cycles.

Nitrogen exists, in order of decreasing oxidation state, as nitrate (NO_3^--N), nitrite (NO_2^--N), ammonia (NH_3-N) and organic nitrogen. Organic nitrogen and ammonia are often measured together and are termed *kjeldahl* nitrogen after the test used to determine their concentration. Phosphorous exists primarily in the form of phosphate, either soluble or suspended.

Human Pathogens

The most significant pathogens present in untreated wastewater can be classified into four groups: bacteria, protozoa, helminths, and viruses (USEPA, 1992). For a human to be infected with one of these organisms, a number of steps must occur, beginning with the presence of an organism in sufficient numbers and a long enough exposure to elicit infection. Because it is technically difficult and expensive to measure all potential pathogens found in stormwater, it has been common practice (as with wastewater) to use a surrogate measure of fecal contamination. The most common indicator of the presence of human pathogens has been the presence of either total coliforms or fecal coliform bacteria. More recently, only *Escherichia coli* and enterococci have been used in recreational waters. Several features of these coliforms make them useful as surrogates for pathogenic contamination: they occur naturally in the feces of warm-blooded animals in higher concentrations than the pathogenic organisms; they are easily and definitively detected; their presence is usually positively correlated with fecal contamination; and they usually respond similarly to environmental conditions and treatment processes as do the pathogenic bacteria (Ogden, 1994).

While the presence of such coliforms indicate some fecal contamination, they are not considered a fool-proof indicator of the presence of pathogenic organisms (Field, 1993). Many bather illnesses are not related to enteric bacteria but to other organisms that cause infections of the skin, eye, ear, nose, and upper respiratory systems. There is a need for epidemiological studies that relate illness to specific pathogens and attempt to trace the source of such pathogens. One such study was conducted by Ogden (1994) to determine the actual impacts of combined sewer overflows for Sacramento County, California. A retrospective public health risk assessment of sickness reports among certain outdoor occupations was done and matched to probable overflow dates. Such data can conclude if there is a correlation from cases reported or found in the database. However, it is more difficult to determine the "no impact" hypothesis, because the data are not exact.

Oxygen Demand

Dissolved oxygen is a necessary component of aquatic life. Decomposition of organic matter by microorganisms depletes the available supply of oxygen, particularly in slower-moving waters (COG, 1987). The direct test for oxygen is the dissolved oxygen test (DO). Indirect methods include the biochemical oxygen demand (BOD) and chemical oxygen demand (COD) tests. Neither of these tests has been proven to be effective in correlating or predicting DO problems and fish kills. The BOD test measures the oxygen utilized, during a specified time period, normally 5 days, for the biochemical degradation of organic material and the oxygen used to oxidize inorganic material. The COD test measures all the oxidizable matter present in the fluid through the use of a strong chemical reaction. It oxidizes even some materials that would not normally provide natural oxygen demands. Correlations with BOD can be established in wastewaters or fairly uniform waters from a specific source. Variability in stormwater runoff often makes this correlation difficult and inaccurate.

Sediment

Sediments may be the most destructive pollutant. Urban development can increase the amount of sediment entering streams many-fold. Because, in some cases, the stream cannot transport the increased sediment load, sediment settles out, filling ponds, wetlands, and reservoirs, smothering benthic communities. It inhibits visual feeders from seeing prey, clogs fish gills, reduces spawning and juvenile survival rates, carries toxics and trace metals, and makes water supplies more difficult to treat. The scouring of sediment from banks and beds destroys habitat and adjacent vegetation. Sediments are measured in terms of total suspended solids, settleable solids, turbidity, and dissolved solids. Volatile solids are those that burn off at 550°C and are organic in nature (Standard Methods, 1989).

Oil and Grease

Scheuler (COG, 1987) and Stenstrom, et al. (1984) provide excellent summaries and analysis of the oil and grease problem in urban runoff. The actual impacts are not easily measurable but are visually apparent in the oily sheen on urban runoff and the oily muck found in catch basins and at outfalls. Oil and grease compounds can be toxic at concentrations found in urban stormwater. In testing, oil and grease is any substance that is soluble in trichlorotriflouroethane and not a specific compound or chemical family. Thus, it is defined by the test done to identify it.

Other pollutants are also of importance in urban runoff sampling, including a list of the priority pollutants, chlorides from road salting, thermal impacts, trash, and floatables.

Pollution Sources

Urban pollution accumulation and washoff into streams and ponds is the summation of many, seemingly unimportant, individual activities and physical processes. Thus, it is called nonpoint. Each typical pollutant has one or several principal sources. These sources introduce the pollutant to the environment. The pollutant is then transported, transformed, deposited, resuspended, biologically taken up, and so on.

Erosion and Sedimentation

Sediment loads from construction sites have been variously reported in the literature to be as high as 50 tons/acre/year. Postconstruction values are several orders of magnitude less than this. Even during postconstruction, the dust and dirt from erosion is the main source of street particulate matter and a main carrier of adsorbed contaminants such as ammonium ions, toxic materials, and phosphates. The most common sources of sediment include natural weathering, construction, urban bare soil erosion, agricultural erosion, litter and dust, highway erosion, streambank, and bed erosion. Erosion and sediment yield are estimated a number of ways, including (Novotny and Chesters, 1981): stream sampling, reservoir sediment surveys, sediment delivery equations, transport functions, and watershed modeling (such as the Opus Model, Smith, 1992). Of these, the well-known Universal Soil Loss Equation for upland erosion is the most common.

Atmospheric Fallout

Dust from dryfall and wetfall is the most important source of pollutants on impervious surfaces. Total average annual dustfall values range from about three to five tons per square mile per month in the Tennessee and Ohio valleys, to 15 to 25 in California and the upper Mississippi, to 30 to 50 in the lower Mississippi and Texas Southwest areas, to over 140 in the Upper Colorado River areas. Table 6-5 gives specific pollution deposition rates for the Washington, DC, area.

It is estimated that about 90% of all the fallout on impervious surfaces eventually makes its way into the drainage system. The sources of this pollution include emissions from industry and vehicles, smoke, wind erosion, volcanoes, coal-burning plants, pesticide drift, naturally occurring forest emissions, and domestic heating. Rainfall in urban areas can contain levels of COD in the 15 to 70 mg/L range, Ammonia-N in the 0.3 to 0.7 mg/L, and lead in the 0.03 mg/L range. Acid rain, due to anthropogenic emissions of sulfur and nitrogen oxides, is a serious problem the world over, caused by rainfall pH in the 4 range. For more information, see Novotny and Chesters, 1981, or NAPAP, 1993.

TABLE 6-5

Average Annual Atmospheric Deposition Rates for the Washington, DC, Area (lbs/ac/yr)

Pollutant	Rural	Suburban	Urban
Total solids	99	155	245
COD	199	133	210
Total nitrogen	19.9	12.8	17.0
Nitrate-N	9.4	5.6	6.8
Ammonia-N	5.5	1.1	1.0
TKN	10.5	7.2	10.2
Total P	0.71	0.50	0.84
Ortho-P	0.28	0.26	0.35
Cadmium	—	0.09	0.003
Copper	—	0.21	0.61
Lead	0.06	0.44	0.53
Iron	—	1.57	5.60
Zinc	0.67	1.35	0.65

Source: Metropolitan Washington Council of Governments, Urban Runoff in the Metropolitan Washington Area — Final Report, prepared for EPA and WRPB, Washington, DC, 1983.

Vehicles

Vehicles cause pollution by abrasion of road surfaces and the subsequent dust, re-suspension of other deposition, and transfer to areas adjacent to roads and curb sections, and inorganic and organic pollutants such as lead, asbestos, oil and grease, copper, chromium, nickel, phosphorus, zinc, and rubber.

Other Human Activities

Other human activities contribute to the pollution problem through normal industrial activities, illegal dumping, illicit connections, household hazardous-waste disposal practices, fertilizer and pesticide applications, pet excretions, leaf and litter, and many more sources.

Most of these pollutants, those which are not directly introduced into the drainage system, accumulate on the road surface, and within a short distance (less than two feet) from the curb. Therefore, many urban pollution studies report pollution loadings and removals in terms of curb length instead of loadings per unit area. There is a strong correlation between urban density and curb miles that can be expressed as (APWA, 1969):

$$CL = 234.5 - 200 * 0.631^{PD} \tag{6.11}$$

or

$$CL = -1.049 + 1.4098\ I + 0.0149\ I^2 \tag{6.12}$$

Where CL = curb length density (ft/acre); PD = population density (persons/acre); and I = impervious percentage (half impervious equals 50).

Refuse accumulation along curbs is relatively rapid. Sartor, Boyd, and Agardy (1974) found accumulations in eight American cities from 48 grams per curb mile per day for residential and 66 for multifamily to 69 and 127 for commercial and industrial, respectively.

While the typical street dirt accumulation has more than 80% of its volume of sizes greater than 100 μm, the pollution particulate sizes range much smaller. Pollution

TABLE 6-6

Pollutants Associated with Street Dirt and Litter

Constituent	Residential (mg/L)	Industrial (mg/L)	Commercial (mg/L)	Transportation (mg/L)[a]
BOD_5[b]	9.16	7.50	8.33	2.3
COD[b]	20.82	35.71	19.44	54
Volatile[b] Solids[b]	71.67	53.57	77.00	51
TKN[b]	1.666	1.392	1.111	0.156
PO_4-P[b]	0.916	1.214	0.833	0.61
NO_3-N[b]	0.050	0.064	0.500	0.079
Pb[c]	1.468	1.339	3.924	12
Cr[c]	0.186	0.208	0.241	0.08
Cu[c]	0.095	0.055	0.126	0.12
Ni[c]	0.022	0.059	0.059	0.19
Zn[c]	0.397	0.283	0.506	1.5
Total Coliforms (No./g)[b]	160,000	82,000	110,000	—
Fecal Coliforms (No./g)[b]	16,000	4,000	5,900	925

[a] Shaheen, D.G., USEPA No. 600/2-75-004, Washington, DC, 1975.
[b] Sartor, J.D., Boyd, G.B., and Agardy, F.J., J. WPCF, Vol. 46, 1974.
[c] Amy, G. et al., USEPA No. 44019-75-004, Washington, DC, 1975.

associated with the dust and dirt fraction of general curb dirt and litter can be significant. Sartor, Boyd, and Agardy (1974) found 77% of COD, 57% of BOD_5, 59% of TKN, and 51% of metals having particulate sizes less than 100 µm. Table 6-6 gives typical values in milligrams per gram of total solids collected from street curb sections for various land uses.

Impacts on Receiving Waters

The impacts of urban stormwater on receiving waters are not as well understood as that of point sources. Urban runoff differs from point source discharges in that it is intermittent and of short duration, highly variable from storm to storm, can be highly but variably diluted as it enters receiving waters, and carries a relatively high load of total suspended solids. Therefore, comparison of concentration limits and other numeric standards, low-flow criteria, lethal dose testing, and other point source methods of analysis do not apply readily to stormwater analysis. Wet-weather criteria have been studied by EPA and proposed by several authors, though a standard has never been adopted. The impacts on receiving waters can be divided into three categories (Mancini and Plummer, 1986):

- Short-term changes in water quality during and immediately after a storm event
- Long-term impacts associated with contaminants adsorbed to settled sediment or nutrients in lakes and ponds with long detention times
- Scour and deposition of bed and banks

The NURP study adopted a three-level definition of problems to assist in evaluating urban stormwater impacts (USEPA, 1983):

- Impairment or denial of beneficial use
- Water-quality criterion violation
- Local public perception

Based on these criteria, NURP concluded that the impacts of urban runoff are highly site-specific. It is clear from various studies that, if the ambient water quality of a stream, lake, pond, or estuary is already near a threshold level, the boost given by a slug of urban stormwater runoff could push it beyond that threshold (Bastian, 1986). Frequent exceedances of EPA's heavy metal criteria are common, though actual impairment of aquatic life from the criteria violations was not observed. Of the heavy metals, copper, lead, and zinc are the most prevalent. Copper was concluded to be a significant threat to aquatic life in the Southern and Southeastern United States. Generally speaking, organic pollutants are not a typical urban runoff hazard, though freshwater intakes in the vicinity of urban runoff locations should be tested for such pollutants. This is not to say that in specific cases where there are spills, dumping, or misuse, they cannot cause problems. The physical aspects of urban runoff cause as much or more habitat destruction and species diversity limitation as do the chemical aspects. Pollution adsorption to sediments and possible re-suspension is a suspected, though unmeasured, problem. Coliform bacteria are present in great numbers in all urban runoff. Nutrient and bacterial loadings can severely impact urban lakes and ponds in terms of trophic state and recreational use impairment. A survey of fisheries experts indicated that turbidity, temperature, and nutrient loadings are the three most important urban stormwater impacts on fisheries habitat (Heaney, 1988).

It is clear that urban stormwater managers are interested in the mitigation of real, discernible impacts of urban stormwater runoff: aquatic organisms, aesthetics, and human health and safety. Measurement of constituents such as BOD, COD, and TOC, and application of numeric criteria are a step away from these real impacts toward a more easily measured standard. A further step away is the development of simplistic design criteria (e.g., capture of the first half-inch). These steps away from assessing real impacts have been necessary because of the combined need to do something immediate to stop the rapid degradation of urban waters and the general lack of understanding of the details of cause and effect (Huber, 1988). However, there will be (and is in some states) an eventual shift in consideration of urban stormwater impacts from an emphasis on numerical criteria and simple design criteria toward site-specific, risk-based, and biologically based criteria and standards (Huber, 1988; Heaney, 1988).

Land-Use Baseline Data

Actual sampling information can be compared to baseline information to test for inconsistencies and to, perhaps, avoid needless sampling if initially sampled values compare with average values. The NURP, and other, data provide such a basis. Because of the variability of measurements within storms, among different storms at one site, and among sites, it was desirable to use a measure that tended to reduce this variability somewhat. The measure of the magnitude of urban runoff pollution chosen is termed the event mean concentration (EMC). It is defined as the total constituent mass discharge divided by the total runoff volume for a given storm event. It is defined as the event load divided by the event flow volume. It is calculated as:

$$\text{EMC} = \left[\sum_{i-1}^{n} V_i C_i / \sum_{i-1}^{n} V_i \right] \quad (6.13)$$

where V_i = flow volume per time increment i (liters); C_i = average concentration for time period i (mg/L). $V_i C_i$ is the mass of pollution for a given time increment i. The summation of all the $V_i C_i$ values gives the total event loading. The summation of all the V_i values

TABLE 6-7

Example EMC Calculation for TSS

Average Discharge Rate (L/sec)	Measured Concentration (C_i) (mg/L)	Incremental Runoff Volume (V_i) (L)	Incremental Mass of Pollutant (mg)
8	25	4,800.00	120,000.00
16	145	9,600.00	1,392,000.00
45	180	27,000.00	4,860,000.00
60	120	36,000.00	4,320,000.00
48	80	28,800.00	2,304,000.00
35	45	21,000.00	945,000.00
22	20	13,200.00	264,000.00
12	23	7,200.00	165,600.00
5	25	3,000.00	75,000.00
Totals		150,600.00	14,445,600.00

gives the total event flow volume. The division of the two gives the event mean concentration (EMC).

Table 6-7 demonstrates this calculation for a runoff event sampling TSS. Assume that the sampling increment is 10 min. Then, dividing the total mass of pollutant by the total volume of discharge gives EMC = 14,445,600 mg/150,600 L = 95.9 mg/L.

With few exceptions, EMCs were found to not vary significantly for similar land uses from site to site for the same constituent and were found to be distributed lognormally. Therefore, measures of central tendency (median and mean) and scatter (standard deviation, coefficient of variation) could be calculated, as well as expected values at any frequency of occurrence, by using the natural logarithmic transformation of the raw data. Standard statistical tests and sampling theory can also be used on the lognormally distributed data.

EMCs also did not vary with storm volume. That is, over experienced storm depths or duration, the mean concentration of the pollutant over the storm event did not vary significantly with runoff volume. What this implies, and what has been shown by various authors, is that the total volume of pollution (as opposed to the concentration) will vary directly with the volume of runoff. And, because the volume of runoff varies directly with the amount of impervious area, the total pollution volume varies directly with impervious area.

Based on these findings, certain calculations of pollutant loading become simplified. After taking the natural logarithm of each of N observations, the statistics can be calculated as follows:

Mean $\qquad M = (x_i)/N \qquad$ (6.14)

Standard deviation $\qquad S = [(x_i - M)^2/(N - 1)]^{1/2} \qquad$ (6.15)

Coefficient of variation $\qquad C = S/M \qquad$ (6.16)

Median $\qquad m = M/(1 + C^2)^{1/2} \qquad$ (6.17)

Expected value $\qquad X = \exp(M + ZS) \qquad$ (6.18)

where M = natural logarithm of the mean value of the EMC observations; x = natural logarithm of an individual EMC observation; N = number of observations; S = standard deviation of the logarithms of the observations; C = coefficient of variation of the logarithm of the observations; m = median of the natural logarithm of the observations; Z = the standard normal probability from probability table for normal distribution found in any statistics textbook; and X = expected value (nonlogarithm) of the logarithms of the observations.

It should be noted that, for standard deviation calculations, the following identity is helpful:

$$\Sigma(x_i - M)^2 = \Sigma x_i^2 - (\Sigma x_i)^2/N \qquad (6.19)$$

While the median value is of concern for more acute pollution situations, the mean value is of value when considering long-term chronic or cumulative pollution. Conversion between the two can be obtained from Equation (6.17). Table 6-8 gives median event mean concentrations for various land uses for a number of common constituents from several sources. The measured coefficient of variation from the NURP data is also presented to facilitate calculations of mean values. While the analysis of the NURP information did not prove a statistical relationship between land use and EMC (except for open areas), the differences in values for different constituents, for different land uses, are instructional and have been distinguished in use (SQTF, 1993).

An example conversion between the median and the mean can be done for the commercial/industrial value for BOD_5. The median value is 9.3 mg/L, and the coefficient of variation is 0.31. To compute annual loadings, the mean is desired. Recall that these are the log transformed values in the equations. Therefore, from Equation (6.17):

$$m = M/(1 + C^2)^{1/2}$$

$$\ln(9.3) = M/(1 + 0.31^2)^{1/2}$$

$$M = 2.3347$$

$$e^{0.988} = 10.33 \text{ mg/l} = EMC$$

The total annual loading of pollution can then easily be approximated by multiplying the mean concentration in the runoff by the total annual runoff as follows (SMFT, 1993):

$$L = 0.2266 * EMC * [0.15 + 0.75\ I] * P * A \qquad (6.20)$$

where L = pollution loading (lbs/year); EMC = mean event mean concentration (mg/L); I = fraction of impervious area (acres); P = annual rainfall (in); and A = watershed area (acres).

For the example above, the annual loading of BOD_5 for an area of 10 acres, 35% impervious and an annual rainfall of 45 inches is:

$$L = 0.2266 * 10.33 * [0.15 + 0.75\ (0.35)] * 45 * 10 = 434 \text{ lbs}$$

A refinement of this method, termed the Simple Method, is given in Chapter 13.

Urban Runoff Quality Data Collection

The importance of proper sampling is paramount because the decisions that will be based on samples can be little more accurate than the sample itself (Wullschleger, Zanoni, and

TABLE 6-8
Median Event Mean Concentrations and Coefficients of Variation for Various Land Uses

Pollutant	Residential Median	Residential CV	Mixed/NURP Urban Site Median[d] Median	Mixed/NURP Urban Site Median[d] CV/90%[e]	Commercial/Industrial Median	Commercial/Industrial CV	Highways[c] Mean	Open/Nonurban Median	Open/Nonurban CV
BOD (mg/L)	10.0	0.41	7.8/9[d]	0.52/15[e]	9.3	0.31	9.7/24[f]	8.0[a]	—
COD (mg/L)	73	0.55	65/65[d]	0.58/140[e]	57	0.39	130-360/14.76[f]	40/51[a]	0.78
TSS (mg/L)	101	0.96	67/100[d]	1.14/300[e]	69	0.85	150-450	70/216[a]	2.92
Total lead (μg/L)	144	0.75	114/144[d]	1.35/350[e]	104	0.68	500-1300/960[f]	30/0.0[a]	1.52
Total copper (μg/L)	33	0.99	27/34[d]	1.32/93[e]	29	0.81	70-180/103[f]	0.0[a]	—
Total zinc (μg/L)	135	0.84	154/160[d]	0.78/500[e]	226	1.07	300-750/410[f]	195/0.0[a]	0.66
Cadmium							0-30		
TKN (mg/L)	1.9	0.73	1.29/1.5[d]	0.50/3.30[e]	1.18	0.43	1.78/2.996[f]	0.965 / 0.610[b] / 1.36[a]	1.00
NO$_2$ + NO$_3$ (mg/L)	0.736	0.83	0.558/0.68[d]	0.67/1.75[e]	0.572	0.48	0.830/1.14[f]	0.543 / 0.730[a]	0.91
Total P (mg/L)	0.383	0.69	0.263/0.33[d]	0.75/0.70[e]	0.201	0.67	0.440/0.79[f]	0.121 / 0.150[b] / 0.230[a]	1.66
Soluble P (mg/L)	0.143	0.46	0.056/0.12[d]	0.75/0.21[e]	0.080	0.71	0.170	0.026 / 0.040[b] / 60[a]	2.11

[a] From Stormwater Quality Task Force, State of California, California Stormwater Best Management Practice Handbooks — Municipal, March 1993. (No coefficient of variation.)

[b] From MWCOG, 1987. Virginia hardwood forest (no coefficient of variation).

[c] Highway data from Stormwater Quality Task Force, State of California, California Stormwater Best Management Practice Handbooks — Municipal, March 1993, and FHWA, 1990.

[d] NURP Median EMC values for combined urban sites. From U.S. Environmental Protection Agency, EPA PB 84-18552, December 1983, Table 6-7.

[e] NURP ninetieth percentile values for combined urban sites. From U.S. Environmental Protection Agency, EPA PB 84-18552, December 1983, Table 6-7.

[f] From Smith, D.L. and Lord, B.N., Trans. Res. Rec., 1279, 1990. Results from six street sites.

General Source: From U.S. Environmental Protection Agency, EPA PB 84-18552, December 1983.

Hansen, 1976). Monitoring takes place within the stormwater collection system, at its outfalls, or in the receiving waters. It is either storm-related or storm event independent. This section will provide some basic data and information on urban stormwater sampling.

Site Selection

Within-system sampling must take place at a defined point source that 40 CFR 122.2 defines for urban purposes as any pipe, ditch, channel, tunnel, conduit, well, or discrete fissure. The sites should be easily accessible in adverse weather conditions, not inviting to vandals (private backyards work well for this), not subject to flow interruptions or backwater conditions, and physically close enough to the normal location of the sampling technician to meet start-of-sampling criteria, if first flush grab samples are necessary. The site must also represent only that land use or other phenomena that is to be measured. Commingling flows from several land sources will lead to misleading results. However, when other options are not feasible, known single-source concentrations can be used and tested against a site with several land uses and weighting used.

Receiving stream sites must clearly represent a particular watershed and reach of stream. The site should reflect the runoff from a particular area or region of interest and be sited to fairly represent this site. The limits of the stream reach for which the gauge serves as a representative sample should be determined approximately by assessing land-use homogeneity in the tributaries above and below the gauge site and the distance upstream for which the reach flow is consistent.

Safety considerations should be paramount, including use of the appropriate equipment, safety gear, the provision for two persons at all sites with any safety risk at all, and ready or constant communications at other sites. Sampling in manholes normally requires training in confined space entry.

Storm Event Sampling

EPA has established guidelines for storm event discharge characterization sampling under the NPDES permit application. The storm depth must be greater than 0.1 inch accumulation to ensure adequate flow is generated. It must be preceded by a minimum 72 h dry period to allow for pollution buildup before the storm. And, the rainfall depth and duration should be representative (defined as not varying from the mean by more than 50% in either category).

It cannot be known for sure if the event being sampled is characteristic until after all the criteria have been met. However, use of National Weather Service or private real-time radar information and quick mobilization should allow enough time to set up and sample prior to the rainfall–runoff event beginning. Automated samplers can be used to avoid constant mobilization preparedness. They can be programmed to trigger based on flow or rainfall. Automated samplers cannot be used in all cases though, because some constituents have short holding times or the automated sampling tubing can be a source of contamination from sample to sample. Grab samples must be used in these cases.

Sample Types

There are many types of samples common to urban stormwater quality programs, including sediment sampling, fish tissue sampling, water quality sampling, biological assessment, etc. Water quality sampling includes measurement of the physical (temperature, pH, turbidity, etc.) and chemical (toxics, nitrogen, etc.) conditions of the water. Water quality sampling consists of two basic types of samples: grab samples and composite samples (EPA, 1992). Either of these sample types can be collected manually or by automated equipment. Grab samples are discrete snapshots of the water quality at an instant in time (usually less than 5 to 15 min). Composite samples are a mixture of samples taken in discrete intervals over the storm event.

Details on sampling technique, preservation, quality control, and handling can be found in EPA (1976, 1992a, 1992b, 1993) as well as standard laboratory analysis manuals such as Standard Methods (1989). The development of complex sampling programs for receiving waters is beyond the scope of this text. However, the principles provided previously concerning sampling objectives should be employed in such programs.

References

Agricultural Research Service, Field Manual for Research in Agricultural Hydrology, Handbook No. 224, USDA, Washington, DC, 1962.

American Association of State Highway and Transportation Officials, *Model Drainage Manual*, Washington, DC, 1991.

American Public Works Association, Water Pollution Aspects of Urban Runoff, FWPCA (EPA), WP-20-15, 1969.

Amy, G., Pitt, R., Singh, R., Bradford, W.L., and LaGraff, M.B., Water Quality Management Planning for Urban Runoff, USEPA No. 44019-75-004, Washington, DC, 1975.

Arnell, V., et al.,, Review of rainfall data application for design and analysis, *Water Sci. Tech.*, Vol. 116, Copenhagen, 1984.

Barfield, B.J., Warner, R.C., and Haan, C.T., *Applied Hydrology and Sedimentology for Disturbed Areas*, Oklahoma Technical Press, 1981.

Bastian, R.K., Potential impacts on receiving waters, *Urban Runoff Quality*, Proc. Eng. Found. Conf., Henniker, NH, Urbonas, B. and Roesner, L.A., Eds., ASCE, 1986.

Benson, M.A., Spurious correlation in hydraulics and hydrology, *ASCE J. Hydraulics*, 91, HY4, July 1965.

Bowker, A.H. and Lieberman, G.J., *Engineering Statistics*, Prentice Hall, New York, 1972.

Brater and King, *Handbook of Hydraulics for the Solution of Hydraulic Engineering Problems*, 6th ed., McGraw-Hill, New York, 1976.

Bureau of Reclamation, Water Measurement Manual, 2nd ed., U.S. Department of Interior, Denver, CO, 1981.

Burnham, M. and Davis, D., Accuracy of Computed Water Surface Profiles, U.S. Army Corps of Engineers, Hydrologic Engineering Center, Davis, California, December 1986.

Cheremisinoff, N., Flow measurement and instrumentation, *Natl. Environ. J.*, November/December 1991.

Chow, V.T., *Handbook of Applied Hydrology*, McGraw-Hill, New York, 1964.

Clark, I., *Practical Geostatistics*, Applied Science Publishers, London, 1979.

Clemmens, A.J., Bos, M.G., and Replogle, J.A., RBC broad-crested weirs for circular sewers and pipes, *J. Hydrology*, 68, 349–368, 1984b.

Clesceri, L.S., Greenberg, A.E., and Trussell, R.R., Eds., Standard Methods for the Examination of Water and Wastewater, APHA-AWWA-WPCF, 17th ed., 1989.

Court, A., Reliability of hourly precipitation data, *J. Geophys. Res.*, Washington, DC, December 1960.

Davies, P.H., Toxicology and chemistry of metals in urban runoff, in *Urban Runoff Quality*, Proc. Eng. Found. Conf., June 23–27, Henniker, NH, Urbonas, B. and Roesner, L.A., Eds., 1986.

Dean, J.D. and Snyder, W.M., Temporally and areally distributed rainfall, *ASCE J. Irrig. and Drain.*, 103, IR2, June 1977.

Driscoll, E.D., Lognormality of point and nonpoint source pollution concentrations, in *Urban Runoff Quality*, Proc. Eng. Found. Conf., June 23–27, Henniker, NH, Urbonas, B. and Roesner, L.A., Eds., 1986.

Driscoll, E.D., Shelley, P.E., and Strecker, E.W., Pollutant Loadings and Impacts from Highway Stormwater Runoff, Vol. I, FHWA-RD-88-006, April 1990.

Field, R., Use of Coliform as an Indicator of Pathogens in Storm-Generated Flows, Proc. 6th Int. Conf. on Urban Storm Drainage, Niagara, Ontario, Canada, September 1993.

Fulford, J.M. and Sauer, V.B., Comparison of Velocity Interpolation Methods for Computing Open Channel Discharge, Selected Papers in Hydrologic Science, USGS WSP 2290, January 1986.

Gilbert, R.O., *Statistical Methods for Environmental Pollution Monitoring*, Van Nostrand Reinhold, New York, 1987.

Gill, M.A., Flow measurement by triangular broad-crested weir, *Water Power and Dam Construction*, 37, 8, 1985.

Gilman, C.S., Rainfall, in *Handbook of Applied Hydrology*, Chow, V.T., Ed., McGraw-Hill, New York, 1964, chap. 9.

Gottschalk, L. and Jutman, T., Calculation of Areal Means of Meteorological Variables for Watersheds, Proc. of Den 7ende Nordiske Hydrologiske Konferense, Forde, Norway, 1983.

Gray, D.M., *Handbook on the Principles of Hydrology*, National Research Council of Canada, 1970.

Greenville, South Carolina, Manning's n Manual, City/County of Greenville, 1992.

Gwinn, E.R. and Parsons, D.A., *Discharge equations for HS, H, and HL flumes*, ASCE J. Hydraul. Div., 102, HY1, January 1976.

Gumbel, E.J., Spurious correlation and its significance to physiology, *J. of Am. Stat. Assoc.*, June 1926.

Heaney, J.P., Cost-effectiveness and urban storm-water quality criteria, in *Urban Runoff Quality Controls*, Proc. Eng. Found. Conf., Potosi, MO, Roesner, L.A., Urbonas, B., and Sonnen, M.B., Eds., ASCE, 1988.

Herricks, E.E., Disciplinary integration: the solution, in *Urban Runoff Quality*, Proc. of Eng. Found. Conf., June 23–27, Henniker, NH, Urbonas, B. and Roesner, L.A., Eds., 1986.

Hershfield, D.M., Rainfall Frequency Atlas of the United States, TP 40, U.S. Department of Commerce, Weather Bureau, 1961.

Hershfield, D.M., On the Spacing of Rain Gauges, IASH and WMO Symp., Design of Hydrometeorological Syst., Quebec City, Canada, 1965.

Hirsch, R.M. et al., Statistical analysis of hydrologic data, in *Handbook of Hydrology*, Maidment, D.R., Ed., McGraw Hill, New York, 1993, chap. 17.

Hirsch, R.M. and Gilroy, E.J., Methods of fitting a straight line to data: examples in water resources, *Water Res. Bull.*, 20, 5 October 1984.

Hosking, J.R.M., Wallis, J.R., and Wood, E.F., Estimation of the generalized extreme value distribution by the method of probability weighted moments, *Technometrics*, 27, 3, 1985.

Huber, W.C., Technological, hydrological, and BMP basis for water-quality criteria and goals, in *Urban Runoff Quality Controls*, Proc. Eng. Found. Conf., Potosi, MO, by Roesner, L.A., Urbonas, B., and Sonnen, M.B., Eds., ASCE, 1988.

Huff, D., *How to Lie with Statistics*, W. W. Norton & Co., New York, 1954.

Huff, F.A., Radar Analysis of Urban Effects on Rainfall, 17th Conf. on Radar Meteorol., Am. Meteorol. Soc., Boston, MA, 1975.

Huff, F.A., Spatial and Temporal Correlation of Precipitation in Illinois, Illinois State Water Survey, Cir. 141, Urbana, IL, 1979.

Interagency Advisory Committee on Water Data, Guidelines for Determining Flood Flow Frequency, Bull. 17B U.S. Department of Int., USGS, Reston, VA, 1982.

Kelway, P.S., A scheme for assessing the reliability of interpolated rainfall estimates, *J. Hydrology*, 21, 1974.

Kilpatrick, F.A., Use of flumes in measuring discharges at gaging stations, *Surface Water Techniques*, Book 1, USGS, Washington, DC, 1965, chap. 16.

Kitanidis, P.K., Geostatistics, in *Handbook of Hydrology*, Maidment, D.R., Ed., McGraw Hill, New York, 1993, chap. 19.

Loftis, J.C., McBride, G.B., and Ellis, J.C., Considerations of scale in water quality monitoring and data analysis, *Water Res. Bull.*, 27, 2, April 1991.

Ludwig, J.H. and Ludwig, R.G., Design of Palmer-Bowlus flumes, *Sewage and Industrial Wastes*, 23, 9, September 1951.

Mancini, J.L. and Plummer, A.H., Urban runoff and water quality criteria, in *Urban Runoff Quality*, Proc. Eng. Found. Conf., Henniker, NH, Urbonas, B. and Roesner, L.A., Eds., ASCE, 1986.

McGuiness, J.L., Accuracy of estimating watershed mean rainfall, *J. Geophys. Res.*, 68, 1963.

McTrans Center, University of Florida, Gainesville, FL.

Metropolitan Washington Council of Governments, Urban Runoff in the Metropolitan Washington Area — Final Report, prepared for EPA and WRPB, Washington, DC, 1983.

National Acid Precipitation Assessment Program, 1992 Report to Congress, Washington, DC, June 1993.

Nguyen, V.T.V., Rouselle, J., and McPherson, M.B., Evaluation of areal versus point rainfall with sparse data, *Canadian J. Civ. Eng.*, 8, 2, 1981.

Niemczynowicz, J., Areal intensity-duration frequency curves for short term rainfall events, *Nordic Hydrology*, 13, 4, 1982.

Niemczynowicz, J., Storm tracking using raingauge data, *J. Hydrology*, 93, 135–152, 1987.

Niemczynowicz, J., Moving Storms as an Areal Input to Runoff Simulation Models, Int. Symp. on Urban Hydrology and Municipal Eng., Markham, June 15, 1988.

Novotny, V. and Chesters, G., *Handbook of Nonpoint Pollution*, Van Nostrand Reinhold, New York, 1981.

Ogden Environmental and Energy Services, Inc., Storm Water Sampling Guide, Fairfax, VA, 1992.

Ogden Environmental and Energy Services, Inc., Retrospective Study Report Public Health Assessment for Outflows from the Combined Sewer System, San Francisco, CA, 1994.

Reinelt, L.E., Horner, R.R., and Mar, B.W., Nonpoint source pollution monitoring program design, *J. Water Res. Plann. and Dev.*, ASCE, 114, 3, 1988.

Robinson, A.R., Trapezoidal Flumes for Measuring Flow in Irrigation Channels, ARS 41-140, Agricultural Research Service, USDA, 1968.

Robinson, A.R. and Chamberlain, A.R., Trapezoidal flumes for open channel flow measurement, *Trans. ASAE*, 3, 2, 1960.

Salas, J.D., Analysis and modelling of hydrologic time series, in *Handbook of Hydrology*, Maidment, D.R., Ed., McGraw Hill, New York, 1993, chap. 19.

Sartor, J.D., Boyd, G.B., and Agardy, F.J., Water pollution aspects of street surface contaminants, *J. WPCF*, 46, 1974.

Shaheen, D.G., Contribution of Urban Roadway Usage to Water Pollutions, USEPA No. 600/2-75-004, Washington, D.C., 1975.

Shilling, W., A Quantitative Assessment of Uncertainties in Stormwater Modeling, Proc. Int. Conf. Urban Storm Drain., Gothenburg, Sweden, 1984.

Simanton, J.R. and Osborn, H.B., Reciprocal-distance estimate of point rainfall, *ASCE J. Hydraulics*, 106, HY7, July 1980.

Skogerboe, G.V., Bennett, R.S., and Walker, W.R., Generalized discharge relations for cutthroat flumes, *ASCE J. Irrig. and Drain.*, 98, IR4, December 1974.

Skogerboe, G.V. et al., Design and Calibration of Submerged Open Channel Flow Measurement Structures — Part 3, Cutthroat Flumes, Utah Water Research Laboratory, College of Engineering, Utah State University, Logan, UT, Rpt. WG 31-4, April 1967.

Smith, D.L. and Lord, B.N., Highway water quality control — summary of 15 years of research, *Trans. Research Record*, No. 1279, 1990.

Smith, J.A., Precipitation, in *Handbook of Hydrology*, Maidment, D.R., Ed., McGraw-Hill, New York, 1993, chap. 3.

Smith, R.E., Opus: An Integrated Simulation Model for Transport of Nonpoint-Source Pollutants at the Field Scale, ARS-98, Agricultural Research Service, USDA, July 1992.

Soil Conservation Service, A Method for Estimating Volume and Rate of Runoff in Small Watersheds, SCS TP 149, January 1968.

Soil Conservation Service, Guide for the Use of Technical Release No. 55 — Urban Hydrology, SCS, Albany, NY, 1977.

Sonnen, M.B., Guidelines for the Monitoring of Urban Runoff Quality, EPA-600/2-83-124, November 1983.

Sonnen, M.B., Review of data needs and collection technology, in *Urban Runoff Quality*, Proc. of Eng. Found. Conf., June 23–27, Henniker, NH, Urbonas, B. and Roesner, L.A., Eds., 1986.

Stedinger, J.R., Vogel, R.M., and Foufoula-Georgiou, E., Frequency analysis of extreme events, in *Handbook of Hydrology*, Maidment, D.R., Ed., McGraw Hill, New York, 1993, chap. 18.

Stormwater Quality Task Force, State of California, California Stormwater Best Management Practice Handbooks — Municipal, March, 1993.

Terstriep, M.L., Design of Data Collection Systems, Urban Runoff Pollution, NATO ASI Series, Torno, Marsalek and Desbordes Eds., 1986.

Thiessen, A.H., Precipitation for large areas, *Monthly Weather Review*, 39, 1082–1084, July 1911.

Thorpe, W.R., Rose, C.W., and Simpson, R.W., Areal interpolation of rainfall with a double Fourier series, *J. Hydrology*, 42, 1979.

U.S. Department of the Interior, Bureau of Reclamation, Design of Small Dams, 1977.

U.S. Environmental Protection Agency, Methodology for the Study of Urban Storm Generated Pollution and Control, EPA-600/2-76-145, 1976.

U.S. Environmental Protection Agency, Results of the Nationwide Urban Runoff Program, Vol 1 — Final Report, EPA PB 84-18552, December 1983.

U.S. Environmental Protection Agency (USEPA), in association with the U.S. Agency for International Development, Manual: Guidelines for Water Reuse, EPA/625/R-92/004, September 1992.

U.S. Environmental Protection Agency, Office of Water, NPDES Storm Water Sampling Guidance, EPA 833-B-92-001, 1992a.

U.S. Environmental Protection Agency, Guidance Manual for the Preparation of Part 2 of the NPDES Permit Applications for Discharges from Municipal Separate Storm Sewer Systems, EPA 833-B-92-002, 1992b.

U.S. Environmental Protection Agency, Investigations of Inappropriate Pollutant Entries into Storm Drainage Systems, EPA/600/R-92/238, January 1993.

U.S. Geological Survey, National Handbook of Recommended Methods for Water-Data Acquisition, Office of Water Data Coordination, Reston, VA, 1984.

Valiela, D. and Whitfield. P.H., Monitoring strategies to determine compliance with water quality objectives, *Water Res. Bull.*, 25, 1, February 1989.

Vanleer, B.R., The California pipe method of water measurement, *Eng. News Record*, August 3, 1922 and August 21, 1924.

Vogel, J.L. and Huff, F.A., Mesoscale Analysis of Urban Related Storms, Proc. of Natl. Symp. for Hydrologic Modeling, Am. Geo. Union, Davis, CA, 1975.

Vogel, J.L., Potential urban rainfall prediction measurement system, *Water Sci. Tech.*, 16, Copenhagen, 1984.

Wanielista, M.P. et al., Stormwater Management Manual, University of Central Florida, Orlando, 1981.

Wells, E.A. and Gotaas, H.B., Design of venturi flumes in circular conduits, *Trans. ASCE*, 123, 1958.

Wilson, W.T., Discussion of precipitation at Barrow, Alaska, *Trans. Am. Geophys. Union*, Vol. 35, pp. 206–207, 1954.

Wullschleger, R.E., Zanoni, A.E., and Hansen, C.A., Methodology for the Study of Urban Generated Pollution and Control, EPA-600/2-76-145, August 1976.

Yevjevich, V., *Stochastic Processes in Hydrology*, Water Resources Publications, Ft. Collins, CO, 1972a.

Yevjevich, V., *Probability and Statistics in Hydrology*, Water Resources Publications, Ft. Collins, CO, 1972b.

Appendix A — Sources of Data

Following is a list of data sources that might prove useful for stormwater management analysis and design.

Principle Hydrology Data Sources

- Meteorological Data
    National Oceanography and Atmospheric Agency (NOAA)
    Climatic Data Center
    Asheville, North Carolina 28801
- Regional and local flood studies
- U.S. Geological Survey regional and any site studies
- Surveyed high-water marks and site visits by local agencies
- Hydrology data from others (see below)

Principle Watershed Data Sources

- U.S. Geological Survey maps ("Quad" sheets)
    U.S. Geological Survey
    Rocky Mountain Mapping Center
    Stop 504
    Denver Federal Center
    Denver, Colorado 80225
    (303) 236-5829
- EROS aerial photographs
    U.S. Geological Survey
    EROS Data Center
    Sioux Falls, South Dakota 57198
    (605) 594-6151
- U.S. Geological Survey local offices
- State and local maps and aerial photos
- State geological maps
- Soil Conservation Service and BLM soils maps
- County soils maps
- Site visits
- Watershed data from others

Principle Site Data Sources

- Local agency files of aerial drainage surveys

- Local agency files for existing facilities
- Site visits by local agency
- Field or aerial surveys from others (see below)

Principle Regulatory Data Sources

- Federal Flood Plain delineations and studies
    Federal Emergency Management Agency
    Flood Map Distribution Center
    6930 (A-F) San Tomas Road
    Baltimore, Maryland 21227-6227
    Watts 71-800-638-6620
- State floodplain delineations and studies
- FHWA design criteria and practices
    Federal Highway Administration
    U.S. Department of Transportation
    400 Seventh Street SW
    Washington, DC 20590
- State laws
- Local ordinances and master plans
- Local agency policy statements
- Corps of Engineers Section 404 permit program (see Environmental below)
- U.S. Coast Guard
- U.S. Environmental Protection Agency (EPA) (see Environmental below)
- State EPAs (see Environmental below)
- Federal Register
    Superintendent of Documents
    U.S. Printing Office
    Washington, DC 20402
    (202) 783-3238

Principle Environmental Data Sources

- U.S. Environmental Protection Agency data and studies
- Corps of Engineers data and studies
- U.S. Geological Survey water quality data
- State water quality data
- Environmental statements prepared by other federal, state, and local agencies as well as private parties
- Environmental data from others (see below)

Data Availability and Collection

Principle Demographic, Economic, and Political Data Sources

- Local agency files for existing facilities
- Local agency plans for proposed facilities
- Local agency field or aerial surveys
- Site visits by local agency
- Local agency planning, budgeting, and scope documents
- Internal local agency reports, memorandums, minutes, and verbal communications

Other Data Sources

- U.S. Bureau of Reclamation (USBR)
    U.S. Bureau of Reclamation Center
    Denver, Colorado 80225
    (303) 236-8098
- Regional and State U.S. Bureau of Land Management (BLM)
- Regional U.S. Environmental Protection Agency (EPA)
- Regional U.S. Federal Emergency Management Agency (FEMA)
- Regional and State U.S. Fish and Wildlife Service (USFWS)
- Regional and State U.S. Forest Service (USFS)
- Regional and State U.S. Soil Conservation Service (SCS)
- Regional and State U.S. Corps of Engineers (COE)
- Regional U.S. Coast Guard (USCG)
- Regional and State U.S. Geological Survey (USGS)
- Regional and State Federal Highway Administration (FHWA)
- National Weather Service (NWS)
- National Marine Fisheries Service (NMFS)
- National Oceanic and Atmospheric Administration (NOAA)
- Any state counterparts to the above federal agencies
- Any local counterparts to the above federal agencies
- State or local irrigation, drainage, flood control, and watershed districts
- Any Indian councils
- Municipal governments
- Any planning districts
- Any regional water quality control boards
- Any river basin compacts, commissions, committees, and authorities
- Tennessee Valley Authority
- Private citizens
- Private industry

Appendix B — Field Investigation Form and Checklist

Form 1
Field Visit Investigation Form

Date:_____ Project:_____ By:_____
Structure Type_____ Pier Type_____
Size or Span_____ Skew _____
# of Barrels or Spans_____ Inlet_____
Clear Ht_____ Outlet_____
Abut Types_____ % Grade of Road_____
Inlet Type_____ % Grade of Stream_____
Existing Wtwy Cover_____
Overflow Begins @ El._____ Length of Overflow_____
                                Check for Debris_____
Max AHW (ft)_____ Check for Ice_____
Reason:_____ Side Slopes_____
_____ Height of Banks_____

Up or Downstream Restriction:
Outlet Channel, Base_____

Manning's n Value:
Type of Material in Stream_____
Ponding_____

Check Bridges Upstream and Downstream
Check Land Use Upstream and Downstream

Survey Required?   Yes____   No____

Remarks:

Form 2
Hydraulic Survey Field Inspection Check List

I. <u>General Project Data</u>

1. Project Number:_____ 2. County:_____
3. Road Name:_____
4. Site Name:_____, Station_____ M.P._____
5. Site Description: ( ) Cross Drain, ( ) Irrigation, ( ) Storm Drain, ( ) Long. Encroach, ( ) Channel,( ) Other _____
6. Survey Source: ( ) Field, ( ) Aerial, ( ) Other_____
7. Date Survey Received:_____, From_____

Data Availability and Collection 205

    8. Site Inspected by_____ on _____

II. <u>Office Preparation For Inspection</u>

1. Reviewed:
   Aerial Photos - ( ) Yes, Photo #'s _____, ( ) None Available
   Mapping/Maps - ( ) Yes, Map #'s _____, ( ) None Available
   Reports - ( ) Yes, ( ) No, ( ) None Available at this time
   Municipal Permanent File - ( ) Yes, ( ) No, ( ) No file data found

2. Special Requirements and Problems Identified for Field Checking:
   - ( ) Hydrologic Boundary - obtain hydrologic channel geometry
   - ( ) Adverse Flood History - obtain HW Marks/dates/eye witness
   - ( ) Irrigation Ditch - obtain several Water Right depths
   - ( ) Permits Req'd - ( ) COE ( ) Dam, ( ) Coast Guard, ( ) State/Local
   - ( ) Other_____
   - ( ) Adverse Channel Stability and Alignment History - Check for headcutting, bank caving, braiding, increased meander activity
   - ( ) Structure Scour - check flow alignment, scour at culvert outlet or evidence of bridge scour
   - ( ) Obtain bed/bank material samples at _____

III. <u>Field Inspection</u>
   (The following details obtained at the site are annotated on the Drainage Survey)

1. Survey appears correct: - ( ) Yes, ( ) Apparent errors are: _____
   which were resolved by: _____

2. Flooding Apparent? - ( ) No, ( ) Yes, HW marks obtained, ( ) Yes but HW marks not obtained because _____

3. Do all Floods Reach Site? - ( ) Yes, ( ) No and details obtained, ( ) No but details not obtained because _____

4. Do Floodwaters Enter Irrigation Ditch ? - ( ) N/A, ( ) No, ( ) Yes and details obtained, ( ) Yes but details not obtained because _____

5. Hydrologic Ch. Geom. obtained? - ( ) Yes, ( ) No because _____

6. Channel Unstable? - ( ) No , ( ) Yes because of ( ) headcutting observed and ( ) amount/location obtained, ( ) bank caving, ( ) braiding, ( ) increased meander activity, ( ) Other _____

7. Structure Scour in Evidence? - ( ) No, ( ) Minor, ( ) Yes and ( ) obtained bed/bank samples and ( ) noted any flow alignment problems, ( )Yes and ( ) bed/bank material samples not obtained and ( ) flow alignment not noted because _____

8. Irrigation facility? - ( ) No, ( ) Yes and several water right related depths obtained, ( ) Yes and No water right related depths obtained because _____
_____

9. Manning's "n" obtained? - ( ) Yes, ( ) No because _____
_____

10. Property damage due to BW? - ( ) No ( ) Yes and elevation/property type checked, ( ) Yes but elevation/property type not obtained because _____
_____

11. Environmental Hazards Present? - ( ) No, ( ) Yes, details obtained, ( ) Yes, details not obtained because _____
_____

12. Ground Photos Taken? - ( ) Upstream floodplain and all property, ( ) Downstream floodplain and all property, ( ) Site looking from downstream, ( ) Site looking from upstream, ( ) Channel Material w/scale, ( ) Evidence of channel instability, ( ) Evidence of scour, ( ) Existing structure inlet/outlet, ( ) Other__
_____

13. Effective drainage area visually verified? ( ) Yes, ( ) No because _____
_____

IV. <u>Post Inspection Survey Annotation</u>

1. Section II Findings annotated on survey? - ( ) Yes, ( ) No and see section attached (attach typed explanation by site station and site name, and check list section and number).

2. Survey Originals and check lists forwarded to ( ) Municipality's Roadway Unit, _____ ea, site's, and the ( ) Public Works or Engineering, _ _____ea. site's for hydraulic design.

_____/s/_____
(Designer Making Inspection)

7
Urban Hydrology

7.1 Introduction

Hydrology is generally defined as a science dealing with the interrelationship between water on and under the Earth and in the atmosphere. For the purpose of this chapter, hydrology will deal with estimating flood magnitudes, volumes, and time distributions as the result of precipitation. In the design of stormwater management facilities, floods are usually considered in terms of peak runoff or discharge in cubic feet per second (cfs) and hydrographs as discharge per time. For structures that are designed to control volume of runoff, like detention storage facilities, or where flood routing through culverts is used, then the entire discharge hydrograph will be of interest.

There are a large number of hydrologic procedures covering topics such as snow melt, precipitation, water supply, infiltration, drought, evapotranspiration, etc. In basic urban hydrology, procedures commonly used are reduced to:

- Precipitation and losses
- Peak flow determination
- Flow hydrograph or volume determination
- Hydrograph routing and combining
- Storage routing

Chapter 10 and Chapter 11 cover hydrograph routing and storage routine, respectively. This chapter covers the other three major topics.

The analysis of the peak rate of runoff, volume of runoff, and time distribution of flow is fundamental to the design of drainage facilities. Errors in the estimates will result in a structure that is undersized and causes stormwater management problems, or oversized and costs more than necessary. On the other hand, it must be realized that any hydrologic analysis is only an approximation. The relationship between the amount of precipitation on a drainage basin and the amount of runoff from the basin is complex, and too little data are available on the factors influencing the rural and urban rainfall–runoff relationship to expect exact solutions.

The type and source of information available for hydrologic analysis will vary from site to site, and it is the responsibility of the designer to determine what information is available and applicable to a particular analysis.

Hydrologic analysis methods vary from simple formulas (e.g., Rational Method, Berkely Ziegler) to unit hydrograph methods (e.g., SCS Method, Santa Barbara Method) to regression equations based on local data (e.g., U. S. Geological Survey equations) to complex continuous simulation computer models (e.g., Hydrocomp Simulation Model, EPA SWMM Model). Each of these methods has its place in municipal engineering for the analysis and design of different stormwater management facilities. No one method is optimum for all analysis and design, but several of the methods have severe limitations that should be accounted for. Recently, some municipalities and states passed ordinances and legislation that further restricts the use of some hydrologic methods.

The purpose of this chapter is to present several hydrologic methods that have been widely used in municipalities throughout the United States. Other methods may also be applicable, and methods specifically developed for particular locations will, in most cases, produce better results than general methods that have not been verified for local use. Because the number of hydrologic methods available is much too large to be covered in one chapter, the interested reader is encouraged to seek other sources for documentation of other methods and the availability of computer programs to assist in using these methods.

7.2 Concept Definitions

Following are definitions of concepts that will be important for understanding different hydrologic methods. These concepts will be used throughout the remainder of this book in dealing with different aspects of hydrologic studies.

Antecedent moisture condition Antecedent soil moisture conditions are the soil moisture conditions of the watershed at the beginning of a storm. These conditions affect the volume of runoff generated by a particular storm event. Notably, they affect the peak discharge only in the lower range of flood magnitudes — say, below about the ten-year event threshold. Antecedent moisture has a rapidly decreasing influence on runoff, as the frequency of a flood event increases.

Depression storage frequency Depression storage is the natural depressions within a watershed that store runoff. Generally, after the depression storage is filled, runoff will commence. Frequency is the number of times a flood of a given magnitude can be expected to occur on an average over a long period of time. Frequency analysis is then the estimation of peak discharges for various recurrence intervals. Another way to express frequency is with probability. Probability analysis seeks to define the flood flow with a probability of being equaled or exceeded in any year. For example, a 25-year flood has the probability of occurrence of once every 25 years on the average, or a 4% chance of occurrence in any given year.

Hydraulic roughness Hydraulic roughness is a composite of the physical characteristics that influence the flow of water across the Earth's surface, whether natural or channelized. It affects the time response of a watershed and drainage channel, as well as the channel storage characteristics.

Hydrograph The hydrograph is a graph of the time distribution of runoff from a watershed.

Hyetograph The hyetograph is a graph of the time distribution of rainfall over a watershed.

Urban Hydrology

Infiltration Infiltration is a complex process of allowing runoff to penetrate the ground surface and flow through the upper soil surface. The infiltration curve is a graph of the time distribution at which this occurs.

Interception Storage of rainfall on foliage and other intercepting surfaces during a rainfall event is called interception storage.

Lag time The lag time is defined as the time from the centroid of the excess rainfall to the peak of the runoff hydrograph.

Peak discharge The peak discharge, sometimes called peak flow, is the maximum rate of flow of water passing a given point during or after a rainfall event or snowmelt.

Rainfall excess After interception, depression storage, and infiltration have been satisfied, if there is excess water available to produce runoff, this is the rainfall excess.

Stage The stage of a channel is the elevation of the water surface above some elevation datum.

Time of concentration The time of concentration is the time required for water to flow from the most hydraulically remote point of the basin to the location being analyzed. Thus, the time of concentration is the maximum time for water to travel through the watershed, which is not always the maximum distance from the outlet, to any point in the watershed.

Unit hydrograph A unit hydrograph is the direct runoff hydrograph resulting from a rainfall event that has a specific temporal and spatial distribution and that lasts for a specific duration of time (thus, there could be a 5-, 10-, 15-minute, etc., unit hydrograph for the same drainage area). The ordinates of the unit hydrograph are such that the volume of direct runoff represented by the area under the hydrograph is equal to one inch of runoff from the drainage area.

7.3 Hydrologic Design Policies

For all hydrologic analysis, the following factors should be evaluated and included when they will have a significant effect on the final results.

Drainage Basin Characteristics

Size and shape
Slope
Groundcover and land use
Geology
Soil types
Surface infiltration
Ponding and storage
Watershed development potential
Antecedent moisture design conditions

History of urban development
Other characteristics

Stream Channel and Conveyance System Characteristics

Geometry and configuration
Natural controls
Artificial controls
Channel modifications
Aggradation–degradation
Debris
Manning's n
Slope
Connectedness of impervious areas
Age, condition, and type of structures
Other characteristics

Floodplain Characteristics

Slope
Vegetation
Alignment
Storage
Location of structures and development
Obstructions to flow
Other characteristics

Meteorological Characteristics

Time rate, geographical distribution, and amounts of precipitation
Historical flood heights
Other characteristics

Many hydrologic methods are available. The methods included in this chapter are recommended for general municipal use, and the circumstances for their use are included in the description of each method. Engineers and hydrologists will often argue the virtues of one hydrologic method over another, but most methods will produce acceptable results if used within the limitations of the method and calibrated or verified for local use. Calibration is probably much more important than the method selected and will be further discussed later in this chapter. This is true because most methods have factors or parameters that can be adjusted within a wide range to reflect special situations or conditions. Sound engineering judgment is still necessary, because urban hydrology is still as much art as science.

Municipalities should accept only those hydrologic methods that have been proven to produce acceptable results for their area, and complete source documentation is submitted or available for approval. Consistent application of similar methods is important.

The methods in this chapter were selected for use based on several considerations, including the following:

- Verification of their accuracy in duplicating local hydrologic estimates in municipalities throughout the United States and for a range of design storms
- Availability of equations, nomographs, and public-domain computer programs
- Use and familiarity with the methods by local municipalities and consulting engineers
- Applicability to a wide range of geographical locations with different hydrologic characteristics

It should be remembered that no method is universally applicable, and that attempting to apply it in situations in which its basic assumptions or underlying bases are violated will compromise every method's accuracy. Recognition of circumstances that may render a particular method inappropriate is important. It is also important to make an estimate of the relative accuracy of the predictions of the methods used, through sensitivity analysis or a less formal recognition of approximations and judgments employed. Urban drainage calculations are often performed with little or no true flow data; use hydrology, which is a rough approximation; employ parameters with much leeway for judgment; and use assumptions on the stability of land use and flow conveyance conditions that are almost always violated.

7.4 Design Frequency and Risk

Risk

Because it is not economically or often physically feasible to design stormwater management structures for the maximum runoff a watershed is capable of producing, a design frequency is usually established. This use of a frequency implies an assumed risk of the failure of the structure, or it is overwhelming with an event of greater magnitude than that used for design. This area of consideration is often called risk-based analysis. While use of risk-based analysis is not common in urban stormwater design, knowledge of the risks and uncertainties involved is appropriate for any type of design and analysis.

The frequency with which a given flood can be expected to occur is the reciprocal of the probability or chance that the flood will be equaled or exceeded in a given year. The probability of occurrence of an event (say a flood of magnitude equal to or greater than $X1$) is expressed as:

$$P\{X1\} = N1/N \qquad (7.1)$$

where $N1$ is the number of occurrences of a flood greater than or equal to a certain event, and N is the total population of observations. P is always between zero and one. The return period is then the reciprocal:

$$Tr = 1/P\{X1\} \tag{7.2}$$

For example, the probability of a 5-year flood being equaled or exceeded in any one year has a probability of occurrence of 20% (0.2). Alternately, it would occur, on the average, once every five years for any given stream. In the same way, a 5-year rainfall would tend to occur once every five years for any given spot on the ground. Across a municipality, there may be many 5-year or greater storms each year, because each spot in the city would tend to experience a 5-year or greater storm every five years. The probability of nonoccurrence of X1 is:

$$P\{not\ X1\} = 1-P\{X1\} = 1-1/Tr \tag{7.3}$$

The probability of nonoccurrence of an event X1 in n years is:

$$P\{not\ X1^n\} = (1-1/Tr)^n \tag{7.4}$$

Risk (R) is defined as the probability that an event X1 will occur in n years and is given as:

$$R = 1-(1-1/Tr)^n \tag{7.5}$$

Or rearranged:

$$Tr = 1/[1-(1-R)^{1/n}] \tag{7.6}$$

These equations provide the designer an estimate of the reasonableness of the design parameters. For example, what is the risk that the capacity of a culvert designed for a 25-year flood will be equaled or exceeded in the first ten years? Using Equation (7.5), the risk is calculated to be about 33% as:

$$R = 1-(1-1/Tr)^n = 1-(1-1/25)^{10} = 0.335$$

Or, what should be the design flood return period (recurrence interval) to reduce the risk of overtopping of a roadway to 5% within the 50-year design life of a highway culvert? Using Equation (7.6), the return period is:

$$Tr = 1/[1-(1-R)^{1/n}] = 1/[1-(1-.05)^{1/50}] = 975 \text{ years}$$

A flood with a recurrence interval of 975 years would be considered excessive for normal design purposes and may still not protect the designer from liability, even if it could be accurately estimated (Lewis, 1992).

Risk-Based Analysis

All estimates of design and economic parameters have inherent error and variability due to the complex physical, social, and economic situations for even the most simple hydrologic and hydraulic analyses and projects. The designer typically develops a "most likely" estimate of a certain design parameter (for example, 10-year storm rainfall or Manning's roughness coefficient) and then uses sensitivity analysis to test the impact of variability in the parameter estimate on the final solution. However, there is nothing but a gut feeling " on the actual likelihood of the parameter being correct or its actual likely range. Engineering judgment is relied on for all such answers.

Risk-based analysis was developed to attempt to estimate the actual amount of risk and uncertainty inherent in any design and to treat each key variable and parameter in probabilistic terms. It can seek specifically to quantify risk and uncertainty, taking into account the vagaries of the physical environment, the limitations of understanding of physical phenomena involved, cost and willingness to pay, institutional rigidities, and social or political risk (Shabman, 1985). It can be used to quantify the likelihood of physical and economic success of a project (COE, 1993). Risk-based design is often taking the place of more conventional analysis, especially in larger federal flood-control projects.

Risk is an estimated chance of an occurrence (such as flooding). Uncertainty is the error associated with key parameters or functions used in computing economic or reliability estimates. For flood-control projects, economic variables of interest include damage relationships, structure and contents values, structure elevations and types, and flood warning and evacuation times and effectiveness. Uncertainty arises due to errors in sampling, measurement, estimation and forecasting, and modeling. For hydrologic and hydraulic analysis, stage and discharge are of prime importance. Uncertainty in discharge is due to short or nonexistent flood records, inaccurate rainfall–runoff modeling, and inaccuracy in known flood flow regulation, where it exists. Stage uncertainty comes from errors and unknowns in roughness, geometry, debris accumulation, ice effects, sediment effects, and others (COE, 1993).

An example of the difference in design approach is the determination of design height for a levee. Normal design approaches would be to determine the design flood, develop a backwater profile, design the levee to contain the design flood, and add freeboard. Enlightened design would also plan to account for storms larger than the design storm by controlling the location of overtopping to reduce damage to the maximum extent practicable. Risk-based design takes a different approach. The end goal of risk-based design is to have a levee with a known reliability and performance. For example, it may be designed to contain the 100-year flood with a 95% reliability and the 500-year flood with a 50% reliability. It also seeks to manage events exceeding the design. The levees may look the same, but the design for the risk-based designed levee included a determination of the chances of failure, even for floods less than the design flood. This is not always comforting to local citizens, but it is more realistic.

The difference is represented in design by adding a probability density function around each estimated parameter. For example, each point on a simple stage discharge curve has a bell-shaped or other shaped probability density function built in. The most likely stage for a discharge of, say, 100 cfs is elevation 520. But, there is a 20% chance the elevation will be above 523, and so on.

Detailed discussion of risk-based analysis beyond this introduction is beyond the scope of this book. Readers are encouraged to explore the emerging literature on the subject or obtain information from the Army Corps of Engineers, Hydrologic Engineering Center in Davis, California.

Design Frequency

A drainage facility should be designed to accommodate a discharge with a given return period(s). The design should be such that the backwater (termed the "headwater") caused by the structure for the design storm does not:

- Increase the flood hazard significantly for property
- Dangerously overtop a street or highway
- Exceed a certain depth on a street or highway embankment

Different design storm frequencies or return periods are important for different types of studies or structure sizes and uses:

- Large hydraulic structures are often designed for infrequent floods such as some fraction of the Probable Maximum Flood (PMF) when considering structural integrity, and slightly less remote events when designing outlet structures (100-year or 500-year events).
- Major urban conveyance systems are often designed for the 100-year flood, particularly if they fall under the purview of the National Flood Insurance program administered by FEMA. The range of design frequencies is generally from the 25-year to 100-year events. Studies on flood damages have shown that if the 25-year and more frequent storms are controlled, the majority of the damages in urban flooding will be avoided (Johnson, 1985). The practice for some major structures, such as highway bridges, is to design to pass the 50-year peak flow. Some agencies make allowance for debris buildup partial blockage (St. Paul District, 1985). Often, channels are designed for a flow much less than the 100-year event but should pass the 100-year event in the floodplain without structural damage to adjacent properties.
- Smaller urban conveyance systems, storm drains, and feeder streams (often termed the *convenience system*) are designed for a range of flows generally centering on the 5-year to 10-year storms. Although many municipalities allow for gutter and inlet design for design storms less than the 5-year storm, benefit cost analysis has shown that the 5-year storm is often the most cost-effective standard to use for storm drain design (Casamayor and Rodgers, 1980).
- Concern for erosion control in channels usually centers around the 2-year to 5-year storms. This frequency of storm is often called the *channel-forming* or *dominant* discharge. Most stable streams form naturally to pass this range of floods within their banks. Often, partial duration frequency flow values are important here, because the number of storms is also important (Chow, 1964).
- Pollution calculations are concerned with frequent storms on the order of fractions of a year. This is because these storms are numerous and, cumulatively, convey the major portion of pollution to and through streams. Again, partial duration values are important here, because the number of storms is important.

Following are some design frequency ranges that are used by many municipalities across the United States. These frequencies should be increased or decreased to account for local conditions, specific site conditions, importance of the stormwater management facility, and other factors important to the proposed facilities and adjacent development.

Cross Drainage

Cross-drainage facilities transport storm runoff under roadways. For many municipalities, such drainage facilities are designed to accommodate a 25-year flood, though designs from the 10-year through the 100-year are not uncommon. Assuming a 25-year storm is chosen, the peak flows and hydrographs used for cross-drainage design should be based on fully developed land use conditions. Thus, the cross drainage should be designed so that the roadway is not overtopped for all floods that are equal to or less than the 25-year frequency event. Thus, if a storm drainage system crosses under a roadway, then a 25-year flood must be routed through the system to show that the roadway will not be overtopped by this event. The excess storm runoff from events larger than the 25-year storm may be allowed to inundate the roadway or may be stored in areas other than on the roadway until the drainage system can accommodate the additional runoff. The final design should be checked using the 100-year flood to be sure structures are not flooded or increased damage does not occur to the highway or adjacent property for this design event.

Storm Drains

Storm drains are often designed to accommodate a 10-year flood. Frequency ranges from as low as the 5-year storm to as high as the 25-year storm are common. The design should be such that the storm runoff does not:

- Increase the flood hazard significantly for property
- Encroach onto the street or highway so as to cause a significant hazard
- Limit traffic, emerging vehicles, or pedestrian movements to an unreasonable extent

Based on these design criteria, a design involving temporary street or road inundation is acceptable practice for flood events greater than the design event but not for floods that are equal to or less than the design event.

Inlets

Inlets to storm drain systems should be designed for a 5- or 10-year flood depending on the roadway type and other factors at the site. Some cities allow for a much more frequent storm design for inlets (such as the 2-year storm) while sizing the storm drain, which carries the flow from the inlets for the 5- or 10-year storm. See Chapter 8 for more details on such designs.

Detention and Retention Storage Facilities

All storage facilities should be designed to provide sufficient storage and release rates to accommodate a range of design storm events. The 2- and 10-year design storm events are common and cover the range of normal flooding, although some areas are using the 25-year storm as the design storm. Larger regional ponds may be designed to accommodate events up to the 100-year storm. The design should be such that the storm runoff does not:

- Increase the flood hazard significantly for adjacent, upstream, or downstream property as defined in the drainage ordinance
- Cause any safety hazards associated with the facility

Emergency spillway facilities should be provided to accommodate the 100-year storm. The final design should be checked to be sure that the downstream flood peaks for the storage discharge hydrographs have not increased for the 2- and 10-year floods. Dam safety requirements may dictate emergency spillway size.

Review Frequency

After sizing a stormwater management facility using a design event, a review frequency flood event should be used, such as the 100-year flood. This is done to ensure that there are no unexpected flood hazards inherent in the proposed facilities. In some cases, a flood event larger than the 100-year flood should be used to ensure the safety of the drainage structure and downstream development.

7.5 Hydrologic Procedure Selection

In performing hydrologic calculations, it is important to have a systematic approach and to know the following:

- The types and accuracy of answers needed
- The available methodologies to provide necessary information
- The available data and information
- The assumptions inherent in and the limitations on selected methods

A general approach for hydrologic calculations could be outlined as follows:

Step 1: Determine requirements (e.g., peak flow, hydrograph, etc.) and accuracy and select a design procedure.
Step 2: Collect necessary data.
Step 3: Identify design storm criteria and develop the design storm or rainfall.
Step 4: Compute time of concentration or other lag times required.
Step 5: Determine rainfall excess if appropriate to the methodology.
Step 6: Compute peak rate of runoff or flood hydrograph.
Step 7: Perform detention storage or channel routing, if appropriate.
Step 8: Estimate or test sensitivity to engineering judgments and data error ranges. Adjust approach as appropriate.
Step 9: Document all estimates and calculations in detail.

Streamflow measurements for determining a flood frequency relationship at a site are usually unavailable; in such cases, it is accepted practice to estimate peak runoff rates and hydrographs using statistical, empirical, or physically based formula methods. In general, results from using several methods should be compared, not averaged. If hydrologic procedures have been developed for the local area, they will probably produce the best results and should be used. The accuracy of general hydrologic procedures can be greatly increased for a local area if they are calibrated for use within a specific area. The next section in this chapter discusses calibration.

A consideration of peak runoff rates for design conditions is sometimes adequate for conveyance systems such as storm drains or open channels. However, if the design must include flood routing (e.g., storage basins or complex conveyance networks), a flood hydrograph is required. Many municipal ordinances now require that discharges from stormwater management facilities be evaluated through a portion of the downstream drainage system. For such an analysis, a hydrograph and channel routing are necessary. Although the development of runoff hydrographs (typically more complex than estimating peak runoff rates) is usually accomplished using computer programs, some methods are adaptable to nomographs or other desktop procedures.

Following are some of the commonly used categories or types of methods for generating peak flow estimates and hydrographs for rural and urban watersheds. Some of these methods are available in the computer program HYDRAIN Drainage Design System and HEC-1. The hydrologic model within the HYDRAIN system is called HYDRO.

Analysis of Stream Gauge Data

Where reliable stream gauge data are available, they can be used to develop peak discharges and (less often) hydrographs. Data may be available from the regional U.S. Geological Survey (USGS), Army Corps of Engineers, the local municipality, or other agencies. These data can be used to generate accurate hydrologic estimates.

A large number of types of analysis of gauge data are available in the literature. For peak flow estimates, Log Pearson Type III frequency distribution analysis is considered to be one of the most reliable methods for estimating flood frequency relationships, though often data manipulation and adjustment are necessary (Water Resources Council, 1981).

Rules of thumb have been developed by the USGS relating the frequency estimate required and the length of record desirable as follows:

Design Return Period (Years)	Desired Period of Record (Years)
10	10
25	15
50	20
100	25

Unfortunately, many urban streams are not gauged, and such direct discharge data are seldom available. In these cases, data may be available for similar areas and can be transferred to the watershed in question through one of several methods of regional frequency analysis. Such methods are found in standard hydrology textbooks. Alternately, the gauged watershed could be modeled and the final modeling methodology transferred to the new watershed.

If urban development has been occurring in the watershed throughout the gauged period, allowance must be made to adjust the record for such development. Many methods can be found in the literature for such adjustments (Sarma, Delleur, and Rao, 1969; Dunne and Leopold, 1978; Rantz, 1971; Gundlach, 1978). While data on actual storms are scarce, information indicates that for every 1% increase in impervious area in an urban setting, the peak flow increases between 1 and 1.25%. The former value is for more infrequent storms in the 100-year return period range and the latter for return periods in the range of the 2-year storm. Development has less impact on the larger storms due to the fact that, for these storms, the ground is normally saturated during the storm event, and the impact of impervious area is lessened.

If urban regression equations are available for the areas in question, which have impervious area as one of the parameters, the impact of development can be explicitly determined by differentiating the regression equation with respect to impervious area. This gives the change in peak flow for a change in impervious area. It may also be necessary

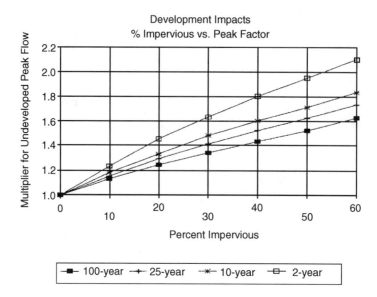

FIGURE 7-1
Impacts of urban development on peak flow. (From McCuen, R., Introduction to Chapter 9, ASCE TC on Hydrology, 1993.)n Hydrology, 1993. Reproduced by permission of ASCE.)

to consider a change in lag time or time of concentration in this analysis, because these times normally shorten considerably with increases in impervious area.

Heggen (1982) provides a nomographic method for using the SCS methodology for estimating changes in runoff volume for known changes in the SCS curve number. Based on an analysis of a number of methods for determining the impacts of development, a general relationship can be derived as depicted in Figure 7-1 (McCuen, 1993). Using this figure, the change in peak flow can be estimated in relation to undeveloped flows.

Regression Equations

If specific gauge data are not available, peak flow, and sometimes whole hydrographs, can be calculated by using regression equations developed empirically for specific geographic regions. The equations are normally in the form of a log-log formula, where the dependent variable would be the peak flow or flow volume for a given frequency, and the independent variables may be parameters such as area, impervious area, slope, channel geometry, and other meteorological, physical, or site-specific data. Regression equations are available in some areas for urban and rural streams, large and small. However, close attention should be paid to limits on data parameters and geographical applicability. The average error of estimate should be assessed as well as sensitivity to specific parameters. Regression results are often used for assistance in calibration as a reality check, or for preliminary estimating in master planning.

A complete, though simulated, flood hydrograph can be computed using regression methods by applying lag time, obtained from the proper regression equation, and peak discharge of a specific recurrence interval, to a dimensionless hydrograph. The coordinates of the runoff hydrograph can be computed by multiplying lag time by the time ratios and peak discharge by the discharge ratios developed for different areas. Such a method has been developed and is being used by some municipalities in Georgia (Inman, 1986), and has been developed or verified by the USGS for other locations. Care must be taken that the definitions for lag time are equivalent, as different agencies sometimes use different definitions.

Rational Method

A number of theoretically derived peak flow methods exist in the literature. Most provide a physically reasonable mathematical relationship between key parameters and peak flow, much like a regression equation. However, unlike regression equations, the methods are meant to be used in general application across a wide range of geography and climate. The parameters are then estimated or calculated using local knowledge, standard values, and engineering judgment. Great errors can occur when these methods are applied in ways or situations contrary to the assumptions inherent in the equations, or when parameters are incorrectly estimated.

The well-known rational method is one such general-use methodology. It provides peak runoff rates for small urban and rural watersheds, often limited to 100 acres or much less, but is best suited to urban storm drain systems. The rational method should be used with caution if the time of concentration exceeds 30 min. Rainfall is a necessary input. Many municipalities will not allow the use of the rational method to estimate hydrographs or for detention storage designs. The rational method will be discussed in detail later in this chapter.

Unit Hydrograph Methods

Most standard hydrology books contain a derivation of the linear theory of hydrologic systems culminating in the unit hydrograph theory. A unit hydrograph is essentially the runoff, distributed correctly in time, from a unit of excess rainfall falling in a certain predetermined time period and applied uniformly over a watershed or subbasin. Thus, the 5-min unit hydrograph is the runoff hydrograph from, say, 1 in of excess rainfall falling uniformly over a 5-min interval. Determination of an actual runoff hydrograph from a storm event (termed *convolution*) is accomplished when measured 5-min blocks of rainfall (termed a rainfall hyetograph) are multiplied by the ordinates of this unit hydrograph, shifted in time by 5-min steps and added together. Use of unit hydrographs involves the determination of excess rainfall through the use of some sort of initial loss and infiltration methodology.

A number of unit hydrograph types are available in the literature. Two of the most common are described briefly here.

Synder's Unit Hydrograph

This method was developed (Snyder, 1938) for the Appalachian area watersheds ranging from 10 to 10,000 square miles. It has been applied to watersheds across the United States by the Corps of Engineers and is one of the methods found in the popular HEC-1 program. It provides a means of generating a synthetic unit hydrograph. It relies on the calculation of lag time and peak flows through two relationships involving area, length measurements, and estimated parameters. Because it does not define the total hydrograph shape, other relationships must be used with the Snyder method for such a definition. For example, HEC-1 uses the Clark relationship for such a definition along with empirically developed estimates of the hydrograph widths at the 50 and 75% of peak levels (HEC, 1990). Details of the method can be found in Chow (1964).

SCS Synthetic Unit Hydrograph

The Soil Conservation Service developed a family of hydrologic procedures, one of which is a synthetic unit hydrograph procedure. It has been widely used for developing rural and urban hydrographs. The unit hydrograph used by the SCS method is based upon an analysis of a large number of natural unit hydrographs from a broad cross section of geographic locations and hydrologic regions. Rainfall is a necessary input. This method is discussed in detail later in this chapter.

Continuous Simulation Models

All of the methods described above are event models, which use rainfall as the input and flood peaks and hydrographs as the output. This is because most municipal stormwater management facilities are designed for a specified flood event, and rainfall data are readily available throughout the country. A consequence of this approach is that the complex interaction between rainfall and resulting storm runoff must be estimated, and only one or two flood events are used in the design. In contrast, continuous simulation models such as the EPA Stormwater Management Model (SWMM) or the Hydrocomp Simulation Model attempt to represent the entire hydrologic system on the computer so as to simulate the natural system. In this way, the model simulates the runoff process including interception, infiltration, overland flow, channel flow, etc. This simulation is over a long period of time and is continuous so that both flood events and low-flow events are simulated. If accurately simulated, the models will provide information on particular aspects of the runoff process, such as antecedent moisture, which is important when estimating flood peaks and hydrographs.

Although these models have been available since the early 1970s, their application for municipal use has been limited. This probably results from the large amount of data needed, and, until recently, the difficulty in using the models. Today, user-friendly continuous simulation models are available, and in time, their use for municipal stormwater management analysis and design will increase. The use of these models is especially useful for water-quality simulation, storm movement analysis, and complex designs and analyses. Due to the complexity of these models and their present limited application, no further discussion of continuous simulation models will be presented in this book. The interested reader should consult the extensive literature related to these models and documentation that is readily available from the model developers and software distributors.

Summary

When sufficient streamflow data are available, the findings from a Log Pearson III method can be used for hydrological estimates. Adjustments or regionalization must often be performed to make the data useful.

If available, regression equations and regression analysis are an acceptable method of estimating peak flows and hydrographs for rural and urban watersheds. These equations and analysis have been shown to be accurate, reliable, and easy to use as well as to provide consistent findings when applied by different hydraulic engineers (Newton and Herin, 1982). Regression equations are used to relate such things as the peak flow or some other flood characteristic at a specified recurrence interval to the watershed's physiographic, hydrologic, and meteorological characteristics.

The major problem with using stream gauge data or regression equations is their lack of availability within many municipalities. Also, where available, limitations of using the equations related to the scope of the database used for regression analysis limits the use in many locations. It is common that these equations can only be used for large or undeveloped watersheds, while most municipal hydrologic studies deal with small urbanized watersheds and site developments. If regression equations are available for specific municipalities, the associated documentation should be used to determine their applicability and how to apply the equations.

The following discussion will focus on three of the most popular hydrologic methods used within most municipalities: the Rational Method, the SCS Unit Hydrograph Method, and the Santa Barbara Unit Hydrograph Method. Other unit hydrograph methods are available, such as the Snyder Synthetic Unit Hydrograph, and, for some areas, local unit hydrographs have been developed, such as the Colorado Unit Hydrograph Method. Those interested in the Snyder Synthetic Unit Hydrograph should consult Chow (1964) or the AASHTO Model Drainage Manual, and for those municipalities where unit hydrograph methods have been developed, associated documentation should be used.

7.6 Calibration

Calibration is a process of varying the parameters or coefficients of a hydrologic method so that it will estimate peak discharges and hydrographs consistent with local rainfall and streamflow data. Most hydrologic procedures used for stormwater management facility design contain general equations that have been developed for large geographic areas. These procedures cannot be expected to take into account local hydrologic conditions, and, as a result, unless calibrated, they may not produce acceptable estimates for analysis and design.

Figure 7-2 shows a hydrograph resulting from flow data as compared to a hydrograph resulting from using a noncalibrated and calibrated hydrologic procedure. It can be seen that the calibrated hydrograph, although not exactly duplicating the hydrograph from streamflow data, is a much better representation of the streamflow hydrograph than the noncalibrated hydrograph.

The accuracy of the hydrologic estimates will have a major effect on the design of stormwater management facilities. Although it might be argued that one hydrologic

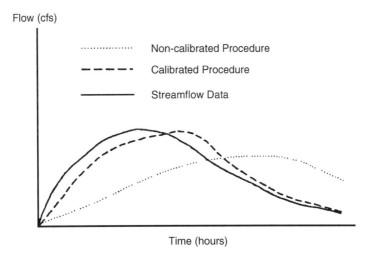

FIGURE 7-2
Hydrograph calibration.

procedure is more accurate than another, practice has shown that all of the methods discussed in this chapter can, if calibrated and not used beyond their intended purposes, produce acceptable results consistent with observed or measured events. What should be emphasized is the need to calibrate the method for local and site-specific conditions. This calibration process can result in much more accurate and consistent estimates of peak flows and hydrographs.

The calibration process can vary depending on the data or information available for a local area. Following are some general steps in the calibration process:

1. If streamflow data are available for an area, the hydrologic procedures can be calibrated to these data. The process would involve generating peak discharges and hydrographs for different input conditions (e.g., slope, area, antecedent soil moisture conditions) and comparing these results to the gauged data. Changes in the procedures would then be made to improve the estimated values as compared to the measured values. It is typically a good idea to first match flow volumes through adjustment of losses, then match peaks and timing. Also, it is best to change one parameter at a time and to measure sensitivity at each change.

   Note: When changing hydrologic procedures, care should be exercised to be sure that the basic theory of the hydrologic procedures is not violated by changes made during the calibration process.

2. After changing the variables or parameters in the hydrologic procedure, the results should be checked against another similar gauged stream or another portion of the streamflow data that were not used for calibration.

3. If some local agency developed procedures or equations (for example, a local regression equation for lag time or rainfall losses) for an area based on streamflow data, general hydrologic procedures can be calibrated to these local procedures. In this way, the general hydrologic procedures can be used for a greater range of conditions (e.g., land uses, watershed size, slope).

4. The calibration process should be undertaken only by personnel highly qualified in hydrologic procedures and design. There are many pitfalls. For example, the use of too long a time step duration may give an artificially low peak flow, simply because two time adjacent steps straddled the peak and did not capture it.

5. Should it be necessary to use unreasonable values for variables in order for the model to produce reasonable results, then the model should be considered suspect and its use carefully considered (e.g., having to use terrain variables that are obviously dissimilar to the geographic area in order to calibrate to measured discharges or hydrographs). However, the skilled modeler can often modify parameters to mimic conditions in nature not available in the model. For example, for small watersheds where absolute timing is not critical, artificially extending the lag time can mimic inadvertent storage without resorting to a series of minor detention routings behind each undersized culvert within a subbasin.

7.7 Precipitation and Losses

Stormwater management facilities are normally designed based on some flood frequency. Floods and their corresponding frequencies are not often available. Therefore, most synthetic

TABLE 7-1

Empirical Factors for Converting
Partial–Duration Series to Annual Series

Return Period	Conversion Factor
2-year	0.88
5-year	0.96
10-year	0.99

hydrologic procedures use rainfall and rainfall frequency as the basic input, and derive flood estimates. To do this, it is commonly assumed that, for example, the 10-year rainfall will produce the 10-year flood. Depending on antecedent soil moisture conditions, and other hydrologic parameters, for a specific storm, this may or may not be true.

Other methods, mostly regression-type equations and gauge analysis, rely on direct estimation of flood magnitudes without resorting to the use of rainfall.

Intensity–Duration–Frequency Curves

Rainfall data are available for most geographic areas. Such data can be obtained from the National Weather Service publications or from local sources, if available. The two most common sources of such information for use in the 37 states east of the 105th meridian for durations from 5 min to 24 h and return periods from 1 to 100 years are Technical Paper 40 (Hershfield, 1961) and NWS HYDRO-35 (NOAA, 1977). Other technical papers and local or regional reports, as well as digital data and information, are available from the National Oceanic and Atmospheric Administration, National Climatic Data Center, Asheville, North Carolina (phone: 704-259-0682), for other areas of the country and less frequent events. Information on the Probable Maximum Precipitation required for some dam safety studies is found in HMR-51 and HMR-52 (NOAA, 1978, 1982).

Because, TP40 and HYDRO-35 are now quite dated, it may be advisable, when generating a new rainfall intensity-duration-frequency (IDF) curve for an area, to obtain the latest rainfall information from NOAA in digital form and develop new IDF curves. From these publications or new data, IDF curves can be developed for the commonly used design frequencies. These IDF curves then become the basic input for the hydrologic procedures discussed in this chapter.

The procedure for developing an IDF curve for a particular return period from TP40 and HYDRO-35 is as follows:

Step 1: For periods 1 h and greater in duration, obtain point rainfall information by interpolating between curves from the maps in TP40. For periods 1 h and less, obtain information in a similar way from HYDRO-35.

Step 2: Because the values given are for partial duration series, they must be converted for most uses to annual series data by multiplying the map-derived information by factors obtained from Table 7-1.

Step 3: Plot family of curves with duration on the horizontal axis and intensity on the vertical axis and perform minor manual smoothing to remove irregularities.

IDF data for any city can be easily fit by an equation of the following form:

$$i = a/(t+b)^c \tag{7.7}$$

TABLE 7-2

Rainfall Intensity Data — Fayetteville, North Carolina

Storm Duration		Rainfall Intensity (in/h)					
Hours	Minutes	2	5	10	25	50	100
0	5	5.17	6.19	6.93	8.02	8.87	9.72
	10	4.33	5.26	5.94	6.91	7.68	8.44
	15	3.70	4.51	5.10	5.95	6.62	7.28
	30	2.58	3.35	3.87	4.62	5.20	5.78
1		1.67	2.24	2.63	3.17	3.58	4.00
2		0.93	1.25	1.47	1.78	2.01	2.25
3		0.68	0.92	1.08	1.31	1.49	1.66
6		0.42	0.57	0.68	0.82	0.93	1.04
12		0.24	0.34	0.40	0.48	0.55	0.61
24		0.14	0.19	0.23	0.28	0.32	0.35

where i = the rainfall intensity (in/hr); t = the duration (min); and a, b, and c are curve-fitting parameters.

By taking the log of Equation (7.7), the equation becomes:

$$\log i = \log a - c \log (t+b) \tag{7.7a}$$

This is the form of the equation for a straight line with c being the slope and $\log a$ being the y-intercept. By assuming trial values for b, a straight line can be fit through plots of $\log (t + b)$ versus $\log i$. This can be done automatically using the linear regression fitting functions found in most common spreadsheet software.

Often, c is set equal to 1, simplifying the equation without losing great accuracy. Adding the term Td to the numerator, where T is the recurrence interval and d is a curve fitting parameter, can approximate all IDF curves for a particular location.

Table 7-2 shows an example of rainfall intensity data available for the Fayetteville, North Carolina, area. Similar data can be obtained for other municipalities.

Rainfall Durations and Time Distributions

When using the design storm method, the urban stormwater designer must decide which design storm to use. Earlier sections talked about the design frequency applicable for the sizing of various components of the drainage system. But, the choice of frequency is only part of the storm selection process. All design storms are made up of four main components: frequency, duration, rainfall distribution, and depth (assuming stationary storms). The storm frequency, depth, and duration are interrelated through use of the intensity–duration–frequency curves.

Duration

The chosen duration of the storm is often determined by the municipality's design criteria or by the type of design being performed. A rule of thumb often used for minor system peak flow design in urbanized areas using the rational method is to use the storm duration that is equal to the time of concentration of the site to the calculation point. However, it can be easily shown that this may not yield the highest peak, if a portion of the site is grassy and the rest is paved. Using only the paved portion with its higher runoff and shorter time of concentration (and thus higher rainfall intensity) often gives the highest peak flow.

Urban Hydrology

TABLE 7-3

IDF Information for the 10-Year Storm

Time (min)	5	10	15	30	60
Depth (in)	0.512	0.867	1.125	1.588	1.967
Intensity (in/hr)	6.145	5.201	4.498	3.176	1.967

In developing a runoff hydrograph using unit hydrograph methods, a storm duration significantly longer than the time of concentration is warranted. For example, it can be shown that the duration of the storm when using SCS methods for unit hydrograph development and a number of pulses of rainfall in the rainfall hyetograph, that the duration of the storm must be almost twice the time of concentration. For detention design, the duration of the storm should be that which yields the highest storage requirement. This then becomes a function of the relative sizes of the pond, watershed, and outlet, but will, in any case, be much longer than the duration necessary for simple peak runoff estimation.

Distribution

The appropriate time distribution of the design rainfall has been a subject of some speculation among urban stormwater designers. It should be remembered that the objective is to produce a runoff event of a particular frequency. It is normally assumed that a rainfall of a particular frequency will, on average, produce a runoff of the denticle frequency, though each storm may not at all reproduce its associated frequency of runoff. While rainfall depths for a given duration have an associated calculated frequency, rainfall distributions generally do not. Thus, a lesser amount of rain, but falling in an atypical way, may produce a higher peak runoff than more rain depth but falling in a more uniformly distributed manner. This distribution selection becomes more important as the size of the watershed decreases and the imperviousness increases. Larger or less impervious watersheds have the opportunity to dampen pulses of rainfall and smooth the runoff hydrographs through attenuation and runoff hydrograph timing combinations.

There are probably a dozen methods used around the United States to distribute rainfall, and others in other countries. Four commonly used methods to distribute rainfall (other than a uniform distribution) are illustrated by an example for a 1-h storm with 5-min time increments.

Example Information

A Midwestern city has IDF curve information for the 10-year storm as shown in Table 7-3. Other method-specific data will be given for each approach.

The "Balanced-Storm Approach"

This approach is described in COE (1982) and in the popular HEC-1 computer software manual (HEC, 1990). In this method, the storm depth at any duration of storm is equal to the depth from the IDF curve. Therefore, the total storm depth in the first 5 min, from this example, is 0.512 in. The total depth after 10 min is 0.867 in. The incremental depth from 5 to 10 min is equal to the difference between the two or 0.867 - 0.512 = 0.355 in. Continuing this process and using linear interpolation or curve fitting for the 5-min increments for which there is no value leads to 12 incremental rainfall depths for each 5 min. These depth increments are then arranged so that the most intense increment is at the center of the storm. The second most intense is placed before it. The third is placed after it, the forth at the front, and so on, staggering the placement of all the increments

TABLE 7-4

Balanced Storm Example

Time (min)	Intensity (in/hr)	Depth (in)	Increment (in)	Arranged (in)	Accum. (in)
0	0	0	0	0	0
5	6.144	0.512	0.512	0.039	0.039
10	5.200	0.867	0.355	0.054	0.094
15	4.497	1.124	0.258	0.079	0.172
20	3.955	1.318	0.194	0.119	0.291
25	3.525	1.469	0.150	0.194	0.485
30	3.175	1.588	0.119	0.355	0.840
35	2.886	1.684	0.096	0.512	1.352
40	2.643	1.762	0.079	0.258	1.610
45	2.436	1.827	0.065	0.150	1.760
50	2.258	1.882	0.054	0.096	1.856
55	2.103	1.928	0.046	0.065	1.921
60	1.967	1.967	0.039	0.046	1.967

until all 12 are placed alternately in front or at the end of the storm. Table 7-4 illustrates this procedure for the example given. The last two columns are the final storm increments and accumulated depth S-curve.

The balanced storm method has the advantage in that it is relatively independent of storm duration. It will always maximize the rainfall depth for any duration for a given frequency. That is, due to the nesting of the depths, any duration chosen will yield the depth of rainfall for that frequency; it is balanced. This has certain advantages when working with different sized areas with one storm. It has the disadvantage in that it is a theoretical frequency distribution, not observed in nature, and gives the most intense storm possible for any given frequency and duration. And, in fact, comparing its S-curve to that of a series of S-curves derived by Huff (discussed below) indicates that the distribution has a less-than-one-in-ten chance of occurring in nature. This has led to a concern that it may overpredict storm peaks for smaller urban areas, in which attenuation and timing effects do not dampen the intense rainfall. It is the method used by HEC-1 when PH records are used (HEC, 1990).

SCS 24-Hour Storm

The Soil Conservation Service developed several semidimensionless 24-h storms that are considered typical of different locations within the United States. Section 7.9 in this chapter covers the different storm types, S-curve appearance, and geographic applicability. The SCS storm is semidimensionless in that the horizontal axis (X-axis) of the storm S-curve covers a time period of 24 h, while the vertical axis (Y-axis) is a dimensionless ratio of the storm accumulation to that time divided by the storm total depth. The SCS 24-h storm is a generalization of the balanced storm approach and is therefore also theoretical in nature. It was developed by taking the IDF curve information from a number of municipalities and developing an average representative curve. Thus, it shares all the advantages and disadvantages of the balanced-storm approach.

One additional disadvantage of the SCS storm approach is that it is cumbersome and inaccurate to use for short duration storms, owing to the general unavailability of detailed tables of values for short time increments. The SCS storm method was derived for larger areas, where longer time increments are applicable (SCS, 1986). The general procedure for storms less than 24 h is to take the central portion of the SCS curve equal to the duration

TABLE 7-5

SCS 24-Hour Storm Method Example

SCS Hour (h)	SCS Value of P/P24	Time (min)	Interp. Rainfall (in)	Increment. Rainfall (in)
11.5	0.2833	0	0.000	0.000
		5	0.276	0.276
		10	0.551	0.276
		15	0.827	0.276
		20	1.103	0.276
		25	1.378	0.276
12.0	0.6632	30	1.654	0.276
		35	1.706	0.052
		40	1.758	0.052
		45	1.810	0.052
		50	1.863	0.052
		55	1.915	0.052
12.5	0.7351	60	1.967	0.052

necessary. So, for example, for a 1-h storm, the time period from 11.5 to 12.5 h would be used. The total rainfall for 1 h (1.967 in) would be proportioned in a manner equivalent to the SCS curve. Because the total portion of the curve is 0.4518 (0.7351 − 0.2833), and 84% of that occurs between 11.5 and 12 h of the curve, 84% of the 1.967 in for a 1-h storm total should also occur in the first half-hour. Table 7-5 shows typical calculations for the example. Note that the actual resultant storm occurs in two blocks. This is due to the fact that the available table has time increments of only 0.5 h, and linear interpolation was used for the 5-min values.

Huff Distributions

Huff (1967) looked at 261 storms in a 400-square-mile network during the period 1955–1966 in East-Central Illinois and assigned typical distributions to them according to whether they tended to have the bulk of the rainfall in the first, second, third, or fourth quartile. These "Huff" curves are applicable to areas of 50 to 400 square miles. The time distributions were expressed as percentages of total duration and total rainfall accumulation into a series of S-curves. For each of the quartile groupings, the storms were analyzed, and a family of curves was derived, one for each probability interval of accumulated rainfall plotted versus time. For the 90% curve, for example, at each point on the curve, 90% of the storms had accumulated at least the given percentage of rainfall by that point in the storm's total duration. Table 7-6 gives the 50% (median) time distributions for each of the four-quartile storms versus percentage of storm duration. Table 7-7 gives first-quartile storm information for the 10%, 50%, and 90% storms. Later, Huff and Vogel (1976) used Chicago data from 471 storms with depths greater than 0.5 in to derive point data distributions following the same procedures as the East-Central Illinois data. Huff (1990) compared the two sets of data to derive curves for areas from 0 to 10 square miles and 10 to 50 square miles. These values are given in Table 7-8.

According to Huff, it is recommended to use first- or second-quartile storms for durations less than 12 h, and fourth-quartile storms for durations greater than 24 h. For most design purposes, the median curves are probably the most applicable. They are more established than the more extreme curves at the 10 and 90% levels, which typify unusual storm distribution conditions. The storm distributions are thought to be applicable to all areas with storm climatology similar to the Midwest. Some have thought they should be

TABLE 7-6

Median Time Distributions of Heavy Storm Rainfall in Areas of 50 to 400 Square Miles

Cumulative Percent of Storm Time	First-quartile Storm	Second-quartile Storm	Third-quartile Storm	Forth-quartile Storm
5	8	2	2	2
10	17	4	4	3
15	34	8	7	5
20	50	12	10	7
25	63	21	12	9
30	71	31	14	10
35	76	42	16	12
40	80	53	19	14
45	83	64	22	16
50	86	73	29	19
55	88	80	39	21
60	90	86	54	25
65	92	89	68	29
70	93	92	79	35
75	95	94	87	43
80	96	96	92	54
85	97	97	95	75
90	98	98	97	92
95	99	99	99	97

Source: Huff, F.A., State of Illinois, Water Survey, ISWS/CIR-173/90, Circular 173, 1990.

TABLE 7-7

Median Time Distributions of Areal Mean Rainfall in First-quartile Storms at 10%, 50%, and 90% Probability Levels

Cumulative Percent of Storm Time	Cumulative Percent of Storm Rainfall for Given Storm Probability		
	10%	50%	90%
5	24	8	2
10	50	17	4
15	71	34	13
20	84	50	28
25	89	63	39
30	92	71	46
35	94	76	49
40	95	80	52
45	96	83	55
50	97	86	57
55	98	88	60
60	98	90	63
65	98	92	67
70	99	93	72
75	99	95	76
80	99	96	82
85	99	97	89
90	99	98	94
95	99	99	97

Source: Huff, F.A., State of Illinois, Water Survey, ISWS/CIR-173/90, Circular 173, 1990.

TABLE 7-8

Median Time Distributions of Heavy Storm Rainfall at a Point and on Areas of 10 to 50 Square Miles

Cumulative Percent of Storm Time	Cumulative Percent of Point Storm Rainfall for a Given Storm Type				Cumulative Percent of Storm Rainfall on Areas of 10 to 50 Square Miles for a Given Storm Type			
5	16	3	3	2	12	3	2	2
10	33	8	6	5	25	6	5	4
15	43	12	9	8	38	10	8	7
20	52	16	12	10	51	14	12	9
25	60	22	15	13	62	21	14	11
30	66	29	19	16	69	30	17	13
35	71	39	23	19	74	40	20	15
40	75	51	27	22	78	52	23	18
45	79	62	32	25	81	63	27	21
50	82	70	38	28	84	72	33	24
55	84	76	45	32	86	78	42	27
60	86	81	57	35	88	83	55	30
65	88	85	70	39	90	87	69	34
70	90	88	79	45	92	90	79	40
75	92	91	85	51	94	92	86	47
80	94	93	89	59	95	94	91	57
85	96	95	92	72	96	96	94	74
90	97	97	95	84	97	97	96	88
95	98	98	97	92	98	98	98	95

Source: Huff, F.A., State of Illinois, Water Survey, ISWS/CIR-173/90, Circular 173, 1990.

applicable wherever the SCS Type II storm is applicable, because, according to SCS records, these areas indicate similar storm distributions.

The Huff distribution has the advantage in that it is derived from real observed data for heavy storms. Therefore, the distribution in time should give a realistic picture of a typical storm. It should be realized that, for the median storm, half the storms observed would have steeper (and half milder) S-curve accumulations.

The Huff distributions have some significant disadvantages for the smaller watersheds. Huff used 30-min data for his observations and thus missed the short-time rainfall burst information inherent in 1-min data. Thus, for longer storm durations with short time increments, the short duration bursts will be missed. If the method is used for these short times, the actual rainfall accumulation for a short time period may be greater or less than the IDF curve would indicate for the same duration of storm. Second, graphical information has been shown to be difficult to read, interpolate accurately, and use (Ruthroff and Bodtmann, 1976). Thus, information in this section has been expressed in Huff's tabular format to facilitate consistent use. Third, because the distribution is duration independent, strict duration criteria must be set or the designer should be required to choose the duration that maximizes the peak, because different durations of the same storm can yield significantly different peak flows. Last, the method used combined a number of storms in each quartile, averaging them at each percentage accumulation interval. This also tends to dampen the variation effects of individual storms, producing an averaged smooth curve unlike that found in nature. Huff's separating the data into quartiles and probabilities somewhat compensates for this and allows more flexibility in its use.

Table 7-9 illustrates the application of the Huff method to the example problem for the first- and second-quartile storms. The basic procedure is to scale the total storm duration into percentages of the total duration and find the dimensionless accumulation of the total storm rainfall from the appropriate table. Multiplying this by the total storm depth gives

TABLE 7-9
Example Problem Using Huff Distributions

	Huff — 1st Quartile		Huff — 2nd Quartile	
Time	Accumulation	Increments	Accumulation	Increments
0	0.00	0.00	0.00	0.00
5	0.54	0.54	0.13	0.13
10	0.91	0.37	0.26	0.14
15	1.18	0.28	0.43	0.17
20	1.36	0.18	0.70	0.27
25	1.50	0.14	1.08	0.37
30	1.61	0.11	1.38	0.30
35	1.68	0.07	1.56	0.18
40	1.74	0.07	1.69	0.13
45	1.81	0.07	1.79	0.10
50	1.88	0.07	1.86	0.07
55	1.92	0.04	1.92	0.06
60	1.97	0.05	1.97	0.05

the accumulation up to that point in the storm. Subtracting subsequent values gives the incremental values of the storm. No arrangement is necessary, because the shape of the S-curve gives the arrangement.

Yen and Chow's Method

Yen and Chow (1980) evaluated hourly data from over 7400 storms from Urbana, Illinois, Boston, Massachusetts; and Elizabeth City, New Jersey, to develop a typical dimensionless rainfall distribution for use in basins under 20 square miles. From these, they developed a rainfall hyetograph in the shape of a triangle. Figure 7-3 depicts this hyetograph.

In this procedure, Tp is the time to peak, Td is the total storm duration, and I is the average hourly rainfall rate. From a known total storm depth and duration, the only unknown to totally define the hyetograph is the value of Tp/Td. Based on a further analysis of many thousand rainfall gauges around the United States, Yen and Chow determined typical values of Tp/Td (Yen and Chow, 1983) (Figure 7-4).

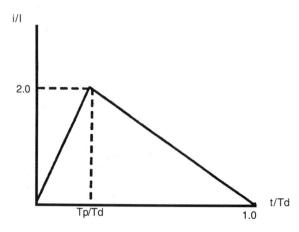

FIGURE 7-3
Yen and Chow's dimensionless hyetograph.

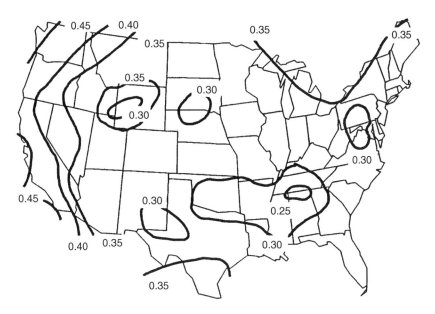

FIGURE 7-4
Tp/Td values. (After Yen, B.C. and Chow, V.T., Local Design Storm Vol. II, Federal Highway Administration, Report Number FHWA/RD-82-064, 1983.)

Applying Yen and Chow's method to the example problem yields the following information:

$$Tp/Td = 0.32$$

$$I = 1.967 \text{ in}$$

$$Td = 60 \text{ min}$$

Since $i/I = 2.0$, the peak instantaneous rainfall rate (i) is 3.934 in/h, and Tp is 19.2 min. Then, 5-min blocks of rainfall are derived from the triangular shape by calculating the area under the triangle for each 5 min. The storm depicted in Table 7-10 results.

TABLE 7-10
Yen and Chow's Method

Time	Accumulated Rain	Incremented Rain
0	0	0
5	0.043	0.043
10	0.171	0.128
15	0.384	0.213
20	0.681	0.297
25	0.983	0.301
30	1.244	0.261
35	1.465	0.221
40	1.646	0.181
45	1.786	0.141
50	1.887	0.1
55	1.947	0.06
60	1.967	0.02

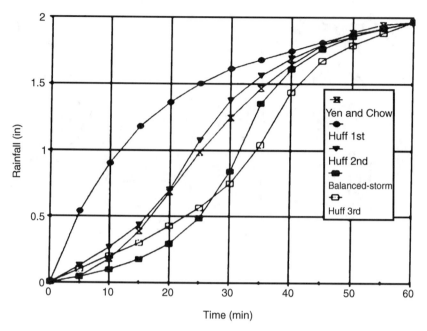

FIGURE 7-5
Comparison of the rainfall methods.

The different storm S-curves are plotted for comparison in Figure 7-5. The Huff third-quartile storm is also plotted to compare it to the balanced storm approach. Notice that the balanced storm has the steepest curve of any of the methods given. The Yen and Chow method, for this example, is roughly equivalent to Huff's second-quartile storm.

7.8 Rational Method

The rational method is a simple formula for the estimation of peak flow rates from small urban and rural watersheds. Its form is a simple ratio between runoff and rainfall rates. It is often recommended for estimating the design storm peak runoff for small urban and rural watersheds in drainage system design. Often, a size limit of less than 100 acres is used, and some regulations prohibit its use for areas larger than 20 acres (South Carolina, 1992). The method is best suited for estimating peak discharges for inlet and storm drainage system design.

Rational Formula

The rational formula estimates the peak rate of runoff at any location in a watershed as a function of the drainage area, runoff coefficient, and mean rainfall intensity for a duration equal to the time of concentration (the time required for water to flow from the most remote point of the basin to the location being analyzed). The rational formula is expressed as follows:

$$Q = CIA \tag{7.8}$$

where Q = maximum rate of runoff (cfs); C = runoff coefficient representing a ratio of runoff to rainfall; I = average rainfall intensity for a duration equal to the time of concentration (in/h); and A = drainage area contributing to the design location (acres).

Note that the units of the rational formula are not consistent, as the left side is in cfs and the right side is in acre-inches per hour. However, the conversion from one set of units to the other is approximately equal to one, thus allowing for the discrepancy.

The rational method, while first introduced in 1889 (Kuichling), is derived from earlier work in America and England. Even though it has frequently come under criticism for its simplistic approach, no other drainage design method has received such widespread use. The equation itself is simply a mass transfer equation: Inflow equals outflow minus losses. It is completely accurate for *totally* impervious surfaces (assuming accurate rainfall). But the "c" factor must account for *every* factor that reduces peak flow. This does not imply that it produces accurate hydrologic estimates. However, many systems have been designed with the rational method that have been tested by storms and found not to be grossly undersized. However, adequate data have not been collected with which to test the validity of the method in a variety of situations. It may, in fact, over- or underdesign storm systems. Opinions in the literature are mixed. With the increased complexity of urban storm drainage systems and the need for hydrograph routing downstream from stormwater management facilities, the use of the rational method is being severely restricted in many municipalities. Proper application of the method and recognition of instances where standard application of the rational method does not apply are still the governing principles.

For example, in some situations, it can be demonstrated that taking the peak from only the directly connected impervious area of a development yields a higher peak flow estimate than taking the runoff from a combination of the grassy and paved areas, even though the area considered is smaller. Or, in backwater situations, the ability of the pipe to handle the calculated peak flow is reduced, even though the peak may be calculated correctly.

Characteristics and Limits of the Rational Method

When using the rational method, some precautions should be considered in order to correctly apply the method:

1. The first step in applying the rational method is to obtain a good topographic map and define the boundaries of the drainage area in question. A field inspection of the area should also be made to determine if the natural drainage divides have been altered.
2. In determining the C value (land use) for the drainage area, thought should be given to future changes in land use that might occur during the service life of the proposed facility that could result in an inadequate drainage system.
3. Because the rational method uses a composite C value for the entire drainage area, if the distribution of land uses within the drainage basin will affect the results of hydrologic analysis, then the basin should be divided into two or more subdrainage basins for analysis.
4. Restrictions to the natural flow, such as highway crossings and dams that exist in the drainage area, should be investigated to see how they affect the design flows. Also, the effects of upstream detention facilities may be taken into account.
5. The charts, graphs, and tables included in this section are given to assist the designer in applying the rational method. The designer should use good engi-

neering judgment in applying these design aids and should make appropriate adjustments when specific site characteristics dictate that these adjustments are or are not appropriate.

Characteristics of, or assumptions inherent in, the rational method may limit its use for analysis of stormwater management facilities. A more detailed discussion can be found in Rossmiller (1980) or Chow (1964). Following are some of these characteristics:

The rate of runoff resulting from any rainfall intensity is a maximum when the rainfall intensity lasts as long or longer than the time of concentration. This assumption limits the size of the drainage basin that can be evaluated by the rational method. For large drainage areas, the time of concentration can be so large that constant rainfall intensities for such long periods do not occur, and shorter, more intense rainfalls can produce larger peak flows. Further, in semiarid and arid regions, storm cells are relatively small with extreme intensity variations, thus making the rational method inappropriate for large watersheds. There is even some question about whether the time of concentration can be measured or is significant (Schaake, Geyer, and Knapp, 1967).

The frequency of peak discharges is the same as that of the rainfall intensity for the given time of concentration. Frequencies of peak discharges depend on rainfall frequencies, antecedent moisture conditions in the watershed, and the response characteristics of the drainage system. For small and largely impervious areas, rainfall frequency is the dominant factor. For larger drainage basins, the physical characteristics of the drainage basin control are the dominant factor. For drainage areas with few impervious surfaces, antecedent moisture conditions usually govern, especially for rainfall events with a return period of 10 years or less.

The fraction of rainfall that becomes runoff (C) is independent of rainfall intensity or volume. The assumption is reasonable for impervious areas, such as streets, rooftops, and parking lots. For pervious areas, the fraction of runoff varies with rainfall intensity and the accumulated volume of rainfall (Schaake, Geyer, and Knapp, 1967). Thus, the art necessary for application of the rational method involves the selection of a coefficient that is appropriate for the storm, soil, and land-use conditions. Many guidelines and tables have been established, but seldom have they been supported with empirical evidence.

The peak rate of runoff is sufficient information for the design. Modern drainage practice often includes detention of urban storm runoff to reduce the peak rate of runoff downstream. No recommended hydrograph is available from the rational method to route through detention storage facilities. Lack of a reliable hydrograph severely limits the use of the rational method for the evaluation of design alternatives available in urban and, in some instances, rural drainage design. Other hydrologic methods discussed in this chapter can be used to generate hydrographs.

One of the greatest drawbacks of the rational method is that its results are not usually replicable from application to application. A formula with the simplicity of the rational method lends itself to wide variations in application, judgment of the parameters, and use of various adjustment procedures (Rossmiller, 1980). Many municipalities have identified one or several adjustments considered important for their local environment. Complex adjustments, which yield only minor changes in the final peak flow estimate, are usually not warranted (except, perhaps, for legal reasons). Despite all of these concerns, the rational method remains a mainstay of hydrologic design.

Urban Hydrology

The rational method depends on four variables: time of concentration, rainfall, area, and the C factor. The results of using the rational formula to estimate peak discharges are sensitive to the variables used. Careful selection of the variables and recognition of unusual circumstances is important. Following is a discussion of the different variables used in this method.

Time of Concentration

The time factor used in the rational method is simply a period within the total storm duration during which the maximum average rainfall intensity occurs. It is sometimes called the *rainfall intensity averaging time*. Popular application of the rational method uses the time of concentration (t_c) for each design point within the drainage basin as the rainfall intensity averaging time.

The time of concentration is the time required for water to flow from the hydraulically most remote point of the drainage area to the point under investigation. The duration of rainfall is then set equal to the time of concentration and is used to estimate the design average rainfall intensity (I). The time of concentration used in the design consists of an inlet time to the point where the runoff enters the storm drain or channel inlet plus the time of flow in a closed conduit or open channel to the design point.

Two common errors should be avoided when calculating the time of concentration. First, in some cases, runoff from a portion of the drainage area that is highly impervious may result in a greater peak discharge than would occur if the entire area were considered. In these cases, adjustments can be made to the drainage area by disregarding those areas where flow time is too slow to add to the peak discharge. Sometimes, it is necessary to estimate several different times of concentration to determine the design flow that is critical for a particular application. Second, when designing a drainage system, the overland flow path is not necessarily the same before and after development and grading operations have been completed. Selecting overland flow paths in excess of 100 ft in urban areas and 300 ft in rural areas should be done only after careful consideration.

Figure 7-6 can be used to estimate inlet time for design conditions that do not involve complex drainage conditions. For each drainage area, the distance is determined from the

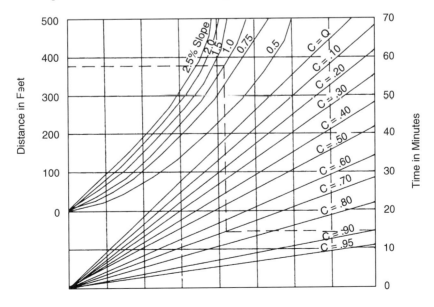

FIGURE 7-6
Overland time of flow. (From Airport Drainage, Federal Aviation Administration, 1965.)

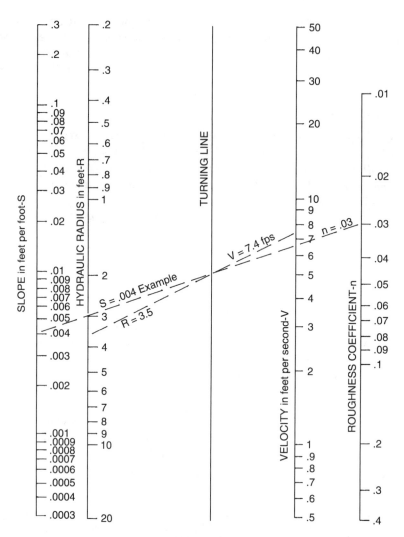

FIGURE 7-7
Manning formula for travel time. (From AASHTO Model Drainage Manual, 1991.)

inlet to the most remote point in the tributary area. From a topographic map, the average slope is determined for the same distance. The runoff coefficient (C) is determined by the procedure described in a subsequent section of this chapter.

To obtain the total time of concentration, the pipe or open channel flow time must be calculated and added to the inlet time. After first determining the average flow velocity in the pipe or channel (using the Manning formula, for example), the travel time is obtained by dividing velocity into the pipe or channel length. Velocity can be estimated by using the nomograph shown in Figure 7-7. Note: time of concentration cannot be less than 5 min.

A large number of other methods is available for estimating inlet time. They fall into categories based on their form and independent variables. The travel time estimation method used for the SCS method could be substituted for the above procedure or used to check the values obtained from the charts and nomographs described for use with the rational method. The SCS travel time estimation method is described under the SCS Unit Hydrograph method in this chapter.

Urban Hydrology

TABLE 7-11

Values of N_k for the Kerby Formula

N_k	Surface Type
0.02	Smooth impervious surface
0.10	Smooth bare packed soil, free of stones
0.20	Poor grass, cultivated row crops, or moderately rough bare surfaces
0.40	Pasture or average grass cover
0.60	Deciduous timberland
0.80	Conifer timberland, deciduous timberland with deep forest litter or dense grass cover

Kirpich (1940) developed a formula for inlet time based on data from six steeply sloped agricultural and forested watersheds in West Tennessee as:

$$t_c = 0.0078 \, (L/(S^{0.5}))^{0.77} \tag{7.9}$$

where t_c = time of concentration (min); L = length of travel (ft); and S = slope of the flow path from the most remote part of the basin to the calculation point divided by the horizontal distance between the two points (ft/ft).

Equation (7.9) is most applicable for natural basins with well-defined channels, bare-earth overland flow, or flow in mowed channels. Rossmiller (1980) gives adjustment factors for other conditions based on literature values:

- For general overland flow and flow in natural grassed channels, multiply t_c by 2.
- For concrete or asphalt surfaces, multiply t_c by 0.4.
- For concrete channels, multiply t_c by 0.2.

Kerby (1959) developed an estimate of the overland flow portion of the inlet time as:

$$t_c = 0.83(N_k L/(S^{0.5}))^{0.467} \tag{7.10}$$

where t_c = time of concentration (min); L = straight line length from the top of the basin to the point of a defined channel (ft); S = slope calculated similar to the Kirpich equation (ft/ft); and N_k = a coefficient of roughness given in Table 7-11.

Rainfall Intensity

The rainfall intensity (I) is the average rainfall rate in inches/hour for a duration equal to the time of concentration for a selected return period. Once a particular return period has been selected for design and a time of concentration calculated for the drainage area, the rainfall intensity can be determined from intensity–duration–frequency curves. Figure 7-8 gives an example of IDF curves for Raleigh, North Carolina. Rational method calculations are made easier by fitting a curve of the form given in Equation (7.7) to the IDF information.

Drainage Area

The choice of drainage area may, on the surface, appear simple. However, there are a number of details to consider, such as:

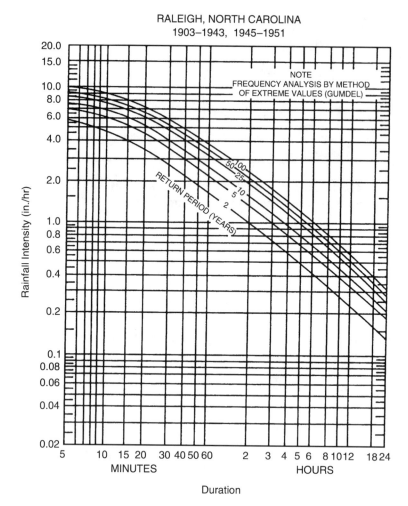

FIGURE 7-8
Intensity duration curve — Raleigh, North Carolina. (From Weather Bureau Tech. Paper 25.)

- Existing and planned flow directions in key areas such as along lot lines, at street intersections, flow bypass points, etc.
- Impacts of regrading on flow direction and speed
- Impacts of drainage structures on various flow frequencies
- Use of pervious and impervious areas in the area calculation, or only directly connected impervious areas
- Use of several or one land-use composite calculation
- Flow directions and amounts for flows in excess of the design storm

Runoff Coefficient

The runoff coefficient (C) is the variable of the rational method least susceptible to precise determination and requires judgment and understanding on the part of the designer. While engineering judgment will always be required in the selection of runoff coefficients, a typical coefficient represents the integrated effects of many drainage basin parameters. The

TABLE 7-12

Frequency Factors for Rational Formula

Recurrence Interval (Years)	C_f
25	1.1
50	1.2
100	1.25

TABLE 7-13

Hydrologic Soils Groups for Orange County, North Carolina

Series Name	Hydrologic Groups	Series Name	Hydrologic Groups
Altavista	C	Herndon	B
Appling	B	Hiwassee	B
Cecil	B	Iredell	D
Chewacla	C	Lignum	C
Congaree	B	Louisburg	B
Creedmoor	C	Orange	D
Enon	C	Tatum	C
Georgeville	B	Vance	C
Goldston	C	Wedowee	D

Source: Soil Conservation Service, Technical Release No. 55, Washington, DC, 1986.

following discussion considers C-factor modifications based on the effects of design frequency, soil groups, land use, and average land slope. The coefficients given in Table 7-15 are applicable for storms of 5-year to 10-year frequencies. Less frequent, higher intensity storms will require modification of the coefficient, because infiltration and other losses have a proportionally smaller effect on runoff (Wright-McLaughlin, 1969). The adjustment of the rational method for use with major storms can be made by multiplying the right side of the rational formula by a frequency factor C_f. The rational formula now becomes:

$$Q = CC_f IA \tag{7.11}$$

The C_f values are listed in Table 7-12. The product of C_f times C should not exceed 1.0.

Three methods for determining the runoff coefficient are presented based on soil groups and land slope (Tables 7-13 and 7-14), land use (Table 7-15), and a composite coefficient for complex watersheds (Table 7-16).

Table 7-14 gives the recommended coefficient (C) of runoff for pervious surfaces by selected hydrologic soil groupings and slope ranges. From this table, the C values for

TABLE 7-14

Recommended Coefficient of Runoff for Pervious Surfaces by Selected Hydrologic Soil Groupings and Slope Ranges

Slope	A	B	C	D
Flat (0–1%)	0.04–0.09	0.07–0.12	0.11–0.16	0.15–0.20
Average (2–6%)	0.09–0.14	0.12–0.17	0.16–0.21	0.20–0.25
Steep	0.13–0.18	0.18–0.24	0.23–0.31	0.28–0.38

Source: Storm Drainage Design Manual, Erie and Niagara Counties Regional Planning Board.

TABLE 7-15

Recommended Coefficient of Runoff Values for Various Selected Land Uses

Description of Area	Runoff Coefficients
Business:	
Downtown areas	0.70–0.95
Neighborhood areas	0.50–0.70
Residential:	
Single-family areas	0.30–0.50
Multiunits, detached	0.40–0.60
Multiunits, attached	0.60–0.75
Suburban	0.25–0.40
Residential (1/2-acre lots or more)	0.30–0.45
Apartment-dwelling areas	0.50–0.70
Industrial:	
Light areas	0.50–0.80
Heavy areas	0.60–0.90
Parks, cemeteries	0.10–0.25
Playgrounds	0.20–0.40
Railroad yard areas	0.20–0.40
Unimproved areas	0.10–0.30

Source: U.S. Department of Transportation, Federal Highway Administration, Hydraulic Engineering Circular No. 19, 1984.

nonurban areas such as forest land, agricultural land, and open space can be determined. Soil properties influence the relationship between runoff and rainfall, because soils have differing rates of infiltration. Infiltration is the movement of water through the soil surface into the soil. Based on infiltration rates, the Soil Conservation Service (SCS) divided soils into four hydrologic soil groups as follows:

Group A: Soils having a low runoff potential due to high infiltration rates (These soils consist primarily of deep, well-drained sands and gravels.)

Group B: Soils having a moderately low runoff potential due to moderate infiltration rates (These soils consist primarily of moderately-deep to deep, moderately well- to well-drained soils, with moderately fine to moderately coarse textures.)

Group C: Soils having a moderately high runoff potential due to slow infiltration rates (These soils consist primarily of soils in which a layer exists near the surface that impedes the downward movement of water or soils with moderately-fine to fine texture.)

Group D: Soils having a high runoff potential due to very slow infiltration rates (These soils consist primarily of clays with high swelling potential, soils with permanently high water tables, soils with a claypan or clay layer at or near the surface, and shallow soils over nearly impervious parent material.)

As an example, a list of soils for Orange County, North Carolina, and their hydrologic classification are presented in Table 7-13.

As the slope of the drainage basin increases, the selected C value should also increase. This is caused by the fact that, as the slope of the drainage area increases, the velocity of overland and channel flow will increase, allowing less opportunity for water to infiltrate the ground surface. Thus, more of the rainfall will become runoff from the drainage area.

Urban Hydrology

TABLE 7-16

Coefficients for Composite Runoff Analysis

Surface	Runoff Coefficients
Street:	
Asphalt	0.70–0.95
Concrete	0.80–0.95
Drives and walks	0.75–0.85
Roofs	0.75–0.95

Source: U.S. Department of Transportation, Federal Highway Administration, Hydraulic Engineering Circular No. 19, 1984.

It is often desirable to develop a composite runoff coefficient based on the percentage of different types of surface in the drainage area. Composites can be made with Tables 7-14 and 7-15. At a more detailed level, composites can be made with Table 7-14 and the coefficients with respect to surface type given in Table 7-16. The composite procedure can be applied to an entire drainage area or to typical sample blocks as a guide to selection of reasonable values of the coefficient for an entire area.

It should be remembered that the rational method assumes that all land uses within a drainage area are uniformly distributed throughout the area. If it is important to locate a specific land use within the drainage area, then another hydrologic method should be used, with which hydrographs can be generated and routed through the drainage system.

Perhaps the most complex formulation for C factor estimation is given by Rossmiller (1980) as follows:

$$C = 7.2(10)^{-7} CN^3 Cr(Cf)^{Cs}(Cn)^{Ci}(Cm) \tag{7.12}$$

where $Cr = RI^{0.05}$; $Cf = (0.01CN)^{0.6}$; $Cs = (-S)^{0.2}$; $Cn = (0.001CN)^{1.48}$; $Ci = 0.15$-i; $Cm = ((Imp+1)/2)^{0.7}$; CN = SCS curve number; RI = recurrence interval (yrs); S = average watershed slope (%) (4% = 4); i = rainfall intensity (in/h); Imp = watershed imperviousness as a decimal (20% = 0.20).

The equation takes into account, respectively, frequency, slope, rainfall intensity, modifier for slope factor, modifier for intensity factor, and surface roughness.

Example Problem — Rational Method

Following is an example problem that illustrates the application of the rational method to estimate peak discharges. Preliminary estimates of the maximum rate of runoff are needed at the inlet to a culvert for a 25-year and 100-year return period. From a topographic map and field survey, the area of the drainage basin upstream from the point in question was measured to be 18 acres. In addition, the following data were measured:

- Average overland slope = 2.0%
- Length of overland flow = 50 ft
- Length of main basin channel = 1300 ft
- Slope of channel = 0.018 ft/ft = 1.8%
- Roughness coefficient (n) of the channel was estimated to be 0.090
- From existing land-use maps, land use for the drainage basin was estimated to be:

- Residential (single family) = 80%
- Graded - sandy soil, 3% slope = 20%
- From existing land-use maps, the land use for the overland flow area at the head of the basin was estimated to be:
  - Lawn = sandy soil, 2% slope = 100%
- A runoff coefficient (C) for the overland flow area is determined from Table 7-15 to be 0.15.
- From Figure 7-6 with an overland flow length of 50 ft, slope of 2%, and a C of 0.15, the overland flow time is 10 min. Channel flow velocity is determined from Figure 7-7 to be 3.5 ft/s ($n = 0.090$, $R = 1.97$, and $S = .018$). Therefore,
  - Flow Time = (1300 ft)/[(3.5 ft/s)(60 s/min)] = 6.2 min and t_c = 10+6.2 = 16.2 min = 16 min
- From Table 7-2 with duration equal to 16 min (values obtained by linear interpolation between values for 15 and 30 min):
  - I_{25} (25-year return period) = 5.86 in/h
  - I_{100} (100-year return period) = 7.18 in/h

A weighted runoff coefficient (C) for the total drainage area is determined in the following table by utilizing the values from Table 7-15.

Land Use	(1) Percent of Total Land Area	(2) Runoff Coefficient	(3) Weighted Runoff Coefficient[a]
Residential (single family)	0.80	0.50	0.40
Graded area	0.20	0.30	0.06
Total weighted runoff coefficient			0.46

[a] Column 3 equals column 1 multiplied by column 2.

From the rational method equation:

$$Q_{25} = C_f CIA = 1.1 \times .46 \times 5.86 \times 18 = 53 \text{ cfs}$$

$$Q_{100} = C_f CIA = 1.25 \times .46 \times 7.18 \times 18 = 74 \text{ cfs}$$

These are the estimates of peak runoff for a 25-year and 100-year design storm for the given drainage basin.

7.9 SCS Hydrologic Methods

The grouping of methods known as the SCS methodology were derived and modified throughout the last half-century (Rallison, 1980). The techniques developed by the U.S. Soil Conservation Service for calculating rates of runoff require the same basic data as the rational method: drainage area, runoff factor, time of concentration, and rainfall. The SCS method, however, is more sophisticated in that it also considers the time distribution of the rainfall, the initial rainfall losses to interception and depression storage, and an infiltration rate that decreases during the course of a storm. With the SCS method, the direct runoff can be calculated for any storm, real or fabricated, by subtracting infiltration and other losses from the rainfall to obtain the precipitation excess. Details of the methodology can be found in the *SCS National Engineering Handbook, Section 4.*

The SCS method can be used to find peak flow, outflow hydrographs, or unit hydrographs for use with rainfall excess hyetographs also generated from a modification of the SCS method. The SCS method includes the following basic equations and concepts:

- Runoff volume can be found using the basic SCS rainfall–runoff equation.
- Curve numbers used in determination of rainfall losses are developed based on soils, land uses, and antecedent moisture assumptions within the drainage area.
- Standard dimensionless storm distributions are used based on geographic location throughout the United States.
- A lag time is determined based on calculation of the time of concentration to the study point (similar to the rational method).

From some or all of this basic information, peaks, hydrographs, storms, and unit hydrographs can be found:

- Peak flows can be estimated from the SCS peak flow methods.
- Total outflow hydrographs can be developed through the SCS tabular method or any of a number of computer models utilizing the SCS unit hydrograph.
- Using one of the rainfall distributions and known rainfall totals, total and excess rainfall amounts are determined using the SCS rainfall–runoff relationship.
- Using the curvilinear (or triangular approximation) dimensionless unit hydrographs, dimensional unit hydrographs are developed for the drainage area based on drainage area and lag time calculations.

Basic Equations and Concepts

The following discussion outlines the equations and basic concepts used in the SCS method.

Rainfall–Runoff Equation

A relationship between accumulated rainfall and accumulated runoff was derived by SCS from experimental plots for numerous soils and vegetative cover conditions. Data for land-treatment measures, such as contouring and terracing, from experimental watersheds were included. The equation was developed mainly for small rural watersheds for which only daily rainfall and watershed data are ordinarily available. It was developed from recorded storm data that included the total amount of rainfall in a calendar day but not its distribution with respect to time. The SCS runoff equation is therefore primarily a method of estimating direct runoff from 24-hour or 1-day storm rainfall, though shorter durations are routinely used. The basic proportionality relationship, the center of the methodology, is as follows:

$$F/S = Q/(P-I_a) \tag{7.13}$$

where Q = accumulated direct runoff (in); P = accumulated rainfall (potential maximum runoff) (in); I_a = initial abstraction including surface storage, interception, and infiltration prior to runoff (in); S = potential maximum retention (in); and F = retained rainfall volume (in).

By substituting $F = (P - I_a) - Q$ into Equation (7.13), the following rainfall–runoff relationship emerges:

$$Q = (P - I_a)^2 / (P - I_a) + S \qquad (7.14)$$

A relationship between I_a and S was developed from experimental watershed data, which removes the necessity for estimating I_a for common usage. The empirical relationship used in the SCS runoff equation is as follows:

$$I_a = 0.2S \qquad (7.15)$$

Research has shown that the value of $0.2S$ is not correct for all circumstances. There is much scatter in the data. Others have suggested $0.1S$ better fits even the SCS data. Other values can be used for $0.2S$, though care must be taken to use equations for which this assumption has not been included. Bosznay (1989) avoided the need for the $0.2S$ assumption (and the use of curve numbers) by seeking to determine S directly and plotting Q directly as a function of S for various assumed values of $(P - I_a)$.

Substituting $0.2S$ for I_a in Equation (7.14), the SCS rainfall–runoff equation becomes:

$$Q = (P - 0.2S)^2 / (P + 0.8S) \qquad (7.16)$$

SCS has also defined a physically based curve number such that:

$$CN = 1000/(10+S) \text{ or} \qquad (7.17)$$

$$S = 1000/CN - 10 \qquad (7.18)$$

Curve numbers vary between about 40 and 99: the higher the curve number, the more runoff per unit of rainfall. The curve numbers reflect all physical aspects of the land use and antecedent moisture within the basin. Curve numbers are discussed in more detail below.

Figure 7-10 shows a graphical solution of Equation (7-16) that enables the precipitation excess from a storm to be obtained if the total rainfall and watershed curve number are known. For example, 4.8 in of direct runoff would result if 6.5 in of rainfall occurs on a watershed with a curve number of 85.

The family of curves in Figure 7-10 can be collapsed to a single curve through manipulation of Equation (7.16) as follows (Hawkins, 1979):

$$Q/S = (P/S - 0.2)^2 / (P/S + 0.8) \qquad (7.19)$$

It has been stated that this equation has little basis in physical reality, but it seems to work well for a variety of applications throughout the United States, except in portions of the arid Southwest (Hjelmfelt, 1991). For very dry climates, the SCS methodology has often not appeared applicable and often gives too high a runoff for large rainfall events. In these conditions, reliance may be placed on regression-type equations available for much of the area (Daly, 1981).

Boughton (1987) demonstrated that the equation is a special case of a spatially varied, saturation overflow model and has derived curves similar in form to those in Figure 7-9

Urban Hydrology

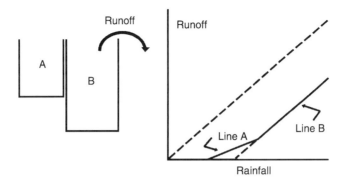

FIGURE 7-9
Spatial variability explanation of SCS curves. (After Boughton, W.C., *J. Irrig. and Drain. Eng.*, ASCE, 113, 3, August, 1987. Reproduced by permission of ASCE.)

from purely physical reasoning. Figure 7-9 illustrates this reasoning. The watershed is considered to be a series of "buckets," each of which produces runoff after it has become filled. The buckets are of different sizes. Each size range of buckets begins to produce runoff at a different point in the storm. When all the buckets are full, runoff equals rainfall, and the rainfall–runoff line in Figure 7-9 approaches 45 degrees. For the simple case of two buckets as illustrated in the figure, line A results when bucket A is full and begins to produce runoff. When bucket B is full, the runoff begins to equal the rainfall, and the whole watershed is producing runoff.

For a real watershed, there are millions of "buckets" of all different sizes. Some runoff begins almost immediately. The watershed response line curves slowly toward the 45-degree dotted line as more and more rainfall occurs. The line shifts to the right for drier initial conditions, for greater initial losses assumed, or for more porous soil. For many different soil and land use conditions (curve numbers), a family of curves would result, eventuating in a figure similar to Figure 7-10, the SCS solution to the runoff equation.

Hjelmfelt (1980) showed that the equation is an effective "frequency transformer"; transforming rainfall to runoff of the same frequency. Regardless of any controversy or questions surrounding the method, it has received wide attention and application across the United States and foreign countries.

Curve Number

In hydrograph applications, runoff is often referred to as rainfall excess or effective rainfall — all defined as the amount by which rainfall exceeds the capability of the land to infiltrate or otherwise retain the rainwater. The principal physical watershed characteristics affecting the relationship between rainfall and runoff are land use, land treatment, soil types, and land slope.

Land use is the watershed cover, and it includes agricultural and nonagricultural uses. Items such as type of vegetation, water surfaces, roads, roofs, etc., are all part of the land use. Land treatment applies mainly to agricultural land use, and it includes mechanical practices such as contouring or terracing and management practices such as rotation of crops.

The SCS method uses a combination of soil conditions and land use (groundcover) to assign a runoff factor to an area. These runoff factors, called runoff curve numbers (CN), indicate the runoff potential of an area when the soil is not frozen. The higher the CN, the higher the runoff potential.

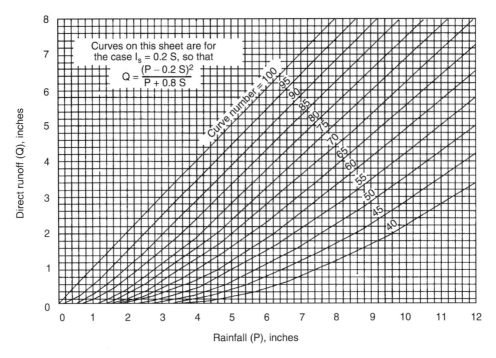

FIGURE 7-10
SCS solution of the runoff equation. [From U.S. Department of Agriculture, Soil Conservation Service, Engineering Division, Urban Hydrology for Small Watersheds, Technical Release 55 (TR-55), 1986.]

Soil properties influence the relationship between rainfall and runoff by affecting the rate of infiltration. The SCS divided soils into four hydrologic soil groups based on infiltration rates (Groups A, B, C, and D). These groups were previously described for the rational method.

Consideration should be given to the effects of urbanization on the natural hydrologic soil group. If heavy equipment can be expected to compact the soil during construction, or if grading will mix the surface and subsurface soils, appropriate changes should be made in the soil group selected. Also, runoff curve numbers vary with the antecedent soil moisture conditions, defined as the amount of rainfall occurring in a selected period preceding a given storm. In general, the greater the antecedent rainfall, the more direct runoff there is from a given storm. A 5-day period is used as the minimum for estimating antecedent moisture conditions. Antecedent soil moisture conditions also vary during a storm; heavy rain falling on a dry soil can change the soil moisture condition from dry to average to wet during the storm period.

A series of tables related to runoff factors are provided here. The first tables (Tables 7-18 to 7-21) give curve numbers for various land uses. These tables are based on an average antecedent moisture condition, i.e., soils that are neither very wet nor very dry when the design storm begins. Curve numbers should be selected only after a field inspection of the watershed and a review of zoning and soil maps.

Several factors, such as the percentage of impervious area and the means of conveying runoff from impervious areas to the drainage system, should be considered in computing CN for urban areas. For example, do the impervious areas connect directly to the drainage system, or do they outlet onto lawns or other pervious areas where infiltration can occur? The curve number values given in Table 7-18 are based on directly connected impervious areas. An impervious area is considered directly connected if runoff from it flows directly into the drainage system. It is also considered directly connected if runoff from it occurs as concentrated shallow flow that runs over pervious areas and then into a drainage system. It is possible that curve number values from urban areas could be reduced by not directly connecting impervious surfaces to the drainage system.

Antecedent Moisture Conditions

Most of the curve numbers given are for what might be termed normal conditions. Drier conditions will work to increase the amount of available storage, effectively sliding the curve in Figure 7-10 to the right (i.e., lower curve number), while wetter conditions will use up some of the available storage, resulting in a higher curve number for a specific storm. Table 7-22 gives conversion factors to convert average curve numbers to wet and dry curve numbers. The SCS defined three wetness conditions. Table 7-23 gives a common definition of the antecedent conditions for the three classifications.

If S is treated as a random variable, then it has been shown (Hjelmfelt, 1991) that the dry and wet conditions correspond roughly to the 10 and 90% exceedance frequency. Or, they correspond roughly to the lower and upper envelope CN curves in a plot of actual rainfall versus runoff values. The normal moisture condition would plot through the middle of the scatter of points for any given watershed.

The actual CN for any storm event can be calculated from the quadratic formulation of the rainfall–runoff equation (Hawkins, 1979):

$$CN = 100/(1 + 1/2(P+2Q-(4Q^2 + 5PQ)^{1/2})) \qquad (7.20)$$

The following discussion will give some guidance for adjusting curve numbers for different types of impervious areas. Imperviousness can be estimated by laying a grid over an aerial photograph and selecting a minimum of 200 sampling points at grid intersections (Cochran, 1963).

Connected Impervious Areas

Urban CNs given in Table 7-18 were developed for typical land-use relationships based on specific assumed percentages of impervious area. These CN values were developed on the following assumptions:

1. Pervious urban areas are equivalent to pasture in good hydrologic condition.
2. Impervious areas have a CN of 98 and are directly connected to the drainage system.

Some assumed percentages of impervious area are shown in Table 7-18.

If all of the impervious area is directly connected to the drainage system, but the impervious area percentages or the pervious land use assumptions in Table 7-18 are not applicable, use Figure 7-11 to compute a composite CN. For example, Table 7-18 gives a CN of 70 for a 1/2-acre lot in hydrologic soil group B, with an assumed impervious area of 25%. However, if the lot has 20% impervious area and a pervious area CN of 61, the composite CN obtained from Figure 7-11 is 68. The CN difference between 70 and 68 reflects the difference in percent impervious area.

Unconnected Impervious Areas

Runoff from these areas is spread over a pervious area as sheet flow. To determine CN when all or part of the impervious area is not directly connected to the drainage system,

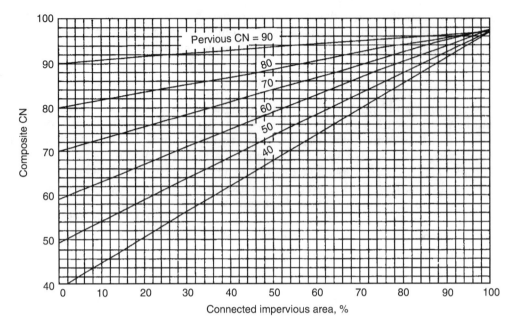

FIGURE 7-11
Composite CN with connected impervious areas.

(1) use Figure 7-12 if the total impervious area is less then 30% or (2) use Figure 7-11 if the total impervious area is equal to or greater than 30%, because the absorptive capacity of the remaining pervious areas will not significantly affect runoff.

When impervious area is less than 30%, obtain the composite CN by entering the right half of Figure 7-12 with the percentage of total impervious area and the ratio of total unconnected impervious area to total impervious area. Then, move left to the appropriate

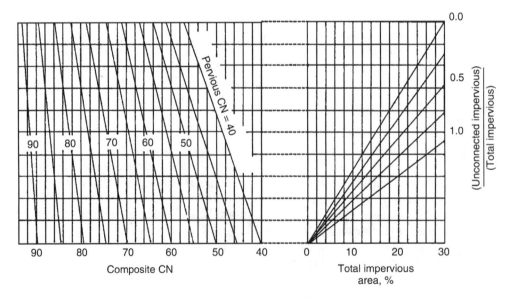

FIGURE 7-12
Composite CN with unconnected impervious areas (total impervious area less than 30%). U.S. Department of Agriculture, Soil Conservation Service, Engineering Division, Urban Hydrology for Small Watersheds, Technical Release 55 (TR-55), 1986.

TABLE 7-17

Composite Curve Number Calculations

(1)	(2)	(3)	(4)	(5)
Land use	Curve number	Area	% of total area	Composite curve number (column 2 × column 4)

pervious CN and read down to find the composite CN. For example, for a 1/2-acre lot with 20% total impervious area (75% of which is unconnected) and pervious CN of 61, the composite CN from Figure 7-12 is 66. If all of the impervious area was connected, the CN (from Figure 7-11) would be 68.

Composite Curve Numbers

When a drainage area has more than one land use, a composite curve number can be calculated and used in the analysis. It should be noted that, when composite curve numbers are used, the analysis does not take into account the location of the specific land uses but sees the drainage area as a uniform land use represented by the composite curve number.

Composite curve numbers for a drainage area can be calculated by entering the required data into a table such as the one presented in Table 7-17.

Rainfall

The SCS method is based on a 24-h rainfall distribution, which has a shape dependent on the location within the United States. These storm time distributions have been prepared by the SCS from analysis of numerous rainfall records. Figures 7-13 and 7-14 show the distributions presently available for the SCS method. Table 7-24 gives the tabular values for the SCS Types I, IA, II, and III distribution.

Shorter durations can be used by taking the most steeply sloped portion of the S-curve corresponding to the desired duration and distributing the rainfall accordingly. For example, a 6-h storm would take the center 6-h portion [hours/values 9 (0.1467) through 15 (0.8538)] of the 24-h rainfall distribution.

Calculation of Lag Time and Time of Concentration

There are many ways to consider definitions of lag time and time of concentration. SCS warns against a simplified consideration of each of them as simply a calculation of a "theoretical velocity of a segment of water moving through a hydraulic system" (SCS, 1985). The lag time (T_l) is really a weighted time of concentration for each segment of the watershed. In hydrograph analysis, it is the time from the center of mass of the rainfall to the peak of the outflow. (Note that the USGS defines lag time in most of its regression equations differently as the time from the center of mass of the rainfall to the center of mass of the outflow. This will lead to errors in calculation if methods are mixed, unless the time is adjusted based on the shape of the outflow hydrograph.) Time of concentration (T_c) is defined as the time of travel for water falling on the hydraulically most distant point in the watershed to the outlet.

The average slope within the watershed together with the overall length and retardance of overland flow are the major factors affecting the runoff rate through the watershed. However, differing storage characteristics for different storm events, seasonal changes, and urbanization will affect the lag time. Thus, average values can sometimes lead to gross errors. Nevertheless, without adequate rainfall–runoff information to estimate basin lag time, regression or synthetic methods are the only available recourse.

TABLE 7-18

Runoff Curve Numbers for Urban Areas[a]

Cover Description Cover Type and Hydrologic Condition	Average Percent Impervious Area[b]	Curve Numbers for Hydrologic Soil Groups			
		A	B	C	D
Fully developed urban areas (vegetation established)					
Open space (lawns, parks, golf courses, cemeteries, etc.)[c]					
Poor condition (grass cover <50%)		68	79	86	89
Fair condition (grass cover 50 to 75%)		49	69	79	84
Good condition (grass cover >75%)		39	61	74	80
Impervious areas:					
Paved parking lots, roofs, driveways, etc. (excluding right-of-way)		98	98	98	98
Streets and roads:					
Paved; curbs and storm drains (excluding right of way)		98	98	98	98
Paved; open ditches (including right of way)		83	89	92	93
Gravel (including right of way)		76	85	89	91
Dirt (including right of way)		72	82	87	89
Western desert urban areas:					
Natural desert landscaping (pervious areas only)		63	77	85	88
Artificial desert landscaping (impervious weed barrier, desert shrub with 1- to 2-in sand or gravel mulch and basin borders)		96	96	96	96
Urban districts:					
Commercial and business	85	89	92	94	95
Industrial	72	81	88	91	93
Residential districts by average lot size:					
1/8 acre or less (townhouses)	65	77	85	90	92
1/4 acre	38	61	75	83	87
1/3 acre	30	57	72	81	86
1/2 acre	25	54	70	80	85
1 acre	20	51	68	79	84
2 acres	12	46	65	77	82
Developing urban areas:					
Newly graded areas (pervious areas only, no vegetation)		77	86	91	94
Idle lands (CNs are determined using cover types similar to those in Table 7-20)					

[a] Average runoff condition, and $I_a = 0.2S$.

[b] The average percent impervious area shown was used to develop the composite CNs. Other assumptions are as follows: impervious areas are directly connected to the drainage system, impervious areas have a CN of 98, and pervious areas are considered equivalent to open space in good hydrologic condition. If the impervious area is not connected, the SCS method has an adjustment to reduce the effect.

[c] CNs shown are equivalent to those of pasture. Composite CNs may be computed for other combinations of open space cover type.

Source: Soil Conservation Service, Technical Release No. 55, Washington, DC, 1986.

TABLE 7-19

Cultivated Agricultural Land[a]

Cover Description Cover Type	Treatment[b]	Hydrologic Condition[c]	Curve Numbers for Hydrologic Soil Group			
			A	B	C	D
Fallow	Bare soil	—	77	86	91	94
	Crop residue cover (CR)	Poor	76	85	90	93
		Good	74	83	88	90
Row crops	Straight row (SR)	Poor	72	81	88	91
		Good	67	78	85	89
	SR + CR	Poor	71	80	87	90
		Good	64	75	82	85
	Contoured (C)	Poor	70	79	84	88
		Good	65	75	82	86
	C + CR	Poor	69	78	83	87
		Good	64	74	81	85
	Contoured and terraced (C & T)	Poor	66	74	80	82
		Good	62	71	78	81
	C&T + CR	Poor	65	73	79	81
		Good	61	70	77	80
	Small grain SR	Poor	65	76	84	88
		Good	63	75	83	87
	SR + CR	Poor	64	75	83	86
		Good	60	72	80	84
	C	Poor	63	74	82	85
		Good	61	73	81	84
	C + CR	Poor	62	73	81	84
		Good	60	72	80	83
	C&T	Poor	61	72	79	82
		Good	59	70	78	81
	C&T + CR	Poor	60	71	78	81
		Good	58	69	77	80
	Close-seeded SR or broadcast	Poor	66	77	85	89
		Good	58	72	81	85
	Legumes or C	Poor	64	75	83	85
	Rotation	Good	55	69	78	83
	Meadow C&T	Poor	63	73	80	83
		Good	51	67	76	80

[a] Average runoff condition, and $I_a = 0.2S$.
[b] Crop residue cover applies only if residue is on at least 5% of the surface throughout the year.
[c] Hydrologic condition is based on a combination of factors that affect infiltration and runoff, including (a) density and canopy of vegetative areas, (b) amount of year-round cover, (c) amount of grass or close-seeded legumes in rotations, (d) percent of residue cover on the land surface (good > 20%), and (e) degree of roughness.

Poor: Factors impair infiltration and tend to increase runoff.
Good: Factors encourage average and better than average infiltration and tend to decrease runoff.

Source: Soil Conservation Service, Technical Release No. 55, Washington, DC, 1986.

TABLE 7-20

Other Agricultural Lands[a]

Cover Description Cover Type	Hydrologic Condition	Curve Numbers for Hydrologic Soil Group			
		A	B	C	D
Pasture, grassland, or range-continuous forage for grazing[b]	Poor	68	79	86	89
	Fair	49	69	79	84
	Good	39	61	74	80
Meadow — continuous grass, protected from grazing and generally mowed for hay	—	30	58	71	78
Brush — brush–weed–grass mixture, with brush the major element[c]	Poor	48	67	77	83
	Fair	35	56	70	77
	Good[d]	30	48	65	73
Woods — grass combination (orchard or tree farm)[e]	Poor	57	73	82	86
	Fair	43	65	76	82
	Good	32	58	72	79
Woods[f]	Poor	45	66	77	83
	Fair	36	60	73	79
	Good[d]	30	55	70	77
Farmsteads — buildings, lanes, driveways, and surrounding lots	—	59	74	82	86

[a] Average runoff condition, and $I_a = 0.2S$.
[b] Poor: <50% groundcover or heavily grazed with no mulch; Fair: 50 to 75% groundcover and not heavily grazed; Good: >75% groundcover and lightly or only occasionally grazed.
[c] Poor: <50% groundcover; Fair: 50 to 75% groundcover; Good: >75% groundcover.
[d] Actual curve number is less than 30; use CN = 30 for runoff computations.
[e] CNs shown were computed for areas with 50% grass (pasture) cover. Other combinations of conditions may be computed from CNs for woods and pasture.
[f] Poor: forest litter, small trees, and brush are destroyed by heavy grazing or regular burning; Fair: woods are grazed but not burned, and some forest litter covers the soil; Good: woods protected from grazing, litter and brush adequately cover soil.

Source: Soil Conservation Service, Technical Release No. 55, Washington, DC, 1986.

TABLE 7-21

Arid and Semiarid Rangelands[a]

Cover Type	Hydrologic Condition[b]	A[c]	B	C	D
Herbaceous — mixture of grass, weeds, and low-growing brush, with brush the minor element	Poor		80	87	93
	Fair		71	81	89
	Good		62	74	85
Oak-aspen — mountain brush mixture of oak brush, aspen, mountain mahogany, bitter brush, maple, and other brush	Poor		66	74	79
	Fair		48	57	63
	Good		30	41	48
Pinyon-juniper — pinyon, juniper, or both; grass understory	Poor		75	85	89
	Fair		58	73	80
	Good		41	61	71
Sagebrush with grass understory	Poor		67	80	85
	Fair		51	63	70
	Good		35	47	55
Desert shrub — major plants include saltbush, greasewood, creosote-bush, blackbrush, bursage, paloverde, mesquite, and cactus	Poor	63	77	85	88
	Fair	55	72	81	86
	Good	49	68	79	84

[a] Average runoff condition, and $I_a = 0.2S$.
[b] Poor: <30% groundcover (litter, grass, and brush overstory); Fair: 30 to 70% groundcover; Good: >70% groundcover.
[c] Curve numbers for Group A have been developed only for desert shrub.

Source: Soil Conservation Service, Technical Release No. 55, Washington, DC, 1986.

TABLE 7-22

Conversion from Average Antecedent Moisture Conditions to Dry and Wet Conditions

CN for Average Conditions	Corresponding CNs for Dry	Corresponding CNs for Wet
100	100	100
95	87	98
90	78	96
85	70	94
80	63	91
75	57	88
70	51	85
65	45	82
60	40	78
55	35	74
50	31	70
45	26	65
40	22	60
35	18	55
30	15	50
25	12	43
15	6	30
5	2	13

Source: SCS, TP-149, 1973.

TABLE 7-23

Rainfall Groups for Antecedent Soil Moisture Conditions During Growing and Dormant Seasons

Antecedent Condition	Growing Season Five-Day Antecedent Rainfall	Dormant Season Five-Day Antecedent Rainfall
Dry[a]	Less than 1.4 inches	Less than 0.5 inches
Average[b]	1.4 to 2.1 inches	0.5 to 1.1 inches
Wet[c]	Over 2.1 inches	Over 1.1 inches

[a] An optimum condition of watershed soils, where soils are dry but not to the wilting point, and when satisfactory plowing or cultivation takes place.
[b] The average case for annual floods.
[c] When a heavy rainfall, or light rainfall and low temperatures, have occurred during the five days previous to a given storm.

Source: SCS, TP-149, 1973.

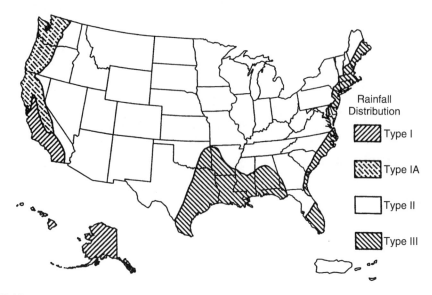

FIGURE 7-13
Approximate geographic boundaries for SCS rainfall distributions. [From U.S. Department of Agriculture, Soil Conservation Service, Engineering Division, Technical Release 55 (TR-55), 1986.]

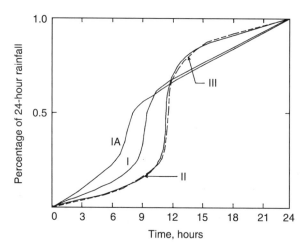

FIGURE 7-14
SCS 24-h rainfall distributions. (From U.S. Department of Agriculture, Soil Conservation Service, Engineering Division, Urban Hydrology for Small Watersheds, Technical Release 55 (TR-55), 1986.

Travel Time and Time of Concentration Empirical Equation

Lag (T_l) can be considered as a weighted time of concentration and is related to the physical properties of a watershed, such as area, length, and slope. The SCS derived the following empirical relationship between lag and time of concentration:

$$T_l = 0.6\ T_c \tag{7.21}$$

The SCS equation to estimate lag is:

$$T_l = (l^{0.8}\ (S + 1)^{0.7}) / (1900\ Y^{0.5}) \tag{7.22}$$

Urban Hydrology

TABLE 7-24

Ordinates of the SCS Type I, IA, II, Type III Precipitation Distributions

Storm Time (Hours)	Precipitation Ratio Type				Storm Time (Hours)	Precipitation Ratio			
	I	IA	II	III		I	IA	II	III
0.0	0.000	0.000	0.0000	0.000					
0.5	0.008	0.010	0.0053	0.005	12.5	0.705	0.683	0.7351	0.702
1.0	0.017	0.022	0.0108	0.010	13.0	0.727	0.701	0.7724	0.751
1.5	0.026	0.036	0.0164	0.015	13.5	0.748	0.719	0.7989	0.785
2.0	0.035	0.051	0.0223	0.020	14.0	0.767	0.736	0.8179	0.811
2.5	0.045	0.067	0.0284	0.026	14.5	0.784	0.753	0.8380	0.830
3.0	0.055	0.083	0.0347	0.032	15.0	0.800	0.769	0.8538	0.848
3.5	0.065	0.099	0.0414	0.037	15.5	0.816	0.785	0.8676	0.867
4.0	0.076	0.116	0.0483	0.043	16.0	0.830	0.800	0.8801	0.886
4.5	0.087	0.135	0.0555	0.050	16.5	0.844	0.815	0.8914	0.895
5.0	0.099	0.156	0.0632	0.057	17.0	0.857	0.830	0.9019	0.904
5.5	0.122	0.179	0.0712	0.065	17.5	0.870	0.844	0.9115	0.913
6.0	0.125	0.204	0.0797	0.072	18.0	0.882	0.858	0.9206	0.922
6.5	0.140	0.233	0.0887	0.081	18.5	0.893	0.871	0.9291	0.930
7.0	0.156	0.268	0.0984	0.089	19.0	0.905	0.884	0.9371	0.939
7.5	0.174	0.310	0.1089	0.102	19.5	0.916	0.896	0.9446	0.948
8.0	0.194	0.425	0.1203	0.115	20.0	0.926	0.908	0.9519	0.957
8.5	0.219	0.480	0.1328	0.130	20.5	0.936	0.920	0.9588	0.962
9.0	0.254	0.520	0.1467	0.148	21.0	0.946	0.932	0.9653	0.968
9.5	0.303	0.550	0.1625	0.167	21.5	0.955	0.944	0.9719	0.973
10.0	0.515	0.577	0.1808	0.189	22.0	0.965	0.956	0.9777	0.979
10.5	0.583	0.601	0.2042	0.216	22.5	0.974	0.967	0.9836	0.984
11.0	0.624	0.623	0.2351	0.250	23.0	0.983	0.978	0.9892	0.989
11.5	0.654	0.644	0.2833	0.298	23.5	0.992	0.989	0.9947	0.995
12.0	0.682	0.664	0.6632	0.500	24.0	1	1	1	1

Source: U.S. Department of Agriculture, Soil Conservation Service, Engineering Division, Urban Hydrology for Small Watersheds, Technical Release 55 (TR-55), 1986.

where T_1 = lag (h); l = length of mainstream to farthest divide (ft); Y = average slope of watershed (%); $S = 1000/CN - 10$; and CN = SCS curve number.

The lag time can be corrected for the effects of urbanization by using Figures 7-15 and 7-16. The amount of modifications to the hydraulic flow length usually must be determined from topographic maps or aerial photographs following a field inspection of the area. The modification to the hydraulic flow length not only includes pipes and channels but also the length of flow in streets and driveways.

In small urban areas (less than 2000 acres), a curve number method can be used to estimate the time of concentration from watershed lag. In this method, the lag for the

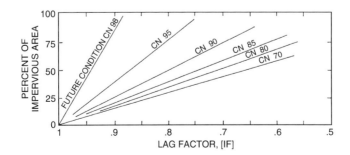

FIGURE 7-15
Factors for adjusting lag for impervious areas within the watershed.

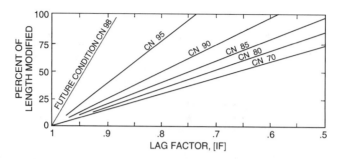

FIGURE 7-16
Factors for adjusting lag for main channel hydraulic improvement. (From HEC-19.)

runoff from an increment of excess infall can be considered as the time between the center of mass of the excess rainfall increment and the peak of its incremental outflow hydrograph.

This equation and the urbanization correction factors should be tested to determine if satisfactory results are obtained for local conditions.

After the lag time is adjusted for the effects of urbanization, Equation (7.21) can be used to calculate the time of concentration for use in the SCS method.

If lag and time of concentration equations have been developed for use within your municipality, these equations can be substituted for Equations (7.21) and (7.22) for use with the SCS Unit Hydrograph method. Also, variations of the Manning's equation can be used to estimate travel time and time of concentration for the SCS Unit Hydrograph method. The following describes the procedures and equations for using the Manning's equation.

Travel Time and Time of Concentration — Manning's Equation

In using the Manning's equation, travel time (T_t) is the time it takes water to travel from one location to another within a watershed, through the various components of the drainage system. Time of concentration (T_c) is computed by summing all the travel times for consecutive components of the drainage conveyance system from the hydraulically most distant point of the watershed to the point of interest within the watershed. Following is a discussion of related procedures and equations (SCS, 1986).

Water moves through a watershed as sheet flow, shallow concentrated flow, open channel flow, or some combination of these. The type that occurs is a function of the conveyance system and is best determined by field inspection. The travel time is the ratio of flow length to flow velocity:

$$T_t = L/(3600V) \qquad (7.23)$$

where T_t = travel time (hr); L = flow length (ft); V = average velocity (ft/s); and 3600 = conversion factor from seconds to hours.

The time of concentration is the sum of T_t values for the various consecutive flow segments along the path extending from the hydraulically most-distant point in the watershed to the point of interest.

$$T_c = T_{t1} + T_{t2} + \dots T_{tm} \qquad (7.24)$$

where T_c = time of concentration (hr), and m = number of flow segments.

TABLE 7-25

Roughness Coefficients (Manning's n) for Sheet Flow

Surface Description	n
Smooth surfaces (concrete, asphalt, gravel, or bare soil)	0.011
Fallow (no residue)	0.05
Cultivated soils:	
Residue cover <20%	0.06
Residue cover >20%	0.17
Grass:	
Short grass prairie	0.15
Dense grasses[a]	0.24
Bermudagrass[b]	0.41
Range (natural)	0.13
Woods:[c]	
Light underbrush	0.40
Dense underbrush	0.80

[a] The n values are a composite of information by Engman (1986).
[b] Includes species such as weeping lovegrass, bluegrass, buffalo grass, blue grama grass, and native grass mixtures.
[c] When selecting n, consider cover to a height of about 0.1 ft. This is the only part of the plant cover that will obstruct sheet flow.

Source: U.S. Department of Agriculture, Soil Conservation Service, Engineering Division, Technical Release 55 (TR-55), 1986.

Sheet flow is flow over plane surfaces, and it usually occurs in the headwater of streams. With sheet flow, the friction value (Manning's n) is an effective roughness coefficient that includes the effect of raindrop impact; drag over the plane surface; obstacles such as litter, crop ridges, and rocks; and erosion and transportation of sediment (see Table 7-25). These n values are for very shallow flow depths of about 0.1 ft or so. For sheet flow of less than 300 ft, use Manning's kinematic solution (Overton and Meadows, 1976) to compute T_t:

$$T_t = [0.42 \, (nL)^{0.8} / (P_2)^{0.5}(S)^{0.4}] \tag{7.25}$$

where T_t = travel time (min); n = Manning roughness coefficient; L = flow length (ft); P_2 = 2-year, 24-h rainfall = 3.6 in; S = slope of hydraulic grade line (land slope) (ft/ft).

This simplified form of the Manning's kinematic solution is based on the following:

- Shallow steady uniform flow
- Constant intensity of rainfall excess (rain available for runoff)
- Rainfall duration of 24 h
- Minor effect of infiltration on travel time

Another approach is to use the kinematic wave equation. For details on using this equation, consult the publication by Regan (1971).

After a maximum of 300 ft (100 ft in urban areas), sheet flow usually becomes shallow concentrated flow. The average velocity for this flow can be determined from Figure 7-17, in which average velocity is a function of watercourse slope and type of channel.

Average velocities for estimating travel time for shallow concentrated flow can be computed using Figure 7-17 or the following equations. These equations can also be used for slopes less than 0.005 ft/ft.

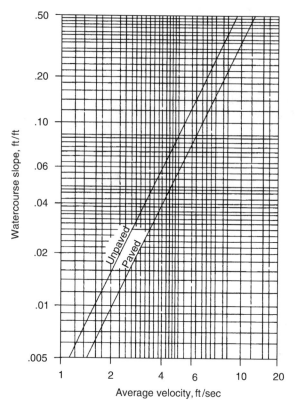

FIGURE 7-17
Average velocities — shallow concentrated flow. (From U.S. Department of Agriculture, Soil Conservation Service, Engineering Division, Urban Hydrology for Small Watersheds, Technical Release 55 (TR-55), 1986.)

Unpaved $\qquad V = 16.1345(S)^{0.5}$ (7.26)

Paved $\qquad V = 20.3282(S)^{0.5}$ (7.27)

where V = average velocity (ft/s), and S = slope of hydraulic grade line (watercourse slope) (ft/ft).

These two equations are based on the solution of Manning's equation with different assumptions for n (Manning's roughness coefficient) and r [hydraulic radius (ft)]: for unpaved areas, n is 0.05, and r is 0.4; for paved areas, n is 0.025, and r is 0.2.

After determining average velocity using Figure 7-17 or Equations (7.26) or (7.27), use Equation (7.23) to estimate travel time for the shallow concentrated-flow segment.

Open channels are assumed to begin where surveyed cross-sectional information has been obtained, where channels are visible on aerial photographs, or where blue lines (indicating streams) appear on U.S. Geological Survey (USGS) quadrangle sheets. Manning's equation or water surface profile information can be used to estimate average flow velocity. Average flow velocity is usually determined for bank-full elevation.

Manning's equation $\qquad V = [1.49 \, (r)^{2/3} \, (s)^{1/2}] / n$ (7.28)

where V = average velocity (ft/s); r = hydraulic radius (equal to a/p_w) (ft); a = cross-sectional flow area (ft²); p_w = wetted perimeter (ft); s = slope of the hydraulic grade line (ft/ft); and n = Manning's roughness coefficient for open channel flow.

After average velocity is computed using Equation (7.28), T_t for the channel segment can be estimated using Equation (7.23). Velocity in channels should be calculated from the Manning equation. Cross sections from all channels that have been field-checked should be used in the calculations. This is particularly true of areas below dams or other flow-control structures.

Sometimes it is necessary to estimate the velocity of flow through a reservoir or lake at the outlet of a watershed. This travel time is normally small and can be assumed to be zero. If the travel time through the reservoir or lake is important to the analysis, then the hydrograph should be routed through the storage facility. A reservoir can have an impact in reducing peak flows that can be accounted for by routing.

The following are some limitations of this procedure that should be taken into account:

- Manning's kinematic solution should not be used for sheet flow longer than 300 ft (100 ft in urban areas). Some areas have recommended shorter distances. (*Georgia Stormwater Management Manual* recommends 50 feet in urban areas.)
- In watersheds with storm drains, carefully identify the appropriate hydraulic flow path to estimate T_c. Storm drains generally handle only a small portion of a large event. The rest of the peak flow travels by streets, lawns, and so on, to the outlet.
- A culvert or bridge can act as a reservoir outlet if there is significant storage behind it. This requires detailed storage routing procedures for analysis.

Drainage Area

The drainage area of a watershed is determined from topographic maps and field surveys. For large drainage areas, it might be necessary to divide the area into subdrainage areas to account for major land-use changes, to account for the location of stormwater management facilities, to obtain analysis results at different points within the drainage area, and to route flows to points of interest. Also, a field inspection of the existing or proposed drainage systems should be made to determine if the natural drainage divides have been altered.

Procedures, Tables, and Figures

The following sections provide procedures for rainfall excess, peak flow calculation, hydrograph calculation, and development of a SCS unit hydrograph. The procedures herein were taken directly or developed from the SCS Technical Release 55 (TR-55), and the *National Engineering Handbook*, Chapter 4 (NEH4), which present simplified procedures to calculate storm runoff volume, peak rate of discharges, and hydrographs.

SCS Peak Discharges

The SCS peak discharge method is applicable for estimating the peak runoff rate from watersheds with homogeneous land uses. The following method is based on the results of computer analyses performed using TR-20, "Computer Program for Project Formulation — Hydrology" (SCS, 1983). The peak discharge equation is as follows:

$$Q_p = q_u A Q F_p \tag{7.29}$$

where Q_p = peak discharge (cfs); q_u = unit peak discharge (cfs/mi²/in); A = drainage area (mi²); Q = runoff (in); and F_p = pond and swamp adjustment factor.

The input requirements for this method are as follows:

- T_c — hours
- Drainage area — mi²
- Rainfall distribution type
- 24-hour design rainfall
- CN value
- Pond and swamp adjustment factor (If pond and swamp areas are spread throughout the watershed and are not considered in the T_c computation, an adjustment is needed.)

Computations

Computations for the peak discharge method, using a Type II rainfall distribution, proceed as follows:

1. The 24-h rainfall depth is determined from the following table for the selected return frequency.

Frequency	24-h Rainfall
2-year	3.34 in
5-year	4.62 in
10-year	5.48 in
25-year	6.67 in
50-year	7.59 in
100-year	8.50 in

2. The runoff curve number, CN, is estimated from Table 7-18, and direct runoff, Q, is calculated using Equation (7.29).
3. The CN value is used to determine the initial abstraction, I_a, from Table 7-27, and the ratio I_a/P is then computed. (P = accumulated rainfall or potential maximum runoff.)
4. The watershed time of concentration is computed and is used with the ratio I_a/P to obtain the unit peak discharge, q_u, from Figure 7-20. If the ratio I_a/P lies outside the range shown in Figure 7-20, the limiting values or another peak discharge method should be used.
5. The pond and swamp adjustment factor, F_p, is estimated from Table 7-26.
6. The peak runoff rate is computed using Equation (7.29).

Limitations

The accuracy of the peak discharge method is subject to specific limitations, including the following:

1. The watershed must be hydrologically homogeneous and describable by a single CN value.
2. The watershed can have only one main stream, or, if more than one, the individual branches must have nearly equal time of concentrations.
3. Hydrologic routing cannot be considered.

TABLE 7-26

Adjustment Factor F_p for Pond and Swamp Areas that are Spread Throughout the Watershed

Pond and Swamp Areas (%)	F_p
0.0	1.00
0.2	0.97
1.0	0.87
3.0	0.75
5.0	0.72

Source: U.S. Department of Agriculture, Soil Conservation Service, Engineering Division, Technical Release 55 (TR-55), 1986.

TABLE 7-27

I_a Values for Runoff Curve Numbers

Curve Number	I_a (in)	Curve Number	I_a (in)
40	3.000	70	0.857
41	2.878	71	0.817
42	2.762	72	0.778
43	2.651	73	0.740
44	2.545	74	0.703
45	2.444	75	0.667
46	2.348	76	0.632
47	2.255	77	0.597
48	2.167	78	0.564
49	2.082	79	0.532
50	2.000	80	0.500
51	1.922	81	0.469
52	1.846	82	0.439
53	1.774	83	0.410
54	1.704	84	0.381
55	1.636	85	0.353
56	1.571	86	0.326
57	1.509	87	0.299
58	1.448	88	0.273
59	1.390	89	0.247
60	1.333	90	0.222
61	1.279	91	0.198
62	1.226	92	0.174
63	1.175	93	0.151
64	1.125	94	0.128
65	1.077	95	0.105
66	1.030	96	0.083
67	0.985	97	0.062
68	0.941	98	0.041
69	0.899		

Source: U.S. Department of Agriculture, Soil Conservation Service, Engineering Division, Technical Release 55 (TR-55), 1986.

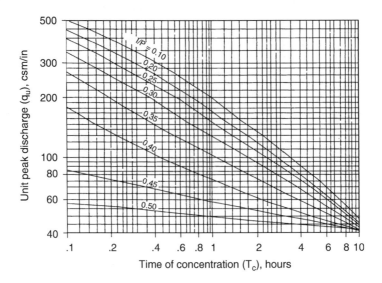

FIGURE 7-18
SCS Type I unit peak discharge graph. [From U.S. Department of Agriculture, Soil Conservation Service, Engineering Division, Urban Hydrology for Small Watersheds, Technical Release 55 (TR-55), 1986.]

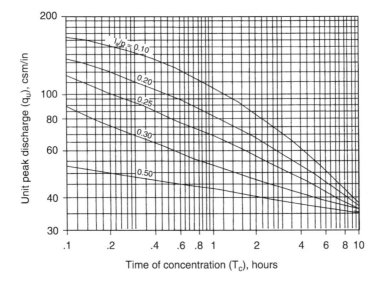

FIGURE 7-19
SCS Type IA unit peak discharge graph. [From U.S. Department of Agriculture, Soil Conservation Service, Engineering Division, Urban Hydrology for Small Watersheds, Technical Release 55 (TR-55), 1986.]

4. F_p applies only to areas located away from the main flow path.
5. Accuracy is reduced if the ratio I_a/P is outside the range given in Figures 7-18 through 7-21.
6. The weighted CN value must be greater than or equal to 40 and less than or equal to 98.
7. The same procedure should be used to estimate pre- and postdevelopment time of concentration when computing pre- and postdevelopment peak discharge.
8. The watershed time of concentration must be between 0.1 and 10 h.

Urban Hydrology

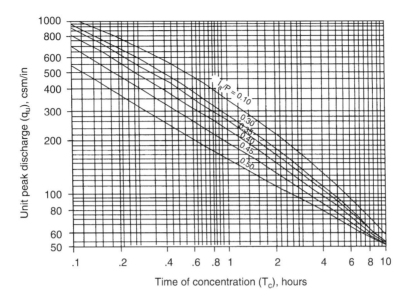

FIGURE 7-20
SCS Type II unit peak discharge graph. [From U.S. Department of Agriculture, Soil Conservation Service, Engineering Division, Urban Hydrology for Small Watersheds, Technical Release 55 (TR-55), 1986.]

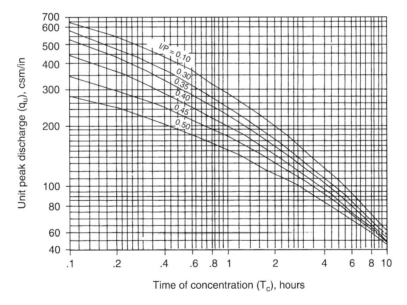

FIGURE 7-21
SCS Type III unit peak discharge graph. [From U.S. Department of Agriculture, Soil Conservation Service, Engineering Division, Urban Hydrology for Small Watersheds, Technical Release 55 (TR-55), 1986.]

SCS Peak Discharge Example

Compute the 25-year peak discharge for a 50-acre wooded watershed that will be developed as follows:

- (1) Forest land — poor cover (hydrologic soil group B) = 10 acres
- (2) Forest land — poor cover (hydrologic soil group C) = 10 acres

- (3) Townhouse residential (hydrologic soil group B) = 20 acres
- (4) Industrial development (hydrologic soil group C) = 10 acres
- Other data include: percentage of pond and swamp area = 0

1. Calculate rainfall excess:
   a. The 25-year, 24-h rainfall is 6.67 in.
   b. The composite weighted runoff coefficient is as follows:

Dev. #	CN	% Total	Area	Composite CN
1	55	0.20	10 ac	11.0
2	70	0.20	10 ac	14.0
3	85	0.40	20 ac	34.0
4	91	0.20	10 ac	18.2
Total		1.00	50 ac	77.2 (use 77)

   c. From Figure 7-10, $Q = 4.0$ in.

2. Calculate time of concentration:

Segment	Type of Flow	Length	Slope (%)
1	Overland n = 0.45	70 ft	2.0
2	Shallow channel	750 ft	1.7
3	Main channel[a]	1100 ft	0.20

   [a] For the main channel, n = .025, width = 10 ft, depth = 2 ft, rectangular channel.

   a. Segment 1 — Travel time from Equation (7.25) with $P_2 = 3.34$ in

   $$T_t = [0.42(0.45 \times 60)^{.8}] / [(3.34)^{.5} (.035)^{.4}]$$

   $$T_t = 12.3 \text{ min}$$

   b. Segment 2 — Travel time from Figure 7-17 and Equation (7.23)

   $$V = 2.6 \text{ ft/s [from Figure 7-17 or Equation (7.23)]}$$

   $$T_t = 750 / 60 (2.6) = 4.8 \text{ mi}$$

   c. Segment 3 — Using Equations (7.28) and (7.23)

   $$V = (1.49/.025) (1.43)^{.67} (.002)^{.5} = 3.4 \text{ ft/s}$$

   $$T_t = 1100 / 60 (3.4) = 5.4 \text{ min}$$

   $$T_c = 12.3 + 4.8 + 5.4 = 22.5 \text{ min } (.38 \text{ h})$$

3. Calculate I_a/P for CN = 77 (from Step 1), $I_a = 0.597$ (Table 7-27):

   $$I_a/P = (.597 / 6.67) = .09$$

a. (Note: Use $I_a/P = 0.10$ to facilitate use of Figures 7-18 through 7-21. Straight line interpolation could also be used.)
4. Estimate unit discharge q_u from Figure 7-20 = 600 cfs
5. Calculate peak discharge with $F_p = 1$ using Equation (7.29):

$$Q_{25} = 600 \, (50/640) \, (4.0) \, (1) = 188 \text{ cfs}$$

SCS Hydrograph Generation

In addition to estimating the peak discharge, the SCS method can be used to estimate the entire hydrograph. The Soil Conservation Service has developed a tabular hydrograph procedure that can be used to generate the hydrograph for small drainage areas. The tabular hydrograph procedure uses unit discharge hydrographs, which have been generated for a series of time of concentrations. SCS has developed tables that give the unit discharges (csm/in) for different times of concentration for the four SCS rainfall distributions (USDA, 1986). Straight-line interpolation can be used for times of concentrations and travel times between the values given in the tables. Because these tables do not account for reservoir routing through storage facilities and are more difficult to use if the drainage area is not homogeneously developed, their application within most municipal applications is limited. Due to this limitation and the availability of more appropriate hydrologic methods for municipal application, the tables and details of their use are not presented in this book but can be obtained from the SCS publication.

For drainage areas that are not homogeneous, where hydrographs need to be generated from subareas and then routed and combined at a point downstream, the designer is referred to the procedures outlined by the SCS in the 1986 version of TR-55 available from the National Technical Information Service in Springfield, Virginia. The catalog number for TR-55, Urban Hydrology for Small Watersheds, is PB87-101580. These procedures involve some channel and reservoir routing assumptions that may affect the accuracy of the hydrologic estimates. What would be more appropriate for most municipalities' hydrologic studies, which involve channel and reservoir routing, would be to use one of the many hydrologic computer models available (e.g., HEC-1, TR-55).

Rainfall Excess Using SCS Methods

Calculation of rainfall excess using the SCS methodology involves a step-wise procedure. Combining Equations (7.13) and (7.14) yields:

$$F = S(P - I_a)/(P - I_a + S) \tag{7.30}$$

Differentiating each side with respect to time yields a loss rate (dF/dt) and a rainfall intensity (dP/dt). Using a difference approximation yields:

$$(F_{i+1} - F_i)/d = S_2/(P_A - I_a + S)^2 * (P_{i+1} - P_i)/d \tag{7.31}$$

$$P > I_a$$

where F_i = loss at time step i; F_{i+1} = loss at time step $i + 1$; d = time step; P_A = average rainfall during time step = $(P_i + P_{i+1})/2$; P_i = rainfall rate at time step i; and P_{i+1} = rainfall rate at time step $i + 1$.

The rainfall excess can then be calculated for each time step (a necessary step in the application of unit hydrographs) through the use of a table with column headings of time, cumulative rainfall, cumulative abstractions ($I_a + F$) from Equation (7.30), cumulative excess rainfall, and excess rainfall hyetograph.

The steps in the development of a rainfall excess hyetograph are as follows:

1. *Determine the SCS curve number for the area.* Divide the area into suitable subbasins as appropriate. Determine the antecedent moisture condition necessary for the design. While some agencies require use of wet conditions for less frequent storm designs, comparison of this approach with gauge data often indicates an overprediction of peak flows (especially when using longer duration storms). Remember that the goal is to reproduce a storm event to test the drainage system, not to develop an overly conservative hypothetical storm with a nominal frequency that is significantly different from its actual event frequency.

2. *Determine the frequency and duration of the design storm.* For smaller watersheds, use of a long duration is not warranted or, in some instances, realistic. For smaller, more impervious watersheds, a shorter, more intense storm will sometimes yield a higher peak flow (Aron and Kibler, 1990). This is especially true if there is a lot of connected imperviousness that may act almost like a separate watershed. The median storm duration east of the Mississippi is about 6 h. Choice of a duration somewhat longer than the time of concentration is usually sufficient.

3. *Calculate S, I_a, T_c, T_l, and d from the equations in the unit hydrograph section.* Make sure d is within the limits set above. It is convenient to round it to the nearest minute and to see if a multiple of 60 will fit within the limits of 0.2 and 0.25 T_l.

4. *Distribute the total rainfall according to the proportions from the proper SCS S-curve.* The difference between the end and the beginning of the storm duration interval chosen is the total portion of the S-curve contained in the storm duration. This can be done by hand using Figure 7-14 or Table 7-24. Then, for each time increment, a proportion can be set up as:

$$P_i/P = R_i/R \qquad (7.32)$$

where P_i = the increment of rainfall at time interval d for use with the unit hydrograph; R_i = the increment of the S-curve at time interval d; and R = total S-curve rainfall dimensionless rainfall at the storm duration

5. *Set up a calculation table (see example) below.* The table should have column headings for time, S-curve accumulated values (R), accumulated rainfall (P), initial abstraction (I_a), infiltration and other losses (F), rainfall excess (Q), and hyetograph values. This format works well with spreadsheet applications. Aaron (1992) provides a linear filter to correct for certain defects in the infiltration assumptions inherent in this SCS methodology.

Rainfall Excess Example

Step 1: Curve Number = 70

Step 2: 10-year, 6-h storm for Charlotte, North Carolina, = 3.67 in; SCS Type II rainfall distribution

TABLE 7-28

Example Problem Solution Table

(1) Time Ord. i (h)	(2) Cumulative Precipitation R_i (in)	(3) SCS Init. P_i (in)	(4) Cumulative Abs. Ia_i (in)	(5) Cumulative Losses F_i (in)	(6) Cumulative Excess Q_i (in)	(7) Cumulative Excess Hyetograph (in)
1.0	0.147	0.000	0.000	0.000	0.00	0.00
1.1	0.150	0.016	0.016	0.000	0.00	0.00
1.2	0.153	0.033	0.033	0.000	0.00	0.00
1.3	0.156	0.049	0.049	0.000	0.00	0.00
1.4	0.159	0.066	0.066	0.000	0.00	0.00
1.5	0.163	0.082	0.082	0.000	0.00	0.00
1.6	0.166	0.101	0.101	0.000	0.00	0.00
1.7	0.170	0.120	0.120	0.000	0.00	0.00
1.8	0.173	0.139	0.139	0.000	0.00	0.00
1.9	0.177	0.158	0.158	0.000	0.00	0.00
2.0	0.181	0.177	0.177	0.000	0.00	0.00
2.1	0.185	0.201	0.201	0.000	0.00	0.00
2.2	0.190	0.226	0.226	0.000	0.00	0.00
2.3	0.195	0.250	0.250	0.000	0.00	0.00
2.4	0.200	0.274	0.274	0.000	0.00	0.00
2.5	0.204	0.298	0.298	0.000	0.00	0.00
2.6	0.210	0.331	0.331	0.000	0.00	0.00
2.7	0.217	0.363	0.363	0.000	0.00	0.00
2.8	0.223	0.395	0.395	0.000	0.00	0.00
2.9	0.229	0.427	0.427	0.000	0.00	0.00
3.0	0.235	0.459	0.459	0.000	0.00	0.00
3.1	0.245	0.509	0.509	0.000	0.00	0.00
3.2	0.254	0.559	0.559	0.000	0.00	0.00
3.3	0.264	0.609	0.609	0.000	0.00	0.00
3.4	0.274	0.659	0.659	0.000	0.00	0.00
3.5	0.283	0.709	0.709	0.000	0.00	0.00
3.6	0.359	1.103	0.858	0.232	0.01	0.01
3.7	0.435	1.498	0.858	0.557	0.08	0.07
3.8	0.511	1.892	0.858	0.833	0.20	0.12
3.9	0.587	2.286	0.858	1.072	0.36	0.16
4.0	0.663	2.681	0.858	1.279	0.54	0.19
4.1	0.678	2.755	0.858	1.316	0.58	0.04
4.2	0.692	2.830	0.858	1.351	0.62	0.04
4.3	0.706	2.905	0.858	1.386	0.66	0.04
4.4	0.721	2.979	0.858	1.419	0.70	0.04
4.5	0.735	3.054	0.858	1.452	0.74	0.04
4.6	0.743	3.093	0.858	1.469	0.77	0.02
4.7	0.750	3.131	0.858	1.486	0.79	0.02
4.8	0.757	3.170	0.858	1.502	0.81	0.02
4.9	0.765	3.209	0.858	1.519	0.83	0.02
5.0	0.772	3.248	0.858	1.535	0.85	0.02
5.1	0.778	3.275	0.858	1.546	0.87	0.02
5.2	0.783	3.303	0.858	1.557	0.89	0.02
5.3	0.788	3.330	0.858	1.568	0.90	0.02
5.4	0.794	3.358	0.858	1.579	0.92	0.02
5.5	0.799	3.385	0.858	1.590	0.94	0.02
5.6	0.803	3.407	0.858	1.599	0.95	0.01
5.7	0.807	3.428	0.858	1.607	0.96	0.01
5.8	0.811	3.450	0.858	1.616	0.98	0.01
5.9	0.816	3.471	0.858	1.624	0.99	0.01
6.0	0.820	3.493	0.858	1.632	1.00	0.01
6.1	0.823	3.512	0.858	1.640	1.01	0.01
6.2	0.827	3.531	0.858	1.647	1.03	0.01

TABLE 7-28 *(Continued)*

Example Problem Solution Table

(1) Time Ord. i (h)	(2) Cumulative Precipitation R_i (in)	(3) SCS Init. P_i (in)	(4) Cumulative Abs. Ia_i (in)	(5) Cumulative Losses F_i (in)	(6) Cumulative Excess Q_i (in)	(7) Cumulative Excess Hyetograph (in)
6.3	0.831	3.550	0.858	1.654	1.04	0.01
6.4	0.834	3.569	0.858	1.661	1.05	0.01
6.5	0.838	3.588	0.858	1.668	1.06	0.01
6.6	0.841	3.604	0.858	1.674	1.07	0.01
6.7	0.844	3.621	0.858	1.681	1.08	0.01
6.8	0.847	3.637	0.858	1.687	1.09	0.01
6.9	0.851	3.654	0.858	1.693	1.10	0.01
7.0	0.854	3.670	0.858	1.699	1.11	0.01

Step 3: $A = 700$ acres; $T_c = 40$ min (calculated); $T_l = 24$ min; $T_p = 27$ min; $d = 6$ min (equals $0.25\ T_l$); $S = 1000/70 - 10 = 4.29$ in; and $I_a = 0.2S = 0.0858$ in

Steps 4 and 5: The time actually started in hour 9 of the SCS Type 2 curve. A description of each column follows:

Time ordinate, in this example, starts with hour 1.0 and goes through hour 7.0 for a total of 6 h. The time increment is 0.1 h (6 min) equal to d.

The total dimensionless time period is hour 15.0 minus hour 9.0 from the SCS Type II storm distribution table or $0.8538 - 0.1467 = 0.7071$. Each dimensionless rainfall increment was linearly interpolated at 0.1 h intervals between the half-hour interval values given in the table.

The precipitation increments were calculated from Equation (7.32) as follows:

$$P_i = 3.67 * ((R_i - R_{i-1})/0.7071) + P_{i-1}$$

Note that some rounding has occurred in the table presentation that did not take place in the spreadsheet calculation.

The initial abstraction was deducted from the precipitation until the total of 0.2S or 0.0858 in was obtained.

Values were taken from a solution of Equation (7.30) using the accumulated values from Column 3.

This is Column 3 minus Column 4 and Column 5.

This value is the incremental excess from Column 6.

Figure 7-22 depicts the accumulated rainfall losses and excess (I_a, F, and P) from Table 7-28. Figure 7-23 shows the excess rainfall hyetograph. The step-like appearance and slight increase in rainfall comes from the half-hour linear interpolation increments and the fact that, while equal values of rainfall are falling for a half hour in 6-min increments, the infiltration (F) is decreasing throughout that time period.

SCS Dimensionless Unit Hydrographs

Unit hydrograph hydrological methods can be employed using SCS methods and information. The SCS methods use, as a basis, a dimensionless unit hydrograph with a shape

Urban Hydrology

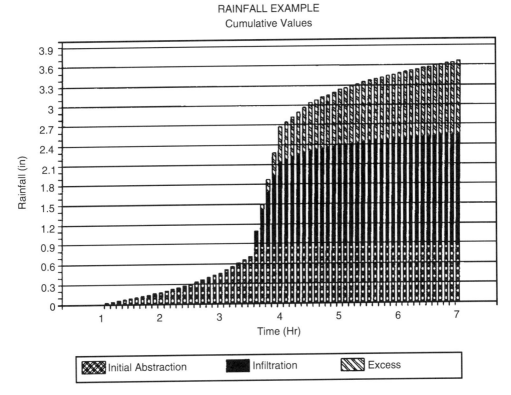

FIGURE 7-22
Rainfall losses and excess values.

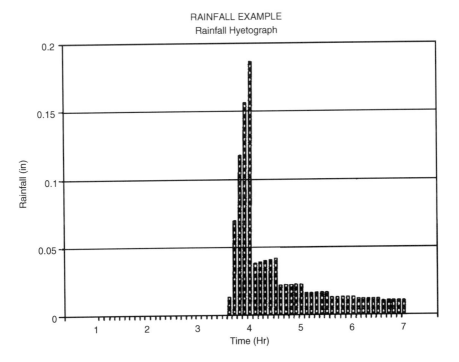

FIGURE 7-23
Rainfall hyetograph.

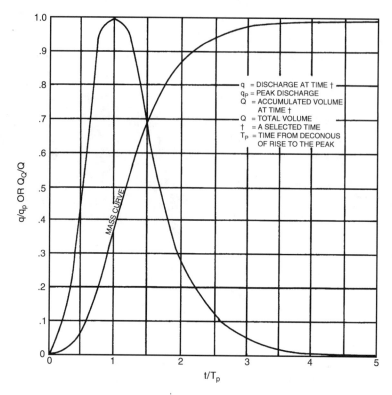

FIGURE 7-24
Dimensionless unit hydrograph and mass curve. (From U.S. Department of Agriculture, Soil Conservation Service, Engineering Division, National Engineering Handbook, Section 4, Hydrology, 1985.)

that is an average compilation of extensive analysis of measured data. Details of the derivation of the unit hydrograph can be found in the SCS National Engineering Handbook, Section 4 (SCS, 1985).

Dimensionless Unit Hydrograph Discussion

Figure 7-24 depicts the average dimensionless curvilinear unit hydrograph, while the discharge ratios are tabulated in Table 7-29. The horizontal axis is a ratio of the time to the time-to-peak (T_p), while the vertical axis is a ratio of discharge to the peak discharge (q_p) or accumulated volume to total flow volume (Q).

It can be seen from this figure that the time base is approximately five times the time-to-peak, and about 3/8 of the total volume occurred prior to the time of the peak.

The triangular hydrograph is a practical approximation of the curvilinear unit hydrograph. Its geometric makeup can easily be described mathematically, which makes it useful in the process of estimating discharge rates and produces results that are sufficiently accurate for most stormwater management facility designs. Figure 7-25 depicts the triangular dimensionless unit hydrograph and the curvilinear hydrograph. The areas under the rising limbs of the two hydrographs are the same, as is the total volume. The SCS developed the following equation to estimate the peak rate of discharge for the dimensionless unit hydrograph:

$$q_p = 484 \, A(Q/T_p) \tag{7.33}$$

TABLE 7-29

Ratios for Dimensionless Unit Hydrograph and Mass Curve

Time Ratios (t/T_p)	Discharge Ratios (q/q_p)	Mass Curve Ratios (Q_a/Q)
0.0	0.000	0.000
0.1	0.030	0.001
0.2	0.100	0.006
0.3	0.190	0.017
0.4	0.310	0.035
0.5	0.470	0.065
0.6	0.660	0.107
0.7	0.820	0.163
0.8	0.930	0.228
0.9	0.990	0.300
1.0	1.000	0.375
1.1	0.990	0.450
1.2	0.930	0.522
1.3	0.860	0.589
1.4	0.780	0.650
1.5	0.680	0.705
1.6	0.560	0.751
1.7	0.460	0.790
1.8	0.390	0.822
1.9	0.330	0.849
2.0	0.280	0.871
2.2	0.207	0.908
2.4	0.147	0.934
2.6	0.107	0.953
2.8	0.077	0.967
3.0	0.055	0.977
3.2	0.040	0.984
3.4	0.029	0.989
3.6	0.021	0.993
3.8	0.015	0.995
4.0	0.011	0.997
4.5	0.005	0.999
5.0	0.000	1.000

Source: U.S. Department of Agriculture, Soil Conservation Service, Engineering Division, *National Engineering Handbook*, Section 4, Hydrology, 1985.

where q_p = peak rate of discharge (cfs); A = area (mi^2); Q = storm runoff volume (in); T_p = time to peak (hr); and 484 includes appropriate conversion factors for the units.

Additional relationships (in consistent units) that can be derived from a geometric consideration of the SCS dimensionless unit hydrographs are:

$$T_L = 0.6\, T_c \tag{7.34}$$

$$T_p = d/2 + T_L \tag{7.35}$$

$$T_b = 2.67\, T_p \text{ (triangular hydrograph)} \tag{7.36}$$

where T_L = watershed lag time (h), and T_b = time of base (h).

According to SCS, d must be between 0.2 and 0.25 T_p. If $d = 0.2\, T_p$, then:

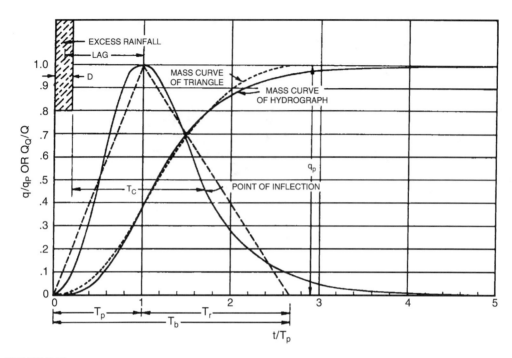

FIGURE 7-25
Dimensionless curvilinear unit hydrograph and equivalent triangular hydrograph. (From U.S. Department of Agriculture, Soil Conservation Service, Engineering Division, National Engineering Handbook, Section 4, Hydrology, 1985.)

$$d = 1.11\ T_p = 0.133\ T_c \qquad (7.37)$$

$$T_p = 0.667\ T_c \qquad (7.38)$$

where d = rainfall time increment (hr).

The constant 484, or peak rate factor (PR), is valid for the average SCS dimensionless unit hydrograph. Any change in the dimensionless unit hydrograph reflecting a change in the percent of volume under the rising side would cause a corresponding change in the shape factor associated with the triangular hydrograph and, therefore, a change in the constant 484. This constant has been known to vary from about 600 in steep terrain to 300 in flat terrain.

Characteristics of the dimensionless hydrograph vary with the size, shape, and slope of the tributary drainage area. The most significant characteristics affecting the dimensionless hydrograph shape are the basin lag and the peak discharge for a given rainfall. Basin lag in this method is defined as the time from the center of mass of rainfall excess to the hydrograph peak. Steep slopes, compact shape, and an efficient drainage network tend to make lag time short and peaks high; flat slopes, elongated shape, and an inefficient drainage network tend to make lag time long and peaks low for the same rainfall. Any change in the 484 factor would also need to be reflected in a change in the shape of the hydrograph to ensure a unit of volume remains under the curve. Special hydrographs have been developed for specific applications.

Meadows, Morris, and Spearmen (1992) provide a methodology for the modification of the SCS curvilinear hydrograph for other peak rate factors. For the curvilinear hydrograph, a two-parameter gamma function was developed that fits the curvilinear form closely:

Urban Hydrology

$$q = q_p * ((t/tp) * e^{(1-(t/tp))})^{(N-1)} \qquad (7.39)$$

N can be found from the fit equation:

$$N = 0.8679 * e^{(0.00353*PR)} \qquad (7.40)$$

where PR is the peak rate factor chosen (484 being the normal value for an N of 4.7). The procedure then is to choose a shape factor and calculate q_p and tp. For various values of t/tp, q and, finally, q/q_p can be calculated.

The dimensionless unit hydrograph with peak rate factor of 484 can also be fitted with an exponential equation as follows:

$$q/q_p = t/T_p^{3.5} * e^x \qquad (7.41)$$

where $x = -3.5*(t/T_p - 1)$.

Unit Hydrograph Applications

Steps in the application of the unit hydrograph method are as follows:

1. Measure basin area and parameters necessary to calculate the time of concentration.
2. Calculate t_p and d from the relationship between lag time, time of concentration, and time step for rainfall increments. Ensure the time step for rainfall increments falls in the acceptable range.
3. Calculate q_p from Equation (7.33) for a peak rate factor of 484 or from Equation (7.39) for a chosen peak rate factor setting t equal to t_p.
4. Use Figure 7-24, Table 7-29, or the Meadows equation (7.39) to determine the flow rates for various time increments equal to d to define the unit hydrograph.
5. Convolute the hydrograph using the excess rainfall hyetograph generated using the SCS or other appropriate rainfall methods.

Unit Hydrograph Example

The rainfall hyetograph from the previous example is used with an SCS unit hydrograph to develop a runoff hydrograph. Figure 7-26 illustrates the triangular and curvilinear hydrographs as well as the accumulated volume of outflow S-curve. The first few iterations of the convolution process are given in Table 7-30 and are illustrated in Figure 7-27.

Step 1: $A = 700$ acres $= 1.09$ sq. mi.

Step 2: $T_c = 40$ min (calculated); $T_l = 24$ min; $T_p = 27$ min $= 0.45$ h; $d = 6$ min (equals $0.25\ T_l$)

Step 3: $Q = 1.0$ for a unit hydrograph

$$q_p = 484AQ/T_p = (484)(1.09)(1.0)/(0.45) = 1172 \text{ cfs}$$

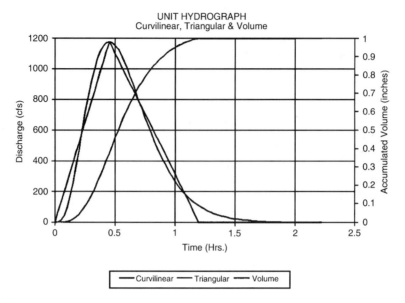

FIGURE 7-26
Unit hydrographs and accumulated runoff volume.

TABLE 7-30

Unit Hydrograph Convolution Example

Rainfall Hyetograph →		0.01	0.07	0.12	0.16	0.19
Curvilinear Unit Hydrograph						
Time (h)	Flow (cfs)	Flow (cfs)		Total Hydrograph		
0.0	0.00	0.00	0.00			
0.1	92.59	1.23	1.23	0.00		
0.2	481.25	12.84	6.39	6.46	0.00	
0.3	913.92	56.60	12.13	33.56	10.91	0.00
0.4	1149.24	150.13	15.25	63.73	56.70	14.44	0.00
0.5	1152.95	295.47	15.30	80.14	107.68	75.06	17.29
0.6	1002.67	461.52	13.31	80.40	135.41	142.54	89.86
0.7	790.11	566.14	10.49	69.92	135.85	179.24	170.65
0.8	579.26	575.34	7.69	55.10	118.14	179.82	214.59
0.9	401.91	510.49	5.33	40.40	93.09	156.38	215.28
1.0	266.99	410.27	3.54	28.03	68.25	123.23	187.22
1.1	171.23	306.12	2.27	18.62	47.35	90.34	147.53
1.2	106.67	215.66	1.42	11.94	31.46	62.68	108.16
1.3	64.85	145.16	0.86	7.44	20.18	41.64	75.05
1.4	38.62	94.16	0.51	4.52	12.57	26.71	49.85
1.5	22.59	59.24	0.30	2.69	7.64	16.64	31.97
1.6	13.01	36.33	0.17	1.58	4.55	10.12	19.92
1.7	7.39	21.80	0.10	0.91	2.66	6.02	12.11
1.8	4.15	12.84	0.06	0.52	1.53	3.52	7.21
1.9	2.30	7.44	0.03	0.29	0.87	2.03	4.22
2.0	1.27	4.25	0.02	0.16	0.49	1.15	2.43
2.1	0.69	2.39	0.01	0.09	0.27	0.65	1.38
2.2	0.37	1.34	0.00	0.05	0.15	0.36	0.77
2.3	0.00	0.73	0.00	0.03	0.08	0.20	0.43
2.4	0.39	0.00	0.04	0.11	0.24		
2.5	0.19	0.00	0.60	0.13			
2.6	0.07	0.00	0.07				
2.7	0.00	0.00					

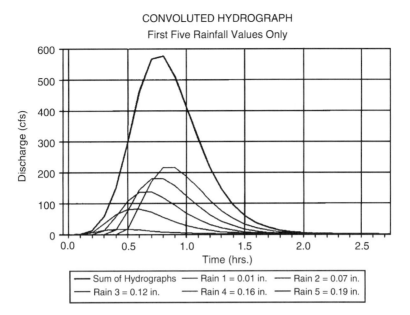

FIGURE 7-27
Convoluted hydrograph — first five values.

7.10 Santa Barbara Urban Hydrograph Method

The Santa Barbara Urban Hydrograph Method (SBUH) was developed by the Santa Barbara County Flood Control and Water Conservation District in an attempt to develop a complete urban hydrograph from rainfall that was simple, understandable, and easy to apply (Stubchaer, 1975). It works by converting design storm incremental excess rainfall depths into instantaneous unit hydrographs and routing them through an imaginary reservoir, with a time delay equal to the basin time of concentration (Puget Sound, 1992). This imaginary reservoir mimics the effects of lag time on the instantaneous runoff hydrograph.

Thus, two steps are used to synthesize the runoff hydrograph — computing the instantaneous hydrograph and routing it to calculate the runoff hydrograph. All of this is accomplished efficiently and easily using a tabular form of rainfall excess and routing that can be accomplished using a spreadsheet. The original approach calculated runoff from pervious (assuming 100% runoff) and impervious areas using Horton's infiltration approach combined with an antecedent precipitation index (API) approach to establish the initial wetness of the soil. More recent applications of the SBUH involve application of the SCS Curve Number (CN) method for rainfall excess calculation. This SBUH approach is included as an option in almost every popular commercial hydrologic model in current use.

The SBUH method applies the curve numbers selected to the SCS equations, in order to compute soil absorption and precipitation excess from the rainfall hyetograph. Each time step of this process generates one block of an instantaneous hydrograph with the same duration. The instantaneous hydrograph is then routed through an imaginary reservoir with a time delay equal to the basin time of concentration. The end product is the runoff hydrograph for that land segment.

The abstraction of runoff, S, is computed from the CN using Equation (7.18). Using the abstraction value and precipitation for the given time step, the runoff depth, D, per unit area is calculated as follows:

$$D(t) = (p(t) - .2(S)2\ )/(p(t) + .8(S)) \qquad (7.42)$$

where $p(t)$ = precipitation for the time increment (in), and S = defined in Equation (7.18)
The total runoff, $R(t)$, for the time increment is computed as follows:

$$R(t) = D(t) - D(t-1) \qquad (7.43)$$

The instantaneous hydrograph, $I(t)$, in cfs, at each time step, dt, is computed as follows:

$$I(t) = 60.5\ R(t)\ A/dt \qquad (7.44)$$

where A = area (acres), and dt = time interval (min).
The runoff hydrograph, $Q(t)$, is then obtained by routing the instantaneous hydrograph, $I(t)$, through an imaginary reservoir with a time delay equal to the time of concentration of the drainage basin. The following equation, derived from the basic finite difference form of the level pool routing equation, estimates the routed flow, $Q(t)$:

$$Q(t+1) = Q(t) + w[I(t) + I(t+1) - 2Q(t)\ ] \qquad (7.45)$$

where $w = dt/(2Tc + dt)$ and Tc = time of concentration for the drainage basin.
An example taken from the Washington State Department of Ecology illustrates the process using a spreadsheet (Washington State Department of Ecology, 2001).

Example: 10-Acre Site — Existing Conditions

Input Data

- Area = 10 acres
- P = 2.9 in
- dt = 10 min
- CN = 74
- S = 3.513514 in
- 0.2S = 0.70 in
- Tc = 73 min
- w = 0.064103
- No impervious area

Output Table — 24-hour Storm (Table 7-31)

- Column (1) = Time increment
- Column (2) = Time (min)
- Column (3) = Type IA storm distribution — 24-h storm with 10-min time increments for this example

Urban Hydrology

TABLE 7-31

(1) Time Increment	(2) Time (minute)	(3) Rainfall Distrib. (fraction)	(4) Incre. Rainfall (inches)	(5) Accumul. Rainfall (inches)	(6) PERVIOUS Accum. Runoff (inches)	(7) PERVIOUS Incre. Runoff (inches)	(8) IMPERVIOUS Accum. Runoff (inches)	(9) IMPERVIOUS Incre. Runoff (inches)	(10) Total Runoff (inches)	(11) Instant Flowrate (cfs)	(12) Design Flowrate (cfs)
1	0	0	0	0	0	0	0	0	0	0.0	0.0
2	10	0.004	0.012	0.012	0.000	0.000	0.000	0.000	0.000	0.0	0.0
3	20	0.004	0.012	0.023	0.000	0.000	0.000	0.000	0.000	0.0	0.0
4	30	0.004	0.012	0.035	0.000	0.000	0.000	0.000	0.000	0.0	0.0
5	40	0.004	0.012	0.046	0.000	0.000	0.000	0.000	0.000	0.0	0.0
6	50	0.004	0.012	0.058	0.000	0.000	0.001	0.001	0.000	0.0	0.0
7	60	0.004	0.012	0.070	0.000	0.000	0.004	0.002	0.000	0.0	0.0
8	70	0.004	0.012	0.081	0.000	0.000	0.007	0.003	0.000	0.0	0.0
9	80	0.004	0.012	0.093	0.000	0.000	0.011	0.004	0.000	0.0	0.0
10	90	0.004	0.012	0.104	0.000	0.000	0.015	0.005	0.000	0.0	0.0
11	100	0.004	0.012	0.116	0.000	0.000	0.020	0.005	0.000	0.0	0.0
12	110	0.005	0.015	0.131	0.000	0.000	0.027	0.007	0.000	0.0	0.0
13	120	0.005	0.015	0.145	0.000	0.000	0.035	0.008	0.000	0.0	0.0
14	130	0.005	0.015	0.160	0.000	0.000	0.044	0.008	0.000	0.0	0.0
15	140	0.005	0.015	0.174	0.000	0.000	0.053	0.009	0.000	0.0	0.0
16	150	0.005	0.015	0.189	0.000	0.000	0.062	0.009	0.000	0.0	0.0
17	160	0.005	0.015	0.203	0.000	0.000	0.072	0.010	0.000	0.0	0.0
18	170	0.006	0.017	0.220	0.000	0.000	0.084	0.012	0.000	0.0	0.0
	180			0.238	0.000		0.097	0.013			
				0.255							
	1130		0.012								
115	1140	0.004	0.012	2.552	0.638	0.007	2.322	0.012	0.007	0.4	
116	1150	0.004	0.012	2.564	0.644	0.007	2.334	0.012	0.007	0.4	0.4
117	1160	0.004	0.012	2.575	0.651	0.007	2.346	0.012	0.007	0.4	0.4
118	1170	0.004	0.012	2.587	0.658	0.007	2.357	0.012	0.007	0.4	0.4
119	1180	0.004	0.012	2.598	0.664	0.007	2.369	0.012	0.007	0.4	0.4
120	1190	0.004	0.012	2.610	0.671	0.007	2.380	0.012	0.007	0.4	0.4
121	1200	0.004	0.012	2.622	0.678	0.007	2.392	0.012	0.007	0.4	0.4
122	1210	0.004	0.012	2.633	0.685	0.007	2.403	0.012	0.007	0.4	0.4
123	1220	0.004	0.012	2.645	0.691	0.007	2.415	0.012	0.007	0.4	0.4
124	1230	0.004	0.012	2.656	0.698	0.007	2.426	0.012	0.007	0.4	0.4
125	1240	0.004	0.012	2.668	0.705	0.007	2.438	0.012	0.007	0.4	0.4
126	1250	0.004	0.012	2.680	0.712	0.007	2.449	0.012	0.007	0.4	0.4
127	1260	0.004	0.012	2.691	0.719	0.007	2.461	0.012	0.007	0.4	0.4
128	1270	0.004	0.012	2.703	0.726	0.007	2.472	0.012	0.007	0.4	0.4
129	1280	0.004	0.012	2.714	0.732	0.007	2.484	0.012	0.007	0.4	0.4
130	1290	0.004	0.012	2.726	0.739	0.007	2.496	0.012	0.007	0.4	0.4
131	1300	0.004	0.012	2.738	0.746	0.007	2.507	0.012	0.007	0.4	0.4
132	1310	0.004	0.012	2.749	0.753	0.007	2.519	0.012	0.007	0.4	0.4
133	1320	0.004	0.012	2.761	0.760	0.007	2.530	0.012	0.007	0.4	0.4
134	1330	0.004	0.012	2.772	0.767	0.007	2.542	0.012	0.007	0.4	0.4
135	1340	0.004	0.012	2.784	0.774	0.007	2.553	0.012	0.007	0.4	0.4
136	1350	0.004	0.012	2.796	0.781	0.007	2.565	0.012	0.007	0.4	0.4
137	1360	0.004	0.012	2.807	0.788	0.007	2.576	0.012	0.007	0.4	0.4
138	1370	0.004	0.012	2.819	0.795	0.007	2.588	0.012	0.007	0.4	0.4
139	1380	0.004	0.012	2.830	0.803	0.007	2.599	0.012	0.007	0.4	0.4
140	1390	0.004	0.012	2.842	0.810	0.007	2.611	0.012	0.007	0.4	0.4
141	1400	0.004	0.012	2.854	0.817	0.007	2.623	0.012	0.007	0.4	0.4
142	1410	0.004	0.012	2.865	0.824	0.007	2.634	0.012	0.007	0.4	0.4
143	1420	0.004	0.012	2.877	0.831	0.007	2.646	0.012	0.007	0.4	0.4
144	1430	0.004	0.012	2.888	0.838	0.007	2.657	0.012	0.007	0.4	0.4
145	1440	0.004	0.012	2.900	0.845	0.007	2.669	0.012	0.007	0.4	0.4

- Column (4) = Column (3) * P
- Column (5) = Accumulated sum of Column (4)
- Column (6) = If $(P < 0.2S) = 0$, If $(P > 0.2S) = [$Column (5) $- 0.2S]^2/[$Column (5) $+ 0.8S]$, where the PERVIOUS AREA S value is used
- Column (7) = Column (6) of the present step - Column (6) of the previous step
- Column (8) = Same as Column (6) except use IMPERVIOUS AREA S value
- Column (9) = Column (8) of the present step - Column (8) of the previous step

- Column (10) = (PERVIOUS AREA/TOTAL AREA) * Column (7) + (IMPERVIOUS AREA/TOTAL AREA) * Column (9)
- Column (11) = [60.5 * Column (10) * Total Area]/dt, where dt = 10 or 60 min
- Column (12) = Column (12) of previous time step + w * {[Column (11) of previous time step + Column (11) of present time step] - [2 * Column (12) of previous time step]}, where w = routing constant = $dt/(2Tc + dt)$ = 0.0641

Example: 10-Acre Site — Developed Conditions

Input Data

- Area = 10 acres
- P = 2.9 in
- dt = 10 min
- PERVIOUS AREA: Area = 6.1 acres CN = 89
- Pervious area (soil compacted by equipment):
  - A = 6.1
  - CN = 89
  - S = 1.235955 in
  - $0.2S$ = 0.25 in
  - Tc = 73 min
  - w = 0.064103
- Impervious area:
  - Area = 3.9 acres
  - CN = 98
  - S = 0.204082
  - $0.2S$ = 0.04
  - Tc = 28 min
  - w = 0.151515

Output Table — 24-Hour Storm

- Column (1) = Time increment
- Column (2) = Time (min)
- Column (3) = Type IA Storm Distribution — 24-h storm with 10-min time increments
- Column (4) = Column (3) * P
- Column (5) = Accumulated sum of Column (4)
- Column (6) = If $(P < 0.2S)$ = 0, if $(P > 0.2S)$ = [Column (5) - $0.2S]^2$/[Column (5) + $0.8S$], where the PERVIOUS AREA S value is used
- Column (7) = Column (6) of the present step - Column (6) of the previous step
- Column (8) = Same as Column (6), except use IMPERVIOUS AREA S value

TABLE 7-32

(1)	(2)	(3)	(4)	(5)	(6)	(7)	(8)	(9)	(10)	(11)	(12)
					PERVIOUS		IMPERVIOUS				
Time Increment	Time (minute)	Rainfall Distrib. (fraction)	Incre. Rainfall (inches)	Accumul. Rainfall (inches)	Accum. Runoff (inches)	Incre. Runoff (inches)	Accum. Runoff (inches)	Incre. Runoff (inches)	Total Runoff (inches)	Instant Flowrate (cfs)	Design Flowrate (cfs)
1	0	0	0	0	0	0	0	0	0	0.0	0.0
2	10	0.004	0.012	0.012	0.000	0.000	0.000	0.000	0.000	0.0	0.0
3	20	0.004	0.012	0.023	0.000	0.000	0.000	0.000	0.000	0.0	0.0
4	30	0.004	0.012	0.035	0.000	0.000	0.000	0.000	0.000	0.0	0.0
5	40	0.004	0.012	0.046	0.000	0.000	0.000	0.000	0.000	0.0	0.0
6	50	0.004	0.012	0.058	0.000	0.000	0.001	0.001	0.000	0.0	0.0
7	60	0.004	0.012	0.070	0.000	0.000	0.004	0.002	0.001	0.1	0.0
8	70	0.004	0.012	0.081	0.000	0.000	0.007	0.003	0.001	0.1	0.0
9	80	0.004	0.012	0.093	0.000	0.000	0.011	0.004	0.002	0.1	0.0
10	90	0.004	0.012	0.104	0.000	0.000	0.015	0.005	0.002	0.1	0.1
11	100	0.004	0.012	0.116	0.000	0.000	0.020	0.005	0.002	0.1	0.1
12	110	0.005	0.015	0.131	0.000	0.000	0.027	0.007	0.003	0.2	0.1
13	120	0.005	0.015	0.145	0.000	0.000	0.035	0.008	0.003	0.2	0.1
14	130	0.005	0.015	0.160	0.000	0.000	0.044	0.008	0.003	0.2	0.1
15	140	0.005	0.015	0.174	0.000	0.000	0.053	0.009	0.003	0.2	0.2
16	150	0.005	0.015	0.189	0.000	0.000	0.062	0.009	0.004	0.2	0.2
105	1040	0.004	0.012	2.436	1.399	0.010	2.207	0.012	0.011	0.6	0.7
106	1050	0.004	0.012	2.448	1.409	0.010	2.219	0.012	0.011	0.6	0.7
107	1060	0.004	0.012	2.459	1.419	0.010	2.230	0.012	0.011	0.6	0.7
				2.471	1.429						
			0.012	2.482							
	1090		0.012								
111	1100	0.004	0.012	2.506	1.460	0.010	2.276	0.012	0.011	0.6	0.6
112	1110	0.004	0.012	2.517	1.470	0.010	2.288	0.012	0.011	0.6	0.6
113	1120	0.004	0.012	2.529	1.480	0.010	2.299	0.012	0.011	0.6	0.6
114	1130	0.004	0.012	2.540	1.490	0.010	2.311	0.012	0.011	0.6	0.6
115	1140	0.004	0.012	2.552	1.500	0.010	2.322	0.012	0.011	0.6	0.6
116	1150	0.004	0.012	2.564	1.510	0.010	2.334	0.012	0.011	0.6	0.6
117	1160	0.004	0.012	2.575	1.521	0.010	2.346	0.012	0.011	0.6	0.6
118	1170	0.004	0.012	2.587	1.531	0.010	2.357	0.012	0.011	0.6	0.6
119	1180	0.004	0.012	2.598	1.541	0.010	2.369	0.012	0.011	0.6	0.6
120	1190	0.004	0.012	2.610	1.551	0.010	2.380	0.012	0.011	0.6	0.6
121	1200	0.004	0.012	2.622	1.562	0.010	2.392	0.012	0.011	0.6	0.6
122	1210	0.004	0.012	2.633	1.572	0.010	2.403	0.012	0.011	0.7	0.6
123	1220	0.004	0.012	2.645	1.582	0.010	2.415	0.012	0.011	0.7	0.6
124	1230	0.004	0.012	2.656	1.592	0.010	2.426	0.012	0.011	0.7	0.7
125	1240	0.004	0.012	2.668	1.603	0.010	2.438	0.012	0.011	0.7	0.7
126	1250	0.004	0.012	2.680	1.613	0.010	2.449	0.012	0.011	0.7	0.7
127	1260	0.004	0.012	2.691	1.623	0.010	2.461	0.012	0.011	0.7	0.7
128	1270	0.004	0.012	2.703	1.633	0.010	2.472	0.012	0.011	0.7	0.7
129	1280	0.004	0.012	2.714	1.644	0.010	2.484	0.012	0.011	0.7	0.7
130	1290	0.004	0.012	2.726	1.654	0.010	2.496	0.012	0.011	0.7	0.7
131	1300	0.004	0.012	2.738	1.664	0.010	2.507	0.012	0.011	0.7	0.7
132	1310	0.004	0.012	2.749	1.675	0.010	2.519	0.012	0.011	0.7	0.7
133	1320	0.004	0.012	2.761	1.685	0.010	2.530	0.012	0.011	0.7	0.7
134	1330	0.004	0.012	2.772	1.695	0.010	2.542	0.012	0.011	0.7	0.7
135	1340	0.004	0.012	2.784	1.706	0.010	2.553	0.012	0.011	0.7	0.7
136	1350	0.004	0.012	2.796	1.716	0.010	2.565	0.012	0.011	0.7	0.7
137	1360	0.004	0.012	2.807	1.726	0.010	2.576	0.012	0.011	0.7	0.7
138	1370	0.004	0.012	2.819	1.737	0.010	2.588	0.012	0.011	0.7	0.7
139	1380	0.004	0.012	2.830	1.747	0.010	2.599	0.012	0.011	0.7	0.7
140	1390	0.004	0.012	2.842	1.758	0.010	2.611	0.012	0.011	0.7	0.7
141	1400	0.004	0.012	2.854	1.768	0.010	2.623	0.012	0.011	0.7	0.7
142	1410	0.004	0.012	2.865	1.778	0.010	2.634	0.012	0.011	0.7	0.7
143	1420	0.004	0.012	2.877	1.789	0.010	2.646	0.012	0.011	0.7	0.7
144	1430	0.004	0.012	2.888	1.799	0.010	2.657	0.012	0.011	0.7	0.7
145	1440	0.004	0.012	2.900	1.810	0.010	2.669	0.012	0.011	0.7	0.7

- Column (9) = Column (8) of the present step - Column (8) of the previous step
- Column (10) = (PERVIOUS AREA/TOTAL AREA) * Column (7) + (IMPERVIOUS AREA/TOTAL AREA) * Column (9)
- Column (11) = [60.5 * Column (10) * Total Area]/dt, where dt = 10 or 60 min
- Column (12) = Column (12) of previous time step + w * {[Column (11) of previous time step + Column (11) of present time step] - [2 * Column (12) of previous time step]}, where w = routing constant = $dt/(2Tc + dt)$ = 0.0641

7.11 Water-Quality Volume and Peak Flow

Water-Quality Volume Calculation

The Water-Quality Volume (WQ_v) is the treatment volume required to remove a significant percentage of the stormwater pollution load. As an example, the procedures developed for the Georgia Stormwater Management Manual (GSWMM, 2001), are based on an 80% removal of the average annual postdevelopment total suspended solids (TSS) load. This is achieved by intercepting and treating a portion of the runoff from all storms and all the runoff from 85% of the storms that occur on average during the course of a year. For other locations, different removal policies might be more appropriate depending on local conditions and the level of pollution removal desired.

The water-quality treatment volume is calculated by multiplying the total inches of rainfall to be included in the water-quality volume (P) by the volumetric runoff coefficient (R_v) and the site area. R_v is defined as:

$$R_v = 0.05 + 0.009(I) \tag{7.46}$$

where I = percent of impervious cover (%).

Therefore, WQ_v is calculated using the following formula:

$$WQ_v = \frac{PR_vA}{12} \tag{7.47}$$

where WQ_v = water-quality volume (acre-feet); P = total rainfall to be included (in); R_v = volumetric runoff coefficient; A = total drainage area (acres); and WQ_v can be expressed in inches simply as $P(R_v) = Q_{wv}$.

Water-Quality Volume Peak Flow Calculation

The peak rate of discharge for the water-quality design storm is needed for the sizing of offline diversion structures, such as for sand filters and infiltration trenches. Choosing an arbitrary storm using the rational method and conventional SCS methods has been found to sometimes underestimate the volume and rate of runoff for rainfall events less than 2 in. This discrepancy in estimating runoff and discharge rates can lead to situations in which a significant amount of runoff bypasses the treatment practice due to an inadequately sized diversion structure and leads to the design of undersized bypass channels.

The following procedure can be used to estimate peak discharges for small storm events. It relies on the water-quality volume and the SCS simplified peak flow discharge procedures from Section 7.9. A brief description of the calculation procedure is presented below.

Step 1: Using WQ_v, a corresponding curve number (CN) is computed utilizing the following equation:

$$CN = 1000/[10 + 5P + 10Q_{wv} - 10(Q_{wv}^2 + 1.25\, Q_{wv}P)^{1/2}] \tag{7.48}$$

where P = rainfall (in); Q_{wv} = water-quality volume, in (PR_v)

Step 2: Once a CN is computed, the time of concentration (t_c) is computed.

Urban Hydrology

Step 3: Using the computed CN, t_c, and drainage area (A), in acres; the peak discharge (Q_{wq}) for the water-quality storm event is computed based on the SCS peak discharge procedures from Section 7.9, Figures 7-18 to 7-21. Use the appropriate rainfall distribution type.

Read initial abstraction (Ia), compute I_a/P
Read the unit peak discharge (q_u) for appropriate tc
Using WQ_v, compute the peak discharge (Q_{wq})

$$Q_{wq} = q_u * A * Q_{wv} \tag{7.49}$$

where Q_{wq} = the water-quality peak discharge (cfs); q_u = the unit peak discharge (cfs/mi_/inch); A = drainage area (mi²); and Q_{wv} = water-quality volume, in inches (PR_v).

Example Problem

Using the following data and information, calculate the water-quality volume and the water-quality peak flow.

Site area = 50 acres

Impervious area = 18 acres

Time of concentration (t_c) = 0.34 h

Total rainfall to be included in water-quality volume (P) = 1.5 in

Calculate water-quality volume (WQ_V)

Compute volumetric runoff coefficient, R_v

$$R_V = 0.05 + (0.009) = 0.05 + (0.009)(18/50 \times 100\%) = 0.37$$

Compute water-quality volume, WQ_v

$$WQ_v = P(R_V)(A)/12 = 1.5(.37)(50)/12 = 2.31 \text{ acre-feet}$$

Calculate water-quality peak flow

Compute runoff volume in inches, Q_{wv}:

$$Q_{wv} = P\ R_v = 1.5 * 0.37 = 0.56 \text{ in}$$

Computer curve number:

$$CN = 1000/[10 + 5P + 10Q - 10(Q_{wv}^2 + 1.25\ Q_{wv}\ P)^{1/2}]$$

$$CN = 1000/[10 + 5*1.5 + 10*0.56 - 10(0.56^2 + 1.25*0.56*1.5)^{1/2}] = 87.5$$

$$t_c = 0.34 \text{ (computed previously)}$$

$$S = 1000/CN - 10 = 1000/87.5 - 10 = 1.43 \text{ in}$$

$$0.2S = I_a = 0.29 \text{ in}$$

$$I_a/P = 0.29/1.5 = 0.193$$

Find q_u: from Figure 7-20, for $I_a/P = 0.193$, and $q_u = 600$ cfs/mi²/in

Compute water-quality peak flow: $Qwq = q_u * A * Q_{wv} = 600 * 50/640 * 0.56 = 26.3$ cfs

7.12 Channel Protection Volume Estimation

The SCS peak discharge procedures from Section 7.9 can be used for estimation of the channel protection volume (CP_v) for storage facility design. This method should not be used for standard detention design calculations. See Chapter 11 for detention design.

Basic Approach

For CP_v estimation, using Figures 7-18 to 7-21, the unit peak discharge (q_U) can be determined based on I_a/P and time of concentration (t_C). Knowing q_U and T (extended detention time, typically 24 h), the q_O/q_I ratio (peak outflow discharge/peak inflow discharge) can be estimated from Figure 7-33.

Using the following equation from SCS (1986), V_S/V_r can be calculated:

$$V_S/V_r = 0.682 - 1.43\,(q_O/q_I) + 1.64\,(q_O/q_I)^2 - 0.804\,(q_O/q_I)^3 \tag{7.50}$$

where V_S = required storage volume (acre-feet); V_r = runoff volume (acre-feet); q_O = peak outflow discharge (cfs); and q_I = peak inflow discharge (cfs).

The required storage volume can then be calculated by:

$$V_S = \frac{(V_S/V_r)(Q_d)(A)}{12} \tag{7.51}$$

where V_S and V_r are defined above; Q_d = the developed runoff for the design storm (in); and A = total drainage area (acres).

While the TR-55 shortcut method reports to incorporate multiple stage structures, experience has shown that an additional 10 to 15% storage is required when multiple levels of extended detention are provided.

Example Problem

Compute the 100-year peak discharge for a 50-acre wooded watershed, which will be developed as follows:

- Forest land — good cover (hydrologic soil group B) = 10 acres
- Forest land — good cover (hydrologic soil group C) = 10 acres
- 1/3 acre residential (hydrologic soil group B) = 20 acres

Urban Hydrology

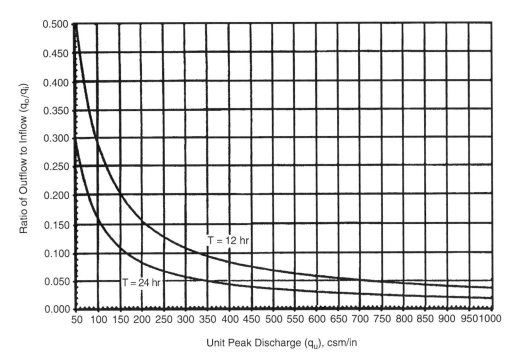

FIGURE 7-28
Detention time versus discharge ratios. (From MDE, 1998.)

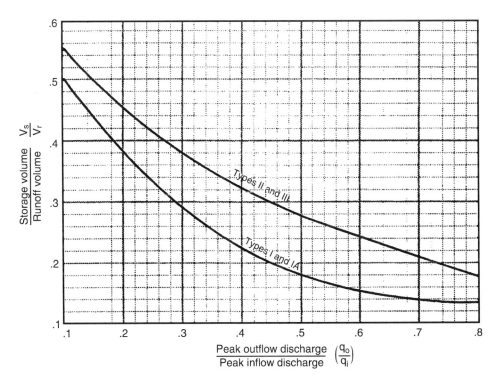

FIGURE 7-29
Approximate detention basin routing for rainfall types I, IA, II, and III.

- Industrial development (hydrological soil group C) = 10 acres

Other data include the following:

- Total impervious area = 18 acres
- % of pond and swamp area = 0
- Local policy states that the total rainfall (P) to be included in the channel protection volume is 1.3 in.
- Use Type II SCS rainfall distribution.

Computations

- Calculate rainfall excess.
  - For this location, the 100-year, 24-h rainfall is 6.91 in (0.288 in/h x 24 h — from local IDF curves).
  - For this location, the 1-year, 24-h rainfall is 2.98 in (.124 in/h x 24 h — from local IDF curves).
  - Composite weighted runoff coefficient is:

Development #	Area (acre)	% Total	CN	Composite CN
1	10	0.20	55	11.0
2	10	0.20	70	14.0
3	20	0.40	72	28.8
4	10	0.20	91	18.2
Total	50	1.00		72

* From Equation (7.16), Q (100-year) = 3.75 in, and Q_d (1-year developed) = 0.80 in.

- Calculate time of concentration.

The hydrologic flow path for this watershed = 1890 ft.

Segment	Type of Flow	Length (ft)	Slope (%)
1	Overland n = 0.24	40	2.0
2	Shallow channel	750	1.7
3	Main channel[a]	1100	0.50

[a] For the main channel, n = 0.06 (estimated), width = 10 ft, depth = 2 ft, rectangular channel.

Segment 1 — Travel time, from Equation (7.25) with P_2 = 3.60 in (0.15 x 24 — from local IDF curves).

$$T_t = [0.42(0.24 \times 40)^{0.8}] / [(3.60)^{0.5} (.020)^{0.4}] = 6.46 \text{ min}$$

Segment 2 — Travel time, from Figure 7-17 or Equation (7.26).

$$V = 2.1 \text{ ft/sec [from Equation (7.26)]}$$

Urban Hydrology 285

$$T_t = 750 / 60 (2.1) = 5.95 \text{ min}$$

Segment 3 — Using Equation (7.28)

$$V = (1.49/.06)(1.43)^{0.67}(.005)^{0.5} = 2.23 \text{ ft/sec}$$

$$T_t = 1100 / 60 (2.23) = 8.22 \text{ min}$$

$$t_c = 6.46 + 5.95 + 8.22 = 20.63 \text{ min } (0.34 \text{ h})$$

- Calculate I_a/P for $C_n = 72$, $I_a = 0.778$ (Table 7-27).

$$I_a/P = (.778 / 6.91) = 0.113$$

(Note: In using Figure 7-20, straight-line interpolation is used.)

- Unit discharge q_u (100-year) from Figure 7-20 = 630 csm/in.
- For the 1-year, $I_a/P = 0.778/2.98 = 0.261$, and from Figure 7-20 q_u (1-year) = 570 csm/in.
- Calculate peak discharge with $F_p = 1$ using Equation (7.29).

$$Q_{100} = 630 (50/640)(3.75)(1) = 185 \text{ cfs}$$

Calculate water-quality volume (WQ_v).

Compute runoff coefficient, R_v.

$$R_V = 0.50 + (IA)(0.009) = 0.50 + (18)(0.009) = 0.21$$

Compute water-quality volume, WQ_v.

$$WQ_v = P(R_V)(A)/12 = 1.3(.21)(50)/12 = 1.14 \text{ acre-feet}$$

- Calculate channel protection volume ($CP_v = V_S$).

Knowing q_u (1-year) = 570 csm/in from Step 3, and T (extended detention time of 24 h, which is stated in local policies), find q_O/q_I from Figure 7-28.

$$q_O/q_I = 0.03$$

For a Type II rainfall distribution and using Equation (7.50),

$$V_S/V_r = 0.682 - 1.43 (q_O/q_I) + 1.64 (q_O/q_I)^2 - 0.804 (q_O/q_I)^3$$

$$V_S/V_r = 0.682 - 1.43 (0.03) + 1.64 (0.03)^2 - 0.804 (0.03)^3 = 0.64$$

Therefore, stream channel protection volume with Q_d (1-year developed) = 0.80 in, from Step 1, is

$$CP_v = V_S = (V_S/V_r)(Q_d)(A)/12 = (0.64)(0.80)(50)/12 = 2.13 \text{ acre-feet}$$

7.13 Water Balance Calculations

Water balance calculations help determine if a drainage area is large enough, or has the right characteristics, to support a permanent pool of water during average or extreme conditions. When in doubt, a water balance calculation may be advisable for retention pond and wetland design.

The details of a rigorous water balance are beyond the scope of this book. However, a simplified procedure is described herein that will provide an estimate of pool viability and point to the need for more rigorous analysis. Water balance can also be used to help establish planting zones in a wetland design.

Basic Equations

Water balance is defined as the change in volume of the permanent pool resulting from the total inflow minus the total outflow (actual or potential):

$$\Delta V = \Sigma I - \Sigma O \tag{7.52}$$

where Δ = change in; V = pond volume (ac-ft); Σ = sum of; I = inflows (ac-ft); and O = outflows (ac-ft).

The inflows consist of rainfall, runoff, and baseflow into the pond. The outflows consist of infiltration, evaporation, evapotranspiration, and surface overflow out of the pond or wetland. Equation (7.52) can be changed to reflect these factors.

$$\Delta V = P + Ro + Bf - I - E - Et - Of \tag{7.53}$$

where P = precipitation (ft); Ro = runoff (ac-ft); Bf = baseflow (ac-ft); I = infiltration (ft); E = evaporation (ft); Et = evapotranspiration (ft); and Of = overflow (ac-ft).

Rainfall (P)

Monthly values are commonly used for calculations of values over a season. Rainfall is then the direct amount that falls on the pond surface for the period in question. When multiplied by the pond surface area (in acres), it becomes acre-feet of volume. As an example, Table 7-33 shows monthly rainfall rates for Atlanta, Georgia, based on a 30-year period of record at Hartsfield-Atlanta International Airport.

Runoff (R_o)

Runoff is equivalent to the rainfall for the period times the efficiency of the watershed, which is equal to the ratio of runoff to rainfall. In lieu of gauge information, Q/P can be estimated one of several ways. The best method would be to perform long-term simulation modeling using rainfall records and a watershed model. Two other methods have been proposed.

Equation (7.46) gives a ratio of runoff to rainfall volume for a particular storm. If it can be assumed that the average storm that produces runoff has a similar ratio, then the R_v value can serve as the ratio of rainfall to runoff. Not all storms produce runoff in an urban setting. Typical initial losses (often called "initial abstractions") are normally taken between 0.1 and 0.2 in. Thus, a factor (F) should be applied to the calculated R_v value to account

Urban Hydrology

TABLE 7-33

Water Balance Values for Atlanta, Georgia

	January	February	March	April	May	June	July	August	September	October	November	December
Precipitation (ft)	0.40	0.40	0.48	0.36	0.36	0.30	0.42	0.31	0.29	0.25	0.32	0.36
Turf Evapotranspiration (ft)	0.06	0.07	0.15	0.27	0.44	0.56	0.61	0.56	0.41	0.25	0.11	0.06

	Annual
Precipitation (ft)	4.25
Turf Evapotranspiration (ft)	3.55

Source: Ferguson, B. and Debo, T.N., *On-Site Stormwater Management*, 1990, and http://www.griffin.peachnet.edu/.

for storms that produce no runoff. Equation (7.54) reflects this approach. Total runoff volume is then simply the product of runoff depth (Q) times the drainage area to the pond.

$$Q = FPR_v \quad (7.54)$$

where Q = runoff volume (in); F = factor; P = precipitation (in); and R_v = volumetric runoff coefficient [see Equation (7.46)].

As an example, Ferguson (1996) performed simulation modeling in an attempt to quantify an average ratio on a monthly basis. For the Atlanta, Georgia, area, he developed the following equation:

$$Q = 0.235P/S^{0.64} - 0.161 \quad (7.55)$$

where P = precipitation (in); Q = runoff volume (in); and S = potential maximum retention (in) [see Equation (7.16).

Baseflow (Bf)

Most stormwater ponds and wetlands have little, if any, baseflow, as they are rarely placed across perennial streams. If so placed, baseflow must be estimated from observation or through theoretical estimates. Methods of estimation and baseflow separation can be found in most hydrology textbooks.

Infiltration (I)

Infiltration is a complex subject and cannot be covered in detail here. The amount of infiltration depends on soils, water-table depth, rock layers, surface disturbance, the presence or absence of a liner in the pond, and other factors. The infiltration rate is governed by the Darcy equation as:

$$I = Ak_hG_h \quad (7.56)$$

TABLE 7-34

Saturated Hydraulic Conductivity

Material	Hydraulic Conductivity	
	in/hr	ft/day
ASTM crushed stone no. 3	50,000	100,000
ASTM crushed stone no. 4	40,000	80,000
ASTM crushed stone no. 5	25,000	50,000
ASTM crushed stone no. 6	15,000	30,000
Sand	8.27	16.54
Loamy sand	2.41	4.82
Sandy loam	1.02	2.04
Loam	0.52	1.04
Silt loam	0.27	0.54
Sandy clay loam	0.17	0.34
Clay loam	0.09	0.18
Silty clay loam	0.06	0.12
Sandy clay	0.05	0.10
Silty clay	0.04	0.08
Clay	0.02	0.04

Source: Ferguson, B. and Debo, T.N., *On-Site Stormwater Management*, 1990, and http://www.griffin.peachnet.edu/.

where I = infiltration (ac-ft/day); A = cross-sectional area through which the water infiltrates (ac); K_h = saturated hydraulic conductivity or infiltration rate (ft/day); and G_h = hydraulic gradient = pressure head/distance.

G_h can be set equal to 1.0 for pond bottoms and 0.5 for pond sides steeper than about 4:1. Infiltration rate can be established through testing, though not always accurately. As a first estimate, Table 7-34 can be used.

Evaporation (E)

Evaporation is from an open lake water surface. Evaporation rates are dependent on differences in vapor pressure, which, in turn, depend on temperature, wind, atmospheric pressure, water purity, and shape and depth of the pond. It is estimated or measured in a number of ways that can be found in most hydrology textbooks. Pan evaporation methods are also used. A pan coefficient of 0.7 is commonly used to convert the higher pan value to the lower lake values.

Table 7-35 gives pan evaporation rate distributions for a typical 12-month period based on pan evaporation information from five stations in and around Georgia. Figure 7-30 depicts a map of annual free water surface (FWS) evaporation averages for Georgia based on a National Oceanic and Atmospheric Administration (NOAA) assessment done in 1982. FWS evaporation differs from lake evaporation for larger and deeper lakes but can be

TABLE 7-35

Evaporation Monthly Distribution

January	February	March	April	May	June	July	August	September	October	November	December
3.2%	4.4%	7.4%	10.3%	12.3%	12.9%	13.4%	11.8%	9.3%	7.0%	4.7%	3.2%

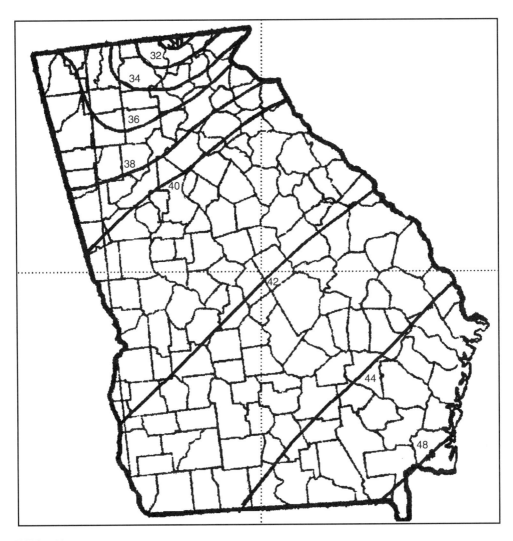

FIGURE 7-30
Average annual free water surface evaporation (in inches). (From NOAA and the Army Corps of Engineers, Hydrometeorological Report Number 52, Application of the Probable Maximum Precipitation Estimates — U.S. East of the 105th Meridian, August, 1982.)

used as an estimate of it for the type of structural stormwater ponds and wetlands being designed in Georgia. Total annual values can be estimated from this map and distributed according to Table 7-35.

Evapotranspiration (Et)

Evapotranspiration consists of the combination of evaporation and transpiration by plants. The estimation of Et for crops in Georgia is well documented and has become standard practice. However, for wetlands, the estimating methods are not documented, and there are no consistent studies to assist the designer in estimating the demand wetland plants would put on water volumes. Literature values for various places in the United States vary around the free water surface lake evaporation values. Estimating Et becomes only important when wetlands are being designed and emergent vegetation covers a significant portion of the pond surface. In these cases, conservative estimates of lake evaporation should be compared to crop-based *Et* estimates and a decision made. Crop-based *Et* estimates can be obtained from typical hydrology textbooks or from the Web sites mentioned above.

Overflow (Of)

Overflow is considered excess runoff, and, in water balance design, is either not considered because the concern is for average values of precipitation, or is considered lost for all volumes above the maximum pond storage. Obviously, for long-term simulations of rainfall–runoff, large storms would play an important part in pond design.

Example Problem

A 26-acre site in Augusta, Georgia, is being developed along with an estimated 0.5-acre surface area pond. There is no baseflow. The desired pond volume to the overflow point is 2 acre-feet. Will the site be able to support the pond volume? From the basic site data, we find that the site is 75% impervious with sandy clay loam soil.

- From Equation (7.46), $R_v = 0.05 + 0.009 (75) = 0.73$. With the correction factor of 0.9, the watershed efficiency is 0.65.
- The annual lake evaporation from Figure 7-31 is about 42 in.
- For a sandy clay loam, the infiltration rate is $I = 0.34$ ft/day (Table 7-34).
- From a grading plan, it is known that about 10% of the total pond area is sloped greater than 1:4.
- Monthly rainfall for Augusta was found from the Web site provided above.

Table 7-36 shows summary calculations for this site for each month of the year.

TABLE 7-36
Summary Information for Austin Acres

1		J	F	M	A	M	J	J	A	S	O	N	D
2	Days/mo	31	28	31	30	31	30	31	31	30	31	30	31
3	Precipitation (in)	4.05	4.27	4.65	3.31	3.77	4.13	4.24	4.5	3.02	2.84	2.48	3.40
4	Evap Dist.	3.2%	4.4%	7.4%	10.3%	12.3%	12.9%	13.4%	11.8%	9.3%	7.0%	4.7%	3.2%
5	Ro (ac-ft)	5.70	6.01	6.55	4.66	5.31	5.82	5.97	6.34	4.25	4.00	3.49	4.79
6	P (ac-ft)	0.17	0.18	0.19	0.14	0.16	0.17	0.18	0.19	0.13	0.12	0.10	0.14
7	E (ac-ft)	0.06	0.08	0.13	0.18	0.22	0.23	0.23	0.21	0.16	0.12	0.08	0.06
8	I (ac-ft)	5.01	4.52	5.01	4.85	5.01	4.85	5.01	5.01	4.85	5.01	4.85	5.01
9													
10	Balance (ac-ft)	0.81	1.59	1.61	-0.23	0.24	0.92	0.91	1.31	-0.63	-1.01	-1.33	-0.13
11	Running Balance (ac-ft)	0.81	2.00	2.00	1.77	2.00	2.00	2.00	2.00	1.37	0.36	0.00	0.00

Explanation of Table:

1. Months of year
2. Days per month
3. Monthly precipitation
4. Distribution of evaporation by month from Table 7-33
5. Watershed efficiency of 0.65 times the rainfall and converted to acre-feet
6. Precipitation volume directly into pond equals precipitation depth times pond surface area divided by 12 to convert to acre-feet

Urban Hydrology

7. Evaporation equals monthly percent of 42 in from line 4 converted to acre-feet
8. Infiltration equals infiltration rate times 90% of the surface area plus infiltration rate times 0.5 (banks greater than 1:4) times 10% of the pond area converted to acre-feet
9. Lines 5 and 6 minus lines 7 and 8
10. Accumulated total from line 10, keeping in mind that all volume above 2 acre-feet overflows and is lost in the trial design

It can be seen that, for this example, the pond has potential to go dry in winter months. This can be remedied in a number of ways, including compacting the pond bottom, placing a liner of clay or geosynthetics, and changing the pond geometry to decrease surface area.

TABLE 7-37

Augusta Precipitation Information

Station: (90495) AUGUSTA_WSO_AIRPORT					From Year=1961 to Year=1990				Missing Data: 0.0%		
		Total Precipitation					Snow		# Days Precip		
	Mean	High–Yr		Low–Yr		1-Day Max	Mean	High–Yr		=>.10	=>.50 =>1.
Ja	4.05	8.91	87	0.75	81	2.78	25/1978	0.3	2.3	88	7	3 1
Fe	4.27	7.67	61	0.69	68	3.50	5/1985	1.0	14.0	73	7	3 1
Ma	4.65	11.92	80	0.88	68	5.31	10/1967	0.0	1.1	80	7	3 1
Ap	3.31	8.43	61	0.60	70	2.71	27/1961	0.0	0.0	0	5	2 1
Ma	3.77	9.61	79	1.57	87	4.44	31/1981	0.0	0.0	0	7	2 1
Jn	4.13	8.84	89	0.68	84	2.95	12/1964	0.0	0.0	0	6	3 1
Jl	4.24	11.43	67	1.02	87	3.67	21/1979	0.0	0.0	0	8	2 1
Au	4.50	11.34	86	0.65	80	5.95	29/1964	0.0	0.0	0	6	3 1
Se	3.02	9.51	75	0.31	84	4.55	20/1975	0.0	0.0	0	5	2 1
Oc	2.84	14.82	90	0.01	63	5.32	12/1990	0.0	0.0	0	4	2 1
No	2.48	7.76	85	0.57	73	3.43	22/1985	0.0	0.0	0	4	2 1
De	3.40	8.65	81	0.96	80	2.89	16/1970	0.0	0.4	71	6	3 1
An	44.66	66.04	64	32.96	78	5.95	29/08/64	1.4	14.4	73	71	30 13
Wi	11.71	20.26	87	5.62	86	3.50	5/02/85	1.3	14.4	73	20	8 3
Sp	11.73	19.93	84	4.00	85	5.31	10/03/67	0.0	1.1	80	18	8 3
Su	12.87	24.89	64	7.08	80	5.95	29/08/64	0.0	0.0	0	20	9 4
Fa	8.35	18.50	90	1.96	84	5.32	12/10/90	0.0	0.0	0	12	5 2

7.14 Downstream Hydrologic Assessment

The purpose of the overbank flood-protection and extreme flood-protection criteria is to protect downstream properties from flood increases due to upstream development. These criteria require the designer to control peak flow at the outlet of a site such that postdevelopment peak discharge equals predevelopment peak discharge. It has been shown that, in certain cases, this does not always provide effective water-quantity control downstream from the site and may actually exacerbate flooding problems downstream. The reasons for this have to do with (1) the timing of the flow peaks, and (2) the total increase in volume of runoff. Further, due to a site's location within a watershed, there may be little reason to require overbank flood control from a particular site. This section outlines a

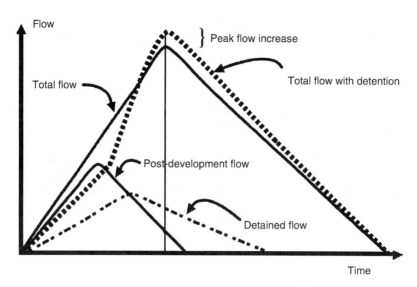

FIGURE 7-31
Detention timing example. (From GSWMM, 2001.)

suggested procedure for determining the impacts of postdevelopment stormwater peak flows and volumes on downstream flows that a community may require as part of a developer's stormwater management site plan.

Reasons for Downstream Problems

Flow Timing

If water-quantity control (detention) structures are indiscriminately placed in a watershed and changes to the flow timing are not considered, the structural control may actually increase the peak discharge downstream. The reason for this may be seen in Figure 7-31. The peak flow from the site is reduced appropriately, but the timing of the flow is such that the combined detained peak flow (the larger dashed triangle) is actually higher than if no detention were required. In this case, the shifting of flows to a later time brought about by the detention pond actually makes the downstream flooding worse than if the postdevelopment flows were not detained.

Increased Volume

An important impact of new development is an increase in the total runoff volume of flow. Thus, even if the peak flow is effectively attenuated, the longer duration of higher flows due to the increased volume may combine with downstream tributaries to increase the downstream peak flows.

Figure 7-32 illustrates this concept. The figure shows the pre- and postdevelopment hydrographs from a development site (Tributary 1). The postdevelopment runoff hydrograph meets the flood-protection criteria (i.e., the postdevelopment peak flow is equal to the predevelopment peak flow at the outlet from the site). However, the post-development combined flow at the first downstream tributary (Tributary 2) is higher than predevelopment combined flow. This is because the increased volume and timing of runoff from the developed site increases the combined flow and flooding downstream. In this case, the detention volume would have to have been increased to account for the

Urban Hydrology

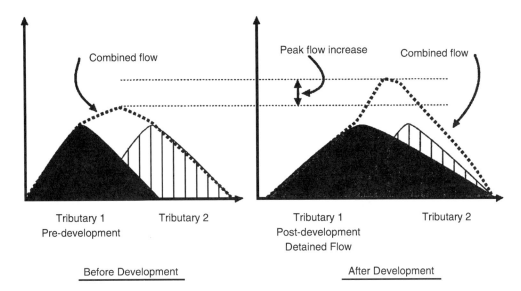

FIGURE 7-32
Effect of increased postdevelopment runoff volume with detention on a downstream hydrograph. (From GSWMM, 2001)

downstream timing of the combined hydrographs to mitigate the impact of the increased runoff volume.

The Ten-Percent Rule

A criterion that has been adopted as a flexible and effective approach for ensuring that stormwater quantity detention ponds actually attempt to maintain predevelopment peak flows throughout the system downstream is the ten-percent rule. The ten-percent rule recognizes the fact that a structural control providing detention has a zone of influence downstream, where its effectiveness can be felt. Beyond this zone of influence, the structural control becomes relatively small and insignificant compared to the runoff from the total drainage area at that point. Based on studies and master planning results for a large number of sites, that zone of influence is considered to be the point where the drainage area controlled by the detention or storage facility comprises 10% of the total drainage area (see page 94). For example, if the structural control drains 10 acres, the zone of influence ends at the point where the total drainage area is 100 acres or greater.

Typical steps in the application of the ten-percent rule are as follows:

- Determine the target peak flow for the site for predevelopment conditions.
- Using a topographic map, determine the lower limit of the zone of influence (10% point).
- Using a hydrologic model, determine the predevelopment peak flows and timing of those peaks at each tributary junction beginning at the pond outlet and ending at the next tributary junction beyond the 10% point.
- Change the land use on the site to postdevelopment, and rerun the model.
- Design the structural control facility such that the overbank flood protection postdevelopment flow does not increase the peak flows at the outlet and the determined tributary junctions.

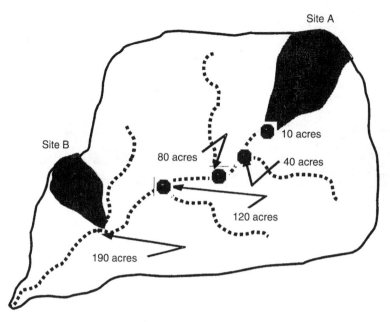

FIGURE 7-33
Example of the ten-percent rule. (From GSWMM, 2001.)

- If it does increase the peak flow, the structural control facility must be redesigned, or one of the following options must be considered:
  - Control of the overbank flood volume may be waived by the local authority, saving the developer the cost of sizing a detention basin for overbank flood control. In this case, the ten-percent rule saved the construction of an unnecessary structural control facility that would have been detrimental to the watershed flooding problems. In some communities, this situation may result in a fee being paid to the local government in lieu of detention. That fee would go toward alleviating downstream flooding or making channel or other conveyance improvements.
  - Work with the local government to reduce the flow elevation through channel or flow conveyance structure improvements downstream.
  - Obtain a flow easement from downstream property owners to the 10% point.

Even if the overbank flood protection requirement is eliminated, other requirements will need to be addressed (e.g., the water-quality treatment, channel protection, any infiltration requirements, and extreme flood protection criteria).

Example Problem

Figure 7-33 illustrates the concept of the ten-percent rule for two sites in a watershed.

Site A is a development of 10 acres, all draining to a wet extended detention stormwater pond. The overbank flooding and extreme flood portions of the design are going to incorporate the ten-percent rule. Looking downstream at each tributary, in turn, it is determined that the analysis should end at the tributary marked "80 acres." The 100-acre (10%) point is between the 80-acre and 120-acre tributary junction points.

The assumption is that, if there is no peak flow increase at the 80-acre point, then there will be no increase through the next stream reach downstream through the 10% point (100 acres) to the 120-acre point. The designer constructs a simple HEC-1 (or other appropriate) model of the 80-acre area using existing conditions in subwatersheds for each tributary. Key detention structures existing in other tributaries must be modeled. An approximate curve number is used, because the *actual* peak flow is not key for initial analysis; only the increase or decrease is important. The accuracy in curve number determination is not as significant as an accurate estimate of the time of concentration. Because flooding is an issue downstream, the pond is designed (through several iterations) until the peak flow does not increase at junction points downstream to the 80-acre point.

Site B is located downstream at the point where the total drainage area is 190 acres. The site is only 6 acres. The first tributary junction downstream from the 10% point is the junction of the site outlet with the stream. The total 190 acres is modeled as one basin, with care taken to estimate the time of concentration for input into the hydrologic model of the watershed. The model shows that a detention facility, in this case, will actually *increase* the peak flow in the stream and thus may not be required.

References

Aaron, G., Adaptations of Horton and SCS infiltration equations to complex storms, *J. Irrig. and Drain. Eng.*, ASCE, 118, 2, March/April 1992.

Aron, G. and Kibler, D.F., Pond sizing for rational formula hydrographs, *AWRA Water Resources Bull.*, 26, 2, April 1990.

Bosznay, M., Generalization of SCS curve number method, *J. Irrig. and Drain. Eng.*, ASCE, 115, 1, February 1989.

Boughton, W.C., Evaluation of partial areas of watershed runoff, *J. Irrig. and Drain. Eng.*, ASCE, 113, 3, August 1987.

Casamayor, J.E., and Rodgers, J.R., Cost Analysis of 2-year, 5-year and 10-year Stormwater Protection, Hydraulics Technical Group, San Antonio, TX, Texas Section, ASCE, 1980.

Chow, V.T., Ed., *Handbook of Hydrology*, McGraw-Hill, New York, 1964.

Cochran, W.G., *Sampling Techniques*, John Wiley and Sons, New York, 1963.

Daly, M., Discussion of empirical investigation of curve number technique, *J. Hydraulic Div.*, ASCE, HY5, May, 651, 1981.

Dunne, T. and Leopold, L.B., *Water in Environmental Planning*, W.H. Freeman, San Francisco, 1978.

Federal Highway Administration, HYDRAIN Documentation, 1991.

Ferguson, B., Estimation of direct runoff in the Thornthwaite water balance, *Prof. Geographer*, 263–271, October 1996.

Ferguson, B. and Debo, T.N., *On-Site Stormwater Management*, 1990.

Georgia Stormwater Management Manual, Volume 2, Technical Handbook, August 2001.

Gundlach, D.L., Adjustment of Peak Discharge Rates for Urbanization, U.S. Army Corps of Engineers Hydrologic Engineering Center, Tech. Paper Number 54, 1978.

Hawkins, R.H., Runoff curve numbers from partial area watersheds, *J. Irrig. and Drain. Eng.*, ASCE, 105, IR4, December 1979.

Hershfield, D.M., Rainfall Frequency Atlas of the United States, Technical Paper 40, 1961.

Hjelmfelt, A.T., Investigation of curve number procedure, *J. Hydraulic Div.*, ASCE, 117, 6, June 1991.

Hjelmfelt, A.T., An empirical investigation of the curve number technique, *J. Hydraulic Div.*, ASCE, 106, 9, September 1980.

Huff, F.A., Time Distributions of Heavy Rainstorms in Illinois, State of Illinois, Water Survey, ISWS/CIR-173/90, Circular 173, 1990.

Huff, F.A., Time distribution of rainfall in heavy storms, *Water Resources Res.*, 3, 4, 1967.

Huff, F.A. and Vogel, J.L., Hydrometeorology of Heavy Storms in Chicago and Northeastern Illinois, Illinois State Water Survey Report of Investigation 82, 1976.

Hydrologic Engineering Center, HEC-1 Flood Hydrograph Package, Davis, California, September 1990.

Hydrologic Engineering Center, Hydrologic Analysis of Ungauged Watersheds Using HEC-1, Davis, California, Training Document Number 15, April 1982.

Hydrologic Engineering Center, U.S. Army Corps of Engineers, HEC-1 Flood Hydrograph Package, 1990.

Johnson, W.K., Significance of location in computing flood damage, *ASCE J. Water Res. Plann. and Manage.*, 111, 1, 65–81, January 1985.

Kerby, W.S., Time of concentration for overland flow, *Civil Eng.*, 29, March 1959.

Kirpich, Z.P., Time of concentration from small agricultural watersheds, *Civil Eng.*, 10, 6, June 1940.

Kuichling, E., The relation between the rainfall and the discharge of sewers in populous districts, *Trans. Am. Soc. of Civ. Eng.*, 20, 1–56, 1889.

Lewis, G.L., Jury verdict: frequency versus risk-based culvert design, *ASCE J. Water Res. Plann. Manage.*, 118, 2, 166–185, March/April 1992.

Maryland Department of the Environment, Maryland Stormwater Design Manual, Volumes I and II, prepared by the Center for Watershed Protection, 2000.

McCuen, R., Introduction to Chapter 9, ASCE TC on Hydrology, DRAFT, 1993.

McCuen, R.H., *A Guide to Hydrologic Analysis Using SCS Methods*, Prentice-Hall, New York, 1982.

Meadows, M.E., Morris, K.B., and Spearmen, W.E., Improved Runoff Estimation Methods and Single Outlet Detention Performance Curves Applicable to Urban Watersheds in the Midlands of South Carolina, Supplemental Rep., Department of Civil Engineering, University of South Carolina and South Carolina Land Resources Conservation Commission, March 1992.

NOAA, Five- to 60-Minute Precipitation Frequency for the Eastern and Central United States, NOAA Technical Report NWS 33, 1982.

NOAA, Five- to 60-Minute Precipitation Frequency for the Eastern and Central United States, NOAA Tech. Memo. NWS HYDRO-35, 1977.

NOAA and the Army Corps of Engineers, Hydrometeorological Report Number 52, Application of the Probable Maximum Precipitation Estimates — U.S. East of the 105th Meridian, August 1982.

NOAA and the Army Corps of Engineers, Hydrometeorological Report Number 51, Probable Maximum Precipitation Estimates, U.S. East of the 105th Meridian, June 1978.

Overton, D.E. and Meadows, M.E., *Storm Water Modeling*, Academic Press, New York, 1976, pp. 58–88.

Pitt, R., Small Storm Hydrology, unpublished report, Department of Civil Engineering, University of Alabama at Birmingham, 1994.

Puget Sound, Stormwater Manual for the Puget Sound Basin, February 1992.

Rantz, S.E., Suggested Criteria for Hydrologic Design of Storm-Drainage Facilities in the San Francisco Bay Region, California, USGS Open File Report, Menlo Park, CA, 1971.

Regan, R.M., A Nomograph Based on Kinematic Wave Theory for Determining Time of Concentration for Overland Flow, Report Number 44, Civil Engineering Department, University of Maryland at College Park, 1971.

Rossmiller, R.L., The Rational Formula Revisited, Proc. Int. Symp. on Urban Storm Runoff, University of Kentucky, Lexington, KY, July 28–31, 1980.

Ruthroff, C.L. and Bodtmann, W.F., Computing derivatives from equally spaced data, *J. Applied Meteorol.*, 15, 11, November 1976.

Sarma, P.G.S., Delleur, J.W., and Rao, A.R., A Program in Urban Hydrology, Part II, Technical Report Number 99, Purdue University, Water Resources Research Center, 1969.

Schaake, J.C., Geyer, J.C., and Knapp, J.W., Experimental examination of the rational method, *J. Hydraulics Div.*, ASCE, 93, HY6, 353–370, November 1967.

Schueler, T.R., Controlling Urban Runoff, Washington Metropolitan Water Resources Planning Board, 1987.

Shabman, L., Water project design and safety: prospects for use of risk analysis in public sector organizations, in *Risk-Based Decision Making in Water Resources*, ASCE Conf. Proc., November 3–5, 1985, pp. 16–29.

Snyder, F.F., Synthetic unit hydrographs, *Trans. Am. Geophysical Union*, 19, Part 1, 447–454, 1938.
South Carolina Stormwater Management and Sediment Reduction Regulations, South Carolina Land Resources Conservation Commission, Division of Engineering, Columbia, South Carolina, 1992.
Stubchaer, J.M., The Santa Barbara Urban Hydrograph Method, Proc. Nat. Symp. on Urban Hydrology and Sediment Control, University of Kentucky, Lexington, July 28–31, 1975, pp. 131–141.
U.S. Army Corps of Engineers, St. Paul District, Analysis for the Determination of Percentage of Plugging for Hydraulic Analysis of Bridges, St. Paul, MN, Hydraulic Design Section.
U.S. Department of Agriculture, Soil Conservation Service, Engineering Division, Urban Hydrology for Small Watersheds, Technical Release 55 (TR-55), 1986.
U.S. Department of Agriculture, Soil Conservation Service, Engineering Division, National Engineering Handbook, Section 4, Hydrology, 1985.
U.S. Department of Transportation, Federal Highway Administration, Hydrology, Hydraulic Engineering Circular Number 19, 1984.
Washington State Department of Ecology, Stormwater Management Manual for Western Washington, Volume III — Hydrologic Analysis and Flow Control Design/BMPs, August 2001.
Water Resources Council Bulletin 17B, Guidelines For Determining Flood Flow Frequency, 1981.
Wright-McLaughlin Engineers, Urban Storm Drainage Criteria Manual, Vol. I and II, prepared for the Denver Regional Council of Governments, Denver, CO, 1969.
Yen, B.C. and Chow, V.T., Local Design Storm Vol. II, Federal Highway Administration, Report Number FHWA/RD-82-064, 1983.
Yen, B.C. and Chow, V.T., Design hyetographs for small drainage structures, *ASCE J. Hydraulics*, 106, HY6, June 1980.

8

Storm Drainage Systems

8.1 Introduction

Storm drainage facilities collect stormwater runoff and convey it away from structures and through the roadway right-of-way in a manner that adequately drains sites and roadways and minimizes the potential for flooding and erosion to properties. Storm drainage facilities consist of curbs, gutters, storm drains, channels, ditches, and culverts. The placement and hydraulic capacities of storm drainage structures and conveyances should be designed to take into consideration damage to adjacent property and to secure as low a degree of risk of traffic interruption by flooding as is consistent with the importance of the road, the design traffic service requirements, and available funds.

8.2 Concept Definitions

The following are definitions of concepts important in storm drain analysis and design as used in this chapter.

Bypass Flow that bypasses an inlet on grade and is carried in the street or channel to the next inlet downgrade. Inlets can be designed to allow a certain amount of bypass. Also, inlets may be designed to allow a certain amount of bypass for one design storm and larger or smaller amounts for other design storms.

Combination inlet A drainage inlet usually composed of a curb opening and a grate inlet.

Curb-opening inlet A drainage inlet consisting of an opening in the roadway curb.

Drop inlet A drainage inlet with a horizontal or nearly horizontal opening.

Equivalent cross slope An imaginary continuous cross slope having conveyance capacity equal to that of the given compound cross slope.

Flanking inlet Inlets placed upstream and on either side of an inlet at the low point in a sag vertical curve. The purpose of these inlets is to intercept debris as the slope decreases and to act in relief of the inlet at the low point.

Frontal flow The portion of the flow that passes over the upstream side of a grate.

Grate inlet A drainage inlet composed of a grate in the roadway section or at the roadside in a low point, swale, or channel.

Grate perimeter The sum of the lengths of all sides of a grate, except that any side adjacent to a curb is not considered a part of the perimeter in weir flow computations.

Gutter That portion of the roadway section adjacent to the curb that is utilized to convey storm runoff water. It may include a portion or all of a traveled lane, shoulder, or parking lane, and a limited width adjacent to the curb may be of different materials and have a different cross slope.

Hydraulic grade line The hydraulic grade line is the locus of elevations to which the water would rise in successive piezometer tubes if the tubes were installed along a pipe run.

Inlet efficiency The ratio of flow intercepted by an inlet to total flow in the gutter.

Pressure head Pressure head is the height of a column of water that would exert a unit pressure equal to the pressure of the water.

Scupper A vertical hole through a bridge deck for the purpose of deck drainage. Sometimes, a horizontal opening in the curb or barrier is called a scupper.

Side-flow interception Flow that is intercepted along the side of a grate inlet, as opposed to frontal interception.

Slottted drain inlet A drainage inlet composed of a continuous slot built into the top of a pipe that serves to intercept, collect, and transport the flow.

Splash-over Portion of the frontal flow at a grate that skips or splashes over the grate and is not intercepted.

Spread The width of flow measured laterally from the roadway curb.

Velocity head Velocity head is a quantity proportional to the kinetic energy of flowing water expressed as a height or head of water.

8.3 Pavement Drainage

Design Steps

There are many details to consider in the design and specification of storm drain systems. ASCE Manuals of Engineering Practice (1960, 1982, 1983) as well as other trade and vendor publications provide construction and specification details beyond the scope of this text. Typical steps in the hydraulic design of a storm drain system include the following:

1. *Design problem and design criteria specification* — Determine the design type to be done and the design criteria to be followed for the specific locality. Where criteria are not adequate or complete, establish internal criteria using other sources including those given in this text.

2. *System drainage area definition and preliminary layout* — Based on street layout or other criteria, identify the total drainage area to be handled (including off-site drainage). With reference to design criteria, lay out the preliminary drainage routes. Consider bypass flow locations for storms in excess of design. If parcels or lots are involved, lay out bypass locations on parcel boundaries where possible.

Look for an ability to introduce environmental features and for integration with other neighborhood amenities. Locate quantity and quality control structures.

3. *Field and office data collection* — Make a field visit to determine feasibility of preliminary layout and to look for site-specific problems such as rock outcrops, large trees, environmentally sensitive areas, utility locations, unmapped structures, etc.

4. *System layout* — Perform final layout in coordination with others on the design team to ensure minimization of design conflicts and an ability to take advantage of any multiobjective use opportunities of the drainage system. Locate all ditches, inlets, manholes, mains, laterals, culverts, etc. Determine flow direction from grading plan or map of site, and mark with arrows on site map.

5. *Hydrologic calculations* — Outline drainage area for each inlet or ditch start. Develop flow estimates for the design frequency throughout the system.

6. *Street flow* — Develop flow and spread calculations for streets or roadways and determine permissible maximum spread. Section 8.5 provides details of spread calculations, and Section 8.6 provides information for gutter flow.

7. *Inlet spacing and layout* — Beginning at the upstream end, determine the location of the upstream inlet based on trial drainage area delineations. Size inlet, and calculate bypass. Continue downstream, locating inlets and determining bypass amounts. Accommodate special situations and point sources of inflow. Seek to visualize the flow of water to determine potential problems. For example, placing reliance on an inlet at the end of a downhill cul-de-sac often allows excess water to bypass the inlet into a residential driveway or parking lot if it should become clogged. Extra inlets should be placed at all street crossings to eliminate flow bypass. Sections 8.6 through 8.9 give details for inlet design.

8. *Pipe sizing and layout* — Beginning at the upstream end, size pipes for the calculated flow and the given design criteria. Sometimes pipe size criteria are different from the street flow criteria. Velocity should not appreciably decrease at inlets. Pipe sizes should never decrease downstream, even for flow on a steeper slope. Take great care in allowing flow to enter inlet boxes from laterals and tapping culverts with pipes larger than 50% of the minimum box dimension. Provide for maintenance and other utility spacing minimums. Minimum pipe sizes may apply in some locations. Sections 8.10 and 8.11 provide details of pipe sizing and hydraulic gradient calculations. For ditches, the additional consideration of erosion and bank protection is a concern. Smooth flow lines and gradual transitions should be planned.

9. *Hydraulic gradeline* — Follow design criteria for hydraulic gradeline calculations in pipes. Begin at the downstream end using a suitable flow depth in the receiving stream if a free outfall cannot be assumed. Sections 8.10 and 8.11 provide details of hydraulic gradeline calculations.

Design Factors

Design factors to be considered for drainage calculations include the following:

- Return period
- Spread
- Inlet types and spacing

- Longitudinal slope
- Cross slope
- Curb and gutter sections
- Roadside and median channels
- Bridge decks
- Shoulder gutter
- Median/median barriers
- Storm drains
- Detention storage
- Cost
- Erosion

Following is a summary discussion of each of these factors. Most factors are covered in detail in later sections of this chapter. Where appropriate, some typical municipal criteria are given.

Return Period

The design storm return period for pavement drainage should be consistent with the frequency selected for other components of the drainage system. See Section 8.5 for further discussion of design storm return period (frequency) criteria.

Spread

For multilane curb and gutter or guttered roadways with no parking, it is not practical to avoid travel lane flooding when grades are flat (0.2 to 1%). However, flooding should never exceed the lane adjacent to the gutter (or shoulder) for design conditions. Municipal bridges with curb and gutter should also use this criterion. For single-lane roadways, at least 8 ft of roadway should remain unflooded for design conditions.

Inlet Types and Spacing

Inlet types should be selected from locally available inlets. Municipalities may want to specify the manufacturer and specific inlet types that are acceptable, and where performance information is available for engineering analysis and design. Inlets should be located or spaced in such a manner that the design curb flow does not exceed the spread limitations. Flow should not be allowed to cross intersecting streets unless approved by the municipal engineering department. Some percentage of lockage should be planned for, if appropriate.

Longitudinal Slope

A minimum longitudinal gradient is more important for a curbed pavement, because it is susceptible to stormwater spread. Flat gradients on uncurbed pavements can lead to a spread problem if vegetation is allowed to build up along the pavement edge.

Desirable gutter grades should not be less than 0.3% for curbed pavements, and a minimum of 0.2% in flat terrain. Minimum grades can be maintained in flat terrain by use of a sawtooth profile.

To provide adequate drainage in sag vertical curves, a minimum slope of 0.3% should be maintained within 50 ft of the level point in the curve. This is accomplished where the length of the curve divided by the algebraic difference in grades is equal to or less than 167. Special gutter profiles should be developed to maintain a minimum slope of 0.2% up to the inlet. Although ponding is not usually a problem at crest vertical curves, on extremely flat curves a similar minimum gradient should be provided to facilitate drainage.

Cross Slope

The design of pavement cross slope is often a compromise between the need for reasonably steep cross slopes for drainage and relatively flat cross slopes for driver comfort. The USDOT, FHWA (1979) reports that cross slopes of 2% have little effect on driver effort in steering, especially with power steering, or on friction demand for vehicle stability. Use of a cross slope steeper than 2% on pavements with a central crown line is not desirable. In areas of intense rainfall, a somewhat steeper cross slope may be necessary to facilitate drainage. In such areas, the cross slope could be increased to 2.5%.

When three or more lanes are inclined in the same direction on multilane pavements, it is desirable that each successive pair of lanes, or the portion of the outside lanes from the first two lanes from the crown line, have an increased slope. The two lanes adjacent to the crown line should be pitched at the normal slope, and successive lane pairs, or portions of the outside lanes, should be increased by about 0.5 to 1%. Where three or more lanes are provided in each direction, the maximum pavement cross slope should be limited to 4%.

Median areas should not be drained across traveled lanes, and shoulders should generally be sloped to drain away from the pavement, except with raised, narrow medians.

Curb and Gutter Sections

Curbing at the outside edge of pavements is normal practice for low-speed, urban highway facilities. Gutters may be 1.0 ft to 3.5 ft wide. Standard curb and gutter has a width of 1.5 ft, and may be integral with the curb. Gutters are on the same cross slope as the pavement on the high side and depressed with a steeper cross slope on the low side, usually 1 in per ft. Typical practice is to place curbs at the outside edge of shoulders or parking lanes on low-speed facilities. The gutter width may be included as a part of the parking lane. Shoulder gutter is not required adjacent to barrier walls on high fills.

Curbed highway sections are relatively inefficient at conveying water, and the area tributary to the gutter section should be kept to a minimum to reduce the hazard from water on the pavement. Where practicable, the flow from major areas draining toward curbed highway pavements should be intercepted by channels as appropriate.

Roadside and Median Channels

Roadside channels are commonly used with uncurbed roadway sections to convey runoff from the highway pavement and from areas that drain toward the highway. Due to right-of-way limitations, roadside channels cannot be used on most urban arterials. They can be used in cut sections, depressed sections, and other locations where sufficient right-of-way is available, and driveways or intersections are infrequent.

It is preferable to slope median areas and inside shoulders to a center swale, to prevent drainage from the median area from running across the pavement. This is particularly important for high-speed facilities and for facilities with more than two lanes of traffic in each direction. If used, temporary storage in shallow medians must be carefully engineered to handle high-intensity rainfall.

Bridge Decks

Drainage of bridge decks is similar to other curbed roadway sections. It is often less efficient, because cross slopes are flatter, parapets collect large amounts of debris, and small drainage inlets on scuppers have a higher potential for clogging by debris. Because of the difficulties in providing and maintaining adequate deck drainage systems, gutter flow from roadways should be intercepted before it reaches a bridge. In many cases, deck drainage must be carried several spans to the bridge end for disposal.

Short, continuous-span bridges, particularly overpasses, may be built without inlets. The water from the bridge roadway should be carried downslope by open or closed chutes near the end of the bridge structure. Some type of bridge end drainage must be provided at all bridges.

Zero gradients and sag vertical curves should be avoided on bridges. The minimum desirable longitudinal slope for bridge deck drainage should be 0.2%. When bridges are placed at a vertical curve, and the longitudinal slope is less than 0.2%, the gutter spread should be checked to ensure a safe, reasonable design.

Scuppers are the recommended method of deck drainage, because they can reduce the problems of transporting a relatively large concentration of runoff in an area of generally limited right-of-way. However, the use of scuppers should be evaluated for site-specific concerns. Scuppers should not be located over embankments, slope protection, navigation channels, driving lanes, or railroad tracks. Runoff collected and transported to the end of the bridge should generally be collected by inlets and down drains, although sod flumes may be used for extremely minor flows in some areas. Downspouts, where required, should be made of rigid corrosion-resistant material not less than 6 in in least dimension and should be provided with cleanouts.

The details of deck drains should be such as to prevent the discharge of drainage water against any portion of the structure or on moving traffic below, and to prevent erosion at the outlet of the downspout. Deck drainage may be connected to conduits leading to stormwater outfalls at ground level. Overhanging portions of concrete decks should be provided with a drip bead or notch. Water in a roadway gutter section should be intercepted prior to the bridge.

The placement of bridges over environmentally sensitive areas should be avoided if possible. Where not possible, precautions should be taken to minimize the adverse environmental impact caused by roadway runoff from bridge decks. Precautions include restricting the use of scuppers and downspouts and directing water through overland treatment systems, detention ponds, or designed wetlands prior to entering the water course (EPA, 1993; Versar, 1985; Woodward-Clyde, 1989). For situations where traffic under the bridge or environmental concerns prevent the use of scuppers, grated bridge drains could also be used.

Shoulder Gutters

Shoulder gutters may be appropriate to protect fill slopes from erosion caused by water from the roadway pavement. Shoulder gutter is required on fill slopes higher than 20 ft — standard slopes are 2:1. It is also required on fill slopes higher than 10 ft — standard slopes are 6:1 and 3:1 if the roadway grade is greater than 2%. In areas where permanent vegetation cannot be established, a shoulder gutter is required on fill slopes higher than 10 ft regardless of the grade. Inspection of the existing/proposed site conditions and contact with maintenance and construction personnel should be made by the designer to determine if vegetation will survive.

Shoulder gutters may be appropriate at bridge ends, where concentrated flow from the bridge deck would otherwise run down the fill slope. This section of gutter should be long enough to include the transitions. Shoulder gutters are not required on the high side of superelevated sections or adjacent to barrier walls on high fills.

Median/Median Barriers

Weepholes are often used to prevent ponding of water against barriers (especially on superelevated curves). In order to minimize flow across traveled lanes, it is preferable to collect the water into a subsurface system connected to the main storm drain system.

Storm Drains

Storm drains are used to convey water from the inlets to an acceptable outlet. Storm drains comprise a drainage system usually consisting of one or more pipes connecting two or more drop inlets. Cross storm drains hydraulically designed to function as a culvert or culverts are an exception to that statement. Storm drains should have adequate capacity so that they can accommodate runoff that enters the system. The storm drainage system for sag vertical curves should have a higher level of flood protection to decrease the depth of ponding on the roadway and bridges.

Storm drains should be designed to protect the roadway from flooding at the appropriate return period. Reserve capacity should be available at critical locations, such as vertical curve sags and at bridge approaches. Where feasible, the storm drains should be designed to avoid existing utilities. Attention should be given to the storm drain outfalls to ensure that the potential for erosion is minimized.

Detention Storage

Reduction of peak flows can be achieved by the storage of runoff in detention basins, storm drains, swales and channels, and other detention storage facilities. Stormwater is then released to the downstream conveyance facility at a reduced flow rate. The concept should be considered for use in highway drainage design, where existing downstream conveyance facilities are inadequate to handle peak flow rates from highway storm drainage facilities, where environmental concerns are present, where the highway would contribute to increased peak flow rates and aggravate downstream flooding problems, and as a technique to reduce the right-of-way, construction, and operation costs of outfalls from highway storm drainage facilities.

Costs

The cost of drainage is neither incidental nor minor on most roads. Careful attention to requirements for adequate drainage and protection of the roadway from stormwater in all phases of location and design will prove to be effective in reducing costs in construction and maintenance. Unless drainage is properly accommodated, maintenance costs will be unduly high.

Construction costs can be minimized through proper layout and design. For example, it is cost effective to avoid deep cuts, rock blasting, numerous junctions, utility relocation, and pumping of stormwater. Storm drain design should anticipate upstream development that will use the system and seek to accommodate such development in system sizing.

8.4 Stormwater Inlet Overview

The primary aim of drainage design is to limit the amount of water flowing along the gutters or ponding at the sags to quantities that will not interfere with the passage of traffic for the design frequency. This is accomplished by placing inlets at such points and at such intervals to intercept flows and control spread.

In this chapter, guidelines are given for evaluating roadway features and design criteria as they relate to gutter and inlet hydraulics and storm drain design. Procedures for performing gutter flow calculations are based on a modification of Manning's equation. Inlet capacity calculations for grate and combination inlets are based on information contained in HEC-12, USDOT, FHWA (1984) and HEC-22, FHWA (2001). Storm drain design is based on the use of the rational formula.

Drainage inlets are sized and located to limit the spread on traffic lanes to tolerable widths for the design storm. Because grates may become blocked by trash accumulation, curb openings, or combination inlets with both grate and curb openings, are advantageous for urban conditions. Grate inlets and depressions of curb-opening inlets should be located outside the through-traffic lanes. Inlet grates should safely accommodate bicycle and pedestrian traffic where appropriate.

Inlets should be selected, sized, and located to prevent silt and debris carried in suspension from being deposited on the traveled way where the longitudinal gradient is decreased.

Inlets at vertical curve sags in the roadway grade should also be capable of limiting the spread to tolerable limits. The width of water spread on the pavement should not be substantially greater than the width of spread encountered on continuous grades. At high discharges, this can only be accomplished by the use of inlets just upstream of the sag inlet on either or both sides of the sag. These additional inlets, often referred to as flanking inlets, serve to pick up the runoff before it reaches the sag and also limit the spread in the event that the sag inlet is clogged by debris. Where there is a danger of damage to adjacent property by runoff overtopping the curb in a sag, flanking inlets should be used, and the location should be checked to ensure that the curb is not overtopped due to insufficient inlet capacity.

Inlets should be located so that concentrated flow and heavy sheet flow will not cross traffic lanes. Where pavement surfaces are warped, as at cross streets or ramps, surface water should be intercepted just before the change in cross slope. Also, inlets should be located just upgrade of pedestrian crossings.

Inlets should be placed upstream of locations where the pavement cross slope reverses, such as on the high side of superelevated horizontal curves, to avoid concentrated flows crossing the roadway. Special care should be given to inlet placement to ensure adequate capacity at bridge approaches and at sag vertical curves, where ponding deeper than the curb height could occur.

The design of a drainage system for a roadway traversing an urbanized region is generally a more complex problem than for roadways traversing sparsely settled rural areas. This is often due to the following:

- The wide roadway sections, flat grades, both in longitudinal and transverse directions, shallow water courses, absence of side channels
- The more costly property damages that may occur from ponding of water or from flow of water through built-up areas

- The fact that the roadway section must carry traffic, but also act as a channel to convey the water to a disposal point (Unless proper precautions are taken, this flow of water along the roadway will interfere with or possibly halt the passage of highway traffic.)

There are four stormwater inlet categories:

- Curb-opening inlets
- Grated inlets
- Combination inlets
- Multiple inlets

In addition, inlets may be classified as being on a continuous grade or in a sump. *Continuous grade* refers to an inlet located on the street with a continuous slope past the inlet with water entering from one direction. The *sump* condition exists when the inlet is located at a low point and water enters from both directions.

Although design storm criteria differ from one municipality to another, the 2-year, 5-year, and 10-year design storms are the most frequent design storms used for stormwater inlet design. In addition to design storm criteria, following are some typical spread limit criteria used by municipalities:

- A maximum spread of 6 ft is allowed in a travel lane.
- For a street with a valley gutter, another foot for the gutter is allowed, with a total maximum spread of 7 ft.
- For a street with a standard 2 ft, 6 in curb and gutter, an additional 2 ft is allowed, with a total maximum spread of 8 ft from the face of the curb.

8.5 Design Frequency and Spread

Following are some recommended criteria for design storm frequency and spread. These criteria should be analyzed and modified to fit local conditions.

The design storm frequency for pavement drainage should be consistent with the frequency selected for other components of the drainage system. However, for a full-shoulder bridge condition, the spread should be contained within the shoulder for a 10-year design storm.

For multilane curb and gutter, or guttered roadways with no parking, it is not practical to avoid travel lane flooding when longitudinal grades are flat (0.2 to 1%). However, flooding should never exceed the lane adjacent to the gutter (or shoulder) for design conditions. Municipal bridges with curb and gutter should also use this criterion. For single-lane roadways, at least 8 ft of roadway should remain unflooded for design conditions.

The major consideration for selecting a design frequency and spread is highway classification. Ponding should be minimized on the traffic lanes of high-speed, high-volume highways, where it is not expected by the public to occur, whereas for local streets, some ponding is common.

TABLE 8-1

Frequency and Spread Design Criteria

Road Classification	Design Frequency	Design Spread
High volume		
<45 mph	10-year	Shoulder + 3 ft
>45 mph	10-year	Shoulder
Sag point	50-year	Shoulder + 3 ft
Collector		
<45 mph	10-year	1/2 driving lane
>45 mph	10-year	Shoulder
Sag point	10-year	1/2 driving lane
Local streets		
Low ADT	5-year	1/2 driving lane
High ADT	10-year	1/2 driving lane
Sag point	10-year	1/2 driving lane

Source: From U.S. Department of Transportation, Federal Highway Administration, Hydraulic Engineering Circular No. 22, August 2001.

Highway speed is another major consideration, because at speeds greater than 45 miles per hour, even a shallow depth of water on the pavement can cause hydroplaning. Design speed is recommended for use in evaluating hydroplaning potential. When the design speed is selected, consideration should be given to the likelihood that legal posted speeds may be exceeded. It is clearly unreasonable to provide the same level of protection for low-speed facilities as for high-speed facilities.

Other considerations include inconvenience, hazards, and nuisances to pedestrian traffic, and buildings adjacent to roadways that are located within the splash zone. These considerations should not be minimized and, in some locations (such as commercial pedestrian areas), may assume major importance.

Table 8-1 shows recommended design criteria for frequency and spread.

8.6 Gutter Flow Calculations

The following form of Manning's equation can be used to evaluate gutter flow hydraulics:

$$Q = [0.56 / n] S_x^{5/3} S^{1/2} T^{8/3} \tag{8.1}$$

where Q = gutter flow rate (cfs); n = Manning's roughness coefficient; S_x = pavement cross slope (ft/ft); S = longitudinal slope (ft/ft); and T = width of flow or spread (ft).

A nomograph for solving Equation (8.1) is presented in Figure 8-1. Manning's n values for various pavement surfaces are given in Table 8-2.

Uniform Cross Section

The nomograph in Figure 8-1 is used with the following procedures to find gutter capacity for uniform cross slopes:

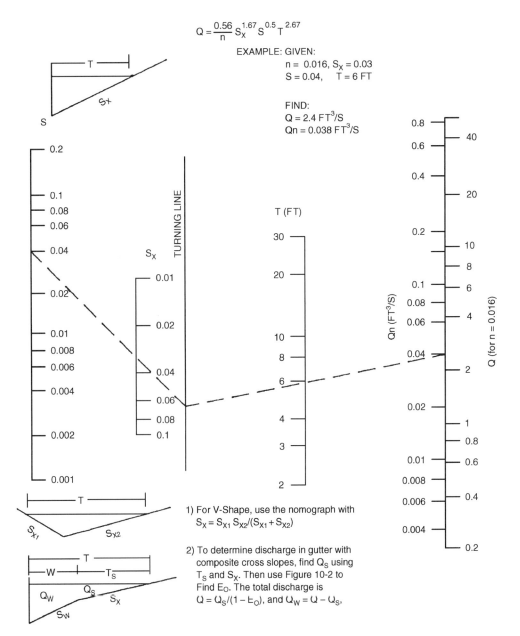

FIGURE 8-1
Flow in triangular gutter sections. (From American Association of State Highway and Transportation Officials, Model Drainage Manual, Washington, DC, 1991.)

Condition 1— Find spread, given gutter flow.
1. Determine input parameters, including longitudinal slope (S), cross slope (S_x), gutter flow (Q), and Manning's n.
2. Draw a line between the S and S_x scales and note where it intersects the turning line.
3. Draw a line between the intersection point from Step 2 and the appropriate gutter flow value on the capacity scale. If Manning's n is 0.016, use Q from Step 1; if not, use the product of Q and n.

TABLE 8-2

Manning's n Value for Street and Pavement Gutters

Type of Gutter or Pavement	Range of Manning's n
Concrete gutter, troweled finish	0.012
Asphalt pavement:	
Smooth texture	0.013
Rough texture	0.016
Concrete gutter with asphalt pavement:	
Smooth	0.013
Rough	0.015
Concrete pavement:	
Float finish	0.014
Broom finish	0.016
For gutters with small slopes, where sediment may accumulate, increase above values of n by	0.002

Note: Estimates are by the Federal Highway Administration.

Source: From USDOT, FHWA, HDS-3 (1961).

4. Read the value of the spread (T) at the intersection of the line from Step 3 and the spread scale.

Condition 2 — Find gutter flow, given spread.

1. Determine input parameters, including longitudinal slope (S), cross slope (S_x), spread (T), and Manning's n.
2. Draw a line between the S and S_x scales, and note where they intersect the turning line.
3. Draw a line between the intersection point from Step 2 and the appropriate value on the T scale. Read the value of Q or Qn from the intersection of that line on the capacity scale.
4. For Manning's n values of 0.016, the gutter capacity Q from Step 3 is selected. For other Manning's n values, the gutter capacity times n (Qn) is selected from Step 3 and divided by the appropriate n value to give the gutter capacity.

Composite Gutter Sections

Figure 8-2 in combination with Figure 8-1 can be used to find the flow in a gutter with width (W) less than the total spread (T). Such calculations are generally used for evaluating composite gutter sections or frontal flow for grate inlets.

Figure 8-3 provides a direct solution of gutter flow in a composite gutter section. The flow rate at a given spread or the spread at a known flow rate can be found from this figure. Figure 8-3 involves a complex graphical solution of the equation for flow in a composite gutter section. Typical of graphical solutions, extreme care in using the figure is necessary to obtain accurate results.

Condition 1 — Find spread, given gutter flow.

1. Determine input parameters, including longitudinal slope (S), cross slope (S_x), depressed section slope (S_w), depressed section width (W), Manning's n, gutter flow (Q), and a trial value of the gutter capacity above the depressed section (Q_s).

Storm Drainage Systems

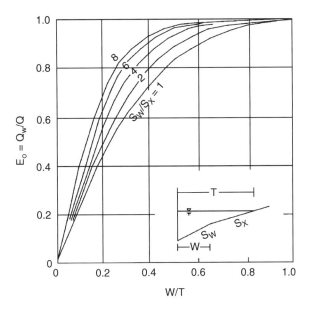

FIGURE 8-2
Ratio of frontal flow to total gutter flow. (From American Association of State Highway and Transportation Officials, Model Drainage Manual, Washington, DC, 1991.)

2. Calculate the gutter flow in W (Q_w), using the following equation:

$$Q_w = Q - Q_s \tag{8.2}$$

3. Calculate the ratios Q_w/Q or E_o and S_w/S_x and use Figure 8-2 to find an appropriate value of W/T.
4. Calculate the spread (T) by dividing the depressed section width (W) by the value of W/T from Step 3.
5. Find the spread above the depressed section (T_s) by subtracting W from the value of T obtained in Step 4.
6. Use the value of T_s from Step 5 along with Manning's n, S, and S_x to find the actual value of Q_s from Figure 8-1.
7. Compare the value of Q_s from Step 6 to the trial value from Step 1. If values are not comparable, select a new value of Q_s and return to Step 1.

Condition 2 — Find gutter flow, given spread.

1. Determine input parameters, including spread (T), spread above the depressed section (T_s), cross slope (S_x), longitudinal slope (S), depressed section slope (S_w), depressed section width (W), Manning's n, and depth of gutter flow (d).
2. Use Figure 8-1 to determine the capacity of the gutter section above the depressed section (Q_s). Use the procedure for uniform cross slopes (Condition 2), substituting T_s for T.
3. Calculate the ratios W/T and S_w/S_x, and, from Figure 8-2, find the appropriate value of E_o (the ratio of Q_w/Q).
4. Calculate the total gutter flow using the following equation:

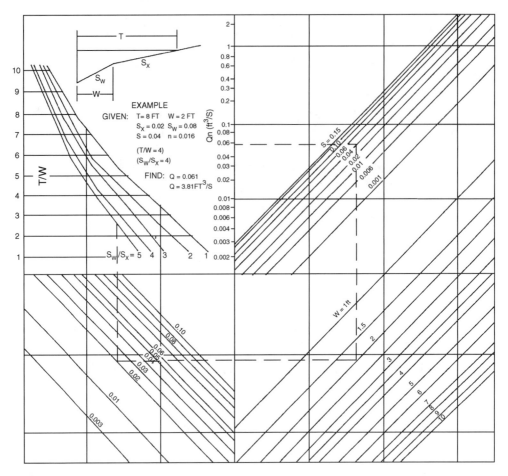

FIGURE 8-3
Flow in composite gutter sections. (From American Association of State Highway and Transportation Officials, Model Drainage Manual, Washington, DC, 1991.)

$$Q = Q_s / (1 - E_o) \qquad (8.3)$$

where Q = gutter flow rate (cfs); Q_s = flow capacity of the gutter section above the depressed section (cfs); and E_o = ratio of frontal flow to total gutter flow (Q_w/Q).

5. Calculate the gutter flow in width (W), using Equation (8.2).

Example Problems
EXAMPLE 1

Given:

$T = 8$ ft
$n = 0.015$
$S_x = 0.025$ ft/ft
$S = 0.01$ ft/ft

Find:

1. Flow in gutter at design spread
2. Flow in width ($W = 2$ ft) adjacent to the curb

Solution:

1. From Figure 8-1, $Qn = 0.03$

$$Q = Qn/n = 0.03/0.015 = 2.0 \text{ cfs}$$

2. $T = 8 - 2 = 6$ ft

$(Qn)_2 = 0.014$ (Figure 8-1) (flow in 6 ft width outside of width W)

$$Q = 0.014/0.015 = 0.9 \text{ cfs}$$

$$Q_w = 2.0 - 0.9 = 1.1 \text{ cfs}$$

Flow in the first 2 ft adjacent to the curb is 1.1 cfs and 0.9 cfs in the remainder of the gutter.

EXAMPLE 2

Given:

$T = 6$ ft
$S_w = 0.0833$ ft/ft
$T_s = 6 - 1.5 = 4.5$ ft
$W = 1.5$ ft
$S_x = 0.03$ ft/ft
$n = 0.014$
$S = 0.04$ ft/ft

Find:

Flow in the composite gutter

Solution:

1. Use Figure 8-1 to find the gutter section capacity above the depressed section.

$$Q_s n = 0.038$$

$$Q_s = 0.038/0.014 = 2.7 \text{ cfs}$$

Calculate $W/T = 1.5/6 = 0.25$ and

$$S_w/S_x = 0.0833/0.03 = 2.78$$

Use Figure 8-2 to find $E_o = 0.64$
Calculate the gutter flow using Equation (8.3):

$$Q = 2.7/(1 - 0.64) = 7.5 \text{ cfs}$$

Calculate the gutter flow in width, W, using Equation (8.2):

$$Q_w = 7.5 - 2.7 = 4.8 \text{ cfs}$$

8.7 Grate Inlet Design

Inlets are drainage structures utilized to collect surface water through grate or curb openings and convey it to storm drains or direct outlet to culverts. Grate inlets subject to traffic should be bicycle-safe and be load-bearing adequate. Appropriate frames should be provided.

Inlets used for the drainage of highway surfaces can be divided into three major classes:

1. Grate inlets — These inlets include grate inlets consisting of an opening in the gutter covered by one or more grates, and slotted inlets consisting of a pipe cut along the longitudinal axis with a grate of spacer bars to form slot openings.
2. Curb-opening inlets — These inlets are vertical openings in the curb covered by a top slab.
3. Combination inlets — These inlets usually consist of a curb-opening inlet and a grate inlet placed in a side-by-side configuration, but the curb opening may be located partly upstream of the grate.

In addition, where significant ponding can occur, in locations such as underpasses and in sag vertical curves in depressed sections, it is good engineering practice to place flanking inlets on each side of the inlet at the low point in the sag. The flanking inlets should be placed so that they will limit spread on low-gradient approaches to the level point and act in relief of the inlet at the low point if it should become clogged or if the design spread is exceeded.

The design of grate inlets will be discussed in this section, curb inlet design in Section 8.8, and combination of inlets in Section 8.9.

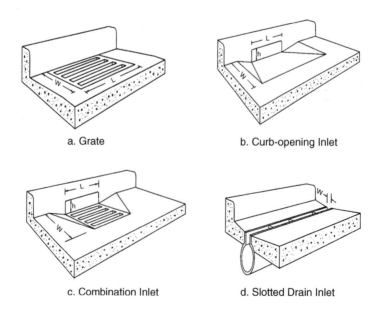

a. Grate

b. Curb-opening Inlet

c. Combination Inlet

d. Slotted Drain Inlet

TABLE 8-3

Grate Debris-Handling Efficiencies

Rank	Grate	Longitudinal Slope	
		(0.005)	(0.04)
1	CV - 3-1/4 - 4-1/4	46	61
2	30 - 3-1/4 - 4	44	55
3	45 - 3-1/4 - 4	43	48
4	P - 1-7/8	32	32
5	P - 1-7/8 - 4	18	28
6	45 - 2-1/4 - 4	16	23
7	Reticuline	12	16
8	P - 1-1/8	9	20

Source: From U.S. Department of Transportation, Federal Highway Administration, Drainage of Highway Pavements, Hydraulic Engineering Circular No. 12, 1984.

Grate Inlets on Grade

The capacity of an inlet depends upon its geometry and the cross slope, longitudinal slope, total gutter flow, depth of flow, and pavement roughness. The depth of water next to the curb is the major factor in the interception capacity of gutter inlets and curb-opening inlets. At low velocities, all of the water flowing in the section of gutter occupied by the grate, called frontal flow, is intercepted by grate inlets, and a small portion of the flow along the length of the grate, termed side flow, is intercepted. On steep slopes, only a portion of the frontal flow will be intercepted if the velocity is high or the grate is short and splash-over occurs. For grates less than 2 ft long, intercepted flow is small.

Agencies and manufacturers of grates have investigated inlet interception capacity. For inlet efficiency data for various sizes and shapes of grates, refer to USDOT/FHA (1984) and inlet grate capacity charts prepared by grate manufacturers.

A parallel bar grate is the most efficient type of gutter inlet; however, when crossbars are added for bicycle safety, the efficiency is greatly reduced. Where bicycle traffic is a design consideration, the curved vane grate and the tilt bar grate are recommended for their hydraulic capacity and bicycle safety features. They also handle debris better than other grate inlets, but the vanes of the grate must be turned in the proper direction.

Where debris is a problem, consideration should be given to debris-handling efficiency rankings of grate inlets from laboratory tests in which an attempt was make to qualitatively simulate field conditions. Table 8-3 presents the results of debris-handling efficiencies of several grates.

The ratio of frontal flow to total gutter flow, E_o, for straight cross slope is expressed by the following equation:

$$E_o = Q_w/Q = 1 - (1 - W/T)^{2.67} \tag{8.4}$$

where Q = total gutter flow (cfs); Q_w = flow in width W, cfs; W = width of depressed gutter or grate (ft); and T = total spread of water in the gutter (ft).

Figure 8-2 provides a graphical solution of E_o for depressed gutter sections or straight cross slopes.

The ratio of side flow, Q_s, to total gutter flow is:

$$Q_s/Q = 1 - Q_w/Q = 1 - E_o \tag{8.5}$$

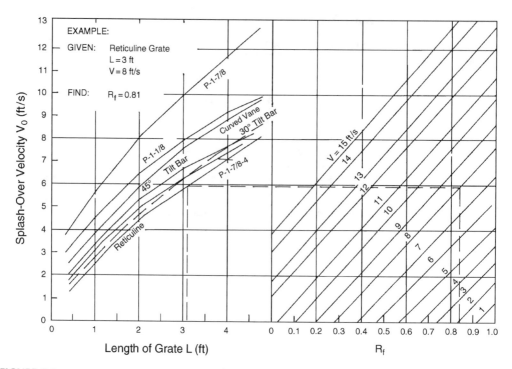

FIGURE 8-4
Grate inlet frontal flow interception efficiency. (From U.S. Department of Transportation, Federal Highway Administration, Drainage of Highway Pavements, Hydraulic Engineering Circular No. 12, 1984.)

The ratio of frontal flow intercepted to total frontal flow, R_f, is expressed by the following equation:

$$R_f = 1 - 0.09 (V - V_0) \quad (8.6)$$

where V = velocity of flow in the gutter (ft/s), and V_o = gutter velocity where splash-over first occurs (ft/s).

This ratio is equivalent to frontal flow interception efficiency. Figure 8-4 provides a solution of Equation (8.6) that takes into account grate length, bar configuration, and gutter velocity at which splash-over occurs. The gutter velocity needed to use Figure 8-4 is total gutter flow divided by the area of flow.

The ratio of side flow intercepted to total side flow, R_s, or side flow interception efficiency, is expressed by:

$$R_s = 1 / [1 + (0.15 V^{1.8} / S_x L^{2.3})] \quad (8.7)$$

where L = length of the grate (ft).

Figure 8-5 provides a solution to Equation (8.7).

The efficiency, E, of a grate is expressed as:

$$E = R_f E_o + R_s (1 - E_o) \quad (8.8)$$

The interception capacity of a grate inlet on grade is equal to the efficiency of the grate multiplied by the total gutter flow:

Storm Drainage Systems

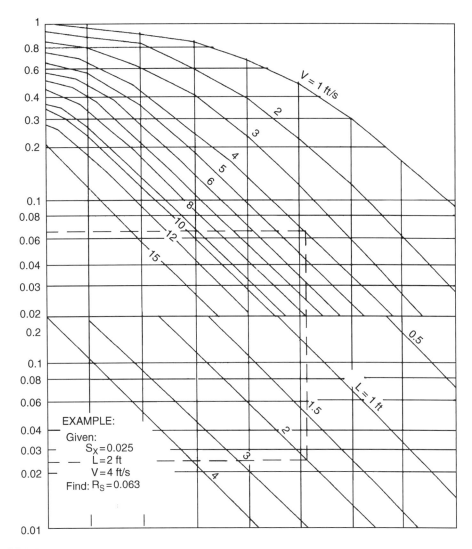

FIGURE 8-5
Grate inlet side flow interception efficiency. (From U.S. Department of Transportation, Federal Highway Administration, Drainage of Highway Pavements, Hydraulic Engineering Circular No. 12, 1984.)

$$Q_i - EQ = Q[R_f E_o + R_s(1 - E_o)] \tag{8.9}$$

Example Problem

The following example illustrates the use of this procedure.

Given:

$W = 2$ ft
$T = 8$ ft
$S_x = 0.025$ ft/ft
$S = 0.01$ ft/ft

$E_o = 0.69$
$Q = 3.0$ cfs
$V = 3.1$ ft/s
Gutter depression = 2 in

Find:

Interception capacity of:

1. A curved vane grate
2. A reticuline grate 2-ft long and 2-ft wide

Solution:

1. From Figure 8-4 for curved vane grate, $R_f = 1.0$
2. From Figure 8-4 for reticuline grate, $R_f = 1.0$
3. From Figure 8-5 $R_s = 0.1$ for both grates.
4. From Equation (8.9)

$$Q_i = 3.0[1.0 * 0.69 + 0.1(1 - 0.69)] = 2.2 \text{ cfs}$$

For this example, the interception capacity of a curved vane grate is the same as that for a reticuline grate for the sited conditions.

Grate Inlet in Sag

A grate inlet in a sag operates as a weir up to a certain depth dependent on the bar configuration and size of the grate, and as an orifice at greater depths. For a standard gutter inlet grate, weir operation continues to a depth of about 0.4 foot above the top of grate and, when depth of water exceeds about 1.4 ft, the grate begins to operate as an orifice. Between depths of about 0.4 and 1.4 ft, a transition from weir to orifice flow occurs.

The capacity of grate inlets operating as a weir is:

$$Q_i = CPd^{1.5} \qquad (8.10)$$

where P = perimeter of grate, excluding bar widths and the side against the curb (ft); C = 3.0; and d = depth of water above grate (ft).

The capacity of grate inlets operating as an orifice is:

$$Q_i = CA(2gd)^{0.5} \qquad (8.11)$$

where $C = 0.67$ orifice coefficient; A = clear opening area of the grate (ft^2); and $g = 32.2$ ft/s.

Figure 8-6 is a plot of Equations (8.10) and (8.11) for various grate sizes. The effects of grate size on the depth at which a grate operates as an orifice is apparent from the chart. Transition from weir to orifice flow results in interception capacity less than that computed by either the weir or the orifice equation. This capacity can be approximated by drawing in a curve between the lines representing the perimeter and net area of the grate to be used.

Storm Drainage Systems

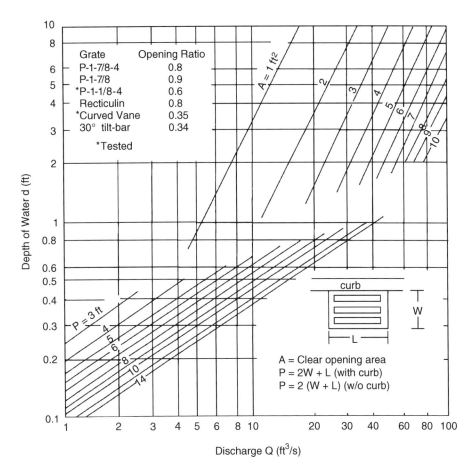

FIGURE 8-6
Grate inlet capacity in sump conditions. (From U.S. Department of Transportation, Federal Highway Administration, Drainage of Highway Pavements, Hydraulic Engineering Circular No. 12, 1984.)

Example Problem

The following example illustrates the use of Figure 8-6.

Given:

A symmetrical sag vertical curve with equal bypass from inlets upgrade of the low point, allow for 50% clogging of the grate.

Q_b = 3.6 cfs
Q = 8 cfs, 10-year storm (design)
Q_b = 4.4 cfs
Q = 11 cfs, 25-year storm (check)
T = 10 ft, design
S_x = 0.05 ft/ft
$d = TS_x$ = 0.5 ft

Find:

Grate size for design Q and depth at curb for check Q. Check spread at S = 0.003 on approaches to the low point.

Solution:

1. From Figure 8-6, a grate must have a perimeter of 8 ft to intercept 8 cfs at a depth of 0.5 ft. Some assumptions must be made regarding the nature of the clogging in order to compute the capacity of a partially clogged grate. If the area of a grate is 50% covered by debris so that the debris-covered portion does not contribute to interception, the effective perimeter will be reduced by a lesser amount than 50%. For example, if a 2-ft × 4-ft grate is clogged so that the effective width is 1 ft, then the perimeter, P = 1 + 4 + 1 = 6 ft, rather than 8 ft, the total perimeter, or 4 ft, half of the total perimeter. The area of the opening would be reduced by 50% and the perimeter by 25%. Therefore, assuming 50% clogging along the length of the grate, a 4 × 4, a 2 × 6, or a 3 × 5 grate would meet requirements of an 8-ft perimeter 50% clogged.
2. Assuming that the installation chosen to meet the design is a double 2 × 3 ft grate for 50% clogged conditions:

$$P = 1 + 6 + 1 = 8 \text{ ft}$$

3. For 10-year flow: d = 0.5 ft (from Figure 8-6)
4. For 25-year flow: d = 0.6 ft (from Figure 8-6),

$$T = 12.0 \text{ ft}$$

5. At the check flow rate, ponding will extend 2 ft into a traffic lane if the grate is 50% clogged in the manner assumed.
6. The American Society of State Highway and Transportation Officials (AASHTO) geometric policy recommends a gradient of 0.3% within 50 ft of the level point in a sag vertical curve.
7. Check T at S = 0.003 for the design and check flow:
8. Q = 3.6 cfs, T = 8.2 ft (10-year storm) (Figure 8-1)
9. Q = 4.4 cfs, T = 9 ft (25-year storm) (Figure 8-1)
10. Thus, a double 2 × 3-ft grate 50% clogged is adequate to intercept the design flow at a spread that does not exceed design spread, and spread on the approaches to the low point will not exceed design spread. However, the tendency of grate inlets to clog completely warrants consideration of a combination inlet, or curb-opening inlet in a sag where ponding can occur, and flanking inlets on the low-gradient approaches.

8.8 Curb Inlet Design

Curb Inlets on Grade

Following is a discussion of the procedures for the design of curb inlets on grade that is followed by curb inlets in a sump.

Storm Drainage Systems

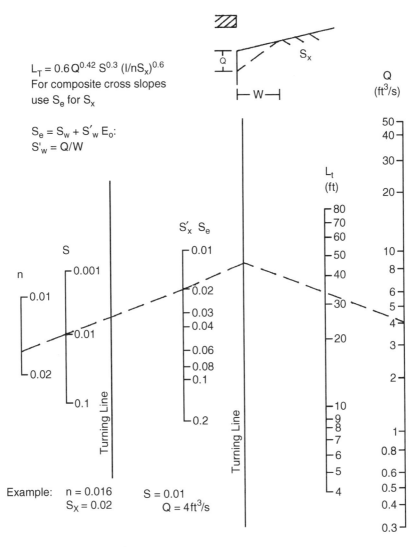

FIGURE 8-7
Curb-opening and slotted drain inlet length for total interception. (From U.S. Department of Transportation, Federal Highway Administration, Drainage of Highway Pavements, Hydraulic Engineering Circular No. 12, 1984.)

Curb-opening inlets are effective in the drainage of highway pavements, where flow depth at the curb is sufficient for the inlet to perform efficiently. Curb openings are relatively free of clogging tendencies and offer little interference to traffic operation. They are a viable alternative to grates in many locations, where grates would be in traffic lanes or would be hazardous for pedestrians or bicyclists. The length of curb-opening inlet required for total interception of gutter flow on a pavement section with a straight cross slope is determined using Figure 8-7. The efficiency of curb-opening inlets shorter than the length required for total interception is determined using Figure 8-8.

The length of inlet required for total interception by depressed curb-opening inlets or curb openings in depressed gutter sections can be found by the use of an equivalent cross slope, S_e, in the following equation:

$$S_e = S_x + S'_w E_o \tag{8.12}$$

where E_o = ratio of flow in the depressed section to total gutter flow; S'_w = cross slope of the gutter measured from the cross slope of the pavement (ft/ft); $S_x = (a/12W)$ (ft/ft); and a = gutter depression (in).

It is apparent from examination of Figure 8-7 that the length of curb opening required for total interception can be significantly reduced by increasing the cross slope or the equivalent cross slope. The equivalent cross slope can be increased by use of a continuously depressed gutter section or a locally depressed gutter section.

Steps for using Figures 8-7 and 8-8 in the design of curb inlets on grade are given below:

1. Determine the following input parameters:
   a. Cross slope = S_x (ft/ft)
   b. Longitudinal slope = S (ft/ft)
   c. Gutter flow rate = Q (cfs)
   d. Manning's n = n
   e. Spread of water on pavement = T (ft) from Figure 8-1
2. Enter Figure 8-7 using the two vertical lines on the left side labeled n and S. Locate the value for Manning's n and longitudinal slope and draw a line connecting these points and extend this line to the first turning line.
3. Locate the value for the cross slope (or equivalent cross slope) and draw a line from the point on the first turning line through the cross slope value, and extend this line to the second turning line.
4. Using the far right vertical line labeled Q, locate the gutter flow rate. Draw a line from this value to the point on the second turning line. Read the length required from the vertical line labeled L_T.
5. If the curb-opening inlet is shorter than the value obtained in step 4, Figure 8-8 can be used to calculate the efficiency. Enter the x-axis with the L/L_T ratio and draw a vertical line upward to the E curve. From the point of intersection, draw a line horizontally to the intersection with the y-axis and read the efficiency value.

Example Problem

Given:

S_x = 0.03 ft/ft

n = 0.016

S = 0.035 ft/ft

S'_w = 0.083

Q = 5 cfs

Find:

1. Q_i for a 10-ft curb-opening inlet
2. Q_i for a depressed 10-ft curb-opening inlet with a = 2 in, W = 2 ft, T = 8 ft (Figure 8-1)

Storm Drainage Systems

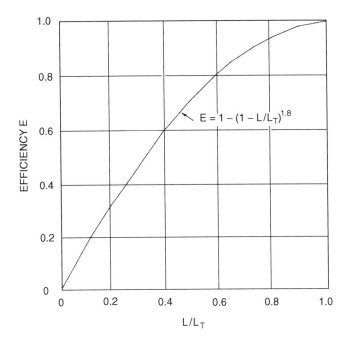

FIGURE 8-8
Curb-opening and slotted drain inlet interception efficiency. (From U.S. Department of Transportation, Federal Highway Administration, Drainage of Highway Pavements, Hydraulic Engineering Circular No. 12, 1984.)

Solution:

1. From Figure 8-7, $L_T = 41$ ft

$$L/L_T = 10/41 = 0.24$$

1. From Figure 8-8, $E = 0.39$

$$Q_i = EQ = 0.39 \times 5 = 1.95 \text{ cfs} = 2 \text{ cfs}$$

2. $Qn = 5.0 \times 0.016 = 0.08$ cfs

$$S_w/S_x = (0.03 + 0.083)/0.03 = 3.77$$

$$T/W = 3.5 \text{ (from Figure 8-3)}$$

$$T = 3.5 \times 2 = 7 \text{ ft}$$

$$W/T = 2/7 = 0.29 \text{ ft}$$

$$E_o = 0.72 \text{ (from Figure 8-2)}$$

$$\text{Therefore, } S_e = S_x + S'_w E_o$$

$$S_e = 0.03 + 0.083(0.72) = 0.09$$

From Figure 8-7, $L_T = 23$ ft

$$L/L_T = 10/23 = 0.43$$

From Figure 8-8, $E = 0.64$

$$Q_i = 0.64 \times 5 = 3.2 \text{ cfs}$$

The depressed curb-opening inlet will intercept 1.6 times the flow intercepted by the undepressed curb opening and over 60% of the total flow.

Curb Inlets in Sump

For the design of a curb-opening inlet in a sump location, the inlet operates as a weir to depths equal to the curb-opening height, and as an orifice at depths greater than 1.4 times the opening height. At depths between 1.0 and 1.4 times the opening height, flow is in a transition stage.

The capacity of curb-opening inlets in a sump location can be determined from Figure 8-9, which accounts for the operation of the inlet as a weir and as an orifice at depths greater than 1.4 h. This figure is applicable to depressed curb-opening inlets, and the depth at the inlet includes any gutter depression. The height (h) in the figure assumes a vertical orifice opening (see sketch in Figure 8-9). The weir portion of Figure 8-9 is valid for a depressed curb-opening inlet when $d < (h + a/12)$.

The capacity of curb-opening inlets in a sump location with a vertical orifice opening but without any depression can be determined from Figure 8-10. The capacity of curb-opening inlets in a sump location with other than vertical orifice openings can be determined by using Figure 8-11.

Steps for using Figures 8-9, 8-10, and 8-11 in the design of curb-opening inlets in sump locations are given below:

1. Determine the following input parameters:
   a. Cross slope = S_x (ft/ft)
   b. Spread of water on pavement = T (ft) from Figure 8-1
   c. Gutter flow rate = Q (cfs) or dimensions of curb-opening inlet [L (ft) and H (in.)]
   d. Dimensions of depression if any [a (in) and W (ft)]
2. To determine discharge given the other input parameters, select the appropriate Figure (8-9, 8-10, or 8-11, depending whether the inlet is in a depression and if the orifice opening is vertical).
3. To determine the discharge (Q), given the water depth (d), locate the water depth value on the y-axis and draw a horizontal line to the appropriate perimeter (p), height (h), length (L), or width times length (hL) line. At this intersection, draw a vertical line down to the x-axis and read the discharge value.
4. To determine the water depth given the discharge, use the procedure described in step 3 except enter the figure at the value for the discharge on the x-axis.

Storm Drainage Systems

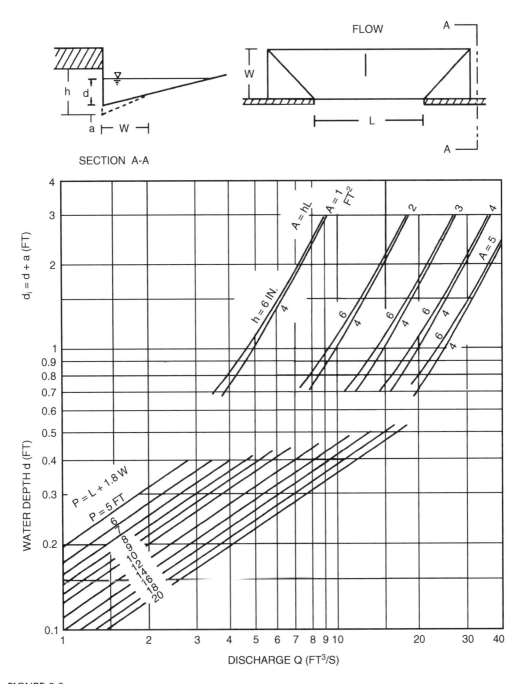

FIGURE 8-9
Depressed curb-opening inlet capacity in sump locations. (From U.S. Department of Transportation, Federal Highway Administration, Urban Drainage Design Manual, Hydraulic Engineering Circular No. 22, August 2001.)

326 *Municipal Stormwater Management, Second Edition*

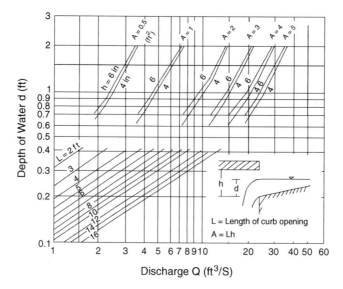

FIGURE 8-10
Curb-opening inlet capacity in sump locations. (From U.S. Department of Transportation, Federal Highway Administration, Urban Drainage Design Manual, Hydraulic Engineering Circular No. 22, August 2001.)

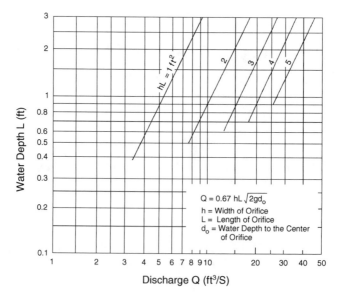

FIGURE 8-11
Curb-opening inlet orifice capacity — inclined and vertical orifice throats. (From U.S. Department of Transportation, Federal Highway Administration, Urban Drainage Design Manual, Hydraulic Engineering Circular No. 22, August 2001.)

Storm Drainage Systems

Example Problem

Given:

Curb-opening inlet in a sump location

a. L = 5 ft
b. h = 5 in

1. Undepressed curb opening

$$S_x = 0.05 \text{ ft/ft}$$

$$T = 8 \text{ ft}$$

2. Depressed curb opening

$$S_x = 0.05 \text{ ft/ft}$$

$$a = 2 \text{ in.}$$

$$W = 2 \text{ ft, } T = 8 \text{ ft}$$

Find:

Discharge Q_i

Solution:

1. d = TS_x = 8 x 0.05 = 0.4 ft
   a. d < h
   b. From Figure 8-10, Q_i = 3.8 cfs
2. d = 0.4 ft
   a. h + a/12 = (5 + 2/12)/12 = 0.43 ft
   b. Because d < 0.43, the weir portion of Figure 8-9 is applicable (lower portion of the figure)
   c. P = L + 1.8W = 5 + 3.6 = 8.6 ft
   d. From Figure 8-9, Q_i = 5 cfs
   e. At d = 0.4 ft, the depressed curb-opening inlet has about 30% more capacity than an inlet without depression.

8.9 Combination Inlets

On a continuous grade, the capacity of an unclogged combination inlet with the curb opening located adjacent to the grate is approximately equal to the capacity of the grate inlet alone. Thus, capacity is computed by neglecting the curb-opening inlet, and the design procedures should be followed based on the use of Figures 8-4, 8-5, and 8-6.

All debris carried by stormwater runoff that is not intercepted by upstream inlets will be concentrated at the inlet located at the low point, or sump. Because this will increase the probability of clogging for grated inlets, it is generally appropriate to estimate the capacity of a combination inlet at a sump by neglecting the grate inlet capacity. Assuming complete clogging of the grate, Figures 8-9, 8-10, and 8-11 for curb-opening inlets should be used for design.

8.10 Energy Losses in a Pipe System

Following are the equations needed to calculate the energy losses for the hydraulic grade line calculations for storm drainage systems.

Friction Losses

Energy losses from pipe friction may be determined by rewriting the Manning equation.

$$S_f = [Qn/(1.486\ AR^{2/3})]^2 \tag{8.13}$$

Then, the headlosses due to friction may be determined by the following formula:

$$H_f = S_f L \tag{8.14}$$

where H_f = friction headloss (ft); S_f = friction slope (ft/ft); and L = length of outflow pipe (ft).

Velocity Headlosses

From the time stormwater first enters the storm drain system at the inlet until it discharges at the outlet, it will encounter a variety of hydraulic structures such as inlets, manholes, junctions, bends, contractions, enlargements, and transitions that will cause velocity headlosses. Velocity losses may be expressed in a general form derived from the Bernoulli and Darcy-Weisback equations.

$$H = KV^2/2g \tag{8.15}$$

where H = velocity headloss (ft); K = loss coefficient for the particular structure; V = velocity of flow (ft/s); and g = acceleration due to gravity 32.2 ft/s².

Entrance Losses

Following are the equations used for entrance losses for beginning flows:

$$H_{tm} = V^2/2g \tag{8.16}$$

TABLE 8-4
Values of K for Change in Direction of Flow in Lateral

K	Degree of Turn (in Junction)
0.19	15
0.35	30
0.47	45
0.56	60
0.64	75
0.70	90 and greater

Note: K values for other degree of turns can be obtained by interpolating between values in Table 8-4.

$$H_e = KV^2/2g \qquad (8.17)$$

where H_{tm} = terminal (beginning of run) loss (ft); H_e = entrance loss for outlet structure (ft); and K = 0.5 (assuming square-edge).

Other terms are defined above.

Junction Losses — Incoming Opposing Flows

The headloss at a junction, H_{j1}, for two almost equal and opposing flows meeting head on with the outlet direction perpendicular to both incoming directions is considered the total velocity head of outgoing flow:

$$H_{j1} = (V_3^2)\,(\text{outflow})/2g \qquad (8.18)$$

where H_{j1} = junction losses (ft).
Other terms are defined above.

Junction Losses — Changes in Direction of Flow

When main storm drainpipes or lateral lines meet in a junction, velocity is reduced within the chamber and specific head increases to develop the velocity needed in the outlet pipe. The sharper the bend (approaching 90°), the more severe this energy loss becomes. When the outlet conduit is sized, determine the velocity and compute headloss in the chamber by the following formula:

$$H_b = K(V^2)\,(\text{outlet})/2g \qquad (8.19)$$

where H_b = bend headloss (ft), and K = junction loss coefficient.
Table 8-4 lists the values of K for various junction angles.

Junction Losses — Several Entering Flows

The computation of losses in a junction with several entering flows utilizes the principle of conservation of energy. For a junction with several entering flows, the energy content of the inflows is equal to the energy content of outflows plus additional energy required by the collision and turbulence of flows passing through the junction. The total junction losses at the sketched intersection are as follows:

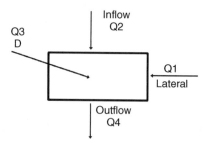

The following equation can be used to calculate junction losses:

$$H_{j2} = [(Q_4 V_4^2) - (Q_1 V_1^2) - (Q_2 V_2^2) + (K Q_1 V_1^2)] / (2 g Q_4)] \qquad (8.20)$$

where H_{j2} = junction losses (ft); Q = discharges (cfs); V = horizontal velocities (ft/s) (V_3 is assumed to be zero); g = acceleration due to gravity, 32.2 ft/s; and K = bend loss factor.

Where subscript nomenclature is as follows: Q_1 = 90° lateral (cfs); Q_2 = straight-through inflow (cfs); Q_3 = vertical dropped-in flow from an inlet (cfs); Q_4 = main outfall = total computed discharge (cfs); and V_1, V_2, V_3, V_4 are the horizontal velocities of foregoing flows, respectively in feet per second.

Assume: $H_b = K(V_1^2)/2g$ for change in direction. Also, no velocity head of an incoming line is greater than the velocity head of the outgoing line. Water surface of inflow and outflow pipes in junction to be level.

When losses are computed for any junction condition for the same or a lesser number of inflows, the above equation will be used with zero quantities for those conditions not present. If more directions or quantities are at the junction, additional terms will be inserted with consideration given to the relative magnitudes of flow and the coefficient of velocity head for directions other than straight through.

The final step in designing a storm drain system is to check the hydraulic grade line (HGL) as described in the next section of this chapter. Computing the HGL will determine the elevation, under design conditions, to which water will rise in various inlets, manholes, junctions, etc.

In Figure 8-12 is a summary of energy losses that should be considered. Following this, in Figure 8-13 is a sketch showing the proper and improper use of energy losses in developing a storm drain system.

8.11 Storm Drains

After the tentative locations of inlets, storm drains, and outfalls with tailwaters have been determined and the inlets sized, the next logical step is the computation of the rate of discharge to be carried by each storm drain and the determination of the size and gradient of pipe required to accommodate this discharge. This is done by proceeding in steps from upstream of a line to downstream to the point at which the line connects with other lines or the outfall, whichever is applicable. The discharge for a run is calculated, the storm drain serving that discharge is sized, and the process is repeated for the next run downstream.

$$H_{tm} = \frac{V^2}{2g}$$

TERMINAL JUNCTION LOSSES
(at beginning of run)
Where g = gravitational constant
2.2 feet per second
per second

$$H_e = 0.5 \frac{V^2}{2g}$$

ENTRANCE LOSSES
(for structure at beginning of run)
Assuming square edge

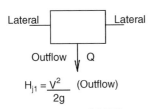

$$H_{j1} = \frac{V^2}{2g} \quad \text{(Outflow)}$$

JUNCTION LOSSES

Use only where flows are identical to above, otherwise use H_{J2} Equation.

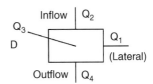

$$H_{J2} = \frac{Q_4 V_4^2 - Q_1 V_1^2 - Q_2 V_2^2 + K Q_1 V_1^2}{2g Q_4}$$

JUNCTION LOSSES
(After FHWA)

Total losses to include H_{J2} plus losses for change in direction of less than 90 (H_b).

Where K = Bend loss factor
Q_3 = Vertical dropped-in flow from an inlet
V_3 = Assumed to be zero

FRICTION LOSS (H_f)

$$H_f = S_f \times L$$

Where H_f = friction head
S_f = friction slope
L = length of conduit

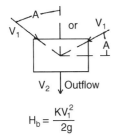

$$H_b = \frac{KV_1^2}{2g}$$

BEND LOSSES
(changes in direction of flow)

Where K	Degree of Turn in Junction
0.19	15
0.35	30
0.47	45
0.56	60
0.64	75
0.70	90

$$S = \left(\frac{Q_n}{1.486 \, AR^{\frac{2}{3}}}\right)$$

Where Q = discharge of conduit
n = Mannings coefficient of roughness (use 0.013 for R.C. Pipes)
A = area of conduit
R = hydraulic radius of conduit (D/4 for round pipe)

TOTAL ENERGY LOSSES AT EACH JUNCTION
$H_T = H_{tm} + H_e + H_{j1}$ or $H_{j2} + H_b + H_f$

FIGURE 8-12
Summary of energy losses. (From American Association of State Highway and Transportation Officials, Model Drainage Manual, Washington, DC, 1991.)

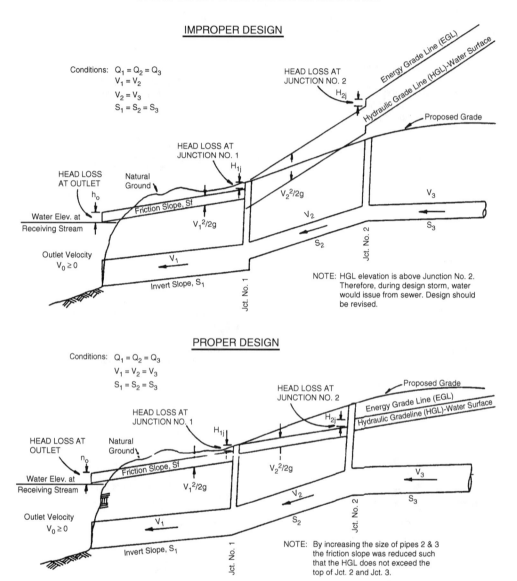

FIGURE 8-13
Energy and hydraulic grade lines for storm drains under constant discharge. (From American Association of State Highway and Transportation Officials, Model Drainage Manual, Washington, DC, 1991.)

It should be recognized that the rate of discharge to be carried by any particular section of storm drain is not necessarily the sum of the inlet design discharge rates of all inlets above that section of pipe, but as a general rule is somewhat less than this total. It is useful to understand that the time of concentration is most influential, and, as the time of concentration grows larger, the proper rainfall intensity to be used in the design grows smaller.

For ordinary conditions, storm drains should be sized on the assumption that they will flow full or practically full under the design discharge but will not be placed under pressure head. The Manning formula is recommended for capacity calculations.

Storm Drainage Systems

The standard recommended maximum and minimum slopes for storm drains should conform to the following criteria:

1. The maximum hydraulic gradient should not produce a velocity that exceeds 20 feet per second.
2. The minimum desirable physical slope should be 0.5% or the slope, which will produce a velocity of 2.5 ft per second when the storm drain is flowing full, whichever is greater.

Systems should generally be designed for nonpressure conditions. When hydraulic calculations do not consider minor energy losses such as expansion, contraction, bend, junction, and manhole losses, the elevation of the hydraulic gradient for design flood conditions should be at least 1.0 foot below the ground elevation. As a general rule, minor losses should be considered when the velocity exceeds 6 feet per second (lower if flooding could cause critical problems). If all minor energy losses are accounted for, it is usually acceptable for the hydraulic gradient to reach the gutter elevation or the bottom of the casting (Nashville, 1988).

Formulas for Gravity and Pressure Flow

The most widely used formula for determining the hydraulic capacity of storm drain pipes for gravity and pressure flows is the Manning formula, and it is expressed by the following equation:

$$V = \frac{[1.486 R^{2/3} S^{1/2}]}{n} \tag{8.21}$$

where V = mean velocity of flow (ft/s); R = the hydraulic radius (ft) — defined as the area of flow divided by the wetted flow surface or wetted perimeter (A/WP); S = the slope of hydraulic grade line (ft/ft); and n = Manning's roughness coefficient.

In terms of discharge, the above formula becomes:

$$Q = [1.486\, A R^{2/3} S^{1/2}]/n \tag{8.22}$$

where Q = rate of flow (cfs); A = cross-sectional area of flow (ft²).

For pipes flowing full, the above equations become:

$$V = [0.590\, D^{2/3} S^{1/2}]/n \tag{8.23}$$

$$Q = [0.463\, D^{8/3} S^{1/2}]/n \tag{8.24}$$

where D = diameter of pipe (ft).

The Manning's equation can be written to determine friction losses for storm drainpipes as:

$$H_f = [2.87\, n^2 V^2 L]/[S^{4/3}] \tag{8.25}$$

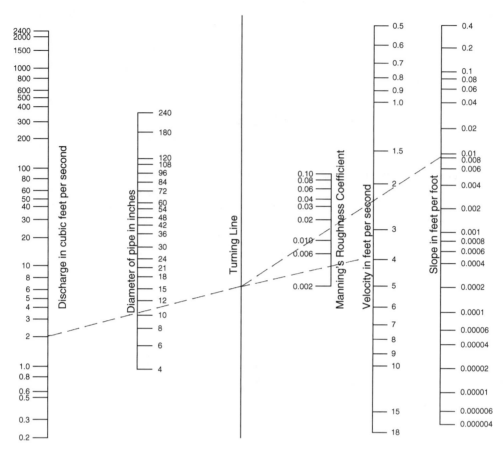

FIGURE 8-14
Nomograph for solution of Manning's formula for flow in storm drains. (From American Association of State Highway and Transportation Officials, Model Drainage Manual, Washington, DC, 1991.)

$$H_f = [29\ n^2LV^2]/[(R^{4/3})(2g)] \tag{8.26}$$

where H_f = total headloss due to friction (ft); n = Manning's roughness coefficient; D = diameter of pipe (ft); L = length of pipe, ft; V = mean velocity (ft/s); R = hydraulic radius (ft); g = acceleration of gravity, 32.2 (ft/s). The nomograph solution of Manning's formula for full flow in circular storm drain pipes, which is shown on Figures 8-14 through 8-16 and Figure 8-17, has been provided to solve the Manning's equation for part-full flow in storm drains.

Saatci (1990) provides a direct solution for depth and velocity for the partial flow situation if discharge (Q), slope (S), and Manning's n are known. The procedure, originally derived for metric units but converted for application here, lends itself to spreadsheet calculations for storm sewer design. The following equations, derived from Manning equations and geometric considerations, are used.

$$K = 0.673\ Qn\ D^{-8/3}\ S^{-1/2} \tag{8.27}$$

$$\theta = 3\pi/2\ \{1- [1- (\pi K)^{1/2}]^{1/2}\}^{1/2} \tag{8.28}$$

$$h/D = 1/2\ [1 - \cos(\theta/2)] \tag{8.29}$$

Storm Drainage Systems

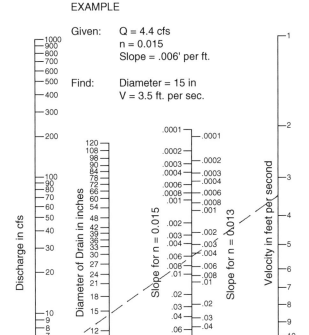

FIGURE 8-15
Nomograph for computing required size of circular drain, flowing full, n = 0.013 or 0.015. (From American Association of State Highway and Transportation Officials, Model Drainage Manual, Washington, DC, 1991.)

$$A = D^2 [(\theta - \sin \theta)/8] \tag{8.30}$$

where Q = the discharge (cfs); K = a constant; h = the depth (ft); D = the pipe diameter (ft); n = Manning's roughness; S = the pipe slope (ft/ft); and θ = the central angle.

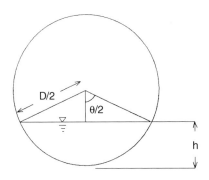

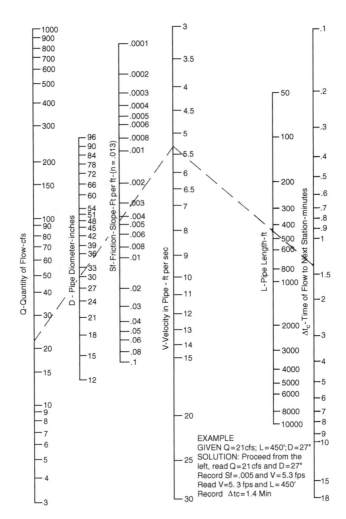

FIGURE 8-16
Concrete pipe flow nomograph. (From American Association of State Highway and Transportation Officials, Model Drainage Manual, Washington, DC, 1991.)

Equation (8.28) is fit to the curves of K versus 2. This method is valid for values of 2 from 0 to 265 degrees. The procedure is illustrated by an example. A 2.5-ft diameter pipe has a flow of 2 cfs, an *n* value of 0.013 and a slope of 0.0022. Find the depth and velocity of flow.

From Equation (8.27):

$$K = 0.673 Q n D^{-8/3} S^{-1/2} = (0.673)(2)(0.013)(2.5)^{-8/3}(0.0022)^{-1/2} = 0.0324$$

Substituting *K* into Equation (8.28) gives:

$$2 = 3\pi/2 \{1- [1-(\pi K)^{1/2}]^{1/2}\}^{1/2} = 3\pi/2 \{1-[1-(\pi\, 0.0324)^{1/2}]^{1/2}\}^{1/2} =$$

$$1.9702 \text{ rad} = 112.9 \text{ deg which is less than 265 deg (OK).}$$

Substituting into Equation (8.29) gives:

Storm Drainage Systems

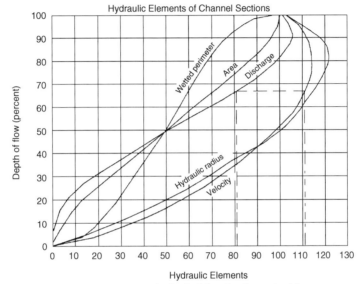

FIGURE 8-17
Values of various elements of circular section for various depths of flow. (From American Association of State Highway and Transportation Officials, Model Drainage Manual, Washington, DC, 1991.)

$$h/D = 1/2 \,[1 - \cos(\theta/2)]$$

$$h/D = 1/2 \,[1 - \cos(112.9/2)] = 0.224$$

Then, $h = (0.224)(2.5) = 0.560$ ft. The flow area is calculated from Equation (8.30) as:

$$A = D^2\,[(\theta - \sin\theta)/8] = (2.5)^2\,[(1.9702 - \sin 112.9)/8] = 0.875 \text{ ft}^2$$

And, the velocity is

$$Q/A = 2/0.875 = 2.286 \text{ ft/s}.$$

Hydraulic Grade Line

The total energy in a pipe is equal to:

$$V^2/2g + P/\gamma + Z \tag{8.31}$$

where P = Pressure head (lb/ft²); γ = specific weight of water (62.4 lb/ft³); V = average velocity of flow (ft/s); g = acceleration of gravity (32.2 ft/s²); and Z = elevation relative to some datum (ft).

Each term in this equation has the dimension of length. The quantity $P/\gamma + Z$ establishes the elevation of the hydraulic grade line. For nonpressure flow, the term P/γ is just the flow depth. The hydraulic grade line is the elevation to which water would rise if not for the fact that it is contained in a pipe under pressure.

The general design procedure is to establish a downstream control elevation. From that elevation, calculations proceed upstream from junction to junction or manhole to manhole. At the lower end of each junction, the pipe friction losses from the downstream section, expressed in terms of feet of loss, are added to the downstream hydraulic gradeline elevation. At the upstream end of each junction, the minor losses through the junction are added. If the transition is to a ditch section, the losses added are the entrance losses to the pipe system. This then establishes the starting elevation of the hydraulic gradeline for the next upstream leg of the calculations.

Hydraulic Grade Line Design Procedure

The headlosses are calculated beginning from the downstream control point to the first junction, and the procedure is repeated for the next junction. The computation for an outlet control may be tabulated on Figure 8-18 using the following procedure:

1. Enter in Column 1 the station for the junction immediately upstream of the outflow pipe. Hydraulic grade line computations begin at the outfall and are worked upstream taking each junction into consideration.
2. Enter in Column 2 the outlet water surface elevation if the outlet will be submerged during the design storm, or 0.8 diameter plus invert out elevation of the outflow pipe, whichever is greater.
3. Enter in Column 3 the diameter (D_o) of the outflow pipe.
4. Enter in Column 4 the design discharge (Q_o) for the outflow pipe.
5. Enter in Column 5 the length (L_o) of the outflow pipe.
6. Enter in Column 6 the friction slope (S_f) in ft/ft of the outflow pipe. This can be determined by using the following formula:

$$S_f = Q^2/K^2 \tag{8.32}$$

where S_f = friction slope (ft/ft).

$$K = [1.486 \, AR^{2/3}]/n, \text{ ft}^3$$

7. Multiply the friction slope (S_f) in Column 6 by the length (L_o) in Column 5 and enter the friction loss (H_f) in Column 7. On curved alignments, calculate curve losses by using the formula $H_c = 0.002 \, (\Delta)(V_o^2/2g)$, where Δ = angle of curvature in degrees and add to the friction loss.
8. Enter in Column 8 the velocity of the flow (V_o) of the outflow pipe.
9. Enter in Column 9 the contraction loss (H_o) by using the formula $H_o = [0.25 \, (V_o^2)]/2g$, where g = 32.2 ft/s².

Storm Drainage Systems

STATION	HYDRAULIC GRADE LINE						JUNCTION LOSS												Inlet	Rim. Elev.
	Outlet Water Surf. Elev.	D_o	Q_o	L_o	S_{f_o}	H_f	V_o	H_o	Q_1	V_1	Q_1V_1	$\dfrac{V_1^2}{2g}$	H_1	ANGLE	H_Δ	H_t	1.3 H_t	0.5 H_t	FINAL H	Water Surf. Elev.	
(1)	(2)	(3)	(4)	(5)	(6)	(7)	(8)	(9)	(10)	(11)	(12)		(13)	(14)	(15)	(16)	(17)	(18)	(19)	(20)	(21)

$H_1 = 0.35 \dfrac{V_1^2}{2g}$ $H_o = 0.25 \dfrac{V_o^2}{2g}$ $H_\Delta = K \dfrac{V_1^2}{2g}$

90°K = 0.70	50°K = 0.47	20°K = 0.16
80°K = 0.66	40°K = 0.38	15°K = 0.10
70°K = 0.61	30°K = 0.28	
60°K = 0.55	25°K = 0.22	

FINAL $H = H_f + H_t$

$H_t = H_o + H_i + H_\Delta$

FIGURE 8-18
Hydraulic grade line computation form. (From American Association of State Highway and Transportation Officials, Model Drainage Manual, Washington, DC, 1991.)

10. Enter in Column 10 the design discharge (Q_i) for each pipe flowing into the junction. Neglect lateral pipes with inflows of less than 10% of the mainline outflow. Inflow must be adjusted to the mainline outflow duration time before a comparison is made.

11. Enter in Column 11 the velocity of flow (V_i) for each pipe flowing into the junction (for exception, see Step 10).

12. Enter in Column 12 the product of Q_i times V_i for each inflowing pipe. When several pipes inflow into a junction, the line producing the greatest Q_i times V_i product is the line that will produce the greatest expansion loss (H_i). (For exception, see Step 10).

13. Enter in Column 13 the controlling expansion loss (H_i) using the formula $H_i = [0.35\,(V_i^2)]/2g$.

14. Enter in Column 14 the angle of skew of each inflowing pipe to the outflow pipe (for exception, see Step 10).

15. Enter in Column 15 the greatest bend loss (H_Δ) calculated by using the formula $H_\Delta = [KV_i^2]/2g$, where K = the bend loss coefficient corresponding to the various angles of skew of the inflowing pipes.

16. Enter in Column 16 the total headloss (H_t) by summing the values in Column 9 (H_o), Column 13 (H_i), and Column 15 (H_Δ).

17. If the junction incorporates adjusted surface inflow of 10% or more of the mainline outflow, i.e., drop inlet, increase H_t by 30% and enter the adjusted H_t in Column 17.
18. If the junction incorporates full diameter inlet shaping, such as standard manholes, reduce the value of H_t by 50% percent and enter the adjusted value in Column 18.
19. Enter in Column 19 the final H, the sum of H_f and H_t, where H_t is the final adjusted value of the H_t.
20. Enter in Column 20 the sum of the elevation in Column 2 and the final H in Column 19. This elevation is the potential water surface elevation for the junction under design conditions.
21. Enter in Column 21 the rim elevation or the gutter flow line, whichever is lowest, of the junction under consideration in Column 20. If the potential water surface elevation exceeds the rim elevation or the gutter flow line, whichever is lowest, adjustments are needed in the system to reduce the elevation of the hydraulic grade line (H.G.L.).
22. Repeat the procedure starting with Step 1 for the next junction upstream.

Figure 8-18 can be used to summarize the hydraulic grade line computations.

All storm drains should be designed such that velocities of flow will not be less than 2.5 feet per second at design flow or lower, with a minimum slope of 0.5%. For flat flow lines, the general practice is to design components so that flow velocities will increase progressively throughout the length of the pipe system. Upper reaches of a storm drain system should have flatter slopes than the slopes of lower reaches. Progressively increasing slopes keep solids moving toward the outlet and deter settling of particles due to steadily increasing flow streams.

The slopes are calculated by the modified Manning formula (term previously defined):

$$S = [(nV)^2]/[2.208 \, R^{4/3}] \qquad (8.33)$$

If downstream drainage facilities are undersized for the design flow, an above- or belowground detention structure may be needed to reduce the possibility of flooding. Using larger-than-needed storm drainpipe sizes and restrictors to control the release rates at manholes and junction boxes in the storm drain system can provide the required storage volume. The same design criteria for sizing the detention basin is used to determine the storage volume required in the system. Figure 8-19 can be used to summarize the storm drain design computations.

Example Problem

The following example will illustrate the hydrologic calculations needed for storm drain design using the rational formula (see Chapter 7 for rational method description and procedures). Figure 8-20 shows a hypothetical storm drain system that will be used in this example. Table 8-5 is a tabulation of the data needed to use the rational equation to calculate inlet flow rate for the seven inlets shown in the system layout.

Table 8-6 shows the data and results of the calculations needed to determine the design flow rate in each segment of the hypothetical storm drain system.

COMPUTED ________ DATE ________

CHECKED ________ DATE ________

ROUTE ________
SECTION ________
COUNTY ________

STATION			DRAINAGE AREA "A" (ACRES)			"A" "C"		FLOW TIME MIN.						VELOCITY (f.p.s.)		INVERT ELEV.			
FROM	TO	LENGTH FEET	INCREMENT	TOTAL	RUNOFF COEFFICIENT "C"	INCREMENT	TOTAL	TO UPPER END	IN SECTION	RAINFALL INTENSITY "I" In/hour	TOTAL RUNOFF C.I.A. = Q (c.f.s.)	DIAMETER PIPE inches	CAPACITY FULL c.f.s.	FLOWING FULL	DESIGN FULL	UPPER END	LOWER END	MANHOLE INVERT DROP	SLOPE OF DRAIN ft/ft

FIGURE 8-19
Storm drain computation form. (From American Association of State Highway and Transportation Officials, Model Drainage Manual, Washington, DC, 1991.)

8.12 Environmental Design Considerations

Roadway Pollution

In 1972, Sartor and Boyd developed data on the accumulation of pollutants in streets and roads. Well over 75% of all pollutants accumulate within three to five feet of the gutter. The types of pollutants coming off roads include solids, metals, nutrients, oil and grease, bacteria, and other pollutants. Specific sources include vehicle-related pollutants and myriad other contributors. Some of the more common are as follows (Kobinger, 1982):

- Exhaust emissions containing nitrogen oxides, hydrocarbons, carbon monoxide, metals, and salts that can settle over the municipality
- Particulants from pavement wear and windblown deposits
- Nutrients and pesticides from roadside maintenance programs
- Sediment from bank and shoulder erosion
- Hydrocarbons from leaks and spills
- Bacteria from fecal deposition and animal carcasses
- Thermal impacts from solar heating of pavement
- A number of trace metals from rust, tire wear, brake lining wear, engine wear, plating deterioration, etc.
- Salts from deicing activities.

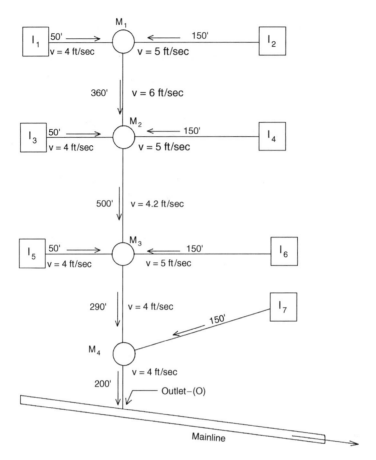

FIGURE 8-20
Hypothetical storm drain system layout. (From American Association of State Highway and Transportation Officials, Model Drainage Manual, Washington, DC, 1991.)

TABLE 8-5

Hydrologic Data

Inlet[a]	Drainage Area	Time of Concentration	Rainfall Intensity	Runoff Coefficient	Inlet Flow[b] (cfs)
1	2.0	8	6.3	0.9	11.3
2	3.0	10	5.9	0.9	15.8
3	2.5	9	6.1	0.9	13.6
4	2.5	9	6.1	0.9	13.6
5	2.0	8	6.3	0.9	11.3
6	2.5	9	6.1	0.9	13.6
7	2.0	8	6.3	0.9	11.3

[a] Inlet and storm drain system configuration are shown in Figure 8-20.
[b] Calculated using the rational equation (see Chapter 7).

Studies have also shown that actual physical impacts of these pollutants are site specific and hard to quantify. Acute (first flush) and chronic impacts have been reviewed in the literature. Sediment sampling from commercial and residential areas indicated the deposition of metals and PCBs downstream from outfall points (Shaheen, 1975). Many of the metals go through several changes in species, making them less available for later uptake. Average density of traffic (ADT) is the greatest predictor of pollution, with a value of

TABLE 8-6

Storm Drain System Calculations

Storm Drain Segment	Drainage Area (acres)	Time of Concentration (min)	Rainfall Intensity (in/h)	Runoff Coefficient	Design Flow Rate (cfs)
I_1-M_1	2.0	8.0	6.3	0.9	11.3
I_2-M_1	3.0	10.0	5.9	0.9	15.8
M_1-M_2	5.0	10.5	5.8	0.9	25.9
I_3-M_2	2.5	9.0	6.1	0.9	13.6
I_4-M_2	2.5	9.0	6.1	0.9	13.6
M_2-M_3	10.0	11.5	5.6	0.9	50.2
I_5-M_3	2.0	8.0	6.3	0.9	11.3
I_6-M_3	2.5	9.0	6.1	0.9	13.6
M_3-M_4	14.5	13.5	5.3	0.9	68.6
I_7-M_4	2.0	8.0	6.3	0.9	11.3
M_4-O	16.5	14.7	5.1	0.9	75.4

30,000 being the normal dividing point between a rural and urban roadway. Specific land use and traffic type also heavily influences the pollutant type and intensity.

A number of state DOTs have instituted programs to investigate stormwater pollution from their highways, most notably California, Florida, Washington, and Virginia. The Federal Highway Administration has an ongoing program of study and research in this area and has published a number of reports on the subject, most notably three separate studies: (1) characterizing the pollution sources and types, (2) developing possible best management practices (BMPs) and (3) developing a design and analysis methodology. Attempts have been made to predict pollutant discharge using physically based modeling and regression equations. In-stream aquatic sampling has been used as a screening technique to prioritize management programs with some success in Washington State (Horner and Mar, 1985).

BMP measures to treat highway runoff fall into several categories, including source control, postdeposition treatment, and postrunoff treatment. Source controls include traffic regulation, litter control, and chemical use controls. Postdeposition treatment includes street cleaning, debris removal, spill cleanup, and curb or barrier elimination. Postrunoff treatment includes vegetative controls, wet detention, infiltration, and wetlands.

Design of streets to take into account sensitive wetland or other water areas can have a great long-term effect in reduction of pollutants. In many situations, there is ample room along road shoulders or within interchange areas to develop highway controls. At the least, vegetative controls and erosion protection can be built into many designs. Some specific structural BMPs include:

- Velocity controls
- Direct discharge reduction
- Retention and infiltration ponds
- Infiltration/exfiltration trenches
- Streets without curbs
- Grassed swales
- Vegetated filter strips and overland flow
- Screened catch basins
- Street designs for sumped catch basins
- Exfiltrating curb and gutter

- Coalescing plate oil/water separator
- Oil/grit separators
- First flush devices
- Porous pavement (parking)
- Vegetated pavement blocks (parking)

There are also some low-cost nonstructural operations, maintenance, and management measures that can be effective and can be incorporated into ongoing programs throughout a municipality. EPA has recommended that municipalities look at snow maintenance, vegetation management, road repair, transportation planning, structural retrofitting, catchbasin cleaning, litter control, and targeted street sweeping for general pollution reduction.

Retrofitting streets during roadway rehabilitation is also an option in design. It is recommended that specific screening and studies be undertaken prior to investing in more costly retrofitting of structural measures. Certain rules-of-thumb have been developed for such screening, and sampling techniques have been researched and demonstrated (Versar, 1985). Detailed design guidelines are available (Woodward-Clyde, 1989).

Structural Roadway BMP Design Overview

Design of highway and roadway BMPs is similar to general BMP design in that pollutant loads, runoff volumes, and impacts to receiving water body must be estimated; a BMP must be sized to achieve a particular goal or objective; and the details of maintenance, placement, materials, etc., are worked out. Highway impacts on water bodies immediately downstream are normally short-term and acute in nature. Therefore, exceedance frequency type design is normally appropriate. Woodward-Clyde (1989) provides a detailed design procedure, while Versar (1989) provides specific BMP design guidelines for vegetative control, wet ponds, infiltration systems, and wetlands. Relatively ineffective structural BMPs include sumped catch basins and detention ponds. Performance of these can be improved somewhat with proper design and a strong maintenance program. See the Structural Best Management Practices, Chapter 13, for more details on BMP design.

Vegetative measures — Include grassed channels and overland flow buffer strips. The effectiveness is dependent on the density of the stand and the amount of flow short-circuiting the strip or flowing with depth in the channel. Vegetative BMPs reduce nutrients through soil and vegetative processes; breakdown hydrocarbons by bacterial degradation; filter suspended solids; cause settling of suspended solids through velocity reduction; and complex metals through soil adsorption and biological assimilation.

Retention basins — Performance of wet ponds ranges from poor to excellent depending on the size of the pond relative to the drainage area. The primary removal mechanism is sedimentation, though soluble nutrients reduction often takes place through biological uptake. This can be enhanced with the construction of a bench along the shore for emergent vegetation growth. Often, a partnership can be worked out wherein the roadway and other development share the pond for pollution treatment.

Infiltration systems — Infiltration systems temporarily store roadway runoff, allowing it to infiltrate into the ground. This type of system can be effective in the removal of pollutants, though construction quality control and maintenance practices are essential for its overall effectiveness. Pretreatment of runoff for removal of oil and sediment greatly lengthens the life of an infiltration system (Galli, 1992a, 1992b). Low water tables (3 ft from trench or pond bottom) and permeable soil (more than 1/2" per hour) are requirements for the use of such a system.

Wetlands — Wetlands are really a combination of the above methods with a thriving and complex ecosystem, a wider range of chemical reactions, and dense growth of marsh plants. Removal capabilities are high for many types of pollutants. Constructed wetlands for the purpose of stormwater treatment may not have the complex ecological system of a natural or enhanced wetland (MWCOG, 1992). The three important components in wetland creation are water, soil, and vegetation (Hammer, 1992). Sufficient baseflow must be available to maintain water levels that may limit urban applications to locations where joint use can be made of the location. The soils in natural wetlands are hydric in nature and anaerobic, which leads to a wide potential for chemical reaction within the soil. Wetland vegetation is capable of surviving up to 5 days inundated. In areas where cold weather shortens the growing season, pollution removal may be drastically reduced or even negative. A wide variety of plants are recommended to provide a balance in chemical and biological reaction. Maintenance costs in the first few years are relatively high until the marsh vegetation is established.

References

American Association of State Highway and Transportation Officials, Model Drainage Manual, Washington, DC, 1991.
American Society of Civil Engineers, Manual of Practice No. 37, Design and Construction of Sanitary and Storm Sewers, ASCE, 1960.
American Society of Civil Engineers, Manual of Practice No. 60, Gravity Sanitary Sewer Design and Construction, ASCE, 1982.
American Society of Civil Engineers, Manual of Practice No. 62, Existing Sewer Evaluation and Rehabilitation, ASCE, 1983.
Bellevue, Wasahington, Standards and Specifications, chap. 4, Storm Drainage and Streams, Drawing No. 34, 1988.
Environmental Protection Agency, Guidance Specifying Management Measures for Sources of Nonpoint Pollution in Coastal Waters, EPA Office of Water, 840-B-92-002, January 1993.
Environmental Protection Agency, Detailed Guidance for Part 2 Municipal NPDES Permit Application., 1993.
Federal Aviation Administration, Manual on Airport Drainage, Advisory Circular 150-5320-5B, Washington, DC, GPO, 1984.
Federal Highway Administration, HYDRAIN Documentation, 1991.
Galli, F.J., Preliminary Analysis of the Performance and Longevity of Urban BMPs installed in Prince George's County, Maryland, Department of Env. Resources, Prince George's County, 1992a.
Galli, F.J., Analysis of Urban BMP Performance and Longevity in Prince George County, Maryland,, Metor. Wash. COG, August 1992b.
GKY & Assoc., HYDRAIN: Integrated Drainage Design Computer System, Springfield, VA, 1990.
Horner, R.R. and Mar, B.W., Assessing the Impacts of Operating Highways on Aquatic Ecosystems, Trans. Res. Bd. Rpt. 1017, 1985.
Kobinger, N.P. Sources and Migration of Highway Runoff Pollutants (Four Volumes), Rpt. No. FHWA/RD-84/057-060, Fed. Hwy. Admin., Washington, DC, 1982.
Metropolitan Washington Council of Governments, A Current Assessment of Urban Best Management Practices, March 1992.

Saatci, A., Velocity and depth of flow calculations in partially filled pipes, *ASCE Journal of Environmental Engineering*, 116, No. 6, November/December 1990.

Sartor, J.D. and Boyd, G.B., Water Pollution Aspects of Street Surface Contaminants, EPA-R2-72-081, November 1972.

Shaheen, D.G., Contributions of Urban Roadway Usage to Water Pollution, Report EPA-600/2-75-004, April 1975.

U.S. Department of Transportation, Federal Highway Administration, FHWA-RD-79-30, 31, 1979.

U.S. Department of Transportation, Federal Highway Administration, Drainage of Highway Pavements, Hydraulic Engineering Circular No. 12, 1984.

U.S. Department of Transportation, Federal Highway Administration, Urban Drainage Design Manual, Hydraulic Engineering Circular No. 22, August 2001.

Versar, Inc., Springfield, Virginia, Management Practices for Mitigation of Highway Stormwater Runoff Pollution Volumes 1–4, 1985.

Woodward-Clyde Consultants, Pollutant Loadings and Impacts from Highway Stormwater Runoff, Volumes 1–4, Design Procedure, 1989.

Appendix A — Metric Design Figures

Following are the metric figures for Chapter 8 related to storm drainage systems. Note that there are no metric figures for Figures 8-12, 8-18, and 8-20, because metric conversion would not change these figures.

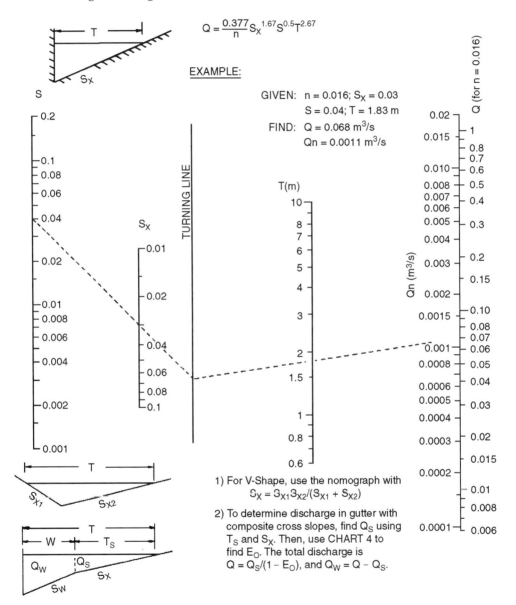

FIGURE 8A-1
Flow in triangular gutter sections.

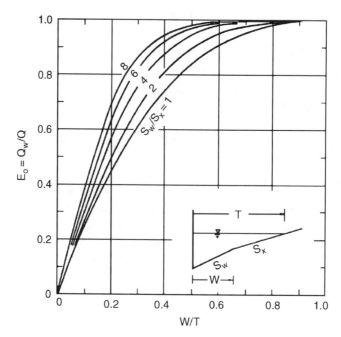

FIGURE 8A-2
Ratio of frontal flow to total gutter flow.

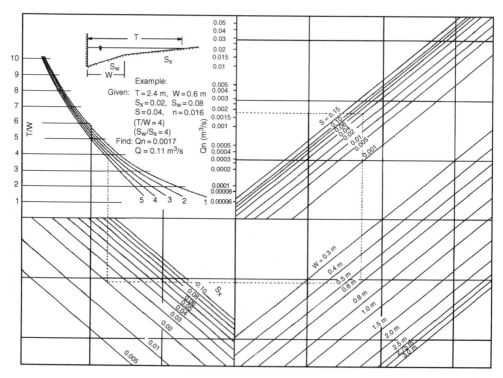

FIGURE 8A-3
Flow in composite gutter sections.

Storm Drainage Systems

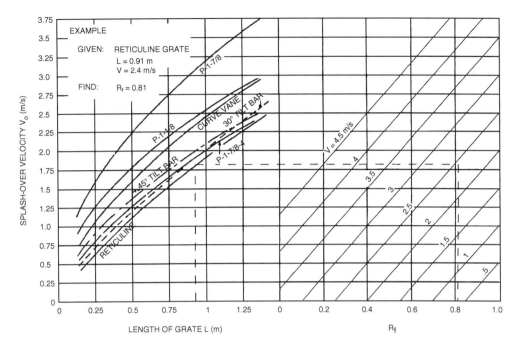

FIGURE 8A-4
Grate inlet frontal flow interception efficiency.

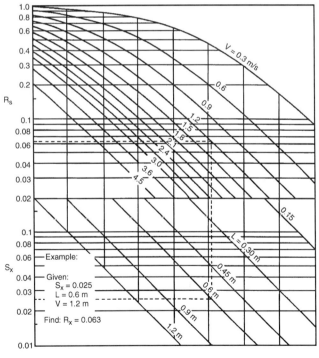

FIGURE 8A-5
Grate inlet side flow interception efficiency.

350 Municipal Stormwater Management, Second Edition

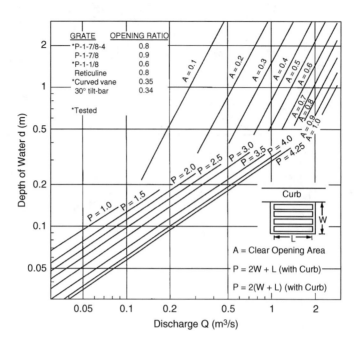

FIGURE 8A-6
Grate inlet capacity in sump conditions.

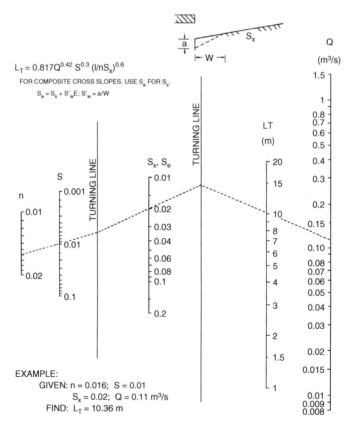

FIGURE 8A-7
Curb-opening and slotted drain inlet length for total interception.

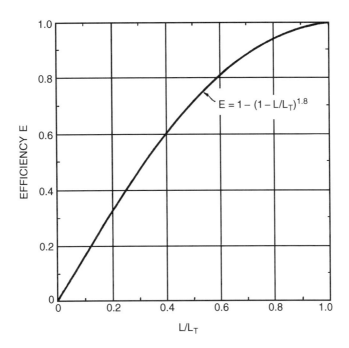

FIGURE 8A-8
Curb-opening and slotted drain inlet interception efficiency.

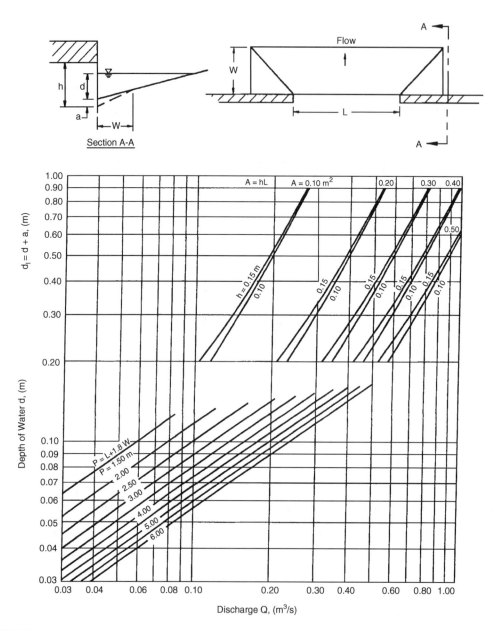

FIGURE 8A-9
Depressed curb-opening inlet capacity in sump locations.

Storm Drainage Systems

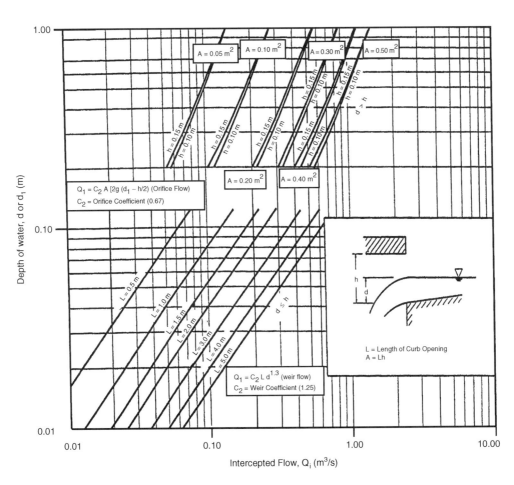

FIGURE 8A-10
Curb-opening inlet capacity in sump locations.

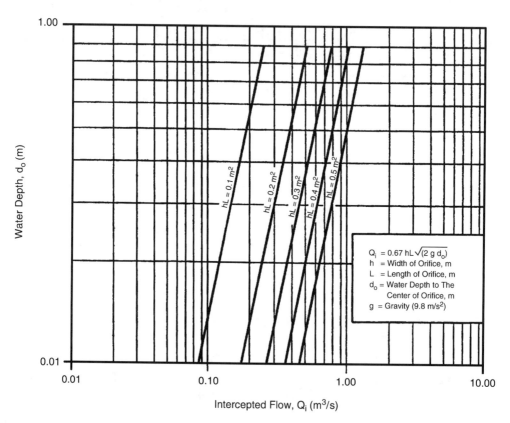

FIGURE 8A-11
Curb-opening inlet orifice capacity — inclined and vertical orifice throats.

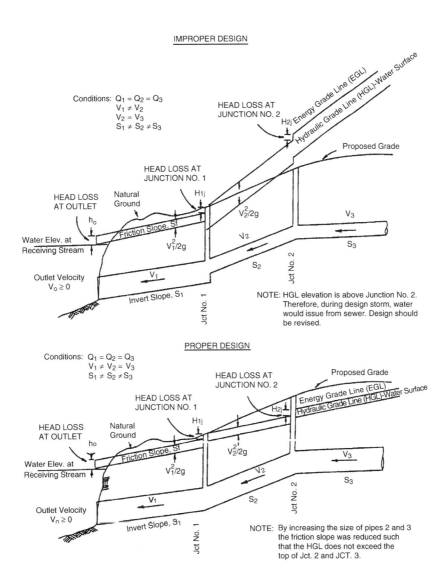

FIGURE 8A-13
Energy and hydraulic grade lines for storm drains under constant discharge.

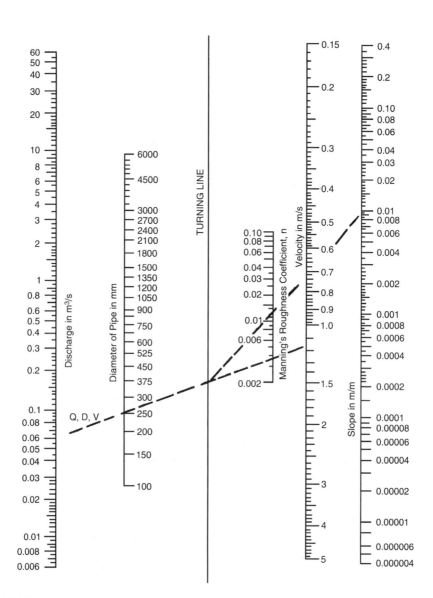

FIGURE 8A-14
Nomograph for solution of Manning's formula for flow in storm drains.

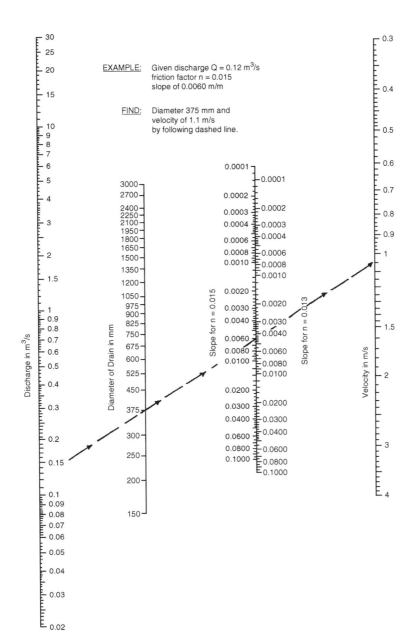

FIGURE 8A-15
Nomograph for computing required size of circular drain, flowing full $n = 0.013$ or 0.015.

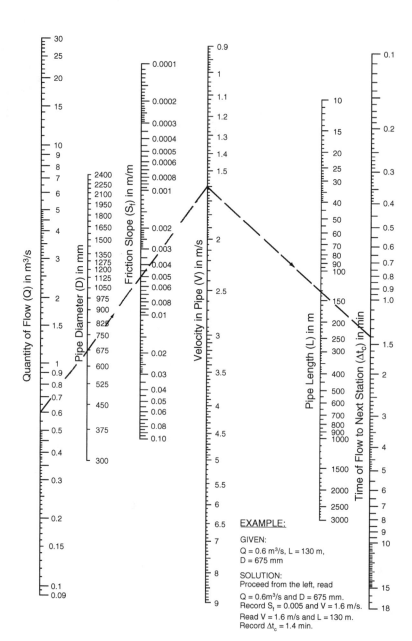

FIGURE 8A-16
Concrete pipe flow nomograph.

Storm Drainage Systems

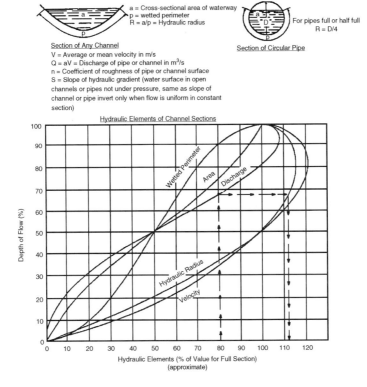

FIGURE 8A-17
Values of various elements of circular section for various depths of flow.

Storm Drain Computation Sheet

Station		Length (m)	Drainage Area A (ha)		Runoff Coefficient C	A × C		Flow Time (min)		Rainfall Intensity I (mm/h)	Total Runoff 0.0278 CIA = Q (m³/s)	Diameter Pipe (mm)	Capacity Full (m³/s)	Velocity (m/s)		Invert Elev.		Manhole Invert Drop	Slope of Drain (m/m)
From	To		Increment	Total		Increment	Total	To Upper End	In Section					Flowing Full	Design Flow	Upper End	Lower End		

FIGURE 8A-19
Storm drain computation form.

9
Design of Culverts

9.1 Introduction

A culvert is defined as the following:

- A structure used to convey surface runoff through embankments.
- A structure that is usually designed hydraulically to take advantage of submergence to increase hydraulic capacity.
- A structure, as distinguished from bridges, that is usually covered with embankment and is composed of structural material around the entire perimeter, although some are supported on spread footings with the streambed serving as the bottom of the culvert.
- A structure that is 20 ft or less in centerline length between extreme ends of openings for multiple boxes. However, a structure designed hydraulically as a culvert is treated in this chapter, regardless of its total length.
- For economy and hydraulic efficiency, culverts should be designed to operate with the inlet submerged during flood flows, if conditions permit.
- Cross-drains are those culverts and pipes that are used to convey runoff from one side of a highway to another.

Culvert flow is one of the most complex forms of hydraulics an engineer will encounter. Easy-to-understand nomographs and "cookbook" procedures can belie the reality that control of flow in a culvert can shift dramatically and unpredictably between inlet control (weir or orifice), barrel control, and outlet control, causing relatively sudden rises in headwater and, in the worst case in larger culverts, structural damage or failure due to sudden pressure surges. Certain types of culverts on certain slopes can exhibit a looped rating curve depending on whether the flow is rising or falling. Therefore, the most critical aspect of culvert design is to determine stable and predictable performance for all expected flow levels. When the type of flow is known, the well-known equations for orifice, weir, or pipe flow and backwater profiles can be applied to determine the relationships between head and discharge (Blaisdell, 1966). The good news is that modern culvert design nomographs, computer programs, and instructions are based on sound theory and extensive laboratory and field studies, and can be relied on in all but the most uncommon situations.

The design of a culvert is influenced by cost, hydraulic efficiency, purpose, and the topography at the proposed culvert site. Thus, physical data must be integrated with engineering and economic considerations. The information contained in this chapter should give the designer the ability to design culverts taking into account the factors that influence their design and selection.

The primary purpose of a culvert is to convey surface water, but, properly designed, it may also be used to restrict flow and reduce downstream storm runoff peaks. In addition to the hydraulic function, a culvert must also support the embankment and roadway for traffic conveyance and protect the traveling public and adjacent property owners from flood hazards to the extent practicable and in a reasonable and prudent manner.

Primary considerations for the final selection of any drainage structure are that its design be based upon appropriate hydraulic principles, economy, and minimized effects on adjacent property by the resultant headwater depth and outlet velocity. The allowable headwater elevation is that elevation above which damage may be caused to adjacent property and a highway. It is this allowable headwater depth that is the primary basis for sizing a culvert.

To ensure safety during major flood events, access and egress routes to developed areas should be checked for a more infrequent flood than the design flood. Often, the 100-year flood is checked to determine if these streets will provide safe access for emergency vehicles and local residents. Thus, an analysis of how the culvert will function during a less-frequent flood is an important part of culvert design. Some municipalities have different culvert design standards for streets of different classifications.

At many sites, a bridge or a culvert will fulfill the structural and hydraulic requirements. The structural choice should be based on risk of property damage, construction and maintenance cost, traffic safety, environmental considerations, risk of failure, and aesthetic considerations.

Performance curves should be developed for all culverts for evaluating the hydraulic capacity of a culvert for various headwaters. These will display the consequence of high-flow rates at the site and any possible hazards. Sometimes a small increase in flow rate can affect a culvert design. If only the design peak discharge is used in the design, the designer cannot assess what effects increases in the estimated design discharge will have on the culvert design.

For culverts with significant headwater storage, the site should be treated as a detention design, and flow should be routed using techniques found in Chapter 11.

9.2 Concept Definitions

The following are definitions of concepts that will be important for understanding different culvert design procedures. These concepts are used throughout the remainder of this book in dealing with different aspects of culvert designs.

Critical depth Critical depth can best be illustrated as the depth at which water flows over a weir, this depth being attained automatically, where no other backwater forces are involved. Critical depth is the depth at which the specific energy of a given flow rate is at a minimum. For a given discharge and cross-section geometry, there is only one critical depth. Appendixes A and B at the end of this chapter give a series of critical-depth figures for the different culvert shapes.

Cross-drainage culverts Cross-drainage culverts extend under a roadway and transport runoff from one side of a roadway and discharge the runoff at the other side of the roadway.

Energy grade line The energy grade line represents the total energy at any point along the culvert barrel.

Free outlets Free outlets are outlets with a tailwater equal to or lower than critical depth. For culverts having free outlets, lowering of the tailwater has no effect on the discharge or the backwater profile upstream of the tailwater.

Hydraulic grade line The hydraulic grade line is the depth to which water would rise in vertical tubes connected to the sides of the culvert barrel. In full flow, the energy grade line and the hydraulic grade line are parallel lines separated by the velocity head, except at the inlet and the outlet.

Improved inlets Flared, improved, or tapered inlets indicate a special entrance condition that decreases the amount of energy needed to pass the flow through the inlet and, thus, increases the capacity of culverts at the inlet.

Invert Invert refers to the inside bottom of the culvert.

Normal flow Normal flow occurs in a channel reach when the discharge, velocity, and depth of flow do not change throughout the reach. The water surface profile and channel bottom slope will be parallel. This type of flow will be approximated in a culvert operating on a steep slope, provided the culvert is sufficiently long.

Soffit Soffit refers to the inside top of the culvert. The soffit is also referred to as the crown of the culvert.

Steep and mild slope A steep-slope culvert operation is where the computed critical depth is greater than the computed uniform depth. A mild-slope culvert operation is where critical depth is less than uniform depth.

Submerged inlets Submerged inlets are those inlets having a headwater greater than about 1.2 times the diameter of the culvert or barrel height.

Submerged outlets Partially submerged outlets are outlets with tailwater that is higher than critical depth and lower than the height of the culvert. Submerged outlets are outlets having a tailwater elevation higher than the soffit (crown) of the culvert.

Uniform flow Uniform flow is flow in a prismatic channel of constant cross section having a constant discharge, velocity, and depth of flow throughout the reach. This type of flow will exist in a culvert operating on a steep slope, provided the culvert is sufficiently long.

9.3 Culvert Design Steps and Criteria

Design Steps

Determine and Analyze Site Characteristics

Site characteristics include the generalized shape of the highway embankment, bottom elevations and cross sections along the streambed, the approximate length of the culvert, and the allowable headwater elevation. In determining the allowable headwater elevation, roadway elevations and the elevation of upstream property should be considered. The consequences of exceeding the allowable headwater elevation should be evaluated and kept in mind throughout the design process.

Culvert design is actually a trial-and-error procedure, because the length of the barrel cannot be accurately determined until the size is known, and the size cannot be precisely determined until the length is known. In most cases, however, a reasonable estimate of length will be accurate enough to determine the culvert size.

A field visit is essential to determine any site characteristics that might affect the culvert design and the effect the culvert might have on the natural drainage system.

Perform Hydrologic Analysis

Outline the drainage area above the culvert site. Develop flow estimates for the design frequencies, and determine if sufficient storage will be available at the culvert inlet to justify routing flows through the culvert for design purposes. Design frequencies should include the design flow rate and a larger flood (check flood), which is usually the 100-year flow rate. The probable accuracy of the estimate should be kept in mind as the design proceeds. The accuracy is dependent on the method used to define the flow rate, the available data on which it is based, etc. If routing through the culvert is needed, a hydrograph must be developed.

Perform Outlet Control Calculations and Select Culvert

These calculations are performed before inlet control calculations in order to select the smallest feasible barrel that can be used without the required headwater elevation in outlet control exceeding the allowable headwater elevation. The full-flow outlet control performance curve for a given culvert (size, inlet edge, shape, material) defines its maximum performance. Therefore, the inlet improvements beyond the beveled edge or changes in inlet invert elevation will not reduce the required outlet control headwater elevation. This makes the outlet control performance curve an ideal limit for improved inlet design. The results of these calculations should be the outlet control performance curve. In addition to considering the allowable headwater elevation, the velocity of flow at the exit to the culvert should be checked to determine if downstream erosion problems would be created.

Perform Inlet Control Calculations for Conventional and Beveled-Edge Culvert Inlets

Perform the inlet control calculations to develop the inlet control performance curve to determine if the culvert design selected will be on inlet or outlet control for the design, and check flood frequencies. A fall may be incorporated upstream of the culvert to increase the flow through the culvert.

Perform Throat-Control Calculations for Side- and Slope-Tapered Inlets (Optional)

The same concepts are involved here as with a conventional or beveled-edge culvert design.

Analyze the Effect of Falls on Inlet Control Section Performance (Optional)

The purpose of this step is to determine if having a fall before the inlet of the culvert would increase the capacity of the culvert, and if a fall can be justified from a cost perspective and site characteristics.

Design Side- and Slope-Tapered Inlets (Optional)

Side- and slope-tapered inlets can be used to significantly increase the capacity of many culvert designs. Develop performance curves based on side- and slope-tapered inlets,

Design of Culverts

and determine from a cost perspective and site characteristics if such a design would be justified.

Complete File Documentation

As outlined in Chapter 16, complete the documentation file for the final design selected.

Engineering and Technical Design Criteria

The design of a culvert should take into account many different engineering and technical aspects at the culvert site and adjacent areas. The following design criteria should be considered for all culvert designs as applicable:

- Engineering aspects
  - Flood frequency
  - Velocity limitations
  - Buoyancy protection
- Site criteria
  - Length and slope
  - Debris control
  - Ice buildup
- Design limitations
  - Headwater limitations
  - Tailwater conditions
  - Storage — temporary or permanent
- Design options
  - Culvert inlets
  - Inlets with headwalls
  - Wingwalls and aprons
  - Improved inlets
  - Material selection
  - Culvert skews
  - Culvert sizes and shapes
- Related designs
  - Weep holes
  - Outlet protection
  - Erosion and sediment control
  - Environmental considerations
  - Safety considerations

Some culvert designs are relatively simple, involving a straightforward determination of culvert size and length. Other designs are more complex, where structural, hydraulic, environmental, or other considerations must be evaluated and provided for in the final design. The designer must incorporate personal experience and judgment to determine which criteria must be evaluated and how to design the final culvert installation. Following is a discussion of each of the above criteria as it relates to culvert siting and design.

Flood Frequency

The appropriate flood frequency for determining the flood-carrying capacity of a culvert is dependent on the roadway at the crossing and the level of risk associated with failure of the culvert crossing, as well as on the level of risk associated with increasing the flood hazard to upstream (backwater) or downstream (redirection of floodwaters) property. It is recommended that culverts be designed to accommodate the following minimum flood frequencies:

- All cross-drainage culverts — 25-year frequency
- Other culverts — 10-year frequency

Future development of contributing watersheds and floodplains that have been zoned or delineated should be considered in determining the design flood frequency. In addition, the 100-year frequency storm should be routed through all culverts to be sure structures are not flooded or increased damage does not occur to the roadway or adjacent property for this design event.

An economic or risk analysis (Young, Taylor, and Costello, 1970) may justify a design to pass floods greater than those noted above, where potential damage to adjacent property, to human life, or heavy financial loss due to flooding as determined by a flood assessment or analysis, commensurate with the site. Even designing for an infrequent flood may not protect the designer or owner from potential liability should the culvert backwater onto private property at any flow level (Lewis, 1992). Risk concepts may be used in the design.

Also, in compliance with the National Flood Insurance Program, it is necessary to consider the 100-year frequency flood at locations identified as being special flood-hazard areas. This does not necessitate that the culvert be sized to pass the 100-year flood, provided the capacity of the drainage system including the culvert plus flow bypassing the culvert is sufficient to accommodate the 100-year flood without raising the water surface elevation 1 ft, or if raised greater than 1 ft, mitigation is provided. The designer should review the municipal floodway regulations for more information related to floodplain regulations and designs.

Velocity Limitations

Minimum and maximum velocities should be considered when designing a culvert. The maximum velocity should be consistent with channel stability requirements at the culvert outlet. As outlet velocities increase, the need for channel stabilization at the culvert outlet increases. If velocities exceed permissible velocities for the various types of nonstructural outlet lining material available, the installation of structural energy dissipators is appropriate.

The maximum allowable velocity for corrugated metal pipe is 10 ft/sec. There is no specified maximum allowable velocity for reinforced concrete pipe, but outlet protection should be provided where discharge velocities will cause erosion problems.

A minimum velocity of 2.5 ft/sec when the culvert is flowing partially full is recommended to ensure a self-cleaning condition during partial depth flow. When velocities below this minimum are anticipated to cause unacceptable sedimentation, the installation of a sediment trap upstream of the culvert should be considered. For streams carrying heavy sediment loads, higher velocities and lower headwater depths may be required to ensure sediment passage and avoid accumulation of sediment near the culvert entrance.

Buoyancy Protection

Headwalls, endwalls, slope paving, or other means of anchoring to provide buoyancy protection should be considered for all flexible culverts. Buoyancy is more serious with steepness of the culvert slope, depth of the potential headwater (debris blockage may increase), flatness of the upstream fill slope, height of the fill, large culvert skews, or mitered ends.

Length and Slope

Because the length of the culvert will affect the capacity of culverts on outlet control, the length should be kept to a minimum, and existing facilities should not be extended without determining the decrease in capacity that will occur. In addition, the culvert length and slope should be chosen to approximate existing topography, and, to the degree practicable, the culvert invert should be aligned with the channel bottom and the skew angle of the stream, and the culvert entrance should match the geometry of the roadway embankment.

Debris Control

In designing debris control structures, it is recommended that the publication Hydraulic Engineering Circular No. 9 titled "Debris — Control Structures" (FHA, 1971) be consulted. Debris control should be considered:

- Where experience or physical evidence indicates the watercourse will transport a heavy volume of controllable debris
- For culverts located in mountainous or steep regions
- For culverts that are under high fills
- Where cleanout access is limited. However, access must be available to clean out the debris-control device.

Ice Buildup

Ice buildup should be mitigated as necessary by:

- Increasing the culvert height 1 ft above the total of the maximum observed ice buildup plus any winter flow depth
- Increasing the culvert width to encompass the observed channel's static ice width plus 10%, where appropriate, to prevent property damage

Headwater Limitations

The allowable headwater elevation is determined from an evaluation of land use upstream of the culvert and the proposed or existing roadway elevation. Headwater is the depth of water above the culvert invert at the entrance end of the culvert. In general, the constraint that gives the lowest allowable headwater elevation establishes the criteria for the hydraulic calculations.

The following criteria related to headwater should be considered:

- The allowable headwater for design frequency conditions should allow for or consider the following upstream controls:
  - Reasonable freeboard
  - Upstream property damage
  - Elevations established to delineate floodplain zoning
  - Low point in the road grade that is not at the culvert location
  - Ditch elevation of the terrain that will permit flow to divert around culvert
  - Following are some recommended HW/D design criteria:
    - For drainage facilities with cross-sectional area equal to or less than 30 sq ft - HW/D = to or <1.5
    - For drainage facilities with cross-section area greater than 30 sq ft - HW/D = to or <1.2
- The headwater should be checked for the 100-year flood to ensure compliance with floodplain management criteria, and, for most facilities, the culvert should be sized to maintain flood-free conditions on major thoroughfares for one half-lane of two-lane facilities and one lane of multilane facilities.
- The maximum acceptable outlet velocity should be identified. The headwater should be set to produce acceptable velocities, or stabilization or energy dissipation should be provided where acceptable velocities are exceeded.

After determining the allowable headwater elevation, tailwater elevation, and approximate length, invert elevations must be established. Scour can be minimized if the culvert has the same slope as the channel. In addition, the flow conditions and velocity in the channel upstream from the culvert should be investigated to determine if scour would occur.

If there is insufficient headwater elevation available to convey the required discharge, it will be necessary to use a larger culvert, lower the inlet invert, use an irregular cross section, use an improved inlet if in inlet control, use multiple barrels, or use a combination of these measures. If the inlet invert is lowered, special consideration must be given to scour and sedimentation at the entrance.

Tailwater Conditions

The hydraulic conditions downstream of the culvert site must be evaluated to determine a tailwater depth for a range of discharges. At times, there may be a need for calculating backwater curves to establish the tailwater conditions. If the culvert outlet is operating with a free outfall, the critical depth and equivalent hydraulic grade line should be determined. Tailwater elevations can determine whether a culvert will operate with a free outfall or under submerged conditions. For culverts that discharge to an open channel, the stage–discharge curve for the channel must be determined. See Chapter 10 in this book.

If an upstream culvert outlet is located near a downstream culvert inlet or other control, the headwater elevation of the downstream control may establish the design tailwater depth for the upstream culvert. If the culvert discharges to a lake, pond, or other major water body, the expected high-water elevation of the particular water body may establish the culvert tailwater.

Design of Culverts

Storage — Temporary or Permanent

If storage is being assumed upstream of the culvert, consideration should be given to:

- The total area of flooding
- The average time that bankfull stage is exceeded for the design flood up to 48 h in rural areas or 6 h in urban areas
- Availability of the storage area for the life of the culvert through the purchase of right-of-way or easement

Culvert Inlets

Selection of the type of inlet is an important part of culvert design, particularly with inlet control. Hydraulic efficiency and cost can be significantly affected by inlet conditions. The inlet coefficient K_e is a measure of the hydraulic efficiency of the inlet, with lower values indicating greater efficiency. All methods described in this chapter, directly or indirectly, use inlet coefficients. Recommended inlet coefficients are given in Table 9-1.

Following are some considerations that should be used to determine the type of inlet to be used for a particular installation.

Projecting Inlets or Outlets

- Extend beyond the embankment of the roadway
- Have low construction cost
- Are susceptible to damage during roadway maintenance and automobile accidents
- Have poor hydraulic efficiency for thin materials
- Should be mitered to fill slope
- Are used predominantly with metal pipe
- Include anchoring inlet to concrete slope paving to strengthen the weak leading edge

Headwalls with Bevels

- Increase the efficiency of metal pipe
- Provide embankment stability and erosion protection
- Provide protection from buoyancy
- Shorten the required structure length
- Reduce maintenance damage

Improved Inlets

- Should be considered for culverts that will operate in inlet control.
- Can increase the hydraulic performance of the culvert but may also add to the total culvert cost. Therefore, they should only be used if practicable.

TABLE 9-1

Inlet Coefficients

Type of Structure and Design of Entrance	Coefficient K_e
Pipe, Concrete	
Projecting from fill, socket end (groove-end)	0.2
Projecting from fill, square cut end	0.5
Headwall or headwall and wingwalls:	
Socket end of pipe (groove-end)	0.2
Square-edge	0.5
Rounded [radius = 1/12(D)]	0.2
Mitered to conform to fill slope	0.7
*End-section conforming to fill slope	0.5
Beveled edges, 33.7° or 45° bevels	0.2
Side- or slope-tapered inlet	0.2
Pipe, or Pipe-Arch, Corrugated Metal	
Projecting from fill (no headwall)	0.9
Headwall or headwall and wingwalls square-edge	0.5
Mitered to fill slope, paved or unpaved slope	0.7
End-section[a] conforming to fill slope	0.5
Beveled edges, 33.7° or 45° bevels	0.2
Side- or slope-tapered inlet	0.2
Box, Reinforced Concrete	
Headwall parallel to embankment (no wingwalls):	
Square-edged on three edges	0.5
Rounded on three edges to radius of [1/12(D)] or beveled edges on three sides	0.2
Wingwalls at 30° to 75° to barrel	
Square-edged at crown	0.4
Crown edge rounded to radius of [1/12(D)] or beveled top edge	0.2
Wingwalls at 10° or 25° to barrel	
Square-edged at crown	0.5
Wingwalls parallel (extension of sides)	
Square-edged at crown	0.7
Side- or slope-tapered inlet	0.2

[a] End section conforming to fill slope, made of either metal or concrete, are the sections commonly available from manufacturers. From limited hydraulic tests, they are equivalent in operation to a headwall in inlet and outlet controls. Some end sections, incorporating a closed taper in their design, have superior hydraulic performance.

Source: From Federal Highway Administration, Hydraulic Design of Improved Inlets for Culverts, Hydraulic Engineering Circular No. 13, 1972.

Commercial End Sections

- Are available for corrugated metal and concrete pipe
- Retard embankment erosion and incur less damage from maintenance
- May improve projecting metal pipe entrances by increasing hydraulic efficiency, reducing the accident hazard, and improving appearance
- Are hydraulically equal to a headwall but can be equal to a beveled or side-tapered entrance if a flared, enclosed transition takes place before the barrel

Inlets with Headwalls

Headwalls may be used for a variety of reasons:

- Increasing the efficiency of the inlet
- Providing embankment stability
- Providing embankment protection against erosion
- Providing protection from buoyancy
- Shortening the length of the required structure

The relative efficiency of the inlet depends on the pipe material. Headwalls are usually required for all metal culverts, and where buoyancy protection is necessary. Corrugated metal pipe in a headwall is essentially square-edged with an inlet coefficient of about 0.5. For tongue-and-groove or bell-and-concrete pipe, little increase in hydraulic efficiency is realized by adding a headwall.

Wingwalls and Aprons

Wingwalls are used where the side slopes of the channel adjacent to the entrance are unstable, or where the culvert is skewed to the normal channel flow. Little increase in hydraulic efficiency is realized with the use of normal wingwalls, regardless of the pipe material used and, therefore, the use should be justified for other reasons. Wingwalls can be used to increase hydraulic efficiency if designed as a side-tapered inlet.

If high headwater depths are to be encountered, or the approach velocity in the channel will cause scour, a short channel apron should be provided at the toe of the headwall. This apron should extend at least one pipe diameter upstream from the entrance, and the top of the apron should not protrude above the normal streambed elevation.

Improved Inlets

Where inlet conditions control the amount of flow that can pass through the culvert, improved inlets can greatly increase the hydraulic performance at the culvert. For these designs, refer to Section 9.8, which describes the design of improved inlets.

Material Selection

The material selection should consider replacement cost and difficulty of construction as well as traffic delay. The material selected should be based on a comparison of the total cost of alternate materials over the design life of the structure, which is dependent upon the following:

- Durability (service life)
- Structural strength
- Hydraulic roughness
- Bedding conditions
- Abrasion and corrosion resistance
- Watertightness requirements

There are a number of different types of culvert pipe materials on the market, a number of variations within each basic material type. There is significant competition among pipe manufacturers concerning which material is superior generally and for specific applications. Cost and longevity data from other than unbiased sources is not always trustworthy, as myriad assumptions that go into the estimates can slant the results in any direction.

Generally, material costs are comparable for the smaller diameters among the three major categories: concrete, steel, and plastic, with installation costs being higher (normally in the range of 10 to 30%) for plastic and steel because of the greater bedding care that must be taken due to the pipe's flexibility. As the size increases above 36 in diameter, concrete pipe begins to be more expensive, by as much as 40% for 72-in pipes. This factor is balanced by the fact that concrete pipes have an expected life of 60 to 100 years or more, while steel pipes have an expected life of 25 to 75 years, depending heavily on coating type(s) and thickness, the abrasive nature of the sediments, the acidity of the flow and soil, the use of street salts, groundwater resistivity and chemical content, and the soil conductivity. Plastic pipes have not had a track record to estimate longevity, though it is generally thought to fall between concrete and steel pipes. Fire hazards may be an issue with plastic pipe.

Bituminous coatings and pavings are often used with corrugated metal pipe to prolong life by providing resistance to corrosion and abrasion. When coated, the layer should be a minimum of 0.05-in thick. Culverts are hot- or cold-dipped. Bottom-paved pipe normally has the bottom 1/4 covered with a minimum thickness of 1/8 in. above the corrugations. Pipe arches have the bottom 40% covered. Polymer coatings and fiber bonding are also used and should be applied in accordance with AASHTO and ASTM standards. Corrosion is repaired with aluminum-filled epoxy or aluminum-filled and fibered asphaltic coatings.

On balance, most municipalities require the use of concrete pipe for storm drains and sewers, and for culverts placed in critical areas or within the public right-of-way. Often, other types of pipe are allowed for less-important applications such as driveways or low-traffic parking lot entrances. For larger pipes or pipes on slopes too steep to allow easy concrete pipe installation, steel pipes are sometimes specified. The basic pipe materials and applicable testing standards from various sources are given in Table 9-2. Table 9-3 gives recommended Manning's n values for different materials used for culverts.

Culvert Skews

Culvert skews should not exceed 45 degrees as measured from a line perpendicular to the roadway centerline without approval from the municipal engineering department.

Culvert Sizes and Shapes

The culvert size and shape selected should be based on engineering and economic criteria related to site conditions. The following minimum sizes should be used to avoid maintenance problems and clogging, and provide sufficient capacity:

- 24 in for interstate system
- 18 in for other systems
- 12 in for a side-drain or drive

Land-use requirements can dictate a larger or different barrel geometry than required for hydraulic considerations. It is also recommended to use only arch or oval shapes if required by hydraulic limitations, site characteristics, structural criteria, or environmental criteria.

TABLE 9-2

Culvert Pipe Material Types and Standards

Pipe Material	Standards
Aluminum-alloy structural plate	ASTM B790, B746 — culvert manufacture ASTM B209 — alloy type
Galvanized Structural plate	ASTM A761, A796 — culvert manufacture ASTM A444 — pipe material AASHTO M243 — field coatings of asphaltic mastic or tar base material for shallow buried pipe
Corrugated metal pipe	Joined with rivet lap joint construction (annular corrugations) or continuous lock or welded seam (helical corrugations) and wrapped in nonwoven filter fabric or O-ring gaskets Aluminum alloy corrugated pipe fabrication — ASTM B745, ASTM B209 — alloy material Aluminum-coated steel type-2 fabrication — ASTM A760, with ASTM A819 coupling bands Galvanized steel fabrication — ASTM A760, with ASTM A444 material Installation according to AASHTO Standard Specifications of Highway Bridges Section, 12 and 23, ASTM A796, A798
High-density polyethylene pipe (HDPE)	Male and female ends with gasketed joints, ASTM D3212, or external coupling band joints in accordance with AASHTO standards Precast fittings (wyes, tees, etc.) not normally accepted in lieu of precast storm sewer manholes Corrugated HDPE — manufacture AASHTO M294, cell class minimum 324420C, ASTM D3350 with flexibility factor less than 0.095 Ribbed HDPE — manufacture ASTM F894, cell class minimum 334433C, ASTM D3350 Smoothwall HDPE — manufacture ASTM F714, cell class minimum 35434C, ASTM D3350 Installation according to AASHTO Standard Specifications of Highway Bridges Section 18, ASTM D2321.
Polyvinyl chloride pipe (PVC)	Storm sewer pipe, profile wall should be the bell-and-spigot type with elastomeric joints and smooth inner walls in accordance with AASHTO M304, minimum cell class 12454C or 12364C, ASTM D1784 Smooth wall PVC manufacture ASTM F679 or AASHTO M278 with same cell class minimum Precast fittings (wyes, tees, etc.) not normally accepted in lieu of precast storm sewer manholes Installation according to AASHTO Standard Specifications of Highway Bridges Section 18, ASTM D2321.
Reinforced concrete pipe (RCP)	Bell- or groove-and-spigot or tongue pipe required with rubber gasket, ASTM C443 Mastic type joints are acceptable for noncritical pipes if joints are wrapped with a 1-ft wide nonwoven geotextile fabric Material class III, IV, or V, ASTM C76 Elliptical RCP minimum class HE-II, ASTM C507 Installation according to AASHTO Standard Specifications of Highway Bridges Sections 17 and 28
Reinforced concrete box sections	Material standard, ASTM C789 with compressive strength tested prior to shipping Precast box sections, ASTM C850 Joints should be smooth to allow for a continuous line, sealed with butyl rubber or asphaltic mastic, and wrapped with a 1-ft wide nonwoven geotextile fabric Steel reinforcement minimum outside cover of 1 in extending into the male and female ends; steel wire should be between 1/2 in and 2 in from the ends of the pipe

TABLE 9-3

Manning's n Values

Type of Conduit	Wall and Joint Description	Manning's n
Concrete pipe	Good joints, smooth walls	0.011–0.013
	Good joints, rough walls	0.014–0.016
	Poor joints, rough walls	0.016–0.017
	Badly spalled	0.015–0.020
Concrete box	Good joints, smooth finished walls	0.014–0.018
	Poor joints, rough, unfinished walls	0.014–0.018
Corrugated metal pipes and boxes annular corrugations	2 2/3 by 1/2 in corrugations	0.027–0.022
	6 by 1 in corrugations	0.025–0.022
	5 by 1 in corrugations	0.026–0.025
	3 by 1 in corrugations	0.028–0.027
	6 by 2 in structural plate	0.035–0.033
	9 by 2 1/2 in structural plate	0.037–0.033
Corrugated metal pipes	2 2/3 by 1/2 in corrugated 24-inch plate width	0.024–0.012
Helical corrugations, full circular-flow spiral rib metal pipe	3/4 by 3/4 in recesses at 12-in spacing, good joints	0.012–0.013
High-density polyethylene (HDPE)		
	Corrugated smooth liner	0.009–0.015
	Corrugated	0.018–0.025
Polyvinyl chloride (PVC)	0.009–0.011	

Note: For further information concerning Manning n values for selected conduits, consult Hydraulic Design of Highway Culverts, Federal Highway Administration, HDS No. 5, p. 163.

Source: From Federal Highway Administration, Hydraulic Design of Highway Culverts, Hydraulic Design Series No. 5, 1985.

Multiple barrel culverts should fit within the natural dominant channel with minor widening of the channel so as to avoid conveyance loss through sediment deposition in some of the barrels. Multiple barrel culverts should be avoided where:

- The approach flow is high velocity, particularly if supercritical (these sites require a single barrel or special inlet treatment to avoid adverse hydraulic jump effects)
- Irrigation canals or ditches are present unless approved by the canal or ditch owner
- Fish passage is required unless special treatment is provided to ensure adequate low flows (commonly, one barrel is lowered)

Weep Holes

Weep holes are sometimes used to relieve uplift pressure. Filter materials should be used in conjunction with the weep holes in order to intercept the flow and prevent the formation of piping channels. The filter materials should be designed as underdrain filter so that it will not become clogged and so that piping cannot occur through the pervious material and the weep hole. Plastic woven filter cloth would be placed over the weep hole in order to keep the pervious material from being carried into the culvert. If weep holes are used to relieve uplift pressure, they should be designed in a manner similar to underdrain systems.

Outlet Protection

See Chapter 12 for information on the design of outlet protection. In general, scour holes at culvert outlets provide efficient energy dissipators. As such, outlet protection for the selected culvert design flood should be provided only where the outlet scour-hole depth computations indicate:

- The scour hole will undermine the culvert outlet
- The expected scour hole may cause costly property damage
- The scour hole causes a nuisance effect (most common in urban areas)
- The scour hole blocks fish passage
- The scour hole will restrict land-use requirements

Erosion and Sediment Control

For erosion and sediment control, the use of silt boxes, brush silt barriers, temporary silt-fence and filter cloth, and check dams may be appropriate. These measures should be utilized as necessary during construction to minimize pollution of streams and damages to wetlands.

Environmental Considerations

In addition to controlling erosion, siltation, and debris at the culvert site, care must be exercised in selecting the location of the culvert site. Environmental considerations are an important aspect of culvert design. Where compatible with good hydraulic engineering, a site should be selected that will permit the culvert to be constructed to cause the least impact on the stream or wetlands. This selection must consider the entire site, including any necessary lead channels.

Safety Considerations

Traffic should be protected from culvert ends as follows:

- Small culverts (30 in diameter or less) should use an end section or a sloped headwall.
- Culverts greater than 30 in diameter should receive one of the following treatments:
  - Extended to the appropriate clear zone distance per AASHTO Roadside Design Guide
  - Safety treated with a grate if the consequences of clogging and causing a potential flooding hazard are less than the hazard of vehicles impacting an unprotected end (If a grate is used, an open area should be provided between the bars of 1.5 to 3.0 times the area of the culvert entrance.)
  - Shielded with a traffic barrier if the culvert is large, cannot be extended, has a channel that cannot be safely traversed by a vehicle, or has a significant flooding hazard with a grate
- Periodically inspect each site to determine if safety problems exist for traffic or for the structural safety of the culvert and embankment.

- Grating, if used for human safety, should be designed at an angle to cause the flow to force someone up above the culvert and not against the grate. See discussion in Chapter 11.

9.4 Culvert Flow Controls and Equations

An exact theoretical analysis of culvert flow is extremely complex because the following is required:

- Analyzing nonuniform flow with regions of gradually varying and rapidly varying flow
- Determining how the flow type changes as the flow rate and tailwater elevations change
- Applying backwater and drawdown calculations
- Applying energy and momentum balance
- Applying the results of hydraulic model studies
- Determining if hydraulic jumps occur, and if they are inside or downstream of the culvert barrel

The design procedures contained in this chapter are for the design of culverts for a constant discharge, considering inlet and outlet control. Generally, the hydraulic control in a culvert will be at the culvert outlet if the culvert is operating on a mild slope. Entrance control usually occurs if the culvert is operating on a steep slope.

For outlet control, the headlosses due to tailwater and barrel friction are predominant in controlling the headwater of the culvert. The entrance will allow the water to enter the culvert faster than the backwater effects of the tailwater, and barrel friction will allow it to flow through the culvert.

For inlet control, the entrance characteristics of the culvert are such that entrance headlosses are predominant in determining the headwater of the culvert. The barrel will carry water through the culvert more efficiently than the water can enter the culvert. Proper culvert design and analysis requires checking for inlet and outlet control to determine which will govern particular culvert designs. For more information on inlet and outlet control, see the Federal Highway Administration publication titled "Hydraulic Design of Highway Culverts," HDS-5 (1985).

Design Equations

Figure 9-1 illustrates the terms and dimensions used in the culvert headwater equations.

There are many combinations of conditions that may classify a particular culvert's hydraulic operation. By consideration of a succession of parameters, the designer may arrive at the appropriate calculation procedure. The most common types of culvert operations for any barrel type are classified as follows.

Critical Depth — Outlet Control

The entrance is unsubmerged ($HW < 1.5\ D$), the critical depth is less than uniform depth at the design discharge ($d_c < d_u$), and the tailwater is less than or equal to critical depth

Design of Culverts

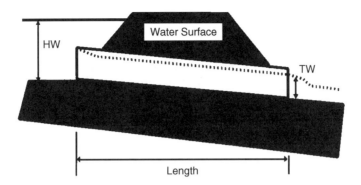

FIGURE 9-1
Culvert terms and dimensions.

($TW < d_c$). This condition is a common occurrence where the natural channels are on flat grades and have wide, flat floodplains. The control is critical depth at the outlet.

$$HW = d_c + V_c^2/(2g) + H_e + H_f - SL \qquad (9.1)$$

where HW = headwater depth (ft); d_c = critical depth (ft); V_c = critical velocity (ft/s); g = 32.2 ft/sec^2; H_e = entrance headloss (ft); H_f = friction headloss (ft); S = slope of culvert (ft/ft); and L = length of culvert (ft).

Tailwater Depth — Outlet Control

The entrance is unsubmerged ($HW < 1.5\,D$), the critical depth is less than uniform depth at design discharge ($d_c < d_u$), and TW is greater than critical depth ($TW > d_c$) and is less than D ($TW < D$). This condition is a common occurrence where the channel is deep, narrow, and well defined. The control is tailwater at the culvert outlet. The outlet velocity is the discharge divided by the area of flow in the culvert at tailwater depth.

$$HW = TW + V^2/(2g) + H_e + H_f - SL \qquad (9.2)$$

where HW = headwater depth (ft); TW = tailwater at the outlet (ft); V = velocity based on tailwater depth (ft/s); g = 32.2 ft/sec^2; H_e = entrance headloss (ft); H_f = friction headloss (ft); S = slope of culvert (ft/ft); and L = length of culvert (ft).

Tailwater Depth > Barrel Depth — Outlet Control

This condition will exist if the critical depth is less than uniform depth at the design discharge ($d_c < d_u$) and TW depth is greater than D ($TW > D$), or if the critical depth is greater than the uniform depth at the design discharge ($d_c > d_u$) and TW is greater than ($SL + D$), [$TW > (SL + D)$]. The HW may or may not be greater than 1.5 D, though often it is greater. If critical depth of flow is determined to be greater than the barrel depth (only possible for rectangular culvert barrels), then this operation will govern. Outlet velocity is based on full flow at the outlet.

$$HW = H + TW - SL \qquad (9.3)$$

where HW = headwater depth (ft); H = total head loss of discharge through culvert (ft); TW = tailwater depth (ft); and SL = culvert slope times length of culvert (ft).

Tailwater < Barrel — Outlet Control

The entrance is submerged ($HW > 1.5\,D$), and the tailwater depth is less than D ($TW < D$). Normally, the designer should arrive at this type of operation only after previous consideration of the operations depth covered when the critical depth, tailwater depth, or slug flow controls the flow in outlet control conditions. On occasion, it may be found that ($HW > 1.5\,D$) for the three previously outlined conditions, but ($HW < 1.5\,D$) for Equation (9.4). If so, the higher HW should be used. Outlet velocity is based on critical depth, if TW depth is less than critical depth. If TW depth is greater than critical depth, outlet velocity is based on TW depth.

$$HW = H + P - SL \quad (9.4)$$

where HW = headwater depth (ft); H = total headloss of discharge through culvert (ft); P = empirical approximation of equivalent hydraulic grade line (ft), $P = (d_c + D)/2$ if TW depth is less than critical depth at design discharge, if TW is greater than critical depth, then $P = TW$; and SL = culvert slope times length of culvert (ft).

Tailwater Insignificant — Inlet Control

The entrance for this condition may be submerged or unsubmerged, the critical depth is greater than uniform depth at the design discharge ($d_c > d_u$), TW depth is less than SL (tailwater elevation is lower than the upstream flowline). Tailwater depth with respect to the diameter of the culvert is inconsequential as long as the above conditions are met. This condition is a common occurrence for culverts in rolling or hilly country. The control is critical depth at the entrance for HW values up to about $1.5\,D$. Control is the entrance geometry for HW values over about $1.5\,D$. HW is determined from empirical curves in the form of nomographs that are discussed later in this chapter. If TW is greater than D, outlet velocity is based on full flow at the outlet. If TW is less than D, outlet velocity is based on uniform depth for the culvert.

Inlet or Outlet Control

For slug flow operation, the entrance may be submerged or unsubmerged, critical depth is greater than uniform depth at the design discharge ($d_c > d_u$), TW depth is greater than ($SL + d_c$) (TW elevation is above the critical depth at the entrance), and TW depth is less than $SL + D$ (TW elevation is below the upstream soffit). TW depth with respect to D is inconsequential as long as the above conditions are met. This condition is a common occurrence for culverts in rolling or hilly country. The control for this type of operation may be at the entrance or the outlet, or control may transfer back and forth between the two (commonly called slug flow). For this reason, it is recommended that HW be determined for entrance control and outlet control and the higher of the two determinations be used. Entrance control HW is determined from the inlet control nomographs, and outlet control HW is determined by Equations (9.3) or (9.4) or the outlet control nomographs.

If TW depth is less than D, outlet velocity should be based on TW depth. If TW depth is greater than D, outlet velocity should be based on outlet full flow.

9.5 Design Procedures

The following design procedure provides a convenient and organized method for designing culverts for constant discharge, considering inlet and outlet control. There are two

Design of Culverts

procedures presented for designing culverts: (1) the manual use of inlet and outlet control nomographs and (2) the use of the HY8 culvert analysis microcomputer program.

It is recommended that the HY8 culvert analysis microcomputer program be used for culvert design, because it will allow the designer to easily develop performance curves, rather than only examine one design situation. The HY8 program uses the theoretical basis for the nomographs to size a culvert. In addition, this system can evaluate improved inlets, route hydrographs, consider road overtopping, and evaluate outlet streambed scour. By using water surface profiles, this procedure is more accurate in predicting backwater effects and outlet scour. The following will outline the design procedures for use of the nomographs. The use of the computer model is described in Section 9.11.

Tailwater Elevations

In some cases, culverts fail to perform as intended because of tailwater elevations high enough to create backwater. The problem is more severe in areas where gradients are flat, and, in some cases, in areas with moderate slopes. Thus, as part of the design process, the normal depth of flow in the downstream channel at discharges equal to those being considered should be computed.

If the tailwater computation leads to water surface elevations below the invert of the culvert exit, there are obviously no problems; if elevations above the culvert invert are computed, the culvert capacity will be somewhat less than assumed. The tailwater computation can be simple, and on steep slopes requires little more than the determination of a cross section downstream, where normal flow can be assumed, and a Manning equation can be calculated. (See Chapter 10 for more information on open-channel analysis.) Conversely, with sensitive flood-hazard sites, if the slopes are flat, or natural and man-made obstructions exist downstream, a water surface profile analysis reaching beyond these obstructions may be required.

Use of Inlet- and Outlet-Control Nomographs

The use of nomographs requires a trial-and-error solution. The solution is quite easy and provides reliable designs for many applications. It should be remembered that velocity, hydrograph routing, roadway overtopping, and outlet scour require additional, separate computations beyond what can be obtained from the nomographs.

Figures 9-2 and 9-3 show examples of inlet-control and outlet-control nomographs that can be used to design concrete pipe culverts. For culvert designs not covered by these nomographs, refer to the complete set of nomographs given in Appendixes A and B at the end of this chapter. Following is the design procedure that requires the use of inlet- and outlet-control nomographs:

Step 1: List design data
  Q = discharge (cfs)
  L = culvert length (ft)
  S = culvert slope (ft/ft)
  K_e = inlet loss coefficient
  V = velocity (ft/s)
  TW = tailwater depth (ft)
  HW = allowable headwater depth for the design storm (ft)

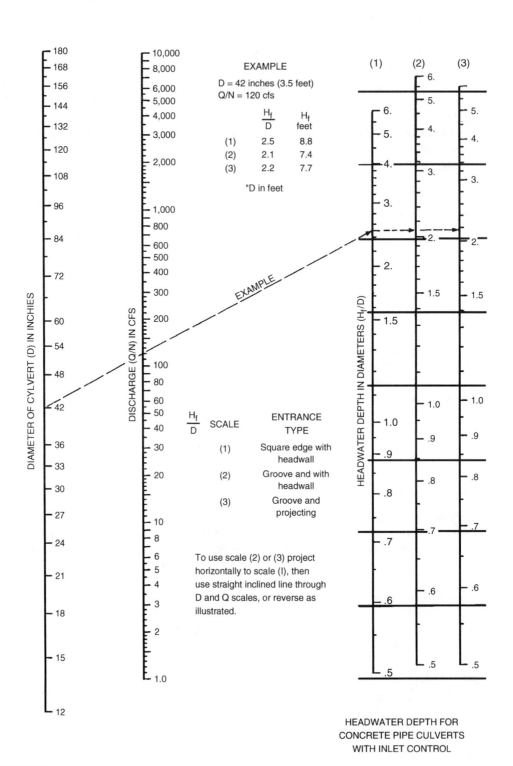

FIGURE 9-2
Inlet-control nomograph. (From FHWA, 1973.)

Design of Culverts

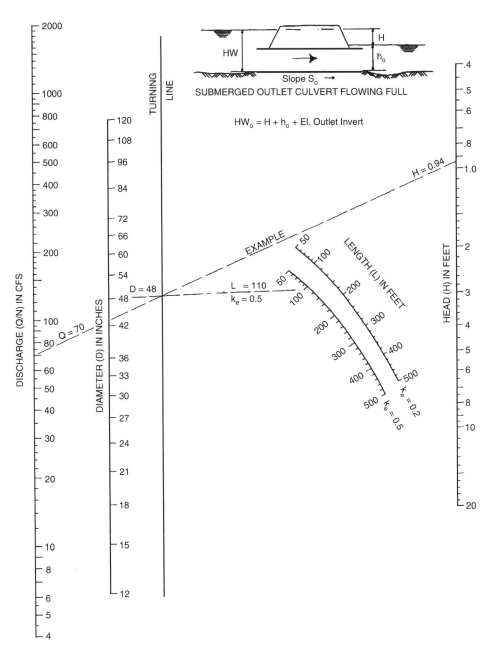

FIGURE 9-3
Outlet-control nomograph. (From Bureau of Public Road, 1963.)

Step 2: Determine trial culvert size by assuming a trial velocity 3 to 5 ft/s and computing the culvert area, $A = Q/V$. Determine the culvert diameter (inches).

Step 3: Find the actual HW for the trial-size culvert for inlet and outlet control.

For *inlet control*, enter inlet-control nomograph with D and Q and find HW/D for the proper entrance type.

Compute HW, and, if too large or two small, try another culvert size before computing HW for outlet control.

For *outlet control*, enter the outlet-control nomograph with the culvert length, entrance loss coefficient, and trial culvert diameter.

To compute HW, connect the length scale for the type of entrance condition and culvert diameter scale with a straight line, pivot on the turning line, and draw a straight line from the design discharge through the turning point to the headloss scale H. Compute the headwater elevation HW from the following equation:

$$HW = H + h_o - LS \qquad (9.5)$$

where $h_o = 1/2$ (critical depth + D), or tailwater depth, whichever is greater.

Step 4: Compare the computed headwaters and use the higher HW nomograph to determine if the culvert is under inlet or outlet control. If outlet control governs and the HW is unacceptable, select a larger trial size and find another HW with the outlet-control nomographs. Because the smaller size of culvert had been selected for allowable HW by the inlet-control nomographs, the inlet control for the larger pipe need not be checked.

Step 5: Calculate exit velocity and expected streambed scour to determine if an energy dissipator is needed. A performance curve for any culvert can be obtained from the nomographs by repeating the steps outlined above for a range of discharges that are of interest for that particular culvert design. A graph is then plotted of headwater versus discharge with sufficient points so that a curve can be drawn through the range of interest. These curves are applicable through a range of headwater, velocities, and scour depths versus discharges for a length and type of culvert. Usually, curves with length intervals of 25 to 50 ft are satisfactory for design purposes. Such computations are made much easier by available computer programs.

To complete the culvert design, roadway overtopping should be analyzed. A performance curve showing the culvert flow as well as the flow across the roadway is a useful analysis tool. Rather than using a trial-and-error procedure to determine the flow division between the overtopping flow and the culvert flow, an overall performance curve can be developed.

The overall performance curve can be determined as follows:

Step 1: Select a range of flow rates and determine the corresponding headwater elevations for the culvert flow. The flow rates should fall above and below the design discharge and cover the entire flow range of interest. Inlet- and outlet-control headwaters should be calculated.

Step 2: Combine the inlet- and outlet-control performance curves to define a single performance curve for the culvert.

Step 3: When the culvert headwater elevations exceed the roadway crest elevation, overtopping will begin. Calculate the equivalent upstream water surface depth above the roadway (crest of weir) for each selected flow rate. Use these water surface depths and Equation (9.6) to calculate flow rates across the roadway.

$$Q = C_d L (HW)^{1.5} \qquad (9.6)$$

where Q = overtopping flow rate (cfs); C_d = overtopping discharge coefficient; L = length of roadway (ft); and HW = upstream depth, measured from the roadway crest to the water surface upstream of the weir drawdown (ft).

Step 4: See Figure 9-4 for guidance in determining a value for C_d. For more information on calculating overtopping flow rates, see pages 39–42 in HDS No. 5 (FHA, 1985).

Step 5: Add the culvert flow and the roadway overtopping flow at the corresponding headwater elevations to obtain the overall culvert performance curve.

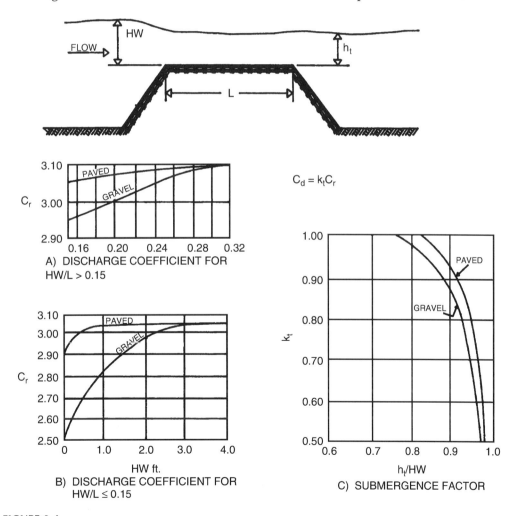

FIGURE 9-4
Discharge coefficient for roadway overtopping. (From Federal Highway Administration, Hydraulic Design of Highway Culverts, Hydraulic Design Series No. 5, 1985.)

Storage Routing

A significant storage capacity behind a highway embankment attenuates a flood hydrograph. Because of the reduction of the peak discharge associated with this attenuation, the required capacity of the culvert, and its size, may be reduced considerably. If significant storage is anticipated behind a culvert, the design should be checked by routing the design hydrographs through the culvert to determine the discharge and stage behind the culvert. Routing procedures are outlined in HDS No. 5 (FHA, 1985). Section 9.10 in this chapter gives some basic information on flood routing and culvert design.

9.6 Culvert Design Example

The following example problem illustrates the procedures to be used in designing culverts using the nomographs. The example problem is as follows: Size a culvert given the following design conditions, which were determined by physical limitations at the culvert site and hydraulic procedures described elsewhere in this book.

Input Data

- Discharge for 10-year flood = 70 cfs
- Discharge for 100-year flood = 176 cfs
- Allowable H_w for 10-year discharge = 4.5 ft
- Allowable H_w for 100-year discharge = 7.0 ft
- Length of culvert = 100 ft
- Natural channel invert elevations — inlet = 15.50 ft, outlet = 15.35 ft
- Culvert slope = 0.0015 ft/ft
- Tailwater depth for 10-year discharge = 3.0 ft
- Tailwater depth for 100-year discharge = 4.0 ft
- Tailwater depth is the normal depth in downstream channel
- Entrance type = groove end with headwall

Step 1: Assume a culvert velocity of 5 ft/s
   Required flow area = 70 cfs/5 ft/s = 14 ft² (for the 10-year flood).

Step 2: The corresponding culvert diameter is about 48 in.
   This can be calculated by using the formula for area of a circle:
   Area = $(3.14\ D^2)/4$ or $D = $ (Area times $4/3.14)^{0.5}$
   Therefore: $D = ((14\text{ sq ft} \times 4)/3.14)^{0.5} \times 12\text{ in/ft} = 50.7$ in

Step 3: A grooved-end culvert with a headwall is selected for the design. Using the inlet-control nomograph (Figure 9-2), with a pipe diameter of 48 in and a discharge of 70 cfs; read a HW/D value of 0.93.

Step 4: The depth of headwater (HW) is $(0.93) \times (4) = 3.72$ ft, which is less than the allowable headwater of 4.5 ft.

Design of Culverts

Step 5: The culvert is checked for outlet control by using Figure 9-3.

With an entrance loss coefficient K_e of 0.20, a culvert length of 100 ft, and a pipe diameter of 48 in, an H value of 0.77 ft is determined. The headwater for outlet control is computed by the equation:

$$HW = H + h_o - LS$$

For the tailwater depth lower than the top of culvert, $h_o = T_w$ or 1/2 (critical depth in culvert + D), whichever is greater.

$$h_o = 3.0 \text{ ft or } h_o = 1/2 (2.55 + 4.0) = 3.28 \text{ ft}$$

The headwater depth for outlet control is:

$$HW = H + h_o - LS$$

$$HW = 0.77 + 3.28 - (100) \times (0.0015) = 3.90 \text{ ft}$$

Step 6: Because HW for outlet control (3.90 ft) is greater than the HW for inlet control (3.72 ft), outlet control governs the culvert design.

Thus, the maximum headwater expected for a 10-year recurrence flood is 3.90 ft, which is less than the allowable headwater of 4.5 ft.

Step 7: The performance of the culvert is checked for the 100-year discharge. The allowable headwater for a 100-year discharge is 7 ft; critical depth in the 48-in diameter culvert for the 100-year discharge is 3.96 ft.

For outlet control, an H value of 5.2 ft is read from the outlet-control nomograph. The maximum headwater is:

$$HW = H + h_o - LS$$

$$HW = 5.2 + 4.0 - (100) \times (0.0015) = 9.05 \text{ ft}$$

This depth is greater than the allowable depth of 7 ft; thus, a larger size culvert must be selected.

Step 8: A 54-in diameter culvert is tried and found to have a maximum headwater depth of 3.74 ft for the 10-year discharge and of 6.97 ft for the 100-year discharge. These values are acceptable for the design conditions.

Step 9: Estimate outlet exit velocity. Because this culvert is on outlet control and discharges into an open channel downstream, the culvert will be flowing full at the flow depth in the channel. Using the 100-year design peak discharge of 176 cfs and the area of a 54-in or 4.5-ft diameter culvert, the exit velocity will be: $Q = VA$. Therefore:

$$V = 176 / (\pi(4.5)^2)/4 = 11.8 \text{ ft/s}.$$

With this high velocity, some energy dissipator may be needed downstream from this culvert for streambank protection. It will first be necessary to compute a scour-hole depth and then decide if protection is needed. See Chapter 12 for design procedures related to energy dissipators.

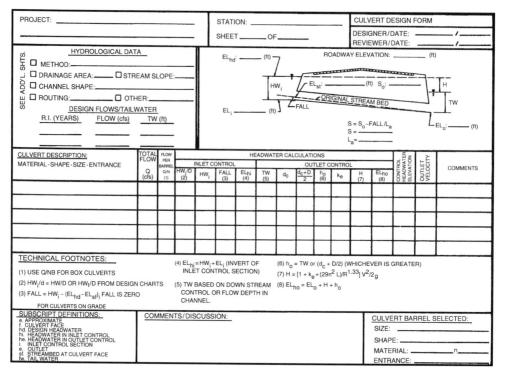

FIGURE 9-5
Culvert design calculation form. (From Federal Highway Administration, Hydraulic Design of Highway Culverts, Hydraulic Design Series No. 5, 1985.)

Step 10: The designer should check minimum velocities for low-frequency flows if the larger storm event (100-year) controls culvert design. Note: Figure 9-5 provides a convenient form to organize culvert design calculations.

9.7 Long-Span Culvert

Long-span culverts are better defined on the basis of structural design aspects than on the basis of hydraulic considerations. According to the AASHTO Specifications for Highway Bridges, long-span structural plate structures: (1) exceed certain defined maximum sizes for pipes, pipe-arches, and arches, or (2) may be special shapes of any size that involve a long radius of curvature in the crown or side plates. Special shapes include vertical and horizontal ellipses, underpasses, low- and high-profile arches, and inverted pear shapes. Generally, the spans of long-span culverts range from 20 to 40 ft.

Long-span culverts depend on interaction with the earth embankment for structural stability. Therefore, proper bedding and selection and compaction of backfill are of utmost importance. For multiple-barrel structures, care must be taken to avoid unbalanced loads during backfilling. Anchorage of the ends of long-span culverts is required to prevent flotation or damage due to high velocities at the inlet. This is especially true for mitered inlets. Severe miters and skews are not recommended.

Long-span culverts generally are hydraulically short (low length to equivalent diameter ratio) and flow partly full at the design discharge. The same hydraulic principles apply

Design of Culverts

to the design of long-span culverts as to other culverts. However, due to their large size and variety of shapes, it is possible that design nomographs are not available for the barrel shape of interest. For these cases, dimensionless inlet control design curves have been prepared. For these nomographs and design curves, consult the publication, HDS No. 5 (FHA, 1985).

For outlet control, backwater calculations are usually appropriate, because design headwaters exceeding the crowns of these conduits are rare. The bridge design techniques of HDS No. 1 (FHA, 1978) are appropriate for the design of most long-span culverts.

9.8 Design of Improved Inlets

A culvert operates in inlet or outlet control. As previously discussed under outlet control, headwater depth, tailwater depth, entrance configuration, and barrel characteristics influence a culvert's capacity. The entrance configuration is defined by the barrel cross-sectional area, shape, and edge condition, while the barrel characteristics are area, shape, slope, length, and roughness.

The flow condition for outlet control may be full or partly full for all or part of the culvert length. The design discharge usually results in full flow. Inlet improvements in these culverts reduce the entrance losses, which are only a small portion of the total headwater requirements. Therefore, only minor modifications of the inlet geometry that results in little additional cost are justified.

In inlet control, only entrance configuration and headwater depth determine the culvert's hydraulic capacity. Barrel characteristics and tailwater depth are of no consequence. These culverts usually lie on relatively steep slopes and flow only partly full. Entrance improvements can result in full, or nearly full, flow, thereby increasing culvert capacity significantly.

Figure 9-6 illustrates the performance of a 30-in circular culvert in inlet control with three commonly used entrances: thin-edged projecting, square-edged, and groove-edged.

Improved inlets include inlet geometry refinements beyond those normally used in conventional culvert design practice. Several degrees of improvements are possible, including bevel-edged, side-tapered, and slope-tapered inlets.

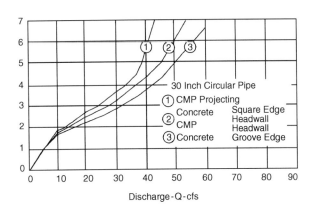

FIGURE 9-6
Performance curves — inlet control. (From Federal Highway Administration, Hydraulic Design of Highway Culverts, Hydraulic Design Series No. 5, 1985.)

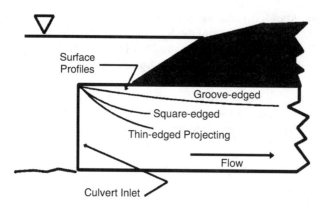

FIGURE 9-7
Schematic flow contractions for conventional culvert inlets.

It is clear that inlet type and headwater depth determine the capacities of many culverts. For a given headwater, a groove-edged inlet has a greater capacity than a square-edged inlet, which, in turn, outperforms a thin-edged projecting inlet.

The performance of each inlet type is related to the degree of flow contraction. A high degree of contraction requires more energy, or headwater, to convey a given discharge than a low degree of contraction. Figure 9-7 shows schematically the flow contractions of the three inlet types.

Bevel-Edged Inlet

The first degree of inlet improvement is a beveled edge (Figure 9-8). The bevel is proportioned based on the culvert barrel or face dimension and operates by decreasing the flow contraction at the inlet. A bevel is similar to a chamfer, except that a chamfer is smaller and is generally used to prevent damage to sharp concrete edges during construction.

Adding bevels to a conventional culvert design with a square-edged inlet increases culvert capacity by 5 to 20%. The higher increase results from comparing a bevel-edged inlet with a square-edged inlet at high headwaters. The lower increase is the result of comparing inlets with bevels, with structures having wingwalls of 30 to 45 degrees.

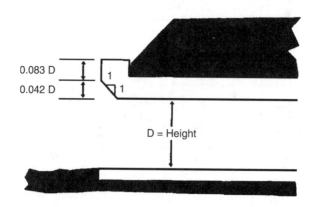

FIGURE 9-8
Bevel-edged inlet.

Design of Culverts

Although the bevels referred to in this book are plane surfaces, rounded edges, which approximate the bevels, are also acceptable.

As a minimum, bevels should be used on all culverts that operate in inlet control, conventional and improved inlet types. The exception to this is circular concrete culverts, where the socket end performs much the same as a beveled edge.

Culverts flowing in outlet control cannot be improved as much as those in inlet control, but the entrance loss coefficient, k_e, is reduced from 0.5 for a square edge to 0.2 for beveled edges. It is recommended that bevels be used on all culvert entrances if little additional cost is involved.

Side-Tapered Inlet

The second degree of improvement is a side-tapered inlet. This inlet has an enlarged face area with the transition to the culvert barrel accomplished by tapering the sidewalls. The inlet face has the same height as the barrel, and its top and bottom are extensions of the top and bottom of the barrel. The intersection of the sidewall tapers and barrel is defined as the throat section. If a headwall and wingwalls are going to be used at the culvert entrance, side-tapered inlets should add little if any to the overall cost, while significantly increasing hydraulic efficiency.

The side-tapered inlet provides an increase in flow capacity of 25 to 40% over that of a conventional culvert with a square-edged inlet. Whenever increased inlet efficiency is needed, or when a headwall and wingwalls are planned to be used for a culvert installation, a side-tapered inlet should be considered.

Slope-Tapered Inlet

A slope-tapered inlet is the third degree of improvement. Its advantage over the side-tapered inlet without a depression is that more head is available at the inlet. This is accomplished by incorporating a fall in the enclosed entrance section. The slope-tapered inlet can have over a 100% greater capacity than a conventional culvert with square edges. The degree of increased capacity depends largely upon the amount of fall available. Because this fall may vary, a range of increased capacities is possible.

Side- and slope-tapered inlets should be used in culvert design when they can economically be used to increase the inlet efficiency over a conventional design. For a complete discussion of tapered inlets, including figures and illustrations, see FHA, 1985, pages 65–93.

Improved Inlet Performance

Table 9-4 compares the inlet-control performance of the different inlet types. The top part of the table shows the increase in discharge that is possible for a headwater depth of 8 ft. The bevel-edged inlet, side-tapered inlet, and slope-tapered inlet show increases in discharge over the square-edged inlet of 16.7, 30.4, and 55.6%, respectively. It should be noted that the slope-tapered inlet incorporates only a minimum fall. Greater increases in capacity are often possible if a larger fall is used. The bottom part of the table depicts the reduction in headwater that is possible for a discharge of 500 cfs. The headwater varies from 12.5 ft for the square-edged inlet to 7.6 ft for the slope-tapered inlet. This is a 39.2% reduction in required headwater.

TABLE 9-4

Inlet Control Performance of Different Inlet Types

Comparison of Inlet Performance at Constant Headwater for 6 ft × 6 ft Concrete Box Culvert			
Inlet Type	Headwater (ft)	Discharge (cfs)	% Improvement
Square-edge	8.0	336	0.0
Bevel-edge	8.0	392	16.7
Side-tapered	8.0	438	30.4
Slope-tapered[a]	8.0	523	55.6

Comparison of Inlet Performance at Constant Discharge for 6 ft × 6 ft Concrete Box Culvert			
Inlet Type	Discharge (cfs)	Headwater (ft)	% Improvement
Square-edge	500	12.5	0.0
Bevel-edge	500	10.1	19.2
Side-tapered	500	8.8	29.6
Slope-tapered[a]	500	7.6	39.2

[a] Minimum fall in inlet = D/4 = 6/4 = 1.5 ft

Source: From Federal Highway Administration, Hydraulic Design of Improved Inlets for Culverts, Hydraulic Engineering Circular No. 13, 1972.

9.9 Design Procedures for Bevel-Edged Inlets

This section will outline the procedures and figures to use when incorporating bevel-edged inlets in the design of culverts. For the design of side- and slope-tapered inlets, consult the detailed design criteria and example designs outlined in "Hydraulic Design of Highway Culverts" (in FHA, 1985).

Inlet-control figures for culverts with beveled edges are included in Appendixes A and B at the end of this chapter. The figures for bevel-edged inlets are used for design in the same manner as the conventional inlet-design nomographs discussed earlier. For box culverts, the dimensions of the bevels to be used are based on the culvert dimensions. The top bevel dimension is determined by multiplying the height of the culvert by a factor. The side bevel dimensions are determined by multiplying the width of the culvert by a factor. For a 1:1 bevel, the factor is 1/2 in/ft. For a 1.5:1 bevel, the factor is 1 in/ft.

For example, the minimum bevel dimensions for a 8 ft × 6 ft box culvert with 1:1 bevels would be as follows:

Top bevel = d = 6 ft × 1/2 in/ft = 3 in
Side bevel = b = 8 ft × 1/2 in/ft = 4 in

For a 1.5:1 bevel, computations would result in d = 6 in and b = 8 in.

Design Figure Limits

The improved inlet design figures are based on research results from culvert models with barrel width, B, to depth, D, ratios of from 0.5:1 to 2:1.

For box culverts with more than one barrel, the figures are used in the same manner as for a single barrel, except that the bevels must be sized on the basis of the total clear opening rather than on individual barrel size. For example, in a double 8 ft by 8 ft box culvert:

Top bevel is proportioned based on the height of 8 ft, which results in a bevel of 4 in for the 1:1 bevel and 8 in for the 1.5:1 bevel.

Side bevel is proportioned based on the clear width of 16 ft, which results in a bevel of 8 in for the 1:1 bevel and 16 in for the 1.5:1 bevel.

The ratio of the inlet face area to the barrel area remains the same as for a single-barrel culvert. Multibarrel pipe culverts should be designed as a series of single-barrel installations because each pipe requires a separate bevel.

For multibarrel installations exceeding a 3:1 width-to-depth ratio, the side bevels become excessively large when proportioned on the basis of the total clear width. For these structures, it is recommended that the side bevel be sized in proportion to the total clear width, B, or three times the height, whichever is smaller. The top bevel dimension should always be based on the culvert height. The shape of the upstream edge of the intermediate walls of multibarrel installations is not as important to the hydraulic performance of a culvert as the edge condition of the top and sides. Therefore, the edges of these walls may be square, rounded with a radius of one-half their thickness, chamfered, or beveled. The intermediate walls may also project from the face and slope downward to the channel bottom to help direct debris through the culvert.

It is recommended that skewed inlets not be used for multiple-barrel installations, as the intermediate wall could cause an extreme contraction in the downstream barrels. This would result in underdesign due to a greatly reduced capacity. Skewed inlets should be avoided whenever possible, and should not be used with side- or slope-tapered inlets.

9.10 Flood Routing and Culvert Design

Flood routing through a culvert is a practice that evaluates the effect of temporary upstream ponding caused by the culvert's backwater. By not considering flood routing, it is possible that the findings from culvert analyses will be conservative. If the selected allowable headwater is considered acceptable without flood routing, then costly overdesign of the culvert and outlet protection may result, depending on the amount of temporary storage involved.

There are many ramifications associated with culvert flood routing:

- Ownership or easement of the upstream property may be required.
- A perceived loss of a subjective safety factor may arise.
- Credibility, in court as well as in technical negotiations, may be affected.
- Evaluation of environmental concerns may be required.
- Realistic assessments of potential flood hazards may need to be made.
- Estimation of sediment problems may be necessary.

Ignoring temporary storage effects on reducing the selected design flood magnitude by assuming that this provides a factor of safety is not recommended. This practice results in inconsistent factors of safety at culvert sites, as it is dependent on the amount of temporary storage at each site. Further, with little or no temporary storage at a site, the factor of safety would be unity, thereby precluding a factor of safety. If a factor of safety is desired, it is essential that flood-routing practices be used to ensure that consistent and defensible factors of safety are used at all culvert sites.

Improved hydrology methods or changed watershed conditions are factors that can cause an older, existing culvert to be inadequate. A culvert analysis that relies on findings that ignore any available temporary storage may be misleading. A flood-routing analysis may show that what was thought to be an inadequate existing culvert is, in fact, adequate.

Often, existing culverts require replacement due to corrosion or abrasion. This can be costly, particularly where a high fill is involved. A less costly alternative is to place a smaller culvert inside the existing culvert. A flood-routing analysis may, where there is sufficient storage, demonstrate that this is acceptable in that no increase in flood hazard results.

With legal proceedings or in resolving conflicting design findings, it is essential that credible and defensible practices be used. By ignoring flood routing where significant storage occurs, findings may be discredited. With legal proceedings, claims of design negligence may result, depending on the nature of the case.

With culvert flood routing, a more realistic assessment can be made where environmental concerns are important. The temporary time of upstream ponding can be easily identified. This allows environmental specialists to assess whether such ponding is beneficial or harmful to localized environmental features such as fisheries, beaver ponds, wetlands, and uplands.

Potential flood hazards increase whenever a culvert increases the natural flood stage. Some of these hazards can conservatively be assessed without flood routing. However, some damages associated with culvert backwater are time-dependent and thus require an estimate of depth versus duration of inundation. Some vegetation and commercial crops can tolerate longer periods of inundation than others, and to greater depths. Such considerations become even more important when litigation is involved.

Complex culvert sediment deposition (silting) problems require the application of a sediment routing practice. This practice requires a time–flood discharge relationship, or hydrograph. This flood hydrograph must be coupled to a flood–discharge and sediment–discharge relationship in order to route the sediment through the culvert site.

There are situations in which culvert sizes and velocities obtained through flood routing will not differ significantly from those obtained by designing to the selected peak discharge and ignoring any temporary upstream storage. This occurs when:

- There is no significant temporary pond storage available (as in deep incised channels).
- The culvert must pass the design discharge with no increase in the natural channel's flood stage.
- Runoff hydrographs last for long periods of time, such as with snowmelt runoff or irrigation flows.

Design Procedure

The design procedure for flood routing through a culvert is the same as for reservoir routing (see Chapter 11). The site data and roadway geometry are obtained and the hydrology analysis completed to include estimating a hydrograph. Once this essential information is available, the culvert can be designed.

Flood routing through a culvert can be time-consuming. It is recommended that the HY8 computer program be used, as it contains software that quickly routes floods through a culvert to evaluate an existing culvert (review), or to select a culvert size that satisfies given criteria (design). However, the designer should be familiar with the culvert flood routing design process. This familiarization is necessary in order to do the following:

- Recognize and test suspected software malfunctions
- Circumvent any software limitations
- Flood-route manually where the software is limited
- Understand and discuss culvert flood routing in a credible manner

9.11 HY8 Culvert Analysis Microcomputer Program

The HY8 culvert analysis microcomputer program will perform the calculation for the following:

- Culvert analysis (including independent, multiple-barrel sizing)
- Hydrograph generation
- Hydrograph routing
- Roadway overtopping
- Outlet scour estimates

The HY8 culvert analysis microcomputer program has several user-friendly features that permit easy data entry, editing, and comparison of several design alternatives. Data are entered by selecting options on a menu or by entering numeric data at prompts. These data are periodically summarized in tables. Any incorrect entry can be changed, and design variations can be quickly analyzed. Another feature of the HY8 culvert analysis microcomputer program is that plots of irregular cross sections, channel rating curves, and culvert performance curves can be obtained if the terminal has graphics capabilities. For an example of using HY8 for culvert design, see AASHTO (1998),

References

American Association of State Highway and Transportation Officials, Highway Drainage Guidelines, 1982.
American Association of State Highway and Transportation Officials, Model Drainage Manual, 1992, 1998.
American Concrete Pipe Association, Culvert Durability Study, No. 02-902, 1982.
Blaisdell, F.W., Flow in culverts and related design philosophies, *Proc. ASCE J. Hy.*, 92, HY2, March 1966.
City of Indianapolis, Indiana, Specifications for Construction of Stormwater Drainage Improvements, Department of Capital Asset Management, April 1994.
Federal Highway Administration, Debris-Control Structures, Hydraulic Engineering Circular No. 9, 1971.
Federal Highway Administration, Hydraulic Design of Improved Inlets for Culverts, Hydraulic Engineering Circular No. 13, 1972.
Federal Highway Administration, Hydraulics of Bridge Waterways, Hydraulic Design Series No. 1, 1978.
Federal Highway Administration, Hydraulic Design of Highway Culverts, Hydraulic Design Series No. 5, 1985.

Federal Highway Administration, HY8 Culvert Analysis Microcomputer Program Applications Guide, Hydraulic Microcomputer Program HY8, 1987.

HYDRAIN Culvert Computer Program (HY8), Available from McTrans Software, University of Florida, Gainesville.

Lewis, G.L., Jury verdict: frequency versus risk-based culvert design, *ASCE J. Water Res. Plann. and Manage.*, 118, 2, March/April 1992, pp. 166–185.

U.S. Department of Interior, Design of Small Canal Structures, 1983.

Young, G.K., Taylor, R.S., and Costello, L.S., Evaluation of the Flood Risk Factor in the Design of Box Culverts, Rpt. No. FHWA-RD-74-11, Sept. 1970.

Appendix A

Nomographs and Charts — English Units*

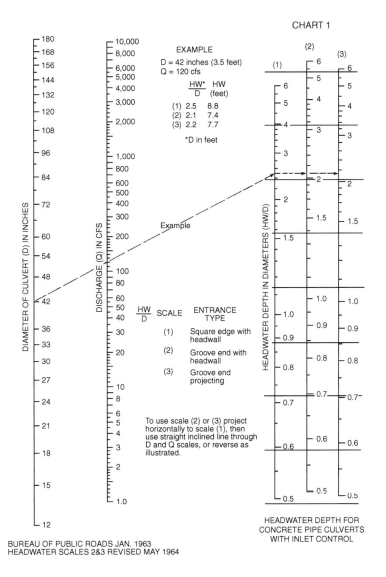

* *Source:* American Association of State Highway and Transportation Officials, Model Drainage Manual, 1992, 1998.

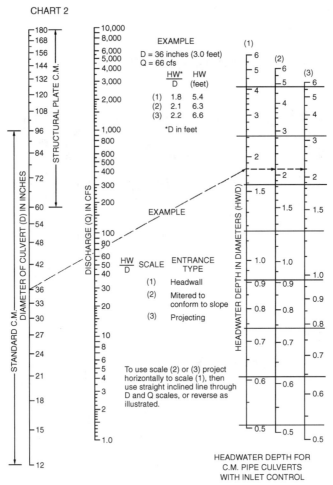

Design of Culverts

CHART 3

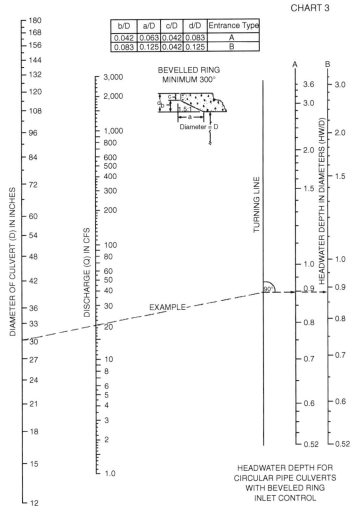

HEADWATER DEPTH FOR
CIRCULAR PIPE CULVERTS
WITH BEVELED RING
INLET CONTROL

FEDERAL HIGHWAY ADMINISTRATION MAY 1973

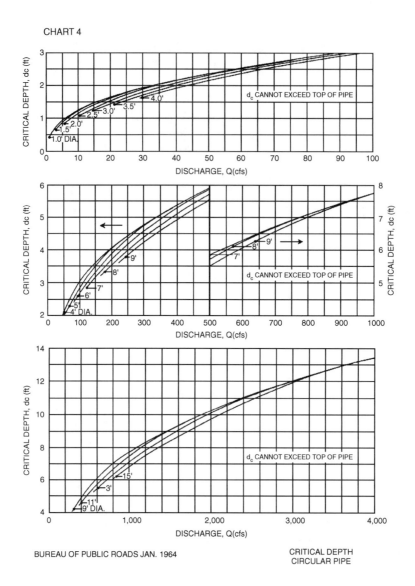

BUREAU OF PUBLIC ROADS JAN. 1964

CRITICAL DEPTH
CIRCULAR PIPE

Design of Culverts

CHART 5

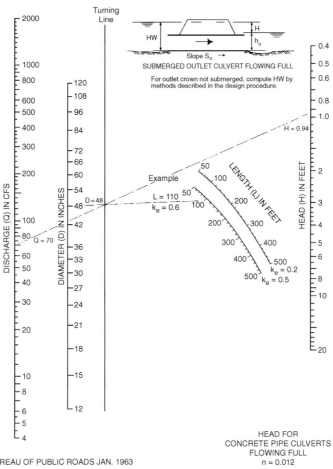

HEAD FOR
CONCRETE PIPE CULVERTS
FLOWING FULL
n = 0.012

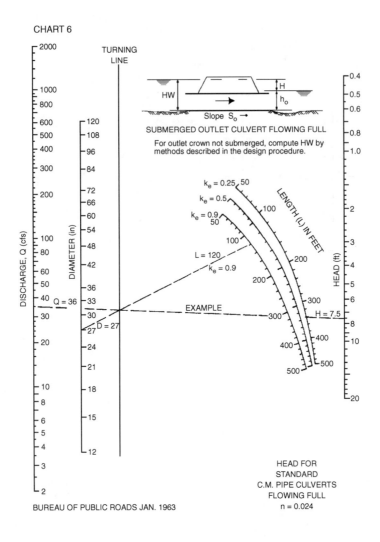

Design of Culverts

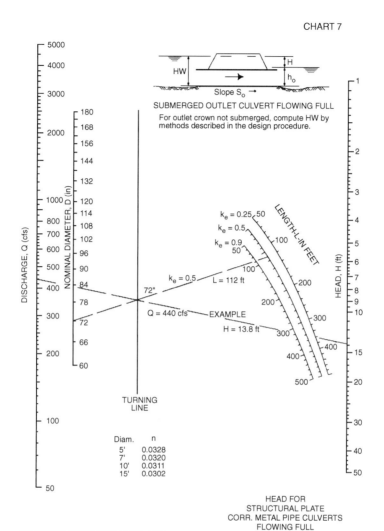

CHART 7

HEAD FOR
STRUCTURAL PLATE
CORR. METAL PIPE CULVERTS
FLOWING FULL
$n = 0.0328$ TO 0.0302

BUREAU OF PUBLIC ROADS JAN. 1963

CHART 8

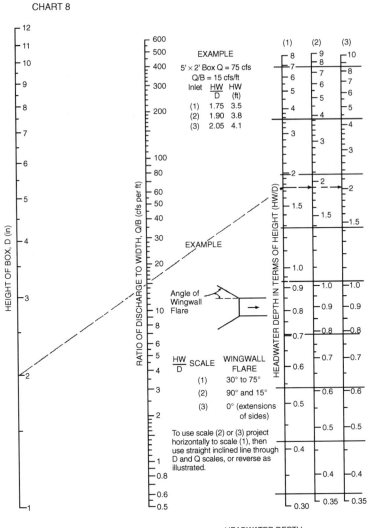

HEADWATER DEPTH
FOR BOX CULVERTS
WITH INLET CONTROL

BUREAU OF PUBLIC ROADS JAN. 1963

Design of Culverts

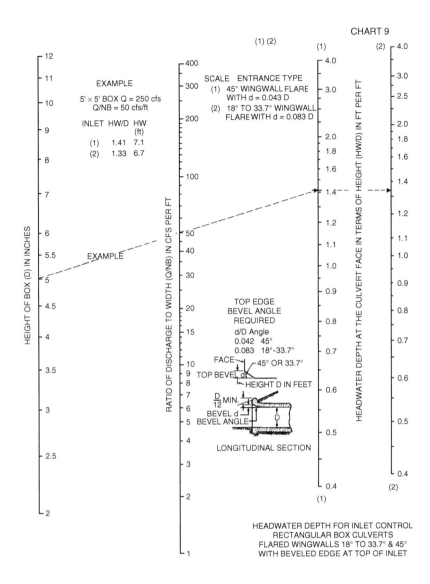

CHART 9
HEADWATER DEPTH FOR INLET CONTROL
RECTANGULAR BOX CULVERTS
FLARED WINGWALLS 18° TO 33.7° & 45°
WITH BEVELED EDGE AT TOP OF INLET

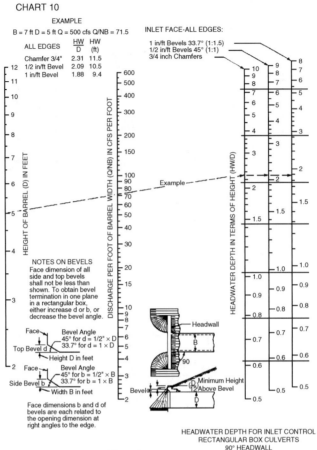

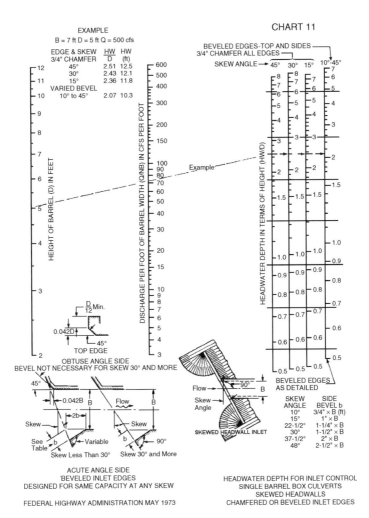

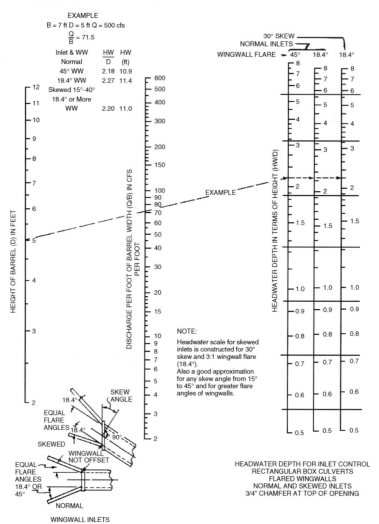

Design of Culverts

CHART 13

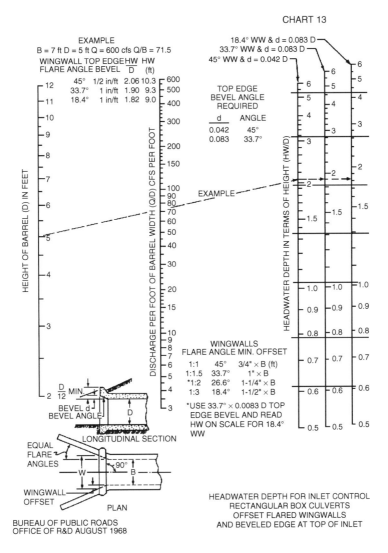

HEADWATER DEPTH FOR INLET CONTROL
RECTANGULAR BOX CULVERTS
OFFSET FLARED WINGWALLS
AND BEVELED EDGE AT TOP OF INLET

BUREAU OF PUBLIC ROADS
OFFICE OF R&D AUGUST 1968

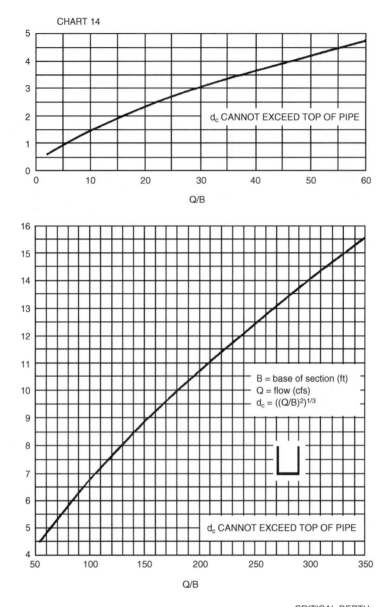

BUREAU OF PUBLIC ROADS JAN. 1963

CRITICAL DEPTH
RECTANGULAR SECTION

Design of Culverts

CHART 15

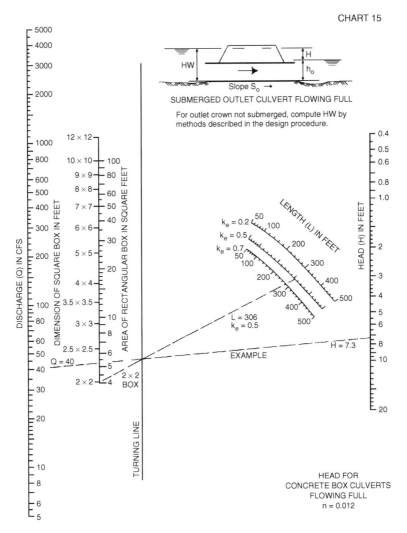

SUBMERGED OUTLET CULVERT FLOWING FULL

For outlet crown not submerged, compute HW by methods described in the design procedure.

HEAD FOR
CONCRETE BOX CULVERTS
FLOWING FULL
n = 0.012

BUREAU OF PUBLIC ROADS JAN. 1963

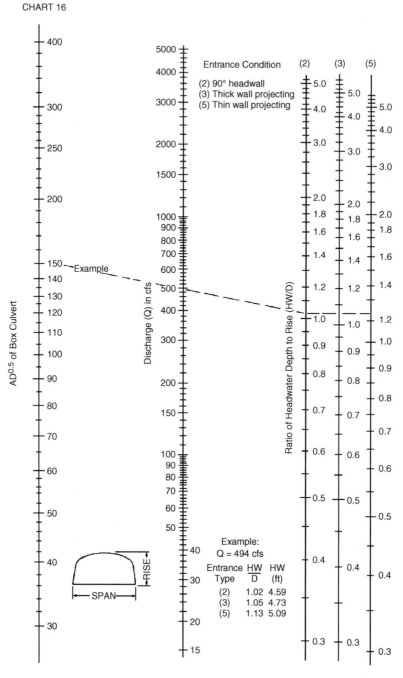

CHART 16

HEADWATER DEPTH
FOR C.M. BOX CULVERTS
RISE/SPAN < 0.3
WITH INLET CONTROL

Nomographs adapted from material furnished by
Kaiser Aluminum and Chemical Corporation.

Design of Culverts

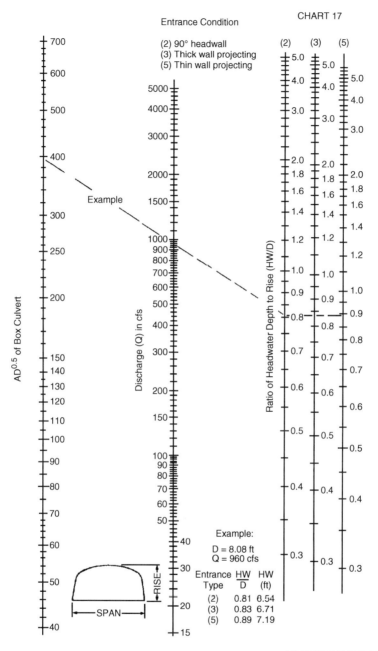

CHART 17

HEADWATER DEPTH
FOR C.M. BOX CULVERTS
0.3 ≤ RISE/SPAN < 0.4
WITH INLET CONTROL

Nomographs adapted from material furnished by Kaiser Aluminum and Chemical Corporation. Duplication of this nomograph may distort scale.

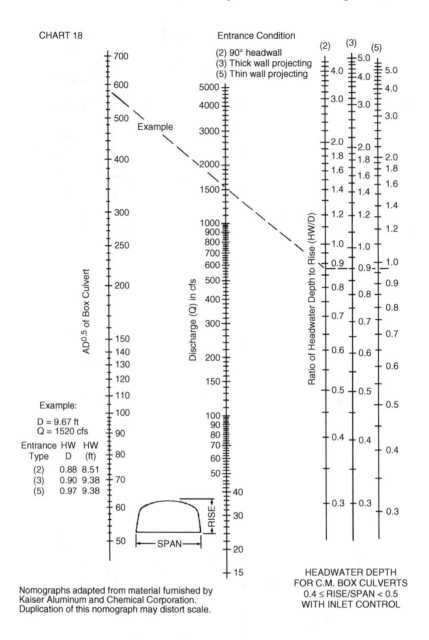

Design of Culverts

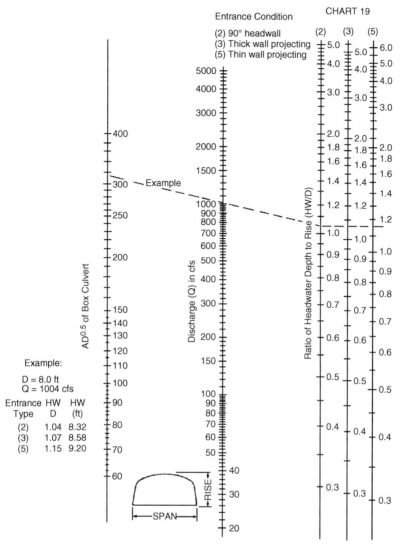

HEADWATER DEPTH
FOR C.M. BOX CULVERTS
0.5 ≤ RISE/SPAN
WITH INLET CONTROL

414 Municipal Stormwater Management, Second Edition

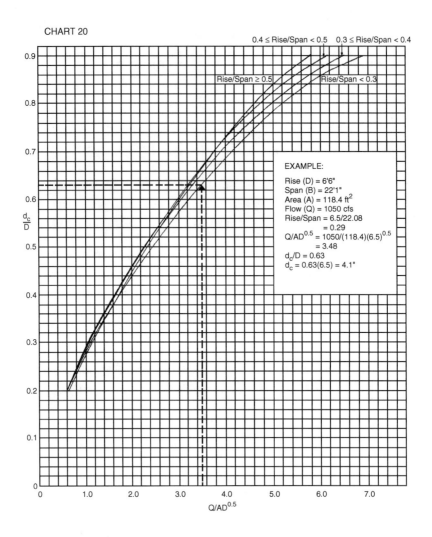

Design of Culverts

CHART 21

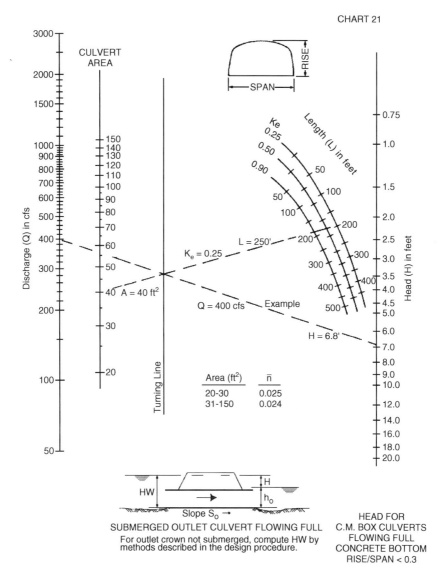

HEAD FOR
C.M. BOX CULVERTS
FLOWING FULL
CONCRETE BOTTOM
RISE/SPAN < 0.3

Nomographs adapted from material furnished by
Kaiser Aluminum and Chemical Corporation.
Duplication of this nomograph may distort scale.

CHART 22

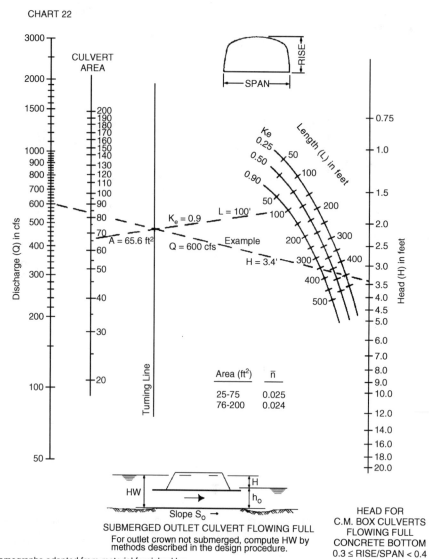

HEAD FOR
C.M. BOX CULVERTS
FLOWING FULL
CONCRETE BOTTOM
$0.3 \leq \text{RISE/SPAN} < 0.4$

Nomographs adapted from material furnished by
Kaiser Aluminum and Chemical Corporation.
Duplication of this nomograph may distort scale.

Design of Culverts

CHART 23

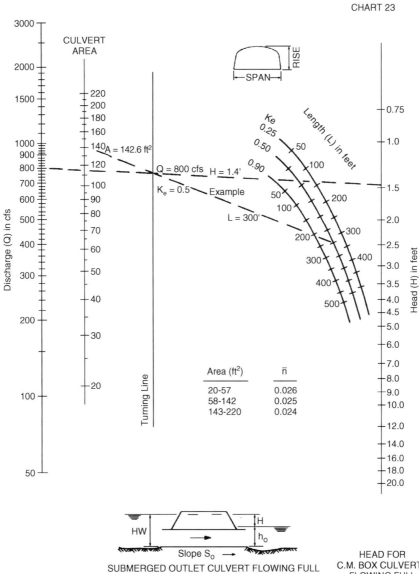

HEAD FOR
C.M. BOX CULVERTS
FLOWING FULL
CONCRETE BOTTOM
$0.4 \leq \text{RISE/SPAN} < 0.5$

Nomographs adapted from material furnished by
Kaiser Aluminum and Chemical Corporation.
Duplication of this nomograph may distort scale.

CHART 24

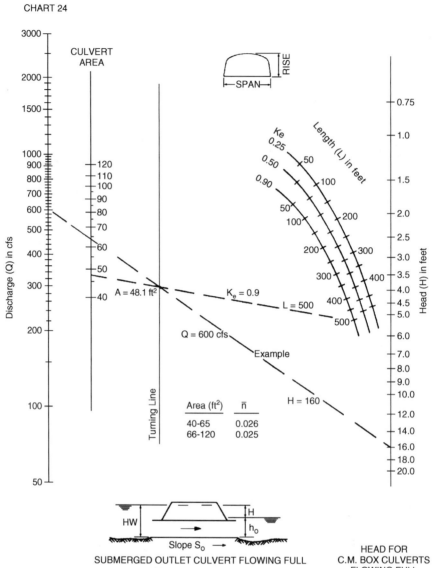

SUBMERGED OUTLET CULVERT FLOWING FULL

HEAD FOR
C.M. BOX CULVERTS
FLOWING FULL
CONCRETE BOTTOM
$0.5 \leq$ RISE/SPAN

Nomographs adapted from material furnished by
Kaiser Aluminum and Chemical Corporation.
Duplication of this nomograph may distort scale.

Design of Culverts

CHART 25

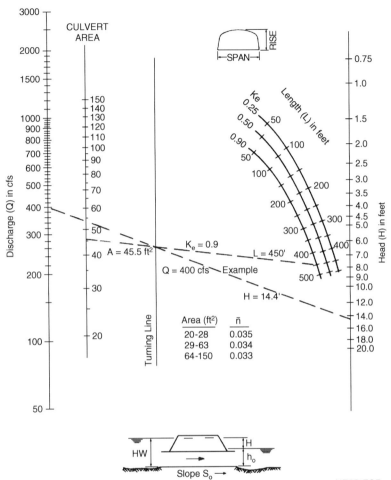

SUBMERGED OUTLET CULVERT FLOWING FULL

Nomographs adapted from material furnished by
Kaiser Aluminum and Chemical Corporation.
Duplication of this nomograph may distort scale.

HEAD FOR
C.M. BOX CULVERTS
FLOWING FULL
CORRUGATED METAL BOTTOM
0.3 < RISE/SPAN

420

CHART 26

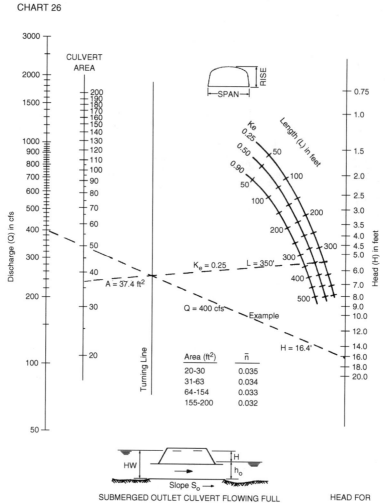

SUBMERGED OUTLET CULVERT FLOWING FULL

Nomographs adapted from material furnished by
Kaiser Aluminum and Chemical Corporation.
Duplication of this nomograph may distort scale.

HEAD FOR
C.M. BOX CULVERTS
FLOWING FULL
CORRUGATED METAL BOTTOM
$0.4 \leq$ RISE/SPAN < 0.5

Design of Culverts

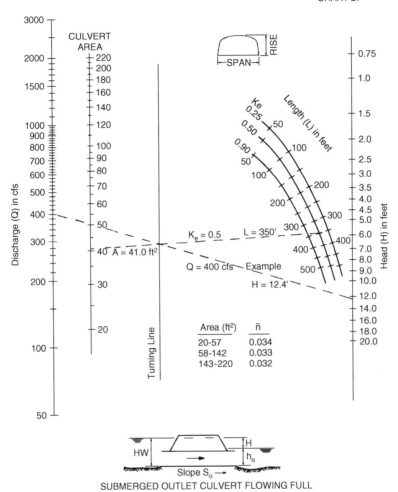

CHART 27

SUBMERGED OUTLET CULVERT FLOWING FULL

Nomographs adapted from material furnished by
Kaiser Aluminum and Chemical Corporation.
Duplication of this nomograph may distort scale.

HEAD FOR
C.M. BOX CULVERTS
FLOWING FULL
CORRUGATED METAL BOTTOM
0.4 ≤ RISE/SPAN < 0.5

422   Municipal Stormwater Management, Second Edition

CHART 28

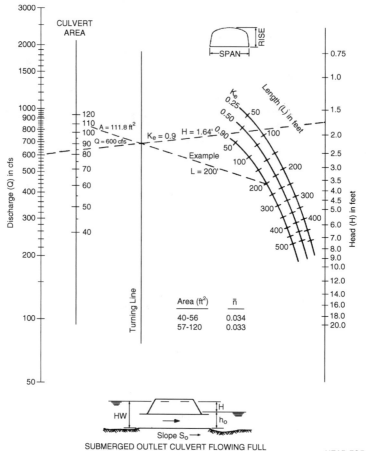

HEAD FOR
C.M. BOX CULVERTS
FLOWING FULL
CORRUGATED METAL BOTTOM
0.5 ≤ RISE/SPAN

Nomographs adapted from material furnished by
Kaiser Aluminum and Chemical Corporation.
Duplication of this nomograph may distort scale.

Design of Culverts

CHART 29

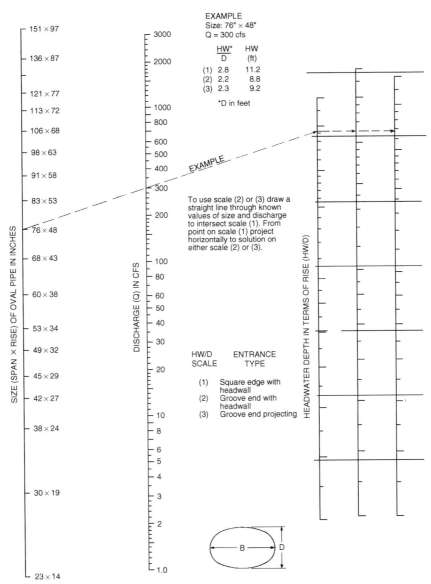

HEADWATER DEPTH FOR
OVAL CONCRETE PIPE CULVERTS
LONG AXIS HORIZONTAL
WITH INLET CONTROL

BUREAU OF PUBLIC ROADS JAN. 1963

424

CHART 30

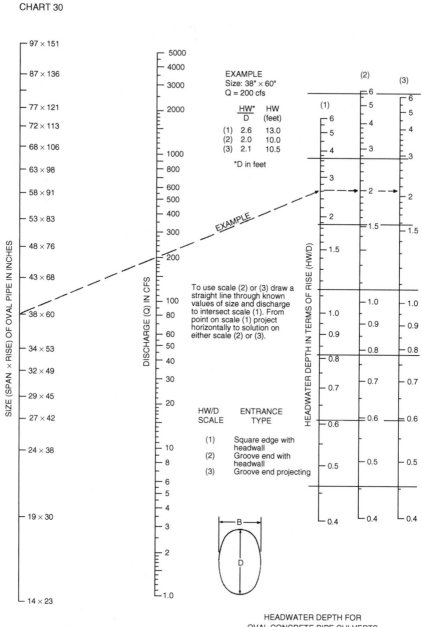

HEADWATER DEPTH FOR
OVAL CONCRETE PIPE CULVERTS
LONG AXIS VERTICAL
WITH INLET CONTROL

BUREAU OF PUBLIC ROADS JAN. 1963

Design of Culverts

CHART 31

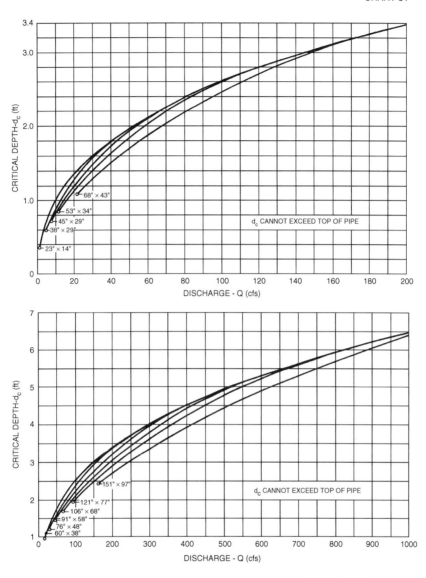

BUREAU OF PUBLIC ROADS JAN. 1964

CRITICAL DEPTH
OVAL CONCRETE PIPE
LONG AXIS HORIZONTAL

CHART 32

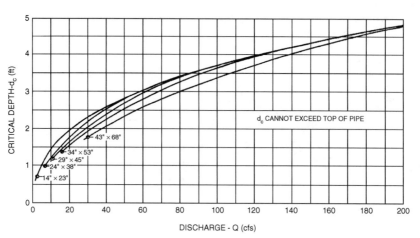

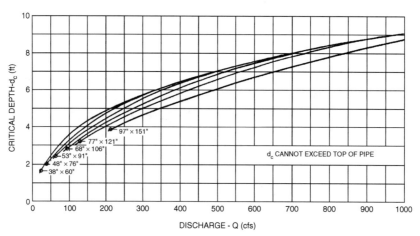

BUREAU OF PUBLIC ROADS JAN. 1964

CRITICAL DEPTH
OVAL CONCRETE PIPE
LONG AXIS VERTICAL

Design of Culverts

CHART 33

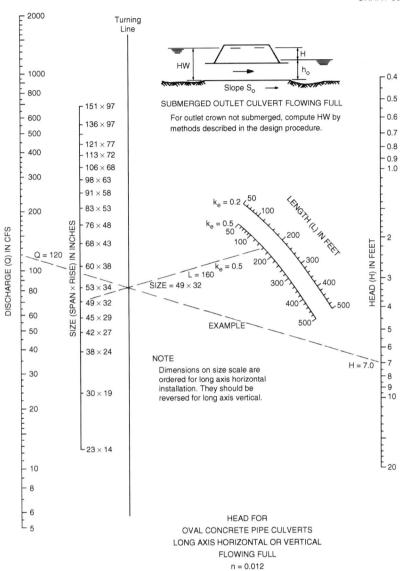

HEAD FOR
OVAL CONCRETE PIPE CULVERTS
LONG AXIS HORIZONTAL OR VERTICAL
FLOWING FULL
n = 0.012

BUREAU OF PUBLIC ROADS JAN. 1963

CHART 34

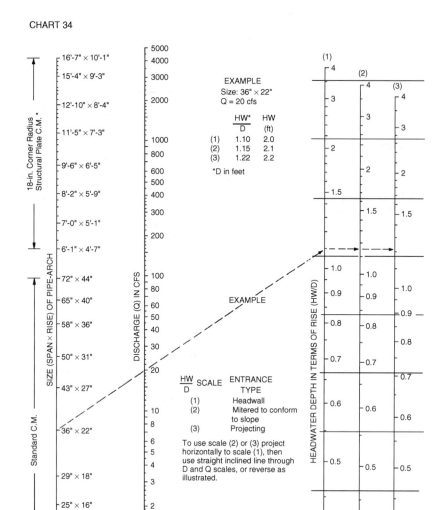

HEADWATER DEPTH FOR
C.M. PIPE-ARCH CULVERTS
WITH INLET CONTROL

* Additional sizes not dimensioned are listed in fabricator's catalog.

BUREAU OF PUBLIC ROADS JAN. 1963

Design of Culverts

CHART 35

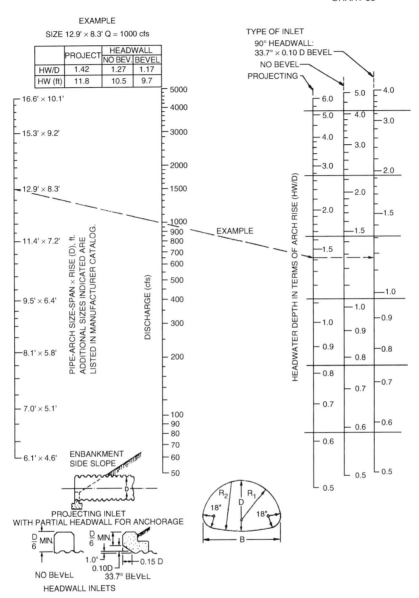

HEADWATER DEPTH FOR INLET CONTROLS
STRUCTURAL PLATE PIPE-ARCH CULVERTS
18 in. RADIUS CORNER PLATE
PROJECTING OR HEADWALL INLET
HEADWALL WITH OR WITHOUT EDGE BEVEL

BUREAU OF PUBLIC ROADS
OFFICE OF R&D JULY 1968

430

CHART 36

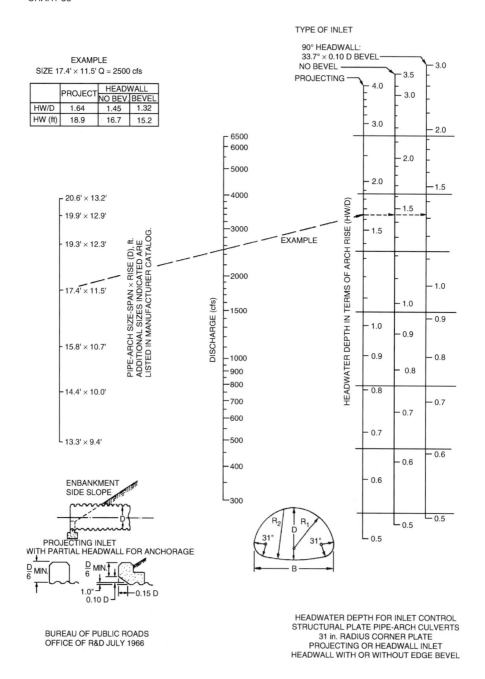

HEADWATER DEPTH FOR INLET CONTROL
STRUCTURAL PLATE PIPE-ARCH CULVERTS
31 in. RADIUS CORNER PLATE
PROJECTING OR HEADWALL INLET
HEADWALL WITH OR WITHOUT EDGE BEVEL

Design of Culverts

CHART 37

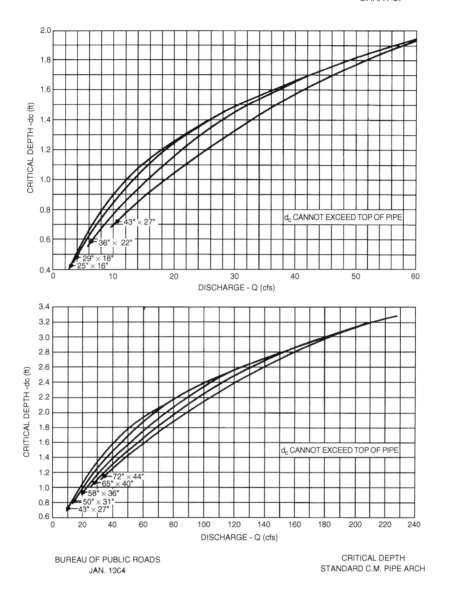

BUREAU OF PUBLIC ROADS
JAN. 1964

CRITICAL DEPTH
STANDARD C.M. PIPE ARCH

CHART 38

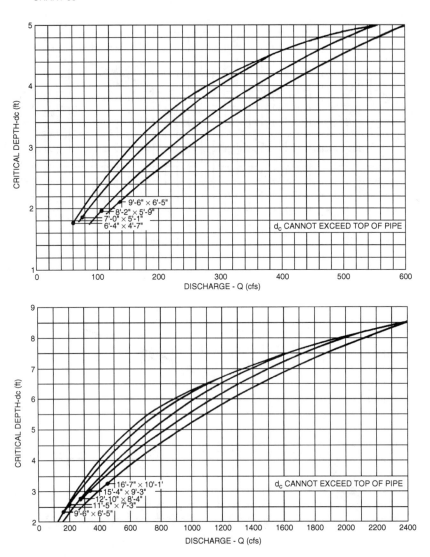

BUREAU OF PUBLIC ROADS
JAN. 1964

CRITICAL DEPTH
STRUCTURAL PLATE
C.M. PIPE ARCH
18 in. CORNER RADIUS

Design of Culverts

CHART 39

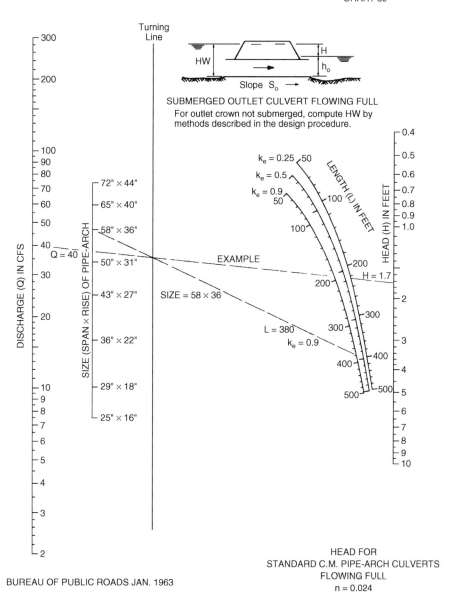

HEAD FOR
STANDARD C.M. PIPE-ARCH CULVERTS
FLOWING FULL
n = 0.024

BUREAU OF PUBLIC ROADS JAN. 1963

434

CHART 40

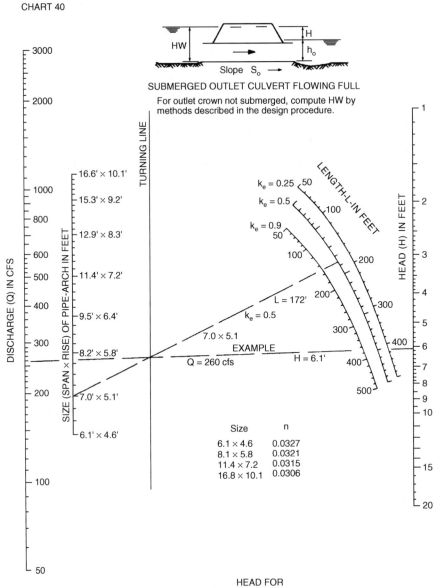

HEAD FOR
STRUCTURAL PLATE
C.M. PIPE ARCH CULVERTS
18 in. CORNER RADIUS
FLOWING FULL
n = 0.0327 TO 0.0306

BUREAU OF PUBLIC ROADS JAN. 1963

Design of Culverts

CHART 41

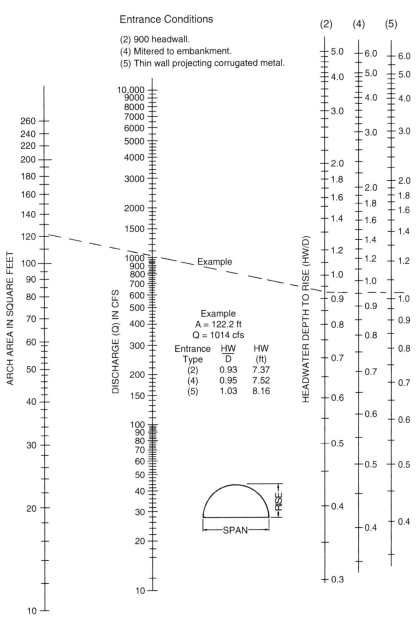

HEADWATER DEPTH
FOR C.M. ARCH CULVERTS
$0.3 \leq$ RISE/SPAN < 0.4
WITH INLET CONTROL

Nomographs adapted from material furnished by Kaiser Aluminum and Chemical Corporation. Duplication of this nomograph may distort scale.

CHART 42

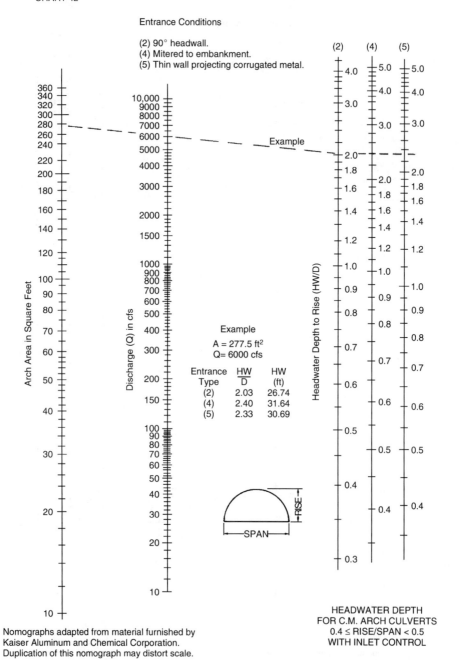

HEADWATER DEPTH
FOR C.M. ARCH CULVERTS
0.4 ≤ RISE/SPAN < 0.5
WITH INLET CONTROL

Nomographs adapted from material furnished by
Kaiser Aluminum and Chemical Corporation.
Duplication of this nomograph may distort scale.

Design of Culverts

CHART 43

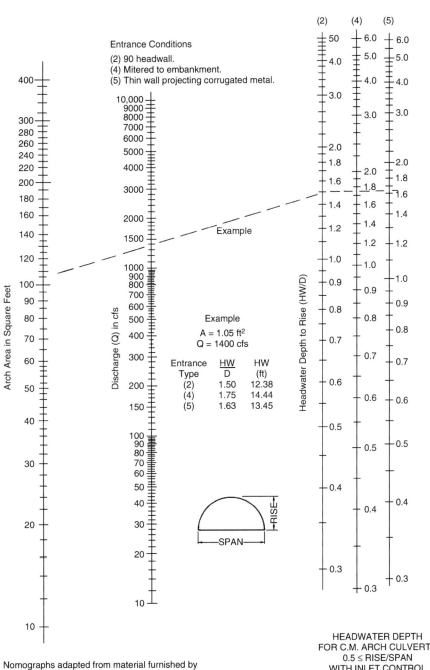

HEADWATER DEPTH
FOR C.M. ARCH CULVERTS
0.5 ≤ RISE/SPAN
WITH INLET CONTROL

438 *Municipal Stormwater Management, Second Edition*

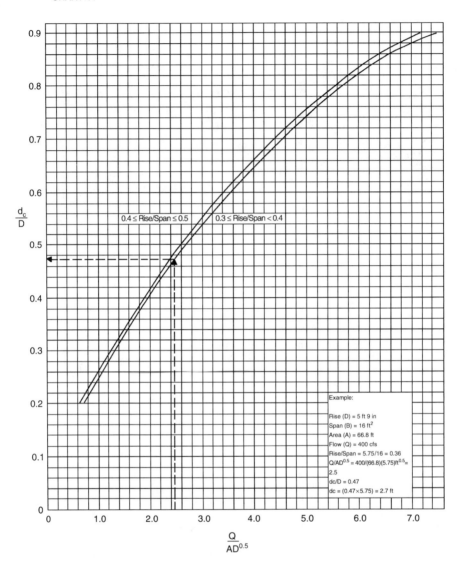

CHART 44

DIMENSIONLESS CRITICAL DEPTH CHART
FOR C.M. ARCH CULVERTS

Design of Culverts

CHART 45

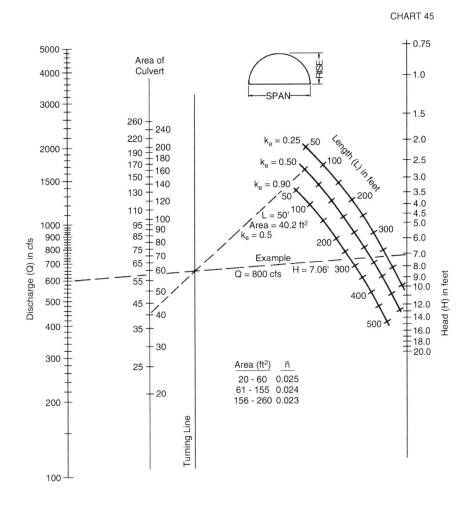

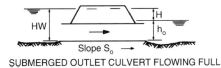

SUBMERGED OUTLET CULVERT FLOWING FULL

Nomographs adapted from material furnished by
Kaiser Aluminum and Chemical Corporation.
Duplication of this nomograph may distort scale.

HEAD FOR
C.M. ARCH CULVERTS
FLOWING FULL
CONCRETE BOTTOM
$0.3 \leq \text{RISE/SPAN} < 0.4$

CHART 46

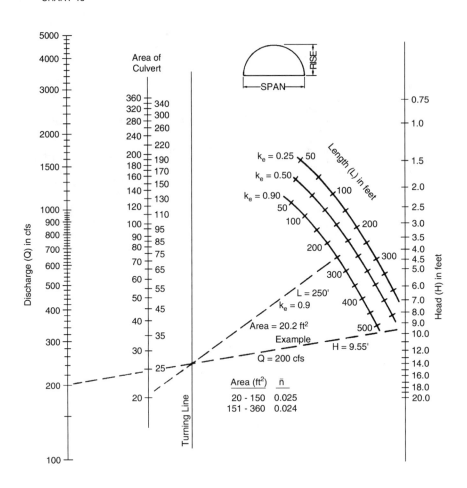

HEAD FOR
C.M. ARCH CULVERTS
FLOWING FULL
CONCRETE BOTTOM
$0.4 \leq RISE/SPAN < 0.5$

Nomographs adapted from material furnished by
Kaiser Aluminum and Chemical Corporation.
Duplication of this nomograph may distort scale.

Design of Culverts

CHART 47

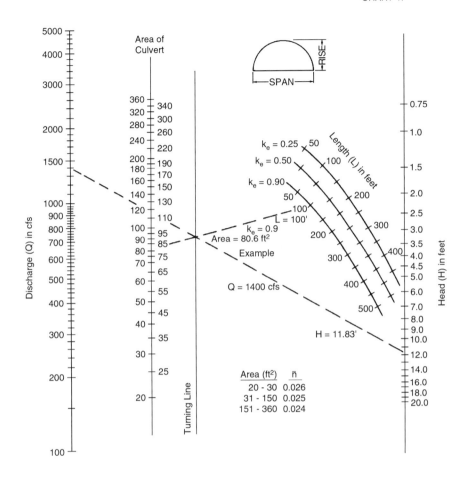

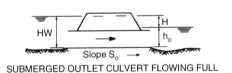

SUBMERGED OUTLET CULVERT FLOWING FULL

HEAD FOR
C.M. ARCH CULVERTS
FLOWING FULL
CONCRETE BOTTOM
$0.5 \leq$ RISE/SPAN

Nomographs adapted from material furnished by
Kaiser Aluminum and Chemical Corporation.
Duplication of this nomograph may distort scale.

CHART 48

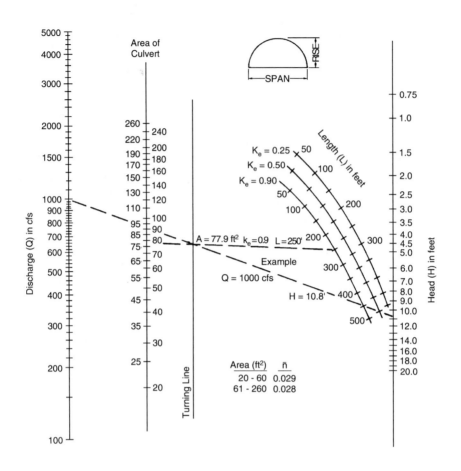

SUBMERGED OUTLET CULVERT FLOWING FULL

HEAD FOR
C.M. ARCH CULVERTS
FLOWING FULL
EARTH BOTTOM ($n_b = 0.022$)
$0.3 \leq$ RISE/SPAN < 0.4

Nomographs adapted from material furnished by
Kaiser Aluminum and Chemical Corporation.
Duplication of this nomograph may distort scale.

Design of Culverts

CHART 49

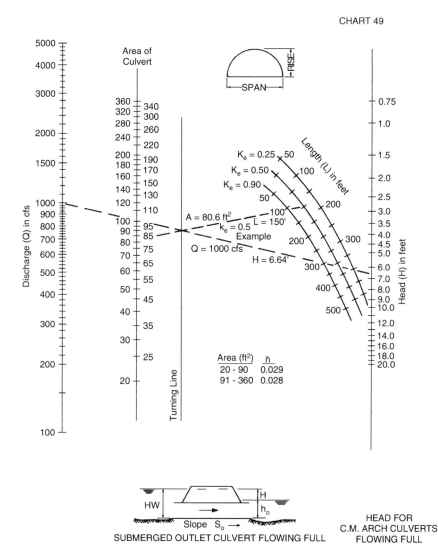

HEAD FOR
C.M. ARCH CULVERTS
FLOWING FULL
EARTH BOTTOM ($n_b = 0.022$)
$0.4 \leq \text{RISE/SPAN} < 0.5$

Nomographs adapted from material furnished by
Kaiser Aluminum and Chemical Corporation.
Duplication of this nomograph may distort scale.

444	Municipal Stormwater Management, Second Edition

CHART 50

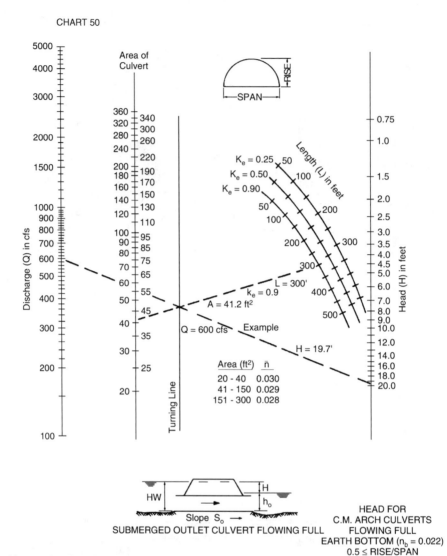

HEAD FOR
C.M. ARCH CULVERTS
FLOWING FULL
EARTH BOTTOM ($n_b = 0.022$)
$0.5 \leq$ RISE/SPAN

Nomographs adapted from material furnished by
Kaiser Aluminum and Chemical Corporation.
Duplication of this nomograph may distort scale.

Design of Culverts

CHART 51

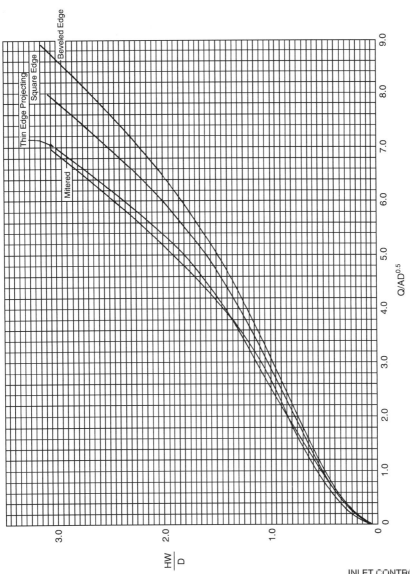

INLET CONTROL
HEADWATER DEPTH FOR
CIRCULAR OR ELLIPTICAL
STRUCTURAL PLATE
C.M. CONDUITS

CHART 52

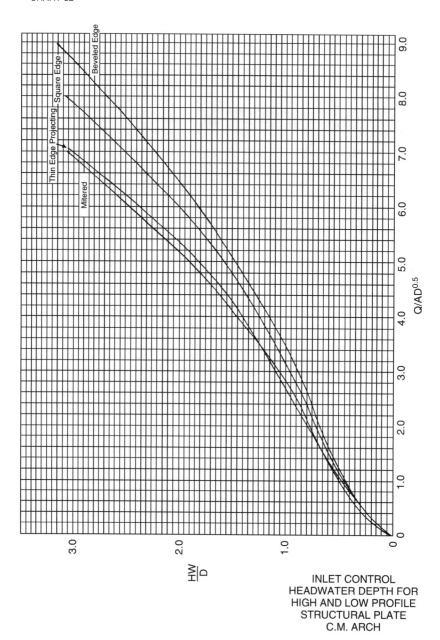

INLET CONTROL
HEADWATER DEPTH FOR
HIGH AND LOW PROFILE
STRUCTURAL PLATE
C.M. ARCH

Design of Culverts

CHART 53

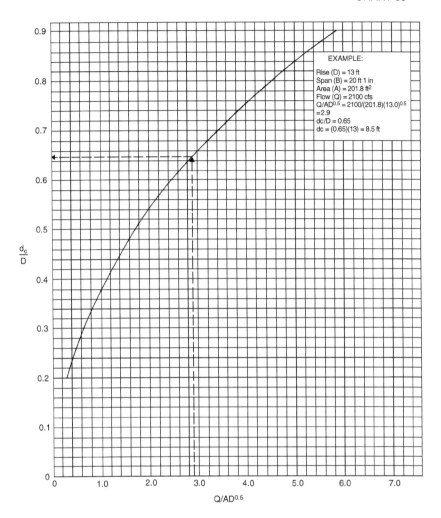

DIMENSIONLESS CRITICAL DEPTH CHART
FOR STRUCTURAL PLATE
ELLIPSE LONG AXIS HORIZONTAL

CHART 54

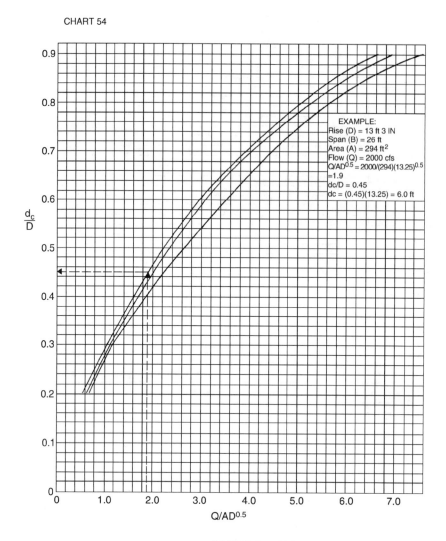

DIMENSIONLESS CRITICAL DEPTH CHART
FOR STRUCTURAL PLATE
LOW-AND HIGH-PROFILE ARCHES

Design of Culverts

CHART 55

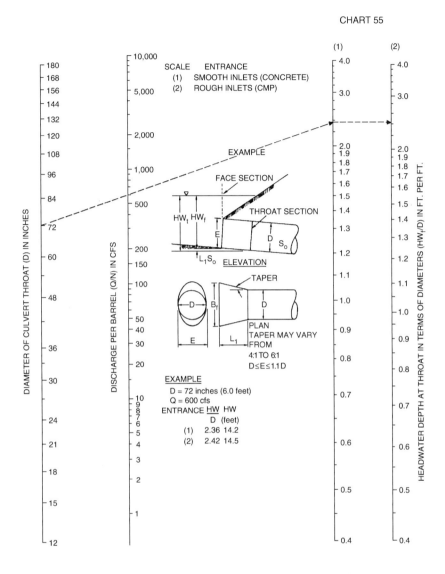

THROAT CONTROL
FOR SIDE-TAPERED INLETS
TO PIPE CULVERT
(CIRCULAR SECTION ONLY)

CHART 56

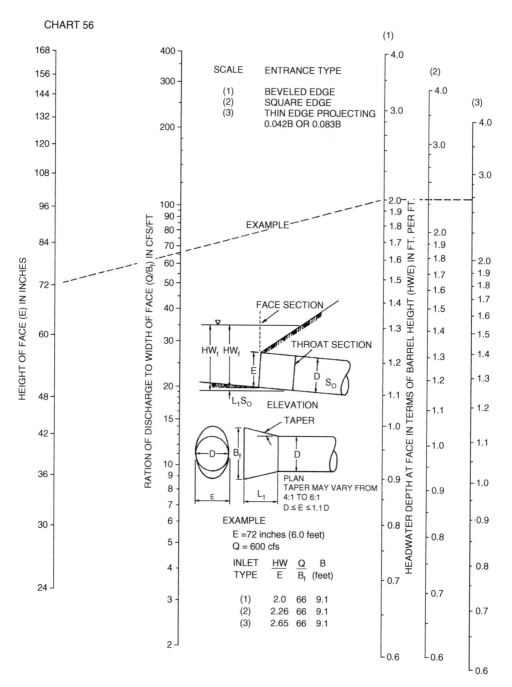

FACE CONTROL FOR SIDE-TAPERED
INLETS TO PIPE CULVERTS
(NON-RECTANGULAR SECTIONS ONLY)

Design of Culverts

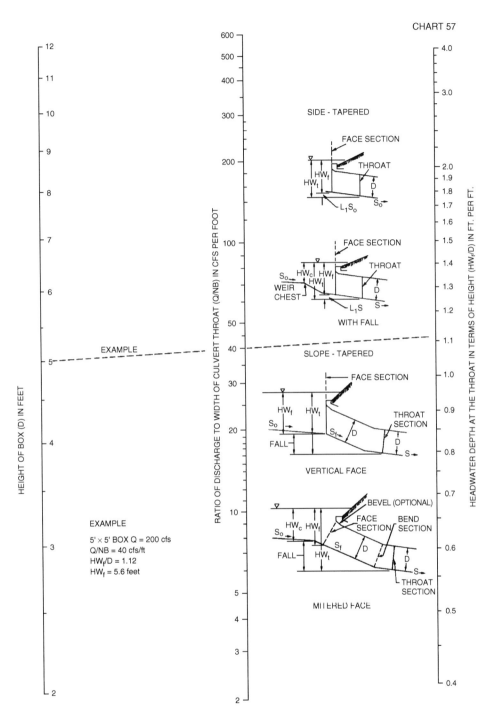

CHART 57

THROAT CONTROL FOR
BOX CULVERTS WITH
TAPERED INLETS

CHART 58

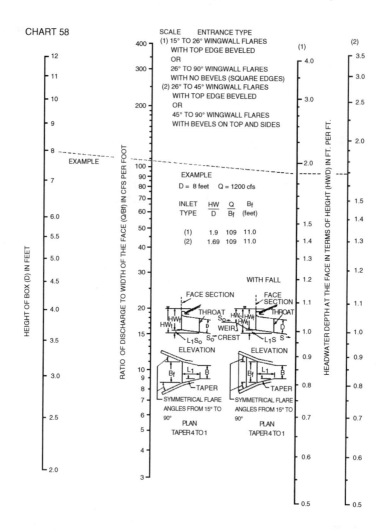

FACE CONTROL FOR BOX CULVERTS
WITH SIDE-TAPERED INLETS

Design of Culverts

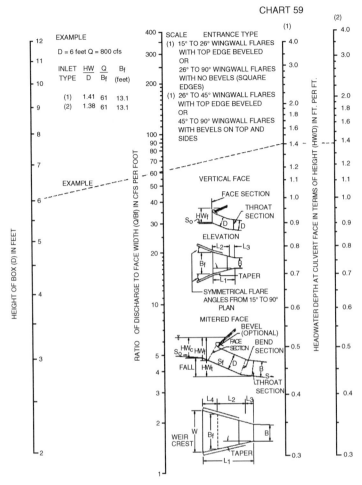

CHART 59

FACE CONTROL FOR BOX CULVERTS
WITH SLOPE TAPERED INLETS

CHART 60

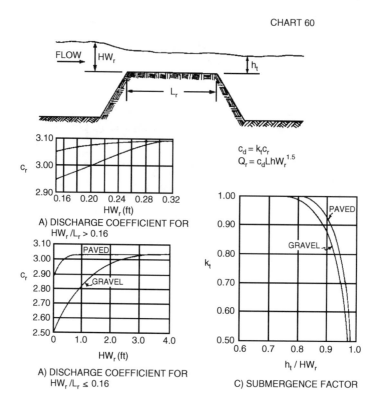

DISCHARGE COEFFICIENTS FOR
ROADWAY OVERTOPPING

Appendix B

Nomographs and Charts — Metric Units*

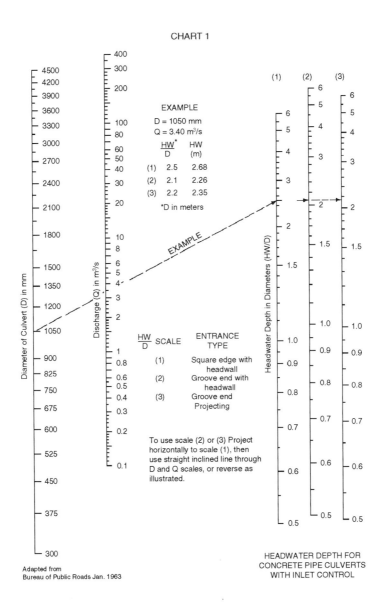

CHART 1

HEADWATER DEPTH FOR CONCRETE PIPE CULVERTS WITH INLET CONTROL

Adapted from Bureau of Public Roads Jan. 1963

---

* *Source:* American Association of State Highway and Transportation Officials, Model Drainage Manual, 1992, 1998.

CHART 2

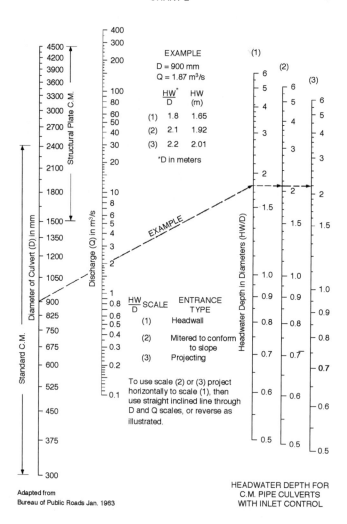

HEADWATER DEPTH FOR
C.M. PIPE CULVERTS
WITH INLET CONTROL

Adapted from
Bureau of Public Roads Jan. 1963

Design of Culverts

CHART 3

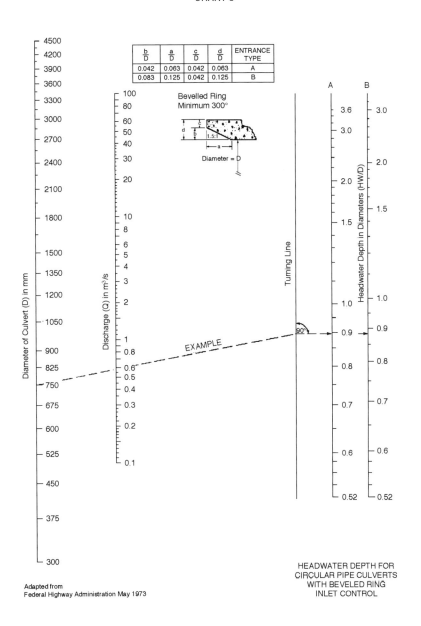

HEADWATER DEPTH FOR
CIRCULAR PIPE CULVERTS
WITH BEVELED RING
INLET CONTROL

Adapted from
Federal Highway Administration May 1973

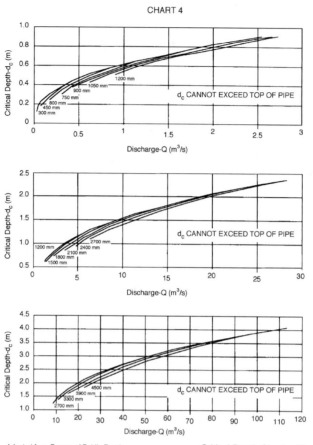

Adapted from Bureau of Public Roads

Critical Depth-Circular Pipe

Design of Culverts

CHART 5

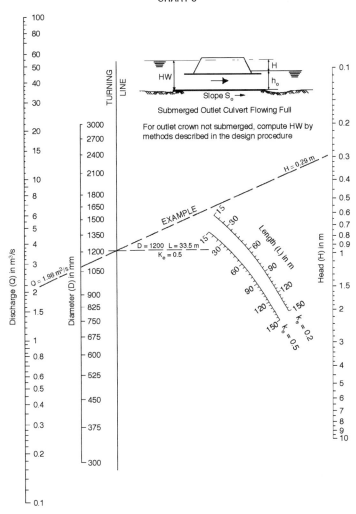

HEAD FOR
CONCRETE PIPE CULVERTS
FLOWING FULL
n = 0.012

Adapted from
Bureau of Public Roads Jan. 1963

460 Municipal Stormwater Management, Second Edition

CHART 6

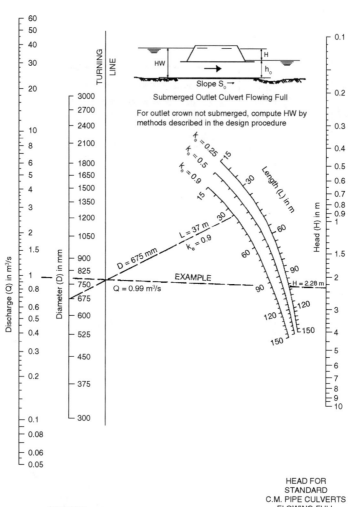

HEAD FOR
STANDARD
C.M. PIPE CULVERTS
FLOWING FULL
n = 0.024

Adapted from
Bureau of Public Roads Jan. 1963

Design of Culverts

CHART 7

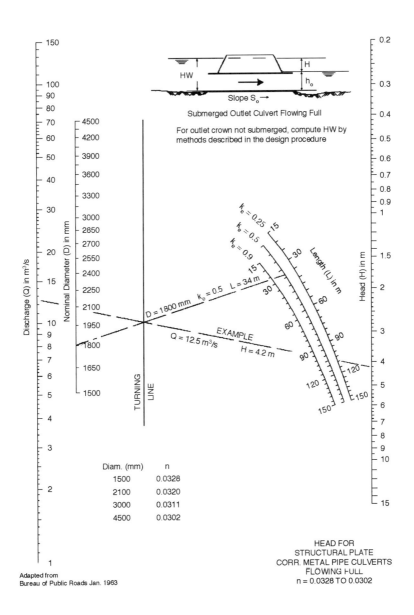

HEAD FOR
STRUCTURAL PLATE
CORR. METAL PIPE CULVERTS
FLOWING FULL
n = 0.0328 TO 0.0302

Adapted from
Bureau of Public Roads Jan. 1963

462

CHART 8

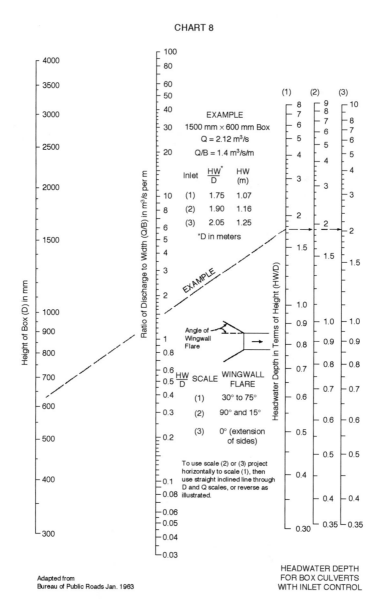

HEADWATER DEPTH
FOR BOX CULVERTS
WITH INLET CONTROL

Adapted from
Bureau of Public Roads Jan. 1963

Design of Culverts

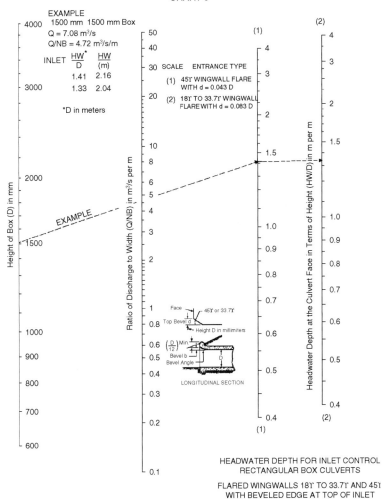

CHART 9

HEADWATER DEPTH FOR INLET CONTROL
RECTANGULAR BOX CULVERTS
FLARED WINGWALLS 18° TO 33.7° AND 45°
WITH BEVELED EDGE AT TOP OF INLET

464 Municipal Stormwater Management, Second Edition

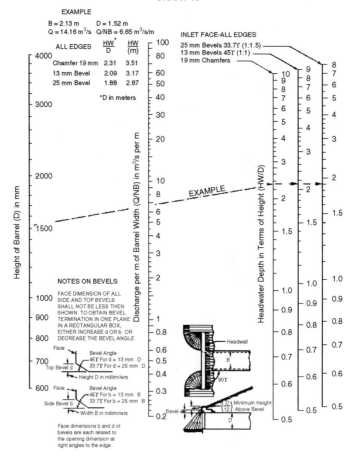

CHART 10

HEADWATER DEPTH FOR INLET CONTROL
RECTANGULAR BOX CULVERTS
90° HEADWALL
CHAMFERED OR BEVELED INLET EDGES

Design of Culverts

CHART 11

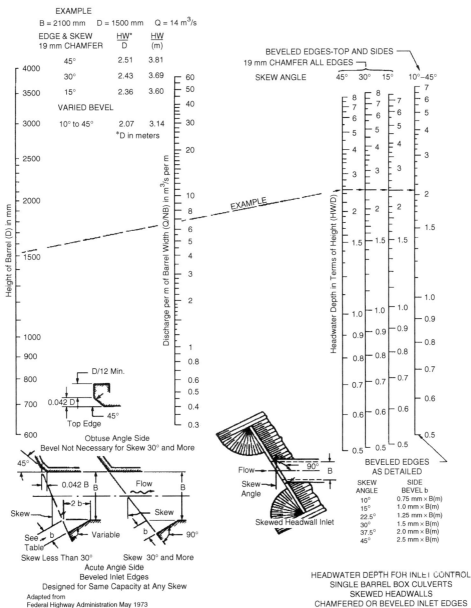

HEADWATER DEPTH FOR INLET CONTROL
SINGLE BARREL BOX CULVERTS
SKEWED HEADWALLS
CHAMFERED OR BEVELED INLET EDGES

466

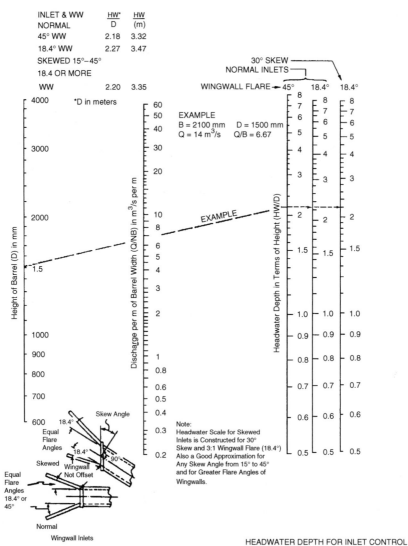

CHART 12

HEADWATER DEPTH FOR INLET CONTROL
RECTANGULAR BOX CULVERTS
FLARED WINGWALLS
NORMAL AND SKEWED INLETS
19 mm CHAMFERED AT TOP OF OPENING

Design of Culverts

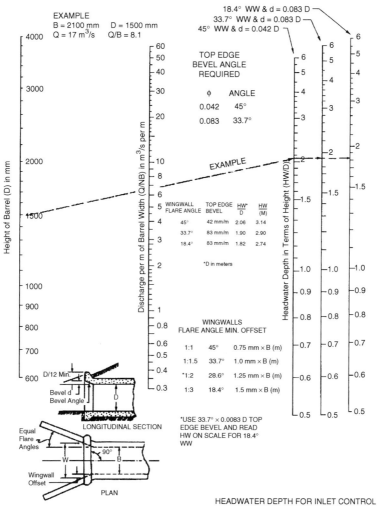

CHART 13

HEADWATER DEPTH FOR INLET CONTROL
RECTANGULAR BOX CULVERTS
OFFSET FLARED WINGWALLS
AND BEVELED EDGE AT TOP OF INLET

CHART 14

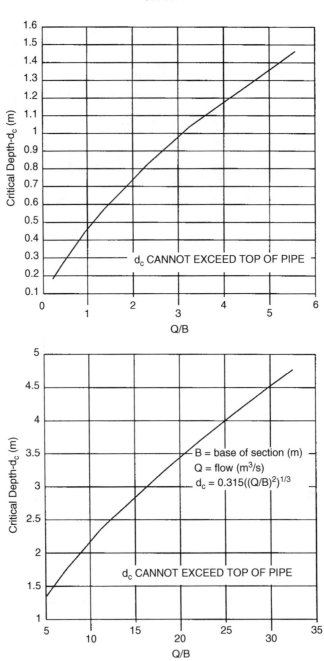

Adapted from Bureau of Public Roads

Critical Depth-Rectangular Section

Design of Culverts

CHART 15

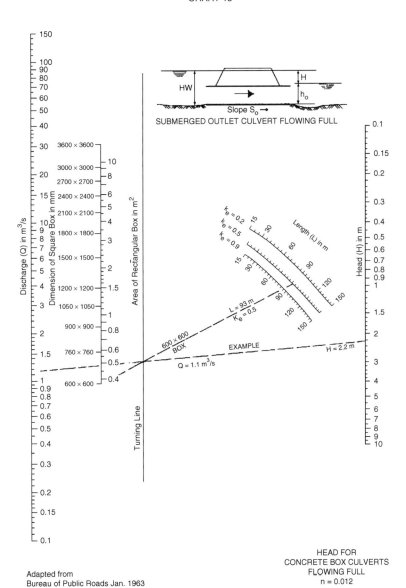

HEAD FOR
CONCRETE BOX CULVERTS
FLOWING FULL
n = 0.012

Adapted from
Bureau of Public Roads Jan. 1963

CHART 16

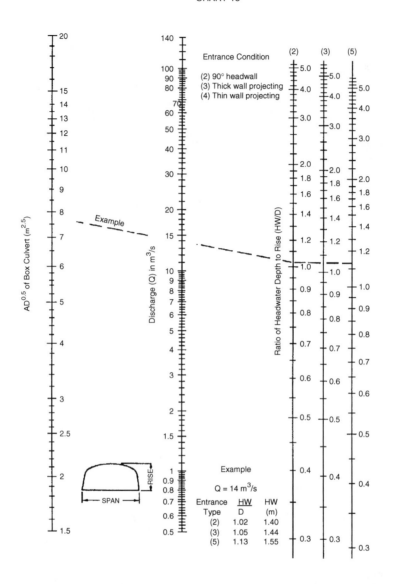

Adapted from
Kaiser Aluminium and Chemical Corporation

HEADWATER DEPTH
FOR C.M. BOX CULVERTS
RISE/SPAN < 0.3
WITH INLET CONTROL

Design of Culverts

CHART 17

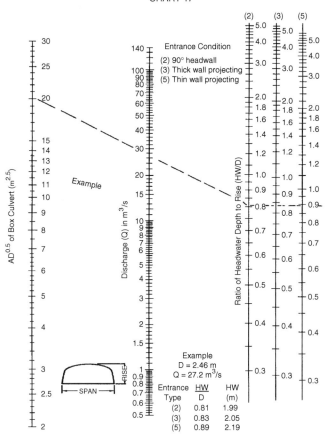

HEADWATER DEPTH
FOR C.M. BOX CULVERTS
$0.3 \leq \text{RISE/SPAN} < 0.4$
WITH INLET CONTROL

Adapted from
Kaiser Aluminium and Chemical Corporation

CHART 18

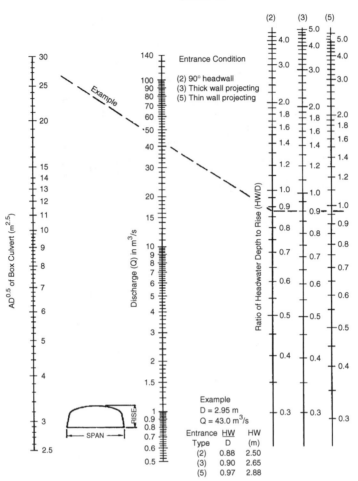

HEADWATER DEPTH
FOR C.M. BOX CULVERTS
$0.4 \leq$ RISE/SPAN < 0.5
WITH INLET CONTROL

Adapted from
Kaiser Aluminium and Chemical Corporation

Design of Culverts

CHART 19

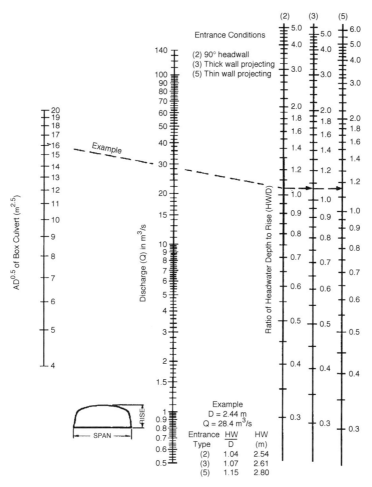

HEADWATER DEPTH
FOR C.M. BOX CULVERTS
0.5 ≤ RISE/SPAN
WITH INLET CONTROL

Adapted from
Kaiser Aluminium and Chemical Corporation

CHART 20

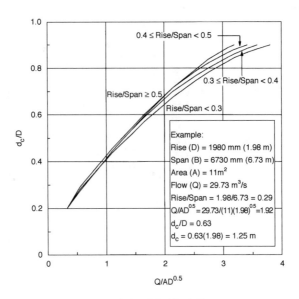

Dimensionless Critical Depth Chart
for Corrugated Metal Box Culverts

Design of Culverts

CHART 21

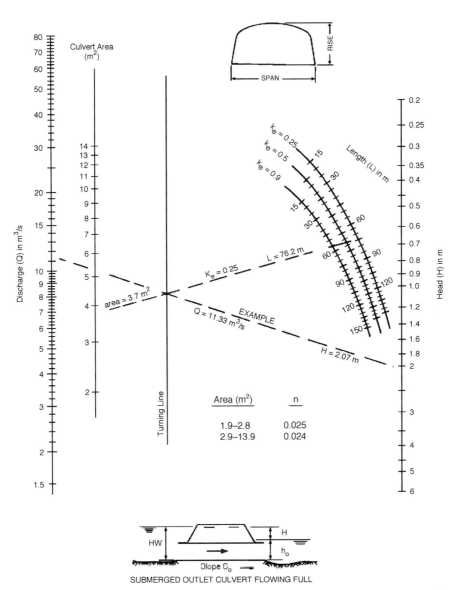

HEAD FOR
C.M. BOX CULVERTS
FLOWING FULL
CONCRETE BOTTOM
RISE/SPAN < 0.3

Adapted from
Kaiser Aluminium and Chemical Corporation

476 Municipal Stormwater Management, Second Edition

CHART 22

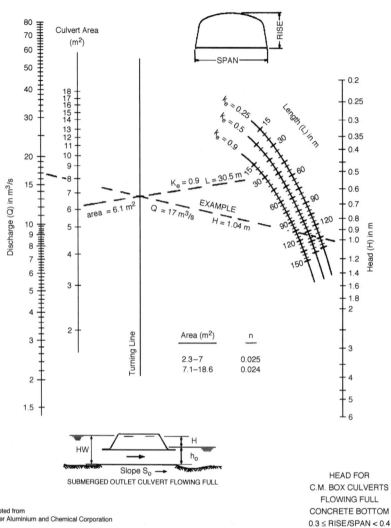

HEAD FOR
C.M. BOX CULVERTS
FLOWING FULL
CONCRETE BOTTOM
$0.3 \leq \text{RISE/SPAN} < 0.4$

Adapted from
Kaiser Aluminium and Chemical Corporation

Design of Culverts

CHART 23

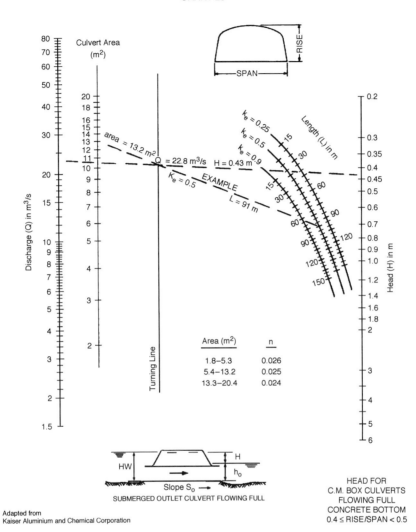

HEAD FOR
C.M. BOX CULVERTS
FLOWING FULL
CONCRETE BOTTOM
$0.4 \leq$ RISE/SPAN < 0.5

Adapted from
Kaiser Aluminium and Chemical Corporation

478

CHART 24

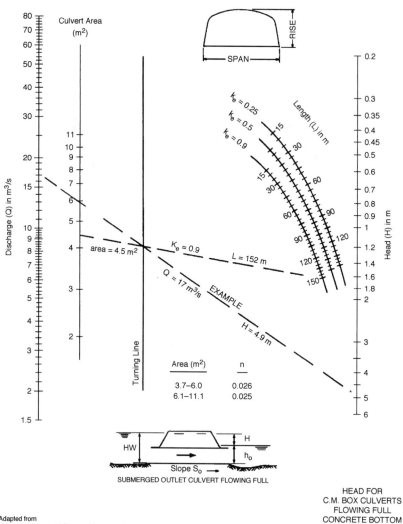

SUBMERGED OUTLET CULVERT FLOWING FULL

Adapted from
Kaiser Aluminium and Chemical Corporation

HEAD FOR
C.M. BOX CULVERTS
FLOWING FULL
CONCRETE BOTTOM
0.5 ≤ RISE/SPAN

CHART 25

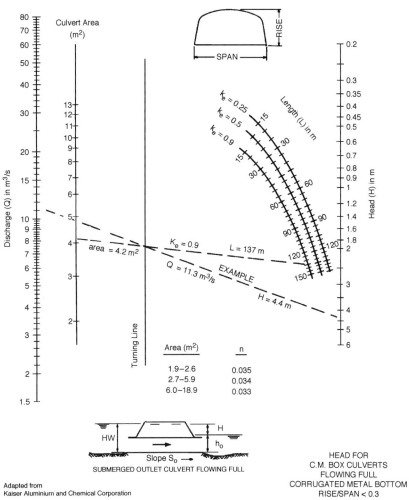

HEAD FOR
C.M. BOX CULVERTS
FLOWING FULL
CORRUGATED METAL BOTTOM
RISE/SPAN < 0.3

Adapted from
Kaiser Aluminium and Chemical Corporation

480 Municipal Stormwater Management, Second Edition

CHART 26

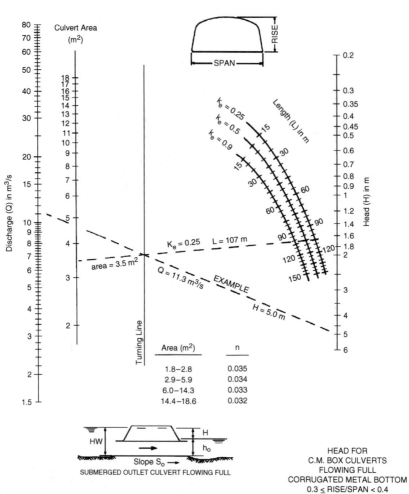

HEAD FOR
C.M. BOX CULVERTS
FLOWING FULL
CORRUGATED METAL BOTTOM
$0.3 \leq \text{RISE/SPAN} < 0.4$

Adapted from
Kaiser Aluminium and Chemical Corporation

Design of Culverts

CHART 27

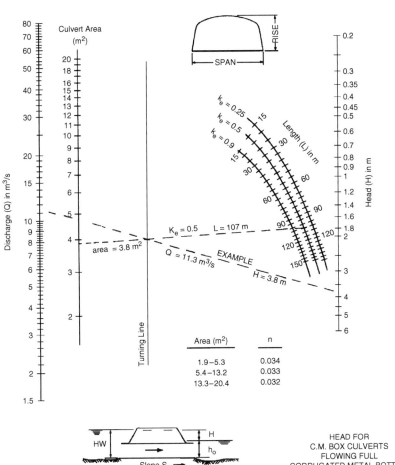

HEAD FOR
C.M. BOX CULVERTS
FLOWING FULL
CORRUGATED METAL BOTTOM
$0.4 \leq$ RISE/SPAN < 0.5

Area (m²)	n
1.9–5.3	0.034
5.4–13.2	0.033
13.3–20.4	0.032

Adapted from
Kaiser Aluminium and Chemical Corporation

CHART 28

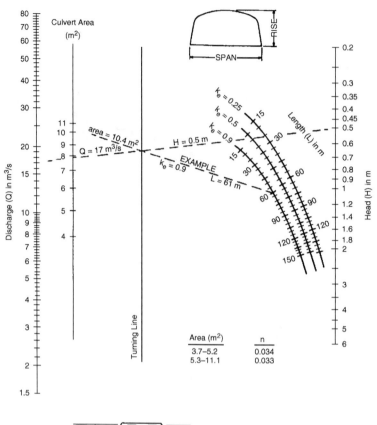

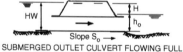
SUBMERGED OUTLET CULVERT FLOWING FULL

Adapted from
Kaiser Aluminum and Chemical Corporation

HEAD FOR
C.M. BOX CULVERTS
FLOWING FULL
CORRUGATED METAL BOTTOM
0.5 ≤ RISE/SPAN

Design of Culverts

CHART 29

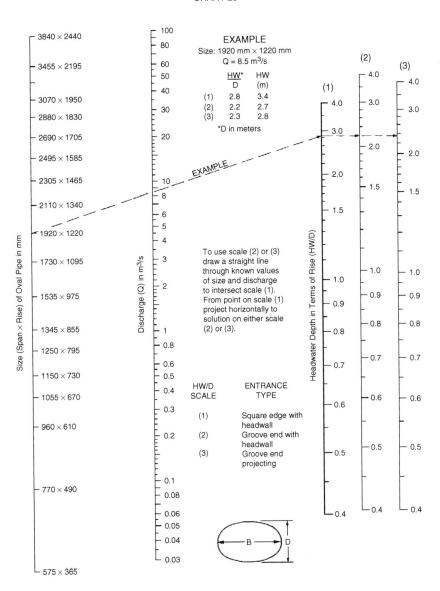

Adapted from
Bureau of Public Roads Jan. 1963

HEADWATER DEPTH FOR
OVAL CONCRETE PIPE CULVERTS
LONG AXIS HORIZONTAL
WITH INLET CONTROL

CHART 30

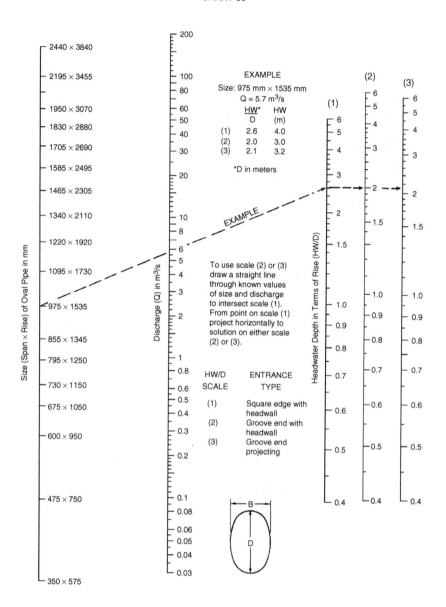

HEADWATER DEPTH FOR
OVAL CONCRETE PIPE CULVERTS
LONG AXIS VERTICAL
WITH INLET CONTROL

Adapted from
Bureau of Public Roads Jan. 1963

Design of Culverts

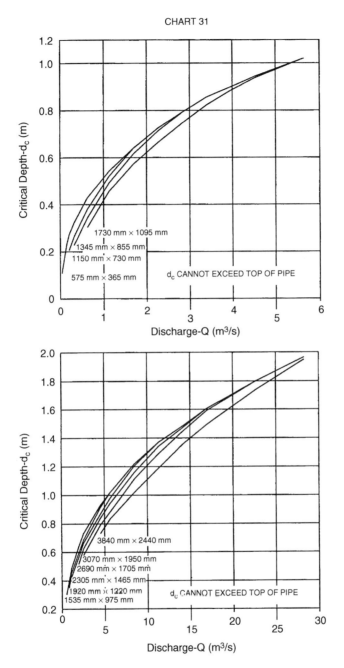

Adapted from Bureau of Public Roads

Critical Depth-Oval Concrete Pipe
Long Axis Horizontal

CHART 32

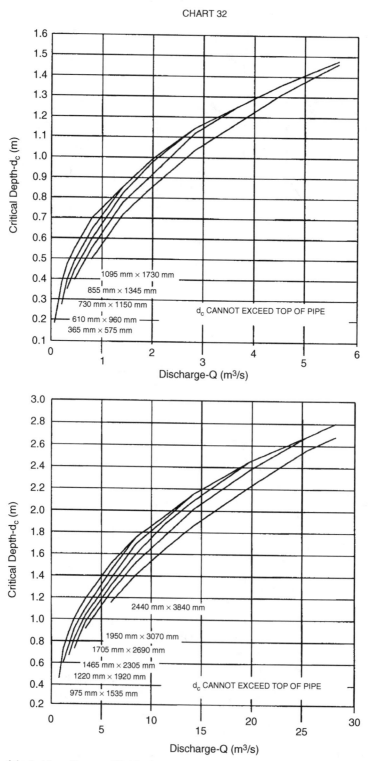

Adapted from Bureau of Public Roads

Critical Depth-Oval Concrete Pipe
Long Axis Vertical

Design of Culverts

CHART 33

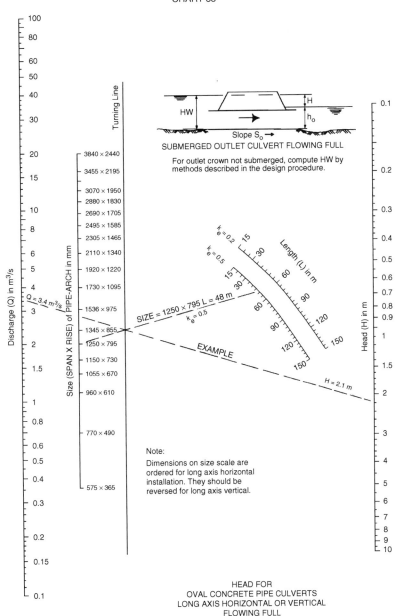

HEAD FOR
OVAL CONCRETE PIPE CULVERTS
LONG AXIS HORIZONTAL OR VERTICAL
FLOWING FULL
n = 0.012

Adapted from
Bureau of Public Roads Jan. 1963

488 Municipal Stormwater Management, Second Edition

CHART 34

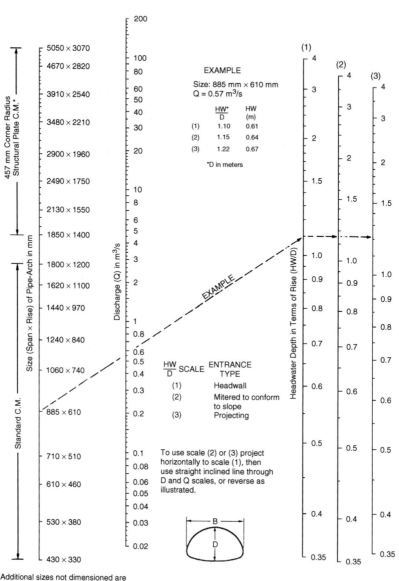

HEADWATER DEPTH FOR
C.M. PIPE-ARCH CULVERTS
WITH INLET CONTROL

Design of Culverts

CHART 35

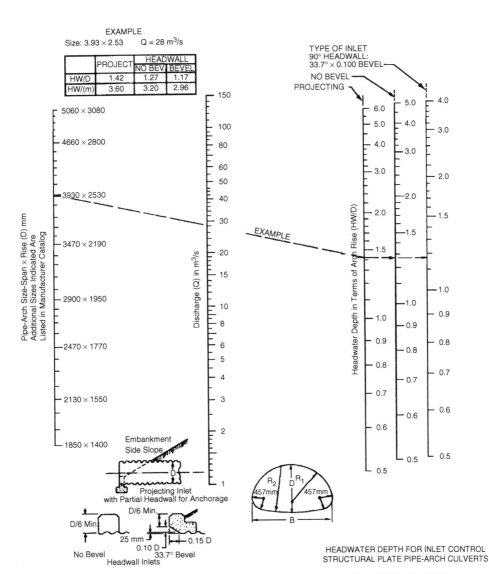

HEADWATER DEPTH FOR INLET CONTROL
STRUCTURAL PLATE PIPE-ARCH CULVERTS

457 mm RADIUS CORNER PLATE
PROJECTING OR HEADWALL INLET
HEADWALL WITH OR WITHOUT EDGE BEVEL

Adapted from
Bureau of Public Roads Office of R&D
July 1968

CHART 36

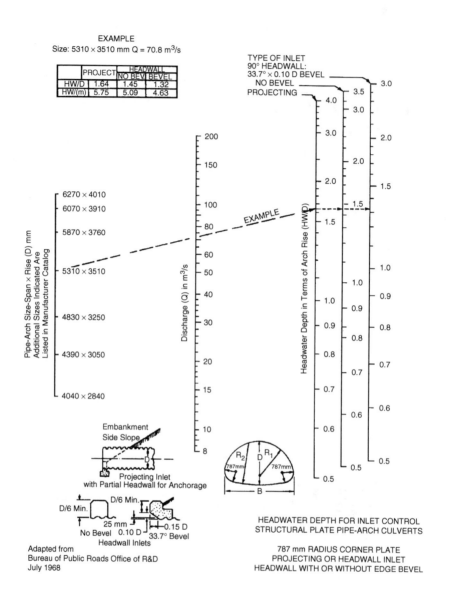

Design of Culverts

CHART 37

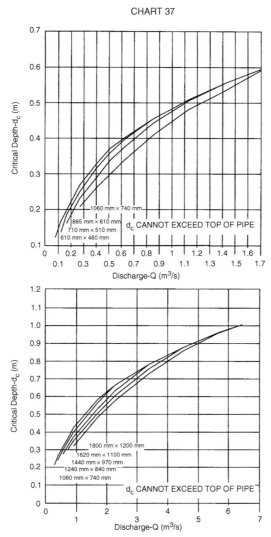

Adapted from Bureau of Public Roads

Critical Depth-Standard C.M. Pipe Arch

CHART 38

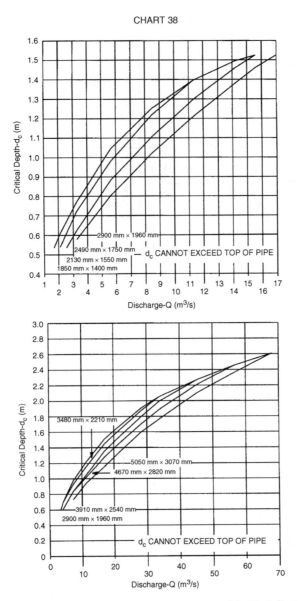

Adapted from Bureau of Public Roads

Critical Depth-Structural Plate C.M. Pipe
Arch, 457 mm Corner Radius

Design of Culverts

CHART 39

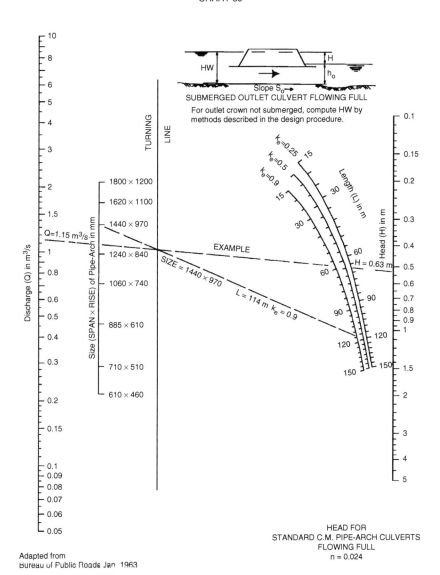

HEAD FOR
STANDARD C.M. PIPE-ARCH CULVERTS
FLOWING FULL
n = 0.024

Adapted from
Bureau of Public Roads Jan 1963

494 Municipal Stormwater Management, Second Edition

CHART 40

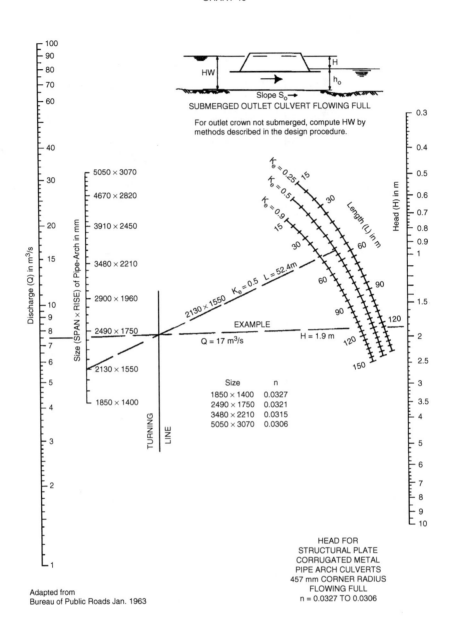

HEAD FOR
STRUCTURAL PLATE
CORRUGATED METAL
PIPE ARCH CULVERTS
457 mm CORNER RADIUS
FLOWING FULL
n = 0.0327 TO 0.0306

Adapted from
Bureau of Public Roads Jan. 1963

Design of Culverts

CHART 41

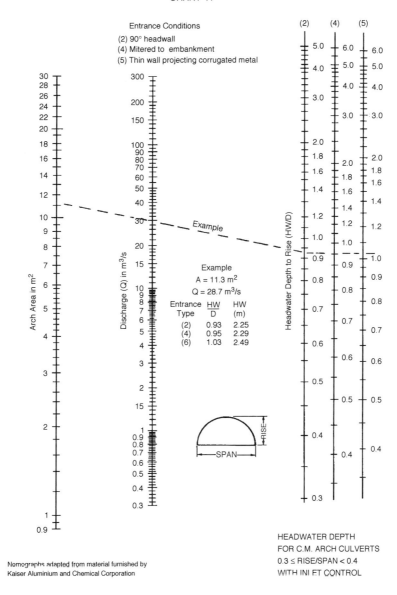

HEADWATER DEPTH
FOR C.M. ARCH CULVERTS
$0.3 \leq \text{RISE/SPAN} < 0.4$
WITH INLET CONTROL

Nomographs adapted from material furnished by
Kaiser Aluminium and Chemical Corporation

496 Municipal Stormwater Management, Second Edition

CHART 42

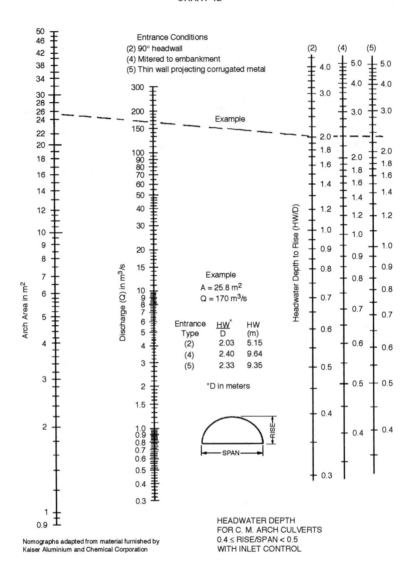

HEADWATER DEPTH
FOR C. M. ARCH CULVERTS
$0.4 \leq \text{RISE/SPAN} < 0.5$
WITH INLET CONTROL

Nomographs adapted from material furnished by
Kaiser Aluminium and Chemical Corporation

Design of Culverts

CHART 43

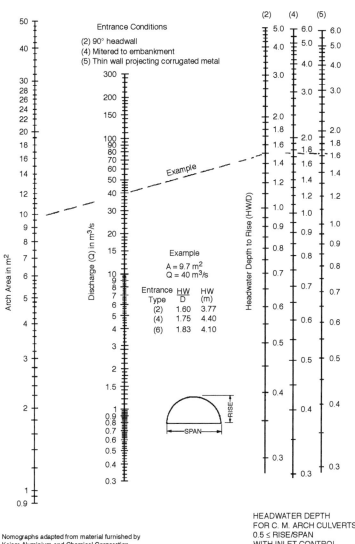

HEADWATER DEPTH
FOR C. M. ARCH CULVERTS
$0.5 \leq$ RISE/SPAN
WITH INLET CONTROL

Nomographs adapted from material furnished by
Kaiser Aluminium and Chemical Corporation

CHART 44

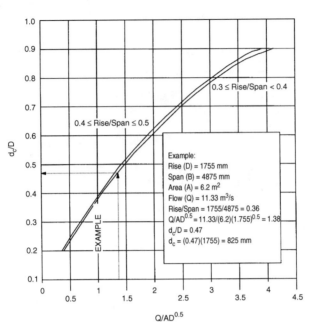

Dimensionless Critical Depth Chart
for Corrugated Metal Arch Culverts

Design of Culverts

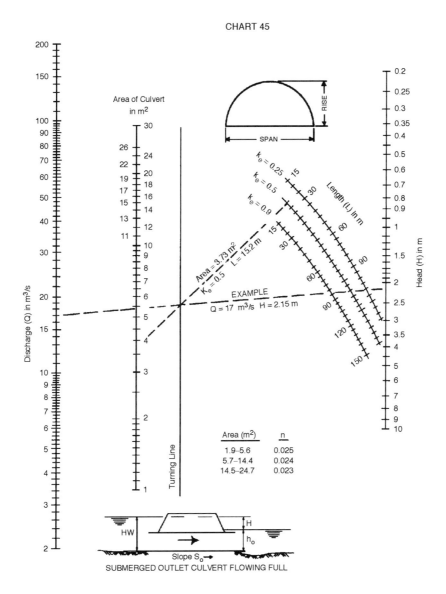

CHART 45

HEAD FOR
C. M. ARCH CULVERTS
FLOWING FULL
CONCRETE BOTTOM
$0.3 \leq RISE/SPAN < 0.4$

Adapted from material furnished by
Kaiser Aluminium and Chemical Corporation

CHART 46

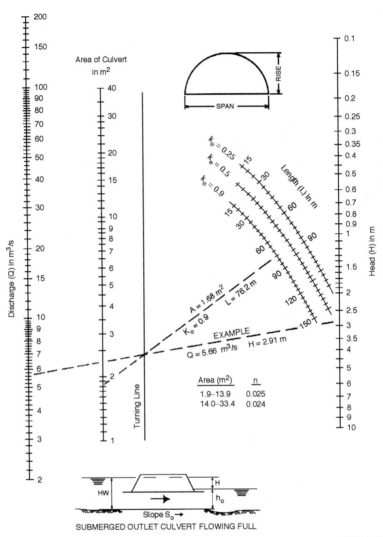

SUBMERGED OUTLET CULVERT FLOWING FULL

HEAD FOR
C. M. ARCH CULVERTS
FLOWING FULL
CONCRETE BOTTOM
0.4 ≤ RISE/SPAN < 0.5

Adapted from material furnished by
Kaiser Aluminium and Chemical Corporation

Design of Culverts

CHART 47

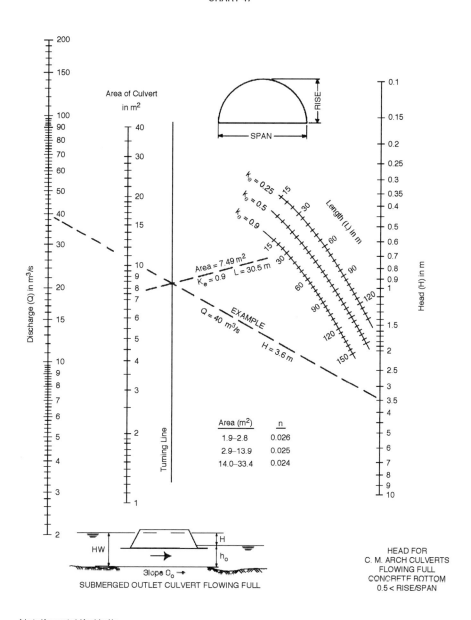

HEAD FOR
C. M. ARCH CULVERTS
FLOWING FULL
CONCRETE BOTTOM
0.5 < RISE/SPAN

Adapted from material furnished by
Kaiser Aluminium and Chemical Corporation

CHART 48

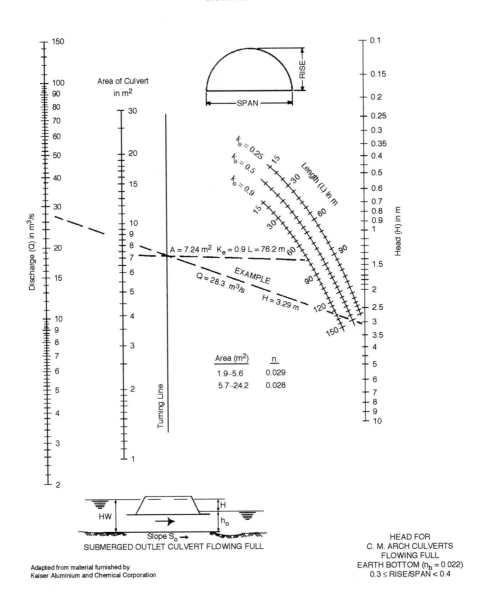

HEAD FOR
C. M. ARCH CULVERTS
FLOWING FULL
EARTH BOTTOM ($n_b = 0.022$)
$0.3 \leq$ RISE/SPAN < 0.4

Adapted from material furnished by
Kaiser Aluminium and Chemical Corporation

Design of Culverts

CHART 49

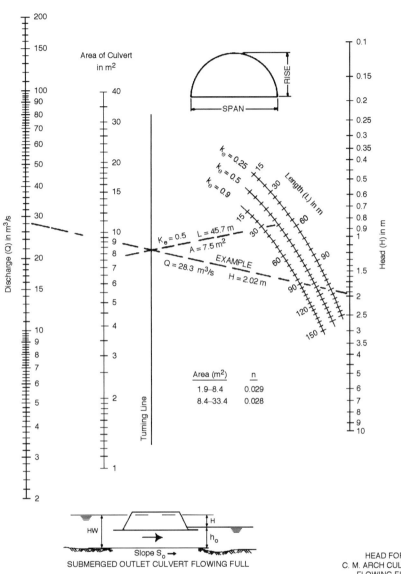

HEAD FOR
C. M. ARCH CULVERTS
FLOWING FULL
EARTH BOTTOM ($n_b = 0.022$)
$0.4 \leq \text{RISE/SPAN} < 0.5$

Adapted from material furnished by
Kaiser Aluminium and Chemical Corporation

504

CHART 50

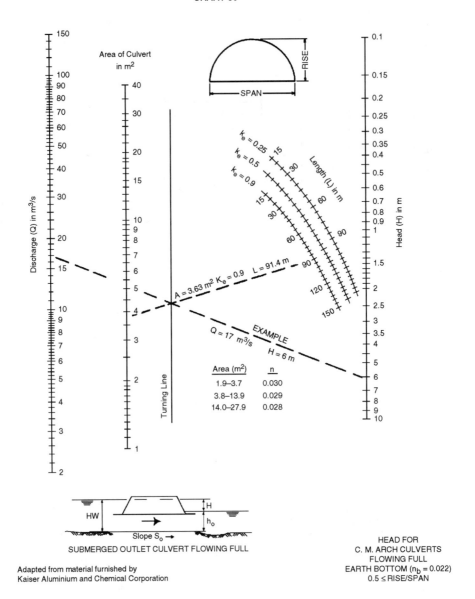

HEAD FOR
C. M. ARCH CULVERTS
FLOWING FULL
EARTH BOTTOM ($n_b = 0.022$)
$0.5 \leq$ RISE/SPAN

Adapted from material furnished by
Kaiser Aluminium and Chemical Corporation

Design of Culverts

CHART 51

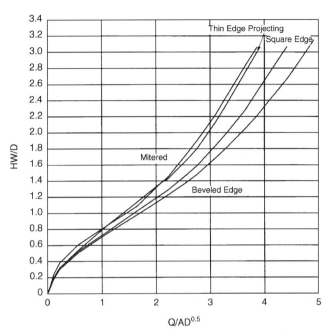

Inlet Control-Headwater Depth for Circular or Elliptical
Structural Plate Corrugated Metal Conduits

CHART 52

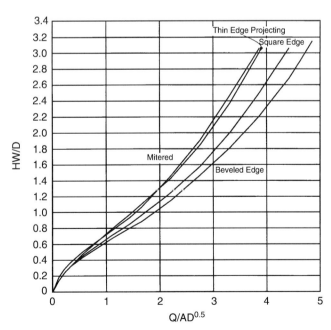

Inlet Control-Headwater Depth for High and Low Profile
Structural Plate Corrugated Metal Arch

CHART 53

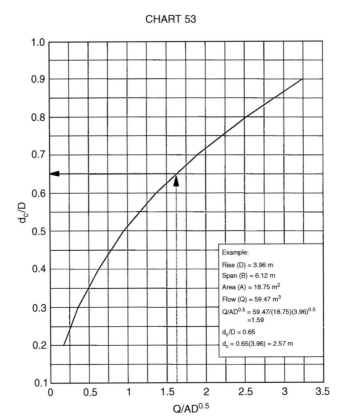

Dimensionless Critical Depth Chart for Structural Plate
Ellipse Long Axis Horizontal

Design of Culverts

CHART 54

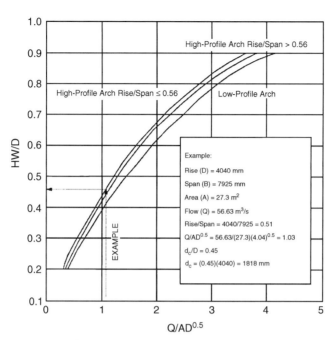

Dimensionless Critical Depth Chart for Structural Plate
Low- and High-Profile Arches

CHART 55

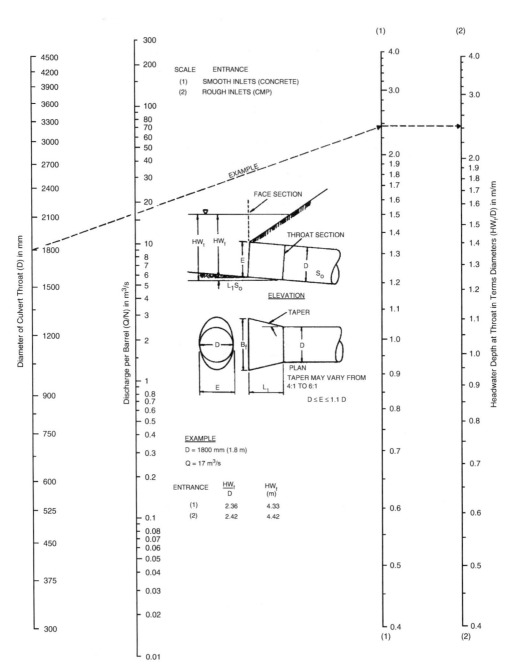

THROAT CONTROL
FOR SIDE-TAPERED INLETS TO PIPE CULVERT
(CIRCULAR SECTION ONLY)

Design of Culverts

CHART 56

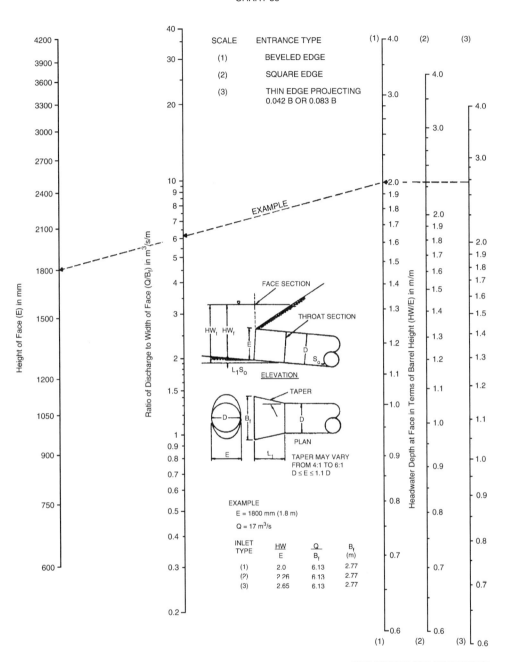

FACE CONTROL FOR SIDE-TAPERED
INLETS TO PIPE CULVERTS
(NON-RECTANGULAR SECTIONS ONLY)

CHART 57

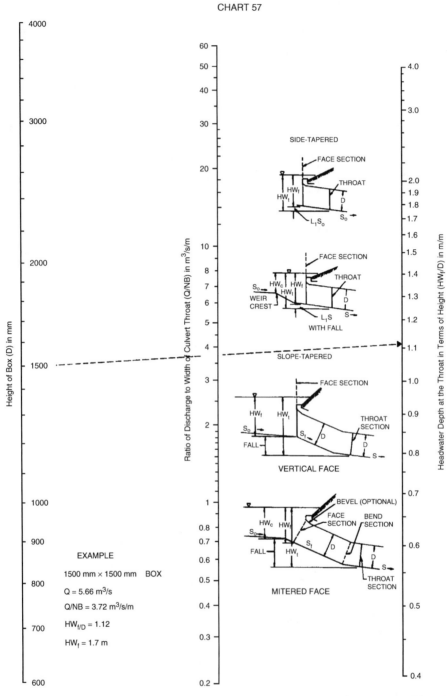

THROAT CONTROL FOR BOX CULVERTS WITH TAPERED INLETS

Design of Culverts

CHART 58

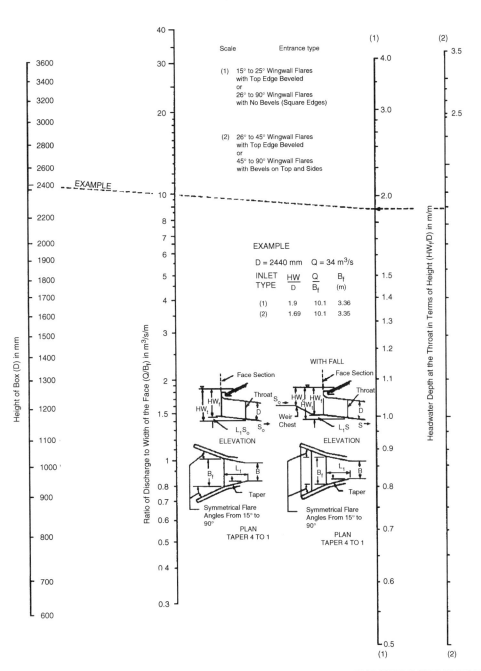

FACE CONTROL FOR BOX CULVERTS
WITH SIDE-TAPERED INLETS

CHART 59

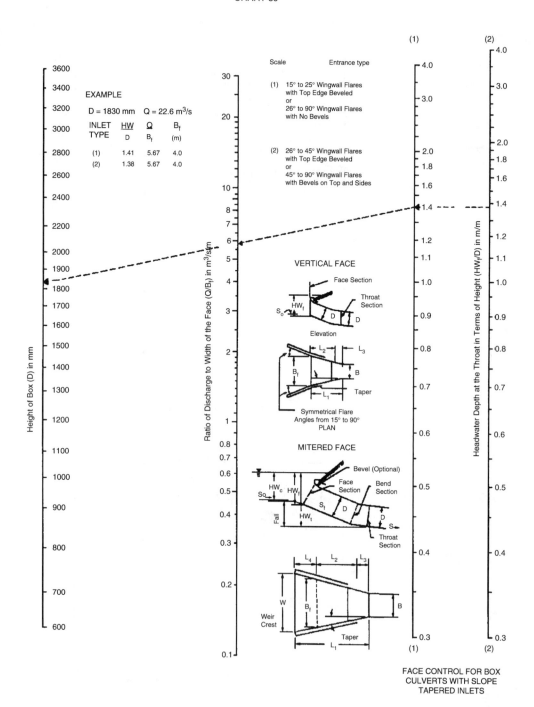

FACE CONTROL FOR BOX
CULVERTS WITH SLOPE
TAPERED INLETS

Design of Culverts

CHART 60

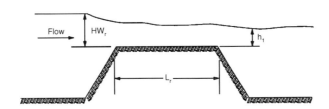

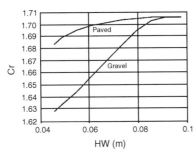

A) Discharge Coefficient for $HW_r/L_r > 0.15$

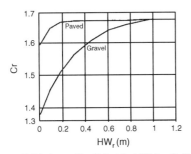

B) Discharge Coefficient for $HW_r/L_r \leq 0.15$

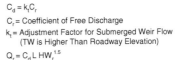

$C_d = k_t C_r$
C_r = Coefficient of Free Discharge
k_t = Adjustment Factor for Submerged Weir Flow
(TW is Higher Than Roadway Elevation)
$Q_r = C_d L \, HW_r^{1.5}$

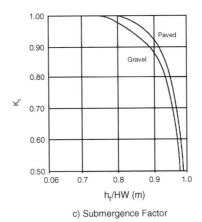

c) Submergence Factor

Discharge Coefficients for Roadway Overtopping

10

Open Channel Design

10.1 Introduction

A consideration of open channel hydraulics is an integral part of projects in which artificial channels and improvements to or analysis of natural channels are a primary concern. The design of channels in urban environments is complicated by a number of factors including changing hydrology due to watershed land use changes, near bank encroachments, numerous channel crossings and modifications, intermittent upstream construction with sediment surges, and difficult or unknown environmental considerations.

In many municipalities, urban stream corridors have become unattractive, abandoned fields full of weeds and trash. However, many other municipalities have made the decision to consider urban streams a resource rather than a nuisance and have taken steps to make their corridors attractive to wildlife and man. While it is nearly impossible to have a pristine and fully natural stream in an urban environment, it is possible to have an attractive and functional urban stream. Examples abound of greenways, bike paths, walking trails, boardwalk wetlands, multiobjective parks and recreational areas — even urban salmon spawning areas. These examples take advantage of the value in transit that stormwater runoff provides as runoff moves from one location to another location in the drainage system.

But the beginning of any channel design or modification is to understand the hydrology and hydraulics of the stream and watershed. This chapter emphasizes procedures for performing uniform flow calculations that aid in the selection or evaluation of appropriate channel linings, depths, and grades for natural or man-made channels. Allowable velocities are provided, along with procedures for evaluating channel capacity using Manning's equation. Additional sections discuss greenways, stable channel design, and other environmental considerations.

10.2 Design Criteria

Channel Types

The three main classifications of open-channel types according to channel linings are vegetative, flexible, and rigid. Vegetative linings include grass with mulch, sod, lapped sod, and a variety of bio-engineered systems. Rock riprap and some forms of flexible man-made linings or gabions are examples of flexible linings, while rigid linings are generally concrete or rigid block. Often combinations are used (e.g., toe stone with vegetated upper banks).

Vegetative Linings

Vegetation, where practical, is the most desirable lining for an artificial channel. It stabilizes the channel body, consolidates the soil mass of the bed, checks erosion on the channel surface, provides habitat, and controls the movement of soil particles along the channel bottom. Conditions under which vegetation may not be acceptable, however, include but are not limited to the following:

- Standing or continuous flowing water
- Lack of the regular maintenance necessary to prevent domination by taller or woody vegetation
- Lack of nutrients and inadequate topsoil
- Excessive shade
- High velocities

Proper seeding, mulching, and soil preparation are required during construction to assure establishment of a healthy growth of grass. Soil testing should be performed, and the results should be evaluated by an agronomist or local agricultural extension office to determine soil treatment requirements for pH, nitrogen, phosphorus, potassium, and other factors. In many cases, temporary erosion-control measures (such as jute or straw matting or spray tacking substances) are required to provide time for the seeding to establish a viable vegetative lining (UDFCD, 1991). Sodding should be staggered to avoid seams in the direction of flow. Lapped or shingle sod should be staggered and overlapped by approximately 25%, with the top overlap portion facing downstream to prevent sod rollup during high flows. Staked sod is usually necessary only for use on steeper slopes to prevent sliding. Newer vegetated grid and net-mesh systems proport to hold vegetation in place at realtively high velocities, and have a considerable track record in roadside applications.

Flexible Linings

Rock riprap, including rubble, is the most common type of flexible lining. It presents a rough surface that can dissipate energy and mitigate increases in erosive velocity. These linings are usually less expensive than rigid linings and have self-healing qualities that reduce maintenance. Depending on the soil underneath, they may require use of filter fabric or gravel layers and allow the infiltration and exfiltration of water. The growth of grass and weeds through the lining may present maintenance problems, though vegetation can help stabilize the channel if it is not woody and of large enough diameter to cause turbulence and riprap displacement. The use of a flexible lining may be restricted where right-of-way is limited, because the higher roughness values create larger cross sections, or due to environmental considerations.

Rigid Linings

Rigid linings are generally constructed of concrete and used where smoothness offers a higher capacity for a given cross-sectional area. Higher velocities, however, create the potential for scour at channel lining transitions. A rigid lining can be destroyed by flow undercutting or eroding under the lining, soil support loss at improperly constructed or maintained joints, channel headcutting, erosion along the tops of channels from improperly accounted side flow, or the buildup of hydrostatic pressure behind the rigid surfaces.

Filter fabric may be required to prevent soil loss through pavement cracks. Soil cement can also be used profitably in areas where sand materials low in organic materials are available (Reese, 1987). Such channels in Arizona have withstood high velocities of sediment-laden flow without abrasion or structural failure.

Experience in Albuquerque, New Mexico, and the Corps of Engineers in Southern California have produced design tips (Blair, 1984; USCOE, 1978) for concrete channels:

- Eliminate contraction joints by using continuous reinforcing.
- Use new joint sealants such as ethylene vinyl acetate foam or two-part urethane.
- Use a staggered or stepped joint (see Blair, 1984) to eliminate joint failure.
- Use 0.3 to 0.5% steel longitudinally and 0.2 to 0.27% transversely and place it in the upper portion of the middle third of thickness (longitudinal bar on top).
- Use side ditches along tops of channels to intercept sheet flow and convey it to chutes or dip inlets.
- Form a low flow or trickle channel in the bottom of the channel (see UDFCD, 1991).
- Use 7- to 8-in. concrete thickness.
- Use underdrains of 6-in. perforated pipe in gravel or coarse sand when experiencing temporary or permanent high-water tables.
- Limit joints and weep holes when possible.
- Use expansion joints only when abutting other structures.
- Ensure proper compaction of soils under and adjacent to the concrete.

Rigid Low-Flow Channels

Under continuous base flow conditions, when a vegetative lining alone would not be appropriate, a small concrete (normally V-shaped) pilot or trickle channel could be used to handle the continuous low flows. Vegetation could then be maintained for handling larger flows. The trickle channel allows the remainder of the channel to remain dry and easy to maintain; reduces erosion caused by a meandering low-flow channel; and allows for sediment transport at low flows due to the higher velocities in the trickle channel (UDFCD, 1991). Sometimes, rock-lined channels are used for trickle channels; however, they are more aesthetically pleasing but require more maintenance and can encourage sediment deposition. Rock imbedded in concrete can obtain the best of both designs, but at greater cost. A low-flow pipe has also been used effectively if it is designed with a minimum diameter of 24 in., maintains a minimum of 3 ft/s velocity at half-full conditions, and provides access manholes every 300 to 500 ft.

General Design Criteria

In general, the following criteria should be used for open-channel design and analysis:

- Channels with bottom widths greater than 10 ft should be designed with a minimum bottom cross slope of 12 to 1.
- Low-flow and high-flow sections should be considered in the design of channels with large cross sections (Q > 50 cfs).
- Channel side slopes should be stable throughout the entire length, and side slope should depend on the channel material. A maximum of 3:1 is usually allowed for vegetation and 2:1 for riprap, unless otherwise justified by calculations.

- Superelevation of the water surface at horizontal curves should be accounted for by increased freeboard.
- Trapezoidal or parabolic cross sections are preferred to triangular shapes.
- If relocation of a stream channel is unavoidable, the cross-sectional shape, meander pattern, roughness, sediment transport capacity, and slope should conform to the existing conditions insofar as practicable. Some means of energy dissipation may be necessary when existing conditions cannot be duplicated.
- Streambank stabilization should be provided, when appropriate, as a result of any stream disturbance, and should include upstream and downstream banks as well as the local site.
- Open-channel drainage systems are often sized to handle a 10-year design storm. The 100-year design storm should be routed through the channel floodplain system to determine if the 100-year plus applicable freeboard flood elevation restrictions are exceeded, structures are flooded, or flood damages increased.
- Where complex sections are used, the low-flow channel is often sized to handle the dominant discharge (normally between the 2-year and 5-year flow), while the high-flow channel handles less frequent flows depending on overbank flooding circumstances. Low-flow channels are also designed to handle base flows in larger streams.
- Sediment transport requirements must be considered for conditions of flow below the design frequency. A low-flow channel component within a larger channel can reduce maintenance by improving sediment transport in the channel.

Channel Transitions

The following criteria should be considered at channel transitions in order for the channel system to operate as designed:

- Transition to channel sections should be smooth and gradual.
- A straight line connecting flow lines at the two ends of the transition should not make an angle greater than 12.5 degrees with the axis of the main channel.
- Transition sections should be designed to provide a gradual transition to avoid turbulence and eddies.
- Energy losses in transitions should be accounted for as part of the water surface profile calculations.
- Scour downstream from rigid-to-natural and steep-to-mild slope transition sections should be accounted for through velocity-slowing and energy-dissipating devices.

10.3 Hydraulic Terms and Equations

Design analysis of natural and artificial channels proceeds according to the basic principles of open-channel flow (Chow, 1970; Henderson, 1966). The basic principles of fluid mechanics — continuity, momentum, and energy — can be applied to open-channel flow with the additional complication that the position of the free surface is usually one of the

Open Channel Design

unknown variables. The determination of this unknown is one of the principle problems of open-channel flow analysis, and it depends on quantification of the flow resistance. Natural channels display a much wider range of roughness values than artificial channels.

Flow Classification

The classification of one-dimensional open-channel flow can be summarized as follows.

I. Steady flow
   1. Uniform flow
   2. Nonuniform flow
      a. Gradually varied flow
      b. Rapidly varied flow
II. Unsteady flow
   1. Unsteady uniform flow (rare)
   2. Unsteady nonuniform flow
      a. Gradually varied unsteady flow
      b. Rapidly varied unsteady flow

The steady uniform flow case and the steady nonuniform flow case are the fundamental types of flow treated in municipal engineering hydraulics. Rapidly varying flow is handled through special design procedures such as inlet or weir equations.

Hydraulic Terms and Definitions

Total Energy Head
The total energy head is the specific energy head plus the elevation of the channel bottom with respect to some datum. The locus of the energy head from one cross section to the next defines the energy grade line.

Specific Energy
Specific energy (E) is defined as the energy head relative to the channel bottom. If the channel is not too steep (slope less than 10%), and the streamlines are nearly straight and parallel (so that the hydrostatic assumption holds), the specific energy E becomes the sum of the depth and velocity head:

$$E = y + \alpha (V^2/2g) \tag{10.1}$$

where y = depth (ft); α = kinetic energy correction coefficient; V = mean velocity (ft/s); and g = gravitational acceleration (32.2 ft/s²).

The kinetic energy correction coefficient is taken to have a value of one for turbulent flow in prismatic channels but may be significantly different from one in natural channels.

Kinetic Energy Coefficient
As the velocity distribution in a river varies from a maximum at the design portion of the channel to essentially zero along the banks, the average velocity head, computed as $V^2/2g$

for the stream at a section, does not give a true measure of the kinetic energy of the flow, but rather normally underpredicts the actual value. A weighted average value of the kinetic energy is obtained by multiplying the average velocity head, above, by a kinetic energy coefficient, α, defined as:

$$\alpha = [\Sigma(q_i v_i^2)/(QV^2)] = [\Sigma(a_i v_i^3)/(QV^3)] \tag{10.2}$$

where v_i = average velocity in subsection i (ft/s); q_i = discharge in same subsection i (cfs); a_i = flow area of subsection i (ft²); Q = total discharge in river (cfs); and V = average velocity in river at section or Q/A (ft/s).

Kolupaila (1956) has given typical values of α for regular channels between 1.1 and 1.2 (with an average value of 1.15) and for natural channels between 1.15 and 1.5 (with an average value of 1.3). For natural channels with flooded overbanks, the value can range between 1.5 and 2.0 with an average value of 1.75.

Steady and Unsteady Flow

A steady flow is where the discharge passing a given cross section is constant with respect to time. The maintenance of steady flow requires that the rates of inflow and outflow be constant and equal. When the discharge varies with time, the flow is unsteady.

Uniform Flow and Nonuniform Flow

A nonuniform flow is one in which the velocity and depth vary in the direction of motion, while they remain constant in uniform flow. Uniform flow can occur only in a prismatic channel, which is a channel of constant cross section, roughness, and slope in the flow direction; however, nonuniform flow can occur in a prismatic channel or in a natural channel with variable properties.

Gradually Varied and Rapidly Varied

A nonuniform flow, in which the depth and velocity change gradually enough in the flow direction that vertical accelerations can be neglected, is referred to as a gradually-varied flow; otherwise, it is considered to be rapidly varied.

Froude Number

The Froude number is an important dimensionless parameter in open-channel flow. It represents the ratio of inertia forces to gravity forces and is defined by:

$$F = V/(gd)^{.5} \tag{10.3}$$

where V = mean velocity = Q/A (ft/s); g = acceleration of gravity (ft/s²); d = hydraulic depth = A/B (ft); A = cross-sectional area of flow (ft²); and B = channel top width at the water surface (ft).

This expression for Froude number applies to any single-section channel of nonrectangular shape.

Critical Flow

The variation of specific energy with depth at a constant discharge shows a minimum in the specific energy at a depth called critical depth at which the Froude number has a value

Open Channel Design

of one. Critical depth is also the depth of maximum discharge, when the specific energy is held constant.

Subcritical Flow

Depths greater than critical occur in subcritical flow, and the Froude number is less than one. In this state of flow, small water surface disturbances can travel upstream and downstream, and the control is always located downstream.

Supercritical Flow

Depths less than critical depth occur in supercritical flow, and the Froude number is greater than one. Small water surface disturbances are always swept downstream in supercritical flow, and the location of the flow control is always upstream.

Hydraulic Jump

Hydraulic jumps occur at abrupt transitions from supercritical to subcritical flow in the flow direction. There are significant changes in depth and velocity in the jump, and energy is dissipated. For this reason, the hydraulic jump is often employed to dissipate energy and control erosion at stormwater management structures.

Equations

The following equations are those most commonly used to analyze open channel flow.

Manning's Equation

For a given channel geometry, slope, and roughness, and a specified value of discharge Q, a unique value of depth occurs in steady uniform flow. It is called the normal depth. The normal depth is used to design artificial channels in steady, uniform flow and is computed from Manning's equation:

$$Q = [(1.49/n)AR^{2/3}S^{1/2}] \qquad (10.4)$$

where Q = discharge (cfs); n = Manning's roughness coefficient; A = cross-sectional area of flow (ft²); R = hydraulic radius = A/P (ft); P = wetted perimeter (ft); and S = channel slope (ft/ft).

The selection of Manning's n is generally based on observation; however, considerable experience is essential in selecting appropriate n values. See Section 10.4 for a discussion of Manning's n values and selection tables.

If the normal depth computed from Manning's equation is greater than critical depth, the slope is classified as a mild slope, while, on a steep slope, the normal depth is less than critical depth. Thus, uniform flow is subcritical on a mild slope and supercritical on a steep slope.

In channel analysis, it is often convenient to group the channel properties in a single term called the channel conveyance K:

$$K = (1.49/n)AR^{2/3} \qquad (10.5)$$

and then Manning's equation can be written as:

$$Q = KS^{1/2} \qquad (10.6)$$

The conveyance represents the carrying capacity of a stream cross section based upon its geometry and roughness characteristics and is independent of the streambed slope.

The concept of channel conveyance is useful when computing the distribution of overbank flood flows in the stream cross section and the flow distribution through the opening in a proposed stream crossing. Manning's equation should not be used for determining highwater elevations in a bridge opening.

Continuity Equation

The continuity equation is the statement of conservation of mass in fluid mechanics. For the special case of steady flow of an incompressible fluid, it assumes the simple form:

$$Q = A_1 V_1 = A_2 V_2 \qquad (10.7)$$

where Q = discharge (cfs); A = flow cross-sectional area (ft²); and V = mean cross-sectional velocity (ft/s) (which is perpendicular to the cross section).

The subscripts 1 and 2 refer to successive cross sections along the flow path. The continuity equation can be used with Manning's equation to obtain the steady uniform flow velocity as:

$$V = Q/A = [(1.49/n)R^{2/3}S^{1/2}] \qquad (10.8)$$

Energy Equation

The energy equation expresses conservation of energy in open-channel flow expressed as energy per unit weight of fluid, which has dimensions of length and is therefore called energy head. The energy head is composed of potential energy head (elevation head), pressure head, and kinetic energy head (velocity head). These energy heads are scalar quantities that give the total energy head at any cross section when added. Written between an upstream open-channel cross section designated 1 and a downstream cross section designated 2, the energy equation is as follows:

$$h_1 + \alpha_1(V_1^2/2g) = h_2 + \alpha_2(V_2^2/2g) + h_L \qquad (10.9)$$

where h_1 and h_2 are the upstream and downstream stages, respectively (ft); α = kinetic energy correction coefficient; V = mean velocity (ft/s); and h_L = headloss due to local cross-sectional changes (minor loss) as well as boundary resistance (ft).

The stage h is the sum of the elevation head z at the channel bottom and the pressure head, or depth of flow y, i.e., $h = z + y$. The terms in the energy equation are illustrated graphically in Figure 10-1. The energy equation states that the total energy head at an upstream cross section is equal to the energy head at a downstream section plus the intervening energy headloss. The energy equation can be applied only between two cross sections, at which the streamlines are nearly straight and parallel, so that vertical accelerations can be neglected.

10.4 Manning's *n* Values

The Manning's *n* value is an important variable in open-channel flow computations. Variation in this variable can significantly affect discharge, depth, and velocity estimates.

Open Channel Design 523

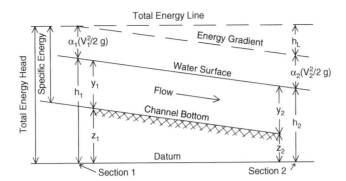

FIGURE 10-1
Terms in the energy equation. (From FHWA, 1990.)

Because Manning's n values depend on many different physical characteristics of natural and man-made channels, care and good engineering judgment must be exercised in the selection process.

The following general factors should be considered when selecting the value of Manning's n:

1. The physical roughness of the bottom and sides of the channel can affect n values. Fine particle soils on smooth, uniform surfaces result in relatively low values of n. Coarse materials, such as gravel or boulders, or pronounced surface irregularity, cause higher values of n.

2. The value of n depends on the height, density, type of vegetation, and how the vegetation affects the flow through the channel reach. The n value will increase in the spring and summer, as vegetation grows and foliage develops, and diminish in the fall, as the dormant season approaches.

3. Channel shape variations, such as abrupt changes in channel cross sections or alternating small and large cross sections, will require somewhat larger n values than normal. These variations in channel cross section become particularly important if they cause the flow to meander from side to side.

4. A significant increase in the value of n is possible if severe meandering occurs in the alignment of a channel. Meandering becomes particularly important when frequent changes in the direction of curvature occur with relatively small radii of curvature.

5. Active channel erosion or sedimentation will tend to increase the value of n, because these processes may cause variations in the shape of a channel. The potential for future erosion or sedimentation in the channel should also be considered.

6. Obstructions such as log jams or deposits of debris will increase the value of n. The level of this increase will depend on the number, type, and size of obstructions.

7. To be conservative, it is better to use a higher resistance for capacity calculations and a lower resistance for stability calculations. Sensitivity studies may be important.

8. Due to floodplain vegetation, n may vary with depth of flow.

All of these factors should be considered with respect to type of channel, degree of maintenance, seasonal requirements, and other considerations, as a basis for making a determination of an appropriate design n value. The probable condition of the channel when the design event is anticipated should be considered. Values representative of a freshly constructed channel are rarely appropriate as a basis for design capacity calculations.

Recommended Manning's n values for channels that are either excavated or dredged and natural are given in Table 10-1. Recommended Manning's n values for artificial channels with rigid, unlined, temporary, and riprap linings are given in Table 10-2. For more information, refer to the publication "Guide for Selecting Manning's Roughness Coefficients for Natural Channels and Flood Plains" (USDOT, 1984).

Natural Channels

There will be times when the determination of Manning's n value will be difficult in compound channels or for special circumstances. The partitioning of roughness among its several components is not new and has been used for sediment transport for years. Cowan (1956) developed a method for the partitioning of the various components that make up Manning's n as follows:

$$n = (n_0 + n_1 + n_2 + n_3 + n_4) m_5 \tag{10.10}$$

where n = Manning's roughness coefficient for natural channel; n_0 = coefficient associated with lining material type; n_1 = coefficient associated with degree of irregularity; n_2 = coefficient associated with variations of the channel cross section; n_3 = coefficient associated with channel obstructions; n_4 = coefficient associated with channel vegetation; and m_5 = coefficient associated with channel meandering.

Table 10-3 gives pertinent information for the determination of each of the coefficients. For channels flowing in alluvial materials, the roughness and flow rate are interdependent and are determined by the bed grain size distribution and the bedforms of the alluvium (ripples, dunes, antidunes, etc.). There have been many analyses to determine roughness and stage–discharge relationships. For example, see ASCE (1977) and Simons and Senturk (1976).

10.5 Manning's n Handbook

Given the importance of the Manning's n value in open-channel flow computations, it is important to obtain the n value that accurately represents the resistance to flood flows in local channels and floodplains. Because there is not an exact method of selecting the n value, a range of n values is often given for different channel and floodplain conditions with limited guidelines for final n value selection. To veteran engineers, this means the exercise of sound engineering judgment and experience in selecting a value for n for a specific location within the drainage system; for beginners, it can be no more than a guess, and different individuals will obtain different results. Although there are publications available that give recommended n values for particular streams and rivers contained in the publication, these examples often do not represent local conditions of interest to the design engineer. One approach to assist local design professionals in selecting a Manning's n value for channels and floodplain areas within the municipality is to develop a Manning's n Handbook with recommended n values for channels and floodplain areas depicting the different streams and floodplains within the municipality (Greenville, South Carolina, 1992).

TABLE 10-1

Uniform Flow Values of Roughness Coefficient (n)

Type of Channel and Description	Minimum	Normal	Maximum
Excavated or Dredged			
a. Earth, straight and uniform	0.016	0.018	0.020
1. Clean, recently completed	0.018	0.022	0.025
2. Clean, after weathering	0.022	0.025	0.030
3. Gravel, uniform section, clean	0.022	0.027	0.033
b. Earth, winding and sluggish			
1. No vegetation	0.023	0.025	0.030
2. Grass, some weeds	0.025	0.030	0.033
3. Dense weeds and plants in deep channels	0.030	0.035	0.040
4. Earth bottom and rubble sides	0.025	0.030	0.035
5. Stony bottom and weedy sides	0.025	0.035	0.045
6. Cobble bottom and clean sides	0.030	0.040	0.050
c. Dragline-excavated or dredged			
1. No vegetation	0.025	0.028	0.033
2. Light brush on banks	0.035	0.050	0.060
d. Rock cuts			
1. Smooth and uniform	0.025	0.035	0.040
2. Jagged and irregular	0.035	0.040	0.050
e. Channels not maintained, weeds and brush uncut			
1. Dense weeds, high as flow depth	0.050	0.080	0.120
2. Clean bottom, brush on sides	0.040	0.050	0.080
3. Same, highest stage of flow	0.045	0.070	0.110
4. Dense brush, high stage	0.080	0.100	0.140
Natural Streams			
Minor Streams (Top Width at Flood Stage <100 ft)			
a. Streams on plain			
1. Clean, straight, full stage, no rifts or deep pools	0.025	0.030	0.033
2. Same as above, but with more stones and weeds	0.030	0.035	0.040
3. Clean, winding, some pools and shoals	0.033	0.040	0.045
4. Same as above, but with some weeds and some stones	0.035	0.045	0.050
5. Same as above, lower stages, more ineffective slopes and sections	0.040	0.048	0.055
6. Same as 4, but with more stones	0.045	0.050	0.060
7. Sluggish reaches, weedy, deep pools	0.050	0.070	0.080
8. Very weedy reaches, deep pools, or floodways with heavy stand of timber and underbrush	0.075	0.100	0.150
b. Mountain streams, no vegetation in channel, banks usually steep, trees and brush along banks submerged at high stages			
1. Bottom: gravels, cobbles, few boulders	0.030	0.040	0.050
2. Bottom: cobbles with large boulders	0.040	0.050	0.070
Floodplains			
a. Pasture, no brush			
1. Short grass	0.025	0.030	0.035
2. High grass	0.030	0.035	0.050
b. Cultivated area			
1. No crop	0.020	0.030	0.040
2. Mature row crops	0.025	0.035	0.045
3. Mature field crops	0.030	0.040	0.050
c. Brush			
1. Scattered brush, heavy weeds	0.035	0.050	0.070
2. Light brush and trees, in winter	0.035	0.050	0.060
3. Light brush and trees, in summer	0.040	0.060	0.080
4. Medium to dense brush, in winter	0.045	0.070	0.110
5. Medium to dense brush, in summer	0.070	0.100	0.160
d. Trees			
1. Dense willows, summer, straight	0.110	0.150	0.200

TABLE 10-1 *(Continued)*

Uniform Flow Values of Roughness Coefficient (n)

Type of Channel and Description	Minimum	Normal	Maximum
2. Cleared land, tree stumps, no sprouts	0.030	0.040	0.050
3. Same as above, but with heavy growth of sprouts	0.050	0.060	0.080
4. Heavy stand of timber, a few down trees, little undergrowth, flood stage below branches	0.080	0.100	0.120
5. Same as above, but with flood stage reaching branches	0.100	0.120	0.160
Major Streams (Top Width at Flood Stage >100 ft); n Value is Less than that for Minor Streams of Similar Description, Because Banks Offer Less Effective Resistance			
a. Regular section with no boulders or brush	0.025	—	0.060
b. Irregular and rough section	0.035	—	0.100

Source: From Chow, V.T., Ed., *Open Channel Hydraulics*, McGraw Hill, New York, 1959.

TABLE 10-2

Manning's Roughness Coefficients for Artificial Channels — n

| Lining Category | Lining Type | Depth Range | | |
		0–0.5 ft	0.5–2.0 ft	>2.0 ft
Rigid	Concrete	0.015	0.013	0.013
	Grouted riprap	0.040	0.030	0.028
	Stone masonry	0.042	0.032	0.030
	Soil cement	0.025	0.022	0.020
	Asphalt	0.018	0.016	0.016
Unlined	Bare soil	0.023	0.020	0.020
	Rock cut	0.045	0.035	0.025
Temporary[a]	Woven paper net	0.016	0.015	0.015
	Jute net	0.028	0.022	0.019
	Fiberglass roving	0.028	0.022	0.019
	Straw with net	0.065	0.033	0.025
	Curled wood mat	0.066	0.035	0.028
	Synthetic mat	0.036	0.025	0.021
Gravel riprap	1-in D50	0.044	0.033	0.030
	2-in D50	0.066	0.041	0.034
Rock riprap	6-in D50	0.104	0.069	0.035
	12-in D50	—	0.078	0.040

[a] Some "temporary" linings become permanent when buried.

Note: Values listed are representative values for the respective depth ranges. Manning's roughness coefficients, n, vary with the flow depth.

Source: From U.S. Department of Transportation, Federal Highway Administration, Design of Stable Channels with Flexible Linings, Hydraulic Engineering Circular No. 15, Washington, DC, 1986.

The determination of the n values for the streams and floodplains contained in a local Manning's n Handbook could be determined by using Equation (10.10) and Table 10-3 and extensive field studies to arrive at values for the variables in the equation. Another approach would be to obtain velocity and other field measurements within the example streams and floodplains and calculate the appropriate n value using Equation (10.8). In either case, developing Manning n values for local conditions should improve the accuracy of the open-channel flow computations and provide a valuable service to local engineers and designers.

Open Channel Design

TABLE 10-3

Cowan's Coefficients for Computing Manning's n

Material	Specific Condition	Value
Bank material	Earth	0.020
(n_0)	Rock cut	0.025
	Fine gravel	0.024
	Coarse gravel	0.028
Degree of irregularity	Smooth	0.000
(n_1)	Minor	0.005
	Moderate	0.010
	Severe	0.020
Cross-section variations	Gradual	0.000
(n_2)	Alternating some	0.005
	Alternating much	0.010 to 0.015
Obstructions	Negligible	0.000
(n_3)	Minor	0.010 to 0.015
	Appreciable	0.020 to 0.030
	Severe	0.040 to 0.060
Vegetation	Low	0.005 to 0.010
(n_4)	Medium	0.010 to 0.025
	High	0.025 to 0.050
	Very high	0.050 to 0.100
Meandering	Minor	1.000
(m_5)	Appreciable	1.150
	Severe	1.300

Source: From Chow, V.T., Ed., *Open Channel Hydraulics*, McGraw Hill, New York, 1959.

10.6 Best Hydraulic Section

The cost of an open channel, other than environmental or regulatory costs, is comprised of (1) the acquisition of land, (2) excavation, and (3) lining. Land cost varies with the top width of the channel, excavation cost with the cross-sectional area, and lining with the wetted perimeter. In cases where land cost predominates, the most economical design is a buried facility. In other cases, it can be shown that the most economical design is one comprised of the so-called "best hydraulic section." For a given discharge, slope, and channel roughness, maximum velocity implies minimum cross-sectional area. From Manning's equation, if velocity is maximized and area is minimized, wetted perimeter will also be minimized. The best hydraulic section, therefore, simultaneously minimizes area and wetted perimeter.

The best of all prismatic sections is the semicircle. Similarly, the best rectangular section is half of a square (i.e., width equal twice the depth), and the best triangular section is an isosceles right triangle, etc. Of course, the semicircular section would be impractical because of the labor involved in excavation and forming the lining. Because rectangular and triangular sections are special cases of trapezoidal sections, it follows that the best hydraulic section for trapezoidal sections is the one of primary practical interest.

Given that the desired side slope, M to one, has been selected for a given channel, the minimum wetted perimeter (P) exists when:

$$P = 4y(1 + M^2)^{0.5} - 2My \tag{10.11}$$

(Figure 10-2 shows a definition of variables.)

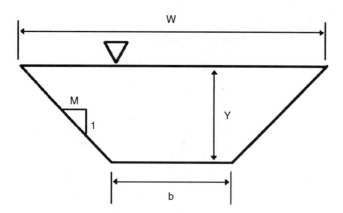

FIGURE 10-2
Trapezoidal channel — definition of variables.

From the geometry of the channel cross section and the Manning equation, design equations can be developed for determining the dimensions of the best hydraulic section for a trapezoidal channel.

The depth of the best hydraulic section is defined by:

$$y = C_M (Qn/(S^{1/2}))^{3/8} \tag{10.12}$$

where

$$C_M = \frac{[(k + 2(M^2 + 1)^{1/2})^{2/3}]^{3/8}}{[1.49(k + M)^{5/3}]^{3/8}} \tag{10.13}$$

The associated bottom width is:

$$B = ky \tag{10.14}$$

The cross-sectional area of the resulting channel is:

$$A = By + My^2 \tag{10.15}$$

Table 10-4 lists values of C_M and K for various values of M.

10.7 Uniform Flow Calculations

Following is a discussion of the equations that can be used for the design and analysis of open-channel flow. Manning's equation, presented in three forms below, is recommended for evaluating uniform flow conditions in open channels:

$$v = (1.49/n) R^{2/3} S^{1/2} \tag{10.16}$$

$$Q = (1.49/n) A R^{2/3} S^{1/2} \tag{10.17}$$

Open Channel Design

TABLE 10-4
Values of C_M and k for Determining Bottom Width and Depth of Best Hydraulic Section

M	C_M	k	Remarks
0/1	0.790	2.00	Vertical Sides
0.5/1	0.833	1.236	
0.577/1	0.833	1.155	60° Side Slopes
1.0/1	0.817	0.828	
1.5/1	0.775	0.606	
2.0/1	0.729	0.472	
2.5/1	0.688	0.385	
3.0/1	0.653	0.325	
3.5/1	0.622	0.280	
4.0/1	0.595	0.246	
5.0/1	0.522	0.198	
6.0/1	0.518	0.166	
8.0/1	0.467	0.125	
10.0/1	0.430	0.100	
12.0/1	0.402	0.083	

$$S = [Qn/(1.49 \, A \, R^{2/3})]^2 \tag{10.18}$$

where v = average channel velocity (ft/s); Q = discharge rate for design conditions (cfs); n = Manning's roughness coefficient; A = cross-sectional area (ft²); R = hydraulic radius A/P (ft); P = wetted perimeter (ft); and S = slope of the energy grade line (ft/ft).

For prismatic channels, in the absence of backwater conditions, the slope of the energy grade line, water surface, and channel bottom are equal. Area, wetted perimeter, hydraulic radius, and channel top width for standard channel cross sections can be calculated from geometric dimensions.

Direct Solution

When the hydraulic radius, cross-sectional area, and roughness coefficient and slope are known, discharge can be calculated directly from Equation (10.17). The slope can be calculated using Equation (10.18) when the discharge, roughness coefficient, area, and hydraulic radius are known.

Nomographs for obtaining direct solutions to Manning's equation are presented in Figures 10-4 and 10-5. Figure 10-3 provides a general solution for the velocity form of Manning's equation, while Figure 10-4 provides a solution of Manning's equation for trapezoidal channels.

General Solution Nomograph

The following steps are used for the general solution using the nomograph in Figure 10-3:

1. Determine open-channel data, including slope in ft/ft, hydraulic radius in ft, and Manning's n value.
2. Connect a line between the Manning's n scale and slope scale and note the point of intersection on the turning line.
3. Connect a line from the hydraulic radius to the point of intersection obtained in Step 2.
4. Extend the line from Step 3 to the velocity scale to obtain the velocity in ft/s.

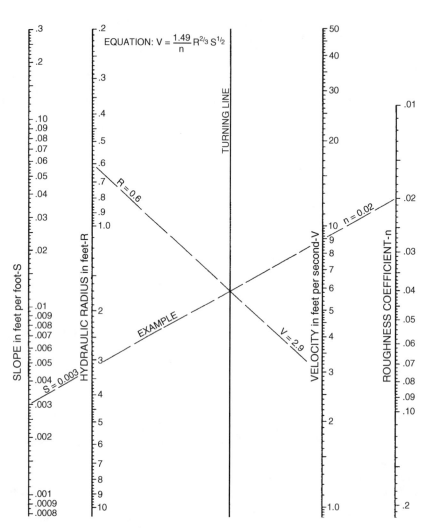

FIGURE 10-3
Nomograph for the solution of Manning's equation. (From USDOT, FHWA, HDS-3, 1961.)

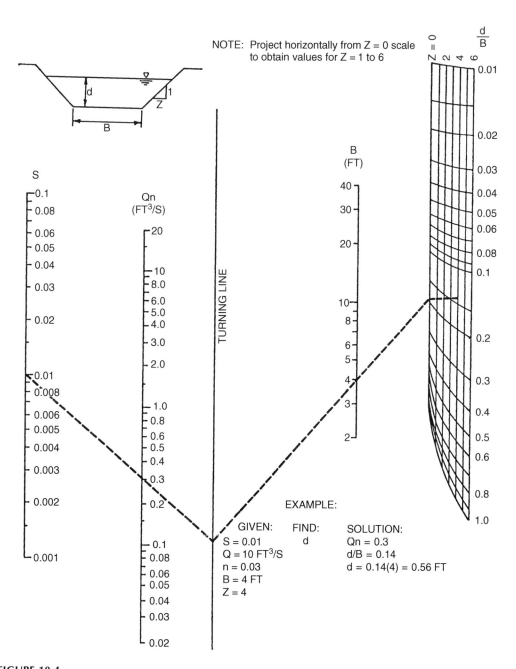

FIGURE 10-4
Solution of Manning's equation for trapezoidal channels. (From U.S. Department of Transportation, Federal Highway Administration, Design of Stable Channels with Flexible Linings, Hydraulic Engineering Circular No. 15, Washington, DC, 1986.)

Trapezoidal Solution Nomograph

The trapezoidal channel nomograph solution to Manning's equation in Figure 10-4 can be used to find the depth of flow if the design discharge is known or the design discharge if the depth of flow is known:

1. Determine input data, including slope in ft/ft, Manning's n value, bottom width in ft, and side slope in ft/ft.
2. Given the design discharge, do the following:
   a. Find the product of Q times n, connect a line from the slope scale to the Qn scale, and find the point of intersection on the turning line.
   b. Connect a line from the turning point from Step 2a to the b scale, and find the intersection with the $z = 0$ scale.
   c. Project horizontally from the point located in Step 2b to the appropriate z value, and find the value of d/b.
   d. Multiply the value of d/b obtained in Step 2c by the bottom width b to find the depth of uniform flow, d.
3. Given the depth of flow, do the following:
   a. Find the ratio d divided by b, and project a horizontal line from the d/b ratio at the appropriate side slope, z, to the $z = 0$ scale.
   b. Connect a line from the point located in Step 3a to the b scale and find the intersection with the turning line.
   c. Connect a line from the point located in Step 3b to the slope scale and find the intersection with the Qn scale.
   d. Divide the value of Qn obtained in Step 3c by the n value to find the design discharge, Q.

Trial-and-Error Solution

A trial-and-error procedure for solving Manning's equation is used to compute the normal depth of flow in a uniform channel when the channel shape, slope, roughness, and design discharge are known. For purposes of the trial-and-error process, Manning's equation can be arranged as:

$$AR^{2/3} = (Qn)/(1.49\ S^{1/2}) \qquad (10.19)$$

where A = cross-sectional area (ft); R = hydraulic radius (ft); Q = discharge rate for design conditions (cfs); n = Manning's roughness coefficient; and S = slope of the energy grade line (ft/ft).

To determine the normal depth of flow in a channel by the trial-and-error process, trial values of depth are used to determine A, P, and R for the given channel cross section. Trial values of $AR^{2/3}$ are computed until the equality of Equation (10.19) is satisfied such that the design flow is conveyed for the slope and selected channel cross section. Graphical procedures for simplifying trial-and-error solutions are presented in Figure 10-5 for trapezoidal channels.

1. Determine input data, including design discharge, Q, Manning's n value, channel bottom width, b, channel slope, S, and channel side slope, z.

Open Channel Design

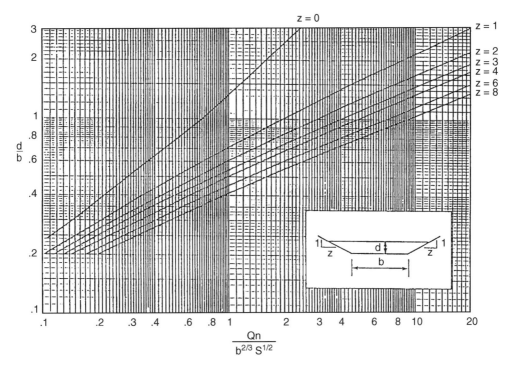

FIGURE 10-5
Trapezoidal channel capacity. (From Storm Water Management Manual, Nashville, 1988.)

2. Calculate the trapezoidal conveyance factor using the equation:

$$K_T = (Qn)/(b^{8/3}S^{1/2}) \qquad (10.20)$$

where K_T = trapezoidal open channel conveyance factor; Q = discharge rate for design conditions (cfs); n = Manning's roughness coefficient; b = bottom width (ft); and S = slope of the energy grade line (ft/ft).

3. Enter the x-axis of Figure 10-5 with the value of K_T calculated in Step 2, and draw a line vertically to the curve corresponding to the appropriate z value from Step 1.
4. From the point of intersection obtained in Step 3, draw a horizontal line to the y-axis and read the value of the normal depth of flow over the bottom width, d/b.
5. Multiply the d/b value from Step 4 by b to obtain the normal depth of flow.

Irregular Channels

The calculation of uniform flow in irregular channels, either in shape or roughness, is a common problem for urban stormwater designers. There are several techniques to handle it, depending on the accuracy required and the shape of the channel (US COE, 1970; Chow, 1959).

Average Roughness

For channels with simple concave geometry that have different roughnesses, for example, on the bed and banks, an equivalent Manning's n can be calculated and applied to the whole channel (Chow, 1959). If the channel is broken into N subsections, then the equivalent Manning's n value can be expressed as:

$$n_e = (P_1 n_1^x + P_2 n_2^x + \ldots + P_N n_N^x)^y / P^y \qquad (10.21)$$

where n_e = equivalent Manning's n and x, y = coefficients depending on basic assumptions.

Other variables are as defined previously. For the assumption that the mean velocity in each subsection is equal to each other and the mean velocity in the channel, $x = 3/2$ and $y = 2/3$. For the assumption that the sum of the forces resisting the flow equals the total force resisting the flow, $x = 2$ and $y = 1/2$. For this type of channel, using a single Manning's equation and an equivalent Manning's n is preferable to using a subdivided channel, as the subdivision process leads to an overestimation of channel conveyance by as much as 7 to 10% (Garbrecht and Brown, 1991; Bensen and Dalrymple, 1967).

Irregular Channel Calculations

For more complex convex channel shapes (flow with overbanks) or width-to-depth ratios greater than 10, uniform flow can be calculated by breaking the channel into N subsections, projecting vertical lines from the perimeter break point upward to the surface. Then, for each subsection, i:

$$v_i = (K_i / A_i) S^{1/2} \qquad (10.22)$$

where v_i = mean velocity in subsection i (ft/s); K_i = conveyance in subsection i (ft²) [see Equation (10.5)]; and A_i = flow area in subsection i (ft²).

And, from the continuity equation:

$$Q = VA = \Sigma(v_i A_i) = \Sigma(K_i) S^{1/2} \qquad (10.23)$$

where it should be noted that K_i, the sectional conveyance, equals $(1.49/n_i) A_i R_i^{2/3}$.

When computing the hydraulic radius of the subsections, the water depth common to the two adjacent subsections is not counted as wetted perimeter. The procedure then is to guess at a depth of flow; calculate conveyance at each subsection for the various flow areas, wetted perimeters, and Manning's n values; and then find the total discharge. By plotting a stage–discharge curve, the depth for the known Q can be readily determined through interpolation. This method lends itself to spreadsheet calculations. The slope is normally assumed equal in each section.

Alternately, a backwater computer model can be used to quickly determine depth of flow by inputting the complex section and simply changing the vertical values by the product of slope times distance and moving the section an arbitrary distance upstream for several cross sections.

10.8 Critical-Flow Calculations

Critical depth depends only on the discharge rate and channel geometry. The general equation for determining critical depth is:

TABLE 10-5

Critical Depth for Uniform Flow in Selected Channel Types

Channel Type	Semiempirical Equation[a] for Estimating Critical Depth	Range of Applicability
Rectangular[b]	$d_c = [Q^2/(gb^2)]^{1/3}$	N/A
Trapezoidal[b]	$d_c = 0.81[Q^2/(gz^{0.75}b^{1.25})]^{0.27} - b/30z$	$0.01 < 0.5522Q/b^{2.5} < 0.4$ For $0.5522Q/b^{2.5} < 0.1$, use rectangular equa.
Triangular[b]	$d_c = [(2Q^2)/(gz^2)]^{1/5}$	N/A
Circular[c]	$d_c = 0.325(Q/D)^{2/3} + 0.083D$	$0.3 < d_c/D < 0.9$
General[d]	$(A^3/T) = (Q^2/g)$	N/A

[a] Assumes uniform flow with the kinetic energy coefficient equal to 1.0
[b] Reference: French, R.H., *Open Channel Hydraulics*, McGraw Hill, New York, 1985.
[c] Reference: USDOT, FHWA, HDS-4, 1965
[d] Reference: Brater and King, 1976

Note: d_c = critical depth (ft); Q = design discharge (cfs); g = acceleration due to gravity (32.2 ft/s²); b = bottom width of channel (ft); z = side slopes of a channel (horizontal to vertical); D = diameter of circular conduit (ft); and T = top width of water surface (ft).

$$Q^2/g = A^3/T \tag{10.24}$$

where Q = discharge rate for design conditions (cfs); g = acceleration due to gravity (32.2 ft/s²); A = cross-sectional area (ft²); and T = top width of water surface (ft).

A trial-and-error procedure is needed to solve Equation (10.24).

Semiempirical equations (as presented in Table 10-5) can be used to simplify trial-and-error critical-depth calculations. The following equation from Chow (1959) is used to determine critical depth with the critical flow section factor, Z:

$$Z = Q/g^{0.5} \tag{10.25}$$

where Z = critical flow section factor; Q = discharge rate for design conditions (cfs); and g = acceleration due to gravity (32.2 ft/s²).

The following guidelines are presented for evaluating critical-flow conditions of open-channel flow:

1. A normal depth of uniform flow within about 10% of critical depth is unstable (relatively large depth changes are likely for small changes in roughness, cross-sectional area, or slope) and should be avoided in design, if possible.
2. If the velocity head is less than one-half the mean depth of flow, the flow is subcritical.
3. If the velocity head is equal to one-half the mean depth of flow, the flow is critical.
4. If the velocity head is greater than one-half the mean depth of flow, the flow is supercritical.
5. If an unstable critical depth cannot be avoided in design, the least favorable type of flow should be assumed for the design.

The Froude number, *Fr*, calculated by the following equation, is useful for evaluating the type of flow conditions in an open channel:

$$Fr = v/(gA/T)^{0.5} \tag{10.26}$$

where Fr = Froude number (dimensionless); v = velocity of flow (ft/s); g = acceleration of gravity (32.2 ft/s^2); A = cross-sectional area of flow (ft^2); and T = top width of flow (ft).

If Fr is greater than 1.0, flow is supercritical; if it is under 1.0, flow is subcritical. Fr is 1.0 for critical flow conditions.

In compound channels (i.e., with both a high- and low-flow channel) or with flow in the main channel and overbanks, there can be more than one depth at which critical depth will occur. There is no readily accepted method for analysis of this phenomenon. However, more details can be found in Petryk and Grant (1978), Blalock and Strumm (1981), Schoellhamer, Peters, and Larock (1985), and Chaudhry and Bhallamudi (1988).

10.9 Vegetative Design

A two-part procedure, adapted from Chow (1959), Temple et al., (1987) and Temple (1979) and presented below, is recommended for the design of temporary and vegetative channel linings. Part 1, the *design stability* component, involves determining channel dimensions for low vegetative retardance conditions, using Class D as defined in Table 10-6. Part 2, the *design capacity* component, involves determining the depth increase necessary to maintain capacity for higher vegetative retardance conditions, using Class C as defined in Table 10-6. If a temporary lining is to be used during construction, vegetative retardance Class E should be used for the design stability calculations.

If the channel slope exceeds 10%, or a combination of channel linings will be used, additional procedures not presented below are required. References include HEC-15 (USDOT, FHWA, 1986) and HEC-14 (USDOT, FHWA, 1982).

Design Stability

The following are the steps for design stability calculations:

1. Determine appropriate design variables, including discharge, Q, bottom slope, S, cross-section parameters, and vegetation type.
2. Use Table 10-15 to assign a maximum velocity, v_m, based on vegetation type and slope range.
3. Assume a value of n and determine the corresponding value of vR from the n versus vR curves in Figure 10-6. Use retardance Class D for permanent vegetation and E for temporary construction. The methods of Kouwen and Li (1980) and Kouwen (1988) can be used for vegetal Manning's n values for channels with tall weeds. Resistance in wetlands areas can be obtained from Kadlec (1990). Alternately, n can be calculated by the following equation (Gwinn and Ree, 1980; Green and Garton, 1983):

$$n = \exp\{[0.01329C(\ln R_v)^2] - [0.09543C(\ln R_v)] + [0.2971C] - 4.16\} \quad (10.27)$$

where $R_v = vR/v$; v = channel velocity (ft/s); R = channel hydraulic radius (ft); v = kinematic viscosity of water (ft^2/s); and C = 10.0, 7.643, 5.601, 4.436, and 2.876 for retardance classes A, B, C, D, and E, respectively.

4. Calculate the hydraulic radius using the equation:

Open Channel Design

TABLE 10-6

Classification of Vegetal Covers to Degrees of Retardancy

Retardance	Cover	Condition
A	Weeping Lovegrass	Excellent stand, tall (average 30 in)
	Yellow Bluestem Ischaemum	Excellent stand, tall (average 36 in)
B	Kudzu	Very dense growth, uncut
	Bermuda grass	Good stand, tall (average 12 in)
	Native grass mixture, little bluestem, bluestem, blue gamma, other short- and long-stem Midwest grasses	Good stand, unmowed
	Weeping lovegrass	Good stand, tall (average 24 in)
	Laspedeza sericea	Good stand, not woody, tall (average 19 in)
	Alfalfa	Good stand, unmowed (average 11 in)
	Weeping lovegrass	Good stand, unmowed (average 13 in)
	Kudzu	Dense growth, uncut
	Blue gamma	Good stand, uncut (average 13 in)
C	Crabgrass	Fair stand, uncut (10–48 in)
	Bermuda grass	Good stand, mowed (average 6 in)
	Common lespedeza	Good stand, uncut (average 11 in)
	Grass–legume mix[a]	Good stand, uncut (6–8 in)
	Centipede grass	Very dense cover (average 6 in)
	Kentucky bluegrass	Good stand, headed (6–12 in)
D	Bermuda grass	Good stand, cut to 2.5 in
	Common lespedeza	Excellent stand, uncut (average 4.5 in)
	Buffalo grass	Good stand, uncut (3–6 in)
	Grass–legume mix[b]	Good stand, uncut (4–5 in)
	Lespedeza sericea	After cutting to 2 in (good before cutting)
E	Bermuda grass	Good stand, cut to 1.5 in
	Bermuda grass	Burned stubble

[a] Summer (orchard grass, redtop, Italian ryegrass, and common lespedeza).
[b] Fall, spring (orchard grass, redtop, Italian ryegrass, and common lespedeza).

Note: Covers classified were tested in experimental channel and were green and generally uniform.

Source: From U.S. Department of Transportation, Federal Highway Administration, Design of Stable Channels with Flexible Linings, Hydraulic Engineering Circular No. 15, Washington, DC, 1986.

$$R = (vR)/v_m \tag{10.28}$$

where R = hydraulic radius of flow (ft); vR = value obtained from Figure 10-6 in Step 2; and v_m = maximum velocity from Step 2 (ft/s).

5. Use the following form of Manning's equation to calculate the value of vR:

$$vR = (1.49\, R^{5/3}\, S^{1/2})/n \tag{10.29}$$

where vR = calculated value of vR product; R = hydraulic radius value from Step 4 (ft); S = channel bottom slope (ft/ft); and n = Manning's n value assumed in Step 3.

6. Compare the vR product value obtained in Step 5 to the value obtained from Figure 10-6 for the assumed n value in Step 3. If the values are not reasonably close, return to Step 3 and repeat the calculations using a new assumed n value.

7. For trapezoidal channels, find the flow depth using Figures 10-4 or 10-5, as described in Section 10.7. The depth of flow for other channel shapes can be evaluated using the trial-and-error procedure described in Section 10.7.

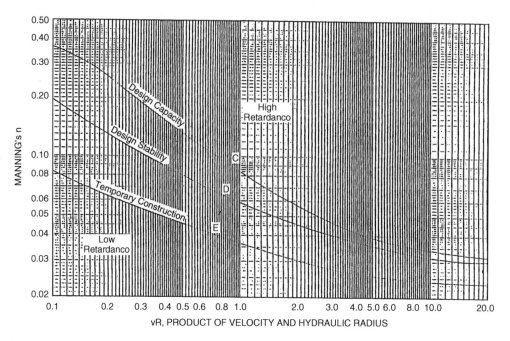

FIGURE 10-6
Manning's n values for vegetated channels. (From USDA, TP-61, 1947.)

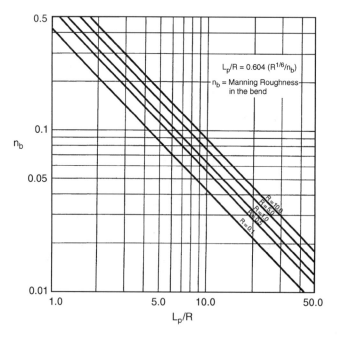

FIGURE 10-7
Protection length, L_p, downstream of channel bend. (From U.S. Department of Transportation, Federal Highway Administration, Design of Stable Channels with Flexible Linings, Hydraulic Engineering Circular No. 15, Washington, DC, 1986.)

Open Channel Design

If bends are considered, calculate the length of downstream protection, L_p, for the bend using Figure 10-7. Provide additional protection, such as gravel or riprap in the bend and extending downstream for length, L_p.

Design Capacity

The following are the steps for design capacity calculations:

1. Assume a depth of flow greater than the value from Step 7 above, and compute the waterway area and hydraulic radius.
2. Divide the design flow rate, obtained using appropriate procedures from Chapter 7, by the waterway area from Step 1 to find the velocity.
3. Multiply the velocity from Step 2 by the hydraulic radius from Step 1 to find the value of vR.
4. Use Figure 10-6 to find a Manning's n value for retardance Class C based on the vR value from Step 3.
5. Use Manning's equation [Equation (10.16)] or Figure 10-3 to find the velocity using the hydraulic radius from Step 1, Manning's n value from Step 4, and appropriate bottom slope.
6. Compare the velocity values from Steps 2 and 5. If the values are not reasonably close, return to Step 1 and repeat the calculations.
7. Add an appropriate freeboard to the final depth from Step 6. Generally, 20% is adequate.
8. If bends are considered, calculate superelevation of the water surface profile at the bend using the equation:

$$\text{delta } d = (v^2 T)/(g R_c) \tag{10.30}$$

where delta d = superelevation of water surface profile due to bend (ft); v = average velocity from Step 6 (ft/s); T = top width of flow (ft); g = acceleration of gravity (32.2 ft/s²); and R_c = mean radius of the bend (ft).

9. Add freeboard consistent with the calculated delta d.

Erosion Control

Practice has shown that complete protection of grassed channels from erosion is costly (Wright-McLaughlin, 1969). It is far better to provide reasonable erosion-free design with the intention to take additional erosion-control measures and corrective steps after the first year of operation. However, the use of erosion-control cut-off walls at regular intervals in a grassed channel is often desirable. Such cut-off walls will safeguard a channel from serious erosion in case of a large storm flow prior to the grass developing a good root system.

Erosion-control cut-off walls are usually of reinforced concrete, approximately 8-in thick and from 18 in to 2 ft deep, extending across the entire bottom of the channel. They can be shaped to fit a slightly sloped bottom to help direct water to the trickle channel. Grass will not grow under bridges, and, therefore, the erosion tendency is large. A cut-off wall and the use of riprap at the downstream edge of a bridge is good practice in some alluvial bed environments.

At bends in the channel, special erosion-control measures may be needed. However, once a good growth of grass is established and, if the design velocities, depths, and curvatures are satisfactory, normally, erosion at bends will not be a problem.

In maintaining the appropriate channel slope, the designer may find it necessary to use frequent drops. Erosion tends to occur at the edges immediately downstream of a drop even though it may be only 6 to 18 in high (Wright-McLaughlin, 1969). Proper use of riprap and gabions at these drops may be necessary.

10.10 Approximate Flood Limits

For small streams and tributaries not included in the floodplain studies, analysis may be required to identify the 100-year flood elevation and to evaluate floodplain encroachment as required. For such cases, when the designer can demonstrate that a complete backwater analysis is unwarranted, approximate methods may be used.

100-Year Flood Elevation

A generally accepted method for approximating the 100-year flood elevation is outlined in the following steps (Powell, James, and Jones, 1980):

1. Divide the stream or tributary into reaches that may be approximated using average slopes, cross sections, and roughness coefficients for each reach.
2. Estimate the 100-year peak discharge for each reach using an appropriate hydrologic method from Chapter 7.
3. Compute normal depth for uniform flow in each reach using Manning's equation for the reach characteristics from Step 1 and peak discharge from Step 2.
4. Use the normal depths computed in Step 3 to approximate the 100-year flood elevation in each reach. The 100-year flood elevation is then used to delineate the floodplain. This approximate method is based on several assumptions, including the following:
    a. A channel reach is accurately approximated by average characteristics throughout its length.
    b. The cross-sectional geometry, including area, wetted perimeter, and hydraulic radius, of a reach may be approximated using typical geometric properties that can be used in Manning's equation to solve for normal depth.
    c. Uniform flow can be established, and backwater effects are negligible between reaches.
    d. Expansion and contraction effects are negligible.

As indicated, the approximate method is based on a number of restrictive assumptions that may limit the accuracy of the approximation and applicability of the method. The designer is responsible for appropriate application of this method.

Open Channel Design

Setback Limits

After the 100-year flood elevation and floodplain are established, floodway setback limits may be approximated by requiring conveyance of the encroached section, including any allowable flood elevation increases, to equal the conveyance of the nonencroached section. From Manning's equation, the conveyance is given as follows:

$$K = (1.49/n) \, A \, R^{2/3} \tag{10.31}$$

$$Q = KS^{1/2} \tag{10.32}$$

where K = channel conveyance; A = cross-sectional area (ft²); R = hydraulic radius A/P (ft); P = wetted perimeter (ft); Q = discharge (cfs); and S = slope of the energy grade line (ft/ft).

The following procedure may be used to approximate setback limits for a stream:

1. Divide the stream cross section into segments for which the geometric properties may be easily solved, and estimate a Manning's n value for each segment.
2. Compute the area, hydraulic radius, and conveyance [Equation (10.31)] of each segment for the encroached and nonencroached segments.
3. Sum the conveyance for each cross-sectional segment to obtain the total conveyance for the encroached and nonencroached conditions.
4. Set the total conveyance of the encroached cross section equal to the total conveyance of the nonencroached section, and solve for the allowable encroachment by trial and error.

This method for approximating the allowable encroachment is based on the assumptions that the 100-year flood elevation has been established or can be approximated and that the energy grade line of the encroached and nonencroached sections remains unchanged. The accuracy of the results obtained using this method may be highly subject to the accuracy of the flood elevation used. In addition, because the method assumes no change in the energy grade line, the method should not be used near bridges or similar contraction–expansion areas.

For typical natural channel cross sections, the procedure may result in an equality that is difficult to solve for the allowable encroachment dimensions. Morris (1984) provides a series of dimensionless graphs that are solved for the allowable encroachment as a percentage of the nonencroached overbank width. These graphs are based on an allowable flood elevation increase of 1 ft and assume a symmetrical cross section with triangular overbanks and equal encroachment on both overbanks. The limitations listed above for the general procedure also apply.

Because of the simplifying assumptions required, this approximate method will have limited applicability. Generally, only very small streams will satisfy the assumptions, and the designer should use extreme caution to avoid misapplication.

10.11 Uniform Flow — Example Problems

Following are some example problems using the direct solution of the Manning's equation for irregular channels and for grassed lined channels designed for stability and capacity.

Direct Solution of Manning's Equation

Use Manning's equation to find the velocity, v, for an open channel with a hydraulic radius value of 0.6 ft, an n value of 0.020, and slope of 0.003 ft/ft. Solve using Figure 10-4:

1. Connect a line between the slope scale at 0.003 and the roughness scale at 0.020, and note the intersection point on the turning line.
2. Connect a line between that intersection point and the hydraulic radius scale at 0.6 ft, and read the velocity of 2.9 ft/s from the velocity scale.

Irregular Channel (Example 1)

A channel consists of four subsections with the properties contained in Table 10-7. A discharge of about 3500 cfs flows in the channel. Trial depths were developed, and all parameters per subsection were computed based on the trial depth. The discharge was calculated and compared to the desired discharge. This was continued until the values matched within a set tolerance. Based on this table, the maximum depth was found to be about 5.0 ft.

Grassed Channel Design Stability (Example 2)

A trapezoidal channel is required to carry 50 cfs at a bottom slope of 0.015 ft/ft. Find the channel dimensions required for design stability criteria (retardance Class D) for a grass mixture.

1. From Table 10-16, the maximum velocity, v_m, for a grass mixture with a bottom slope less than 5% is 4 ft/s.
2. Assume an n value of 0.035, and find the value of vR from Figure 10-6.

$$vR = 5.4$$

3. Use Equation (10.28) to calculate the value of R:

$$R = 5.4/4 = 1.35 \text{ ft}$$

TABLE 10-7

Example 1

	Section 1	Section 2	Section 3	Section 4
Top width, ft	25	75	15	75
Average Depth, ft	2.5	4.5	3	1
Area, ft²	62.50	337.50	45.00	75.00
Wetted Perimeter, ft	25.5	75	15.1	75
Hydraulic Radius, ft	2.45	4.50	2.98	1.00
Manning n	0.045	0.035	0.035	0.050
Conveyance	3762	39,181	3969	2235
Total conveyance		49,147		
Total area, ft²		520		
Discharge (slope = 0.005), cfs		3489		
Mean velocity, ft/s		6.71		

Open Channel Design

4. Use Equation (10.29) to calculate the value of vR:

$$vR = [1.49\,(1.35)^{5/3}\,(0.015)^{1/2}]/0.035 = 8.60$$

5. Because the vR value calculated in Step 4 is higher than the value obtained from Step 2, a higher n value is required, and calculations are repeated. The results from each trial of calculations are presented below:

Assumed n Value	vR (Figure 10-6)	R [Equation (10.28)]	vR [Equation (10.29)]
0.035	5.40	1.35	8.60
0.038	3.8	0.95	4.41
0.039	3.4	0.85	3.57
0.040	3.2	0.80	3.15

Select $n = 0.040$ for stability criteria.

6. Use Figure 10-4 to select channel dimensions for a trapezoidal shape with 3:1 side slopes.

$$Qn = (50)\,(0.040) = 2.0$$

$$S = 0.015$$

For $b = 10$ ft, $d = (10)\,(0.098) = 0.98$ ft

For $b = 8$ ft, $d = (8)\,(0.14) = 1.12$ ft

Select: $b = 10$ ft, such that R is approximately 0.80 ft

$z = 3$
$d = 1$ ft
$v = 3.9$ ft/s (equation 10.8)\
$Fr = 0.76$ (equation 10.26)
Flow is subcritical

Design capacity calculations for this channel are presented in Example 3 below.

Grassed Channel Design Capacity (Example 3)

Use a 10-ft bottom width and 3:1 side-slopes for the trapezoidal channel sized in Example 2 and find the depth of flow for retardance Class C:

1. Assume a depth of 1.0 ft and calculate the following (see Figure 10-6):

$$A = (b + zd)\,d = [10 + (3)\,(1)]\,(1)$$

$$A = 13.0 \text{ ft}^2$$

$$R = \{[b + zd]\,d\}/\{b + [2d\,(1 + z^2)^{0.5}]\}$$

$$R = \{[10 + (3)(1)] (1)\}/\{10 + [(2)(1)(1 + 3^2)^{0.5}]\} = 0.796 \text{ ft}$$

2. Find the velocity.

$$v = Q/A = 50/13.0 = 3.85 \text{ ft/s}$$

3. Find the value of vR.

$$vR = (3.85)(0.796) = 3.06$$

4. Using the vR product from Step 3, find Manning's n from Figure 10-6 for retardance Class C.

$$n = 0.047$$

5. Use Figure 10-3 or Equation (10.16) to find the velocity for $S = 0.015$, $R = 0.796$, and $n = 0.047$.

$$v = 3.34 \text{ ft/s}$$

6. Because 3.34 ft/s is less than 3.85 ft/s, a higher depth is required and calculations are repeated. Results from each trial of calculations are presented below:

Assumed Depth (ft)	Area (ft²)	R (ft)	Velocity Q/A (ft/s)	vR	Manning's n (Figure 10-6)	Velocity [Equation (10.16)]
1.0	13.00	0.796	3.85	3.06	0.047	3.34
1.05	13.81	0.830	3.62	3.00	0.047	3.39
1.1	14.63	0.863	3.42	2.95	0.048	3.45
1.2	16.32	0.928	3.06	2.84	0.049	3.54

7. Select a depth of 1.1 with an n value of 0.048 for design capacity requirements. Add at least 0.2 ft for freeboard to give a design depth of 1.3 ft. Design data for the trapezoidal channel are summarized as follows:

Vegetation lining = grass mixture, $v_m = 4$ ft/s
$Q = 50$ cfs
$b = 10$ ft, $d = 1.3$ ft, $z = 3$, $S = 0.015$ ft/ft
Top width = (10) + (2)(3)(1.3) = 17.8 ft
n (stability) = 0.040, $d = 1.0$ ft, $v = 3.9$ ft/s, Froude number = 0.76 [Equation (10.26)]
n (capacity) = 0.048, $d = 1.1$ ft, $v = 3.45$ ft/s, Froude number = 0.64 [Equation (10.26)]

10.12 Gradually Varied Flow

In reality, all flow varies gradually or rapidly. The most common occurrence of gradually varied flow in storm drainage is the backwater created by some downstream control culverts, bridges, storm drain inlets, channel constrictions or the drawdown from a weir,

Open Channel Design

dam or other overflow point. For these conditions, the flow depth will be different from normal depth in the channel, and the water surface profile should be computed using backwater techniques.

Many computer programs are available for computation of backwater curves. The most general and widely used one-dimensional flow programs are HEC-2, developed by the U.S. Army Corps of Engineers (1982), and Bridge Waterways Analysis Model (WSPRO), developed for the Federal Highway Administration (FHA, 1989). These programs can be used to compute water surface profiles for natural and artificial channels.

For prismatic channels, the backwater calculation can be computed manually using the direct step method, as presented by Chow (1959). For an irregular nonuniform channel, the standard step method is recommended,, although it is a more tedious and iterative process. The use of HEC-2 is recommended for standard step calculations.

Cross sections for water surface profile calculations should be normal to the direction of flow. The number of sections required will depend on the irregularity of the stream and floodplain. In general, a cross section should be obtained at each location where there are significant changes in stream width, shape, or vegetal patterns. Sections should usually be no more than four to five channel widths apart or 100 ft apart for ditches or streams, and 500 ft apart for floodplains, unless the channel is very regular.

Direct Step Method

The direct step method is limited to prismatic channels. A form for recording the calculations described below is presented in Table 10-8 (Chow, 1959).

1. Record the following parameters across the top of Table 10-8:

   Q = design flow (cfs)

   n = Manning's n value

   S_o = channel bottom slope (ft/ft)

   α = energy coefficient

   y_c = critical depth (ft)

   y_n = normal depth (ft)

2. Using the desired range of flow depths, y, recorded in column 1, compute the cross-sectional area, A, the hydraulic radius, R, and average velocity, v, and record results in columns 2, 3, and 4, respectively.
3. Compute the velocity head, $\alpha v^2/2g$, in ft, and record the result in column 5.
4. Compute specific energy, delta E, in ft, by summing the velocity head in column 5 and the depth of flow in column 1. Record the result in column 6.
5. Compute the change in specific energy, delta E, between the current and previous flow depths, record the result in column 7 (not applicable for row 1).
6. Compute the friction slope using the equation:

$$S_f = (n^2 v^2)/(2.22 R^{4/3}) \tag{10.33}$$

where S_f = friction slope (ft/ft); n = Manning's n value; v = average velocity (ft/s); and R = hydraulic radius (ft).

Record the result in column 8.

TABLE 10-8

Water Surface Profile Computation Form for the Direct Step Method

Location _____

Q = _____ n = _____ S$_o$ = _____ n = _____ k$_o$ = _____ y$_c$ = _____ y$_n$ = _____

Station (1)	z (2)	y (3)	A (4)	R (5)	V (6)	nv²/2g (7)	H (8)	$\bar{S}_1$ (9)	S$_1$ (10)	Δx (11)	h$_1$ (12)	h$_e$ (13)	H (14)
1.													
2.													
3.													
4.													
5.													
6.													
7.													
8.													
9.													
10.													
11.													
12.													
13.													
14.													
15.													
16.													
17.													
18.													
19.													
20.													
21.													
22.													

Notes: _____

Note: Q = 400 cfs; n = 0.025; S$_o$ = 0.0016 ft/ft; α = 1.10; y$_c$ = 2.22 ft; and y$_n$ = 3.36 ft.

Source: From Chow, V.T., Ed., *Open Channel Hydraulics*, McGraw Hill, New York, 1959.

7. Determine the average of the friction slope between this depth and the previous depth (not applicable for row 1). Record the result in column 9.
8. Determine the difference between the bottom slope, S$_o$, and the average friction slope, $\bar{S}_f$, from column 9 (not applicable for row 1). Record the result in column 10.
9. Compute the length of channel between consecutive rows or depths of flow using the equation:

$$\text{delta } x = \text{delta } E/(S_o - S_f) = \text{Col. 7/Col. 10} \qquad (10.34)$$

where delta x = length of channel between consecutive depths of flow (ft); delta E = change in specific energy (ft); S_o = bottom slope (ft/ft); and S_f = friction slope (ft/ft).

Record the result in column 11.

10. Sum the distances from the starting point to give cumulative distances, x, for each depth in column 1, and record the result in column 12.

Standard Step Method

The standard step method is a trial-and-error procedure applicable to natural and prismatic channels. The step computations are arranged in tabular form, as shown in Table 10-9 and described below (Chow, 1959):

Open Channel Design

TABLE 10-9

Water Surface Profile Computation Form for the Standard Step Method

y (1)	A (2)	R (3)	$R^{4/3}$ (4)	V (5)	$\alpha V^2/2g$ (6)	E (7)	ΔE (8)	S_f (9)	$\bar{S}_f$ (10)	$S_0 - \bar{S}_f$ (11)	Δx (12)	x (13)
5.00	150.00	3.54	5.40	2.667	0.1217	5.1217	—	0.000370				
4.80	142.08	3.43	5.17	2.819	0.1356	4.9356	0.1861	0.000433	0.000402	0.001198	155	155
4.60	134.32	3.31	4.94	2.979	0.1517	4.7517	0.1839	0.000507	0.000470	0.001130	163	318
4.40	126.72	3.19	4.70	3.156	0.1706	4.5706	0.1811	0.000598	0.000553	0.001047	173	491
4.20	119.28	3.08	4.50	3.354	0.1925	4.3925	0.1781	0.000705	0.000652	0.000948	188	679
4.00	112.00	2.96	4.25	3.572	0.2184	4.2184	0.1741	0.000850	0.000778	0.000822	212	891
3.80	104.88	2.84	4.02	3.814	0.2490	0.0490	0.1694	0.001020	0.000935	0.000665	255	1146
3.70	101.38	2.77	3.88	3.948	0.2664	3.9664	0.0826	0.001132	0.001076	0.000524	158	1304
3.60	97.92	2.71	3.78	4.085	0.2856	3.8856	0.0808	0.001244	0.001188	0.000412	196	1500
3.55	96.21	2.68	3.72	4.158	0.2958	3.8458	0.0398	0.001310	0.001277	0.000323	123	1623
3.50	94.50	2.65	3.66	4.233	0.3067	3.8067	0.0391	0.001382	0.001346	0.000254	154	1777
3.47	93.48	2.63	3.63	4.278	0.3131	3.7831	0.0236	0.001427	0.001405	0.000195	121	1898
3.44	92.45	2.61	3.59	4.326	0.3202	3.7602	0.0229	0.001471	0.001449	0.000151	152	2050
3.42	91.80	2.60	3.57	4.357	0.3246	3.7446	0.0156	0.001500	0.001486	0.000114	137	2187
3.40	91.12	2.59	3.55	4.388	0.3292	3.7292	0.0154	0.001535	0.001518	0.000082	188	2375

Note: $Q = 400$ cfs; $n = 0.025$; $S_0 = 0.0016$ ft/ft; $\alpha = 1.10$; $y_c = 2.22$ ft; and $y_n = 3.36$ ft.

Source: From Chow, V.T., Ed., *Open Channel Hydraulics*, McGraw Hill, New York, 1959.

1. Record the following parameters across the top of Table 10-9:

    Q = design flow (cfs)

    n = Manning's n value

    S_o = channel bottom slope (ft/ft)

    α = energy coefficient

    k_e = eddy head loss coefficient (ft)

    y_c = critical depth (ft)

    y_n = normal depth (ft)

2. Record the location of the measured channel cross sections and the trial water surface elevation, z, for each section in columns 1 and 2. The trial elevation will be verified or rejected based on computations of the step method.
3. Determine the depth of flow, y, based on trial elevation and channel section data. Record the result in column 3.
4. Using the depth from Step 3 and section data, compute the cross-sectional area, A, in ft², and hydraulic radius, R, in ft. Record results in columns 4 and 5.
5. Divide the design discharge by the cross-sectional area from Step 4 to compute the average velocity, v, in ft/s. Record the result in column 6.
6. Compute the velocity head, $\alpha v^2/2g$, in ft, and record the result in column 7.
7. Compute the total head, H, in ft, by summing the water surface elevation, z, in column 2 and velocity head in column 7. Record the result in column 8.
8. Compute the friction slope, S_f, using Equation (10.33), and record the result in column 9.
9. Determine the average friction slope, $\bar{S}_f$, between the sections in each step (not applicable for row 1). Record the result in column 10.

10. Determine the distance between sections, delta x, and record the result in column 11.
11. Multiply the average friction slope, $\overline{S}_f$, (column 10), by the reach length, delta x (column 11), to give the friction loss in the reach, h_f. Record the result in column 12.
12. Compute the eddy loss using the following equation:

$$h_e = (k_e v^2)/2g \qquad (10.35)$$

where h_e = eddy headloss (ft); k_e = eddy headloss coefficient, ft (for prismatic and regular channels, $k_e = 0$; for gradually converging and diverging channels, $k_e = 0$ to 0.1 or 0.2; for abrupt expansions and contractions, $k_e = 0.5$); v = average velocity (ft/s) (column 6); and g = acceleration due to gravity (32.2 ft/s²).

13. Compute the elevation of the total head, H, by adding the values of h_f and h_e (columns 12 and 13) to the elevation at the lower end of the reach, which is found in column 14 of the previous reach. Record the result in column 14.
14. If the value of H computed above does not agree closely with that entered in column 8, a new trial value of the water surface elevation is used in column 2, and calculations are repeated until agreement is obtained. The computation may then proceed to the next step or section reported in column 1.

10.13 Gradually Varied Flow — Example Problems

Direct Step Method

Use the direct step method to compute a water surface profile for a trapezoidal channel using the following data:

$Q = 400$ cfs
$B = 20$ ft
$z = 2$
$S = 0.0016$ ft/ft
$n = 0.025$
$\alpha = 1.10$

A dam backs up water to a depth of 5 ft immediately behind the dam. The upstream end of the profile is assumed to have a depth 1% greater than normal.

Results of calculations, as obtained from Chow (1959), are reported in Table 10-10. Values in each column of the table are briefly explained below.

1. Depth of flow, in ft, arbitrarily assigned values ranging from 5 to 3.4 ft.
2. Water area, in ft², corresponding to the depth, y, in column 1.
3. Hydraulic radius, in ft, corresponding to y in column 1.
4. Mean velocity, in ft/s, obtained by dividing 400 cfs by the water area in column 2.
5. Velocity head, in ft, calculated using the mean velocity from column 4 and an α value of 1.1.

TABLE 10-10
Direct Step Method Results for Example

Station (1)	Z (2)	y (3)	A (4)	V (5)	$\alpha V^2/2g$ (6)	H (7)	R (8)	$R^{4/3}$ (9)	S_f (10)	$\bar{S}_f$ (11)	Δx (12)	h_f (13)	h_e (14)	H (15)
0 + 00	605.000	5.00	150.00	2.667	0.127	605.122	3.54	5.40	0.000370	—	—	—	—	605.122
1 + 55	605.048	4.80	142.08	2.819	0.1356	605.184	3.43	5.17	0.000433	0.000402	155	0.062	0	605.184
3 + 18	605.109	4.60	134.32	2.979	0.1517	605.261	3.31	4.92	0.000507	0.000470	163	0.077	0	605.261
4 + 91	605.186	4.40	126.72	3.156	0.1706	605.357	3.19	4.70	0.000598	0.000553	173	0.096	0	605.357
6 + 79	605.286	4.20	199.28	3.354	0.1925	605.479	3.08	4.50	0.000705	0.000652	188	0.122	0	605.479
8 + 91	605.426	4.00	112.00	3.572	0.2184	605.644	2.96	4.25	0.000850	0.000778	212	0.165	0	605.644
11 + 46	605.633	3.80	104.38	3.814	0.2490	605.882	2.84	4.02	0.001020	0.000935	255	0.238	0	605.882
13 + 04	605.786	3.70	101.38	3.948	0.2664	606.052	2.77	3.88	0.001132	0.001076	158	0.170	0	606.052
15 + 00	605.999	3.60	97.92	4.085	0.2856	606.285	2.71	3.78	0.001244	0.001188	196	0.233	0	606.285
16 + 23	606.146	3.55	96.21	4.158	0.2958	606.442	2.68	3.72	0.001310	0.001277	123	0.157	0	606.442
17 + 77	606.343	3.50	94.50	4.233	0.3067	606.650	2.65	3.66	0.001382	0.001346	154	0.208	0	606.650
18 + 98	606.507	3.47	93.48	4.278	0.3131	606.820	2.63	3.63	0.001427	0.001405	121	0.170	0	606.820
20 + 50	606.720	3.44	92.45	4.326	0.3202	607.040	2.61	3.59	0.001471	0.001449	152	0.220	0	607.040
21 + 87	606.919	3.42	91.80	4.357	0.3246	607.244	2.60	3.57	0.001500	0.001486	137	0.204	0	607.244
23 + 75	607.201	3.40	91.12	4.388	0.3292	607.530	2.59	3.55	0.001535	0.001518	188	0.286	0	607.530

6. Specific energy, E, in ft, obtained by adding the velocity head in column 5 to the depth of flow in column 1.
7. Change of specific energy, delta E, in ft, equal to the difference between the E value in column 6 and that of the previous step.
8. Friction slope, S_f, computed by Equation (10.33), with $n = 0.025$, v as given in column 4, and R as given in column 3.
9. Average friction slope between the steps, $\overline{S}_f$, equal to the arithmetic mean of the friction slope computed in column 8 and that of the previous step.
10. Difference between the bottom slope, S_o, 0.0016 and the average friction slope, $\overline{S}_f$, in column 9.
11. Length of the reach, delta x, in ft, between the consecutive steps computed by Equation (10.34) or by dividing the value of delta E in column 7 by the value of $S_o - \overline{S}_f$ in column 10.
12. Distance from the section under consideration to the dam site. This is equal to the cumulative sum of the values in column 11 computed for previous steps.

Standard Step Method

Use the standard step method to compute a water surface profile for the channel data and stations considered in the previous example. Assume the elevation at the dam site is 600 ft.

Results of the calculations, after Chow (1959), are reported in Table 10-11. Values in each column of the table are briefly explained below:

1. Section identified by station number such as "station 1 + 55." The locations of the stations are fixed at the distances determined in the previous example to compare the procedure with that of the direct step method.
2. Water surface elevation, z, at the station. A trial value is first entered in this column; this will be verified or rejected on the basis of the computations made in the remaining columns of the table. For the first step, this elevation must be given or assumed. Because the elevation of the dam site is 600 ft and the height of the dam is 5 ft, the first entry is 605.00 ft. When the trial value in the second step has been verified, it becomes the basis for the verification of the trial value in the next step, and the process continues.
3. Depth of flow, y, in ft, corresponding to the water surface elevation in column 2. For instance, the depth of flow at station 1 + 55 is equal to the water surface elevation minus the elevation at the dam site minus the distance from the dam site times bed slope.

$$605.048 - 600.00 - (155)(0.0016) = 4.80 \text{ ft}$$

4. Water area, A, in ft², corresponding to y in column 3.
5. Hydraulic radius, R, in ft, corresponding to y in column 3.
6. Mean velocity, v, equal to the discharge, 400 cfs, divided by the water area in column 4.
7. Velocity head, in ft, corresponding to the velocity in column 6 and an α value of 1.1.

Open Channel Design

TABLE 10-11
Standard Step Method Results for Example

Station (1)	Z (2)	y (3)	A (4)	V (5)	αV²/2g (6)	H (7)	R (8)	R^(4/3) (9)	S_f (10)	$\bar{S}_f$ (11)	Δx (12)	h_f (13)	h_e (14)	H (15)
0 + 00	605.00	5.00	150.00	2.667	0.1217	605.122	3.54	5.40	0.000370	—	—	—	—	605.122
1 + 55	605.048	4.80	142.08	2.819	0.1356	605.184	3.43	5.17	0.000433	0.000402	155	0.062	0	605.184
3 + 18	605.109	4.60	134.32	2.979	0.1517	605.261	3.31	4.92	0.000507	0.000470	163	0.077	0	605.261
4 + 91	605.186	4.40	126.72	3.156	0.1706	605.357	3.19	4.70	0.000598	0.000553	173	0.096	0	605.357
6 + 79	605.286	4.20	119.28	3.354	0.1925	605.479	3.08	4.50	0.000705	0.000652	188	0.122	0	605.479
8 + 91	605.426	4.00	112.00	32.572	0.2184	605.644	2.96	4.25	0.000850	0.000778	212	0.165	0	605.644
11 + 46	605.633	3.80	104.88	3.814	0.2490	605.882	2.84	4.02	0.001020	0.000935	255	0.238	0	605.882
13 + 04	605.786	3.70	101.38	3.948	0.2664	606.052	2.77	3.88	0.001132	0.001076	158	0.170	0	606.052
15 + 00	605.999	3.60	97.92	4.085	0.2856	606.285	2.71	3.78	0.001244	0.001188	196	0.233	0	606.285
16 + 23	606.146	3.55	96.21	4.158	0.2958	606.442	2.68	3.72	0.001310	0.001277	123	0.157	0	606.442
17 + 77	606.343	3.50	94.50	4.233	0.3067	606.650	2.65	3.66	0.001382	0.001346	154	0.208	0	606.650
18 + 98	606.507	3.47	93.48	4.278	0.3131	606.820	2.63	3.63	0.001427	0.001405	121	0.170	0	606.820
20 + 50	606.720	3.44	92.45	4.326	0.3202	607.040	2.61	3.59	0.001471	0.001449	152	0.220	0	607.040
21 + 87	606.919	3.42	91.80	4.357	0.3246	607.244	2.60	3.57	0.001500	0.001486	137	0.204	0	607.244
23 + 75	607.201	3.40	91.12	4.388	0.3292	607.530	2.59	3.55	0.001535	0.001518	188	0.286	0	607.530

8. Total head, H, equal to the sum of z in column 2 and the velocity head in column 7.
9. Friction slope, S_f, computed by Equation (10.33), with $n = 0.025$, v from column 6, and R from column 5.
10. Average friction slope through the reach, $\bar{S}_f$, between the sections in each step, approximately equal to the arithmetic mean of the friction slope just computed in column 9 and that of the previous step.
11. Length of the reach between the sections, delta x, equal to the difference in station numbers between the stations.
12. Friction loss in the reach, h_f, equal to the product of the values in columns 10 and 11.
13. Eddy loss in the reach, h_e, equal to zero.
14. Elevation of the total head, H, in ft, computed by adding the values of h_f and h_e in columns 12 and 13 to the elevation at the lower end of the reach, which is found in column 14 of the previous reach. If the value obtained does not agree with that entered in column 8, a new trial value of the water surface elevation is assumed until agreement is obtained. The value that leads to agreement is the correct water surface elevation. The computation may then proceed to the next step.

10.14 Hydraulic Jump

A hydraulic jump can occur when flow passes rapidly from supercritical to subcritical depth. The evaluation of the hydraulic jump should consider the high energy loss and erosive forces that are associated with the jump. For rigid-lined facilities such as pipes or concrete channels, the forces and the change in energy can affect the structural stability or the hydraulic capacity. For grass-lined channels, unless the erosive forces are controlled, serious damage can result. Control of jump location is usually obtained by check dams or grade control structures that confine the erosive forces to a protected area. Flexible material such as riprap or rubble usually affords the most effective protection.

The analysis of the hydraulic jump inside storm drains must be approximate, because of the lack of data for circular, elliptical, or arch sections. The jump can be approximately located by intersecting the energy grade line of the supercritical and subcritical flow reaches. The primary concerns are whether the pipe can withstand the forces, which may separate the joint or damage the pipe wall, and whether the jump will affect the hydraulic characteristics. The effect on pipe capacity can be determined by evaluating the energy grade line, taking into account the energy lost by the jump. In general, for Froude numbers less than 2.0, the loss of energy is less than 10%. French (1985) provides semiempirical procedures to evaluate the hydraulic jump in circular and other nonrectangular channel sections.

For long box culverts with a concrete bottom, the concerns about jump are the same as for storm drains. However, the jump can be adequately defined for box culverts and drains and for spillways using the jump characteristics of rectangular sections.

The relationship between variables for a hydraulic jump in rectangular sections can be expressed as:

$$d_2 = -(d_1/2) + [(d_1^2/4) + (2v_1^2 d_1/g)]^{1/2} \tag{10.36}$$

where d_2 = depth below jump (ft); d_1 = depth above jump (ft); v_1 = velocity above jump (ft/s); and g = acceleration due to gravity (32.2 ft/s²).

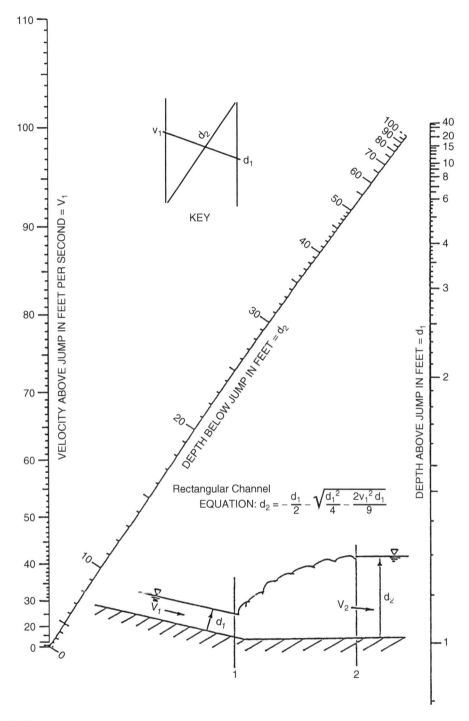

FIGURE 10-8
Nomograph for solving the rectangular channel hydraulic jump equation. (From U.S. Dept. Of Interior, 1973.)

A nomograph for solving Equation (10.36) is presented in Figure 10-8. Additional details on hydraulic jumps can be found in HEC-14 (US DOT, 1982), Chow (1959), Peterska (1978), and French (1985).

10.15 Channel Bank Protection

The protection of the channel bed and banks is both an art and a science. Rarely will the designer find a situation that includes a straight uniform channel with uniform soils and no extenuating circumstances. Therefore, the designer must choose among a wide variety of possible bank protection methods to meet the demands of space restrictions, rapid transitions, cost limitations, soil variability, aesthetic demands, environmental and habitat concerns, regulatory demands, and much more. Several of the more common types of bank protection used are described below. For all but the generic types, the designer should consult additional literature to ensure a complete understanding of the limitations and details of design method.

Streambanks subject to erosion are protected by stabilizing eroding soils, planting vegetation, covering the banks with various materials, or building structures to deflect stream currents away from the bank. Placement and type of bank protection vary, depending on the cause of erosion, environmental objectives, and cost. Table 10-12 identifies streambank protection measures appropriate for different problems, and ranks them according to overall environmental benefits.

Riprap Design

The following procedure is based on results and analysis of laboratory and field data (Maynord, 1987; Reese, 1984, 1988). This procedure applies to riprap placement in natural and prismatic channels and has the following assumptions and limitations:

- Minimum riprap thickness equal to d_{100}
- Value of d_{85}/d_{15} less than 4.6

TABLE 10-12

Streambank Protection Measures

Problems	Appropriate Protection Measures
Toe erosion and upper bank failure	1. Earth core dikes with vegetation
	2. Stone toe protection with vegetation
	3. Cribwalls with vegetation
	4. Gabions
	5. Bulkheads
Local scour	1. Woody vegetation
	2. Grass
	3. Riprap and cribwalls with vegetation
	4. Conventional riprap
	5. Gabions
General scour	1. Woody vegetation
	2. Grass
	3. Riprap with vegetation
	4. Conventional riprap
	5. Gabions
Mass failure	1. Rock toe protection with vegetation
	2. Cribwalls with vegetation
	3. Bulkheads
Overbank runoff and piping	1. Flow diversion
	2. Drop inlets
	3. Drop structures

- Froude number less than 1.2
- Side slopes up to 2:1
- Safety factor of 1.2
- Maximum velocity less than 18 ft/s

If significant turbulence is caused by boundary irregularities, such as installations near obstructions or structures, this procedure is not applicable.

Guidance for riprap design in a typical small stream urban setting is from a combination of sources including Reese (1984, 1988), Maynord (1987), and guidance from several federal agencies. Equation 10.37 gives the D_{50} size of stone (in inches) for riprap placed in a channel with average velocity V and depth d.

$$D_{50} = 0.0136\ V^3/d^{0.5}K^{1.5} \tag{10.37}$$

K is the side slope correction factor and can be found from Equation (10.38). It should be used for all side slope placement on slopes steeper then 1V:4H. For other placement, K can be taken equal to one (1.0). θ is equal to the bank angle with the horizontal (e.g., a 1V:3H slope has a θ value of 18.43 degrees). The method assumes a typical riprap angle of repose of 39 degrees. If stone is very flat or very rounded, the 0.396 term should be replaced by the $\sin^2$ of the estimated riprap angle of repose.

$$K = [1 - (\sin^2\theta/0.396)]^{0.5} \tag{10.38}$$

Equation (10.37) is based on a safety factor of 1.2 and a stone specific weight of 165 lbs/ft³. For situations other than a uniform straight channel, the D_{50} size from Equation (10.37) should be multiplied by a stability correction factor found below and used in Equation (10.39).

Condition	Stability Factor (SF)
Uniform flow; straight or mildly curving reach [curve radius/channel topwidth ($R_c/T > 30$)]; little impact from wave action and floating debris; little uncertainty in design parameters	1.0–1.2
Gradually varied flow; moderate bend curvature ($30 > R_c/T > 10$); moderate impact from waves or debris; moderate uncertainty in design parameters	1.3–1.6
Approaching rapidly varied flow; sharp bend curvature ($10 > R_c/T$); significant impact from waves or debris; high flow turbulence; significant uncertainty in design parameters	1.6–2.0

$$C_{SF} = (SF/1.2)^{1.5} \tag{10.39}$$

If the rock density is significantly different from 165 lbs/ft³, the D_{50} size found in Equation (10.37) should be multiplied by a specific weight correction factor (C_w) found in Equation (10.40). S_w is the specific weight of the stone.

$$C_w = [102.6 / (S_w - 62.4)]^{1.5} \tag{10.40}$$

The riprap layer thickness should be a minimum of D_{100} and the D_{85}/D_{15} value should be less than 4.6. Stone should be angular in shape. Riprap should be placed so as not to be flanked by the flow. The end of the protected section should be keyed into the bank to prevent scouring failure. For riprap blanket thicknesses greater than D_{100}, the following reductions in D_{50} stone size are allowed:

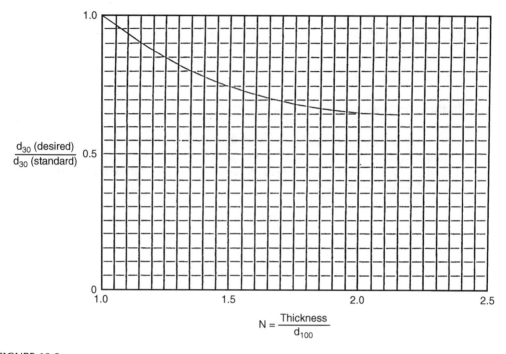

FIGURE 10-9
Riprap lining thickness adjustment for d_{85}/d_{15} = 2.0 to 2.3. (From Maynord, S.T., Stable Riprap Size for Open Channel Flows, Ph.D. dissertation, Colorado State University, Fort Collins, CO, 1987.)

- For blanket thickness equal to 1.5 D_{100} the D_{50} size can be reduced 25%.
- For blanket thickness equal to 2.0 D_{100} the D_{50} size can be reduced 40%.

Channel design must account for riprap thickness in channel excavation. If the complete channel is to be designed as a riprap channel, the riprap controls the velocity, and an iterative process between riprap size and average channel velocity due to riprap roughness is necessary. Channel roughness for riprap-lined channels can be evaluated from (D_{50} in feet):

$$n = 0.0395 \, (D_{50})^{1/6} \qquad (10.41)$$

The stone weight for the selected diameter can be calculated from:

$$W = 0.5236 \, S_w \, D_{50}^3 \qquad (10.42)$$

where W = stone weight (lbs), and S_w = specific weight of the stone (lb/ft³).

Normally, an apron of riprap is placed and keyed into the channel bottom. Figure 10-10 gives details of the placement by several methods. Gradation of riprap material should meet the following rules for stone weight (W) for the upper and lower limit curves for gradation (USCOE, 1970):

- W_{50} lower limit not less than size indicated by procedure above
- W_{100} lower limit ≥ W_{50} lower limit
- Upper limit W_{100} ≤ 5 lower limit W_{50}
- Lower limit W_{15} ≥ 1/16 W_{100}

Open Channel Design

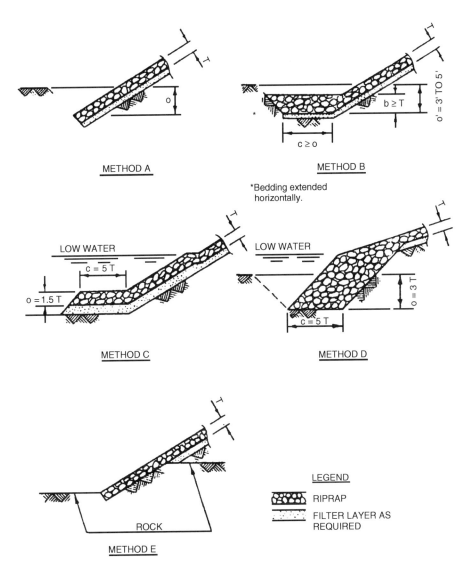

FIGURE 10-10
Typical riprap placement. (From COE, 1971.)

The relationship between the weight (W in pounds) of the stone and its equivalent spherical diameter (D in feet) is:

$$D = (6W/\pi\gamma_s)^{1/3} \qquad (10.43)$$

where γ_s = the specific weight of the stone (about 165 lbs/ft^3).

The methods in Figure 10-10 should be used under the following circumstances:

- Method A — Excavation made in the dry extend below the bed to anticipated scour depth
- Method B — If toe excavation is wet, the horizontal rock should be extended horizontally at a depth of 3 to 5 ft for medium-sized channels, thickness "b" not less than the layer thickness, $c \geq a$

- Method C — For underwater placement with low velocities, toe at existing channel bottom, and a = 1.5 thickness, and c = 5 thickness.
- Method D — Underwater placement with expected erosion, a thickened rock toe should be placed in trench, a = 3 thickness, and c = 5 thickness
- Method E — Channel bottom is in rock, the layer should be keyed into the rock

Riprap Design Example

A natural channel has a calculated velocity of 8.5 ft/s and a depth of 3 ft. Riprap is to be placed on the outside of a bend on a 1V:3H side slope. Moderate turbulence is expected (1.5 safety factor assumed). The side-slope correction factor can be found from:

$$K = [1- (Sin^2\theta/0.396)]^{0.5} = [1- (Sin^2 18.43/0.396)]^{0.5} = 0.865$$

Then, the D_{50} is:

$$D_{50} = 0.0136\ V^3/d^{0.5}K^{1.5} = 0.0136\ (8.5)^3/(3)^{0.5}(0.865)^{1.5} = 6\ \text{inches}$$

Applying the stability correction factor of 1.5:

$$C_{SF} = (SF/1.2)^{1.5} = (1.5/1.2)^{1.5} = 1.397$$

Then, the corrected D_{50} size is:

$$1.397 * 6 = 8.4\ \text{inches (use 8 1/2 inches)}$$

The design is then finished by determining the available gradation, use of filter cloth, placement means, thickness, extent, and end treatment.

Grouted Riprap

Grouted riprap has advantages over loose riprap in that it can be placed in steeper areas, is more stable for low-flow channels, can take advantage of smaller stone sizes, and reduces weed growth and trash accumulation. Because it is no longer a flexible lining, it loses the self-healing properties of dumped stone. Grout should contain air entrainment, have a 28-day strength of at least 2400 psi and a high slump (5 to 7 in) in order to penetrate at least 2 ft into the riprap layer (UDFCD, 1991). Grout penetration may be accomplished by rodding, vibrating, or pumping. The fines from a typical gradation of riprap can be removed to allow for better penetration. The appearance can be improved by leaving the outer rocks exposed (similar to exposed aggregate). Weep holes should be provided to allow for reduction in lift forces and hydrostatic pressure buildup behind the stone.

Gabions and Rock Mattresses

Wire-enclosed rock mattresses and blocks (gabions) can be used in most places where concrete would otherwise be specified, because riprap would be unstable. Gabions are used routinely for vertical channel banks, rock fill drops and weirs, mattress-type bank and shore protection, and other miscellaneous water resource uses (Jacobs, 1984). However, gabion wire is susceptible to chemical degradation from high-sulfate soils or acidic

Open Channel Design

TABLE 10-13

U.S. Gabion Dimensions Maccaferri (8 × 10 mesh)

Letter Code	Length (ft)	Width (ft)	Height (ft)	Number of Diaphragms	Capacity (yd³)	Min. Rock Dimensions (in.)
A	6	3	3	1	2.0	4
B	9	3	3	2	3.0	4
C	12	3	3	3	4.0	4
D	6	3	1.5	1	1.0	4
E	9	3	1.5	2	1.5	4
F	12	3	1.5	3	2.0	4
G	6	3	1	1	0.66	4
H	9	3	1	2	1.0	4
I	12	3	1	3	1.33	4

Slope (Reno) Mattresses (6 × 8 mesh)						
Letter Code	Length (ft)	Width (ft)	Thickness (in)	Number of Cells	Capacity (yd³)	Area (yd²)
Q	9	6	6	3	1.0	6
R	12	6	6	4	1.33	8
T	9	6	9	3	1.5	6
U	12	6	9	4	2.0	8

Source: From Maccaferri, Maccaferri Gabions for River Training, Earth Control and Soil Conservation, Bologna, Italy, 1990.

pollution (plastic-coated wire can reduce this impact) and from the abrasive action of cobbles, rocks, and angular sand transported in the channel. The normal service life of wire-enclosed rock is considered to be 15 years if there are no extraordinary circumstances to shorten this expectancy (UDFCD, 1991; Stephenson, 1979; Bekaert, 1979). Maintenance requirements include periodic inspection to patch broken or cut wires. Vandalism can be a problem with wire-enclosed rock. In some urban areas, mattresses should be buried under several inches of grassed soil to protect them.

Tests conducted on the stability of such mattresses found that the thickness of the wire-enclosed rock could be one-third that of riprap while maintaining the same resistance to erosion, and that the stability of the rocks was twice that of nonenclosed rocks (Simons et al., 1984). Field experience from studies by the U.S. Army and in a few sites in Wyoming also prove the same finding. Many of the failures of gabions come from bedding failures, not of the mattress. Standard sizes of gabions are given in Table 10-13.

Another type of wire enclosed member is termed a "rock sausage" or "sack gabion," due to its round and elongated shape. Originally pioneered centuries ago by the Chinese (in bamboo and rope cages), rock sausages have found wide use in the Far East. Little design information is available (Posey, 1973).

Considerations for design of mattresses of wire-enclosed rock (similar to the proprietary Reno Mattress, Maccaferri, 1976, 1979) for placement in channels in lieu of tested local information and experience (Simons et al., 1984) are:

- Manning roughness can be found from:

$$n_b = d_{90}^{1/6}/31.69 \tag{10.44}$$

where n_b = Manning roughness coefficient, and d_{90} = size fraction for which 90% is finer (ft).

- Bed, maximum bank, and critical and bank critical shear stress are found from (for a chosen rock median diameter):

$$\tau = \gamma DS \qquad (10.45)$$

$$\tau_m = 0.75\tau \qquad (10.46)$$

$$\tau_c = 0.1 \, (\gamma_s - \gamma) \, d_m \qquad (10.47)$$

$$\tau_s = \tau_c [\, 1 - (\sin^2 \Theta / \sin^2 \phi)]^{0.5} \qquad (10.48)$$

where τ = bed shear stress (lbs per ft^2); τ_m = maximum shear on the banks (lbs per ft^2); τ_c = critical shear stress (lbs per ft^2); τ_s = critical shear stress on banks (lbs per ft^2); γ = specific weight of water (62.4 lbs/ft^3); D = flow depth (ft); d_m = median rock diameter (ft); S = channel slope (ft/ft); Θ = bank slope,(degrees); and ϕ = angle of repose of rock (about 41 degrees).

- For bed placement, compare τ with τ_c. For bank placement, compare τ_m with τ_s. If the former is greater than the latter, there will be some displacement or deformation. If the former is less than 20% greater than the latter, there will only be deformation, the mattresses will remain stable.
- Mattress thickness can be computed from:

$$T_m = 6.67 \, \tau_c - 16.67 \qquad (10.49)$$

where T_m = mattress thickness (in).

- For velocities over 15 to 25 ft/s, consider grouting the mattresses with sand asphalt mastic rather than using a considerably thicker mattress. Percentages of sand, filler, and bitumen are about 70:15:15, with slightly more sand and bitumen for underwater placement. Application rates are 20 to 35 lbs/ft^2 for consolidating and 35 to 50 for sealing. (Maccaferri, 1976). Grouting modifies the Manning n value to 0.016.
- For smaller velocities, use a geotextile filter under the mattresses. For larger velocities, design a gravel filter for placement under the mattresses. Geotextiles are often used in combination with granular material to provide drainage (Jacobs, 1984; UDFCD, 1991). Filter fabrics can tear at steeper slopes and should be limited to placement on 2.5:1 slopes. In fine silts and clays, clogging of the pores can also take place, and a granular filter is recommended. Compute the velocity at the mattress-filter and filter-bank interfaces using Manning equation with n = 0.02 for filter fabric and n = 0.025 for gravel.

$$V_f = 1.486/n_f \, (d_m/2)^{2/3} \, S^{1/2} \qquad (10.50)$$

$$V_b = 0.5 \, V_b \qquad (10.51)$$

where V_f = velocity at mattress-filter interface (ft/s), and V_b = velocity at filter-bank soil interface (ft/s).

- Compare with erosive velocity of bank soil from the permissible velocity method above or, for noncohesive and cohesive soils, respectively:

$$V_e = 1.67 \, d_s^{1/2} \tag{10.52}$$

$$V_e = 24.92 \, \tau_e^{1/2} \tag{10.53}$$

τ_e can be computed from maximum permissible velocities given earlier or from Figure 10-11.

- If $V_b < V_e$ use geotextile filter fabric.
- If $V_b > V_e$ design a gravel filter with a thickness that is not less than 6 to 9 in using the following gradation:
  - d50 (filter)/d50 (base) < 40
  - 5 < d15 (filter)/d15 (base) < 40
  - d15 (filter)/d85 (base) < 5
- The filter thickness can be computed from the following equation [Use 6 in or L (in the units of d_m), whichever is greater]:

$$L = 4d_m[1 - (V_e/V_f)^2] \tag{10.54}$$

Soil Cement

Soil cement has been used successfully and cost effectively in a number of applications throughout the United States and the world. It has been especially successful in arid climates, where the naturally sandy soil, low in organic content, and wide unstable streams are conducive for soil cement use. Soil cement is a material formed by blending,

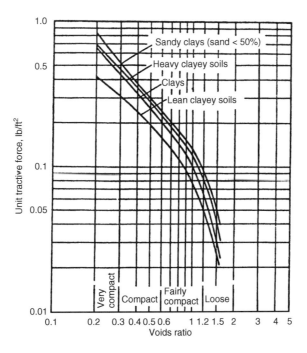

FIGURE 10-11
Permissible unit tractive force for channel in cohesive material from USSR data. (From USSR, Maximum Permissible Velocity in Open Channels, Hydro-Technical Construction, Number 5, Moscow, 1936.)

compacting, and curing a mixture of soil/aggregate, Portland cement, and water and allowing it to harden in place (Reese, 1987).

The type of soils conducive for use of soil cement has the following characteristics (PCA, 1975):

- Sands with at least 55% passing No. 4 sieve
- No material retained on the No. 2 sieve
- Material in the amount of 5 to 35% passing the No. 200 sieve
- Plasticity index (PI) less than 8 [Pima County recommends 3 (1986)]
- Less than 2% organic material

Violation of the fines standard may require more cement to make up the strength loss. Violation of the plasticity standard will require hydrated lime to improve friability. Soils with too much organic material can be treated with calcium chloride to hasten hydration. Gravel content can be increased to improve abrasion resistance for bank protection in high sediment transport situations. Pima County, AZ, has experienced long-duration flows with high sediment content and velocities over 20 ft/s against soil-cement bank protection without noticeable loss of material. Other instances of abrasion and wear have been investigated with mixed results (Reese, 1987). The most common cause of failure was flanking and undermining of the soil-cement monolithic structure. Soil-cement emplacements for bank protection should be tied well into the bank and toed well into the soil below expected scour depths. A Manning n value of 0.022 is normally assigned to soil-cement, trimmed and with a smooth appearance.

Cement content by dry weight is normally in the 7 to 14% range (bags of cement weigh 94 pounds) (PCA, 1976). ASTM cement types I and II are normally used with type V used for high-sulfate content situations in the soil or water (>0.2% in soil and >1500 ppm in the water). Compressive strength tests from ASTM are available for soil-cement as D1632, D1633, and D2901 as well as wet-dry testing and freeze-thaw testing. Density of soil-cement can normally be taken as 120 lbs per cubic foot.

Soil-cement for riverbank protection and dam construction is normally made from soil excavated and mixed on site and placed in lifts by road construction equipment (Nussbaum and Colley, 1971). Lifts are kept moist between passes and can be dusted with cement to improve bonding lift to lift. The relationship between lift height, slope, and width is given by:

$$W = T\,(S^2 + 1)^{1/2}\,SL \tag{10.55}$$

where W = width of each lift (ft); T = thickness of the layer perpendicular to the bank slope (ft); S = bank slope horizontal (SH:1V); and L = thickness of each compacted lift (ft).

Most lifts are 8 ft wide due to construction equipment limitations. Lifts as little as 4 ft wide would be structurally stable if they could be placed easily with modified equipment. Slope is dictated by structural failure if the slope is not steep enough and soil failure if it is too steep. Many bank placements are at 1:1 slopes. Nowatzki and Reely (1986) and Nussbaum and Colley (1971) provide details of slope stability testing. Lift thicknesses are between 4 and 8 in. Figure 10-12 shows a typical placement of this type.

Facings of soil-cement are generally 6 to 12 in thick. Pima County specified 500 psi compressive strength plus 2% additional cement for a series of channels that underwent record flooding without loss. Compaction up to 3:1 slopes can be accomplished with a bulldozer with street pads, or a roller compactor can be let down the side of the channel by rope and winch.

Open Channel Design

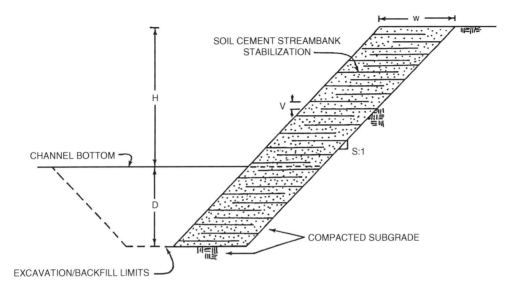

FIGURE 10-12
Typical soil cement placement. (From Reese, A.J., Use of Soil-Cement for Bank Protection, DRAFT ETL 1110-2, U.S. Army Corps of Engineers, Waterways Experiment Station, March 1987.)

Bioengineering

Soil bioengineering is an applied science combining mechanical, biological, and ecological concepts to construct living structures for bank and slope protection. Biological methods have been used in the United States for a number of years but without the scientific background afforded by the bioengineering science. Soil bioengineering, pioneered in Europe, is gaining popularity in the United States and has been used for various applications of bank and slope protection (Sotir and Gray, 1989; Gray and Leiser, 1982; Schiechtl, 1980).

Bioengineering methods use structural support to hold live plantings in place, while the root structure grows and the plants are established. This is done through the use of sprigging, live crib walls (see Figure 10-13 for a typical example), cut brush layers, live fascines, live stakes, and other methods.

Advantages of soil bioengineering include the strength of intertwining root masses, self-healing properties, natural appearance, habitat enrichment, faster recovery of ecosystems after disturbance, and resistance to slope failure. Disadvantages include labor-intensive installations, need for stability control until roots are established, and dependence on ability of materials to root and grow. Many of these disadvantages can be countered with use of native vegetation, engineering stability analyses, skilled designers, and the integration of structural members with the plantings as necessary.

Detailed knowledge of plant species (grasses, legumes, shrubs, and trees) and ability to withstand inundation in bank and terrace zones is necessary for this method to be successful (Logan, 1979). Often, an interdisciplinary team is helpful. Nurseries need notice if they will be counted on to supply various types of vegetation stock. Willows can be cut from other streambanks in the area or supplied by maintenance crews. Planting by phase is often done to facilitate the best mixes of vegetation, availability periods, and growing seasons. In any case, close monitoring is necessary until the growth is established and assured.

Description	Notes
Six foot willow switches are wired together to form a mat which is then secured to the bank by stakes, fascines, poles or rock fill. The toe of the slope is reinforced with a brushlayer anchored by a live fascine or rip-rap. Mattress should lie perpendicular to the water.	1. The mat establishes a complete cover for the bank. Entire layer must be slightly covered with soil or fill. 2. Captures sediment and rebuilds bank. Plants provide long term durability and erosion control. 3. Establishes dense riparian growth. Allows for invasion of surrounding riparian vegetation. 4. An economical technique which provides protection soon after it is established. 5. Applicable on banks with a uniform gradient not exceeding 2:1. The toe of the slope is reinforced with a brushlayer anchored by a live fascine. 6. Mattress should lie perpendicular to the water. 7. Site should have moderate water level fluctuations.

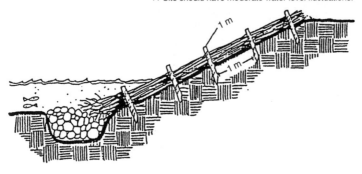

FIGURE 10-13
Typical live crib wall.

Manufactured Bank-Protection Methods

There is a wide variety of manufactured bank-protection products. They include mats and coverings, crib membranes, filter cloths, blocks and articulated concrete matting, rigid grid blocks, structural wall modules of concrete and various materials, sand-filled mattresses, concrete-filled mattresses, etc.

Manufactured materials have the advantage of uniformity, consistency, targeted strengths for specific applications, ability for steep or vertical banks, nice appearance in developed urban settings, tested stability, and technical support from suppliers. Disadvantages include unit cost, labor-intensive installation, and the possibility of lack of availability should additional or matching materials be needed and the supplier is no longer in business.

10.16 Physical Stability of Channels

Overview

In addition to being an integral component of storm drainage systems, natural, or undisturbed, streams possess numerous environmental benefits that are derived from the diversity and stability of the stream and its associated riparian ecosystem. Urbanization is associated with processes that disturb natural streams, increasing velocity and volume of runoff, resulting in increased erosion and sediment problems within the stream system. The purpose of this section is to provide some guidelines for ensuring the stability of streams and for designing drainage projects and streambank protection measures with environmental benefits. There are detailed textbooks on this subject. This section will only serve to familiarize the reader with concepts and concerns.

Channel Response to Change

Channels in natural and constructed states are not naturally stable and unchanging. All channels attempt to adjust to the ever-changing combination of flows, sediment loads, topography, climate, and bed and bank conditions imposed on them. Any change in these variables will result in some channel modification. If a channel is designed without due regard for a more stable configuration, that channel will be stable only as long as rigid bank linings and bed stabilizers hold it in check or channel dredging removes excess sediment. Channels have three degrees of freedom or three ways of becoming unstable: width changes, depth changes, and planiform or horizontal instability (horizontal migration, for example). There are many detailed treatments of stream sediment transport and stable channel design from a river engineering approach and a geomorphic approach, and a detailed treatment is beyond the scope of this book. See, for example, SLA (1982), ASCE (1977), Simons and Senturk (1976), USCOE (1989), or Schumm (1977). This section will summarize some general ways to consider and recognize stream stability problems and considerations for stable channel design, and it will refer to other sources of detailed information.

Despite almost 8 years since the publication of the first edition of this book, progress in the ability to predict the impacts of urban development of urban stream channels is still in its infancy (Bledsoe, 2001). Individual basic physical principles are understood well enough, but streams in urban and urbanizing environments exhibit a diverse array of adjustments and thresholds in response to the imposed changes in hydrology, sediment delivery, and water quality. When the complexities of biology are added, predicting ecological changes is an even more daunting task (Herricks, 2001; Yoder, Miller, and White, 2000).

Most authorities on channel erosion indicate that the eventual form and magnitude of channel adjustment is a direct function of stream power (specific gravity of water times flow and energy slope) per unit channel area, as it is applied to the most susceptible bank or bed reaches. Thus, a channel with a cobble bottom but erodible banks would widen rather than deepen, while a channel with stiff clay banks and a silty-clay bed will most likely incise with bank failure resulting from oversteepening of banks and mass wasting or rotational slump rather than shear-induced transport from bank materials. To the extent that stream power (discharge, specifically) increases with development, we can expect channel changes. For example, it can be shown, in most states, through direct application of the state's peak flow regression equations, that normal urban development will cause a discharge frequency shift, wherein the 10-year storm peak occurs every 2 years, and the 100-year peak flow occurs every ten years. If streams tend to carry the 2- to 5-year storm at bankfull (Schumm, 1977), then a simple application of the Manning's equation will show stream enlargement to handle the increased flows. However, these channel changes might be influenced by myriad other variables, including modification of the flow duration curve, geomorphic history, changes in sediment inflow, stream temperature, vegetation loss, near-bank disturbance, etc. Any direct connection between urban development (i.e., imperviousness) and channel change is only a trend, not a predictive equation (Bledsoe, 2001).

Studies by a number of authors since the 1980s have led to an understanding that simple control of peak flow will not control channel erosion. Rather, in some cases, it is the *duration* of bankfull flows, often called the dominant discharge, that matters, not the actual large peak flow magnitude. The reasoning is that larger flows move out of banks with little actual increase in shear stress on bed and banks (except in some alluvial streams in the arid West). It is the shear stress over time that causes erosion in many channel types. This apparent reality has led local communities and researchers to derive modified design criteria, beyond peak flow controls. MacRae (1997) postulates an approach that relies on a zero net change in boundary sheer stress. The State of Georgia (ARC, 2002) reduces the 2-year flow to a small percentage of its normal peak through extended detention. King County, WA, (Sovern and MacDonald, 2001) adopted a criteria that attempts to adjust the postdevelopment flow duration curve to more closely match the predevelopment curve.

What is of eventual concern is an answer to the related questions of: What will the stream look like after urban development and how can we control avoidable impacts to arrive at a final stream form and character that meets goals, objectives, or regulatory requirements? What should our goals and objectives be for this stream? Is it possible to have a stream of rural characteristics in an urban setting? What does a "nice" urban stream look like? How elastic is our stream in its ability to spring back from temporary impacts? How malleable is the stream in adjusting to new character in the face of impacts (Westman, 1985)? And finally, what other things besides pure science are necessary for a stream to be restored or preserved (Riley, 1997)?

We generally know that the following set of characteristics indicates a stream at risk for instability and response magnitude (Bledsoe, 2001):

- High specific stream power relative to the most erodible channel boundary
- Capacity limited — fine bed material, especially sand
- Little or no grade control or hard points
- Low density of root volume in the banks
- Noncohesive, fine grained, uniform grained, sparsely vegetated banks
- Large ratio of large woody debris size to channel width, causing local scouring
- Incised channel allowing little floodplain energy dissipation
- Near an energy threshold associated with abrupt changes in planiform or incision initiation
- Flashy flows resulting in bank saturation
- Low roughness of form and vegetation
- Floodplain susceptible to chutes, cutoffs, and avulsions
- Steep bank angles

There are myriad simple tools and classification schemes (Rosgen, 1994; USCOE, 1994) to help us categorize, assess, and understand urban streams. Probably no tool can substitute for experience in similar streams and comparing the stream in question to other streams in the same watershed or area. Below are some different methods to assess and design stable stream channels. The designer or assessor should look at several methods and apply a good bit of judgment before making final form and character decisions.

Proportionality Approaches

Lane (1955), using methods from fluvial geomorphology, provided a simple proportion that allows for a qualitative "feel" for channel response to changes.

Open Channel Design

$$QS \propto Q_s D_{50} \qquad (10.56)$$

where Q = discharge; S = channel slope; Q_s = sediment discharge; and D_{50} = median grain size.

In a proportion, if one side increases or decreases, the other must do likewise. Consider, for example, the response of a stream or channel to upstream construction and greatly increased sediment yields. Q_s increases, but Q is constant. Therefore, to maintain the proportionality, the slope (S) must increase, and D_{50}, the median sediment size transported, must decrease. The result is channel aggradation (sedimentation) near the inflow point of the sediment until such time as the slope is steep enough to transport the excess sediment.

Urban development increases Q. To adjust, S must decrease or Q_s or D_{50} or both must increase. The result is channel degradation (erosion), widening, deepening (depending on the relative soils and vegetation in bed and banks), and eventual slope flattening. If a sinuous channel is straightened, the slope is increased, and a response similar to flow increase will normally result. Uncontrolled urban development normally shortens channels and increases discharge.

If a large retention pond is built in line with a stream carrying significant sediment, the result is channel degradation below the dam due to decreased sediment transport (Q_s) and the responding slope decrease. If a main channel is dredged to make it deeper, the base level is dropped for each of the tributaries. The result of this base level lowering is an increased gradient (S) in the lower portions of the tributaries. To balance this impact, Q_s increases, inducing headcutting or rapid erosion of the tributaries.

Simons and Senturk (1976) give a compilation of channel response proportions based on the work and experience of many authors:

- Depth, D, is proportional to discharge, Q.
- Channel width, W, is proportional to both Q and Q_s.
- Channel shape, expressed as width to depth ratio (W/D), is directly proportional to Q_s.
- Sinuosity, the measure of tendency to meander, is proportional to valley slope, S_v, and inversely proportional to Q_s.

The best way to avoid instability problems in urban stream channels and to maximize environmental benefits is to maintain streams in as natural a condition as possible, and, when channel modification is necessary, to avoid altering channel dimensions, channel alignment, and channel slope as much as possible. When channel modification is necessary, the following set of guidelines should be followed to minimize erosion problems and maximize environmental benefits:

- Avoid channel enlargement whenever possible.
- When channels must be enlarged, avoid streambed excavation that would significantly increase streambed slope or streambank height.
- When channel bottom widths are increased more than 25%, provide for a low-flow channel to concentrate flows during critical low-flow periods.
- Avoid channel realignment whenever possible.

Assessing Stream Stability

Whether existing erosion on streams needs to be repaired and protected from additional erosion depends on several factors, including type, extent, severity, and location of the

TABLE 10-14

General Conditions for Stable and Eroding Banks

Characteristics	Stable Bank	Eroding Bank
Bank slope	Not vertical; may be compound with vegetated berm at toe	Often vertical or near vertical; may have moss or sod or other failed material at toe
Bank cover	May have variety of vegetation	General absence of vegetation growing on slope, including ferns or moss
Trees	Often has trees growing on bank or on the bed at toe	Standing live or dead trees inside the bank line, often leaning toward channel, fallen trees may obstruct flow
Bankline	Relatively uniform or smoothly curing	Irregular, sometimes with scalloped appearance
Sediment	Sediment located in bars, bars may be partially stabilized by vegetation, especially along bank toe	Entire bed may be covered with sediment; bars not stabilized

erosion. Streambanks that appear to be well-vegetated and stable usually are, but often, banks that appear to be unstable and eroding may be relatively stable. Differentiating between stable and eroding banks involves observing and evaluating evidence (that may at times be contradictory) about boundary conditions. General conditions frequently associated with stable and eroding banks are summarized in Table 10-14 (Nunnally and Keller, 1979).

Judging the severity of streambank erosion is a somewhat subjective task that may involve the amount or the rate of erosion, relative location of the problem, and even the type of erosion involved. Erosion is a natural process, and several feet of lateral streambank erosion in a bend, or meander, might be no cause for concern. On the other hand, if the erosion threatens a culvert installation, it might be considered severe. Table 10-15 provides typical basic guidance for establishing severity, in Midwestern and Eastern climates, but it should be used with considerable discretion.

Various modifications to streams have good and adverse physical impacts. Table 10-16 gives a summary of these.

Stable Channel Design Approaches

There are four main approaches to consider stability in channel design: maximum permissible velocity approach, tractive stress approach, sediment transport approach, and geomorphic or regime equations and observations (USDA, 1977). A detailed treatment of all of these methods is beyond the scope of this book. The USDA (1977) or Simons and

TABLE 10-15

Degree of Erosion Severity

Degree of Erosion	Characteristics
Stable	Little or no evidence of erosion; if eroding banks are present, they are small in extent (less than 5 linear feet or 50 square feet), and rates are modest (less than 1/2 foot per year); greater erosion may be tolerated at bends if it causes no associated problems.
Moderate	Extent of problem or rate of erosion exceeds criteria for stable class, but is less than severe
Severe	Erosion covers large area of bank and is occurring at a rate in excess of one foot per year or a rate that is unacceptable for safety, environmental, or economical reasons.

TABLE 10-16

Consideration of Channel Modifications

Positive Impacts	Negative Impacts
Reservoirs, Ponds and Detention	
Flood flow and stage reduction at flood flow	Loss of land
Downstream damage reduction	Filling of reservoir with sediment
Source of water for multiuses	Loss of wetlands
Recreation uses	Modification of stream flow regime
	Conveyance encroachment and risk
	Water quality changes, thermal impacts
	Evaporation losses in arid climates
	Elimination in fish spawning and movement
Levees and Flood Walls	
No flooding from exterior until exceedance flows reached	May induce flooding up- and downstream of site
Protection of property values	Potential of sudden large loss if exceeded
	Interior flooding and pumping needed
	Maintenance and observation
Channelization and Channel Widening	
Flood stage reductions for all flows	Potential impact on fish spawning and habitat
Local damage reduction through project reach	Changed sediment transport
	Increased maintenance
	Induced flooding downstream if loss of floodplain storage
Diversions	
Flood stage reductions for all flows	Increased maintenance
Local damage reduction in project reach	Induced flooding downstream if loss of floodplain storage
Nonstructural Efforts	
Individual structures protected	High residual damage to infrastructure
Few environmental impacts	Emergency responses required for specific events
Low cost	Individual maintenance required

Source: From U.S. Army Corps of Engineers, Hydrologic Engineering Analysis Concepts for Cost-Shared Flood Damage Reduction Studies, EP 1110-2-6005, December, 1992.

Senturk (1976) should be consulted for details of the first three applications and Rosgen (1994) for the latter. Use of any of these approaches should be tempered with engineering judgment and experience. There is no substitute for understanding the stream types in the area and for knowing what has happened to similar streams when they have been designed, impacted by development, or modified. The sediment transport method will not be covered in this book, because the level of skill involved in understanding the methods and the complexities of sediment transport are beyond the needed skills of the normal urban stormwater manager.

Maximum Permissible Velocity Method

The final design of artificial open channels should be consistent with the velocity limitations for the selected channel lining. Maximum velocity values for selected lining categories for bare-earth channels are presented in Table 10-17. Seeding and mulch should be used only when the design value does not exceed the allowable value for bare soil. Velocity limitations for vegetative linings are reported in Table 10-18. Vegetative lining calculations

TABLE 10-17

Maximum Velocities for Comparing Lining Materials

Material	Clear Water	Water with Colloidal Silt	Water with Noncolloidal Silt, Sand, or Gravel
Fine sand (colloidal)	1.5	2.5	1.5
Sand loam (noncolloidal)	1.45	2.5	2.0
Silt loam (noncolloidal)	2.0	3.0	2.0
Alluvial silt (noncolloidal)	2.0	3.5	2.0
Alluvial silt (colloidal)	3.75	5.0	3.0
Firm loam	2.5	3.5	2.25
Volcanic ash	2.5	3.5	2.0
Fine gravel	2.5	5.0	3.75
Stiff clay (very colloidal)	3.75	5.0	3.0
Graded loam to cobbles (noncolloidal)	3.75	5.0	5.0
Graded silt to cobbles (colloidal)	3.75	5.0	3.0
Coarse gravel	4.0	6.0	6.5
Cobbles and shingles	5.0	5.5	6.5
Shales and hard pans	6.0	6.0	5.0

Source: From Fortier, S. and Scoby, F.C., Permissible Canal Velocities, Trans. ASCE, Vol. 89, pp. 940–956, 1926. Reproduced by permission of ASCE.

are presented in Section 10.9, and riprap procedures are presented in Section 10.16. If lining costs are expensive, there are methods available to design channels to minimize such costs (Trout, 1982; Loganathan, 1991).

For D_{75} grain sizes greater than or equal to 1 mm, which act as discrete particles, the permissible velocities for both sediment-laden water and sediment-free water can be found

TABLE 10-18

Maximum Velocities for Vegetative Channel Linings

Vegetation Type	Slope Range (%)[a]	Maximum Velocity[b] (ft/s)	
		Erosion-Resistant Soils	Easily Eroded Soils
Bermuda grass	0–5	8	6
	5–10	7	5
	>10	6	4
Kentucky bluegrass	0–5	7	5
Buffalo grass	5–10	6	4
	>10	5	3
Grass mixture	0–5[a]	5	4
	5–10	4	3
Lespedeza sericea[a]	0–5[c]	3.5	2.5
Kudzu, alfalfa			
Annuals[d]	0–5	3.5	2.5
Sod		4.0	4.0
Lapped sod		5.5	5.5

[a] Do not use on slopes steeper than 10%, except for side-slope in combination channel.

[b] Use velocities exceeding 5 ft/s only where good stands can be established and maintained.

[c] Do not use on slopes steeper than 5%, except for side-slope in combination channel.

[d] Annuals — used on mild slopes or as temporary protection until permanent covers are established.

Source: From U.S. Department of Agriculture, Soil Conservation Service (SCS), Handbook of Channel Design for Soil and Water Conservation, TP-61, 1954.

from using Equations (10.57) and (10.58) fit to USDA (1977) information. Sediment-laden water is defined as having fine suspensions of sediment carried in the flow at concentrations greater than 20,000 ppm by weight. Flows with concentrations of fine sediment less than 1000 ppm by weight are considered sediment free. Permissible velocities for flows in between can be found by linear interpolation between the two values found in Equations (10.38) and (10.39).

$$V_s = 2.6356 \, D_{75}^{0.30596} \quad D_{75} > 0.4 \text{ mm} \qquad (10.57)$$

$$V_c = 1.4653 \, D_{75}^{0.37990} \quad D_{75} > 2.0 \text{ mm} \qquad (10.58)$$

where V_s = sediment-laden water permissible velocity (ft/s); V_c = sediment-free water permissible velocity (ft/s); and D_{75} = sediment particle size for which 75% of particles by weight are smaller (mm).

For sizes less than the indicated limits that act as discrete particles, use 2.0 ft/s for the maximum permissible velocity without adjustments. For other sizes and types of soil, the following adjustments should be made:

- Adjust all channel types for both alignment and depth.
- Adjust channels with soils that behave as discrete particles for side slope.
- Apply a frequency adjustment factor for storms with a design frequency that is less frequent than the 10% flood (more than 10-year return period) and the bed or bank materials do not operate as discrete particles.
- Apply a density correction to channels for all soil types except clean sands and gravels with less than 5% passing the #200 sieve.

For deeper channels, the velocity can be multiplied by 1.1 for 4-ft depth and 1.4 for 18-ft depth. Linear interpolation between these depths is acceptable. If the channel is curved, the permissible velocity should be multiplied by a factor based on a measured radius of curvature (to the channel centerline) and channel width. Table 10-19 gives appropriate values.

For bank materials that behave as discrete particles (noncolloidal), an adjustment in permissible velocity should be made based on side-slope of the bank. Table 10-20 gives appropriate values. Frequency factor corrections for channels with bed or bank materials

TABLE 10-19

Adjustment Factors for Channel Alignment

Curve Radius/Top Width	Correction Factor
16	1.00
14	0.99
12	0.96
10	0.93
8	0.89
6	0.81
5	0.70

Source: From U.S. Department of Agriculture, Soil Conservation Service (SCS), Design of Open Channels, Tech. Release 25, October 1977.

TABLE 10-20

Adjustment Factors for Channel Side-Slope

Cotangent of Slope Angle (x:1)	Correction Factor
1.5	0.50
2.0	0.72
2.5	0.82
3.0	0.86
4.0	0.90
>4.0	1.00

Source: From U.S. Department of Agriculture, Soil Conservation Service (SCS), Design of Open Channels, Tech. Release 25, October 1977.

that do not act as discrete particles, and when the design discharge is less frequent than the 10% storm, can be made using the following equation:

$$C_f = 1.705 - 0.3085 \ln (F) \tag{10.59}$$

where C_f = factor to multiply by permissible velocity, and F = flood frequency in percent chance (100-year storm = 1.0).

Density corrections should be made for all soils except clean sands and gravels with less than 5% material passing the #200 sieve. Corrections can be made on the basis of void ratio of the soil using the following equations:

SM, SC, GM and GC soils $\qquad C_d = -0.617e + 1.417 \qquad (10.60)$

CL and ML soils $\qquad C_d = -0.567e + 1.480 \qquad (10.61)$

CH and MH soils $\qquad C_d = -0.358e + 1.361 \qquad (10.62)$

where C_d = correction factor multiplied by the permissible velocity; e = soils void ratio = $G(62.4/\Gamma_d) - 1$; G = specific gravity = specific weight of material/specific weight of water; and Γ_d = dry density of solid material and voids (lb/ft³).

Application of this method involves determination of design discharge and approximation of sediment concentration. Based on the bed and bank materials and the channel configuration, determine if the permissible velocity method is applicable. Compare the design velocities with the permissible velocities. If the design velocities are greater than the permissible velocities, either reconfigure the channel or provide bed and bank protection as applicable.

Tractive Stress Method

Channels become unstable when the ability of the material in the bed and banks is overcome by the shearing force of the water. The theoretical point shear stress of the water on the boundary of an infinitely wide channel can be expressed as follows:

$$\tau = 62.4 \, D \, S_t \tag{10.63}$$

where τ = average boundary shear stress (lbs/ft²); D = flow depth (ft); S_t = friction slope due to particle resistance (ft/ft); and 62.4 is the specific weight of water (lbs/ft³).

S_t is calculated by partitioning the friction loss among the various components that go into Manning's n. The base value of Manning's n, based only on particle size, n_t, can be found from:

$$n_t = D_{75}^{1/6}/39 \qquad (10.64)$$

D_{75} is in inches in this equation. Then, S_t can be found from a proportion as:

$$S_t = (n_t/n)^2 \, S \qquad (10.65)$$

where n and S are the channel total friction slope and Manning's n.

The maximum shear force on the bed of the channel is about 0.97τ, while the maximum on the side slope of a trapezoidal channel is about 0.75τ. Shear coefficients can get as high as 2.5 on the outside of sharp bends (Ippen and Drinker, 1962). An approximate correction for the actual maximum shear stress on the downstream outer bank of a curved reach can be obtained from the empirical equation:

$$\tau_b = [-0.133 \, (R_c/W) + 2.224] \, \tau \qquad (10.66)$$

where τ_b = shear stress in the bend (lbs/ft²); R_c = bend radius to channel centerline (ft); and W = top width (ft).

For turbulent flood conditions in an urban stream, the ability of a noncohesive soil particle (in the size range $0.25'' < D_{75} < 5.0''$) to resist movement depends on its density, position, and size. A critical shear stress for particles on the bed can be defined for these conditions as:

$$\tau_c = 0.4 \, D_{75} \qquad (10.67)$$

where τ_c = critical shear stress (lbs/ft²), and D_{75} = sediment particle size for which 75% of particles by weight are smaller, inches ($0.25'' < D_{75} < 5.0''$).

If the particle or stone is on a side slope of angle Θ and its natural angle of repose (the angle of a dumped pile of such stone) is Φ, then the critical shear stress for a particle on a side slope is given as:

$$\tau_{ss} = \tau_c \, [(z^2 - \cot^2\Phi)/(1+z^2)]^{1/2} = \tau_c \, K_{ss} \qquad (10.68)$$

where τ_{ss} = critical shear stress on a side slope (lbs/ft²); Θ = side slope angle (degrees); and Φ = natural angle of repose of sediment (degrees).

The angle of repose (Φ) for particles in this size range can be found from the set of equations below for descriptive shape factors of the particles:

Very rounded	$\Phi = 33.82 - 2.09 \, D_{75} + 21.83 \log (D_{75})$	(10.69)
Moderately rounded	$\Phi = 34.96 - 1.83 \, D_{75} + 18.95 \log (D_{75})$	(10.70)
Slightly rounded	$\Phi = 35.14 - 0.76 \, D_{75} + 12.80 \log (D_{75})$	(10.71)
Slightly angular	$\Phi = 37.63 - 1.71 \, D_{75} + 15.22 \log (D_{75})$	(10.72)
Moderately angular	$\Phi = 39.05 - 1.73 \, D_{75} + 13.59 \log (D_{75})$	(10.73)
Very angular	$\Phi = 40.14 - 1.51 \, D_{75} + 11.11 \log (D_{75})$	(10.74)

where Φ = natural angle of repose of sediment (degrees), and D_{75} = sediment particle size for which 75% of particles by weight are smaller, inches ($0.25" < D_{75} < 5.0"$).

If the density is different from 160 lb/ft³, the critical tractive stress should be multiplied by:

$$T = (\Gamma_s - 62.4)/97.6 \tag{10.75}$$

where T = correction factor, and Γ_s = specific weight of particle (lb/ft³).

For grain sizes finer than $D_{75} = 0.25"$, the TP-25 (USDA, 1977) or EM 1110-2-1418 (USCOE, 1994) references should be consulted and the Shields incipient movement diagram found in those texts should be used.

For channels with substantial inflows of sediment, a minimum velocity or shear stress to avoid sediment deposition may be as important as a maximum to avoid erosion (USCOE, 1994). For meandering channels and at the outside of bends, erosion may occur even if the average velocity is within bounds. An allowable shear stress must be paired with other geometric data to actually define a stable channel, because the shear stress can be satisfied by a wide range of width, depth, and slope combinations. It is necessary to supplement this approach with relations for the other variables derived from existing conditions or other sources.

A basic design approach is as follows (USCOE, 1994):

- Determine average velocities or shear stresses in cross sections for a range of design flows.
- Match the velocity–discharge curve as much as possible from an existing channel in a redesigned channel by controlling cross section, slope, and roughness.
- Base calculations on an average section or divide the stream into reaches of similar geometry.
- Determine the incipient motion return frequency from a flow–frequency and a stage–velocity curve and ensure channel stability at a dominant or channel forming discharge.

Regime Equations for Channel Proportions

Regime methods are based on the development of empirical equations from observed stream data. These methods can yield unrealistic answers when blindly applied to streams not similar to the observed streams of the database. Because regime equations do not directly account for physical processes in their regressions and are not based on physical principles that can then be relied upon and extrapolated with more comfort, they have been often criticized. However, they have the advantage of being based on true observations rather than theory and, thus, bring a certain comfort level.

Determination of stable channel widths or width-to-depth ratios can be difficult. Flow area functions are generally more accurate. Local measurements on similarly situated channels are the best source of data. Regime equations can be useful in helping to determine channel proportions when other information is not available and the channel does not have rigid boundaries.

Regime equations are developed from observations of data and are often combined with physical relationships. They can be helpful in stability analysis on urban streams and rivers as the designer considers the potential impacts of stream development and modifications. They typically apply to alluvial channels like those found most often in the West and Southwest. They can be applied to any channel with a boundary that contains a layer

Open Channel Design

of loose material of the same type that is moved along the bed (termed the *bed load* or *bed material load*). There are a large number of regime equations, and, because they are primarily empirical in nature, they cannot be readily applied beyond the conditions for which they were derived. Some of the better examples and discussions from the literature include Lacey (1930), Blench (1966), Ackers and White (1973), Hey, Bathurst, and Thorne (1982), Ackers (1972), White, Bettess, and Paris (1982), Chien (1955, 1957), Inglis (1948), and Rosgen (1994).

Simons and Albertson Equations

The equations provided below were developed by Simons and Albertson and apply generally to alluvial channels, though most of the data were derived in the West. According to Simons and Albertson, the Froude number F should be less than 0.3 for stable channels in alluvium. F is expressed as $V/(gD)^{1/2}$.

$$D = 1.23\ R \quad (R \text{ from 1 to 7 feet}) \tag{10.76}$$

$$D = 2.11 + 0.934\ R \quad (R \text{ from 7 to 12 feet}) \tag{10.77}$$

$$W = 0.9\ P \tag{10.78}$$

$$W = 0.92\ T - 2.0 \tag{10.79}$$

C_1–C_5 **Coefficients by Channel Type**

	A	B	C	
$P = C_1\ Q^{0.512}$	3.30	2.51	2.12	(10.80)
$R = C_2\ Q^{0.361}$	0.37	0.43	0.51	(10.81)
$A = C_3\ Q^{0.873}$	1.22	1.08	1.08	(10.82)
$V = C_4\ (R^2 S)^{1/3}$	13.9	16.1	16.0	(10.83)
$W/D = C_5\ Q^{0.151}$	6.5	4.3	3.0	(10.84)

where P = wetted perimeter (ft); R = hydraulic radius (ft); A = flow area (ft²); V = mean channel velocity (ft/s); D = mean depth (ft); W = mean channel width (ft); T = channel top width (ft); C_i = coefficients by different channel types defined as: A = sand bed and sand banks, B = sand bed and cohesive banks, C = cohesive bed and banks (cohesive $PI > 7$).

Gravel Bed Equations

The following set of equations is based on hydraulic geometry relationships and is most applicable to gravel-bed streams. These equations are based on data on 70 streams in the Alberta, Canada, area (Hey, Bathurst, and Thorne, 1982; Bray, 1975):

$$B = 2.38\ Q_2^{0.527} \tag{10.85}$$

$$A = 0.632\ Q_2^{0.860} \tag{10.86}$$

$$d = 0.266 \, Q_2^{0.333} \tag{10.87}$$

$$v = 1.58 \, Q_2^{0.140} \tag{10.88}$$

and additionally:

$$B/d = 8.95 \, Q_2^{0.194} \tag{10.89}$$

$$S = 0.0354 \, Q_2^{-0.342} \tag{10.90}$$

Planiform estimates can be made. Slopes falling above the line indicated by: $S = 0.06 \, Q_2^{-.44}$ will tend to be braided rather than meandering.

Analytical Methods for Channel Proportions

A number of investigators have proposed a hybrid approach to stable channel design that involves simultaneous solution of the governing equations. Each of these methods considers discharge, sediment transport, and bed material composition as independent variables, and width, depth, and slope as dependent variables (USCOE, 1994). With three variables, three equations are required. The designer chooses among several equations relating sediment transport and resistance and then uses a relationship between the independent variables and width. Five or six methods gained popularity among major channel designers, though each had severe limitations on data trials, and applicable range. See USCOE (1994) for a further discussion. One method, graphical in nature, will be discussed in greater detail due to its ease of use.

Neill Method

Neill (1984) developed a hybrid approach to regime analysis combining physical considerations and regime supplements. His methods have been discussed in detail by the Corps of Engineers (USCOE, 1994) which should be referenced for further information on derivation. His approach can be used for trapezoidal channel design or stability checking. It works best for channels in the low transport range and those that are neither aggrading nor degrading actively. High sediment transport can cause the slope to increase greatly, and adjustments in the procedure for this are only approximate. Braided or multichanneled streams would have a higher slope and be wider than indicated. The method should not be applied to these types of streams (USCOE, 1990).

The basic approach is to (1) determine a stable slope, (2) determine a channel width, (3) determine a flow depth, and (4) check the results against other stable channel methods. It is normally an iterative procedure.

The design discharge should be one that can remain stable. Therefore, using design discharges above about the 10-year flood will result in a channel too wide and flat to be stable and result in submeandering within the channel bottom. If a larger capacity is desired, Neill recommends using the greater of 10-year or 50% of the design discharge and providing overbank storage and high-flow channels for the remainder of the design flow.

Channel slope is based on critical tractive force and regime relationships blended together. Figure 10-14 gives the chart for slope selection based on dominant discharge. The slope value should be multiplied by 1.7 for moderate sediment transport and 3.0 for high sediment transport. Moderate transport includes active bedload several times per year, channel bars, and some basin gullying of sandy soil. High transport includes active

Open Channel Design

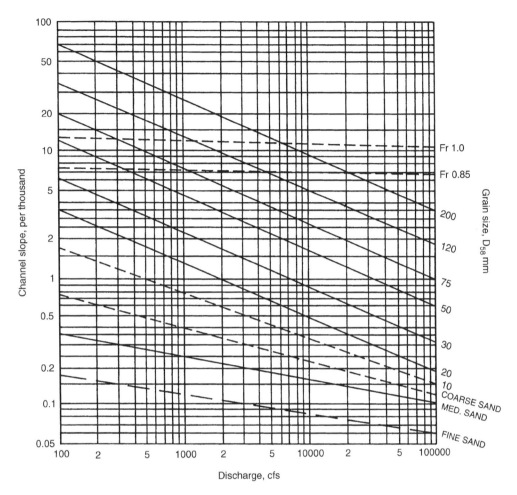

FIGURE 10-14
Slopes for stable channels. (From Neill, C., Hydraulic Design of Stable Flood Control Channels — Draft Guidelines for Preliminary Design, Seattle District COE, Northwest Hydraulic Consultants, January 1984.)

bed for a substantial portion of the year, frequent bars and stream splitting or braiding, and generally sheet and rill erosion in the basin or caving banks.

Channel width is based on regime concepts alone. It is rough, and its results should be compared to other local streams. The width equation is as follows:

$$W = C_w Q^{1/2} \qquad (10.91)$$

where W = width (ft); Q = dominant discharge (cfs); and C_w = 1.6 for high bank resistance, 2.1 for moderate bank resistance, and 2.7 for low bank resistance.

Channel depth is taken from Figure 10-15 on a preliminary basis. More accurate determination of channel cross-sectional geometry can be obtained through the use of various applicable resistance equations. This figure can be used for preliminary selection and quick comparison of alternatives and different Q values.

Checking of the relationships can be done through the use of three charts prepared by Neill for allowable mean velocity (Figure 10-16), allowable shear stress on cohesive materials (Figure 10-17), and limiting slope (Figure 10-18).

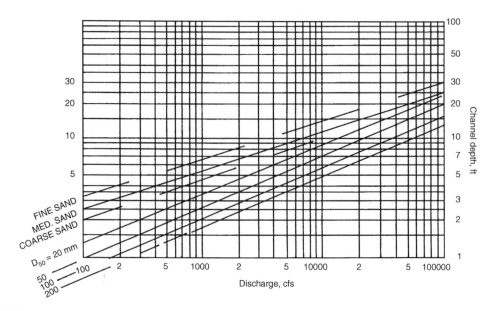

FIGURE 10-15
Preliminary estimate of channel depth. (From Neill, C., Hydraulic Design of Stable Flood Control Channels — Draft Guidelines for Preliminary Design, Seattle District COE, Northwest Hydraulic Consultants, January 1984.)

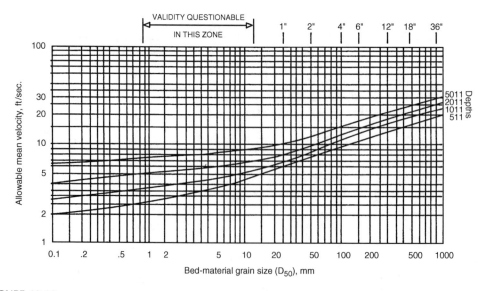

FIGURE 10-16
Allowable mean velocity for stable channels. (From Neill, C., Hydraulic Design of Stable Flood Control Channels — Draft Guidelines for Preliminary Design, Seattle District COE, Northwest Hydraulic Consultants, January 1984.)

Open Channel Design

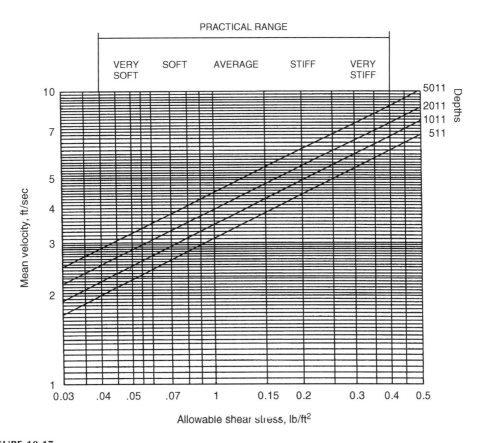

FIGURE 10-17
Allowable mean velocity and shear stress for cohesive soils and stable channels. (From Neill, C., Hydraulic Design of Stable Flood Control Channels — Draft Guidelines for Preliminary Design, Seattle District COE, Northwest Hydraulic Consultants, January 1984.)

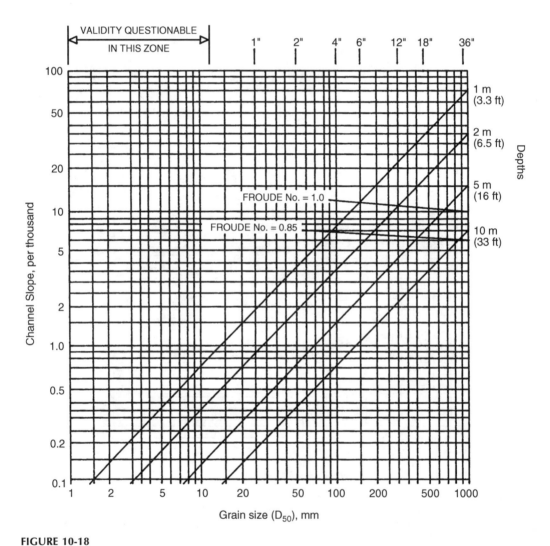

FIGURE 10-18
Allowable limiting slope for stable channels. (From Neill, C., Hydraulic Design of Stable Flood Control Channels — Draft Guidelines for Preliminary Design, Seattle District COE, Northwest Hydraulic Consultants, January 1984.)

Open Channel Design

Stable Channel Design Examples

Permissible Velocity Example

A natural stream channel is to be modified to convey the 2% chance flood (50-year return period). Hydraulic analysis indicates a trapezoidal channel with 2:1 side slopes, a 40 ft bottom width, a design flow depth of 8.7 ft, and design velocity of 5.45 ft/s. The material is glacial outwash, clean sandy gravel soil with a D_{75} of 2.25 in. Sediment yield estimates and visual inspection of the channel indicate the transport concentration is less than about 500 ppm. The channel will be excavated as a straight reach with one bend of 600 ft radius of curvature. Determine the allowable velocity and reach stability.

1. The D_{75} size in mm is 57.15. From Equation (10.92) for clear flow conditions:

$$V_c = 1.4653\ D_{75}^{\ 0.3799} \quad D_{75} > 2.0\ mm \qquad (10.92)$$

$$V_c = 1.4653\ (57.15)^{0.3799}$$

$$V_c = 6.8\ ft/s$$

2. Correction factors are applied:
   Depth correction factor = 1.22
   Side slope correction factor = 0.72
   Alignment correction factor
     Curve radius/topwidth = (600)/74.8 = 8.02
     Correction factor = 0.89
   Frequency and density correction factors do not apply
3. Straight reach permissible velocity
   = (6.8)(1.22) = 8.31 ft/s bed velocity
   = (6.8)(1.22)(0.72) = 5.88 ft/s side slope
4. Curved reach permissible velocity
   = (8.31)(0.89) = 7.40 ft/s bed velocity
   = (5.88)(0.89) = 5.24 ft/s side slope

Because the curved reach side slope permissible velocity is less than the actual mean velocity (5.24 versus 5.45 ft/s), there should be a check on bank protection methods that are available. Because the values are so close, it may be appropriate to construct the channel with vegetated protection and watch the bend to determine its stability.

Note that, for the sediment-laden condition, the permissible velocity increases to 9.08 ft/s uncorrected and 7.1 ft/s in the curved reach on the side slope. This increase is due to the fact that the sediment load acts as an exchange mechanism for the bed and banks, depositing and scouring particles. It may also deposit a protective coating of colloidal clay, effectively armoring the channel from normal erosion.

Tractive Stress Design Example

A channel is to be constructed in an urban subdivision. The design flow is 262 cfs at a depth of 3.5 ft and velocity of 3.23 ft/s. The slope of the energy grade line and the channel, assuming uniform flow, is 0.0026 ft/ft. The bottom width of the channel is 18 ft with side slopes of 1 1/2:1. The mature channel total Manning's n value is expected to be 0.045. The channel is going to be lined with angular gravel (GM) with a D_{75} of 0.90 in (22.9 mm). Determine the channel's stability.

1. Because $D_{75} > 0.25$ in, the large-particle method can be applied. The particle resistance factor then is:

$$n_t = D_{75}^{1/6}/39 \tag{10.93}$$

$$= (0.9)^{1/6}/39 = 0.0252$$

The friction slope due to the gravel is:

$$S_t = (n_t/n)^2 S = (0.0252/0.045)^2 (0.0026 = 0.00082)$$

And point shear stress is:

$$\tau = 62.4 \, D \, S_t = (62.4)(3.5)(0.00082) = 0.179 \text{ lbs/ft}^2$$

2. The shear on the side slope and bed is taken as 75 and 97%, respectively, as:

$$\tau \text{ (side slope)} = 0.75 \, \tau = 0.134 \text{ lbs/ft}^2$$

$$\tau \text{ (bed)} = 0.97 \, \tau = .174 \text{ lbs/ft}^2$$

3. The shear in the bend can be calculated from the ratio of the radius of curvature and the bottom width as: Radius/bottom width = 150/18 = 8.33

$$\tau_b = [-0.133 \, (R_c/W) + 2.224] \, \tau \tag{10.94}$$

$$\tau_b = [-0.133 \, (8.33) + 2.224] \, (.174) = 0.194 \text{ lbs/ft}^2 \text{ for the bed}$$

$$\tau_b = [-0.133 \, (8.33) + 2.224] \, (.134) = 0.150 \text{ lbs/ft}^2 \text{ for the side slope*}$$

4. The allowable tractive stress can be found from:

$$\tau_c = 0.4 \, D_{75} = (0.4)(0.90) = 0.36 \text{ lbs/ft}^2 \text{ for the bed}$$

Adjusting for the side slope:

$$\Phi = 40.14 - 1.51 \, D_{75} + 11.11 \log (D_{75}) \tag{10.95}$$

$$= 40.14 - 1.51 \, (0.90) + 11.11 \log (0.90 = 38.3 \text{ deg})$$

$$\tau_{ss} = \tau_c \, [(z^2 - \cot^2\Phi)/(1+z^2)]^{1/2} \tag{10.96}$$

$$= 0.36 \, [(1.5^2 - \cot^2 38.3)/(1+1.5^2)]^{1/2} = 0.161 \text{ lbs/ft}^2$$

Comparing actual and allowable, the channel will be stable for bed and side slope.

---

* Some designers do not calculate a separate shear for side slopes in bends.

Open Channel Design

Regime Equation Example

A type-B channel is to be designed to convey 600 cfs at bankfull with 2:1 side slopes and $n = 0.022$.

1. $P = 2.51 \, Q^{0.512} = 2.15 * 600^{0.512} = 66.4$ ft
2. $R = C_2 \, Q^{0.361} = 0.43 * 600^{0.361} = 4.33$ ft
3. $A = C_3 \, Q^{0.873} = 1.08 * 600^{0.873} = 288$ ft² (equals PR)
4. $V = Q/A = 600/288 = 2.08$ ft/s
5. $D = 1.23R = 1.23 * 4.33 = 5.3$ ft
6. Froude number $= V/[gd]^{1/2} = 2.08/[32.2 * 5.3]^{1/2} = 0.159 < 0.3$
7. $W = 0.9 \, P = 59.76$ ft; $W = 0.92 \, T - 2.0 \, T = 67.1$ ft.
   For 2:1 side slope bottom width $B = 67.1 - (4)(5.3) = 45.9$ ft
8. Find the regime equation channel slope needed from $V = C_4 \, (R^2 S)^{1/3} = 16.1 \, [(4.33)^2 S]^{1/3} = 2.08$ ft/s Then: $S = 0.000114$ ft/ft.
9. Assuming Manning's equation, find the slope necessary to provide conveyance capacity for the channel dimensions given.

$$V = 2.08 = 1.49/n \, R^{2/3} \, S^{1/2} = 1.49/0.022 \; 4.33^{2/3} \, S^{1/2}$$

$$S = 0.00013 \text{ difference of 14\% in slope}$$

One can choose slope or one in the middle, depending on the topography of the site. If a slope flatter than the Manning's slope is provided, the channel conveyance needs to be increased to account for this.

Neill Relations

Given a dominant discharge of 500 cfs, S (natural) = 0.008, n = 0.025, D_{50} = 0.5 mm and cross-sectional shape to be trapezoidal with side slopes of 2:1, find stable channel dimensions and slope. Low transport condition applies (Neill, 1984).

Checking Figure 10-14, a slope of 0.006 is indicated. From the width relationships for moderate bank resistance, the width is found to be 47 ft. A rough approximation of depth from Figure 10-15 is 1.65 ft. Based on these values, the Manning equation is solved, giving a velocity of about 6.45 ft/s. Checking this against Figure 10-16 gives a satisfactory result. Checking against Figure 10-18 gives a slope in about the right range, though the chart does not go to such low depths.

It should be noted that conventional design might have chosen a greater depth and thus a higher velocity and narrower channel. While this channel would have carried the flow adequately, it would probably have needed significant bank protection and constant maintenance.

Stream Structures

Grade Control Structures

Grade control structures are used to prevent streambed degradation. This is accomplished in two ways. First, the structures provide local base levels that prevent bed erosion and subsequent slope increases. Second, some structures provide controlled dissipation of energy between the upstream and downstream sides of the structure. Structure choice depends on the type of existing or anticipated erosion, cost, and environmental objectives.

The best source of design guidance is the National Engineering Handbook, Section 11, "Drop Spillways," and Section 14, "Chute Spillways" and UDFCD (1991). All grade control consists of a control section, an adjacent protection section, and an energy dissipation section.

The general design procedure for grade control structures is to:

- Determine the total fall through the design reach
- For the selected channel, determine the stable channel gradient
- Determine the amount of fall to be controlled by the grade control or drop structures (from the tailwater of the upstream structure to the head on the next structure downstream should be a stable slope)
- Select the size, location, and type of structures to be used

Grade control design will be improved if entrance walls are rounded, end walls are straight rather than curved or flared, the exit channel is larger than the control section, a preformed scour hole is supplied if the exit channel is 1.5 times the control section crest length, and the riprap section length downstream should be 100 times the depth of flow (Biedenharn, 1987).

Sills

A sill (or stabilizer) is a structure that extends across a channel and has a surface that is flush with the channel invert or that extends a foot or two above the invert. Because sills are intended to prevent scouring of the bed, they should be placed close enough together to control the energy gradeline and prevent scour between structures.

Drop Structures, Chutes, and Flumes

Drop structures provide for a vertical drop in the channel invert between the upstream and downstream sides, whereas chutes and flumes provide for a more gradual change in invert elevation. Because of the high energies that must be dissipated, preformed scour holes or plunge pools are required below these structures. When the ratio of the drop to critical depth in the channel is less than one, the CIT-type drop structure is often used (Vanoni and Pollak, 1959; Murphy, 1967). When it is greater than one, the SAF type is normally used (USACE WES, 1988). Figures 10-19 and 10-20 illustrate these two types of drop structures. The other most common type of structure used extensively in the rural Southeast is a sheet pile type shown in Figure 10-21 (Little and Murphy, 1982). A simple drop structure can be formed from a culvert section as illustrated in Figure 10-22 (Biedenharn, 1987). Figure 10-23 shows a typical gabion control structure for a trapezoidal channel. The equation of discharge is:

$$q = 2.63 \, H^{1.62} \qquad (10.97)$$

where q = unit discharge (cfs/ft) and H = total head on crest (ft).

This type of structure shows submergence at a submergence ratio of about 80%. It will be stable as long as the unit discharge is less than the limit given as follows:

$$q < 38.5[(H_c + h)/H_c]^{-3.5} \qquad (10.98)$$

where H_c = crest height (ft), and h = tailwater height relative to the crest (ft).

Sills may be notched at the thalweg location to concentrate low flows to improve aquatic habitat and water quality, or for aesthetic reasons. In highly visible locations, sills extending

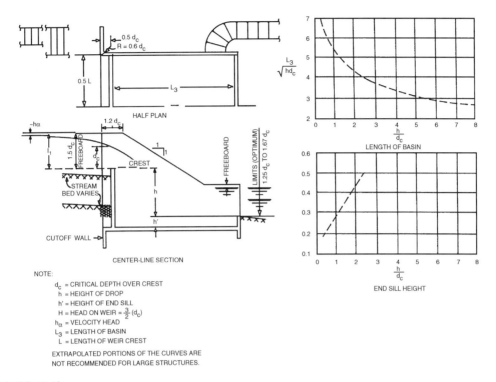

FIGURE 10-19
CIT type drop structure. (From Vanoni, V.A. and Pollak, R.E., Experimental Design of Low Rectangular Drops for Alluvial Flood Channels, Rpt. E-82, California Institute of Technology, Pasadena, CA, 1959. With permission.)

above the channel invert may be constructed of, or faced with, materials such as natural stone that create an attractive appearance. Sills may be modified to allow for passage of boats or fish, if desired.

10.17 Environmental Considerations for Channels

Floodplains and surface water resources provide many environmentally related benefits to a local urban municipality. Natural areas, ponds and lakes, and linear parks can be sources of municipal recreation. Habitat areas and flood-control areas can be in and adjacent to water sources. Water quality can be improved using the overbank and wetland areas adjacent to streams and rivers. This section covers both instream and floodplain environmental features and activities.

Instream Mitigation Features

A consideration of the environmental impacts of channel projects is, or should be, an important part of every channel design and alteration project. When properly designed, channels can maintain flood-control objectives and habitat preservation or restoration objectives. This type of consideration has not always been the case. Thaxton and Sneed (1982) reviewed the use of environmental features in Corps of Engineers projects, and

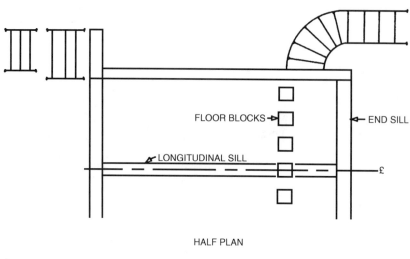

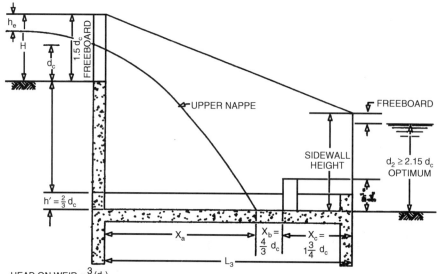

NOTE:
- H = HEAD ON WEIR = $\frac{3}{2}(d_c)$
- h_e = VELOCITY HEAD
- d_2 = TAIL WATER DEPTH
- d_c = CRITICAL DEPTH OVER CREST
- h = HEIGHT OF DROP
- h' = HEIGHT OF END SILL
- L_3 = LENGTH OF STILLING BASIN = $X_a + X_b + X_c$
- X_a = HORIZONTAL DISTANCE FROM CREST TO INTERSECTION OF UPPER NAPPE AND STILLING BASIN FLOOR
- X_b = HORIZONTAL DISTANCE FROM INTERSECTION OF UPPER NAPPE AND STILLING BASIN FLOOR TO UPSTREAM FACE OF FLOOR BLOCKS
- X_c = HORIZONTAL DISTANCE FROM UPSTREAM FACE OF FLOOR BLOCKS TO END OF STILLING BASIN

FIGURE 10-20
SAF type drop structure. (From U.S. Army Corps of Engineers, Waterways Experiment Station (WES), Hydraulic Design Criteria, Vicksburg, MS, 1988.)

Open Channel Design

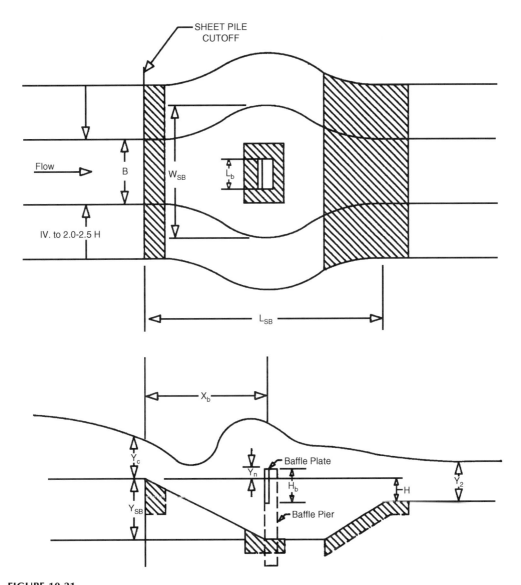

FIGURE 10-21
Low drop grade control structure. (From Little, W.C. and Murphy, J.B., Model Study of Low Drop Grade Structures, ASCE J. Hydraul., 108, HY10, October 1982. Reproduced by permission of ASCE.)

FIGURE 10-22
Culvert section drop structure.

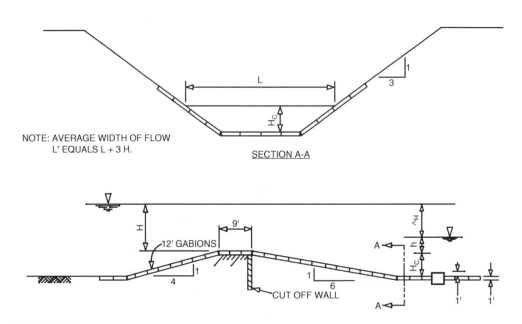

FIGURE 10-23
Typical gabion trapezoidal channel control structure. (From U.S. Army Corps of Engineers, Gabion Channel Control Structures, ETL 1110-2-194, 1974.)

McCarley et al. reviewed the use for flood-control projects (1990). Both found some successful but not widespread integration of environmental features in the past. The Corps of Engineers surveyed hundreds of environmental measurement and assessment techniques in an attempt to find quantitative measurement techniques for impact and improvement of streams and rivers (Henderson, 1982).

The beneficial and adverse environmental impacts of various types of channel projects have been detailed by Nunnally and Shields (1985) and are summarized in Table 10-21.

Open Channel Design

TABLE 10-21

Possible Adverse Environmental Impacts of Channel-Related Projects

Negative Environmental Impacts	Positive Environmental Impacts
Channel Brush Snagging	
Reduction of macroinvertebrate food supply	Aesthetic improvements
Reduction of fish habitat	Recreational improvements
Channel Bank Clearing	
Reduced shade, thermal impacts	Limited improvement of aesthetics
Destroyed terrestrial habitat	Limited recreational improvement
Reduced aquatic habitat	
Reduced food supply to food chain base	
Increased in-channel photosynthesis, shift toward autotrophic state	
Reduced bank stability, increased erosion	
Channel Excavation	
Same as clearing	Can show landscaped aesthetic appeal
Channel instability, erosion and sedimentation	Can improve some types of recreation
Physical destruction and reduction of aquatic habitat	
Can reduce recreational fishing	
Inadvertent wetlands draining	
Channel Paving	
Same as clearing and excavating	Possible improved aesthetics if original channel was very poor
Reduced aesthetic appeal	
Temperature fluctuations	
Loss of all habitat	
Block access if sides are steep and smooth	
Streambank Protection	
Loss of near bank habitat	Reduced erosion
Recreation can be impaired if access or use are reduced	Possible improved aesthetics for urban environment
Normally impaired aesthetics	Recreation can be improved if access is increased
Water quality effects same as clearing	Stable banks provide habitat for small creatures and macroinvertebrates
Terrestrial habitat reduction	Projecting bank protection can create instream habitat and slack water habitat
Same aquatic habitat as clearing	
Detention Control and Drops	
Blockage of fish passage	Sediment reduction if flow is reduced
Natural sediment reduction	Can be aesthetically pleasing
Can block recreational use	
Can cause hazard	
Levees and Floodwalls	
Creation of drier floodplain conditions	Can provide recreational overlooks, trails, access points, etc.
Modified flow regime may impact spawning	
Floodplain habitat reduction due to converted land use	
Normally aesthetically dull	

Impacts are interrelated in complex ways but can be categorized as follows: aesthetics, recreation, water quality, terrestrial habitat, and aquatic habitat. They can also be seen as primary, secondary, and tertiary. For example, a channel-straightening project has the primary effect of rapid bed and bank erosion, the secondary effect of reduced water quality due to increased suspended sediment levels, and the tertiary effect of impacts on recreation and aquatic habitat.

General environmental guidelines employed in successful channel projects include the following (COE, 1985):

- Minimize structural modifications and channel alterations
- Pay close attention to geomorphic and stable channel analysis, because channel instability is one of the largest impacts
- Lay out channel and modifications in great detail, showing all existing and created habitat and habitat (such as specific trees) to be saved

In addition, the following aspects should be considered:

- Low-flow channels are subchannels constructed within main channels. They can concentrate flow for biological habitat restoration, recreation, aesthetics, and water-quality benefits.
- High-flow channels are used to pass larger flood flows. Benefit can be accrued if only one side of the channel is excavated — single-side construction.
- Natural channels have a periodic meander with natural pools and riffles. An artificially but correctly sized meander, pool and riffle sequence, will help the channel to remain stable and provide fish habitat. Five to seven channel widths is the normal spacing between pools.
- Flow-through design for high flows to enter cutoff and oxbow lakes preserves habitat and allows for flood-control diversions.
- Instream habitat structures include sills, large rocks, ledges and overhangs, deflectors, bank-embedded pipes with large holes cut in them, etc.
- Paved channels can be improved if fish are present by providing cover, tree shade, and resting places.
- Construction timing and scheduling for nonsensitive seasons is important, especially in areas that experience fish migration.
- For migrating fish, culvert outlets should have resting pools, low entrance jumps, and sufficient flows. Baffles can reduce velocities. One municipality has even installed lights for especially long culverts to be used during the spawning season.
- Interior drainage areas and borrow pits can be redesigned as ponds for habitat as well as flood control.

Riprap and the Environment

Riprap has not been considered anything but a net negative in the popular literature concerning stream environmental health. The actual data and studies tell a different story. In nearly every study of macroinvertebrate health and diversity, those streams that have enhanced habitat diversity through the use of well thought out riprap schemes have increased stream health (Shields et al., 1995). Often riprap is mixed with soil-vegetation mixtures to attain the best of both worlds – the interstices within the riprap and its bank stabilization capabilities and the habitat and aesthetically pleasing visual feel of vegetation. Macroinvertebrate health and diversity compare favorably with vegetated channels, and the combination of both is best of all. On the contrary, those streams that have been stripped of natural vegetation and covered with uniform riprap have suffered habitat and environmental deterioration (see Figures 10-24 and 10-25).

Open Channel Design

FIGURE 10.24
Riprap vegetation combination for bank protection.

FIGURE 10.25
After vegetation establishment.

There are generally three scales to consider with respect to riprap design and environmental consideration: (1) micro-scale, (2) meso-scale, and (3) macro-scale. At the micro-scale level the diversity of size distribution of the riprap, the actual size, and the stability of the riprap substrate are key factors in creation of macroinvertebrate habitat (Minshall, 1984). Interstices between rocks create natural protected areas for macroinvertebrates, while deposition of finer particles in some interstices creates a secondary substrate. Micro-scale effects for fish health and diversity are mixed depending on the relative need of the species for the slowing effects of dense vegetation vs. the diversity of flow created by rough riprap banks (Shields et al., 1995).

At the meso-scale, about equal to channel-width scale, planned riprap structures can create significant fish habitat. Most designers are familiar with fish boulders. However, other types of structures including groins, dikes, irregular placement, rearing benches, bendway weirs, and spur dikes provide habitat superior to continuous revetment and often times rivaling vegetated banks.

At the macro-scale continuous revetment and the confinement of stream to narrow channels has led to deterioration of habitat and loss of diversity and health. Impacts include increased velocity and depth, loss of diversity, cutoff of floodplains, etc.

On balance there are several ways riprap can be used effectively, all including retention of diversity in riprap placement and mixture of riprap with vegetative treatments.

Environmental Consideration of Floodplains

Surface waters, floodplains, and the watersheds draining to them must be seen as one interrelated ecological system. One part of that system is the open channels and the immediate floodplain areas. These areas can provide the following (FEMA, 1986):

- Water resource values, including natural moderation of flood flows, water-quality benefits, and groundwater recharge
- Living resource values such as animal habitat, plant habitat, and human–nature interaction areas
- Cultural resource values such as archeological sites, historical sites, scientific, cultural and recreational sites, as well as areas for agriculture, forestry, and aquaculture

Floodplain Environmental Activities

Examples of management of floodplains for enhancement and preservation of natural values are included in Table 10-22 (FEMA, 1986).

Greenway Planning

The use of linear parks and greenways in urban floodplain and channel management is gaining wide public acceptance around the United States. Any program seeking to enhance the environment of open channels within a local urban area could benefit from a greenway program. Greenway programs are often combined with an overall stream restoration effort.

The National Parks Service provides a successful federally assisted program for stream corridor restoration, Rivers and Trails Conservation Assistance Program, though many municipalities find it easier and faster to work independently. Other state conservation interests can be found in the River Conservation Directory (1990). Other guidance can be

TABLE 10-22

Environmental Management of Floodplains

1. Natural Flood Storage and Conveyance
   Minimize floodplain filling
   Maintain wetlands
   Minimize grading and regrading to minimize soil compaction
   Relocate nonconforming structures out of the floodplain
   Return sites to natural contours
   Preserve natural drainage when constructing infrastructure (such as utility lines, bridges, sanitary lines) in the floodplain
   Prevent intrusion into sensitive ecosystem areas
2. Water Quality Maintenance
   Maintain wetland and vegetative buffers
   Use sound agricultural practices
   Control urban runoff quality
   Minimize erosion around streams and lakes
   Restrict sources of potential toxics and pathogens from the floodplain
   Restrict transportation of toxics and pathogens from routes over or adjacent to sensitive waterways
   Plan for emergency response to spills
3. Groundwater Recharge
   Require pervious surfaces near floodplains for low-traffic areas
   Design for runoff infiltration
   Dispose of waste materials in noncontaminating ways
4. Living Resources
   Identify and protect wildlife habitat and ecologically sensitive areas from destruction
   Require topsoil protection programs
   Restrict wetland drainage
   Reestablish floodplain ecosystems
   Minimize tree cutting and vegetation removal
   Design floodgates and seawalls for natural flow
5. Cultural Resources
   Provide public access to and near water resources
   Locate and preserve historical and cultural resources and integrate with recreational areas
6. Agricultural Resources
   Minimize soil erosion on cropped areas
   Control chemical and fertilizer use
   Strengthen "carrot"-type incentive programs locally
   Minimize irrigation return flows
   Eliminate feedlot operations and cattle access to streambanks
   Discourage new agricultural production areas
   Encourage sound agricultural practices in productive areas remote from flood problems
7. Aquacultural Resources
   Construct impoundments that minimize impact on natural drainage
   Limit exotic species not natural to the area
   Discourage large-scale mechanical operations
   Use caution in animal waste disposal
8. Forestry Resources
   Control clear cutting
   Encourage enforcement of state and federal laws
   Include fire management in overall management plans
   Require erosion control on all logging operations

Source: From Federal Emergency Management Agency, A Unified National Program for Floodplain Management, FEMA 100, March 1986.

found in the *Riverworkbook* (National Park Service, 1988), the Illinois Stream Preservation Handbook (State of IL, 1983), and Greenways (Flink and Searns, 1993). The steps in the development of a corridor restoration include identification of issues, resources, and problems; development of public involvement program; setting of goals; consideration of alternatives; setting a plan of action; and taking action.

Eugster (1988), Flanagan (1991) and the National Park Service (1983) lay out greenway planning steps once the planning team is put together:

- Define role and function of the proposed greenway.
- Determine project goals and orientation including early land acquisition/donation and development of regulations for land use practices.
- Initiate the project through approaching each stakeholder with the proposal and "selling it."
- Involve the public in hearings, workshops, etc.
- Assess resources and land value, often through the use of GIS and a point rating overlay system.
- Analyze the issues including resource use conflicts and public attitudes.
- Explore regulatory and administrative alternatives to take advantage of programs and possible financial resources.
- Assess the political situation.
- Develop an implementation strategy.

Stream Channel Preservation and Restoration

We would be remiss in a book on urban stormwater management not to discuss, however briefly, stream channel restoration and preservation. This topic fills books of several hundred pages and cannot be but briefly surveyed here (Interagency Group, 2000). There are many good general and specific topical references on the subject, several of which appear in the reference list at the end of this chapter.

The term restoration, in its purest sense, means the reestablishment of predisturbance aquatic functions and related physical, chemical, and biological characteristics. Restoration is a holistic process not achieved through the isolated manipulation of individual elements. In the urban setting, the definition becomes a bit more clouded. Is it possible to have a pristine rural stream in the midst of an urban environment? Many practitioners would say no. What then is the fall-back position for urban streams? There is no easy answer and certainly no general rule. However, the holistic master planning process described in Chapter 14 can lead a group of stakeholders to defining the values and characteristics of a particular stream to be preserved or restored within the confines of the existing development, and can define the limits on watershed development to preserve key stream character and habitat.

Restoration Guiding Principles

There are certain generally accepted guiding principles in the restoration of streams, urban or rural (EPA, 2001). Below is a list of the most common and widely held.

Preserve and Protect Aquatic Resources

Existing, relatively intact ecosystems are the keystone for conserving biodiversity. They provide the biota and other natural materials needed for the recovery of impaired systems. Thus, restoration does not replace the need to protect aquatic resources in the first place.

Restore Ecological Integrity

Restoration should reestablish insofar as possible the ecological integrity of degraded aquatic ecosystems, including biological and structural integrity. Its key ecosystem processes, such as nutrient cycles, succession, water levels and flow patterns, and the dynamics of sediment erosion and deposition, fluctuate naturally within a natural range of variability.

Restore Natural Structure

Urban-related alteration of channel form or other physical characteristics often leads to other problems such as habitat degradation, changes in flow regimes, and siltation. Restoring the original site morphology and other physical attributes is then an essential first step to the success of other aspects of a restoration project.

Restore Natural Function

Structure and function are closely linked in urban stream corridors. Reestablishing the appropriate natural structure can bring back beneficial functions. It is essential to identify what functions should be present, and make missing or impaired functions priorities in the restoration.

Work within the Watershed and Broader Landscape Context

Restoration requires a design based on the entire watershed, not just the part of the water body that may be the most degraded site. Activities throughout the watershed can have adverse effects on the aquatic resource that is being restored. A localized restoration project may not be able to change what goes on in the whole watershed, but it can be designed to better accommodate watershed effects.

Understand the Potential of the Watershed

A watershed has the capacity to become only what its physical and biological setting will support. Establishing restoration goals for a water body requires knowledge of the historical range of conditions that existed on the site prior to degradation and what future conditions might be. Anticipate future changes.

Address Ongoing Causes of Degradation

Restoration efforts are likely to fail if the sources of degradation persist. Therefore, it is essential to identify the causes of degradation and eliminate or remediate ongoing stresses wherever possible. It is important in the urban setting to identify the institutional root causes of stream degradation and seek to change those as well. The Five Whys methodology of Chapter 1 can be used for such an analysis.

Develop Clear, Achievable, and Measurable Goals

Goals direct implementation and provide the standards for measuring success. The chosen goals should be achievable ecologically and socioeconomically, given the available resources and the extent of community support for the project.

Use a Reference Site

Reference sites are areas that are comparable in structure and function to the proposed restoration site before it was degraded. In an urban setting, reference sites can also mean other urban streams that have retained or attained the character and values desired in the stream in question. As such, reference sites may be used as models (with due regard for the uniqueness of each stream) for restoration projects, as well as a yardstick for measuring the progress of the project.

Involve the Skills and Insights of a Multidisciplinary Team

Restoration can be a complex undertaking that integrates a wide range of disciplines including ecology, aquatic biology, hydrology and hydraulics, geomorphology, engineering, planning, communications, and social science.

Design for Self-Sustainability

Long-term viability is best assured by minimizing the need for continuous maintenance of the site and overdependence on man's intervention. To do so requires establishing a native ecology with due regard for changed and changing conditions that may push an ecosystem past a threshold of no return. An ecosystem in good condition is more likely to have the ability to adapt to changes.

Use Passive Restoration, when Appropriate

Often simply reducing or eliminating the sources of degradation and allowing recovery time will be enough to allow the site to naturally regenerate or move to a new plateau of structural and biological stability. Many times, there are reasons for restoring a water body as quickly as possible, but there are other situations when immediate results are not critical. It is important to note that, while passive restoration relies on natural processes, it is still necessary to analyze the site's recovery needs and determine whether time and natural processes can meet them.

Use Natural Fixes and Bioengineering Techniques, where Possible

Bioengineering is a method of construction combining live plants with dead plants or inorganic materials to produce living, functioning systems to prevent erosion, control sediment and other pollutants, and provide habitat. Bioengineering techniques can often be successful for erosion control and bank stabilization, flood mitigation, and even water treatment.

Monitor and Adapt where Changes are Necessary

Every combination of watershed characteristics, sources of stress, and restoration techniques is unique and, therefore, restoration efforts may not proceed exactly as planned. Adapting a project to at least urbanization-induced change or new information is normal. Monitoring, therefore, is crucial. This process of monitoring and adjustment is known as adaptive management.

Restoration Practices

Brown (2000) examined 24 different types of stream restoration practices and included over 450 individual practice installations. The practice types were broadly classified into four practice groups based on their intended restoration objective: bank protection, grade control, flow deflection/concentration, and bank stabilization. Each practice was evaluated in the field according to four simple visual criteria: structural integrity, function, habitat

TABLE 10-23

Stream Restoration Measures

Bank Protection Group	Flow Deflection/Concentration Group
Imbricated riprap	Wing deflectors
Rootwad revetment	Single-wing deflectors
Boulder revetments	Double-wing deflectors
Single-boulder revetment	Log vane
Double-boulder revetment	Rock vane/J-rock vane
Large-boulder revetment	Cut-off sill
Placed rock	Linear deflector
Lunkers	
A-jacks	

Grade Control Group	Bank Stabilization/Bioengineering Group
Rock vortex weir	Vegetative/bioengineering practices
Rock cross vane	Coir fiber log
Step pool	Live fascine
Log drop/V-log drop	Brush mattress
	Bank regarding

enhancement, and vegetative stability. He found that most of the 24 practices remained stable after an average of 4 years — yet, a majority did not fully achieve their intended purpose: that of establishing or enhancing effective habitat.

Table 10-23 indicates these 24 practices, which serve as a fairly comprehensive summary of the most common types of practices for stream restoration.

Some of Brown's findings are especially telling when attempts are made to apply rural understandings to changing urban streams (Rosgen, 1994; Caraco, 2000). It was not the design specifications that caused most of the failures but rather a lack of understanding of the stream morphology, installing practices where channel conditions were inappropriate, or poor construction practice. It is difficult to predict stable stream channel geometry in urban streams, and unless the geometry is correct from the start, any subsequent channel adjustment can and will cause practice failure. Most of these projects attempted to create a natural (e.g., predisturbance) type channel morphology in an unnatural, disturbed watershed.

Each practice type has a relatively narrow range of stream conditions for which it is best suited. In a rapidly changing or poorly understood urban environment, it is easy to exceed one, or more, thresholds for a particular practice, resulting in poor performance or outright failure. Most often, practitioners fail to recognize the increased shear stresses (and associated scour) in the vicinity of structures, the depth of scour in just a few events, the natural tendency for urban channels to enlarge, and the force water exerts on any body on which it impinges. His study found that the key factors for practice success were a good understanding of stream processes and an accurate assessment of current and future stream channel conditions.

Restoration practices should not only concentrate instream but throughout the whole floodplain corridor and beyond. Urban stream assessment must also include at least a cursory watershed assessment — is the flow regime changing?, what is happening in terms of sediment production, pollution generation, temperature flux, riparian intrusion, land use changes, etc.? Often, the best progress toward stream restoration, and especially preservation, is to change land uses and look at retrofits that can reduce the forces pushing the stream to, and over, a threshold.) (Opportunities to use the stream corridor for multiple uses abound and are often economically viable. (Flink and Searns, 1993; Interagency Group, 2000.)

Stream Health Assessment

There are many approaches to assessing urban streams from physical, chemical, and biological perspectives (Galli, 1996; Rosgen, 1994; USDA, 1999; USEPA, 1999; Hunter, 1991; Knutson and Naef, 1997; McCarley et al., 1990; Steedman, 1998). Physical approaches range from geomorphologic to attempting to estimate stresses and predict shape and form changes due to imposed shear stresses and flow duration curve changes. Ecological integrity is measured through sampling of the macroinvertebrate community, fish, periphyton, and habitat assessment. Stream form is compared to known types with measured characteristics and predictions made of stable form. In most cases, reference streams are used for examples and standard setting.

It should be stated at the outset that even experienced practitioners are poor at being able to predict a final stable stream form and ecological relationships in an urban setting, especially when the stream boundary is mobile. Part of the reason is that urban streams (actually, all streams and most of nature as well) tend toward thresholds. Imposed stresses, whether ecological, chemical, velocity, sediment, temperature, etc., can be handled by the stream system up to a threshold. At that point, a rapid degradation of the steam occurs. Stream research generally indicates that certain zones of stream quality exist, most notably at about 10% impervious cover, where sensitive stream elements are lost from the system (Brown, 2000). A second threshold appears to exist at around 25 to 30% impervious cover,

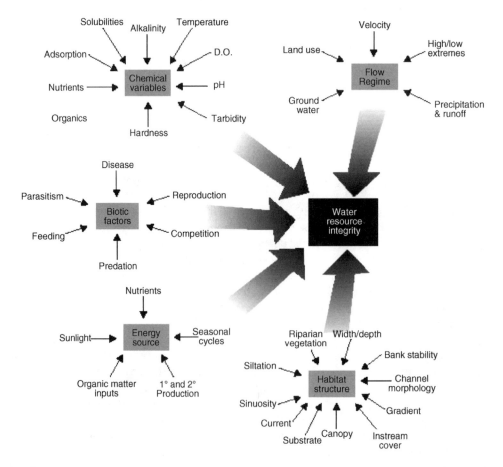

FIGURE 10-26
Factors that influence the integrity of streams. (From USDA, 1998. With permission.)

where most indicators of stream quality consistently shift to a poor condition (e.g., diminished aquatic diversity, water quality, and habitat conditions).

Stream assessments must take into account parameters of importance to the goals and objectives of the stream. For example, if trout habitat is important, then the parameters must include such things as (Hunter, 1991) vegetation type, overhang and shade percentage, bank angle, height and overhang, active channel width, wet width, depth and velocity, sinuosity, gradient, discharge, substrate composition and embeddedness, pool-to-riffle ratio, instream cover and structures, temperature, and macroinvertebrate types and density.

One of the better stream assessment techniques is the Rapid Stream Assessment Technique (RSAT) developed for Montgomery County by the Washington COG (Galli, 1996). It represents a synthesis of many of the more robust and highly technical techniques applied in a commonsense way. While the specifics of the technique are not applicable everywhere (it is limited to non-limestone, smaller Piedmont streams), the approach is.

Six categories of assessment are used: channel stability, scouring and sediment deposition, physical instream habitat, water quality, riparian habitat conditions, and biological indicators. The collection of data is rigorous in that over 30 parameters are measured at 400 ft transects along the stream. Other data are gathered as well. Each of the six general categories is divided into excellent, good, fair, and poor categories with a range of about zero at the low end to seven to eleven at the high end.

The USDA also developed an assessment tool with the key parameters indicated in Figure 10-26.

References

Ackers, P. and White, W.R., Sediment transport: new approach and analysis, *ASCE J. Hydraul. Div.*, 99, HY11, November 1973.
Ackers, P., River regime: research and application, *J. Inst. of Water Eng.*, 26, 5, 1972.
American Society of Civil Engineers, Sedimentation Engineering, ASCE Manuals and Reports on Engineering Practice No. 54, 1977.
Atlanta Regional Commission, The Georgia Stormwater Manual, 2002.
Bekaert, Gabions, Terra Aqua Corp., Riverdale Station, NY, 1979.
Bensen, M.A. and Dalrymple, T., USGS Techniques of Water Resources Investigations, Book 3, 1967.
Biedenharn, D.S., Grade Control Structures, U.S. Army Corps of Engineer's Training Course Notes, USAEWES, November 1987.
Blair, L.A., Trapezoidal Concrete Flood Control Channels in Albuquerque: Lessons Learned, Proc. 1984 Int. Symp. Urban Hydrology, Hydraulics and Sediment Control, Lexington, KY, 1984.
Blalock, M.E. and Strumm, T.W., Minimum specific energy in compound channel, *ASCE J. Hydrology*, 107, 699–717, 1981.
Bledsoe, B., Relationships of Stream Responses to Hydrologic Changes, Proc. ASCE Eng. Found. Conf.: *Linking Stormwater BMP Designs and Performance to Receiving Water Impact Mitigation*, Urbonas, B., Ed., August 19–24, Snowmass Village, CO, 2001, pp. 127–144.
Blench, T., Mobile-Bed Fluviology, Department of Tech. Svc., University of Alberta, Edmonton, Canada, 1966.
Bray, D.I., Representative discharges for gravel-bed rivers in Alberta, CA, *J. Hydrology*, 27, 1975.
Brown, K.B., Urban Stream Restoration Practices: An Initial Assessment, The Center For Watershed Protection, Elliot City, MD, 2000, www.cwp.org.
Caraco, D., The dynamics of urban stream channel enlargement, *Watershed Protection Techniques*, The Center for Watershed Protection, Elliot City, MD, 3, 3, 729–734, 2000.

Chaudhry, M.H. and Bhallamudi, S.M., Computation of critical depth in symmetrical compound channels, *J. Hydraul. Res.*, 26, 4, 1988.
Chien, N., A Concept of Lacey's Regime Theory, Proc. ASCE, Irrig. and Drain. Div., February 1955.
Chien, N., A Concept of the Regime Theory, ASCE Trans., paper 2884, with discussions, 1957.
Chow, V.T., Ed., *Open Channel Hydraulics*, McGraw Hill, New York, 1959.
Cowan, W.L., Estimating hydraulic roughness coefficients, *Agric. Eng.*, 37, 7, 1956.
Denver Urban Drainage and Flood Control District, Urban Storm Drainage Criteria Manual, with changes dated 1991.
Denver Urban Drainage and Flood Control District, Flood Hazard News, September 1980.
EPA, Stream Restoration Principles, Watershed Ecology Team, U.S. EPA Office of Wetlands, Oceans and Watersheds, 2001, http://www.epa.gov/owow/wetlands/restore/principles.html#list.
Eugster, J.G., Steps in state and local greenway planning, *Nat. Wetlands Newsletter*, September – October 1988.
Federal Emergency Management Agency, A Unified National Program for Floodplain Management, FEMA 100, March 1986.
Federal Highway Administration, Bridge Waterways Analysis Model (WSPRO), Users Manual, FHWA IP-89-027, 1989.
Flanagan, R.D., Multi-purpose Planning for Greenway Corridors, ASFPM Natl. Conf. Handout, 1991.
Flink, C.A. and Searns, R.M., *Greenways — A Guide to Planning, Design and Development*, Island Press, Washington, DC, 1993.
Fortier, S. and Scoby, F.C., Permissible Canal Velocities, *Trans. ASCE*, Vol. 89, pp. 940–956, 1926.
French, R.H., *Open Channel Hydraulics*, McGraw Hill, New York, 1985.
Galli, J., Final Technical Memorandum: Rapid Stream Assessment Technique (RSAT) Field Methods, Department of Environmental Programs, Metropolitan Washington Council of Governments. Washington, DC, 1996.
Garbrecht, J. and Brown, G.O., Calculation of total conveyance in natural channels, *ASCE J. Hydraul.*, 117, 6, June 1991.
Gray, D.T. and Leiser, A.T., *Biotechnical Slope Protection and Erosion Control*, Van Nostrand Reinhold, New York, 1982.
Green, J.E.P. and Garton, J.E., Vegetation Lined Channel Design Procedures, Trans. ASAE, 1983.
Gwinn, W.R. and Ree, W.O., Maintenance Effects on the Hydraulic Properties of a Vegetation-Lined Channel, Trans. ASAE, 1980.
Harza Engineering Company, Storm Drainage Design Manual, prepared for the Erie and Niagara Counties Regional Planning Board, Harza Engineering Company, Grand Island, NY, 1972.
Henderson, J.E., Handbook of Environmental Quality Measurement and Assessment: Methods and Techniques, IR E-82-2, USAEWES, 1982.
Herricks, E., Observed Stream Responses to Changes in Runoff Quality, Proc. ASCE Eng. Found. Conf.: Linking Stormwater BMP Designs and Performance to Receiving Water Impact Mitigation, Urbonas, B., Ed., August 19–24, Snowmass Village, CO, 2001, pp. 145–157.
Hey, R.D., Bathurst, J.C., and Thorne, C.R., *Gravel-Bed Rivers*, John Wiley & Sons, NY, 1982, chap. 19.
Hunter, C.J., *Better Trout Habitat, Montana Land Reliance*, Island Press, Washington, DC, 1991.
Inglis, C.C., Historical Note on Empirical Equations, Developed by Engineers in India for Flow of Water in Sand and Alluvial Channels, Proc. Int. Assoc. for Hydrology Res., Second Meeting, Stockholm, 1948.
Interagency Floodplain Management Review Committee, Sharing the Challenge: Floodplain Management into the 21st Century, Executive Office of the President, Washington, DC, June 1994.
Interagency Group Stream Corridor Restoration: Principles, Processes and Practices, 2000, http://www.usda.gov/stream_restoration/newgra.html.
Ippen, A.T. and Drinker, P.A., Boundary Shear Stress in Curved Trapezoidal Channels, Proc. ASCE, HY5, September 1962.
Jacobs, E.L., Gabions: Flexible Solutions for Urban Erosion, Proc. Symp. Urban Hydrology, Hydraul., and Sed. Cont., Lexington, KY, 1984.
Kadlec, R.H., Overland flow in wetlands: vegetation resistance, *ASCE J. Hydraul. Eng.*, 116, 5, 1990.
Knutson, K.L. and Naef, V.L., Recommendations for Washington's Priority Habitats, Washington Department of Fish and Wildlife, Olympia, WA, 1997.

Kolupaila, K., Methods of determination of the kinetic energy factor, *The Port Engineer*, Calcutta, India, 5, 1, January 1956.

Kouwen, N., Field estimation of the biomechanical properties of grass, *J. Hydraul. Res.*, 26, 5, 1988.

Kouwen, N. and Li, R., Biomechanics of vegetative channel linings, *ASCE J. Hydraul.*, 106, HY6, June 1980.

Lacey, G., Stable Channels in Alluvium, Proc. Inst. for Civ. Eng., 229, 259–292, 1930.

Lane, E.W., The Importance of Fluvial Morphology in Hydraulic Engineering, Proc. ASCE, 21, 745, 1955.

Little, W.C. and Murphy, J.B., Model study of low drop grade structures, *ASCE J. Hydraul.*, 108, HY10, October 1982.

Logan, Vegetation and Mechanical Systems, USDA Forest Service, Missoula, MT, February 1979.

Loganathan, G.V., Optimal design of parabolic canals, *ASCE J. Irrig. and Drain.*, 117, 5, 1991.

Maccaferri, Flexible Linings in Reno Mattress and Gabions for Canals and Canalized Water Courses, Bologna, Italy, 1985.

Maccaferri, Maccaferri Gabions Channeling Works, Bologna, Italy, 1989.

Maccaferri, Maccaferri Gabions for River Training, Earth Control and Soil Conservation, Bologna, Italy, 1990.

MacRae, C.R., Experience from morphological research on Canadian streams: Is control of the two-year frequency runoff event the best basis for stream channel protection, in: *Effects of Watershed Development and Management on Aquatic Ecosystems*, Proc. ASCE Engineering Foundation Conference, Roesner, L., Ed., August 4–9, 1996, Snowmass Village, CO, 1997, pp. 145–162.

Maynord, S.T., Stable Riprap Size for Open Channel Flows, Ph.D. dissertation, Colorado State University, Fort Collins, CO, 1987.

McCarley, R.W. et al., Flood-Control Channel National Inventory, MP HL-90-10, USAEWES, October 1990.

Minshall, G.W., Aquatic insect-substratum relationship, in *The Ecology of Aquatic Insects*, Resh, V.H. and Rosenberg, D.M., Eds., Praeger Publishers, New York, 1984, chap. 12.

Morris, J.R., A method of estimating floodway setback limits in areas of approximate study, in *Proc. 1984 Int. Symp. Urban Hydrology, Hydraul. and Sediment Control*, Lexington, KY: University of Kentucky, 1984.

Murphy, T.E., Drop Structures for the Gering Valley Project, Scotsbluff County, NE, USAEWES Tech. Rpt. 2-760, February 1967.

National Park Service, Greenway Planning, U.S. Department of the Interior, National Park Service, Mid-Atlantic Regional Office, September 1983.

National Park Service, Riverworkbook, U.S. Department of the Interior, National Park Service, Mid-Atlantic Regional Office, 1988.

National Park Service, River Conservation Directory, Department of the Interior, National Association for State River Conservation Programs, 1990.

Neill, C., Hydraulic Design of Stable Flood Control Channels — Draft Guidelines for Preliminary Design, Seattle District COE, Northwest Hydraulic Consultants, January 1984.

Nowatzki, E.A. and Reely, B.T., Effects of Fly Ash Content on Strength and Durability Characteristics of Pantano Soil-Cement Mixes, Department of Civil Engineering, University of Arizona, Tucson, AZ, 1986.

Nunnally, N.R. and Keller, E., Use of Fluvial Processes to Minimize Adverse Effects of Stream Channelization, Water Res. Inst., UNC, July 1979.

Nunnally, N.R. and Shields, F.D., Incorporation of Environmental Features in Flood Control Projects, TR E-85-3, USAEWES, 1985.

Nussbaum, P.J. and Colley, B.E., Dam Construction and Facing with Soil-Cement, Portland Cement Assoc., R&D Bull., 1971.

Ogden Environmental, A Nationwide Inventory of Stream Corridor Planning Programs, Ogden Environmental, Nashville, TN, 1993.

Peterska, A.J., Hydraulic Design of Stilling Basins and Energy Dissipators, Engineering Monograph No. 25, U.S. Department of Interior, Bureau of Reclamation, Washington, DC, 1978.

Petryk, S. and Grant, E.U., Critical flow in rivers with flood plains, *ASCE J. Hydraul.*, 583–594, May 1978.

Pima County Department of Transportation and Flood Control Distribution, Soil-Cement Applications and Use in Pima County for Flood Control Projects, Tucson, AZ, 1986.
Portland Cement Association (PCA), Soil-Cement Slope Protection for Embankments: Planning and Design, 1975.
Portland Cement Association (PCA), Soil-Cement for Water Control — Laboratory Tests, 1976.
Posey, C.J., Stability of Rock Sausages, University of Connecticut Institute of Water Resources, Rpt. No. 19, November 1973.
Powell, R.F., James, L.D., and Jones, D.E., Approximate method for quick flood plain mapping, *ASCE J. Water Resour.*, 106, WR1, March 1980.
Reese, A.J., Riprap sizing, four methods, in *Proc. ASCE Conf. Water Resour. Dev., Hydraul. Div., ASCE,* Schreiber, D.L., Ed., 1984.
Reese, A.J., Use of Soil-Cement for Bank Protection, DRAFT ETL 1110-2, U.S. Army Corps of Engineers, Waterways Experiment Station, March 1987.
Reese, A.J., Nomographic Riprap Design, miscellaneous paper HL 88-2. Vicksburg, MS: U.S. Army Engineers, Waterways Experiment Station, 1988.
Riley, A.L., *Restoring Streams in Cities*, Island Press, Washington, DC, 1997.
Rosgen, D., A Classification of Natural Rivers, CATENA 22, 1994, pp. 169–199.
Rosgen, D., Applied River Morphology, Wildland Hydrology, Pogosa Springs, CO, 1994.
Schiechtl, H., Bioengineering for Land Reclamation and Conservation, University of Alberta Press, Edmonton, Alberta, Canada, 1980.
Schoellhamer, D.H., Peters, J.C., and Larock, B.E., Subdivision Froude number, *ASCE J. Hydraul.* 111, 1099–1104, 1985.
Schueler, T.R., The importance of imperviousness, *The Practice of Watershed Protection*, Schueler, T.R. and Holland, H.K., Eds., published by the Center for Watershed Protection, Ellicott City, MD, 2000.
Schumm, S.A., *The Fluvial System*, John Wiley & Sons, NY, 1977.
Shields, F.D., Cooper, C.M., and Testa, S., Towards a greener riprap: environmental considerations from microscale to macroscale, *Proc. River, Coastal and Shoreline Protection Erosion Control Using Riprap and Armourstone,* Thorne, C.R., Abt, S.R., Barends, F.B., Maynord, S.T., and Pilarezyk, K.W., Eds., John Wiley & Sons, New York, 1995.
Simons, D.B. and Albertson, M.L., Uniform Water Conveyance Channels in Alluvial Material, Tran. ASCE, Vol. 128, Part 1, Paper No. 3399, 1963.
Simons, D.B. and Senturk, F., *Sediment Transport Technology*, Water Resources Press, Ft. Collins, CO, 1976.
Simons, Li & Assoc. (SLA), Engineering Analysis of Fluvial Systems, SLA, Ft. Collins, CO, 1982.
Sotir, R.B. and Gray, D.H., Fill Slope Repair Using Soil Bioengineering Systems, Public Works, December 1989.
Sovern, D.T. and MacDonald, A., Can instream integrity be obtained through on-site controls?, Proc. ASCE Eng. Found. Conf.: *Linking Stormwater BMP Designs and Performance to Receiving Water Impact Mitigation*, Urbonas, B., Ed., August 19–24, Snowmass Village, CO, 2001, pp. 47–59.
State of Illinois, Stream Preservation Handbook, IDOT, 1983.
Steedman, R.J., Modification and assessment of an index of biotic integrity to quantify stream quality in Southern Ontario, *Canadian J. Fish. and Aquatic Sci.*, 45, 492–501, 1988.
Stephenson, D., Rockfill and Gabions for Erosion Control, Civil Eng. in South Africa, September 1979.
Temple, D.M., Tractive Force Design of Vegetated Channels, Proc. ASAE-CSAE Joint Meeting, University of Manitoba, Winnipeg, Canada, June 1979.
Temple, D.M. et al., Stability Design of Grass-Lined Open Channels, USDA ARS Handbook # 667, 1987.
Thackston, E.L. and Sneed, R.B., Review of Environmental Consequences of Waterway Design, TR E-82-4, USAEWES, 1982.
Trout, T.J., Channel design to minimize lining material costs, *ASCE J. Irrig. and Drain.*, 108, IR4, December 1982.
U.S. Army Corps of Engineers, Hydraulic Design of Flood Control Channels, EM 1110-2-1601, 1970.
U.S. Army Corps of Engineers, Gabion Channel Control Structures, ETL 1110-2-194, 1974.

U.S. Army Corps of Engineers, Design Criteria — Paved Concrete Flood Control Channels, ETL 1110-2-236, June 1978.
U.S. Army Corps of Engineers, Environmental Engineering for Flood Control Channels, DRAFT, 1985.
U.S. Army Corps of Engineers, Waterways Experiment Station (WES), Hydraulic Design Criteria, Vicksburg, MS, 1988.
U.S. Army Corps of Engineers, Stability of Flood Control Channels, DRAFT, 1990.
U.S. Army Corps of Engineers, Hydrologic Engineering Analysis Concepts for Cost-Shared Flood Damage Reduction Studies, EP 1110-2-6005, December 1992.
U.S. Army Corps of Engineers, Engineering and Design: Channel Stability Assessment for Flood Control Projects, EM 1110-2-1418, Department of the Army, Washington, DC, October 1994.
U.S. Department of Agriculture, Soil Conservation Service (SCS), Handbook of Channel Design for Soil and Water Conservation, TP-61, 1954.
U.S. Department of Agriculture, Soil Conservation Service (SCS), Design of Open Channels, Tech. Release 25, October 1977.
U.S. Department of Agriculture (USDA), Stream Visual Assessment Protocol, USDA Natural Resource Conservation Service, National Water and Climate Center Technical Note 99-1, 1999.
U.S. Department of Transportation, Federal Highway Administration, Design Charts for Open Channel Flow, Hydraulic Design Series No. 3, Washington, DC, 1973.
U.S. Department of Transportation, Federal Highway Administration, Hydraulic Design of Energy Dissipators for Culverts and Channels, Hydraulic Engineering Circular No. 14, Washington, DC, 1982.
U.S. Department of Transportation, Federal Highway Administration, Guide for Selecting Manning's Roughness Coefficients for Natural Channels and Flood Plains, FHWA-TS-84-204, Washington, DC, 1984.
U.S. Department of Transportation, Federal Highway Administration, Design of Stable Channels with Flexible Linings, Hydraulic Engineering Circular No. 15, Washington, DC, 1986.
U.S. Department of Transportation, Federal Highway Administration, Urban Drainage Design Manual, Second Edition, Hydraulic Engineering Circular No. 22, Washington, DC, 2001.
U.S. Environmental Protection Agency (USEPA), Rapid Bioassessment Protocols for Use in Streams and Wadeable Rivers: Periphyton, Benthic Macroinvertebrates and Fish, Second Edition, Office of Water, Washington, DC, EPA/841-B-99-002, 1999.
USSR, Maximum Permissible Velocity in Open Channels, Hydro-Technical Construction, Number 5, Moscow, 1936.
Vanoni, V.A. and Pollak, R.E., Experimental Design of Low Rectangular Drops for Alluvial Flood Channels, Rpt. E-82, California Institute of Technology, Pasadena, CA, 1959.
Westman, W.E., *Ecology, Impact Assessment and Environmental Planning*, John Wiley and Sons, NY, 1985.
White, W.R., Bettess, R., and Paris, E., Analytical approach to river regime, *ASCE J. Hydraul.*, 108, HY10, 1982.
Wright-McLaughlin Engineers, Urban Storm Drainage Criteria Manual, Vol. 2, prepared for the Denver Regional Council of Governments, Wright-McLaughlin Engineers, Denver, CO, 1969.
Yoder, C.O., Miller, R.J., and White, D., Using Biological Criteria to Assess and Classify Urban Streams and Develop Improved Landscape Indicators, Ohio EPA, Proc. USEPA Nat. Conf. on Tools for Urban Resour. Manage. and Prot., February 7–10, Chicago, IL, 2000, pp. 56–82.

11

Storage and Detention Facilities

11.1 Introduction

The traditional design of storm drainage systems has been to collect and convey storm runoff as rapidly as possible to a suitable location, where it can be discharged. As areas urbanize, this type of design may result in major drainage and flooding problems downstream. The impacts of urban development include faster and higher peaks, more flow volume, higher velocities, higher temperatures, lower base flows during nonstorm conditions, and greater levels of pollution. This chapter deals with flood flow peak reduction, timing strategies, and design. A later chapter will address stormwater quality (Chapter 15). Because municipalities use different design storms for detention design, examples are given in this chapter based on different design storm events (e.g., 2-year, 10-year, 25-year, and 100-year). The principles and procedures are the same for any design storm that is selected.

11.2 Uses and Types of Storage Facilities

Under favorable conditions, the temporary storage of some of the storm runoff can decrease downstream flows and the cost of the downstream conveyance system. Detention storage facilities can range from small facilities contained in parking lots or other on-site facilities to large lakes and reservoirs. They can be:

- Single purpose or dual purpose
- On-line or off-line
- Regional or site specific
- Temporary or permanent
- To reduce runoff or delay runoff
- Single objective or multiobjective
- Single outlet or multioutlet
- Above ground or underground
- Wet ponds or dry detention

- Integrated with the surroundings or separate from them
- Structural in nature or developed from natural materials

This chapter provides general design criteria for detention and retention storage facilities, as well as procedures for performing preliminary and final sizing and reservoir routing calculations.

It should be noted that the location of storage facilities is very important, as it relates to the effectiveness of these facilities to control downstream flooding. Small facilities will only have minimal flood-control benefits, and these benefits will quickly diminish as the flood wave travels downstream. Multiple storage facilities located in the same drainage basin will affect the timing of the runoff through the conveyance system, which could decrease or increase flood peaks in different downstream locations. Thus, it is important for the designer to design storage facilities as a stormwater management structure that controls runoff from a defined area and interacts with other stormwater management structures within the drainage basin. Effective stormwater management must be coordinated on a regional or basin-wide planning basis. The municipality should encourage and participate in such planning.

Urban stormwater storage facilities are often referred to as detention or retention facilities. For the purposes of this chapter, detention facilities are those that are designed to reduce the peak discharge and only detain runoff for some short period of time. These facilities are designed to completely drain after the design storm has passed. Recharge basins are a special type of detention facility designed to drain into the groundwater table. Retention facilities are designed to contain a permanent pool of water while discharging runoff at a controlled rate. Because most of the basic design procedures are the same for detention and retention facilities, the term *storage facilities* will be used in this chapter to include detention and retention facilities. If special procedures are needed for detention or retention facilities, these will be specified.

Routing calculations needed to design storage facilities, although not extremely complex, are time consuming and repetitive. To assist with these calculations, there are many available reservoir routing computer programs. Storage facilities should be designed and analyzed using reservoir routing calculations. An approximate hand-calculation method is presented in this chapter that closely approximates reservoir routing for smaller facilities. For small facilities in highly impervious areas, with accurate target flow reduction criteria, approximate methods may be appropriate. Sometimes retention facilities are designed with no outlet. For this type of facility, a long-term water balance is important rather than routing.

Uses of Storage Facilities

The use of storage facilities for stormwater management has increased dramatically in recent years. The benefits of storage facilities can be divided into two major control categories of quality and quantity.

Controlling the quantity of stormwater using storage facilities can provide the following potential benefits:

- Prevent or reduce peak runoff rate increases caused by urban development
- Mitigate downstream drainage capacity problems
- Recharge groundwater resources
- Reduce or eliminate the need for downstream outfall improvements

Control of stormwater quality using storage facilities offers the following potential benefits:

- Decrease downstream channel erosion (with proper design) through velocity control and flow reduction
- Reduce pollution loading through deposition, chemical reaction, and biological uptake mechanisms
- Improve base flow conditions
- Show aesthetic and ecological habitat benefits at multiobjective sites
- Control sediment deposition
- Improve water quality through stormwater filtration (wet ponds only)

The objectives for managing stormwater quantity by storage facilities are typically based on limiting peak runoff rates to match one or more of the following values:

- Historic rates for specific design conditions (i.e., postdevelopment peak equals predevelopment peak for a particular frequency of occurrence)
- Nonhazardous discharge capacity of the downstream drainage system
- A specified value for allowable discharge set by a regulatory jurisdiction

Types of Storage Facilities

Table 11-1 gives a number of applications of ways to reduce or delay runoff. It is provided to encourage the urban stormwater manager to "think outside the box" when it comes to stormwater management design. Urban development will have impacts. Streams in urban areas will never again be rural streams. But, these streams do not have to become eyesores or sterile linear stretches of eroded urban blight. Urban stormwater managers are in a position to bring about a change in thinking about how precipitation is handled from the time it first hits a surface until it leaves the municipal jurisdiction. Just as pollution in urban settings is the accumulated product of many minor acts of careless or overt polluting, so urban flooding and erosion is the result of many minor (and a few major) development choices. Many municipalities and counties have learned how to handle the problem and to bring about the changes necessary to consider rain as a resource and surface streams as positive neighborhood enhancements.

One other application of detention that is little considered is the spill-over pond placed along a stream or river. In many cases detention ponds are designed to capture and treat the runoff from the site when the parcel is located adjacent to a stream where, due to timing considerations, the water should be allowed to release ahead of the peak flow coming downstream from above. But should detention just be waived? Another approach is to require that a spill-over pond be designed. Such a pond is built in the floodplain with a side-channel spillway set at an elevation wherein the peak flow from the stream slows over the spillway — thus shaving the peak directly from the stream flood flow. As the stream flow subsides, the stored water in the detention pond flows out by gravity on the downstream end of the pond. In this way the peak flow on the stream is directly reduced. The capture volume can be set at the volume that would have been required had normally designed detention been mandated.

TABLE 11-1

Advantages and Disadvantages of Measures for Reducing or Delaying Runoff

Measure	Advantages	Disadvantages
Cisterns and covered ponds	Alternate uses for the water stored Occupy small areas Land above has alternate uses Water conservation	Expensive to install Not large capacity Restricted maintenance access
Open space and grassed areas	Aesthetically pleasing Cost effective Pollution removal	Land availability Public acceptance problems if not well done Safety issues
Blue-green storage	Multipurpose capabilities High public acceptance Aesthetically pleasing Often cost effective	Difficulty finding suitable sites Maintenance special difficulties May be safety hazard
Rooftop ponding	Runoff delay Cooling effect on building Possible fire protection	Structural loading Clogging possibility if trees overhang Freezing problems during winter Roof leakage
Surface ponds	Ability to control large areas Can be aesthetically pleasing Multipurpose capability Aquatic habitat provider Can increase land value Pollution reduction	Requires large areas Possible pollution and eutrophication Pest breeding area Can become urban eyesore Safety hazard Maintenance problems possible
Increased roughness on roof — ripples and gravel	Runoff delay	Cost and structural loading
Porous pavement, gravel and paving blocks	Runoff reduction potential Groundwater recharge Gravel may be cheaper Pollution reduction	Cost and maintenance Clogging or earth compaction possibility Groundwater pollution Frost heave Grass and weeds grow through
Grassed channels and ditches, vegetative strips	Runoff delay Some runoff reduction Aesthetically pleasing Pollution reduction	Land loss Maintenance increase
Ponding on impervious areas	Runoff delay Pollution reduction	Restricts other uses when raining Damage due to wetness and freeze-thaw Dirt and debris collection in depressions
Infiltration devices	Runoff reduction Groundwater recharge Little evaporation loss Pollution reduction Water conservation	Clogging Initial expense Groundwater contamination
"Microlandscaping," xeriscaping, terracing, flow routing, urban forestry	Runoff reduction and delay Aesthetically pleasing Pollution reduction Water conservation	More expensive to design, construct, and maintain

11.3 Design Criteria

Storage may be concentrated in large basin-wide or regional facilities, or distributed throughout an urban drainage system. Possible dispersed or on-site storage may be developed in depressed areas in parking lots, road embankments and freeway interchanges, parks and other recreation areas, and small lakes, ponds, and depressions within urban developments. The utility of any storage facility depends on the amount of storage, its location within the system, and its operational characteristics. An analysis of such storage facilities should consist of comparing the design flow(s) at a point or points downstream of the proposed storage site with and without storage. In addition to the design flow, other flows in excess of the design flow that might be expected to pass through the storage facility should be included in the analysis (i.e., 100-year flood). The design criteria for storage facilities should include:

- Release rate
- Storage volume
- Grading and depth requirements
- Safety considerations and landscaping
- Environmental impacts and multiobjective use
- Outlet works
- Location

Commonly, control structure release rates are designed to approximate predeveloped peak runoff rates for the 2-year through 10-year storms, with emergency overflow capable of handling the 100-year discharge. The point of this comparison is at the outlet of the pond, at the property boundary, or at some point downstream, where the developed property makes up some proportion of the total drainage to that point. Design calculations are required to demonstrate that runoff from the 2- and 10-year design storms are controlled. If so, runoff from intermediate storm return periods can be assumed to be adequately controlled. Multistage control structures may be required to control runoff from the 2- and 10-year storms. Other municipalities' basic criteria concentrate on the 2-year and 25-year storms or even control of the 100-year storm. This has been done in an effort to reduce downstream effects of urban development (caused by the increase of volume of flow as much as the peak) to an extent that often cannot be accomplished when the criteria is to meet predeveloped conditions. Recently, some communities began using the 25-year storm to design outlet control structures when different runoff volumes are controlled for other purposes (i.e., water quality, channel stability, infiltration), see Georgia Stormwater Management Manual (2001).

If sedimentation during construction causes loss of detention volume, design dimensions should be restored before completion of the project, as-built requirements or post-construction surveys are often used to satisfy this requirement. For detention facilities, all detention volume should be drained within 72 h. For a detailed discussion of the many types of detention and design specifications for many types of outlet works, see Stahre and Urbonas (1990).

Following is a discussion of the general grading and depth criteria for typical storage facilities followed by criteria related to detention and retention facilities.

General Design Criteria

The construction of storage facilities usually requires excavation or placement of earthen embankments to obtain sufficient storage volume. Vegetated embankments should be less than 20 ft in height and should have side slopes no steeper than 3:1 (horizontal to vertical). Riprap-protected embankments should be no steeper than 2:1. Geotechnical slope stability analysis is recommended for embankments greater than 10 ft in height and is mandatory for embankment slopes steeper than those given above. Procedures for performing slope stability evaluations can be found in most soil engineering textbooks, including those by Spangler and Handy (1982) and Sowers and Sowers (1970).

A minimum freeboard of 1 ft above the 100-year design storm high-water elevation is often provided for impoundment depths between about 5 to 20 ft. Smaller impoundments may or may not have freeboard requirements. Impoundment depths greater than 20 ft are subject to the requirements of the Safe Dams Act (see Section 11.4) unless the facility is excavated to this depth.

Other considerations when setting depths include flood elevation requirements, public safety, land availability, land value, present and future land use, water table fluctuations, soil characteristics, maintenance requirements, and required freeboard. Aesthetically pleasing features are also important in urbanizing areas.

In designing the detention, the designer should gain information on some or all of the following:

- Design policy, standards, and criteria
- Hydrologic characteristics of the tributary basin or area
- Future development potential and resulting hydrology
- Downstream conditions
- Risks of design flow exceedance
- Precision and accuracy requirements
- Local practice in design types and methods
- Regulatory and legal requirements
- Backwater impacts
- Low-flow requirements
- The aquatic ecosystem of the stream and potential detention pond impacts
- Dual purpose requirements
- Constructability and value engineering
- Availability of materials and land
- Physical suitability of the site utilities, easements, pedestrian access, building location, open-space reservation, etc.
- Soils suitability
- Groundwater tables and locations of wells and other potable water supplies
- Special factors such as archeological sites, historical sites, endangered species, wetlands, rock formations, etc.
- Maintenance requirements
- Cost and schedule limitations
- Aesthetic and recreational requirements
- Human interaction with the site
- Institutional, social, and political concerns

Specific Detention Design Criteria

Areas above the normal high-water elevations of storage facilities should be sloped at a minimum of 5% toward the facilities to allow drainage and to prevent standing water. Careful finish grading is required to avoid creation of upland surface depressions that may retain runoff. The bottom area of storage facilities should be graded toward the outlet to prevent standing water conditions. A minimum 2% bottom slope is recommended. A low-flow or pilot channel constructed across the facility bottom from the inlet to the outlet is recommended to convey low flows and prevent standing water conditions. Often, a sediment collection forebay is provided with easy maintenance access.

Specific Retention Design Criteria

The maximum depth of permanent storage facilities will be determined by site conditions, design constraints, climatic conditions, and environmental needs. In general, if the facility provides a permanent pool of water, a depth sufficient to discourage growth of weeds, without creating undue potential for anaerobic bottom conditions, should be considered. A depth of 6 to 8 ft is generally reasonable unless fishery requirements dictate otherwise. Aeration may be required in permanent pools to prevent anaerobic conditions. Where aquatic habitat is required, the cognizant wildlife experts should be contacted for site-specific criteria relating to such things as depth, habitat, and bottom and shore geometry. In some cases, a shallow bench along the perimeter is constructed to encourage emergent vegetation growth to enhance the pollution reduction capabilities, safety, or aesthetics of the pond.

Outlet Works Design Criteria

Outlet works selected for storage facilities typically include a principal spillway and an emergency overflow and must be able to accomplish the design functions of the facility. Outlet works can take the form of combinations of drop inlets, pipes, weirs, and orifices. The principal spillway is intended to convey the design storm(s) without allowing flow to enter an emergency outlet. For large storage facilities, selecting a flood magnitude for sizing the emergency outlet should be consistent with the potential threat to downstream life and property if the facility embankment were to fail. The minimum flood to be used to size the emergency outlet for a more major system pond is the 100-year flood. For smaller ponds, in some municipalities, lesser floods are used for the emergency overflow spillway. The sizing of a particular outlet works should be based on results of hydrologic routing calculations.

Outlet works are designed to fit a specific design standard or to meet a specific need. Often, the standard for stormwater quantity control (see Chapters 13 and 15 for a discussion of dual quality–quantity designs) is to control several different storm frequencies with one pond. Sometimes a specific range of frequencies is of particular interest, or a certain location within the watershed must be specifically protected. All outlet works should strive to meet stated goals while considering safety, aesthetics, maintenance minimization, longevity, debris handling, cost, and simplicity of design.

Location Design Criteria

Stahre and Urbonas (1990) classify storage as source control or downstream control. Source control consists of a larger number of smaller basins located high in the drainage or conveyance system. They might include local percolation, injection or infiltration basins, smaller inlet control basins, and on-site detention. Downstream controls are normally on- or off-line, larger, open pond-type basins or large dry detention ponds. Source controls have the advantage in that they are lower in cost to build, can be targeted to a certain area or pollution source and type, and are flexible in design types. Maintenance and regulation are more difficult and costly for many source controls providing the same level of protection as fewer and larger downstream controls, and source controls are not normally effective for controlling flooding very far downstream. Downstream controls offer greater opportunities for larger-scale multiobjective site use. On the other hand, downstream controls require coordination for funding and development (perhaps master planning), require more upfront capital funding, may have greater environmental impacts in the forms of abandonment of reaches upstream from the pond, and impact on flow interruption and riparian wetlands. Thus, there is a trade-off between source and downstream controls. Usually, through master planning, a mix of both is found to be most appropriate.

In addition to controlling the peak discharge from the outlet works, storage facilities will change the timing of the entire hydrograph. If several storage facilities are located within a particular basin, it is important to determine what effects a particular facility may have on combined hydrographs in downstream locations. Great cost savings can be realized by combining detention with channel improvements, oversizing detention in one area to compensate for no detention in another, and using different criteria for the sizing of the ponds based on the needs of the site or watershed (Hartigan, 1986; Hartigan and George, 1989; Jones, 1990; State of Maryland, 1986, 1987; James, Bell, and Leslie, 1987; Tyrpak, 1990; McCuen and Moglen, 1988).

Detention can be located within floodplains and still effectively control flooding through the use of timing calculations. In this situation, the flood peak coming down the stream rarely coincides with local on-site flooding. It is often advantageous to allow the on-site water to pass with simple erosion control and a properly sized conveyance system. Then, locate the detention pond to skim the peak from the oncoming flood hydrograph through the use of a side-channel weir or a simple flow-through depression along the banks (Smith and Cook, 1987).

For all storage facilities, channel routing calculations should proceed downstream to a suitable location (normally to the point where the controlled land area is less than 10% of the total drainage to that point). At this point, the effect of the hydrograph routed through the proposed storage facility on the downstream hydrograph should be assessed and shown not to have detrimental effects on downstream areas.

11.4 Safe Dams Act

National responsibility for the promotion and coordination of dam safety lies with the Federal Emergency Management Agency (FEMA). In addition, state agencies have some responsibility for administration of the provisions of the Federal Dam Safety Act. Under the federal regulations, a dam is an artificial barrier that does or may impound water and that is 20 ft or greater in height or has a maximum storage volume of 30 acre-feet or more. Detailed engineering requirements are given in the regulations for new dams, and these regulations should be consulted for all engineering requirements. A number of exemptions are allowed from the Safe Dams Act, and the cognizant state office should be contacted to resolve questions.

TABLE 11-2

Size Categories for Dam Classification

Category	Storage (acre-feet)	Height (ft)
Small	30 to <1000	20 to <41
Intermediate	1000 to 50,000	41 to 100
Large	>50,000	>100

Dams are classified as new or existing, by hazard potential, and by size. Hazard potential categories are listed below.

Category 1

Category 1 dams are located where failure would probably result in:

- Loss of human life
- Excessive economic loss due to damage of downstream properties
- Public hazard
- Public inconvenience due to loss of impoundment and damage to roads or any public or private utilities

Category 2

Category 2 dams are located where failure may damage downstream private or public property, but such damage would be relatively minor and within the general financial capabilities of the dam owner. Public hazard or inconvenience due to loss of roads or any public or private utilities would be minor and of short duration. Chances of loss of human life would be possible but remote.

Category 3

Category 3 dams are located where failure may damage uninhabitable structures or land, but such damage would probably be confined to the dam owner's property. No loss of human life would be expected.

Dams are also classified as small, intermediate, or large depending on their storage capacity and height. Table 11-2 gives the criteria for this classification.

11.5 General Design Procedure for Storage Routing

The following data will be needed to complete storage design and routing calculations:

- Inflow hydrograph for all selected design storms
- Stage–storage curve for proposed storage facility (see Figure 11-1 for an example)
- Stage–discharge curve for all outlet control structures (see Figure 11-2 for an example)

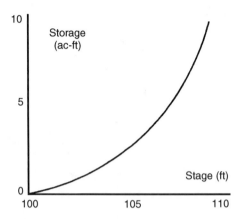

FIGURE 11-1
Example stage–storage curve.

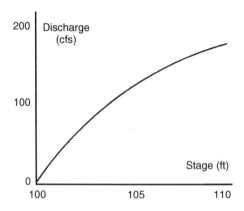

FIGURE 11-2
Example stage–discharge curve.

Using these data, a design procedure is used to route the inflow hydrograph through the storage facility with different basin and outlet geometry until the desired outflow hydrograph is achieved.

Stage–Storage Curve

A stage–storage curve defines the relationship between the depth of water and storage volume in a reservoir. The data for this type of curve are usually developed using a topographic map and the double-end area, frustum of a pyramid, prismoidal formulas, or circular conic section. The double-end area formula is expressed as:

$$V_{1,2} = [(A_1 + A_2)/2]d \qquad (11.1)$$

where $V_{1,2}$ = storage volume (ft³) between elevations 1 and 2; A_1 = surface area at elevation 1 (ft²); A_2 = surface area at elevation 2 (ft²); and d = change in elevation between points 1 and 2 (ft).

The frustum of a pyramid is expressed as:

Storage and Detention Facilities

$$V = [d/3][A_1 + (A_1 \times A_2)^{0.5} + A_2]/3 \qquad (11.2)$$

where V = volume of frustum of a pyramid (ft^3). Other terms are as defined above.
The prismoidal fosrmula for trapezoidal basins is expressed as:

$$V = LWD + (L + W)ZD^2 + 4/3\, Z^2 D^3 \qquad (11.3)$$

where V = volume of trapezoidal basin (ft^3); L = length of basin at base (ft); W = width of basin at base (ft); D = depth of basin (ft); and Z = side-slope factor, ratio of horizontal to vertical.
The circular conic section formula is:

$$V = 1.047\, D\, (R_1^2 + R_2^2 + R_1 R_2) \qquad (11.4)$$

$$V = 1.047\, D\, (3R_1^2 + 3ZDR_1 + Z_2 D^2) \qquad (11.5)$$

where R_1 and R_2 = bottom and surface radii of the conic section (ft); D = depth of basin (ft); and Z = side-slope factor, ratio of horizontal to vertical.

Often, the stage–storage curve can take the form of a power curve, where storage is a function of a constant times stage to some power. This can be easily done through the use of computerized spreadsheets by taking the logarithm and fitting a straight line to sets of area-elevation or volume-elevation points. In fitting a dam to an existing site, areas or volumes can be determined by planimeter. For areas where detention will be excavated, one of the formulas above can be used, and size, depth, shape, and side-slope adjusted to obtain the necessary storage volume.

Stage–Discharge Curve

A stage–discharge curve defines the relationship between the depth of water and the discharge or outflow from a storage facility. A typical storage facility has two spillways: principal and emergency. The principal spillway is designed with a capacity sufficient to convey the design flood without allowing flow to enter the emergency spillway. A pipe culvert, weir, or other appropriate outlet can be used for the principal spillway or outlet. The emergency spillway is sized to provide a bypass for floodwater during a flood that exceeds the design capacity of the principal spillway. This spillway should be designed taking into account the potential threat to downstream life and property if the storage facility were to fail. The stage–discharge curve should take into account the discharge characteristics of the principal and emergency spillways.

General Procedure

Following are the steps in a general procedure for using the above data in the design of storage facilities:

Step 1: Compute inflow hydrograph for runoff from the 2-, 10-, and 100-year design storms using the procedures outlined in Chapter 7 (or use the local design criteria). Both pre- and postdevelopment hydrographs are required for the 2- and 10-year design storms. Only the postdevelopment hydrograph is required for runoff from the 100-year design storm.

Step 2: Perform preliminary calculations to evaluate detention storage requirements for the hydrographs from Step 1 (see Section 11.7). If storage requirements are satisfied for runoff from the 2- and 10-year design storms, runoff from intermediate storms is assumed to be controlled.

Step 3: Determine the physical dimensions necessary to hold the estimated volume from Step 2, including freeboard. The maximum storage requirement calculated from Step 2 should be used. From the selected shape, determine the maximum depth in the pond.

Step 4: Select the type of outlet and size the outlet structure. The estimated peak stage will occur for the estimated volume from Step 2. The outlet structure should be sized to convey the allowable discharge at this stage.

Step 5: Perform routing calculations using inflow hydrographs from Step 1 to check the preliminary design using the storage routing equations. If the routed post-development peak discharges from the 2- and 10-year design storms exceed the predevelopment peak discharges, or if the peak stage varies significantly from the estimated peak stage from Step 4, then revise the estimated volume and return to Step 3.

Step 6: Consider emergency overflow from runoff due to the 100-year (or specified) design storm and established freeboard requirements.

Step 7: Evaluate the downstream effects of detention outflow to ensure that the routed hydrograph does not cause downstream flooding problems. The exit hydrograph from the storage facility should be routed through the downstream channel system until a confluence point is reached. Example criteria for locating the confluence point are where the drainage area being analyzed represents 10% of the total drainage area (Debo and Reese, 1992).

Step 8: Evaluate the control structure outlet velocity and provide channel and bank stabilization if the velocity will cause erosion problems downstream.

This procedure can involve a significant number of reservoir routing calculations to obtain the desired results. Because there are almost limitless methods for design of outlet structures, no generic guidance will cover every situation. But, for outlet structures that must control multiple storm frequencies, there are several approaches. Normally, for the 2-year and 10-year storm scenarios, the 10-year storm volume will determine the size of the pond, while the 2-year storm peak control need will determine the size of a minimum outlet structure. Thus, the pond is sized for the 10-year storm. Then, the 2-year storm is routed through the pond, and the 2-year outlet structure is sized. The depth of the storage when the 2-year storm is routed through the pond then normally becomes the elevation of the invert or weir for the 10-year storm outlet. The 10-year storm is then routed through both outlets.

11.6 Outlet Hydraulics

The outlet is the heart of the detention and retention design. The outlet meters the water through the pond at a controlled rate and in a controlled way. There are many types and combinations of outlet works. Most of these outlet works consist of a combination of weirs and orifices in conjunction with outlet pipes or conduits. If culverts are used as outlet works or HW/D criteria are not met, such that orifices act as culverts, procedures presented in Chapter 9 should be used to develop stage–discharge data. Combinations of various components and special use outlet works are discussed in this chapter.

It is important to obtain accuracy in the design of outlet works through proper estimation of discharge coefficients, submergence, rating curves, control shifts, etc. But, it should be remembered that the accuracy of hydrologic estimates is often poor (McCuen, 1983), and, therefore, overzealous optimization of detention is hardly warranted (ASCE, 1985).

The decision to use off-line storage versus on-line storage will affect the type of outlet structure used and its size. Nix and Tsay (1988) have shown that the use of off-line storage can reduce the size of the basin required by as much as 20 to 40%. This kind of storage can be used, for example, with a channel diversion such as a side-channel weir to skim higher flows into the detention structure at its upstream end for later release at the detention structure's downstream end. The structure then lies parallel to the flow channel.

Combination Outlets

Combinations of weirs, pipes, and orifices can be put together to provide a variable control stage–discharge curve suitable for control of multiple storm flows. They are generally of two types: shared outlet control and separate outlet controls. Shared outlet control is typically a number of individual outlet openings, weirs, or drops at different elevations on a riser pipe or box that all flow to a common larger conduit or pipe. Separate outlet controls are less common and normally consist of a single opening through the dam of a detention facility in combination with an overflow spillway for emergency use.

The most common shared outlet control practice is to place a low-flow outlet at the base of a riser pipe. The riser pipe can then serve as an emergency overflow spillway and could also have openings, slots, or perforations in it for variable discharge by stage. In this case, separate stage–discharge curves are computed for each element of the outflow and for each type of possible control. The curve that gives the highest stage for a given flow is expected to control. The total stage–discharge curve is the upper stage envelope of all possible shared outlet stage–discharge curves. Figure 11-3 illustrates this practice, and Figure 11-4 illustrates the various curves and the controlling curve. For the case of the overflow riser pipe diagrammed in Figure 11-3, the control can follow a three-step process.

- Initially, as the flow reaches and begins to overflow the riser pipe, weir flow controls, and an equation of the weir form is applicable.
- As the stage rises, the flow begins to interfere with itself around the circular opening, and eventually, the flow control transitions to that of orifice flow.
- If the flow is great and the outlet pipe is small in comparison, the barrel of the pipe may become the controlling factor, and the outlet may operate as a barrel control or outlet control culvert.

Perforated Risers

To extend detention times, risers can be designed with perforations, though clogging potential is high for such designs. In that case, the top of the riser is left open with a trash rack and vortex plate attached. The flow through the perforations acts like an orifice, with the head measured from the center of the opening to the water surface. Holes should normally be a minimum of three diameters, center-to-center, apart. If the pipe is thick, the discharge will be that of a short tube with appropriate adjustment in discharge coefficient.

618   *Municipal Stormwater Management, Second Edition*

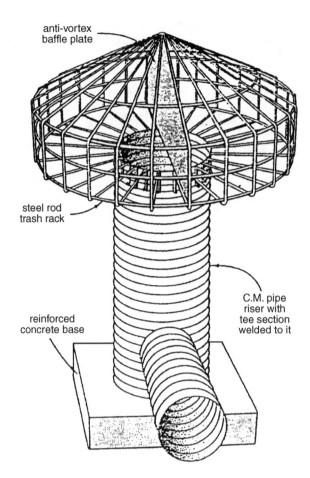

FIGURE 11-3
Corrugated metal pipe riser with trash rack and baffle.

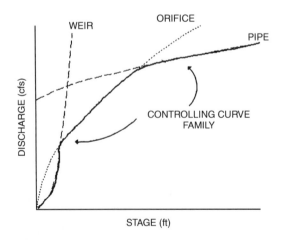

FIGURE 11-4
Stage–discharge curve for riser type detention pond.

Storage and Detention Facilities 619

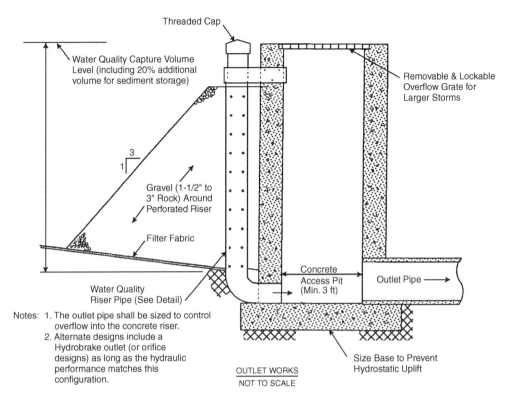

FIGURE 11-5
Typical perforated riser outlet design. (From Denver Urban Drainage and Flood Control District (UDFCD)., Urban Drainage Criteria Manual — Vol. 3", Denver UDFCD, 1992.)

If it is corrugated metal pipe, it will act as an orifice. Clogging can be reduced with the use of a jacket of gravel and wire mesh around the pipe. Various designs are provided by MWCOG (1987), UDFCD (1992), and Washington State Ecology (1992). Figure 11-5 illustrates this practice, and Table 11-3 gives hole spacing and sizes.

There is some question concerning the accuracy of using a simple orifice equation for such designs (MWCOG, 1987). McEnroe, Steichen, and Schweiger (1988) conducted experiments with perforated risers and developed an equation to predict flows through the perforations in the riser. The outlet capacity must be greater than the riser for the equation to retain validity.

$$Q = C_h [2A_h/3h_h][(2g)^{0.5}(h^{3/2})] \tag{11.6}$$

where Q = discharge through all the riser holes (cfs); C_h = discharge coefficient given as 0.611 by the author; A_h = area of all the holes in the perforated section (ft²); h_h = distance from $d/2$ below the centerline of the lowest hole to $d/2$ above the centerline of the highest hole (where d is the vertical) spacing between holes (ft); h = head on riser pipe measured from $d/2$ below the centerline of the lowest hole (ft).

Two-Way Drop Inlets

A two-way drop inlet is a combination inlet/outlet structure to a conduit or channel in which the flow discharges over one of two weirs, which are parallel to the axis of the outlet channel or conduit. Figure 11-6 gives details of the design and construction (USCOE,

TABLE 11-3
Maximum Number of Perforated Columns

Riser Diameter (in)	Hole Diameter (in)			
	1/4"	1/2"	3/4"	1"
4	8	8	—	—
6	12	12	9	—
8	16	16	12	8
10	20	20	14	10
12	24	24	18	12

Hole Diameter (in)	Area of Hole (in^2)
1/8"	0.013
1/4"	0.049
3/8"	0.110
1/2"	0.196
5/8"	0.307
3/4"	0.442
7/8"	0.601
1"	0.785

Source: From Denver Urban Drainage and Flood Control District (UDFCD)., Urban Drainage Criteria Manual Vol. 3", Denver UDFCD, 1992.)

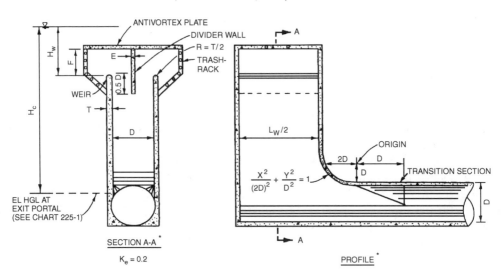

FIGURE 11-6
Two-way drop inlet. (From COE, 1988.)

1988). As in other combination inlets, the flow transitions through three stages. The U.S. Army Corps of Engineers recommends that, for larger applications, the inlet be designed to avoid the unstable, slug flow–orifice flow regime. The basic procedure is to assume a weir crest length and to plot each of the rating curves and ensure orifice flow never controls. The conduit control curve is plotted with two friction factors, high and low, to establish a sensitivity range. When a satisfactory plot is achieved, the antivortex plate is set 1 ft above the pool elevation required to establish conduit controlled flow. The following design information applies (SCS, 1965):

- Initially determine the minimum pond water surface elevation to set the weir crest elevations.
- Determine the conduit invert, diameter, and length from flow calculations.
- Determine, from structural considerations, dimensions T and E (Figure 11-6).
- Weir flow discharge coefficient has been determined to be 3.8 when the weir is rounded to a radius of half the thickness of the weir walls (see Figure 11-6).

Orifice flow calculations are done using the following equations:

$$Q = C_o A_o (2gH_w)^{1/2} \tag{11.7}$$

$$A_o = (1/2) L_w (D-E) \tag{11.8}$$

where Q = discharge (cfs); C_o = orifice discharge coefficient; A_o = area of the orifice (ft²); H_w = head on the weir (ft); D = conduit diameter (ft); E = separation wall thickness (ft); L_w = length of the weir (ft); and C_o can be determined from:

$$C_o = C_1[(E/D)^{0.083}][(L_w/2d)^{-0.2934}] \tag{11.9}$$

$$C_1 = -15.6993 (T/D)^2 + 11.3136 (T/D) - 0.2032 \tag{11.10}$$

where T = thickness of the wall (ft).

Conduit control is developed when full pressure flow exists throughout the inlet structure and is determined from:

$$Q = A(2gH_c/K)^{0.5} \tag{11.11}$$

where A = conduit cross-sectional area (ft²); g = gravitational constant (ft/s²); H_c = head on pipe equal to the difference between the upper pool elevation and the hydraulic gradeline elevation at the exit portal (ft) (use the outlet pipe center elevation for freeflow conditions and tailwater elevation for submerged conditions); and K = total loss coefficient on velocity head (i.e. $KV^2/2g$) equal to entrance loss coefficient (0.2 for this design) plus friction loss in pipe plus 1.0 exit loss.

Drop-type inlets can also be placed at different elevations, as in Figure 11-7, to gain control over several different flow frequencies (Rossmiller, 1982). In this figure, the lower inlet pipe was sized to control the 2-year flow, while the two drop inlets were sized for the 5- through 10-year and 25- through 100-year flows, respectively.

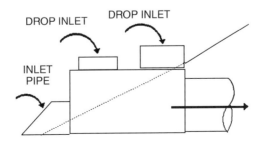

FIGURE 11-7
Multiple drop inlets.

Weirs

A weir is a type of a head-area flow control used to control outflow from storage facilities, open channels, pipes, etc. The edge or surface over which the water flows is called the crest. The overflowing sheet of water is termed the nappe. When the weir is thin and beveled on the downstream side such that a sharp upstream edge exists, the nappe springs clear of the weir (except for very low heads). This is called a sharp-crested weir. If not, it is called a broad-crested weir (or sometimes "weirs not sharp-crested"). Under the two categories of sharp-crested and broad-crested weirs, there are many subcategories based on weir shape. Sharp-crested weirs are used for flow measurement, while broad-crested weirs, due to greater durability, are often used in conjunction with flow structures. Sharp-crested weirs are quite susceptible to floating debris and sediment deposition in the slack water area upstream from the weir. This deposition will eventually modify the approach conditions, requiring an adjustment for approach velocity or, better, cleaning out of the approach channel or pool.

For a weir of a proper size and shape, with steady state flow conditions, and weir-to-pool height relationship, critical flow will exist near the overflow point. Thus, only one depth of flow can exist for any discharge, largely independent of channel roughness or other uncontrollable circumstances. Discharges are calculated by measuring the depth of flow from the crest of the weir vertically upward to the water surface elevation at a point upstream from the weir, where the drawdown effects of the accelerated flow are negligible — about three times the head on the weir.

Sharp-Crested Weirs

Commonly used shapes for sharp-crested weirs are rectangular, trapezoidal (or Cipolletti), and triangular (or V-notched). The sharp-crested weirs consist of a thin plate, 1/8 to 1/4 in thick, with a bevel on the downstream side and the upstream side sharpened. The crest plate should be vertical and the upstream side smooth. The connection to the channel should be watertight. The lower nappe of the weir should be vented with a pipe through the side wall if the weir is suppressed. To lower the approach velocity factor, the height of the weir from the channel bottom should be at least twice the maximum expected head and the upstream flow area eight times the nappe cross-sectional area, for a distance 15 to 20 times the head on the crest (Grant, 1989). The weir height should be over 1 ft. The approach should be straight at least 20 times the maximum expected head and have as little slope as possible. Submerged flow conditions should be investigated, and the weir crest elevation set to avoid submergence. Details on weir construction can be obtained from USBR (1974, 1978).

The standard generic weir flow equation is:

$$Q = CLH^n \qquad (11.12)$$

where Q = discharge (cfs); H = head above weir crest excluding velocity head (ft); L = horizontal effective weir length (ft); and n = 1.5 for nonvariable flow width weirs such as rectangular, and other values for other weirs.

Submerged Weir Correction

If the water surface downstream of the weir rises to the point that there is not free ventilation of the underside of the nappe, with free fall distance less than about twice the head on the weir, the discharge may increase due to the formation of a partial vacuum

Storage and Detention Facilities

under the nappe. If the tailwater rises to a point equal to the crest elevation, the weir is considered submerged, and an incomplete vertical contraction occurs. When the tailwater height (H_t), measured relative to the crest elevation above the crest, rises to about 2/3 that of the head on the crest (H), the discharge becomes appreciably affected. A correction for submergence for any of the shapes of weirs discussed below can be given by the Villemonte equation (Villemonte, 1947):

$$Q_s = Q_f(1 - (H_2/H_1)^n)^{0.385} \quad (11.13)$$

where Q_s = submergence flow (cfs); Q_f = free flow (cfs); H_1 = upstream head above crest (ft); H_2 = downstream head above crest (ft); and n = 1.5 for non-variable flow width weirs and other values for other weirs.

Approach Velocity Correction

If the conditions for neglecting approach velocity are violated it may be prudent to include an estimate of the approach velocity as a correction to the calculated discharge. The approach velocity, V, is simply the discharge divided by the cross sectional area of the approach channel. The approach velocity head is then:

$$h_v = V^2/2g \quad (11.14)$$

where V = approach velocity (ft/s); g = acceleration of gravity (32.2 ft/s²); and h_v = approach velocity head (ft).

The corrected head above the weir is then:

$$H_v = [(H+h_v)^{3/2} - h^{3/2}]^{2/3} \quad (11.15)$$

where H_v = corrected head on weir (ft); h_v = approach velocity head (ft); and H = head above weir crest excluding velocity head (ft).

Sharp-Crested Rectangular Weirs

Sharp-crested rectangular weirs with no end contractions and with two end contractions are illustrated in Figures 11-8 and 11-9. The discharge equation coefficient in Equation (11.12) for a vertical sharp-crested rectangular weir configuration is given by Rehbock as quoted in Rouse (1949):

$$C = 3.237 + 0.428(H/H_c) + 0.0175/H \quad (11.16)$$

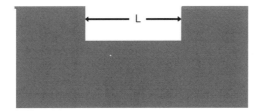

FIGURE 11-8
Sharp-crested weir (two end contractions).

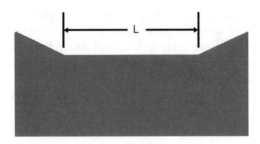

FIGURE 11-9
Sharp-crested weir (no end contractions).

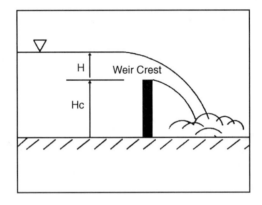

FIGURE 11-10
Sharp-crested weir and head.

where H_c = height of weir crest above channel bottom (ft), and H = head above weir crest, excluding velocity head (ft).

An average value of this equation, 3.33, is often used in practice and in flow measurement manuals. This equation is accurate for H/H_c ratios less than about 10. For H/H_c ratios greater than about 15, the weir acts as a sill, and the flow equation becomes the following (Chow, 1959):

$$C = 5.68 \,(1 + H_c/H)^{1.5} \tag{11.17}$$

When the ends of the weir do not coincide with the sides of the approach pool or channel, the flow from the sides of the channel approaches the weir opening at an angle. The momentum of this flow carries it past the opening edge and causes a contraction of the flow as it corners through the opening and turns downstream. If the weir ends coincide with the pool or channel sides, this contraction is suppressed, and more flow can pass through the weir opening. Provision must be made for ventilation of the underside of the weir. The correction for contracted sides for rectangular weirs is given in Equation (11.19). Corrections are built into the discharge coefficients of other weir equations. For the contraction to be complete, the distance from the walls to the weir ends must be greater than twice the head on the weir.

$$L = L' - 0.1NH \tag{11.18}$$

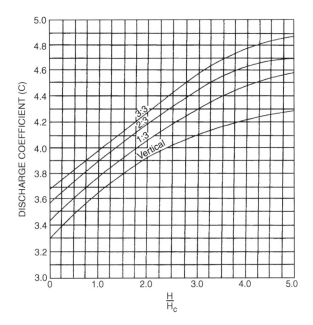

FIGURE 11-11
Discharge coefficients for sharp-edged suppressed rectangular weirs.

where L' = measured horizontal weir length (ft); N = number of contracted sides [$N = 2$ for both sides contracted; $N = 1$ for one side contracted; $N = 0$ for no contraction (suppressed weir)]; and H = head above weir crest, excluding velocity head (ft).

Rantz (1982) provides an alternate way to obtain the discharge coefficient and the contraction correction coefficient. Figure 11-11 gives the variation of the discharge coefficient with sloped approach conditions and for various values of H/H_c. Figure 11-12 gives corrections (k_c) for various ratios of contraction ratios (b/B) and H/H_c ratios, where b is the measured weir length, and B is the channel width. If the rectangular weir is set between two rounded vertical abutments instead of sharp edges, a further correction is made to increase the discharge (Rantz, 1982).

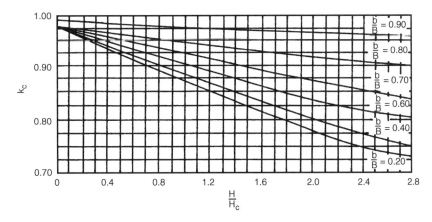

FIGURE 11-12
Adjustment factor k_c for contracted rectangular sharp-edged weirs.

Assume the radius of curvature of the abutment is r. The following rule applies:

for: $r/b = 0$ correction factor = k_c

for: $0 \leq r/b \leq 0.12$ correction factor = interpolated   (11.19)

for: $r/b > 0.12$ correction factor = $(1+k_c)/2$

The height of the weir crest above the bottom of the approach channel, and the distance from the sides of the notches to the sides of the channel, should each be more than twice the head on the crest. The weir should not be used for heads less than 0.2 ft or for heads greater than 1/2 the crest length (USBR, 1974). Discharge coefficients lose accuracy for crest lengths less than 1 ft, and triangular weirs should be used instead (Grant, 1989).

Trapezoidal and Cipolletti Weirs

Trapezoidal weirs of all dimensions have been fabricated. In all cases but the special case of the Cipolletti weirs, the discharge is assumed to be a combination of a rectangular weir and a triangular weir. However, the discharge coefficients have to be experimentally determined in the field using flow meter measurements at various stages. For the special case of the Cipolletti weir, the side slopes of the trapezoidal section are 4V:1H. This slightly greater area, for the same weir length, in the bottom of the trapezoid, b, is designed to counteract the contraction effects and make the Cipolletti weir act as if it were a fully suppressed rectangular weir of length of b. The equation of discharge for the Cipolletti weir is:

$$Q = 3.367 \, L \, H^{1.5} \tag{11.20}$$

Considering the velocity of approach, the formula becomes (USBR, 1974):

$$Q = 3.367 \, L \, (H + 1.5h_v)^{1.5} \tag{11.21}$$

where h_v = approach velocity (ft/s).

The height of the weir crest above the bottom of the approach channel and the distance from the sides of the notches to the sides of the channel should each be more than twice the head on the crest. The weir should not be used for heads less than 0.2 ft or for heads greater than 1/3 the crest length (USBR, 1974).

V-Notch Weirs

The V-notch weir is particularly suited for small flows, as it concentrates the low flow in the center, allowing the nappe to spring free from the crest. The minimum distance from the sides of the weir to the channel banks should be twice the head on the weir. Also, the minimum distance from the point of the notch to the channel bottom should be more than twice the head on the weir (USBR, 1974). The 90 degree V-notch weir can handle flows up to about 10 cfs, easily.

The discharge through a V-notch weir can be calculated from the following equation (Brater and King, 1976):

$$Q = 2.5 \tan(\theta/2) H^{2.5} \tag{11.22}$$

where Q = discharge (cfs); θ = angle of v-notch (degrees); and H = head on apex of notch (ft).

Much more data are available on standard 90-degree and 120-degree notches than other size openings, the wider angle accommodating a much wider range of flows than the narrower angle (Leupold and Stevens, 1991). For these, the standard equations can more accurately be taken, respectively, as (Rantz, 1982):

$$Q = 2.47 \, H^{2.50} \quad (90°) \tag{11.23}$$

and

$$Q = 4.35 \, H^{2.50} \quad (120°) \tag{11.24}$$

Broad-Crested Weirs

The equation generally used for the broad-crested weir is Equation (11.12). The typical form the weir takes is a rectangular block in the bottom of the open channel. Rounding of the upstream edge of the rectangular block improves the flow efficiency and renders the weir less sensitive to sediment deposition and weir height effects. If R is the radius of curvature and P the height of the weir, ratios of R/P less than 0.094 cause no significant reduction in the C factor. Rounding of the upstream edge in the R/P range from 0.094 up to 0.25 causes a gradual increase in the C factor for all values of H/P (H is the head on the crest). Rounding above the R/P ratio of 0.25 has no further effect on the C factor (Ramamurthy et al., 1988). The effects of an inclined upstream face can be found in Clemmens et al. (1984) and Replogle (1978). Information on C values as a function of sharp-edged (upstream edge) rectangular weir crest breadth and head is given in Table 11-4.

TABLE 11-4
Broad-Crested Weir Coefficient C Values as a Function of Weir Crest Breadth and Head

Measured Head, H[a] (ft)	Breadth of the Crest of Weir (ft)									
	0.50	0.75	1.00	1.50	2.00	2.50	3.00	4.00	5.00	10.00	15.00
0.2	2.80	2.75	2.69	2.62	2.54	2.48	2.44	2.38	2.34	2.49	2.68
0.4	2.92	2.80	2.72	2.64	2.61	2.60	2.58	2.54	2.50	2.56	2.70
0.6	3.08	2.89	2.75	2.64	2.61	2.60	2.68	2.69	2.70	2.70	2.70
0.8	3.30	3.04	2.85	2.68	2.60	2.60	2.67	2.68	2.68	2.69	2.64
1.0	3.32	3.14	2.98	2.75	2.66	2.64	2.65	2.67	2.68	2.68	2.63
1.2	3.32	3.20	3.08	2.86	2.70	2.65	2.64	2.67	2.66	2.69	2.64
1.4	3.32	3.26	3.20	2.92	2.77	2.68	2.64	2.65	2.65	2.67	2.64
1.6	3.32	3.29	3.28	3.07	2.89	2.75	2.68	2.66	2.65	2.64	2.63
1.8	3.32	3.32	3.31	3.07	2.88	2.74	2.68	2.66	2.65	2.64	2.63
2.0	3.32	3.31	3.30	3.03	2.85	2.76	2.27	2.68	2.65	2.64	2.63
2.5	3.32	3.32	3.31	3.28	3.07	2.89	2.81	2.72	2.67	2.64	2.63
3.0	3.32	3.32	3.32	3.32	3.20	3.05	2.92	2.73	2.66	2.64	2.63
3.5	3.32	3.32	3.32	3.32	3.32	3.19	2.97	2.76	2.68	2.64	2.63
4.0	3.32	3.32	3.32	3.32	3.32	3.32	3.07	2.79	2.70	2.64	2.63
4.5	3.32	3.32	3.32	3.32	3.32	3.32	3.32	2.88	2.74	2.64	2.63
5.0	3.32	3.32	3.32	3.32	3.32	3.32	3.32	3.07	2.79	2.64	2.63
5.5	3.32	3.32	3.32	3.32	3.32	3.32	3.32	3.32	2.88	2.64	2.63

[a] Measured at least 2.5H upstream of the weir.

Source: From Brater, E.F. and King, H.W., *Handbook of Hydraulics*, 6th ed., McGraw Hill, New York, 1976.

TABLE 11-5

Broad-Crested Weir Submergence Discharge Reduction as a Function of Head Ratio

H_2/H_1	R/P=0.0 Q_s/Q	R/P = 0.125 Q_s/Q	R/P = 1.00 Q_s/Q
0.80	0.95	0.98	1.00
0.85	0.87	0.90	0.96
0.90	0.77	0.80	0.88
0.95	0.48	0.60	0.70

Source: Ramamurthy, R.S., Tim, U.S., and Rao, M.V.J., Characteristics of square-edged and round-nosed broad-crested weirs, ASCE J. Irrig. Drain. Eng., 114, 1, 1988. Reproduced by permission of ASCE.

Submergence Effects

Submergence effects cause a reduction in the discharge coefficient. Table 11-5 gives approximate values of reduction in flow for a range of R/P values. H_2 and H_1 are the downstream and upstream heads with reference to the crest elevation, respectively, and Q_s is the submerged flow, while Q is the unsubmerged flow.

Triangular Channel Broad-Crested Weirs

Broad-crested weirs with upstream ends suitably rounded can be used. Gill (1985) collected data from many authors and slightly modified equations for discharge coefficients. The discharge coefficient is independent of vertex angle over the range tested. His equations are as follows:

$$Q = 0.64\, C_d\, (0.4\, g)^{1/2}\, H^{2.5}\, \tan(\theta/2) \tag{11.25}$$

and

$$C_d = [(H/L) - 0.0718]^{0.032} \tag{11.26}$$

where Q = discharge (cfs); C_d = discharge coefficient; g = acceleration of gravity (32.2 ft/s^2); H = static head on crest (ft); θ = bottom vertex angle of channel; and L = stream-wise length of weir crest (ft).

Ogee Shapes

The shape of the nappe over a flat broad-crested weir is often unstable, leading to a varying flow rate for the same head. In an attempt to correct for this deficiency, the ogee-shaped spillway was developed. The purpose of the ogee crest is to structurally mimic the falling stream of flow over a sharp-crested weir and support the nappe at suitable pressures over a range of flows. Extensive model testing has been accomplished for this ogee shape. A complete description of the design procedure for this type of crest can be found in Reese and Maynord (1987), USCOE (1988), and USBR (1977). The latest procedure is to fit the upstream quadrant with an equation in the form of an ellipse and the downstream quadrant with another equation of the falling free-flowing nappe.

Special Weirs

A variety of specially shaped sharp-edged weirs have been developed for different purposes. Most have attempted to provide proportional or linear stage–discharge curves for irrigation, sediment control, and flood control purposes, as well as flow measurement (Rao and Chandrasekaran, 1977; Murthy and Pillai, 1977, 1978a, 1978b; Murthy and Giridhar, 1989; Swamee et al., 1991; Sandvik, 1985). For erosion control or settling design, the velocity can be controlled, as any function of depth, to meet any relationship. Different shapes can be superimposed on each other to achieve almost any physically possible desired effect.

One type of weir is the inverted triangle weir with a discharge relationship given in Equation (11.27). The limits of the equation are $0.22D_t \le H \le 0.94D_t$. Flows less than that are termed base flows and cannot be approximated by this linear relationship.

$$Q = 0.4481\ C_d / g\ W_t\ D_t^{1/2}\ (H_t - 0.0817 D_t) \tag{11.27}$$

where Q = discharge (cfs); C_d = discharge coefficient taken as 0.615; g = acceleration of gravity (32.2 ft/s²); W_t = bottom width of triangle (ft); D_t = height of triangle (ft); and H_t = depth of flow in triangular weir (ft).

The proportional weir (or "Sutro" weir, though the Sutro shape is only one of many similar shapes) is distinguished from other control devices by having a linear head–discharge relationship achieved by allowing the discharge area to vary nonlinearly with head. Although more complex to design and construct, a proportional weir may significantly reduce the required storage volume for a given site for detention design. Design equations for proportional weirs are as follows (Sandvik, 1985; Murthy and Pillai, 1978; Rouse, 1949; Pratt, 1914):

$$Q = 4.97\ a^{0.5}\ b(H - a/3) \tag{11.28}$$

$$x/b = 1 - (1/3.17)\ (\arctan (y/a)^{0.5}) \tag{11.29}$$

where Q = discharge (cfs).

Dimensions a, b, h, x, and y are shown in Figure 11-13.

Dimension a is assumed and, for design discharges, b is calculated until reasonable dimensions are determined. Using these values and Equation (6.27), the coordinates of the weir are found.

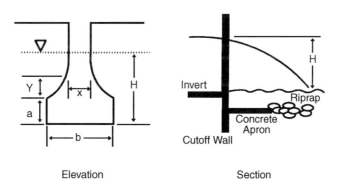

FIGURE 11-13
Proportional weir dimensions.

Side-Channel Weirs

Side-channel spillways are located along channel banks with the spillway crest parallel to the alignment of the channel. As the water level in the channel or stream rises above the crest, the excess is diverted into the side channel. This type of structure works well in urban settings for flood diversions, skimming surface waters or pollution from streams, irrigation purposes, or for detention ponds located along channels, CSO or other storm sewer overflows or in floodplains. In conjunction with debris or surface pollution removal, a floating boom is often placed diagonally across the channel (USBR, 1978).

While the flow from a side weir is complex and dependent on a large number of factors, according to the USBR (1978), for normal purposes where flow is relatively tranquil, a standard weir equation with discharge coefficient will suffice. The exit channel must be set to allow for free flow over the weir without submergence. Overtopping of the weir walls should be prevented with freeboard. In closed conduit situations, open channels with faster flows, or, where a grate is needed for accuracy, the complexities of side-channel weir design must be dealt with. The water surface profile encountered along the lateral side weir varies as does the discharge over the weir. The normal approach is to determine the discharge per unit length of weir, attempting to account for a large number of factors, and then integrate along the weir using a momentum or constant specific energy approach. Various authors approached the problem for different types of weirs and channel shapes: Hager (1987) — general discussion of factors; Uyumaz and Smith (1991) — design procedure for circular and rectangular channels; Uyumaz (1992) — triangular channels; and Cheong (1991) or Ramamurthy, Tim, and Carballada (1986) — trapezoidal channels.

Metcalf & Eddy (1972) provide a description of the three types of situations encountered in side-channel weir design. Figure 11-14 shows possible water surface profiles for the following: (a) steeply sloping supercritical flow situation, (b) subcritical flow situation

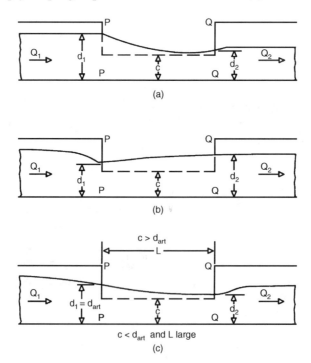

FIGURE 11-14
Side-channel weir water surface profiles. (From Metcalf & Eddy, Inc., *Wastewater Engineering*, McGraw-Hill, New York, 1972. With permission.)

Storage and Detention Facilities 631

with the crest elevation above critical depth, and (c) subcritical flow situation with the crest elevation below critical depth. In (a) and (c), the water surface is falling along the axis of the weir crest. In (b), it is actually rising, if small amounts of water are withdrawn. If the ratio of the height of the weir to the specific energy is less than about 0.6, a falling water surface profile is likely (Metcalf & Eddy, 1972). Approximate solutions for the falling water surface profile and rising water surface profile, respectively, are:

$$Q_w = 0.67 \, L^{0.72} \, E_w^{1.645} \tag{11.30}$$

$$Q_w = 3.32 \, L^{0.83} \, E_w^{1.67} \tag{11.31}$$

Based on the work of Akers (1957), Metcalf & Eddy (1972), De Marchi (1934), and Uyumaz and Smith (1991), the following design procedure can be used with the assumption that specific energy is constant along the weir (found to be true within ±3%).

The variables used in the following design procedure are as follows:

B, D = Channel width or diameter (ft)
c, p = Height of weir above channel (ft)
m = Discharge coefficient of weir
$h_{1,2}$ = Up- and downstream heads on the weir (ft)
D = Hydraulic depth, cross-sectional area/free surface width (ft)
$d_{1,2}$ = Depths of flow up- and downstream (ft)
E = Specific energy relative to the channel bottom (ft)
E_w = Specific energy relative to the weir crest elevation (ft)
$F_{1,2}$ = Up- and downstream Froude numbers at ends of weir
L = Length of weir (ft)
$L_{1,2}$ = Length of weir in rising water surface case (ft)
n_2 = Ratio of h_1/h_2
$Q_{1,2}$ = Discharge up- and downstream (cfs)
$V_{1,2}$ = Velocity up- and downstream (ft/s)
$y_{1,2}$ = Flow depths up- and downstream ends of weir (ft)
$\alpha_{1,2}$ = Velocity coefficient up- and downstream
α' = Pressure-head correction
θ = Channel slope angle
ϕ = Varied flow function

The normal situation is to have an existing or designed channel with a known flow and a desired reduction in that flow, or flow level, to achieve some goal. The side weir is then designed to remove the desired amount of flow from the channel when it is flowing at the design flow.

1. Calculate the upstream depth, velocity, and flow type.
   Upstream depth is calculated from the known flow and flow geometry using the Manning equation.
   Upstream velocity is calculated from the known flow and flow geometry using the Manning equation.

Calculate the upstream Froude number. According to Metcalf & Eddy (1979), it may be taken as 1.2 upstream and 1.4 downstream.

$$F_1 = V_1 / (gD_1\cos\theta/\alpha_1) \tag{11.32}$$

If the Froude number is greater than 1, the flow is supercritical (upper profile in Figure 11-14). If it is less than 1, the flow is subcritical (the lower two water surface profiles in Figure 11-14).

2. Compute critical depth in the channel and set the trial weir height.

By setting the Froude number equal to 1 in Equation (11.32), critical depth is determined.

Set the weir height above the channel bottom. For subcritical flow, if the weir height is set above critical depth, the water surface will be rising. By setting the weir height below critical depth, the water surface will be falling. For supercritical flow, the water surface will be falling. Another check is given by Metcalf & Eddy (1979): if the ratio of the weir height to the specific energy is less than about 0.6, the water surface is falling.

Note: There may be a number of factors that go into setting the weir height, including low-flow maintenance restrictions, physical geometry of the channel, need for tail channel slope or free overfall, need to minimize weir length, etc.

3. Compute the specific energy.

The specific energy relative to the weir crest elevation may be computed as (α' is taken as 1.0 upstream and 0.95 downstream):

$$E_w = \alpha(V^2/2g) + \alpha'(d-c) \tag{11.33}$$

The head on the weir is defined as the depth of flow minus the weir height in feet.

4. Find the weir length.

Compute c/E_w.

Find the weir length from (Akers, 1957):

$$L = 2.03\ B\ (5.28 - 2.63\ c/E_w) \tag{11.34}$$

Note: In this equation, the ratio of the up- and downstream heads on the weir is set to 10.0

5. Check length for flow target (optional).

Solve for the downstream head on the weir. The flow goes through critical depth at the upstream end. At critical flow:

$$\alpha V_1^2/2g = \alpha' h_1/2 \tag{11.35}$$

Then, the specific energy is:

$$E_w = \alpha' h_1/2 + \alpha' h_1 \tag{11.36}$$

$$h_1 = E_w/(1.5\alpha') \text{ and } h_2 = 0.1\ h_1$$

Storage and Detention Facilities 633

Substitute h_2 into the specific energy equation to find V_2.

From the continuity equation $Q = VA$, find Q_2 and compare it to the target.

For the rising water case, the determination of weir length and flow over the weir is handled in a different way. The problem becomes one of calculating the channel discharge at the beginning and ending of an assumed weir length, the difference being the discharge over the side weir. Knowing conditions at section 1 (Q, y_1), ϕ_2 and thus Q_2 can be found. Alternately, a Q_2 value can be set and the L value determined. See Chow (1959), Subramanya and Awasthy (1972), Metcalf & Eddy (1972), and Uyumaz and Smith (1991) for more details. The general equations are:

$$L_{1,2} = 3/2\, B/m\, (\phi_2 - \phi_1) \tag{11.37}$$

$$\phi = [(2E - 3c)/(E - c)][(E - y)/(y - c)]^{0.5} - 3\sin^{-1}[(E-y)/(E-c)]^{0.5} \tag{11.38}$$

The discharge coefficient is given as:

for rectangular channels $F_1 < 1$

$$m = [(2/3)0.611][1 - (3F_1^2/F_1^2+2)]^{0.5} \tag{11.39}$$

for rectangular channels $F_1 > 2$

$$m = 2/3\,(0.036 - 0.008\, F_1) \tag{11.40}$$

Discharge coefficients for other types of channels and weirs should be taken from the above referenced literature.

Example of Side-Channel Weir

A rectangular channel is 10 ft wide and flows 5 ft deep. The slope is 0.001 and Manning's n is taken as 0.014. A side-channel weir is desired to reduce the depth to about 3.7 ft.

- From the Manning equation, the upstream normal depth flow characteristics are:

$$Q_1 = 310 \text{ cfs}$$

$$V_1 = 6.20 \text{ ft/s}$$

- The desired flow characteristics downstream from the weir are:

$$Q_2 = 206 \text{ cfs}$$

$$V_2 = 5.56 \text{ ft/s}$$

- The Froude number in the approaching flow is:

$$F = V/(gD\cos\theta/\alpha)^{1/2} = (6.20)/[(32.2 * 5 * 1)/1.2]^{1/2} = 0.035 \text{ (subcritical)}$$

- At critical depth F = 1 and:

$$1 = Q/(B\ D_c^{3/2}\ g^{1/2}) = 310/(10 * D_c^{3/2} * 32.2^{1/2})\ \text{therefore,}\ D_c = 3.1\ \text{ft}$$

- Set weir height below critical depth. (After several trials, it is set at 1.2 ft.)
- Specific energy is:

$$E_w = \alpha V^2/2g + \alpha'(y\text{-}c) = (1.2 * 6.2)^2/(2 * 32.2) + (5\text{-}1.2) = 4.516\ \text{ft}$$

$$c/E_w = 1.2/4.516 = 0.266\ \text{(falling water surface profile)}$$

- Weir length is calculated as:

$$L = 2.03\ B\ (5.28 - 2.63\ c/E_w) = 2.03*10 * (5.28 - 2.63*0.266) = 93\ \text{ft}$$

- Checking flow target, h_1 is at critical flow upstream, where depth equals twice the velocity head, and substituting into the specific energy equation for the velocity head:

$$h_2 = 0.1\ h_1 = (0.1)*(2/3)*(4.516) = 0.301\ \text{ft}$$

$$E_w = \alpha V^2/2g + \alpha'(h_2) = (1.4* V_2)^2/(2*32.2)+(0.95*0.301) = 4.516\ \text{ft}$$

$$V_2 = 13.95\ \text{ft/s}$$

$$y_2 = h_2 + c = 0.301 + 1.2 = 1.501\ \text{ft}$$

$$Q_2 = AV = (10)(1.501)(13.95) = 209\ \text{cfs. OK}$$

Orifice Meters and Nozzles

Any constriction in the flow in a pressure conduit or pipe will cause a velocity increase and a pressure reduction. A number of methods under the general headings of orifice meter or flow nozzle are available on the market, the thin-plate orifice being the most commonly used orifice meter. Flow nozzles range from a venturi-like insert at one end to a thin nozzle approaching an orifice plate at the other.

Flow nozzles are more expensive than orifice plates and are sensitive to upstream conditions. Twenty or more diameters upstream to the nearest bend or fitting are recommended. Some nozzles are designed to fit on the end of a freely discharging pipe flowing full or partially full. Figure 11-15 shows a typical flow nozzle.

The orifice is the least expensive of the pressure-reducing flow devices for pipe full-flow. It is normally a thin plate with a hole cut in it clamped at a pipe flange or joint. It also produces the greatest headloss across the orifice and is the most sensitive to upstream disturbances. Flows heavy in debris or sediment will often reduce the accuracy, although eccentric orifices with the opening at the pipe invert reduce this somewhat. Orifices are used commonly for detention and retention pond outlet works design.

The basic discharge equation for the orifice is as follows:

$$Q = C_o A (2gH)^{0.5} \qquad (11.41)$$

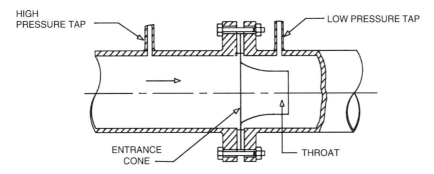

FIGURE 11-15
Typical flow nozzle. (After Shelley and Kirkpatrick, 1975.)

where Q = discharge (cfs); A = cross-sectional area of pipe (ft^2); g = acceleration due to gravity (32.2 ft/s^2); C_o = orifice discharge coefficient; and H = head from center of pipe to the water surface (ft).

For a single orifice, as illustrated in Figure 11-16, the orifice discharge can be determined using Equation (11.41). If the orifice discharges as a free outfall, then the effective head is measured from the center of the orifice to the upstream (headwater) surface elevation. If the orifice discharge is submerged, then the effective head is the difference in elevation of the headwater and tailwater surfaces as shown in Figure 11-17.

The discharge coefficient varies with orifice geometry and edge conditions, and its value is quite sensitive when not a simple square-edge orifice. The discharge coefficient is the product of the velocity coefficient and the contraction coefficient. The velocity coefficient is obtained experimentally by dividing the actual velocity at the *vena contracta* (the smallest area of the contracting jet emerging from the orifice) by the theoretical velocity. The value is on the order of 0.94 to 0.99 (Nelson, 1983). The contraction coefficient is the ratio of the flow area at the *vena contracta* to the area of the orifice. Typical values are in the range of 0.61 to 0.67. However, any number of factors may reduce or suppress the full contraction, giving a significantly higher discharge coefficient, including rounding the edges, providing a short tube downstream, or placing the orifice such that the orifice edge coincides with the wall, top, or bottom of the entrance area.

For sharp square-edge entrance conditions, the coefficient, C_o, can be taken to be 0.6 for a wide range of sizes for circular and rectangular orifices. For orifices with a sharp edge, but having a length of pipe discharging to the atmosphere downstream, the discharge coefficient is taken as 0.8 (Bauer et al., 1969). If part of the boundary is suppressed due to improving the inlet through beveling, rounding, or placement at the boundary, an empirical correction can be applied to the coefficient as follows (USBR, 1974):

$$C_s = 1 + 0.15r \qquad (11.42)$$

where C_s = correction coefficient multiple of C_o, and r = ratio of the suppressed portion of the perimeter of the orifice to the total orifice perimeter.

See Bauer et al. (1969) for a discussion of combinations of conditions.

If the approach velocity is significant, the head in the discharge equation must be corrected by adding to it the velocity head of the approaching flow ($V^2/2g$). Pipes smaller than 12 in may be analyzed as a submerged orifice if H/D is greater than about 1.5. For other conditions, the opening operates as a weir up to a certain point, and culvert analysis with inlet conditions is appropriate. When in doubt, construct two stage–discharge curves, one for each condition, and use the one yielding the higher stage for a given flow (Barfield,

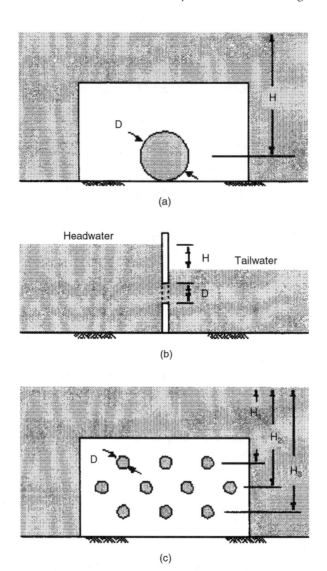

FIGURE 11-16
Orifice definitions. (From Georgia Stormwater Management Manual, Volume 2, Technical Handbook, First Edition, August, 2001.)

Warner, and Haan, 1981). For rectangular orifices with low head, a corrected equation taking head into account is:

$$Q = 2/3\ C_o\ W\ (2g)^{0.5}\ (H_2^{3/2} - H_1^{3/2}) \qquad (11.43)$$

where W = width of the rectangular orifice (ft); H_1 = distance from top of orifice to water surface (ft); and H_2 = distance from bottom of orifice to water surface (ft)

Variable gates have also been used as orifices. For a discussion of various types of variable gates and discharge coefficients see USCOE (1987), Bauer et al. (1969), Rantz (1982), and USBR (1974).

Flow through multiple orifices, such as the perforated plate shown in Figure 11-16(c), can be computed by summing the flow through individual orifices. For multiple orifices

Storage and Detention Facilities 637

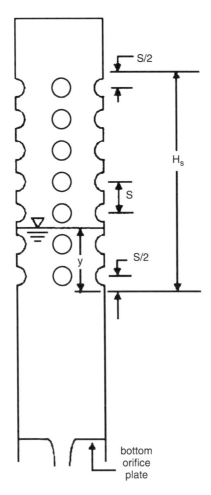

FIGURE 11-17
Perforated riser. (From Georgia Stormwater Management Manual, Volume 2, Technical Handbook, First Edition, August, 2001.)

of the same size and under the influence of the same effective head, the total flow can be determined by multiplying the discharge for a single orifice by the number of openings.

Perforated orifice plates for the control of discharge can be of any size and configuration. However, the Denver Urban Drainage and Flood Control District developed standardized dimensions that have worked well. Table 11-6 gives appropriate dimensions. The vertical spacing between hole centerlines is always 4 in.

For rectangular slots, the height is normally 2 in, with variable width. Only one column of rectangular slots is allowed.

Figure 11-18 provides a schematic of an orifice plate outlet structure for a wet ED pond showing the design pool elevations and the flow-control mechanisms.

A special kind of orifice flow is a perforated riser, as illustrated in Figure 11-17. In the perforated riser, an orifice plate at the bottom of the riser, or in the outlet pipe just downstream from the elbow at the bottom of the riser, controls the flow. It is important that the perforations in the riser convey more flow than the orifice plate so as not to become the control. Referring to Figure 11-17, a shortcut formula has been developed to estimate the total flow capacity of the perforated section (McEnroe, 1988):

TABLE 11-6

Circular Perforation Sizing

Hole Diameter (in)	Minimum Column Hole Centerline Spacing (in)	Flow Area per Row (in²)		
		1 Column	2 Columns	3 Columns
1/4	1	0.05	0.1	0.15
5/16	2	0.08	0.15	0.23
3/8	2	0.11	0.22	0.33
7/16	2	0.15	0.3	0.45
1/2	2	0.2	0.4	0.6
9/16	3	0.25	0.5	0.75
5/8	3	0.31	0.62	0.93
11/16	3	0.37	0.74	1.11
3/4	3	0.44	0.88	1.32
13/16	3	0.52	1.04	1.56
7/8	3	0.6	1.2	1.8
15/16	3	0.69	1.38	2.07
1	4	0.79	1.58	2.37
1 1/16	4	0.89	1.78	2.67
1 1/8	4	0.99	1.98	2.97
1 3/16	4	1.11	2.22	3.33
1 1/4	4	1.23	2.46	3.69
1 5/16	4	1.35	2.7	4.05
1 3/8	4	1.48	2.96	4.44
1 7/16	4	1.62	3.24	4.86
1 1/2	4	1.77	3.54	5.31
1 9/16	4	1.92	3.84	5.76
1 5/8	4	2.07	4.14	6.21
1 11/16	4	2.24	4.48	6.72
1 3/4	4	2.41	4.82	7.23
1 13/16	4	2.58	5.16	7.74
1 7/8	4	2.76	5.52	8.28
1 15/16	4	2.95	5.9	8.85
2	4	3.14	6.28	9.42
Number of columns refers to parallel columns of holes				
Minimum steel plate thickness		1/4"	5/16"	3/8"

Source: Urban Drainage and Flood Control District, Denver, CO,

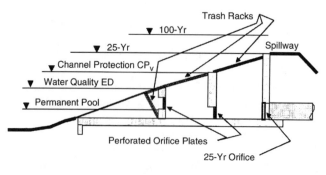

FIGURE 11-18
Schematic of orifice plate outlet structure. (From Georgia Stormwater Management Manual, Volume 2, Technical Handbook, First Edition, August 2001.)

Storage and Detention Facilities 639

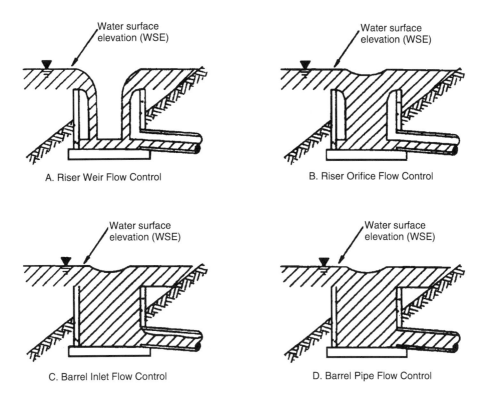

FIGURE 11-19
Riser flow diagrams. (From Virginia Department of Conservation and Recreation, Virginia Stormwater Management Handbook, 1999.)

$$Q = C_p(2A_p/3H_s)(2g^{0.5})(H^{3/2}) \tag{11.44}$$

where Q = discharge (cfs); C_p = discharge coefficient for perforations (normally 0.61); A_p = cross-sectional area of all the holes (ft²); and H_s= distance from $S/2$ below the lowest row of holes to $S/2$ above the top row (ft).

The hydraulic analysis of the design must take into account the hydraulic changes that will occur as depth of storage changes for the different design storms. As shown in Figure 11-19, as the water passes over the rim of a riser, the riser acts as a weir. However, when the water surface reaches a certain height over the rim of a riser, the riser will begin to act as a submerged orifice. The designer must compute the elevation at which this transition from riser weir flow control to riser orifice flow control takes place for an outlet where this change in hydraulic conditions will change. Also note in Figure 11-19 that, as the elevation of the water increases further, the control can change from barrel inlet flow control to barrel pipe flow control. Figure 11-20 shows another condition, where weir flow can change to orifice flow that must be taken into account in the hydraulics of the rating curve, as different design conditions result in changing water surface elevations.

Drop Inlet Boxes

Figure 11-21 illustrates three different types of box-type inlet structures for control of multiple discharges (after Rossmiller, 1982).

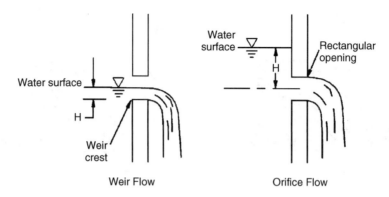

FIGURE 11-20
Weir and orifice flow. (From Virginia Department of Conservation and Recreation, Virginia Stormwater Management Handbook, 1999.)

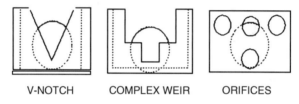

FIGURE 11-21
Inlet boxes for detention control.

In the first of these, a V-notch weir is used as the controlling mechanism. As flow increases, the ability to pass that flow also increases. The spread of the V-notch is sized to pass flow at the correct rate for several design flows. The flow through the notch needs to be adjusted for submergence effects based on the inlet head on the conduit needed to pass the flow through the V-notch. At some point, the flow through the notch will not be sufficient to pass the required discharge due to submergence. At that point, the top of the box elevation is set, and flow begins to pass over the box and through the conduit.

The second of the figures is a complex weir. In this case, a more narrow weir lower in elevation is used for the more frequent storm. A wider weir is then placed at the maximum elevation of the more frequent flood to handle necessary attenuation for less-frequent storms. The 100-year storm is then passed over the top of the first two weirs and the flow box. The stage–discharge curve is calculated as the combination of the inner lower weir, and the head on it, and the two pieces of the upper outer weir, and the lesser stage on it. The inner weir is not contracted, while the outer weir has end contractions as appropriate to the design. The flow coefficient is approximate, because this type of weir has not been extensively model tested. This type of weir also works well as a spill-through-type opening through a dam.

The third type of box-type outlet consists of multiple orifices at different elevations passing through the box walls. The orifices are placed such that the lower orifice(s) control a more frequent flow, while the upper orifice invert is placed at an elevation to begin to operate for a less-frequent design storm. There is significant trial and error in the placement of orifices. It should also be remembered that orifice flow does not begin until the orifice is suitably submerged. Until that time, the orifice will act as a weir, and culvert design equations would be applicable, though calculations for such low flows are rarely warranted.

Rooftop Detention

Rooftop detention is often a viable option for larger commercial and industrial facilities. Water weighs 62.4 pounds per cubic foot. Six inches of standing water places a loading of 31.2 pounds per square foot. This is roughly equivalent to the snow loading requirement in many municipalities, though most building codes are uniformly silent on rooftop detention. Most municipalities limit storage depth on rooftops to about 3 in. To help prevent clogging, designs often include the provision for overtopping into an unrestricted downspout, and periodic inspection. Roof deflection may need to be considered in storage calculations (APWA, 1981). Often, roofs for detention storage are level or have a pitch of 0.25 in/ft.

Figure 11-22 gives the detail of a typical roof drain restrictor device. When water reaches the top of the ring, it spills into the center and through the strainer section. Minimum hole spacing is 2 in on center. Discharge capacity for this type of drain per set of holes (half hole plus full hole above, 1" in diameter) in cfs is:

- 0.0022 cfs for 1.5 in depth
- 0.0046 cfs for 2.0 in depth
- 0.0114 cfs for 2.5 in depth
- 0.0137 cfs for 3.0 in depth
- 0.0156 cfs for 3.5 in depth
- 0.0179 cfs for 4.0 in depth

For this type of restrictor, the 6-in. diameter collar has 9 sets of holes, and the 12 in. has 18 sets. Other devices offer inverted parabolic openings for a linear depth inflow relationship, loading distribution pads built into the device, sumped designs, and durable plastic or aluminum construction.

Design and analysis are normally fairly simple, often being more of a mass balance than a true routing. Water is stored as a wedge based on roof slope toward the drain line. Simple triangular, graphical, or chainsaw routing, modified rational methods (Aaron and Kibler, 1990; Curry, 1974), or even approximate probability methods (Meredith, Middleton, and Smith, 1990) can provide reasonable results for this type of application, provided storm volumes are correct; however, errors can be encountered that may result in inadequate designs and protection of downstream facilities (see later discussion of the modified

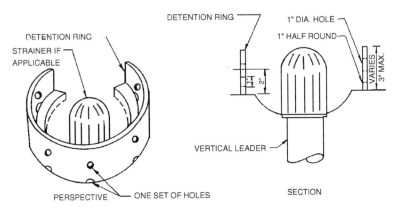

FIGURE 11-22
Typical roof drain restrictor device. (From Bellevue, Washington, Specifications Manual — Chapter 4 Storm Drainage and Streams, Bellevue, WA, 1988.)

rational method). Tables of required storage and allowable release rates based on rooftop area can be prepared for a specific site and rainfall depth. An example of this type of analysis is given below. Care should be taken in gaining a clear understanding of assumptions in such a mass-balance method and accounting for any possible underpredictions of resulting storage volumes. These examples are given for illustration purposes, and the authors do not recommend the use of these methods. The following examples show the large discrepancy that can be obtained from using approximate methods, and thus, routing is recommended for the design of all storage and detention facilities.

Frequent inspection is one of the keys to successful rooftop detention. When rooftops are near, trees leaves can clog the outlet devices allowing water to stand for long periods of time; sometimes leading to roof leaks (Poertner, 1974).

Example Rooftop Design

A rooftop with an area of 1.19 acres (200 ft by 260 ft) is to be designed to hold 3 in of water (maximum) for the 25-year storm. Assuming virtually all the rainfall will run off, a factor of 0.95 is used as a C factor and A is 1.19. The maximum allowable discharge from the roof is 1.5 cfs. Following most municipal practices, the minimum time of concentration is taken as 5 min.

Assuming the roof is level with parapets around the sides, the total allowable storage of the rooftop is: $260 * 200 * 3/12 = 13{,}000$ ft^3.

Modified Rational — Mass Balance Method

For any duration, the rainfall is constant and equal to the intensity given in the IDF curve. This volume of rainfall accumulates on the roof at the rate indicated in Table 11-7. A release rate, assumed constant for simplicity, is then determined to limit the accumulated volume to the maximum allowable.

Application of the rainfall intensity information is:

$$Q = CIA = 0.95 * I * 1.19 = 1.13\ I$$

TABLE 11-7

Modified Rational–Mass Balance For Rooftop Storage Example

(1) Duration (hr)	(2) Intensity (in/hr)	(3) Inflow (cfs)	(4) Volume (cfs-hr)	(5) Accumulated Volume (ft^3)	(6) Average Outflow Rate (cfs)	(7) Outflow Volume (cfs-hr)	(8) Storage Volume (ft^3)
0	0.00	0.00	0.00	0.00	0.0	0.00	0.00
0.083	8.00	9.04	0.75	2700.00	1.5	0.12	2268.00
0.167	6.60	7.46	1.25	4500.00	1.5	0.25	3600.00
0.25	5.50	6.22	1.56	5616.00	1.5	0.38	4248.00
0.50	4.00	4.52	2.26	8136.00	1.5	0.75	5436.00
0.75	3.33	3.76	2.82	10,160.00	1.5	1.13	6102.00
1	2.66	3.01	3.01	10,836.00	1.5	1.50	5436.00
2	1.68	1.90	3.80	13,680.00	1.5	3.00	2880.00
3	1.28	1.45	4.35	15,660.00	1.5	4.50	-540.00
6	0.80	0.90	5.40	19,440.00	1.5	9.00	-12,960.00
12	0.46	0.52	6.24	22,464.00	1.5	18.00	-42,336.00

Table 11-7 gives information on application of the modified rational or mass-balance method for this problem. For the outflow rate of 1.5 cfs, the storage volume required is 6102 cubic feet at a storm duration of about 45 min.

Using this method and trial outflow rates, the lowest possible average outflow rate was determined to meet the requirements for less than 13,000 ft^3 of storage. That value was 0.30 cfs for a storm duration of about 6 h and a storage of 12,960 ft^3.

Columns 1 and 2 show the rainfall intensity–duration–frequency curve information for the 25-year storm for this municipality. Column 3 shows the results of the multiplication of Column 2 by 1.13. Column 4 shows the product of Columns 1 and 3, and Column 5 is obtained by multiplying Column 4 by 3600. Column 6 displays a trial or target discharge rate, and Column 7 shows it converted to a volume by multiplying by Column 1. Column 8 shows the difference between Columns 4 and 7 multiplied by 3600.

An average trial outflow rate of 0.30 cfs will handle the storm volume at the maximum 3 in of depth. As mentioned previously, this design method, although used by some municipalities, is somewhat unrealistic in that the average outflow will vary with head on the outlet, and, thus, storage is somewhat underpredicted. Because of the unreliability of this method, some municipalities (i.e., within the Atlanta, Georgia, metropolitan area) will not accept storage designs based on this method.

Drywells for Roof Drains

A more expensive alternate for roof detention is to take roof water to a drywell. Figure 11-23 shows an example. Drywells have the advantage of pollution reduction as well, but may have maintenance difficulties if not well designed and discharging onto the surface or through well graded soils or gravel. Overflow Ys should be placed at the ground level and cleanouts supplied in case of clogging.

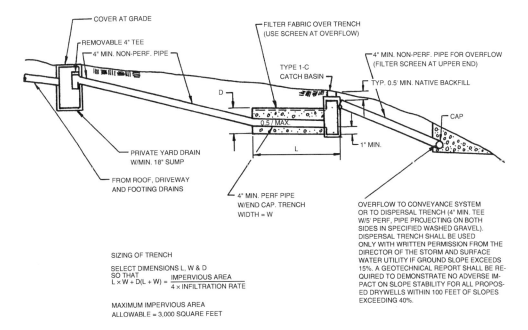

FIGURE 11-23
Typical drywell. (From Bellevue, Washington, Specifications Manual — Chapter 4 Storm Drainage and Streams, Bellevue, WA, 1988.)

Detention Chambers and Pipes

For tight locations on smaller sites (typically areas with high land costs), it may be cost effective to store water in a chamber or oversized pipe or in an inlet chamber. For details, see Urbonas and Stahre (1993).

Parking Lot Detention

Parking lot detention can be used when other sites are not available. Parking lot detention has been shown to be more effective than rooftop detention in reducing peak flow in intense urban developments (McCuen, 1975). Obviously, such detention sites will share the same space as parked vehicles. Detention sites in parking lots should not be located where normal traffic or pedestrians will pass through. Maximum depths should be less than 7 to 12 in and grades less than 7% (Poertner, 1974). Maximum ponding should occur no more frequently than every two years, and sites should drain in 30 min or less. A 25% freeboard is often added to the site.

Outlets should be as vandal (and owner) proof as possible. Figure 11-24 illustrates a type of parking lot detention outlet approved for Bellevue, Washington (1988). In this case, the grate is sized for discharge restriction. Poertner (1974) describes a similar outlet device, but with the addition of an orifice plate seated in the riser collar below the grate. Flow restriction is then not the responsibility of the grate (which acts as a trash rack) but of the orifice below the grate. Columbus, Ohio, uses a similar design, but with a lower orifice plate.

Other outlets include simple curb openings or curb inlets discharging directly to a ditch or stream. All outlets should provide for emergency overflows. An alternative to storing water on the lot is to construct the lot with grassy swales or gravel-filled trenches as dividers and store water in the swales or trenches.

Porous Pavement

An alternative, discussed in more detail in Chapter 13, is porous pavement. It has a rather large storage capacity and, when considering the cost of conventional pavement

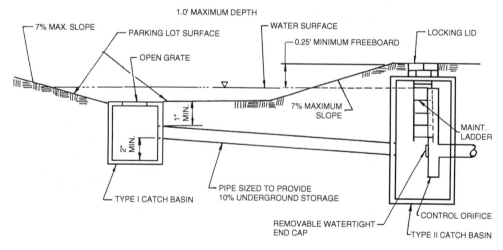

FIGURE 11-24
Typical parking lot ponding basin overflow. (From Bellevue, Washington, Specifications Manual — Chapter 4 Storm Drainage and Streams, Bellevue, WA, 1988.)

and the drainage facilities, is often more cost effective to install than conventional pavement and drainage facilities. Maintenance includes sweeping with vacuum assist. For low-volume traffic areas with well-drained subsoils, it can be a cost-effective solution. Water storage in porous pavement for a combined surface and base thickness of 10 in is about 2.4 in. For a combined surface and base thickness of 25 in, it is nearly 7 in of rainfall (Poertner, 1974). Exfiltration systems can be employed with porous pavement systems to improve performance and to allow for the acceptance of off-site drainage (MWCOG, 1987).

Pipes and Culverts

Discharge pipes are often used as outlet structures for stormwater control facilities. The design of these pipes can be for single or multistage discharges. A reverse-slope underwater pipe is often used for water-quality or channel-protection outlets.

Pipes smaller than 12-in diameter may be analyzed as a submerged orifice as long as H/D is greater than 1.5. Note: For low-flow conditions when the flow reaches and begins to overflow the pipe, weir flow controls. As the stage increases, the flow will transition to orifice flow.

Pipes greater than 12-in diameter should be analyzed as a discharge pipe with headwater and tailwater effects taken into account. The outlet hydraulics for pipe flow can be determined from the outlet control culvert nomographs and procedures given in Chapter 9, or by using Equation (11.45) (NRCS, 1984).

The following equation is a general pipe flow equation that is derived through the use of the Bernoulli and continuity principles:

$$Q = a[(2gH) / (1 + k_m + k_p L)]^{0.5} \qquad (11.45)$$

where Q = discharge (cfs); A = pipe cross-sectional area (ft^2); g = acceleration of gravity (ft/s^2); H = elevation head differential (ft); k_m = coefficient of minor losses (use 1.0); k_p = pipe friction coefficient = $5087 n^2/D^{4/3}$; and L = pipe length (ft).

Combination Outlets

Combinations of orifices, weirs, and pipes can be used to provide multistage outlet control for different control volumes within a storage facility (i.e., water-quality volume, channel-protection volume, infiltration volume, overbank flood-protection volume, and extreme flood-protection volume).

They are generally of two types of combination outlets: shared outlet control structures and separate outlet controls. Shared outlet control is typically a number of individual outlet openings (orifices), weirs, or drops at different elevations on a riser pipe or box that all flow to a common larger conduit or pipe. Figure 11-25 shows an example of a riser designed for a wet ED pond. The orifice plate outlet structure in Figure 11-18 is another example of a combination outlet.

Separate outlet controls are less common and may consist of several pipe or culvert outlets at different levels in the storage facility that are either discharged separately or are combined to discharge at a single location.

The use of a combination outlet requires the construction of a composite stage–discharge curve (as shown in Figure 11-26) suitable for control of multiple storm flows. The design of multistage combination outlets is discussed later in this section.

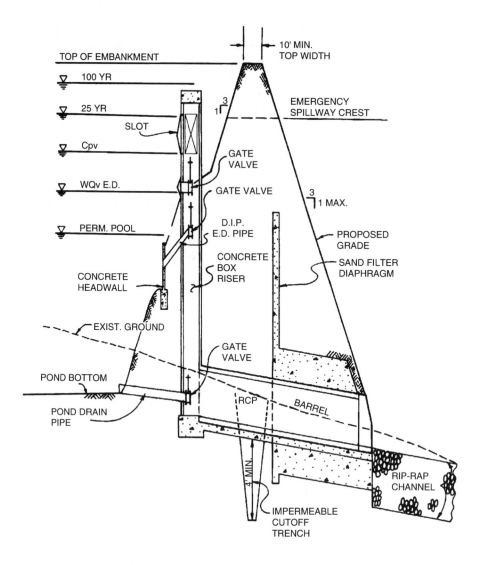

FIGURE 11-25
Schematic of combination outlet structure. (From Georgia Stormwater Management Manual, Volume 2, Technical Handbook, First Edition, August 2001.)

Multistage Outlet Design Procedure

A combination outlet such as a multiple orifice plate system or multistage riser is often used to provide adequate hydraulic outlet controls for the different design requirements (e.g., water quality, channel protection, overbank flood protection, infiltration requirements, and extreme flood protection) for stormwater ponds, stormwater wetlands, and detention-only facilities. Separate openings or devices at different elevations are used to control the rate of discharge from a facility during multiple design storms.

A design engineer may be creative to provide the most economical and hydraulically efficient outlet design possible in designing a multistage outlet. Many iterative routings are usually required to arrive at a minimum structure size and storage volume that

Storage and Detention Facilities

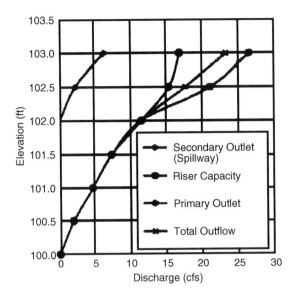

FIGURE 11-26
Composite stage–discharge curve. (From Georgia Stormwater Management Manual, Volume 2, Technical Handbook, First Edition, August 2001.)

provides proper control. The stage–discharge table or rating curve is a composite of the different outlets that are used for different elevations within the multistage riser.

Below are the steps for designing a multistage outlet. Note that, if a structural control facility will not control one or more of the required storage volumes (WQ_v, CP_v, Q_{p25}, and Q_f), then that step in the procedure is skipped.

Control of water quality, channel protection, control of the 25-year storm event, and extreme flood control are included in the following steps. Other volume controls could be added and different storm events used depending on local regulations and site requirements.

Step 1: *Determine stormwater control volumes* — Using the procedures from Chapter 7, estimate the required storage volumes for water-quality treatment (WQ_v), channel protection (CP_v), overbank flood control (Q_{p25}), and extreme flood control (Q_f).

Step 2: *Develop stage–storage curve* — Using the site geometry and topography, develop the stage–storage curve for the facility in order to provide sufficient storage for the control volumes involved in the design.

Step 3: *Design water quality outlet* — Design the water quality extended detention (WQ_v-ED) orifice using either Method 1 or Method 2 outlined in Section 11-7. If a permanent pool is incorporated into the design of the facility, a portion of the storage volume for water quality will be above the elevation of the permanent pool. The outlet can be protected using a reverse slope pipe, a hooded protection device, or another acceptable method.

Step 4: *Design channel protection outlet* — Design the stream channel protection extended detention outlet (CP_v-ED) using either method from Section 11-7. For this design, the storage needed for channel protection will be stacked on top of the water-quality volume storage elevation determined in Step 3. The total stage–discharge rating curve at this point will include water quality control orifice and the outlet used for stream channel protection. The outlet should be protected in a manner similar to that for the water-quality orifice.

Step 5: *Design overbank flood protection outlet* — The overbank protection volume is added above the water-quality and channel-protection storage. Establish the Q_{p25} maximum water surface elevation using the stage–storage curve, and subtract the CP_v elevation to find the 25-year maximum head. Select an outlet type and calculate the initial size and geometry based upon maintaining the predevelopment 25-year peak discharge rate. Develop a stage–discharge curve for the combined set of outlets (WQ_v, CP_v, and Q_{p25}). This procedure is repeated for control (peak flow attenuation) of the 100-year storm (Q_f).

Step 6: *Check performance of the outlet structure* — Perform a hydraulic analysis of the multistage outlet structure using reservoir routing to ensure that all outlets will function as designed. Several iterations may be required to calibrate and optimize the hydraulics and outlets that are used. Also, the structure should operate without excessive surging, noise, vibration, or vortex action at any stage. This usually requires that the structure have a larger cross-sectional area than the outlet conduit.

Step 7: *Size the emergency spillway* — It is recommended that all stormwater impoundment structures have an emergency spillway. An emergency spillway provides a degree of safety to prevent overtopping of an embankment if the primary outlet or principal spillway should become clogged or otherwise inoperative. The 100-year storm should be routed through the outlet devices and emergency spillway to ensure the hydraulics of the system will operate as designed.

Step 8: *Design outlet protection* — Design necessary outlet protection and energy dissipation facilities to avoid erosion problems downstream from outlet devices and emergency spillway(s). See Chapter 12, for more information.

Step 9: *Perform buoyancy calculations* — Perform buoyancy calculations for the outlet structure and footing. Flotation will occur when the weight of the structure is less than or equal to the buoyant force exerted by the water.

Step 10: *Provide seepage control* — Seepage control should be provided for the outflow pipe or culvert through an embankment. The two most common devices for controlling seepage are (1) filter and drainage diaphragms and (2) antiseep collars.

11.7 Extended Detention Outlet Design

Extended detention orifice sizing is required in design applications that provide extended detention for downstream volume control (e.g., channel protection or the ED portion of the water-quality volume). In such cases, an extended detention orifice or reverse slope pipe can be used for the outlet. For example, for a structural control facility providing water-quality extended detention and channel-protection control (wet ED pond, micropool ED pond, and shallow ED wetland), there will be a need to design two outlet orifices — one for the water-quality control outlet and one for the channel-protection drawdown.

(The following procedures are based on the water quality outlet design procedures included in the Virginia Stormwater Management Handbook, 1999. In these procedures, only water quality and channel protection are included, other volume controls could be added depending on local requirements.)

The outlet hydraulics for peak control design (overbank flood protection and extreme flood protection) is usually straightforward in that an outlet is selected that will limit the peak flow to some predetermined maximum. Because volume and the time required for water to exit the storage facility are not usually considered, the outlet design can easily be calculated and routing procedures used to determine if quantity design criteria are met.

In an extended detention facility for water-quality treatment or downstream channel protection, however, the storage volume is detained and released over a specified amount of time (e.g., 24 h). The release period is a brim drawdown time, beginning at the time of peak storage of the water-quality volume until the entire calculated volume drains out of the basin. This assumes that the brim volume is present in the basin prior to any discharge. In reality, however, water is flowing out of the basin prior to the full or brim volume being reached. Therefore, the extended detention outlet can be sized using one of the following methods:

1. Use the maximum hydraulic head associated with the storage volume and maximum flow, calculate the orifice size needed to achieve the required drawdown time, and route the volume through the basin to verify the actual storage volume used and the drawdown time.
2. Approximate the orifice size using the average hydraulic head associated with the storage volume and the required drawdown time.

These two procedures are outlined in the examples below and can be used to size an extended detention orifice for water quality and channel protection.

Example Outlet Design

Method 1: Maximum Hydraulic Head with Routing

A wet ED pond sized for the required water-quality volume will be used here to illustrate the sizing procedure for an extended-detention orifice.

Given the following information, calculate the required orifice size for water-quality design.

Given:

Water-Quality Volume (WQ_v) = 0.76 ac ft = 33,106 ft^3

Maximum Hydraulic Head (H_{max}) = 5.0 ft (from stage versus storage data)

Step 1: Determine the *maximum* discharge resulting from the 24-h drawdown requirement. It is calculated by dividing the water-quality volume (or channel-protection volume) by the required time to find the average discharge, and then multiplying by two to obtain the maximum discharge.

$$Q_{avg} = 33{,}106 \text{ ft}^3 / (24 \text{ h})(3600 \text{ s/h}) = 0.38 \text{ cfs}$$

$$Q_{max} = 2 * Q_{avg} = 2 * 0.38 = 0.76 \text{ cfs}$$

Step 2: Determine the required orifice diameter by using the orifice equation (11.41) and Q_{max} and H_{max}:

$$Q = CA(2gH)^{0.5}, \text{ or } A = Q / C(2gH)^{0.5}$$

$$A = 0.76 / 0.6[(2)(32.2)(5.0)]^{0.5} = 0.071 \text{ ft}^3$$

Determine pipe diameter from $A = 3.14d^2/4$, then $d = (4A/3.14)^{0.5}$:

$$D = [4(0.071)/3.14]^{0.5} = 0.30 \text{ ft} = 3.61 \text{ in}$$

Use a 3.6-in diameter water-quality orifice.

Routing the water-quality volume of 0.76 ac ft through the 3.6-in water-quality orifice will allow the designer to verify the drawdown time as well as the maximum hydraulic head elevation. The routing effect will result in the actual drawdown time being less than the calculated 24 h. Judgment should be used to determine whether the orifice size should be reduced to achieve the required 24 h or if the actual time achieved will provide adequate pollutant removal.

Method 2: Average Hydraulic Head and Average Discharge

Using the data from the previous example, use Method 2 to calculate the size of the outlet orifice.

Given:

Water-Quality Volume (WQ$_v$) = 0.76 ac ft = 33,106 ft^3

Average Hydraulic Head (h$_{avg}$) = 2.5 ft (from stage versus storage data)

Step 1: Determine the average release rate to release the water quality volume over a 24-h time period.

$$Q = 33{,}106 \text{ ft}^3 / (24 \text{ h})(3600 \text{ s/h}) = 0.38 \text{ cfs}$$

Step 2: Determine the required orifice diameter by using the orifice equation (11.41) and the average head on the orifice:

$$Q = CA(2gH)^{0.5}, \text{ or } A = Q / C(2gH)^{0.5}$$

$$A = 0.38 / 0.6[(2)(32.2)(2.5)]^{0.5} = 0.05 \text{ ft}^3$$

Storage and Detention Facilities

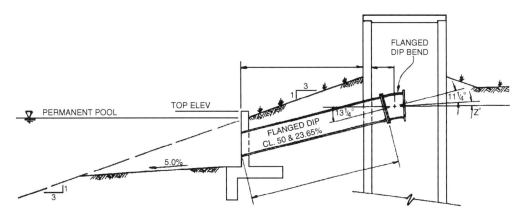

FIGURE 11-27
Reverse slope pipe outlet. (From Georgia Stormwater Management Manual, Volume 2, Technical Handbook, First Edition, August 2001.)

Determine pipe diameter from $A = 3.14 r^2 = 3.14 d^2/4$, then $d = (4A/3.14)^{0.5}$:

$$D = [4(0.05)/3.14]^{0.5} = 0.252 \text{ ft} = 3.03 \text{ in}$$

Use a 3-in diameter water-quality orifice.

Use of Method 1, utilizing the maximum hydraulic head and discharge and routing, results in a 3.6-in diameter orifice (though actual routing may result in a changed orifice size), and Method 2, utilizing average hydraulic head and average discharge, results in a 3.0-in diameter orifice.

Extended-Detention Outlet Protection

Small low-flow orifices such as those used for extended-detention applications can easily clog, preventing the structural control from meeting its design purpose(s) and potentially causing adverse impacts. Therefore, extended-detention orifices need to be adequately protected from clogging. There are a number of different anticlogging designs, including the following:

- The use of a reverse slope pipe attached to a riser for a stormwater pond or wetland with a permanent pool (see Figure 11-27) (The inlet is submerged 1 ft below the elevation of the permanent pool to prevent floatables from clogging the pipe and to avoid discharging warmer water at the surface of the pond.)
- The use of a hooded outlet for a stormwater pond or wetland with a permanent pool (see Figures 11-28 and 11-29)
- Internal orifice protection through the use of an overperforated vertical stand pipe with 1/2-in orifices or slots that are protected by wirecloth and a stone filtering jacket (see Figure 11-30)
- Internal orifice protection through the use of adjustable gate valves to achieve an equivalent orifice diameter

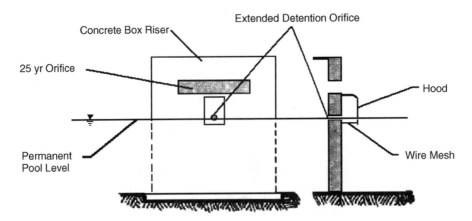

FIGURE 11-28
Hooded outlet. (From Georgia Stormwater Management Manual, Volume 2, Technical Handbook, First Edition, August 2001.)

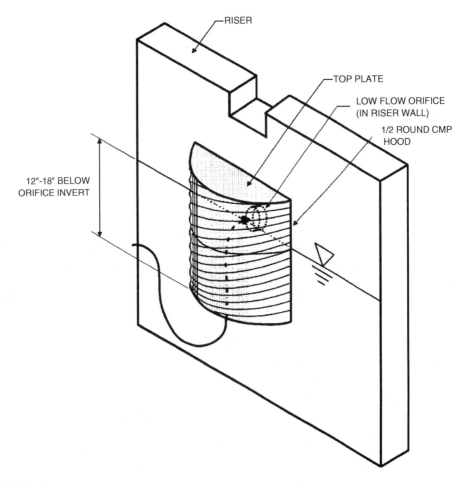

FIGURE 11-29
Half-round CMP orifice hood. (From Georgia Stormwater Management Manual, Volume 2, Technical Handbook, First Edition, August 2001.)

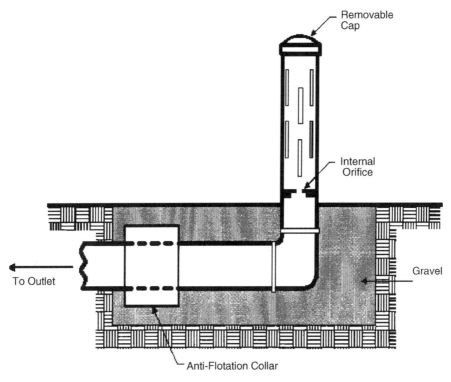

FIGURE 11-30
Internal control for orifice protection. (From Georgia Stormwater Management Manual, Volume 2, Technical Handbook, First Edition, August 2001.)

11.8 Preliminary Detention Calculations

A preliminary estimate of the storage volume required for peak flow attenuation may be obtained from a simplified design procedure that replaces the actual inflow and outflow hydrographs with the standard triangular shapes shown in Figure 11-31.

The required storage volume may be estimated from the area above the outflow hydrograph and inside the inflow hydrograph, expressed as:

$$V_S = 0.5 tb(Q_i - Q_o) \qquad (11.46)$$

where V_S = storage volume estimate (ft^3); Q_i = peak inflow rate (cfs); Q_o = peak outflow rate (cfs); and tb = duration of storage facility inflow (sec).

Any consistent units may be used for Equation (11.46).

Alternative Method

An alternative preliminary estimate of the storage volume required for a specified peak flow reduction can be obtained by the following regression equation procedure (Wycoff and Singh, 1976):

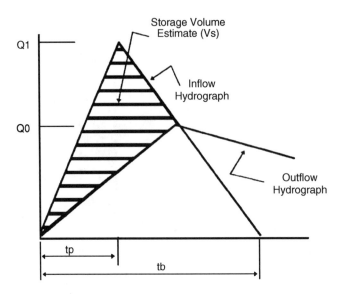

FIGURE 11-31
Triangular shaped hydrographs (preliminary analysis).

1. Determine input data, including the allowable peak outflow rate, Q_o, the peak flow rate of the inflow hydrograph, Q_i, the time base of the inflow hydrograph, t_b, and the time-to-peak of the inflow hydrograph, t_p.
2. Calculate a preliminary estimate of the ratio V_S/V_r using the input data from Step 1 and the following equation:

$$V_S/V_r = [1.291(1-Q_o/Q_i)^{0.753}]/[(t_b/t_p)^{0.411}] \qquad (11.47)$$

where V_S = volume of storage (in); V_r = volume of runoff (in); Q_o = outflow peak flow (cfs); Q_i = inflow peak flow (cfs); t_b = time base of the inflow hydrograph (hr) (determined as the time from the beginning of rise to a point on the recession limb where the flow is 5% of the peak.); and t_p = time to peak of the inflow hydrograph (hr).
3. Multiply the volume of runoff, V_r, times the ration V_s/V_r calculated in Step 2 to obtain the estimated storage volume V_s.

Peak Flow Reduction

A preliminary estimate of the potential peak flow reduction for a selected storage volume can be obtained by the following procedure:

1. Determine the following:
   Volume of runoff, V_r
   Peak flow rate of the inflow hydrograph, Q_i
   Time base of the inflow hydrograph, t_b
   Time to peak of the inflow hydrograph, t_p
   Storage volume, V_S

Storage and Detention Facilities 655

2. Calculate a preliminary estimate of the potential peak flow reduction for the selected storage volume using the following equation (Singh, 1976):

$$Q_o/Q_i = 1 - 0.712(V_S/V_r)^{1.328}(t_b/t_p)^{0.546} \tag{11.48}$$

where Q_o = outflow peak flow (cfs); Q_i = inflow peak flow (cfs); V_S = volume of storage (in); V_r = volume of runoff (in); t_b = time base of the inflow hydrograph (hr) (determined as the time from the beginning of rise to a point on the recession limb where the flow is 5% of the peak.); and t_p = time to peak of the inflow hydrograph (hr).

3. Multiply the peak flow rate of the inflow hydrograph, Q_i, times the potential peak flow reduction calculated from Step 2 to obtain the estimated peak outflow rate, Q_o, for the selected storage volume.

11.9 Routing Calculations

The following procedure is used to perform routing through a reservoir or storage facility (Puls Method of storage routing):

Step 1: Develop an inflow hydrograph, stage–discharge curve, and stage–storage curve for the proposed storage facility. Example stage–storage and stage–discharge curves are shown in Figures 11-1 and 11-2.

Step 2: Select a routing time period, Δt, to provide at least five points on the rising limb of the inflow hydrograph.

Step 3: Use the storage–discharge data from Step 1 to develop storage characteristics curves that provide values of $S \pm (O/2)\Delta t$ versus stage. An example tabulation of storage characteristics curve data is shown in Table 11-8.

Step 4: For a given time interval, I_1 and I_2 are known. Given the depth of storage or stage, H_1, at the beginning of that time interval, $S_1 - (O_1/2)\Delta t$ can be determined from the appropriate storage characteristics curve (Figure 11-32).

TABLE 11-8
Storage Characteristics

(1) Stage(ft)	(2) Storage[a] (ac-ft)	(3) Discharge[b] (cfs)	(4) Discharge[b] (ac-ft/hr)	(5) S-(O/2)Δt (ac-ft)	(6) S+(O/2)Δt (ac-ft)
100	0.05	0	0	0.05	0.05
101	0.3	15	1.24	0.20	0.40
102	0.8	35	2.89	0.56	1.04
103	1.6	63	5.21	1.17	2.03
104	2.8	95	7.85	2.15	3.45
105	4.4	143	11.82	3.41	5.39
106	6.6	200	16.53	5.22	7.98
107	10	275	22.73	8.11	11.89

[a] Obtained from the Stage–Storage Curve.
[b] Obtained from the Stage–Discharge Curve.
Note: t = 10 min = 0.167 h and 1 cfs = 0.0826 ac-ft/hr.

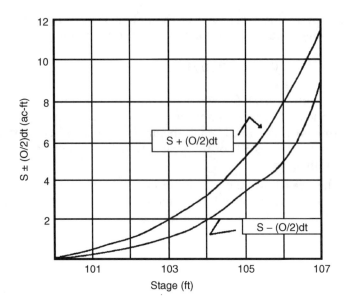

FIGURE 11-32
Storage characteristic curve.

Step 5: Determine the value of $S_2 + (O_2/2)\,\Delta t$ from the following equation:

$$S_2 + (O_2/2)\,\Delta t = [S_1 - (O_1/2)\,\Delta t] + [(I_1 + I_2)/2\,\Delta t] \tag{11.49}$$

where S_2 = storage volume at time 2 (ft³); O_2 = outflow rate at time 2 (cfs); Δt = routing time period (sec); S_1 = storage volume at time 1 (ft³); O_1 = outflow rate at time 1 (cfs); I_1 = inflow rate at time 1 (cfs); and I_2 = inflow rate at time 2 (cfs).

Other consistent units are equally appropriate.

Step 6: Enter the storage characteristics curve at the calculated value determined in Step 5 of $S_2 + (O_2/2)\Delta t$ and read off a new depth of water, H_2.

Step 7: Determine the value of O_2, which corresponds to a stage of H_2 determined in Step 6, using the stage–discharge curve.

Step 8: Repeat Steps 1 through 7 by setting new values of I_1, O_1, S_1, and H_1 equal to the previous I_2, O_2, S_2, and H_2, and using a new I_2 value. This process is continued until the entire inflow hydrograph has been routed through the storage facility.

11.10 Example Problem

This example demonstrates the application of the methodology presented in this chapter for the design of a typical detention storage facility. Example inflow hydrographs and associated peak discharges for both pre- and postdevelopment conditions are assumed to have been developed using hydrologic methods from Chapter 7.

For this example, the storage facilities are designed for runoff from both the 2- and 10-year design storms, and an analysis is done using the 100-year design storm runoff to ensure that the structure can accommodate runoff from this storm without damaging

TABLE 11-9

Example Runoff Hydrographs

	Predevelopment Runoff		Postdevelopment Runoff	
(1)	(2)	(3)	(4)	(5)
Time (Hours)	2-Year (cfs)	10-Year (cfs)	2-Year (cfs)	10-Year (cfs)
0.0	0	0	0	0
0.1	18	24	38	50
0.2	61	81	125	178
0.3	127	170	190 > 150	250 > 200
0.4	150	200	125	165
0.5	112	150	70	90
0.6	71	95	39	50
0.7	45	61	22	29
0.8	30	40	12	16
0.9	21	28	7	9
1.0	13	18	4	5
1.1	10	15	2	3
1.2	8	13	0	1

adjacent and downstream property and structures. Example peak discharges from the 2- and 10-year design storm events are as follows:

- Predeveloped 2-year peak discharge = 150 cfs
- Predeveloped 10-year peak discharge = 200 cfs
- Postdevelopment 2-year peak discharge = 190 cfs
- Postdevelopment 10-year peak discharge = 250 cfs

Because the postdevelopment peak discharge must not exceed the predevelopment peak discharge, the allowable design discharges are 150 and 200 cfs for the 2- and 10-year storms, respectively. Example runoff hydrographs are shown in Table 11-9. Inflow durations from the postdevelopment hydrographs are about 1.2 and 1.25 h, respectively, for runoff from the 2- and 10-year storms.

Preliminary Volume Calculations

Preliminary estimates of required storage volumes are obtained using the simplified method outlined in Section 11.8. For runoff from the 2- and 10-year storms, the required storage volumes, V_S, are computed using Equation (11.46):

$$V_S = 0.5 T_i (Q_i - Q_o)$$

2-year storm: $V_S = [0.5(1.2 \times 3600)(190 - 150)]/43{,}560 = 1.98$ acre-feet

10-year storm: $V_S = [0.5(1.25 \times 3600)(250 - 200)]/43{,}560 = 2.58$ acre-feet

Design and Routing Calculations

Stage–discharge and stage–storage characteristics of a storage facility that should provide adequate peak flow attenuation for runoff from the 2- and 10-year design storms are presented in Table 11-10. The storage–discharge relationship was developed by requiring

TABLE 11-10

Stage-Discharge-Storage Data

(1) Stage (ft)	(2) Q (cfs)	(3) S (acre-feet)	(4) $S_1 - (O/2)\Delta t$ (acre-feet)	(5) $S_1 + (O/2)\Delta t$ (acre-feet)
0.0	0	0.00	0.00	0.00
0.9	10	0.26	0.30	0.22
1.4	20	0.42	0.50	0.33
1.8	30	0.56	0.68	0.43
2.2	40	0.69	0.85	0.52
2.5	50	0.81	1.02	0.60
2.9	60	0.93	1.18	0.68
3.2	70	1.05	1.34	0.76
3.5	80	1.17	1.50	0.84
3.7	90	1.28	1.66	0.92
4.0	100	1.40	1.81	0.99
4.5	120	1.63	2.13	1.14
4.8	130	1.75	2.29	1.21
5.0	140	1.87	2.44	1.29
5.3	150	1.98	2.60	1.36
5.5	160	2.10	2.76	1.44
5.7	170	2.22	2.92	1.52
6.0	180	2.34	3.08	1.60
6.4	200	2.58	3.41	1.76
6.8	220	2.83	3.74	1.92
7.0	230	2.95	3.90	2.00

the preliminary storage volume estimates of runoff for both the 2-and 10-year design storms to be provided when the corresponding allowable peak discharges occurred. Storage values were computed by solving the broad-crested weir equation for head, H, assuming a constant discharge coefficient of 3.1, a weir length of 4 ft, and no tailwater submergence. The capacity of storage relief structures was assumed to be negligible.

Storage routing was conducted for runoff from the 2- and 10-year design storms to confirm the preliminary storage volume estimates and to establish design water surface elevations. Routing results using the stage–discharge–storage data given in Table 11-10 and the stage-discharge-storage data given in Table 11-10, and 0.1-h time steps, are given in Tables 11-11 and 11-12 for runoff from the 2- and 10- year design storms, respectively. The preliminary design provides adequate peak discharge attenuation for both the 2- and 10-year design storms.

For the routing calculations, the following equation was used:

$$S_2 + (O_2/2)\,\Delta t = [S_1 - (O_1/2)\,\Delta t] + [(I_1 + I_2)/2\,\Delta t]$$

Also, Column 6 = Column 3 + Column 5

Because the routed peak discharge is lower than the maximum allowable peak discharges for both design storm events, the weir length could be increased or the storage decreased. If revisions are desired, routing calculations must be repeated.

Although not shown for this example, runoff from the 100-year storm should be routed through the storage facility to establish freeboard requirements and to evaluate emergency overflow and stability requirements. In addition, the preliminary design provides hydraulic details only. Final design should consider site constraints such as depth of water, side slope stability and maintenance, grading to prevent standing water, and provisions for public safety.

Storage and Detention Facilities 659

TABLE 11-11

Storage Routing for the 2-Year Storm

(1) Time (h)	(2) Inflow (cfs)	(3) [(I₁ + I₂)]/2 (acre-ft)	(4) H₁ (ft)	(5) S₁ − (O₁/2)Δt (acre-ft) (6)−(8)	(6) S₂ + (O₂/2) Δt (acre-ft) (3) + (5)	(7) H (ft)	(8) Outflow (cfs)
0.0	0	0.00	0.00	0.00	0.00	0.00	0
0.1	38	0.16	0.00	0.00	0.16	0.43	3
0.2	125	0.67	0.43	0.10	0.77	2.03	36
0.3	190	1.30	2.03	0.50	1.80	4.00	99
0.4	125	1.30	4.00	0.99	2.29	4.80	130 < 150 OK
0.5	70	0.81	4.80	1.21	2.02	4.40	114
0.6	39	0.45	4.40	1.12	1.57	3.60	85
0.7	22	0.25	3.60	0.87	1.12	2.70	55
0.8	12	0.14	2.70	0.65	0.79	2.02	37
0.9	7	0.08	2.08	0.50	0.58	1.70	27
1.0	4	0.05	1.70	0.42	0.47	1.03	18
1.1	2	0.02	1.30	0.32	0.34	1.00	12
1.2	0	0.01	1.00	0.25	0.26	0.70	7
1.3	0	0.00	0.70	0.15	0.15	0.40	3

TABLE 11-12

Storage Routing for the 10-Year Storm

(1) Time (h)	(2) Inflow (cfs)	(3) [(I₁ + I₂)]/2 (acre-ft)	(4) H₁ (ft)	(5) S₁ − (O₁/2)Δt (acre-ft) (6)−(8)	(6) S₂ + (O₂/2)Δt (acre-ft) (3) +(5)	(7) H (ft)	(8) Outflow (cfs)
0.0	0	0.00	0.00	0.00	0.00	0.00	0
0.1	50	0.21	0.21	0.00	0.21	0.40	3
0.2	178	0.94	0.40	0.08	1.02	2.50	49
0.3	250	1.77	2.50	0.60	2.37	4.90	134
0.4	165	1.71	4.90	1.26	2.97	2.97	173 < 200 OK
0.5	90	1.05	5.80	1.30	2.35	4.00	137
0.6	50	0.58	4.95	1.25	1.83	4.10	103
0.7	29	0.33	4.10	1.00	1.33	3.10	68
0.8	16	0.19	3.10	0.75	0.94	2.40	46
0.9	9	0.10	2.40	0.59	0.69	1.90	32
1.0	5	0.06	1.90	0.44	0.50	1.40	21
1.1	3	0.03	1.40	0.33	0.36	1.20	16
1.2	1	0.02	1.20	0.28	0.30	0.90	11
1.3	0	0.00	0.90	0.22	0.22	0.60	6

Downstream Effects

An estimate of the potential downstream effects (i.e., increased peak flow rate and recession time) of detention storage facilities may be obtained by comparing hydrograph recession limbs from the predevelopment and routed postdevelopment runoff hydrographs. Example comparisons are shown in Figure 11-33 for the 10-year design storms.

Potential effects on downstream facilities should be minor when the maximum difference between the recession limbs of the predeveloped and routed outflow hydrographs is less than about 20%. As shown in Figure 11-33, the example results are well below 20%; downstream effects can thus be considered negligible and downstream flood routing omitted.

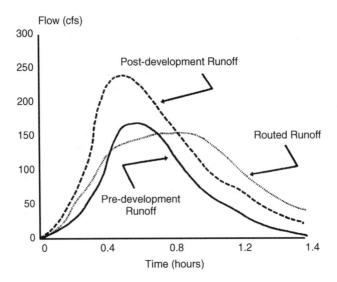

FIGURE 11-33
Runoff hydrographs.

11.11 "Chainsaw Routing" Technique for Spreadsheet Application

Overview

The chainsaw routing procedure is a shortcut method of routing runoff hydrographs through storage facilities to approximate the outflow hydrograph (Malcom, 1987). The same information is required for the chainsaw routing procedure as the storage indication method, as follows:

- Inflow hydrographs for all design storms
- Stage–storage curve for the proposed facility
- Stage–discharge curve for all outlet control structures

In this method, the actual routing procedure has been simplified to allow the computations to be performed by hand or in a standard spreadsheet program. The normal routing equation is derived as follows:

$$dS/dt = I - O \qquad (11.50)$$

which can be simplified as:

$$(S_j - S_i)/\Delta t = (I_i + I_j)/2 - (O_i + O_j)/2 \qquad (11.51)$$

The chainsaw method simplifies it as:

$$\Delta S_{i,j} = (I_i - O_j)\, \Delta t_{i,j} \qquad (11.52)$$

where S = storage (ft³); I = inflow (cfs); O = outflow (cfs); Δt = time step (sec); and i,j = time step subscripts j following i in time.

In this method, the approximation is not considered a trapezoid but a parallelogram with $S_{i,j}$ representing storage across two time steps. In order to ensure accuracy, the computation interval should be less than or equal to 0.1 of the time to peak of the inflow hydrograph.

Stage–Storage

The stage–storage function represents the relation of accumulated storage volume to elevation within the basin. This relation can be expressed as a graph or as a function. However, writing the relation as a function is most beneficial for spreadsheet applications in the chainsaw routing method. The source of the data for the stage–storage function is typically a site plan or topographic map that illustrates the contours of the area proposed to be used for detention storage.

Apart from fitting a curve to a prism, discussed in a previous section, one method of writing a stage–storage function for natural basin shapes is accomplished by assuming that the stage–storage function can be approximated by a power relationship of the form:

$$S = aH^m \tag{11.53}$$

where a = constant in the stage–storage power equation, and m = coefficient in the stage–storage power equation.

The steps for developing a stage–storage function for a natural area using the planimetered end-area method are as follows:

1. Determine the elevations of interest within the storage volume of the detention basin and list them in increasing order in a spreadsheet or table. The units should be in feet.
2. Planimeter or measure the contour elevations for all stages within the detention basin storage volume and enter the areas in a column of a spreadsheet or table. The units should be in square feet.
3. Compute the incremental storage volume by using the average-end area method (upper contour area plus the lower contour area divided by two and multiplied by the difference in elevation).
4. Compute the accumulated volume by adding each incremental volume calculated in Step 3.
5. Compute the relative stage by setting the bottom elevation of the pond equal to zero and adjusting all subsequent elevations appropriately.
6. Calculate the natural logarithm of the accumulated volume in Step 4. These values are assigned the variable name S_i.
7. Calculate the natural logarithm of the stage listed in Step 5. These values are assigned the variable name H_i. Plot Step 5 versus Step 6.
8. Calculate the exponent m in Equation (11.53) by fitting a representative straight line through the points. Select two points on the line. The exponent m is then computed as:

$$m = [\ln(S_2 / S_1)] / [\ln(H_2 / H_1)] \tag{11.54}$$

9. Compute the variable a with Equation (11.55).

$$a = S_2 / H_{s2}^m \qquad (11.55)$$

Chainsaw Routing

The steps to be completed for the chainsaw routing are accomplished through a spreadsheet with the following columns:

Column 1: Time increment, seconds less than or equal to one-tenth of the time to peak of the inflow hydrograph

Column 2: The inflow hydrograph can be generated using any of the appropriate methods described in Chapter 7 of this book or can be automatically generated using the procedure in the next section with an equation fit to the SCS shape

Column 3: Storage in cubic feet

Column 4: Stage computed from the storage in Column 3 and the computed stage–storage relationship

Column 5: Total outflow based on the stage–discharge relationship set in succeeding columns

Columns 6–?: Other stage–discharge relationships for the various portions of the outlet structure, such as orifice and overflow weir, etc.

In order to begin the computations, the spreadsheet or table must be initialized by performing the following steps:

1. Set the initial inflow and outflow equal to zero.
2. Set the initial stage equal to the invert elevation of the outlet spillway or outlet pipe that would be at the pond surface for a wet pond.
3. Set the initial storage volume to its rightful value based on the starting stage. For a dry pond, the value would be set to zero.

Refer to Table 11-13 for an example of several rows from a typical routing. Computations then take a time step from time i to time j. The example will follow the computations from time 28 min to time 32 min:

Step 1: Column 3 — The change in storage for time i-j is computed as the inflow from time step i minus the outflow from time step i times the time increment.

TABLE 11-13

Example Section of Chainsaw Routing Table

1	2	3	4	5	Outflow Devices	
Time (min)	Inflow (cfs)	Storage (ft³)	Stage (ft)	Outflow (cfs)	Culvert (cfs)	Weir (cfs)
28	328	103,103	5.97	130	115	15
32	358	150,623	6.69	142	117	25
36	368	202,463	7.32	151	120	31

Source: After Malcom, H.R., Elements of Urban Stormwater Design, North Carolina State University, 1987.

Storage and Detention Facilities

This increment of storage is added to the storage total in Column 3 for a new total storage at time step j.

$$(328-130) \text{ cfs} * 4 \text{ min} * 60 \text{ sec/min} = 47{,}520 \text{ ft}^3$$

$$47{,}520 + 103{,}103 = 150{,}623 \text{ ft}^3$$

Step 2: Column 4 — At time step j, the new storage is calculated from the stage–storage relationship solved for stage.

$$\text{new stage} = 6.69 \text{ ft from stage–storage curve}$$

Step 3: Columns 6–7 — Calculate the new discharge through each portion of the detention facility outlet works based on the stage in Column 4.

$$\text{discharges} = 117 \text{ cfs (culvert) and } 25 \text{ cfs (weir)}$$

Step 4: Column 5 — Sum the total discharges from Columns 6 through the end for the new total discharge at time step j.

$$117 + 25 = 142 \text{ cfs outflow}$$

Then, begin Step 1 again for the next time step and so on through the routing.

Numerical Instability

In some cases (when the stage–discharge relation has high outflows relative to the stage–storage relation with low storage values), this method will result in unstable results. This becomes evident when the outflow exceeds the inflow, perhaps to the degree that negative storage will be computed. This error can be corrected by reinitializing the row at which this instability occurs, with the following procedure.

1. Set outflow (Column 5) equal to inflow (Column 2).
2. Set stage (Column 4) equal to outflow (Column 5) using the stage–discharge function.
3. Set storage (Column 3) equal to stage (Column 4), by using the stage–storage function.

11.12 Modified Rational Method Detention Design

The modified rational method is actually a collection of methods used in different parts of the country, each employing the rational method to calculate a peak flow and then, after some fashion, building an outflow hydrograph around it. This hydrograph is then compared to a certain release rate or release hydrograph, and the required storage volume is determined graphically or numerically in tabular form.

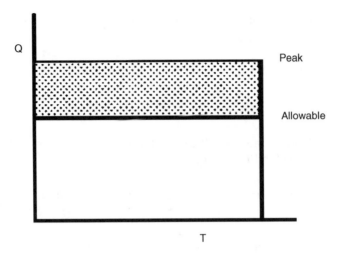

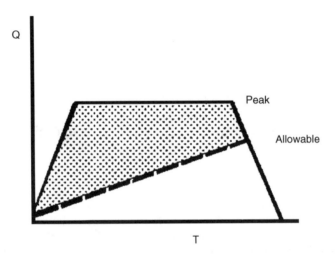

FIGURE 11-34
The mass-balance and graphical hydrograph methods for modified rational method detention design.

Figure 11-34 depicts two common ways to perform a modified rational method design of detention. The top diagram in the figure is termed the mass-balance method. The bottom diagram can be called the graphical hydrograph method. The cross-hatched portion is the estimated storage volume. The first is the most common method used in the Eastern United States. The second is slightly more conservative and generally gives more realistic answers when compared to fully routed applications.

In the mass-balance method, shown in the top half of Figure 11-34, both a constant inflow hydrograph and constant release rate are used. It basically assumes a constant rainfall rate for a certain duration, neglecting the flow buildup to peak. The total volume of inflow at any time is equal to 60CIAt, where C, I, and A are defined in the rational method, and t is time in consistent units of measure. The release rate is a constant that can be expressed as R.

In the graphical hydrograph method, depicted in the bottom half for Figure 11-34, the inflow hydrograph from the watershed accumulates until the time of concentration is reached. Then, it levels out until the storm duration is reached. The recession limb is assumed to be similar to the rising limb. The total inflow volume under each of the inflow

Storage and Detention Facilities

hydrographs for any duration is equal, though the volume is accumulated at different overall rates.

The required storage volume equation for the mass-balance method's volume is:

$$V = 60*(CiAt - Rt) \tag{11.56}$$

where V = the required volume of the pond (ft³); C = the postdevelopment C factor; I = the rainfall intensity from the IDF curve (in/h); R = the allowable release rate (cfs); and t = the storm duration to maximize the volume (min).

The required storage volume equation for the graphical hydrograph method is:

$$V = 60*[CiAt - R(t+t_c)/2] \tag{11.57}$$

where all variables are defined as before, except t_c is the postdevelopment time of concentration for the watershed in minutes. It can be seen that, for a storm duration equal to the time of concentration ($t = t_c$), the two methods yield identical volume estimates. For durations longer than the time of concentration, the second method yields a larger volume.

The normal method of solution is to set up a table for calculation of volume over a range of durations similar to Table 11-7. That duration that maximizes the required storage volume is chosen. It is normally a trial-and-error procedure. However, a closed-form solution can be found that gives the required durations and volumes without trial and error by using the following method. Both methods can be transformed to solve directly for critical duration or maximum storage volume by:

1. Substituting the equation:

$$I = a/(t+b) \tag{11.58}$$

   where I = the rainfall intensity (in/h); t = the time (min); and a and b are curve-fitting constants for rainfall intensity (I)

2. Differentiating the resulting volume equation with respect to time
3. Setting the result equal to zero and solving for the critical duration
4. Substituting the expression for critical duration found in Step 3 into the original volume equation to find the maximum volume

The values of a and b in Equation (11.58) are easily found by curve-fitting Equation (11.58). Most IDF curves can be accurately expressed by such an equation for durations less than 1 h. The logarithm of each side of the equation is taken as:

$$\log(I) = \log(a) - \log(t+b) \tag{11.59}$$

This is an equation of a straight line with slope equal to minus one and intercept equal to $\log a$. The y value is $\log(I)$ and the x value is $\log(t + b)$. The values of a and b are found using any standard spreadsheet's linear regression function. The table below illustrates the procedure. Column 1 is the time and Column 2 is the intensity value from the municipality's IDF curve. Column 3 adds the trial value of b indicated at the top of the table to the time. Columns 4 and 5 are the logarithms of Columns 3 and 2, respectively, and are the X and Y values the spreadsheet performs the regression on. Column 5 is the dependent variable. Columns 6 and 7 are optional, showing the accuracy. The procedure is to use trial values of b until the slope (from the regression output) of the regression line is close to -1.0 (in this case, -1.0001). The value of a can then be calculated from the regression output. An alternate graphical method is found in Froehlich (1993) for sites without developed IDF curves.

When this is done, the critical storm duration and maximum velocity for the mass-balance method are found from Equations (5) and (6), respectively. The critical duration and maximum volume are found for the graphical hydrograph method in Equations (7) and (8), respectively. Note that the volume from the graphical hydrograph method is always greater than the mass balance. This difference is in the 20% range in most cases.

$$T = [(ACab)/r]^{0.5} - b \qquad (11.60)$$

$$V_M = 60*[CaA - 2(CabAR)^{1/2} + Rb] \qquad (11.61)$$

$$T = [(2ACab)/r]^{0.5} - b \qquad (11.62)$$

$$V_M = 60*[CaA - (2CabAR)^{1/2} + R/2\,(b-t_c)] \qquad (11.63)$$

where t = the critical storm duration (min); a = municipal specific constant in the rainfall equation; b = municipal specific constant in the rainfall equation; C = fully developed runoff factor in the rational method equation; A = area (acres); R = release rate (cfs); V_M = maximum storage volume (ft^3); and t_c = fully developed time of concentration (min).

Actual application in a municipality that has a predetermined release rate criteria (such as predeveloped conditions with a C factor of 0.1) can be developed graphically by dividing the volume equation by the area (A), giving a storage volume per unit area and a release rate, effectively canceling A from the equation. Then, a simple plot of unit volume versus C factor can be developed for the design storm and the required volume read directly, for curves of constant time of concentration. Equation (11.63) would become:

$$V_{MA} = 60*[Ca - (2CabR_A)^{1/2} + R_A/2\,(b-t_c)] \qquad (11.64)$$

where V_{MA} = the critical volume per unit drainage area (ft^3/acre), and R_A = the release rate per unit drainage area (cfs/acre).

There has been much discussion about the proper use of the modified rational method for detention design. Discussion centers on the question of whether the method properly sizes ponds or for what limits it is applicable. Some engineers have found that it undersizes ponds when compared to other methods. Some have said it works well for smaller ponds with watershed size limits given in the range of from 2 acres on the low end to about 100 acres on the high end. There are several reasons for the undersizing statements.

The SCS method is often compared as a standard. Depending on how the SCS method is employed, it can show an undersize problem or it can show that the modified rational method works well for smaller basins. If, for example, a rational method C factor of 0.2 is used for predevelopment conditions, the resulting target discharge for detention reduction will be significantly higher than the corresponding predevelopment peak predicted using the SCS method for many of the predevelopment land uses. For a 10-year storm, a C factor of 0.2 corresponds roughly to an SCS curve number in the low 60s. However, the range of curve numbers for existing soils and cover conditions without development range from about 30 to as high as 89, depending on cover, soils, and slope. If a high C factor is chosen but a low target value is set for the SCS method in comparison, the modified rational method will greatly underpredict the required volume. Because the SCS method was derived originally for undeveloped land use conditions, it is probably more accurate across a range of conditions than an arbitrary C factor. The answer is to lower the predevelopment C factor estimate to correspond to cover, slope, storm frequency, and soils conditions inherent in the SCS method. Table 7-7, in Chapter 7, can give guidance for estimating a more accurate undeveloped C factor. Forest and cultivated land will be at the low end of the range, meadow and lawns in the middle, and poorly covered pasture at the high end of the range.

Due to its theoretical worst-case rainfall distribution, the SCS method has been thought to often overpredict volume and peak flows as imperviousness increases. As a small watershed runoff hydrograph moves from being generated by an infiltration-dominated pervious surface to a mass-transfer-dominated impervious surface, the basic assumptions of the rational method become more and more true. A block of rainfall is translated into runoff with little loss. If the SCS method is set, through reasonable adjustment of the curve number and time of concentration, to mimic peak flows predicted by the rational method, for the more impervious cases, the volumes required begin to match more closely. In a series of tests for a Midwestern city, the modified rational method consistently sized ponds adequately when compared to the SCS method when the predevelopment and postdevelopment outflow peaks were matched, and the SCS outflow hydrograph was routed through the modified rational designed pond.

If the required amount of peak flow reduction increases the volume in the runoff, hydrograph tails become increasingly important. The modified rational method does not reflect this in its averaging period approach and then tends to underpredict somewhat. Also, if a constant discharge is used for the modified rational method, the discharge will be too large for a portion of the period and will also tend toward underprediction.

The same considerations apply for comparison of the modified rational method and routing using hydrographs generated using the Huff distribution. When the peak outflows from the Huff distribution and the rational method are approximately equal, the volumes also tend to be approximately equal. But, unlike the SCS method, when the durations of the Huff storm are lengthened, the peak decreases but the volume increases. If this duration storm is chosen, the modified rational method will underpredict the required volume (according to the Huff method) by between 20 and 60%.

On balance, the authors recommend that if, the modified rational method is to be used, it should be used with caution and be limited to highly impervious developments under about 5 acres for normal detention design. For extended detention times, methods that can account for longer storms and actual rainfall distributions should be used.

Example Problem

A 2-acre site is to be developed as a restaurant. The predevelopment C factor is estimated to be 0.1, and the postdevelopment C factor is estimated to be 0.85. A 5-min time of concentration is chosen. The local municipality requires that detention be provided such that the pre- and postdevelopment 10-year storms outflow peaks match. The IDF curve and a and b values are given in Table 11-14.

The predevelopment peak flow is calculated to be:

$$Q = CIA = 0.1*9.69*2 = 1.94 \text{ cfs} = R$$

Using the mass-balance method, the required volume is:

$$V_M = 60*[CaA - 2(CabAR)^{1/2} + Rb]$$

$$= 60*[0.85*222.37*2 - 2*(0.85*222.37*18.48*2*1.94)^{1/2} + 1.94*18.48] = 10,860 \text{ ft}^3$$

TABLE 11-14

Example Curve Fitting

			b = 18.48; a = 222.4; slope = –1			
1	2	3	4	5	6	7
t (min)	I actual (in/h)	t + b (min)	ln(t + b) (X)	ln(I) (Y)	i pred. (in/h)	dif
5	9.692	23.48	3.156	2.271	9.468	–0.23
10	7.767	28.48	3.349	2.05	7.805	0.039
15	6.527	33.48	3.511	1.876	6.64	0.113
30	4.504	48.48	3.881	1.505	4.585	0.081
60	2.88	78.48	4.363	1.058	2.832	–0.05

Using the graphical hydrograph method, the required volume is:

$$V_M = 60 \cdot [CaA - (2CabAR)^{1/2} + R/2\,(b-t_c)]$$

$$= 60 \cdot [0.85 \cdot 222.37 \cdot 2 - (2 \cdot 0.85 \cdot 222.37 \cdot 18.48 \cdot 2 \cdot 1.94)^{1/2} + 1.94/2(18.48-5)]$$

$$= 13{,}590 \text{ ft}^3$$

Using the unit area formulation of the graphical hydrograph method gives:

$$V_{MA} = 60 \cdot [Ca - (2CabR_A)^{1/2} + R_A/2\,(b-t_c)]$$

$$= 60 \cdot [0.85 \cdot 222.37 - (2 \cdot 0.85 \cdot 222.37 \cdot 18.48 \cdot 1.94/2)^{1/2} + (1.94/2)/2 \cdot (18.48-5)]$$

$$= 6{,}795 \text{ ft}^3/\text{acre}$$

which is the actual storage volume divided by the area.

11.13 Hand-Routing Method for Small Ponds

There are many approximate methods for estimating the volume of storage required for detention design, including graphical estimation procedures and a number of variations involving the rational method. Each of these methods attempts to approximate the results that would be generated through a full storage-indication routing described in previous sections without actually performing the routing. The problem with these methods is that they often underpredict the detention volume required, because the shortened triangular design storm does not contain sufficient volume.

The method provided here was developed based on the work of Horn (1987) and graphically approximates the routing of the volume of flow from any storm using a hydrograph shape equivalent to the SCS dimensionless unit hydrograph and approximated by the equations below. It will provide a volume that is slightly greater than the volume generated by unit hydrograph methods with convoluted outflow hydrographs.

Limitations

This method is approximately equivalent to a standard routing method and can be used with the following limitations:

Storage and Detention Facilities 669

- The method is subject to the limitations of the SCS unit hydrograph methodology and shape.
- The method should be limited to smaller applications, in which a single basin is modeled with the SCS dimensionless unit hydrograph shape.
- The method requires the use of a single outlet with a continuous stage–discharge curve that can be expressed as a power function of the form:

$$S = bH^n \qquad (11.65)$$

where b = constant in power equation for either weir or orifice flow as (Lk_w) for weir flow or ($a_o k_o (2g)^{1/2}$) for orifices, and n = the coefficient in the stage–discharge equation which is equal to 0.5 for orifice flow and 1.5 for weir flow.

- Outflow must begin when inflow begins (no dead storage capacity) and fits the power equation from the start of outflow.
- Stage–storage can be expressed as a power function of the form:

$$S = aH^m \qquad (11.66)$$

where a = constant in the stage–storage power equation, and m = coefficient in the stage–storage power equation.

Basic Approach Overview

Given that the inflow hydrograph can be approximated by the SCS dimensionless unit hydrograph and that both the stage–storage and stage–discharge functions can be expressed as simple power functions, the following dimensionless substitutions can be made to the hydrologic continuity equation used for storage routing and dimensionless parameters developed.

$$\alpha = m/n \qquad (11.67)$$

$$K = a/b^\alpha \qquad (11.68)$$

$$N_r = (KQ_i^{\alpha-1})/tP \qquad (11.69)$$

$$R = Q_o/Q_i \qquad (11.70)$$

N_r is called the "routing number," and R is called the "attenuation ratio," which is the peak flow reduction target. Figures 11-35 and 11-36 show relationships between N_r and R for spillway outlets or orifice outlets for a range of values of α. These figures constitute the routing in this method. For a known attenuation ratio, the routing number can be read directly from the graph. Alternately, for a known N_r the attenuation ratio (R) can be read.

For design, the basic approach is to use a known attenuation ratio target and determine N_r. From N_r, a known stage–storage function, known t_p, known Q_i, and K can be determined. From a known K, b can be found, and thus, orifice size or weir length.

For analysis, the process works in reverse. From known physical site information, N_r is calculated. From Figures 11-35 or 11-36, the attenuation ratio is read, and the outflow peak value is calculated.

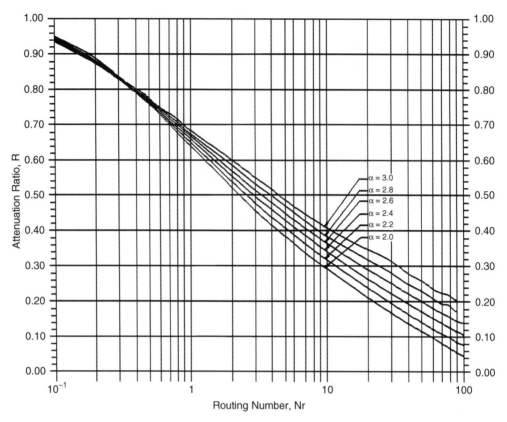

FIGURE 11-35
Attenuation ratio for orifices. (From Horn, D.R., *ASCE J. Hydraul. Eng.*, 113, 11, 1441–1450, 1987. With permission.)

Emergency Spillway Approximation

Most municipalities require that all detention be able to pass safely some infrequent postdevelopment peak flow through an emergency spillway. If this spillway is to be the only spillway for the detention facility, then the orifice or weir and embankment or wall height need to be sized for this flow as well as the 2-year and 10-year or other design controlled peaks. If an additional overflow weir is to be added to the minor storage facility, an approximate method will normally give adequate results.

The basic procedure is to distribute the additional flow from the 50-year peak inflow between the previously sized orifice or weir and an overflow spillway. No routing is done, and no peak attenuation is assumed. It is assumed that the crest of the emergency overflow spillway is set at an elevation equal to the stage (H) of the peak 10-year storm outflow.

For orifice or weir flow for the main outlet and weir flow for the overflow emergency spillway, the combined flow equation is:

$$Q_d = b*[(H+\Delta H)^n - (H)^n] + k_{ew} * L_{ew} * \Delta H^{1.5} \quad (11.71)$$

where Q_d = difference between the 50-year peak postdevelopment inflow and the orifice or weir outflow (Q_p) at peak stage (H) (i.e., the flow to be proportioned between the orifice or weir and the emergency overflow spillway) (cfs); k_{ew} = emergency overflow spillway

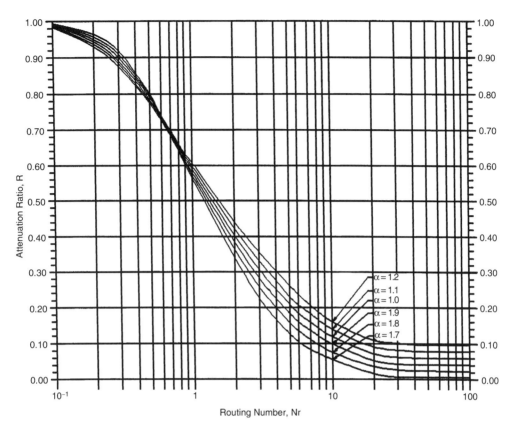

FIGURE 11-36
Attenuation ratio for weirs. (From Horn, D.R., *ASCE J. Hydraul. Eng.*, 113, 11, 1441–1450, 1987. With permission.)

discharge coefficient; L_{ew} = emergency overflow spillway effective length (ft); and ΔH = head on the emergency overflow spillway measured from H upward (ft).

And, b and n are the constant and coefficient in the stage–discharge equations for orifice or weir flow of the principal outlet for the detention pond as described previously.

Details of Approach

For design, the following steps should be accomplished to size a small detention pond. By putting this procedure in a spreadsheet, the design can be done quite easily and fast.

1. *Data input* — Calculate or estimate C factors, curve numbers, and times of concentration for the pre- and postdevelopment conditions. Determine location and approximate size limits of pond, vertically and horizontally. Choose an orifice or weir type and discharge coefficient (k_o or k_w).

2. *Basic calculations* — Calculate peak flows for the design and emergency spillway discharges using rational or another suitable method. Calculate runoff volume from SCS curve number equations. Calculate time-to-peak for the design storm using equations in Chapter 7 for the SCS dimensionless unit hydrograph.

3. *Volume estimate and detention layout* — Estimate design storm storage volume required from the following equation:

$$V_s = 1.39 * 60 * t_p * (Q_i - Q_o) \qquad (11.72)$$

Estimate reasonable peak design storm stage at peak outflow using the design storm predevelopment peak flow (or other municipal design criteria) as the flow target and the weir or orifice equation, as appropriate. With known volume estimate and stage at peak outflow, perform pond layout and develop the stage–storage power function.

4. *Find routing number* — For the design storm, calculate R and α from the equations above. Enter appropriate figure for orifice or spillway flow (Figures 11-35 and 11-36, respectively) to read the routing number (N_r).

5. *Find preliminary weir or orifice size and other data* — From the routing number, calculate K. From K, calculate b and, finally, preliminary orifice or weir size. From orifice or weir size and peak outflow target, calculate the actual head and storage from the stage–discharge and stage–storage equations. If the weir or orifice size is to be rounded to a next available size, proceed through Steps 6 and 7. If not, go to Step 8 to size the emergency spillway.

6. *Recalculate routing number (Optional)* — From a chosen (rounded) weir or orifice size for the design storm flow, calculate b and K. Finally, calculate a new routing number (N_r).

7. *Recalculate R and other data (Optional)* — Enter appropriate figure for orifice or spillway flow (Figures 11-35 and 11-36, respectively) with the new routing number (N_r), and find the attenuation number (R). From R, calculate new Q_o and make sure it is less than the target. If so, calculate new stage and storage for this peak outflow for the design storm flow. This is the final size and stage data.

8. *Size emergency spillway* — Using the equation for the emergency spillway sizing, input an assumed ΔH, and calculate the emergency overflow spillway length (L_{ew}). Often, 1 ft is used for ΔH, though site constraints and safety considerations should determine the assumed head.

Design Example — Sizing of a Small Pond for Orifice Flow

A detention pond is to be sized for a small commercial development. The drainage area to the pond is 2.0 acres Other data are as given. The design criteria for this community is that the 10-year storm peak for postdevelopment should be equal to the 10-year storm peak for predevelopment using a C factor of 0.3 for predevelopment conditions. A 50-year storm must be used to size an emergency spillway. A 6-h storm is to be used for design purposes.

Step 1 — Input Data

Area = 2 acres

$k_o = 0.6$

C predevelopment = 0.2

C postdevelopment = 0.8

CN predevelopment = 62

CN postdevelopment = 85

t_c predevelopment = 5 min

Storage and Detention Facilities

t_c postdevelopment = 5 min
i_{10} = 7.03 in/h
P_{6hr} = 3.72 in
i_{50} = 9.00 in/h

Step 2 — Basic Calculations

Rational method peak flow and SCS method for volume and time-to-peak using a given rainfall IDF curve:

$$Q_{10} \text{ (predevelopment)} = CiA = 0.2*7.03*2 = 2.81 \text{ cfs}$$

$$Q_{10} \text{ (postdevelopment)} = CiA = 0.8*7.03*2 = 11.25 \text{ cfs}$$

$$Q_{50} \text{ (postdevelopment)} = CiA = 0.8*9.00*2 = 14.40 \text{ cfs}$$

The runoff volume is estimated from the SCS equations (see Chapter 7):

$$S = 1000/CN - 10 = 1000/85 - 10 = 1.765$$

$$Q_v = (P - 0.2S)^2/(P+0.8S)$$

$$= (3.72 - 0.2*1.765)^2/(3.72 + 0.8*1.765) = 2.21 \text{ in}$$

The time to peak is estimated from the volume equation for the SCS unit hydrograph:

$$Q_v * 43560/12 * A = 1.39 * 60 * t_p * Q_{10}$$

$$2.21 * 43560/12 * 2 = 1.39 * 60 * t_p * 11.25$$

$$t_p = 17.10 \text{ min} = 1026 \text{ seconds}$$

Step 3 — Preliminary Estimates

Based on a triangular hydrograph approximation, a first estimate of volume (using the 10-year storm peaks and volume) required is:

$$V = 1.39 * 60 * t_p * (Q_i - Q_o) = 1.39*60*17.10*(11.25-2.81) = 12{,}036 \text{ ft}^3$$

Based on site characteristics, the basin should produce a storage volume of about 12,000 cubic feet at a depth of about 5 ft. Using a simple spreadsheet program, a basin with bottom width of 25 ft and length of 55 ft with side-slope of 2.5:1 was chosen. A curve fit to these data yielded values for the stage–storage equation of:

$$a = 1535$$

$$m = 1.35$$

Step 4 — Routing Number
For an orifice, $n = 0.5$ (for a weir it is 1.5)

$$\alpha = m/n = 1.35/0.5 = 2.7$$

$$R = Q_o/Q_i = 2.81/11.25 = 0.25$$

From Figure 11-36, $N_r = 34$.

Step 5 — Orifice Size and Other Data
From Equation (11.69), K is calculated (t_p in seconds), and then the orifice size is backed out:

$$K = N_r * t_p / Q_i^{\alpha-1} = 34 * 17.10 * 60 / (11.25^{2.7-1}) = 570$$

$$b = (a/K)^{n/m} = (1535/570)^{0.37} = 1.44$$

$$a_o = b/(8.02 * k_o) = 1.44/(8.02*0.6) = 0.30 \text{ ft}^2$$

$$D = [(4/\pi) * a_o]^{1/2} = 0.617 \text{ ft} = 7.4 \text{ in}$$

In this case, it was desired to use a standard available size of 7.5 in. To check this, the calculations proceed backward. This is also the procedure used to check an existing pond.

Step 6 (Optional) — Recalculate Routing Number

$$a_o = \pi * (D^2/4) = 0.307 \text{ ft}^2$$

$$b = a_o k_o (2g)^{1/2} = 1.48$$

$$\alpha = m/n = 1.35/0.5 = 2.7$$

$$K = a/b^\alpha = 1535/1.48^{2.7} = 533$$

$$N_r = KQ_I^{\alpha-1}/t_p = (533*11.25^{1.7})/(17.10*60) = 31.81$$

Step 7 (Optional) — Recalculate R and Other Data
From Figure 11-36, $R = 0.255$.

$$Q_o = RQ_I = 0.255 * 11.25 = 2.87 \text{cfs (Target 2.81 cfs) (OK)}$$

$$H = 1/(2g) * [Q_o/(k_o a_o)]^2 = 1/(2g) * [2.87/(0.6*0.307)]^2 = 3.91 \text{ ft}$$

$$S = 1535 H^{1.35} = 9672 \text{ ft}^3$$

Step 8 — Emergency Spillway Approximation
Assume a ΔH of 1 ft and a k_{ew} of 3.0

$$Q_d = b*[(H+\Delta H)^n - (H)^n] + k_{ew} * L_{ew} * \Delta H^{1.5}$$

$$14.4 - 2.87 = 1.48 * [(3.91 + 1)^{0.5} - (3.91)^{0.5}] + 3 * L_{ew} * 1^{1.5}$$

$$L_{ew} = 3.7 \text{ ft (OK)}$$

Method Comparison

It should be noted that, for this problem, the required storage using the graphical hydrograph method (discussed in the modified rational section) is 8090 ft^3. The required volume by routing SCS method hydrographs through the designed pond using HEC-1 is 8620 ft^3. Figure 11-37 shows a comparison of the inflow hydrographs among the three methods when the peak flows are kept equal for pre- and postdevelopment conditions.

The fully routed hydrograph is the most realistic of the three in terms of convolution of the hydrograph for the design storm and routing through the pond. In this example, all three methods give equivalent volumes. This is not always the case. The hand-routing method is more sensitive to large or small reductions in peak flow and may tend to underpredict required volumes for great peak flow reductions and overpredict required volumes for small peak flow reductions. Therefore, it is recommended that it be used in the mid-range of the R values given in the figures.

Outflow hydrographs using the Huff rainfall distribution (discussed in Chapter 7) vary depending on the duration of storm chosen. If a duration is chosen to maximize the peak outflow, the resulting volume required for storage is often equivalent or slightly less than that required for modified rational method design. However, if longer durations are chosen, the peak flows will decrease, but the required storage volumes will increase. In this case, it is not uncommon, using the Huff distribution, for the required volume to be 20 to 60% higher than that necessary for the modified rational method.

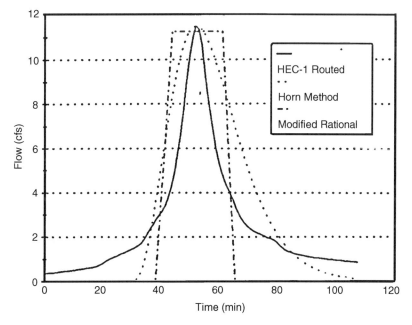

FIGURE 11-37
Comparison of three detention design methods.

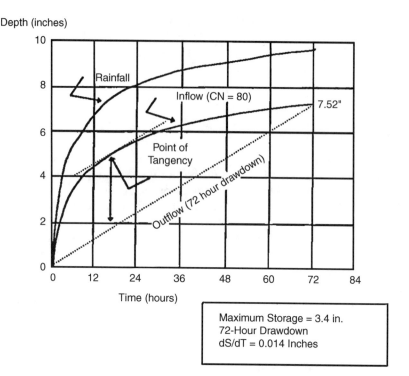

FIGURE 11-38
Mass-routing curve.

11.14 Land-Locked Retention

Watershed areas that drain to a central depression with no positive outlet (playa lakes) are typical of many topographic areas, including karst topography, and can be evaluated using a mass-flow routing procedure to estimate flood elevations. Although this procedure is fairly straightforward, the evaluation of storage facility outflow is a complex hydrogeologic phenomenon that requires good field measurements and a thorough understanding of local conditions. Because outflow rates for flooded conditions are difficult to calculate, field measurements are desirable.

The steps in the procedure presented below for the mass-routing procedure are illustrated by the example graph given in Figure 11-38.

Step 1: Obtain cumulative rainfall data for the 100-year frequency, 10-day duration design event from Figure 11-39.

Step 2: Calculate the cumulative inflow to the land-locked retention facility using the rainfall data from Step 1 and runoff procedure from Chapter 7. Plot the mass inflow to the retention facility.

Step 3: Develop the facility outflow from field measurements of hydraulic conductivity, taking into consideration worst-case water table conditions. Hydraulic conductivity should be established using *in situ* test methods, then results compared to observed performance characteristics of the site. Plot the mass outflow as a straight line with a slope corresponding to worst-case outflow in inches/hour.

Storage and Detention Facilities 677

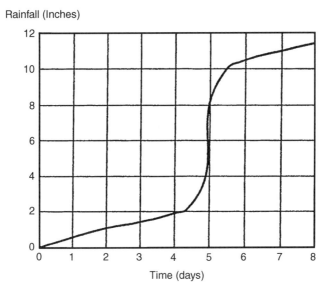

FIGURE 11-39
Cumulative rainfall data for 100-year, 10-day design storm. (Note: Local data should be used to develop this figure.)

Step 4: Draw a line tangent to the mass inflow curve from Step 2, which has a slope parallel to the mass outflow line from Step 3.

Step 5: Locate the point of tangency between the mass inflow curve of Step 2 and the tangent line drawn for Step 4. The distance from this point of tangency and the mass outflow line represents the maximum storage required for the design runoff.

Step 6: Determine the flood elevation associated with the maximum storage volume determined in Step 5. Use this flood elevation to evaluate flood-protection requirements of the project. The zero volume elevation should be established as the normal wet season water surface or water table elevation or the pit bottom, whichever is highest.

Step 7: If runoff from the project area discharges into a drainage system tributary to the land-locked depression, detention storage facilities are required to comply with the predevelopment discharge requirements for the project. Unless the storage facility is designed as a retention facility, including water budget calculations, environmental needs, and provisions for preventing anaerobic conditions, relief structures should be provided to prevent standing water conditions.

11.15 Retention Storage Facilities

The use of retention storage facilities that have a permanent pool (wet ponds) is often discouraged because of the extensive maintenance that is sometimes required. Provisions for weed control and aeration for prevention of anaerobic conditions should be considered. Also, facilities should not be built that have the potential of becoming nuisances or health hazards. Note, wet ponds are required where water-quality problems are to be addressed.

Water budget calculations are required for all permanent pool facilities and should consider performance for average annual conditions. The water budget should consider all significant inflows and outflows including, but not limited to, rainfall, runoff, infiltration, exfiltration, evaporation, and outflow.

Average annual runoff may be computed using a weighted runoff coefficient for the tributary drainage area multiplied by the average annual rainfall volume. Infiltration and exfiltration should be based on site-specific soils testing data. Evaporation may be approximated using the mean monthly pan evaporation or free water surface evaporation data appropriate for the given facility location.

11.16 Retention Facility Example Problem

A shallow retention facility with an average surface area of 3 acres and a bottom area of 2 acres is planned for construction at the outlet of a 100-acre watershed. The watershed is estimated to have a postdevelopment runoff coefficient of 0.3. Site-specific soils testing indicates that the average infiltration rate is about 0.1 in/h. Determine for average annual conditions if the facility will function as a retention facility with a permanent pool.

1. From rainfall records, the average annual rainfall is about 50 in.
2. From local data, the mean annual evaporation is 35 in.
3. The average annual runoff is estimated as:

    Runoff = (0.3) (50 in) (100 acres) = 1500 acre-in

4. The average annual evaporation is estimated as:

    Evaporation = (35 in) (3 acres) = 105 acre-in

5. The average annual infiltration is estimated as:

    Infiltration = (0.1 in/h) (24 h/day) (365 days/year) (2 acres)

    Infiltration = 1752 acre-in

6. Neglecting facility outflow and assuming no change in storage, the runoff (or inflow) less evaporation and infiltration losses is:

    Net Budget = 1500 - 105 - 1752 = -357 acre-in

    Thus, the proposed facility will not function as a retention facility with a permanent pool.

7. Revise pool design as follows:

    Average surface area = 2 acres and bottom area = 1 acre

8. Recompute the evaporation and infiltration:

Storage and Detention Facilities

$$\text{Evaporation} = (35)(2) = 70 \text{ acre-in}$$

$$\text{Infiltration} = (0.1)(24)(365)(1) = 876 \text{ acre-in}$$

9. The revised runoff less evaporation and infiltration losses is:

$$\text{Net Budget} = 1500 - 70 - 876 = +554 \text{ acre-in}$$

The revised facility is assumed to function as a retention facility with a permanent pool.

11.17 Construction and Maintenance Considerations

An important step in the design process is identifying whether special provisions are warranted to properly construct or maintain proposed storage facilities. To assure acceptable performance and function, storage facilities that require extensive maintenance are discouraged. The following maintenance problems are typical of urban detention facilities, and facilities should be designed to minimize these problems (see Poertner, 1974; MWCOG, 1987; and APWA, 1981 for more details.):

- Weed growth
- Grass and vegetation maintenance
- Sedimentation control
- Bank deterioration
- Standing water or soggy surfaces
- Mosquito control
- Blockage of outlet structures
- Litter accumulation
- Maintenance of fences and perimeter plantings

Proper design should focus on the elimination or reduction of maintenance requirements by addressing the potential for problems to develop. Following are some examples:

- Weed growth and grass maintenance may be addressed by constructing side slopes that can be maintained using available power-driven equipment, such as tractor mowers.
- Sedimentation may be controlled by constructing traps to contain sediment for easy removal, or low-flow channels to reduce erosion and sediment transport.
- Bank deterioration can be controlled with protective lining or by limiting bank slopes.
- Standing water or soggy surfaces may be eliminated by sloping basin bottoms toward the outlet, constructing low-flow pilot channels across basin bottoms from the inlet to the outlet, or by constructing underdrain facilities to lower water tables.
- In general, when the above problems are addressed, mosquito control will not be a major problem.

- Outlet structures should be selected to minimize the possibility of blockage (i.e., very small pipes tend to block quite easily and should be avoided).
- Finally, one way to deal with the maintenance associated with litter and damage to fences and perimeter plantings is to locate the facility for easy access, where this maintenance can be conducted on a regular basis.

11.18 Protective Treatment

Protective treatment may be required to prevent entry to facilities that present a hazard to children and, to a lesser extent, all persons. Fences may be required for detention areas where one or more of the following conditions exist:

- Rapid stage increases would make escape practically impossible where small children frequent the area.
- Water depths exceed 2.5 ft for more than 24 h or are permanently wet and have side slopes steeper than 4:1.
- A low-flow watercourse or ditch passing through the detention area has a depth greater than 5 ft or a flow velocity greater than 5 ft/s.
- Side slopes equal or exceed 1.5:1.

Guards or grates may be appropriate for other conditions, and debris accumulation may become a problem. In some cases, it may be advisable to fence the watercourse or ditch rather than the detention area. Fencing should be considered for dry retention areas with design depths in excess of 2.5 ft for 24 h, unless the area is within a fenced, limited-access facility.

Trash Racks and Safety Grates

Trash racks and safety grates serve several functions:

- They trap larger debris well away from the entrance to the outlet works, where they will not clog the critical portions of the works.
- They trap debris in such a way that relatively easy removal is possible.
- They keep people and large animals out of confined conveyance and outlet areas.
- They provide a safety system whereby persons caught in them will be stopped prior to the very high velocity flows immediately at the entrance to outlet works, and persons will be carried up and onto the outlet works, allowing for an ability to climb to safety.
- Well-designed trash racks can have an aesthetically pleasing appearance.

When designed well, trash racks serve these purposes without interfering significantly with the hydraulic capacity of the outlet (or inlet, in the case of conveyance structures) (ASCE, 1985; Allred-Coonrod, 1991). The location and size of the trash rack depends on a number of factors, including headlosses through the rack, structural convenience, safety, and size of outlet. Well-designed trash racks can also have an aesthetically pleasing appearance.

Storage and Detention Facilities 681

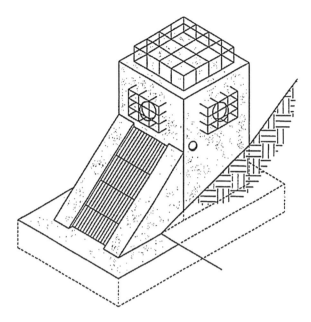

FIGURE 11-40
Example of various trash racks used on a riser outlet structure. (From Virginia Department of Conservation and Recreation, Virginia Stormwater Management Handbook, 1999.)

An example of trash racks used on a riser outlet structure is shown in Figure 11-40. The inclined vertical bar rack is most effective for lower stage outlets. Debris will ride up the trash rack as water levels rise. This design also allows for removal of accumulated debris with a rake while standing on top of the structure.

Trash racks at entrances to pipes and conduits should be sloped at about 3H:1V to 5H:1V to allow trash to slide up the rack with flow pressure and rising water level — the slower the approach flow, the flatter the angle. Rack opening rules-of-thumb abound in the literature. Figure 11-41 gives opening estimates based on outlet diameter (UDFCD, 1992). Judgment should be used in that an area with higher debris (e.g., a wooded area) may require more opening space.

The bar opening space for small pipes should be less than the pipe diameter. For larger diameter pipes, openings should be 6 in or less. Collapsible racks have been used in some places if clogging becomes excessive or a person becomes pinned to the rack. Alternately, debris for culvert openings can be caught upstream from the opening by using pipes placed in the ground or a chain safety net (USBR, 1978; UDFCD, 1991). Racks can be hinged on top to allow for easy opening and cleaning.

The control for the outlet should not shift to the grate. And, the grate should not cause the headwater to rise above planned levels. Therefore, headlosses through the grate should be calculated. A number of empirical loss equations exist, though many have difficult-to-estimate variables. Three will be given to allow for comparison. Metcalf & Eddy (1972) give the following equation (based on German experiments) for losses. Grate openings should be calculated assuming a certain percentage blockage as a worst case to determine losses and upstream head. Often, 40 to 50% is chosen as a working assumption (Perham, 1987). Abt et al. (1992) studied effects of blockage in supercritical flow and found that, at a 3H:1V or flatter slope, debris tended to be pushed up and off the grate and that localized flooding could occur for a 40% blockage.

$$H_g = K_{g1} (w/x)^{4/3} (V_u^2/2g) \sin \theta_g \tag{11.73}$$

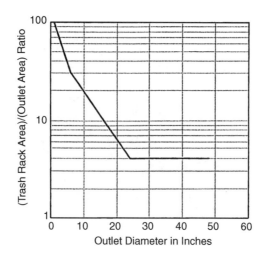

FIGURE 11-41
Minimum rack size versus outlet diameter. (From Denver Urban Drainage and Flood Control District (UDFCD)., Urban Drainage Criteria Manual — Vol. 3", Denver UDFCD, 1992. With permission.)

where H_g = headloss through grate (ft); K_{g1} = bar shape factor: 2.42 — sharp-edged rectangular, 1.83 — rectangular bars with semicircular upstream faces, 1.79 — circular bars, and 1.67 — rectangular bars with semicircular up- and downstream faces; w = maximum cross-sectional bar width facing the flow (in); x = minimum clear spacing between bars (in); V_u = approach velocity (ft/s); and θ = angle of the grate with respect to the horizontal (degrees).

The Corps of Engineers (USCOE, 1988) developed curves for trash racks based on similar and additional tests. These curves are for vertical racks, but, presumably, they can be adjusted in a manner similar to the previous equation through multiplication by the sine of the angle of the grate with respect to the horizontal.

$$H_g = K_{g2} V_u^2/2g \qquad (11.74)$$

where K_{g2} is defined from a series of fit curves as:

- Sharp-edged rectangular (length/thickness = 10)

$$K_{g2} = 0.00158 - 0.03217 A_r + 7.1786 A_r^2$$

- Sharp-edged rectangular (length/thickness = 5)

$$K_{g2} = -0.00731 + 0.69453 A_r + 7.0856 A_r^2$$

- Round-edged rectangular (length/thickness = 10.9)

$$K_{g2} = -0.00101 + 0.02520 A_r + 6.0000 A_r^2$$

- Circular cross section

$$K_{g2} = 0.00866 + 0.13589 A_r + 6.0357 A_r^2$$

And, A_r is the ratio of the area of the bars to the area of the grate section.

- Creager and Justin (1950) give:

$$K_{g2} = 1.45 - 0.45 A_o - A_r^2 \qquad (11.75)$$

where A_r = ratio of net open area (gross grate area less bar area) divided by gross area of the grate.

References

Aaron, G. and Kibler, D.F., Pond sizing for rational formula hydrographs, *Water Resour. Bull.*, 26, 2, 1990.

Akers, P., A Theoretical Consideration of Side Weirs in Stormwater Overflows, *Proc. Inst. of Civ. Eng.*, 6, 2, 1957.

Allred-Coonrod, J.E., Safety Grates in Open Channels, MS Prof. Paper, University of New Mexico, Albuquerque, 1991.

American Public Works Association, Urban Stormwater Management, Special report, 49, 1981.

American Society of Civil Engineers, Stormwater Detention Outlet Control Structures, Task Committee on the Design of Outlet Control Structures, ASCE, New York, NY, 1985.

Bat, S.R., Brisbane, T.E., Frick, D.M., and McKnight, C.A., Trash rack blockage in supercritical flow, *ASCE J. Hydraul. Eng.*, 118, 12, December 1992.

Bauer, W.J., Louie, D.S., and Voordin, W.L., Basic hydraulics, Section 2, in *Handbook of Applied Hydraulics*, Davis, C.V. and Sorenson, K.E., Eds., McGraw-Hill, New York, 1969.

Bellevue, Washington, Specifications Manual — Chapter 4 Storm Drainage and Streams, Bellevue, WA, 1988.

Brater, E.F. and King, H.W., *Handbook of Hydraulics*, 6th ed., McGraw Hill, New York, 1976.

Cheong, H., Discharge coefficient of lateral diversion from trapezoidal channel, *ASCE J. Irrig. and Drain. Eng.*, 117, 4, 1991.

Chow, C.N., *Open Channel Hydraulics*, McGraw Hill, New York, 1959.

Clemmens, A.J., Replogle, J.A., and Bos, M.G., Rectangular measuring flumes for lined and earthen channels, *ASCE J. Irrig. Drain. Eng.*, 110, 2, 1984a.

Creager, W.P. and Justin, J.D., *Hydroelectric Handbook*, 2nd ed., John Wiley & Sons, New York, 1950.

Curry, L., Relationship of Rational and Unit Graph Methods in Retention Basin Design, Proc. Natl. Symp. on Urban Rainfall and Runoff and Sediment Control, University of Kentucky, 1974.

De Marchi, G., Essay on the performance of lateral weirs, *L'Energia Electricia*, Milan, Italy, 11, 11, November, 1934; as cited in Uyumaz, A. and Smith, R.H., Design procedure for flow over side weir, *ASCE J. Irrig. and Drain. Eng.*, 117, 1, 1991.

Debo, T.N. and Reese, A.J., Determining Downstream Analysis Limits for Detention Facilities, Proceedings for NOVATECH 92, Int. Conf. Innovative Technol in the Domain of Urban Stormwater Drain., Lyon, France, 1992.

Denver Urban Drainage and Flood Control District (UDFCD)., Urban Drainage Criteria Manual — Vol. 2", Denver UDFCD, 1969 with changes 1991.

Denver Urban Drainage and Flood Control District (UDFCD)., Urban Drainage Criteria Manual — Vol. 3", Denver UDFCD, 1992.

Froehlich, D.C., Short-duration-rainfall intensity equations for drainage design, *ASCE J. Irrig.and Drain. Eng.*, 119, 5, 1993.

Georgia Stormwater Management Manual, Volume 2, Technical Handbook, First Edition, August 2001.

Grant, D.M., *ISCO Open Channel Flow Measurement Handbook*, 3rd ed., 1989.

Hager, W.H., Lateral outflow over side weirs, *ASCE, J. Hydraul. Eng.*, 113, 4, 1987.

Hartigan, J.P., Regional BMP Masterplans, Proc. ASCE Eng. Found. Conf., *Urban Runoff Technology*, Henniker, NH, 1986.

Hartigan, J.P. and George, Optimizing the Performance of a Regional Stormwater Detention System, ASCE Civ. Eng. Convention and Expo., 1989.

Horn, D.R., Graphic estimation of peak flow reduction in reservoirs, *ASCE J. Hydraul. Eng.*, 113, 11, 1441–1450, 1987.
James, W.P., Bell, J.F., and Leslie, D.L., Size and location of detention storage, *ASCE J. Water Res. Plann. and Manage.*, 113, 1, 1987.
Jones, J.E., Multipurpose Stormwater Detention Ponds, Public Works, 1990.
Leupold & Stevens, Inc., *Steven Water Resources Data Book*, 5th ed., Leupold & Stevens, Beaverton, OR, 1991.
Malcom, H.R., Elements of Urban Stormwater Design, North Carolina State University, 1987.
McCuen, R.H., Design Accuracy of Stormwater Detention Basins, Tech. report, Department of Civil Engineering, University of Maryland, 1983.
McCuen, R.H., Flood Runoff from Urban Areas, Completion report. A-025-Md, University of Maryland, 1975.
McCuen, R.H. and Moglen, G.E., Multicriterion stormwater management methods, *ASCE J. Water Res. Plann. and Manage.*, 114, 4, 1988.
McEnroe, B.M., Steichen, J.M., and Schweiger, R.M., Hydraulics of Perforated Riser Inlets for Underground-Outlet Terraces, Trans. ASAE, Vol. 31, No. 4, 1988.
Meredith, D.D., Middleton, A.C., and Smith, J.R., Design of detention basins for industrial sites, *ASCE J. Water Res. Plann. and Manage.*, 116, 4, 1990.
Metcalf & Eddy, Inc., *Wastewater Engineering*, McGraw-Hill, New York, 1972.
Metropolitan Government of Nashville and Davidson County, Stormwater Management Manual — Vol. 2 Procedures. Prepared by AMEC, Inc. (formerly The Edge Group) and CH2M Hill, July 1988.
Metropolitan Washington Council of Governments (MWCOG), Controlling Urban Runoff, 1987.
Murthy, K.K. and Giridhar, D.P., Inverted V-notch: practical proportional weir, *ASCE J. Irrig. & Drain. Eng.*, 115, 6, December 1989.
Murthy, K.K. and Pillai, K.G., Some aspects of quadratic weirs, *ASCE J. Hydraulics*, 103, 9, September 1977.
Murthy, K.K. and Pillai, K.G., Design of constant accuracy linear proportional weir, *ASCE J. Hydraulics*, 104, 4, April 1978a.
Murthy, K.K. and Pillai, K.G., Modified proportional V-notch weirs, *ASCE J. Hydraulics*, 104, 5, May 1978b.
Nelson, S.B., Water engineering, in *Standard Handbook for Civil Engineers*, Merritt, F.S., Ed., McGraw-Hill, New York, 1983, chap. 21.
Nix, S.J. and Tsay, T., Alternative strategies for stormwater detention, Water Res. Bull., 24, 3, June 1988.
NRCS, Engineering Field Manual for Conservation Practices, Soil Conservation Service, Engineering Division, Washington, DC, 1984.
Perham, R.E., Floating Debris Control: A Literature Review, U.S. Army CREL, Hanover, NH, 13, 1987.
Poertner, H.G., Practices in the Detention of Urban Stormwater Runoff, American Public Works Assoc., OWWR Proj. C-3380, 1974.
Pratt, E.A., Another proportional flow-sutro weir, *Eng. News-Record*, 72, 9, 1914.
Ramamurthy, A.S., Tim, U.S., and Carballada, L.B., Lateral weirs in trapezoidal channels, *ASCE J. Irrig. and Drain. Eng.*, 112, 2, 1986.
Ramamurthy, R.S., Tim, U.S., and Rao, M.V.J., Characteristics of square-edged and round-nosed broad-crested weirs, *ASCE J. Irrig. Drain. Eng.*, 114, 1, 1988.
Rantz, S.E., Measurement and Computation of Streamflow, 2 Vols., U.S. Geological Survey Water Supply Paper 2175, 1982.
Rao, N.S. and Chandrasekaran, D., Proportional weirs as velocity controlling devices, *ASCE J. Hydraulics*, 103, 6, June 1977.
Reese, A.J. and Maynord, S.T., Design of spillway crests, *ASCE J. Hydraul. Eng.*, 113, 4, 1987.
Replogle, J.A., Flumes and Broad Crested Weirs: Mathematical Modelling and Laboratory Ratings, Flow Measurement of Fluids, North-Holland Publishing Co., Amsterdam, 1978, pp. 321–328.
Rossmiller, R.L., Outlet Structure Hydraulics for Detention Facilities, Proc. Int. Symp. on Urban Hydrology, Hydraul. and Sediment Control, University of Kentucky, Lexington, KY, 1982.
Rouse, H., *Engineering Hydraulics*, John Wiley & Sons, New York, 1949.

Sandvik, A., Proportional weirs for stormwater pond outlets, *Civil Engineering, ASCE*, pp. 54–56, March 1985.
Shelley, P.E. and Kirkpatrick, G.A., Sewer Flow Measurement — a State-of-the-Art Assessment, EPA 600/2-75-027, Washington, DC, 1975.
Smith P.H. and Cook, J.S., Stormwater Management Detention Pond Design within Floodplain Areas, Trans. Res. Record 1017, 1987.
Soil Conservation Service, Hydraulics of Two-way Covered Risers, TR No. 29, 1965.
Sowers, G.B. and Sowers, G.F., *Introductory Soil Mechanics and Foundations*, 3rd ed., MacMillan, New York, 1970.
Spangler, M.G. and Handy, R.L., *Soil Engineering*, 4th ed., Harper & Row, New York, 1982.
Stahre, P. and Urbonas, G., *Stormwater Detention*, Prentice Hall, New York, 1990.
State of Maryland, The Effects of Alternative Stormwater Management Design Policy on Detention Basins, Department of Environ., Sed. and Stormwater Div., 1986.
State of Maryland, Design Procedures for Stormwater Management Detention Structures, Department of Environ., Sed. and Stormwater Div., 1987.
Swamee, P.K. et al., Alternate linear weir design, *ASCE J. Irrig. and Drain. Eng.*, 117, 3, May/June 1991.
Tyrpak, K.A., Individualized Water Detention Saves Money, Public Works, January 1990.
Urban Drainage and Flood Control District (UDFCD), Criteria Manual, Denver, CO, 1999.
U.S. Army Corps of Engineers, Hydraulic Design Criteria, USAE Waterways Experiment Station, Vicksburg, MS, 1988.
U.S. Bureau of Reclamation, Design of Small Canal Structures, 1978.
U.S. Bureau of Reclamation, Water Measurement Manual, 1974, http://www.usbr.gov/wrrl/fmt/wmm/
U.S. Department of the Interior, Bureau of Reclamation, Design of Small Canal Structures, 1978.
Uyumaz, A., Side weir triangular channel, *ASCE J. Irrig. and Drain. Eng.*, 118, 6, 1992.
Uyumaz, A. and Smith, R.H., Design procedure for flow over side weir, *ASCE J. Irrig. and Drain. Eng.*, 117, 1, 1991.
Villemonte, J.R., Submerged weir discharge studies, *Eng. News Record*, December 25, 1947.
Virginia Department of Conservation and Recreation, Virginia Stormwater Management Handbook, 1999.
Washington State Department of Ecology, Stormwater Management Manual for the Puget Sound Basin, 1992.
Wycoff, R.L. and Singh, U.P., Preliminary hydrologic design of small flood detention reservoirs, *Water Resour. Bull.*, 12, 2, 337–349, 1976.

12

Energy Dissipation

12.1 Introduction

The failure or damage of many culverts and detention basin outlet structures can be traced to unchecked erosion. Erosive forces at work in the natural drainage network are often increased by the construction of highways or other urban developments. Interception and concentration of overland flow and constriction of natural waterways inevitably results in an increased erosion potential. To protect the culvert and adjacent areas, it is sometimes necessary to employ an energy dissipating device.

An energy dissipator is any device designed to protect downstream areas from erosion by reducing the velocity of flow to acceptable limits. Unlike bank protection, which reduces the channel's susceptibility to erosion, energy dissipaters reduce the erosive force of the flow. These devices cover a wide range in complexity and cost, and the particular type selected will depend on the assessment of the erosion hazard. This assessment includes determining the ability of the natural channel to withstand erosive forces and the scour potential represented by the superimposed flow conditions. The purpose of this chapter is to aid in selecting and designing an energy dissipator capable of controlling erosive velocities that could damage channels and stream banks.

Generally, energy dissipators should be employed whenever the velocity of flow leaving a stormwater management facility exceeds the erosion velocity of the downstream channel system. Several standard energy dissipator designs are in use, including hydraulic jump, forced hydraulic jump, impact basins, drop structures, stilling wells, and riprap.

12.2 Recommended Energy Dissipators

Erosion problems at culvert or detention basin outlets are common. Determination of the flow conditions, scour potential, and channel erosion resistance should be standard procedure for all designs. Two types of scour can occur in the vicinity of culvert and other outlets: general channel degradation and local scour. Channel degradation may proceed in a fairly uniform manner over a long length, or may be evident in one or more abrupt drops progressing upstream with every runoff event. The abrupt drops, referred to as headcutting, can be detected by location surveys or by periodic maintenance following construction.

Local scour is the result of high-velocity flow at the culvert outlet, but its effect extends only a limited distance downstream. The highest outlet velocities will be produced by long, smooth-barrel culverts and channels on steep slopes. At most sites, these cases will require protection of the outlet. However, protection is also often required for culverts and channels on mild slopes. For these culverts flowing full, the outlet velocity will be the critical velocity with low tailwater and the full barrel velocity for high tailwater.

Standard practice has been to use the same treatment at the culvert entrance and exit, however, this does not always make sense. It is important to recognize that the inlet is designed to improve culvert capacity, improve structural stability, and reduce headloss, while the outlet structure should provide a smooth flow transition back to the natural channel or into an energy dissipator.

For many designs, the following outlet protection and energy dissipators provide sufficient protection at a reasonable cost.

- Riprap apron
- Riprap outlet basins
- Baffled outlets

This chapter will focus primarily on these measures. The reader is referred to the Federal Highway Administration Hydraulic Engineering Circular No. 14 titled, Hydraulic Design of Energy Dissipators for Culverts and Channels (1983), for the design procedures of the other energy dissipators.

For special conditions that involve steep chutes or pipes in tight physical surroundings, with few debris problems or where debris can be caught on trash grates, an internal dissipator may be needed. The stilling well energy dissipator developed by the Corps of Engineers may be applicable (USCOE, 1988). For this dissipator, energy dissipation is accomplished by flow expansion in the stilling well, impact on the well, and by momentum loss by the upward redirection of the flow. Interested designers should consult the Corps of Engineers' publication for further details.

12.3 Design Criteria

The dissipator type selected for a site must be appropriate to the location. In this chapter, the terms internal and external are used to indicate the location of the dissipator in relationship to the culvert. An external dissipator is located outside of the culvert, and an internal dissipator is located within the culvert barrel.

Dissipator Type Selection

Internal dissipators are used where:

- The scour hole at the culvert outlet is unacceptable
- The right of way is limited
- Debris is not a problem
- Moderate velocity reduction is needed

Natural scour holes are used where:

- Undermining of the culvert outlet will not occur, or it is practicable to be checked by a cutoff wall.
- The expected scour hole will not cause costly property damage.
- There is no nuisance effect.

External dissipators are used where:

- The outlet scour hole is not acceptable.
- A moderate amount of debris is present.
- The culvert outlet velocity (V_o) is moderate, Fr < 3.

Stilling basins are used where:

- The outlet scour hole is not acceptable.
- Debris is present.
- The culvert outlet velocity (V_o) is high, Fr > 3.

Ice Buildup

If ice buildup is a factor, it should be mitigated by:

- Sizing the structure to not obstruct the winter low flow
- Using external dissipators

Debris Control

Debris control should be designed using FHWA Hydraulic Engineering Circular No. 9, Debris-Control Structures (FHA, 1964), and should be considered:

- Where cleanout access is limited
- If the dissipator type selected cannot pass debris

Flood Frequency

The flood frequency used should be the flood frequency used for the culvert design or a greater frequency, if justified by:

- Low risk of failure of the crossing or development
- Substantial cost savings
- Limited or no adverse effect on the downstream channel
- Limited or no adverse effect on downstream development

Maximum Culvert Exit Velocity

The culvert exit velocity should be consistent with the maximum velocity in the natural channel or should be mitigated by using:

- Normal channel stabilization
- Special riprap transition below the stilling basins
- Energy dissipation

Tailwater Relationship

The hydraulic conditions downstream should be evaluated to determine a tailwater depth and the maximum velocity for a range of discharges:

- Open channels (see Chapter 10) should be evaluated.
- A lake, pond, or large water body should be evaluated using the high-water elevation that has the same frequency as the design flood for the culvert, if events are known to occur concurrently (statistically dependent). If statistically independent, evaluate the joint probability of flood magnitudes and use a likely combination. Use a reasonable worst case.
- Tidal conditions should be evaluated using the mean high tide, but they should be checked using low tide.

Material Selection

The material selected for the dissipator should be based on a comparison of the total cost over the design life of alternate materials and should not be made using first cost as the only criteria. This comparison should consider replacement cost, maintenance cost, the difficulty of construction, and traffic and other delays.

Culvert Outlet Type

In choosing a dissipator, the selected culvert end treatment has the following implications:

- Culvert ends that are projecting or mitered to the fill slope offer no outlet protection.
- Headwalls provide embankment stability and erosion protection. They provide protection from buoyancy and reduce damage to the culvert.
- Commercial end sections add little cost to the culvert and may require less maintenance, retard embankment erosion, and incur less damage from maintenance.
- Aprons do not reduce outlet velocity, but, if used, they should extend at least one culvert height downstream. They should not protrude above the normal streambed elevation.
- Wingwalls are used where the side slopes of the channel are unstable, where the culvert is skewed to the normal channel flow, to redirect outlet velocity, or to retain fill.

Safety Considerations

Traffic should be protected from external energy dissipators by locating them outside the appropriate "clear zone" distance per AASHTO Roadside Design Guide or shielding them with a traffic barrier.

Weep Holes

If weep holes are used to relieve uplift pressure, they should be designed in a manner similar to underdrain systems.

12.4 Design Procedure

Data Needs

The following data may be needed for the design and evaluation of energy dissipation facilities:

- Culvert and other terminal outlet structures
    - Design capacity
    - Outlet velocity
    - Type of control
    - Length
    - Barrel slope
    - Tailwater
    - Outlet depth
    - Froude number
- Channel
    - Capacity
    - Allowable velocity
    - Bottom slope
    - Debris and bedload
    - Cross-section dimensions
    - Soil plasticity index
    - Normal depth
    - Saturated shear strength
    - Average velocity
- Allowable scourhole dimensions, based on site conditions
    - Depth (h_s)
    - Length (L_s)
    - Width (W_s)
    - Volume (V_s)

TABLE 12-1

Suggested Outlet Protection Type Based on Froude Number and Velocity

Outlet Protection	Fr ≤ 2.5	Fr between 2.5 and 4.5	Fr ≥ 4.5 and V < 50[a] V ≥ 50[a]
Riprap apron	X		
Riprap outlet basin	X		
Baffled outlet	X[b]	X[b]	X[b]

[a] Velocity is based on the energy to be dissipated. Theoretically, the dissipation velocity can be calculated using the equation:

$$V = (2gh)^{0.5} \qquad (12.1)$$

where V = theoretical dissipation velocity (ft/s); g = acceleration due to gravity (32.2 ft/s²); and h = energy head to be dissipated (ft) (can be approximated as the difference between channel invert elevations at the inlet and outlet).

[b] Practical application requires that $1 \leq Fr \leq 9$.

Note: For more outlet protection types within the categories given above, refer to the FHWA Hydraulic Engineering Circular No. 14 for design procedures.

Procedure Outline

Following are the steps in the procedure that is generally applicable for outlet protection facilities for pipe and culvert type outlets:

Step 1: Compute local scourhole dimensions with the procedure in Section 12.5. A nonerodible layer (e.g., bedrock) may limit scour hole depth but only slightly affect scour hole width and length.

Step 2: Compare the local scour hole dimensions from Step 1 to the allowable scour hole dimensions from site data. If the allowable dimensions are exceeded, outlet protection is required.

Step 3: If outlet protection is required, choose an appropriate type. Suggested outlet protection facilities and applicable flow conditions (based on Froude number and dissipation velocity) are presented in Table 12-1. When outlet protection facilities are selected, appropriate design flow conditions and site-specific factors affecting erosion and scour potential, construction cost, and long-term durability should be considered. Consult the Federal Highway Administration Hydraulic Engineering Circular No. 14 (FHA, 1983) for other energy dissipators if the ones given in Table 12-1 are not applicable.

Step 4: Design downstream transition section if flow velocity leaving the end sill or end of riprap apron is higher than allowable.

Step 5: If outlet protection is not provided, energy dissipation will occur through formation of a local scour hole. A cutoff wall will be needed at the discharge outlet to prevent structural undermining. The wall depth should be slightly greater than the computed scour hole depth, h_s. The scour hole should then be stabilized. If the scour hole is of such size that it will present maintenance, safety, or aesthetic problems, other outlet protection will be needed.

Step 6: Evaluate the downstream channel stability and provide appropriate erosion protection if channel degradation is expected to occur.

Following is a discussion of applicable conditions for each outlet protection measure:

Riprap aprons may be used when the outlet Froude number (Fr) is less than or equal to 2.5. In general, riprap aprons prove economical for transitions from culverts to overland sheet flow at terminal outlets but may also be used for transitions from culvert sections to stable channel sections. Stability of the surface at the termination of the apron should be considered.

Riprap outlet basins may also be used when the outlet Fr is less than or equal to 2.5. They are generally used for transitions from culverts to stable channels. Because riprap outlet basins function by creating a hydraulic jump to dissipate energy, performance is impacted by tailwater conditions.

Baffled outlets have been used with outlet velocities up to 50 ft/sec. Practical application typically requires an outlet Froude number between 1 and 9. Baffled outlets may be used at terminal outlet and channel outlet transitions. They function by dissipating energy through impact and turbulence and are not significantly affected by tailwater conditions.

12.5 Local Scourhole Estimation

Estimates of erosion at culvert outlets must consider factors such as discharge, culvert diameter, soil type, duration of flow, and tailwater depth. In addition, the magnitude of the total erosion can consist of local scour and channel degradation. Blaisdell has also developed a design procedure for plunge pools from culverts that should be consulted when culverts discharge at some height above a streambed (Blaisdell and Anderson, 1988a,b, 1991).

Empirical equations for estimating the maximum dimensions of a local scourhole are presented in Table 12-2. These equations are based on test data obtained as part of a study conducted at Colorado State University (USDOT, FHWA, HEC-14, 1983). A form for recording the following local scourhole computations is presented in Figure 12-1. Following are the calculation steps:

1. Prepare input data, including:

   Q = design discharge (cfs)

   For circular culvert, D = diameter (in)

   For other shapes, use the equivalent depth $Y_e = (A/2)^{0.5}$ (A = cross-sectional area)

   t = time of scour (min)

   V_o = outlet mean velocity (ft/s)

   $$\tau_c = 0.0001 \, (S_v + 180) \tan (30 + 1.73 \, PI) \tag{12.2}$$

   where τ_c = critical tractive shear stress (lb/in²); S_v = saturated shear strength (lb/in²); and PI = plasticity index from Atterburg limits.

   The time of scour should be based on a knowledge of peak flow duration. As a guideline, a time of 30 min is recommended. Tests indicate that approximately 2/3 to 3/4 of the maximum scour occurs in the first 30 min of the flow duration.

2. Based on the channel material, select the proper scour equations and coefficients from Table 12-2.

TABLE 12-2

Experimental Coefficients For Culvert Outlet Scour

Material	Nominal Grain Size d_{50} (mm)	Scour Equation	Depth (h_s)				Width (W_s)				Length (L_s)				Volume (V_s)			
			α	β	θ	α_s	α	β	θ	α_e	α	β	θ	α_e	α	β	θ	α_e
Uniform sand	0.20	V-1 or V-2	2.72	0.375	0.10	2.79	11.73	0.92	0.15	6.44	16.82	0.71	0.125	11.75	203.36	2.0	0.375	80.71
Uniform sand	2.0	V-1 or V-2	1.86	0.45	0.09	1.76	8.44	0.57	0.06	6.94	18.28	0.51	0.17	16.10	101.48	1.41	0.34	79.62
Graded sand	2.0	V-1 or V-2	1.22	0.85	0.07	0.75	7.25	0.76	0.06	4.78	12.77	0.41	0.04	12.62	36.17	2.09	0.19	12.94
Uniform gravel	8.0	V-1 or V-2	1.78	0.45	0.04	1.68	9.13	0.62	0.08	7.08	14.36	0.95	0.12	7.61	65.91	1.86	0.19	12.15
Graded gravel	8.0	V-1 or V-2	1.49	0.50	0.03	1.33	8.76	0.89	0.10	4.97	13.09	0.62	0.07	10.15	42.31	2.28	0.17	32.82
Cohesive sandy clay 60% sand, PI 15	0.15	V-1 or V-2	1.86	0.57	0.10	1.53	8.63	0.35	0.07	9.14	15.30	0.43	0.09	14.78	79.73	1.42	0.23	61.84
Clay, PI 5–16	Various	V-3 or V-4	0.86	0.18	0.10	1.37	3.55	0.17	0.07	5.63	2.82	0.33	0.09	4.48	0.62	0.93	0.23	2.48

Equations

V-1. *For Circular Culverts.* Cohesionless material or the 0.15 mm cohesive sand clay

$$\left[\frac{h_s}{D}, \frac{W_s}{D}, \frac{L_s}{D}, \text{ or } \frac{V_s}{D^3}\right] = \alpha\left(\frac{Q}{\sqrt{g}D^{5/2}}\right)^\beta \left(\frac{t}{t_o}\right)^\theta$$

where t_o = 316 min.

V-2. *For Other Culverts Shapes.* Same material as above.

$$\left[\frac{h_s}{y_e}, \frac{W_s}{y_e}, \frac{L_s}{y_e}, \text{ or } \frac{V_s}{y_e^3}\right] = \alpha\left(\frac{Q}{\sqrt{g}D^{5/2}}\right)^\beta \left(\frac{t}{t_o}\right)^\theta$$

where t_o = 316 min.

V-3. *For Circular Culverts.* Cohesive sandy clay with PI = 5–16

$$\left[\frac{h_s}{D}, \frac{W_s}{D}, \frac{L_s}{D}, \text{ or } \frac{V_s}{D^3}\right] = \alpha\left(\frac{\rho V^2}{\tau_c}\right)^\beta \left(\frac{t}{t_o}\right)^\theta$$

where t_o = 316 min.

V-4. *For Other Culvert Shapes.* Cohesive sandy clay with PI = 5–16

$$\left[\frac{h_s}{y_e}, \frac{W_s}{y_e}, \frac{L_s}{y_e}, \text{ or } \frac{V_s}{y_e^3}\right] = \alpha_e\left(\frac{\rho V^2}{\tau_c}\right)^\beta \left(\frac{t}{t_o}\right)^\theta$$

where t_o = 316 min.

Source: From Federal Highway Administration, Hydraulic Design of Energy Dissipators for Culverts and Channels, Hydraulic Engineering Circular No. 14, 1983.

Energy Dissipation

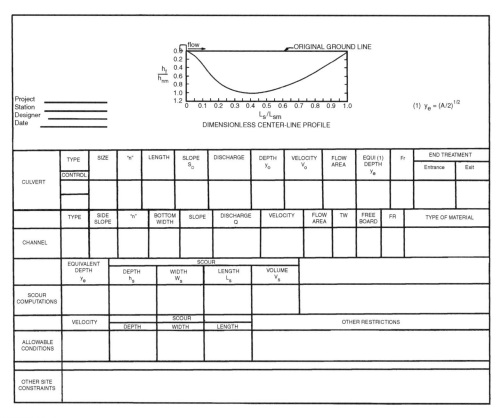

FIGURE 12-1
Culvert, channel, scour and other site data form. (From Federal Highway Administration, Hydraulic Design of Energy Dissipators for Culverts and Channels, Hydraulic Engineering Circular No. 14, 1983.)

3. Using the results from the equations selected in Step 2, compute the following scourhole dimensions:

Depth (h_s)
Length (L_s)
Width (W_s)
Volume (V_s)

Observations indicate that a nonerodible layer at a depth less than h_s below the pipe outlet affects only scourhole depth. The width and length may be computed using the equations in Table 12-2.

12.6 Riprap Aprons

A flat riprap apron can be used to prevent erosion at the transition from a circular or box culvert outlet to a natural channel. Protection is provided primarily by having sufficient length and flare to dissipate energy by expanding the flow. Riprap aprons are appropriate when the culvert outlet Fr is less than or equal to 2.5.

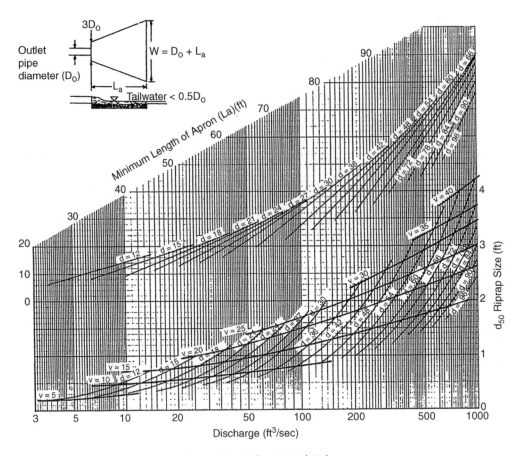

FIGURE 12-2
Riprap apron under minimum tailwater conditions. (From U.S. Department of Agriculture, Soil Conservation Service, Standards and Specifications for Soil Erosion and Sedimentation in Developing Areas, Washington, DC, 1975.)

Design Procedure

The design procedure presented in this section is taken from USDA, SCS (1975). Two sets of curves, one for minimum and one for maximum tailwater conditions, are used to determine the apron size and the median riprap diameter, d_{50}. If tailwater conditions are unknown, or if both minimum and maximum conditions may occur, the apron should be designed to meet criteria for both. Although the design curves are based on round pipes flowing full, they can be used for partially full pipes and box culverts. The design procedure consists of the following steps:

Step 1: If possible, determine tailwater conditions for the channel. If tailwater is less than one-half the discharge flow depth (pipe diameter if flowing full), minimum tailwater conditions exist, and the curves in Figure 12-2 apply. Otherwise, maximum tailwater conditions exist, and the curves in Figure 12-3 should be used.

Step 2: Determine the correct apron length and median riprap diameter, d_{50}, using the appropriate curves from Figures 12-2 and 12-3. If tailwater conditions are uncertain, find the values for minimum and maximum conditions, and size the apron as shown in Figure 12-4.

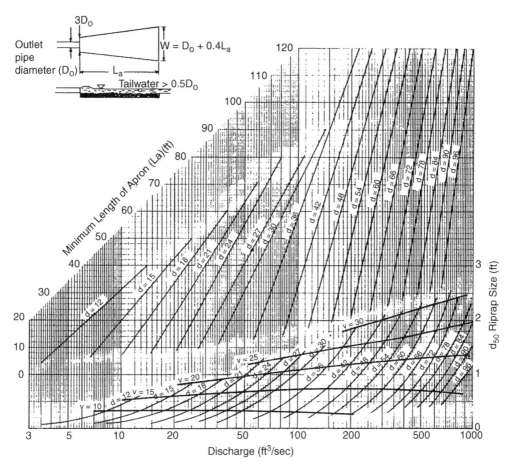

FIGURE 12-3
Riprap apron under maximum tailwater conditions. (From U.S. Department of Agriculture, Soil Conservation Service, Standards and Specifications for Soil Erosion and Sedimentation in Developing Areas, Washington, DC, 1975.)

For pipes flowing full: Use the depth of flow, d, which equals the pipe diameter, in feet, and design discharge, in cfs, to obtain the apron length, L_a, and median riprap diameter, d_{50}, from the appropriate curves.

For pipes flowing partially full: Use the depth of flow, d, in feet, and velocity, V, in ft/s. On the lower portion of the appropriate figure, find the intersection of the d and V curves, then find the riprap median diameter, d_{50}, from the scale on the right. From the lower d and V intersection point, move vertically to the upper curves until intersecting the curve for the correct flow depth, d. Find the minimum apron length, L_a, from the scale on the left.

For box culverts: Use the depth of flow, d, in feet, and velocity, V, in ft/s. On the lower portion of the appropriate figure, find the intersection of the d and V curves, then find the riprap median diameter, d_{50}, from the scale on the right. From the lower d and V intersection point, move vertically to the upper curve until intersecting the curve equal to the flow depth, d. Find the minimum apron length, L_a, using the left scale.

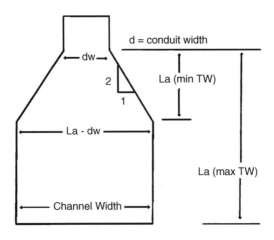

FIGURE 12-4
Riprap apron schematic for uncertain tailwater conditions. (From U.S. Department of Agriculture, Soil Conservation Service, Standards and Specifications for Soil Erosion and Sedimentation in Developing Areas, Washington, DC, 1975.)

Step 3: If tailwater conditions are uncertain, the median riprap diameter should be the larger of the values for minimum and maximum conditions. The dimensions of the apron will be as shown in Figure 12-4. This will provide protection under either of the tailwater conditions.

Design Considerations

The following items should be considered during riprap apron design:

1. The maximum stone diameter should be 1.5 times the median riprap diameter.

$$d_{max} = 1.5 \times d_{50}$$

   d_{50} = the median stone size in a well-graded riprap apron.

2. The riprap thickness should be 1.5 times the maximum stone diameter or 6 in, whichever is greater.

   Apron thickness = $1.5 \times d_{max}$

   (Apron thickness may be reduced to $1.5 \times d_{50}$ when an appropriate filter fabric is used under the apron.)

3. The apron width at the discharge outlet should be at least equal to the pipe diameter or culvert width, d_w. Riprap should extend up both sides of the apron and around the end of the pipe or culvert at the discharge outlet at a maximum slope of 2:1 and a height not less than the pipe diameter or culvert height, and it should taper to the flat surface at the end of the apron.

4. If there is a well-defined channel, the apron length should be extended as necessary so that the downstream apron width is equal to the channel width. The sidewalls of the channel should not be steeper than 2:1.

5. If the ground slope downstream of the apron is steep, channel erosion may occur. The apron should be extended as necessary until the slope is gentle enough to prevent further erosion.

Energy Dissipation

6. The potential for vandalism should be considered if the rock is easy to carry. If vandalism is possible, the rock size may be increased or the rocks held in place using concrete or grout.

Example Problems

Example 1 — Riprap Apron Design for Minimum Tailwater Conditions

A flow of 280 cfs discharges from a 66-in pipe with a tailwater of 2 ft above the pipe invert. Find the required design dimensions for a riprap apron.

1. Minimum tailwater conditions = $0.5\, d_o$ = 0.5166 in = 0.5 (5.5 ft). Therefore, $0.5\, d_o$ = 2.75 ft.
2. Because TW = 2 ft, use Figure 12-2 for minimum tailwater conditions.
3. From Figure 12-2, the apron length, L_a, and median stone size, d_{50}, are 38 ft and 1.2 ft, respectively.
4. The downstream apron width equals the apron length plus the pipe diameter:

$$W = d + L_a = 5.5 + 38 = 43.5 \text{ ft}$$

5. Maximum riprap diameter is 1.5 times the median stone size:

$$1.5\,(d_{50}) = 1.5\,(1.2) = 1.8 \text{ ft}$$

6. Riprap depth = $1.5\,(d_{max}) = 1.5\,(1.8) = 2.7$ ft.

Example 2 — Riprap Apron Design for Maximum Tailwater Conditions

A concrete box culvert 5.5 ft high and 10 ft wide conveys a flow of 600 cfs at a depth of 5.0 ft. Tailwater depth is 5.0 ft above the culvert outlet invert. Find the design dimensions for a riprap apron.

1. Compute $0.5\, d_o = 0.5\,(5.0) = 2.5$ ft.
2. Because TW = 5.0 ft is greater than 2.5 ft, use Figure 12-3 for maximum tailwater conditions.

$$V = Q/A = [600/(5)(10)] = 12 \text{ ft/s}$$

3. In Figure 12-3, at the intersection of the curve, d_o = 60 in, V = 12 ft/s, and d_{50} = 0.4 ft. Reading up to the intersection with d = 60 in, find L_a = 40 ft.
4. Apron width downstream = $d_w + 0.4\, L_a = 10 + 0.4\,(40) = 26$ ft.
5. Maximum stone diameter = $1.5\, d_{50} = 1.5\,(0.4) = 0.6$ ft.
6. Riprap depth = $1.5\, d_{max} = 1.5\,(0.6) = 0.9$ ft.

12.7 Riprap Basin Design

Outlet velocities from culverts and other drainage structures are one of the primary indicators of erosion potential. Different culvert and outlet designs should be investigated

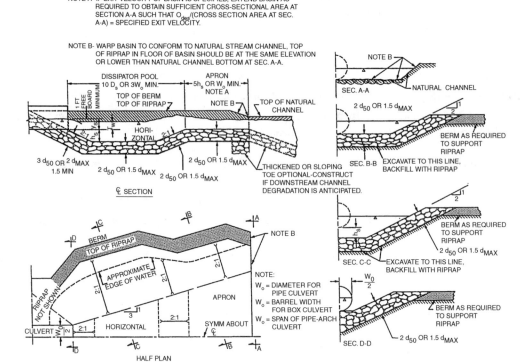

FIGURE 12-5
Details of riprap outlet basin. (From Federal Highway Administration, Hydraulic Design of Energy Dissipators for Culverts and Channels, Hydraulic Engineering Circular No. 14, 1983.)

to limit the outlet velocity, but the degree of velocity reduction is, in most cases, limited. The continuity equation, $Q = VA$, can be utilized to compute channel exit velocities and culvert velocities. One method to reduce the exit velocities from outlets is to install a riprap basin. A riprap outlet basin is a preshaped scour hole lined with riprap that functions as an energy dissipator by forming a hydraulic jump.

General details of the basin recommended in this chapter are shown in Figure 12-5. Principal features of the basin are the following:

1. The basin is preshaped and lined with riprap of median size (d_{50}).
2. The floor of the riprap basin is constructed at an elevation of h_s below the culvert invert. The dimension h_s is the approximate depth of scour that would occur in a thick pad of riprap of size d_{50} if subjected to design discharge. The ratio of h_s to d_{50} of the material should be between 2 and 4.
3. The length of the energy dissipating pool is 10 x h_s or 3 x W_o, whichever is larger. The basin overall length is 15 x h_s or 4 x W_o, whichever is larger.

Design Procedure

Following are the steps in the procedure for the design of riprap basins:

Step 1: Estimate the flow properties at the brink (outlet) of the culvert. Establish the outlet invert elevation such that $TW/y_o \leq 0.75$ for the design discharge.

Energy Dissipation 701

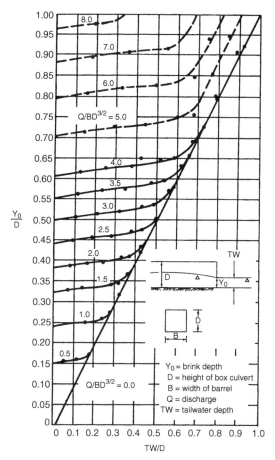

FIGURE 12-6
Dimensionless rating curves for outlets on horizontal and mild slopes — rectangular curves. (From Federal Highway Administration, Hydraulic Design of Energy Dissipators for Culverts and Channels, Hydraulic Engineering Circular No. 14, 1983.)

Step 2: For subcritical flow conditions (culvert on mild or horizontal slope), utilize Figures 12-6 or 12-7 to obtain y_o/D, then obtain V_o by dividing Q by the wetted area associated with y_o. D is the height of a box culvert. If the culvert is on a steep slope, V_o will be the normal velocity obtained by using the Manning equation for appropriate slope, section, and discharge.

Step 3: For channel protection, compute the Froude number for brink conditions with $y_e = (A/2)^{1.5}$. Select d_{50}/y_e appropriate for locally available riprap (usually the most satisfactory results will be obtained if $0.25 < d_{50}/y_e < 0.45$). Obtain h_s/y_e from Figure 12-8, and check to see that $2 < h_s/d_{50} < 4$. Recycle computations if h_s/d_{50} falls out of this range.

Step 4: Size basin as shown in Figure 12-5.

Step 5: Where allowable, dissipator exit velocity is specified:

Determine the average normal flow depth in the natural channel for the design discharge.

Extend the length of the energy basin (if necessary) so that the width of the energy basin at section A-A (Figure 12-5) times the average normal flow depth

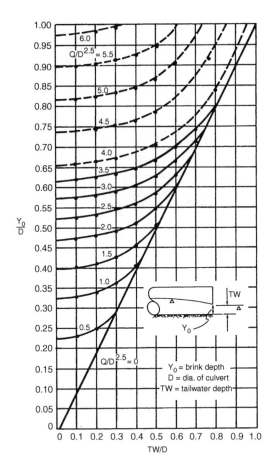

FIGURE 12-7
Dimensionless rating curves for outlets on horizontal and mild slopes — circular curves. (From Federal Highway Administration, Hydraulic Design of Energy Dissipators for Culverts and Channels, Hydraulic Engineering Circular No. 14, 1983.)

in the natural channel is approximately equal to the design discharge divided by the specified exit velocity.

Step 6: In the exit region of the basin, the walls and apron of the basin should be warped (or transitioned) so that the cross section of the basin at the exit conforms to the cross section of the natural channel. Abrupt transition of surfaces should be avoided to minimize separation zones and eddies.

Step 7: If high tailwater is a possibility and erosion protection is necessary for the downstream channel, the following design procedure is suggested:

Design a conventional basin for low tailwater conditions in accordance with the instructions above.

Estimate centerline velocity at a series of downstream cross sections using the information shown in Figure 12-9.

Shape downstream channel and size riprap using Figure 12-12 and the stream velocities obtained above.

For materials, construction techniques, and design details, consult the Federal Highway publication HEC-11 titled, Use of Riprap for Bank Protection (1967).

Energy Dissipation

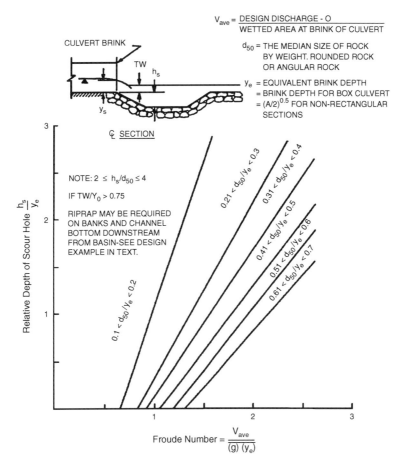

FIGURE 12-8
Relative depth of scour hole versus Froude number at brink of culvert with relative size of riprap as a third variable. (From Federal Highway Administration, Hydraulic Design of Energy Dissipators for Culverts and Channels, Hydraulic Engineering Circular No. 14, 1983.)

Design Considerations

Riprap basin design should include consideration of the following:

1. The dimensions of a scourhole in a basin constructed with angular rock can be approximately the same as the dimensions of a scourhole in a basin constructed of rounded material when rock size and other variables are similar.

2. When the ratio of tailwater depth to brink depth, TW/y_o, is less than 0.75 and the ratio of scour depth to size of riprap, h_s/d_{50}, is greater than 2.0, the scourhole should function efficiently as an energy dissipator. The concentrated flow at the culvert brink plunges into the hole, a jump forms against the downstream extremity of the scourhole, and flow is generally well dispersed as it leaves the basin.

3. The mound of material formed on the bed downstream of the scourhole contributes to the dissipation of energy and reduces the size of the scourhole; that is, if the mound from a stable scoured basin is removed and the basin is again subjected to design flow, the scourhole will enlarge.

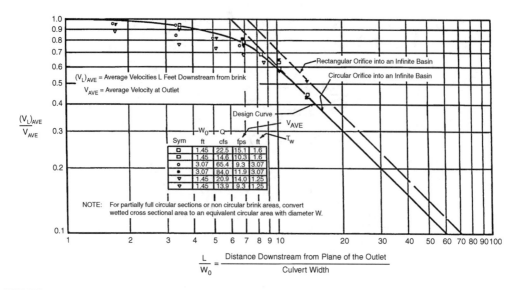

FIGURE 12-9
Distribution of centerline velocity for flow from submerged outlets to be used for predicting channel velocities downstream from culvert outlet, where high tailwater prevails. (From Federal Highway Administration, Hydraulic Design of Energy Dissipators for Culverts and Channels, Hydraulic Engineering Circular No. 14, 1983.)

4. For high tailwater basins (TW/y_o greater than 0.75), the high velocity core of water emerging from the culvert retains its jet-like character as it passes through the basin and diffuses, similarly to a concentrated jet diffusing in a large body of water. As a result, the scour hole is much shallower and generally longer. Consequently, riprap may be required for the channel downstream of the rock-lined basin.

5. It should be recognized that there is a potential for limited degradation to the floor of the dissipator pool for rare-event discharges. With the protection afforded by the $2(d_{50})$ thickness of riprap, the heavy layer of riprap adjacent to the roadway prism, and the apron riprap in the downstream portion of the basin, such damage should be superficial.

6. See Standards in HEC No. 11 (FHA, 1967) for details on riprap materials and use of filter fabric.

7. Stability of the surface at the outlet of a basin should be considered using the methods for open-channel flow as outlined in Chapter 10 of this book.

Example Problems

Following are some example problems to illustrate the design procedures outlined above.

Example 1

Given:

Box culvert 8 ft by 6 ft
Design discharge $Q = 800$ cfs
Supercritical flow in culvert

Energy Dissipation

Normal flow depth = brink depth
$y_o = 4$ ft
Tailwater depth $TW = 2.8$ ft

Find:

Riprap basin dimensions for these conditions

Solution: Definitions of terms in Steps 1–5 can be found in Figures 12-5 and 12-8.

1. $y_o = y_e$ for rectangular section; therefore, with y_o given as 4 ft, $y_e = 4$ ft
2. $V_o = Q/A = 800/(4 \times 8) = 25$ ft/s
3. Froude Number = Fr = $V/(g \times y_e)^{.5}$ ($g = 32.3$ ft/s²)
   Fr = $25/(32.2 \times 4)^{.5} = 2.20 < 3.0$ OK
4. $TW/y_e = 2.8/4.0 = 0.7$
   Therefore, $TW/y_e < 0.75$ OK
5. Try $d_{50}/y_e = 0.45$; $d_{50} = 0.45 \times 4 = 1.80$ ft
   From Figure 12-8, $h_s/y_e = 1.6$; $h_s = 4 \times 1.6 = 6.4$ ft; $h_s/d_{50} = 6.4/1.8 = 3.6$ ft; $2 < h_s/d_{50} < 4$ OK
6. $L_s = 10 \times h_s$ (L_s = length of energy dissipator pool)
   $L_s = 10 \times 6.4 = 64$ ft
   L_s min = $3 \times W_o = 3 \times 8 = 24$ ft
   Therefore, use $L_s = 64$ ft
   $L_B = 15 \times h_s$ (L_B = overall length of riprap basin)
   $L_B = 15 \times 6.4 = 96$ ft
   L_B min = $4 \times W_o = 4 \times 8 = 32$ ft
   Therefore, use $L_B = 96$ ft
7. Thickness of riprap:
   On the approach = $3 \times d_{50} = 3 \times 1.8 = 5.4$ ft
   Remainder = $2 \times d_{50} = 2 \times 1.8 = 3.6$ ft
   Other basin dimensions are designed according to details shown in Figure 12-5.

Example 2

Given:

Same design data as Example 1, except $TW = 4.2$ ft.
The downstream channel can tolerate only 7 ft/s discharge.

Find:

Riprap basin dimensions for these conditions

Solutions: Note — High tailwater depth, $TW/y_o = 4.2/4 = 1.05 > 0.75$.

1. Use the riprap basin designed in Example 1, with $d_{50} = 1.8$ ft; $h_s = 6.4$ ft; $L_s = 64$ ft; and $L_B = 96$ ft.
2. Design riprap for downstream channel. Utilize Figure 12-9 for estimating average velocity along the channel. Compute equivalent circular diameter D_e for brink area from:

$$A = 3.14 D_e^2/4 = y_o \times W_o = 4 \times 8 = 32 \text{ ft}^2$$

$$D_e = ((32 \times 4)/3.14)^{.5} = 6.4 \text{ ft}$$

$$V_o = 25 \text{ ft/s (from Example 1)}$$

3. Set up the following table:

L/D$_e$ (Assume) (D$_e$ = W$_o$)	L (Compute) ft	V$_L$/V$_o$ (Figure 12-9)	V$_1$ ft/s	Rock Size d$_{50}$ (ft) (Figure 12-12)
10	64	0.59	14.7	1.4
15[a]	96	0.37	9.0	0.6
20	128	0.30	7.5	0.4
21	135	0.28	7.0	0.4

[a] L/W$_o$ is on a logarithmic scale so interpolations must be made logarithmically.

4. Riprap should be at least the size shown but can be larger. As a practical consideration, the channel can be lined with the same size rock used for the basin. Protection must extend at least 135 ft downstream from the culvert brink. Channel should be shaped and riprap installed in accordance with details in HEC No. 11 (1967).

Example 3

Given:

6 ft diameter CMC
Design discharge $Q = 135$ cfs
Slope channel $S_o = 0.004$ ft/ft
Manning's $n = 0.024$
Flow is subcritical
Tailwater depth $TW = 2.0$ ft
Normal depth in pipe for $Q = 135$ cfs is 4.5 ft
Normal velocity is 5.9 ft/s

Find:

Riprap basin dimensions for these conditions

Solution: Following are the steps in the solution:

1. Determine y_o and V_o.

From Figure 12-7, $y_o/D = 0.45$.

$Q/D^{3/2} = 135/6^{3/2} = 1.53$.

$TW/D = 2.0/6 = 0.33$.

$y_o = 0.45 \times 6 = 2.7$ ft.

$TW/y_o = 2.0/2.7 = 0.74$.

$TW/y_o < 0.75$ OK

Brink area (A) for $y_o/D = 0.45$

  For $y_o/D = d/D = 0.45$ and $A/D^2 = 0.3428$

  Therefore, $A = 0.3428 \times 6^2 = 12.3$ ft²

  $V_o = Q/A = 135/12.3 = 11.0$ ft/s

2. For Froude number calculations at brink conditions, $y_e = (A/2)^{1/2} = (12.3/2)^{1/2} = 2.48$ ft.
3. Froude number = Fr = $V_o/(32.2 \times y_e)^{1/2} = 11/(32.2 \times 2.48)^{1/2} = 1.23 < 3$ OK
4. For most satisfactory results - $0.25 < d_{50}/y_e < 0.45$

    Try $d_{50}/y_e = 0.25$; $d_{50} = 0.25 \times 2.48 = 0.62$ ft

    From Figure 12-8, $h_s/y_e = 0.75$; therefore, $h_s = 0.75 \times 2.48 = 1.86$ ft

    Check: $h_s/d_{50} = 1.86/0.62 = 3$; $2 < h_s/d_{50} < 4$ OK
5. $L_s = 10 \times h_s = 10 \times 1.86 = 18.6$ ft or $L_s = 3 \times W_o = 3 \times 6 = 18$ ft; therefore, use $L_s = 18.6$ ft.

    $L_B = 15 \times h_s = 15 \times 1.86 = 27.9$ ft, or $L_B = 4 \times W_o = 4 \times 6 = 24$ ft; therefore, use $L_B = 27.9$ ft, $d_{50} = 0.62$ ft, or $d_{50} = 8$ in.

Other basin dimensions should be designed in accordance with details shown in Figure 12-5.

When using the design procedure outlined in this chapter, it is recognized that there is some chance of limited degradation of the floor of the dissipator pool for rare-event discharges. With the protection afforded by the 3 x d_{50} thickness of riprap on the approach and the 2 x d_{50} thickness of riprap on the basin floor and the apron in the downstream portion of the basin, the damage should be superficial.

Figure 12-10 is provided as a convenient form to organize and present the results of riprap basin designs.

12.8 Baffled Outlets

The baffled outlet is a boxlike structure with a vertical hanging baffle and an end sill, as shown in Figure 12-11. Energy is dissipated primarily through the impact of the water striking the baffle and, to a lesser extent, through the resulting turbulence. This type of outlet protection has been used with outlet velocities up to 50 ft/sec and with Froude numbers from 1 to 9. Tailwater depth is not required for adequate energy dissipation, but a tailwater will help smooth the outlet flow. Denver Urban Drainage District has developed a design modification that allows the basin to drain during low flows (UDFCD, 1991).

		TW	y_e	(1) TW/y_e	d_{50}/y_e	d_{50}	h_s/y_e	h_s	(2) h_s/d_{50}
LOW TW TW/$y_e \leq 0.75$									
	V ALLOWABLE	L/D_e (3)	L		V_{ave}/V_L	V_L			
HIGH TW TW/$y_e > 0.75$									

Length of Pool = Larger of $10h_s$ or $3W_o$ = _____ ft.
Length of Apron = $5h_s$ or W_o = _____ ft.
Thickness of Approach = $3d_{50}$ or $2d_{max}$ = _____ ft.
Thickness of Remainder of Basin = $2d_{50}$ or $1.5d_{max}$ = _____ ft.

(1) TW/$y_e \leq 0.75$ for Low TW Design
(2) $2 < h_s/d_{50} < 4$
(3) $D_e = [4A/\pi]^{1/2}$

FIGURE 12-10
Riprap basins. (From Federal Highway Administration, Hydraulic Design of Energy Dissipators for Culverts and Channels, Hydraulic Engineering Circular No. 14, 1983.)

Design Procedure

The following design procedure is based on physical modeling studies summarized from the U.S. Department of Interior (1978). The dimensions of a baffled outlet as shown in Figure 12-11 should be calculated as follows:

1. Determine input parameters, including:

    h = energy head to be dissipated (ft) (can be approximated as the difference between channel invert elevations at the inlet and outlet)

    Q = design discharge (cfs)

    V = theoretical velocity (ft/s) = $2gh$

    $A = Q/V$ = flow area (ft²)

    $d = A^{0.5}$ = flow depth entering the basin (ft) (assumes square jet)

    Fr = $V/(gd)^{0.5}$ = Froude number, dimensionless

2. Calculate the minimum basin width, W, in ft, using the following equation:

$$W/d = 2.88 Fr^{0.566} \tag{12.3}$$

    where W = minimum basin width (ft); d = depth of incoming flow (ft); and Fr = $V/(gd)^{0.5}$ = Froude number, dimensionless.

    The limits of the W/d ratio are from 3 to 10, which corresponds to Froude numbers 1 and 9. If the basin is much wider than W, flow will pass under the baffle, and energy dissipation will not be effective.

3. Calculate the other basin dimensions as shown in Figure 12-11, as a function of W. Standard construction drawings for selected widths are available from the U.S. Department of the Interior (1978).

Energy Dissipation

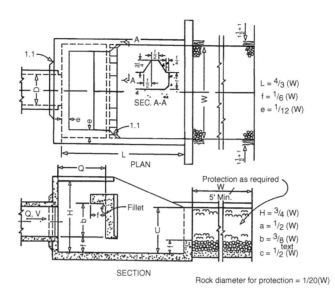

FIGURE 12-11
Schematic of baffled outlet. (From U.S. Department of the Interior, 1978.)

4. Calculate required protection for the transition from the baffled outlet to the natural channel based on the outlet width. A riprap apron should be added of width W, length L (or a 5-ft minimum), and depth $f = (W/6)$. The side slopes should be 1.5:1, and median rock diameter should be at least $W/20$.

5. Calculate the baffled outlet invert elevation based on expected tailwater. The maximum distance between expected tailwater elevation and the invert should be $b + f$, or some flow will go over the baffle with no energy dissipation. If the tailwater is known and fairly controlled, the baffled outlet invert should be a distance, $b/2 + f$, below the calculated tailwater elevation. If tailwater is uncontrolled, the baffled outlet invert should be a distance, f, below the downstream channel invert.

6. Calculate the outlet pipe diameter entering the basin assuming a velocity of 12 ft/s flowing full.

7. If the entrance pipe slopes downward, the outlet pipe should be turned horizontal for at least 3 ft before entering the baffled outlet.

8. If the upstream and downstream ends of the pipe will be submerged, provide an air vent of diameter approximately 1/6 the pipe diameter near the upstream end to prevent pressure fluctuations and possible surging flow conditions.

Example Problem

A cross-drainage pipe structure has a design flow rate of 150 cfs, a head, h, of 30 ft, and a tailwater depth, TW, of 3 ft above ground surface. Find the baffled outlet basin dimensions and inlet pipe requirements.

1. Compute the theoretical velocity from:

$$V = (2gh)^{0.5} = [2(32.2 \text{ ft/s}^2)(30 \text{ ft})]^{0.5} = 43.95 \text{ ft/s}$$

This is less than 50 ft/s, so a baffled outlet is suitable.

2. Determine the flow area using the theoretical velocity as follows:

$$A = Q/V = 150 \text{ cfs}/43.95 \text{ ft/s} = 3.41 \text{ ft}^2$$

3. Compute the flow depth using the area from Step 2:

$$d = (A)^{0.5} = (3.41 \text{ ft}^2)^{0.5} = 1.85 \text{ ft}$$

4. Compute the Froude number using the results from Steps 2 and 3:

$$Fr = V/(gd)^{0.5} = 43.95 \text{ ft/s}/[(32.2 \text{ ft/s}^2)(1.85 \text{ ft})]^{0.5} = 5.7$$

5. Determine the basin width using Equation (12.3) with the Froude number from Step 4.

$$W = 2.88 \text{ } dFr^{0.566} = 2.88 \text{ } (1.85) \text{ } (5.7)^{0.566}$$

$$W = 14.27 \text{ ft (minimum)}$$

Use 14 ft, 4 in as design width.

6. Compute the remaining basin dimensions (as shown in Figure 12-11):
   $L = 4/3 \text{ } (W) = 19.1 \text{ ft}$
   Use $L = 19 \text{ ft, 2 in}$
   $f = 1/6 \text{ } (W) = 2.39 \text{ ft}$
   Use $f = 2 \text{ ft, 5 in}$
   $e = 1/12 \text{ } (W) = 1.19 \text{ ft}$
   Use $e = 1 \text{ ft, 3 in}$
   $H = 3/4 \text{ } (W) = 10.75 \text{ ft}$
   Use $H = 10 \text{ ft, 9 in}$
   $a = 1/2 \text{ } (W) = 7.17 \text{ ft}$
   Use $a = 7 \text{ ft, 2 in.}$
   $b = 3/8 \text{ } (W) = 5.38 \text{ ft}$
   Use $b = 5 \text{ ft, 5 in}$
   $c = 1/2 \text{ } (W) = 7.17 \text{ ft}$
   Use $c = 7 \text{ ft, 2 in.}$

   Baffle opening dimensions would be calculated from f (Figure 12-11).

7. Basin invert should be at $b/2 + f$ below tailwater, or (5 ft, 5 in)/2 + 2 ft, 5 in. = 5.125 ft. Use 5 ft 2 in; therefore, invert should be 2 ft, 2 in. below ground surface.

8. The riprap transition from the baffled outlet to the natural channel should be 14 ft, 4 in long by 14 ft, 4 in wide by 2 ft, 5 in deep ($W \times W \times f$). Median rock diameter should be of diameter $W/20$, or about 9 in.

9. Inlet pipe diameter should be sized for an inlet velocity of about 12 ft/s. $(3.14d)^2/4 = Q/V$; $d = [(4Q)/3.14V)]^{0.5} = [(4(150 \text{ cfs})/3.14(12 \text{ ft/s})]^{0.5} = 3.99 \text{ ft}$. Use 48-in pipe. If a vent is required, it should be about 1/6 of the pipe diameter or 8 in.

Energy Dissipation

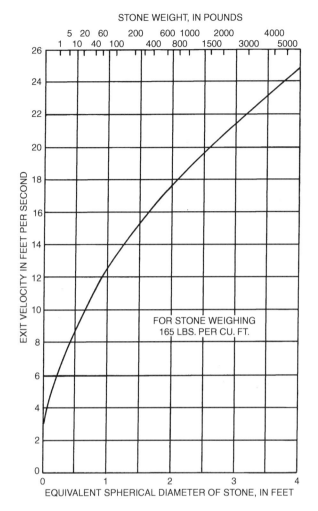

FIGURE 12-12
Riprap size for use downstream of energy dissipater. (From Searcy, J.K., Use of Riprap for Bank Protection, Federal Highway Administration, Washington, DC, 1967.)

12.9 Downstream Channel Transitions

Some energy dissipators result in exit velocity and depth near critical conditions. This flow condition rapidly adjusts to the downstream or natural channel regime; however, critical velocity may be sufficient to cause erosion problems requiring protection adjacent to the basin exit. Figure 12-12 provides an estimate of the riprap size recommended for use downstream of energy dissipators.

The Corps of Engineers (USCOE, 1988) developed an estimate of riprap size required below stilling basins in moderately and highly turbulent situations. The estimate has proven effective in model studies and in the field. The equation based on the work of Isbash is:

$$D_{50} = (V^2/2g)\,(1/C^2)\,[\gamma_w / (\gamma_s - \gamma_w)] \tag{12.4}$$

and

$$W_{50} = (\pi D_{50}^3/6) \gamma_s \qquad (12.5)$$

where W_{50} = weight of stone for which 50% of material by weight contains stone of lesser weight (lbs); D_{50} = spherical diameter of stone having weight of W_{50} (ft); V = average velocity of flow over the end sill of the stilling basin (ft/s); g = acceleration of gravity (ft/s^2); γ_w = specific weight of water (62.4 lbs/ft^3); γ_s = specific weight of stone, normally about 165 (lbs/ft^3); and C = coefficient equal to 0.86 for riprap just downstream from stilling basins, and 1.2 for riprap dumped as a temporary closure dam in a channel

It should be noted that the USBR performed a series of studies on stone stability downstream from various sites. Their design curve for stability is identical to the high-turbulence value (0.86) in the USCOE equation (Peterka, 1978). The curve provided by the FHWA (Figure 12-12) is identical to the low-turbulence value (1.2) in the COE equation. It is, therefore, recommended that the COE high-turbulence values be used for all situations of high-velocity flow and for all larger structures, while the FHWA curve (or the low turbulence COE values) can be used downstream from smaller structures with low-turbulence. The COE low turbulence values are nearly equivalent to the riprap sizing method given in Chapter 10, but without the adjustment capability.

The gradation and thickness of the stone downstream from stilling basins is recommended to fit the following criteria (LL means lower limit, UL means upper limit) (HDC, 1988):

- UL W_{50} should satisfy thickness requirements specified below
- LL W_{100} < 2 LL W_{50}
- UL W_{100} < 5 LL W_{50} and satisfy thickness requirements below
- LL W_{15} > 1/16 UL W_{100}
- UL W_{15} < UL W_{50} and satisfy any graded stone filter requirements
- Bulk volume of stone < W_{15} should not exceed the void volume in the riprap
- W_0 to W_{25} may be used in place of W_{15} to better meet available stone gradations
- Thickness of the stone should be the greater of 2 $D_{50\,max}$ or 1.5 $D_{100\,max}$

The riprap should be extended downstream to a point where the velocity has slowed enough for nonerosive conditions in the natural channel. The end of the riprap should be toed into the banks and bed. The extent of the protection downstream has not been settled with theory or experiment. Rice and Kadavy (1993), based on tests for SAF type stilling basins (baffle blocks with end sill), provide a preliminary direct relation for the downstream length of riprap as:

$$L_s = 4.5\, Fr\, D_1 \qquad (12.6)$$

where L_s = length of riprap section (ft); Fr = Froude number based on velocity and depth at basin entrance; and D_1 = depth at basin entrance (ft).

The Denver Urban Drainage and Flood District uses a rule developed by considering an expanding jet as (UDFCD, 1991):

$$L_t = [1 / (2\tan\theta)]\, [(A_t / Y_t) - W] \qquad (12.7)$$

where L_t = length of protection (ft); W = conduit width or diameter (ft); $A_t = Q/V$ = design discharge/allowable downstream velocity (ft^2); Y_t = tailwater depth (ft); and θ =

Energy Dissipation

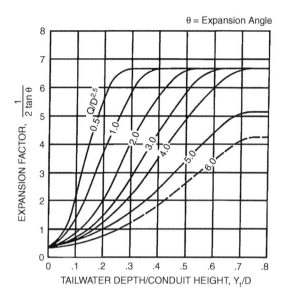

FIGURE 12-13
Expansion factor for circular conduits. (From Denver Urban Drainage and Flood Control District (UDFCD), Urban Storm Drainage Criteria Manual, 1969 with changes 1991.)

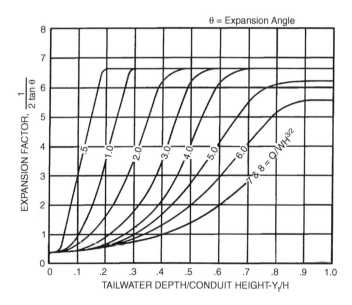

FIGURE 12-14
Expansion factor for rectangular conduits. (From Denver Urban Drainage and Flood Control District (UDFCD), Urban Storm Drainage Criteria Manual, 1969 with changes 1991.)

expansion angle of discharging flow, from Figures 12-13 and 12-14 for circular and rectangular conduits.

For multiple conduits, see UDFCD (1991). If unreasonable results are calculated, a rule-of-thumb is:

$$3 \text{ times conduit diameter or height} < L < 10 \text{ times diameter or height}$$

12.10 Energy Dissipator Computer Model

The HYDRAIN Culvert Analysis microcomputer program HY8 consists of three main options:

1. Culvert analysis
2. Hydrograph generation
3. Routing

The culvert analysis portion of the HYDRAIN model includes a submodel for the design and analysis of energy dissipators. Although the model is specifically designed for energy dissipators associated with culverts, by modifying the culvert discharge data, energy dissipators for other stormwater management facilities (i.e., weirs from detention facilities, open channels) can also be designed using this model.

The model has several user-friendly features that permit easy data entry, changing input data to determine the effects on the energy dissipator design, and output of summary design information. Very little input data are required, which makes the model very easy to use. The design procedures included in the model follow the design procedures used in the Federal Highway Publication — HEC-14 (FHA, 1983). This publication was also used to develop most of the concepts and design procedures included in this chapter.

The energy dissipator model includes the design and analysis of many dissipators that will probably not be used within municipal stormwater management applications. Some of the dissipators were designed for large flood-control structures and not the usual urban applications. The scour hole estimation procedure and the design of riprap basins are two of the more useful options within the model. Information about this model can be obtained from the Federal Highway Administration or as part of the HYDRAIN Culvert Computer Program (HY8) available from McTrans (FHA, 1987; HYDRAIN).

References

Blaisdell, F.W. and Anderson, C.L., A comprehensive generalized study of scour at cantilevered pipe outlets — background, *J. Hydraul. Res.*, 26, 4, 1988a.

Blaisdell, F.W. and Anderson, C.L., A comprehensive generalized study of scour at cantilevered pipe outlets — results, *J. Hydraul. Res.*, 26, 5, 1988b.

Blaisdell, F.W. and Anderson, C.L., Plunge pool energy dissipator, *ASCE J. Hydraul. Eng.*, 117, 3, 1991.

Denver Urban Drainage and Flood Control District (UDFCD), Urban Storm Drainage Criteria Manual, 1969 with changes 1991.

Federal Highway Administration, Debris-Control Structures, Hydraulic Engineering Circular No. 9, 1964.

Federal Highway Administration, Use of Riprap for Bank Protection, Hydraulic Engineering Circular No. 11, 1967.

Federal Highway Administration, Hydraulic Design of Energy Dissipators for Culverts and Channels, Hydraulic Engineering Circular No. 14, 1983.

Federal Highway Administration, HY8 Culvert Analysis Microcomputer Program Applications Guide, Hydraulic Microcomputer Program HY8, 1987.

HYDRAIN Culvert Computer Program (HY8), Available from McTrans Software, University of Florida, Gainesville, FL.

Peterka, A.J., Hydraulic Design of Stilling Basins and Energy Dissipators, Eng. Mono. No. 25, USBR, 1978.
Rice, C.E. and Kadavy, K.C., Protection against scour at SAF stilling basins, *ASCE J. Hydraul. Eng.*, 119, 1, January, 1993.
Searcy, J.K., Use of Riprap for Bank Protection, Federal Highway Administration, Washington, DC, 1967.
U.S. Army Corps of Engineers, Hydraulic Design Criteria (HDC), USAEWES, 1988.
U.S. Department of Agriculture, Soil Conservation Service, Standards and Specifications for Soil Erosion and Sedimentation in Developing Areas, Washington, DC, 1975.

13

Structural Best Management Practices

13.1 Introduction

This chapter discusses structural Best Management Practices (BMPs). This topic is a rapidly evolving art and science. In part, this is due to new regulatory pressures brought about by federal, state, and local regulations. But there is an additional factor, and that is the growing realization that the impacts of urban development on stream systems go far beyond simple pollution loading into ecological and habitat impacts, and that no single structural BMP in a single location can handle mitigation of these impacts. When these other aspects are considered, the design of BMPs becomes much more complex and much more uncertain (Urbonas, 2001).

In almost every municipal engineering project, the designer is involved in surface water environmental issues such as:

- Adverse pollution increases due to general development
- Specific permitting for industry
- Wetland encroachment
- Endangered species or flora
- Ecological balance issues
- Greenway, park, or buffer requirements
- Stream filling or dredging
- Nonpoint source pollution or maximum stream load issues
- Erosion control
- Structural and nonstructural controls
- Land-use regulations and limitations
- Mitigation efforts or environmental design criteria
- Aesthetics

In dealing with these and similar situations, it is good practice to use a multidisciplined team approach because of the complex and interrelated problems involved. The team may involve biologists and ecologists as well as landscape architects and engineers. In general, the municipal engineer's role is to determine the hydrology-, hydraulic-, and pollution-related effects and significance on sensitive surface waters and recommend BMPs to be used to mitigate many of the detrimental effects.

There are many BMP variations and differing (and sometimes conflicting) design criteria. Newer BMP designs, many available only commercially, are pushed for a variety of higher technology applications. Many are yet untested over long periods and diverse applications. Data that exist are often not able to be compared to one another. Due to the complex nature of surface water environmental engineering, this chapter will address only the more reliable and accepted structural BMPs. Procedures for other, more complex and experimental practices may be gleaned from the relevant literature and Internet sites. It is believed that the structural BMPs provided in this chapter will address a high percentage of the structurally treated surface water environmental problems to be encountered in municipal stormwater management applications. Efforts continue to address some of these uncertainty issues through database development, research, and controlled testing (Strecker, Quigley, and Urbonas, 2000; Schueler and Caraco, 2001; Chocat, Barraud, and Alfakif, 2001; Lawrence, 2001; Mikkelsen et al., 2001). Additional information appears on the Internet and in conference settings almost daily, so the reader is advised to seek these sources for additional information.

13.2 Surface Waters, Pollution, and Mitigation Measures

Surface Water Drivers for BMP Usage

The surface waters addressed in this chapter include streams or rivers, ponds or lakes, and wetlands. The characteristics of sensitive surface waters are related to such things as geographic location, hydrology, climatology, size, chemical and physical quality, ecology, species present, etc. Normally, it is a combination of factors that makes the specific waters a unique resource. The interrelationship of these factors for sensitive surface waters determines the water's functional values. These values include such things as flood control, wildlife and aquatic habitat, aesthetics, recreation, groundwater recharge, scenic and wild rivers designation, etc. However, all surface waters have unique or valuable characteristics, and should be considered for restoration or preservation during the development or redevelopment process. These stream characteristics are normally reflected in the designated use standards propagated by each state for its surface waters. These standards carry chemical, sediment, physical, and often ecological criteria (Swietlik, 2001).

Various regulatory or planning programs or requirements may influence what kind of controls must be used, including drinking water source protection, TMDL establishment, NPDES permits for stormwater management, endangered species recovery plans, etc. More and more states are establishing BMP design or capture criteria on a mandatory or voluntary basis. The U.S. EPA will establish nationwide standards or at least general criteria through its effluent limitation program for the construction and ultimate development industries — an indirect method of making during- and postconstruction-BMP use mandatory for all developments (USEPA, 1999a, 2001). Unless there has been a numerical standard established for stormwater runoff through a TMDL, the normal practice is to follow local, regional, or statewide design criteria for BMPs and BMP programs in compliance with NPDES stormwater permits or other regulatory controls. Within this context, certain BMP runoff volume or depth capture criteria or design standards are stipulated and implemented through legislation, criteria manuals, local ordinances, master plans, overlay districts, and policies. In any case, BMPs are fast becoming part of every new development or redevelopment in the United States.

Pollution Dynamics

The overall design of structural BMPs is dependent on an understanding of a complex set of man-induced and natural physical, chemical, and biological forces acting on the environment and introducing pollutants to receiving waters. Additionally, dryfall from atmospheric deposition and the settling of pollutants originally as aerosols is a major source of pollution source (Stolzenbach et al., 2001). Pollutants are not the only impact of urban stormwater runoff, and chemicals in runoff are not the only problem. Other impacts of urbanization, including changes in stream flow, geomorphic changes, and degradation of habitat, also become critically important and targets of BMP mitigation. We know that there are physical as well as chemical impacts of urban development, and that the physical changes within a receiving water system are every bit as important to consider and mitigate as the chemical impacts. To the extent that structural BMPs can mitigate physical impacts of urban development, they are considered here, though the ultimate impact of urban development and the BMPs designed to mitigate that impact on ecological systems is still in its exploratory stages. While the designer rarely needs to consider many of these forces directly, it is appropriate to summarize them here. Chapter 15 provides more information on these impacts and a more detailed discussion of nonstructural programs within the context of a watershed-wide treatment train.

Pollution Accumulation

Conventional urban pollution studies assert that the surface load of pollutants for pervious and impervious areas accumulates between storms. This load eventually approaches an upper limit, or steady state condition. As the surface load accumulates to a particular level for a given land use, the rate of removal by the described processes approximates the deposition rate, and the surface load reaches its upper limit or steady state. This upper limit is a function of the rate at which the pollutants are deposited on the basin's surface and of the dry-weather processes that tend to remove the load. These dry-weather processes include wind removal, air currents generated by automotive and pedestrian traffic, adsorption, and decay or oxidation of chemicals from one state to another.

Urban Hot Spots

Various studies have indicated that certain land uses, termed *urban hot spots*, accumulate and export much greater amounts of petroleum-based hydrocarbons, including more exotic priority pollutants such as highly toxic polycyclic aromatic hydrocarbons (PAH) (Schueler and Shepp, 1992, cited in Schueler, 1994). The levels of such pollutants are from several times, to an order of magnitude, higher than the levels found in runoff from typical residential streets and parking areas, and can often be considered toxic. The land uses most associated with elevated levels of these pollutants include:

- Gas/fueling stations
- Vehicle maintenance areas
- Vehicle washing/steam-cleaning areas
- Auto recycling facilities
- Outdoor material storage areas
- Loading and transfer areas
- Landfills
- Construction sites

- Industrial sites
- Industrial rooftops

Use of specialized structural and nonstructural controls for these areas is often included in a local community's BMP program. See Chapter 15 for more information.

Pollution Movement Forces

When rainfall occurs, the surface accumulated pollution load is subject to a series of physical processes that dictates which particles and how much of the particulate load can be entrained into the surface flow at a given time, and how many constituents are dissolved. For a given depth and velocity, the surface flow has a certain energy for moving dissolved and suspended material. This potential is a function of the velocity of the moving water, the specific gravity of the water, its acidity and temperature, and the amount of potential being used to overcome the friction of the overland flow surface.

In order for a given particle to be entrained into the flow, the flow must have sufficient potential to overcome the effects of gravity on the particle and the friction, adhesion, cohesion, and other forces that resist its lateral movement. As particles are entrained into the flow, the water begins to use up its excess energy. When the water reaches the point where its suspended and dissolved load are using all of the potential, an equilibrium is established, where particles begin to settle out of the flow at a rate equal to that at which they are being entrained.

If the velocity or volume of flow becomes greater by adding more water, changes in surface friction, or increases in surface slope, then more potential is created and thus more surface load can be entrained. Similarly, if the velocity becomes lower, then the potential decreases and part of the load must settle out. This latter case is what is normally used in BMP design. By slowing the velocity through the use of ponds, filter strips, or check dams, the potential of the water is decreased, and the particulates begin to settle out of the water. This lowering of the water velocity also provides protection of downstream channels against streambank degradation, bridge scour, and other physical changes. Filtration can also physically block pollutants from movement between the interstices of the granular or matrix material. Chemical or biological changes may also reduce (or increase) the amount of pollutants in a flow.

Pollution Availability and Fate

The total storm washoff load is dependent on the amount of pollution available and the magnitude of the force available to carry it away. The factors then include total depth of rainfall, the duration of the rainfall, the period of time that has elapsed since the last significant precipitation, the pollution accumulation rate, the residual load on the catchment surface at the end of that previous rainfall event, and any intervening chemical or biological changes.

As a storm event begins, and velocities of surface runoff are relatively low, the portion of the surface load that is first susceptible to being washed off will be that portion that requires the least energy to entrain. This load is made up of highly soluble and very fine sediments. The term "first flush" refers to this pollutant load that is readily washed off of a catchment's surface. Several nationwide studies, such as NURP (USEPA, 1983), have documented this occurrence. Estimates are that as much as 40 to 90% of the total pollutant load may be removed during the first half inch of rainfall runoff. Snowmelt pollution can also be important in Northern climates. See Oberts (1994) for a detailed discussion.

TABLE 13-1

Summary of Urban Stormwater Pollutants

Constituents	Effects
Sediments — Suspended solids, dissolved solids, turbidity	Stream turbidity Habitat changes Recreation/aesthetic loss Contaminant transport Filling of lakes and reservoirs
Nutrients — Nitrate, nitrite, ammonia, organic nitrogen, phosphate, total phosphorus	Algae blooms Eutrophication Ammonia and nitrate toxicity Recreation/aesthetic loss
Microbes — Total and fecal coliforms, fecal streptococci viruses, *E.Coli*, enterocci	Ear/intestinal infections Shellfish bed closure Recreation/aesthetic loss
Organic matter — Vegetation, sewage, other oxygen-demanding materials	Dissolved oxygen depletion Odors Fish kills
Toxic pollutants — Heavy metals (cadmium, copper, lead, zinc), organics, hydrocarbons, pesticides/herbicides	Human and aquatic toxicity Bioaccumulation in the food chain
Thermal pollution	Dissolved oxygen depletion Habitat changes
Trash and debris	Recreation/aesthetic loss

Source: From Atlanta Regional Commission (ARC), Georgia Stormwater Manual, 2002.

One almost never knows enough about the chemical–biological process to be totally comfortable with a diagnosis of a pollution problem. But great good can be accomplished with even a modicum of knowledge of pollution sources and transport mechanisms. Table 13-1 gives a summary of commonly found pollutants and their primary impacts on the environment. Table 13-2 gives an overview of pollution removal mechanisms.

TABLE 13-2

Fate of Pollutants by Management Measures

Pollutant	Vegetative Detention Controls	Infiltration Basin	System	Wetlands
Heavy metals	Filtering Settling	Adsorption Filtration	Adsorption Settling	Adsorption
Toxic organics	Adsorption Settling Biodegradation Volatilization	Adsorption Biodegradation Biodegradation Volatilization	Adsorption Settling	Adsorption
Nutrients	Bioassimilation	Bioassimilation	Adsorption	Bioassimilation
Solids	Filtering Settling	Settling	Adsorption	Adsorption
Oil and grease	Adsorption Settling	Adsorption Settling	Adsorption	Adsorption
BOD	Biodegradation	Biodegradation	Biodegradation	Biodegradation
Pathogens	Not applicable	Settling	Filtration	Not applicable

Source: From Maestri, B. et al., Managing Pollution from Highway Stormwater Runoff, Transportation Research Board, National Academy of Science, Transportation Research Record Number 1166, 1988.

The ability of a pollutant to migrate to receiving waters from its source is a function of its:

- Chemical nature
- Physical–chemical properties (such as water solubility and vapor pressure)
- Tendency to adsorb organic matter or sediment
- Interception by any mitigation or control measures

Of the major migration processes, the key mechanism for transporting a pollutant originating from urbanization is a combination of sorption and settling. This is due to many of the runoff constituents being in a particulate form. This is further enhanced as soluble organic chemicals and heavy metals tend to adsorb to suspended sediments, which in turn will settle, given sufficient time and a suitable environment. Once settled, these pollutants will be most commonly transformed through the biological action of degradation, assimilation by microbes or by rooted vegetation, and through a complex set of chemical reactions. Resuspension and other processes can release chemicals, making some pollutants biologically available once again, and moving pollutants through the system along with the transported bed and suspended sediment load.

Stressors Associated with Urban Runoff

Stormwater programs are evolving to the point that ecological health of streams is becoming a predominant goal of urban stormwater programs (Swietlik, 2001). Ecological health is defined as a balanced, integrated, adaptive community of organisms having a species composition, diversity, and functional organization comparable to that of the natural habitat within the region (USEPA, 1990). There may be room in the future for different definitions that take into account the realities of nearly unavoidable urban impacts. But the fact remains that BMP programs (structural and nonstructural) must include these other impacts if they are to be seen as successful.

Biological assessments have become increasingly important tools for managing water quality to meet the goals of the Clean Water Act (CWA). These methods, which use measurements of aquatic biological communities, are particularly important for evaluating the impacts of chemicals for which there are no water-quality standards, and for nonchemical stressors such as flow alteration, siltation, and invasive species. Typically, a reference stream is located and an analysis is done. The biological/ecological health of the reference stream is taken as a standard for other streams within the same ecoregion, or region of similar ecology (Barbour et al., 1999; Cloak and Bicknell, 2001). Newer methods, more in keeping with habitat needs and wet-weather conditions, are being developed and are expected to come into common usage over the next five to ten years (Herricks, 2001; Swietlik, 2001).

Although biological assessments are critical tools for detecting impairment, they do not identify the cause or causes of the impairment. Stressor analysis then comes into play.

There are many different kinds of stressors that cause change in streams — and it is these stressors that may need to be mitigated through the use of BMPs. A slight modification of Karr (1986), Stribling et al. (2001), and USEPA (2000) results in a list of categories of potential stressors as:

- Biotic interactions — disease, reproduction, competition, predation, feeding, and parasitism
- Energy sources — nutrient availability, organic matter inputs, primary and secondary food production, seasonal patterns, sunlight

- Chemical variables — metals, nutrients, dissolved oxygen, chemical solubilities, organics, alkalinity, temperature, hardness, turbidity, other toxics
- Habitat structure — sinuosity, current, riparian vegetation, substrate, instream cover, gradient, channel morphology, bank stability, canopy, bank steepness/height and overhang, large woody debris, in-stream structures
- Flow regime — volume, watershed characteristics, runoff velocity, flow duration and extremes, precipitation, groundwater

The Stressor Identification process (SI) is prompted by biological assessment data indicating that a biological impairment has occurred. The general SI process entails critically reviewing available information, forming possible stressor scenarios that might explain the impairment, analyzing those scenarios, and producing conclusions about which stressor or stressors are causing the impairment.

After stressors have been identified, potential methods for reducing or removing these stressors are investigated. Typically, that means developing a program or activities that use structural BMPs and nonstructural programs and practices. Such analysis is in its infancy but is gaining credence throughout the country and can be expected to greatly impact the planning and use of BMPs in the future. Focus of some of the more recent advances in BMP concepts, such as low impact development, are targeted at reducing these ecological impacts on stream environments (Schueler and Caraco, 2001; Clar and Rushton, 2001; PGC, 1999). Also, the vast body of literature in the areas of smart growth, zero impact development, sustainable development, conservation development, better site design, green infrastructure, and such concepts all tout an ability to reduce urban impacts on the ecological system.

One conclusion to draw from all this discussion is that, more than ever, structural BMPs are simply one facet of an overall set of programs, activities, structures, and design approaches which together allow for development and life to go on but with a reduced impact on the environment. There *will be* impact.

Overall Approach: Structural and Nonstructural BMPs

Thus, keeping pollution from the drainage system is done through a combination of nonstructural BMPs or source-control measures and structural controls. A basic principle, derived from early studies and conceptual analysis of stormwater pollution, is to attempt to control the introduction of, or remove, pollutants as close to the source as possible. Through a series of treatment opportunities located all along the runoff flow path (termed a generalized "treatment train" — see Section 13.8 and Chapter 15 for more information), pollution introduced into the system is captured, accumulated, and held in place until removed through some sort of maintenance program. Additionally, other adverse impacts may be mitigated through BMPs.

It should be remembered that structural BMPs are only one part of an overall BMP program. The most effective way to maintain clean surface waters is to eliminate the sources of pollution, not to remove pollution once it has gotten into the system. Thus, each municipality (and many sites) should also be implementing a baseline BMP program that includes nonstructural educational, regulatory, and programmatic BMPs. Chapter 15 gives general criteria for the assessment of any type of BMP, as well as a procedure for the development of an overall stormwater quality management program.

Close attention to the nature of site design and layout can go a long way toward reducing pollution. For example, landscaping and terracing provide natural buffers, rainfall holding, and infiltration areas. Use of low-maintenance, locally derived vegetation reduces the need for pesticides and fertilizer. Directing roof and driveway runoff across grassy areas provides a natural filtration and infiltration. Clustering designs, zoning density limitations, critical area identification and protection, and joint regional design planning can all work toward an inherent lowering of pollution loading levels.

Nonstructural programs tend to be built into the design of the site through the use of pollution-reducing site design elements, or are programs of pollution reduction. For programmatic nonstructural BMPs, the typical steps include:

- Identification of the pollutant of concern
- Assessment of the vectors by which the pollutant enters the system
- Selection of points within the process where the pollutant-generating activity can most effectively be modified or eliminated
- Development of interception strategies and ways to measure success
- Development of implementation steps, responsibilities, and funding

Thus, for example, the prime source of citizen dumping of used motor oil is in the mind, attitude, and habits of the dumper. Education, reminders, and making appropriate behavior relatively easy are key. Reduction of dumping of used motor oil can be targeted through a combination of a storm drain stenciling program, convenient used-oil turn-in locations, public education, a reporting hotline, point-of-sale pamphlets, comprehensive recycling programs, financial rewards for recycling, strong enforcement, and other activities. For each type of pollutant and point of entry, there can be developed a nonstructural program attacking it at its source. With each program, there logically should also be an attempt to assess the severity and impacts of the problem, the effectiveness of the program, and the value of the investment in pollution reduction. Chapter 15 provides a long list of nonstructural and structural BMPs.

Despite nonstructural measures, some pollution will enter streams. Nonstructural programs will not eliminate all pollution, and, thus, structural measures may be necessary. Once in the system, pollution is much more difficult to remove or abate. Pollutants are normally removed from surface waters through some combination of structural BMPs through the basic mechanisms of:

- Physical removal
- Settling
- Filtering
- Chemical reaction
- Biochemical transformation

All structural measures to remove pollutants from the drainage system use a combination of these basic and other ancillary mechanisms.

After other design modifications (such as smart growth and low-impact development concepts), conventional BMPs of the type described herein will commonly suffice for most urban site and small watershed development projects. For most pollutants, nonpoint pollution discharges from frequent minor storms are more critical than discharges during infrequent major storms. First-flush conditions result in relatively high pollutant concentrations (analogous to raw sewage) during the initial, relatively low discharge, stages of stormwater runoff. This can induce shock loading and a short-term degradation of the water quality of receiving waters. First-flush effects are attributed primarily to the washoff of particles from paved areas.

Pollution management should commonly rely on controls for minor storms having a recurrence interval of 1 year or less. As such, management techniques that isolate first-flush, or smaller, discharges take advantage of the smaller required storage capacities for these discharges. Placing several BMPs in line (site-based treatment train) can produce dramatic increases in pollutant removal efficiency.

BMPs and Disease Vectors

There is much concern about the unintended consequences of BMPs. There can be all sorts of adverse impacts of BMPs: temperature fluxes from heated pond water suddenly released in a slug, infiltrated toxics entering the groundwater, release and concentration of metals in the heated soup of a retention pond, release of decayed vegetation in oxygen demand and nutrients, etc. A more recent concern has been the unintended provision of breeding grounds for mosquitoes and associated diseases such as West Nile virus, yellow fever, and malaria (Metzger et al., 2002; Kluh et al., 2002).

Of the 200 species of mosquitoes in the U.S. (and more than 3000 worldwide) all need water to complete their lifecycle from eggs to larvae to pupae to adult. This cycle can take as little as two weeks in many species. Some species (floodwater species) actually prefer to lay eggs in moist soil subject to flooding — for example, an extended detention pond or floodplain overbank area. While early mosquito elimination practices concentrated on the use of highly toxic pesticides and draining wetlands, current practices concentrate on use of environmentally friendly larvicides, larva eating fish, and habitat control or elimination of unwanted standing water.

Several types of BMPs contain ideal breeding grounds for mosquitoes including retention ponds, commercial type units with standing water in sumps, standing water in the interstices of riprap energy dissipaters, underground sand filter basins, long-term extended detention, the multi-chambered treatment train, and similar BMPs. In addition, poorly maintained BMPs tend to clog or create scour holes where water stands.

Local stormwater managers should consider forming an alliance with local vector control agencies and staffs and cross-training their own BMP inspectors. Maintenance manuals, guidance, and agreements should contain language concerning the elimination of unintended mosquito breeding areas and the maintenance of BMP features designed to inhibit mosquito breeding.

A several-year study sponsored by CALTRANS in 1998–2000 looked at hundreds of BMP sites by survey and visit and sought retrofit opportunities in several as a pilot study. At times designs to keep mosquitoes from breeding are at odds with other design needs. For example, shallow safety or emergent vegetation benches in retention ponds inhibit mosquito eating fish from getting to the mosquito larvae. But the steeper slopes and deeper ponds necessary are a safety concern and reduce the ability of vegetative uptake of nutrients. Key recommendations for BMP design include (Metzger et al., 2002):

- Ensure BMPs drain within 72 hours to cut off the cycle prior to its completion.
- Ensure gravity flow throughout the whole BMP.
- Loose riprap with deeper interstices should be avoided in favor of grouted riprap. Ensure free drainage of all systems, and avoid pumped systems where possible.
- Pipes and pipe systems should be designed to drain fully, keeping in mind locations where sediment will build up, or should be sealed from mosquito entrance.
- Sumped systems should be sealed leaving no openings more than $1/16$ in.
- Dewatering systems should be designed for easy access and use.
- Submerge inlets and outlets to reduce the water surface available for breeding, or provide fine mesh netting at clear water effluent pipes.

- Permanent pools should be sized with quality and quantity to support mosquito eating fish such as *Gambusia affinis*.
- Ensure pool water surfaces and other BMP designs are accessible to vector control crews, and coordinate with local crews and agencies, including access roads and lightweight or counterbalanced access hatches.
- Concrete liners should be used wherever vegetation is not necessary to reduce unwanted growth.

13.3 Estimation of Pollutant Concentration and Loads

There are many methods for estimating the concentration and loading of pollutants to surface waters. Physically based models attempt to mimic the accumulation and removal of pollutants as well as the chemical reactions within the receiving streams. More empirical models rely on general data and information on pollution concentrations in surface runoff and then predict pollution through an estimation of surface runoff volumes. Regression equations use significant variables to predict loadings of various constituents based on data sets. This type of model can be used with little or no data but is very rough in its estimates. Such models are less effective for "what if" analysis that may extend the situation beyond the limits of databases. They are not accurate in prediction of acute or shock loadings. Physically based models require substantial data for calibration over the range of expected conditions but can be very effective when data exists and can simulate the most important physical, biological, and chemical aspects of the problem.

Pollutant Loads

USEPA (1983) determined that, based on the sampling done during the National Urban Runoff Program (NURP), there are certain pollutants that may be typically found in urban stormwater. More detail is contained in Chapter 6 of this book. Some of the conventional pollutants show up in significant concentrations in most samples, notably the metals, but most others were present in measurable quantities in less than 15% of the samples. Many of these constituents are related to automotive traffic or industrial activities, while others are characteristic of fertilizing and insect-control practices. Common pollutants found in studies include coliform bacteria, total suspended sediment (TSS), total phosphorus (TP), total nitrogen, (TN), 5-day biochemical oxygen demand (BOD5), chemical oxygen demand (COD), total copper (TCu), total lead (TPb), total zinc (TZn), and oil and grease.

There are numerous sources of these pollutants. Table 13-3 summarizes some of the more important ones for key pollutant types (EPA, 1999a).

Event Mean Concentrations

NURP was designed and executed under the auspices of USEPA in the late 1970s and early 1980s. Its main goal was to provide reliable data and information characterizing runoff from urban sites (USEPA, 1983). Twenty-eight sites were monitored from across the United States. While there were some differences in the objectives and procedures of the sites, a common base of information emerged. Later, sampling data from municipalities with NPDES permits generally confirmed NURP's findings, though there is always variability when pollutant monitoring is concerned.

TABLE 13-3

Summary of Pollutant Sources

Contaminant	Contaminant Sources
Sediment and floatables	Streets, lawns, driveways, roads, construction activities, atmospheric deposition, drainage channel erosion
Pesticides and herbicides	Residential lawns and gardens, roadsides, utility right-of-ways, commercial and industrial landscaped areas, soil washoff
Organic materials	Residential lawns and gardens, commercial landscaping, animal wastes
Metals	Automobiles, bridges, atmospheric deposition, industrial areas, soil erosion, corroding metal surfaces, combustion processes
Oil and grease/hydrocarbons	Roads, driveways, parking lots, vehicle maintenance areas, gas stations, illicit dumping to storm drains
Bacteria and viruses	Lawns, roads, leaky sanitary sewer lines, sanitary sewer cross-connections, animal waste, septic systems
Nitrogen and phosphorus	Lawn fertilizers, atmospheric deposition, automobile exhaust, soil erosion, animal waste, detergents

Source: Typical loadings of pollutants by land-use type are provided in Horner, R.R., Skupien, J.J., Livingston, E.H., and Shaver, E.H., Fundamentals of Urban Runoff Management: Technical and Institutional Issues, Terrene Institute and U.S. Environmental Protection Agency, Washington DC, 1994.

Because of the variability of measurements within storms, among different storms at one site, and among sites, it was desirable to use a measure that tended to reduce this variability somewhat. The measure of the magnitude of urban runoff pollution chosen is termed the event mean concentration (EMC). EMC is defined as the total constituent mass discharge divided by the total runoff volume for a given storm event. With few exceptions, EMCs were found to not vary significantly for similar land uses from site to site for the same constituent and were found to be distributed lognormally. This is not to say that the variation of pollution concentration within an outflow hydrograph does not change. Studies in Austin, TX (1990), have shown that the more impervious an area is, the more pronounced is the first-flush effect and the concentration of pollutants in the first flush.

Because EMCs are distributed lognormally, measures of central tendency (median and mean) and scatter (standard deviation, coefficient of variation), as well as expected values at any frequency of occurrence, could be calculated by using the logarithmic transforma-

TABLE 13-4

Typical Loadings of Pollutants from Runoff by Urban Land Use (lb/ac/year)

Land Use	TSS	TP	TKN	NH_3	NO_2	BOD	COD	Pb	Zn	Cu
Commercial	1000	1.5	6.7	1.9	3.1	62	420	2.7	2.1	0.4
Parking	400	0.7	5.1	2.0	2.9	47	270	0.8	0.8	0.04
HDR	420	1	4.2	0.8	2.0	27	170	0.8	0.7	0.03
MDR	190	0.5	2.5	0.5	1.4	13	72	0.2	0.2	0.14
LDR	10	0.04	0.03	0.02	0.1	NA	NA	0.01	0.04	0.01
Freeway	880	0.9	7.9	1.5	4.2	NA	NA	4.5	2.1	0.37
Industrial	860	1.3	3.8	0.2	1.3	NA	NA	2.4	7.3	0.5
Park	3	0.03	1.5	NA	0.3	NA	2	0.0	NA	NA
Construction	6000	80	NA	NA	NA	NA	NA	NA	NA	NA

HDR: High-Density Residential, MDR: Medium-Density Residential, LDR: Low-Density Residential.

NA: Not available; insufficient data to characterize loadings.

Source: From Horner, R.R., Skupien, J.J., Livingston, E.H., and Shaver, E.H., Fundamentals of Urban Runoff Management: Technical and Institutional Issues, Terrene Institute and U.S. Environmental Protection Agency, Washington DC, 1994.

Pollutant	Units	Residential		Mixed		Commercial		Open/Non-Urban	
		Median	COV	Median	COV	Median	COV	Median	COV
BOD	mg/l	10	0.41	7.8	0.52	9.3	0.31	—	—
COD	mg/l	73	0.55	65	0.58	57	0.39	40	0.78
TSS	mg/l	101	0.96	67	1.14	69	0.85	70	2.92
Total Lend	µg/l	144	0.75	114	1.35	104	0.68	30	1.52
Total Copper	µg/l	33	0.99	27	1.32	29	0.81	—	—
Total Zinc	µg/l	135	0.84	154	0.78	226	1.07	195	0.66
Total Kjeldahl Nitrogen	µg/l	1900	0.73	1288	0.50	1179	0.43	965	1.00
Nitrate + Nitrite	µg/l	736	0.83	558	0.67	572	0.48	543	0.91
Total Phosphorus	µg/l	383	0.69	263	0.75	201	0.67	121	1.66
Soluble Phosphorus	µg/l	143	0.46	56	0.75	80	0.71	26	2.11

FIGURE 13-1
NURP data EMC summary.

tion of the raw data. Standard statistical tests and sampling theory can also be used on the lognormally distributed data. See Chapter 6 for more information on EMC values.

In selecting a method for estimation of potential washoff loads for a particular site, it is often decided to use methods that estimate washoff loads by land-use type. Total loadings are then determined based on event mean concentrations of pollutants and runoff volumes. Table 13-4 gives typical EMCs for various land uses based on a number of sources. This information should be compared to local information, when available. Initial data from a number of municipalities throughout the East and Midwest indicate that, other than lead, most constituents did not vary significantly from the NURP information. The reduction in lead is thought to be based on the use of lead-free gasoline since the NURP data were collected. Arid western regions tend to have higher EMC values, probably due to the longer inter-event dry periods (CWP, 2000). Figure 13-1 gives the information found in the NURP study.

Nationwide Regression Equations Method

Reconnaissance studies of urban storm runoff loads commonly require preliminary estimates of mean seasonal or mean annual loads of chemical constituents at sites where little or no storm runoff or concentration data are available (Colyer and Yen, 1983). To make preliminary estimates, a regional regression analysis can be used to relate observed mean seasonal or mean annual loads at sites where data are available for physical, land-use, or climate characteristics. The result of a major study by the U.S. Geological Survey and the U.S. Environmental Protection Agency resulted in the development of regression equations that can be used to estimate mean loads for chemical oxygen demand, suspended solids,

TABLE 13-5
Water Quality Nationwide Regression Equations

Dependent Variable (W)	Regression Constant (a)	SQRT(DA) (b)	IA (c)	MAR (d)	MJT (e)	X2 (f)	Bias Correction Factor (BCF)
COD	1.1174	2.0069	0.0051				1.298
SS	1.5430	1.5906		0.0264	−0.0297		1.521
DS	1.8449	2.5468			−0.0232		1.251
TN	−0.2433	1.6383	0.0061			−0.4442	1.345
AN	−0.7282	1.6123	0.0064	0.0226	−0.0210	−0.4345	1.277
TP	−1.3884	2.0825		0.0234	−0.0213		1.314
DP	−1.3661	1.3955					1.469
CU	−1.4824	1.8281			−0.0141		1.403
PB	−1.9679	1.9037	0.0070	0.0128			1.365
ZN	−1.6302	2.0392	0.0072				1.322

Dependent Variable (W)	N	Degrees of Freedom	R^2 (Note 1)	SE in Average Prediction Error (ASEP)			
				Logs-γ (Note 2)	Logs (Note 3)	−%	+%
COD	59	56	0.53	0.3020	0.311	−51	105
SS	47	43	0.43	0.412	0.433	−63	171
DS	13	10	0.61	0.310	0.349	−55	123
TN	41	37	0.49	0.345	0.363	−57	131
AN	51	45	0.49	0.316	0.337	−54	119
TP	51	47	0.65	0.303	0.316	−52	107
DP	28	26	0.20	0.372	0.388	−59	144
CU	30	27	0.41	0.361	0.381	−58	140
PB	56	52	0.46	0.353	0.368	−57	133
ZN	34	31	0.59	0.310	0.326	−53	128

Notes: R^2 is the proportion of variance in W explained by the sample regression; SE is the standard error of the model; ASEP is given in log units and percent; the conversion from logs to percent is: ASEP(-percent) = $100[10^{-ASEP(logs)} - 1]$; in the model is W = $10^{[a+bSQRT(DA)+cIA+dMAR+eMJT+fX2]}$BCF. W is the mean load, in pounds, associated with a runoff event; here are shown regression coefficients for indicated explanatory variables.

Source: Tasker, D.T. and Driver, N.E., Nationwide regression models for predicting urban runoff water quality at unmonitored sites, *Water Resour. Bull.*, 24, 5, October 1988.

dissolved solids, total nitrogen, total ammonia plus nitrogen, total phosphorous, dissolved phosphorous, total copper, total lead, and total zinc (Tasker and Driver, 1988).

The Nationwide Regression Equations Method (NRE) is based on the assumption that explanatory variables, such as drainage area, basin imperviousness, mean annual rainfall, mean-minimum January temperature, and general land-use categories, can explain regional variation in annual or seasonal storm loads. Coefficients for the regression equations were estimated by a generalized-least-squares (GLS) regression method that accounts for cross correlation and differences in reliability of sample estimates between sites. Table 13-5 presents the 10 regression equations that can be used to predict mean loads for different constituents at unmonitored sites with drainage areas ranging from 0.015 to 1 square mile. The following characteristics are used in the equations.

1. Total contributing drainage area (DA), in square miles
2. Impervious area (IA), as a percent of total contributing drainage area
3. An indicator variable (X1) that is 1, if residential land use (LUR) plus nonurban land use (LUN) exceeds 75% of the total contributing area, and is 0 otherwise

TABLE 13-6

Range of Variables Used in Regression Equations

Dependent Variable	DA		Variables IA		MAR		MJT	
	Min.	Max.	Min.	Max.	Min.	Max.	Min.	Max.
COC	0.019	0.070	4	100	8.38	62.00	3.2	58.7
SS	0.019	0.707	4	100	8.38	49.38	3.2	50.1
DS	0.020	0.450	19	99	10.24	37.61	11.4	35.8
TN	0.019	0.830	4	100	11.83	62.00	3.2	58.7
AN	0.019	0.707	4	100	8.38	62.00	3.2	58.7
TP	0.190	0.830	4	100	8.38	62.00	3.2	58.7
DP	0.020	0.707	5	99	8.38	46.18	10.8	35.8
CU	0.014	0.830	6	99	8.38	62.00	15.3	58.7
PB	0.019	0.830	4	100	8.38	62.00	3.2	58.7
ZN	0.019	0.830	13	100	8.38	62.00	11.4	58.7

4. An indicator variable (X2) that is 1, if commercial land use (LUC) plus industrial land use (LUI) exceeds 75% of the total contributing drainage area, and is 0 otherwise
5. Mean annual rainfall (MAR), in inches
6. Mean-minimum January temperatures (MJT), in degrees Fahrenheit

In general, the equations given in Table 13-6 should not be used to estimate mean storm loads at sites whose characteristics are much beyond the range of values shown in Table 13-5. There has been some work directed at extending these equations for larger drainage areas (Tasker, Gilroy, and Jennings, 1990). This work uses a scheme of summing predictions for small areas to make load predictions for large urban basins. The summing scheme assumes that the mean annual load for a large basin can be computed as the sum of mean annual loads for subbasins. Although this method involves many tedious calculations, a computer model is available to make most of the necessary calculations. This method has not at this time been verified by observed data.

Example Application (Tasker and Driver, 1988)

To compute the mean annual load of total nitrogen (TN), in pounds, at a 0.5 mi² basin that is 90% residential with impervious area of 30% and in a region where the mean number of storms per year is 79, first compute the mean load for a storm, W, using the appropriate equation from Table 13-3.

$$W = 10^{(X)} * 1.345$$

where $X = [a + b(DA^{1/2}) + cIA + fX2]$; $X = [-0.2433 + 1.6383(0.5)^{1/2} + 0.0061(30) - 0.4442(0)]$; and $W = 16.9$ pounds.

The mean annual load can be calculated by multiplying W by 79, the average number of storms per year, to yield a mean annual load of $TN = 79(16.9) = 1335$ lb. The 90% confidence interval for this example is calculated as follows:

Given:

$n - p = 41 - 4 = 37$ (degrees of freedom — Table 13-3)

$\alpha = 0.1$ (90% confidence/0.1 level of significance)

$t\alpha/2, n - p = 1.64$ (from statistical table — value of z score for two-tailed tests at 0.10 level of significance)

$x_i = (X1\ DA^{1/2}\ X2\ IA)$ (vector of basin characteristics)

$x_i = (1\ .707\ 0\ 30)$

From Table 13-5, U = the following:

0.0346	−0.0449	0.00472	−0.00033
−0.0449	0.1031	0.0130	0.00012
0.00472	0.0130	0.00421	−0.00035
−0.0003	0.00012	−0.00035	0.0000073

$\gamma^2 = (0.345)^2 = 0.119$ (From Table 13–3)
BCF = 1.345 (From Table 13–3)

Formulas:

The fit of the regression models may be measured by the R^2 value (Table 13-5), which is the fraction of variance in W explained by the model. A measure of how good the models are for prediction at unmonitored sites is the average variance of prediction. This statistic is computed by averaging over the sites the estimated variance of prediction at a site used in the regression. The variance of prediction at site V_{pi} is estimated by (Tasker and Driver, 1988):

$$V_{pi} = \gamma^2 + x_i U x'_i \quad (13.1)$$

A 100 (1 - α) confidence interval for the true mean load for a storm, W_i, at a particular unmonitored site i can be computed by:

$$\frac{1}{T}\frac{W_i}{1.345} < W < \frac{W_i}{1.345}T \quad (13.2)$$

where W_i is the regression estimate for one of the models in Table 13-3 and

$$T = 10^X \text{ where } X = (t(\alpha/2, n-p)V_{pi})^{1/2} \quad (13.3)$$

in which $t(\alpha/2, n - p)$ is the critical value of the t-distribution for $n - p$ degrees of freedom and is tabulated in many statistical texts. The variance-covariance matrices, U, needed to calculate V_{pi} in T for each of the 10 regression equations are given in Table 13-7.

Although the calculations needed to use these formulas can be tedious, a computer program is available to make most of the necessary computations (Tasker, Gilroy, and Jennings, 1990). This interactive computer program calculates the expected value of mean load for a storm, for the 10 constituents, given the appropriate basin characteristics along with the 90% confidence interval and standard deviation of the load.

Calculate:

$$V_{pi} = \gamma^2 + x_i U x'_i = 0.119 + x_i U x'_i = 0.119 + 0.014 = 0.133$$

$$T = 10^X \text{ where } X = (t(\alpha/2, n-p)V_{pi})^{1/2}$$

$$T = 10^X \text{ where } X = [1.64(0.133)^{1/2}] = 3.96$$

$$\frac{1}{T}\frac{W_i}{1.345} < W < \frac{W_i}{1.345}T$$

TABLE 13-7

Variance-Covariance Matrices (U)

COD	Constant	SQRT (DA)	IA	
Constant	1.9363E-02	−2.716E-02	−1.682E-04	
SQRT (DA)	−2.716E-02	6.4332E-02	9.8363E-05	
IA	−1.682E-04	9.8363E-05	2.7996E-06	
SS	**Constant**	**SQRT (DA)**	**MAR**	**MJT**
Constant	1.2799E-01	−4.440E-02	−3.779E-03	1.0304E-03
SQRT (DA)	−4.440E-02	1.2989E-01	2.8962E-05	−2.991E-04
MAR	−3.779E-05	2.8962E-05	1.4849E-04	−6.608E-05
MJT	1.0304E-03	−2.991E-04	−6.608E-05	7.3585E-05
DS	**Constant**	**SQRT (DA)**	**MJT**	
Constant	5.6262E-02	−5.360E-02	−1.248E-03	
SQRT (DA)	−5.360E-02	3.9703E-01	−3.337E-03	
MJT	−1.248E-03	−3.337E-03	9.9534E-05	
TN	**Constant**	**SQRT (DA)**	**X2**	**IA**
Constant	3.4590E-02	−4.493E-02	4.7196E-03	−3.324E-04
SQRT (DA)	−4.493E-02	1.0309E-01	1.3013E-02	1.1757E-04
X2	4.7196E-03	1.3013E-02	4.2067E-02	−3.484E-04
IA	−3.324E-04	1.1757E-04	−3.484E-04	7.2720E-06
AN	**Constant**	**SQRT (DA)**	**X2**	
Constant	1.0696E-01	−5.283E-02	1.8916E-02	
SQRT (DA)	−5.283E-02	1.0073E-01	7.7172E-03	
X2	1.8916E-02	7.7172E-03	3.8246E-02	
IA	−4.887E-04	1.2527E-04	−3.384E-04	
MAR	−2.605E-03	2.7698E-04	−5.631E-04	
MJT	1.2411E-03	3.6202E-05	3.8766E-04	
AN	**IA**	**MAR**	**MJT**	
Constant	−4.887E-04	−2.605E-03	1.2411E-03	
SQRT (DA)	1.2527E-04	2.7698E-04	3.6202E-05	
X2	−3.8246E-02	−5.631E-04	3.8766E-04	
IA	6.5176E-06	9.2175E-06	−6.013E-06	
MAR	9.2175E-06	9.3696E-05	−5.564E-05	
MJT	−6.013E-06	−5.564E-05	4.4696E-05	
TP	**Constant**	**SQRT (DA)**	**MAR**	**MJT**
Constant	5.5400E-02	−2.589E-02	−1.660E-03	6.7787E-04
SQRT (DA)	−2.589E-02	6.1221E-02	5.5267E-05	8.9042E-05
MAR	−1.660E-03	5.5267E-05	7.0813E-05	−4.022E-05
MJT	6.7787E-04	8.9042E-05	−4.022E-05	3.3482E-05
DP	**Constant**	**SQRT (DA)**		
Constant	3.7200E-02	−9.503E-02		
SQRT (DA)	−9.503E-02	2.8924E-01		
CU	**Constant**	**SQRT (DA)**	**MJT**	
Constant	6.5351E-02	−6.876E-02	−1.122E-03	
SQRT (DA)	−6.876E-02	1.3005E-01	5.8628E-04	
MJT	−1.122E-03	5.8628E-04	3.0108E-05	
PB	**Constant**	**SQRT (DA)**	**IA**	**MAR**
Constant	7.1623E-02	−4.938E-02	−3.567E-04	−9.865E-04
SQRT (DA)	−4.938E-02	9.1551E-02	2.3984E-04	1.3275E-04
PB	**Constant**	**SQRT (DA)**	**IA**	**MAR**

TABLE 13-7 *(Continued)*

Variance-Covariance Matrices (U)

IA	−3.567E-04	2.3984E-04	4.2164E-06	1.7693E-06
MAR	−9.865E-04	1.3275E-04	1.7693E-06	2.5135E-05

ZN	Constant	SQRT (DA)	IA
Constant	0.4020E-01	−0.4700E-01	−0.3564E-03
SQRT (DA)	−0.4700E-01	0.9232E-01	0.2412E-03
IA	−0.3564E-03	0.2412E-03	0.4873E-05

Note: Values are in scientific notation, i.e., −1.9363E-02 = 0.019363.

$$\frac{1}{3.96}\frac{16.9}{1.345} < W < \frac{16.9}{1.345}3.96$$

$$3.2 < W < 49.8$$

Therefore, a 90% confidence interval for $W = 16.9$ is (3.2, 49.8).

The Simple Method

The following method of computing pollutant loadings is referred to as the simple method and is adapted from earlier EPA reports as modified by Schueler (MWCOG, 1987). The simple method provides a quick and reasonable estimate of pollutant loadings with a minimal amount of required data. The simple method is best for small urban watersheds (<640 acres), when only stormwater runoff and pollutant load estimates are desired, when there is a need for quick and reasonable load estimates, when only percent imperviousness and runoff pollutant concentrations are available, and when only planning level estimates are needed (Ohrel, 2000). It should not be used when baseflow runoff carries high pollutant loads, when there are large watersheds (>640 acres), when there are nonurban land uses (e.g., construction sites, industrial, areas, rural development, agricultural uses), as reliable C values are unavailable, and there is ambiguity about watershed's percent imperviousness.

The basic loading equation is:

$$L = RO * C * A * 2.72 \tag{13.4}$$

where L = pollutant washoff load (lbs); RO = runoff depth (ft); C = event mean concentration of pollutant (mgl); A = drainage area (acres); and 2.72 = units conversion constant.

The runoff depth, RO, is computed from the following formula:

$$RO = (P * Pj * Rv)/12 \tag{13.5}$$

where P = rainfall depth (in); Pj = fraction of events producing runoff; Rv = volume runoff coefficient; and 12 = units conversion constant.

To use this method, the following steps should be taken:

Step 1: Estimate rainfall depth (P) for the time interval or period of interest. Annual values often range from about 30 to 60 in in the United States. Seasonal and even individual storm values can be input. Storm durations should be appropriate to the type of BMP being tested.

TABLE 13-8

Land-Use Definitions

Land Use	% Imp	Description
Wooded	1	Wooded land
Pasture	1	Meadow, pasture, or open land
Crops	1	Agriculture, row crops
Rural Residence	6	Single-family, 1 dwl/2 + ac
Low Residence	12	Single-family, 1 dwl/ac
Med Residence	30	Single-family, 2 + dwl/ac; Duplex
High Residence	60	Multi-family, apartment/townhouse
Commerce	75	Light commercial with landscaping
Industry	90	Light manufacturing, warehousing

Source: Ogden, 1990.

Step 2: Rv is similar to the rational method C factor, but it relates volumes of runoff, not rates. Determination of a volume-of-runoff coefficient (Rv) is based on an equation developed by the Metropolitan Washington Council of Governments (1987) after Driscoll (1983). Regression of computed runoff coefficients and measured imperviousness at 44 small urban catchments resulted in the following equation:

$$Rv = 0.05 + 0.009 * I \qquad (13.6)$$

where I is the percent imperviousness of the watershed (use 50 for 50%) obtained from site plans or aerial photographs.

Rv does not have a component for baseflow. For smaller or more densely developed watersheds, or for pollutants not found in appreciable quantities in baseflow, this is not important. If baseflow is important, a separate pollutant loading for baseflow can be computed given knowledge of baseflow amounts and mean pollutant concentrations sampled from baseflow. An approximate method is given in COG (1987). Percent impervious values can be estimated from Table 13-8 for different land-use categories though local measurements are best (Ogden, 1990).

Step 3: Rv is adjusted by the fraction of rainfall from the many smaller storms that produce no runoff (Pj) and were underrepresented in NURP studies. For an individual storm, use a value of 1 to avoid double counting. Washington, DC, rainfall analysis suggests that about 90% of the storms produce runoff (COG, 1987), and a value of 0.9 can be used for Pj.

Step 4: Calculate the total pollutant load from Equation (13.4) for a given event mean concentration for the pollutant of interest.

Individual storm pollution estimates are very approximate.

Acute or "Shock" Pollutant Loading Estimates

The U.S. Department of Transportation, based on earlier work, developed detailed procedures and information dealing with the characterization of individual storm runoff pollutant loads from a statistical perspective from streets and highways, and the prediction of water-quality impacts they cause. The procedures developed are based on monitoring data from 993 individual storm events at 31 highway runoff sites in 11 states. The federal document provides a step-by-step procedure for computing the estimated impacts on water quality of a stream or lake that receives highway runoff (FHWA, 1990). The basic approach is to determine statistics on storms and runoff for a particular area and compute exceedance frequencies. Guidance is provided for evaluating whether or not a water-quality problem will result, and the degree of pollution control required to mitigate impacts to acceptable levels. The federal document provides many worksheets, maps, and detailed examples to assist the user in the application of the concepts. This statistical methodology is applicable to many types of acute loading problems, though EMC and other parameters may need to be modified for application to other than roadway drainage. In addition, recently, highways have been shown to be a source of polynuclear aromatic hydrocarbons (PAH), a highly toxic pollutant released from asphalt on hot sunny days.

Simulation Models

In instances where a municipality has a significant amount of historical data for the drainage areas served by storm drain outfalls, including historical precipitation data and receiving water concentration and flow data, the municipality may elect to use more complex dynamic models to derive pollutant loads and to analyze the effects of discharges on receiving waters. Normally, these more complex models are used because of the significance of the resource being protected. Details on the basic theory and types of such models can be found in a number of sources including a summary by Huber (1986). New and specialized models appear from time to time, as well as pre- and postprocessors for existing models.

Dynamic models have an additional benefit over steady-state models in that dynamic models determine the entire discharge concentration–frequency distribution over time. Therefore, dynamic models take into consideration the inherent variability of data associated with urban stormwater discharges, including concentration, flow rate, and runoff volume. This would enable the modeler to examine the effects of stormwater discharges on receiving water quality in terms of the runoff pollutograph.

Another benefit of using a dynamic model is that it allows the modeler to consider a multitude of "what-if" scenarios. For example, when sufficient historical data are available, the modeler could consider the benefits and risks associated with alternative BMP strategies.

For purposes of computing pollutant loadings, a number of models are available, including EPA's Stormwater Management Model (SWMM), BASINS, QUAL2E, MOUSE, SUNOM, and the Hydrologic Simulation Program (HSPF); U.S. Army Corps of Engineers' Storage, Treatment, Overflow, Runoff Model (STORM); and Illinois State Water Survey's Model QILLUDAS (or Auto-QI); and others. An application of the simple EMC methodology to pollution loading is SLAMM (Pitt and Voorhees, 1995). It includes abilities to input various types of structural and nonstructural practices and to perform stochastic analysis of pollution loads.

13.4 BMP Design Concepts

Basis for Design Criteria for Structural BMPs

The design of structural BMPs for pollutant removal is essentially a matter of mass balance of pollutants, in many ways similar to a typical pond level-pool routing equation with one additional term: mass inflow - mass outflow ± mass creation/destruction =

accumulated mass. The mass inflow for a time period is a function of the flow volume times the mean pollutant concentration. The mass created or destroyed is a function of the chemical, biological, or physical changes that take place within the BMP. For any given storm, there can be an actual net export of pollutants from the BMP due to release of pollutants in some way bound within the BMP (often in the sediments or biomass). The goal then is to design a BMP to, on average, provide sufficient pollution reduction to meet a mass outflow target, standard, or goal.

The difficulties in the design become apparent when the complexities of urban rainfall–runoff sequences are compounded with the great number of pollutant types and concentration variability (even within a single storm) and with the myriad chemical, biological, and physical reactions that can take place among the various constituents. In addition, the pollutant removal rates of BMPs vary considerably even among carefully constructed, well maintained and similar BMPs at contiguous locations. The result in application is often derived rules-of-thumb or simplifications for certain localities based on statistical or continuous simulation analysis of rainfall and runoff, assumptions about pollutant concentrations, and estimated pollutant removal rates (Nix et al., 1988). Adjustment of theory is left to monitoring and performance observation.

The rules-of-thumb for design purposes are often based on capture and treatment of a certain depth of rainfall over the watershed draining to the BMP or based on runoff from all or only the impervious areas of a watershed. Rainfall and runoff in urban stormwater pollution calculations are normally related using a rational method-type C factor (or its volume equivalent R_v) because of the simplicity, the fact that much of the NURP analysis was done in a similar way, the many unknowns in the analysis, and the wide variability of rainfall–runoff produced when using the initial loss estimates of SCS TR-55 methods for small storms (Harrington, 1987). Therefore, for example, capture of the first inch of rainfall can equate to capture of the first half-inch of runoff given an appropriate C factor for the watershed.

The size of event or depth of runoff to be captured and treated depends on several factors, including the pollution target, loading rate, type of BMP, its location (online or offline), etc. The total pollution removed is a function of the portion captured and the part of the captured portion that is actually removed prior to discharge from the BMP back to receiving waters. For example, treating 1/2 in of runoff totally through infiltration may be equivalent to treating 1 in of runoff partially through wet pond removal mechanisms. Or, diverting some small volume of flow through an offline filtration system may be more effective than capturing the total flow for a lesser period through an online extended detention system.

For practical purposes, it can be assumed that the percent of pollution mass captured is proportional to the percent of flow volume captured. Several authors have attempted to demonstrate, using various methods, that capture of about 0.25 to 0.5 in of runoff will capture in the range of 80 to 90% of the annual runoff volume and, thus, that percentage of the pollution (Wanielista and Yousef, 1992; Urbonas, Guo, and Tucker, 1990; Roesner, Burgess, and Aldrich, 1989). The reason is, of course, that the vast majority of the storms in any given year only produce a small amount of runoff. If the area experiences a pronounced first-flush phenomena, then the pollution capture may be in that order.

Another approach used in some locations uses site-based rainfall records. Measures designed to control storms producing less than 1 in (in more arid climates) to about 1 to 1.5 in of rainfall will control nonpoint pollution discharges for over 90% of the storms each year. For example, Figure 13-2 illustrates capture and treatment depth requirements for the 90th percentile storm for Athens, GA. Use of this criteria would result in a design standard for the capture and treatment of the runoff from the first 1.4" of rainfall. Figure

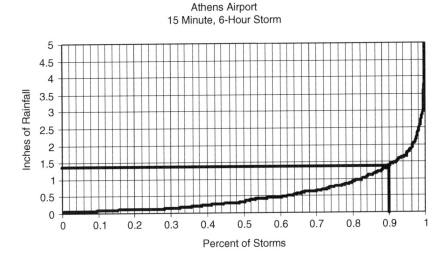

FIGURE 13-2
Storm-based capture criteria — Athens, GA.

13-3 illustrates the impact of arid climate on the 90% capture depth, where the corresponding value is about 0.7 in for Reno, NV.

The first-flush phenomena is much less pronounced for areas of lower imperviousness. Studies have shown that areas with less than 50% impervious will typically show little first flush. They also show that, the higher the imperviousness, the more depth of capture is necessary to achieve the same result. A sliding scale should be specified based on storm runoff from a certain rainfall depth, not a constant depth (Chang and Souer, 1990).

Specific analyses to determine these capture volumes for any location are done using continuous simulation of a long rainfall record or statistical analysis of the rainfall record. A detailed generic method for planning level statistical analysis in areas without developed rules-of-thumb is given by Driscoll (Woodward-Clyde, 1986) and illustrated by

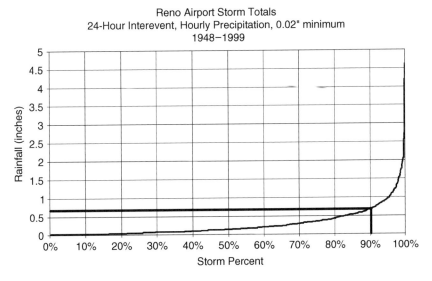

FIGURE 13-3
Storm-based capture criteria — Reno, NV.

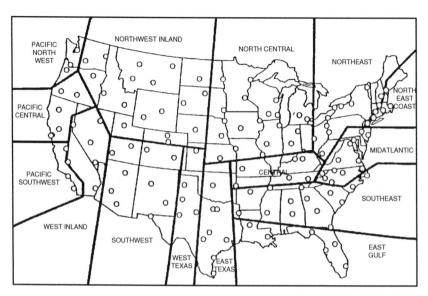

FIGURE 13-4
Fifteen rainfall zones in the United States. (From Driscoll, E.D., Palhegyi, G.E., Strecker, E.W., and Shelley, P.E., Analysis of Storm Event Characteristics for Selected Rainfall Gauges Throughout the United States, Woodward-Clyde Consultants, prepared for U.S. EPA, Washington, DC, 1989.)

Urbonas and Stahre (1993). Nix, Heaney, and Huber (1988) illustrate application of continuous simulation to derive removal efficiencies for trial BMP sizes.

Quite often, capture is based on mean annual or mean storm statistics. Driscoll et al. (1989) provide such statistics for the United States in 15 rainfall zones. Information is for all storms with depths greater than 0.1 in. Depths less than that were considered to produce no runoff. Figure 13-4 and Table 13-9 present a summary of this information for use in

TABLE 13-9

Annual and Individual Storm Statistic Typical Values for 15 Rainfall Zones in the United States

Rainfall Zone	Annual Storm Depth (in)		Annual No. of Storms		Storm Separation (h)		Duration (h)		Intensity (in/h)		Volume (in)	
	Avg.	CV	Avg.	CV	Avg.	CV	Avg.	CV	Avg.	CV	Avg.	CV
Northeast	34.6	0.18	70	0.13	126	0.94	11.2	0.81	0.067	1.23	0.50	0.95
Northeast, coastal	41.4	0.21	63	0.12	140	0.87	11.7	0.77	0.071	1.05	0.66	1.03
Mid-Atlantic	39.5	0.18	62	0.13	143	0.97	10.1	0.84	0.092	1.20	0.64	1.01
Central	41.9	0.19	68	0.14	133	0.99	9.2	0.85	0.097	1.09	0.62	1.00
North Central	29.8	0.22	55	0.16	167	1.17	9.5	0.83	0.087	1.20	0.55	1.01
Southeast	49.0	0.20	65	0.15	136	1.03	8.7	0.92	0.122	1.09	0.75	1.10
East Gulf	53.7	0.23	68	0.17	130	1.25	6.4	1.05	0.178	1.03	0.80	1.19
East Texas	31.2	0.29	41	0.22	213	1.28	8.0	0.97	0.137	1.08	0.76	1.18
West Texas	17.3	0.33	30	0.27	302	1.53	7.4	0.98	0.121	1.13	0.57	1.07
Southwest	7.4	0.37	20	0.30	473	1.46	7.8	0.88	0.079	1.16	0.37	0.88
West, inland	4.9	0.43	14	0.38	786	1.54	9.4	0.75	0.055	1.06	0.36	0.87
Pacific Southwest	10.2	0.42	19	0.36	476	2.09	11.6	0.78	0.054	0.76	0.54	0.98
Northwest, inland	11.5	0.29	31	0.23	304	1.43	10.4	0.82	0.057	1.20	0.37	0.93
Pacific Central	18.4	0.33	32	0.25	265	2.00	13.7	0.80	0.048	0.85	0.58	1.05
Pacific Northwest	35.7	0.19	71	0.15	123	1.50	15.9	0.80	0.035	0.73	0.50	1.09

Source: From Driscoll, E.D., Palhegyi, G.E., Strecker, E.W., and Shelley, P.E., Analysis of Storm Event Characteristics for Selected Rainfall Gauges Throughout the United States, Woodward-Clyde Consultants, prepared for U.S. EPA, Washington, DC, 1989.

TABLE 13-10

Settling Velocities for Typical Urban Runoff Pollutants

Size Fraction	Percent of Particle Mass in Urban Runoff	Average Settling Velocity (ft/h)
1	0–20	0.03
2	20–40	0.3
3	40–60	1.5
4	60–80	7.0
5	80–100	65.0

Source: From Woodward-Clyde Consultants, Methodology for Analysis of Detention Basins for Control of Urban Runoff Quality, prepared for EPA Office of Water, Washington, DC, 1986.

design of extended detention basins and retention basins, where mean storm volume criteria are called for. This local rainfall analysis has been done for many places in the United States, resulting in tables of surface area and drainage area relationships for extended detention ponds with permanent pools (for example, Professional Engineers of North Carolina, 1988, or Washington State, 1992).

CV is the coefficient of variation, and an indication of the scatter of the data, it is defined as the standard deviation divided by the mean. The volume runoff coefficient can be found in the discussion on the simple method. The runoff volume, given the rainfall volume from the table above, is simply the rainfall volume times the volume runoff coefficient.

For designs using settling velocities, typical settling velocity values for urban particulate pollutants are given in Table 13-10 (Woodward-Clyde, 1986).

Unified Sizing Criteria

One method, originated at the Center for Watershed Protection and articulated in several state design manuals including those of Georgia, Maryland, and New York, is the concept of a "unified sizing criteria." This concept states that, for many types of structural BMPs, or BMPs and other structure combinations, there is an ability to size them to mitigate the major impacts of urban development as articulated in five major ways:

- Remove pollutants through capturing and treating a water-quality volume (WQ_v)
- Protect the stability of downstream channels through overdetention of a channel protection volume (CP_v)
- Ensure appropriate recharge of groundwater by infiltrating a recharge volume (RE_v) equal to, or some percentage of, the predevelopment recharge volume
- Protect downstream properties from a fairly frequent nuisance flood, often taken as the 10-year (Q_{p10}) or 25-year (Q_{p25}) storms
- Protect floodplain developments and floodplains from adverse impacts from the extreme flood, often taken as the 100-year flood event (Q_f)

While these conditions take on slightly different details from state to state or application to application, they are all generally similar. Some states omit one or more of the criteria, or handle them in different ways. For example, it might be that the extreme flood is handled through floodplain preservation rather than peak flow reduction. Table 13-11 shows typical requirements.

The idea then is to design a set of structures, and other site designs, that can mitigate all of these potential impacts according to the standards stipulated. Figure 13-5 shows an

TABLE 13-11

Summary of the Stormwater Sizing Criteria for the Unified Sizing Method

Sizing Criteria	Description
Water quality	Treat the runoff from a certain percentage of storms (80 to 95% typically) that occur in an average year; for the Southeast, this may equate to over an inch of rainfall and, for the arid West, less than an inch; the actual number is derived from rainfall data or taken as a rule of thumb if a specific percentage is not specified
Channel protection	Provide extended detention of the (typically 1- or 2-year) storm event released over a period of 24 h to reduce bankfull flows and protect downstream channels from erosive velocities and unstable conditions
Recharge volume	Fraction of water-quality volume depending on hydrologic soil group
Overbank flood protection	Provide peak discharge control of the (10-, 25-year, etc.) storm event such that the postdevelopment peak rate does not exceed the predevelopment rate to reduce overbank flooding
Extreme flood protection	Evaluate the effects of the (100-year typically) storm event on the stormwater management system, adjacent property, and downstream facilities and property; manage the impacts of the extreme storm event through detention controls and/or floodplain management

outlet structure for a wet pond that has been designed to handle these impacts. There is a permanent pool to allow for recharge and partial water-quality treatment. An additional volume, discharging through a reverse slope pipe to reduce clogging potential, handles the remainder of the water-quality volume. A 24-h extended detention volume is stored above the water-quality volume for the 1-year storm released over a 24-h period for channel protection. A riser contains an orifice to handle the 25-year storm, and the 100-year storm is handled by a weir structure.

This unified design criteria is often combined with better site design techniques (CWP, 1998) and even a crediting mechanism wherein the developer reduces the required water-quality treatment volume if environmentally conscious site design practices are provided (such as buffers, grassy swales, etc.). The items listed below summarize this three-prong attack (ARC, 2002).

Stormwater Better Site Design

The first step in addressing stormwater management begins with the site planning and design process. The goal of better site design is to reduce the amount of runoff and pollutants that are generated from a development site and provide for some nonstructural on-site treatment and control of runoff by implementing a combination of approaches collectively known as stormwater better site design practices. These include maximizing the protection of natural features and resources, developing a site design that minimizes impact, reducing overall site imperviousness, and utilizing natural systems for stormwater management.

Unified Stormwater Sizing Criteria

An integrated set of design criteria for stormwater quality and quantity management that address the entire range of hydrologic events is needed. These criteria allow the site engineer to calculate the stormwater control volumes required for water quality, downstream channel protection, and overbank and extreme flood protection.

Stormwater Credits for Better Site Design

A set of stormwater credits can be used to provide developers and site designers an incentive to implement better site design practices that can reduce the volume of storm-

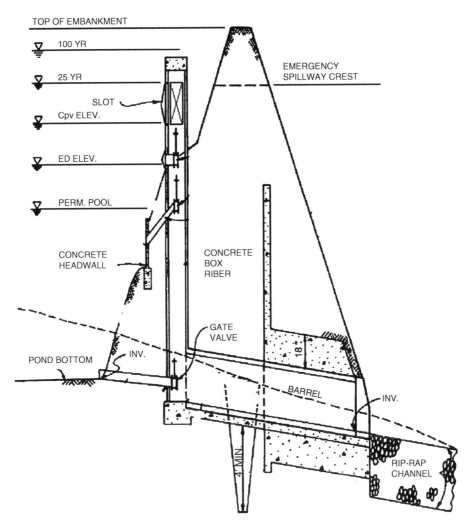

FIGURE 13-5
Outlet structure for a unified criteria wet pond.

water runoff and minimize the pollutant loads from a site. The credit system directly translates into cost savings to the developer by reducing the size of structural stormwater control and conveyance facilities.

13.5 Overview of Structural BMP Types

Basic Structural BMP Types

Structural BMPs can be classified by the predominant removal mechanism into the four major categories of detention basins, filtration devices, vegetative filtration, and special devices.

- *Detention systems*, as the term implies, hold runoff for a period of time, allowing for settling of the solid pollutants. Other removal mechanisms, which are generally of lesser importance, include filtration through vegetation, infiltration, evapotranspiration, and biological and chemical transformation. Basic types include extended detention, retention ponds, and constructed wetlands.

- *Filtration systems* remove pollutants through the natural filtering process of soil and reduction in runoff volume. Infiltration devices generally transport runoff to groundwater. Filtration (often called exfiltration) devices filter water through an engineered layer of soil and then discharge it to the drainage system. Specific devices include infiltration trenches or basins, sand filters, peat filters, exfiltrating storm drain systems and trenches, porous pavement, Washington DC sand filters, Delaware sand filters, Austin first-flush basins, bioretention areas, and variations.

- *Vegetative filtration systems* operate through the contact of runoff with vegetation. They tend to be used for on-site controls. Pollutant removal mechanisms include sedimentation and vegetal filtration, infiltration, evapotranspiration, and biological uptake. Devices include buffers or filter strips with flow spreaders, grassed swales (with or without check dams), and other devices.

- *Specialty devices* tend to be manufactured devices or more highly engineered devices, which have a variety of removal mechanisms and are used on smaller sites, industrial-type sites, or for special applications. They can include such structures as swirl concentrators, oil and grit separators, chemical treatment devices, water-quality inlets, coalescing filters, sumped catch basins, various catch basin inserts, floating curtain devices, and others.

Table 13-12 lists the structural stormwater control practices commonly used in many metropolitan areas. These structural controls are recommended for use in a wide variety of applications. A detailed discussion of most of the these controls, as well as design criteria and procedures, are included later in this chapter.

The following measures are used for stormwater quantity control and not stormwater quality control. For more information, see Chapter 11.

Structural Control	Description
Dry detention/dry extended detention basins	Dry detention basins and dry extended detention (ED) basins are surface facilities intended to provide for the temporary storage of stormwater runoff to reduce downstream water-quantity impacts
Multipurpose detention areas	Multipurpose detention areas are site areas used for one or more specific activities, such as parking lots and rooftops, which are also designed for the temporary storage of runoff
Underground detention	Underground detention tanks and vaults are an alternative to surface dry detention for space-limited areas where there is not adequate land for a dry detention basin or multipurpose detention area

13.6 Selection of Structural Management Measures

Introduction

There are many detailed and readily available discussions of structural best management practices. New data and variations on design criteria and structural variations are often being published. In addition, new devices and modifications on older devices now in experimental stages become available on a regular basis. The reader interested in general

TABLE 13-12

General Application Structural Controls

Structural Control	Description
Stormwater Detention Systems	
Stormwater ponds • Dry extended detention pond • Wet pond • Wet extended detention pond • Micropool extended detention pond • Multiple pond systems	Stormwater ponds are constructed stormwater retention basins that have a permanent pool (or micropool) of water; runoff from each rain event is detained and treated in the pool
Stormwater wetlands • Shallow wetland • Extended detention shallow wetland • Pond/wetland systems • Pocket wetland	Stormwater wetlands are constructed wetland systems used for stormwater management; stormwater wetlands consist of a combination of shallow marsh areas, open water, and semiwet areas above the permanent water surface
Wetland systems • Submerged gravel wetlands	Submerged gravel wetlands systems use wetland plants in a submerged gravel or crushed rock media to remove stormwater pollutants; these systems should only be used in mid- to high-density environments where the use of other structural controls may be precluded; the long-term maintenance burden of these systems is uncertain
Stormwater Filtration Systems	
Bioretention areas	Bioretention areas are shallow stormwater basins or landscaped areas that utilize engineered soils and vegetation to capture and treat stormwater runoff; runoff may be returned to the conveyance system or allowed to partially exfiltrate into the soil
Sand filters • Austin sand filter • Surface sand filter • Perimeter sand filter	Sand filters are multichamber or surface application structures designed to treat stormwater runoff through filtration, using a sand bed as its primary filter media; filtered runoff may be returned to the conveyance system or allowed to partially exfiltrate into the soil
Infiltration trench	An infiltration trench is an excavated trench filled with stone aggregate used to capture and allow infiltration of stormwater runoff into the surrounding soils from the bottom and sides of the trench
Other filtering practices • Organic filter • Underground sand filter	Organic filters are surface sand filters where organic materials such as a leaf compost or peat/sand mixture serve as the filter media; these media may be able to provide enhanced removal of some contaminants, such as heavy metals; given their potentially high maintenance requirements, they should only be used in environments that warrant their use Underground sand filters are sand filter systems located in an underground vault. These systems should only be considered for extremely high density or space-limited sites.
Infiltration basin	While, in theory, infiltration basins provide excellent pollutant removal capabilities, the reality is that infiltration basins have historically experienced high rates of failure due to clogging associated with poor design, construction, and maintenance; in addition, because many areas have soils with high clay content, the infiltration basin has limited applicability; they also could have an unacceptably high maintenance burden; in some areas with high infiltration capacity, pollution of groundwater may be a major concern; thus, they are *not* recommended

TABLE 13-12 *(Continued)*

General Application Structural Controls

Structural Control	Description
Vegetative Filtration Systems	
Enhanced swales • Dry swale • Wet swale/wetland channel	Enhanced swales are vegetated open channels explicitly designed and constructed to capture and treat stormwater runoff within dry or wet cells formed by check dams or other means
Biofilters • Filter strip • Grass channel	Filter strips and grass channels provide biofiltering of stormwater runoff as it flows across the grass surface; however, by themselves, these controls usually cannot meet the performance goals; consequently, filter strips and grass channels should be used only as a pretreatment measure or as part of a treatment train approach
Specialty Devices	
Hydrodynamic devices • Gravity (oil-grit) separator	Hydrodynamic controls use the movement of stormwater runoff through a specially designed structure to remove target pollutants; they are typically used on smaller impervious commercial sites and urban hot spots; these controls typically do not meet water-quality removal performance goals and therefore should be used only as a pretreatment measure and as part of a treatment approach
Porous surfaces • Porous concrete • Modular porous paver systems	Porous surfaces are permeable pavement surfaces with an underlying stone reservoir to temporarily store surface runoff before it infiltrates into the subsoil; porous concrete is the term for a mixture of course aggregate, Portland cement, and water that allows for rapid infiltration of water; modular porous paver systems consist of open void paver units laid on a gravel subgrade; both porous concrete and porous paver systems provide water-quality and quantity benefits, but have high workmanship and maintenance requirements, as well as high failure rates
Alum treatment	Alum treatment provides for the removal of suspended solids from stormwater runoff entering a wet pond by injecting liquid alum into storm drain lines on a flow-weighted basis during rain events; alum treatment should be only considered for large-scale projects where high water quality is desired
Porous asphalt	Porous asphalt surfaces are easily clogged by clays, silts, and oils, resulting in a potentially high maintenance burden to maintain the effectiveness of this structural control; further, summer heat in some locations can cause the asphalt to melt, destroying the porous properties of the surface
Media filter inserts	Media filter inserts such as catch basin inserts and filter systems are easily clogged and require a high degree of regular maintenance and replacement to achieve the intended water-quality treatment performance, and should not be used for areas of new development or redevelopment; these structural controls may serve a potential use in stormwater retrofitting
Proprietary systems • Commercial stormwater controls	Proprietary controls are manufactured structural control systems available from commercial vendors designed to treat stormwater runoff and/or provide water-quantity control. Proprietary systems often can be used on small sites and in space-limited areas, as well as in pretreatment applications; however, proprietary systems are often more costly than other alternatives, may have high maintenance requirements, and often lack adequate independent performance data

and detailed design discussions of BMPs should consult any of the following: EPA (1983), MWCOG (1987, 1992), California Stormwater Task Force (1993a), WA State (1992), Austin Texas (1988), Northern Virginia (1992), Southwest Florida Watershed Management District (1986, 1990), Denver Urban Drainage and Flood Control District (1993), Maestri et al. (1988), Minnesota PCA (1989), Massachusetts Department of Environmental Protection (1997), USEPA (1999a), Atlanta Regional Commission, (2002), and local or state guidance, or specific BMP studies from the literature. Many states, regional authorities, and local governments have published BMP manuals on the Internet (see Chapter 15 for several site listings) including New York, Maryland, Massachusetts, Idaho, Texas, Vermont, and Washington, with more appearing almost monthly. It should be noted that there are at least five or six other major sources of pollution where structural BMPs are applicable (as well as nonstructural BMPs) that are adequately covered elsewhere (including EPA guidance documents and their Web site) and will not be treated in this book:

- Soil erosion during construction is an important source of pollution of receiving waters. Structural erosion control measures are well covered in state erosion control manuals such as North Carolina (1991), Virginia (1992) or Georgia (undated), and EPA publications (EPA, 1992a).
- Pollution from nonurban nonpoint sources (e.g., agriculture, silvaculture) may greatly affect the incoming quality of surface waters to the urban environment and the ability of the municipality to meet its own targets and goals. Such nonpoint measures are covered in detail in USEPA (1993) and numerous other references and recent Web sites.
- Pollution from industrial facilities can also be a source of standard and non-standard urban pollution. There are also certain municipal activities considered to be equivalent to industrial activities requiring NPDES permits (such as vehicle maintenance facilities, landfills, airports, etc.). Development of industrial BMP plans is covered in detail in EPA publications (EPA, 1992b) and in state and local guidance (such as California Stormwater Task Force, 1993b), and on Web site manuals targeting specific industries.
- Illicit connections and illegal dumping can also be important sources of pollution. The permanent connection to the drainage system is normally illegal without an NPDES permit. Structural controls have been used to treat the stormwater prior to discharge or in lieu of discharge, though the controls are similar to those for other industrial and commercial uses (EPA, 1993).
- Combined sewer and sanitary sewer overflows are handled in different evolving federal programs. Combined sewer overflows are a major source of urban pollution during wet weather and a major source of cost for treatment at overtaxed wastewater plants. At times, the interaction between separating or managing combined sewers and fixing leaking sanitary sewers causes stormwater problems, as drainage flow paths are cut off, and storm flows find other routes downhill.

Any BMP has unique capabilities and persistent limitations. Additionally, each site imposes limitations and each overall watershed location requires consideration of the stormwater and other management objectives for that watershed. Selection of BMPs is always a balance between competing needs, physical limitations, sociological or institutional constraints, financial constraints, and pollution removal needs.

The longevity of some BMPs is limited to such a degree that their widespread use is encouraged only under certain circumstances. Of particular concern are the infiltration practices, such as basins, trenches, and porous pavement. The poor longevity of these

BMPs is attributable to a number of factors: lack of pretreatment, poor construction practices, application to infeasible sites, lack of regular maintenance, and, in some cases, fundamental difficulties in basic design. Very often, the lifespans of BMPs can be increased to acceptable lengths if local communities adopt enhanced designs and commit to strong maintenance and inspection programs.

BMP options are adaptable to most regions of the country, though there may be extra efforts required for extremely arid regions of the West and colder climates of the North. In these regions, conventional BMP designs may need to be refined to account for high evaporation rates during the growing season or subfreezing and snow melt conditions as well as other issues.

No single BMP option can be applied to all development situations, and all BMP options require careful site assessment prior to design. Pond options are applicable to the widest range of development situations, but typically require a minimum drainage area. On the other hand, infiltration practices often have more limited applications and require field verification of soils, water tables, slope, and other factors.

Several BMPs can have significant secondary environmental impacts, although the extent and nature of these impacts is uncertain and site-specific. Pond systems, which offer reliable pollutant removal and longevity, tend to be associated with the greatest number and strongest degree of secondary environmental impacts. Careful site assessment and design are often required to prevent stream warming, natural wetland destruction, and riparian habitat modification. Infiltration of polluted stormwater to drinking water supplies would also be of concern, and would limit use of such devices to nonwater supply area applications.

Relatively limited cost data exist to aid in the assessment of the comparative cost-effectiveness of urban BMP options. Presently, the selection of BMPs is based more on longevity, feasibility, and local design factors than on comparative cost. A couple of more recent summaries of cost information provide a good starting place to search the literature (Raghavan, Koustas, and Liao, 2001; Brown and Schueler, 1997). It is expected that construction costs for all BMPs will increase in the future due to the enhanced designs needed for more reliable pollutant removal and longevity, and to handle ecological impacts of urban development. Costs may also increase in response to increasingly complex permitting requirements. Maintenance costs may rival construction costs over the design life of many BMPs; however, many jurisdictions currently do not have very active BMP maintenance programs.

Several fundamental uncertainties still exist with respect to urban BMPs and need to be resolved through basic research. These uncertainties include the toxicity of residuals trapped within BMPs; the interaction of groundwater and BMPs (both ponds and infiltration); impacts of different kinds of BMPs on downstream ecology and channel stability; the ability of BMPs to remove dissolved pollutants; the ability for microscale BMPs to be maintained by private property owners; and the long-term performance of urban BMPs (Urbonas, 2001).

General BMP Pollutant Removal Effectiveness

Structures designed without specific consideration of stormwater quality improvement are generally ineffective for pollutant removal, except for removal of the largest particles. Following are some examples of such measures:

- *Sumped catch basins* — Catch basins should be considered only as temporary storage locations for pollutant loads. Sediment and adhered pollution may settle out during the recession limb of a storm hydrograph, but the next storm will mobilize and flush out such sediments, leaving a new deposit. Cleaning is seldom

routine or sufficiently frequent. The first flush from catch basins is often more septic than raw sewage.
- *Dry detention basins* — While a municipality may consider dry detention basins effective for stormwater quantity management, a detention time of even a few hours is insufficient to permit settling of the smaller fractions of the suspended materials associated with pollutants.
- *Temporary filtration systems* — Filtration systems (straw bales, sand bags, filter cloth fences, gravel filters) are used extensively as temporary sediment control measures during construction and when establishing a new vegetative cover. However, the finer solids are not effectively trapped, and runoff pollution control levels are low.

Structural stormwater controls are engineered facilities intended to treat stormwater runoff and mitigate the effects of increased stormwater runoff peak rate, volume, and velocity due to urbanization. Though each of these structural controls provides pollutant removal capabilities, the relative capabilities vary between structural control practices and for different pollutant types.

Pollutant removal capabilities for a given structural stormwater control practice are based on a number of factors, including the physical, chemical, and biological processes that take place in the structural control and the design and sizing of the facility. In addition, pollutant removal efficiencies for the same structural control type and facility design can vary widely, depending on the tributary land use and area, incoming pollutant concentration, rainfall pattern, time of year, maintenance frequency, and numerous other factors.

There is some discussion on whether pollution removal efficiency measured in terms of percentage removal of pollution from influent is the best approach (Strecker, Quigley, and Urbonas, 2000; Winer, 2000). For example, even the most rudimentary BMP could attain a very high TSS removal if mudflow entered the control, even though the effluent would still have a high TSS loading. On the other hand, the most comprehensively designed site treatment train could achieve little removal if nearly clear water were entering the system. However, given the current state of the art, the ability to determine actual effluent concentrations is limited to those sites with extensive monitoring and may be reserved for only the most stringent numerical limited situations.

To assist the designer in evaluating the relative pollutant removal performance of the various structural control options, Table 13-13 provides *median* removal efficiencies for most of the standard control practices included in this chapter, as well as standard deviations in the data set for many sets of controls. Pollutant removal capabilities can vary significantly for several reasons, including design variations, monitoring error, natural scatter in pollution loading, seasonal or climatic factors, and differences in incoming effluent (EPA, 1999a).

Design pollutant reduction percentages should be derived from sampling data, design comparisons, and professional judgment. A structural control design may be capable of exceeding these performances, however, the values in the table are reasonable median values that can be assumed to be achieved for most older BMPs. When the structural control is sized, designed, constructed, and maintained in accordance with recommended specifications and practices in this chapter, higher removals can be anticipated. The data can be interpreted and applied in the following ways:

- If the design of the structure does not (or has not) rigidly follow(ed) the criteria below, then pollutant removal of 80% TSS and other removals cannot be assured. In that case, median values or below should be used. The designer can access

TABLE 13-13

Pollutant Removal Efficiencies for Structural Stormwater Controls

	TSS	TP	Sol P	TN	NO$_x$[b]	Cu	Zn
Stormwater Dry Ponds							
Quantity control pond	3	19	0	5	9	10	5
Dry extended detention pond	61	20	−11	31[a]	−2[a]	29[a]	29[a]
Stormwater Wet Ponds							
Wet pond	79	49	62	32	36	58	65
Wet ED pond	80	55	67	35	63	44	69
Multi-pond system	91	76	69	N/D	87	N/D	N/D
Group median ± 1 std. dev.	80 ± 27	51 ± 21	66 ± 27	33 ± 20	43 ± 39	57 ± 22	66 ± 22
Stormwater Wetlands							
Shallow marsh	83	43	29	26	73	33	42
Extended detention wetland	69	39	32	56	35	N/D	−74
Pond/wetland system	71	56	43	19	40	58[a]	56
Submerged gravel wetland[a]	83	64	−10	19	81	21	55
Group median ± 1 std. dev.	76 ± 43	49 ± 36	36 ± 45	30 ± 34	67 ± 54	40 ± 45	44 ± 40
Pocket wetland[a]	76	50	N/D	N/D	N/D	N/D	40
Filtration Practices							
Bioretention[a]	N/D	65	N/D	49	16	97	95
Surface sand filter	87	59	−17[a]	32	−13	49	80
Perimeter sand filter[a]	79	41	68	47	−53	25	69
Organic filter	88	61	30[a]	41[a]	−15	66[a]	89
Underground sand filter[a]	58	45	21	5	−87	32	56
Group median ± 1 std. dev.	86 ± 23	59 ± 38	3 ± 46	38 ± 16	−14 ± 47	49 ± 26	88 ± 17
Infiltration Practices							
Infiltration trench[a]	100	42	100	42	82	N/D	N/D
Porous pavement[a]	95	65	10	83	N/D	N/D	99
Group median ± 1 std. dev.	95	70 ± 37	85.3[a]	51 ± 24	82[a]	N/D	99[a]
Vegetative Filtration Systems							
Ditches (no WQ design features)	31	−16	−25[a]	−9	24[a]	14[a]	0[a]
Grass channel[a]	68	29	40	N/D	−25	42	45
Dry swale[a]	93	83	70	92	90	70	86
Wet swale[a]	74	28	−31	40	31	11	33
Group median (w/o ditches) ± 1 std. dev.	81 ± 14	34 ± 33	38 ± 46	84[a]	31 ± 49	51 ± 40	71 ± 36
Specialty Devices							
Oil-grit separator[a]	−8	−41	40	N/D	47	−11	17
Stormceptor®[a]	25	19	21	N/D	6	30	21

[a] Indicates fewer than five data points.
[b] Nitrate and nitrite nitrogen.

Note: The performance of specific proprietary commercial devices and systems must be provided by the manufacturer and should be verified by independent third-party sources and data.

the national BMP database (referenced below) to compare the design to others and help determine probable removal that way.

- If the design of the structure follows the guidance in this chapter, then, for most of the structures (with the exception of dry detention), 80% TSS removal can be assumed to be achieved.

Where the pollutant removal capabilities of an individual structural stormwater control are not deemed sufficient for a given site application, additional controls may be used in

TABLE 13-14

Pathogen, Carbon, and Hydrocarbon Pollutant Removal Efficiencies for Structural Stormwater Controls

	Bacteria[b]	Organic Carbon[c]	Hydrocarbons
Stormwater wet ponds	70	43	81[a]
Stormwater dry ponds	78[a]	25	N/D
Stormwater wetlands	78[a]	18	85[a]
Filtering practices	37	54	84[a]
Water quality swales	−25[a]	69[a]	62[a]
Ditches (no WQ design features)	5	18	N/D

[a] Indicates fewer than five data points.
[b] Includes fecal streptococci, fecal coliform, E. coli, and total coliform.
[c] Includes BOD, COD and TOC removal data

series in a treatment train approach. More details on using structural stormwater controls in series are provided below.

For additional information and data on the ever-changing range of pollutant removal capabilities for various structural stormwater controls, the reader is referred to the National Pollutant Removal Performance Database (2nd ed., Winer, 2000) and the National Stormwater Best Management Practices (BMP) Database at www.bmpdatabase.org.

Bacteria and other pathogen removals, and organic carbon and hydrocarbon removals are given by Winer (2000) and are summarized by structural group in Table 13-14.

Structural BMP Screening

Outlined below is a screening process for structural stormwater controls. This process is intended to assist the site designer and design engineer in the selection of the most appropriate structural controls for a development site, and provides guidance on factors to consider in their location.

In general, the following four criteria should be evaluated in order to select the appropriate structural control(s) or group of controls for a development:

- Stormwater treatment suitability
- Water-quality performance
- Site applicability
- Implementation considerations

In addition, for a given site, the following factors should be considered, and any specific design criteria or restrictions need to be evaluated:

- Physiographic factors
- Soils
- Special watershed or stream considerations

Finally, environmental regulations that may influence the location of a structural control on site, or may require a permit, need to be considered.

The following steps provide a selection process for comparing and evaluating various structural stormwater controls using two screening matrices and a list of location and permitting factors. These tools are provided to assist the design engineer in selecting the subset of structural controls that will meet the stormwater management and design objectives for a development site or project.

The tables provided are taken from the Georgia Stormwater Manual for which the authors were principals in its development. Thus, the tables should be locally modified for such things as minimum watershed drainage area, etc. For example, watershed area for a wet pond must ensure the ability to maintain the pond elevation. When in doubt, a water balance (see Chapter 7) should be accomplished.

Step 1: Overall Applicability

Through the use of Table 13-15, the site designer evaluates and screens the overall applicability of the full set of structural controls, as well as the constraints of the site in question. The following are the details of the various screening categories and individual characteristics used to evaluate the structural controls.

Stormwater Management Suitability

The first columns of the table examine the capability of each structural control option to provide water-quality treatment, downstream channel protection, overbank flood protection, and extreme flood protection; other suitable factors could be added for local application. A blank entry means that the structural control cannot or is not typically used to meet a particular criterion. This does not necessarily mean that it should be eliminated from consideration, but rather is a reminder that more than one structural control may be needed at a site (e.g., a bioretention area used in conjunction with dry detention storage). The following are given as example suitability criteria, which are often selected.

- *Ability to treat the water-quality volume* (WQ_v) — This indicates whether a structural control provides treatment of a selected water-quality volume (WQ_v).
- *Ability to provide channel protection* (CP_v) — This indicates whether the structural control can be used to provide the extended detention of a selected channel-protection volume (CP_v). The presence of a check mark indicates that the structural control can be used to meet CP_v requirements. A star indicates that the structural control may be sized to provide channel protection in certain situations, for instance, on small sites.
- *Ability to provide overbank flood protection* (Q_{p25}) — This indicates whether a structural control can be used to meet the overbank flood-protection criteria. The presence of a check mark indicates that the structural control can be used to provide peak reduction of a design storm event.
- *Ability to provide extreme flood protection* (Q_f) — This indicates whether a structural control can be used to meet the extreme flood protection criteria. The presence of a check mark indicates that the structural control can be used to provide peak reduction of the 100-year storm event.

Relative Water-Quality Performance

The second group of columns in Table 13-15 provides an overview of the pollutant removal performance of each structural control option, when designed, constructed, and maintained according to approved criteria and specifications.

- *Ability to provide TSS and sediment removal* — This column indicates the capability of a structural control to remove sediment in runoff to a level acceptable to the local municipality.

Structural Best Management Practices

TABLE 13-15
Structural Control Screening Matrix — Overall Applicability

Structural Control Category	Structural Control	Stormwater Treatment Suitability				Water Quality Performance*				Site Applicability					Implementation Considerations			
		Water Quality	Channel Protection	Overbank Flood Protection	Extreme Flood Protection	TSS/ Sediment Removal Rate	Nutrient Removal Rate (TP/TN)	Bacteria Removal Rate	Hotspot Application	Drainage Area (acres)	Space Req'd (% of tributary imp. Area)	Site Slope	Minimum Head Required	Depth to Water Table	Residential Subdivision Use	High Density/ Ultra-Urban	Capital Cost	Maintenance Burden
Stormwater ponds	Wet pond	✓	✓	✓	✓				✓	25 min**					✓		Low	Low
	Wet ED pond	✓	✓	✓	✓	80%	50%/30%	70%	✓	10 min**	2–3%	15% max	6 to 8 ft	2 ft, if hotspot or aquifer	✓		Low	Low
	Micropool ED pond	✓	✓	✓	✓				✓	25 min**					✓		Low	Moderate
	Multiple ponds	✓	✓	✓	✓				✓						✓		Low	Low
Stormwater wetlands	Shallow wetland	✓	✓	✓	✓				✓						✓		Moderate	Moderate
	Whallow ED wetland	✓	✓	✓	✓	80%	40%/30%	70%	✓	25 min	3–5%	8% max	3 to 5 ft	2 ft, if hotspot or aquifer	✓		Moderate	Moderate
	Pond/wetland	✓	✓	✓					✓				6 to 8 ft		✓		Moderate	Moderate
	Pocket wetland	✓	✓						✓	5 min			2 to 3 ft	below WT	✓	✓	Moderate	High
Bioretention	Bioretention areas	✓	★			80%	60%/50%	Insuff. data	✓	5 max**	5%	6% max	5 ft	2 ft	✓	✓	Moderate	Moderate
Sand filters	Surface sand filter	✓	★						✓	10 max**		6% max	5 ft	2 ft		✓	High	High
	Perimeter sand filter	✓	★			80%	50%/25%	40%	✓	2 max**	2–3%		2 to 3 ft			✓	High	High
Infiltration	Infiltration trench	✓	★			80%	60%/60%	90%		5 max	2–3%	6% max	1 ft	4 ft	✓	✓	High	High
Enhanced swales	Dry swale	✓	★			80%	50%/50%	Insuff. data	✓	5 max		4% max	3 to 5 ft	2 ft	✓		Moderate	Low
	Wet swale	✓	★			80%	25%/40%	Insuff. data	✓	5 max	10–20%		1 ft	below VT	✓		High	Low

Note: ✓ = meets suitability criteria; ★ = Can be incorporated into the structural control in certain situations.
* Pollutant removal rates are average removal efficiencies for design purposes.
** Smaller area acceptable with adequate water balance and anti-clogging device.
*** Drainage area can be larger in some instances.

Source: Georgia Stormwater Management Manual, 2001.

- *Ability to provide nutrient treatment* — This column indicates the capability of a structural control to remove the nutrients nitrogen and phosphorus in runoff, which may be of particular concern with certain downstream receiving waters.
- *Ability to provide bacteria removal* — This column indicates the capability of a structural control to remove bacteria in runoff. This capability may be of particular focus in areas with public beaches, shellfish beds, or to meet water regulatory quality criteria under the Total Maximum Daily Load (TMDL) program.
- *Ability to accept hot spot runoff* — This last column indicates the capability of a structural control to treat runoff from designated hot spots. Hot spots are land uses or activities with higher potential pollutant loadings. Examples of hot spots might include gas stations, convenience stores, marinas, public works storage areas, vehicle service and maintenance areas, commercial nurseries, and auto recycling facilities. A check mark indicates that the structural control may be used on a hot spot site, however, it may have specific design restrictions. Please see the specific design criteria of the structural control for more details.

Site Applicability

The third group of columns in Table 13-15 provides an overview of the specific site conditions or criteria that must be met for a particular structural control to be suitable. In some cases, these values are recommended values or limits that can be exceeded or reduced with proper design or depending on specific circumstances.

- *Drainage area* —This column indicates the approximate minimum or maximum drainage area that is considered suitable for the structural control practice. If the drainage area present at a site is slightly greater than the maximum allowable drainage area for a practice, some leeway can be permitted if more than one practice can be installed. The minimum drainage areas indicated for ponds and wetlands should not be considered inflexible limits, and may be increased or decreased depending on water availability (baseflow or groundwater), the mechanisms employed to prevent outlet clogging, or design variations used to maintain a permanent pool (e.g., liners).
- *Space required (space consumed)* —This comparative index expresses how much space a structural control typically consumes at a site in terms of the approximate area required as a percentage of the area draining to the control.
- *Slope* — This column evaluates the effect of slope on the structural control practice. Specifically, the slope restrictions refer to how flat the area where the facility is installed must be and how steep the contributing drainage area or flow length can be.
- *Minimum head* — This column provides an estimate of the minimum elevation difference needed at a site (from the inflow to the outflow) to allow for gravity operation within the structural control.
- *Water table* — This column indicates the minimum depth to the seasonally high water table from the bottom or floor of a structural control.

Implementation Considerations

The last group of columns of Table 13-15 provide additional considerations for the applicability of each structural control option.

- *Residential subdivision use* — This column identifies whether or not a structural control is suitable for typical residential subdivision development (not including high-density or ultra-urban areas).
- *Ultra-urban* — This column identifies those structural controls that are appropriate for use in very high-density (ultra-urban) areas, or areas where space is a premium.
- *Construction cost* — The structural controls are ranked according to their relative construction cost per impervious acre treated as determined from cost surveys.
- *Maintenance* — This column assesses the relative maintenance effort needed for a structural stormwater control, in terms of three criteria: frequency of scheduled maintenance, chronic maintenance problems (such as clogging), and reported failure rates. It should be noted that all structural controls require routine inspection and maintenance.

Step 2: Specific Criteria

Table 13-16 provides an overview of various specific design criteria and specifications, or exclusions for a structural control that may be present due to a site's general physiographic character, soils, or location in a watershed with special water resources considerations.

Physiographic Factors

Three general factors to consider are low-relief, high-relief, and karst terrain. Note: Other local physiographic factors may also be important in some locations. Special geotechnical testing requirements may be needed in karst areas. The local reviewing authority should be consulted to determine if a project is subject to terrain constraints.

- Low-relief areas need special consideration, because many structural controls require a hydraulic head to move stormwater runoff through the facility.
- High relief may limit the use of some structural controls that need flat or gently sloping areas to settle out sediment or to reduce velocities. In other cases, high relief may impact dam heights to the point that a structural control becomes infeasible.
- Karst terrain can limit the use of some structural controls as the infiltration of polluted waters directly into underground streams found in karst areas may be prohibited. In addition, ponding areas may not reliably hold water in karst areas.

Soils

The key evaluation factors are based on an initial investigation of the NRCS hydrologic soils groups at the site. Note that more detailed geotechnical tests are usually required for infiltration feasibility and during design to confirm permeability and other factors.

Special Watershed or Stream Considerations

The design of structural stormwater controls is fundamentally influenced by the nature of the downstream water body that will be receiving the stormwater discharge. Consequently, designers should determine any use classifications that have been developed for the watershed in which their project is located prior to design. In addition, the designer should consult with the appropriate review authority to determine if the development project is subject to additional structural control criteria as a result of an adopted local watershed plan or special provision.

TABLE 13-16
Structural Control Screening Matrix — Specific Criteria (General Application Controls)

Structural Control Category	Physiographic Factors				Special Watershed Considerations					
	Low Relief	High Relief	Karst	Soils	Trout Stream	High-Quality Stream	Acquifer Protection	Reservoir Protection	Shellfish/Beach
Stormwater ponds	Limit maximum normal pool depth to about 4 ft (dugout) Providing pond drain can be problematic	Embankment heights restricted	Require poly or clay liner Max ponding depth Geotechnical tests	"A" soils may require pond liner "B" soils may require infiltration testing	Limit use of due to thermal impacts Limit ED to 12 hr Offline design and provide shading	Evaluate stream warming	May require liner if "A" soils are present Pretreat hotspots 2 to 4 ft separation distance from water table		Moderate bacteria removal Design for waterfowl prevention Provide 48 hr ED for max coliform dieoff
Stormwater wetlands		Embankment heights restricted	Require poly liner Geotechnical tests	"A" soils may require pond liner	Limit use of due to thermal impacts Offline design and provide shading	Evaluate for stream warming	May require liner if "A" soils are present Pretreat hotspots 2 to 4 ft separation distance from water table		Provide 48 hr ED for max coliform dieoff
Bioretention and sand filters	Several design variations will likely be limited by low head		Use poly liner or impermeable membrane to seal bottom	Clay or silty soils may require pretreatment	Evaluate for stream warming		Needs to be designed with no exfiltration (i.e., outflow to groundwater)		Moderate to high coliform removal
Infiltration	Minimum distance to water table of 2 ft	Maximum slope of 6% Trenches must have flat bottom	Generally not allowed	Infiltration rate > 0.5 in./hr			Maintain safe distance from wells and water table No hotspot runoff	Maintain safe distance from bedrock and water table Pretreat runoff	Maintain safe distance from water table
Enhanced swales	Generally feasible; however, slope <1% may lead to standing water in dry swales	Often infeasible if slopes are 4% of greater					Hotspot runoff must be adequately treated	Hotspot runoff must be adequately treated	Poor coliform removal

Source: Georgia Stormwater Management Manual, 2001.

In some cases, higher pollutant removal or environmental performance is needed to fully protect aquatic resources and human health and safety within a particular watershed or receiving water. Therefore, special design criteria for a particular structural control or the exclusion of one or more controls may need to be considered within these watersheds or areas. Examples of important watershed factors to consider include:

- *Primary trout streams* — Cold- and cool-water streams have habitat qualities capable of supporting trout and other sensitive aquatic organisms. Therefore, the design objective for these streams is to maintain habitat quality by preventing stream warming, maintaining natural recharge, preventing bank and channel erosion, and preserving the natural riparian corridor. Some structural controls can have adverse downstream impacts on cold-water streams, and their design may need to be modified or use restricted.
- *High–quality streams* — These streams may also possess high–quality cool-water or warm-water aquatic resources or endangered species. The design objectives are to maintain habitat quality through the same techniques used for cold-water streams, with the exception that stream warming is not as severe of a design constraint. These streams may also be specially designated by local authorities.
- *Wellhead protection* — Areas that recharge existing public water supply wells present a unique management challenge. The key design constraint is to prevent possible groundwater contamination by preventing infiltration of hot spot runoff. At the same time, recharge of unpolluted stormwater is encouraged to maintain flow in streams and wells during dry weather.
- *Reservoir or drinking water protection* — Watersheds that deliver surface runoff to a public water supply reservoir or impoundment are a special concern. Depending on the treatment available at the water intake, it may be necessary to achieve a greater level of pollutant removal for the pollutants of concern, such as bacteria pathogens, nutrients, sediment, or metals. One particular management concern for reservoirs is ensuring that stormwater hot spots are adequately treated so that they do not contaminate drinking water.
- *Swimming/shellfish* — Watersheds that drain to public swimming waters or shellfish harvesting areas require a higher level of stormwater treatment to prevent closings caused by bacterial contamination from stormwater runoff. In these watersheds, structural controls should be explicitly designed to maximize bacteria removal.

Step 3: Location and Permitting Considerations

In the last step, a site designer assesses the physical and environmental features at the site to determine the optimal location for the selected structural control or group of controls. Table 13-17 provides a condensed summary of current restrictions as they relate to common site features that may be regulated under local, state, or federal law. These restrictions fall into one of three general categories:

- Locating a structural control within an area that is expressly prohibited by law
- Locating a structural control within an area that is strongly discouraged, and is allowed only on a case by case basis (Local, state, and federal permits must be

TABLE 13-17

Location and Permitting Checklist

Site Feature	Location and Permitting Guidance
Jurisdictional wetland (waters of the U.S.) U.S. Army Corps of Engineers Section 404 Permit	Jurisdictional wetlands should be delineated prior to siting structural control. Use of natural wetlands for stormwater quality treatment is contrary to the goals of the Clean Water Act and should be avoided. Stormwater should be treated prior to discharge into a natural wetland. Structural controls may also be *restricted* in local buffer zones, although they may be utilized as a nonstructural filter strip (i.e., accept sheet flow). Should justify that no practical upland treatment alternatives exist. Where practical, excess stormwater flows should be conveyed away from jurisdictional wetlands.
Stream channel (waters of the U.S) U.S. Army Corps of Engineers Section 404 Permit	All waters of the U.S. (streams, ponds, lakes, etc.) should be delineated prior to design. Use of any waters of the U.S. for stormwater quality treatment is contrary to the goals of the Clean Water Act and should be avoided. Stormwater should be treated prior to discharge into waters of the U.S. In-stream ponds for stormwater quality treatment are highly discouraged. Must justify that no practical upland treatment alternatives exist. Temporary runoff storage preferred over permanent pools. Implement measures that reduce downstream warming.
100-year floodplain Local stormwater review authority	Grading and fill for structural control construction is generally discouraged within the ultimate 100-year floodplain, as delineated by FEMA flood insurance rate maps, FEMA flood boundary and floodway maps, or more stringent local floodplain maps. Floodplain fill cannot raise the floodplain water surface elevation by more than a tenth of a foot.
Stream buffer Check with appropriate review authority whether stream buffers are required	Consult local authority for stormwater policy. Structural controls are discouraged in the streamside zone (within 25 ft or more of streambank, depending on the specific regulations). There may be specific additional requirements by related local laws and regulations.
Utilities Local review authority	Call appropriate agency to locate existing utilities prior to design. Note the location of proposed utilities to serve development. Structural controls are discouraged within utility easements or rights of way for public or private utilities.
Roads Local DOT, DPW, or state DOT	Consult local DOT or DPW for any setback requirement from local roads. Consult DOT for setbacks from state-maintained roads. Approval must also be obtained for any stormwater discharges to a local or state-owned conveyance channel.
Structures Local review authority	Consult local review authority for structural control setbacks from structures. Recommended setbacks for each structural control group are provided in the performance criteria in this manual.
Septic drain fields Local health authority	Consult local health authority. Recommended setback is a minimum of 50 ft from drain field edge.
Water wells Local health authority State laws and restrictions related to site features	Typically 100-ft setback for stormwater infiltration. Typically 50-ft setback for all other structural controls. Consult location and permitting guidance from appropriate state agencies.

obtained, and the applicant will need to supply additional documentation to justify locating the stormwater control within the regulated area.)

- Structural stormwater controls must be set back a fixed distance from the site feature

Structural Best Management Practices

This checklist is intended only as a general guide to location and permitting requirements as they relate to siting of stormwater structural controls. Consultation with the appropriate regulatory agency is the best strategy.

Example Application

A 20-acre institutional area (e.g., church and associated buildings) is being constructed in a dense urban area within a metropolitan area. The impervious coverage of the site is 40%. The site drains to an urban stream that is highly impacted from hydrologic alterations (accelerated channel erosion). The stream channel is deeply incised, consequently, flooding is not a problem. The channel drains to an urban river that is tributary to a phosphorus-limited drinking water reservoir. Low-permeability soils limit infiltration practices.

- *Objective:* Avoid additional disruptions to receiving channel and reduce pollutant loads for sediment and phosphorus to receiving waters.
- *Target removals:* Provide stormwater management to mitigate for accelerated channel incision and reduce loadings of key pollutants by the following:
  - Sediment: 80%
  - Phosphorus: 40%
- *Activity/runoff characteristics:* The proposed site is to have large areas of impervious surface in the form of parking and structures. However, there will be a large contiguous portion of turf grass proposed for the front of the parcel that will have a relatively steep slope (approximately 10%) and will drain to the storm drain system associated with the entrance drive. Stormwater runoff from the site is expected to exhibit fairly high sediment levels and seasonally high phosphorus levels (due to turf grass management).

Table 13-18 lists the results of the selection analysis using Tables 13-15 and 13-16 described previously.

While there is a downstream reservoir to consider, there are no special watershed factors or physiographic factors that preclude the use of any of the practices from the structural control list. However, due to the size of the drainage area, most stormwater ponds and wetlands are removed from consideration. In addition, the site's impermeable soils remove an infiltration trench from being considered. Due to the need to provide overbank flood control as well as channel protection storage, a micropool ED pond will likely be needed, unless some downstream regional storage is available to control the overbank flood.

To provide additional pollutant removal capabilities in an attempt to better meet the target removals, bioretention, surface sand filters, and perimeter sand filters can be used to treat the parking lot and driveway runoff. The bioretention provides some removal of phosphorus while improving the aesthetics of the site. Surface sand filters provide higher phosphorus removal at a comparable unit cost to bioretention but are not as aesthetically pleasing. The perimeter sand filter is a flexible, easy-to-access practice (but at higher cost) that provides good phosphorus removal and additionally high oil- and grease-trapping ability.

The site drainage system can be designed so that the bioretention and sand filters drain to the micropool ED pond for redundant treatment. Vegetated dry swales could also be used to convey runoff to the pond, which would provide pretreatment. Pocket wetlands and wet swales were eliminated from consideration due to potential for nuisance conditions. Underground sand filters could also be used at the site; however, cost and aesthetic considerations were significant enough to eliminate them from consideration.

TABLE 13-18

Sample Structural Control Selection Matrix

General Application Structural Control Alternative	Stormwater Treatment Suitability	Site Applicability	Implementation Considerations	Physiographic Factors/Soils	Special Watershed Considerations	Other Issues
Wet pond	✓	X				
Wet ED pond	✓	X				
Micropool ED pond	✓	✓	✓	✓	None	
Multiple ponds	✓	X				
Shallow wetland	✓	X				
ED shallow wetland	✓	X				
Pocket wetland	✓	✓	✓	✓	None	Odor/mosquitoes
Infiltration trench	✓[1]	✓	✓	X	None	Aesthetics
Surface sand filter	✓[1]	✓[2]	✓	✓	None	Higher cost
Perimeter SF	✓[1]	✓[2]	✓	✓	None	
Bioretention	✓[1]	✓[2]	✓	✓	None	
Dry swale	✓[1]	✓[2]	✓	✓	None	
Wet swale	✓[1]	✓[2]	✓	✓	None	Odor/mosquitoes

[1] Only when used with another structural control that provides water-quantity control.
[2] Can treat a portion of the site.

13.7 Online Versus Offline Structural Controls

Introduction

Structural stormwater control systems are designed to be online or offline. Online facilities are designed to receive, but not necessarily control or treat, the entire runoff volume up to the design event. Online structural controls must be able to handle the entire range of storm flows.

Offline facilities, on the other hand, are designed to receive only a specified flow rate through the use of a flow regulator (i.e., diversion structure, flow splitter, etc.). Flow regulators are typically used to divert the water-quality volume (WQ_v) to an offline structural control sized and designed to treat and control the WQ_v. After the design runoff flow has been treated and controlled, it can be returned to the conveyance system. Figure 13-6 shows an example of an offline sand filter and an offline enhanced dry swale.

Flow Regulators

Flow regulation to offline structural stormwater controls can be achieved by:

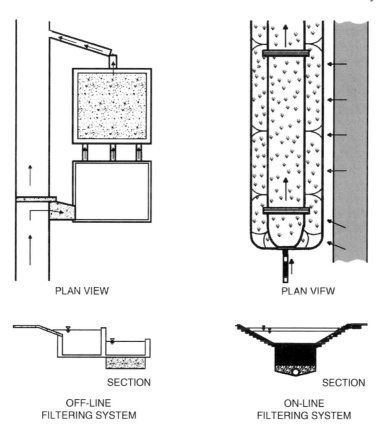

FIGURE 13-6
Example of online versus offline structural controls. (From Center for Watershed Protection (CWP), Design of Stormwater Filtering Systems, prepared for Chesapeake Research Consortium, Edgewater, MD, Center for Watershed Protection, Ellicott City, MD, 1996.)

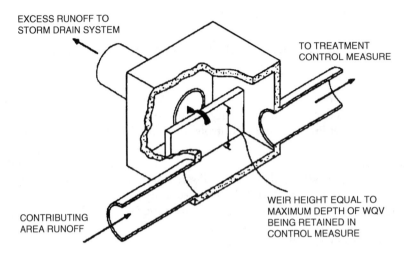

FIGURE 13-7
Pipe interceptor diversion structure. (From City of Sacramento, California, Guidance Manual for On-Site Stormwater Quality Control Measures, Department of Utilities, 2000.)

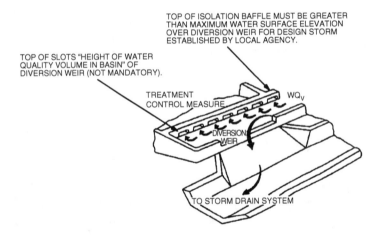

FIGURE 13-8
Surface channel diversion structure. (From City of Sacramento, California, Guidance Manual for On-Site Stormwater Quality Control Measures, Department of Utilities, 2000.)

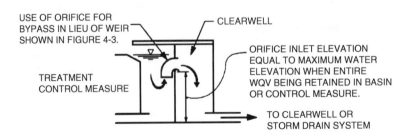

FIGURE 13-9
Outlet flow regulator. (From City of Sacramento, California, Guidance Manual for On-Site Stormwater Quality Control Measures, Department of Utilities, 2000.)

- Diverting the water-quality volume or other specific maximum flow rate to an offline structural stormwater control
- Bypassing flows in excess of the design flow rate

Flow regulators can be flow splitter devices, diversion structures, or overflow structures. A number of examples are shown in Figures 13-7 through 13-9.

13.8 Regional Versus On-Site Stormwater Management

Introduction

Using individual, on-site structural stormwater controls for each development is the typical approach for controlling stormwater quantity and quality. The developer finances the design and construction of these controls and, initially, is responsible for all operation and maintenance.

A potential alternative approach is for a community to install a few strategically located regional stormwater controls in a subwatershed rather than require on-site controls (see Figure 13-10). For this discussion, regional stormwater controls are defined as facilities designed to manage stormwater runoff from multiple projects and properties through a local jurisdiction-sponsored program, where the individual properties may assist in the financing of the facility, and the requirement for on-site controls is eliminated or reduced.

Advantages and Disadvantages of Regional Stormwater Controls

Regional stormwater facilities are significantly more cost-effective, because it is normally easier and less expensive to build, operate, and maintain one large facility than several small ones. Regional stormwater controls are generally better maintained than individual site controls, because they are large, highly visible and typically the responsibility of the local government. In addition, a larger facility can pose less of a safety hazard than numerous small ones, because it is more visible and easier to secure.

There are also several disadvantages to regional stormwater controls. In many cases, a community must provide capital construction funds for a regional facility, including the costs of land acquisition. However, if a downstream developer is the first to build, that

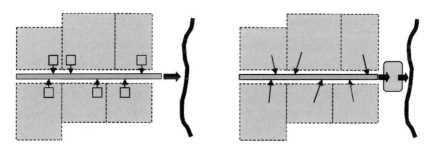

Structural Controls on Each Development Site

Regional Structural Stormwater Control

FIGURE 13-10
On-site versus regional stormwater management.

person could be required to construct the facility and later be compensated by upstream developers for the capital construction costs and annual maintenance expenditures. Conversely, an upstream developer may have to establish temporary control structures if the regional facility is not in place before construction. Maintenance responsibilities generally shift from the homeowner or developer to the local government when a regional approach is selected. The local government would need to establish a stormwater utility or some other program to fund and implement stormwater control. Finally, a large in-stream facility can pose a greater disruption to the natural flow network and is more likely to affect wetlands within the watershed.

Below are summarized some of the pros and cons of regional stormwater controls.

Advantages of Regional Stormwater Controls

The following are advantages of regional stormwater controls:

- *Reduced construction costs* — Design and construction of a single regional stormwater control facility can be far more cost-effective than numerous individual on-site structural controls.
- *Reduced operation and maintenance costs* — Rather than multiple owners and associations being responsible for the maintenance of several stormwater facilities on their developments, it is simpler and more cost-effective to establish scheduled maintenance of a single regional facility.
- *Higher assurance of maintenance* — Regional stormwater facilities are far more likely to be adequately maintained, as they are large and have a higher visibility, and are typically the responsibility of the local government.
- *Maximum utilization of developable land* — Developers would be able to maximize the utilization of the proposed development for the purpose intended by minimizing the land normally set aside for the construction of stormwater structural controls.
- *Retrofit potential* — Regional facilities can be used by a community to mitigate existing developed areas that have insufficient or no structural controls for water quality and quantity, as well as provide for future development.
- *Other benefits* — Well-sited regional stormwater facilities can serve as a recreational and aesthetic amenity for a community.

Disadvantages of Regional Stormwater Controls

The following are some disadvantages of regional stormwater controls:

- *Location and siting* — Regional stormwater facilities may be difficult to site, particularly for large facilities or in areas with existing development.
- *Capital costs* — The community must typically provide capital construction funds for a regional facility, including the costs of land acquisition.
- *Maintenance* — The local government is typically responsible for the operation and maintenance of a regional stormwater facility.
- *Need for planning* — The implementation of regional stormwater controls requires substantial planning, financing, and permitting. Land acquisition must be in place ahead of future projected growth.

- *Water-quality, channel, and flood protection* — Without on-site water-quality and channel protection, regional controls do not protect smaller streams upstream from the facility from degradation and streambank erosion and flood protection.
- *Ponding impacts* — Upstream inundation from a regional facility impoundment can eliminate floodplains, wetlands, and other habitat.
- *Mimicking nature* — Nature rarely uses regional controls. The normal course is to handle rainfall initially close to where it falls through a series of microscale treatment facilities (low spots that catch and infiltrate water, sheet flow through forest and meadow, etc.).

Important Considerations for the Use of Regional Stormwater Controls

If a community decides to implement a regional stormwater control, then it must ensure that the conveyances between the individual upstream developments and the regional facility can handle the design peak flows and volumes without causing adverse impact or property damage. Full-buildout conditions in the regional facility drainage area should be used in the analysis.

In addition, unless the system consists of completely man-made conveyances (i.e., storm drains, pipes, concrete channels, etc.), then on-site structural controls for water-quality and downstream channel protection will need to be required for all developments within the regional facility's drainage area. Federal water-quality provisions do not allow the degradation of water bodies from untreated stormwater discharges, and it is U.S. EPA policy to not allow regional stormwater controls that would degrade stream quality between the upstream development and the regional facility. Further, without adequate channel protection, aquatic habitats and water quality in the channel network upstream of a regional facility may be degraded by streambank erosion if they are not protected from bankfull flows and high velocities.

Based on these concerns, both the U.S. EPA and the U.S. Army Corps of Engineers have expressed opposition to in-stream regional stormwater control facilities. In-stream facilities should be avoided if possible and will likely be permitted on a case-by-case basis only. It is important to note that siting and designing regional facilities should ideally be done within a context of a stormwater master planning or watershed planning to be effective.

13.9 Using Structural Stormwater Controls in Series

Combined Measures

Although the management measures mentioned here can be used individually to remove pollutants, it may be desirable to consider combining two or more of these measures. Examples might include the following:

- Use of infiltration wells in a detention basin to increase pollutant removal while concurrently decreasing the long-term storage requirements
- Use of overland flow through vegetative filters (strips and channels) and wetlands so as to filter suspended sediments from runoff upstream from an infiltration basin or trench
- Use of wetland marshes adjacent to retention ponds
- Use of oil and grit separators to remove larger sizes of grit and oil and grease on particular sites upstream from regional basins

Combining two or more management measures can often have advantages. Combinations of measures may:

- Increase the operational life of a given BMP
- Increase pollutant removal effectiveness
- Overcome any site-limiting factors

Wet detention ponds should be used in combination with vegetative filters to provide sediment removal before the runoff can enter an infiltration system or wetland. Additionally, infiltration basins can be used to provide for the temporary or permanent storage of runoff.

Wetlands should be used only in combination with vegetative filters and detention, not with infiltration. Typically, wetlands should receive inflow from vegetated conveyance facilities and a wet detention pond. Any planned discharge from a wetland should be into a vegetated conveyance facility. Wetlands should not precede an infiltration system, as accumulated sediment and decaying matter may clog the infiltration mechanism. It must be recognized that conditions favorable to wetlands (high water table, impervious soils, etc.) are unfavorable for infiltration measures.

Many types of structural BMPs also provide, or can be designed to provide, flood-control capabilities. Retention ponds, extended detention, infiltration basins, and porous pavement are of particular benefit for flood control, at least for the more frequent nuisance floods. Additionally, some older detention facilities or interior drainage pumping sump areas can be retrofitted to maximize pollution removal without severely limiting flood-control capabilities. In some municipalities, flood-control design criteria had been overly restrictive, providing excess basin volume. Through a modification of such criteria, retrofitting for water quality could be easily accomplished.

Recently, structural BMPs have taken on the added responsibility of providing channel protection by storing a design volume of runoff for an extended time before discharging into the receiving drainage system. The result has been stabilization of downstream streambanks, providing protection for the system as storm runoff travels downstream.

Stormwater Treatment Trains

The minimum stormwater management standards are an integrated planning and design approach with components that work together to limit the adverse impacts of urban development on downstream waters and riparian areas. This approach is sometimes called a stormwater "treatment train." A more detailed discussion on the five-part watershed treatment train is given in Chapter 15. This chapter considers only the third and fourth portions of the five-part treatment train, those parts where structural controls more readily apply. The treatment train is illustrated in Figure 13-11.

As shown in Figure 13-11, this treatment train can be thought of as having five major components:

1. *Education and prevention programs* begin in the minds of potential polluters to educate and change habits and practices. These might include used-oil recycling programs, sweeping practices, homeowner use of certain toxic chemicals, fertilizer or gardener programs, etc.
2. *Runoff and load generation* is the first physical line of defense against pollution. It begins with the development of land-use change policies that reduce the inherent impact of development on streams. There are a wide number of practices (see Chapter 1) in the United States and abroad that include modifying land-use

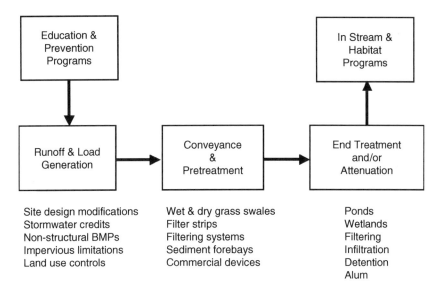

FIGURE 13-11
Generalized stormwater treatment train.

and zoning codes and development practices to reduce impervious areas and to build in pollution reducing features.

3. *Conveyance and pretreatment* is the initial designed stormwater system. It includes a wide array of commercial devices used in small drainage situations, low-impact development approaches (PGC, 1999), pollution-reducing structural conveyance BMPs (swales, linear wetlands, buffers, exfiltrating pipe systems, etc.), and small-site BMPs (bioretention, infiltration trenches, dry wells, rain barrels, etc.). The idea is to intercept and remove pollutants at the entry to the minor stormwater system or within the first several reaches of the conveyance system.

4. *End-of-pipe treatment and attenuation* refers to the wide array of structural BMP devices that remove pollution from stormwater, including ponds, wetlands, infiltration, filtration, channel practices, and special devices. Often, these BMPs are designed for multiple uses in controlling water-quality and quantity impacts, as well as recharge of groundwater and channel protection.

5. *In-stream and habitat programs* refer to an array of practices to restore and protect streams, including instream structures, stream assessment, bank treatment and stabilization, riparian corridor preservation or restoration, channel modifications, fish habitat, etc.

The combinations of structural stormwater controls are limited only by the need to employ measures of proven effectiveness and meet local regulatory and physical site requirements. Figures 13-12 through 13-14 illustrate the application of the treatment train concept for a moderate-density residential neighborhood, a small commercial site, and a large shopping mall site.

In Figure 13-12, rooftop runoff drains over grassed yards to backyard grass channels. Runoff from front yards and driveways reaches roadside grass channels. Finally, all stormwater flows drain to a micropool ED stormwater pond.

A gas station and convenience store are depicted in Figure 13-13. In this case, the decision was made to intercept hydrocarbons and oils using a commercial gravity (oil-grit) separator located on the site prior to draining to a perimeter sand filter for removal of finer

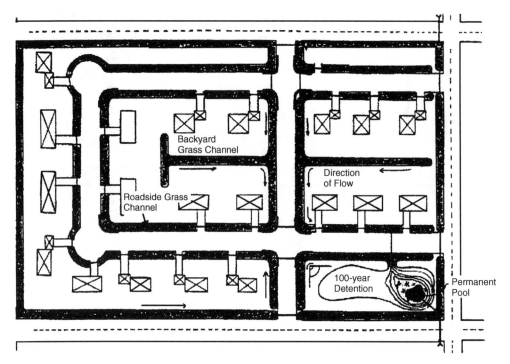

FIGURE 13-12
Example treatment train — residential subdivision (3.1.6-3). (Adapted from NIPC, 2000.)

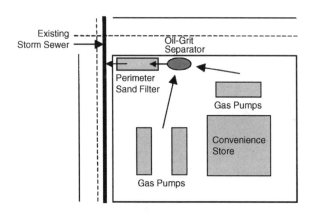

FIGURE 13-13
Example treatment train — small commercial development (3.1.6-4).

particles and TSS. No stormwater control for channel protection is required, as the system drains to the municipal storm drain pipe system. Overbank and extreme flood protection are provided by a regional stormwater control downstream.

Figure 13-14 shows an example treatment train for a commercial shopping center. In this case, runoff from rooftops and parking lots drains to depressed parking lot, perimeter grass channels, and bioretention areas. Slotted curbs are used at the entrances to these swales to better distribute the flow and to settle out the very coarse particles at the parking lot edge for sweepers to remove. Runoff is then conveyed to a wet ED pond for additional pollutant removal and channel protection. Overbank and extreme flood protection are provided through parking lot detention.

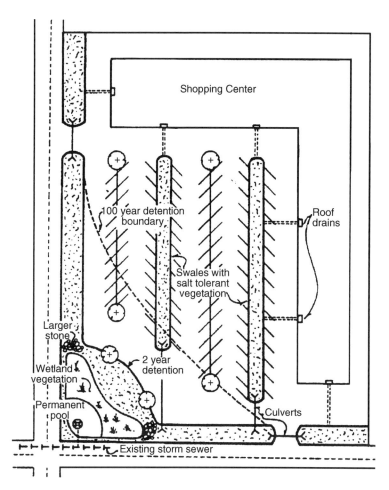

FIGURE 13-14
Example treatment train — commercial development (3.1.6-5). (From NIPC, 2000.)

For all of the examples, other stormwater-related design criteria could be included, such as groundwater recharge and infiltration requirements.

Calculation of Pollutant Removal for Structural Controls in Series

For two or more structural stormwater controls used in combination, it is often important to have an estimate of the pollutant removal efficiency of the treatment train. Pollutant removal rates for structural controls in series are not additive. For pollutants in particulate form, the actual removal rate (expressed in terms of percentage of pollution removed) varies directly with the pollution concentration and sediment size distribution of runoff entering a facility.

For example, a stormwater pond facility will have a much higher pollutant removal percentage for very turbid runoff than for clearer water. When two stormwater ponds are placed in series, the second pond will treat an incoming particulate pollutant load very different from the first pond. The upstream pond captures the easily removed larger sediment sizes, passing on an outflow with a lower concentration of TSS but with a higher relative proportion of finer particle sizes. Hence, the removal capability of the second pond for TSS, when expressed in terms of capture efficiency, is considerably less than for the first pond. Recent findings suggest that the second pond in series can provide as little as half the removal efficiency of the upstream pond.

To estimate the pollutant removal rate of structural controls in series, a method is used in which the removal efficiency of a downstream structural control is reduced to account for the pollutant removal of the upstream control(s). The following steps are used to determine the pollutant removal:

- For each drainage area, list the structural controls in order, upstream to downstream, along with their expected average pollutant removal rates from Table 13-7 for the pollutants of concern.
- Apply the following equation for calculation of approximate total accumulated pollution removal for controls in series:

Final Pollutant Removal = (Total load * Control1 removal rate) + (Remaining load * Control 2 removal rate) + ... for other Controls in series

Example Application

TSS is the pollutant of concern. Due to urban hot spot concerns for oil and grease, a commercial device is inserted that has only a 20% sediment removal rate. A stormwater pond is designed at the site outlet. A second stormwater pond is located downstream from the first one in series. What is the total TSS removal rate? The following information is given:

Control 1 (Commercial Device) = 20% TSS removal
Control 2 (Stormwater Pond 1) = 80% removal (use 1.0 _ design removal rate)
Control 3 (Stormwater Pond 2) = 40% TSS removal (use 0.5 s design removal rate)

Then, applying the controls in order and working in terms of units of TSS starting at 100 units:

For Control 1: 100 units of TSS _ 20% removal rate = 20 units removed
  100 units − 20 units removed = 80 units of TSS remaining
For Control 2: 80 units of TSS _ 80% removal rate = 64 units removed
  80 units − 64 units removed = 16 units of TSS remaining
For Control 3: 16 units of TSS _ 40% removal rate = 6 units removed
  16 units − 6 units removed = 10 units of TSS remaining = 90% removal

13.10 Routing with WQ_v Removed

When offline structural controls such as bioretention areas, sand filters, and infiltration trenches capture and remove the water-quality volume (WQ_v), downstream structural controls do not have to account for this volume during design. That is, the WQ_v may be subtracted from the total volume that would otherwise need to be routed through the downstream structural controls.

From a calculation standpoint, this would amount to removing the initial WQ_v from the beginning of the runoff hydrograph, thus creating a notch in the runoff hydrograph. Because most commercially available hydrologic modeling packages cannot handle this

Structural Best Management Practices

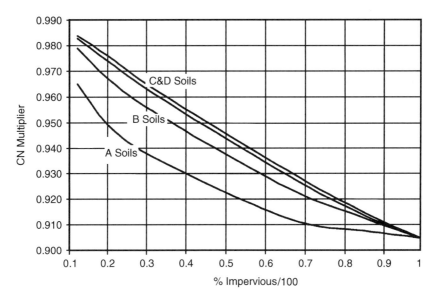

FIGURE 13-15
Curve number adjustment factor.

type of action, the following method has been created to facilitate removal from the runoff hydrograph of approximately the WQ_v:

- Enter the horizontal axis on Figure 13-15 with the impervious percentage of the watershed and read upward to the predominant soil type (interpolation between curves is permitted).
- Read left to the CN multiplier.
- Multiply the curve number for the subwatershed that includes the water quality basin by this CN multiplier – this provides a smaller curve number.

The difference in curve number will generate a runoff hydrograph that has a volume less than the original volume by an amount approximately equal to a typical WQ_v. This method should be used only for bioretention areas, filter facilities, and infiltration trenches, where the drawdown time is less than or equal to 24 h.

For example, a site design employs an infiltration trench for the WQ_v and has a curve number of 72, is B type soil, and has an impervious percentage of 70%; the factor from Figure 13-15 is 0.921. The curve number to be used in calculation of a runoff hydrograph for the quantity controls would be $(72 \times 0.92) = 66$.

13.11 Specific Structural BMP Design Guidance

Overview

Following are design considerations for representative designs for some of the more effective management measures that can be used for water-quality control in urban areas. Each section gives typical recommended specifications, operation and maintenance requirements, and performance standards. Again, it should be noted that climatic and site-specific conditions may lead to major modifications of each of these sets of criteria. The criteria are given to illustrate the considerations necessary for BMP design and are not meant to be an exhaustive or restrictive treatment. There are many variations on the basic device types to accommodate specific site restrictions or take advantage of site opportunities. The following BMPs are included in this section:

- Detention ponds
  - Extended dry detention basins
  - Retention ponds
  - Constructed wetlands
  - Bioretention areas
- Infiltration and filtration
  - Austin first-flush filtration basin
  - Sand filters
  - Infiltration trenches
  - Porous pavement
- Vegetative filtration
  - Filter strips and flow spreaders
  - Grassed swales
- Special devices
  - Water quality inlets
  - Oil/grit separators
  - Catch basin inserts

It is beyond the scope of this chapter to describe all structural BMPs that can be used to control the water-quality aspects of urban stormwater runoff. These 13 BMPs were selected to represent the four predominant removal mechanisms. Descriptions of other structural BMPs can be obtained from the literature cited at the end of this chapter. For additional information on the BMPs included in Table 13-13, see the Georgia Stormwater Management Manual (2002) or the Maryland Stormwater Manual (2000).

There are also modifications for many of the structural BMPs for arid- and for cold-weather applications. Water scarcity, higher sediment loads, effluent-dominated streams, need for flood control and the importance of groundwater recharge greatly influence BMPs in arid areas. Cold weather requires consideration of freezing of pipes and outlets, increased storage volume for spring thaw or ice buildup, seasonal usage or circulation, weather-tolerant plants or seasonal plant management, salt tolerance, and the lack of wintertime infiltration or filtration. See CWP (2000) and CWP (1997) or MCES (2001) for arid- and cold-weather applications, respectively.

Overall Design Approach

See Chapter 17 for a discussion of the overall design approach for site development and integration of stormwater site design concepts and BMPs into site design. There are many approaches to the design of BMPs. Figure 13-16 illustrates the approach used by the Washington Department of Ecology (2001). Figure 13-17 shows the unified design criteria approach favored by Georgia.

Structural Best Management Practices

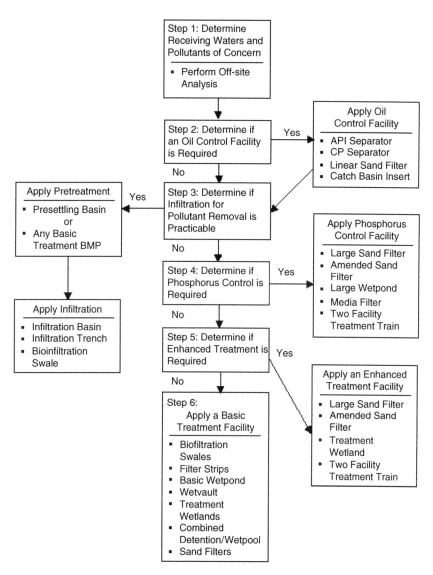

FIGURE 13-16
WDE overall design approach for BMPs.

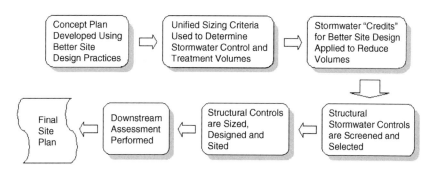

FIGURE 13-17
GSM overall design approach.

Perhaps the common thread that runs through all of the design approaches, when given, is obtaining a clear understanding of regulatory requirements (federal, state, and local), looking at less-expensive source controls and site design approaches first, planning structural controls as part of a treatment train approach, and fitting them into the site design early in the concept plan process, and making arrangements for long-term maintenance according to local practice.

Extended Dry Detention Basin

An extended dry detention basin is a detention basin with increased runoff residence time sufficient to remove settleable pollutants to acceptable levels of concentration. There are two basic types of extended detention ponds. The first is normally dry and drains completely in a specified period of time. The second maintains a permanent pool with the extended detention volume above the permanent pool and really is a combination of a wet pond and extended detention (Schueler and Helfrich, 1988). The second type will be covered under wet ponds.

Extended dry ponds are used where lack of water or other multiuse considerations preclude the use of wet ponds or constructed wetlands. Extended ponds have been used for recreational fields, tennis courts, parking lots, roof tops, nature areas, etc. They are considered dry, though wetland plants can grow and pools and muddy areas can appear due to settling or poor grading (UDFCD, 1999). Often, a cement low-flow channel is put in the bottom of the structure to attempt to drain the floor, and channel trickle flows through the structure.

They also handle flood-control objectives more economically than do wet ponds, because they do not have extra storage volume set aside for a permanent pool. The flood storage volume is set aside above the pollution-control volume, a 40-h drawdown time is normally recommended, and a multistage riser arrangement is used to handle the outflow (UDFCD, 1999). They are generally targeted where particulate removal is the focus. Extended ponds can be readily converted from sediment traps used during construction (UDFCD, 1992). They are also applicable where soil with poor infiltration characteristics exists, because they do not depend on infiltration for proper operation. They are also often used downstream from other water-quality controls in a treatment train approach (ARC, 2002).

Disadvantages of extended dry ponds include difficulty in keeping some outlet types unclogged, difficulty to mow if wet, rapid accumulation of sediments, swampy eyesore possibility if not maintained, and modification of flow regime if inline. Maintenance is estimated to be 3 to 5% of construction cost annually (MWCOG, 1992).

Dry extended detention ponds are the least-expensive stormwater treatment practice, on a cost per unit area treated. The construction costs associated with these facilities range considerably. One recent study evaluated the cost of all pond systems (Brown and Schueler, 1997). Adjusting for inflation, the cost of dry extended detention ponds can be estimated with the following equation:

$$C = 12.4V^{0.760} \tag{13.7}$$

where C = construction, design, and permitting cost; and V = volume needed to control the 10-year storm (cubic feet).

Figure 13-18 shows a schematic of a typical pond design. Figure 13-19 shows a picture of such a pond. Other designs include vaults and tanks for underground applications and myriad aesthetic and pollutant-removal enhancing features such as micropools and marshes.

Structural Best Management Practices

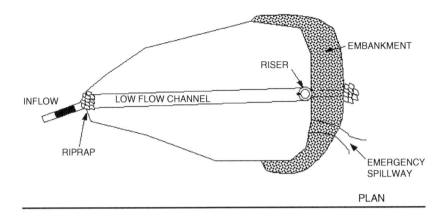
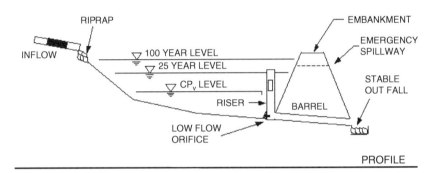

FIGURE 13-18
Extended detention pond.

FIGURE 13-19
Extended detention pond picture.

Erosion of the banks and the pond bottom have been a problem for this type of design, and special care should be taken to provide vegetation that can stand frequent inundation and a noneroding low-flow channel. Structural bank protection could be provided in lieu of vegetation, though aesthetic requirements may make it costly.

Dry extended detention ponds can also be useful stormwater retrofits and have two primary applications as a retrofit design. In many communities, detention basins have been designed for flood control in the past. It is possible to modify these basins to incorporate features that encourage-water quality control and channel protection. It is also possible to construct new dry extended detention ponds in open areas of a watershed to capture existing drainage, or create them above a road crossing or culvert.

In arid and semiarid regions, some modifications may be needed to conserve scarce water resources. Any landscaping plans should prescribe drought-tolerant vegetation wherever possible. In addition, the wet forebay may be replaced with an alternative form of pretreatment, such as a dry sediment chamber. In regions that have distinct wet and dry seasons, regional extended detention ponds can be designed to act as a recreational area, such as a ball field, during the dry season (CWP, 2000).

In cold climates, some additional design features are needed to treat the spring snowmelt. One such modification is to increase the volume available for detention to help treat this relatively large runoff event. In some cases, dry ponds may be an option as a snow storage facility to promote some treatment of plowed snow. If a pond is used to treat road runoff, or is used for snow storage, landscaping should incorporate salt-tolerant species. Finally, sediment removal from the forebay may need to be completed more frequently than in warmer climates to account for sediment deposited as a result of road sanding (CWP, 1997).

Pond Sizing

It has been shown by several authors that detention of all storms more frequent than about the 6-month storm for a period of at least 24 h can affect a long-term suspended solid (TSS) removal rate by about 80%. However, if ponds are designed with a single orifice to detain the larger design storm, the 24-h period smaller storms will not be detained adequately. Also, the smaller particles attract pollutants more readily than the larger but are settled less easily. Flocculation of the smaller particles may provide better results than expected. To provide at least 24 h of detention over the complete spectrum of small storms, one of two approaches is used (California Stormwater Task Force, 1993; Schueler, 1987; Grizzard et al., 1986; Stanley, 1994):

- The pond is designed to detain the larger storms for a period approaching 40 h (in the eastern part of the United States), resulting in about a 24-h time for the smaller storms
- More than one outlet orifice is used with one placed at the pool bottom for the smaller storms and a second placed at a mid-height level for the larger storms, resulting in a 24-h drawdown time for all storm volumes

For wet ponds, there are analytical methods to estimate pollution reduction. For ponds without a permanent pool below the extended detention volume, there is no comparable generally accepted analytic method (Urbonas and Stahre, 1993). Most localities have developed rules of thumb for the pond-sizing volume for dry extended detention ponds or ponds with small wet ponds. These rules are designed to remove a high percentage of suspended solids and often provide a volume-of-basin to volume-of-mean annual runoff event ratio (VB/VR) of 1.0 to 2.5. Common design volume criteria include the following:

- First-flush (half-inch) runoff volume should be detained for 24 h
- The runoff volume produced by a 1.0 in storm should be detained 24 h
- Detain the runoff from the 6-month, 24-h storm for 24 h
- Some multiple of mean runoff volume (see retention pond design for runoff information)
- Design storm — 1- to 2-year storm
- The design volume from the 2-year, 24-h storm (Idaho DEQ,2001).

Northern Virginia (1992) provides rules for pond size for a 48-h drawdown time as follows:

$$S_a = 4375 \, C - 875 \text{ or } 32.25 \, I_\% \tag{13.8}$$

where S_a = storage per controlled acre (ft^3); C = rational method C factor; and $I_\%$ = upstream watershed percent impervious (%).

Sizing of the pond dimensions is often a trial-and-error procedure fitting the required volume to the site. Generally, wider and shallower ponds are more efficient than deeper ponds for the same volume (California Stormwater Task Force, 1993). Distance between inlet and outlet should be maximized, though this is not critical, because most of the settling takes place during quiescent periods (Driscoll, 1986).

A TR-55 shortcut method (Harrington, 1987; MDE, 2000; ARC, 2000) is illustrated and explained in Chapter 7 along with an example. This method is recommended in lieu of actual routing calculations, though routing and even long-term simulation is preferred where the size and importance of the pond warrants the effort. Chapter 11 gives routing details.

Outlets

Physically, the extended detention time is provided by an outlet structure with a small opening(s). There are several ways to accomplish the outlet control arrangement. A preferred method when possible is a micropool at the outlet with a reverse slope pipe. A micropool at the outlet can prevent resuspension of sediment and outlet clogging. This is often paired with a trash screen to protect blockage of overflow pipes and openings (Denver UDFCD, 2001; ARC, 2001). A reverse slope pipe draws from below the permanent pool extending in a reverse angle up to the riser, and determines the water elevation of the micropool. Because these outlets draw water from below the level of the permanent pool, they are less likely to be clogged by floating debris. Other methods include gravel packet perforated risers, perforated risers with an internal orifice for control, negatively sloped orifice pipes, V-notched weirs, concrete block structures with multiple openings, concrete manhole riser with submerged internal orifice plate/pipe, etc. Southwest Florida Management District recommends clogging protection for all outlets less than 6 in cross-sectional area, 2 in dimension or 20 degrees for V-notch weirs (Southwest Florida, 1990).

GKY (1989) provides an equation for single-orifice outlet design for ponds where surface area does not vary greatly with depth (larger ponds with steeper side-slopes or any size with vertical side-slopes).

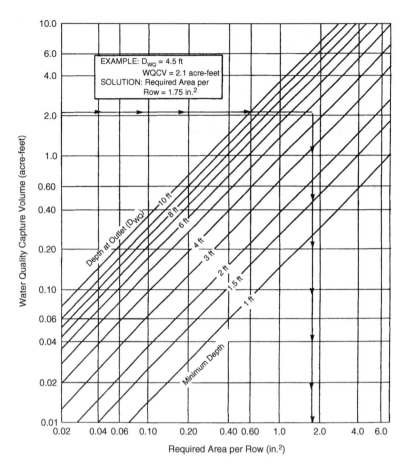

FIGURE 13-20
Water-quality outlet sizing — extended detention basin with 40-h drawdown time of the capture volume. (After UDFCD, 1992, 1994.)

$$A = [24A(H - Ho)^{0.5}]/[3600CT(2g)^{0.5}] \qquad (13.9)$$

where a = orifice area (in^2); A = average surface area of pond (ft^2); C = discharge coefficient; T = drawdown time (h); g = gravitational acceleration (ft/s^2); H = full pond elevation (ft); and Ho = elevation when pond is empty (ft).

If a perforated riser is used for drawdown control without an internal orifice (i.e., the perforations provide the control) use Figure 13-20 for perforation opening requirements (C value of 0.65 assumed) per row. The rows are spaced vertically 4 in on center. Use a gravel pack around the riser and keep perforations larger than about 3/4 to 1 in. Use of filter fabric against the riser has been shown to clog (California Stormwater Task Force, 1993). This may limit this type of outlet structure to larger pond areas.

A much smaller orifice can be used if it is placed inside the riser pipe downstream from the perforated section or inside a riser manhole in a sump-like condition. Figure 13-21 gives a schematic of this arrangement.

Figures 13-22 and 13-23 demonstrate a newer outlet design that incorporates a trash rack and orifice plate combination. While this type of outlet works best with a micropool at the outlet, it can also be used for a dry extended detention pond, though maintenance of the trash rack will need to be done to insure passage of the flow. For more information on this outlet including multistorm variations, see UDFCD (1999).

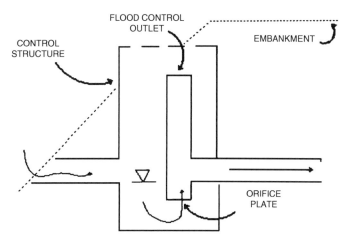

FIGURE 13-21
Extended detention basin with internal orifice.

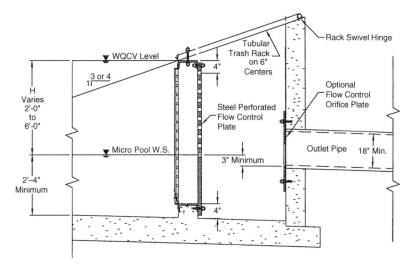

FIGURE 13-22
Extended detention basin outlet.

Pollutant Removal Information

See Table 13-13 for overall pollutant removal summary information. Soluble pollutant removal rates are low for extended dry detention ponds but can be enhanced with greatly increased detention time, through the use of shallow marshes to increase biological uptake, or through the use of an infiltration device downstream from the outlet orifice (Schueler, 1987; Urbonas and Ruzzo, 1986; ARC, 2002). The lower end of the ranges will be used unless experience indicates otherwise. Freezing conditions can be expected to reduce removal rates by 50%.

Because of the great cost of sediment removal (perhaps 20 to 40% of first cost) (Schueler and Helfrich, 1988), even at intervals of 10 years or more, there is often a paved sediment forebay placed at the entrance of an extended detention pond or a wet pond. Sediment accumulation in the pond proper can be estimated from the simple method for TSS and a suitable trap efficiency or from local data and criteria (Schueler, 1987). The size of the forebay is developed from settling theory and is typically 5 to 10% of the water-quality

FIGURE 13-23
Extended detention basin outlet picture.

control volume (UDFCD, 1999; ARC, 2002). Urbonas and Stahre (1993) recommend forebays with depths of at least 3.5 ft to avoid resuspension, detention times for the average annual runoff rate to be 5 min, and loading rates to be approximately 50 ft/h.

Sediments collected in detention and retention ponds have not been found to be classified as hazardous waste (Carr, Geinopolos, and Zanoni, 1982; Zanoni, 1986; Nightingale, 1987; Stanley, 1994). If they are found to be hazardous, RCRA requires generators of hazardous wastes to monitor and manage them in accordance with specified procedures. All accumulated sediment must be handled and stored in a way that will not affect surface or groundwater. In general, it should be disposed of in a location where it will be stable and not in contact with humans (Minnesota PCA, 1989). For on-site disposal, it should be covered with at least 4 in of topsoil and vegetated. In all cases, sediment must be disposed of in accordance with any applicable solid-waste regulations.

Sediments in the bottoms of wet ponds also accumulate metals in the first couple inches of soil. Nightingale (1987) recommends monitoring for lead content and removing sediments if the lead concentration exceeds 1000 mg/kg of soil, as suggested in a California standard. For a more complete discussion of the issue of contaminated sediments in BMPs, see Cox and Livingston, 1997.

General Design Criteria

- Hydrograph centroid should be delayed through control from a small orifice in the release structure.
- Do not locate on fill if seepage is considered to be great.
- Extended detention design storm should be limited to 1-year runoff events.
- Drawdown time should be at least 24 h average to 40 h for maximum design storm.
- Runoff in excess of a 2-year event should normally be passed through the basin with peak discharge controls only.

Typical Required Specifications

- Pilot channel of paved or concrete material is needed for erosion control (alternately, use turf if little low flow). Size such that any event runoff will overflow the low-flow channel onto the pond floor.
- Side slopes should be no greater than 3:1 if mowed.
- Inlet and outlet should be located to maximize flow length.
- Design for full development upstream of control.
- Riprap protection (or other suitable erosion control means) for the outlet and all inlet structures into the pond is needed.
- One-half ft minimum freeboard above peak stage for top of embankment is necessary for design storm.
- Emergency spillway should be designed to pass the 50-year storm event (must be paved in fill areas).
- Maintenance access should be <15% slope, 10 ft wide.
- Trash racks, filters, or other debris protection is necessary on control.
- Maximum depth should not exceed 10 ft.
- Antivortex plates are necessary.
- Ensure that there is no outlet leakage and use antiseep collars.
- Benchmark for sediment removal.
- Careful finish grading should be specified to ensure slope toward the outlet and no pools and low spots.

Typical Recommended Specifications

- Two-stage design (top stage — dry during the mean storm, bottom stage — inundated during storms less than the mean storm event)
- Top-stage slopes between 2% and 5% and a depth of 2 to 5 ft
- Bottom stage maintained as shallow wetland or pool (6 to 12 in)
- Manage buffer and pond as meadow
- Minimum 25-ft wide buffer around pool
- On-site disposal areas for two sediment removal cycles
- Antiseep collars on barrel of principal spillway
- Impervious soil boundary
- Design as offline pond to bypass larger flows
- Design a sediment settling basin for pretreatment of the larger particles
- Maximum contributing drainage area to be served by a single dry ED basin is 75 acres
- Inflow channels are to be stabilized with flared riprap aprons, or the equivalent (A sediment forebay sized to 0.1 in per impervious acre of contributing drainage should be provided for dry detention and dry ED basins that are in a treatment train with offline water-quality treatment structural controls.)

Sizing Example

An extended detention pond for an area of 50 acres is to be sized for a VB/VR ratio of 2.0. The rational coefficient is 0.6, and the percent imperviousness is 35%. The location is Nashville, Tennessee. Because no local information is available, rainfall statistics will come from Driscoll et al. (1989). The mean annual storm volume is (central area) 0.62 in (V), and the intensity is 0.097 in/h.

- The volume runoff coefficient is calculated from Shelley's equation (13.6) as:

$$Rv = 0.05 + 0.009 * I = 0.05 + 0.009 * 35 = 0.365$$

- VR is then calculated as:

$$VR = V * Rv * A * 43,560/12 = 0.62 * 0.365 * 50 * 43,560/12 = 41,073 \text{ ft}^3$$

- Then, VB is found from:

$$VB = 2.0 \, VR = 82,000 \text{ ft}^3$$

The sediment forebay, if desired, is calculated using Urbonas and Stahre's (1993) criteria:

- Mean annual discharge is:

$$Q = C * i * A = 0.6 * 0.097 * 50 = 2.91 \text{ cfs}$$

- Surface Area = Mean Annual Discharge/Loading Rate

$$A = Q * 43,560/12 * 1/50 = 2.91 * 43,560/12 * 1/50 = 211 \text{ ft}^2$$

- Volume for 5-min detention time:

$$V = Q * 5 * 60 = 2.91 * 5 * 60 = 873 \text{ ft}^3$$

- Depth is found from

$$D = V/A = 873/211 = 4.13 \text{ ft}$$

- For a 2:1 length-to-width ratio, the horizontal dimensions are:

$$2 * W^2 = A$$

$$W = 10.27 \text{ ft}$$

$$L = 20.54 \text{ ft}$$

Typical maintenance needs for a dry pond are shown in Table 13-19.

TABLE 13-19

Typical Maintenance Activities for Dry Detention/Dry ED Basins

Activity	Schedule
Remove debris from basin surface to minimize outlet clogging and improve aesthetics	Annually and following significant storm events
Remove sediment buildup	As needed, based on inspection
Repair and revegetate eroded areas	
Perform structural repairs to inlet and outlets	
Mow to limit unwanted vegetation	Routine

Source: From Denver Urban Drainage and Flood Control District (UDFCD), Drainage Criteria Manual Vol. 3, Denver, CO, 1999.

Retention Pond

A retention (or wet) pond is designed as a permanent pool of water, often with additional flood control and extended detention storage volume available above the permanent pool. Retention ponds generally hold water for release only through evapotranspiration and infiltration, though there are designs that maintain an extended detention volume above the wet pond. Retention ponds have proven to be the most effective common structural BMP type in the Eastern United States, when considering all aspects of cost, performance, and maintenance.

Pond volumes are designed to control storm-related pond overflows to a design standard, so that the contributing drainage area and groundwater is capable of supporting a permanent pool. The pond provides pollutant removal through settling of particulates, infiltration, filtration, and biological uptake of soluble contaminants. To be effective in removing pollutants, there must be sufficient runoff residence time, so most design modifications are conceived to lengthen residence time. Good results can be achieved through direct long-term simulation of the pond using the SWMM, STORM, or other model.

Expected performance can be very good, with proper design, depending upon the basin's size relative to the following five characteristics:

- Watershed area
- Vegetative cover of watershed and pond
- Seasonal effects and variation
- Soil erodibility and infiltration rate
- Storm characteristics

There is some difference of opinion on the requirement for an extended detention pool above the retention pond. There is little evidence to suggest there will be a substantial increase in pollutant removal to justify the additional cost (California Stormwater Task Force, 1993). The Denver Urban Drainage and Flood Control District recommends a drawdown (detention) time of 12 h for the design storm before reaching the permanent pool level poststorm (UDFCD, 1992). Other states or localities have different criteria for the extended detention pool. Some put part of the water-quality volume in the extended detention pool (ARC, 2002).

There are several different variants of stormwater pond design, the most common of which include the wet pond, the wet extended detention pond, and the micropool extended detention pond. In addition, multiple stormwater ponds can be placed in series or parallel to increase performance or meet site design constraints. Below are descriptions of each design variant:

- *Wet pond* – Wet ponds are stormwater basins constructed with a permanent (dead storage) pool of water. Stormwater runoff displaces the water already present in the pool during a storm event. Temporary storage (live storage) can be provided above the permanent pool elevation for larger flows.
- *Wet extended detention (ED) pond* — A wet extended detention pond is a wet pond in which the volume of runoff to be treated is divided between the permanent pool and the extended detention storage provided above the permanent pool. During storm events, runoff is detained above the permanent pool and released over an extended time. The design has similar pollutant removal to a traditional wet pond but consumes less space.
- *Micropool extended detention (ED) pond* — The micropool extended detention pond is a variation of the wet ED pond in which only a small micropool is maintained at the outlet to the pond. The outlet structure is sized to detain a volume of water for a specified period of time. The micropool prevents resuspension of previously settled sediments and also prevents clogging of the low-flow orifice.
- *Multiple pond systems* — Multiple pond systems consist of constructed facilities that provide volume storage in two or more cells. The additional cells can create longer pollutant removal pathways and improve downstream protection.

Figures 13-24 through 13-31 portray schematics and pictures of many of these variations. Each of the schematics provide examples of design for multiple impacts under the unified design criteria concept including water quality, channel protection, nuisance flooding, and extreme flooding. The outlet structures would be more simple, and the volumes

FIGURE 13-24
Wet pond.

FIGURE 13.25
Micropool ED pond.

FIGURE 13.26
Wet ED pond.

FIGURE 13.27
Multipond system.

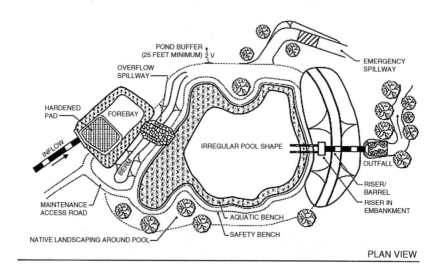
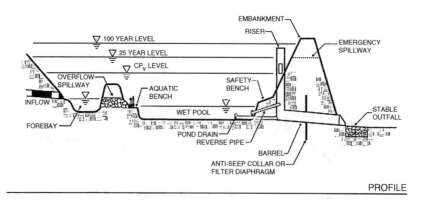

FIGURE 13-28
Wet pond schematic.

Structural Best Management Practices 785

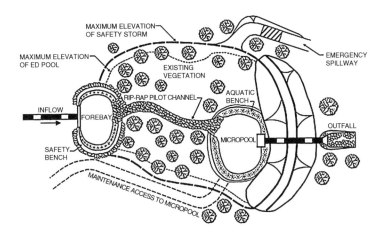

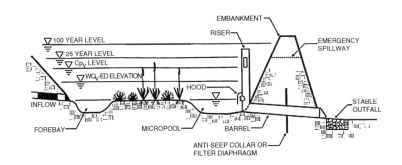

FIGURE 13-29
Micropool ED pond schematic.

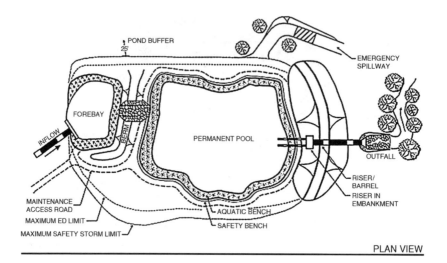
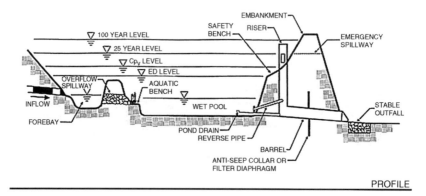

FIGURE 13-30
Wet ED pond schematic.

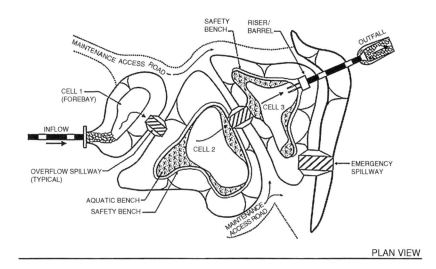
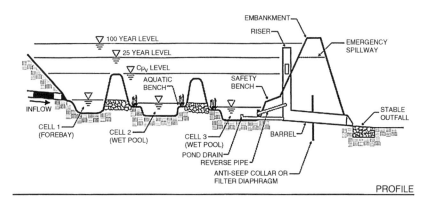

FIGURE 13-31
Multipond schematic.

smaller, without these multiple design objectives. A simpler pond is shown in Figure 13-32 (WDE, 2001).

Wet ponds provide opportunities for multiobjective applications involving recreation and nature areas. They are ideal for large tributary areas. Pondscaping (see Figure 13-33) has become almost an art form in some communities, wherein the BMP becomes a prime landscape feature (Karouna, 1992). Effective arrangements reflected natural and human interaction themes. Recent applications feature pretreatment for oil and grease, flow spreader berms, and lengthened flow paths. It is important to obtain native plants that can propagate and stand frequent inundation without significant maintenance. Whitlow and Harris (1979), among others, provide a guide for flood tolerance of plants in each area of the United States. MDE (2000) provides a landscaping guide for wet ponds and other BMPs.

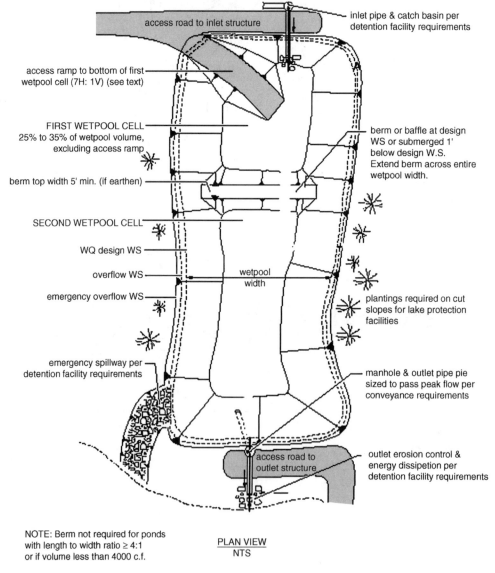

FIGURE 13-32
Basic wet pond schematic.

FIGURE 13-33
Pondscaping.

Emergent vegetation is also used. Approximately 10-ft wide shelves with depths of 1 to 2 ft provide an ideal setting for emergent vegetation and increased pollution removal, though emergent vegetation can work against effective mosquito control. If mosquitoes are a problem, this shelf could be reduced, pond level fluctuated, steeper slopes added, or Gambusia (mosquito fish) added to the pond for control (California Stormwater Task Force, 1993). Gambusia are stocked at the rate of about 1000 fish per pond surface acre. Nonrooted vegetation or rock filters near the outlet can also assist in pollution removal. Consider designing and managing the pond as a fishing lake.

Wet pond costs are generally 25 to 40% higher than conventional detention ponds (MWCOG, 1992). Brown and Schueler (1997) provide an estimate of pond costs as follows:

$$C = 24.5V^{0.705} \tag{13.10}$$

where C = construction, design, and permitting costs, and V = volume in the pond to include the 10-year storm (cubic feet).

Ponds do not consume a large area (typically 2 to 3% maximum of the contributing drainage area). Therefore, the land consumed to design the pond will not be very large. It is important to note, however, that these facilities are generally large. Other practices, such as filters or swales, may be squeezed into relatively unusable land, but ponds need a relatively large continuous area.

Maintenance costs range from 3 to 5% of first costs annually. Wet ponds accumulate sediment under water and thus do not need to be cleaned for aesthetic reasons as often as dry ponds (Hartigan, 1988). The pond muck can have pollutant removal capabilities. See Techniques (1994) for a complete discussion.

Some other possible disadvantages of wet ponds include safety considerations of a pond, floating debris and scum, waterfowl increasing the nutrient loading, and odors. Resuspension and remobilization of pollutants can be a problem if a large storm passes through the pond during unfavorable chemical conditions (UDFCD, 1992). Pond water quality can become quite poor. If a storm displaces this water, the outflow can be more polluted than the inflow, and warmer. Larger ponds also inhibit fish passage, change stream flow regimes, and can destroy wetlands adjacent to the stream. Pond fingerprinting is used to

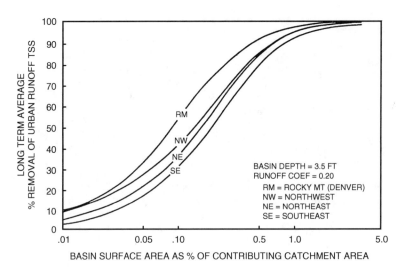

FIGURE 13-34
Wet detention basins size performance relationship. (From USEPA, 1983.)

reduce environmental impacts through flow diversion, pond shaping ("doughnut ponds"), pond series, and other techniques. Additional environmental factors are discussed by Schueler and Galli (1992). See the extended detention pond discussion for brief consideration of bottom sediment contamination.

Pollution Removal Efficiency

Wet ponds can be effective in removal of soluble and particulate fractions of pollution. Estimates of removal rates vary widely in the literature. Approximated removal rates for wet ponds of different designs can be estimated from Table 13-13. Because a particular detention basin may exhibit variable storm-to-storm performance characteristics, the long-term average performance should be considered rather than individual events. Freezing conditions can be expected to reduce removal rates by 50%.

Results from the NURP study (EPA, 1983) provide approximate removal rates for TSS for a design with an average depth of 3.5 ft and an average impervious area upstream of 20% (low-density residential). Figure 13-34 gives this removal information. These curves can be converted to VB/VR curves through the use of the mean depth and a 0.2 runoff coefficient (Hartigan, 1988). It is a reasonable estimate for TSS and for the particulate fractions of other pollutants as well (Driscoll, 1986).

For additional information and data on pollutant removal capabilities for stormwater ponds, see the National Pollutant Removal Performance Database (2nd ed.) available at www.cwp.org and the National Stormwater Best Management Practices (BMP) Database at www.bmpdatabase.org.

Pond Sizing

There are several basic and differing design methodologies for retention ponds that reflect consideration of settling as the major removal mechanism, or a further consideration of biological uptake as an additional important mechanism (Hartigan, 1988; Nix, Heaney, and Huber, 1988; Wanielista and Yousef, 1993). Because many pollutants have an affinity for suspended solids in runoff, design of a wet pond for pollution removal by sedimentation is a logical approach (Randall, 1982). Basins with average retention times on the order of 2 weeks during the biologically active growing season will exhibit better removal rates due to the soluble portion of nutrients becoming available for biological uptake (Hartigan, 1988). The most important factors in this nutrient cycling are phosphorus-loading and decay rates, hydraulic residence time, and mean depth. The two leading models used in the eutrophication modeling of retention ponds are those of Walker (1985, 1987) and Reckhow (1988).

The detailed statistical method of Driscoll (Woodward-Clyde, 1986; Driscoll, 1988) can be used for pond sizing with permanent pools, where there are no other local or state criteria. The necessary rainfall information for this method can be obtained from Driscoll et al. (1989), if local rainfall statistical estimates are not available, which was provided earlier in this chapter. Local rainfall analysis has been done for many places in the United States, resulting in tables of surface area and drainage area relationships for extended detention ponds with permanent pools (for example, Professional Engineers of North Carolina, 1988, or Puget Sound, 1992).

An alternative method has been proposed by Hvitved-Jacobsen, Yousef, and Wanialista (1988). In this method, the rainfall record is analyzed to determine rainfall statistics for a selected inter-event dry period. For example, to ensure at least a 72-h recommended retention period, the rainfall would be analyzed to determine the mean rainfall (and runoff) volume, assuming storms must be separated by dry intervals of at least 72 h. If there is not at least a 72-h separation, then all the precipitation is considered to be one runoff event. By selecting a storm return frequency (such as the 4-month storm) and an inter-event dry period (such as 72 h), the designer controls the number of overflows per year (three on average, in this example) and the average time of retention within the pond. While this is equated to 1.06 in of rainfall in a location in Denmark, the pond must be designed to contain the runoff from 3.25 in of rainfall for a location in Orlando, Florida (Hvitved-Jacobsen, Yousef, and Wanialista, 1988).

Volume requirements differ from place to place and depend on intended retention time within the pond. For example, Washington State Department of Ecology requires that the 6-month, 24-h storm runoff volume be retained for a normal pond, and twice that for a large pond (WDE, 2000). Denver Urban Drainage and Flood Control District requires a different volume depending on type of facility and imperviousness ratio (UDFCD, 1999). New York State, Maryland, and Georgia require retention of a water-quality control volume depending on local or regional rainfall characteristics and a simple runoff relationship. Common pond volume criteria include:

- 0.5 in per impervious acre
- 2.5 times the mean runoff
- Runoff from a certain return period storm
- Runoff from a certain percentage volume frequency storm
- 4.0 times the mean runoff or approximately 2 weeks' retention, whichever is greater

In arid and semiarid climates, these predicted volumes should be considered after decreases due to evaporation and groundwater losses have been accounted for. If there is some question of the pond's ability to hold water, a water balance should be performed, taking into account all losses (see Chapter 7). Infiltration through the pond bottom can be reduced through the use of a clay or plastic liner at and below the permanent water line. These facilities often provide release temperatures significantly higher than the receiving waters. As such, care should be exercised when discharging into cold-water fisheries. Some authors recommend adding volume to account for sediment deposition, up to 25%. This is especially important when construction may occur within the watershed upstream. In one case, a pond was designed for pump back to a treatment plant (Segarra-Garcia and Loganathan, 1993). A sediment forebay sized to 10% of the total pond volume will extend the pond life from a 10- to 20-year cycle to twice that. The forebay should be 4 to 6 ft deep and have exit velocities less than 1 ft/sec (Techniques, 1994b).

Wet ponds typically take up 0.5 to 2% of a watershed area. Calculations of pond volume are similar to those for extended detention when a VB/VR ratio is the design criteria. Two variations on this method are:

$$VR = V * Rv * A * 43{,}560/12 \qquad (13.11)$$

$$VR = V * A_i * 43{,}560/12 \qquad (13.12)$$

where VR = volume of mean annual runoff (in); V = volume of mean annual rainfall (in); Rv = volume runoff coefficient; A = total drainage area (acres); and A_i = total impervious drainage area (acres).

If an extended detention volume is provided above the permanent pool, its outlet calculation is handled similarly to a dry extended detention pond. See Chapter 7 for an approximate routing method for extended detention. One advantage of having a permanent pool is that a negatively sloped pipe or inverted siphon can be used to draw water from below the surface to bleed off the extended detention portion of the pond, reducing clogging. Alternately, the Denver Flood Control District recommends a shorter 12-h drawdown time for such a design, and has provided a figure (Figure 13-35) to determine the required perforated riser area per row similar to the 40-h drawdown figure presented earlier. Table 11-4 gives the arrangement for perforations in the rows 4 in apart vertically on center.

It should be noted that performance surveys of detention ponds designed using state-of-the-art methods still achieved a wide range of removal rates of TSS (Driscoll and Strecker, 1993). Fifty percent of the ponds surveyed in Maryland had achieved removal rates between 60 and 80%, with 25% on either side. Assuming a target rate of 80% TSS removal, 50% of the ponds failed to achieve the target, though 75% were above 60% TSS removal.

Typical Required Specifications

- Minimum length-to-width ratio of 3:1 (preferably expanding outward toward the outlet)
- Irregular shorelines for larger ponds provide visual variety
- Design for full development upstream of control
- Inlet and outlet located to maximize flow length (Use baffles if short circuiting cannot be prevented with inlet-outlet placement.)
- Permanent pool depth: minimum — 2 to 3 ft, maximum — 9 to 10 ft, average — 3 to 6 ft
- Side slopes no greater than 3:1 if mowed
- Riprap protection (or other suitable erosion control means) for the outlet and all inlet structures into the pond (Individual boulders or baffle plates can work for this.)
- Minimum of 25 acres is needed for wet pond and wet ED pond to maintain a permanent pool, 10 acres minimum for micropool ED pond (A smaller drainage area may be acceptable with an adequate water balance and anticlogging device.)

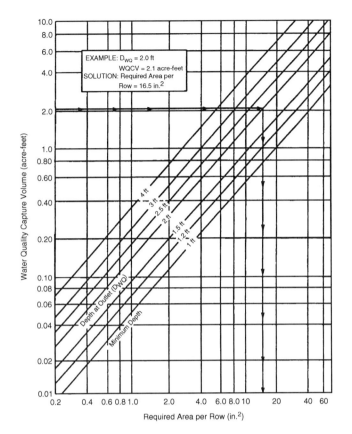

FIGURE 13-35
Water-quality outlet sizing: wet extended detention pond with a 12-h drain time. (UDFCD, 1992, 1994)

- Antiseep collars on barrel of principal spillway
- One-half ft minimum freeboard above peak stage for top of embankment
- Emergency drain, i.e., sluice gate, drawdown pipe; capable of draining within 24 h
- Emergency spillway designed to pass the 50-year storm event
- Bypass greater than 2-year storms
- Space required will be approximately 2 to 3% of the tributary drainage area
- There should be no more than a 15% slope across the pond site
- Minimum elevation difference needed at a site from the inflow to the outflow is 6 to 8 ft
- If used on a site with an underlying water supply aquifer or when treating a hot spot, a minimum separation distance of 2 ft is required between the bottom of the pond and the elevation of the seasonally high water table
- Underlying soils of hydrologic group "C" or "D" should be adequate to maintain a permanent pool (Most group "A" soils and some group "B" soils will require a pond liner.)
- Trash racks, filters, hoods, or other debris control on riser
- Maintenance access (< 15% slope and 10 to 12 ft wide)
- Benchmark for sediment removal

- Paved or concrete channel
- Ponds cannot be located within a stream or any other navigable waters of the U.S., including wetlands, without obtaining a Section 404 permit under the Clean Water Act

Typical Recommended Specifications

- Multiobjective use such as amenities or flood control
- Landscaping management of buffer as meadow
- Design for multifunction as flood control and extended detention
- Minimum length-to-width ratio of 3:1 to 4:1 (preferably wedge shaped)
- Use reinforced concrete instead of corrugated metal
- Sediment forebay for larger ponds (often designed for 5 to 15% of total volume); forebay should have separate drain for dewatering; grass biofilters for smaller ponds
- Consider artificial mixing for small sheltered ponds
- Provision should be made for vehicle access at a 4:1 slope
- Impervious soil boundary to prevent drawdown
- Shallow marsh area around fringe 25 to 50% of area (including aquatic vegetation)
- Safety bench at toe of slope (minimum 10 ft wide)
- Minimum 25-ft wide buffer around pool
- Minimum 100 ft from a private well; if well is down-gradient from a hot spot land use, then the minimum setback is 250 ft
- Minimum 50 ft from a septic system tank/leach field
- The perimeter of all deep pool areas (4 ft or greater in depth) should be surrounded by two benches: safety and aquatic
- For larger ponds, a safety bench extends approximately 15 ft outward from the normal water edge to the toe of the pond side slope
- The maximum slope of the safety bench should be 5%
- An aquatic bench extends inward from the normal pool edge (15 ft on average) and has a maximum depth of 18 in below the normal pool water surface elevation
- Mow embankment and side slopes at least twice a year
- Emergency drain to allow drawdown within 24 h
- On-site disposal areas, for two sediment removal cycles, protected from runoff
- Consider chemical treatment by alum if algal bloom are a problem
- An oil-and-grease skimmer for sites with high production of such pollutants
- For the design of outlet structures for ponds, see Chapter 11

Typical Maintenance Standards for Extended Detention and Wet Ponds

- Sediment to be removed when 20% of storage volume of the facility is filled (design storage volume must account for volume lost to sediment storage)
- Sediment traps should be cleaned when filled

TABLE 13-20

Typical Maintenance Activities for Ponds

Activity	Schedule
Clean and remove debris from inlet and outlet structures; mow side slopes	Monthly
If wetland components are included, inspect for invasive vegetation	Semiannual inspection
Inspect for damage, paying particular attention to the control structure	Annual inspection
Check for signs of eutrophic conditions	
Note signs of hydrocarbon buildup, and remove appropriately	
Monitor for sediment accumulation in the facility and forebay	
Examine to ensure that inlet and outlet devices are free of debris and operational	
Check all control gates, valves, or other mechanical devices	
Repair undercut or eroded areas	As needed
Perform wetland plant management and harvesting	Annually (if needed)
Remove sediment from the forebay	5 to 7 years or after 50% of the total forebay capacity has been lost
Monitor sediment accumulations, and remove sediment when the pool volume has become reduced significantly, or the pond becomes eutrophic	10 to 20 years or after 25% of the permanent pool volume has been lost

Source: From Watershed Management Institute (WMI), Operation Maintenance and Management of Stormwater Management Systems, prepared for U.S. EPA, Office of Water, 1997.

- No woody vegetation should be allowed on the embankment without special design provisions
- Other vegetation over 18 in high should be cut unless it is part of planned landscaping
- Debris should be removed from blocking inlet and outlet structures and from areas of potential clogging
- The control should be kept structurally sound, free from erosion, and functioning as designed
- Periodic removal of dead vegetation should be accomplished
- No standing water is allowed within extended detention pond unless specifically designed for
- An annual inspection is required, reports to be kept by owner
- The site should be inspected and debris removed after every major storm

Table 13-20 provides a summary of maintenance requirements for wet ponds (ARC, 2002).

Alum Treatment System

To enhance the water-quality aspects of wet ponds, an alum treatment system can be added. The process of alum (aluminum sulfate) treatment provides treatment of stormwater runoff from a piped stormwater drainage system entering a wet pond by injecting liquid alum into storm drain lines on a flow-weighted basis during rain events. When added to runoff, liquid alum forms nontoxic precipitates of aluminum hydroxide and aluminum phosphate, which combine with phosphorus, suspended solids, and heavy metals, causing them to be deposited into the sediments of the receiving waters in a stable inactive state.

The alum precipitate formed during coagulation of stormwater can be allowed to settle in receiving water or to collect in small settling basins. Alum precipitates are stable in sediments and will not redissolve due to changes in redox potential or pH under conditions normally found in surface water bodies. Laboratory or field testing may be necessary to verify feasibility and to establish design, maintenance, and operational parameters, such as the optimum coagulant dose required to achieve the desired water-quality goals, chemical pumping rates, and pump sizes.

Construction costs for existing alum stormwater treatment facilities in Florida have ranged from $135,000 to $400,000. The capital construction costs of alum stormwater treatment systems is independent of watershed size and depends primarily on the number of outfall locations treated. Estimated annual operations and maintenance costs for chemicals and routine inspections range from approximately $6500 to $25,000 per year. These costs include chemical, power, manpower for routine inspections, equipment renewal, and replacement costs.

Alum treatment has consistently achieved a 85 to 95% reduction in total phosphorus, 90 to 95% reduction in orthophosphorus, 60 to 70% reduction in total nitrogen, 50 to 90% reduction in heavy metals, 95 to 99% reduction in turbidity and TSS, 60% reduction in BOD, and >99% reduction in fecal coliform bacteria, compared with raw stormwater characteristics.

Alum treatment systems are fairly complex, and design details are beyond the scope of this book. Further information can be obtained from the Internet and by contacting local municipalities and engineers who have designed and implemented successful systems.

Design Example

An example of the design of a wet ED pond is given. It is based on work done by the Center for Watershed Protection for the Georgia Stormwater Manual (2002), which should be referred to for further details and complete design forms, etc. See Chapter 7 for background information on equations and approaches used here.

Base Data

Site Area = Total Drainage Area (A) = 38.0 acres

Measured Impervious Area = 13.8 acres; or I = 13.8/38 = 36.3%

Soils Types: 60% "C", 40% "B"

Zoning: Residential ($1/2$ acre lots)

Hydrologic Data

	Pre	Post
CN	65	78
tc	0.31 h	0.17 h

Phase I — Computation of Preliminary Stormwater Storage Volumes and Peak Discharges
Step 1 — Compute Runoff Control Volumes from the Unified Stormwater Sizing Criteria —

- Compute Runoff Coefficient, Rv

$$Rv = 0.05 + (I)(0.009)$$

$$= 0.05 + (36.3)(0.009) = 0.38$$

- Compute WQ_v
  - Note: Design storm depth is 1.2 in

$$WQv = (1.2")\,(Rv)\,(A)$$

$$= (1.2")\,(0.38)\,(38.0\text{ ac})\,(1\text{ft}/12\text{in})$$

$$= 1.44 \text{ ac-ft}$$

Step 2 — Develop Site Hydrologic and Hydrologic Input Parameters — Note that any hydrologic models using SCS procedures, such as TR-20, HEC-HMS, or HEC-1, can be used to perform preliminary hydrologic calculations.

Condition	Area (ac)	CN	Tc (h)
Predeveloped	38	65	0.31
Postdeveloped	38	78	0.17

Perform preliminary hydrologic calculations:

Condition	Q1-year	Q10-year	Q25-year	Q100-year
Runoff	Inches	Cfs	cfs	cfs
Predeveloped	0.7	22	101	147
Postdeveloped	1.4	67	202	267

Step 3 — Compute Channel Protection Volume (Cpv) — Note that, with design criteria, for stream channel protection, provide 24 h of extended detention for the 1-year event.

- Use the SCS approach to compute channel protection storage volume (see Chapter 7).
  - Initial abstraction (Ia) for CN of 78 is 0.564: [$Ia = (200/CN - 2)$]

$$Ia/P = (0.564)/3.4 \text{ in} = 0.17$$

$$Tc = 0.17 \text{ h}$$

$$qu = 800 \text{ csm/in (Type II Storm)}$$

- Knowing qu and T (extended detention time), find qo/qi for a Type II rainfall distribution.
- Peak outflow discharge/peak inflow discharge (qo/qi) = 0.022.

$$Vs/Vr = 0.683 - 1.43(qo/qi) + 1.64(qo/qi)^2 - 0.804(qo/qi)^3$$

- Where Vs equals channel protection storage (Cpv) and Vr equals the volume of runoff in inches.

$$Vs/Vr = 0.65$$

- Therefore,

$$V_s = C_{pv} = 0.65(1.4")(1/12)(38 \text{ ac}) = 2.9 \text{ ac-ft } (126{,}324 \text{ cubic feet})$$

- Define the average $CP_v\text{-}ED$ release rate.
- The above volume, 2.9 ac-ft, is to be released over 24 h.

$$(2.9 \text{ ac-ft} \times 43{,}560 \text{ ft}^2/\text{ac}) / (24 \text{ hrs} \times 3{,}600 \text{ sec/hr}) = 1.46 \text{ cfs}$$

Step 4 — Compute Overbank Flood-Protection Volume — Note the design criteria of a 25-year storm for overbank flood protection.

- For a Q_{in} of 202 cfs, and an allowable Q_{out} of 101 cfs, and a runoff volume of 552,584 cubic feet (12.69 ac-ft), the V_s necessary for 25-year control is 3.55 ac-ft, under a developed CN of 78. Note that 6.5 in of rain fell during this event, with approximately 4.0 in of runoff.
- While the TR-55 shortcut method reports to incorporate multiple stage structures, experience has shown that an additional 10 to 15% storage is required when multiple levels of extended detention are provided inclusive with the 25-year storm. So, for preliminary sizing purposes, add 15% to the required volume for the 25-year storm.

$$Q_{p\text{-}25} = 3.55 \times 1.15 = 4.1 \text{ ac-ft.}$$

Step 5 — Analyze Safe Passage of 100-Year Design Storm (Q_f) — At final design, provide safe passage for the 100-year event, or detain it, depending on downstream conditions and local policy. Based on field observation and review of local requirements, no control of the 100-year storm is necessary. If control was required, storage estimates would have been made similar to the Q_p volume in the previous substep.

Phase II — Determination if the Development Site and Local Conditions are Appropriate for the Use of a Stormwater Pond

Step 1 — Screen BMPs Using Site-Specific Data — The site area and drainage area to the pond are 38.0 acres. Existing ground at the pond outlet is 919 MSL. Soil-boring observations reveal that the seasonally high water table is at elevation 918. The underlying soils are SC (sandy clay) and are suitable for earthen embankments and to support a wet pond without a liner. The stream invert at the adjacent stream is at elevation 916.

Other site screening aspects listed in Chapter 13 were assessed, and a pond was found to be suitable.

Step 2 — Confirm Local Design Criteria and Applicability — There are no additional requirements for this site.

Phase III — Performance of Final Design Sizing

Step 1 — Determine Pretreatment Volume — Size wet forebay to treat 0.1 in/impervious acre. (13.8 ac) (0.1 in) (1'/12") = 0.12 ac-ft
(Forebay volume is included in WQv as part of permanent pool volume.)

Step 2 — Determine Permanent Pool Volume (and Water-Quality ED Volume) — Size permanent pool volume to contain 50% of WQv:

$$0.5 \times (1.44 \text{ ac-ft}) = 0.72 \text{ ac-ft. (includes 0.12 ac-ft of forebay volume)}$$

Size ED volume to contain 50% of WQv:

$$0.5 \times (1.44 \text{ ac-ft}) = 0.72 \text{ ac-ft}$$

Note that this design approach assumes that all of the ED volume will be in the pond at once. While this will not be the case, because there is a discharge during the early stages of storms, this conservative approach allows for ED control over a wider range of storms, not just the target rainfall.

Step 3 — Determine Pond Location and Preliminary Geometry; Conduct Pond Grading and Determine Storage Available for Permanent Pool and Water Quality Extended Detention — This step involves initially grading the pond (establishing contours) and determining the elevation–storage relationship for the pond. Storage must be provided for the permanent pool (including sediment forebays), extended detention (WQv-ED), Cpv-ED, and 25-year storm, plus sufficient additional storage to pass the 100-year storm with minimum freeboard. An elevation–storage table and curve are prepared using the average area method for computing volumes.

Step 4 — Set Basic Elevations for Pond Structures —

The pond bottom is set at elevation 920.0.

Provide gravity flow to allow for pond drain; set riser invert at 919.5.

Set barrel outlet elevation at 919.0.

Step 5 — Set Water Surface and Other Elevations —

Required permanent pool volume = 50% of WQv = 0.72 ac-ft. From the elevation-storage table, an elevation 924.0 (1.04 ac-ft > 0.72 ac-ft) site can accommodate it, and it allows a small safety factor for fine sediment accumulation — OK

Forebay volume provided in two pools with avg. vol. = 0.08 ac-ft each (0.16 ac-ft > 0.12 ac-ft) OK

Required extended detention volume (WQv-ED) = 0.72 ac-ft. From the elevation–storage table (volume above permanent pool), read elevation 926.0 (0.73 ac-ft > 0.72 ac-ft) OK.

Set ED wsel = 926.0.

Note that total storage at elevation 926.0 = 1.77 ac-ft (greater than required WQv of 1.44 ac-ft).

Step 6 — Compute the Required WQv-ED Orifice Diameter to Release 0.72 ac-ft Over 24 h —

Avg. ED release rate = (0.72 ac-ft)(43,560 ft^2/ac)/(24 h)(3600 sec/h) = 0.36 cfs

Average head = (926.0 - 924.0)/ 2 = 1.0'

Use orifice equation to compute cross-sectional area and diameter:

$Q = CA(2gh)^{0.5}$, for Q = 0.36 cfs h = 1.0 ft; C = 0.6 = discharge coefficient solve for A

$$A = 0.36 \text{ cfs } / [(0.6)((2)32.2 \text{ ft/s}^2)(1.0 \text{ ft}))^{0.5}] \text{ A} = 0.075 \text{ ft}^2,$$
$$A = \pi d2 / 4; \text{ dia.} = 0.31 \text{ ft} = 3.7''$$

Use 4-in pipe with 4-in gate valve to achieve equivalent diameter

Compute the stage–discharge equation for the 3.7" dia. WQv orifice

$$Q_{WQ_v\text{-}ED} = CA(2gh)^{0.5} = (0.6)(0.075 \text{ ft}^2)[((2)(32.2 \text{ ft}/s^2))^{0.5}](h^{0.5}),$$

$$Q_{WQ_v\text{-}ED} = (0.36) h^{0.5}, \text{ where: } h = \text{wsel} - 924.15$$

Note that when calculating head, account for one half of orifice diameter.

Step 7 — Compute Extended Detention Orifice Release Rate(s) and Size(s), and Establish Cpv Elevation —

Set the Cpv pool elevation

Required Cpv storage = 2.9 ac-ft (based on stage–discharge curve)

From the elevation–storage table, read elevation 929 (this includes the WQv)

Set Cpv wsel = 929

Size to release average of 1.46 cfs

Average WQv-ED orifice release rate is 0.66 cfs, based on average head of 3.34' (926 − 924.16 + (929 − 926)/2)

Cpv-ED orifice release = 1.46 - 0.66 = 0.80 cfs

Head = (929 - 926.0)/2 = 1.5'

Use orifice equation to compute cross-sectional area and diameter

$$Q = CA(2gh)^{0.5}, \text{ for } h = 1.5'$$

$$A = 0.80 \text{ cfs} / [(0.6)((2)(32.2'/s^2)(1.5'))^{0.5}]$$

$$A = 0.14 \text{ ft}^2, A = \pi d^2 / 4;$$

$$\text{dia.} = 0.42 \text{ ft} = 5.0"$$

Use 6" pipe with 6" gate valve to achieve equivalent diameter

Compute the stage-discharge equation for the 5.0" dia. Cpv orifice

$$QCpv\text{-}ED = CA(2gh)^{0.5} = (0.6)(0.14 \text{ ft}^2)[((2)(32.2'/s^2))^{0.5}](h^{0.5}),$$

$$QCpv\text{-}ED = (0.67)(h^{0.5}), \text{ where: } h = \text{wsel} - 926.21$$

(Note: account for one half of orifice diameter when calculating head)

Step 8 — Calculate Qp25 (25-Year Storm) Release Rate and Water Surface Elevation and Check 100-Year Overflow for Safety or Detention Depending on Local Design Criteria — In order to calculate the 25-year release rate and water surface elevation, the designer must set up a stage–storage–discharge relationship for the control structure for each of the low-flow release pipes (WQv-ED and Cpv-ED) plus the 25-year storm.

At this point, the design proceeds in a manner similar to design of a standard multiyear detention pond, where each outflow device is sized taking into account the total discharge at the starting elevation. See ARC, 2002 for details.

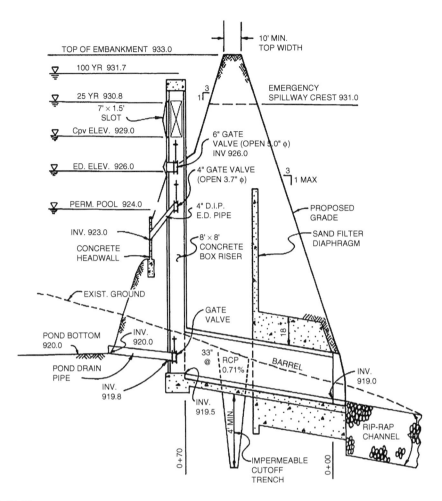

FIGURE 13-36
Example problem spillway structure.

Step 9 — Prepare Vegetation and Landscaping Plan — See MDE (2000) for example landscaping plans and detailed landscaping guidance.

Figures 13-36 and 13-37 give elevation views of the spillway and planned outlet structure for this BMP.

Constructed Stormwater Wetlands

Wetlands can provide a highly effective management measure for mitigation of pollution from runoff, because they have the ability to assimilate large quantities of suspended and dissolved materials from inflow. A basic understanding of wetlands types and function is a necessary prerequisite to the use of wetlands for stormwater pollution control. Mitsch and Gosselink (1986) and Hammer (1989, 1992) provide basic information on wetlands, constructed wetlands for wastewater treatment, and freshwater wetland creation. Schueler (1992) presents a comprehensive consideration of wetland creation with emphasis on the Eastern United States. Later literature amplifies and extends his findings.

Wetlands functions and values include (EPA Region 7, 1992):

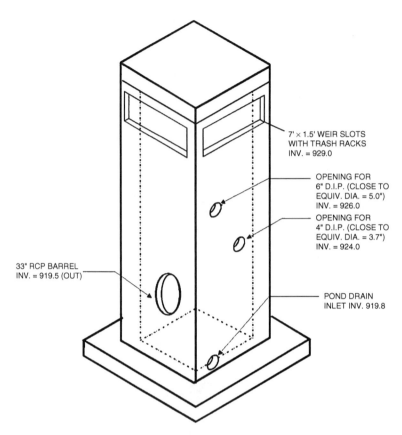

FIGURE 13-37
Example problem outlet structure.

- Flood reduction and conveyance
- Sediment control and reduction
- Habitat for fish, waterfowl, rare and endangered species, and other wildlife
- Enhancement of water quality
- Wastewater treatment
- Groundwater recharge
- Recreation, education, and research
- Food and timber production
- Open space and aesthetic values

The design of wetland management measures is complex, being, generally, a function of "nearly everything." But, the three most important components of wetland creation and function are water, soil, and vegetation. The hydrology of the wetland essentially drives the physical and chemical processes within the wetland. These, in turn, drive the ecosystem character and growth. Water carries the nutrients into the wetland system. It also affects such critical primary variables as pH, temperature, degree of substrate anoxia, and soil salinity. Species composition and richness, primary productivity of the vegetation, organic accumulation, and nutrient cycling are all primarily controlled by the hydrology of the wetland. Thus, the hydrology of the wetland controls its ability to remove pollutants through changes in velocity and flow rate, water depth and fluctuation, detention time, circulation and distribution patterns, seasonal hydrological influences, and groundwater conditions (Livingston, 1988).

Wetlands are an intermediate form between uplands and aquatic systems. Not that they are always transitional in the sense of succession from one to the other, but they share vital features of each. Wetland areas are subject to the random ebb and flow of surface water. As such, they are sensitive to changes in hydrology and may respond to minor changes in flow or hydro-period with major changes in species richness and ecosystem productivity (Mitsch and Gosselink, 1986). Wetlands that are intermittently flooded are different from those that are only intermittently exposed to the drying atmosphere. Saturated soils maintain a different ecosystem than partially saturated systems. The major parameter for the measurement of gross wetland hydrology is the residence time, which is the measure of the average time that water stays in the wetland. The reciprocal of residence time is termed the *turnover rate*.

Wetland soils are termed hydric. This means that the soil is saturated, flooded, or ponded long enough during the growing season to develop anaerobic conditions that favor the growth and regeneration of hydrophytic (adopted to wetland conditions) vegetation (SCS, 1985). If the soils contain less than 12 to 20% organic materials, they are considered mineral soils. More organic material than that, and they are considered organic, or peat, soils (Hammer, 1992). The soil is the basic medium in which many of the wetland chemical transformations take place and in which are stored the primary chemicals for wetland plants (Mitsch and Gosselink, 1986). These two basic types of soils support vegetation in differing ways. Organic soils hold water better, while mineral soils have more minerals biochemically available to plants. Most urban wetlands will be based on mineral soils.

Contrary to thinking that a wetland is a lush vegetated paradise for plants, a wetland environment can be harsh and demanding on vegetation, requiring many adaptations for survival and, often, a narrow range of conditions under which a particular species can survive at all. Nonetheless, there are over 500 species of wetland plants in the United States (Hammer, 1992). A wetland plant is defined as a species that can tolerate or regulate sufficiently to survive up to 5 days inundation during the growing season. These plants are often classified as vascular (having internal structural transport mechanisms) or nonvascular, and as free-floating or rooted. The rooted forms are subdivided into submergent, emergent, and floating-leaved types. Woody species are sometimes classified separately but actually fall under the rooted emergent category, though their roots are rarely under water.

Wetlands plants tend to insulate themselves from the changes the environment brings about. They do this through individual methods and through peat production to stabilize the nutrient and moisture source. All wetland plants have elaborate mechanisms for survival that include structural methods (such as air spaces) and physiological methods (such as modified respiration and photosynthesis methods). Even with this, the limitations imposed by the wetlands environment, along with plant competition, limit the areal extent of species. Zones of plant communities tend to be sharply delineated by abiotic factors such as soils, inundation durations, saturation, and water chemistry (EPA Region 7, 1992). Families of plant species with fairly similar tolerances group together on these steep ecological gradients (Mitsch and Gosselink, 1986). Horner (1988) found some evidence that urban (versus rural) wetlands tended toward more opportunistic and monocultural vegetative stands, probably due to modified hydrology. Virginia planners have divided these zones into aquatic, reed, and riparian zones. Aquatic is almost always inundated, while riparian is normally above the water surface. The reed zone begins slightly above mean low water level and extends into the water to a depth of about 18 in (Northern Virginia, 1992). Shenot (1993) recorded a number of volunteer species at ponds in Maryland dominated by exotic grasses (Graminea), soft rush (Juncus Effusus), and various sedges. Cattails are also notorious volunteers.

Stormwater Wetlands

Stormwater wetlands differ from more natural wetlands in that they are not normally designed with the diversity of species and ecological functions of natural wetlands (MWCOG, 1992). Rather, they are designed to maximize pollution removal. Horner found that a major difference between urban and rural wetlands was simply the amount of trash in urban wetlands detracting from aesthetic appeal (Horner, 1988). Schueler (1992) gives other differences between natural and urban wetlands in the mid-Atlantic region, including groundwater versus runoff-dominated water balance, hydro-period gradual and buffered versus hydro-period based on runoff and is rapid, adjusting wetland boundaries versus controlled boundaries, high diversity maintained by seedbank versus low diversity established by planner, self-maintaining versus man-maintained, complex topography versus simple topography, low sediment supply versus high sediment supply, and high wildlife habitat versus low wildlife habitat. There may be some question about the use of wetlands in arid Western climates due to the possible need for irrigation of the vegetation and that the wet season occurs during times of vegetation dormancy (California Stormwater Task Force, 1993). Many inadvertent riparian wetlands have been created over time with the use of effluent and irrigation return flows. Thus, it is possible that these could be designed as storm flow treatment wetlands as well as constructed for effluent polishing (Hammer, 1989).

Stormwater wetlands are structural practices similar to wet ponds that incorporate wetland plants in a shallow pool. As stormwater runoff flows through the wetland, pollutant removal is achieved by settling, and biological uptake within the practice. Wetlands are among the most effective stormwater practices in terms of pollutant removal, and they also offer aesthetic value. There are several design variations of the stormwater wetland, each design differing in the relative amounts of shallow and deep water and dry storage above the wetland. Schueler (1992) gives four standard wetlands designs used in the mid-Atlantic region: the shallow marsh system, the pond/wetland system, the extended detention wetland system, and the pocket stormwater wetland. Except for the pocket wetland, as the names imply, the major differences are the incorporation of wet pond or extended detention pond designs into the wetland marsh system. Table 13-21 gives characteristic information for the four types. Figures 13-38 and 13-43 provide examples of a typical wetland marsh and pocket wetland.

The pond/wetland system consists of two separate cells — a deep pond leading to a shallow wetland. The pond removes pollutants and reduces the space required for the system.

Pocket wetlands seldom are more than a tenth of an acre in size, and serve development sites of 10 acres or less. Due to their size and unreliable water supply, pocket wetlands do not possess all of the benefits of other wetland designs. Most pocket wetlands have no sediment forebay. Despite many drawbacks, pocket wetlands may be an attractive BMP alternative for smaller development situations. For example, they have been used successfully at culvert outlets and at parking lot exits in Minnesota.

Another type of wetland feature used most often in the West is a wetland bottom channel. They are used downstream from detention ponds to "polish" the effluent and, in their own right, wherever there is an ability for wide and gently sloping channels. Wetland channels are designed with a 2-year flow velocity of less than 2.5 ft/s (n = 0.03) and flow depth of 3 to 5 ft, and 8:1 minimum width-to-depth ratio (UDFCD, 1992, 1999). Grade control or drop structures are used to control the longitudinal slope. See more on this type of BMP under swales, below.

TABLE 13-21
Attributes of Four Stormwater Wetland Systems

Attribute	Design 1: Shallow Marsh System	Design 2: Pond/Wetland System	Design 3: Extended Detention Wetland	Design 4: Pocket Wetland
General description	Flow through a sediment forebay, channels, and micropool near outlet; islands and benches promote circulation; majority is less than 18 in deep	Two-celled pond/wetland system with pond upstream from wetland; hard barrier or other separation between pond and marsh; micropool at outlet; wetland fringe on bench around pool; less area required due to pool pollutant removal	Forebay with low and high marsh zone; whole area floods when runoff occurs; may rise as much as 3 ft with 12- to 24-h drawdown; less space requirement due to vertical storage	Small in size (less than 0.1 acre) with drainage areas less than 10 acres; normally without forebay and less complex; small inlet stilling pool and micropool at outlet with trash rack; fluctuating water levels, may be below groundwater level to support constant moisture
Pollutant removal	Moderate reliable removal of sediments and nutrients	Moderate to high reliable removal of sediments and nutrients	Moderate less reliable removal of sediments and nutrients	Moderate, with possibility of resuspension and groundwater displacement
Land consumption	High	Moderate	Moderate	Moderate but very flexible
Water balance	Dry-weather baseflow recommended as primary source of water supply to wetland	Dry-weather baseflow recommended to maintain water elevations; groundwater not recommended as the	groundwater not recommended as the	Water supply supplied by excavation to groundwater
Deep water cells	Forebay, channels, micropool	Pond, micropool	Forebay, micropool	Micropool, if possible
Outlet configuration	Reversed slope pipe extending from riser, withdrawn about 1 ft below normal pool; pipe and pond drain equipped with gate valve; Alternate is perforated riser emptying into weir box			Broad-crested weir with half-round trash rack and pond drain
Sediment cleanout cycle	Cleanout of forebay every 2 to 5 years	Clean out pond every 10 to 15 years	Cleanout of forebay every 2 to 5 years	Cleanout of wetland every 5 to 10 years, on-site disposal and stockpile mulch
Native plant diversity	High, if complex microtopography is present	High, with sufficient wetland complexity area	Moderate, fluctuating water levels impose physiological constraints	Low to moderate due to small surface area and poor control of water levels
Wildlife habitat potential	High, with complexity and wide buffer (50 ft)	High, with wide buffer; it attracts waterfowl	Moderate, with buffer	Low due to small area and low diversity

Source: From Schueler, T.J., Design of Stormwater Wetland Systems, Metropolitan Washington Council of Governments, Washington, DC, October, 1992.

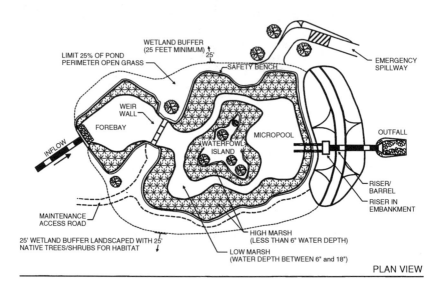

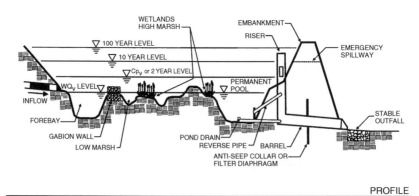

FIGURE 13-38
Shallow marsh wetland schematic. (From Schueler, T.J., Design of Stormwater Wetland Systems, Metropolitan Washington Council of Governments, Washington, DC, October 1992.)

Structural Best Management Practices

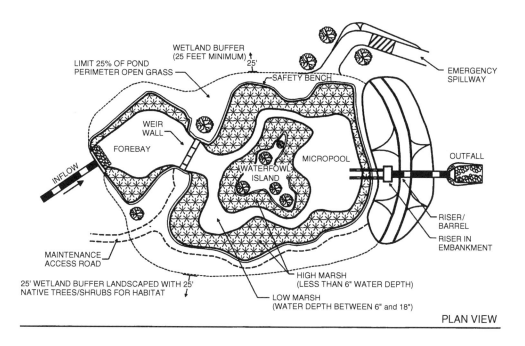

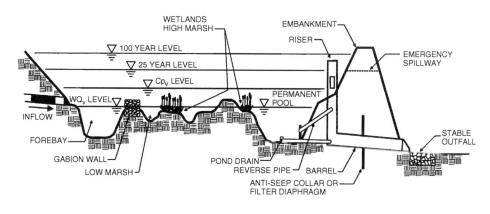

FIGURE 13-39
Example stormwater wetland. (From Atlanta Regional Commission (ARC), Georgia Stormwater Manual, 2002.)

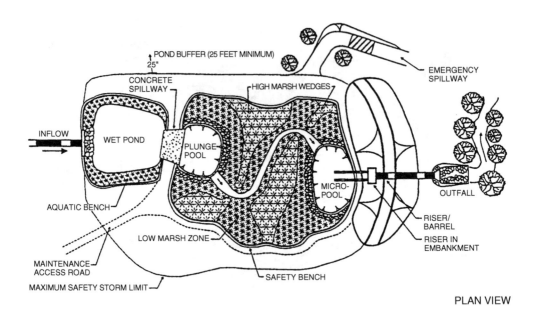

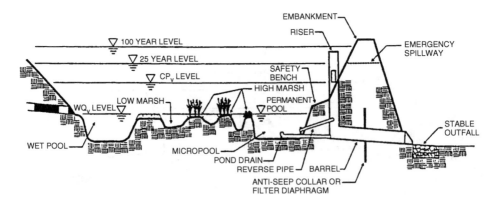

FIGURE 13-40
Pond wetland system schematic. (From Schueler, T.J., Design of Stormwater Wetland Systems, Metropolitan Washington Council of Governments, Washington, DC, October 1992.)

Structural Best Management Practices

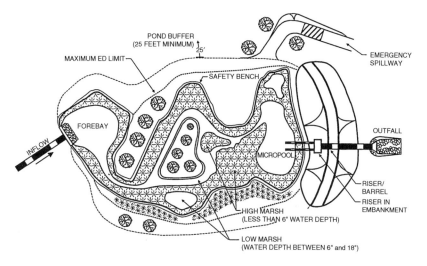

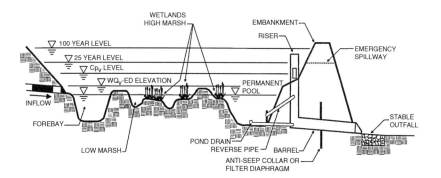

FIGURE 13-41
Extended detention wetland. (From Schueler, T.J., Design of Stormwater Wetland Systems, Metropolitan Washington Council of Governments, Washington, DC, October 1992.)

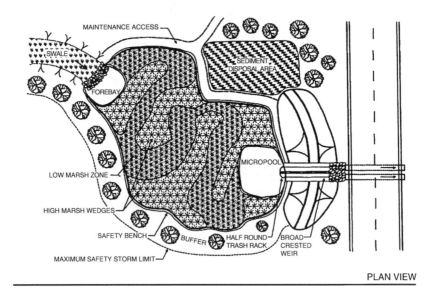

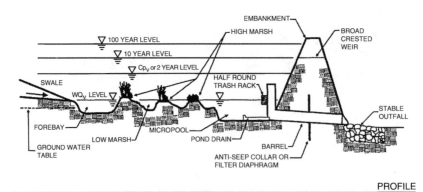

FIGURE 13-42
Typical pocket wetland. (From Schueler, T.J., Design of Stormwater Wetland Systems, Metropolitan Washington Council of Governments, Washington, DC, October 1992.)

FIGURE 13-43
Example pocket wetland.

A final type of wetland is the submerged gravel wetland. The submerged gravel wetland system consists of one or more treatment cells that are filled with crushed rock or gravel, and is designed to allow stormwater to flow subsurface through the root zone of the constructed wetland. In this design, runoff flows through a rock filter with wetland plants at the surface. Pollutants are removed through biological activity on the surface of the rocks, as well as by pollutant uptake of the plants. This practice is fundamentally different from other wetland designs, because, while most wetland designs behave like wet ponds with differences in grading and landscaping, gravel-based wetlands are more similar to a filtering system (CWP, 2001). The outlet from each cell is set at an elevation to keep the rock or gravel submerged. Wetland plants are rooted in the media, where they can directly take up pollutants. In addition, algae and microbes thrive on the surface area of the rocks. In particular, the anaerobic conditions on the bottom of the filter can foster the denitrification process. Although widely used for wastewater treatment in recent years, only a handful of submerged gravel wetland systems have been designed to treat stormwater. Mimicking the pollutant removal ability of nature, this structural control relies on the pollutant-stripping ability of plants and soils to remove pollutants from runoff. Several commercial devices offer a similar approach. Figure 13-44 shows an example of such a BMP.

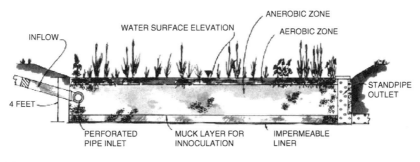

FIGURE 13-44
Submerged gravel wetland system.

Pollution Removal Efficiency

Wetlands can be very effective in removal of soluble and particulate fractions of pollution. But, attention to details in design is vital. Soils, hydrology, and plants must support each other well to achieve high removal rates. Detailed construction supervision is also essential. Estimates of removal rates vary widely in the literature, and many of these are for experimental wetlands. Rough estimates of removal rates for wetlands can be estimated from Table 13-13. The greatest consistency in pollutant removal among the various studies was for BOD, suspended solids, and metals (Livingston, 1988). In the nongrowing season, there may even be nutrient export (especially ammonia and orthophosphorus) (MWCOG, 1992). While metals tend to accumulate in wetlands, nutrients tend to cycle (Horner, 1988). Pollution removal takes place through biological uptake (microbial, algal, vascular plant), sedimentation, volatilization, adsorption, precipitation, and filtration. Rooted vegetation removes nutrients through the soil, while nonrooted vegetation removes it directly from the water.

The pollution removal efficiency of the submerged gravel wetland is similar to a typical wetland. Recent data show a TSS removal rate in excess of 80%. This reflects the settling environment of the gravel media. These systems also exhibit removal of about 60% TP, 20% Tn, and 50% Zn (ARC, 2002). The growth of algae and microbes among the gravel media has been determined to be the primary removal mechanism of the submerged gravel wetland.

Pollutant removal efficiency depends physically on aquatic treatment volume, surface area-to-volume ratio, and surface area-to-watershed ratio. Pollutant removal can be enhanced by maximizing any of these variables and increasing the flow path, providing pretreatment with forebays, and using redundant pollutant removal pathways (Schueler, 1992). Large storms can overwhelm smaller wetlands, greatly reducing the pollutant removal capabilities. Offline design practices, or increasing the size, can circumvent this problem somewhat. Plant harvesting prior to winter dieback may not increase the nutrient removal efficiency. Much of the biomass in wetlands is retained in accumulating bottom sediments or roots. Removal of the particulate portions of nutrients is often lower when baseflow is monitored (Techniques, 1994). Pollutant removal can be enhanced through use of the BMP treatment train concept using a pond or sump prior to the wetland.

Wetlands are a relatively inexpensive stormwater practice. Construction cost data for wetlands are rare, but one simplifying assumption is that they are typically about 25% more expensive than stormwater ponds of an equivalent volume. Wetlands consume about 3 to 5% of the land that drains to them, which is relatively high compared with other stormwater management practices. In areas where land value is high, this may make wetlands an infeasible option. For wetlands, the annual cost of routine maintenance is typically estimated at about 3 to 5% of the construction cost. Alternatively, a community can estimate the cost of the maintenance activities outlined in the maintenance section. Wetlands are long-lived facilities (typically longer than 20 years). Thus, the initial investment in these systems may be spread over a relatively long time period (CWP, 2001).

Wetland Design

Wetland design standards differ somewhat around the country because of the great changes in climate, soils, and vegetation types. Therefore, only general standards and design guidance can be given here. Table 13-22 gives design criteria summaries for the four types of wetlands.

Wetlands should be sized to capture 90% of the total annual stormwater volume. In Washington, DC, this amounts to a 3-month storm and a rainfall depth of 1.25 in over the whole watershed. This is in excess of any dry-weather flow or groundwater seepage. As

TABLE 13-22

Comparative Design Considerations of Four Stormwater Wetland Systems

Design Criterion	Design 1: Shallow Marsh System	Design 2: Pond/Wetland System	Design 3: Extended Detention Wetland	Design 4: Pocket Wetland
Storage allocation	40% pool 60% marsh	70% pool 30% marsh	30% marsh 20% pool 50% extended detention	80% marsh 20% pool
Surface area allocation	40% low marsh 20% deep pool 40% high marsh	45% deep pool 25% low marsh 30% high marsh	20% deep pool 35% low marsh 45% high marsh	10% deep pool 40% low marsh 50% high marsh
Flow path length-to-width ratio	1:1	1:1	1:1	NA
Dry-weather path length-to-width ratio	2:-	2:1	2:1	2:1
Wetland area/watershed area	0.02 minimum	0.01 minimum	0.01 minimum	0.01 minimum
Contributing watershed area	>25 acres with dry-weather flow	>25 acres with dry-weather flow	>10 acres	1 to 10 acres
Forebay	Required	Required	Required	Optional
Micropool	Required	Required	Required	Required
Outlet configuration	Reverse-slope pipe or hooded broad-crested weir	Reverse-slope pipe or hooded broad-crested weir	Reverse-slope pipe or hooded broad-crested weir	Hooded broad-crested weir

Source: From Schueler, T.J., Design of Stormwater Wetland Systems, Metropolitan Washington Council of Governments, Washington, DC, October 1992.

for wet ponds, other rules of thumb abound. Calculations are also similar to extended ponds or wet ponds.

A sediment forebay should be provided and should be at least 10% of the treatment volume and be 4 to 6 ft deep. It should be separated from the main area by a berm or gabion structure. A cattail forebay can be useful if trapping of oil and grease and trash is important. Low-flow channels can cause short circuiting and should be avoided (California Stormwater Taskforce, 1993).

The micropool at the outlet creates sufficient depth for the reverse slope pipe. The reverse slope pipe should be about 1 ft below the permanent pond surface. It can discharge into a stilling well riser that can have openings at higher elevations for flood control. The micropool, like the sediment forebay, should be 10% of the treatment volume and be 4 to 6 ft deep. A drain for the micropool, as well as the wetland, should be placed at the bottom of the micropool.

The maximum depths for most emergent types of vegetation is from 3 to 12 in. Local experience should dictate. The specifications given below are typical for the eastern United States. The high marsh zone is planted zero to 6 in deep, while the low marsh zone is typically 6 to 18 in deep. It must be remembered that, for the extended detention portion of the design, as well as for fringe areas of the other designs, inundation-tolerant species should be selected, because they may become wet 10 to 30 times per year. Vegetation is planted using any of four techniques: using container-grown plant stock, using wetland mulch, broadcasting seeds (extended detention zone), and allowing volunteer growth (Schueler, 1992; Thunhorst, 1993). Transplanting stock is the most reliable method of establishing preferred vegetation if the plants are well taken care of during transport to the site and are preplanted, and if the peat pots are broken open to facilitate root spreading. The use of donor soils as a mulch can be effective as a primary and secondary planting method, should such soils be available. Volunteer growth and broadcast seeding should not normally be relied upon as the primary means of vegetation establishment, except for small pocket wetlands and for fill-in areas between plantings. Seeding should be done on the basis of pure, live seed to avoid purity and germination problems (Gorbisch, 1994). Careful inspection during the first 3 to 5 years will be necessary to reinforce plantings that fail and to control exotic species. Shenot (1993) found four planted species to persist and spread in Maryland: broadleaf arrowhead (Saggitaria latifolia), soft stem bulrush (Scirpus validus), common three square (Scirpus americanus), and pickerelweed (Pontederia cordata). Wild rice showed high persistence. Other specifications are given below. Figure 13-45 gives a shallow marsh planting strategy.

Adjustment of the water balance can be done by slightly raising or lowering the pool elevation, diverting more flow around or through the pond, reducing infiltration with pond bottom liners installed below rooted vegetation, and by changing orifice sizes on outlets.

Permitting

Wetlands permitting has always been a concern. If a wetlands project involves fill into a wetlands, a Section 404 permit will be required. It is not recommended to destroy natural wetlands to create stormwater wetlands, because their basic functions are different. "Fingerprinting" can be used to combine natural and constructed wetlands in a location (Schueler, 1992). Generally, a project will be permitted if there are no practicable alternatives, toxic effluent or water quality standards are not violated, endangered species or critical habitat will not be jeopardized, waters of the United States will not be degraded, and appropriate steps are taken to minimize unavoidable adverse impacts. At present, permitting for the removal of wetlands vegetation from constructed wetlands does not require a Section 404 permit. The EPA Wetlands Protection Hotline can be called for more information at 1-800-832-7828.

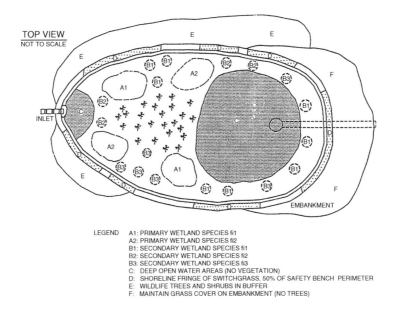

FIGURE 13-45
Shallow marsh planting strategies. (From Schueler, 1987.)

Other laws that may impact activities involving wetlands and 404 permitting include the following:

- The National Environmental Policy Act (NEPA) for required environmental impact statements or environmental assessments
- The Endangered Species Act
- The Fish and Wildlife Coordination Act
- The National Historic Preservation Act
- The Food Security Act of 1985
- The Food, Agriculture, Conservation and Trade Act of 1990
- The Emergency Wetlands Resources Act
- The North American Wetlands Conservation Act
- The Water Resources Development Act
- Historic or archeological preservation concerns
- State and local laws or zoning overlays

Design of a constructed stormwater wetland proceeds through the development stages of feasibility evaluation, concept plan development, pond sizing, pond feature development, pondscaping design and planning, construction and grading, wetland bed preparation, pondscape and vegetation establishment, and postconstruction inspection and maintenance (Schueler, 1992).

Typical Standard Specifications for Constructed Wetlands

Typical Required Specifications

- Inflow of water must be greater than that leaving the basin by infiltration.

- Design must be for an extended detention time of 24 h for the 1-year storm.
- The orifices used for extended detention will be vulnerable to blockage from plant material or other debris that will enter the basin with stormwater runoff. Therefore, some form of protection against blockage should be installed (such as some type of noncorrodible wire mesh). Reverse slope pipes are recommended.
- Surface area of the wetland should account for a minimum of 3% of the area of the watershed draining into it.
- The length-to-width ratio should be at least 2 to 1.
- The elevation difference needed at a site from the inflow to the outflow is 3 to 5 ft, and 2 to 3 ft for pocket wetland.
- If used on a site with an underlying water supply aquifer or when treating a hot spot, a separation distance of 2 ft is recommended between the bottom of the wetland and the elevation of the seasonally high water table; pocket wetland is typically below water table.
- Permeable soils are not well suited for a constructed stormwater wetland without a high water table. Underlying soils of hydrologic group "C" or "D" should be adequate to maintain wetland conditions. Most group "A" soils and some group "B" soils will require a liner. Evaluation of soils should be based upon an actual subsurface analysis and permeability tests.
- A 4- to 6-ft deep micropool must be included in the design at the outlet to prevent the outlet from clogging and resuspension of sediments, and to mitigate thermal effects.
- Maximum depth of any permanent pool areas should generally not exceed 6 ft.
- A soil depth of at least 4 in should be used for shallow wetland basins.
- The deeper area of the wetland should include the outlet structure so that the outflow from the basin is not interfered with by sediment buildup.
- A forebay should be established at the pond inflow points to capture larger sediments, and it shoud be 4 to 6 ft deep. Direct maintenance access to the forebay should be provided with access 15 ft wide minimum and 5:1 slope maximum. Sediment depth markers should be provided.
- If high water velocity is a potential problem, some type of energy dissipation device should be installed.
- The designer should maximize use of pre- and postgrading pondscaping design to create horizontal and vertical diversity and habitat.
- A minimum of three aggressive emergent wetland species (primary species) of vegetation should be established in quantity on the wetland.
- Three additional emergent wetland species (secondary species) of vegetation should be planted on the wetland, although in far less numbers than the two primary species.
- Thirty to 50% of the shallow (12 inches or less) area of the basin should be planted with wetland vegetation.
- Approximately 50 individuals of each secondary species should be planted per acre, set out in 10 clumps of approximately 5 individuals, and planted within 6 ft of the edge of the pond in the shallow area leading up to the pond's edge. They should be spaced as far apart as possible, but there is no need to segregate species to different areas of the wetland.

- Wetland mulch, if used, should be spread over the high marsh area and adjacent wet zones (-6 to +6 in of depth) to depths of 3 to 6 in.
- A minimum 25 ft buffer, for all but pocket wetlands, should be established and planted with riparian and upland vegetation (50 ft buffer if wildlife habitat value required in design).
- Surrounding slopes should be stabilized by planting in order to trap sediments and some pollutants and prevent them from entering the wetland.
- A written maintenance plan should be provided and adequate provision made for ongoing inspection and maintenance, with more intense activity for the first 3 years after construction.
- The wetland should be maintained to prevent loss of area of ponded water available for emergent vegetation due to sedimentation and accumulation of plant material.
- Stormwater wetlands cannot be located within navigable waters of the U.S., including wetlands, without obtaining a Section 404 permit under the Clean Water Act, and any other applicable state permit. In some isolated cases, a wetlands permit may be granted to convert an existing degraded wetland in the context of local watershed restoration efforts.
- If a wetland facility is not used for overbank flood protection, it should be designed as an offline system to bypass higher flows rather than passing them through the wetland system.
- Minimum setback requirements for stormwater wetland facilities are 10 ft from property line, 100 ft from private well (if well is down-gradient from a hot spot land use, then the minimum setback is 250 ft), and 50 ft from a septic system tank/leach field.

Typical Recommended Specifications

- To minimize maintenance as much as possible, it is recommended that wetland basins be installed on stabilized watersheds and not be used for sediment control.
- Complex topography can be maintained by bioengineering methods (such as fascines) or straw bales and geotextile rolls.
- It is recommended that the frequently flooded zone surrounding the wetland be located within approximately 10 to 20 ft from the edge of the permanent pool.
- Soil types conducive to wetland vegetation should be used during construction.
- The wetland should be designed to allow slow percolation of the runoff through the substrate (add a layer of clay for porous substrates).
- The depth of the forebay should be in excess of 3 ft and contain approximately 10% of the total volume of the normal pool.
- As much vegetation as possible and as much distance as possible should separate the basin inlet from the outlet.
- The water should gradually get shallower about 10 ft from the edge of the pond.
- The planted areas should be made as square as possible within the overall design of the wetland, rather than long and narrow.
- The only site preparation necessary for the actual planting (besides flooding the basin) is to ensure that the substrate is soft enough to permit relatively easy insertion of the plants.

- Wetland siting should take into account the location and use of other site features, such as natural depressions, buffers, and undisturbed natural areas, and should attempt to aesthetically fit the facility into the landscape. Bedrock close to the surface may prevent excavation.

Operation and Maintenance Requirements

- A stormwater management easement and maintenance covenant should be required for each facility. The maintenance covenant should require the owner of the wetland to periodically clean the structure.
- Sediment forebays should be cleaned every 2 to 5 years except for pocket wetlands without forebays, which are cleaned after a 6-in accumulation of sediment.
- The ponded water area may be maintained by raising the elevation of the water level in the permanent pond, by raising the height of the orifice in the outlet structure, or by removing accumulated solids by excavation.
- Water levels may need to be supplemented or drained periodically until vegetation is fully established.
- It may be desirable to remove contaminated sediment bottoms or to harvest aboveground biomass and remove it from the site in order to permanently remove pollutants from the wetland.

Table 13-23 provides a summary of maintenance requirements for wetlands (ARC, 2002).

TABLE 13-23

Typical Maintenance Activities for Wetlands

Activity	Schedule
Replace wetland vegetation to maintain at least 50% surface area coverage in wetland plants after the second growing season	One-time activity
Clean and remove debris from inlet and outlet structures	Frequently (3 to 4 times/year)
Mow side slopes	
Monitor wetland vegetation and perform replacement planting as necessary	Semiannual inspection (first 3 years)
Examine stability of the original depth zones and micro-topographical features	Annual inspection
Inspect for invasive vegetation, and remove where possible	
Inspect for damage to the embankment and inlet/outlet structures; repair as necessary	
Note signs of hydrocarbon buildup, and remove appropriately	
Monitor for sediment accumulation in the facility and forebay	
Examine to ensure that inlet and outlet devices are free of debris and operational	
Repair undercut or eroded areas	As needed
Harvest wetland plants that have been choked out by sediment buildup	Annually
Remove sediment from the forebay	5 to 7 years or after 50% of the total forebay capacity has been lost
Monitor sediment accumulations, and remove sediment when the pool volume has become reduced significantly, plants are choked with sediment, or the wetland becomes eutrophic	10 to 20 years or after 25% of the wetland volume has been lost

Source: Adapted from Watershed Management Institute (WMI), Operation Maintenance and Management of Stormwater Management Systems, prepared for U.S. EPA, Office of Water, 1997; and CWP, 1992.

Bioretention Areas

Bioretention is a best management practice (BMP) developed in the early 1990s by engineers and scientists under contract to the Prince George's County, Maryland, Department of Environmental Resources (PGDER) (Bitter and Bowers, 2000). Bioretention utilizes soils and woody and herbaceous plants to remove pollutants from stormwater runoff. Bioretention areas (also referred to as bioretention filters or rain gardens-depending on design) are structural stormwater controls that capture and temporarily store the water quality volume using soils and vegetation in shallow basins or landscaped areas to remove pollutants from stormwater runoff.

Bioretention areas are engineered facilities in which runoff is conveyed as sheet flow to the treatment area, which consists of a grass buffer strip, ponding area, organic or mulch layer, planting soil, and vegetation. An optional sand bed can also be included in the design to provide aeration and drainage of the planting soil. The City of Alexandria, Virginia, modified the bioretention BMP design to include an underdrain within the sand bed to collect the infiltrated water and discharge it to a downstream sewer system. This modification was required because impervious subsoils and marine clays prevented complete infiltration in the soil system. This modified design makes the bioretention area act more as a filter that discharges treated water than as an infiltration device. Design modifications have been tested that include aerobic and anaerobic zones in the treatment area. The anaerobic zone will promote denitrification (USEPA, 1999).

Bioretention should be used in stabilized drainage areas to minimize sediment loading in the treatment area. For example, bioretention is an ideal stormwater management BMP for median strips, parking lot islands, and swales. These areas can be designed or modified so that runoff is diverted directly into the bioretention area or conveyed into the bioretention area by a curb and gutter collection system. Bioretention is usually best used upland from inlets that receive sheet flow from graded areas, and at areas that will be excavated.

There are numerous design applications, both on- and offline, for bioretention areas (Figures 13-46 through 13-51). These include use on single-family residential lots (rain gardens), as offline facilities adjacent to parking lots, along highway and road drainage swales, within larger landscaped pervious areas, and as landscaped islands in impervious or high-density environments.

Bioretention is an excellent stormwater treatment practice due to the variety of pollutant removal mechanisms. Each of the components of the bioretention area is designed to perform a specific function. The grass strip (or grass channel) reduces incoming runoff velocity and filters particulates from the runoff. The ponding area provides for temporary storage of stormwater runoff prior to its evaporation, infiltration, or uptake and provides additional setline capacity. The organic or mulch layer provides filtration as well as an environment conductive to the growth of microorganisms that degrade hydrocarbons and organic material. The planting soil in the bioretention facility acts as a filtration system, and clay in the soil provides adsorption sites for hydrocarbons, heavy metals, nutrients, and other pollutants. Woody and herbaceous plants in the ponding area provide vegetative uptake of runoff and pollutants and also serve to stabilize the surrounding soils. Finally, a sand bed provides for positive drainage and aerobic conditions in the planting soil and provides a final polishing treatment media.

Pollution Removal Efficiency

Table 13-13 provides overall pollutant removal rates. In a situation where a removal rate is not deemed sufficient, additional controls may be put in place at the given site in a series or treatment train approach. These rates are applicable when the facility is sized,

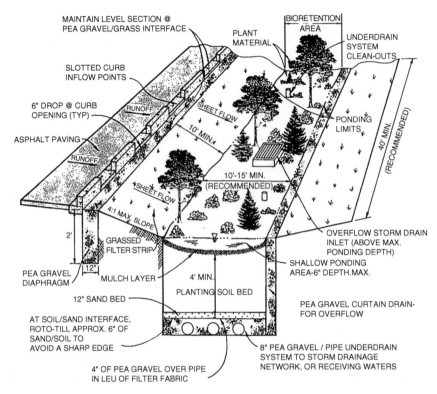

FIGURE 13-46
Schematic of a typical bioretention area.

designed, constructed, and maintained in accordance with the recommended specifications. Undersized or poorly designed bioretention areas can reduce removal performance. Column studies have been done recently to test the ability of bioretention areas to remove pollutants with promising results (Davis et al., 1998; Coffman and Winogradoff, 1999). Bioretention areas appear to remove pollutants at a higher rate than basic sand filters, although this conclusion is based on limited monitoring data. It is thought that the soil filtration of bioretention areas can achieve 60% phosphorus removal and 90% removal of metals and hydrocarbons. More research is needed to confirm whether they also can reliably remove sediment and bacteria, but the soil filtration mechanism used in bioretention should promote high removal rates for these parameters (Schueler, 2000). The combination of plants, soils, and organics rival or surpass the ability of other filter media and methods for nutrients and metals (Schueler, 2000).

For additional information and data on pollutant removal capabilities for bioretention areas, see the National Pollutant Removal Performance Database (2nd ed.) available at www.cwp.org and the National Stormwater Best Management Practices Database at www.bmpdatabase.org.

Design

Bioretention areas are suitable for many types of development, from single-family residential to high-density commercial projects. Bioretention is also well suited for small lots,

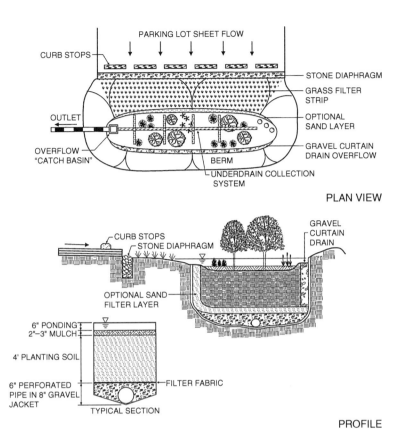

FIGURE 13-47
Schematic of a typical on-line bioretention area. (From Claytor and Schueler, 1996.)

including those of 1 acre or less. Because of its ability to be incorporated in landscaped areas, the use of bioretention is extremely flexible. Bioretention areas are an ideal structural stormwater control for use as roadway median strips and parking lot islands, and are also good candidates for the treatment of runoff from pervious areas, such as on a golf course. Bioretention can also be used to retrofit existing development and stormwater quality treatment capacity (PGDER, 1993).

Typical Required Specifications

- The maximum drainage area should be 5 acres (0.5 to 2 acres preferred).
- Approximately 5% of the tributary impervious area is required (minimum of 200 square feet for small sites).
- All designs except small residential applications should maintain a length-to-width ratio of at least 2:1.
- The maximum recommended ponding depth of the bioretention areas is 6 in.
- No more than a 6% site slope is acceptable.
- Elevation difference needed at the site from the inflow to the outflow is 5 ft.
- A separation distance of 2 ft is recommended between the bottom of the bioretention facility and the elevation of the seasonally high water table.
- Do not allow exfiltration of filtered hot spot runoff into the groundwater.

822 Municipal Stormwater Management, Second Edition

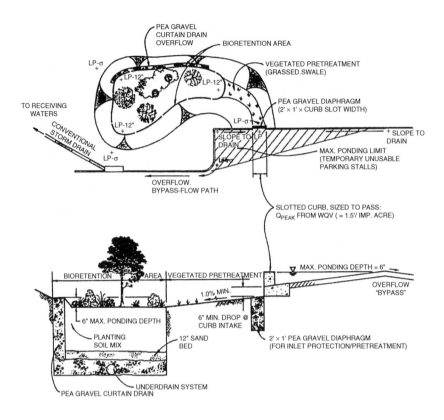

FIGURE 13-48
Schematic of a typical offline bioretention area. (From Claytor and Schueler, 1996.)

FIGURE 13-49
Parking lot bioretention area.

FIGURE 13-50
Parking island bioretention.

FIGURE 13-51
Residential rain garden. (From Center for Watershed Protection.)

- Bioretention systems are designed for intermittent flow and must be allowed to drain and re-aerate between rainfall events. They should not be used on sites with a continuous flow from groundwater, sump pumps, or other sources.
- The planting soil filter bed is sized using a Darcy's Law equation with a filter bed drain time of 48 h and a coefficient of permeability (k) of 0.5 ft/day.
- The planting soil bed must be at least 4 ft in depth. Planting soils should be sandy loam, loamy sand, or loam texture with a clay content ranging from 10 to 25%. The soil must have an infiltration rate greater than 0.27 in/h and a pH between 5.5 and 6.5. In addition, the planting soil should have a 1.5 to 3% organic content and a maximum 500 ppm concentration of soluble salts.

FIGURE 13-52
Bioretention area after a storm.

- A mulch layer should consist of 2 to 4 in of commercially available fine shredded hardwood mulch or shredded hardwood chips.
- The sand bed should be 12 to 18 in thick. Sand should be clean and have less than 15% silt or clay content.
- Pea gravel for the diaphragm and curtain, where used, should be ASTM D 448 size No. 8 (1/8 to 1/2 in).
- The underdrain collection system is equipped with a 6-in perforated PVC pipe in an 8-in gravel layer. The pipe should have 3/8-in perforations, spaced at 6-in centers, with a minimum of four holes per row. The pipe is spaced at a maximum of 10 ft on center, and a minimum grade of 0.5% must be maintained. A permeable filter fabric is placed between the gravel layer and the planting soil bed.

Typical Recommended Specifications

- For online configurations, a grass filter strip with a pea gravel diaphragm is typically utilized as the pretreatment measure. The required length of the filter strip depends on the drainage area, imperviousness, and the filter strip slope.
- For offline applications, grass channel with a pea gravel diaphragm flow spreader is used for pretreatment. The length of the grass channel depends on the drainage area, land use, and channel slope. The minimum grassed channel length should be 20 ft.
- Adequate pretreatment and inlet protection for bioretention systems is provided when all of the following are provided: (a) grass filter strip below a level spreader, or grass channel, (b) pea gravel diaphragm, and (c) an organic or mulch layer.
- Outlet pipe is to be provided from the underdrain system to the facility discharge.
- A dense and vigorous vegetative cover should be established over the contributing pervious drainage area before runoff can be accepted into the facility.
- The bioretention area should be vegetated to resemble a terrestial forest ecosystem, with a mature tree canopy, subcanopy of understory trees, scrub layer, and

TABLE 13-24

Typical Maintenance Activities for Bioretention Areas

Activity	Schedule
Pruning and weeding to maintain appearance	As needed
Mulch replacement when erosion is evident	
Removal of trash and debris	
Inspect inflow points for clogging (offline systems); remove any sediment	Semiannually
Inspect filter strip/grass channel for erosion or gullying; reseed or sod as necessary	
Trees and shrubs should be inspected to evaluate their health and remove any dead or severely diseased vegetation	
The planting soils should be tested for pH to establish acidic levels; if the pH is below 5.2, limestone should be applied; if the pH is above 7.0 to 8.0, then iron sulfate plus sulfur can be added to reduce the pH	Annually
Replace mulch over the entire area	2 to 3 years
Replace pea gravel diaphragm if warranted	

Source: From EPA, 1999.

herbaceous groundcover. Three species each of trees and scrubs are recommended to be planted.

- The tree-to-shrub ratio should be 2:1 to 3:1. On average, the trees should be spaced 8 ft apart. Plants should be placed at regular intervals to replicate a natural forest. Woody vegetation should not be specified at inflow locations.
- After the trees and shrubs are established, the groundcover and mulch should be established.
- Planting recommendations include native plant species over nonnative species, vegetation should be selected based on a specified zone of hydric tolerance, and a selection of trees with an understory of shrubs and herbaceous materials should be provided.

The required planting soil filter bed area is computed using the following equation (based on Darcy's Law):

$$A_f = (WQ_v)(d_f) / [(k)(h_f + d_f)(t_f)] \tag{13.13}$$

where A_f = surface area of ponding area (ft²); WQ_v = water quality volume (or total volume to be captured); d_f = filter bed depth (4 ft min); k = coefficient of permeability or hydraulic conductivity of filter media (ft/day); (typically would use 0.5 ft/day for silt-loam); h_f = average height of water above filter bed (ft); (typically 3 in, which is half of the 6-in ponding depth); and t_f = design filter bed drain time (days) (2.0 days or 48 h is recommended maximum).

More detailed design, soil testing, infiltration testing, and construction specifications can be found at WDE (2001).

Maintenance

Table 13-24 provides a summary of typical maintenance requirements for bioretention areas. Because there is such diversity in types, each will have its own specific requirements.

Design Example

This example focuses on the design of a bioretention facility to meet the water-quality treatment requirements of the site. It is based on work done by the Center for Watershed

Protection for the Georgia Stormwater Manual (ARC, 2002). In general, the primary function of bioretention is to provide water-quality treatment and not large storm attenuation. As such, flows in excess of the water-quality volume are typically routed to bypass the facility or pass through the facility. Where quantity control is required, the bypassed flows can be routed to conventional detention basins (or some other facility, such as underground storage vaults). Under some conditions, channel protection storage can be provided by bioretention facilities.

Base Data

Site Area = Total Drainage Area (A) = 3.0 ac
Impervious Area = 1.9 ac; or I =1.9/3.0 = 63.3%
Soils Type "C"
Hydrologic Data

	Pre	Post
CN	70	88
tc	0.39	0.20

Phase I — Computation of Preliminary Stormwater Storage Volumes and Peak Discharges
Step 1 — Compute Runoff Control Volumes from the Unified Stormwater Sizing Criteria —

Compute water quality volume (WQ_v).
Note that design criteria is capture of runoff from the first 1.2 in of rainfall. See Chapter 7, Section 7-11 for more information.
Compute runoff coefficient (Rv).

$$R_v = 0.05 + (63.3)(0.009) = 0.62$$

Compute *WQv.*

$$WQ_v = (1.2") (R_v) (A) / 12$$

$$= (1.2") (0.62) (3.0ac) (43{,}560 ft^2/ac) (1ft/12in)$$

$$= 8{,}102 \text{ ft}^3$$

Step 2 — Compute Runoff Volumes and Peaks for Other Flows — Include the channel protection volume (assume 1-year, 24-h storm), overbank flood design criteria (assume 25-year storm), and peak flow for 100-year storm. Calculations are not included here.

Phase II — Determination if the Development Site and Local Conditions are Appropriate for the Use of a Stormwater Pond

Step 1 — Screen BMPs Using Site-Specific Data — Existing ground elevation at the facility location is 922.0 ft, mean sea level. Soil-boring observations reveal that the seasonally high water table is at 913.0 ft and underlying soil is silt loam (ML). Adjacent creek invert is at 912.0 ft.

Step 2 — Confirm Local Design Criteria and Applicability — There are no additional local criteria that must be met for this design.

Phase III — Performance of Final Design Sizing

Bioretention areas can be on- or offline. Online facilities are generally sized to receive, but not necessarily treat, the 25-year event. Offline facilities are designed to receive a more or less exact flow rate through a weir, channel, manhole, flow splitter, etc. This facility is situated to receive direct runoff from grass areas and parking lot curb openings and piping for the 25-year event, and *no special flow diversion structure is incorporated.*

Step 1 — Determine Size of Bioretention Ponding/Filter Area —

$$A_f = (WQ_v)(d_f) / [(k)(h_f + d_f)(t_f)]$$

where A_f = surface area of filter bed (ft²); d_f = filter bed depth (ft); k = coefficient of permeability of filter media (ft/day); h_f = average height of water above filter bed (ft); and t_f = design filter bed drain time (days) (48 h is recommended).

$$A_f = (8,102 \text{ ft}^3)(5') / [(0.5'/\text{day})(0.25' + 5')(2 \text{ days})]$$

(With $k = 0.5'/\text{day}$, $h_f = 0.25'$, $t_f = 2$ days)

$$A_f = 7,716 \text{ sq ft}$$

Step 2 — Set Design Elevations and Dimensions of Facility — Based on site, assume a roughly 2 to 1 rectangular shape. Given a filter area requirement of 7716 sq ft, say facility is roughly 65' by 120' (see Figure 13-53). Set top of facility at 921.0 ft, with the berm at 922.0 feet. The facility is 5 ft deep, which will allow 3 ft of freeboard over the seasonally high water table. See Figure 13-54 for a typical section of the facility.

Step 3 — Design Pretreatment — Pretreat with a grass channel, based on guidance provided in the table below. For a 3.0 acre drainage area, 63% imperviousness, and slope less than 2.0%, provide a 90 ft grass channel at 1.5% slope. The value from the table is 30 ft for a 1-acre drainage area.

Pretreatment Grass Channel Guidance for 1.0 Acre Drainage Area

Parameter	≤33% Impervious		Between 34% and 66% Impervious		≥67% Impervious		Notes
Slope	≤2%	≥2%	≤2%	≥2%	≤2%	≥2%	Max slope = 4%
Grass channel min. length (ft)	25	40	30	45	35	50	Assumes a 2 ft bottom width

Source: Adapted from Claytor and Schueler, 1996.

Step 4 — Size Underdrain Area — Base underdrain design on 10% of the A_f or 772 sq ft. Using 6-in perforated plastic pipes surrounded by a 3-ft-wide gravel bed, 10 ft on center (o.c.).

(772 sq ft)/3' per foot of underdrain = 257', *say 260' of perforated underdrain*

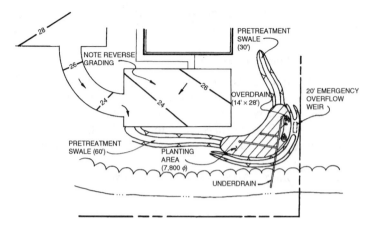

FIGURE 13-53
Plan view of bioretention facility. (From Atlanta Regional Commission, Georgia Stormwater Manual, 2002.)

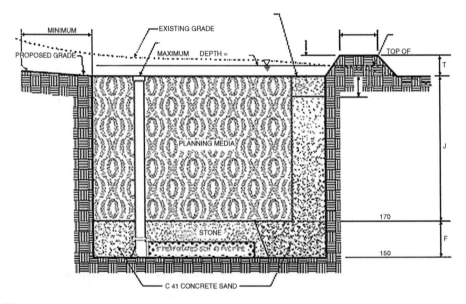

FIGURE 13-54
Typical section of bioretention facility. (Atlanta Regional Commission, Georgia Stormwater Manual, 2002.)

Step 5 — Design Emergency Overflow — To ensure against the planting media clogging, design a small ornamental stone window of 2 to 5-in stone connected directly to the sand filter layer. This area is based on 5% of the A_f or 386 sq ft, say, 14 by 28 ft. See Figures 13-53 and 13-54.

The parking area, curb, and gutter are sized to convey the 25-year event to the facility. Should filtering rates become reduced due to facility age or poor maintenance, an overflow weir is provided to pass the 25-year event. Size this weir with 6 in of head, using the weir equation.

$$Q = CLH^{3/2}$$

where $C = 2.65$ (smooth-crested grass weir); $Q = 19.0$ cfs; and $H = 6$ in.

Solve for L:

$$L = Q / [(C)(H^{3/2})] \text{ or } (19.0 \text{ cfs}) / [(2.65)(.5)^{1.5}] = 20.3' \text{ (say 20')}$$

Outlet protection in the form of riprap or a plunge pool/stilling basin should be provided to ensure nonerosive velocities.

Step 6 — Prepare Vegetation and Landscaping Plan — Choose plants based on factors such as whether native or not, resistance to drought and inundation, cost, aesthetics, maintenance, etc. Select species locations (i.e., on center planting distances) so species will not shade out one another. Do not plant trees and shrubs with extensive root systems near pipe work. A potential plant list is available in your area from a local university, landscape architect, or library.

Sand Filtration Systems

Overview

Sand filters (also referred to as *filtration basins*) are structural stormwater controls that capture and temporarily store stormwater runoff and pass it through a filter bed of sand. Most sand filter systems consist of two-chamber structures. The first chamber is a sediment forebay or sedimentation chamber, which removes floatables and heavy sediments. The second is the filtration chamber, which removes additional pollutants by filtering the runoff through a sand bed and through microbial action in the media. The filtered runoff is typically collected and returned to the conveyance system, though it can also be partially or fully exfiltrated into the surrounding soil in areas with porous soils (ARC, 2002).

Sand filters are distinguished from infiltration devices by the fact that, normally, an outfall handles the effluent after it has been filtered rather than percolating it to groundwater. Thus, they are relatively independent of groundwater depth unless the water table intrudes on a structural feature of the system.

Sand filters are intended primarily for water-quality enhancement. In general, sand filters are preferred over infiltration practices, such as infiltration trenches, when contamination of groundwater with conventional pollutants — BOD, suspended solids, and fecal coliform — is of concern. This usually occurs in areas where underlying soils alone cannot treat runoff adequately or where groundwater tables are high. In most cases, sand filters can be constructed with impermeable basin or chamber bottoms, which help to collect, treat, and release runoff to a storm drainage system or directly to surface water with no contact between contaminated runoff and groundwater (EPA, 1999).

The pollutant removal capability of traditional sand filters may not be high or reliable enough for watershed managers who desire higher levels of nutrient or bacteria removal (Glick, Chang, and Barret, 1998). Consequently, researchers have had a strong interest in testing whether organic media may be a more effective substitute for sand as a filter medium. In this regard, the use of other media has been proposed. Variations include grassed surfaces, gravel pretreatment, plastic screen pretreatment, compost (Lief, 1999), peat (LCRA, 1997), calcitic limestone, or processed steel fibers for phosphorous removal (KCDNR, 1998), stone reservoir storage above the sand layer, grated chambers, and a vertical sand layer with flow moving horizontally. Each design variation has advantages and weaknesses. For example, a filter using organic material may have higher removal rates for some pollutants, but some have suffered in ability to perform hydraulically through increased clogging.

In summary, sand filters have a number of advantages over retention devices and infiltration devices such as the following (Galli, 1992; Schueler, 2000; WDE, 2001):

- The basic sand filter design works well for many small development sites that do not require unusually high pollutant removal requirements. They have a history of successful employment in diverse applications including dry climates and areas with high water table or poor draining soils.

- Pollutant removal is moderate to high for most pollutants. The basic sand filter appears to be capable of removing approximately 80% of incoming sediment, 40% of total phosphorus, and 60% of most metals. Basic sand filter bacteria-removal performance is mixed, and other practices should be considered when bacteria removal is the prime stormwater treatment objective. Sand filters are also consistent nitrate-leakers, and, consequently, may not be a wise choice in coastal watersheds where nitrogen removal is a priority. Likewise, designers working in phosphorus-sensitive watersheds may want to use other media to boost phosphorus removal rates, because sand filters show little ability to remove soluble forms of phosphorus that are most important in reducing eutrophication. Sand filters also have no ability to remove chlorides or dissolved organic carbon (but then again, few other stormwater practices have much capability in this regard).

- Several media appear to be useful when phosphorus removal is the primary stormwater treatment objective. The evidence shows that soil filtration, whether present in bioretention areas or dry swales, can boost phosphorus removal rates to about 60 or 70%. Incorporating calcitic limestone or processed steel fiber amendments within sand filters also appears to improve phosphorus removal, but it remains to be seen whether the cost and loss of hydraulic performance make it worth the effort. The use of peat sand filters is a third strategy, given that they can remove as much as 50% of total phosphorus, but it should be noted that most of the removal was for organic forms of phosphorus that are not as biologically available. Several media demonstrated little or no ability to boost phosphorus removal rates, including zeolites, compost, and pea gravel.

- Sand filters are more resistant to clogging, and easier and cheaper to repair if clogged than infiltration trenches or basins. A pretreatment sedimentation chamber or grassed swale is absolutely essential in the basic sand filter design. Sedimentation storage prior to the filter accounts for much of the observed pollutant removal in the system and helps to reduce the bypass of untreated runoff from these offline practices.

- Sand filters can be easily coupled with flood control BMPs, or in a treatment train situation, which can also be used to pretreat discharge for sediment, or in an offline bypass situation, where only the water-quality volume is routed to the filter.

- Sand filters are quite effective in removing hydrocarbons, which is particularly important for stormwater hot spots. They also do not pose a groundwater contamination threat, making them ideal for urban hotspot applications.

- On a dollar-per-runoff-volume treated basis, sand filters are usually more economical to construct than infiltration devices or oil-and-grit chambers.

Configurations

There are two basic kinds of sand filters: surface sand filters and underground sand filters. Within these general categories, there are a growing number of configurations

DESIGN VARIABLES	Austin Sand Filter Full Sedimentation	Austin Sand Filter Partial Sedimentation	District of Columbia Underground Sand Filter	Delaware Sand Filter	Alexandria Stone Reservoir Trench	Texas Vertical Sand Filter	Peat Sand Filter	Washington Compost Filter System
Applicable Development Situations and Drainage Area	Most sites can serve 1 to 30 acres	Most sites can serve 1 to 30 acres	No more than 10 impervious acres of high urban D.A.	No more than 5 acres of impervious parking lot	2 to 3 acres max. of com-mercial or multi-family	Primarily roadway runoff to date	1 to 50 acres	1 to 50 acres
Filter Bed Profile	18" sand, 4-6 inches of gravel. A layer of sod on the surface of the filter bed is optional.	18" sand, 4-6 inches of gravel. A layer of sod on the surface of the filter bed is optional.	Gravel or Enkadrain screen over 30" of sand	18" of sand	2-4 feet of stone, over 18" of sand and 6" of gravel	Up to 6 feet of sand supported by gabions on either side	Grass on 12" of peat and 2 feet of sand, then gravel	One foot of compost over 8" of rock and gravel
Filter Bed Area (sf/Ia)	100	180	200	360	183	N/A	436	200 ft per cfs
Total Treatment Volume	First 1/2" of runoff with 24 hr. drawdown sediment chamber	First 1/2" of runoff S.C. = 20% of WQV	First flush of runoff (0.3" to 0.5")	First 1" of runoff	First 1/2" of runoff	First 1/2" of runoff	First 1/2" of runoff	N/A
Pretreatment Method	Dry sediment chamber	Dry sediment chamber	3 foot wet micropool plus gravel or geo-textile screen	Shallow wet pool	Wet micropool stone blanket	Dry sediment chamber	Wet micropool	Dry sediment chamber
Pretreatment Volume	sc >> fb	sc ~= fb	sc >> fb	sc = fb	sc < fb	sc >> fb	0.1 acre-inch sc < fb	sc < fb

FIGURE 13-55
Summary data on sand filters.

and variations of sand filters involving filtration media on smaller site scales. These facilities are designed to capture a certain flow volume (normally the first flush), pretreat it for settling of the heaviest particles, and filter the flow through an engineered media to an outflow point.

Some kind of sand filter can be applied to almost any development site. The primary physical requirement is a minimum of two or three feet of head differential existing between the inlet and outlet of the filter bed. This is needed to provide gravity flow through the bed. Otherwise, use of sand filters is only limited by their cost and local maintenance capability. Sand filters are particularly suitable for smaller development sites where other stormwater practices are often not practical. These include infill and ultra-urban developments, small commercial establishments, parking lots, and industrial applications. Care should be exercised in approving sand filters for individual lots and residential developments, as most homeowners lack the incentives or resources to regularly perform needed sand replacement operations. Also, filters should only be installed in stabilized sites with low sediment production.

Figure 13-55 (Schueler, 1994) gives a summary of eight different kinds of sand filters in terms of basic variables and characteristics.

Each sand filter design utilizes a slightly different profile within the filter bed — some with gravel in the bottom, others with concrete sides and limited outlet ports. All have an underdrain system except the vertical sand filter. The required surface area of the filter is usually a direct function of the impervious acreage treated, and varies regionally due to rainfall patterns and local criteria for the volume needed for water-quality treatment

(Schueler, 1994). In addition, designs often differ with respect to the type and volume of pretreatment afforded. The most common form of pretreatment is a wet or dry sedimentation chamber. Pretreatment provides significant pollutant removal and should be considered an integral part of the design (Austin Texas, 1997). Gravel or geotextile screens are sometimes used as a secondary form of protection. The relative volume dedicated to pretreatment versus filtration tends to vary considerably from one area to the next, as shown in Figure 13-55. Nearly all sand filters are constructed offline. Runoff volumes in excess of the water-quality treatment volume must be bypassed to a downstream quantity control structure.

Maintenance

Schueler and the U.S. EPA provide maintenance and longevity tips (Schueler, 1994, 2000; EPA, 1999) for sand filter systems. Regular maintenance is an essential component of the operation of a sand filter. At least once a year, each filter should be inspected after a storm to assess the filtration capacity of the filter bed. Filters should be checked more in applications with high wind-blown sediment loads or where construction tracking of sediment onto paved areas flowing to the basin occurs. Most filters exhibit diminished capacity after a few years due to surface clogging by organic matter, fine silts, hydrocarbons, and algal matter. Maintenance operations to restore the filtration capacity are relatively simple — manual removal of the top few inches of discolored sand followed by replacement with fresh sand. If periodic sand replacement is not conducted, the filter will not be effective. WMI (1997) and others report chronic clogging problems in sand filters installed in residential areas due to lack of maintenance and off-site sediment deposition.

Typically, sand filters begin to experience clogging problems within 3 to 5 years (USEPA, 1999). Accumulated trash, paper, and debris should be removed from the sand filters every 6 months or as necessary to keep the filter clean. A record should be kept of the dewatering times for all sand filters to determine if maintenance is necessary. Corrective maintenance of the filtration chamber includes removal and replacement of the top layers of sand, gravel, and filter fabric that has become clogged. The removed media may usually be disposed in a landfill. The City of Austin tests its waste media before disposal. Sand filter systems may also require the periodic removal of vegetative growth.

A number of techniques are being developed to reduce the frequency of sand replacement or to make the operation more convenient (Schueler, 1994):

- *Surface screen* — Underground sand filters in heavily urbanized areas tend to receive large quantities of trash, litter, and organic detritus. To combat this problem, the District of Columbia specifies the use of a wide-mesh geotextile screen (EnkaDrain 9120) on the surface of the filter bed to trap these materials. During maintenance operations, the screen is rolled up, removed, cleaned, and reinstalled.

- *Careful selection of sod* — Some sand filters that are constructed with a grass cover crop have lost significant filtration capability soon after construction. The clogging is often traced to sod that has an unusually high fraction of fine silts and clays. In other situations, grass roots grow into the sand layer and improve the filtration rate.

- *Limiting use of filter fabric to separate layers* — Often, the loss of filtration capacity occurs where filter fabric is used to separate different layers or media within the filter bed, such as in "sandwich" filters. As a general rule, the less use of filter fabric to separate layers, the better. In many situations, layers of different media

FIGURE 13-56
Scraping polluted surface of sand filter.

can be intergraded together at the boundary (e.g., 50:50 peat/sand) or by a shallow layer of pea gravel.

- *Providing easier access* — During sand replacement operations, heavy and often wet sand must be manually removed from the filter bed. It is surprising that so few designs help a maintenance worker conveniently perform this operation.
- *Pretreatment* — The frequencay of sand replacement can also be reduced by devoting a greater volume to runoff pretreatment in the sedimentation chamber. Several designs provide up to 50% of the total runoff treatment volume in the sedimentation chamber.
- *Visibility and simplicity* — The filter should be visible, i.e., that it be easily recognized as a stormwater practice (so that owners realize what it is) and can be quickly located (so that it can be routinely inspected). This often requires the designer to consider the appearance and aesthetics of the final product so that it does not come to resemble a concrete sandbox. Simplicity is also key. Experience has shown that over complex designs create greater operation and maintenance costs.
- *Imperviousness* — Limit sand filters only to sites that are entirely impervious.

Typical maintenance requirements for all types of sand filters, with suitable consideration for differences in design, are included in Table 13-25. Figure 13-56 shows the accumulation of oily grit on the surface of a perimeter sand filter.

The following sections describe individual sand filters.

Austin First-Flush Filtration Basin

The Austin first-flush filtration basin has been in use in Austin, Texas, for a number of years and, thus, has a proven track record. There are now over a thousand in existence there. It is used when there is insufficient water for a retention pond or natural extended detention pond. It consists, as shown in Figures 13-56 and 13-57 of, in turn, a flow energy dissipator, an extended detention sedimentation chamber for settling larger solids, a flow

TABLE 13-25

Typical Maintenance Activities for Sand Filters

Activity	Schedule
Ensure that contributing area, facility, inlets, and outlets are clear of debris	Monthly
Ensure that the contributing area is stabilized and mowed, with clippings removed	
Remove trash and debris	
Check to ensure that the filter surface is not clogging (also check after moderate and major storms)	
Ensure that activities in the drainage area minimize oil/grease and sediment entry to the system	
If permanent water level is present (perimeter sand filter), ensure that the chamber does not leak, and normal pool level is retained	
Check to see that the filter bed is clean of sediment, and the sediment chamber is not more than 50% full or 6 in, whichever is less, of sediment; remove sediment as necessary	Annually
Make sure that there is no evidence of deterioration, spalling, or cracking of concrete	
Inspect grates (perimeter sand filter)	
Inspect inlets, outlets, and overflow spillway to ensure good condition and no evidence of erosion	
Repair or replace any damaged structural parts	
Stabilize any eroded areas	
Ensure that flow is not bypassing the facility	
Ensure that no noticeable odors are detected outside the facility	
If filter bed is clogged or partially clogged, manual manipulation of the surface layer of sand may be required; remove the top few inches of sand, roto-till or otherwise cultivate the surface, and replace media with sand meeting the design specifications	As needed
Replace any filter fabric that has become clogged	

Source: From Watershed Management Institute (WMI), Operation Maintenance and Management of Stormwater Management Systems, prepared for U.S. EPA, Office of Water, 1997; Pitt, R., Lilbum, M., Nix, S., Durrans, S., and Burian, S., Guidance Manual for Integrated Wet Weather Flow Collection and Treatment Systems for Newly Urbanized Areas, U.S. EPA Office of Research and Development, 1997.

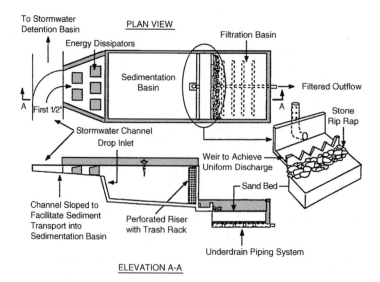

FIGURE 13-57
Austin first-flush basin — full sedimentation design. (From Austin, 1988.)

TABLE 13-26

Average Annual Pollutant Removal Capability of Austin First-Flush Basins

Pollutant	Percent Removal
TSS	75–90%
TP	30–60%
TN	30–50%
TKN	40–60%
TOC	40–60%
BOD	30–50%
Metals	40–80% (depending on metal)
Bacteria	40–70%

Source: From Austin, 1988, 1990.

spreader device, a sand filter, and an underdrain outlet. It depends on a hydraulic head to drive the system and thus requires sufficient drop over the structure to operate.

This type of basin can be used for areas from 1 to 50 acres. Beyond 50 acres, the first-flush criteria becomes harder to justify. Most of the basins are designed as offline basins, though some have flow-through capabilities. Runoff enters the basin by a diversion weir sized to pass the first-flush runoff. The height of the weir is such that the water in the basin backs up to the weir height when the capture volume is achieved. Additional runoff then cannot enter the basin due to the backwater, and flows past and out to a flood control detention structure. Experience has shown this device should only be used in areas not being developed. Even with a forebay, fine sediment can penetrate the filtration media resulting in a need for replacement (Austin, 1988b).

Pollution Removal Efficiency

Pollution removal takes place through settling and filtration. If peat or possibly leaf compost is used as part of the filtration media, the dissolved fraction removal rate will probably improve (California Stormwater Task Force, 1993). Field studies in Austin (1990) indicate the removal estimates shown in Table 13-26, which are similar to other types of surface sand filters (See Table 13-13). Results varied considerably with one site, achieving rates for almost all constituents in the 80% - 90% range, through combined retention with the filter. Negative removal rates can be expected for dissolved solids and nitrate-nitrogen.

Design

The storage volume is based on a first-flush runoff volume from 1/2 inch of rainfall per impervious acre. Included in the volume is runoff from pervious areas that flow to impervious areas. Austin designs utilize infiltration trenches or basins for treatment. The sedimentation basin is designed for either full or partial sedimentation according to the Austin criteria (Austin, 1988b). This results in two separate but related design criteria (see Figures 13-23 and 13-24).

Full-Sedimentation Design

For full sedimentation, the pretreatment basin is designed to hold the full capture volume and release it over a period of 24 h, and is designed to remove particles with diameters of 20 microns and specific gravity of 2.65. A sediment trap can be provided at 10% of the sediment basin volume. The horizontally graded filtration component is designed to achieve drawdown for the design flow. The filtration standard for full-sediment capture upstream is based on a coefficient of permeability of 3.5 ft per day, an average hydraulic head on the filter surface of 3 ft, and a sand depth of 18 in. The equation for the surface area of the sedimentation basin is based on a 10-ft maximum depth:

$$A_s = A_d H / D_p \qquad (13.14)$$

where A_s = minimum surface area of the sedimentation basin (acres); A_d = contributing drainage area (acres); H = depth of runoff captured from A_d (ft); and D_p = pond depth (ft).

The product $A_d H$ is the capture volume in this equation, assuming no pervious area additions. The water-quality volume can be substituted for this product term. Based on the Austin assumptions on size, the minimum surface area can be found by substituting 10 for D_p.

The equation for the surface area of the filtration basin depends on the infiltration rate through the sand. The equation is as follows (cf. Equation 13.13):

$$A_f = [A_d H L] / \{k(h_f + L) t_f \qquad (13.15)$$

where A_f = area of filtration basin (acres); A_d = contributing drainage area (acres); H = depth of runoff captured from A_d (ft); L = sand bed depth (ft) (1.5 ft minimum); k = coefficient of permeability (ft/day) (3.5 for full-sedimentation and 2.0 for partial-sedimentation designs, respectively); h_f = average water depth over the filter surface, may be taken as one-half the distance from the sand surface to the maximum water surface elevation for 20% of the capture volume; and t_f = drawdown time for filtration basin, days (Austin uses 40 h, 1.67 days).

Using the typical and minimum values for the Austin design: $L = 1.5$, $k = 3.5$ for full sedimentation, and 2.0 for partial sedimentation, $h_f = 3$, $t_f = 1.67$, the minimum surface area equation becomes for full sedimentation and partial sedimentation, respectively:

$$A_f = A_d H / 18 \qquad (13.16)$$

$$A_f = A_d H / 10 \qquad (13.17)$$

Partial-Sedimentation Design

The partial-sedimentation basin is designed to settle out only the largest particles. Its volume should be 20% of the capture volume minimum and 3 ft minimum depth. For partial capture, the coefficient of permeability for the filtration section is assumed to be 2 ft per day to allow for the faster clogging of the media. The equations for filtration basin sizing given above apply for this case.

Filter Design — For both designs, the filter is designed with a sand filter or a trench design. The gravel filter design consists of an 18 in minimum layer of sand, geotextile fabric, and a gravel filter surrounding 4-in perforated pipes underlain with another geotextile fabric. The trench design consists of 12 to 18 in of sand with perforated pipes placed in trenches at the bottom. Figures 13-58 and 13-59 give details.

Typical Design Specifications

Full-Sedimentation Basin —

- The basin should be designed to capture the 0.5 in runoff from the 25-year storm and bypass the rest.

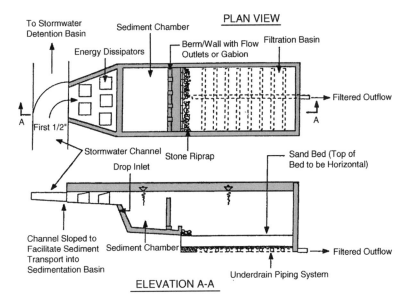

FIGURE 13-58
Austin first-flush basin — partial sedimentation design. L From Austin, 1988.)

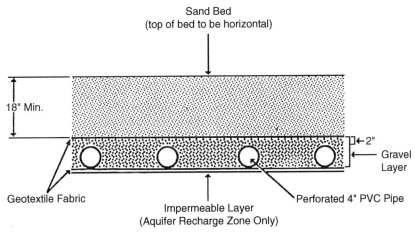

FIGURE 13-59
Gravel filter design. (From Austin, 1988)

- Inlets should discharge water uniformly and at low velocity. Consider an energy dissipator for inlet velocities greater than 3 ft/s.
- Minimum depth of sedimentation basin should be 3 ft.
- Outlets should provide for a drawdown of 24 h minimum. Recommended outlet is a perforated riser pipe with an internal orifice or throttle plate. Outlets should be protected from blockage.
- An impermeable basin liner should be provided.
- Length-to-width ratio should be 2:1 or greater.
- Distance from inlet to outlet should be maximized. Baffles may be necessary.

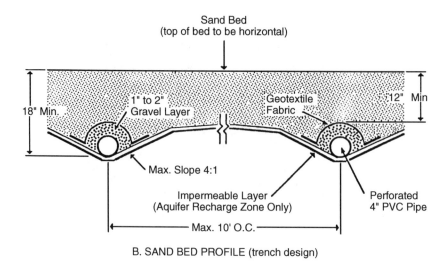

FIGURE 13-60
Trench filter design. (From Austin, 1988.)

- 0.5 ft freeboard should be provided.

Partial-Sedimentation Basin —

- The volume should be a minimum of 20% of the water-quality capture volume.

Filtration Basin —

- A flow spreader should be used on the inlet channels or pipe outfalls to spread the incoming runoff more evenly over the surface of the basin to promote better infiltration.
- The filtration basin storage volume above the sand surface should be 20% of the capture volume minimum.
- The sand bed may be either a gravel filter design or a trench design.
- Sand should be 0.02 to 0.04 in diameter.
- Gravel should be 1/2 to 2 in diameter.
- Drain pipe mains and laterals should be 4-in diameter PVC Schedule 40 pipe with 3/8 in perforations. The pipe should be no more than 10 ft apart and have a grade of no less than 1/8 in per foot. The perforation rows should be no more than 6 in apart. Cleaning access must be provided.
- The surface of the basin should be graded to have a slope close to zero in order to achieve a uniform ponding depth across the entire surface of the basin.
- The side-slopes of the basin should be no steeper than 3:1 (*h:v*) to allow proper vegetative stabilization.
- Inlet channels leading to the basin should be stabilized to prevent incoming runoff velocities from reaching erosive levels and scouring the basin floor (customarily done by riprapping the inlet channels or pipe outfalls).
- Maintenance access should be provided 10 ft wide and less than 4:1 slope.
- On-site sediment disposal should be provided.

Operation and Maintenance Requirements

- The sedimentation basin should be cleaned prior to putting the site into operation.
- Silt will be removed from the sedimentation basin when it reaches a depth of 6 in.
- Debris and trash should be removed when necessary.
- Vegetation growing in the basin will be kept mowed less than 18 in.
- Annual inspection and inspection after major storms, and removal of debris and trash, is required.
- When capture volume drawdown exceeds 60 h or dry drawdown exceeds 96 h, the sedimentation basin needs corrective maintenance.
- Filtration silt removal should be done when it exceeds 1/2 in on the surface.
- Corrective maintenance is required any time the filtration basin does not draw down the capture volume in 36 h after the sedimentation basin has emptied.

Design Sizing Example

A commercial site with 20 acres is to be developed. The total impervious area draining to the Austin first-flush basin is 17 acres. An additional acre of pervious area will drain to the pond. A full filtration design is desired. The capture volume required is 18 acres * 1/2 in = 0.75 acre-ft = 32,670 ft³

> Based on site characteristics, the sedimentation basin depth is 8 ft.
>
> The surface area is: 32,670/8 = 4084 ft²
>
> For a 3:1 length-to-width ratio, the horizontal dimensions are 3 W^2 = 4084.
>
> Therefore:
>
>> Width = 37 ft
>>
>> Length = 110.4 ft
>
> If a sediment trap were installed, its volume would be: V_t = 0.1 * 32,670 = 3267 ft³
>
> For a sand depth of 18 in, 40 h draw down time, k = 3.5, and 3 ft average depth above the filter surface. The surface area of the filtration basin is:
>
> $$A_f = A_d \, H \, L \, / \, k \, (h_f + L) \, t_f$$
>
> = (43,560 * 18 * 0.0417 * 1.5) / (3.5 * (3+1.5) * 1.67) = 1,864 ft²

Check of depth:

> Approximate average inflow rate = 32,670/24 = 1361 ft³/h
>
> Approximate average outflow rate = 32,670/40 = 817 ft³/h
>
> 24-h volume maximum = (1316 - 817)*24 = 11,976 ft³ (37% of capture vol)
>
> Depth = 11,976/1864 = 6.4 ft **OK**

Surface Sand Filter

Surface sand filters are generally used in an offline configuration, where the water-quality volume is diverted to the filter facility through the use of a flow diversion structure and flow splitter. They are a variation on the Austin first-flush structure, the first of this kind.

FIGURE 13-61
Surface sand filter.

Stormwater flows greater than the water quality volume are diverted to other controls or downstream using a diversion structure or flow splitter.

A surface sand filter facility consists of a two-chamber open-air structure, which is located at ground level. The first chamber is the sediment forebay (a.k.a., sedimentation chamber), while the second chamber houses the sand filter bed. Flow enters the sedimentation chamber, where settling of larger sediment particles occurs. Runoff is then discharged from the sedimentation chamber through a perforated standpipe into the filtration chamber. After passing though the filter bed, runoff is collected by a perforated pipe and gravel underdrain system. Figure 13-61 provides a picture of an example surface sand filter, and Figure 13-62 shows typical plan view and profile schematics of a surface sand filter.

Design Information

Much of this design information is taken from ARC (2002), WDE (2001), and Claytor and Schueler (1996). The entire treatment system (including the sedimentation chamber) must temporarily hold at least 75% of the water quality capture volume. Figure 13-63 illustrates the distribution of the treatment volume (0.75 water-quality capture volume) among the various components of the surface sand filter, including:

V_s — Volume within the sedimentation basin

V_f — Volume within the voids in the filter bed

$V_{f\text{-temp}}$ — Temporary volume stored above the filter bed

A_s — Surface area of the sedimentation basin

A_f — Surface area of the filter media

h_s — Height of water in the sedimentation basin

h_f — Average height of water above the filter media

d_f — Depth of filter media

Special inlet structures have been designed for diversion of water into sand filters, and other first-flush BMPs, and are shown below. They should be sized to handle the peak

Structural Best Management Practices   841

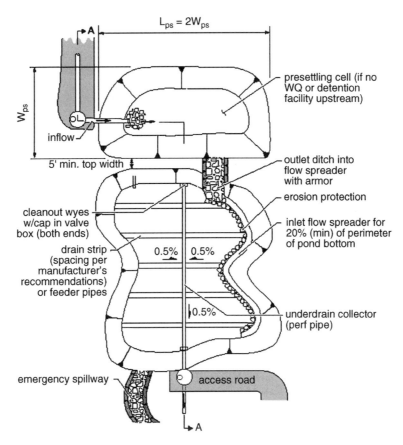

FIGURE 13-62
Typical surface sand filter schematic. (From Atlanta Regional Commission, Georgia Stormwater Manual, 2002.)

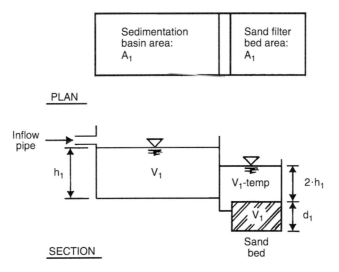

FIGURE 13-63
Surface sand filter volumes. (From Claytor and Schueler, 1996.)

flow from the water-quality capture volume. See Chapter 7, Section 7-11, for information on how to compute this peak flow. There are also proprietary diversion and splitter structures available on the market, many of which fit into ordinary manhole openings.

Pretreatment of runoff in a sand filter system is provided by the sedimentation chamber. The sedimentation chamber must be sized to at least 25% of the computed water-quality capture volume and have a length-to-width ratio of at least 2:1. Inlet and outlet structures should be located at opposite ends of the chamber. Inlets to surface sand filters are to be provided with energy dissipators. Exit velocities from the sedimentation chamber must be nonerosive. The Camp–Hazen equation can be used to compute the required surface area:

$$A_s = - (Q_o/w) * Ln (1-E) \qquad (13.18)$$

where A_s = sedimentation basin surface area (ft²); Q_o = rate of outflow = the capture volume over a 24-h period; w = particle settling velocity (ft/sec); and E = trap efficiency.

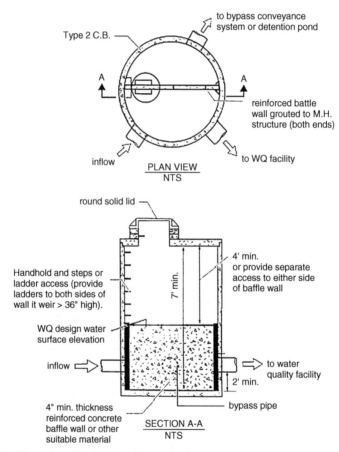

FIGURE 13-64
Flow splitter Option 1. (From Washington State Department of Ecology, Stormwater Management Manual for Western Washington, 2001.)

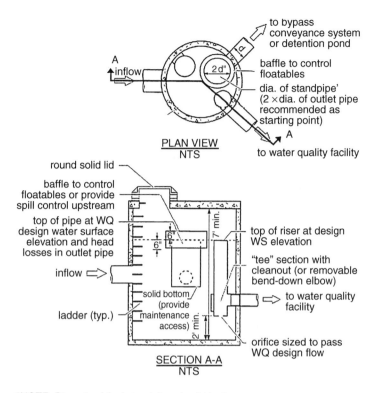

FIGURE 13-65
Flow splitter Option 2. (From Washington State Department of Ecology, Stormwater Management Manual for Western Washington, 2001.)

Assuming:

- 90% sediment trap efficiency (0.9)
- Particle settling velocity (ft/sec) = 0.0033 ft/sec for imperviousness <75%
- Particle settling velocity (ft/sec) = 0.0004 ft/sec for imperviousness ≥75%
- Average of 24-h holding period

Then:

$$A_s = (0.066)(WQ_v) \text{ ft}^2 \text{ for } I < 75\%$$

$$A_s = (0.0081)(WQ_v) \text{ ft}^2 \text{ for } I \geq 75\%$$

Figure 13-68 shows a typical inlet pipe from the sedimentation basin to the filter media basin for the surface sand filter. An outlet pipe is to be provided from the underdrain system to the facility discharge. Due to the slow rate of filtration, outlet protection is generally unnecessary (except for emergency overflows and spillways). An emergency or bypass spillway must be included in the surface sand filter to safely pass flows that exceed the design storm flows. The spillway prevents filter water levels from overtopping the

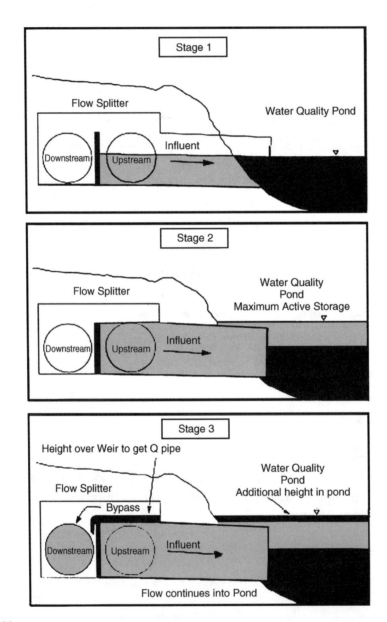

FIGURE 13-66
Flow splitter option 3. (From Ontario Ministry of the Environment, Stormwater Management Planning and Design Manual, Toronto, Canada, 1999.)

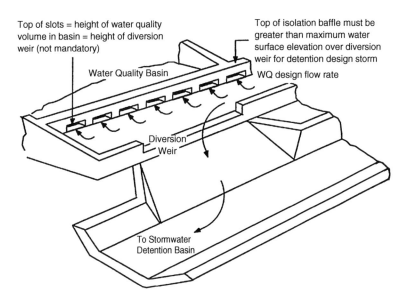

FIGURE 13-67
Flow splitter option 4. (From WDE, 1999.)

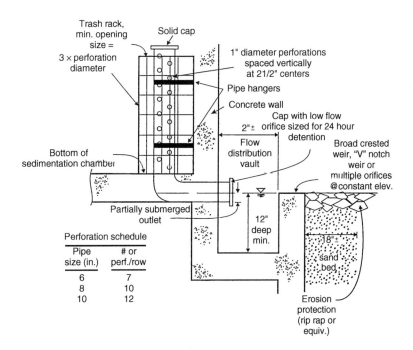

FIGURE 13-68
Surface sand filter perforated standpipe. (From Claytor and Schueler, 1996.)

TABLE 13-27

Sand Medium Specification

U.S. Sieve Number	Percent Passing
4	95–100
8	70–100
16	40–90
30	25–75
50	2–25
100	<4
200	<2

Source: From Washington State Department of Ecology (WDE), Stormwater Management Manual for Western Washington, 2001.

embankment and causing structural damage. The emergency spillway should be located so that downstream buildings and structures will not be impacted by spillway discharges.

The filter area is sized based on the principles of Darcy's Law. A coefficient of permeability (k) of 2 to 3.5 ft/day for sand should be used. The filter bed is typically designed to completely drain in 40 h or less.

The filter area is sized using Equation (13.19) [same as Equation (13.13)]:

$$A_f = (WQ_v)(d_f) / [(k)(h_f + d_f)(t_f)] \tag{13.19}$$

where A_f = surface area of filter bed (ft²); d_f = filter bed depth (typically 18 in, no more than 24 in); k = coefficient of permeability of filter media (ft/day) (use 2 to 3.5 ft/day for sand); h_f = average height of water above filter bed (ft) (1/2 h_{max}, which varies based on site, but h_{max} is typically ≤6 ft); and t_f = design filter bed drain time (days) (1.67 days or 40 h is recommended maximum).

The filter media consists of an 18-in layer of clean washed medium sand (meeting ASTM C-33 concrete sand or per Table 13-27) on top of the underdrain system. Three inches of topsoil are placed over the sand bed. Permeable filter fabric is placed above and below the sand bed to prevent clogging of the sand filter and the underdrain system. Figure 13-69 illustrates typical media cross sections.

The sand in a filter should consist of a medium sand meeting the size gradation (by weight) given in Table 13-27. The contractor should obtain a grain size analysis from the supplier to certify that the No. 100 and No. 200 sieve requirements are met.

The filter bed is equipped with a 6-in perforated PVC pipe (AASHTO M 252) underdrain in a gravel layer. The underdrain must have a minimum grade of 1/8-in/ft (1% slope). Holes should be 3/8-in diameter and spaced approximately 6 in on center. Gravel should be clean washed aggregate with a maximum diameter of 3.5 in and a minimum diameter of 1.5 in with a void space of about 40%. Aggregate contaminated with soil shall not be used. The structure of the surface sand filter may be constructed of impermeable media such as concrete, or through the use of excavations and earthen embankments. When constructed with earthen walls or embankments, filter fabric should be used to line the bottom and side slopes of the structures before installation of the underdrain system and filter media.

Impermeable liners are generally required for soluble pollutants such as metals and toxic organics, and where the underflow could cause problems with structures. Impermeable liners may be clay, concrete, or geomembrane. Clay liners should have a minimum thickness of 12 in and meet the specifications given in WDE (2001). If a geomembrane

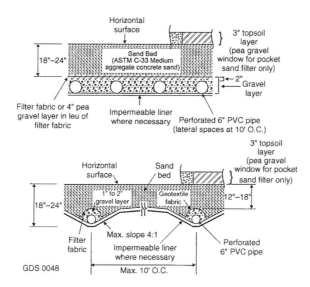

FIGURE 13-69
Typical filter media cross section. (From Claytor and Schueler, 1996.)

liner is used, it should have a minimum thickness of 30 mils and be ultraviolet resistant. The geomembrane liner should be protected from puncture, tearing, and abrasion by installing geotextile fabric on the top and bottom of the geomembrane. Concrete liners may also be used for sedimentation chambers and for sedimentation and sand filtration basins less than 1000 square feet in area. Concrete should be 5-in thick Class A or better and should be reinforced by steel wire mesh.

Adequate access must be provided for all sand filter systems for inspection and maintenance, including the appropriate equipment and vehicles. Access grates to the filter bed need to be included in a perimeter sand filter design. Facility designs must enable maintenance personnel to easily replace upper layers of the filter media. Surface sand filter facilities can be fenced to prevent access. Inlet and access grates to perimeter sand filters may be locked. Surface filters can be designed with a grass cover to aid in pollutant removal and prevent clogging. The grass should be capable of withstanding frequent periods of inundation and drought. More information on design and construction specifications can be found at WDE (2001), ARC (2002), and MCES (2001).

The overall design becomes one of fitting the structure into the site and remaining within the boundaries of sizing limits. Based on experience, 75% of the water-quality capture volume must fit within the three basic compartments within the sand filter: sedimentation chamber, head above the sand bed, and saturated pore spaces within the sand bed as indicated in Equation (13.20).

$$V_{min} = 0.75 \, WQ_v = V_s + V_f + V_{f\text{-temp}} \qquad (13.20)$$

where V_f = water volume within filter bed/gravel/pipe = $A_f * d_f * n$; n = porosity = 0.4 for most applications; $V_{f\text{-temp}}$ = temporary storage volume above the filter bed = $2 * h_f * A_f$; V_s = volume within sediment chamber = $V_{min} - V_f - V_{f\text{-tem}}$; h_s = height in sedimentation chamber = V_s / A_s

Ensure h_s and h_f fit available head and other dimensions still fit — change as necessary in design iterations until all site dimensions fit. A design example will suffice to show the logical process involved in sizing and layout.

Design Example

This example focuses on the design of a surface sand filter to meet the water-quality treatment requirements of the site. It is based on work done by the Center for Watershed Protection for the Georgia Stormwater Manual (ARC, 2002). Channel protection and overbank flood control are not addressed in this example.

Base Data

Site Area = Total Drainage Area (A) = 3.0 acre
Impervious Area = 1.9 acre; or I = 1.9/3.0 = 63.3%
Soils Type "B"
Hydrologic Data

	Pre	Post
CN	57	83
tc	0.36	0.15

Phase 1 — Computation of Preliminary Stormwater Storage Volumes and Peak Discharges

Step 1 — Compute Runoff Control Volumes — Note that design criteria is capture of runoff from the first 1.2 in of rainfall. See Chapter 7, Section 7-11, for more information.

Compute runoff coefficient (R_v)

$$R_v = 0.05 + (63.3)(0.009) = 0.62$$

Compute WQ_v

$$WQ_v = (1.2")(R_v)(A)/12$$

$$= (1.2")(0.62)(3.0\,ac)(43{,}560\,ft^2/ac)(1\,ft/12\,in)$$

$$= 8{,}102\ ft^3 = 0.186\ ac\text{-}ft$$

Step 2 — Compute Runoff Volumes and Peaks for Other Flows — Include the channel protection volume (assume 1-year, 24-h storm), overbank flood design criteria (assume 25-year storm), and peak flow for 100-year storm. Calculations are not included here.

Phase II — Determination if the Development Site and Local Conditions are Appropriate for the Use of a Stormwater Pond

Step 1 — Screen BMPs Using Site-Specific Data — Existing ground elevation at the facililty location is 22.0 ft, mean sea level. Soil-boring observations reveal that the seasonally high water table is at 13.0 ft. Adjacent creek invert is at 12.0.

Step 2 — Confirm Local Design Criteria and Applicability — There are no additional requirements for this site.

Phase III — Performance of Final Design Sizing

Step1 — Compute WQv Peak Discharge (Q_{wq}) and Head —

Water quality volume:

WQ$_v$ previously determined to be 8102 cubic ft.

Step 2 — Determine Available Head (See Figure 13-70) — See Figure 13-70. Low point at parking lot is 23.5. Subtract 2 ft to pass Q$_{25}$ discharge (21.5) and a half foot for channel to facility (21.0). Low point at stream invert is 12.0. Set outfall underdrain pipe 2 ft above stream invert, and add 0.5 ft to this value for drain (14.5). Add to this value 8 in for the gravel blanket over the underdrains, and 18 in for the sand bed (16.67). The total available head is 21.0 - 16.67 or 4.33 ft. Therefore, the average depth, *hf*, is *hf* = 4.33 ft/2, and *hf* = 2.17 ft.

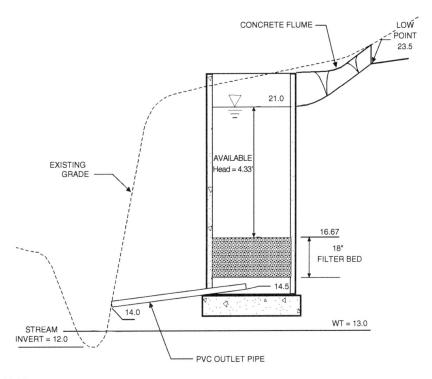

FIGURE 13-70
Inlet elevations and heads for sand filter.

The peak rate of discharge for the water quality design storm is needed for the sizing of offline diversion structures, such as sand filters and grass channels. Conventional SCS methods have been found to underestimate the volume and rate of runoff for rainfall events less than 2 in. This discrepancy in estimating runoff and discharge rates can lead to situations in which a significant amount of runoff bypasses the filtering treatment practice due to an inadequately sized diversion structure and leads to the design of undersized bypass channels. See Chapter 7, Section 7-11, for description of the procedure for estimation of peak flow.

Step 3 — Using the Water-Quality Volume (WQv), Compute the Water-Quality Peak Discharge (Q$_{wq}$) —

$$Q_{wq} = q_u * A * WQ_V$$

where Q_{wq} = the peak discharge (cfs); q_u = the unit peak discharge (cfs/mi_/in); A = drainage area (square miles); and WQ_v = water-quality volume (watershed in).

For this example, the steps are as follows:

Compute modified CN for 1.2 in rainfall

$$P = 1.2''$$

$$Q = WQ_v \div \text{area} = (8{,}102 \text{ ft}^3 \div 3 \text{ ac} \div 43{,}560 \text{ ft}^2/\text{ac} \times 12 \text{ in/ft}) = 0.74''$$

$$CN = 1000/[10+5P+10Q-10(Q^2+1.25*Q*P)^{1/2}]$$

$$= 1000/[10+5*1.2+10*0.74-10(0.74^2+1.25*0.74*1.2)^{1/2}]$$

$$= 95.01$$

Use $CN = 95$

For $CN = 95$ and the $T_c = 0.15$ h, compute the Q_p for a 1.2-in storm. With the $CN = 95$, a 1.2-in storm will produce 0.74 in of runoff. $I_a = 0.105$; therefore, $I_a/P = 0.105/1.2 = 0.088$. From Chapter 2, $q_u = 625$ csm/in, and therefore, $Q_{wq} = (625 \text{ csm/in})(3.0 \text{ ac}/640\text{ac/sq mi})(0.74 \text{ in}) = 2.2$ *cfs*.

Step 4 — Size Flow Diversion Structure: — Size a low-flow orifice to pass 2.2 cfs with 1.5 ft of head using the orifice equation:

$$Q = CA(2gh)^{1/2}; \; 2.2 \text{ cfs} = (0.6)(A)[(2)(32.2 \text{ ft/s}^2)(1.5')]^{1/2}$$

$$A = 0.37 \text{ sq ft} = \pi d^2/4: d = 0.7' \text{ or } 8.5''; \text{ use 9 in}$$

The flow diversion structure is an open-topped rectangular stilling vault with an orifice and pipe leading into the sedimentation chamber, and an overflow weir over the top of the vault for the 25-year storm into a receiving channel, bypassing the sand filter.

The 25-year wsel is set at 23.0. Use a concrete weir to pass the 25-year flow (17.0 cfs) into a grassed overflow channel using the weir equation. Assume 2 ft of head to pass this event. Overflow channel should be designed to provide sufficient energy dissipation (e.g., riprap, plunge pool, etc.) so that there will be nonerosive velocities.

$$Q = CLH^{3/2}$$

$$17 = 3.1 \, (L) \, (2')^{1.5}$$

$L = 1.94$ ft; use $L = 2$ ft - 0 in, which sets flow diversion chamber dimension.

Weir wall elev. = 21.0. Set low-flow invert at 21.0 - [1.5' + (0.5*9''*1ft/12'')] = 19.13.

Step 5 — Size Filtration Bed Chamber —

From Darcy's Law:

$$A_f = WQ_v \, (d_f) / [k \, (h_f + d_f) \, (t_f)]$$

where $d_f = 18$ in; $k = 3.5$ ft/day; $h_f = 2.17$ ft; and $t_f = 40$ h.

$$A_f = (8102 \text{ cubic feet}) (1.5') / [3.5 (2.17' + 1.5') (40 \text{ h}/24 \text{ h/day})]$$

$A_f = 567.7$ *sq ft*; using a 2:1 ratio, say filter is *17' by 34'* (= 578 sq ft)

Step 6 — Size Minimum Sedimentation Chamber — From Camp–Hazen equation, for $I < 75\%$: $A_s = 0.066 \, (WQ_v)$

$$A_s = 0.066 \, (8{,}102 \text{ cubic ft}) \text{ or } 535 \text{ sq ft}$$

Given a width of 17 ft, the length will be 535 ft/17 ft or 31.5 ft (use 17 ft × 32 ft).

Step 7 — Compute V_{min} —

$$V_{min} = \tfrac{3}{4}(WQ_v) \text{ or } 0.75 \, (8{,}102 \text{ cubic feet}) = 6{,}077 \text{ cubic feet}$$

Step 8 — Compute Storage Volumes within Entire Facility and Sedimentation Chamber Orifice Size —

Volume within filter bed (V_f):

$$V_f = A_f \, (d_f) \, (n); \; n = 0.4 \text{ for sand}$$

$$V_f = (578 \text{ sq ft}) \, (1.5') \, (0.4) = 347 \text{ cubic feet}$$

Temporary storage above filter bed ($V_{f\text{-temp}}$):

$$V_{f\text{-temp}} = 2 h_f A_f$$

$$V_{f\text{-temp}} = 2 \, (2.17') \, (578 \text{ sq ft}) = 2{,}509 \text{ cubic feet}$$

Compute remaining volume for sedimentation chamber (V_s):

$$V_s = V_{min} - [\, V_f + V_{f\text{-temp}} \,] \text{ or } 6{,}077 - [347 + 2{,}509] = 3{,}221 \text{ cubic feet}$$

Compute height in sedimentation chamber (h_s):

$$h_s = V_s / A_s$$

(3221 cubic ft)/(17 ft × 32 ft) = 5.9 ft, which is larger than the head available (4.33 ft); increase the size of the settling chamber, using 4.33 ft as the design height (3221 cubic ft)/4.33 ft = 744 sq ft; 744 ft/17 ft yields a length of 43.8 ft (say, 44 ft)

New sedimentation chamber dimensions are 17 ft by 44 ft

With adequate preparation of the bottom of the settling chamber (roto-till earth, place gravel, then surge stone), the bottom can infiltrate water into the substrate. The runoff will enter the groundwater directly without treatment. The stone will eventually clog without protection from settling solids, so use a removable geotextile to facilitate maintenance. Note that there is 2.17 ft of freeboard between the bottom of the recharge filter and the water table. The layout of the sand filter is given in Figure 13-71.

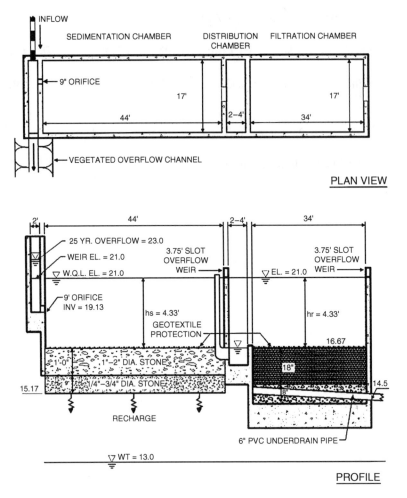

FIGURE 13-71
Layout of sand filter.

Step 9 — Design Inlet Standpipe — Provide perforated standpipe with orifice sized to release volume (within sedimentation basin) over a 24 h period (see Figure 13-72). Average release rate equals 3221 ft³/24 h = 0.04 cfs

Equivalent orifice size can be calculated using orifice equation:

$$Q = CA(2gh)^{1/2}, \text{ where h is average head, or } 4.33'/2 = 2.17'.$$

$$0.04 \text{ cfs} = 0.6*A*(2*32.2 \text{ ft}/s^2*2.17 \text{ ft})^{1/2}$$

$$A = 0.005 \text{ ft}^2 = pD^2/4$$

Therefore, equivalent orifice diameter equals 1 in.

Recommended design is to cap standpipe with low-flow orifice sized for 24-h detention. Overperforate pipe by a safety factor of 10 to account for clogging. Note that the size and number of perforations will depend on the release rate needed to achieve 24 h detention. A multiple orifice stage–discharge relation needs to be developed for the proposed perforation configuration. Standpipe should discharge into a flow distribution chamber prior

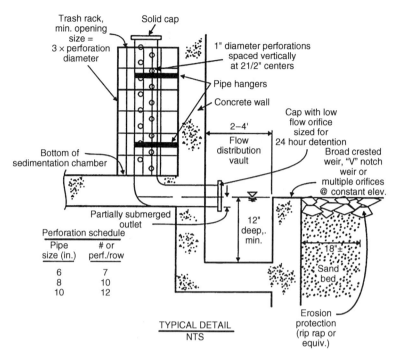

FIGURE 13-72
Inlet structure for sand filter.

to filter bed. Distribution chamber should be between 2 and 4 ft in length and same width as filter bed. Flow distribution to the filter bed can be achieved with a weir or multiple orifices at constant elevation. See Figure 13-72 for standpipe details.

Step 10 — Design Inlets, Pretreatment Facilities, Underdrain System, and Outlet Structures —

Compute overflow weir between sediment chamber and sand filter.

Assume overflow that needs to be handled is equivalent to the 9-in orifice discharge under a head of 3.5 ft (i.e., the head in the diversion chamber associated with the 25-year peak discharge).

$$Q = CA(2gh)^{1/2}$$

$$Q = 0.6(0.44 \text{ ft}^2)[(2)(32.2 \text{ ft/s}^2)(3.5 \text{ ft})]^{1/2}$$

$$Q = 3.96 \text{ cfs, say } 4.0 \text{ cfs}$$

For the overflow from the sediment chamber to the filter bed, size to pass 4 cfs. Weir equation:

$$Q = CLh^{3/2}$$

Assume a maximum allowable head of 0.5 ft.

$$4.0 = 3.1 * L * (0.5 \text{ ft})^{3/2}$$

FIGURE 13-73
Picture of perimeter stormwater filter. (From Atlanta Regional Commission, Georgia Stormwater Manual, 2002.)

$$L = 3.65 \text{ ft}, \text{ Use } L = 3.75 \text{ ft}.$$

Similarly, for the overflow from the filtration chamber to the outlet of the facility, size to pass 4.0 cfs.

Weir equation:

$$Q = CLh^{3/2}$$

Assume a maximum allowable head of 0.5 ft.

$$4.0 = 3.1 * L * (0.5 \text{ ft})^{3/2}$$

$$L = 3.65 \text{ ft}, \text{ Use } L = 3.75 \text{ ft}.$$

Adequate outlet protection and energy dissipation (e.g., riprap, plunge pool, etc.) should be provided for the downstream overflow channel. See Chapter 12.

Linear (Perimeter) Sand Filter

The perimeter sand filter is an enclosed filter system typically constructed just below grade in a vault along the edge of an impervious area such as a parking lot. The system consists of a sedimentation chamber and a sand bed filter. Runoff flows into the structure through a series of inlet grates located along the top of the control.

The linear filter (Figure 13-27) was originally tested in Delaware (Shaver, 1991). It basically intercepts sheet flow coming off a parking area or street and filters it through a recommended 18 in of sand after flowing through a pretreatment chamber, designed like a water-quality inlet. The outfall pipe is recommended to be 6-in diameter or less. A grate over the outfall opening is wrapped in filter fabric.

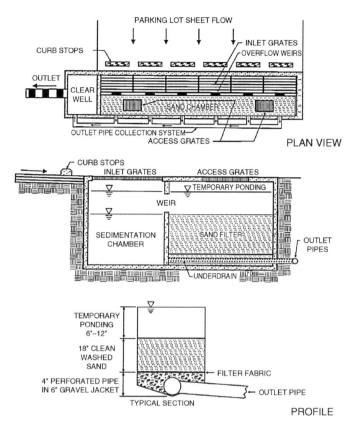

FIGURE 13-74
Perimeter sand filter schematic. (From Atlanta Regional Commission, Georgia Stormwater Manual, 2002.)

Like other sand filters, the perimeter sand filter is deigned primarily as an offline system for stormwater quality (i.e., the removal of stormwater pollutants) and will typically need to be used in conjunction with another structural control to provide downstream channel protection, overbank flood protection, and extreme flood protection, if required.

The entire treatment system (including the sedimentation chamber) must temporarily hold at least 75% of the water-quality capture volume prior to filtration. The sedimentation chamber must be sized to at least 50% of the computed WQ_v. Figure 13-75 illustrates the distribution of the treatment volume (0.75 WQ_v) among the various components of the perimeter sand filter, including:

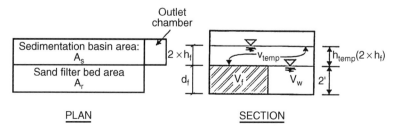

FIGURE 13-75
Volume distribution in a perimeter sand filter. (From Claytor and Schueler, 1996.)

- V_w — Wet pool volume within the sedimentation basin
- V_f — Volume within the voids in the filter bed
- V_{temp} — Temporary volume stored above the filter bed
- A_s — Surface area of the sedimentation basin
- A_f — Surface area of the filter media
- h_f — Average height of water above the filter media ($1/2\ h_{temp}$)
- d_f — Depth of filter media

The filter area is sized based on the principles of Darcy's Law [Equation (13.19)]. A coefficient of permeability (k) of 2 to 3.5 ft/day for sand should be used. The filter bed is typically designed to completely drain in 40 h or less.

The filter media should consist of a 12- to 18-in layer of clean washed medium sand on top of the underdrain system with a sand specification similar to the surface sand filter. Figure 13-69 illustrates a typical media cross section.

The perimeter sand filter is typically equipped with a 4-in perforated PVC pipe underdrain in a gravel layer. The underdrain must have a minimum grade of 1/8 in/ft (1% slope). Holes should be 3/8-in diameter and spaced approximately 6 in on center. A permeable filter fabric should be placed between the gravel layer and the filter media. Gravel should be clean-washed aggregate with a maximum diameter of 3.5 in and a minimum diameter of 1.5 in with a void space of about 40%. Aggregate contaminated with soil should not be used.

Computation for the perimeter sand filter generally follows that of the surface sand filter. Volume calculations change slightly:

- Compute V_f = water volume within filter bed/gravel/pipe = $A_f * d_f * n$
- Where n = porosity = 0.4 for most applications
- Compute V_w = wet pool storage volume $A_s * 2$ ft minimum
- Compute V_{temp} = temporary storage volume = $V_{min} - (V_f + V_w)$
- Compute h_{temp} = temporary storage height = $V_{temp} / (A_f + A_s)$
- Ensure $h_{temp} \geq 2 * h_f$, otherwise decrease h_f and recompute; ensure dimensions fit available head and area and change as necessary in design iterations until all site dimensions fit
- Size distribution slots from sediment chamber to filter chamber

Underground Sand Filter

The underground sand filter is a design variant of the sand filter located in an underground vault designed for high-density land use or ultra-urban applications, where there is not enough space for a surface sand filter or other structural stormwater controls. They are usually used on highly impervious sites of 1 acre or less. The maximum drainage area that should be treated is 5 acres. High groundwater may damage underground structures or affect the performance of filter underdrain systems. There should be sufficient clearance (at least 2 ft is recommended) between the seasonal high groundwater level (highest level of groundwater observed) and the bottom of the sand filter to obtain adequate drainage (WDE, 2001).

The underground sand filter is a three-chamber system (Figures 13-76 through 13-79):

- The initial chamber is a sedimentation (pretreatment) chamber that temporarily stores runoff and utilizes a wet pool to capture sediment. One foot of sediment storage is recommended.

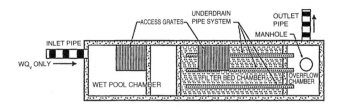
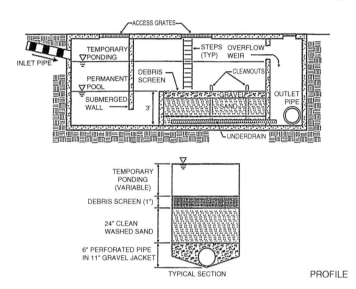

FIGURE 13-76
Schematic of underground "DC" sand filter. (From Maryland Department of the Environment, Maryland Stormwater Design Manual, Volumes I and II, prepared by Center for Watershed Protection, 2000.)

- The sedimentation chamber is connected to the sand filter chamber by a submerged wall that protects the filter bed from oil and trash. It should extend 1 ft above and below the design flow water level and be spaced a minimum of 5 ft horizontally from the inlet. In the event of plugging, passage of flows must be provided for. Access should be provided to both sides of the baffle.
- A maximum of 6 inches between the top of the flow spreader and the top of the sand bed is recommended to reduce sand disturbance. A flow spreader or pipe and manifold system (minimum 8-in pipe) may be used.
- The filter bed is typically 18 to 24 in deep but varies by design type and may have a protective screen of gravel or permeable geotextile on top to limit clogging (King County, 1998).
- At grade, access panels should be provided for the entire length of the sand bed, have lift rings, and weigh no more than 5 tons each (MCES, 2001).
- The sand filter chamber also includes an underdrain system with inspection and cleanout wells. Perforated drain pipes under the sand filter bed extend into a third chamber that collects filtered runoff. Internal diameters of underdrain pipes should be a minimum of 6 in and two rows of $1/2$-in holes spaced 6 in apart longitudinally (maximum), with rows 120 degrees apart (laid with holes downward). Maximum perpendicular distance between two feeder pipes must be 15 ft. All piping is to be schedule 40 PVC or greater wall thickness. Drain

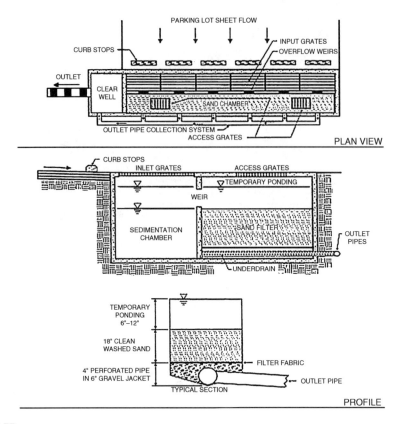

FIGURE 13-77
Delaware sand filter. (From Metropolitan Council Environmental Services, Minnesota Small Sites Urban BMP Manual, Mears Park Center, St. Paul, MN, 2001.)

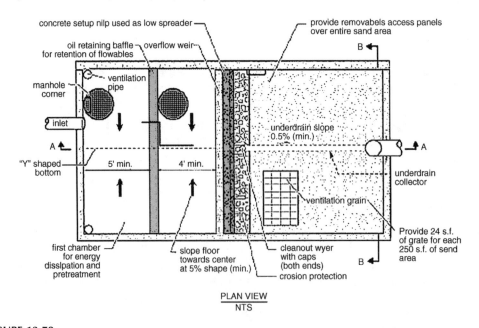

FIGURE 13-78
Washington state sand filter plan view. (From Washington State Department of Ecology, Stormwater Management Manual for Western Washington, 2001.)

Structural Best Management Practices

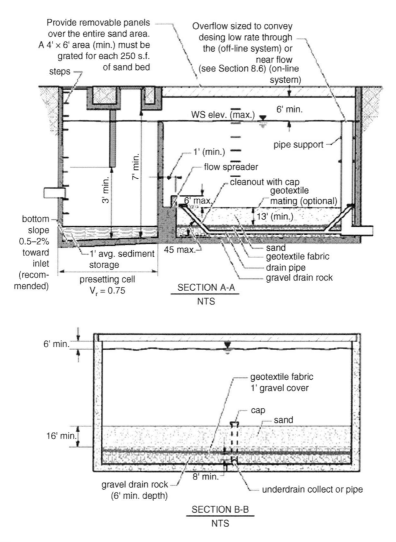

FIGURE 13-79
Washington State sand filter cross sections. (From Washington State Department of Ecology, Stormwater Management Manual for Western Washington, 2001.)

piping could be installed in basin and trench configurations (WDE, 2001; King County, 1998).
- Drain rock should be 0.75 to 1.5 in rock or gravel backfill, washed free of clay and organic material (WDE, 2001).
- To prevent anoxic conditions, a minimum of 24 square feet of ventilation grate must be provided for each 250 square feet of sand bed surface area. Placement at each end is preferred unless the sand bed is small.
- A dewatering valve may be provided just above the sand bed. To facilitate maintenance of the sand filter, an inlet shutoff/bypass valve is recommended.
- Cleanout wyes with caps or junction boxes must be provided at both ends of the collector pipes. Cleanouts must extend to the surface of the filter. A valve box must be provided for access to the cleanouts. Access for cleaning all under-

drain piping should be provided. This may consist of installing cleanout ports, which tee into the underdrain system and surface above the top of the sand bed.
- Flows beyond the filter capacity are diverted through an overflow weir. Underground sand filters are typically constructed offline but can be constructed online. For recommended offline construction, the overflow between the second and third chambers is not included.

Due to its location below the surface, underground sand filters should only be used where adequate inspection and maintenance can be ensured. Thus, adequate maintenance access must be provided to the sedimentation and filter bed chambers. In addition, the underground vault should be tested for watertightness prior to placement of filter layers.

Underground sand filter pollutant removal rates are similar to those for surface and perimeter sand filters.

There are several variations on the underground sand filter, including the "DC" sand filter, Delaware sand filter, and the Washington State sand filter in Figures 13-71, 13-72, and 13-73/13-74, respectively.

The Delaware sand filter contains two chambers and a clearwell. Water enters the first chamber, flows over a series of weirs and into a second chamber containing the filter material. Water ponds in both chambers. Filtered water is collected in gravel and pipe underdrain systems. The DC filter was developed in the 1980s by the District of Columbia. It uses three chambers: a sediment chamber and oil and grease floatables trap, a filter bed chamber, and third clearwell chamber for discharge to the drainage system. A submerged opening connects the first two chambers, providing a seal against floating pollution. Storage is provided in the first two chambers.

Organic Sand Filters

Organic filters can utilize a variety of organic materials as the filtering media. Two typical media bed configurations are the peat/sand filter and compost filter. The peat filter includes an 18-in 50/50 peat/sand mix over a 6-in sand layer and can be optionally covered by 3 in of topsoil and vegetation. The compost filter has an 18-in compost layer. Both variants utilize a gravel underdrain system. Peat-sand filters are an engineered soil/peat layering used for the filtration of the first flush of runoff. They contain hybrid filtration capability consisting of a nutrient-removing cover grass mowed surface, peat layer, and sand (or peat/sand) layer underlain by underdrains set in gravel layers.

Pollutant removal for sand filters is presented in Table 13-13. Relative peat and organic sand filter pollutant removals are discussed in the general sand filter section. Special design modifications, including slower infiltration rates and an extra peat layer on the bottom, have been proposed to enhance nutrient removal (Galli, 1992). Organic filter media, such as peat sand and compost, show some promise in removing higher levels of hydrocarbons and metals, and should be seriously considered for hot spot sites. They do not, however, appear to perform much better than basic sand filters when it comes to removing nutrients. Peat-sand filters have been tested in the Coordinated Anacostia Retrofit Program (CARP) and found to be effective in pollution reduction, especially TP, BOD, and bacteria (Galli, 1990, 1992). Peat is an excellent material for promoting microbial growth and assimilation of nutrients. It also has high water-retention capabilities and low bulk density. Peat-sand filters were originally used for wastewater treatment, and much of the design information currently comes from those sources (Farnham and Brown, 1972; Farnham and Noonan, 1988; Brown and Farnham, 1976). Sand filtration is coming into some prominence for wastewater treatment for small sites. The gradual decomposition of organic media can result in the export of nitrate and soluble phosphorus. Compost filter

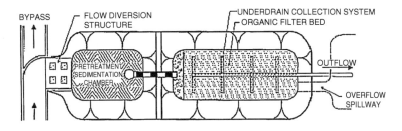

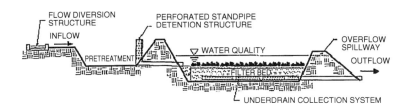

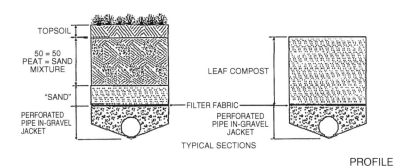

FIGURE 13-80
Peat-sand filter. (From Atlanta Regional Commission, Georgia Stormwater Manual, 2002.)

experience clearly indicates that organic filters are a poor choice if they are likely to encounter dry weather flows (Schueler, 2000; WDE, 2001).

Figure 13-80 shows a typical peat layout and cross section. These layers are placed in filtration beds, similar to the Austin first-flush basin. Thus, there is great flexibility in size and location.

Peat-sand filters are configured like other surface sand filters and should have a pretreatment pool or detention facility similar to a sediment forebay or extended detention pond to remove the larger sediment particles. A flow spreader should be provided to ensure even flow movement over the bed. The filter is normally an offline system with a flow splitter to divert larger flows than the capture volume around the basin. Like other filtration systems, peat-sand filters are restricted to stabilized watersheds. Organic filters are typically used in high-density applications, or for areas requiring an enhanced pollutant removal ability. Maintenance is typically higher than for the surface sand filter facility due to the potential for clogging. In addition, organic filter systems potentially have a higher head requirement than sand filters. The minimum head requirement (elevation difference needed at a site from the inflow to the outflow) for an organic filter is 5 to 8 ft.

Peat is classified into three types: fabric, hematic, and sapric. Fabric is the most fibrous and light. It has a high hydraulic conductivity (up to 140 cm/h) and is brown or yellowish in color. Sapric peat is highly decomposed, nonfibrous, more dense, and has a low hydrau-

FIGURE 13-81
Peat sand filter picture. (From Atlanta Regional Commission, Georgia Stormwater Manual, 2002.)

lic conductivity (down to 0.025 cm/h). It is dark grey to black in color. Hematic is intermediate in all respects.

The peat should be fabric or hematic (not sapric) with less than 30% minerals by weight, and should be shredded to a uniform consistency before use (Tomaseck, Johnson, and Mulloy, 1987). It is recommended that a soil scientist test peat supplies prior to bulk purchase.

Peat is placed in 2- to 4-in lifts, compacted lightly, and raked to ensure good contact with the next lift. It should not be allowed to dry out fully during storage or construction. Granular fine ground 200-mesh Ag-lime calcitic limestone can be mixed into the top 4 to 6 in to enhance P removal.

The sand layer should be clean fine to medium grain (0.1 to 0.5 mm) low phosphate. It is placed below the peat layer to draw the water through the peat layer. Between the peat and sand layers, there is a 4-in 50-50 peat-sand mix to ensure a uniform flow and transition from one layer to the next. The bottom layer is a 6-in thick layer of very fine to medium gravel (2 to 20 mm). Perforated PVC drainpipes are supplied similar to the Austin basin.

A test site in Montgomery County, Maryland, was designed to allow for up to 2 ft (2/3) of the capture volume storage on top of the peat bed. A 1 in/h infiltration rate was proposed to avoid saturation of the peat bed and long surface ponding. The drawdown time was then 24 h.

According to Galli (1991), the original peat basins were sized using one of three proposed rules. The first is to capture the first half-inch from the watershed drainage area. The necessary surface area is then equal to the capture volume divided by the maximum depth allowed on the surface (2 ft in the test site). The second rule is based on a wastewater-like consideration of maximum allowable unit area loading of phosphorus and involves calculating the total runoff per year and the estimated phosphorus concentration. Hard numbers are difficult to derive by this method. The third rule relies only on not exceeding the N- and P-removal capabilities of the surface grasses. Values for a Minnesota study of reed, foxtail, and fescue filter bed vegetation ranged from 210 to 250 lb/acre for N and from 33 to 37 lb/acre P annually (Elling, 1985).

Maintenance requires mowing during the growing season, and removing grass clippings. Sediment and trash removal is important for aesthetic purposes and to keep systems from clogging. Questions remain concerning use of this system, including clogging factors; restoration of the peat layer if clogged; leeway in peat layering and typing; long-term stability, repeatability, and consistency of pollutant removal; ponding depth; best combinations of ponds and beds; and best cover crops (Galli, 1991).

Infiltration Trenches

An infiltration trench (a.k.a. infiltration galley) is a rock-filled trench with no outlet that receives stormwater runoff. Infiltration trenches are one of a number of infiltration devices that accept runoff and exfiltrate it to the groundwater. Stormwater runoff passes through some combination of pretreatment measures, such as a swale and detention basin, and into the trench. There, runoff is stored in the void space between the stones and infiltrates through the bottom and into the soil matrix. The primary pollutant-removal mechanism of this practice is filtering through the soil. Infiltration trenches have select applications. While they can be applied in most regions of the country, their use is sharply restricted by concerns due to common site factors, such as potential groundwater contamination, soils, and clogging. Unlike other controls, infiltration devices can actually reduce the total volume of storm-related runoff flow through groundwater recharge.

Infiltration trenches can be utilized in most regions of the country, with some design modifications in cold and arid climates. In regions of karst (i.e., limestone) topography, these stormwater management practices may not be applied due to concerns of sinkhole formation and groundwater contamination (EPA, 1999).

Trenches are used for on-site type controls and are limited by site characteristics including depth to water table and depth to bedrock. Due to the relatively narrow shape, infiltration trenches can be adapted to many different types of sites and can be utilized in retrofit situations. Unlike some other structural stormwater controls, they can easily fit into the margin, perimeter, or other unused areas of developed sites. A Canadian city developed design specifications to allow for placement of infiltration trenches along roadways (UBC, 2000). However, two features that can restrict their use are the potential of infiltrated water to interfere with existing infrastructure and the relatively poor infiltration of most urban soils (EPA, 1999).

FIGURE 13-82
Infiltration trench example. (From Atlanta Regional Commission, Georgia Stormwater Manual, 2002.)

FIGURE 13-83
Infiltration trench in San Diego, California. (From CALTRANS.)

FIGURE 13-84
Infiltration trench construction.

To protect groundwater from potential contamination, runoff from designated hot spot land uses or activities must not be infiltrated. Infiltration trenches should not be used for manufacturing and industrial sites, where there is a potential for high concentrations of soluble pollutants and heavy metals. In addition, infiltration should not be considered for areas with a high pesticide concentration (ARC, 2002).

Figures 13-82 through 13-84 show typical designs.

Experience has shown that, if sediment is not kept from the trench surface, especially during construction, it will prematurely clog (Galli, 1992b; Currier, 2002). Maintenance of this BMP is critical to keep the surface permeable. Thus, they should be used only in areas where sediment can be kept from the site, mostly highly impervious areas. Failure rates as high as 50% have been experienced in Maryland, Florida, and recently in California (Currier, 2002) due to construction-related clogging, filtration media compaction, lack of pretreatment, high groundwater, lack of maintenance, and poor soils. Other primary concerns with infiltration trenches are metals accumulation and groundwater contamination. Groundwater issues should be investigated if that is the source of water supply, the soil is granular, and intense nonresidential development is present in the drainage area.

Effectiveness can be enhanced through thorough soils investigations prior to design, placing a sand layer at the bottom of the trench, avoiding sites where construction is evident, and Rototilling the trench bottom to preserve infiltration rates (Schueler et al., 1992).

Pollutant removal is accomplished by a complex series of physical and chemical transformations including straining, sorption, trapping, precipitation, and more. Table 13-13 provides estimated removal ranges.

Infiltration Trench Design

Infiltration trenches are excavations typically filled with stone to create an underground reservoir for stormwater runoff (Figure 13-85). This runoff volume gradually exfiltrates through the bottom and sides of the trench into the subsoil and eventually reaches the water table. By diverting runoff into the soil, an infiltration trench not only treats the water-quality volume, but also helps to preserve the natural water balance on a site and can recharge groundwater and preserve baseflow. Due to this fact, infiltration systems are limited to areas with highly porous soils (0.5 in/h percolation rate) and where the water table and bedrock are located at least 3 ft below the bottom of the trench. In addition, infiltration trenches must be carefully sited to avoid the potential of groundwater contamination.

Infiltration trenches are flexible and have been used in a number of ways, including parking lot swales, perimeter strips or median strips, wet wells for downspouts, and neighborhood or industrial park swales (Schueler, 1987). Figure 13-86 shows a typical infiltration trench design.

Oversized pipes and wet vaults have been used in underground applications to hold water for discharge into trenches through perforations. These are often preceded by an oil-grit chamber. Japanese engineers have retrofitted a whole urban area with infiltrating curb and gutter, wet wells for catch basins, and infiltrating (actually, exfiltrating) perforated pipe storm drains (Koyama and Fujita, 1989). This type of design is also common in southern Florida. Typical standard trench designs include three basic types (Schueler, 1987). Commercial systems also exist for smaller site applications (e.g., Infiltrator Systems, Inc., undated)

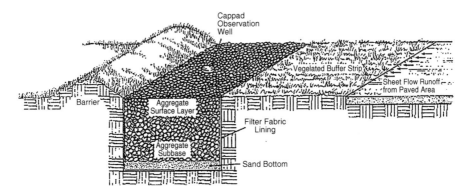

FIGURE 13-85
Typical infiltration trench design. (From Atlanta Regional Commission, Georgia Stormwater Manual, 2002.)

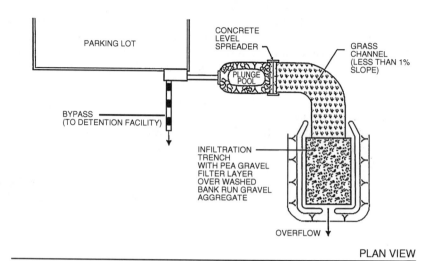

FIGURE 13-86
Typical infiltration trench layout. (From Atlanta Regional Commission, Georgia Stormwater Manual, 2002.)

- *Full exfiltration trench* — Runoff can exit the trench only by exfiltrating through the stone into the underlying soils. The storage volume is based on runoff volume of the 2-year storm.
- *Water quality trench* — The storage volume is based on first-flush volume, either 1/2 in of runoff per impervious acre or 1 in of runoff in contributing area.
- *Partial exfiltration system* — The trench is not designed to rely completely on exfiltration to dispose of the captured runoff volume. A perforated pipe is used to drain part of the volume, being placed beneath or near the top of the trench. The system is not as effective as a full exfiltration system.

Pretreatment minimizes maintenance requirements for trenches. Without such pretreatment, life expectancy is less than 5 years (Schueler et al., 1992; Maryland, 1987). Suspended sediment loads that will clog the trench can be reduced by requiring that the stormwater runoff pass through a sediment forebay, oil-and-grit chamber, or sumped catch basin prior to entering the trench. Grass filter strips have proven ineffective as pretreatment devices. Hydrocarbon loadings (oil and grease) that will clog the filter fabric and sand filter underlying the trench can be reduced by the use of oil-and-grit chambers (when receiving large parking lot and roadway runoff).

TABLE 13-28

Soil Infiltration Estimates

Soil Texture	Minimum Infiltration Rate (in/h)	Estimated Long-Term Infiltration Rate for BMP (in/h)	Effective Water Capacity (in/in)	SCS Hydrologic Soil Group Classification
Sand	8.27	2.0	0.35	A
Loamy sand	2.41	0.5	0.31	A
Sandy loam	1.02	0.25	0.25	B
Loam	0.52	0.13	0.19	B
Silt loam	0.27	N/D	0.17	C

Source: Rawls, W.J., Brakensiek, D.L., and Saxon, K.E., Estimation of soil properties, *Trans. ASAE*, 25, 5, 1982; and Washington State Department of Ecology (WDE), Stormwater Management Manual for Western Washington, 2001.

There are various volume-based design criteria similar to other structural BMPs. The smallest typical volume captured is 0.5 in of runoff from adjoining impervious areas. There are also a number of design procedures used around the country, varying from simple estimates of volumes and infiltration rates to field testing of infiltration and multidimensional consideration of infiltration and groundwater movement (Maryland, 1984; Chescheir, Fipps, and Skaggs, 1990). Certainly, the level of complexity of analysis should be linked to the level of detail of available site data and information. Often, local or regional guidance is available.

Drawdown times should be limited to no more than 40 h. Typical times are more commonly less than 40 h (even as low as 6 h) to allow for a safety factor against clogging. The volume to be infiltrated is determined based on methods given previously. Typical infiltration rates for different types of soils are given in Table 13-28. The long-term rates should be used unless periodic maintenance and any necessary media replacement can be assured.

Soils with infiltration rates less than 0.17 in/h are rarely suitable for infiltration. Soils with infiltration rates greater than 0.3 in/h are recommended. Site materials testing should be done. Typical mandatory *in situ* soils investigations include a falling head percolation test (ASTM D 5084-90). Information is provided by Kent, Washington (2000), Northern Virginia BMP Manual (1992), and WDE (2001). Infiltration basins with pipes as exfiltration devices would be designed slightly differently in that the volume of storage would include a term for the available pipe volume subtracted from the rock volume.

There are several methods used to calculate the actual dimensions of the trench. Washington State has a single-storm volume method and a simulation model method that better accounts for back-to-back storms (WDE, 2001). South Florida Watershed Management District (SFWMD, 1994) offers a different approach for an exfiltrating trench with a perforated pipe in the middle of it laid in partially saturated soil conditions. Through multiple runs of the simulation model, it recently derived specific curves for the region. All the methods involve knowing the necessary water-quality volume, porosity of the soil (often called void space fraction), and timing. Knowing the capture volume, the soil porosity, and the infiltration rate, the trench (designed for sheet flow input through aggregate at the surface) can be sized. Often, the excavation cost or other site constraints set the maximum depth for the trench (i.e., a certain elevation above the seasonally high water table). If not, then a second equation can be developed for the depth. There are then two unknowns: trench bottom area and depth. These can be solved using two alternate expressions for the known water quality capture volume (WQ_v), one in terms of trench volume, and one in terms of flow rate — Equations (13.21) and (13.22).

$$WQ_v = n\, V_{trench} = nA_fD_f \quad (13.21)$$

$$WQ_v = t_f k A_f / 12 \quad (13.22)$$

There is some difference of opinion concerning whether the void space term should be included in the numerator of Equation (13.22), reducing the effective infiltration area in the bottom of the trench. However, especially with a bottom sand layer, the infiltrated water will spread across the whole bottom area of the trench relatively quickly. Plus, infiltration out the sides of the trench is neglected. Solving for the bottom surface area and the trench depth gives:

$$A_f = 12 WQ_v / (kt_f) \quad (13.23)$$

$$D_f = kt_f / (12n) \quad (13.24)$$

where A_f = bottom surface area of trench (ft^2); WQ_v = water-quality capture volume (ft^3); k = percolation rate of surrounding soil (in/h); n = void space fraction (taken between 0.32 and 0.40 for gravel); D_f = depth for infiltration trench (ft); and t_f = drawdown time of filter (h) (local design criteria).

While it is assumed that the flow enters the trench and flows to the trench bottom instantaneously, for design storms of longer duration, it may be advisable to take into account the amount of infiltration during the storm. The Georgia Stormwater Manual (ARC, 2002) takes into account the storm duration as shown in Equation (13-23). The manual recommends a fill time of 2 h for most applications.

$$A_f = WQ_v / (nD_f + kT/12) \quad (13\text{-}23)$$

where T = design storm duration (h) (2 h recommended by ARC, 2002).

The depth calculated is the maximum trench depth. Other site constraints may limit the actual trench depth to something much less. Both methods ignore ponding volume above the trench and any volume within underdrain systems, should the flow be introduced to the gravel fill material through a perforated underdrain system.

Typical Operation and Maintenance Requirements

A stormwater management easement and maintenance covenant should be required for each facility. The maintenance covenant should require the owner of the infiltration trench to periodically clean the structure. The trench should be monitored after every large storm (>1 in in 24 h) for the first year after completion of construction and be monitored quarterly thereafter. Sediment buildup in the top foot of stone aggregate or the surface inlet should be monitored on the same schedule as the recommended observation well.

Detailed maintenance and inspection criteria can be found in Maryland's Inspector's Manual (Maryland, 1985; EPA, 1999). Typical maintenance requirements are summarized in Table 13-29.

Typical Required Specifications

- Use in small drainage areas less than 5 to 10 acres.
- Drain the design volume in no more than 40 h.

TABLE 13-29

Typical Maintenance Activities for Infiltration Trenches

Activity	Schedule
Ensure that contributing area, facility, and inlets are clear of debris	Monthly
Ensure that the contributing area is stabilized	
Remove sediment and oil/grease from pretreatment devices, as well as overflow structures	
Mow grass; filter strips should be mowed as necessary; remove grass clippings	
Check observation wells following 3 days of dry weather; failure to percolate within this time period indicates clogging	Semiannual inspection
Inspect pretreatment devices and diversion structures for sediment buildup and structural damage	
Remove trees that start to grow in the vicinity of the trench	
Replace pea gravel/topsoil and top surface filter fabric (when clogged)	As needed
Perform total rehabilitation of the trench to maintain design storage capacity	Upon failure
Excavate trench walls to expose clean soil	

Source: From EPA, 1999.

- A sediment forebay and grass channel, or equivalent upstream pretreatment, must be provided.
- Trench depths should be between 3 to 8 ft, to provide for easier maintenance. The width of a trench must be less than 25 ft.
- Broader, shallow trenches reduce the risk of clogging by spreading the flow over larger areas for infiltration.
- The bottom slope of a trench should be flat across its length and width to evenly distribute flows, encourage uniform infiltration through the bottom, and reduce the risk of clogging.
- A minimum of one soils log is required for every 50 ft of trench length, and no fewer than 2 soils logs for each proposed trench location. Borings should be taken to a depth of at least 5 ft below the trench depth.
- Each soils log should extend a minimum of 3 ft below the bottom of the trench, describe the SCS series of the soil, the textural class of the soil horizon(s) through the depth of the log, and note any evidence of high groundwater level, such as mottling. In addition, the location of impermeable soil layers or dissimilar soil layers should be determined.
- Soil with minimum infiltration rates of 0.17 in/h or less are not suitable for infiltration trenches. Minimum infiltration rates of 0.5 in/h are required in some municipalities.
- Soils that have a 30% or greater clay content are not suitable for infiltration trenches.
- The use of infiltration systems on fill is not allowed due to the possibility of creating an unstable subgrade.
- A minimum of 2 to 4 ft difference is required between the bottom of the infiltration trench and the groundwater table, and 3 ft to bedrock. If close to these minimums, a groundwater mounding study should be done.
- Site slope must be less than 20%, trench must be horizontal.
- Minimum setback requirements for infiltration trench facilities are 10 ft from property lines, 25 ft from a building foundation, 100 ft from a private well, 1200 ft from a public water supply well, 100 ft from a septic system tank/leach field, 100 ft from surface waters, and 400 ft (100 ft for a tributary) from surface drinking water sources.

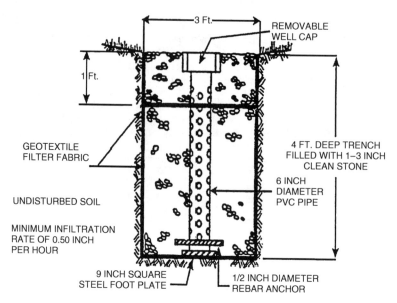

FIGURE 13-87
Observation well details. (From Southeastern Wisconsin Regional Planning Commission, Costs of Urban Nonpoint Source Water Pollution Control Measures, Tech. Report No. 31, 1991.)

- The aggregate material for the trench should consist of a clean aggregate with a maximum diameter of 3 in and a minimum diameter of 1.5 in.
- The aggregate should be graded such that there will be few aggregates smaller than the selected size. For design purposes, void space for these aggregates may be assumed to be in the range of 30 to 40%.
- A 6-in layer of clean, washed sand is placed on the bottom of the trench to encourage drainage and prevent compaction of the native soil, while the stone aggregate is added.
- The aggregate should be completely surrounded with an engineering filter fabric. If the trench has an aggregate surface, filter fabric should surround all of the aggregate fill material except for the top 1 ft.
- An observation well should be installed for every 50 ft of trench length.
- The observation well should consist of perforated PVC pipe, 4 to 6 in diameter, located in the center of the structure, and be constructed flush with the ground elevation of the trench. Figure 13-87 shows details of an observation well.
- The top of the observation well should be capped to discourage vandalism and tampering.
- Bypass larger flows.

Typical Recommended Specifications

- Installation can take place under a swale to increase the storage of the infiltration system.
- Infiltration trenches work well for residential lots, commercial areas, parking lots, and open-space areas.
- Infiltration systems should not be constructed until all construction areas draining to them are fully stabilized.

- An analysis should be made to determine any possible adverse effects of seepage zones when there are nearby building foundations, basements, roads, parking lots, or sloping sites.
- To reduce the potential for costly maintenance and system reconstruction, it is strongly recommended that the trench be located in an open or lawn area, with the top of the structure as close to the ground surface as possible. Infiltration trenches should not be located beneath paved surface, such as parking lots.
- Infiltration trenches are designed for intermittent flow and must be allowed to drain and allow re-aeration of the surrounding soil between rainfall events. They must not be used on sites with a continuous flow from groundwater, sump pumps, or other sources.

Design Example

A 2-acre site, 100% impervious, is to be treated by an infiltration trench. The soils have an infiltration rate of 1.02 in/h, which will be maintained. The seasonal high groundwater table is at a depth of 9.6 ft. The criterion is to stay 3 ft above water table. The trench is to be designed to treat the runoff from 1 in of rainfall (the 6-h, 6-month storm). The soil porosity is 0.40. A 40-h drawdown is the design criteria.

Maximum trench depth is 9.6 − 3 = 6.6 ft.

$$R_v = 0.05 + 0.009 \,\%I = 0.05 + 0.009 * 100 = 0.95$$

The runoff volume to be treated is:

$$WQ_v = 1.0 * R_v * 2/12 = 0.158 \text{ acre-ft} = 6{,}897 \text{ ft}^3$$

Then:

$$V_t = V_w/n = 6{,}897/0.40 = 17{,}243 \text{ ft}^3$$

$$A_f = 12 WQ_v/(kt_f) = (12*6897)/(1.02*40) = 2{,}028 \text{ ft}^2 \text{ minimum surface area}$$

$$D_f = 17{,}243/2{,}028 = 8.5 \text{ ft too deep}$$

Select:

$$D_f = 9.6 - 3 = 6.6 \text{ ft}$$

$$A_f = V_t/H = 17{,}243/6.6 = 2{,}613 \text{ ft}^2$$

Select a width of 10 ft (W) to fit site and minimize groundwater mounding.

$$L = A_f/W = 2{,}613/10 = 261 \text{ ft}$$

Check:

$$LWD_f n = V_w = (261)(10)(6.6)(0.40) = 6{,}890 \text{ ft}^3 \text{ (OK)}$$

Note that this example makes no allowance for a safety factor based on unknowns in the site data and information. To complete the design, a sensitivity analysis should be done by varying the unknown variables over an expected range and assessing the risk of variable choices. Some designers use a safety factor of from 2 to 10. In addition, the pretreatment approach will need to be sized, overflows checked, etc. See ARC (2002) for a detailed design example of an infiltration trench.

Infiltration basins are designed similarly to trenches. They have exhibited high failure rates, often becoming de facto wetlands. Use of infiltration basins should be limited to sites with good to ideal conditions and with strict construction control and maintenance.

Porous Pavement

Porous pavement consists variously of open-graded asphaltic aggregate pavement (gap graded mix or "popcorn" mix), pervious concrete pavement, or concrete or plastic grid paving blocks filled with soil and, normally, vegetated. Porous concrete (also referred to as enhanced porosity concrete, porous concrete, Portland cement pervious pavement, pervious pavement) is a subset of a broader family of pervious pavements. Porous concrete is thought to have a greater ability than porous asphalt to maintain its porosity in hot weather and, thus, may have broader application and will be discussed herein. Although porous concrete has seen growing use since the first edition of this book, there is still very limited practical experience with this measure. According to the U.S. EPA, porous pavement sites have had a high failure rate — approximately 75%. Failure has been attributed to poor design, inadequate construction techniques, soils with low permeability, heavy vehicular traffic, and poor maintenance. Other researchers have found many sites that have been successful, and newer guidance is expected over the next few years.

Porous concrete consists of a specially formulated mixture of Portland cement, uniform, open-graded course aggregate, and water. The concrete layer has a high permeability, often many times that of the underlying permeable soil layer, and allows rapid percolation of rainwater through the surface and into the layers beneath. The void space in porous concrete is in the 15 to 22% range, compared to 3 to 5% for conventional pavements. The permeable surface is placed over a layer of open-graded gravel and crushed stone. The void spaces in the stone act as a storage reservoir for runoff.

Porous concrete is designed primarily for stormwater quality, i.e., the removal of stormwater pollutants. However, it can provide limited runoff quantity control, particularly for smaller storm events. For some smaller sites, trenches can be designed to capture and infiltrate larger volumes in addition to the water-quality capture volume. Porous concrete will need to be used in conjunction with another structural control to provide overbank and extreme flood protection, if required.

Porous concrete has the positive characteristics of volume reduction due to infiltration, groundwater recharge, and an ability to blend into the normal urban landscape relatively unnoticed. It also allows a reduction in the cost of other stormwater infrastructure, a fact that may offset the greater placement cost somewhat. Modifications or additions to the standard design have been used to pass flows and volumes in excess of the water-quality volume, or to increase storage capacity or treatment. These include:

- Placing a perforated pipe near the top of the crushed stone reservoir to pass excess flows after the reservoir is filled
- Providing surface detention storage in a parking lot, adjacent swale, or detention pond with suitable overflow conveyance
- Connecting the stone reservoir layer to a stone-filled trench

- Adding a sand layer and perforated pipe beneath the stone layer for filtration of the water-quality volume
- Placing an underground detention tank or vault system beneath the layers

Porous concrete should not be used in areas that experience high rates of wind erosion, or in drinking water aquifer recharge areas. Pretreatment of sediments, oils, and greases is required to lengthen the service life of the structure. Porous pavement can be effective in flood peak reductions. One study showed peak reductions by as much as 83% through the use of porous pavement (Field, 1985). Freeze and thaw normally does not greatly affect well-designed porous pavement. Many environmental groups tout porous concrete as the answer to impervious problems. It is not that easy. Porous concrete systems are typically used in low-traffic areas such as the following types of applications:

- Parking pads in parking lots
- Overflow parking areas
- Residential street parking lanes
- Recreational trails
- Golf cart and pedestrian paths
- Emergency vehicle and fire access lanes

Slopes should be flat or gentle to facilitate infiltration versus runoff, and the seasonally high water table or bedrock should be a minimum of 2 ft below the bottom of the gravel layer if infiltration is to be relied on to remove the stored volume.

A drawback is the cost and complexity of porous concrete systems compared to conventional pavements. Porous concrete systems require a high level of construction workmanship to ensure that they function as designed. They experience a high failure rate if they are not designed, constructed, and maintained properly.

The function of the porous pavement is such that water infiltrates through the porous upper layer and into a storage reservoir of stone aggregate below. The runoff eventually percolates into the ground or runs out of the stone reservoir through an underdrain collection system. Figures 13-88 through 13-91 show some example applications. Figure 13-92 shows a cross section of a common porous concrete configuration.

FIGURE 13-88
Close-up of porous pavement.

FIGURE 13-89
Porous pavement parking lot.

FIGURE 13-90
Porous pavement street ribbon edge.

Normally, a filter fabric is placed on top of the subbase to keep fine particles from migrating into the aggregate reservoir. Overlap is recommended. To keep polluted runoff from entering the groundwater, sometimes impervious membranes are placed between the reservoir and the subbase. Alternate drain lines are provided to convey the water to an outlet point. Often, stone overflows are provided as a backup to clogging.

Porous pavement has been used for highway and airport paving since about 1947 and 1967, respectively. It has advantages over normal pavement in that it reduces vehicle hydroplaning, retains water, enhances groundwater recharge, and can be cost-effective if curbs and other drainage infrastructure can be avoided (Thelen et al., 1972). It has been effective along roadways as filtration strips (Niemczynowicz, 1990).

While porous pavement can be highly effective in flood control and pollution reduction, it comes at the cost of a high maintenance responsibility. Studies have shown up to a 76% failure rate in Maryland when porous pavement was not protected from off-site sediment,

Structural Best Management Practices

FIGURE 13-91
Porous pavement sidewalk.

not constructed properly or protected during construction, or was resurfaced with nonporous material (Maryland, 1991). Without proper maintenance and construction, the pores in the upper level can become clogged during construction (if unprotected) or within 3 to 5 years. Clogging of the first 1/2 in is sufficient to prevent percolation. European and Japanese experience with such devices has been more positive, indicating the controls may have more potential than early results indicate.

Care during construction must be taken to avoid spills and sediment export to the site. Construction vehicles may not track across the surface. Vehicle washing must not occur on the surface. Premature rolling of the hot mix may also collapse the pores (Diniz, 1980). Routine maintenance measures include sweeping with a vacuum assist sweeper followed by pressure washing several times a year.

Spills must be cleaned immediately to restore function to within 95%. If not cleaned for a longer period, full restoration is impossible (Diniz, 1980). For minor clogging, drilling holes through the surface can be effective. For major system clogging, nothing short of replacement of the surface is required. Alternately, grated openings can be cut in the surface and the water drained directly to the aggregate reservoir and finger drains used, though this is not as effective. Avoid placing in high hydrocarbon-generating areas. Also, do not drain pervious areas to porous pavement.

Pollutant Removal Efficiency

Porous pavement is designed to remove pollutants from atmospheric deposition onto the surface and runoff from areas immediately adjacent to the site. Porous pavement with infiltration removes pollution by the same methods and to the same levels as infiltration trenches

TABLE 13-30

Average Annual Pollutant Removal Capability of Porous Pavement

	Porous Pavement Design Type		
Pollutant	0.5 in of Runoff per Impervious Acre	1.0 inch of Runoff per Impervious Acre	2-Year Design Storm Treatment
TSS	60–80%	80–100%	80–100%
TP	40–60%	40–60%	60–80%
TN	40–60%	40–60%	60–80%
BOD	60–80%	60–80%	80–100%
Bacteria	60–80%	60–80%	80–100%
Metals	40–60%	60–80%	80–100%

Source: From Schueler, 1987; Day, G.E., Smith, D.R., and Bowers, J., Runoff and Pollution Abatement Characteristics of Concrete Grid Pavements, Virginia Water Resources Research Center, Virginia Polytechnic Institute and State University, Rpt. PB82-168469, Bull. 135, October 1981; EPA, 1999; and Atlanta Regional Commission (ARC), Georgia Stormwater Manual, 2002.

and other infiltration devices: adsorption, straining, trapping, and microbial decomposition in the subsoil (Schueler et al., 1992). Studies have shown it can be very effective in removing pollutants if the majority of the pollution is infiltrated to the groundwater (Schueler, 1987; Thelen and Howe, 1978; Urban and Gburek, 1980; Gburek and Urban, 1980; Day, Smith, and Bowers, 1981; Field, Masters, and Singer, 1982; Thelen et al., 1972; Diniz, 1976, 1980).

Table 13-30 presents pollutant removal ranges for three alternate designs of aggregate: half-inch capture from impervious areas, 1-in capture from impervious areas, and 2-year storm exfiltration (erosion control volume too) (Schueler, 1987). See Table 13-13 for general pollutant-removal capabilities.

Pollutant removal can be enhanced by increasing the degree of exfiltration, maximizing the drainage time, and performing the routine maintenance mentioned above.

Porous Pavement Design

Porous pavement is typically composed of four layers (Diniz, 1980; Northern Virginia, 1992; Puget Sound, 1992; ARC, 2002) as shown in Figure 13-92. The aggregate reservoir

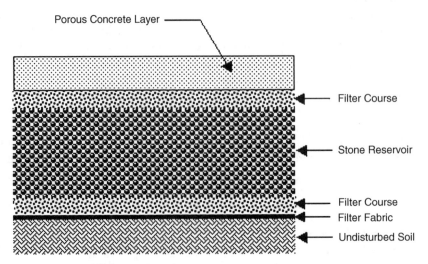

FIGURE 13-92
Cross section for porous pavement.

can sometimes be avoided or minimized if the subgrade is sandy and there is adequate time to infiltrate the necessary runoff volume into the sandy soil without bypassing the water-quality volume. Descriptions of each of the layers is presented below.

Porous Concrete Layer

The porous concrete layer consists of an open-graded concrete mixture usually ranging from depths of 2 to 4 in, depending on required bearing strength and pavement design requirements. Porous concrete can be assumed to contain 18% voids (porosity = 0.18) for design purposes. Thus, for example, a 4-in thick porous concrete layer would hold 0.72 in of rainfall. The omission of the fine aggregate provides the porosity of the porous pavement. To provide a smooth riding surface and to enhance handling and placement, a coarse aggregate of 3/8 in maximum size is normally used. Use coarse aggregate (3/8 to No. 16). See Thelen and Howe (1978) for detailed design criteria of this layer.

Top Filter Layer

This layer consists of 0.5-in diameter crushed stone to a depth of 1 to 2 in. This layer serves to stabilize the porous asphalt layer. It can be combined with the reservoir layer using suitable stone.

Reservoir Layer

The reservoir gravel base course consists of washed, bank-run gravel, 1.5 to 2.5 in in diameter with a void space of about 40%. The void space is calculated from the method of Smith, Rice, and Spelman (1974). Soils should have infiltration capabilities similar to infiltration trenches. For uniform-sized gravel, a voids volume will be in the range of 38 to 46%. For a 40% typical void volume (porosity), a 12-in thick reservoir will hold 4.8 in of water (Thelen et al., 1972).

The thickness of the reservoir layer depends on the greater of storage volume required, frost penetration depth, and bearing load thickness, but typically ranges from 2 to 4 ft. The frost penetration thickness can be found from standard tables available from asphalt manufacturers or associations, and varies from about 1 ft in Tennessee to about 60 in in Maine. The bearing thickness can be determined from Table 13-33 for a known traffic level and California Bearing Ratio (CBR). The layer must have a minimum depth of 9 in. The layer should be designed to drain completely in 48 h and should be designed to store at a minimum the water quality volume (*WQv*).

Bottom Filter Layer

The surface of the subgrade should be a 6-in layer of sand or a 2-in thick layer of 0.5-in crushed stone, and be completely flat to promote infiltration across the entire surface. This layer serves to stabilize the reservoir layer, to protect the underlying soil from compaction, and is the interface between the reservoir layer and the filter fabric covering the underlying soil.

Filter Fabric

It is important to line the entire trench area, including the sides, with filter fabric prior to placement of the aggregate. The filter fabric serves an important function by inhibiting soil from migrating into the reservoir layer and reducing storage capacity. Fabric should be MIRFI # 14 N or equivalent.

Underlying Soil

The underlying soil should have an infiltration capacity of at least 0.5 in/h, but preferably greater than 0.50 in/h, as initially determined from NRCS soil textural classification, and subsequently confirmed by field geotechnical tests. The minimum geotechnical testing is one test hole per 5000 square feet, with a minimum of two borings per facility (taken within the proposed limits of the facility). Infiltration trenches cannot be used in fill soils. Soils at the lower end of this range may not be suited for a full infiltration system. Test borings are recommended to determine the soil classification, seasonal high groundwater table elevation, and impervious substrata, and an initial estimate of permeability. Often, a double-ring infiltrometer test is done at subgrade elevation to determine the impermeable layer, and, for safety, one-half the measured value is allowed for infiltration calculations.

Puget Sound recommends using half the estimated infiltration rate to account for a safety factor for infiltration unknowns (Puget Sound, 1992). Other site limitations are also similar to trenches, including depth to bedrock (2 ft), depth to seasonally high water table (2 to 4 ft), and drainage area (1/4 to 10 acres). An additional consideration, most applicable to the arid West, is to limit applications if wind erosion is expected to blow sediments from adjacent barren areas (Schueler, 1987). If the ground is sloped, the reservoir size must be increased to account for the level water surface within a nonlevel reservoir. Alternately, terraces can be constructed. Drainage times are limited to 24 h in the Puget Sound (Puget Sound, 1992, for the 6-month, 24-h storm) and 72 h in the Washington, DC, area (Schueler, 1987) to avoid anaerobic or septic conditions in the reservoir or freezing of the water.

The minimum surface area of the trench can be determined, in a manner similar to the infiltration trench, from the following equation:

$$A = WQ_v / (n_g d_g + kT/12 + n_p d_p) \qquad (13.24)$$

where A = surface area; WQ_v = water-quality volume (or total volume to be infiltrated); n = porosity (g of the gravel, p of the concrete layer); d = depth or gravel layer (ft) (g of the gravel, p of the concrete layer); k = percolation (in/h); and T = fill time (time for the practice to fill with water) (h).

Additional specifications are found below. More details can be found from Thelen et al. (1972), Schueler (1987), Puget Sound (1993), Maryland (1984), Diniz (1980), ARC (2002), and EPA (1999). A design modeling approach that performs a step-wise routing/flow balance for porous pavement design is presented in Diniz (1976). Table 13-31 gives a typical gradation for the asphaltic aggregate, though variations can be found (Smith, Rice, and Spelman, 1974; Diniz, 1980; Maryland, 1984). Diniz (1980) or Thelen et al. (1978) recommend the asphalt mix gradation given in Table 13-32. Figure 13-93 illustrates placement of porous concrete.

Operation and Maintenance Requirements

A stormwater management easement and maintenance covenant should be required for each facility. The maintenance covenant should require the owner of the porous pavement to periodically clean the structure. Sediment should be kept off of the pavement before, during, and after construction to prevent premature clogging. The surface of the porous pavement should be vacuum swept at least 3 times a year, followed by high pressure jet hosing, to keep the pores free from clogging. Sand or ash should never be applied to porous pavement. Spot clogging of the porous pavement layer can be relieved by drilling $1/4$-in holes through the porous asphalt layer every square foot. The observation well should be monitored several times during the first few months after construction and be monitored quarterly thereafter. Water depth in the well should be measured at 0-, 24-, and

TABLE 13-31

Open-Graded Asphalt Concrete Formulation

Material	Screen	Weight (%)	Volume (%)	Probable Particle Data	
				Width (mm)	Weight (g)
Aggregate	Through 1/2	2.8	2.2	10.7	1.667
	Through 3/8	59.6	46.3	8.0	0.697
	Through #4	17.0	13.3	4.0	0.087
Subtotal coarse aggregate		79.4	61.8		
	Through #8	2.8	2.2	2.0	0.0109
	Through #16	10.4	8.0	1.0	0.00136
	Through 200	1.9	1.5	0.06	0.000294
Asphalt		5.5	10.5		
Air		0	16.0		
Totals		100.0	100.0		

Source: From Maryland, 1984.

TABLE 13-32

Aggregate Gradation Limits for Porous Asphalt Mixes

Sieve Size	Gradation (% by Weight) Puget Sound	Gradation (% by weight) Franklin Institute	Gradation (% by weight) Diniz
1/2"	100	100	100
3/8"	95–100	90–100	90–100
#4	30–50	35–50	35–50
#8	5–15	15–32	15–32
#16	—	0–15	2–15
#200	2–5	0–3	2–15

Source: From Diniz, E.V., Porous Pavement — Phase I, Design and Operational Criteria, USEPA 600/2-80-135, August 1980; Thelen, E. and Howe, L.F., Porous Pavement, Franklin Institute Press, Philadelphia, PA, 1978; and Puget Sound, 1992.

FIGURE 13-93
Porous pavement placement. (From Atlanta Regional Commission, Georgia Stormwater Manual, 2002.)

TABLE 13-33

Minimum Thickness of Porous Paving Top to Subbase

Traffic Group	General Character	California Bearing Ratio			EAL[a]
		15-plus	10–14	6–9	
1	Light traffic	5"	7"	9"	5 or less
2	Med. light traffic (max. 1000 VPD[b])	6"	8"	11"	6–20
3	Medium traffic (max. 3000 VPD)	7"	9"	12"	21–75

[a] EAL = equivalent axle load, equals 18,000 lbs average daily.
[b] VPD = vehicles per day.

Note: CBR values less than 5 were improved using gravel mix to CBR = 6.

Source: From Thelen, E. and Howe, L.F., Porous Pavement, Franklin Institute Press, Philadelphia, PA, 1978.

FIGURE 13-94
Cleaning porous pavement.

48-h intervals after a storm to determine the clearance rate. Table 13-34 summarizes maintenance requirements.

Typical Required Specifications

- Geotechnical investigation should be required prior to design, including a minimum of one soil log (to a depth of 4 ft below the anticipated bottom of the stone reservoir) for each 5000 square feet of infiltration surface area with minimum of three logs per BMP.
- Soil infiltration rate should be greater than 0.27 in/h and clay content less than 30% for partial exfiltration system and greater than 0.52 in/h for full exfiltration system.
- Porous pavement should be designed to exfiltrate a minimum of runoff volume equal to the first 1/2 in of runoff from the contributing impervious area.

TABLE 13-34

Typical Maintenance Activities for Porous Concrete Systems

Activity	Schedule
Initial inspection	Monthly for 3 months after installation
Ensure that the porous paver surface is free of sediment	Monthly
Ensure that the contributing and adjacent area is stabilized and mowed, with clippings removed	As needed, based on inspection
Vacuum-sweep porous concrete surface followed by high-pressure hosing to keep pores free of sediment	Four times a year
Inspect the surface for deterioration or spalling	Annually
Check to make sure that the system dewaters between storms	
Spot clogging can be handled by drilling half-inch holes through the pavement every few feet	Upon failure
Rehabilitation of the porous concrete system, including the top and base course as needed	

- It is only feasible on sites with gentle slopes (less than 5%).
- Design infiltration rate should be equal to 1/2 of the infiltration rate determined from soil textural analysis.
- There should be a minimum of 2 ft of clearance between stone reservoir level and bedrock.
- There should be a minimum of 2 to 4 ft between stone reservoir level and seasonally high water table.
- It should be sited at least 10 ft down-gradient from buildings and 100 ft away from drinking water wells.
- To protect groundwater from potential contamination, runoff from designated hot spot land uses or activities must not be infiltrated. Porous pavement should not be used for manufacturing and industrial sites, where there is a potential for high concentrations of soluble pollutants and heavy metals. In addition, porous pavements should not be considered for areas with a high pesticide concentration.
- Porous pavements are also not suitable in areas with karst geology without adequate geotechnical testing by qualified individuals, and in accordance with local requirements.
- They should not be constructed over fill soils.
- Vegetative strip or diversion berm is required to protect the pavement area from off-site runoff before and during construction.
- If porous pavement areas receive oily runoff from off-site areas, a pretreatment facility should be constructed to remove oil, grit, and sediments before entering the porous pavement.
- Dry subgrade should be covered with engineering filter fabric such as Mirafi #14N or its equivalent on bottom and sides.
- Pavement sections should consist of four layers, as shown in Figure 13-92.
- Stone should be clean, washed stone meeting municipal roadway standards.
- Reservoir base course should consist of 1 to 3 in crushed stone aggregate compacted lightly at the depth required to achieve design storage.
- Filter courses to be $1/2$-in crushed stone aggregate at a 1 to 2 in depth.

- Surface course should be laid in one lift at the design depth with compaction done while the surface is cool enough to resist a 10-ton roller. Only one or two passes are required.
- After final rolling, no vehicular traffic should be permitted on pavement until cooling and hardening, a minimum of 1 day.
- Stone reservoir should be designed to completely drain within a maximum of 2 to 3 days after design storm event.
- Stone reservoir should be designed for a minimum residence time of 12 h.
- The porous pavement site should be posted with signs indicating the nature of the surface and warning against resurfacing, using abrasives, and parking heavy equipment.
- An observation well should be installed on the downslope end of the porous pavement area to monitor runoff clearance rates.
- The observation well should consist of perforated PVC pipe, 4- to 6-in diameter, constructed flush with the ground, with the top of the well capped.

Typical Recommended Specifications

- Use is limited in application to parking lots, service roads, emergency and utility access lanes, and other low-traffic areas.
- Use is limited to sites between 1/4 and 10 acres.
- They should not be constructed near groundwater drinking supplies.
- Heavy equipment should be prevented from compacting the underlying soils before and during construction.
- Other porous pavement types such as concrete lattice blocks, perforated concrete grid slabs, reinforced plastic pavers, etc., placed over a porous subgrade and filled with soil may be used in lieu of the asphalt surface course.
- Signs should be posted identifying the area and specifying maintenance requirements.

Design Example

Data

A 1.5-acre overflow parking area is to be designed to provide water-quality treatment using porous concrete for at least part of the site to handle the runoff from the whole overflow parking area. Initial data shows:

- Borings show depth to water table is 5.0 ft
- Boring and infiltrometer tests show sand-loam with percolation rate (k) of 1.02 in/h
- Structural design indicates the thickness of the porous concrete must be at least 3 in

Water-Quality Volume

$$R_v = 0.05 + 0.009 \, I \text{ (where } I = 100 \text{ percent)}$$

$$= 0.95$$

$$WQ_v = 1.2 \, R_v \, A \, / \, 12 = 1.2 * 0.95 * 1.5/12 \text{ converted to cubic feet from acre-feet}$$

$$= 6{,}207 \text{ cubic feet}$$

Surface Area

A porosity value $n = 0.32$ will be used for the gravel and 0.18 for the concrete layer. All infiltration systems should be designed to fully dewater the entire WQ_v within 24 to 48 h after the rainfall event at the design percolation rate. A fill time of $T = 2$ h can be used for most designs. Choose a depth of gravel pit of 3 ft (including layer under concrete), which fits the site with a 2 ft minimum to water table (other lesser depths could be chosen, making the surface area larger). The minimum surface area of the trench can be determined, in a manner similar to the infiltration trench, from the following equation:

$$A = WQ_v / (n_g d_g + kT/12 + n_p d_p)$$

$$= 6{,}207 / (0.32*3 + 1.02*2/12 + 0.18 * 3/12)$$

$$= 5{,}283 \text{ square feet}$$

Check of drain time:

depth = 3*12 + 3 inches to sand layer = 39 inches @ 1.02 in/hr = 38 hours (ok)

Overflow will be carried across the porous concrete and tied into the drainage system for the rest of the site.

Modular Paving Blocks

Modular blocks are a pavement surface composed of structural units with void areas that are filled with pervious materials such as sand or grass turf. They are installed over a gravel base course that provides storage as runoff infiltrates through the porous paver system into underlying permeable soils.

Modular block paving originated from the use of slotted, flexible steel railways in postwar Europe as a means of temporarily repairing damaged airfields. Later, they were used for temporary parking lots. Eventually, concrete, then plastic, then advanced materials modular blocks evolved (Day, Smith, and Bowers, 1981).

The various porous paver materials available commercially can be separated into the following categories (WDE, 2001):

- *Flexible plastic cellular confinement systems* — Examples include Geoweb, Grasspave 2 and Gravelpave 2, and GRASSYTM PAVERS
- *Molded plastic materials* — Examples include Geoblock and Checkerblock
- *Interlocking concrete blocks* — Examples include UNI Eco-stone and Turfstone
- *Cast-in-place concrete blocks* — These are made with reusable forms to create voids needed for planting grass; they can be reinforced with welded wire mesh to prevent differential settlement (Grasscrete is an example of this technology.)

Paving blocks come in three types: castellated, poured in place, and lattice. Castellated (such as Monoslabs and Checker Blocks) are roughly cast blocks with "castles" cast into the surface for roughness. Lattice blocks (such as Turfblocks and Grasstone) are normally

FIGURE 13-95
Lattice-type paver system.

precast flat concrete slabs with different opening patterns in them. Poured-in-place types (such as Grasscrete) are poured over plastic forms with reinforcing (Field, Masters, and Singer, 1982). Plastic grids or cells are placed over a porous subgrade and then filled with soil. They have some reported advantages over the concrete modular blocks, including an 82% surface permeability (Presto Products, 1990).

Modular pavement is applicable where pavement is desirable or required for low-volume traffic areas. This practice is most applicable for new construction, but it can be used in existing developments to expand a parking area or even to replace existing pavement if that is a cost-effective measure, or for aesthetic reasons. Possible areas for use of these paving materials include the following (WDE, 2001):

- Parking aprons, taxiways, blast pads, and runway shoulders at airports (heavier loads may demand the use of reinforced grid systems)
- Emergency stopping and parking lanes and vehicle crossovers on divided highways
- On-street parking aprons in residential neighborhoods
- Recreational vehicle camping-area parking pads
- Private roads, easement service roads, and fire lanes
- Industrial storage yards and loading zones (heavier loads may demand the use of reinforced grid systems)
- Driveways for residential and light commercial use
- Bike paths, walkways, patios, and swimming pool aprons

Paving systems have become quite complex in the last 5 years, finding application in larger and more prominent developments where reduction of impervious area is important and a credit against a stormwater user fee may apply to help defray the cost of a system. They have a growing place of importance in communities looking to institute impervious reduction measures consistent with smart growth and conservation-type development. Figures 13-97 through 13-99 show three recent applications. Figure 13-100 gives typical approaches to modular block use (ARC, 2002).

Structural Best Management Practices

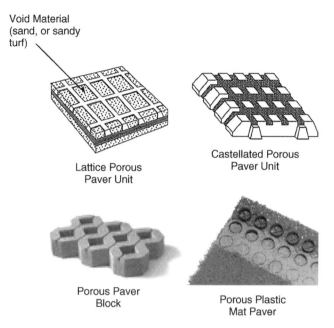

FIGURE 13-96
Various types of modular pavers. (From Atlanta Regional Commission, Georgia Stormwater Manual, 2002.)

FIGURE 13-97
Overflow parking use of a grid system.

FIGURE 13-98
Stadium parking with grid system.

FIGURE 13-99
Pedestal paving for sidewalks.

Paving blocks have the advantage in that they tend to seal less easily than the porous paving. Also, replacing individual blocks is easier than patching porous pavement. However, they tend to be more expensive and less able to resist loadings without misalignment. Consequently, they are often used in overflow parking areas, driveways, sidewalks, trails, or shoulder areas. Horizontal infiltration cutoff walls keep flow from moving just under the block layer downslope and exiting without infiltrating (Urbonas and Stahre, 1993). Pollution-removal rates are generally comparable to those of porous pavement, being dependent on the infiltrated amount and the type and depth of soil materials (Day, Smith, and Bowers, 1981).

A major drawback is the cost and complexity of modular porous paver systems compared to conventional pavements. Porous paver systems require a very high level of construction workmanship to ensure that they function as designed. In addition, there is the difficulty and cost of rehabilitating the surfaces should they become clogged. There-

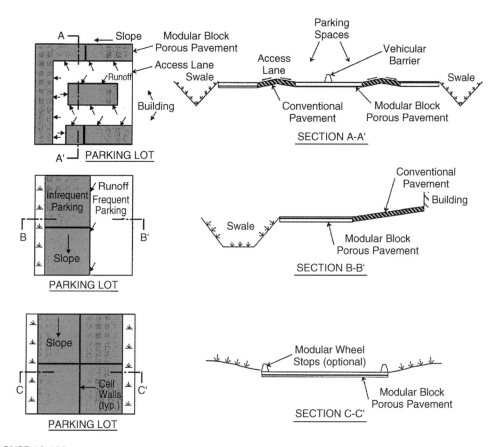

FIGURE 13-100
Traditional uses of modular blocks.

TABLE 13-35

Typical Maintenance Activities for Modular Porous Paver Systems

Activity	Schedule
Ensure that the porous paver surface is free of sediment	Monthly
Check to make sure that the system dewaters between storms	
Ensure that contributing area and porous paver surface are clear of debris	As needed, based on inspection
Ensure that the contributing and adjacent area is stabilized and mowed, with clippings removed	
Vacuum-sweep porous paver surface to keep free of sediment	Typically three to four times a year
Inspect the surface for deterioration or spalling	Annually
Totally rehabilitate the porous paver system, including the top and base course, as needed.	Upon failure

fore, consideration of porous paver systems should include the construction and maintenance requirements and costs.

Table 13-35 provides a summary of maintenance operations for modular block systems.

Typical Design Specifications for Modular Blocks

Modular porous pavers are typically placed on a gravel (stone aggregate) base course. Runoff infiltrates through the porous paver surface into the gravel base course, which acts as a storage reservoir as it exfiltrates to the underlying soil. The infiltration rate of the soils in the subgrade must be adequate to support drawdown of the entire runoff capture volume typically within 24 to 48 h. Special care must be taken during construction to avoid undue compaction of the underlying soils, which could affect the soils' infiltration capability (ARC, 2001).

The design volumes, site limitations, and soil layers are the same as those for conventional porous pavement, with the exception that a Terzaggi filtering layer is placed under the modular block to serve as a bedding and filter layer. Other design criteria are given below.

All installations of modular pavement should be designed and constructed according to the manufacturer's specifications. Facilities using vegetative cover in combination with pavers must be capable of disposing of stored waters within time limits necessary to avoid damage to the ground cover (24 to 36 h for most grasses). Parking areas should avoid extensive ponding for periods exceeding more than an hour or two:

- Modular blocks should typically be used in applications where the pavement receives tributary runoff only from impervious areas. If runoff is coming from adjacent pervious areas, it is important that those areas be fully stabilized to reduce sediment loads and prevent clogging of the block surface.
- Limit impervious tributary area to no more than two to three times the modular block area.
- Modular blocks are not recommended on sites with a slope greater than 2%.
- A minimum of 2 ft of clearance is required between the bottom of the gravel base course and underlying bedrock or the seasonally high groundwater table.
- Modular blocks should be sited at least 10 ft downgrade from buildings and 100 ft away from drinking water wells.
- Size downstream structures as if the surface were 30% impermeable if infiltration is possible, and 65% impermeable if underdrains are necessary due to the poor infiltration qualities of the subbase (UDFCD, 1992).
- Use perforated pipe underdrain in each cell for poorly draining soils.
- Fill modular block with coarse sandy soil or sandy sod with a high infiltration rate (8 in/h recommended for sand-only filling).
- A 1-in top course (filter layer) of sand (ASTM C-33 concrete sand) underlain by filter fabric is placed under the porous pavers and above the gravel base course.
- The filter course layer should be a minimum of 9 in of clean, well-graded stone 1/8 to 3/4 in in diameter. When in doubt of the storage volume, use Equation (13-25) to determine minimum depth.

$$d = V / A * n \qquad (13.25)$$

where d = gravel layer depth (ft); V = water quality volume or total volume to be infiltrated; A = surface area (square feet); and n = porosity (use n= 0.32).

- Install a filter fabric on top of the subbase.

Grassed Swales

Grassed swales (also referred to as enhanced swales, vegetated open channels, or water-quality swales) are one of a number of landscape- and grading-related BMPs that primarily use biofiltration and limited infiltration to remove pollutants. The second major type is a

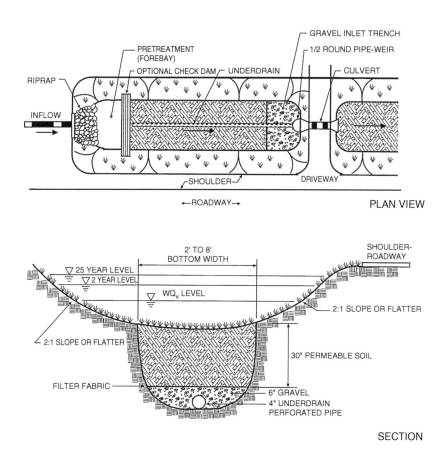

FIGURE 13-101
Example of an engineered dry swale.

filter strip, discussed subsequently. A number of other related practices can be used, including terracing, clustering, groundcover, runoff diversion and redirection, disconnecting of impervious areas, etc. (Bellevue, 1991). In some flat areas, swales are designed essentially as retention basins with sluggish flow (Yousef, Wanielista, and Harper, 1985; Wanielista et al., 1981).

A natural grass channel would differ from the grassed swales discussed in this section in that the natural grass channel would not have an engineered filter media or other design features to enhance pollutant-removal capabilities and, therefore, would have a lower pollutant-removal rate than for a dry or wet swale. Natural grass channels can partially infiltrate runoff from small storm events in areas with pervious soils. When properly incorporated into an overall site design, a natural grass channel can reduce impervious cover, accent the natural landscape, and provide aesthetic benefits.

A grassed swale looks similar to a ditch but is designed to be wider and to maintain flow below the height of the grass vegetation up to a certain design flow. It must be designed and constructed carefully to preserve even distribution of flow across the channel, avoid flow concentration and erosion, and avoid high velocities that will limit water-vegetation contact time and, perhaps, erode the channel or push the vegetation over. There have been problems reported with swales in several states, including standing water and erosion problems.

There are two primary swale designs that have some effectiveness in pollutant removal: the dry swale and the wet swale (or wetland channel). Below are descriptions of these two designs, which are shown in Figures 13-103 and 13-104.

FIGURE 13-102
Example of an engineered wet swale.

- *Dry swale* — The dry swale is a vegetated conveyance channel designed to include a filter bed of prepared soil that overlays an underdrain system. Dry swales are sized to allow the entire water-quality volume to be filtered or infiltrated through the bottom of the swale. Because they are dry most of the time, they are often the preferred option in residential settings.
- *Wet swale (wetland channel)* — The wet swale is a vegetated channel designed to retain water or marshy conditions that support wetland vegetation. A high water table or poorly drained soils are necessary to retain water. The wet swale essentially acts as a linear shallow wetland treatment system, where the water quality volume is retained.

Swales can be used in a variety of developments types: however, they are primarily applicable to residential and institutional areas of low to moderate density, where the impervious cover in the contributing drainage area is relatively small, and along roads and highways. Dry swales are mainly used in moderate- to large-lot residential developments, small impervious areas (parking lots and rooftops), and along rural highways. Wet

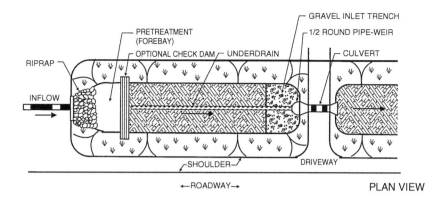
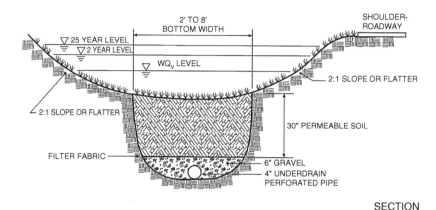

FIGURE 13-103
Schematic of a dry swale. (From Center for Watershed Protection.)

swales tend to be used for highway runoff applications, small parking areas, and in commercial developments as part of a landscaped area.

Because of their relatively large land requirement, swales are generally not used in higher density areas. In addition, wet swales may not be desirable for some residential applications, due to the presence of standing and stagnant water that may create nuisance odor or mosquito problems.

The topography and soils of a site will determine the applicability of the use of one of the two swale designs. Overall, the topography should allow for the design of a swale with sufficient slope and cross-sectional area to maintain nonerosive velocities.

Pollutant-Removal Efficiency

Pollutant-removal efficiency for grassed swales varies widely (Schueler et al., 1992). Some researchers show moderate to high removal rates, while others show little pollutant removal. The reasons for this great variation reflect major differences in design conditions.

Swales appear to be quite effective in the removal of metals and suspended solids, with less long-term effectiveness for nutrients. Runoff from highways indicates a high concentration of metals within the first 50 to 100 ft from the inlet to the swale and within the top 5 cm of soil (Wigington, Randall, and Grizzard, 1986; Yousef, Wanielista, and Harper, 1985). A 200-ft swale in Washington State removed suspended solids and lead by 80% (Wang et al., 1982). Reeves (2000) in a test of a 200-ft channel (which was then reduced to 100 ft) found pollutant removals in the range illustrated in Table 13-36.

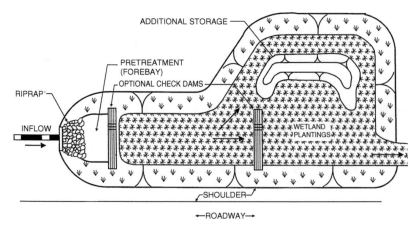

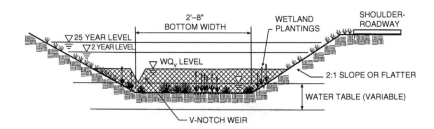

FIGURE 13-104
Schematic of a wet swale. (From Center for Watershed Protection.)

TABLE 13-36

Pollutant-Removal Efficiency

Pollutant	100-ft Biofilter (%)	200-ft Biofilter (%)
Suspended sediment	60	83
Turbidity	60	65
TPH (hydrocarbons)	49	75
Total zinc	16	63
Dissolved zinc	negative	30
Total lead	15	67
Total aluminum	16	63
Total copper	2	46
Total phosphorus	45	29
Bioavailable P	72	40
Nitrate-N	negative	negative
Bacteria	negative	negative

Source: From Reeves, E., Performance and condition of biofilters in the Pacific Northwest, in *Practice of Watershed Protection*, Article 112, Center for Watershed Protection, Ellicott City, MD, 2000.

While it is difficult to distinguish between different designs based on the small amount of available data, grassed channels generally have poorer removal rates than wet and dry swales, although wet swales appear to export soluble phosphorous (Harper, 1988; Koon, 1995; Reeves, 2000). It is not clear why swales export bacteria. One explanation is that bacteria thrive in the warm swale soils. Another is that studies have not accounted for some sources of bacteria, such as local residents walking dogs within the grassed swale (CWP, 2001; EPA, 1999).

The improvement of pollutant removal with check dams or swale blocks included in the design is not known, but some improvement can be surmised, because they tend to hold the water longer in a particular area, bleeding down through a small hole on the lower side of the check dam (Yousef, Wanielista, and Harper, 1985). This allows the sedimentation and infiltration mechanisms to work longer and more effectively. Other factors to improve removal include appropriate length, small slope or a series of terraces, more pervious soil, and any other practice that lengthens the contact time with the vegetation.

The effectiveness of swales can also be enhanced by integrating them into the general design of the area through limiting the amount of directly connected impervious area and through educating the public on low fertilizer maintenance procedures and to catch grass clippings. Table 13-13 gives overall pollutant removal capabilities of swales along with other BMPs.

Both the dry swale and wet swales described above are considered effective in TSS removal. If designed and constructed in accordance with recommended specifications and well maintained, they are presumed to be able to remove 80% of the total suspended solids load in typical urban postdevelopment runoff.

The pollutant removal capability of plain grass channels (as opposed to a dry or wet swale) is not well documented. Based on available data, the TSS pollutant removal for grass channels is in the range of 65%, with a similar range for hydrocarbons, a 20 to 50% range for metals, 25% for TP, and 15% for TN (CWP, 2000).

Dry Swale Design

Low velocity and shallow depth are the key design criteria. Thus, achievement of a low slope and insurance that the flow spreads evenly across the grassed channel are of great importance.

Dry swale channels are sized to store and infiltrate the entire water-quality volume (WQv) with less than 18 in of ponding and allow for full filtering through the permeable soil layer. The maximum ponding time is 48 h, though a 24-h ponding time is more desirable.

The bed of the dry swale on slopes less than about 1.5% consists of a permeable soil layer of at least 30 in in depth, above a 4- to 6-in diameter perforated PVC pipe longitudinal underdrain in a 6-in gravel layer. The underdrain can often be neglected for steeper sloped systems (WDE, 2001). WDE (2001) also recommends a low-flow notch channel flowing out of a concrete stilling basin at the head of the swale for flatter slopes. The soil media should have an infiltration rate of at least 1 ft/day (1.5 ft/day maximum) and contain a high level of organic material to facilitate pollutant removal. A permeable filter fabric is placed between the gravel layer and the overlying soil, or the pipe and gravel are wrapped in fabric.

The channel and underdrain excavation should be limited to the width and depth specified in the design. The bottom of the excavated trench shall not be loaded in a way that causes soil compaction, and scarified prior to placement of gravel and permeable soil. The sides of the channel shall be trimmed of all large roots. The sidewalls will be uniform with no voids and scarified prior to backfilling. The underdrain system should discharge to the storm drainage infrastructure or a stable outfall. See ARC (2001) for a detailed design example for a dry swale.

TABLE 13-37

Criteria for Optimum Swale Performance

Parameter	Optimal Criteria	Minimum Criteria[a]
Hydraulic residence time	9 min	5 min
Average flow velocity	0.9 ft/s	0.9 ft/s
Swale width	8 ft	2 ft
Swale length	200 ft	100 ft
Swale slope	2–6%	1%
Side slope ratio (H:V)	4:1	2:1

[a] Criteria at or below minimum values can be used when compensatory adjustments are made to the standard design. Specific guidance on implementing these adjustments will be discussed in the design section.

Source: From Horner, R.R., Biofiltration for Storm Runoff Water Quality Control, prepared for the Washington State Department of Ecology, Center for Urban Water Resources Management, University of Washington, Seattle, WA, 1993.

Special care must be taken to establish a healthy stand of vegetation tolerant to frequent inundation. See discussions under the wetlands BMP for more information. Irrigation may be a problem in the drier climates. Because the system must work only during the wet season, it may be acceptable to begin to irrigate in time to establish a grass stand just prior to the onset of the wet season (California Stormwater Task Force, 1993). Local information is available from agricultural extension services on types of vegetation. Information on Eastern vegetation can be found in Schueler (1987). Puget Sound recommends grass heights of 6 in and flow depth for design discharges of 5 in (1992). Hydraulic design of grassed swales can be found in Chapter 10, or any of a number of state sediment-control manuals.

Flows larger than the swale design flow need to be checked and accounted for. It may be possible to design a compound channel with the bottom designed as a swale and cutout sides designed to handle the higher flows. It may be extremely difficult to keep the flow from concentrating, channelizing, and meandering. Bottom hardpoints or grade control, flow spreaders, and effective maintenance of the channel and its vegetation are required. Puget Sound recommends grade control every 50 ft and a 200 ft minimal length channel. Designing meandering channels may be necessary to achieve the 200-ft criteria.

Actual, pollutant-removal-based design criteria for the design of a swale is highly dependent on empirical evidence and is more applicable regionally than for all parts of the country with differing hydrology, soils, and vegetation types. Horner (1988) has applied the method of vegetative channel design presented in Chapter 10. This method has been used on the West Coast (Puget Sound, 1992; California Stormwater Task Force, 1993). WDE (2001) requires a 22-min residence time. Others design toward 5 to 10 min. The design criteria for the Puget Sound area is the 6-month, 24-h storm, a length of 200 ft, and a maximum velocity of 1.5 ft/s. See Chapter 7 for calculation of a peak discharge rate for the water quality volume.

Typical steps in the design of a dry or wet swale are given in Horner (1993) as follows:

1. Determine the flow rate to the system
2. Determine the slope of the system
3. Select a swale shape (skip if filter strip design)
4. Determine required channel width
5. Calculate the cross-sectional area of flow for the channel
6. Calculate the velocity of channel flow

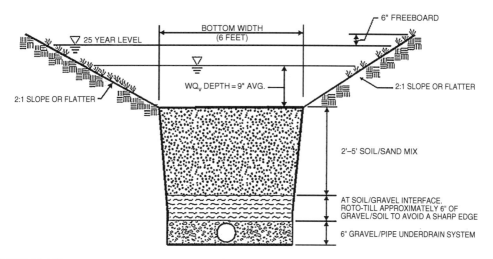

FIGURE 13-105
Typical dry swale trapezoidal cross section. (From ARC, 2001.)

7. Calculate swale length
8. Select swale location based on the design parameters
9. Select a vegetation cover for the swale
10. Check for swale stability

Design the channel using the vegetative method given in Chapter 10. The stability check in that method should be for the largest design flow required to be carried within the banks. Ensure the velocity for the pollution removal design storm is less than the maximum criteria (1.5 ft/s for Puget Sound). Reeves (2000), in studies of dry swales, found Manning's n values to range just under 0.20 and recommends this value for all design. WDE (2001) recommends 0.24 for rarely maintained swales. Other alternate design methods are given by Maryland (1984, 2000), WDE (2001), ARC (2001), and Wanielista and Yousef (1993).

Inlets to enhanced swales must be provided with energy dissipators such as riprap. Pretreatment of runoff in dry and wet swale systems is typically provided by a sediment forebay located at the inlet. The pretreatment volume should be equal to 0.1 in per impervious acre. This storage is usually obtained by providing check dams at pipe inlets and driveway crossings. Enhanced swale systems that receive direct concentrated runoff may have a 6-in drop to a pea gravel diaphragm flow spreader at the upstream end of the control. A pea gravel diaphragm and gentle side slopes should be provided along the top of channels to provide pretreatment for lateral sheet flows. Enhanced swales must be adequately designed to safely pass flows that exceed the design storm flows. Adequate access should be provided for all dry and wet swale systems for inspection and maintenance.

The following topsoil mix should be placed at least 8 in deep:

- Sandy loam 60–90%
- Clay 0–10%
- Composted organic matter, 10–30% (excluding animal waste, toxics)

Use compost amended soil where practicable. Till to at least an 8-in depth. For longitudinal slopes of <2%, use more sand to obtain more infiltration. If groundwater contamination is a concern, seal the bed with clay or a geomembrane liner (WDE, 2001).

Wet Swale

Wet swale channels are sized to retain the entire water-quality volume (WQv) with less than 18 in of ponding at the maximum depth point.

Check dams can be used to achieve multiple wetland cells. V-notch weirs in the check dams can be utilized to direct low-flow volumes. Outlet protection must be used at any discharge point from a wet swale to prevent scour and downstream erosion. Where wet swales do not intercept the groundwater table, a water balance calculation should be performed to ensure an adequate water budget to support the specified wetland species.

If longitudinal slopes are greater than 2%, the wet swale must be stepped so that the slope within the stepped sections averages 2%. Steps may be made of retaining walls, log check dams, or short riprap sections. No underdrain or low-flow drain is required.

A high-flow bypass is required for flows greater than the water-quality design flow to protect wetland vegetation from damage. Unlike grass, wetland vegetation will not quickly regain an upright attitude after being laid down by high flows. New growth, usually from the base of the plant, often taking several weeks, is required to regain its upright form. The bypass may be an open channel parallel to the wet biofiltration swale. Mowing of wetland vegetation is not required. However, harvesting of dense vegetation may be desirable in the fall after plant die-back to prevent the sloughing of excess organic material into receiving waters.

Grass Channels

Grass channels, also termed "biofilters," are typically designed to provide nominal treatment of runoff as well as meet runoff velocity targets for the water-quality design storm. Grass channels are well suited to a number of applications and land uses, including treating runoff from roads and highways and pervious surfaces.

Grass channels differ from the enhanced dry swale design in that they do not have an engineered filter media to enhance pollutant-removal capabilities and, therefore, have a lower pollutant-removal rate than for a dry or wet (enhanced) swale. Grass channels can partially infiltrate runoff from small storm events in areas with pervious soils. When properly incorporated into an overall site design, grass channels can reduce impervious cover, accent the natural landscape, and provide aesthetic benefits.

When designing a grass channel, the two primary considerations are channel capacity and minimization of erosion. Runoff velocity should not exceed 1.0 ft/sec during the peak discharge associated with the water-quality design rainfall event, and the total length of a grass channel should provide at least 5 min of residence time. To enhance water-quality treatment, grass channels must have broader bottoms, lower slopes, and denser vegetation than most drainage channels. Additional treatment can be provided by placing check-dams across the channel below pipe inflows, and at various other points along the channel. Figure 13-106 shows a typical outlet structure at the end of a swale.

Typical Required Specifications: Dry and Wet Swales

- Maximum drainage area is about 5 acres.
- Grassed swales should only convey standing or flowing water following a storm.
- As a BMP, grass swales should be designed for the 6-month, 24-h design storm, water-quality peak flow, or according to local criteria.

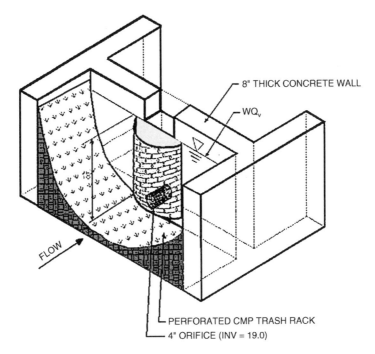

FIGURE 13-106
Typical outlet control structure. (From ARC, 2001.)

- Provide a bypass for high flows if the swale cannot be stable for the 10-year storm. The swale should be able to handle the 2-year storm. Typically, the 10-year storm is passed through the swale with a minimum freeboard.
- Size pretreatment forebays for 0.1 in per impervious acre of drainage.
- Limit to runoff velocities less than 3 ft/sec (less than 1.5 to 2.5 ft/sec optimal).
- Maximum design flow depth to be 1 ft.
- Use a Manning's n value = 0.20 for mowed channels and 0.24 for rarely mowed.
- Swale slopes should not exceed 2.5% without 6- to 12-in drop structures
- Depth at the downstream end of the swale at a drop structure should not exceed 18 in.
- Bottom swale width should be 2 to 8 ft. Dry swales wider than 10 ft should have a divider berm at the inlet.
- Slopes less than 1.5% require an underdrain system, greater than 2.5% require drop structures.
- Install level spreaders (minimum 1-in gravel) at the head and every 50 ft in swales of greater than 4 ft width. Include sediment cleanouts (weir, settling basin, or equivalent) at the head of the swale as needed (WDE, 2001).
- Swale cross section should have side slopes of 3:1 ($h:v$) or flatter.
- Underlying soils should have a high permeability ($fc > 0.5$ in/h).
- Swale area should be tilled before grass cover is established.
- Dense cover of a water-tolerant, erosion-resistant grass should be established over the swale area.
- Wet swale has water table at, or near, the surface.

Typical Recommended Specifications: Dry and Wet Swales

- As a BMP, grassed swales are to be limited to residential or institutional areas where the percentage of impervious area is relatively small.
- A seasonally high water table is to be greater than 2 ft below the bottom of the swale.
- Check dams can be installed in swales to promote additional infiltration. The recommended method is to sink a railroad tie halfway into the swale. Riprap stone should be placed on the downstream side to prevent erosion.
- Maximum ponding time behind the check dam is to be less than 24 h.
- Swale siting should take into account the location and use of the other site features, such as buffers and undisturbed natural areas, and should attempt to aesthetically fit the facility into the landscape.
- A wet swale can be used where the water table is at or near the soil surface, or where there is a sufficient water balance in poorly drained soils to support a wetland plant community.

Typical Operation and Maintenance Procedures

A stormwater management easement and maintenance covenant should be required for each facility. The maintenance covenant should require the owner of the grassed swale to periodically clean the structure. Grass swales should be maintained to keep grass cover dense and vigorous. Maintenance should include periodic mowing, occasional spot reseeding, and weed control. Swale grasses should never be mowed close to the ground. Fertilization of grass swales should be done when needed to maintain the health of the grass, with care not to overapply the fertilizer. Table 13-38 gives a maintenance summary in tabular format.

Filter Strips And Flow Spreaders

Filter strips are uniformly graded and densely vegetated sections of land, engineered and designed to treat runoff and remove pollutants through vegetative filtering and

TABLE 13-38
Typical Maintenance Activities for Enhanced Swales

Activity	Schedule
For dry swales, mow grass to maintain a height of 4 to 6 in; remove grass clippings	As needed (frequent/seasonally)
Inspect grass along side slopes for erosion and formation of rills or gullies, and correct	Annually (semiannually the first year)
Remove trash and debris accumulated in the inflow forebay	
Inspect and correct erosion problems in the sand/soil bed of dry swales	
Based on inspection, plant an alternative grass species if the original grass cover has not been successfully established	
Replant wetland species (for wet swale) if not sufficiently established	
Inspect pea gravel diaphragm for clogging, and correct the problem	
Rototill or cultivate the surface of the sand/soil bed of dry swales if the swale does not draw down within 48 h	As needed
Remove sediment buildup within the bottom of the swale once it has accumulated to 25% of the original design volume	

Source: From Watershed Management Institute, Operation Maintenance and Management of Stormwater Management Systems, prepared for U.S. EPA, Office of Water, 1997; and Pitt, R., Lilbum, M., Nix, S., Durrans, S., and Burian, S., Guidance Manual for Integrated Wet Weather Flow Collection and Treatment Systems for Newly Urbanized Areas, U.S. EPA Office of Research and Development, 1997.

infiltration. Filter strips are best suited to treating runoff from roads and highways, roof downspouts, very small parking lots, and pervious surfaces. They are also ideal components of the outer zone of a stream buffer, or as pretreatment for another structural stormwater control. Filter strips can serve as a buffer between incompatible land uses, be landscaped to be aesthetically pleasing, and provide groundwater recharge in areas with pervious soils. Filter strips may adopt any vegetated form from grassy meadows to forests. The density of the vegetation, the flatness of the surface, the permeability of the soil, and the evenness of the flow spread all help determine the pollution-removal effectiveness (Schueler et al., 1992; Northern Virginia, 1992). Buffers, when engineered, can function as filter strips.

There is significant concentration among stormwater designers on stream corridors and riparian buffer areas. These areas have significant importance in the urban setting and deserve a detailed treatment. This section is restricted to an engineered filter strip for the purposes of pollution removal. The reader is directed to comprehensive treatments of stream corridors, including Riley (1997), Rosgen (1994), FISRWG (1998), and USDA (1998).

Filter strips are commonly used along stream banks (riparian corridor buffers), downstream from agricultural runoff, around area inlets, as pretreatment for other BMPs, and as areas for sheet flow from paved areas to run through. They are often associated with the disconnection of directly connected impervious areas. Landscaping can often take this type of BMP into account almost invisibly through the appropriate use of multifunction flow spreaders and proper structure siting, drainage, and grading. Filter strips and buffers can have multiple uses. However, not all uses are compatible with pollution removal. By their invisibility, they can fall victim to unintended use by property owners, contractors, and even local governments (Heraty, 1993; Cooke, 1991). A typical filter strip is depicted in Figure 13-107.

Filter strips rely on the use of vegetation to slow runoff velocities and filter out sediment and other pollutants from urban stormwater. There can also be a significant reduction in runoff volume for smaller flows that infiltrate pervious soils while contained within the filter strip.

Filter strips must accept stormwater runoff as overland sheet flow in order to effectively filter suspended materials out of the overland flow. In order to function properly, flow entering a filter strip must be spread relatively uniformly over the width of the strip. Filter

FIGURE 13-107
Filter strip.

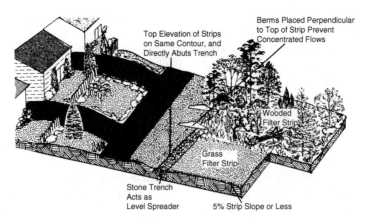

FIGURE 13-108
Schematic of a filter strip. (From Schueler, 1987.)

strip applications should be limited to drainage areas of 5 acres. Even filter strips of 10 to 20 ft in width can have a great effect in removing the coarser particles. In arid climates, they may need to be irrigated to maintain a healthy appearance.

There are two different filter strip designs: a simple filter strip and a design that includes a permeable berm at the bottom. The presence of the berm increases the contact time with the runoff, thus reducing the overall width of the filter strip required to treat stormwater runoff. Filter strips are typically an online practice, so they must be designed to withstand the full range of storm events without eroding.

Aesthetics are important in the design of buffer strips. However, there is some disagreement on the possibility for multiobjective use of buffer areas. Certainly, use of the areas as habitat is seen as permissible, and some authors argue that an undisturbed forest area makes the best buffer due to a greater ability to retain nutrients (Maine, 1992; Schueler, 1987). Filter strips tend to attract urban dwelling edge species such as songbirds and squirrels (Northern Virginia, 1992). But some authors recommend against use as pedestrian walkways or recreational areas. It appears that site-specific factors will dictate whether other uses are permissible. Human interaction with such areas tends to ensure the site is better maintained in terms of trash and erosion control, though there may be some compaction of the soil and wearing away of vegetation if not well designed for human traffic control. Greenways have been effectively used as buffer areas along streams. It may be advisable to plant different vegetation in specific zones depending on inundation, wetness, and other factors (New Jersey, 1989; Karouna, 1992). The use of experimental vegetation such as vetiver grass may improve erosion control and pollutant removal (BSTID, 1993). Shade trees and the allowance for some debris along stream edges will enhance biodiversity.

Pollutant-Removal Efficiency

Pollution-removal mechanisms are the same as those for grassy swales. Filter strips can reliably remove sediment and other settleable solids but are less reliable for soluble nutrients and metals. Pollutant removal is dependent on length, slope, soil permeability, flow volume, and velocity. Two factors work in opposition to each other when compared to swales. The filter strip is normally not as long as a grassy swale, limiting the time of contact between the vegetation and the runoff. But, if designed and maintained correctly, the flow across the buffer strip is in the form of sheet flow moving very slowly across the strip. Table 13-13 gives general estimates of pollution-removal efficiencies for filter strips and buffers.

Various studies conducted primarily for agricultural sites have shown high removal rates for sediments, pesticides, bacteria, and nutrients (Readling, 1992; NCDEHNR, 1991). However, variability is great from site to site, and few of the sites experienced urban stormwater runoff. The values given in Table 13-13 are therefore conservative, expressing the thought that many strips are not well constructed or maintained. Somewhat higher removal rates can be expected for more ideal conditions.

Because the removal rate is directly related to the contact time, it is imperative that the flow be kept from concentrating and short circuiting through the filter strip, typically through the use of a flow spreader (Dillaha et al., 1989). If this happens, most of the width of the strip becomes ineffective, simply lengthening the small feeder stream in its course to the receiving waters. Most strips that have failed to remove target pollutants have failed due to poor design, high slopes, narrow widths, erosion, and short circuiting (Schueler et al., 1992).

There may be site-specific factors that greatly affect the ability of the strip to remove nitrogen and phosphorus. Gilliam and Skaggs (1987) found denitrification to greatly reduce nitrate-nitrogen all within the first few yards of the riparian wetland buffer. High water table may influence the ability for vegetation to uptake soluble nutrients, as does the time of year and phase of the vegetation growth (Schueler, 1987; CWP, 2001).

Often the first few yards of the strip accumulates sediment in the shape of fans or ridges if high loads exist (Gilliam and Skaggs, 1987). Eventually, the flow is diverted to another crossing area, concentrated or diverted around the strip. Also, the turf builds up above the pavement section. At this point, the turf must be stripped and sediment removed and lowered to reestablish the pollution-removal capability.

Design of Filter Strips

As for other types of BMPs, there are several design approaches for filter strips. Often, a simple width rule of thumb is used, matched to a specific design criteria, and empirical study results, and politics. Filter-strip minimums have been established across the country from as little as 25 ft to as much as 300 ft (Palfrey and Bradley, 1982). The use of minimum buffer widths makes it simple to apply the criteria uniformly but may not be as effective as a variable buffer width based on important variables and factors. Minimum buffer widths along streams are given by North Carolina and shown in Table 13-39 (NCDEHNR, 1991).

The Denver Urban Drainage and Flood Control District recommends sizing the buffer such that the runoff from the 2-year storm will load the strip at a rate of 0.05 cfs/linear foot, for a flow of less than 1 in of depth for grass that is at least 2 in high. The flow length is the greater of 8 ft or 0.2 times the flow path from the impervious area upstream from the strip (UDFCD, 1992, 1999). The Puget Sound manual recommends designing filter

TABLE 13-39

Recommended Minimum Widths of Streamside Buffer Zones in North Carolina

Type of Water Body	Percent Slope of Adjacent Lands				
	0–5	6–10	11–20	21–45	46+
Streamside Buffer Zone Width					
Intermittent and perennial	50	50	50	50	50
Perennial trout waters	50	66	75	100	125
Public water supplies	50	100	150	150	200

Source: From North Carolina Department of Environment, Health and Natural Resources (NCDEHNR), An Evaluation of Vegetative Buffer Areas for Water Quality Protection, Raleigh, NC, February 1991.

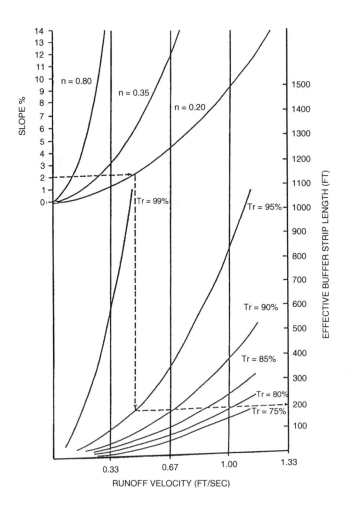

FIGURE 13-109
Effective buffer length determination for selected trap efficiency. (From Wong, S.L. and McCuen, R.H., The Design of Vegetative Buffer Strips for Sediment and Runoff Control, Stormwater Management in Coastal Areas, Tidewater Administration, Maryland Department of Natural Resources, Annapolis, MD, 1982.)

strips similar to swales using the vegetated channel design method given in Chapter 10 (Puget Sound, 1993). Wong and McCuen provide a model for filter-strip design, though rather large values can result (Wong and McCuen ,1982). Figure 13-109 gives a nomograph for calculation of buffer width, given slope, roughness, and target trap efficiency. Yu et al. (1992) recommend 80- to 100-ft widths.

Swift (1986) gives the following equations for buffer width perpendicular to flow for various conditions. His equations are based on earlier work and studies in southern Appalachian sandy loam soils. His equations are designed to filter out particles greater than 0.05 mm.

- Forested buffer with brush barriers 10 ft wide:

$$W = 32 + 0.4 * \% \text{ Slope} \tag{13.25}$$

- Forested buffer without brush barrier:

$$W = 43 + 1.39 * \% \text{ Slope} \tag{13.26}$$

- Grassed buffer width:

$$W = 74 + 1.39 * \% \text{ Slope} + 6.4 * \% \text{ Slope}/10 \qquad (13.27)$$

where W = width (ft), and % Slope = percent slope of buffer area.

The use of buffer strips along roadways is encouraged in Maine (Maine, 1993). Turnouts along the road direct flow off the pavement and across a buffer. They recommend flows less than 1 cfs be released at any one point and recommend a formula for calculating flow:

$$I_a * 0.000139 + P_a * 0.00069 < 1.0 \text{ cfs} \qquad (13.28)$$

where I_a = impervious area (ft²), and P_a = pervious area (ft²).

The actual length perpendicular to the flow depends on the target pollutants to be removed. Phillips (1989) developed two models to predict buffer length based on sedimentation and removal of small sediments and dissolved solids, respectively. The sediment model predicted much narrower strips than did the dissolved nutrient removal model. For the sediment model, slope was the determining factor. For nutrient removal, length of the flow path across the strip was the factor, as it reflected contact time.

Filter strips should be used to treat small drainage areas. Flow must enter the filter strip as sheet flow spread over the width (long dimension normal to flow) of the strip, generally no deeper than 1 to 2 in. As a rule, flow concentrates within a maximum of 75 ft for impervious surfaces and 150 ft for pervious surfaces (CWP, 1996). For longer flow paths, special provision must be made to ensure design flows spread evenly across the filter strip.

Filter strips should be integrated within site designs. Filter strips should be constructed outside the natural stream buffer area whenever possible to maintain a more natural buffer along the streambank. Filter strips should be designed for slopes between 2 and 6%. Greater slopes than this would encourage the formation of concentrated flow. Flatter slopes would encourage standing water.

Filter strips should not be used on soils that cannot sustain a dense grass cover with high retardance. Designers should choose a grass that can withstand relatively high velocity flows at the entrances and wet and dry periods. The filter strip should be at least 15 ft long to provide filtration and contact time for water-quality treatment. Preferable is 25 to 40 ft (where available), though length will normally be dictated by design method. It is desirable to attain at least a 5-min contact time. The top and toe of the slope should be as flat as possible to encourage sheet flow and prevent erosion.

An effective flow spreader is to use a pea gravel diaphragm at the top of the slope (ASTM D 448 size no. 6, 1/8 to 3/8 in). The pea gravel diaphragm (a small trench running along the top of the filter strip) serves two purposes. First, it acts as a pretreatment device, settling out sediment particles before they reach the practice. Second, it acts as a level spreader, maintaining sheet flow as runoff flows over the filter strip. Other flow spreaders include a concrete sill, curb stops, or curb and gutter with sawteeth cut into it.

Ensure that flows in excess of design flow move across or around the strip without damaging it. Often, a bypass channel or overflow spillway with protected channel section is designed to handle higher flows. Pedestrian traffic across the filter strip should be limited through channeling onto sidewalks.

ARC (2002) provides a design approach that allows for the direct calculation of contact time. Maximum discharge loading per foot of filter strip width (perpendicular to flow path) is found using a modified form of the Manning's equation:

$$q = \frac{0.00236}{n}Y^{5/3}S^{1/2} \tag{13.29}$$

where q = discharge per foot of width of filter strip (cfs/ft); Y = allowable depth of flow (in); S = slope of filter strip (%); and n = Manning's n roughness coefficient (use 0.15 for medium grass, 0.25 for dense grass, and 0.35 for very dense Bermuda-type grass).

The minimum length of a filter strip is:

$$W_{fMIN} = \frac{Q}{q} \tag{13.30}$$

where W_{fMIN} = minimum filter strip width perpendicular to flow (ft).

Filter without Berm

Normally, size filter strip (parallel to flow path) for a contact time of 5 min minimum. The equation for filter length is based on the SCS TR55 travel-time equation (SCS, 1986):

$$L_f = \frac{(T_t)^{1.25}(P_{2-24})^{0.625}(S)^{0.5}}{3.34n} \tag{13.31}$$

where L_f = length of filter strip parallel to flow path (ft); T_t = travel time through filter strip (min); P_{2-24} = 2-year, 24-h rainfall depth (in); S = slope of filter strip (%); and n = Manning's n roughness coefficient (use 0.15 for medium grass, 0.25 for dense grass, and 0.35 for very dense Bermuda-type grass).

Filter Strips with Berm

Size outlet pipes to ensure that the bermed area drains within 24 h. Specify grasses resistant to frequent inundation within the shallow ponding limit. Berm material should be of sand, gravel, and sandy loam to encourage grass cover (sand: ASTM C-33 fine aggregate concrete sand 0.02 to 0.04 in, gravel: AASHTO M-43 ½ to 1 in). Size filter strip to contain the WQ_v within the wedge of water backed up behind the berm. Maximum berm height is 12 in.

Filter Strips for Pretreatment

A number of other structural controls, including bioretention areas and infiltration trenches, may utilize a filter strip as a pretreatment measure. The required length of the filter strip depends on the drainage area, imperviousness, and the filter strip slope. Table 13-40 provides sizing guidance for bioretention filter strips for pretreatment.

The use of level spreaders is encouraged wherever there is some question as to the ability of sheet flow to exist, for runoff from parking lots and for all larger areas. Level

TABLE 13-40

Bioretention Filter-Strip Sizing Guidance

Parameter	Impervious Areas				Pervious Areas (Lawns, etc.)			
Maximum inflow approach length (ft)	35		75		75		100	
Filter strip slope (max = 6%)	<2%	>2%	<2%	>2%	<2%	>2%	<2%	>2%
Filter strip minimum length (ft)	10	15	20	25	10	12	15	18

Source: From Claytor and Schueler, 1996.

spreader design criteria can be found in state sediment handbooks (Virginia, 1992; North Carolina, 1991). The Virginia and North Carolina spreader consists of a diversion berm graded to disperse the flow from the end of the berm into a small V-shaped ditch graded to zero slope at the lip. Other types of spreaders include small troughs with V-shaped sawteeth cut into the downstream side, curbs with holes cut into them, gravel-filled trenches graded to zero slope, and curb blocks with openings.

Schueler (1994) and local experience provide guidance to maintain the integrity of buffers during and after construction.

- Planning stage
  - Require buffer limits to be present on all clearing and grading and erosion-control plans.
  - Record all buffer boundaries on official maps and plats.
  - Clearly establish acceptable and unacceptable uses for the buffer by ordinance and through preparing public education and awareness campaign literature.
  - Establish clear vegetation targets and management rules for different lateral zones of the buffer.
  - Design plantings to encourage proper use and discourage traffic in sensitive areas. Anticipate human traffic, and provide walkways in combination with flow spreaders or with multiple openings to allow flow passage.
  - Carefully plan multiple-use opportunities and locations compatible with the buffer's purpose.
  - Provide incentives for owners to protect buffers through perpetual conservation easements rather than deed restrictions.
- Construction stage
  - Do a preconstruction stakeout of buffers to establish the limits of disturbance.
  - Set the limit of disturbance based on the drip line of any forested areas of the buffer to avoid root damage.
  - Conduct preconstruction meetings and education to familiarize contractors with the limit of disturbance locations and buffer limits.
  - Mark the limit of disturbance with a silt fence barrier, signs, or other methods to exclude construction equipment.
- Postdevelopment stage
  - Mark buffer boundaries with permanent signs or attractive barriers in spots.
  - Educate property owners and homeowner associations on the purpose, limits, and allowable uses of the buffer.
  - Conduct periodic walks of the buffer to inspect the condition.
  - Encourage the use of volunteer monitoring groups and enforcement reporting.
  - Quickly reforest or replant bare or eroded areas.

Typical Required Specifications

- Flow spreaders and filter strips should be limited to drainage areas of 5 acres or less.
- Flow must enter the filter strip as sheet flow spread over the width (long dimension normal to flow) of the strips, generally no deeper than 1 to 2 in. As a rule,

flow concentrates within a maximum of 75 ft for impervious surfaces, and 150 ft for pervious surfaces (CWP, 1996). For longer flow paths, special provision must be made to ensure design flows spread evenly across the filter strip.
- Capacity of the spreader should be determined by 10-year peak flow storm.
- The filter strip length (perpendicular to flow) should be calculated based on the 6-month frequency, 24-h duration storm.
- Grade of level spreader should be 0%.
- Release runoff to outlet onto undisturbed stabilized areas in sheet flow and do not allow it to reconcentrate below the structure.
- Filter strips should be designed for slopes between 2 and 6%. Greater slopes than this would encourage the formation of concentrated flow. Flatter slopes would encourage standing water.
- Filter strips should not be used on soils that cannot sustain a dense grass cover with high retardance. Designers should choose a grass than can withstand relatively high-velocity flows at the entrances and during wet and dry periods.
- The filter strip should be at least 15 ft long to provide filtration and contact time for water-quality treatment. Twenty-five ft is preferred (where available), though length will normally be dictated by design method.
- All disturbed areas should be vegetated immediately after construction.
- Filter-strip width should be a minimum of 20 ft.

Typical Recommended Specifications

- The top edge of the filter strip should directly abut the contributing impervious area.
- Runoff water containing high sediment loads should be treated in a sediment-trapping device before release in a flow spreader.
- The top edge of the filter strip should follow the same elevational contour line.
- The spreader lip is to be protected with an erosion-resistant material, such as fiberglass matting or a rigid nonerodible material for higher flows, to prevent erosion and allow vegetation to become established.
- Wooded filter strips are preferred to grassed strips.
- Filter strips should be integrated within site designs.
- Filter strips should be constructed outside the natural stream buffer area, whenever possible, to maintain a more natural buffer along the streambank.
- The top and toe of the slope should be as flat as possible to encourage sheet flow and prevent erosion.
- Ensure that flows in excess of design flow move across or around the strip without damaging it. Often, a bypass channel or overflow spillway with protected channel section is designed to handle higher flows.
- Pedestrian traffic across the filter strip should be limited through channeling onto sidewalks.

Typical Operation and Maintenance Requirements

Maintenance is minimal, consisting of lawn care and mowing for grassed areas, litter removal, erosion control, removal of built-up sediment, releveling and grading, and

TABLE 13-41

Typical Maintenance Activities for Filter Strips

Activity	Schedule
Mow grass to maintain a 2 to 4 in height	Regularly (frequently)
Inspect pea gravel diaphragm for clogging and remove built-up sediment	Annual inspection
Inspect vegetation for rills and gullies, and correct; seed or sod bare areas	(semiannual first year)
Inspect to ensure that grass has established; if not, replace with an alternative species	

Source: From Center for Watershed Protection, Design of Stormwater Filtering Systems, prepared for Chesapeake Research Consortium, Edgewater, MD, Center for Watershed Protection, Ellicott City, MD, 1996.

inspections. Periodic turf or vegetation replacement may be necessary in areas of die-off or excess sediment buildup. A stormwater management easement and maintenance covenant should be required for each facility. The maintenance covenant should require the owner of the filter strip/flow spreader to periodically clean the structure. The flow spreader should be inspected after every rainfall until vegetation is established, and needed repairs should be made promptly. After the area is stabilized, inspections should be made periodically. Vegetation should be kept in a healthy, vigorous condition. The filter strip and flow spreader should be maintained in a manner to achieve sheet flow. Table 13-41 provides a summary of maintenance activities.

Design Example

Basic Data

> Small commercial lot 150 ft deep x 100 ft wide located in Smyrna, Georgia
> Drainage area (A) = 0.34 acres
> Impervious percentage (I) = 70%
> Slope equals 4%, Manning's n = 0.25
> Water-quality control volume based on runoff from 1.2 in

Calculate Maximum Discharge Loading per Foot of Filter Strip Width
Using Equation (13.29):

$$q = 0.00236/0.25 * (1.0)^{5/3} * (4)^{1/2} = 0.019 \text{ cfs/ft}$$

Water-Quality Peak Flow (See Chapter 7)

> Compute the water-quality volume in inches:

$$WQ_v = 1.2 (0.05 + 0.009 * 70) = 0.82 \text{ inches}$$

> Compute modified CN for 1.2-in rainfall (P = 1.2):

$$CN = 1000/[10+5P+10Q-10(Q^2+1.25*Q*P)^{1/2}]$$

$$= 1000/[10+5*1.2+10*0.82-10(0.82^2+1.25*0.82*1.2)^{1/2}]$$

$$= 96.09 \text{ (Use CN = 96)}$$

For $CN = 96$ and an estimated time of concentration (T_c) of 8 min (0.13 h), compute the Q_{wq} for a 1.2-in storm.

From Chapter 7, $I_a = 0.083$, therefore, $I_a/P = 0.083/1.2 = 0.069$.

For a Type II storm (using the limiting values), $q_u = 950$ csm/in, and therefore:

$$Q_{wq} = (950 \text{ csm/in}) (0.34ac/640ac/mi^2) (0.82") = 0.41 \text{ cfs}$$

Minimum Filter Width

Using Equation (13.30),

$$W_{fMIN} = Q/q = 0.41/0.019 = 22 \text{ ft}$$

Because the width of the lot is 100 ft, the actual width of the filter strip will depend on site grading and the ability to deliver the drainage to the filter strip in sheet flow through a pea gravel filled trench.

Filter without Berm

2-year, 24-h storm = 0.17 in/h or 0.17 * 24 = 4.08 in

Use 5-min travel (contact) time.

Using Equation (13.31):

$$L_f = (5)^{1.25} * (4.08)^{0.625} * (4)^{0.5} / (3.34 * 0.25) = 43 \text{ ft}$$

Note that reducing the filter strip slope to 2% and planting a denser grass (raising the Manning n to 0.35) would reduce the filter-strip length to 22 ft.

Filter with Berm

Pervious berm height is 6 in.

Compute the water-quality volume in cubic feet:

$$WQ_v = R_v * 1.2/12 * A = (0.05 + 0.009 * 70) * 1.2/12 * 0.34 = 0.023 \text{ Ac-ft or } 1{,}007 \text{ ft}^3$$

For a berm height of 6 in, the wedge of volume captured by the filter strip is:

$$\text{Volume} = W_f * 1/2 * L_f * 0.5 = 0.25 W_f L_f = 1{,}007 \text{ ft}^3$$

For a maximum width of the filter of 100 ft, the length of the filter would then be 40 ft.

For a 1-ft berm height, the length of the filter would be 20 ft.

Oil/Grit Separators (Water-Quality Inlet, Gravity Separator)

A water-quality inlet is typically a three-stage underground retention system designed to remove heavy particles and hydrocarbons from stormwater runoff. Oil-grit separators are hydrodynamic separation devices designed to remove grit and heavy sediments, oil and grease, debris, and floatable matter from stormwater runoff through gravitational settling and trapping. Gravity separator units contain a permanent pool of water and typically consist of an inlet chamber, separation/storage chamber, a bypass chamber, and an access port for maintenance purposes. Runoff enters the inlet chamber, where heavy sediments and solids drop out. The flow moves into the main gravity separation chamber, where further settling of suspended solids takes place. Oil and grease are skimmed and stored in a waste oil storage compartment for future removal. After moving into the outlet chamber, the clarified runoff is then discharged.

The performance of these systems is based primarily on the relatively low solubility of petroleum products in water and the difference between the specific gravity of water and the specific gravities of petroleum compounds. Gravity separators are not designed to separate other products such as solvents, detergents, or dissolved pollutants. The typical gravity separator unit may be enhanced with a pretreatment swirl concentrator chamber, oil draw-off devices that continuously remove the accumulated light liquids, and flow control valves regulating the flow rate into the unit.

Gravity separators are best used in commercial, industrial, and transportation land uses and are intended primarily as a pretreatment measure for high-density or ultra-urban sites, or for use in hydrocarbon hot spots, such as gas stations and areas with high vehicular traffic. However, gravity separators cannot be used for the removal of dissolved or emulsified oils and pollutants such as coolants, soluble lubricants, glycols, and alcohols.

This type of BMP has been used for many years in industrial applications, rather than in urban stormwater applications, where a known and steady flow can be assured. In industrial applications, great strides have been made in design of grit chambers and in coalescing plate-type devices. When translated to stormwater applications, two basic problems are realized:

- An expectation of removal of pollutants other than grit and oil has been created.
- Stormwater runoff can overwhelm devices designed for steady flow measured in gallons per minute, not widely varied flow measured in cubic feet per second.

When adopted for stormwater flows, a basic mismatch occurs. The conveyance portion of the system is designed for the same flow as the rest of the stormwater system (for example, the 10-year flow), while the storage and treatment portion of the chamber is designed for very low flows. Typical design criteria of capture of 400 cubic feet per impervious acre are equivalent to capture of only the first tenth of an inch of runoff. Runoff residence times average less than 30 min (Schueler, 1994). Ongoing studies have shown that, when the larger flows enter the chamber, they resuspend sediments resulting in little actual capture and treatment. Average sediment depth accumulates to only about 2 in, and varies from storm to storm.

The first problem can be solved by combining this type of application with others designed to remove other target pollution, or by modifying the structure to provide other types of removal mechanisms. The second problem is more difficult to handle, because many of these devices are in-line type structures unable cheaply to shunt excess flows past the device and to an outlet. In all new design cases, this type of device should be designed or retrofitted as an offline system.

Innovative modifications to these devices that improve pollution removal have been developed and are marketed as proprietary commercial devices.

To be cost-effective, water-quality inlets should be limited to capturing runoff from small high-density sites such as maintenance shops, gas stations, and certain industrial areas where high concentrations of oils from small contributing areas are expected. However, gravity separators cannot be used for the removal of dissolved or emulsified oils and pollutants such as coolants, soluble lubricants, glycols, and alcohols. These facilities have a fairly high initial cost and must be frequently cleaned. Cleanout costs can be as high as $1000 to $2000 per site annually, though some municipalities use partial cleanout techniques.

Pollution-Removal Efficiency

True pollution-removal for this type of structure occurs only when the structure is cleaned out. There is some likelihood that the grit and oil from this type of structure is toxic and requires special disposal actions. Thus, many municipalities and industries find themselves in an expensive "Catch 22."

The pollution-removal efficiency has not been reported in the literature and must be inferred. Schueler et al. (1992) and recent MWCOG study results report that sediment removal of coarse materials appears to be moderate, but that resuspension occurs for every moderate storm. However, the types of pollutants found adsorbed to the sediments match those found in receiving water bodies where there are no such devices, indicating that the right target pollutants are being addressed. Different offline-type devices or other experimental modifications should go a long way in improving the pollutant-removal capabilities of the system.

In industrial applications, specification sheets give typical oil-removal rates (relying on coalescing of oil droplets) of 95% for flows of 2 to 3 cfs but infer initial concentrations of 2000 mg/L, much higher than the 2 to 10 mg/L commonly found in urban runoff. But for highly oily runoff, coalescing-type structures have shown an ability to reduce concentrations of oil to the 10 mg/L range (Lettenmaier and Richey, 1985). Studies on sumped catch basins in Switzerland showed that only about 10% of the solids were removed by the basins (Conradin, 1989).

These devices have limited effectiveness as stand-alone treatment of stormwater runoff quality, unless hydrocarbons and grit are the only pollutant of concern. When used with infiltration in the bottom, as suggested in Schueler (1987), there is an expectation of clogging. Hydrocarbons in urban runoff can effectively clog the infiltration capacity of underlying soils, because they tend to attach themselves to particles in the water column and settle to the bottom of the BMP. Three-chamber oil-and-grit devices may remove from 40 to 60% of the hydrocarbons found in parking lot and street runoff if they are cleaned quarterly during wet seasons. Three-chamber oil-and-grit devices may also remove a moderate portion (less than 40%) of the suspended sediment and associated adsorbed pollutants. When combined with the high cost of installation (from $6000 to $10,000 per site) they are not, at present, considered very effective on a unit cost basis. However, there are many types of sites where they are necessary, being one of the only ways to remove their target pollutants. Pollution removal can be enhanced by:

- Maximizing volume in the first two chambers
- Using a coalescing-material-type chamber for oil
- Protecting the orifice between first two chambers with a trash rack
- Extending an inverted elbow at least 3 ft into the pool to improve oil removal
- Performing regular maintenance

Recent testing of gravity separators has shown that they can remove between 40 and 50% of the TSS loading when used in an offline configuration (Curran, 1996; Henry, 1999). Gravity separators also provide removal of debris, hydrocarbons, trash, and other floatables. They provide only minimal removal of nutrients and organic matter.

Structural Best Management Practices

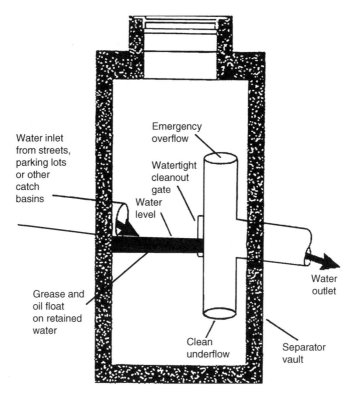

FIGURE 13-110
Example T-spill control catch basin. (From Washington Ecology, 1999.)

Oil/Water Separator Design

There are several types of oil/water separators. The most simple is a minor spill-control separator that consists of a simple sumped catch basin with a T-outlet to remove water from below the surface (Puget Sound, 1992). It is illustrated in Figure 13-110 for comparison purposes only, but it is not recommended for pollutant-removal planning.

Most other water-quality inlets are designed to remove sediment and hydrocarbon loadings before they are conveyed through a storm drain or into an infiltration facility. This is done through settling of the heavier grit and trapping (settling upward) of floatables and oil. Figures 13 111 and 13-112 displays the basic layout of the three-chambered type used in the Washington, DC, area. A modified design that provides another chamber and a submerged orifice for improved oil removal similar to industrial applications has been suggested by Washington, DC, engineers (1992).

This type of oil/grit separator has the following features:

- Chamber one traps particulates and has two 6-in orifices (with trash rack) at midwater depth, through a partition between chambers one and two.
- Chamber two traps floating oil and has an inverted elbow pipe to regulate water levels in chambers one and two.
- Chamber three receives discharge from chamber two, and has an overflow, outfall pipe set with the crown below the horizontal invert of the inverted elbow pipe from chamber two.
- An access manhole should be provided into each chamber.

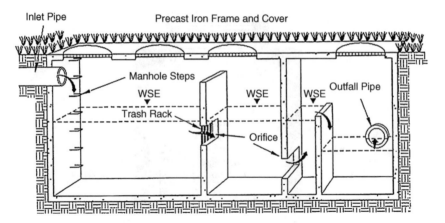

FIGURE 13-111
Example oil-grit separator schematic.

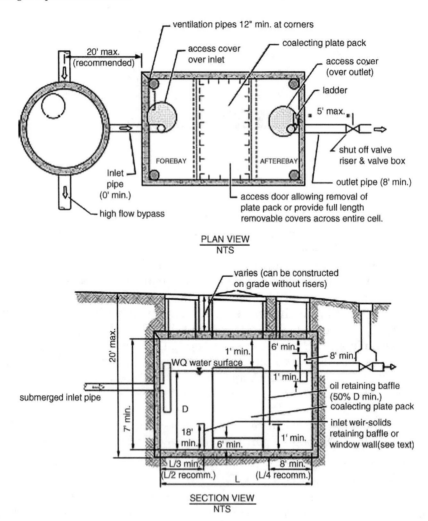

FIGURE 13-112
Example three-chambered oil grit separator schematic. (From Washington State Department of Ecology, Stormwater Management Manual for Western Washington, 2001.)

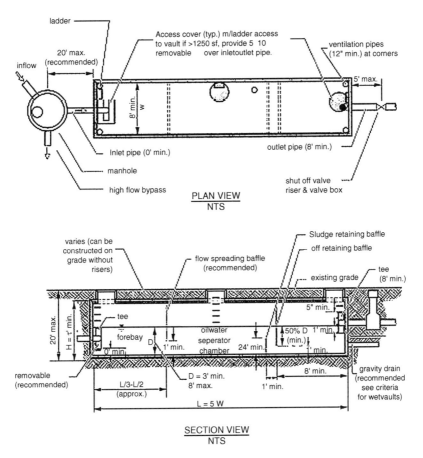

FIGURE 13-113
Coalescing-filter oil-grit separator schematic. (From Washington State Department of Ecology, Stormwater Management Manual for Western Washington, 2001.)

Inlets must be cleaned at least twice each year. The total wet storage should be, as a minimum, 400 cubic feet per impervious acre. These devices apply to areas less than 2 acres.

Another type of separator has been described for use initially in California and has been picked up by other states (WDE, 2001) (Figure 13-113). Sizing of this conventional separator has been approximated based on industrial sizing criteria.

Large droplets of oil rise faster, just like large particles fall faster, according to Stokes' Law. Just as large particles need time to settle, large oil droplets need time to float. The equation for rise rate is:

$$V_p = 1.79 \, (d_p - 1) \, d^2 \times 10^{-8} / n \qquad (13.23)$$

where V_p = rise rate (ft/s); d_p = density of oil (between 0.85 and 0.95) (gm/cc); d = diameter of minimum droplet to be removed (microns); and n = absolute viscosity of the water (poises).

The absolute viscosity of water can be reasonably linearly interpolated between values of 0.0179 poises at 32 degrees F and 0.010 poises at 68 degrees F. The distribution of droplet sizes can be estimated from Table 13-42 assuming a water temperature for removal in the winter or wet season.

Sizing equations based on the American Petroleum Institute procedure (API, 1990) and modified by California Stormwater Task Force (1993) are as follows:

TABLE 13-42

Size and Volume Distribution of Oil Droplets

Drop Size (microns)	Percent by Volume Smaller	Percent by Size Smaller
20	1	4
40	1	6
60	1	12
80	11	47
100	26	67
120	32	80
140	57	88
160	78	95
180	98	98
200	99	99

Source: From California Stormwater Task Force, 1993.

$$D = (Q/RV)^{0.5} \tag{13.24}$$

$$L = VD/V_p \tag{13.25}$$

$$W = Q/(VD) \tag{13.26}$$

where D = depth (3 to 8 ft) (ft); Q = design flow (cfs); R = length-to-width ratio (e.g., 2:1 equals 2); V = allowable horizontal velocity equal to 15 times V_p, but less than 0.05 ft/s; L = length (ft); W = width (2 to 3 times the depth, but less than 20 ft) (ft); and V_p = rise velocity of oil droplet - 0.00055 ft/s recommended by Puget Sound (ft/s).

If the depth exceeds 8 ft, design parallel units, and split the flow. The baffle height-to-depth ratio for top baffles is 0.85, and 0.15 for bottom baffles. The flow distribution baffle is located at 0.1 L from the entrance. Add freeboard. A design flow less than the 6-month discharge should capture sufficient annual volume to achieve a high efficiency. Install a bypass for flows in excess of design. The depth-to-width ratio is normally 0.3 to 0.5 with a width of 6 to 16 ft (Puget Sound, 1992).

The design procedure then is:

- Select a removal efficiency and then a target diameter from Table 13-42 based on estimated influent concentration and effluent target ratio (e.g., 10 mg/L out divided by 60 mg/L in = 83% efficiency) or on design criteria target (e.g., 80%).
- Estimate oil viscosity and water viscosity, and calculate rise rate (V_p).
- Estimate a design flow rate using impervious area and design rainfall rate.
- Calculate V and D to place each within allowable limits. Consider parallel units if depth or velocity limits are exceeded.
- Calculate L and W, and set the baffle heights and locations.
- Complete the design.

There is some question whether a conventional separator can remove particles less than 150 microns efficiently without coalescing the smaller droplets into larger ones, though there are little data on oil droplet size in stormwater. An alternate design criteria has been developed for sizing water-quality inlets using a coalescing plate design. The coalescing process, whereby small droplets are merged with others, forming large droplets on the surfaces of parallel plates made of fiberglass or polypropylene, has been used and is being modified (using, among other things, corrugated filters) for more efficient use in storm drain inlets. The use of coalescing filter media is more commonplace in the West for sites where the oil droplets can be expected to be less than 150 microns (California Stormwater Task Force, 1993).

Packaged coalescing units can be provided for flows up to several cfs. Larger units must be sized by the designer. A number of manufacturers offer coalescing filter materials (or coalescing tubes), so individual specifications and efficiencies must be assessed. The basic equation for coalescing filter design is:

$$A_f = Q/(E_f V_p \text{ Cosine } H) \tag{13.27}$$

where A_f = total surface area of filter (ft^2); H = plate angle measured from horizontal (degrees); V_p = rise velocity of oil droplet (ft/s); and E_f = efficiency of the specific filter (0.35 to 0.95).

Spacing between plates is normally 0.75 to 1.5 in. Add 12 in below the plates for sediment accumulation and 6 to 12 in above for oil accumulation. Add 1 ft of freeboard. Design the forebay for flow distribution and floatable collection. Placement of plates on angles of 45 to 60 degrees allows the solids to slide off the plates to the bottom. Even so, periodic maintenance will be necessary and will be based on local experience, though monthly or poststorm checks are recommended initially. A trash rack with openings smaller than the plates will keep debris out of the plate area. Following are the design steps:

- Collect basic information on available filter types, angles, and efficiencies.
- Calculate V_p and Q as in the procedure for the conventional separator.
- Calculate A_f.
- Select spacing (S) and reasonable width (W) and length (L).
- Calculate number of plates as $N = A_f/LW$.
- Calculate total volume from spacing, number of plates, and plate thickness.
- Complete design and baffle placement.

Typical Required Specifications

The use of gravity (oil-grit) separators should be limited to the following applications:

- Pretreatment for other structural stormwater controls
- High-density, ultra-urban or other space-limited development sites
- Hot-spot areas where the control of grit, floatables, and/or oil and grease are required

Gravity separators are typically used for areas less than 5 acres. It is recommended that the contributing area to any individual gravity separator be limited to 1 acre or less of impervious cover. Gravity separator systems can be installed in almost any soil or terrain. Because these devices are underground, appearance is not an issue, and public safety risks are low. Gravity separators are rate-based devices. This contrasts with most other stormwater structural controls, which are sized based on capturing and treating a specific volume. Gravity separator units are typically designed to bypass runoff flows in excess of the design flow rate. Some designs have built-in high-flow bypass mechanisms. Other designs require a diversion structure or flow splitter ahead of the device in the drainage system. An adequate outfall must be provided.

Oil and water separators should be designed to remove oil and TPH down to 15 mg/L at any time and 10 mg/L on a 24-h average, and produce a discharge that does not cause an ongoing or recurring visible sheen in the stormwater discharge or in the receiving water.

No velocities in the device should exceed the entrance velocity. A trash rack should be included in the design to capture floating debris, preferably near the inlet chamber to prevent debris from becoming oil impregnated. Ideally, a gravity separator design will provide an oil draw-off mechanism to a separate chamber or storage area. Adequate maintenance access to each chamber must be provided for inspection and cleanout of a gravity separator unit. Gravity separator units should be watertight to prevent possible groundwater contamination. The design criteria and specifications of a proprietary gravity separator unit should be obtained from the manufacturer. Further details on the design of separators can be found in WDE (2001) (available on the Internet).

- Separators should be sized for the 6-month, 24-h design storm. Larger storms should not be allowed to enter the separator.
- Separator should be structurally sound and designed for acceptable municipal traffic loadings where subject to traffic loadings.
- Separator should be designed to be watertight.
- Volume of separator should be at least 400 cubic feet per impervious acre tributary to the facility (first two chambers).
- Forebay or first chamber should be designed to collect floatables and larger settleable solids. Its surface area should not be less than 20 square feet per 10,000 square feet of drainage area.
- Horizontal velocity through the separation chamber should be 1 to 3 ft/min or less.
- Separator pool should be at least 4 ft deep.
- Weirs, openings, and pipes should be sized to pass, as a minimum, the storm drain system design storm.
- Manholes should be provided to each chamber to provide access for cleaning.
- The length of the forebay should be 1/3 to 1/2 of the length of the entire separator.
- Screen openings should be about 3/4 in. Include a submerged inlet pipe with a turn-down elbow in the first bay at least 2 ft from the bottom.
- The outlet pipe should be a T, sized to pass the design peak flow and placed at least 12 in below the water surface.
- Use absorbents and skimmers in the afterbay as needed.
- Oil-retaining baffles (top baffles) should be located at least at 1/4 of the total separator length from the outlet and should extend down at least 50% of the water depth and at least 1 ft from the separator bottom. Baffle-height-to-water depth ratios should be 0.85 for top baffles, and 0.15 for bottom baffles.

Recommended Specifications

- Oil-absorbent pads, oil skimmers, or other approved methods for removing accumulated oil should be provided.
- Separator should be located close to the source before pollutants are conveyed to storm drains or other BMPs.

- Use only on sites of less than 1 to 2 acres.
- Provide perforated covers as trash racks on orifices leading from the first to the second chamber.
- The center chamber may contain a coalescing medium to enhance the oil flotation separating process.
- The storm drain inlet in the third chamber should be located above the floor to permit additional settling.
- Stormwater from rooftops and other impervious areas not likely to be polluted with oil should not discharge to the separator.
- Design to bypass flows above 400 cubic feet per acre.

Operation and Maintenance Requirements

A stormwater management easement and maintenance covenant should be required for each facility. The maintenance covenant should require the owner of the separator to periodically clean the structure. Cleaning quarterly should be a minimum schedule, with more intense land uses such as gas stations requiring cleaning as often as monthly. Cleaning should include pumping out wastewater and grit, and having the water processed to remove oils and metals. For coalescing plate separators, the facility should be inspected weekly. Pads are to be replaced when they become clogged and gritty, but no less than annually. Table 13-43 provides a summary of recommended maintenance.

Alum Treatment

The process of alum (aluminum sulfate) injection provides treatment of stormwater runoff from a piped stormwater drainage system entering a wet pond by injecting liquid alum into storm sewer lines on a flow-weighted basis during rain events. When added to runoff, liquid alum forms nontoxic precipitates of aluminum hydroxide [$Al(OH)_3$] and aluminum phosphate [$AlPO_4$], which combine with phosphorus, suspended solids, and heavy metals, causing them to be deposited into the sediments of the receiving waters in a stable, inactive state.

The alum precipitate formed during coagulation of stormwater can be allowed to settle in receiving water or can be collected in small settling basins. Alum precipitates are stable in sediments and will not redissolve due to changes in redox potential or pH under conditions normally found in surface water bodies. Laboratory or field testing may be necessary to verify feasibility and to establish design, maintenance, and operational parameters, such as the optimum coagulant dose required to achieve the desired water-quality goals, chemical pumping rates, and pump sizes.

The use of alum in existing lakes results in immediate and substantial improvements in water clarity and quality, including long-term improvements to benthic communities of the lake. The increased benthic activity has kept floc accumulation depths to less than 1 cm/year. The safety of alum is witnessed by the fact that it has been used for treating

TABLE 13-43

Typical Maintenance Activities for Gravity Separators

Activity	Schedule
Inspect the gravity separator unit	Regularly (quarterly)
Clean out sediment, oil and grease, and floatables, using catch basin cleaning equipment (vacuum pumps); manual removal of pollutants may be necessary	As needed

FIGURE 13-114
Alum treatment example.

potable water since the times of the Romans. The first alum treatment system was designed at Lake Ella in Tallahassee, Florida, in 1986. Since then, there have been 30 to 40 alum treatment systems throughout Florida, and several more in other parts of the United States (Harper, Herr, and Livingston, 1997; Herr and Harper, 1997; Harper and Herr, 1998). Figure 13-115 is a typical schematic for floc collection and disposal.

The practice is applicable to medium- to large-size drainage basins. It is too expensive for small applications. Construction costs for existing alum stormwater treatment facilities in Florida have ranged from $135,000 to $400,000. The capital construction costs of alum stormwater treatment systems is independent of watershed size and depends primarily on the number of outfall locations treated. While these costs seem high, comparative studies done in Florida show that, when land costs are included, alum treatment is often significantly cheaper than wet ponds (Harper and Herr, 1998). Estimated annual operations and maintenance (O&M) costs for chemicals and routine inspections range from approximately $6500 to $25,000 per year. O&M costs include chemicals, power, manpower for routine inspections, and equipment renewal and replacement costs.

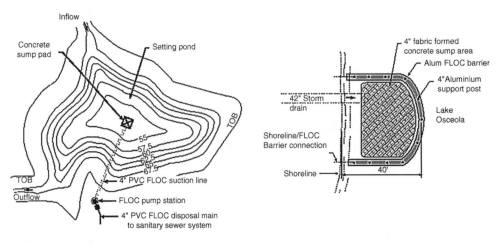

FIGURE 13-115
Floc collection and disposal schematic.

TABLE 13-44

Construction and Operations Costs for Alum Treatment Facilities in Florida

Project	Area Treated (ac)	Construction Cost/System ($)	Estimated Annual O&M Cost ($)	Construction Cost per Area Treated ($/ac)	Annual O&M Cost per Area Treated ($/ac)
Lake Ella	158	200,400	—	1268	—
Lake Dot	305	250,000	—	823	—
Lake Lucerne	272	400,000	16,000	1472	59
Lake Osceola	153	300,000	6500	1959	43
Lake Cannon	490	135,000	13,100	276	27
Channe 2	84	180,000	—	2144	—
Lake Virginia North	64	242,000	—	3769	—
Celebration	158	300,000	25,000	1898	158
Lake Holden	183	292,000	—	1598	—
Lake Tuskawilla	311	242,000	19,627	777	63
Lake Rowena	538	75,000	—	139	—
Lake Mizell	74	300,000	15,389	4049	208
Lake Maggiore (5)	1450	400,000	21,450	1379	74
Webster Avenue	91	130,000	12,397	1423	136
Lake Virginia South	437	288,000	—	659	—
Merritt Ridge	195	201,575	26,298	1033	135
Averages	310	$245,998	$17,307	$1542	$100

Source: From Harper, H.H. and Herr, J., P.E., Alum treatment of stormwater runoff: an innovative BMP for urban runoff problems, Proc. Natl. Conf. on Retrofit Opportunities for Water Resour. Protection in Urban Environ., 1998, http://www.stormwater-resources.com/library.htm#BMPs.

Table 13-44 shows construction and maintenance costs for a number of alum treatment facilities in Florida.

Pollutant-Removal Efficiency

Alum treatment consistently achieved a 85 to 95% reduction in total phosphorus, 90 to 95% reduction in orthophosphorus, 60 to 70% reduction in total nitrogen, 50 to 90% reduction in heavy metals, 95 to 99% reduction in turbidity and TSS, 60% reduction in BOD, and >99% reduction in fecal coliform bacteria, compared with raw stormwater characteristics. Florida lakes that have been treated for several years with alum control have shown remarkable increase in water quality and clarity is shown in Table 13-45.

Design Criteria

Alum treatment systems are fairly complex, and design details are beyond the scope of this book. However, further information can be obtained from the Internet and by contacting local municipalities and engineers who have designed and implemented successful systems. The following are general guidelines for alum treatment systems:

- Injection points should be 100 ft upstream of discharge points.
- Alum concentration is typically 10 micro-g/L.
- Alum treatments systems may need to control pH.
- For new pond design, the required size is approximately 1% of the drainage basin size, as opposed to 10 to 15% of the drainage basin area for a standard detention pond.
- No volume requirement is required when discharging to existing lakes.

Table 13-46 shows typical maintenance requirements.

TABLE 13-45

Comparison of Pre- and Postmodification Water-Quality Characteristics for Typical Alum Stormwater Treatment Systems

Parameter	Units	Lake Ella Before (1974–1985)	Lake Ella After (1/88–5/90)	Lake Dot Before (1986–1988)	Lake Dot After (3/89–8/91)	Lake Osceola Before (6/91–6/92)	Lake Osceola After (2/93–12/96)
# of samples	—	15	11	5	15	12	46
pH	Su	7.41	6.43	7.27	7.17	8.22	7.63
Diss. O2 (1 M)	mg/L	3.5	7.4	6.6	8.8	8.8	8.8
Total N	Ug/L	18.76	417	1545	696	892	856
Total P	Ug/L	232	26	351	24	37	26
BOD	mg/L	41	3.0	16.8	2.7	4.4	3.4
Chlorophyll-a	mg/m^3	180	5.1	55.8	6.3	24.8	21.7
Secchi disc depth	m	0.5	>2.2	<0.8	2.5	1.1	1.2
Dissolved. Al	Ug/L	—	44	—	65	18	51
Florida TSI value	—	98 (Hypereutrophic)	47 (Oligotrophic)	86 (Hypereutrophic)	42 (Oligotrophic)	61 (Eutrophic)	56 (Mesotrophic)
Lake area	—	5.38 ha (13.3 ac)		2.4 ha (5.9 ac)		22.4 ha (55.4 ac)	
Watershed area	—	63.7 ha (57 ac)		123 ha (305 ac)		61.5 ha (153 ac)	
Percent of annual hydraulic inputs treated	%	95		96		9	

Source: From Harper, H.H. and Herr, J., P.E., Alum treatment of stormwater runoff: an innovative BMP for urban runoff problems, Proc. Natl. Conf. on Retrofit Opportunities for Water Resour. Protection in Urban Environ., 1998, http://www.stormwater-resources.com/library.htm#BMPs.

TABLE 13-46

Typical Maintenance Activities for Alum Treatment

Activity	Schedule
Perform routine inspection	Monthly
Monitor water quality and pH	
Perform maintenance of pump equipment, chemical supplies, and delivery system	As needed

Source: From Harper, H.H., Herr, J., P.E., and Livingston, E., Alum Treatment of Stormwater — The First Ten Years: What Have We Learned and Where Do We Go From Here?, 1997, http://www.stormwater-resources.com/library.htm#BMPs.

Proprietary Structural Controls

There are many types of commercially available proprietary stormwater structural controls available for water-quality treatment and quantity control. The array of commercial devices defies logical arrangement but can be categorized into the following:

- *Media filter inserts* — This category includes catch basin inserts (CBI) and offline units, similar to underground sand filters, with a primary pollutant-removal mechanism that consists of filtration through some sort of media. Their target pollutants are floatables, TSS, and oil/hydrocarbons. Some are designed to be used only during construction. Some claim high removal rates for other pollutants, though field data are often lacking.
- *Gravity screen and swirl separator (GSSS) units* — This category is made up of a number of designs that use a combination of screening, flow restriction, baffles, settling, and gravity or vortex separation with settling. Most are offline underground units. They work best for floatables, TSS, and oil.
- *Skimmers* — These are often added to the other units and consist of floating tubes, blocks, or other media made of absorbent materials to soak up oils and other floating pollutants.
- *Constructed wetlands* — These units are designed to be small wetland treatment systems with the hope of providing higher removal rates for nutrients and metals. Some have shown relatively high degrees of success in removal for smaller sites.

Table 13-47 provides summary information for a number of commercially available devices. Web addresses change often but are provided here as a starting point for the would-be purchaser. Inclusion in this table does not constitute an endorsement for any product, and, as always, the designer should be aware of the details of device performance and limitations prior to installation.

Local Acceptance

Many proprietary systems are useful on small sites and space-limited areas where there is not enough land or room for other structural control alternatives. Proprietary systems can often be used in pretreatment applications in a treatment train. However, proprietary systems are often more costly than other alternatives and may have high maintenance requirements. Perhaps the largest difficulty in using a proprietary system is the lack of adequate independent performance data, though this is changing almost monthly. Below are some general guidelines that should be followed before considering the use of proprietary commercial systems.

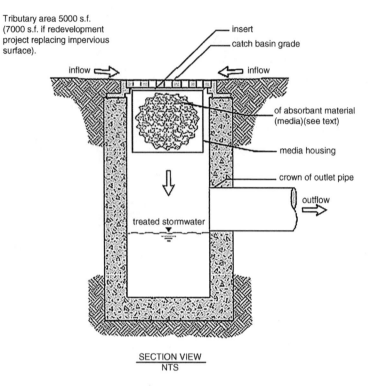

FIGURE 13-116
Typical insert schematic.

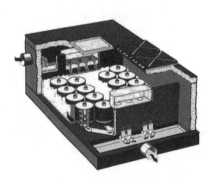

FIGURE 13-117
Typical insert example. (From Stormwater Management, Inc.)

Structural Best Management Practices

FIGURE 13-118
Typical insert example. (From AbTech Industries.)

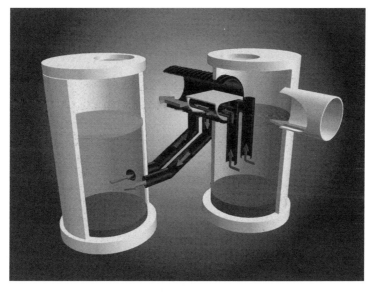

FIGURE 13-119
Typical GSSS example. (From BaySaver, Inc.)

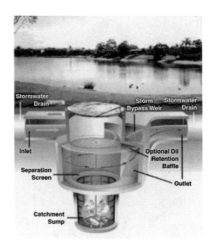

FIGURE 13-120
Typical GSSS example. (From CDS Technologies.)

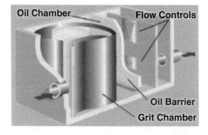

FIGURE 13-121
Typical GSSS example. (From Vortechnics, Inc.)

FIGURE 13-122
Typical GSSS example. (From Stormceptor Corp.)

FIGURE 13-123
Typical GSSS example. (From Best Management Products, Inc.)

FIGURE 13-124
Typical passive skimmer.

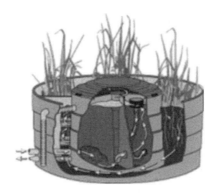

FIGURE 13-125
Typical constructed wetland. (From Storm Treat Systems, Inc.)

FIGURE 13-126
Typical Howland swale.

Several testing protocols are in development or recently published (Minton et al, 2000). The interested reader should search the Internet to find the latest accepted protocol, as they are in a state of flux and negotiation at this writing. However, a commonsense cautionary approach makes sense (WDE, 2001). In order for use, a proprietary system should have a demonstrated capability of meeting the stormwater management goals for which it is being intended. This means that the system must provide:

1. Independent third-party scientific verification of the ability of the proprietary system to meet water-quality treatment objectives and provide water-quantity control as needed.
2. Have a proven record of longevity in the field.
3. Have a proven ability to function in local conditions, where the system will be used (e.g., climate, rainfall patterns, soil types, etc.).

For a propriety system to meet (1) above for water-quality goals, the following monitoring criteria may be established for supporting studies:

- At least 15 storm events must be sampled.
- The study must be independent or independently verified (i.e., may not be conducted by the vendor or designer without third-party verification).
- The study must be conducted in the field, as opposed to laboratory testing.
- Field monitoring must be conducted using standard protocols that require proportional sampling upstream and downstream of the device.
- Concentrations reported in the study must be flow-weighted.
- The propriety system or device must have been in place for at least 1 year at the time of monitoring.

Although local data is preferred, data from other regions can be accepted as long as the design accounts for the local conditions. Local governments may need to submit a proprietary system to further scrutiny based on the performance of similar practices. A poor performance record or high failure rate is valid justification for not allowing the use of a proprietary system or device.

Structural Best Management Practices

TABLE 13-47
Summary Information on Commercial Devices

Manufacturer	Product Name	Description	Performance	Flow Capacity	Maintenance
Media Filter Inserts					
AbTech Industries www.oars97.com 480-874-4000 800-545-8999	Ultra-Urban™ Filter with OARS OnBoard	Designed for use in storm drains; polymer filtration media permanently absorbs oil and grease, while filter captures trash and sediment	Removes 80% of petroleum hydrocarbons; captures 0.45–0.12 yd^3 of debris	0.1 cfs	Replace every 1–3 years; service as needed to remove sediment/debris
Aquashield, Inc. www.aquashieldinc.com 423-870-8888 423-870-1005 fax	Aqua-Guard™ Stormwater Catchbasin Insert	Custom-designed catch basin inserts for drop inlet or curb inlet; sediment/debris collection area followed by removable filter media; constructed out of Stainless Steel and HDPE for durability and long life	Removes 60–80% TSS; certified by CA EPA for removal of 95% dissolved petroleum and oils; removes soluble and insoluble nutrients and dissolved metals	Information not available	Inspection recommended on quarterly basis; vacuum out sediment and replace filter media as needed
Aqua Treatment Systems, Inc. www.gullywasher.com 253-835-9163 800-208-5447 253-835-5477 fax	Gullywasher™ Brand	Trench drain and catch basin inserts designed for litter/oil collection and litter/oil/sediment collection	Information not available	Information not available	Routine inspection — remove wire basket and replace filter pillow as needed
Enviro-Drain, Inc. www.members.aa.net/~filters/ 800-820-1953 360-563-2850	Enviro-Drain™ Stormwater Filter Insert	System is composed of inlet grate, overflow diverter tray, screen unit, and filter tray; filter tray available in assorted mesh sizes	Information not available	0.4 cfs	Maintenance is required at least once a month to clean the screens and replace filters
EP International www.barnsides.com/epinternational/ 305-892-8016	HYDRO-CARTRIDGES	Cartridge equipped with hydro pads that are inserted into old or new inlets	Captures over 90% of petroleum hydrocarbons	Information not available	Information not available
Foss Environmental www.fosscatalog.com 800-909-3677 888-234-3677 fax	STREAMGUARD™ Catch Basins	Catch basin inserts for oil and grease control, sediment control, or trash and debris control	Reduce oil and grease by 93% and sediments by 80%	Information not available	Routine inspection — replace filter pack as needed

TABLE 13-47 (Continued)
Summary Information on Commercial Devices

Manufacturer	Product Name	Description	Performance	Flow Capacity	Maintenance
Media Filter Inserts					
Frostman Environmental, Inc. 920-497-7572	Peat Baskets	Baskets containing peat screenings or peat granules to be placed in catch basins to absorb filter stormwater runoff	Absorbs 90% of petroleum products, 90–98% of heavy metals, and filters 70–90% of heavy sediments	Information not available	Information not available
Hydro Compliance Management, Inc. www.HydroCompliance.com 734-449-8860 800-526-9629	Hydro-Kleen™ Filter Systems	A sedimentation chamber and dual-media filters separate contaminants from stormwater	Hydrocarbons and other contaminates are reduced to nondetect levels	Information not available	Debris removed by vacuum and filters replaced every 4–6 months
KriStar Enterprises, Inc. www.kristar.com 800-579-8819 707-792-4665	Fossil Filter™	Trough structure installed at grate inlets to collect petroleum hydrocarbons; uses Fossil Rock adsorbent material to remove petroleum-based contaminates	Information not available	0.03 cfs/lf	Periodic inspection/maintenance to remove foreign objects; Fossil Rock should be replaced when the surface of the granules is more than 50% coated w/ contaminants, typically a 6-month period
Metro Chem, Inc. www.metrocheminc.com/draind.htm 800-590-2436	Drain Diaper™	Polypropylene fabric insert designed to fit most catch basins; equipped with overflow outlets	Information not available	Up to 300 gpm	Maintenance is required on an as needed basis to replace saturated filters
Pactec, Inc. www.drainpac.com 800-272-2832	DrainPac™	Multilayer filtration insert; can be custom made to conform to any configuration; equipped with a bypass overflow system	Reduces levels of heavy metals and petroleum hydrocarbons to a nondetectable status	0.11–0.31 cfs/ft^2	Information not available

Company	Product	Description	Capacity	Pollutant Removal	Maintenance
Stormwater Management www.stormwatermgt.com 503-240-3393 800-548-4667	StormFilter™	A filter chamber uses radial flow-filter cartridges in a variety of media to treat site-specific pollutants	0.13–1.0 cfs	Removes up to 90% of all solids, 85% of oils and greases, and 91% of solubilized heavy metals	Filter is maintained during the dry season; stormwater management can provide maintenance services
Gravity and Screen Separator Units					
AquaShield, Inc. www.aquashieldinc.com Phone: (423) 870-8888 Fax: (423) 870-1005	Aqua-Filter™	High-flow filtration featuring "Treatment Train" design (i.e., Swirl Concentrator™ for pretreatment followed by filter for polishing)	Information not available	Removes over 80 % TSS and 95% dissolved petroleum and oils; also removes soluble and insoluble nutrients such as phosphorus and nitrogen, and dissolved metals	Easy access, all inspection and maintenance can be performed from surface (no confined entry); AquaShield offers maintenance service package
BaySaver, Inc. www.baysaver.com 301-829-6470 800-229-7283	BaySaver™ Separation System	Two concrete manholes and a HDPE separator unit remove suspended solids, oils, and debris from stormwater runoff	21.8 max cfs	Removes 80% of suspended solids; holds 100–225.6 cf of sediments and 384–868 gal of oil	Removal of pollutants by vacuum truck
Best Management Products Inc. www.bestmp.com 888-354-7585 215-884-2345 215-884-6195 fax	The SNOUT	A hooded outlet cover installed at sumped catch basins to prevent oil and floatables from being drawn downstream	Information not available	Removes 80% of free oils, 95% of floatables, and 50% of suspended solids	Routine inspection and rinsing, flushing the antisiphon vent with water or air to clear
CDS Technologies, Inc. www.cdstech.com.au 407-681-4929 800-848-9955	CDS™ units	Unit filters water through a stainless steel perforated screen while solids sink to bottom	1.1–300 cfs (depending upon unit material)	Removes 80–90% of oils and grease and 100% of particles M the size of the screen opening	Cleaning of sump by vacuum truck 4 times per year; annual inspection of screen surface
Jay R. Smith Mfg. Co./ Environmental Products Group www.jayrsmith.com 334-277-8520 800-767-0466	Ultracept™ Oil/Water Separator	Stainless steel unit skims oil and grease from stormwater; a surge pit accumulates stormwater and pumps it to Ultracept	Information not available	Uses no coalescing plates or filters	Ten-min weekly maintenance; no maintenance contract offered
Practical Best Management LLC www.practicalbestmgmt.com 678-985-2976	CrystalStream Oil/Grit Separator	Stormwater flows into the precast concrete device, where it passes through a series of baffles	Information not available	Captures over 99% of petroleum products and nearly 95% of silt and grit	Maintenance is performed on an as-needed basis, usually every 3 months

TABLE 13-47 (Continued)
Summary Information on Commercial Devices

Manufacturer	Product Name	Description	Performance	Flow Capacity	Maintenance
Gravity and Screen Swirl Separator Units					
Aquashield, Inc. www.aquashieldinc.com 423-870-8888 423-870-1005 fax	Aquaswirl Concentrator	HDPE swirl concentrator with baffle system for oil skimming			Quarterly inspection of the swirl chamber must be performed; typically, annual cleanout is required; the system can be inspected and maintained completely from the surface to eliminate the need for confined space entry
H.I.L. Technology, Inc. www.hil-tech.com	Downstream Defender™	Stormwater spirals down the perimeter of a concrete cylinder, allowing heavier particles to settle out by gravity and the drag forces on the wall and base of the vessel	Removes over 90% of particles greater than 150 microns	0–13.0 cfs (depending on unit size)	Periodical sump vacuum procedure
Stormceptor Corporation www.stormceptor.com 301-762-8361 800-762-4703	Stormceptor™ units	A separation/storage chamber traps solids and oils	Traps up to 80% of inflowing fine and coarse sediment	0.17–2.47 cfs	Inspect on a monthly basis; sediment removed annually by vacuum truck; oil removed when levels surpass 1.0 in
Environment 21, LLC www.envi21.com 716-762-8216	V²B1™ Structural Stormwater Treatment System	Swirl chamber technology combined with vortex design principles remove sand, sediment, metals, and debris	Removes 80% of TSS	3–25.0 cfs	If unit is regularly maintained, only one chamber should need to be pumped
Vortechnics, Inc. www.vortechnics.com 207-878-3662	Vortechs™ Stormwater Treatment System	Hydrodynamic oil-and-grit separator custom-designed to meet local stormwater quality requirements	Removes 80% of annual TSS load; lab and field test data available	1.6–25.0 cfs	Unobstructed access to captured contaminants; Vortechs Systems typically require an annual cleanout

Teichert Precast www.teichert.com 916-386-6964 916-386-8128 fax	Teichert Interceptor	Baffle design, outlet T allows easy effluent sampling, no mechanical equipment	Removes oil droplets with diameters greater than or equal to 150 microns	0.54–1.28 cfs	Inspect throughout rainy season; usually vacuum out once or twice a year
Skimmers AbTech Industries www.oars97.com 480-874-4000 800-545-8999	OARS Passive Skimmer Line Skimmer	Disposable skimmer floats on top of water in a stormwater catch basin and absorbs hydrocarbons	Information not available	Information not available	Information not available
Foss Environmental www.fosscatalog.com 800-909-3677	Passive Skimmer	Floats in sump of catch basins and oil/water separators and absorbs hydrocarbons; absorbent media is contained within a screen pillow	Information not available	Information not available	Information not available
New Pig www.newpig.com 888-HOT-HOGS	Pig Sump Skimmer Pillow	Compact spaghetti-strand absorbent enclosed inside a polyester mesh net w/ attached nylon loop, for use in sumps to absorb petrochemicals only	Absorbs 1.8 gallons per pillow	Information not available	Replace when saturated
Constructed Wetlands Environmental Research Corps www.biofence.com 800-899-2025	Howland Swale™	Four components — siltration trap, pretreatment marsh, detention basin, and vegetated takeoff channel	Removes greater than 80% of TSS	Information not available	Information not available
StormTreat Systems, Inc. www.stormtreat.com 508-778-4449	StormTreat™ Systems	A sedimentation/filtration chamber and constructed contained wetlands remove a broad range of pollutants including bacteria, petroleum hydrocarbons, metals, and nutrients	Meets EPA's recommended 80% removal of TSS	Information not available	Annual inspections and replacement of grit-filter bed; sediment pumping once every 3–5 years using standard septic system pumper; maintenance contracts are available

Source: From CWP, 2000.

Multichambered Treatment Train

Stormwater runoff from paved urban hot spots, particularly automotive service and repair stations, can contain pollutant concentrations 3 to 600 times greater than those found in other urban sources. The higher potential for heavy stormwater pollutant loading becomes apparent when one also considers the multitude of potential hot spots located throughout urban areas. This being the case, it becomes prudent to treat a relatively small amount of runoff at the source, as opposed to allowing contaminated runoff to become part of a much larger volume that may or may not be effectively treated at the end of the pipe (CWP, 2000).

Effective, on-site treatment of stormwater hot spots has been a problem for several reasons. First, most hot spots tend to be small in size and lack adequate space for the installation of typical stormwater management practices such as ponds and wetlands. Second, the use of gravitational settling as a sole pollutant-removal mechanism does not provide sufficient hot spot pollutant removal. Third, infiltration is not an option due to risks of groundwater contamination. Last, the traditional underground approaches using oil-grit separators have not been reported to be effective (Schueler, 1994).

To help solve the hot spot treatment problem, Robert Pitt and his colleagues at the University of Alabama-Birmingham have developed and tested a prototype known as the multichambered treatment train (MCTT) (Pitt, 1996).

It is designed for underground use. Its size varies according to rainfall amount, intensity, and elapsed time between storms, as well as suspended sediment load and desired maintenance regime. Pitt has developed a computer model to aid in the site-specific design. The unit can be sized to contain runoff from various rain events and typically requires between 0.5 and 1.5% of the paved drainage area (Pitt, 2001). Long-term simulations have

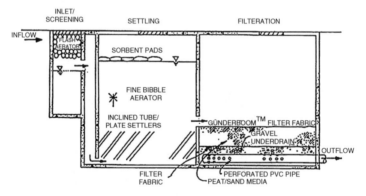

FIGURE 13-127
Multichambered treatment train.

FIGURE 13-128
MCTT Los Angeles CALTRANS site.

TABLE 13-48

Components of the MCTT

Chamber	Component	Description	Function
Inlet	Flash aerator	Small column packing balls with counter current air flow	Removes volatile pollutants and traps trash
	Catch basin sump	Conventional catch basin sump	Traps grit and sand-size particles
Settling	Sorbent pads	Floating absorbent pads	Traps oil and grease
	Fine bubble aerator	Generator-powered fish farm aeration stone	Enhances aeration
	Inclined tube or plate settlers	Plastic tubes 2" x 2', inclined 30–45 degrees, arranged in rows of opposing direction	Increases surfaces area of settling chamber, enhances sedimentation and prevents scour
Filtration	Gunderboom™ filter fabric	Covers top of filter	Reduces channelization, slows infiltration, sorbs oils
	Peat/sand filter media	50/50 mix, at least 12" depth	Removes small and dissolved particles, provides ion exchange
	Filter fabric	Separates peat/sand layer from gravel and pipe layer	Prevents gravel layer from clogging
	Gravel packed underdrain	Perforated PVC pipe and gravel	Provides additional filtration/outlet

been used in the past to size the unit. Costs are anticipated to come down for the design and construction of the MCTT as experience is gained. According to Pitt (2001), the second Wisconsin site cost about the same per acre of treatment as porous pavement.

The MCTT is divided into three main chambers with several options within each chamber (Figure 13-2):

- Stormwater enters the first chamber through a conventional catch basin inlet with litter traps, where the largest particulates are screened out and the bulk of highly volatile materials are removed when they pass over a flash aerator (additional components within each chamber are listed in Table 13-48).
- The stormwater then flows under gravity or is pumped into the settling chamber. Here, settling of fine sediment is enhanced through the use of inclined tube or lamella plate settlers, while floating hydrocarbons and additional volatile compounds are removed by sorbent pad skimmers and bubble diffusers.
- Next, the stormwater flows, or is pumped slowly into, the filtration chamber containing a sand and peat filter bed for final removal of dissolved toxicants. The filter also functions in the partial treatment of runoff that may have bypassed prior chambers in the event of excess stormwater flow. To ensure that the water volume is distributed evenly over the filter bed, a fabric covers the top of the filter. Activated carbon has been added to the Wisconsin test sites to reduce pH changes (Pitt, 2001).

Performance results of the monitoring indicate substantial reductions of total suspended solids, heavy metals, and dissolved and suspended stormwater toxicity from the unit overall. Monitoring data from two full-scale applications of the MCTT in Wisconsin and at CALTRANS appear to confirm that it can achieve consistently high removal rates for solids, nutrients, metals, and two polycyclic aromatic hydrocarbon (Corsi et al., 1999, Barrett, 2001). Removal ranges include 83 to 98% TSS, 60 to 86% COD, 40 to 94% turbidity, 80 to 88% phosphorous, 93 to 96% lead, 90 to 91% zinc, and 65 to 100% for many organic toxicants (Pitt, 2001).

The importance of each chamber depends on the type of pollutants entering the chamber. Many suspended pollutants were removed quite efficiently using just the settling process, whereas the filtration chamber was responsible for further reduction of those same pollutants as well as the additional removal of dissolved pollutants. Suspended solids were reduced somewhat by screening but were almost totally reduced by settling, while filtration was of no consequence. Toxicity was basically unaffected by screening, received slight treatment in the settling chamber, but was reduced significantly by filtration.

Maintenance of the MCTT includes:

- Removal of sediment from the first chamber when depth exceeds 6 in
- Removal and replacement of filter media every three years
- Replacement of sorbent pillows as necessary
- Weekly inspections during rainy periods for trash buildup
- Monthly or quarterly inspections to check for vandalism and other damage

Much of this information was taken from CWP (2000) and Pitt (2001).

References

American Association of State Highway and Transportation Officials, Model Drainage Manual, 1991.
American Petroleum Institute (API), Design and Operation of Oil-Water Separators, Publication 421, 1990.
Atlanta Regional Commission (ARC), Georgia Stormwater Manual, 2002.
Austin, Texas, Inventory of Urban Nonpoint Source Control Practices, Environmental Resources Division, City of Austin, June 1988a.
Austin, Texas, Environmental Quality Criteria Manual, 1988b.
Austin, Texas, Removal Efficiencies of Stormwater Control Structures, May 1990.
Austin, Texas, The First Flush of Runoff and Its Effects on Control Structure Design, June 1990.
Austin, Texas, Evaluation of Nonpoint Source Controls — An EPA/TNRCC Section 319 Grant Report, *Water Quality Report COA-ERM-97- 04*, Drainage Utility Department, Austin, TX, 1997.
Barbour, M.T., Gerritsen, J., Snyder, B.D., and Stribling, J.B., Rapid Bioassessment Protocols for Use in Streams and Wadeable Rivers: Periphyton, Benthic Macroinvertebrates, and Fish, 2nd ed., USEPA, Office of Water, Washington, DC, EPA 841-B-99-002, 1999.
Barrett, M., Performance Summary Report of the Multi-Chambered Treatment Train, prepared for the California Department of Transportation, May 2001.
Board on Science and Technology for International Development (BSTID), Vetiver Grass, National Academy Press, Washington, DC, 1993.
Bellevue, Washington, Standard Drawings, 1988.
Bellevue, Washington, Water Quality Protection for Landscaping Businesses, 1991.
Bitter, S.D. and Bowers, J.K., Bioretention as a water quality best management practice, in *The Practice of Watershed Protection*, Schueler, T.R. and Holland, H.K., Eds., The Center for Watershed Protection, Ellicott City, MD, 2000.
Brown, J.L. and Farnham, R.S., Use of Peat for Wastewater Filtration — Principles and Methods, Proc. 5th Int. Peat Congress, Poznan, Poland, 1976, pp. 349-357.
Brown, W. and Schueler, T., The Economics of Stormwater BMPs in the Mid-Atlantic Region, prepared for Chesapeake Research Consortium, Edgewater, MD, Center for Watershed Protection, Ellicott City, MD, 1997.
Burby, R.J., Kaiser, E.J., Miller, T.L., and Moreau, D.H., *Drinking Water Supplies*, Ann Arbor Science, Ann Arbor, MI, 1983, reprinted 1986.
California Stormwater Task Force, Best Management Practice Handbook — Vol. 1 Municipal, March 1993a.

California Stormwater Task Force, Best Management Practice Handbook — Vol. 2 Industrial/Commercial, March 1993b.

Carr, D.J., Geinopolos, A., and Zanoni, A.E., Characteristics and Treatability of Urban Runoff Residuals, EPA-600/2-82-094, November 1982.

Center for Watershed Protection (CWP), Design of Stormwater Filtering Systems, prepared for Chesapeake Research Consortium, Edgewater, MD, Center for Watershed Protection, Ellicott City, MD, 1996.

Center for Watershed Protection (CWP), Stormwater BMP Design Supplement for Cold Climates, prepared for U.S. EPA Office of Wetlands, Oceans and Watersheds, Washington, DC, 1997.

Center for Watershed Protection (CWP), Better Site Design, Ellicott City, MD, 1998.

Center for Watershed Protection (CWP), Stormwater strategies for arid and semi-arid watersheds, in *The Practice of Watershed Protection*, Schueler, T.R. and Holland, H.K., Eds., Center for Watershed Protection, Ellicott City, MD, 2000a.

Center for Watershed Protection (CWP), Multi-chamber treatment train developed for stormwater hot spots, Technical Note #87 from *Watershed Protection Techniques*, 2, 3, 11–13, in *The Practice of Watershed Protection*, Schueler, T.R. and Holland, H.K., Eds., The Center for Watershed Protection, Ellicott City, MD, 2000b.

Center for Watershed Protection (CWP), Stormwater Fact Sheets — Grass Channel, 2001a.

Center for Watershed Protection (CWP), Stormwater Fact Sheets — Grassed Filter Strip, 2001b.

Center for Watershed Protection (CWP), Stormwater Fact Sheets — Wetlands, 2001c.

Chang, G.J. and Souer, P.C., The First Flush of Runoff and Its Effect on Control Structure Design, Department of Environment and Conservation Services, Austin, TX, 1990.

Chescheir, G.M., Fipps, G., and Skaggs, R.W., Analysis of Stormwater Infiltration Ponds on the North Carolina Outer Banks, North Carolina State University, Report No. 254, September 1990.

Chocat, B., Barraud, S., Alfakif, E., Development of Best Management Practices: Progress in France and Western Europe, Proc. ASCE Eng. Found. Conf.: Linking Stormwater BMP Designs and Performance to Receiving Water Impact Mitigation, Urbonas, B., Ed., August 19–24, Snowmass Village, CO, 2001, pp. 336–353.

City of Sacramento, California, Guidance Manual for On-Site Stormwater Quality Control Measures, Department of Utilities, 2000.

Clar, M. and Rushton, B., Low Impact Development Case Studies, Proc. ASCE Eng. Found. Conf.: Linking Stormwater BMP Designs and Performance to Receiving Water Impact Mitigation, Urbonas, B., Ed., August 19–24, Snowmass Village, CO, 2001, pp. 484–488.

Cloak, D. and Bicknell, J.C., Use of Environmental Indicators to Assess Stormwater Program, Proc. ASCE Eng. Found. Conf.: Linking Stormwater BMP Designs and Performance to Receiving Water Impact Mitigation, Urbonas, B., Ed., August 19–24, Snowmass Village, CO, 2001, pp. 305–315.

Coffman, L. and Winogradoff, D., Bioretention: An Efficient, Cost-Effective Stormwater Management Practice, Proc. Natl. Conf. on Retrofit Opportunities for Water Resour., Protection in Urban Environments, U.S. EPA, Office of Research and Development, EPA/625C-99/001, Washington, DC, 1999, pp. 259–264.

Conradin, F., Study of catch basins in Switzerland, *Urban Stormwater Quality Enhancement*, ASCE, 1989.

Cooke, S. S., Wetland Buffers, Washington Department of Ecology, 1991.

Corsi, S.R., Greb, S.R., Bannerman, R.T., and Pitt, R.E., Evaluation of the Multi-Chambered Treatment Train, USGS Open File Report, 99-270, 1999.

Cox, J.H. and Livingston, E.H., Stormwater Sediments: Hazardous Waste or Dirty Dirt, Fifth Biennial Stormwater Research Conference, SWFWD, 1997.

Currier, B., Infiltration BMPs: CALTRANS Retrofit Pilot Study Experience, University of California, Davis, Center for Water Res. and Environ. Eng., 2002.

Davis, A., Shokouhian, M., Sharma, H., and Minami, C., Optimization of Bioretention for Water Quality and Hydrological Characteristics, Final Report: 01-4-31032, University of Maryland, Department of Civil Engineering, Prince George's County Department of Environmental Resources, Landover, MD, 1998.

Day, G.E., Smith, D.R., and Bowers, J., Runoff and Pollution Abatement Characteristics of Concrete Grid Pavements, Virginia Water Resources Research Center, Virginia Polytechnic Institute and State University, Rpt. PB82-168469, Bull. 135, October 1981.

Denver Urban Drainage and Flood Control District (UDFCD), Drainage Criteria Manual Vol. 3, Denver, CO, 1999.

Denver Urban Drainage and Flood Control District (UDFCD), Flood Hazard News, Vol. 31, No. 1, 2001.

Dillaha, T.A., Reneau, R.B., Mostaghimi, S., and Lee, D., Vegetative Filter Strips for Agricultural Non-Point Source Pollution Control, Trans. ASAE, 1989.

Diniz, E.V., Quantifying the Effects of Porous Pavements on Urban Runoff, Natl. Symp. on Urban Hydrology, Hydraul., and Sediment Control, Lexington, KY, 1976.

Diniz, E.V., Porous Pavement — Phase I, Design and Operational Criteria, USEPA 600/2-80-135, August 1980.

Driscoll, E.D., Detention and retention controls for urban runoff, *Urban Runoff Quality*, ASCE, 1986.

Driscoll, E.D., Long-term performance of water quality ponds, *Urban Runoff Quality Controls*, ASCE, 1988.

Driscoll, E.D. and Strecker, E.W., Assessment of BMPs Being Used in the U.S. and Canada, Proc. 6th Int. Conf. on Urban Storm Drainage, Niagara Falls, Canada, September 1993.

Driscoll, E.D., Palhegyi, G.E., Strecker, E.W., and Shelley, P.E., Analysis of Storm Event Characteristics for Selected Rainfall Gages Throughout the United States, Woodward-Clyde Consultants, prepared for U.S. EPA, Washington, DC, 1989.

Elling, A.E., Managing Vegetation on Peat-Sand Filter Beds for Wastewater Disposal, USDA Forest Service, North Central Forest Experiment Station, Research Note NC-333, 1985.

Enviro-Drain, Informational Brochure, Kirkland, WA, 1994.

Farnham, R.S. and Brown, J.L., Advanced Wastewater Treatment Using Organic and Inorganic Materials, Part 1 — Use of Peat and Peat-Sand Filtration Media, Proc. 4th Int. Peat Congress, Helsinki, Finland, Vol. 4, 1972, pp. 272–286.

Farnham, R.S. and Noonan, T., An Evaluation of Secondary Treatment of Stormwater Inflows in Como Lake, MN, Using a Peat-Sand Filter, Como Lake Restoration, EPA Project. S-005660-02, 1988.

Field, R., Urban runoff: pollution sources, control, treatment, *Wat. Res. Bull.*, AWRA, 21, 2, April 1985.

Field, R., Masters, H., and Singer, M., An overview of porous pavement research, *Wat. Res. Bull.*, AWRA, 18, 2, April 1982.

FISRWG, Stream Corridor Restoration: Principles, Processes, and Practices, The Federal Interagency Stream Restoration Working Group (FISRWG) (15 federal agencies of the U.S. government), GPO Item No. 0120-A; SuDocs No. A 57.6/2:EN 3/PT.653, 1998.

Galli, J., Peat Sand Filters, Coordinated Anacostia Retrofit Program, December 1990.

Galli, J., Peat Sand Filters, *Watershed Restoration Sourcebook*, Metropolitan Washington Council of Governments (MWCOG), Washington, DC, 1992a.

Galli, J., Analysis of Urban BMP Performance and Longevity in Prince George's County Maryland, Metropolitan Washington Council of Governments (MWCOG), Washington, DC, 1992b.

Gburek, W.J. and Urban, J.B., Stormwater Detention and Groundwater Recharge Using Porous Asphalt — Experimental Site, Proc. Int. Symp. on Urban Storm Runoff, University of Kentucky, Lexington, 1980.

Georgia State Soil and Water Conservation Committee, Manual for Erosion and Sediment Control in Georgia, undated.

Ghioto, Singhofen and Assoc., Inc., Stormwater Rule Design Practices and Methods of Calculation, in *Design of Stormwater Facilities*, workshop sponsored by North Carolina Division of Environmental Management and Professional Engineers of North Carolina, 4, May 1988.

Gilliam, J.W. and Skaggs, R.W., Nutrient and sediment removal in wetland buffers, in *Wetland Hydrology, Proc. Natl. Wetland Symp.*, Assoc. of State Wetland Managers, Chicago, IL, September 1987.

GKY, Outlet Hydraulics of Extended Detention Facilities, report for the Northern Virginia Planning Commission, 1989.

Glick, R., Chang, G., and Barret, M., Monitoring and evaluation of stormwater quality control basins, in *Proc. Watershed Manage.: Moving from Theory to Implementation*, Water Environ. Federation Specialty Conf., Denver, CO, 3–6, May 1998.

Gorbisch, E., The do's and don'ts of wetland planning, *Wetland J.*, 6, 1, 1994.
Grizzard, T.L., Randall, C.W., Weand, B.L., and Ellis, K.L., Effectiveness of extended detention ponds, *Urban Runoff Quality*, ASCE, 1986.
Hammer, D.A., Ed., *Constructed Wetlands for Wastewater Treatment*, Lewis Publishers, 1989.
Hammer, D.A., *Creating Freshwater Wetlands*, Lewis Publishers, 1992.
Harper, H. Effects of Stormwater Management Systems on Groundwater Quality, Final Report, Environmental Research and Design, Inc., prepared for Florida Department of Environmental Regulation, Tallahassee, FL, 1988.
Harper, H.H. and Herr, J., P.E., Alum treatment of stormwater runoff: an innovative BMP for urban runoff problems, Proc. Natl. Conf. on Retrofit Opportunities for Water Resour. Protection in Urban Environ., 1998, http://www.stormwater-resources.com/library.htm#BMPs.
Harper, H.H., Herr, J., P.E., and Livingston, E., Alum Treatment of Stormwater — The First Ten Years: What Have We Learned and Where Do We Go From Here?, 1997, http://www.stormwater-resources.com/library.htm#BMPs.
Harrington, B., Feasibility and Design of Wet Ponds to Achieve Water Quality Control, Sediment and Stormwater Division, Maryland Department of the Environment, 1987.
Hartigan, J.P., Basis for design of wet detention basin BMPs, *Urban Runoff Quality Controls*, ASCE, 1988.
Heraty, M., Riparian Buffer Programs, U.S. EPA, 1993.
Herr, P.E. and Harper, H., The Evaluation and Design of an Alum Stormwater Treatment System to Improve Water Quality in Lake Maggiore in St. Petersburg, Florida, 1997, http://www.stormwater-resources.com/library.htm#BMPs.
Herricks, E.E., Observed Stream Responses to Changes in Runoff Quality, Proc. ASCE Eng. Found. Conf.: Linking Stormwater BMP Designs and Performance to Receiving Water Impact Mitigation, Urbonas, B., Ed., August 19–24, Snowmass Village, CO, 2001, pp. 145–157.
Horner, R.R., Biofiltration Systems for Storm Runoff Water Quality Control, Report to Washington State Department of Ecology, 1988a.
Horner, R.R., Long-term effects of urban stormwater on wetlands, *Urban Runoff Quality Controls*, ASCE, 1988b.
Horner, R.R., Biofiltration for Storm Runoff Water Quality Control, prepared for the Washington State Department of Ecology, Center for Urban Water Resources Management, University of Washington, Seattle, WA, 1993.
Horner, R.R., Skupien, J.J., Livingston, E.H., and Shaver, E.H., Fundamentals of Urban Runoff Management: Technical and Institutional Issues, Terrene Institute and U.S. Environmental Protection Agency, Washington DC, 1994.
Huber, W.C., Deterministic Modeling of Urban Runoff Quality, in *Urban Runoff Pollution*, NATO ASI Series, Vol. G10, Springer-Verlag, Berlin, 1986.
Hvitved-Jacobsen, T., Yousef, Y.A., and Wanialista, M.P., Rainfall Analysis for Efficient Detention Ponds, *Urban Runoff Quality Controls*, ASCE, 1988.
Idaho DEQ, Catalog of Stormwater Best Management Practices, 2001, http://www2.state.id.us/deq/water/stormwater_catalog.
Infiltrator Systems, Inc., The Infiltrator, Old Saybrook, CT, undated.
Karouna, N., Native Plant Pondscaping Guide, Watershed Restoration Sourcebook, Metropolitan Washington Council of Governments (MWCOG), April 1992.
Karr, J.R., Fausch, K.D., Angermeier, P.L., Yant, P.R., and Schlosser, I.J., Assessing Biological Integrity in Running Waters, Illinois Natural History Survey, Special Pub. No. 5, 1986.
Kayata, T. and Fujita, S., Pollution Abatement in Tokyo "ESS", Urban Stormwater Quality Enhancement, ASCE, 1989.
Kent, W.A., On Site Infiltration Tests, Development Assistance Brochure #5-10, May 2000.
King County Surface Water Management, Design Manual, September 1998.
King County Department of Natural Resources (KCDNR), Lake Sammamish Water Quality Management Project, Final Report, Washington State Department of Ecology, Seattle, WA, 1998.
Koon, J., Evaluation of Water Quality Ponds and Swales in the Issaquah/East Lake Sammamish Basins, King County Surface Water Management and Washington Department of Ecology, Seattle, WA, 1995.

Kluh, S., Metzger, M.E., Messer, D.F., Hazelrigg, J.E., and Madon, M.B., Stormwater, BMPs, and vectors: The impact of new BMP construction on local public health agencies, *Stormwater Magazine*, 3(2), 40–46, 2002.

Lager, J., Smith, W., Finn, R., and Finnermore, E., Urban Stormwater Management and Technology: Update and Users Guide, U.S. EPA, EPA-600/8-77-014, 1997.

Lawrence, I., Australian Urban Water Best Management Practices Strategic Review, Proc. ASCE Eng. Found. Conf.: Linking Stormwater BMP Designs and Performance to Receiving Water Impact Mitigation, Urbonas, B., Ed., August 19–24, Snowmass Village, CO, 2001, pp. 369–386.

Leif, W., Compost Stormwater Filter Evaluation: Final Report, Snohomish County Department of Public Works. Surface Water Management Division, Everett, WA, 1999.

Lettenmaier, D. and Richey, J., Operational Assessment of a Coalescing Plate Oil/Water Separator, City of Seattle, WA, 1985.

Livingston, E., The Use of Wetlands for Urban Stormwater Management, Urban Runoff Quality Controls, ASCE, 1988.

Lower Colorado River Authority (LCRA), Final Report: Innovative Nonpoint Source Pollution Program for Lake Travis in Central Texas, prepared for Environmental Protection Agency and the Texas Natural Resource Conservation Committee, Contract No. 1900000019, 1997.

Maestri, B. et al., Managing Pollution from Highway Stormwater Runoff, Transportation Research Board, National Academy of Science, Transportation Research Record Number 1166, 1988.

Maine Department of Environmental Protection, Phosphorus Control in Lake Watersheds, September 1992.

Maryland Department of the Environment (MDE), Standards and Specifications for Infiltration Practices, Sediment and Stormwater Administration, February 1984a.

Maryland Department of the Environment (MDE), Standards and Specifications for Porous Pavement, Sediment and Stormwater Administration, February 1984b.

Maryland Department of the Environment (MDE), Inspector's Guidelines Manual for Stormwater Management Infiltration Practices, Sediment and Stormwater Administration, December 1985.

Maryland Department of the Environment (MDE), Minimum Water Quality Objectives and Planning Guidelines for Infiltration Practices, Sediment and Stormwater Administration, April 1986.

Maryland Department of the Environment (MDE), Results of the State of Maryland Infiltration Practices Survey, Sediment and Stormwater Administration, August 1987.

Maryland Department of the Environment (MDE), Water Quality Inlets, Vincent Berg, Sediment and Stormwater Administration, January 1991a.

Maryland Department of the Environment (MDE), Stormwater Infiltration Practices in Maryland: A Second Survey, Sediment and Stormwater Administration, 1991b.

Maryland Department of the Environment (MDE), Maryland Stormwater Design Manual, Volumes I and II, prepared by Center for Watershed Protection (CWP), 2000.

Massachusetts Department of Environmental Protection/Massachusetts Office of Coastal Zone Management, Stormwater Management — Volume One: Stormwater Policy Handbook, and Volume Two: Stormwater Technical Handbook, 1997.

McPherson, J., Water Quality BMPs: Catch Basin Infiltration, presented to the APWA Stormwater Managers Committee, Tacoma, WA, 1992.

Metropolitan Council Environmental Services (MCES), Minnesota Small Sites Urban BMP Manual, Mears Park Center, St. Paul, MN, 2001.

Metropolitan Washington Council of Governments, Controlling Urban Runoff, T. Schueler, Washington, DC, 1987.

Metropolitan Washington Council of Governments, A Current Assessment of Urban Best Management Practices — Techniques for Reducing Nonpoint Source Pollution in the Coastal Zone, Washington, DC, 1992.

Metzger, M.E., Messer, D.F., Beitia, C.L., Myers, C.M., and Kramer, V.L., The dark side of stormwater runoff management: Disease vectors associated with structural BMPs, *Stormwater Magazine*, 3(2), 24–39, 2002.

Mikkelsen, P.S. et al., Best Management Practices in Urban Stormwater Management in Denmark and Sweden , Proc. ASCE Eng. Found. Conf.: Linking Stormwater BMP Designs and Performance to Receiving Water Impact Mitigation, Urbonas, B., Ed., August 19–24, Snowmass Village, CO, 2001, pp. 354–368.

Minnesota Pollution Control Agency, Protecting Water Quality in Urban Areas, Division of Water Quality, 1989.

Minton, G.R., Bucich, P., Blosser, M., Leif, B., Lenhatt, J., Simmler, S., New Stormwater Treatment BMPs: Determining Acceptability to Local Implementing Agencies, Natl. Conf. on Tools for Urban Water Resour. Manage. and Prot. Proc., Chicago, IL, EPA/625/R-00/001, July 2000.

Mitsch, W.J. and Gosselink, J.G., *Wetlands*, Van Nostrand Reinhold, New York, 1986.

New Jersey Department of Environmental Protection, Evaluation and Recommendations Concerning Buffer Zones Around Public Water Supply Reservoirs, Report to Governor Kean, 1989.

Niemczynowicz, S., *Swedish Way to Stormwater Enhancement, Urban SW Quality Enhancement*, Torno, H.C., Ed., ASCE, 1990, pp. 156–168.

Nightingale, H.I., Accumulation of AS, NI, CU and PB in retention and recharge basins soils for urban runoff, *Water Resour. Bull.*, AWRA, 23, 4, 1987.

Nix, S.J., Heaney, J.P., and Huber, W.C., Suspended solids removal in detention basins, *ASCE J. Env. Eng.*, 114, 6, 1988.

North Carolina Sediment Control Commission, Erosion and Sediment Control Planning and Design, North Carolina Department of Natural Resources and Community Development, Division of Land Resources, 1991.

North Carolina Department of Environment, Health and Natural Resources (NCDEHNR), An Evaluation of Vegetative Buffer Areas for Water Quality Protection, Raleigh, NC, February 1991.

North Central Texas Council of Governments, Stormwater Quality Best Management Practices for Residential and Commercial Land Uses, 1st ed., Arlington, TX, July 1993.

Northern Virginia Planning District, Northern Virginia BMP Handbook, Northern Virginia Planning District Commission and Engineers and Surveyors Institute of Northern Virginia, November 1992.

Oberts, G.L., Influence of snowmelt dynamics on stormwater runoff quality, *Techniques for Watershed Protection*, 1, 2, Summer 1994.

Ohrel, R., Simple and complex stormwater load models compared, in *The Practice of Watershed Protection*, Schueler, T.R. and Holland, H.K., Eds., The Center for Watershed Protection, Ellicott City, MD, 2000.

Ontario Ministry of the Environment (OME), Stormwater Management Planning and Design Manual, Toronto, Canada, 1999.

Palfrey, R. and Bradley, E., Buffer Area Study, Maryland Department of Natural Resources, Coastal Research Division, Tidewater Administration, 1982.

Phillips, J. D., An evaluation of factors determining the effectiveness of water quality buffer zones, *J. Hydrology*, 107, 133–145, 1989.

Pitt, R., The Control of Toxicants at Critical Source Areas, The University of Alabama at Birmingham, paper presented at the ASCE/Eng. Found. Conf., Snowbird, UT, August 1996.

Pitt, R., Full-Scale Tests of the Multi-Chambered Treatment Train (MCTT), Proc. ASCE Eng. Found. Conf.: Linking Stormwater BMP Designs and Performance to Receiving Water Impact Mitigation, Urbonas, B., Ed., August 19–24, Snowmass Village, CO, 2001, pp. 529–533.

Pitt, R. and Voorhees, J., Source Loading and Management Model (SLAMM), Seminar Publication: National Conference on Urban Runoff Management: Enhancing Urban Watershed Management at the Local, County, and State Levels, March 30–April 2, 1993; Center for Environmental Research Information, U.S. Environmental Protection Agency, EPA/625/R-95/003, Cincinnati. OH, April 1995, pp. 225–243.

Pitt, R., Lilbum, M., Nix, S., Durrans, S., and Burian, S., Guidance Manual for Integrated Wet Weather Flow Collection and Treatment Systems for Newly Urbanized Areas, U.S. EPA Office of Research and Development, 1997.

Presto Products Company, Geoblock Porous Pavement System: Product Overview, 1990.

Prince Georges County (PGC), Low Impact Development Design Strategies, Department of Environmental Resources, 1999.

Prince George's County Department of Environmental Resources (PGDER), Design Manual for Use of Bioretention, in *Stormwater Management*, Division of Environmental Management, Watershed Protection Branch, Landover, MD, 1993.

Professional Engineers of North Carolina, Design of Stormwater Control Facilities, Workshop Proceedings, May 4, 1988.

Ragahavan, R., Koustas, R.N., and Liao, S., Cost Estimating for Best Management Practices, Proc. ASCE Eng. Found. Conf.: Linking Stormwater BMP Designs and Performance to Receiving Water Impact Mitigation, Urbonas, B., Ed., August 19–24, Snowmass Village, CO, 2001, pp. 196–209.

Randall, C.W., Stormwater Detention Ponds for Water Quality Control, ASCE Conf. on Stormwater Detention Facilities, Henniker, NH, 1982.

Rawls, W.J., Brakensiek, D.L., and Saxon, K.E., Estimation of soil properties, *Trans. ASAE*, 25, 5, 1982.

Readling, K., Flow Spreaders and Buffer Strips, Ogden Environmental BMP Fact Sheet, 1992.

Reckhow, K.H., Empirical models for trophic state in Southeastern U.S. lakes and reservoirs, *Water Resour. Bull.*, 24, 4, 1988.

Reeves, E., Performance and condition of biofilters in the Pacific Northwest, in *Practice of Watershed Protection*, Article 112, Center for Watershed Protection, Ellicott City, MD, 2000.

Riley, A.L., Restoring Streams in Cities, Island Press, Washington, DC, 1997.

Roesner, L.A., Burgess, E.H., and Aldrich, J.A., The hydrology of urban runoff quality, Water Resour. Plann. and Manage., ASCE, 1989.

Rosgen, D., A Classification of Natural Rivers, CATENA 22, 1994, pp. 169–199.

Schueler, T., Performance of stormwater pond and wetland systems, in *Engineering Hydrology*, Kuo, C., Ed., ASCE, 1993.

Schueler, T., Developments in sand filter technology to improve stormwater runoff quality, *Techniques for Watershed Protection*, 1, 2, Summer 1994a.

Schueler, T., Hydrocarbon hotspots in the urban landscape: can they be controlled, *Techniques*, 1, 1, February 1994b.

Schueler, T. and Galli, J., The Environmental Impacts of Stormwater Ponds, Watershed Restoration Sourcebook, Metropolitan Washington Council of Governments, Washington, DC, April 1992.

Schueler, T. and Shepp, D., The Quality of Trapped Sediments and Pool Water within Oil Grit Separators in Suburban Maryland, Washington Council of Governments, 1993.

Schueler, T., Galli, J., Herson, L., Kumble, P., and Shepp, D., Developing Effective BMP Systems for Urban Watersheds, Watershed Restoration Sourcebook, Metropolitan Washington Council of Governments, Washington, DC, April 1992.

Schueler, T.J., Design of Stormwater Wetland Systems, Metropolitan Washington Council of Governments, Washington, DC, October 1992.

Schueler, T.J., Further developments in sand filter technology, in *The Practice of Watershed Protection*, Schueler, T.R. and Holland, H.K., Eds., Center for Watershed Protection, Ellicott City, MD, 2000.

Schueler, T.J. and Caraco, D.S., Prospects for Low Impact Development at Watershed Level, Proc. ASCE Eng. Found. Conf.: Linking Stormwater BMP Designs and Performance to Receiving Water Impact Mitigation, Urbonas, B., Ed., August 19–24, Snowmass Village, CO, 2001, pp. 196–209.

Schueler, T.R. and Helfrich, M., Design of extended detention wet pond systems, *Urban Runoff Quality Controls*, ASCE, 1988.

Segarra-Garcia, R. and Loganathan, V.G., Design of Stormwater Detention Ponds Subject to Random Pollution Loading, Proc. 6th Int. Conf. on Urban Storm Drainage, Niagara Falls, Canada, September 1993.

Shaver, E., Sand Filter Design for Water Quality Treatment, Delaware Department of Natural Resources, 1991.

Shenot, S., An Analysis of Wetland Planting Success at Three SW Management Ponds in Montgomery, MD, M.S. Thesis, University of Maryland, 1993.

Smith, R.W., Rice, J.M., and Spelman, S.R., Design of Open-Graded Asphalt Friction Courses, Federal Highway Administration, FHWA-RD-74-2, January 1974.

South Florida Watershed Management District, Permit Information Manual, Vol. IV, May 1994.

Southeastern Wisconsin Regional Planning Commission (SWRPC), Costs of Urban Nonpoint Source Water Pollution Control Measures, Tech. Report No. 31, 1991.

Southwest Florida Watershed Management District, Permit Information Manual, Vol. I, Revised, March 1990; Vol. IV, 1986.

Stanley, D., An Evaluation of the Pollutant Removal of a Demonstration Urban Stormwater Detention Pond, Albermarle-Pamlico Estuarine Study, Report 94-07, 1994.

Stewart, W., Compost Stormwater Treatment System, W&H Pacific Consultants, Draft Report, Portland, OR, 1992.

Stolzenbach, K.D., Lu, R., Xiong, C., Friedlander, S., and Turco, R., Measuring and Modeling of Atmospheric Deposition on Santa Monica Bay and the Santa Monica Bay Watershed, Institute of the Environment, University of California at Los Angeles, Final Report to the Santa Monica Bay Restoration Project, September 2001.

Strecker, E., Quigley, M.M., and Urbonas, B.R., Determining Urban Stormwater BMP Effectiveness, Proc. Natl. Conf. on Tools for Urban Water Resour. Manage. and Protection, Chicago, IL, February 7–10, USEPA EPA-625-R-00-001, February 2000, pp. 246–267.

Stribling, J.B., Cummins, J.D., Galli, J., Meigs, S., Coffman, L., and Cheng, M., Relating In-stream Biological Condition to BMPs in the Watershed, Proc. ASCE Eng. Found. Conf.: Linking Stormwater BMP Designs and Performance to Receiving Water Impact Mitigation, Urbonas, B., Ed., August 19–24, Snowmass Village, CO, 2001, pp. 287–304.

Swietlik, W. F., Urban Aquatic Life Uses — A Regulatory Perspective, Proc. ASCE Eng. Found. Conf.: Linking Stormwater BMP Designs and Performance to Receiving Water Impact Mitigation, Urbonas, B., Ed., August 19–24, Snowmass Village, CO, 2001, pp. 163–180.

Swift, L.W., Filter strip widths for forestry roads in the Southern Appalachians, *Southern J. Applied Forestry*, 10, 1, 1986.

Tasker, D.T. and Driver, N.E., Nationwide regression models for predicting urban runoff water quality at unmonitored sites, *Water Resour. Bull.*, 24, 5, October 1988.

Techniques, Innovative Leaf Compost System used to Filter Runoff at Small Sites in Northwest, 1, 1, February 1994.

Texas Natural Resource Conservation Commission, Texas Nonpoint Source Book, 2000, www.txnpsbook.org.

Thelen, E. and Howe, L.F., Porous Pavement, Franklin Institute Press, Philadelphia, PA, 1978.

Thelen, E., Grover, W.C., Hoiberg, A.J., and Haigh, T.I., Investigation of Porous Pavements for Urban Runoff Control, The Franklin Institute, Philadelphia, PA, USEPA 11034 DUY 03/72, March 1972.

Thunhorst, G., Wetland Planting Guide for the Northeastern United States, Environmental Concerns, Inc., St. Michaels, MD, 1993.

Tomaseck, M.D., Johnson, G.E., and Mulloy, P.J., Operational problems with a soil filtration system for treating stormwater, *Lake and Reservoir Manage.*, 3, 306–313, 1987.

United States Environmental Protection Agency, Biological Criteria — National Program Guidance for Surface Waters, EPA 440/5-90-004.

United States Environmental Protection Agency, Storm Water Management for Construction Activities, EPA 832-R-92-005, September 1992a.

United States Environmental Protection Agency, Storm Water Management for Industrial Activities, USEPA 832-R-92-006, September 1992b.

United States Environmental Protection Agency — Region 7, Restoring and Creating Wetlands, USEPA, September 1992c.

United States Environmental Protection Agency, Investigation of Inappropriate Pollutant Entries into Storm Drainage Systems, USEPA 600/R-92/238, January 1993.

United States Environmental Protection Agency, Preliminary Data Summary of Urban Storm Water Best Management Practices, USEPA 821-R-99-012, August 1999a.

United States Environmental Protection Agency, Storm Water Technology Fact Sheet: Vegetated Swales, EPA 832-F-99-006, Office of Water, 1999b.

United States Environmental Protection Agency, Storm Water Technology Fact Sheet: Sand Filters, EPA 832-F-99-007, Office of Water, 1999c.

United States Environmental Protection Agency, Storm Water Technology Fact Sheet: Bioretention, EPA 832-F-99-012, Office of Water, 1999d.

United States Environmental Protection Agency, Storm Water Technology Fact Sheet: Infiltration Trench, EPA 832-F-99-019, Office of Water, 1999e.

United States Environmental Protection Agency, Storm Water Technology Fact Sheet: Storm Water Wetlands, EPA 832-F-99-025, Office of Water, 1999f.

United States Environmental Protection Agency, Storm Water Technology Fact Sheet: Modular Treatment Systems, EPA 832-F-99-044, Office of Water, 1999g.

United States Environmental Protection Agency, Stressor Identification Guidance, EPA 822-B-00-025, 2000.

United States Environmental Protection Agency, Draft Data summary for the Construction and Development Industry, Office of Water, Office of Science and Technology, Engineering and Analysis Division, Washington, DC, February 2001.

U.S. Soil Conservation Service, Soils — Hydric Soils of the United States, Natural Bull. No. 430-5-9, Washington, DC, 1985.

University of British Columbia (UBC), Brookswood, Township of Langley, Alternative Stormwater Management, Tech. Bull. #5, October 2000.

Urban, J.B. and Gburek, W.J., Stormwater Detention and Groundwater Recharge Using Porous Asphalt — Initial Results, Proc. Int. Symp. on Urban Storm Runoff, University of Kentucky, Lexington, 1980.

Urbonas, B., Summary of Emergent Stormwater Themes, Proc. ASCE Eng. Found. Conf.: Linking Stormwater BMP Designs and Performance to Receiving Water Impact Mitigation, Urbonas, B., Ed., August 19–24, Snowmass Village, CO, 2001, pp. 1–8.

Urbonas, B. and Ruzzo, W.P., Standardization of Detention Pond Design for Phosphorus Removal, Urban Runoff Pollution, NATO ASI Series Vol. G10, Springer-Verlag, Berlin, 1986.

Urbonas, B. and Stahre, P., *Stormwater Best Management Practices and Detention*, Prentice Hall, Englewood Cliffs, NJ, 1993.

Urbonas, B., Guo, C.Y., and Tucker, L.S., Optimization of Stormwater Quality Capture Volume, Urban Stormwater Quality Enhancement, Proc. Eng. Found. Conf., ASCE, New York, 1990.

USDA, Chesapeake Bay Riparian Handbook, Palone, R.S., Ed., USDA Forest Service Northeastern Area — State and Private Forestry, Morgantown, WV, and Albert H. Todd Chesapeake Bay Program Liaison USDA Forest Service Northeastern Area — State and Private Forestry, Annapolis, MD, May 1997; revised June 1998.

Virginia Department of Conservation and Recreation, Virginia Erosion and Sediment Control Handbook, Division of Soil and Water Conservation, 3rd ed., 1992.

Virginia Department of Conservation and Recreation, Virginia Stormwater Management Handbook, Volumes I and II, 1999.

Walker, W.W., Empirical Methods for Predicting Eutrophication in Impoundments — Report 3: Model Refinements, USAWES Technical Report E-81-9, U.S. Army Corps of Engineers, Vicksburg, MS, 1985.

Walker, W.W., Phosphorus Removal by Urban Runoff Detention Basins, Lake and Reservoir Management: Vol. III, North American Lake Management Society, Washington, DC, 1987.

Wang, T.S., Spyridakis, D.E., Mar, B.W., and Horner, R.R., Transportation, Deposition and Control of Heavy Metals in Highway Runoff, FHWA-WA-RD-39.10, Washington State University, Department of Civil Engineering, Seattle, WA, 1982.

Wanielista, M.P., Yousef, Y.A., Golding, B.L., and Cassagnol, C.L., Stormwater Management Manual, University of Central Florida, October 1981.

Wanielista, M.P. and Yousef, Y.A., *Stormwater Management*, Wiley Interscience, New York, 1993.

Washington, DC, Stormwater Design Handout, City of Washington, Department of Public Works, 1992.

Washington State Department of Ecology, Puget Sound Stormwater Management Manual, February 1992.

Washington State Department of Ecology (WDE), Stormwater Management Manual for Western Washington, 2001.

Watershed Management Institute (WMI), Operation Maintenance and Management of Stormwater Management Systems, prepared for U.S. EPA, Office of Water, 1997.

Watershed Protection Techniques, Pollutant Dynamics of Pond Muck, 1, 2, Summer 1994b.

Whitlow, T.H. and Harris, R.W., Flood Tolerance in Plants: a State-of-the-Art Review, USAEWES Technical Report E-79-2, Vicksburg, MS, 1979.

Wigington, P.J., Randall, C.W., and Grizzard, T.J., Accumulation of selected trace metals in soils of urban runoff drain swales, *Water Resour. Bull.*, 22, 1, February 1986.

Winer, R., National Pollutant Removal Performance Database for Stormwater BMPs, Center for Watershed Protection, U.S. EPA, Office of Science and Technology, 2000.

Wisconsin Department of Natural Resources/University of Wisconsin Extension, The Wisconsin Stormwater Manual, 2000.

Wong, S.L. and McCuen, R.H., The Design of Vegetative Buffer Strips for Sediment and Runoff Control, Stormwater Management in Coastal Areas, Tidewater Administration, Maryland Department of Natural Resources, Annapolis, MD, 1982.

Woodward-Clyde Consultants, Methodology for Analysis of Detention Basins for Control of Urban Runoff Quality, prepared for EPA Office of Water, Washington, DC, 1986.

Woodward-Clyde Consultants, Urban Targeting and BMP Selection, The Terrene Institute, Washington, DC, 1990.

Yousef, Y.A., Wanielista, M.P., and Harper, H.H., Removal of Highway Contaminants by Roadside Swales, Trans. Res. Record 1017, TRB, Washington, DC, 1985.

Zanoni, A.E., Characteristics and treatability of urban runoff residuals, *Water Res.*, 20, 5, 651–659, 1986.

14

Stormwater Master Planning

14.1 The Role of Stormwater Master Planning

Introduction

Stormwater master planning (SWMP) is a term that has come to mean a wide variety of things. The term "master plan" can mean almost anything. When considered herein, the term deals with the technical approach to stormwater solutions over a certain area or in a certain drainage system. This will eliminate the primarily institutional approaches to master planning, where ordinances or financing studies predominate.

A *stormwater master plan* can focus on the following:

- Minor systems and feeder streams in backyards or major systems of streams and rivers within the municipality
- Quantity issues such as flooding and undersizing, or quality issues such as water-quality standards, channel erosion, ecological balance, and riparian corridors
- Existing flooding, erosion, and water-quality problems and their roots, or future and potential problems not yet realized but able to be modeled
- Urban problems, where solutions and space are limited and small-scale integrated solutions abound, or rural problems, where there are many options, land is cheaper, and larger-scale solutions can be planned
- Fixing known problems brought on by neglect, flow increases, and poor planning and design, or avoiding potential problems through modeling and detailed planning, design, and the imposition of regulatory and other institutional controls
- Risk-based type of designs, wherein statistical and probabilistic distributions of estimated variables or ecological risk assessment lead to solutions with estimated probabilities of failure, or deterministic planning where simple designs based on specified criteria lead to construction of solutions
- Plans that are structural in nature, where the solutions consist of brick and mortar (keeping water away from people), or softer, nonstructural solutions consisting of programmatic and regulatory controls (keeping people away from water)
- The plan can be highly technical in nature, generating reams of computer output utilizing the most complex modeling algorithms, or it can be more conceptual in nature, dealing with goal setting, consensus building, and facilitation of citizens

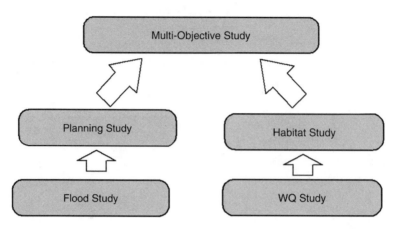

FIGURE 14-1
Basic types of master plans.

- The solutions can be individual site solutions, like a menu, with little regard for the complexities of the whole up- and downstream system or watershed (perhaps using only the rational method), or based on system-wide models that can estimate and portray the interaction of various elements of the plan
- The master plan can be specialized, focusing on one type of problem (lake eutrophication or loss of a specific species), or it can be holistic and integrated, focusing on a blend of factors and objectives

Basic Master Plan Types

Even though there are an infinite number of possible detailed approaches to master planning, there are some general categories of approach or typical types of plans that can be described. For each type, there are typical steps and concerns specific to that type of plan. Because of the great variability in approach and objectives, a detailed discussion of sediment transport, channel stability, and geomorphic analysis will not be covered in this chapter. The reader is referred to Barfield, Warner, and Haan (1985), Simons, Li & Associates (1982), ASCE (1977), Hydraulic Engineering Center (1977b), U.S. COE (1989), or Simons and Senturk (1977) for more information.

Figure 14-1 depicts five basic, or representative, types of master plans. On the left are two levels of complexity for *quantity*-based master plans, and on the right are two levels of complexity for *quality*-based master plans. Holistic watershed planning is the culmination and integration of quality and quantity master planning. Within these five types, there are infinite variations. Each of these basic types, and variations, will be discussed in more detail later in this chapter. But first, it is important to discuss how to arrive at a definition of the type of master plan necessary to attain goals and objectives, and how to determine what those goals and objectives really are.

Keys to Successful Master Planning

In our experience, each of these basic approaches has been termed a "stormwater master plan" (SWMP) or "watershed master plan." Thus, any short chapter on SWMP could not possibly cover all the possibilities and combinations of approaches to develop a stormwater master plan. However, all good master planning follows some well-defined steps, the most important of which are accomplished before a computer model is ever fired up.

There have been thousands of stormwater master plans done, but relatively few have been implemented as planned. A common and obvious reason is lack of funding to accomplish all that has been planned. But, there are deeper issues involved, lack of funding being only the resultant symptom. Several of those issues are discussed here.

The common goal of all master plans is to accomplish something real in a community or watershed. It is not only to come up with good solutions but also to actually implement those solutions. Many master planning efforts fall short of implementation, because they do not plan beyond the end of the final report. The final report (with inches of computer output) is dutifully delivered and presented to political leaders, and then support is sought. By the time it can be fully considered, the momentum has ebbed. The consultant team is out of money, the staff is on to other things, a citizens group has disbanded, and the report is too cumbersome to read and understand without further explanation.

As important as master plans are to any comprehensive stormwater program, by themselves, they will not solve problems or prevent flooding, drainage, or water-quality problems. The master plans represent a blueprint for action that must be taken if these problems are to be solved or prevented. Too often, people see the master plan as the end product and forget that, if the plans are not implemented, little good will result from the completed work. The real work begins when the master plan document is completed.

The written master plan is only a way station on the road to implementation. As such, the process of master planning, of building consensus and support, of involving key stakeholders, of gaining key political backing, is often more important than the final document. Some communities do not even produce a final document, because it is a "living thing," consisting of several or many reports supporting a process.

Another reason master plans fail is that the planners fail to realize that the written plan and computer model are only one component of what it takes to actually implement a master plan. Making the plan successful may involve public education programs, a new ordinance, changes to zoning, and an addendum to the comprehensive plan, funding changes, new staff, new policies, and new procedures. It is these institutional underpinnings (see Chapters 1 and 2 for more on this) that often make or break a master plan. Thus, all but the most simplistic and well-funded master plans must look, however fleetingly, at institutional barriers to success and address those changes as well.

Another reason many master plans fail to live up to expectations is that local municipal government staff have failed to "plan the plan." The next section discusses this in some detail.

14.2 The Master Planning Process: The Scoping Study

Prior to producing a detailed request for proposals or qualifications, a local government should properly set the stage through the use of a master plan action, or scoping, plan. This is a generic process that should be followed to some extent by every master planner for any type of plan. In fact, the scoping process establishes the plan parameters and sets the stage for the scope and request for proposals or qualifications to be developed. The process moves through a thorough needs-and-goals analysis, which assesses all physical problems to be addressed by the master plan and uncovers root causes. The premaster planning process then moves, in turn, through identifying constraints to possible solutions and formation of an appropriate technical approach to find useful solutions.

Step 1 — Make a Needs Analysis

What is the master plan to address? The first step in any master plan is to gain a clear understanding of the actual needs and goals of the municipality. There is an old axiom of sales that states that the customer never tells you what he really needs but what he thinks he needs. That is never truer than in stormwater master planning. To gain a clear understanding of the real needs, a logical and ordered approach must be followed.

A typical program begins when a particular problem becomes apparent. Usually it is flooding, erosion, water quality, or drainage infrastructure problems. The solution normally is to fix the individual problem. However, a quick-fix solution to the apparent problem is rarely a permanent solution, because it rarely addresses the underlying issues and root causes. Chapter 1 of this book laid out an ordered approach to understand the various causes of problems. A master plan should seek to understand the problems at many levels. Even if it cannot address solutions on all the levels, it should seek to call attention to them.

Goals and objectives of the master plan should be carefully determined, or disappointment will invariably result. A master plan is not a magic box out of which instant solutions are pulled. Most problems took a long time to develop and will not be fixed overnight. This becomes even truer as the planner deals with erosion and nonpoint source pollution problems.

Goals are often determined through the use of a staff group, a citizens group, or both. Staff groups are important, because each senior staff member has different needs, goals, and focus. The term stormwater planning means different things to different people. To the municipal administrator, it often means a drainage study to "solve Mrs. Smith's backyard flooding." It might mean "floodplain management" to the community FEMA representative or "water-quality enhancement" to the environmentalist. Whatever its meaning or eventual activities, an ordered understanding of its various components is essential to proper control of stormwater. Developing an approach that, in some way, takes into account the needs of different staff members and the programs they represent, will take longer, but will normally lead to a more satisfactory result and will be better supported across the board.

The use of a citizens group is becoming much more commonplace in master planning, especially if the plan is more than a simple small capital improvement plan. Citizens groups have some significant advantages over simple public meetings, as explained in Chapter 3. A citizens group can have many different uses in master planning, including the following:

- *Political influence* — Members can help, through personal relationships, to sell the master plan concept to political leaders, who often have appointed them.
- *Stakeholder influence* — Members represent key stakeholders groups (environmentalists, home builders, flooded neighbors, etc.) with an ability to present the concepts to their groups and to bring their support to the table.
- *Public educators* — Members are often able to take a slide show about the concept and present it in a variety of settings in nontechnical language.
- *Good ideas* — Often, stakeholders have ideas that the consultant and staff team have not considered.

- *"Guinea pigs"* — If well chosen, the group often serves to represent the reaction of the public to ideas and concepts.
- *"Own" solutions* — Members feel a sense of ownership of solutions they help develop, and they will use their influence to convey this ownership to others, and stick with the process through difficulties.
- *Financial support* — There may be the opportunity to partner with stakeholders in the development of a public–private capital improvement (e.g., a greenway tied to a larger development or commercial and industrial campus).
- *Technical know-how* — Some stakeholder groups contain individuals with unique knowledge of some aspect of science, finance, etc., that can be key to concept development and master plan success.
- *Handle the media* – In some settings, often in smaller towns, stakeholders have personal and business relationships with the media and can help explain concepts and ensure accurate reporting.

The deliverable products should also be defined. Master plans often gather dust on the shelf as implementation funds are sought. When the time comes to actually build the projects, the information may be out of date, as zoning changes and new developments have changed the situation. A master plan should be a dynamic document tied to a mathematical model where appropriate, and that model should be kept up to date periodically or episodically. This can be done quite easily through modern technology that ties mathematical models to geographic information systems (GIS). As changes occur in the field, they are accumulated and then registered in the database. The model is automatically updated and run when appropriate. In this way, master planning is done on the fly as development occurs, changes from assumed conditions are made (there will always be changes), or budgets are modified. Figure 14-2 illustrates this concept. The lighter floodplain is the planned floodplain extent. But an unplanned major development is up for a zoning hearing. Using the "living" model and GIS, the land-use change is quickly input into the land-use layer in the database, and new floodplains are automatically plotted. The darker floodplain shows the impact of the proposed development and opens the door to assessing ways to mitigate the impact. In this case, the analysis took less than half an hour using a skilled operator and an automated GIS-based modeling system.

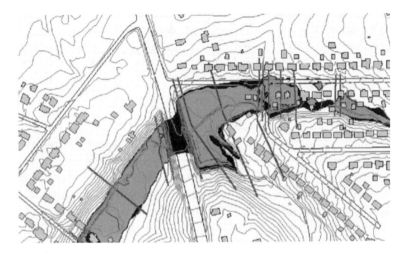

FIGURE 14-2
Modeling tied to GIS.

Without this ability, flooding would have been increased in an unknown way, damaging downstream properties and opening upstream developers (not to mention the municipality) to potential liability. Everybody loses.

One warning is in order. If the technology delivered to a municipality is more than the municipality can keep up, due to lack of training in its use or through burdensome data requirements, it will be less than useless. Several municipalities have spent large sums on complex models tied to GIS systems only to find they were not useful due to data collection procedural problems or levels of sophistication beyond the day-to-day user. This leads directly to Step 2.

Step 2 — Determine Constraints to Possible Solutions

There will always be constraints to solutions to problems. Master plans must be realistic. Expensive structural solutions to flooding problems cannot often be afforded. Key pieces of property may not be available, and the political will may not be there to condemn land even if it is vital to flooding abatement. It is thus important to define the universe in which possible solutions can be found.

This is done through an initial investigation of possible solutions and assessment approaches to the issues of Step 1. The discussion of these topics is often done through a brainstorming process in which different staff elements or citizen representatives discuss ideas. Sometimes, a staff team defines constraints and examples that are then subjected to a citizen review and further discussion. It is important that citizens feel that they are advised but not controlled.

Constraints to any possible solutions fall naturally into several major categories:

- *Technical constraints* to any solution are myriad. One key constraint is lack of necessary data. This is often the case for stream flow and almost always the case for water-quality studies. Another technical constraint may be the ability of the municipal staff to use the tools developed. It makes no sense to recommend a complex watershed model if no one on the staff will be able to keep the model up to date after the departure of some consultant. Still another might be the ultimate use of the plan. Plans to be used daily must have a user-friendly interface; those used once can have a throwaway database. If floodways are required and eventual FEMA updates are envisioned, it is important to specify a model that FEMA accepts and that can generate floodways. With the changing paradigms of stormwater management (see Chapter 1), newer technologies or land-use patterns (e.g., low-impact development concepts, infiltration BMPs, or better site designs) may not be accepted without suitable citizen education and input.
- *Political constraints* are often the most important to the would-be master planner. Some solutions have failed in the past with embarrassing consequences. The planner should think twice before recommending a politically unacceptable approach again, even if it is superior. For example, impact fees may have a bad reputation in some parts of the country, where they have been successfully challenged in court. Detention dams (or "dry dams," as the U.S. Army Corps of Engineers calls them) have a bad reputation in some municipalities where maintenance has proven to be costly. Removing a beloved historical neighborhood to make way for a widened channel may be totally unacceptable. A sense of the political will power is necessary to configure solutions that will be acceptable. This must be tempered with the fact that political climate statements are perceptual factors and often depend on who is making the statement of "fact." Also,

some political constraints will change often: with each election, after a crisis, when a new regulation impacts stormwater, during budget discussions, etc.

- *Social constraints* reflect the values of a given community. For example, channel modification may be seen by a certain municipality as anti-environmental and not allowed. Wetlands may be seen as a positive environmental amenity or as a waste of valuable development land, depending often on the economic condition of the community. Some parts of some municipalities may be off-limits for location of detention facilities or for investing scarce resources. Environmental justice issues can surface as priorities are set for protecting neighborhoods and prioritizing scarce capital resources.

- *Physical constraints* can be such things as soil type, topography, moisture, depth to bedrock, etc. Physical constraints might also include the availability of materials such as sand or riprap. Urban development may constrain certain solutions due to the presence of structures, utility conflicts, or roads.

- *Legal constraints* reflect local, state, and federal restriction of master planning options. Legal constraints are often environmental in nature, such as the necessity of obtaining permits for activities in, or near, a stream. Some municipalities are constrained by state law from exercising very many powers while home-rule states have given strong powers to local municipalities. General case law and perceived liability may restrain municipalities from certain options for fear of a lawsuit or a ruling of a taking of property. Legal boundaries may also constrain some types of solutions that cross outside municipal limits.

- *Financial constraints* almost always limit, to some extent, the types of solutions that are possible. While not all studies are assessed on a strict benefit–cost basis, most projects have to give the perception that they are worth the expense, both initial and ongoing. Budgets for the master plan also usually limit the type of analysis and level of detail. Sometimes suitable options exist, though a wild guess at the front end of a study does not justify rigorous analysis of data based on that guess later in the study. This is often the case with detailed reliance on hydrology developed from little or no data.

Step 3 — Formulate the Technical Approach

The goal of this step is to develop a technical approach that meets the objectives of Step 1 while staying within the bounds set by the constraints of Step 2. This is not always easy, and often cycles through several iterations as the plan progresses.

The goals of the master plan study determine the level of effort and detail involved, as well as the degree of conservatism built into flow estimates. Flood hazard studies may require only an estimate of peak flows. Other quantity-based studies may require the development of design stormwater runoff hydrographs and the timing of these hydrographs with respect to one another. Holistic watershed studies may span the horizon of social and technical considerations, developing goals and objectives far from simple flood control. Thus, there is varying complexity of studies and studies are often done in phases (ARC, 2002; COE, 1992, 1993; CWP, 1998).

The End Product of the Scoping Study

The end product of this master planning scoping study is a clear understanding of what will be done (and not done), how success is measured, what the analysis will be, details of the approach, deliverables, schedule, and cost. The scoping study develops and estimates the probable cost of the plan (like an engineer's estimate in a construction project), and lays out clear tasks for the subsequent RFP.

For more holistic-type master plans, the scoping study becomes part of the overall process and can serve as the initial steps for a citizens and staff group, where vision can be cast and momentum gained. The study can gauge the willingness of political leadership to back any future plan, smoke out stakeholder concerns, and seek staff and private partners. It can also save a municipality the money and embarrassment of beginning a plan that cannot be implemented.

14.3 Computer Model Choice for Master Planning

Choice of a computer model depends on a number of factors, including answers needed, application of information, physical situation, special considerations (such as pressure flow in pipes), available data, and capabilities of the modeler and of the final user. In addition, there may not be a need for a computer model at all. Models are often tied to their ability to represent results graphically through a geographic information system (GIS), automated mapping/facilities management (AM/FM) system, or other means.

A number of models should be considered for use; however, if a model the user is intimately familiar with will suffice, there must be a good reason not to use it. Simply because a model is complex does not necessarily mean it produces better answers. All models depend on verification with or calibration to real data to produce accurate and reliable results. When those data are missing, there is no way to be sure of accuracy. The modeler must be sure that the model used can simulate all the key physical processes (unless it is a "black box" regression-type model) encountered in the watershed. For example, if a hydraulic model cannot determine energy losses through a long culvert, then a supplemental model must be used and the answer inserted into the hydraulic model in the appropriate place.

Physically-based (rather than empirical) models have the advantage in that they can be extended into new situations with some confidence. This is because they are based on physical laws, rather than on fitting some mathematical relationship to a database.

Though there are a number of ways to categorize models (deterministic versus stochastic, lumped versus distributed, continuous versus event, analytical versus empirical, single versus multiple dimension, closed-form versus numerical method, steady versus unsteady, interactive menu-driven versus batch, public versus private), models generally fall into four categories for stormwater master plan studies. Some models span a number of these categories (such as Texas A&M model, SWMM, and SEDIMOT).

1. *Hydraulic models* — Hydraulic models are normally used to determine flow elevations, velocities, distributions, and pressures, given flow rates and boundary characteristics as inputs. Backwater models such as HEC-2 and WSPRO fall into this category, as well as pipe-flow models or special-purpose models such as hydraulic structure design models. The Federal Highway Administration's HY-8 model for culvert design would be an example. Two- and even three-dimensional models exist for more complex problems. Unsteady flow models such as UNET, SWMM, DAMBRK, AD-ICPR, or DWOPER may be necessary for dynamic flow conditions. The leap from simple one-dimensional models to two-dimensional requires a great increase in boundary information input. The more the model

attempts to mimic nature through complex mathematical formulations, the more the modeler must deal with numerical stability as well as trying to obtain accurate answers. For example, the dynamic routing of SWMM's EXTRAN block and the mass conservation in FESWM's two-dimensional finite element network can "blow up" with wrong data, even if it is physically reasonable.

2. *Hydrologic models* — Hydrologic models are normally used to determine flow rates at various points throughout a watershed or basin, given the typical inputs of rainfall, basin characteristics, and basin structure. There are many models that fit this category or can be used for this purpose, such as HEC-1, HEC-HMS, HYDROS, TR-20, PSRM, HYMO, TR-55, ILLUDAS, and more. The leap from single-event to long-term continuous simulation requires much more rainfall data but will most often give more accurate results for such things as flow volumes or interactions of flows in reservoirs or retention ponds.

3. *Water-quality models* — There are a number of water-quality models, each with specific capabilities and drawbacks. Very few water-quality models have been broadly tested and are fully supported. These models fall generally into the categories of statistical models, regression models, and physically-based models. Inputs beyond basin hydrologic inputs include pollutant-loading information or parameters. Several of the most common include SWMM, DR3M, QUAL-2E, STORM, and HSPF. This category includes models that perform such activities as pollutant loading, transport, chemical or biological reaction or interaction, deposition or other treatment, and receiving water-quality impacts. Pollutant-loading estimates are generated based on known constants, rating curves, buildup–washoff relationships, regression equations, sediment transport relationships, or statistical estimates based on data analysis (Huber, 1986). Water-quality models are reviewed in Huber and Heaney (1982) and Whipple et al. (1983). Many newer models and variations appear, though most are variations of older models.

4. *Sediment transport or erosion models* — These models are used to determine channel stability in the short term and long term and to predict sediment yield. Inputs beyond basin hydrology or hydraulic parameters include grain size distribution, transport equation parameters, sediment supply, and a variety of parameters for special functions, depending on the model (such as settling rates for a sediment yield model for use in the design of sediment traps). Common models include SEDIMOT for sediment yield and HEC-6 or TABS for one-dimensional riverine sediment transport.

14.4 Five Basic Types of Master Plans: The Flood Study

The end product of a flood study is the ability to determine, within an acceptable level of accuracy, the flooding elevation (stage) and the probability of occurrence of a flood (frequency) at points along a stream. Sometimes, the distribution of flow over time (the runoff hydrograph) and total volume are important. This is often done for several events and for one or more development scenarios. In addition, there are often one or more alternative solutions explored and tested for existing or anticipated future flooding problems. End products are reports, mapping, models, GIS layers, capital improvement plans, program directions, etc.

Studies vary significantly in scope and complexity. Perhaps the greatest difference in level and type of analysis depends on the relative size of the systems being assessed — major and minor system master plans — and on the amount of remedial or opportunistic design incorporated into the master plan.

Major drainage systems, from the urban stormwater perspective, tend to drain areas larger than about 50 to 100 acres. Sometimes, the regulatory stormwater quality breakpoint of 50 acres (or 36 in. in diameter) is used to help synchronize quantity and quality programs and to provide computation points at major outfalls. In the Eastern United States, a 50-acre break point may also coincide with the upper limit of perennial streams and all the regulatory and practical aspects that implies.

The finer the mesh of subbasins, the greater the cost. For example, changing the subbasin size from 200 acres to 25 acres increases the amount of effort (and cost) several times, depending on the degree of automation in data collection and reduction. Yet, developing a mesh at the 200-acre size limits the usefulness of the master plan to larger streams. Some municipalities, attempting to obtain the best of both worlds, perform basic hydrologic and hydraulic analysis on the larger system and on only selected minor systems with known or anticipated problems. Then, as time goes on, master plans can be developed on a neighborhood basis within the larger mesh, and the results can be connected end-to-end to the overall master plan. This allows the prediction of downstream impacts of smaller site developments without the cost of modeling all subbasins in detail.

Finer system models also must often incorporate a more significant percentage of underground drainage or mixed flow situations. In these cases, the intensity of data collection and potential need for a model that can accurately predict flows, elevations, and pressures within a closed system increases the cost. If there is the potential for backflow, flow exiting the system, or a need for accurate pressure estimates, one of several fully dynamic models should be chosen — those that allow for simultaneous calculation of key parameters throughout the system at different points in time.

Some examples of this first type of master plan include the following:

- A flood hazard potential study to establish Federal Insurance Agency (FIA) floodplain designations to assist in floodplain management and insurance rate setting. Often, the 100-year and 500-year return period flood elevations are developed along a stream. Maps are plotted and used for permitting of new and redeveloped property in and near flood-prone areas. Figure 14-3 illustrates mapping that results from this kind of study.
- A flood-prone area along a major stream may be studied to determine the combination of channel improvements, detention, and home removal or flood proofing that will result in reduced flooding. "What-if" analysis may be done using automated floodplain plotting capabilities and automated hydrologic model flow generation. Figure 14-4 shows an advanced graphic representation of floodplain flooding using an automated process.
- A plot of failure of an upstream dam or detention facility is plotted through an urban area to determine evacuation and timing needs in the event of catastrophic failure or an extreme rain event. Figure 14-5 shows the result of dam failure above Nashville, Tennessee. In this master plan, the floodwave movement and timing is plotted and time-stamped, road overtopping is identified, and flooded structures are noted. Several different failure sequences and types can be modeled coincident with any particular storm throughout the headwaters of the Cumberland River.

Stormwater Master Planning 955

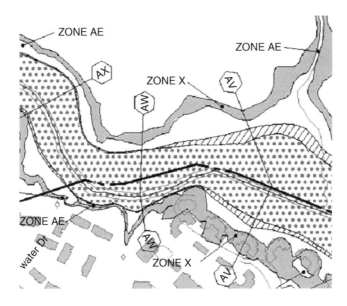

FIGURE 14-3
FIA mapping.

FIGURE 14-4
Automated floodplain plotting. (Courtesy of Amec Earth & Environmental, Inc.)

- Examining the effects of various hydrologic estimates and development scenarios on flooding is another common use. Different hydrologic approaches can yield different results, and a compromise or decision is necessary. Updated land-use or projected use estimates can radically change the floodplain locations and elevations. Figure 14-6 shows such a study done for Charlotte, North Carolina.
- The development of a capital improvement plan for a series of unrelated flooding or infrastructure problems within the neighborhoods of an older section of a community is a common approach for some communities with many smaller problems. Each problem is assessed separately, often using only rational-method

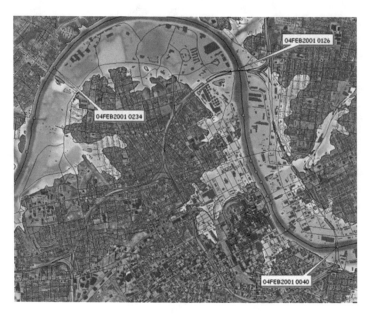

FIGURE 14-5
Dam failure plotting in Nashville, Tennessee. (Courtesy of AMEC Earth & Environmental.)

FIGURE 14-6
Alternate flooding study for Charlotte, North Carolina.(Courtesy of Watershed Concepts, Inc.)

calculations, and preliminary designs are developed and costs are estimated. Off-the-shelf modeling packages are effective for this type of plan. The final plan and costs are then placed into a three-ring binder for use in future capital program planning.
- A SWMM model application to a closed pipe system, or an AD-ICPR application to a system of pipes, channels, and ponds is developed to identify elements of the system that need upsizing, locations for underground storage of excess volume, or the need to change pond configuration to account for backwater effects.

Major Creek Flooding

This type of master plan seeks to correct, or avoid, out-of-bank flooding from major creek systems. Typically, these include creeks that have been studied under the Federal Flood Insurance Program, though, often, the study continues upstream to a point where intermittent or ephemeral stream flow begins. Flood elevation is often key so that a backwater profile is necessary, though, often, an existing FEMA study can be used for lower-budget studies. If benefit–cost analysis is of importance, it is appended to the basic flood study, because the magnitude of flooding is usually great and the cost for mitigation equally great. This is the second type of study and is discussed in the next section. Nonstructural solutions such as relocation, floodproofing, flood insurance, flood warning, etc., often predominate.

Major creek flooding master plans begin by locating flooding areas and by determining information about the flooding, such as depth, frequency, expected annual damages, types of structures, velocities, directly contributing areas, channel characteristics, etc. Typically, an existing condition and future condition(s) scenarios are developed. The development of future conditions is often clouded by uncertainty about development patterns and densities, and the use of flood-control structures. FEMA recently (*Federal Register*, 2001a) made allowance for the plotting of future floodplains on the Federal Insurance Maps as an encouragement to communities to manage floodplains more stringently, though flood insurance rates are still predicated on existing conditions only. In its approach, considering potential future detention or conveyance changes is not allowed due to the uncertainty of actual construction and maintenance of the structures. In other cases, local governments allow consideration of such structures only if they are owned and operated by the local government, and not privately.

Usually, a combination of alternatives is found to be effective in reducing flooding. A frequency such as the 25-year flood is often used for alternative development, because this frequency normally deals with the most cost-effective frequencies for major creek flood problems, though floods of 100-year or greater return period are used for locating floodplain boundaries for safety purposes. Two-dimensional modeling may be necessary for wide and flat floodplain areas where velocities and more exact flood heights are necessary. Unsteady flow modeling may be necessary for calculation of major flood waves during a dam break situation or in steeply sloping terrain. Risk-based analysis is often used to explicitly determine probabilities of failure of possible solutions.

Structural alternatives such as levees and floodwalls, reduced roughness, channelization, cutoffs, diversions, detention and retention, etc., are assessed in combination with non-structural measures. Concern for the environment and for stable channel design should be in the forefront, and environmental features of flood-protection designs should be planned. Environmental enhancement or mitigation measures include such things as high- or low-flow channels, clearing one side only, artificial meanders, artificial wetlands, in-stream habitat structures, etc. Detailed studies have been done for each of these key structural alterations in an attempt to restore ecological function after channel construction (Burch et al., 1984; Hendersen, 1982; Hanson et al., 1985; Nunnally and Shields, 1985; Shields and Palermo, 1982; Thackston and Sneed, 1982; Whitlow and Harris, 1979).

Sometimes these studies also include assessment of channel stability. In these cases, various theories of stream typing, channel degradation, sediment transport and shear stress thresholds, and geomorphology are employed to seek to bring about a stable channel configuration in terms of depth, width, meander ratios, and slopes (Ferguson, 1997; Galli, 1997; Harvey and Watson, 1986; Leopold, Wolman, and Miller, 1964; Ohrel, 2000; Rosgen, 1996; Shields et al., 1995; Schueler and Claytor, 1997; Vannote et al., 1980). Recent thinking assigns the channel-forming flow to the 2- to 5-year flood, because this is the approximate frequency of bankfull flows for most channels that have not been altered by man. If this

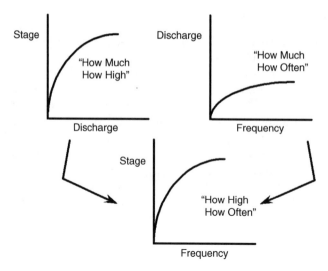

FIGURE 14-7
Basis for the flood assessment study.

flow can be reduced to below the predevelopment flow rate through extended detention, then, as the theory goes, the critical channel-forming shear stress of the 2- to 5-year flow will not be exceeded (ARC, 2002; MacRae, 1993, 1997; Caraco, 2000).

The basis for the flood study is typically the use of hydrologic and hydraulic models or estimates as illustrated in Figure 14-7. The hydrologic model tells us "how much water flows how often." It gives us ability, at the simple end, to estimate a peak flow of a specific return period. Even the rational method is a simple form of a hydrologic model. On the more complex end, a hydrologic model gives us the ability to estimate probabilities of flow and hydrograph shape, volume, and timing for virtually any combination of development scenario and solution strategy. The basic end product is a stage–frequency curve as illustrated in Figure 14-7.

In the absence of sufficient and reliable flow gauge data, the basin-wide hydrologic model relies on physical data to develop estimates of peak flow, and hydrograph shape, volume, and timing. Hydrologic modeling is discussed in more detail below.

A hydraulic model tells us how high and fast the water will flow if we know the flow amount (from the hydrologic model) — the left-hand side of Figure 14-7. At its simplest, it gives us an elevation of flow at a particular point along a stream. If we model a range of flows, perhaps tied to different return frequencies, the end result is a stage–discharge curve.

To obtain the needed information, a frequency–elevation curve, flow is eliminated from the two plots, as indicated in Figure 14-7. When plotted, or interpolated from cross section to cross section, the frequency–elevation curve becomes a two-dimensional plot of floodplain elevation along a stream. More detail on hydraulic modeling is included below.

After objectives and constraints of the plan are obtained, typical technical analysis steps include:

- Data collection
- Preliminary field investigation
- Design storm development
- Hydrologic model development
- Hydraulic model development
- System alternatives development and analysis

Data Collection

This phase determines all available information and ascertains field conditions. Substeps include the following:

1. Review of available documents from, for example, U.S. Geological Survey, U.S. Army Corps of Engineers, Soil Conservation Service, FEMA, other federal and state offices, local records, miscellaneous studies, etc. Determine degree and location of urban development over time from zoning records, annexation records, plat recordings, or other local records and knowledge. Information is also available through the Internet.
2. Obtain historic flow information including discharge–frequency relationships, historic high-water marks, rainfall records, stage–discharge relationships, bridge designs, known cross sections, mapping, etc. Determine whether transfer of information is a viable solution.
3. Plan a hydrologic study. Delineate preliminary subbasins and routing reaches and structures for hydrologic study. Determine whether sufficiently accurate information is available for soils, land use, and topographic modeling. The use of satellite imagery for land use has also become commonplace, as its effective resolution has increased.
4. Plan a hydraulic study. Locate existing or new cross sections and structures requiring specific modeling on maps for later field investigation. Plan the integration of map- and field-derived cross sections and try to anticipate changes in roughness, channel shape and conveyance, overbank storage, etc., for cross-section placement. Remember that cross sections represent average conditions for that reach. Development of strip contour maps for the purposes of hydraulic modeling has now become practical through the use of airborne LIDAR systems. Automated processes greatly accelerate this process.

Preliminary Field Investigation

This phase estimates all hydrologic and hydraulic parameters in the field as well as identifies unusual conditions and allows for the collection of any needed ancillary data peculiar to the study plan. Consider dictating notes and taking digital photographs as you go. Backpacking GPS equipment can help in locating specific features and in determining beginning and end points of observed features and parameters (Figures 14-8 and 14-9). For many structures, X, Y, and Z information must be collected, necessitating a complete infrastructure inventory. Typical substeps include the following:

- Make field estimates of all hydrologic and hydraulic parameters such as roughness coefficients, bridge openings, culvert sizes and inlet types, overtopping circumstances, flooded structure locations, etc.
- Perform a field investigation to ascertain conditions that may impact runoff or hydraulic parameters such as ponding and reservoirs, undersized crossings, debris barriers, recent erosion, new development, hydraulic controls, channelization, location of wetlands or other environmentally sensitive areas, etc.
- Interview local agencies and residents; review newspaper files for historic high-water marks or rainfall information, etc.
- Take pictures of all unusual situations with measurable scale in picture.

FIGURE 14-8
GPS greatly speeds field data collection.

FIGURE 14-9
Digital photos enhance field data collection.

- Finalize cross-section locations, overflow bypass pathways, and subbasin boundaries. Measure cross sections with hand level and tape, or mark survey points if necessary.

Design Storm Development

For event modeling, the development of a design storm is often more art than science. There has been considerable discussion in the literature about proper storm duration, storm movement, distribution of rainfall within storms, adjustment of storm depths for areal distribution, storm types, urban shadow effects, and so on. Each of these factors, and more, may be important in specific instances. For example, storm movement may take on importance if the storm of concern tends to be a fast-moving thunderstorm or frontal system, storm movement is predominantly in one direction, and the basin to be modeled is situated with a long axis parallel to storm movement. It may also be important in mountainous regions, where orographic effects can predominate for the less-frequent events (such as the Big Thompson Colorado flood of the late 1970s and the Fort Collins flooding in the mid 1990s). The adjustment of point rainfall depths for areal distribution will be important for larger watersheds (perhaps greater than 5 to 10 square miles).

However, for other than infrequent event modeling in conjunction with such considerations as dam safety (probably maximum storm development), the modeler should refrain from conjuring a set of theoretical conditions that may have a certain probability of occurrence but whose joint probability would produce an event far more extraordinary than simple frequency analysis would predict.

FIGURE 14-10
Thunderstorm movement and coverage.

The rainfall values given in intensity–duration–frequency (IDF) tables or curves are for point rainfall. If storms are developed over a larger area, these values will overpredict the storm depths and intensities, especially for thunderstorm-type flooding. As shown in Figure 14-10, thunderstorms are generally confined to less than a mile or two in diameter, and switch on and off as they move.

Therefore, the point values must be adjusted downward using an areal adjustment factor. Equation (14.1) has been fit to data and a figure first developed by the National Weather Service (NWS, 1961) by the Corps of Engineers for the HEC-1 computer model (U.S. COE, 1990). Equation (14.2) is a fit of duration versus B_v information provided by the Corps of Engineers.

$$F_r = 1 - B_v * (1 - e^{(-0.015 * A)}) \tag{14.1}$$

$$B_v = 0.35488859 * D^{-0.42722864} \tag{14.2}$$

where F_r = coefficient to adjust point rainfall; B_v = maximum reduction in point rainfall; A = area (mi^2); D = storm duration (h).

Any hydrologic model seeks to provide accurate flow information at all points in the watershed. In reality, for example, the 100-year flow will not occur at all points in the watershed at the same time, and an individual storm will not generate a 100-year flow event throughout a watershed. More typically, the 100-year flow level is achieved as a flood wave moves through a drainage system and then will occur only in part of the watershed. A localized thunderstorm may produce such a storm over a few city blocks, several square miles, or at only a few street intersections. Areas adjacent to these peak flow locations will experience some lesser storm event.

For calibration purposes, rainfall may be set to vary from subbasin to subbasin. Online radar information or multiple gauge locations may be useful for such purposes. Various methods are available for averaging historical storm event rainfall over areas of a watershed. See Chapter 7.

For modeling purposes, it may be appropriate to use a storm of smaller areal extent (and thus a more intense storm) for upland subbasins, and increase the area's extent as computation points lower in the watershed are reached. Many hydrologic models can accomplish this adjustment automatically, generating a number of adjusted storm intensities depending on the drainage area to the computation point. Flow estimates between such points are estimated through transfers according to Equation (14.3).

Accurate and logical estimates of storm volume and peak intensities are important, depending on the type of analysis to be done. For example, a 6-h duration storm may reproduce a frequency distribution of peak flows accurately. However, such frequency distributions of volume are often unknown. In the sizing of retention basins or extended detention basins, storms of longer duration will provide more volume. The choice is often a policy decision. In some cases, continuous simulation of such structures is the most accurate method of design.

Typical steps in the development of a synthetic design storm include the following:

- Obtain rainfall information from local gauge data, the National Weather Service, or other publications. Assess the rainfall to determine intensities, durations, and frequencies. Develop or use intensity–duration–frequency curves as appropriate.
- In conjunction with hydrologic model calibration, choose a design storm frequency and duration.
- Distribute the storm in time using time increments compatible with the watershed size and hydrologic methodology chosen. The SCS distribution methodology given in Chapter 7 can be used for a center-peaked storm. A balanced storm distribution methodology presented by the Corps of Engineers (U.S. COE, 1982) provides a similar distribution. Chapter 7 provides alternate methods for the development of such distributions as first-, second-, or third-quartile storms.
- Consider adjustments for storm movement, orographic effects, areal reduction, or other factors.
- Test the sensitivity of assumptions and choices using the chosen hydrologic model.
- Consider the use of different storms for different purposes.

Hydrologic Model Development

Recall that the normal objective of the hydrologic model is to accurately reproduce flood events, not to concoct a series of rare factors, which together produce a rare event. For example, some planners may combine high antecedent moisture, an adjusted third-quartile storm for late peaking, worst-case storm movement assumptions, fully dense developed conditions, and structural blockage to reproduce, for example, the 100-year event. In fact, the event reproduced is a much rarer occurrence than the 100-year event, because the combination of all the factors must be the joint probability of each independent or semiindependent occurrence. While this may lead to conservative assumptions, it does not reflect reality. It does not reproduce a flood event of known frequency.

A typical hydrologic model, either hand- or computer-generated, accomplishes four functions: runoff hydrograph generation, stream reach routing, pond routing, and combining hydrographs. Runoff hydrograph generation (Chapter 7) includes storm characteristics, losses, and total convoluted hydrograph development. Stream reach routing (Chapter 10) involves moving a stream flood wave from point to point, moving it in time, and changing its shape due to attenuation. Pond routing (Chapter 11) includes development of stage–discharge and stage–storage curves.

For example, in Figure 14-11, a simple watershed is to be modeled. The watershed is first divided into subbasins. The subbasins are indicated on the left of the figure, and the steps to construct a hydrologic model are at the right. Runoff hydrographs are first calculated from subbasins A1 and A2. They are then combined at the tributary junction between the two subbasins and routed downstream to the next calculation point. A runoff hydrograph from subbasin A3 is then calculated and combined with the routed

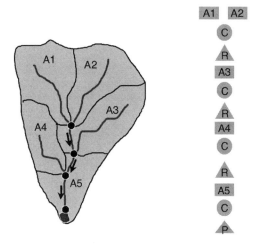

FIGURE 14-11
Basic hydrologic model functions.

hydrograph, and so on. Newer "distributed parameter" models proport to use raster information and calculate across each bit of data.

Hydrologic models may be developed for a number of scenarios. Sometimes a model is developed for specific past conditions to assess its ability to mimic historical storm information. Urbanization adjustments to a frequency record can be made in accordance with information in Chapter 7. Do not forget that urban development greater than about 15 to 20% impervious usually means a much higher incidence of direct connection of paved areas and much shorter lag times. The tendency to use an uncharacteristically long time of concentration converted to a lag time should be avoided. Lag times should be indicative of the basin as a whole and not a single long and unrepresentative flow path. Often one, or more, future development scenarios are developed to allow the planner to evaluate "what-if" questions. Future development normally involves estimates of future land use for a dated planning horizon or for full build-out of a basin given existing or anticipated zoning.

Typical data layers used to determine rainfall losses (and thus runoff amounts) include soils, land use, and basin boundaries, as depicted in Figure 14-12. Routing data include channel shape, slope, roughness, and storage. Timing is controlled through estimation of time of concentration and associated derivations (usually lag time).

Depending on the purpose of the planning, various combinations of structural or nonstructural practices (such as land-use controls) are considered against a developed set of criteria. Usually, the hydrologic and hydraulic models are used in tandem, because a hydrologic model provides stage–frequency information, while a hydraulic model gives stage–discharge data. Some modeling packages combine the two into a seamless package.

Typical steps and considerations in hydrologic modeling are as follows:

- Development of basic data and information
- Development and calibration of runoff parameters
- Delineation of subareas
- Analysis of historic events and development of unit hydrograph parameters
- Development of a typical existing and future conditions model with one or more future development alternatives

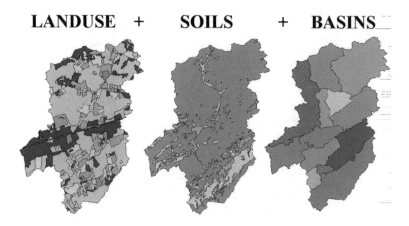

FIGURE 14-12
Basic data layers for hydrologic modeling.

- Analysis with the model of several scenarios (often in conjunction with the hydraulic model)
- Development and testing of recommendations

There is a wide variety of calibration and modeling approaches, depending on available data, complexity of the watershed, detail required, and modeling approach. If peak, discharge frequency, or flood hydrograph information is available in the watershed for historic events, the planner can develop watershed characteristics above the gauge location and seek to calibrate the model to these events. Antecedent moisture is accounted for through suitable curve-number (or other parameter) adjustment. It is usually easier to adjust parameters, within reasonable bounds, for volume matching first. Then, adjust the shape of the runoff hydrograph through adjustments in lag time estimates, in routing parameters, and for inadvertent ponding behind undersized culverts, bridges, etc.

Approximate adjustments for within watershed ponding in swampy areas, behind undersized structures, or through detention can be made through adjustment of the peak flow. This can be accomplished in computer modeling through lengthening of the lag time. There may be different adjustments necessary for different frequency floods. This method avoids breaking the watershed into too many small subbasins, while maintaining suitable accuracy for most applications. However, if subbasin timing of combinations of peaks close to the subbasin outlet is of key importance, this method will delay an individual peak longer than an actual reservoir routing procedure.

The Soil Conservation Service provided a table for such an adjustment for small subbasins in an early version of its TR55 (SCS, 1981). While this version has been superseded, the adjustment factors have been found, through testing by detailed pond routing of many small basins, to be fairly accurate for a first approximation of the peak reducing impacts of many small ponds within basins or for model calibration when there is reason to believe flat swampy areas or storage behind structures plays an important role in peak reduction. Tables 14-1 and 14-2 reproduce two of these SCS adjustment tables. The factors given are multiplied by the predicted peak flow to give a lower, adjusted peak flow. This peak can then be matched through trial-and-error adjustment of the lag time in the subbasin. A fit equation can be easily developed for lag time prediction if a number of minor subbasins are involved. This method should not be used for relatively large basins, or just upstream from key design points.

TABLE 14-1

Adjustment Factors Where Ponding and Swampy Areas are Spread Throughout, or Occur in Central Parts of the Watershed

Percentage of Pond or Swampy Area	Storm Frequency (Years)					
	2	5	10	25	50	100
0.2	0.94	0.95	0.96	0.97	0.98	0.99
0.5	0.88	0.89	0.90	0.91	0.92	0.94
1.0	0.83	0.84	0.86	0.87	0.88	0.90
2.0	0.78	0.79	0.81	0.83	0.85	0.87
2.5	0.73	0.74	0.76	0.78	0.81	0.84
3.3	0.69	0.70	0.71	0.74	0.77	0.81
5.0	0.65	0.66	0.68	0.72	0.75	0.78
6.7	0.62	0.63	0.65	0.69	0.72	0.75
10.0	0.58	0.59	0.61	0.65	0.68	0.71
20.0	0.53	0.54	0.56	0.60	0.63	0.68
25.0	0.50	0.51	0.53	0.57	0.61	0.66

Source: From Soil Conservation Service, Urban Hydrology for Small Watersheds, TR 55, 1981.

TABLE 14-2

Adjustment Factors Where Ponding and Swampy Areas are Located only in Upper Areas of the Watershed

Percentage of Pond or Swampy Area	Storm Frequency (Years)					
	2	5	10	25	50	100
0.2	0.96	0.97	0.98	0.98	0.99	0.99
0.5	0.93	0.94	0.94	0.95	0.96	0.97
1.0	0.90	0.91	0.92	0.93	0.94	0.95
2.0	0.87	0.88	0.88	0.90	0.91	0.93
2.5	0.85	0.85	0.86	0.88	0.89	0.91
3.3	0.82	0.83	0.84	0.86	0.88	0.89
5.0	0.80	0.81	0.82	0.84	0.86	0.88
6.7	0.78	0.79	0.80	0.82	0.84	0.86
10.0	0.77	0.77	0.78	0.80	0.82	0.84
20.0	0.74	0.75	0.76	0.78	0.80	0.82

Source: From Soil Conservation Service, Urban Hydrology for Small Watersheds, TR 55, 1981.

One should not forget that, for the more extreme events, there would be ponding in many places throughout the watershed. Models set up and even calibrated for more frequent events may tend to overpredict peak flows for less-frequent events if areas of significant ponding are ignored. On the other hand, the hundreds of areas of small or minor ponding are normally filled prior to the peak flow and have the effect of simply steepening the rising limb of the hydrograph and lengthening the tail (like a detention pond filling on the rising limb and then overtopping when the peak flow arrives).

If gauge information is available on a similar watershed in the area, or another point in the same watershed, transfer of that information can be accomplished if regression equations are available through a simple ratio:

$$Q_1/(aX_1^b * cY_1^d * \ldots) = Q_2/(aX_2^b * cY_2^d * \ldots) \qquad (14.3)$$

where Q_i = regression predicted discharge or volume, and X_i, Y_i = regression-independent variables such as watershed area, slope, etc.

Each recorded event is assessed to determine the best unit hydrograph and loss rate parameters. Adjustments are made and then tested on other storms, if available, to verify the choices. The eventual goal is to develop good average parameters to use for synthetic event modeling of design storms. Separate sets of parameters may need to be developed for divergent storm frequencies.

Subareas are delineated to balance considerations for selection of subbasins of approximately equal size; the time-step selected for the fastest peaking subbasin; tributary inflow points; stream gauge locations; uniform land use or soil type within a subbasin; the need for computation points; in damage reaches, below pond routing sites; and for future site location considerations.

For larger basins, the development of estimates of baseflow may be important. This is especially true if large inline retention ponds are to be designed, or lakes exist within the system. Special cases may require model changes. This is especially true if the soil is quite porous; quick return flow is known to exist; or karst or rock entrances may influence the infiltration, inflow, and reappearance of flow.

Channel-routing methods and parameters can be determined from gauge data, backwater computations, rating curves, or can be estimated. See Chapter 10 for more information.

Reservoir routing should be done for all major ponding and undersized road crossings with high fill (i.e., potential for significant storage). See Chapter 11 for a discussion of reservoir routing.

The timing of the runoff hydrographs may be significant for flood reduction purposes. Given storm stationarity assumptions, certain subbasins will contribute most directly to flooding in certain locations. These subbasins would be initial targets for flow reduction, because decreases in peak flow in these subbasins would translate most directly to decreases in flooding. Several computer models facilitate such analysis indirectly.

Hydraulic Model Development

For a large open-channel system, hydraulic modeling may take several forms. Hand computations for backwater profiles are almost never done outside of classroom exercises; though it is important to understand the methodologies employed by the computer software used. Normally, it involves developing one-dimensional, steady, gradually varied backwater computations.

Different levels of complexity abound. Commercially available or sole-source modeling packages can combine hydraulic and hydrologic capabilities in a seamless package, but may limit flexibility of application for unusual circumstances. Choice of a hydraulic model depends also on the long-term use of the model. If real-time "what-if" analysis is to be done on a regular basis, then the ability to automate data collection and reduction and to plot floodplains in an accurate automated fashion is a must. Figures 14-13 and 14-14 illustrate two such systems. In Figure 14-13, basic cross sections are automatically taken from digital topographic information, written into a data input file, and then blended (not shown) with surveyed cross-sectional data to form a cross section accurate for modeling based on survey information in the major flow-way, and for plotting tied to the digital information. In Figure 14-14, the model results are plotted on the digital topographic information through a raster interpolation algorithm using a minimization function approach.

It should be understood that there would be conditions when the models discussed above do not apply. For certain applications, the velocity of flow in the transverse direction is as important as in the downstream direction, or the sinuosity of the stream and presence of overbank flow makes accurate modeling with a one-dimensional model inaccurate. In these cases, two-dimensional modeling should be done. Modern modeling packages and

Stormwater Master Planning

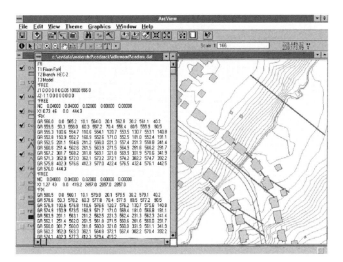

FIGURE 14-13
Automated cross-section development. (Courtesy of Amec Earth & Environmental, Inc.)

FIGURE 14-14
Automated floodplain plotting. (Courtesy of Amec Earth & Environmental, Inc.)

data grid generation and reduction techniques make this approach well within the reach of most experienced modelers. Figure 14-15 shows the automated generation of a finite element grid for modeling of a sinuous channel with several floodplain flow openings. For more flashy systems and for intense short storms, the movement of the floodwave through the system may be important, and an unsteady model should be used.

Typical steps in the development of a backwater model include the following:

- Select and prepare input for cross-section locations.
- Define Manning's n values using a combination of field reconnaissance and analytical calculations.
- Determine flood methods for bridges and define openings, overtopping information, and routing locations.

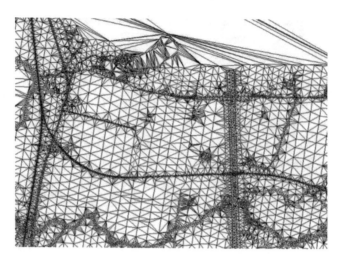

FIGURE 14-15
Two-dimensional grid generation. (Courtesy of Amec Earth & Environmental, Inc.)

- Develop a series of flow profiles for discharges from hydrologic modeling or other information for input for stage–discharge curves for modified puls routing as appropriate.
- Calibrate model to known flow and frequency information, regression equations, or high-water marks for known storms, adjusting cross sections for dead flow areas, Manning's *n* changes, or flow, combining, and timing of peaks.

Spacing, location, and alignment of cross sections must be selected to fully represent the flow condition as a one-dimensional problem. Cross-section spacing depends on changes in roughness, shape, slope, discharge, and abrupt transitions, and locations where calculated information is required. Cross sections should be spaced such that channel conveyance ratio between successive sections is between 0.7 and 1.4. See Chapter 10 for more information on open-channel flow. Conveyance is defined as:

$$k = 1.49/n \; R^{2/3} \, A \qquad (14.4)$$

where n = Manning's n; R = the subsection hydraulic radius (ft); and A = the flow area (ft^2).

Flow lines should cross the centerline of the section at right angles. For high flows, the thalweg of flow tends to shorten and cut across point bars and bends. Flow in overbank areas may be at a different orientation than the main channel. For multiflow profiles, flow lines should be an average of the low- and high-flow lines, or separate sections should be input. Select cross-section computation points to represent the average cross section.

Flow tends to contract through constricted openings at about a 1:1 downstream-to-cross-stream slope. Expanding flow tends to follow more of a 4:1 downstream-to-cross-stream slope. Water outside these constriction and expansion lines does not contribute to the flow area. Flow may actually be upstream in some areas as eddy currents are set up. Elimination of these dead-flow areas should be accomplished by redefining the cross section or by taking advantage of the capabilities of HEC-RAS or other models for encroachment calculations.

Manning's n or other loss coefficients should be appropriate for the flow stage, time of year, and land use. For example, resistance to flow would decrease as flow rises above grass level in the overbank, but would increase significantly as it encounters the leaves of the lower tree limbs. Typically, conservatively high values are chosen for flood height estimates, while conservatively low values are chosen for bank protection design.

The starting point of the backwater profile should be sufficiently far downstream that minor errors in assumed starting elevation would even by the time the profile reaches the design reach. Several starting elevations or methods should be tried to test for model sensitivity if there is no rated control downstream.

Model output should be carefully scrutinized, because results can be generated that are in error but do not stop the program computations (Ogden, 1989; Hoggan, 1989). Check for large changes in top width or conveyance, flow, slope, computed elevation, or velocity from section to section as a start. Channel Froude number should be checked to ensure the flow is in the assumed regime: subcritical or supercritical. The better models will run both profiles. Flow velocities should be examined to see if they are reasonable. Cross sections where critical depth occurs should be noted and checked. If they are few and sparse, look for channel geometry or slope errors. If critical depth occurs at several cross sections in sequence, it is likely that supercritical flow occurs through the reach, and an additional supercritical backwater run may be warranted. Review all bridge sections to ensure flow is reasonable and modeled correctly.

Regulatory floodway determination can be time consuming. Time spent in following a logical approach is usually less than the running of many haphazard trial backwater profile runs. Composite Manning's n values in overbank areas can often give misleading results and should be avoided. The loss of valley storage can actually increase the flows. If, through hydrologic modeling and routing, the flows increase by more than 10% after floodways fringe development is in place, different flow values may be warranted to establish the floodway locations. Floodways may be restricted if anticipated development within a flat floodway fringe area raises the elevation of the flow such that a large area is brought into the floodplain area beyond the actual 100-year nonconstricted limits. Floodway determination generally follows the following steps (U.S. COE, 1978, 1988):

- Check locations for special consideration based on specific land use or zoning. Look for any conditions that might dictate a variation from standard floodway determinations.
- Develop and run natural profiles through the reach.
- Check and eliminate locations for constriction that have excessive velocity or existing erosion.
- Using a model, develop floodway locations for maximum 1-ft surcharge and approximately equal conveyance from both overbanks.
- Check and adjust for excessive velocities, excessive surcharge, smooth alignment (no "balloons"), local width minimums, or other local requirements.

System Alternatives Development and Analysis

There are many methods for developing an analysis scheme for the watershed. The actual procedure chosen depends on the problems identified in Step 1 of the standard master planning procedure. Brainstorming meetings are particularly effective if solutions are not limited to conventional channel improvements or detention.

After a preliminary assessment of the situation and possible solutions on the map, make field reconnaissance of known flooding or erosion locations, taking note of types of structures, relative worth, outbuildings, fences, etc. Look for evidence of geomorphic adjustments of the channel in terms of widening, deepening, large movement of sediment, lateral instability, headcutting in the main channel or tributaries, indications from tree shape and size, etc. Look for evidence of debris problems, structural failure of conveyance, or crossing structures.

Assess the physical possibility of different types of structural solutions, such as channel widening or deepening, levees, possible reservoir or detention locations, high-flow channels, diversions, pumping, etc. Assess nonstructural solutions such as home raising or moving, flood proofing, flood warning, etc. Checking the timing of hydrographs may reveal other locations for reservoirs, which provide indirect peak flow reduction from a side tributary.

After screening likely alternatives, modify the model for each feasible alternative. Modify the outflow–storage relationship for reservoirs and hydraulic geometry for channel adjustments. Various combinations of alternatives may serve to enhance any used alone. Look for opportunities to integrate solutions into the character of the neighborhood or area. Tennis courts and soccer and baseball fields have been used for detention ponds. Greenways, walkways, and linear bike trails work well for floodplain areas. Wetlands can make wonderful strolling areas if boardwalks are provided.

Invariably, if the study progresses to a more formalized design study, costs and benefits will enter the picture, leading to the next type of quantity-based study: cost/benefit analysis.

Minor System Flooding

The minor or "convenience" system is often designed to the 10-year or even more frequent flood. This system is often overwhelmed by intense thundershowers. Although any one site may not be overwhelmed on the average more often than the design frequency, there may be many such flood events in a major urban area over a design frequency time span.

The approach to minor system master plans is to identify and attempt to quantify flooding problems. To do so, this often requires a concerted effort to involve local residents and political leaders. A problem is often not anything more than an inconvenience. Therefore, it is important to determine the design policy and performance level-of-service of the municipality, and prioritize problems based on a professionally driven evaluation system such as life and health, residential structural damage, nonresidential structural damage, utility damage, and nuisance (Debo, Westerich, and Newsom, 1979).

Quantifying these problems involves use of a simple methodology for peak (or sometimes whole hydrograph) flow estimation and a determination of whether the problem is isolated or part of a system-wide deficiency. It is important not to simply pass the problem downstream to the next undersized structure. Solutions are normally straightforward, and cost and materials estimates are often included.

Data collection and analysis can be time consuming and expensive. It is often not worth the expense to analyze the whole of a municipality system. If the system is old, it has probably been tested above its performance capabilities. If an effective complaints system is in existence, those complaints should be the source of determining which parts of the system are analyzed in any detail. Simply because a system backs water up does not constitute a problem per se. It is actually advantageous to store runoff throughout the system in planned locations for economic reasons having to do with pipe size and for pollution- and erosion-reduction reasons.

Some municipalities will use various levels of complexity for analysis of the system. For example, if the rational method was used for the design of systems, it is often applied to much of the system as a first screen. This can be done efficiently using a GIS or AM/FM system. For those systems that (1) exhibit flooding based on this analysis, (2) violate the assumptions in the rational method (such as high tailwater conditions), or (3) are identified through complaint files or inspection reports as problem areas, a more rigorous analysis using hydrograph routing and full dynamic routing can be done for these portions of the drainage system. Typical steps for this type of approach are provided in Table 14-3.

Stormwater Master Planning

TABLE 14-3

Typical Steps in Minor System Flooding Study

Step 1	Determine design criteria and levels-of-service standards for analysis
Step 2	Assess complaints and inspection reports
Step 3	Develop system and runoff information for the rational method or other simplified hydrologic method
Step 4	Perform rational method screening on some or all of the system
Step 5	Prioritize study areas and identify any other areas to be further studied based on other factors
Step 6	Field check sites, interview residents
Step 7	Perform more rigorous analysis, including field data collection
Step 8	Develop and assess alternative solutions
Step 9	Develop cost information and estimates (B/C analysis as appropriate)
Step 10	Prioritize locations for capital or remedial maintenance improvements

FIGURE 14-16
Alternatives analysis for a flood study.

Data collection, if not available, can be performed using automated methods including hand-held dataloggers and automated mapping and facility management systems. Even the modeling can be tied to these systems to provide a complete system design and management package.

Alternative solutions developed under Step 8 can include such things as detention facilities, inlet modifications, pipe upsizing or ditch enlarging, parallel piping, rerouting, etc. (Figure 14-16). Solutions that include allowing ponded water to escape should be checked downstream through the system. It can be shown that the system should be checked to a point downstream where the site draining through the structure comprises about 10% of the total drainage area to that downstream checkpoint (Debo and Reese, 1992). Beyond that point in the system, the increase will normally have negligible impact.

14.5 Five Basic Types of Master Plans: The Cost/Benefit Analysis Master Plan

Another type of master planning builds on a flood study to determine acceptable risks and the associated costs. Using information developed in the flood analysis, economic

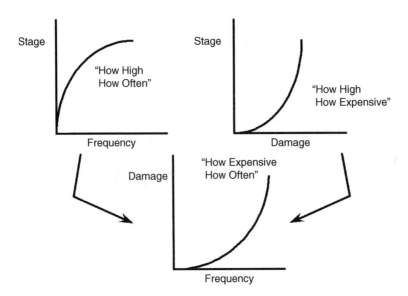

FIGURE 14-17
Benefit-cost analysis basis.

and environmental impacts can be assessed. This initially entails establishing a relation between water surface elevation and associated damage (often referred to as stage–damage curves), as depicted in Figure 14-17. The results of the flood study are depicted on the left of the figure. At the right of the figure is a curve (actually a set of curves, one calculated for each "damage reach") showing stage–damage relationships: how much damage is suffered when the water rises so high.

If the common variable, stage, is eliminated, the result is a damage–frequency curve that provides information on the frequency or risk of suffering certain monetary damages. Based on this relationship, an acceptable level of risk or a balanced level of acceptable damages, is determined, from which design discharges and associated water surface profiles and elevations are established. Acceptable levels of risk might be based upon the likelihood of loss of human life, impacts to residences, impacts to nonresidence structures, or damage to utilities. This information then helps determine the ultimate drainage infrastructure that will be needed to achieve the planning goals. A formal benefit–cost analyses or a more subjective cost-effectiveness approach could be used. There are approximate methods that will yield accurate answers, putting solutions in rank order and approximate absolute ratio values (Ogden, 1988). Decisions may depend on hard-to-quantify features such as integration of a solution site with such things as parks, greenways, or other recreational areas.

Based on the design criteria, preliminary designs can be developed that in turn yield initial cost estimates for the infrastructure. For example, a community might look at different flood-protection strategies along a stream and estimate the costs and flood damage savings for each alternative in an effort to select the most appropriate solution(s) for that community.

The Water Resource Council (1983) laid out typical steps for the development of flood-reduction benefit calculations as given in Table 14-4. These steps serve as guidelines for all agencies planning for floodplain management and design projects. They serve as the basis for a typical major creek master plan for flood control.

Master plans that seek to compare alternative flood-control schemes based on cost objectives will contain, at a minimum, a rudimentary benefit–cost or other economic analysis

TABLE 14-4

Typical Steps in Benefit–Cost Analysis

Step 1	Delineate affected area
Step 2	Determine floodplain characteristics
Step 3	Project activities in affected area
Step 4	Estimate potential land use
Step 5	Project land use
Step 6	Determine existing flood damages
Step 7	Project future flood damages
Step 8	Determine other costs of using floodplain
Step 9	Collect land market value and related data
Step 10	Compute National Economic Development (NED) benefits

Source: From Water Resources Council, Economic and Environmental Principles and Guidelines for Water and Related Land Resource Implementation Studies, Washington, DC, 1983.

(such as cost-effectiveness). Depending on the type of master plan, perform detailed or approximate cost-benefit analyses through the development of stage–damage curves or the use of standard curves found in publications such as Johnson (1985), Arnell (1989), U.S. COE (1985, 1988), or the latest information on depth–damage statistics available from FEMA.

Table 14-5 and Figure 14-18 give generalized residential depth damage information based on U.S. COE data through 2000 (U.S. COE, 2000). Its use is appropriate when no other information exists for nonvelocity zone damages. The water depth in this table is relative to the top of the first finished floor, excluding the basement. The addition of a basement would add approximately 8% to the totals based on older FEMA data (FEMA, 1988), though the curves would probably need to be adjusted slightly. All values given are in terms of percent of the total assessed value of the structure.

Postflood survey information or site-specific appraisals would provide significantly better information than Table 14-4. Losses for nonresidential property should be estimated based on field surveys following techniques given in U.S. COE (1985, 1988), or other available information.

Losses (and therefore benefits) from other than physical flooding damage reduction are more difficult to determine. The magnitude of such losses is not well documented, and accurate estimation may involve the compilation of data from numerous sources. Such costs are not normally large compared to flood damage but can be significant in situations such as a location where flood warning and preparedness occurs more often than actual flood damage. Such "shadow costs" fall into nine categories:

- *Income loss costs* are losses due to loss of income due the halting of the production or delivery of goods and services. Losses are only suffered if they are not recouped by delaying delivery or production or when additional costs are incurred because of a delay. Losses may occur directly due to shutdown of businesses or indirectly due to the closing of roads or shutdown of utilities.
- *Emergency costs* include efforts taken to monitor or forecast flooding; police, fire, or National Guard actions; flood fighting efforts; victim relief and aid costs; and administrative costs.
- *Traffic routing costs* include the costs for added mileage driven by each car rerouted by flooding or its threat, and the time costs for each passenger.
- *Floodproofing costs* are incurred for a prior floodproofing (dry floodproofing) and actions taken during a flood (wet floodproofing).

TABLE 14-5
Residential Depth Percent Damage — Nonvelocity Zone

Flood Depth	1 Story, No Basement		2 Story, No Basement		Split Level		Contents 1 Story, No Basement		Contents 2 Story, No Basement		Contents Split Level	
	Mean	Std Dev	Mean	Std Dev	Mean	Std Dev	Mean	Std Dev	Mean	Std Dev	Mean	Std Dev
-2	0%	0.00%	0%	0.00%	0%	0.00%	0%	0.00%	0%	0.00%	0%	0.00%
-1	2.50%	2.70%	3.00%	4.10%	6.40%	2.90%	2.40%	2.10%	1.00%	3.50%	2.20%	2.20%
0	13.40%	2.00%	9.30%	3.40%	7.20%	2.10%	8.10%	1.50%	5.00%	2.90%	2.90%	1.50%
1	23.30%	1.60%	15.20%	3.00%	9.40%	1.90%	13.30%	1.20%	8.70%	2.60%	4.70%	1.20%
2	32.10%	1.60%	20.90%	2.80%	12.90%	1.90%	17.90%	1.20%	12.20%	2.50%	7.50%	1.30%
3	40.10%	1.80%	26.30%	2.90%	17.40%	2.00%	22.00%	1.40%	15.50%	2.50%	11.10%	1.40%
4	47.10%	1.90%	31.40%	3.20%	22.80%	2.20%	25.70%	1.50%	18.50%	2.70%	15.30%	1.50%
5	53.20%	2.00%	36.20%	3.40%	28.90%	2.40%	28.80%	1.60%	21.30%	3.00%	20.10%	1.60%
6	58.60%	2.10%	40.70%	3.70%	35.50%	2.70%	31.50%	1.60%	23.90%	3.20%	25.20%	1.80%
7	63.20%	2.20%	44.90%	3.90%	42.30%	3.20%	33.80%	1.70%	26.30%	3.30%	30.50%	2.10%
8	67.20%	2.30%	48.80%	4.00%	49.20%	3.80%	35.70%	1.80%	28.40%	3.40%	35.70%	2.50%
9	70.50%	2.40%	52.40%	4.10%	56.10%	4.50%	37.20%	1.90%	30.30%	3.50%	40.90%	3.00%
10	73.20%	2.70%	55.70%	4.20%	62.60%	5.30%	38.40%	2.10%	32.00%	3.50%	45.80%	3.50%
11	75.40%	3.00%	58.70%	4.20%	68.60%	6.00%	39.20%	2.30%	33.40%	3.50%	50.20%	4.10%
12	77.20%	3.30%	61.40%	4.20%	73.90%	6.70%	39.70%	2.60%	34.70%	3.50%	54.10%	4.60%
13	78.50%	3.70%	63.80%	4.20%	78.40%	7.40%	40.00%	2.90%	35.60%	3.50%	57.20%	5.00%
14	79.50%	4.10%	65.90%	4.30%	81.70%	7.90%	40.00%	3.20%	36.40%	3.60%	59.40%	5.40%
15	80.20%	4.50%	67.70%	4.60%	83.80%	8.30%	40.00%	3.50%	36.90%	3.80%	60.50%	5.70%
16	80.70%	4.90%	69.20%	5.00%	84.40%	8.70%	40.00%	3.80%	37.20%	4.20%	60.50%	6.00%

Source: From Federal Emergency Management Agency (FEMA), Flood Insurance Rate Review, 1987.

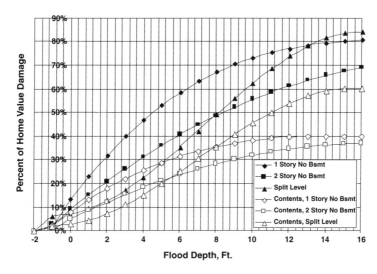

FIGURE 14-18
Stage–damage curve plots.

- *Costs of flood insurance* are incurred for the administration of every policy written for property in a flood-prone area. Costs are in the order of $100 per policy. Specific information is available from the U.S. Army Corps of Engineers for any given year.
- *Temporary relocation and reoccupation costs* include living expenses incurred by flood victims who are forced to find temporary shelter during and after a flood. Reoccupation costs include all opportunity and real costs for arranging the repair of the flooded property.
- *Costs for modified land use* can be incurred when property is not utilized to its full potential because of the threat of flooding.
- *Land market-value reduction costs* are realized when the land prices are depressed due to the location of the parcel in a flood-prone area. These costs are determined through comparison with values of adjacent, nonflood-prone properties.
- *Erosion costs* can be incurred when flooding and associated high velocities erode property or such erosion damages structures.

The actual benefit–cost analysis portion of this type of study follows the steps indicated in Table 14-6.

Damage–frequency curves are developed for each calculation or damage reach. That is, all damages for a certain damage reach are related to the depth at a certain point within the reach. Damage–frequency curves are plotted for all the damage reaches and, often, by damage category. An aggregated curve gives the total damage–frequency curve for the project location. The expected annual damages can be calculated from combining a depth–damage and depth–frequency curve to develop a damage–frequency curve. The expected annual damage is the area under this curve. Figure 14-19 shows the results of such an analysis for a flood-prone neighborhood.

The expected annual damage calculation can be done in tabular form with columns for frequency, frequency interval, damage, average damage for the frequency interval, expected damage for the interval (multiplying the two interval columns), and finally the accumulated damage total, which is the sum of the previous columns. The narrower the interval (i.e., the more values calculated), the more accurate is the procedure mathematically.

TABLE 14-6
Steps in Typical Benefit–Cost Analysis

Existing Condition Costs
Step 1: Delineate the affected area and select calculation reaches
Step 2: Establish elevation–frequency relationships
Step 3: Develop depth–damage relationships
Step 4: Calculate nonphysical depth–damages estimates
Step 5: Calculate damage–frequency relationships
Step 6: Calculate expected annual damages

Future Condition Costs
Step 1: Establish economic and demographic base
Step 2: Project land-use changes
Step 3: Establish new floodplain inventory
Step 4: Establish new elevation–frequency relationships
Step 5: Calculate new nonphysical damages
Step 6: Calculate new equivalent annual damages

Project Costs
Step 1: Calculate project capital and financing costs
Step 2: Calculate project operating and maintenance costs

Benefit Calculation
Step 1: Calculate inundation reduction from structural measures
Step 2: Calculate cost reductions for nonstructural measures
Step 3: Calculate other benefits

B/C Ratio
Step 1: Calculate discounted benefits over project life
Step 2: Calculate discounted costs over project life
Step 3: Ratio discounted benefits and costs

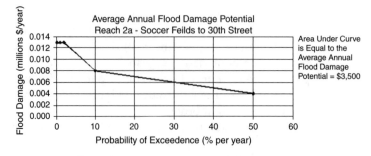

FIGURE 14-19
Benefit cost analysis for home removal.

14.6 Five Basic Types of Master Plans: Stormwater Quality Master Plan

Master planning for stormwater quality is becoming increasingly important in support of myriad regulatory programs and initiatives. For some communities, it is necessary to be able to estimate pollutant loads from stormwater runoff for total maximum daily load input (TMDL), as well as for the expansion of wastewater treatment facilities. For others, the Endangered Species Act leads to a water-quality analysis in support of habitat assessment. Phase I Stormwater NPDES permits under the 1987 Water Quality Act require loading estimates, and Phase I and II require, to some extent, water-quality estimates for new developments. A water-quality master plan can provide the foundation from which to develop broader water-quality assessments. Stormwater quality studies will typically analyze water-quality impacts to receiving waters (and groundwater, particularly in karst regions) and develop structural and nonstructural strategies to reduce or minimize the pollutant loads.

Studies usually involve the development, calibration, and verification of a water-quality model. The level of model sophistication can vary from simple to complex. There are some factors that may mitigate against performing detailed stormwater quality modeling:

- The state-of-the-art in stormwater quality modeling is approximate with most available data, and collecting sufficient calibration data is time and resource consuming.
- It is often difficult to ascertain cause and effect in surface runoff pollution cases, especially if chemical interaction is of concern. Simple models may miss a physical phenomenon of concern, and complex models may have too many adjustments and parameters for which only an educated guess must suffice.
- Methods for treating surface water pollution problems, even if they are well known and quantified, are often only marginally effective, hard to quantify, and expensive to build and maintain.
- Most stormwater quality requirements currently do not involve modeling but a simple design standard (such as a capture criteria).

Despite these drawbacks, stormwater quality modeling is often done on a site-specific or watershed scale. In the future, as the knowledge of the cause and effect of receiving water pollution impacts grows and the knowledge of its mitigation increases, surface quality modeling will grow in prominence. Water-quality modeling is most useful and effective when there are specific pollutants of concern and known impacts on receiving waters are to be mitigated.

The objectives of a typical water-quality study are as follows (Huber, 1986):

- Characterize the urban runoff
- Provide input to receiving water analysis
- Determine effects, sizes, and combinations of control options
- Perform frequency analysis on quality parameters
- Provide input to cost–benefit analysis

A water-quality study normally includes or concentrates on the reduction of surface water pollution to some receiving water. Storms of interest are less than the 2-year storm, because the vast majority of the pollution comes from the many minor storms rather than the few major ones (Roesner, Burgess, and Aldrich, 1990; Livingston, 1986). The study must take into account requirements of the 1987 Water Quality Act. Sometimes, a return period is avoided altogether in favor of a capture criterion tied to a certain depth or percentage of storm capture, or through the use of continuous simulation.

Pollution impacts are acute (short term) and chronic (long term). The response time for pollution impacts in lakes and bays is in the order of weeks to years (Hydroscience, 1979). However, for the type of ponds and streams commonly encountered in urban runoff, the response time for such constituents as dissolved oxygen, bacteria, sediment, or nitrogen is in the order of hours or, at most, days. Statistical or regression-based models are simple to apply and require little calibration. Physically based models have more flexibility in application, can generate pollutographs for single storms more effectively, and can be extended for "what-if" analyses with more confidence.

Sometimes modeling of concentrations and impacts is not strictly necessary. In many cases, the inability to show or quantify direct environmental impairment from stormwater runoff makes the development of designs to mitigate some specific impairment difficult to quantify or justify. Typically, the solution to the cause–effect dilemma is to take one or more steps back from the specific impairment or degradation issue and determine stormwater quality master plan features based on (1) a numerical chemical water-quality standard or designated use basis; (2) a biological indicator or assessment standard; or, most commonly, (3) on a design criteria basis where best management practice (BMP) designs must meet some simple standard (for example, capture and treatment of the first 1/2-in of excess rainfall). If these simplifying steps are appropriate, then detailed water-quality modeling to determine cause and effect is not necessary, and simple estimates of annual or single storm pollution loadings are all that is required. In these cases, a simple loading model is developed, and BMPs are placed within the watershed to reduce these loadings to some target values or percentages. In many cases, stormwater or receiving stream sampling in support of the study or to guide its ultimate application and measure its effectiveness is desirable.

There are then a number of variations of stormwater quality master plans, depending on the results and problems:

- If protection of a lake, bay, or estuary is the main objective, long-term loading estimates and impact assessment is normally done, with phosphorous often being the pollutant of concern. An attempt is made to link these loadings to land use. In these cases, a lake or estuary model is applied in conjunction with the loading estimates to determine long-term impacts of loadings on trophic state. Base flow loadings are also important in these types of studies, with concentration values taken from sampling or standard figures for the region. Alternative land-use or development controls are assessed in conjunction with structural or nonstructural BMPs. Typical of this type of study is a plan to protect a drinking, water supply from the impacts of urban development or agricultural practices.

- If estimations of pollutant thresholds, concentrations, and frequencies in runoff or receiving waters are of concern, the most effective modeling will generate pollutographs for individual storms, or run a continuous simulation for a period to estimate pollutant-removal efficiencies of structural BMPs. The various versions of SWMM (Huber et al., 1981), STORM (HEC, 1977a), DR3M-QUAL (Alley and Smith, 1982), or HSPF (Johanson et al., 1980) are especially effective for this purpose, as is HSPF or its BASINS form (U.S. EPA, 2002). Statistical methods for frequency of pollutant level exceedance can also be used as a pseudocontinuous simulation (Hydroscience, 1979; U.S. EPA, 1983). No stormwater quality model should be used without sufficient model validation data if actual estimates of specific pollutants are to be made (Huber, 1986). Relative changes in pollutant loading can be estimated without site-specific data using generalized values. The models are better at predicting relative changes in unknown values than in predicting the initial values.

- If the design of BMPs to meet simple annual loading reduction targets or design criteria is appropriate, then simple event mean concentration-based annual loading methods can be effective if estimates of pollution reduction for BMP designs can be made. In these cases, annual or seasonal loadings are estimated based on rainfall and runoff volumes. These may be tied to land use for runoff volume and event mean concentrations of pollutants. Then, locations for effective placement of regional or site BMP controls are found and assessed. Solutions include land-use controls, regional BMPs, site development restrictions or design criteria, or other programmatic BMPs such as street sweeping, limitations on herbicide or pesticide usage, maintenance practices, etc. This is probably the most common kind of study performed by the nonspecialist in stormwater quality modeling.

There are a number of simple methods available to make these estimates. The two most popular are the USGS pollution loading regression equation (Tasker and Driver, 1988) and the "simple method" developed by the Washington Metropolitan Council of Governments (Schueler, 1987), and variations of this method. Both of these methods are based on the database generated during the National Urban Runoff Program (NURP) study (U.S. EPA, 1983). These data can be supplemented with local data where available, and a weighted loading or event mean concentration can be estimated. This type of loading calculation lends itself well to GIS and to spreadsheet calculations. The Universal Soil Loss equation (Barfield, Warner, and Haan, 1985) can be used to estimate sediment loss and accumulation, though there is a wide range of sediment loss and delivery depending on the sediment source, type, and delivery mechanisms. Sampling is often the best way to estimate sediment accumulation, or use of regionalized delivery ratio mechanisms tied to watershed area (Schueler, 1987).

BMPs are designed or developed to treat pollution problems with target percent reductions or determinations of the maximum extent practicable for reduction, based on various possible combinations of BMPs. Structural and nonstructural BMPs are used. Nonstructural BMPs are sometimes all that is possible in highly urbanized areas with little room left for structural solutions, although recent advances in microscale commercial devices make retrofitting possible, even if removal of certain kinds of pollutants is still limited.

Regional structural treatment can be most cost-effective and provides the greatest assurance for a long-term commitment to maintenance. Rules of thumb can be developed for first-phase screening analysis of possible sites for structures, using such things as available volume per acre treated or pond surface area related to drainage area.

More recently, master planning tends to involve a treatment train methodology, in which pollution controls are built into the total watershed. See Chapter 13 for further discussion of this concept. When this method is used, there is a greater tendency to employ land-use type controls to limit pollution and quantity of runoff through site design concepts that limit impervious areas and integrate such concepts as grassy swales, buffers, filter strips, and myriad microscale BMPs and other related practices (sometimes termed low-impact development).

Inaccuracies in pollutant load estimation when adequate data are not available make a compelling case for the use of a simple model. For example, a community might develop a simple spreadsheet-based loading model to perform planning level analyses of loadings

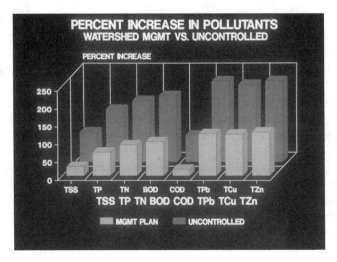

FIGURE 14-20
Pollutant loading model application.

of pollutants, potential removal by stormwater controls, and the impacts of development strategies — or it uses a more complex continuous simulation water-quality model and supporting monitoring to develop a combination of point and nonpoint source loading estimates in support of a watershed assessment or TMDL. The simple method explained in Chapter 13 is often used in a GIS and spreadsheet format for loading calculations. Figure 14-20 shows the end result of an application of a simple loading model to a watershed.

As another example, the better assessment science integrating point and nonpoint sources (BASINS) (U.S. EPA, 2002) model is a system developed to support the development of total maximum daily loads (TMDLs), which require a watershed-based approach that integrates point and nonpoint sources. BASINS can support the analysis of a variety of pollutants at multiple scales, using tools that range from simple to sophisticated. It integrates a geographic information system (GIS), national watershed and meteorologic data, and state-of-the-art environmental assessment and modeling tools into one convenient package.

Originally released in September 1996, BASINS addresses three objectives: (1) to facilitate examination of environmental information, (2) to provide an integrated watershed and modeling framework, and (3) to support analysis of point and nonpoint source management alternatives.

The heart of BASINS is its suite of interrelated components essential for performing watershed and water-quality analysis. These components are grouped into several categories:

- Nationally derived environmental and GIS databases
- Assessment tools (TARGET, ASSESS, and DATA MINING) for evaluating water quality and point source loadings at large or small scales
- Utilities including local data import and management of local water-quality observation data
- Two watershed delineation tools
- Utilities for classifying elevation (DEM), land use, soils, and water-quality data
- An in-stream water-quality model (QUAL2E)
- A simplified GIS-based nonpoint source annual loading model (PLOAD)
- Two watershed loading and transport models (HSPF and SWAT)

- A postprocessor (GenScn) of model data and scenario generator to visualize, analyze, and compare results from HSPF and SWAT
- Many mapping, graphing, and reporting formats for documentation

BASINS' databases and assessment tools are directly integrated within an ArcView GIS environment. By using GIS, a user can fully visualize, explore, and query to bring a watershed to life. The simulation models run in a Windows environment, using data input files generated in ArcView.

14.7 Five Basic Types of Master Plans: The Ecological Study Master Plan

Biological/habitat master planning is similar to a water-quality master plan. However, rather than focusing on water chemistry, the focus is on the aquatic biological communities and supporting habitats. Biological assessments are being implemented on a more frequent basis to assess overall water body health. Biological studies provide the ability to assess acute and long-term effects of nonpoint source impacts to a receiving water in the absence of continuous monitoring data. The resulting data can be used in the design and development of habitat improvement and stream restoration projects, riparian buffers, structural control retrofits, etc.

In order to establish comparative standards for goal and objective setting, ecosystem provinces and regions have been developed nationally. Ecoregions are geographic areas of relative homogeneity in ecological systems or in relationships between organisms and their environment. Within each region, a relatively untouched "reference stream" is determined and assessed using any of a number of bioassessment protocols using biological criteria.

Biological criteria are based on aquatic community characteristics that are measured structurally and functionally. These criteria are used to evaluate the attainment of aquatic life uses. The principal biological evaluation tools used are various indices of biological integrity that apply to fish and macroinvertebrates (U.S. EPA, 1999). The procedure involves the use of a net in the sampling of stream bottoms to collect insects, mollusks, and crustaceans that are collectively called macroinvertebrates. Benthic macroinvertebrates are bottom-dwelling invertebrate organisms easily viewed with the naked eye.

These indices are based on metrics of species richness, trophic composition, diversity, presence of pollution-tolerant individuals or species, abundance of biomass, and the presence of diseased or abnormal organisms. Many states use the results of sampling reference sites to set minimum criteria index scores for use designations in water-quality standards. Figure 14-21 shows ecosystem provinces from which various regions have been derived for each state, and Figure 14-22 shows Level IV ecoregions for Tennessee.

For example, a community may desire to improve the quality and aesthetics of a stream. Biological monitoring and habitat assessment establishes the baseline health of the stream and can be compared to a reference stream in the area. The habitat evaluation can become quite complex and carries significant importance when endangered species are involved. Recent assessment of, for example, the Carolina Heelsplitter Mussel indicates that critical habitat characteristics include: (1) permanent, flowing, cool, clean water; (2) geomorphically stable stream and river channels and banks; (3) pool, riffle, and run sequences within the channel; (4) stable substrata with no more than low amounts of fine sediment; (5) moderate stream gradient; (6) periodic natural flooding; and (7) fish hosts, with adequate living, foraging, and spawning areas (*Federal Register*, 2001b). Thus, these features would

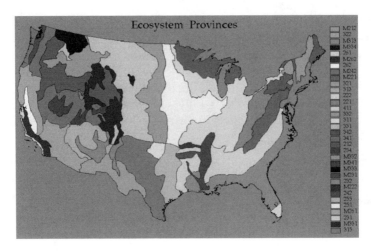

FIGURE 14-21
National ecosystem provinces.

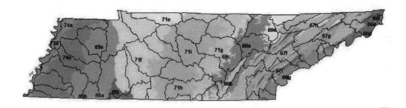

FIGURE 14-22
Level IV ecoregions in Tennessee.

be investigated through field investigations and potential sites for expansion of the habitat of the Heelsplitter would be identified, should range expansion be a goal for the master plan. In some cases, habitat features can be identified through remote sensing and GIS applications. Figure 14-23 shows the results of a multispecies evaluation for San Diego County, California.

This information is assessed to determine causes of impairment (often paired with chemical monitoring), and methods to reduce impairment are investigated. The plan might then include riparian corridor planning, land-use zoning changes, watershed hydrologic response modifications, planned habitat restoration, or any number of actions.

14.8   Five Basic Types of Master Plans: The Holistic Master Plan

Introduction

The comprehensive watershed approach is the most general type of stormwater master planning as well as the most extensive. It has grown out of the overall watershed protection approach first enunciated by the U.S. EPA in the early 1990s and built upon by the U.S. EPA and many federal, state, and local agencies and organizations (U.S. EPA, 1994; MWCOG, 1991; Schueler, 1994; NWC, 1993). Since that time, nearly every federal agency, all state agencies, many local governments, and well over 5000 local watershed nonprofit

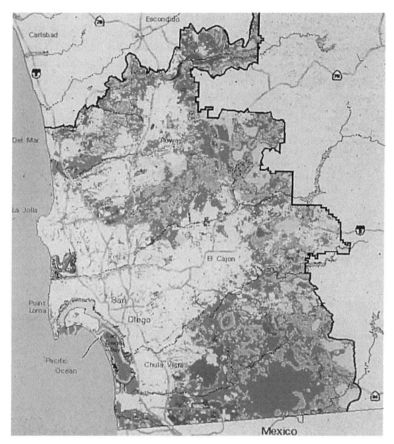

FIGURE 14-23
Multispecies habitat conservation model — San Diego County. (Courtesy of Amec Earth & Environmental, Inc.)

and national organizations have adopted some form of the approach and are applying it at large and small watershed levels. Myriad Web sites, both public and private, describe individual groups, allow for "surfing" a particular watershed, allow access to watershed data, and provide technical background and training to the would-be watershed approach practitioner. It has truly been a revolution under the "water is water is watershed" paradigm described in Chapter 1.

The intent of the comprehensive watershed plan is to assess existing water resources health and to make informed land use and stormwater planning decisions based on the current and projected land-use and development within the targeted watershed and its associated subwatersheds. Watershed-based water-quantity and water-quality goals are typically aimed at maintaining the predevelopment hydrologic and water-quality conditions to the extent practicable through peak discharge control, volume reduction, groundwater recharge, channel protection, and flood protection. In addition, watershed plans may also promote a wide range of additional goals, which include the streambank and stream corridor restoration, habitat protection, protection of historical and cultural resources, enhancement of recreational opportunities, and aesthetic and quality of life issues.

These studies involve a holistic approach to master planning, where hydrology, geomorphology, habitat, water quality, and biological community impacts are analyzed, and solutions are developed. The holistic watershed management approach attempts to provide a framework within which stakeholders from many sectors and viewpoints can work together toward defining and bringing about a long-term environment, which allows reasonable growth and freedom of land use with ecological protection and conservation. Basically, the approach seeks to understand the complex interaction between humans and the ecosystems involved, given the background of climate, topography, and myriad other details of life in the watershed. The idea is to seek to work with nature and find a balance between ecological preservation and conservation and human development pressures.

It is obvious that such an approach can be controversial from the standpoint of land-use controls and development restrictions and ecological risk assessment and compromise. The key to success is balanced local stakeholder involvement and control in development and pursuit of a vision for a certain watershed, along with suitable governmental encouragement, funding, and broad limits or controls. Experience has shown that sound ecological preservation practices can translate into economic development benefits and higher quality of life. But, it is a relearning process for almost everyone.

The process can work at the macroscale, covering large multistate areas, and at the minor urban scale, through the concept of "nesting." Some practitioners recommend that the most effective scale is 2 to 15 square miles to allow for individual stakeholder identification with a particular set of neighborhoods (CWP, 2000). Nesting means planning smaller areas or unique character within a larger overall concept. An urban area can plan its own watershed development and that of areas it may eventually annex, in conjunction with adjacent authorities and regional or state planners. In fact, it makes sense for urban areas to become involved in the planning process to avoid the problems that can result from some outside agency doing it for them and imposing terms.

Basic Approach

The concept is practiced and understood a little differently by each of the major federal agencies, which have a role in some part of watershed stewardship, by each state active in watershed management, and by various agencies, organizations, and local governments. The Soil Conservation Service developed a nine-step process for the planning of its watersheds:

1. Identify all the various problems within the watershed.
2. Develop a set of objectives for the watershed.
3. Inventory the resources within the watershed, including air, soils, water, vegetation, and wildlife.
4. Analyze the resources within the watershed.
5. Develop alternatives to address the problems and opportunities within a watershed.
6. Evaluate and analyze the alternatives.
7. Make decisions concerning the alternatives.
8. Implement the decisions.
9. Reevaluate progress and make adjustments.

Review of many such plans yields a more generic process:

- Phase 1 — Define issues and objectives
  - Identify concerns
  - Involve stakeholders

- Collect and assess data
- Determine critical areas
- Set goals and objectives
- Document everything
* Phase 2 — Develop the plan
  - Assess BMP and management alternatives
  - Develop action plan
  - Build consensus
  - Assign responsibilities
  - Determine measures of progress
* Phase 3 — Implement and evaluate
  - Find funding
  - Set priorities for actions
  - Measure and report progress
  - Review and modify plan
  - Celebrate success

Example: Rapid Watershed Planning

The Center for Watershed Protection developed an eight-step process (CWP, 1998) termed *rapid watershed planning*. As expanded and articulated in the Georgia Stormwater Manual (ARC, 2002), its steps are summarized below.

Step 1 — Identify Initial Goals and Establish a Baseline

Prior to initiating a watershed plan, some broad goals should be identified that define the purpose of the plan initiative, and basic data must be collected to determine a starting point to develop the plan. Information about possible stakeholders, current land use and impervious cover, technical studies (e.g., previous hydrologic/hydraulic studies, floodplain studies, water quality studies, etc.), staffing, and financial resources can help guide the first steps of the plan. Once the broad goals have been identified and defined, specific tasks that may need to be performed include the following.

Task 1: Define Watershed and Subwatershed Boundaries

Most citizens tend to identify with smaller, more neighborhood-oriented subbasins or subwatersheds, perhaps from about 3 to 15 square miles. Sometimes the size can increase if there is a sense of the stream as a well-known unit, such as a popular recreational stream. Therefore, it might be advisable to subdivide larger basins into separate groups with an umbrella oversight group that looks at a larger basin. Table 14-7 shows a typical division of watersheds and subwatersheds arranged by size.

Task 2: Identify Possible Stakeholders

Section 14.2 above describes the many uses of a stakeholder (citizens) group in master planning. In this type of plan, such a group is key to the success of the master plan. So much of the holistic plan is process-driven that a professional facilitator and a strong and involved group are not options. Early on, it is important to identify the partners, or

TABLE 14-7

Description of the Various Watershed Management Units

Watershed Management Unit	Typical Area (Square Miles)	Sample Management Measures
Catchment	0 to 5	Site design measures and structural controls
Subwatershed	5 to 30	Stream classification and management
Watershed	30 to 150	Watershed-based development standards
River basin	Greater than 150	Basinwide planning

stakeholders, that will be involved in some way to make watershed plans happen. Stakeholders might include other categories of nonprofits, environmentalists, developers, small and large business owners, government agencies, flooded homeowners, and neighborhood leaders. Twelve to 15 members are ideal. See Chapter 3 for more information on such a group.

Task 3: Estimate Existing Land Use and Impervious Cover

Estimating existing subwatershed land cover is a recommended baseline task in preparing a watershed plan, because these data can be used in modeling stormwater runoff and estimating pollutant loadings. Land use and impervious cover percentages can be used to initially categorize subwatersheds, help managers set expectations about what can be achieved in each subwatershed, and guide decisions in the watershed.

Task 4: Assemble Historical Monitoring Data in the Watershed

Historical and current monitoring data and stream survey information helps set a numerical baseline from which the master plan will work. Chemical, habitat, ecological, stream flow characteristics, and stream stability information is helpful. There are many sources of data, as indicated in Chapter 6.

Task 5: Assess Existing Mapping Resources

Maps depicting current conditions, including land use, potential pollution sources, problem areas, etc., in each subwatershed, as well as management decisions made during the planning process, are an integral part of the watershed plan. Mapping rounds out the set of baseline information necessary to proceed.

Task 6: Conduct an Audit of Local Watershed Protection Capability

The final element, not strictly data related but institutional, of the watershed baseline is a critical evaluation of the local capability to implement watershed protection tools and management alternatives. This evaluation or audit examines whether existing local programs, regulations, and staff resources are capable of implementing the watershed plan. If not, it identifies key areas that need to be improved.

Step 2 — Set Up a Watershed Management Structure

Establish the institutional organization responsible for the overall management and implementation of the watershed plan. The local government is key to leadership of such an organization. But often, the staff cannot bring about the same results as a nonprofit, neighborhood-based organization. A core set of features is needed to make watershed management structures effective (ARC, 2002):

- Adequate permanent staff to perform facilitation and administrative duties
- A consistent, long-term funding source to ensure a sustainable organization
- All key stakeholders
- A core group of individuals dedicated to the project who have the support of local governmental agencies
- Local ownership of the watershed plan fostered throughout the process
- A process for monitoring and evaluating implementation strategies
- Open communication channels to increase cooperation between organization members

Step 3 — Determine Budgetary Resources Available for Planning

Conduct an analysis to determine what level of staffing, financial, and other resources are available to conduct the plan. Balance the available resources against the estimated cost of developing the plan. Resources may include manpower as well as financial resources. Various agency grants, private nonprofit grants, commercial or industrial participation, local government funding, etc., may all be seen as viable resources.

Step 4 — Project Future Land-Use Change in the Watershed and Its Subwatersheds

Forecast future development, land use, and impervious cover in each subwatershed. In this step, it is recommended that the community forecast future land use and impervious cover based on available planning information such as future land-use plans or master plans for each subwatershed, if possible. Efforts should be made to tie impervious area to land-use projections (helpful in the next step) and to determine how firm the projections are versus how they could be changed and what the appropriate process is. Involving land-use planners and zoning officials may be appropriate as part of the stakeholder group, if one objective of the plan is to influence land-use decisions on behalf of stream quality and conservation.

Step 5 — Fine-Tune Goals for the Watershed and Its Subwatersheds

Use known information about impacts to the watershed, and the goals of larger drainage units (e.g., river basins), to refine and develop goals for the watershed. In addition, determine objectives for each subwatershed to achieve watershed goals. The general goals identified in Step 1 should be added to and modified to reflect the results and inferences of the data collected and analyses performed in Steps 2 through 4.

Goal setting is among the most important steps in watershed planning. It establishes vision and purpose, galvanizes people to action, and provides focus and the ability to measure results. Setting goals must take into account what is important to common citizens. For example, reduction of BOD may be important, but who can identify with such a goal?

To set appropriate and achievable goals, the watershed planning team needs to perform several tasks, including the following.

Task 1: Interpret Goals at the River Basin Level that May Impact the Watershed
Smaller watershed plans nest within larger and statewide efforts. Having parallel or complementary objectives builds synergy.

Task 2: Develop Specific Goals for the Watershed

The goals set at the watershed level are the bottom line of the watershed plan. While these goals may be similar to those developed at the river basin level, they are usually more specific and quantifiable. Examples of watershed goals might include (ARC, 2002):

- Reduce flood damage from current levels
- Reduce pollutant loads from the current level
- Maintain or enhance the overall aquatic diversity in the watershed
- Maintain or improve the current channel integrity in the watershed
- Prevent development in the floodplain
- Allow no net loss of wetlands
- Maintain a connected buffer system throughout the watershed
- Accommodate economic development in the watershed
- Promote public awareness and involvement

Task 3: Assess if Subwatershed Management Objectives Can Be Met with Existing Zoning

One method is to conduct a buildout analysis of current zoning to determine the projected land use and impervious cover in each subwatershed. This analysis can be used to identify which management objectives can be met with existing zoning.

Task 4: Determine if Land-Use Patterns Can Be Shifted among Watersheds

If the current zoning is not compatible with the management objectives, attempts may need to be made to have development shifted to other watersheds or subwatersheds, through zoning map changes or by use of any of a number of conservation tools as indicated in Table 14-8. This is not an easy transition to make, as it intersects basic tenets of land ownership, land use, and associated legal implications.

Step 6 — Develop Watershed and Subwatershed Plans

A watershed plan is a detailed blueprint used to achieve objectives established in the last step. A typical plan may include revised zoning, stormwater design criteria and requirements, potential regional structural stormwater control locations, description of new programs proposed, stream buffer widths, monitoring protocols, and estimates of budget and staff needed to implement the plan. The four tasks needed to establish the watershed plan include the following.

Task 1: Select Watershed Indicators

Indicator monitoring provides timely feedback on how well aquatic resources respond to management efforts. Simple indicators can be selected to track changes in stream geometry, biological diversity, habitat quality, and water quality. The most appropriate indicators will depend largely on the management categories of the individual watersheds.

Task 2: Conduct Watershed-Wide Analyses and Surveys, if Needed

In some situations, a watershed plan may need to incorporate special analyses at the watershed level to supplement basic monitoring and analyses. Other analyses that may be desirable include (ARC, 2002):

TABLE 14-8

Land-Use Planning Techniques

Land Use Planning Technique	Description	Use as a Watershed Protection Measure
Watershed-based	Zoning restrictions specific to a particular watershed or subwatershed	Can be used to protect water resources in a particular watershed and relocate development
Overlay zoning	Superimposes additional regulations or specific development criteria within specific mapped districts	Can require development restrictions or allow alternative site design techniques in specific areas
Impervious overlay zoning	Specific overlay zoning that limits total impervious cover within mapped districts	Can be used to limit potential stormwater runoff and pollutants from a given site or watershed
Performance zoning	Specifies a performance requirement that accompanies a zoning district	Can be used to require additional levels of performance within a watershed or at the site level
Large-lot zoning	Zones land at very low densities	May be used to decrease impervious cover at the site or subwatershed level, but may have an adverse impact on regional or watershed imperviousness and may promote urban sprawl
Transfer of development rights (TDRs)	Transfers potential development from a designated sending area to a designated receiving area	May be used in conjunction with watershed-based zoning to restrict development in areas and encourage development in areas capable of accommodating increased densities
Limiting infrastructure extensions	A conscious decision is made to limit or deny extending infrastructure (such as public sewer, water, or roads) to designated areas to avoid increased development in these areas	May be used as a temporary method to control growth in a targeted watershed or subwatershed; usually delays development until the economic or political climate changes

Source: From Atlanta Regional Commission, Georgia Stormwater Manual, 2002.

- Fishery and habitat sampling
- Stream reconnaissance surveys
- Stormwater structural control performance monitoring
- Bacteria source surveys
- Stormwater outfall surveys
- Detailed wetland identification
- Pollution prevention surveys
- Nutrient budget calculations
- Surveys of potential contaminant source areas
- Hazardous materials surveys
- Stormwater retrofit surveys
- Shoreline littoral surveys
- In-lake monitoring
- Hydrogeologic studies to define surface/groundwater interactions

Task 3: Prepare Subwatershed and Aquatic Corridor Management Maps

Maps that present the plan in a clear, uncomplicated manner are a key product of the subwatershed planning process. Maps range from highly sophisticated GIS maps to simple overlays of USGS quadrangle sheets. Mapping can generally be conducted at two scales, the subwatershed scale and the aquatic corridor scale.

Task 4: Adapt and Apply Watershed Protection Tools

Just as different goals need to be established depending on a watershed's management category, so do the various tools used to protect that resource. For example, while structural stormwater controls are recommended as a component of all management plans, the types of controls used will be different depending on the specific characteristics of a given watershed. Watershed protection tools include (CWP, 1998):

- Land use planning/zoning
- Land acquisition and conservation
- Riparian buffers and greenways
- Better site design techniques
- Structural stormwater controls
- Erosion and sediment control
- Elimination of nonstormwater discharges
- Watershed stewardship programs

Step 7 — Adopt and Implement the Plan

Determine what steps are needed to effectively implement the plan. Implementation of the recommendations of a local watershed management plan can take place through a number of related mechanisms.

In some communities, the watershed or master plan is adopted (often by reference) in its stormwater ordinance and essentially becomes an overlay district, wherein development decisions must follow plan recommendations for various parts of the watershed. In others, it is not mandatory but is referred to when rezoning, and the plans approval decisions are made by staff and zoning boards.

The local long-term capital improvement plan can be derived from the recommendations of the plan. Special assessment districts, fee-in-lieu charges, system development charges, or other funding mechanisms can be established to help pay for specific improvements identified in the plan. Other planning tools, such as comprehensive, corridor, open space, parks, or greenway plans can be modified to incorporate the recommendations of the watershed or stormwater master plan into long-term land-use planning, transportation plans, etc.

Step 8 — Revisit and Update the Plan

Things change. Thus, master planning is an ongoing venture with a long-term perspective and, hopefully, a long-term team in place to oversee the process and implementation. Statewide planning calls for a 5-year cycle, wherein the steps are revisited, often in conjunction with watershed permitting.

Periodically update the plan based on new development in the watershed or results from monitoring data. Each subwatershed or watershed plan should be prepared with a defined management cycle of 5 to 7 years. Individual plans are prepared in an alternating sequence, so that a few are started each year with all plans within a given region or jurisdiction ideally being completed within a 5- to 7-year time span. A management cycle helps balance workloads of watershed staff and managers by distributing work evenly throughout the cycle's time period.

References

Aaron, G. and Laktos, D.R., Penn State Runoff Model — Users Manual, January 1980 version, Pennsylvania State University, January 1980.

Alley, W.M. and Smith, P.E., Multi-Event Urban Runoff Quality Model, USGS Open File Report 82-764, Resten, VA, 1982.

American Society of Civil Engineers, Sedimentation Engineering, ASCE Manuals & Reports No. 54, 1977.

Arnell, N.W., Expected annual damages and uncertainties in flood frequency estimation, *ASCE J. Water Resour. Plann. and Manage.*, 115, 1, January 1989.

Atlanta Regional Commission, Georgia Stormwater Manual, 2002.

Barfield, B.J., Warner, R.C., and Haan, C.T., *Applied Hydrology and Sedimentology for Disturbed Areas*, Oklahoma Technical Press, 1985.

Burch, C.W., Abell, P.R., Stevens, M.A., Dolan, R., and Dawson, B., Environmental Guidelines for Dike Fields, U.S. Army Corps of Engineers, Technical Report E-84-4, September 1984.

Caraco, D.S., Dynamics of urban stream channel enlargement, in The Practice of Watershed Protection, Schueler, T.R. and Holland, H.K., Eds., Center for Watershed Protection, Ellicott City, MD, 2000.

Center for Watershed Protection, Rapid Watershed Planning Handbook, Ellicott City, MD, 1988.

Center for Watershed Protection, Crafting Better Watershed Protection Plans, Article 29, The Practice of Watershed Protection, Schueler, T.R. and Holland, H.K., Eds., Center for Watershed Protection, Ellicott City, MD, 2000.

Debo, T.N., Urban flood damage estimating curves, *J. Hydraul. Div.*, ASCE, HY 10, 1059–1069, October 1982. Reprinted in the *Water Resour J.*, Economic and Social Commission for Asia and the Pacific, United Nations, Bangkok, Thailand, Fall 1983.

Debo, T.N. and Reese, A.J., Determining Downstream Analysis Limits for Detention Facilities, Proc. NOVATECH, Lyon France, November 3–5, 1992.

Debo, T.N., Westerich, D., Newsom, T., Drainage Problem Categorization Study, The Columbus, Georgia, Stormwater Management Program, 1979.

Federal Emergency Management Agency (FEMA), Flood Insurance Rate Review, 1987.

Federal Highway Administration, Bridge Waterways Analysis Model — Research Report, Report No. FHWA/RD-86/108, 1986.

Federal Register, Changes to General Provisions and Communities Eligible for the Sale of Insurance Required to Include Future-Conditions Flood Hazard Information on Flood Maps, 66, 228, 59166–59177, Tuesday, November 27, 2001a.

Federal Register, Endangered and Threatened Wildlife and Plants: Proposed Designation of Critical Habitat for the Carolina Heelsplitter, 66, 133, Wednesday, July 11, 2001b.

Ferguson, B.K., The Alluvial Progress of Piedmont Streams, Proc. of the Eng. Found. Conf., Effects of Watershed Dev. and Manage. on Aquatic Ecosystems, Roesner, L., Ed., Snowbird, UT, 1997.

Galli, J., Development and Application of the RSAT in the Maryland Piedmont, Proc. of the Eng. Found. Conf., Effects of Watershed Dev. and Manage. on Aquatic Ecosystems, Roesner, L., Ed., Snowbird, UT, 1997.

GKY & Assoc., HYDRAIN — Integrated Drainage Design Computer System Version 4.0, Report No. FHWA-RD-92-061, 1992.

Hanson, J.R., Adamus, P.R., Elmer, J.O., and DeWan, T., Environmental Features for Streamside Levee Projects, U.S. Army Corps of Engineers, Technical Report E-85-7, August 1985.

Harvey, M. and Watson, C., Fluvial processes and morphological threshold in incised channel restoration, *Water Resour. Bull.*, June 1986.

Henderson, J.E., Handbook of Environmental Quality Measurement and Assessment: Methods and Techniques, U.S. Army Chief of Engineers, Technical Report E-82-2, May 1982.

Hoggan, D.H., *Computer Assisted Floodplain Hydrology and Hydraulics*, McGraw-Hill, New York, 1989.

Huber, W.C., Modeling urban runoff quality: state-of-the-art, in *Urban Runoff Quality*, ASCE Pub. Proc. Eng. Found. Conf., Henniker, NH, June 23–27, Urbanos, B. and Roesner, L.A., Eds., 1986.

Huber, W.C. and Heaney, J.P., Analyzing Residuals Discharge and Generation from Urban and Nonurban Land Surfaces", in Analyzing Natural Systems, Analysis for Regional Residuals, EPA 600/3-83-046 (NTIS PB83-223321), USEPA, June 1982.

Huber, W.C., Heaney, J.P., Nix, S.J., Dickenson, R.E., and Polmann, D.J., Stormwater Management Model User's Manual, Version III, EPA 600/2-84-109a (NTIS PB84-198432), EPA, November 1981.

Hydrologic Engineering Center, Storage, Treatment, Overflow, Runoff Model, STORM, Users Manual, Generalized Computer Program 723-S8-L7520, Corps of Engineers, Davis, CA, 1977a.

Hydrologic Engineering Center, HEC-6 Scour and Deposition in Rivers and Reservoirs, Users Manual, Hydrologic Engineering Center, Davis, CA, 1977b.

Hydroscience, Inc., A Statistical Method for Assessment of Urban Stormwater Loads — Impacts — Controls, EPA Report 440/3-79-023, EPA, Washington, DC, January 1979.

Johanson, R.C., Imhoff, J.C., Davis, H.H., User's Manual for Hydrologic Simulation Program — Fortran (HSPF), EPA-600/9-80-015, EPA, Athens GA, 1980.

Johnson, W.K., Significance of Location in Computing Flood Damage, *ASCE J. of Water Resour. Plann. and Manage.*, 111, 1, January 1985.

Leopold, L.B., Wolman, M.G., and Miller, J.P., Fluvial Processes in Geomorphology, W.H. Freeman, San Francisco, CA, 1964.

Livingston, E., Stormwater regulatory program in Florida, in *Urban Runoff Quality*, ASCE Pub. Proc. Eng. Found. Conf., Henniker, NH, June 23–27, Urbonas, B. and Roesner, L.A., Eds., 1986.

MacRae, C.R., An Alternative Design Approach for the Control of Instream Erosion Potential in Urbanizing Watersheds, in Conf. Proc. 6th Int. Conf. on Urban Storm Drainage, September 12–17, Niagara Falls, Ontario, Canada, 1993.

MacRae, C.R., Experience from Morphological Research on Canadian Streams, Proc. of the Eng. Found. Conf., Effects of Watershed Dev. and Manage. on Aquatic Ecosystems, Roesner, L., Ed., Snowbird, UT, 1997.

Metropolitan Washington Council of Governments, Watershed Restoration Handbook, Anacostia Restoration Team, Restoring Our Home River: Water Quality and Habitat in the Anacostia, College Park, MD, November 6–7, 1991.

National Watershed Coalition, PL 83-566 — A Holistic Approach to Watershed Management, Proc. 3rd Natl. Watershed Conf., Jackson, MS, May 16–19, 1993.

National Weather Service, Rainfall Frequency Atlas of the United States, Technical Paper 40, U.S. Dept. of Commerce, Washington, DC, 1961.

Nunnally, N.R. and Shields, F.D., Incorporation of Environmental Features in Flood Control Channel Projects, U.S. Army Corps of Engineers, Technical Report E-85-3, May 1985.

Ogden Environmental, Guidelines for Analyzing HEC-1 and HEC-2 Input and Output, 1989.

Ogden, Inc., Master Plan for McCrory Creek, Nashville, TN, for Nashville and Davidson County, TN, 1988.

Ohrel, R., Stream Channel Geometry Used to Assess Land Use Impacts in the Pacific Northwest, in: The Practice of Watershed Protection, Schueler, T.R. and Holland, H.K., Eds., Center for Watershed Protection, Ellicott City, MD, 2000.

Roesner, L.A., Burgess, E.H., and Aldrich, J.A., The hydrology of urban runoff quality management, *Water Resour. Plann. and Manage.*, 764–780, 1990.

Rosgen, D., Applied river morphology, *Wildlife Hydrology*, Pagosa Springs, CO, 1996.

Schueler, T. and Claytor, R., Impervious Cover as an Urban Stream Indicator and Watershed Management Tool, Proc. of the Eng. Found. Conf., Effects of Watershed Dev. and Manage. on Aquatic Ecosystems, Roesner, L., Ed., Snowbird, UT, 1997.

Schueler, T.R., Controlling Urban Runoff, Metropolitan Washington Council of Governments, July 1987.

Schueler, T.R., The Stream Protection Approach, Metropolitan Washington Council of Governments, January 1994.

Shields, F.D. and Palermo, M.R., Assessment of Environmental Considerations in the Design and Construction of Waterway Projects, U.S. Army Chief of Engineers, Technical Report E-82-8, September 1982.

Shields, et al, Experiment in stream restoration, ASCE, *J. Hydraul. Eng.*, June 1995

Simons, D.B. and Senturk, F., Sediment Transport Technology, Water Resources Publications, Ft. Collins, CO, 1977.

Simons, Li & Associates (SLA), Engineering Analysis of Fluvial Systems, SLA, 1982.

Soil Conservation Service, Urban Hydrology for Small Watersheds, TR 55, 1981.

Tasker, G.D. and Driver, N.E., Nationwide regression models for predicting urban runoff water quality at unmonitored sites, *Water Resour. Bull.*, 24, 5, 1091–1101, October 1988.

Thackston, E.L. and Sneed, R.B., Review of Environmental Consequences of Waterway Design and Construction Practices as of 1979, U.S. Army Chief of Engineers, Technical Report E-82-4, April 1982.

U.S. Army Corps of Engineers, Floodway Design Considerations, 1978.

U.S. Army Corps of Engineers, HEC-2 Water Surface Profiles, Hydrologic Engineering Center, 1982a.

U.S. Army Corps of Engineers, Hydrologic Analysis of Ungaged Watersheds Using HEC-1, Training Document No. 15, April 1982b.

U.S. Army Corps of Engineers, Business Depth-Damage Analysis Procedures, Engineer Institute for Water Resources, Research Rpt. 85-R-5, September 1985.

U.S. Army Corps of Engineers, Floodway Determination Using Computer Program HEC-2, Training Doc. No. 5, January 1988a.

U.S. Army Corps of Engineers, National Economic Development Procedures Manual — Urban Flood Damage, Engineer Institute for Water Resources, IWR Report 88-R-2, March 1988b.

U.S. Army Corps of Engineers, Computing Water Surface Profiles with HEC-2 on a Personal Computer, Training Doc. No. 26, September 1988c.

U.S. Army Corps of Engineers, Sedimentation Investigations of Rivers and Reservoirs, EM 1110-2-4000, December 1989.

U.S. Army Corps of Engineers, HEC-1 — Flood Hydrograph Package, September 1990.

U.S. Army Corps of Engineers, Hydrologic Engineering Analysis Concepts for Cost-Shared Flood Damage Reduction Studies, EP 1110-2-6005, December 1992.

U.S. Army Corps of Engineers, Hydrologic Engineering Studies Design, EP 1110-2-6007, February 1993.

U.S. Environmental Protection Agency, Results of the Nationwide Urban Runoff Program, Vol 1., Final Report, Washington, DC, 1983.

U.S. Environmental Protection Agency, The Watershed Protection Approach, EPA840-94-001, Washington, DC, November 1994.

U.S. Environmental Protection Agency, USEPA's Rapid Bioassessment Protocols for Use in Streams and Wadeable Rivers, EPA 841-B-99-002, November 1999.

U.S. Environmental Protection Agency, BASINS 3.0 Fact Sheet, 2002, http://www.epa.gov/OST/BASINS/basinsv3.htm.

Vannote, R.L., Minshall, G.W., Cummins, K.W., Sedell, J.R., and Cushing, C.E., The river continuum concept, *Canadian J. Fisheries and Aquatic Sci.*, 37, 130–137, 1980.

Water Resources Council, Economic and Environmental Principles and Guidelines for Water and Related Land Resource Implementation Studies, Washington, DC, 1983.

Whipple, W., Grigg, N.S., Grizzard, T., Randall, C.W., Shubinski, R.P., and Tucker, L.S., *Stormwater Management in Urbanizing Areas*, Prentice Hall, New York, 1983.

Whitlow, T.H. and Harris, R.W., Flood Tolerance in Plants: A State-of-the-Art Review, U.S. Army Chief of Engineers, Technical Report E-79-2, August 1979.

15

Stormwater Quality Management Programs

15.1 Basic Urban Runoff Quality Understanding

Introduction

The purpose of this chapter is to help you get a practical working knowledge of the area of stormwater quality programs and how to implement them in compliance with NPDES permits and other regulatory programs. After a brief discussion of the background of the subject and basic principles, we will discuss regulatory requirements, NPDES, and establishing and implementing an effective program. Much has been written on the subject, often from a more theoretical perspective. This chapter will summarize such material and focus on methods to actually carry out and pay for the stormwater quality mandates.

The Stormwater Quality Approach — History

In 1972, Congress amended the Clean Water Act (CWA) to prohibit the discharge of any pollutant to waters of the United States from a point source unless the discharge is authorized by an NPDES permit. Initial efforts to improve water quality under the NPDES program primarily focused on reducing pollutants in industrial process wastewater and municipal sewage. These discharge sources were easily identified as being responsible for poor, often drastically degraded, water-quality conditions. As pollution-control measures for industrial process wastewater and municipal sewage were implemented and refined, it became increasingly evident that more diffuse sources of water pollution were also significant causes of water-quality impairment. Specifically, stormwater runoff draining large surface areas, such as agricultural and urban land, was found to be a major cause of water-quality impairment, including the nonattainment of designated beneficial uses. Agricultural runoff was thought to be best addressed, at least initially, through the Section 319 program, primarily through a set of voluntary programs and controls, often encouraged through cost-sharing arrangements. As rural streams continue to fail to meet water-quality standards, this program will transition toward more mandatory controls and requirements. Urban runoff was handled differently.

Urban runoff contains a wide variety of pollutants, often in concentrations and volumes that cause acute and chronic environmental impairment of receiving streams and lakes. Numerous studies have been done over the last 20 years to attempt to quantify the impacts of such runoff and to formulate tools to analyze the impacts and seek to mitigate them.

EPA developed a profile of the impacts of stormwater discharges in 1992 (EPA, 1992a). It was found that between one-third and two-thirds of all designated use impairments for our nation's streams were a result of agricultural and urban runoff. While agricultural nonpoint sources comprise the major source of pollution, urban runoff is also significant.

Impacts on urban receiving waters are varied and often difficult to measure. Because of the transient and intermittent nature of the pollution loading, wastewater measurements and standards do not apply well. The water-quality impacts of stormwater runoff to the receiving water depend on a number of factors, including the magnitude and duration of rainfall events, soil types, time between storms, land-use type and specific activities, illicit connections or illegal dumping, and the ratio of the runoff flow volume to the receiving water flow volumes. While actual impacts of stormwater runoff are often difficult to ascertain, there is a general understanding that urban runoff contains significant levels of pollution comparable to primarily treated effluent from wastewater treatment plants and is a major source of receiving water pollution. In addition to chemical water quality, there are habitat impacts caused by the physical aspects of stormwater runoff: velocity, peak flow, and volume increases.

This lack of data and contradictory approaches to pollution mitigation strategies eventually led to the development of the National Urban Runoff Program (NURP) (USEPA, 1983). Various municipalities and agencies under NURP collected data from many municipal residential, commercial, and light industrial sources in an effort to characterize urban stormwater pollution runoff. A number of trials were made of various approaches to pollution removal, some more clearly successful than others. Understandings gained from the NURP program and related studies and conferences eventually led to an overall approach to stormwater quality management. This approach was codified within the Clean Water Act and amendments under the auspices of the National Pollutant Discharge Elimination System (NPDES).

In 1987, Congress amended the CWA to require implementation, in two phases, of a comprehensive national program for addressing stormwater discharges. The first phase of the program, commonly referred to as Phase I, was promulgated on November 16, 1990 (55 FR 47990). Phase I requires NPDES permits for stormwater discharge from a large number of priority sources, including municipal separate storm sewer systems (MS4s) generally serving populations of 100,000 or more and several categories of industrial activity, including construction sites that disturb 5 or more acres of land. In the intervening period, significant information was gleaned from Phase I cities, additional studies, high-level expert meetings, and general experience. This lead to the Phase II program described below. An EPA report to Congress described the general approach to Phase II as a "cost-effective, flexible approach for reducing environmental harm by stormwater discharges from many point sources of stormwater that are currently unregulated" (USEPA, 1999).

Regulations published on December 9, 1999 (64 FR 68722), finalized the second phase of the stormwater program, expanding the existing program to include discharges of stormwater from smaller municipalities in urbanized areas, other industries, and from construction sites that disturb between 1 and 5 acres of land. Phase II regulations allow certain sources to be excluded from the national program based on a demonstrable lack of impact on water quality. The rule also allows other sources not automatically regulated on a national basis to be designated for inclusion based on increased likelihood for localized adverse impact on water quality. The eventual goal is that the two programs will merge to provide similar requirements. There are to be no additional Phase I communities, regardless of population estimates from the most recent census. Phase II came into effect March 2003, with other urban areas added on the basis of phased watershed permitting or based on subsequent impact analysis by permit agencies. See below for more details on the Phase II program.

15.2 Basic Findings

The research that led to the publication of the Phase I and then Phase II stormwater rules laid the groundwork for approaches taken under those rules. These basic findings and ideas form the foundation on which the stormwater quality approach is built. Other regulatory programs use these basic understandings as a basis for their regulatory approach (e.g., TMDL, ESA, watershed protection, etc.). Since the inception of Phase I, additional research and experience added significant new understandings. Some of the more important aspects of urban stormwater quality will be reviewed prior to discussing how to use these understandings in crafting a stormwater quality management program.

Broad Pollution Categories

The pollutants found in urban runoff include sediment, oxygen-demanding substances, heavy metals, nutrients, bacteria, trash, oil and grease, and other more rare toxic chemicals, pesticides, etc. (APWA, 1991; USEPA, 1990). Pollution that enters stormwater conveyance systems comes from two major categories of sources: dirty stormwater and nonstormwater. Thus, two broad approaches are applicable: (1) how we prevent stormwater from coming in contact with pollutants, and how we remove those pollutants once they mix with stormwater; and (2) how we prevent nonstormwater (in the form of illicit connections and illegal dumping) from entering the stormwater system. All municipal stormwater quality programs must address these two aspects of stormwater pollution, and the basic tenet of the NPDES regulations focuses on these two areas. Table 15-1 summarizes urban stormwater pollution.

Also, as a source category-based approach, under Phase I, polluters to streams in the urban environment were thought of as being divided into four major categories: (1) illicit

TABLE 15-1

Summary of Urban Stormwater Pollutants

Constituents	Effects
Sediments — Suspended solids, dissolved solids, turbidity	Stream turbidity Habitat changes Recreation/aesthetic loss Contaminant transport Filling of lakes and reservoirs
Nutrients — Nitrate, nitrite, ammonia, organic nitrogen, phosphate, total phosphorus	Algae blooms Eutrophication Ammonia and nitrate toxicity Recreation/aesthetic loss
Microbes — Total and fecal coliforms, fecal streptococci viruses, *E.Coli*, enterocci	Ear/intestinal infections Shellfish bed closure Recreation/aesthetic loss
Organic matter — Vegetation, sewage, other oxygen-demanding materials	Dissolved oxygen depletion Odors Fish kills
Toxic pollutants — Heavy metals (cadmium, copper, lead, zinc), organics, hydrocarbons, pesticides/herbicides	Human and aquatic toxicity Bioaccumulation in the food chain
Thermal pollution	Dissolved oxygen depletion Habitat changes
Trash and debris	Recreation/aesthetic loss

connections and illegal dumping, (2) industrial dischargers, (3) construction site runoff, and (4) commercial and residential sites (everybody else). Each of these major sources, types, and categories of pollution is then handled in different ways through a combination of structural and nonstructural practices. Phase II does not overtly use these four categories, though keeping them in mind aids implementation of the Phase II programs described later. While these then become the regulatory categories, the impacts of urban development on stormwater physical, biological, and chemical systems is complex and still not well understood.

Impacts of Urban Development

One issue early in the consideration of urban stormwater runoff was how to define a problem. The NURP study focused on a three-level definition: (1) impairment or denial of beneficial use, (2) water-quality criterion violation, and (3) public perception. However, the issue of whether water-quality criteria or standards are actually applicable to wet-weather discharges and whether the public is the best judge of impairment has not been fully settled to this day. But there is a growing consensus about the ways urban development changes urban-influenced waterbodies, even if we cannot fully agree on what magnitude of change constitutes a problem.

Urban development within a watershed has a direct impact on the amount of runoff versus infiltration, as illustrated in Figure 15-1, and on downstream waters.

The impacts of development on watersheds can be placed into four interrelated primary impact categories (ARC, 2002). A myriad of secondary impacts stem from these four. Any stormwater quality management program (SWQMP) should seek to address, as appropriate, all of these impacts:

- Changes to stream flow
- Changes to stream geometry
- Water-quality impacts
- Degradation of aquatic habitat

Changes to Stream Flow

Urban development alters the hydrology of watersheds and streams by disrupting the natural water cycle. This results in:

- *Increased runoff volumes* resulting in pond and lake flooding and in channel erosion through the increased duration of near bankfull flow periods. In fact,

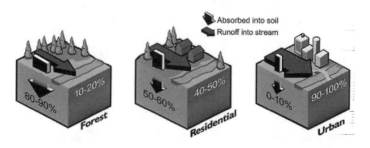

FIGURE 15-1
Land-use changes. (From Atlanta Regional Commission, Georgia Stormwater Manual, 2002.)

some detention programs have been found to exacerbate stream erosion by prolonging the bankfull flow channel-forming discharge, sometimes termed "impulse" erosion, rather than letting it overflow into the floodplain and pass more quickly (Bledsoe, 2001; MacRae, 1993, 1997; Sovern and MacDonald, 2001; Malcolm, 1980).

- *Increased peak runoff discharges* are peak flow increases several times that of pre-development peaks leading to increased flooding. In fact, the actual magnitude of the increase may often be underpredicted due to the tendency of many design approaches to overpredict undeveloped land (predevelopment) peak flows (Strecker, 2001).

- *Greater runoff velocities* are higher flows that lead generally to increased velocities, often overwhelming the delicate balance between sediment inflow and outflow, eroding banks, and causing channel size increase. Sediment transport ability (as measured by shear stress) varies with the third to fifth power of velocity (Simons and Senturk, 1991). Thus, even a relatively small increase in the velocity distribution throughout the year may greatly increase sediment transport capacity (stream power) and long-term erosion.

- *Timing* of peak flows through the normal course of rainfall and storm movement across a watershed is something to which streams have adjusted. Imperviousness, especially directly connected imperviousness, greatly increases the speed of runoff and changes the timing of peak flows, as subbasins join and flow together.

- *Lower dry-weather flows (baseflow)* occur when rainfall cannot soak into the ground and be slowly fed to the stream over the following weeks' or months' baseflow decreases. As illustrated in Figure 15-1, the 80 to 90% of rainfall that infiltrates under a forested scenario decreases dramatically under a paved scenario.

Changes to Stream Geometry

The changes in the rates and amounts of runoff from developed watersheds directly affect the morphology, or physical shape and character, of a city's streams and rivers. Some of the impacts due to urban development include:

- *Stream widening and bank erosion* — Stream channels widen to accommodate and convey the increased runoff and higher stream velocities from developed areas. More frequent small and moderate runoff events undercut and scour the lower parts of the streambank, causing the steeper banks to slump and collapse during larger storms. A stream can widen many times its original size. For example, in Georgia, streams tend to flow at bankfull for storms in the range of the 2-year storm. Figure 15-2, derived from rural and urban regression equations for Georgia, shows that there is a frequency shift when nominal development occurs. The peak flow that a rural stream experienced every 10 years on average, is now experienced every 2 years. And the 100-year flow event now occurs about every 10 years. If the stream channel forms to the 2-year peak flow, then it will tend widen to accommodate the equivalent of the 10-year peak flow prior to development, though, often, a larger change occurs as incision and bank caving cycles are initiated through urban development.

- *Stream degradation* — Another way that streams accommodate higher flows is by downcutting their streambed. Streams will erode the most vulnerable part of the stream perimeter, often the bed. This causes instability in the stream profile, or

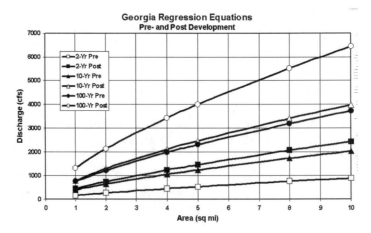

FIGURE 15-2
Georgia regression equations.

elevation along a stream's flow path, which increases velocity and triggers further channel erosion upstream and downstream.

- *Loss of riparian tree canopy* — As streambanks are gradually undercut and slump into the channel, the trees that had protected the banks are undermined and fall into the stream, often creating local scour and more riparian vegetation removal. Also, the natural course of urban development often removes bank vegetation and the protecting effect of root wads.
- *Changes in the channel bed due to sedimentation* — Due to channel erosion and upland sediment supply, deposits in stream as sandbars and other features cover the channel bed, or substrate, with shifting deposits of mud, silt, and sand.

Water-Quality Impacts

Stormwater runoff may contain or mobilize high levels of contaminants, such as sediment, suspended solids, nutrients, heavy metals, pathogens, toxins, oxygen-demanding substances, and floatables. Such contaminants are carried to nearby streams, rivers, lakes, and estuaries. Individually and combined, these pollutants can reduce water quality and threaten one or more designated beneficial uses. Often, an increased volume of runoff or contaminants can lead to violations of applicable state water-quality standards.

Table 15-1 summarizes the key pollutants found in urban stormwater and the kind of impacts that may need to be addressed in an urban stormwater quality management program.

Degradation of Aquatic Habitat

Along with changes in stream hydrology and morphology, the habitat value of streams diminishes due to development in a watershed. Impacts on habitat include:

- *Degradation of substrate habitat structure* — It is now general knowledge that higher and faster flows due to development can scour channels and wash away entire biological communities. Streambank erosion and sediment scour and deposition reduce habitat for many fish species and other aquatic life, while sediment deposits can smother bottom-dwelling organisms and aquatic habitat (Sovern and MacDonald, 2001; Herricks, 2001; Karr et al., 1986; Barbour et al., 1999). An ecological study in New Zealand found that as little as 2 cm of sediment for more than 5 days resulted in almost total mortality of the macroinvertebrate community (Shaver and Hatton, 2001).

- *Loss of pool–riffle structure* — Streams draining undeveloped watersheds often contain pools of deeper, more slowly flowing water that alternate with riffles or shoals of shallower, faster-flowing water. These pools and riffles provide valuable habitat for fish and aquatic insects. As a result of the increased flows and sediment loads from urban watersheds, the pools and riffles disappear and are replaced with more uniform, and often, shallower, streambeds that provide less-varied aquatic habitat.

- *Reduced baseflow* — The makeup of an aquatic community is different from one for an intermittent stream. It is not the other impacts that affect aquatic macroinvertebrates as much as the lack of flowing water in a previously perennial environment during summer dry periods. Reduced baseflow due to increased impervious cover in a watershed and the loss of rainfall infiltration into the soil and water table adversely affect in-stream habitats, especially during periods of drought.

- *Increased stream temperature* — Runoff from warm impervious areas, storage in impoundments, loss of riparian vegetation, and shallow channels can cause an increase in temperature in urban streams. Increased temperatures can reduce dissolved oxygen levels and disrupt the food chain. Certain aquatic species can only survive within a narrow temperature range.

- *Decline in abundance and biodiversity* — When there is a reduction in various habitats and habitat quality, the number and the variety, or diversity, of organisms (wetland plants, fish, macroinvertebrates, etc.) are also reduced. Sensitive fish species and other life forms disappear and are replaced by those organisms that are better adapted to the poorer conditions.

Constant EMC

A useful and somewhat surprising finding from NURP was that the event mean concentration (EMC) of pollutants (the total constituent mass discharge divided by the total runoff volume) was chosen as the best and most useful descriptive statistic for pollution loadings, and that, within the range of data collected, the event mean concentrations were essentially independent of volume of runoff or other factors. In fact, the variability storm to storm was often greater than from site to site. Also, it was found that the variability of EMC values around the mean value follows a log-normal distribution. This allows for various statistical analyses of the probability of pollutant loadings and exceedances.

This constant EMC finding leads directly to the fact that the total loading of wash-off-type pollutants is related primarily to the volume of stormwater and thus is directly dependent on percent imperviousness of the land surface. This leads to simple calculations of pollution loadings based only on estimates of runoff volume and EMC values. Figure 15-3 illustrates typical annual loadings of pollutants from various land-use types for three metals in the summer season. Thus, long-term (chronic) pollutant loadings (and, to a lesser extent, short-term pollution) can be linked directly to land use and rainfall depths. Simple methods of pollution loading calculation, which are basically runoff volume times pollution concentrations for a specific time period, can then be developed and are discussed elsewhere in this text.

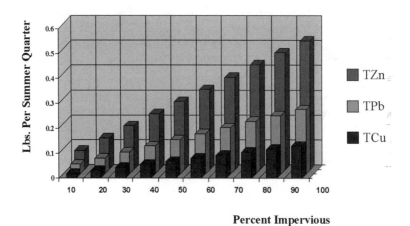

FIGURE 15-3
Total loading of pollutants.

Source Control

Stormwater pollution is the accumulation and aggregation of thousands of individual small sources of pollutants. There is no single end-of-pipe location for stormwater discharges, and the episodic nature, wide variety of pollutants, and potential magnitude of stormwater discharges make end-of-pipe treatment difficult and expensive. While end-of-pipe treatment is acceptable for sewage, it is generally considered that the best way to handle stormwater pollution is to attack the other end, through the use of source controls. In fact, the closer to the source the control can be located, the more effective (and cost effective) it can be (BAS-MAA, 1999). It is often more difficult to remove pollutants from waters once they are there than to prevent them from entering the waters of the United States in the first place.

The source can sometimes be in the minds of polluters. Thus, educational programs and other efforts to change human behavior become important methods for stormwater pollution control. For example, if, through a combination of educational and other related controls, individuals get into the habit of recycling used motor oil rather than dumping it down a catch basin or into a stream, literally millions of quarts of motor oil can be kept from the environment without resorting to a single structural control such as an oil-grit separator.

There are a wide variety of such source controls of a structural nature (such as retention ponds) and of a nonstructural nature (such as zoning ordinances or education initiatives). These controls are often called urban best management practices (BMPs), though other terms are coming into popularity. These practices help form the basis of any control program.

Hot Spots

Most urban stormwater carries relatively few priority pollutants (CWP,2000). But, there are areas of the urban landscape that do not follow this general rule. These areas have been termed *urban hot spots*. Stormwater hot spots are areas of the urban landscape that often produce higher concentrations of certain pollutants, such as hydrocarbons or heavy metals, than are normally found in urban runoff. These areas merit special management and the use of specific pollution prevention activities and structural stormwater controls. Examples of stormwater hot spots include:

- Gas/fueling stations
- Vehicle maintenance areas

- Vehicle washing/steam cleaning
- Auto recycling facilities
- Outdoor material storage areas
- Loading and transfer areas
- Landfills
- Construction sites
- Industrial sites
- Industrial rooftops

The mere presence of priority pollutants and trace metals does not imply an acute toxicity problem, but rather the accumulation of toxics in sediments and ponds over time, and potential for release of these pollutants in toxic concentrations (Pitt and Field, 1990; Schueler and Shepp, 1992). In addition, not all accumulated pollutants are toxic to plants growing in the sediments, to aquatic organisms that live in the sediments, or to higher levels of life that feed on these organisms and may be poisoned due to bioaccumulation of the toxics up through the food chain.

For example, cadmium, copper, lead, and zinc being responsible for aquatic life impacts depends on whether these metals are in toxic/available forms in the sediments. All of these metals tend to be detoxified by sediment sulfides. So, the interaction of these metals with other chemicals drives the actual eventual toxicity. Thus, simple exceedance of a water-quality criterion in a catch basin sump or even in effluent from a small urban hot spot area cannot be taken, per se, as a driver for a more advanced stormwater treatment program. Actual toxicity in the receiving water environment is key, using more than simple chemical analysis including biocriteria and whole effluent toxicity testing.

There are well-established techniques for determining whether a constituent in sediments that occurs in elevated concentrations is a likely cause of aquatic life toxicity or altered organism assemblages, or serves as a significant source of bioaccumulatable chemicals in higher trophic-level organisms (Lee and Jones-Lee, 1999). There are also growing capabilities in what might be termed forensic toxicity, wherein reasons for actual mortality can be ascertained through a scientifically based process of elimination. Ecological risk assessment has also grown to the point where risks of mortality can be quantified if the mechanisms for mortality are understood.

Most urban hot spot treatment programs cannot, and should not, spend the kinds of resources necessary to perform this kind of toxicity testing unless there is seen to be a critical water resource or endangered species needing protection, justifying the expense. Most of the time, a more generalized but targeted treatment program would suffice and may obtain significant pollution reduction for the level of effort.

Specific programs can include the whole array of structural and nonstructural approaches, including such things as education, material handling changes, changes in work practices, substitution of materials, waste disposal, site mapping, prepositioning of spill materials, capture and treatment schemes, inlet stenciling, rerouting drainage, first flush and commercial oil removal devices, and other structural BMPs. One key to the overall effectiveness of structural BMPs is frequent maintenance. There are a number of commercially available devices that purport to treat runoff toxics and hydrocarbons. While a great many of these may, in fact, do what they say, care should be taken to assess them based on independent data from field applications.

Santa Clara Valley (1993) provides the following list of nonstructural approaches to handle hot spot runoff:

- Prevent discharges when changing automotive fluids.
- Use drip pans when working on engines.
- Use special care to prevent leaks from wrecked vehicles.
- Quickly clean up spills of all sizes.
- Keep wastes from entering floor drains and storm drains.
- Use concrete surfaces and roofing over fueling areas to prevent spilled fuel from contact with stormwater.
- Properly store and recycle used batteries.
- Clean parts without using liquid solvents (or use solvent recyclers).
- Capture all metal particles during grinding and finishing operations.
- Properly store and recycle waste oil, antifreeze, and other automotive fluids.
- Select "environmentally friendly" products and control inventory to reduce wastes.
- Keep all working areas inside and away from stormwater.
- Treat all liquid streams from car washing and engine cleaning.
- Train employees on pollution prevention activities for the shop.
- Educate customers on proper recycling and disposal of automotive products.

First Flush and Treatment Volume

Another key aspect of urban stormwater pollution is that for the more impervious areas, higher concentrations of pollutants are generally washed in the first flush of the rainfall and from the many smaller storms, rather than for the few larger storms where dilution is more predominant. This is especially true for smaller, more impervious sites, where first-flush effects can be more readily observed.

Thus, the many smaller storms that occur each year contribute the great majority of pollution to the system. Therefore, control systems designed to treat the first flush and the smaller volumes may be potentially useful, though there is disagreement on whether there actually is a first flush and what volume of control is appropriate (in some northern climates, the first flush is often considered to be the first melt of spring).

For example, Figure 15-4 shows a plot of rainfall from Reno, Nevada, giving total rainfall depth (24-h inter-event dry period) and percent less. From this figure, it can be seen that capture and treatment of runoff from a rainfall depth of about 0.7 in would handle 90% of the storms and a significant proportion of runoff from larger storms. A similar plot from Atlanta, Georgia, would yield a comparable depth of 1.2 in. Treating these relatively small storm volumes results in significant pollution reduction. See Chapters 7 and 13 for more information on water-quality hydrology and BMP sizing.

Many different design concepts have been developed over the years, including capture of 80% TSS, 1/2-in runoff from the 1-in rainfall, 1-in from directly connected impervious areas, etc. The point is that stormwater treatment is a fallback position. The goal is to prevent or mitigate some adverse impact. In wastewater and other point sources, the way to accomplish this is to set some chemical criteria for water quality, allowing discharges below toxicity for indicator species. The thinking is that, if this level of chemical quality can be attained, then whatever impact mitigation is sought can be achieved. But in stormwater, chemical criteria have been found to be almost meaningless. So, we have taken another step back from reality and set capture and treat standards. The hope is that by meeting these standards, we can attain some undefined physical and chemical criteria

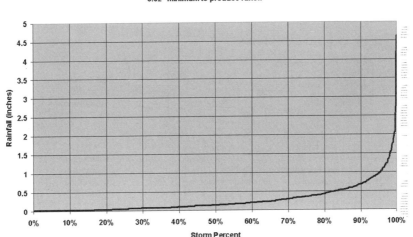

FIGURE 15-4
Reno, Nevada, rainfall data.

that will help attain some impact mitigation (also usually ill-defined for stormwater). These capture criteria are also hard to meet, so many communities have resorted to more of a "black-box" BMP design criteria standard. Using this standard,. the designer meets some structural sizing and design criteria, which is hoped to meet the capture criteria, which is hoped to change some chemical or physical characteristic of the wet-weather discharge, which is hoped to meet some ill-defined impact mitigations strategy.

This is the challenge of urban stormwater quality programs — to regulators and permittees.

Treatment Train Concept

In the intervening years between Phase I of the NPDES permit and Phase II, a clearer understanding of the limitations of any individual BMP application is becoming apparent. Many studies have shown that any specific BMP is limited in a number of ways, including geographic coverage, ability to reduce a specific pollutant of concern, long-term viability and effectiveness, ability to target all the impacts of urban development, and ability to target specific pollution sources. Thus, it is understood that it is only through a combination of BMPs employed within a particular watershed that pollution can be reduced to the maximum extent practicable (Schueler and Caraco, 2001; Strecker, 2001; Harrison, 2001; Sovern and MacDonald, 2001).

As shown in Figure 15-5, this treatment train can be thought of as having five major components:

1. Education and prevention programs begin in the minds of potential polluters to educate and change habits and practices. These might include used oil recycling programs, sweeping practices, homeowner use of certain toxic chemicals, fertilizer or gardener programs, etc. The target is to reduce or eliminate pollution prior to its entering the physical system. Its goal is to change the way individuals and societies function to reduce the generation of pollution into the environment. Broad categories might include education programs, practice changes, and product substitution (Clark, Field, and Pitt, 2001).

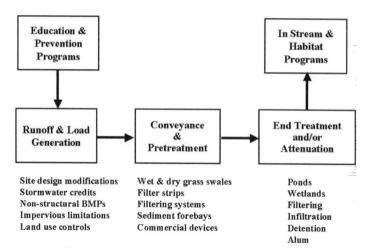

FIGURE 15-5
Watershed treatment train.

2. Runoff and load generation is the first physical line of defense against pollution. It begins with the development of land-use policies that reduce the inherent impact of development on streams. There are a wide number of practices (see Chapter 1) in the United States and abroad that include modifying land-use and zoning codes and development practices to reduce impervious areas and to build in pollution-reducing features. In many cases, the reduced infrastructure requirements actually reduce the development cost of the lots. Crediting and other "carrot" programs provide incentive for developers and landowners to use the methods. Also included is a family of practices designed to remove pollutants prior to rainfall and runoff transport. Street sweeping proved generally ineffective in the NURP study, but review of recent technical advances in sweeper types indicate they may be a significant pollution-reduction mechanism, especially for more arid climates (Deletic et al., 1998; Sutherland and Jelen, 1997). Storm drain stenciling or gluing of a plate on storm drains reminding would-be-polluters that the drain flows to a stream is also popular — the metal or plastic plates last many times longer than the painted-on stencils.

3. Conveyance and pretreatment is the initial designed stormwater system. It includes a wide array of commercial devices used in small drainage situations, low-impact development approaches (PGC, 1999; Shaver and Hatton, 2001), pollution-reducing structural conveyance BMPs (swales, linear wetlands, buffers, exfiltrating pipe systems, etc.), and small-site BMPs [rain gardens (Figure 15-6), infiltration trenches, dry wells, rain barrels, etc.]. The idea is to intercept and remove pollutants at the entry to the minor stormwater system or within the first several reaches of the conveyance system.

4. End-of-pipe treatment and attenuation refers to the wide array of structural BMP devices that remove pollution from stormwater, including ponds, wetlands, infiltration, filtration, channel practices, and special devices. These BMPs are often designed for multiple uses in controlling water-quality and quantity impacts as well as recharge of groundwater and channel protection. Figure 15-7 shows the outlet structure for such a device.

5. In-stream and habitat programs refer to an array of practices to restore and protect streams, including in-stream structures, stream assessment, bank treatment and

FIGURE 15-6
"Rain garden" bioretention area.

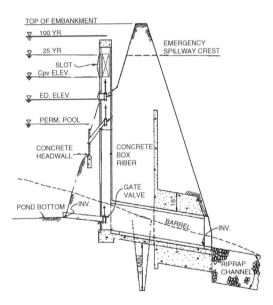

FIGURE 15-7
Multicriteria wet pond outlet structure. (From Atlanta Regional Commission, Georgia Stormwater Manual, 2002.)

stabilization, riparian corridor preservation or restoration, channel modifications, fish habitat, etc. This is the last line of defense against negative stream impact and operates directly on stream needs using a variety of assessment and practice methods (Riley, 1997; Rosgen, 1994). Figure 15-8 shows a citizen group working on a streambank stabilization project.

Water-Quality Standards

The objective of the Clean Water Act (CWA) is to maintain and restore the chemical, physical, and biological integrity of the nation's waters (Swietlick, 2001). These three aspects of water quality, as illustrated in Figure 15-9, together are necessary to achieve ecological integrity.

FIGURE 15-8
Streambank stabilization project.

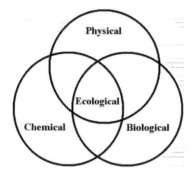

FIGURE 15-9
Ecological integrity.

The U.S. EPA cannot back off from this standard or the other requirements of the CWA. Thus, in the long run, urban waters will need to meet whatever water-quality standard is in force. But do current water-quality standards make sense for stormwater? Most practitioners do not think so and are looking to changes in these standards and the eventual creation of a wet-weather standard more amenable to stormwater management.

The CWA requires states to establish water-quality standards to "protect the public health or welfare" and "enhance the quality of water" [Section 303(c)(2)(A)]. Water-quality standards are to be established for water bodies "taking into consideration their use and value for public water supplies, propagation of fish and wildlife, recreational purposes, and agriculture, industrial, and other purposes, and also taking into consideration their use and value for navigation" [Section 303(c)(2)(A)]. In addition, the CWA establishes the national goal that, wherever attainable, "...water quality provides for the protection and propagation of fish, shellfish, and wildlife and provides for recreation in and on the water..." [Section 101(a)(2)].

EPA's water-quality standards regulations at 40 CFR 131 include several key provisions concerning use designation of water bodies. The regulations at 131.10(j) provide that a state must conduct a use attainability analysis (UAA) whenever the state designates or has designated uses that do not include those specified in CWA Section 101(a)(2), or the state wishes to remove a CWA Section 101(a)(2) use, or to adopt subcategories of uses specified in CWA Section 101(a)(2) that require less stringent criteria. EPA regulations define a UAA as a "structured scientific assessment of the factors affecting the attainment of the use which may include physical, chemical, biological, and economic factors as described in 40 CFR 131.10(g)."

At present, there are chemical-specific criteria, toxicity criteria, biological criteria, contaminated sediment criteria, and physical criteria set by each state based on the overall guidance of the U.S. EPA (Swietlick, 2001). None of these criteria need to take into account economic or technical factors in their development:

- Chemical criteria rely on measurement of the maximum amount of a specific chemical in the water that will prevent acute and chronic effects on aquatic organisms. These were derived from a suite of individual chemical toxicity tests on key organisms.
- Whole effluent toxicity (WET) tests are based on the same overall approach but use the total mix of chemicals in the water to establish impact on organisms.
- Biological criteria are water body response criteria, measuring the response of indicator organisms to the complete set of impacts of urban development on the stream. Biocriteria, in a sense, integrate the impacts of all stressors on a stream over a short period of time within the bodies of the indicator organisms. If the organisms are healthy and diverse, then whatever the combination of stressors, they are not sufficient to adversely impact the stream.
- Contaminated sediment criteria are set for two purposes: protect bottom-dwelling organisms from the toxic effects of pollutants in the sediments, and prevent harmful bioaccumulation effects when these organisms are eaten by other organisms, including man.
- Physical criteria include physical water characteristics such as acidity, temperature, dissolved oxygen, and conductivity.

After much discussion, it was decided that, unlike most NPDES permits, the standard for stormwater permits would be a narrative standard defined as Maximum Extent Practicable. This departure from numerical discharge requirements was to last until 2013, at which time additional information would presumably make the establishment of wet-weather standards potentially possible. This is not to say that numeric criteria for stormwater have not already been developed in isolated cases or that they will not be developed under different regulatory programs (such as TMDL), or as part of an effluent guideline effort or court-ordered compliance.

The primary challenge for state regulators and local governments is to ensure that appropriate water-quality standards have been developed for urban water bodies. There is hope in the future that we might see the development of different designated uses by the states such as an "urban stream habitat use" or "urban stream restoration" (Swietlick, 2001). We will also see over the next decade the wider or more effective use of different stream measurement metrics besides chemical quality, including improved wet weather biocriteria, whole effluent toxicity, and habitat criteria.

15.3 Water Quality Act Overview

The current primary driver for most municipalities engaging in water-quality programs is the Phase I or Phase II NPDES program. Other regulatory programs can also impact a local stormwater program, including the Endangered Species Act, Total Maximum Daily Load Program, Source Water Protection, Wellhead Protection, Coastal Zone Management

Program, and special state, regional, and local protection programs or special requirements. After an overview of the Phase I and Phase II stormwater programs, some of these other programs will be summarized.

Phase I Basic Requirements

The Water Quality Act of 1987, Section 405, amended Section 402 of the Clean Water Act of 1972 to require the U.S. EPA to provide rules and regulations that establish a permit program for stormwater discharges associated with industrial activities and from municipal separate storm drain systems serving populations of 100,000 or more (*Federal Register*, 1988). The first phase of the program, commonly referred to as Phase I, was promulgated on November 16, 1990 (55 FR 47990).

This Act and the promulgated regulations have had a major impact on those municipalities affected. It also required the consideration of smaller municipal separate stormwater systems (MS4s) and the promulgation of regulations to include them at a later date.

The Act requires that dischargers to "waters of the state" (WOS) be subject to the National Pollution Discharge Elimination System (NPDES) permit program. However, the requirements, and the philosophy behind these requirements, outlined in the draft regulation are far-reaching.

The Act, as it applies to municipalities, has three main thrusts:

1. It requires municipalities to effectively prohibit nonstormwater discharges into their publicly owned or operated separate storm drain system.
2. It requires municipalities to control discharge of pollution into their systems to the "maximum extent practicable" (MEP).
3. It defines one system-wide permit rather than permits for each individual discharge point.

The Phase I NPDES permit was in two parts. Part 1 concentrated on identification of and description of the system and dischargers. Part 2 concentrated on a program to reduce pollution discharges.

Part 1 of the permit required identification of known data, structures, outfalls, land uses, water-quality problems, etc. It also required a description of existing program and physical elements such as legal authority, financing, physical characteristics of the area, and controls. There is a dry-weather field screening requirement, wherein any major outfall discharging to waters of the state that shows turbidity, oily sheen, odor, or any obvious degradation or has dry-weather flow is identified and sampled.

Part 2 of the permit had four major aspects:

1. *Program description* entails demonstrating sufficient legal authority, financial capacity, and administrative capability to control pollution, prohibit illicit dumping, and require compliance.
2. *Source characterization* entails identifying all sources of urban stormwater pollution such as landfills, industrial sites that have a pollution problem, etc. It also requires estimation of pollution loadings from all areas of the municipality.
3. *Discharge characterization* entails verification sampling of those low flows under Part 1, which demonstrated problems, as well as representative sampling of from 5 to 10 outfalls representing different land-use types. This sampling is done

during runoff events and is designed to measure first-flush (acute) and long-term (chronic) pollution loadings.

4. *Proposed stormwater management program* entails the complete description of a comprehensive stormwater management program, which can deal effectively with pollutant runoff from the four major categories of dischargers. The program will contain such elements as sampling, inspection, enforcement, BMPs, planning, etc. It will also contain an element to regulate stormwater discharge from industrial sites that discharge into the publicly owned or operated separate stormwater system. This element may have some similarity to the existing industrial pretreatment program.

Municipal Reaction

Unlike most NPDES permits, the one for municipal stormwater has few or no numeric water-quality criteria to meet, few set procedures, forms, or activities, and no true precedent. It is a negotiated permit where, within the confines of broad USEPA guidance and sometimes vague regulatory language, a municipality has the responsibility and the leeway to develop an approach to control the discharge of pollutants to public waters to the "maximum extent practicable" (MEP) and to prohibit nonstormwater from entering into and being discharged from the storm drainage system.

Many municipalities have asked how to get from a sometimes poorly developed urban drainage program to comprehensive stormwater quality management. A review of the literature is not very helpful. While there are reams of information and mounds of case studies on how various BMPs may work or have worked, there is little guidance available for helping a municipality through the process of actually planning and implementing a SWQMP, which employs these BMPs in a cost-effective and prioritized basis.

Concern for adverse impacts of polluted runoff is good stewardship and makes good sense, regardless of legal mandates. Each municipality should, to some appropriate extent, develop and carry out sound policies and programs that meet the objectives of MEP and seek to restore, preserve, and conserve environmental resources.

Phase II Basic Requirements*

EPA has published regulations for permitting of stormwater discharges to waters of the United States under NPDES. These regulations include over 5000 municipal entities under the mandatory designation, and many more under the evaluation and optional designation process.

There are several levels of coverage under this permit, depending on location with respect to recognized urbanized areas and then individual circumstances. You can find a definition of, and list of, urbanized areas at the Census Bureau Web site.

- *Automatic nationwide designation* — Owners or operators of small MS4s located in any incorporated county or place under the jurisdiction of a governmental entity within a census-designated urbanized area were required to develop a local stormwater program. A regulated small MS4 had to apply for coverage within approximately 3 years and 90 days from the date of final regulation (March 2003). Only the portion of a county or Federal Indian Reservation located within an urbanized area is regulated.

---
* 40 CFR Parts 9, 122, 123, and 124, December 8, 1999, pp. 68722–68851.

- *Potential designation/mandatory evaluation by the permitting authority* — Owners or operators of small MS4s located outside of an urbanized area may be designated if they have existing or potential significant water-quality impacts, as determined by criteria set by the permitting authority under a mandatory evaluation. This includes all owners or operators of small MS4s located outside of an urbanized area with a population of at least 10,000 and a population density of at least 1000. A balanced application of the following criteria, on a watershed or other local basis, are recommended: (1) discharge to sensitive waters, (2) high growth or growth potential, (3) high population density, (4) contiguity to an urbanized area, (5) contribution to an exceedance of water-quality standard (including impairment of designated uses), (6) significant contributor of pollutants to the waters of the United States, (7) ineffective control of water-quality concerns by other programs. Must apply within 60 days of notice, unless permitting authority grants a later date.
- *Potential designation/optional evaluation by permitting authority* — Owners or operators of small MS4s located outside of an urbanized area may be designated if they have existing or potential significant water-quality impacts, as determined by criteria set by the permitting authority under an optional evaluation. This includes owners and operators of small MS4s located outside of an urbanized area with a population of less than 10,000 or a density of less than 1000. Similar criteria as the previous category will typically be used for the evaluation.

State and federal facilities (including DOTs, military installations, universities, and correctional facilities) are covered as regulated small MS4s. Several waivers are available, but would not apply to most entities.

A regulated small MS4 must develop, implement, and enforce a stormwater management program designed to reduce the discharge of pollutants from the MS4 to the maximum extent practicable (MEP) and attain water-quality standards.

A regulated small MS4 must submit to the permitting authority, either in the NOI or individual permit, the BMPs to be implemented and the measurable goals for each of the following six minimum control measures:

- Public education and outreach on stormwater impacts
- Public involvement/participation
- Illicit connection and discharge detection and elimination
- Construction site stormwater runoff control
- Postconstruction stormwater management in development/redevelopment
- Pollution prevention/good housekeeping of municipal operations

EPA and authorized states/tribes were encouraged to issue general permits containing most of the details for each municipal program element. Regulated small MS4s may, however, seek to be copermittees on individual municipal stormwater permits.

A municipality's program requirements may possibly be satisfied in four alternative ways:

1. Through **inclusion** in an adjoining municipality's existing stormwater program by way of a NPDES permit modification
2. Through an **agreement** with another governmental or other entity that is already implementing one of the minimum control measures in the same jurisdiction

3. If, in the permit, the permitting authority **recognizes** existing responsibilities among governmental entities for the minimum control measures
4. If, in the permit, the permitting authority **incorporates by reference** a qualifying local, state, or tribal stormwater management program that address one or more of the minimum control measures

Municipalities must evaluate program compliance, the appropriateness of their identified BMPs, and progress toward achievements of their identified measurable goals in annual reports for the first permit term and in years 2 and 4 in subsequent terms. The permitting authority has the discretion to require monitoring, if deemed necessary.

Implementation of BMPs that are consistent with the provisions of the proposed rule constitute compliance with the standard of reducing pollutants to the maximum extent practicable.

Maximum Extent Practicable

Most authorities agree that the state of knowledge about stormwater runoff and its impacts on receiving water did not warrant establishment of wet-weather standards for streams at the time of the Water Quality Act (1987) or at the publication of the Phase II rules more than a decade later. But what should the standard, if any, be? They settled on the term Maximum Extent Practicable.

MS4s are to control stormwater pollution to the maximum extent practicable (MEP). How is that defined? MEP consists of the mix of best management practices (BMPs) and measurable goals that will attain reduction of pollution to the maximum extent practicable to attain water-quality standards. EPA has intentionally not provided a precise definition of MEP to allow maximum flexibility in MS4 permitting. That can be good or bad. The rationale is that MS4s need the flexibility to optimize reductions in stormwater pollutants on a location-by-location basis.

These MEP plans are envisioned to consist of comprehensive stormwater quality management programs (SWQMP) that include management practices, control techniques and systems, design and engineering methods, and other such provisions as the director of the program determines appropriate for control of such pollutants (U.S. EPA, 1992b). Senator Stafford is quoted in the congressional record as stating:

"These are not permits in the normal sense we expect them to be. These are actual programs. These are permits that go far beyond the normal permits we would issue for an industry because they in effect are programs for stormwater management that we would be writing into permits," (*Federal Register*, 1988, p. 49443).

Some guidance on what MEP means is provided by those states that issue a general permit. But, as the U.S. EPA states in its preamble to the final regulations (USEPA, 40 CFR 68754, 1999):

> The pollutant reductions that represent MEP may be different for each small MS4, given the unique local hydrologic and geologic concerns that may exist and the differing possible pollutant control strategies. Therefore, each permittee will determine appropriate BMPs to satisfy each of the six minimum control measures through an evaluative process. EPA envisions application of the MEP standard as an iterative process. MEP should continually adapt to current conditions and BMP effectiveness and should strive to attain water quality standards. Successive iterations of the mix of BMPs and measurable goals will be driven by the objective of assuring maintenance of water quality standards. If, after implementing the six minimum control measures there is still water quality impairment associated with discharges from the MS4, after successive permit

terms the permittee will need to expand or better tailor its BMPs within the scope of the six minimum control measures for each subsequent permit. EPA envisions that this process may take two to three permit terms.

Currently, there are no specific numeric criteria for stormwater discharges (unless established under other regulatory or court-induced programs), and will not be until 2013. MEP is considered a flexible technology-based standard. If you do what you say you are going to do, you are, by definition, in compliance — regardless of the actual water quality. But remember that the Congressionally-mandated goal is to meet water-quality standards (as they are currently defined or may change as newer wet-weather approaches are developed), and EPA plans to negotiate a change in the definition of MEP on the basis of existing or collected monitoring information in each successive permit period. Maybe half of us will be required to monitor receiving waters in the second round. The thumb screws will be tightened until water is fishable, swimable, and whatever else state criteria mandate.

MEP really depends on the consideration of several things as illustrated in Figure 15-10 below:

- Do I have, or can I obtain, the legal authority to carry out the program I am describing?
- Is my technical approach sound in that it is a proven approach, structural or nonstructural, that addresses pollutants of concern in an effective manner?
- Are my defined procedures, policies, staff resources, and equipment appropriate for the level and type of program described?
- Do I have, or can I obtain, dedicated and sufficient funding to support the program I am describing?

Under each of the six minimum controls, it will be necessary to define a program that contains four major considerations: (1) technical approach adequacy and effectiveness, (2) legal authority, (3) financial sufficiency, and (4) administrative and organizational support. In actual analysis, the matrix is greatly expanded to consider every facet of the proposed SWQMP to assure it is funded, staffed, legal, and will actually accomplish environmental or watershed objectives.

Technical adequacy and effectiveness of a planned program can be analyzed under each of the four program areas. Under each area, it will be necessary to specify a priority system

Maximum Extent Practicable	LEGAL	TECHNICAL	ADMINISTRATIVE	FINANCIAL
Public Education				
Public Involvement	✓	✓	✓	✓
Illicit Connection				
Construction	✓		✓	✓
Post-Construction		✓		
Pollution Prevention	✓			

FIGURE 15-10
Matrix definition of MEP.

and a schedule for program development. There are a vast number of possible BMPs that are applicable under each of the categories. It is important to choose only those BMPs (structural and nonstructural) that are in some sense cost-effective for a municipality.

Legal authority must be sufficient to be able to prohibit the discharge of nonstormwater to the municipal system and to require compliance with other measures. Legal authority is based on federal authority and state delegation. Some states allow strong home-rule capabilities, while others may require state legislation to allow municipalities to fully carry out the program mandates. Some states have developed their own stormwater management quality programs and regulations, which must be implemented by municipalities (see, for example, Virginia, 1990). Other states have simply supplied guidance to municipal officials (see, for example, Maine DEP, 1990, 1992).

Local authority in most municipalities rests in the municipal charter, subdivision and zoning ordinances, and other miscellaneous provisions. However, some of the legal references apply only indirectly to stormwater. It may be desirable to revise these portions to more specifically address stormwater quality control management. This will be a dynamic process as the program grows and changes.

Financial sufficiency can be judged on the basis of whether the funding source for the stormwater program is adequate, stable, dedicated, and equitable. Many municipalities are finding that general fund tax-based financing is none of the above and are turning to user-fee-based systems (often termed stormwater utilities). Costs to add a stormwater quality program to a well-established stormwater quantity program could be in the range of a 15 to 30% increase.

So, the approach to defining MEP is to brainstorm a number of BMPs under each of the six minimum controls and seek to define existing and potential answers to each of the four questions. Consider that you may want to do more than the minimum but may not want this "more" to be a regulatory concern (in case it does not happen), so you would attempt to negotiate a more minimal program. On the other hand, you may want the whole program to carry the weight of regulatory oversight and define and negotiate a more comprehensive package.

Measurable Goals

Most readers of the regulations understand that, under its broadest definition, a BMP can be anything that might be expected to reduce or eliminate pollution. But, there is an often overlooked and seemingly innocuous term, "measurable goals," in the definition of MEP. Measurable goals are design objectives or goals that quantify the progress of program implementation and the performance of your BMPs. They are objective markers or milestones that you (and the permitting authority) will use to track the progress and effectiveness of your BMPs in reducing pollutants to the MEP. The thing to remember is that, not only are BMPs enforceable, but so are measurable goals. I do an agreed-to BMP and attain my measurable goal, and I am in compliance. If I perform my negotiated BMP program but do not attain my measurable goal, I am probably not in compliance.

EPA gives several types of measurable goals that may be employed (USEPA, 2002). You can consider developing measurable goals based on one or more of the following general categories:

- *Tracking implementation over time* — Where a BMP is continually implemented over the permit term, a measurable goal can be developed to track how often, or where, this BMP is implemented.

- *Measuring progress in implementing the BMP* — Some BMPs are developed over time, and a measurable goal can be used to track this progress until BMP implementation is completed.
- *Tracking total numbers of BMPs implemented* — Measurable goals also can be used to track BMP implementation numerically, e.g., the number of wet detention basins in place or the number of people changing their behavior due to the receipt of educational materials.
- *Tracking program/BMP effectiveness* — Measurable goals can be developed to evaluate BMP effectiveness, for example, by evaluating a structural BMP's effectiveness at reducing pollutant loadings, or evaluating a public education campaign's effectiveness at reaching and informing the target audience to determine whether it reduces pollutants to the MEP. A measurable goal can also be a BMP design objective or a performance standard.
- *Tracking environmental improvement* — The ultimate goal of the NPDES stormwater program is environmental improvement, which can be a measurable goal. Achievement of environmental improvement can be assessed and documented by ascertaining whether state water-quality standards are being met for the receiving water body or by tracking trends or improvements in water quality (chemical, physical, and biological) and other indicators, such as the hydrologic or habitat condition of the water body or watershed.

Because measurable goals are compliance issues, it is important to define them in ways that can be controlled by the permittee wherever possible. With that as a standard, it is clear that the last two categories suggested by the U.S. EPA can cause problems. In its preamble to the regulations, this potential problem, and its solution, are clarified (40 CFR 68763, 1999):

> Today's rule requires the operator to submit either measurable goals that serve as BMP design objectives or goals that quantify the progress of implementation of the actions or performance of the permittee's BMPs. At a minimum, the required measurable goals should describe specific actions taken by the permittee to implement each BMP and the frequency and the dates for such actions. Although the operator may choose to do so, it is not required to submit goals that measure whether a BMP or combination of BMPs is effective in achieving a specific result in terms of stormwater discharge quality. For example, a measurable goal might involve a commitment to inspect a given number of drainage areas of the collection system for illicit connections by a certain date. The measurable goal need not commit to achieving a specific amount of pollutant reduction through the elimination of illicit connections. Other measurable goals could include the date by which public education materials would be developed, a certain percentage of the community participating in a clean-up campaign, the development of a mechanism to address construction site runoff, and a reduction in the percentage of imperviousness associated with new development projects.

Thus, defining things in ways that can be controlled by the permittee is important. Try to define measurable goals in terms of actions the municipality can take and control, not on the basis of uncontrollable responses from ill-defined citizens or stakeholders, or on the basis of specific but impossible to predict numerical water-quality responses. You would hate to miss a measurable goal of say, 25% response, by one percent and be sued for being out of compliance. This is not to say that measuring these things, citizen response and water-quality improvement, for example, are not important, just that they may best be used as tools to measure progress outside the pressures, and potential legal ramifications, of the compliance format. Thus, under this permit, monitoring and measuring water-quality are

Stormwater Quality Management Programs

generally a parallel component of the program designed to track progress but uncoupled from compliance pressures… at least for now.

EPA strongly recommends that measurable goals include, where appropriate, the following three components:

- The activity, or BMP, to be completed
- A schedule or date of completion
- A quantifiable target to measure progress toward achieving the activity or BMP

15.4 The Six Minimum Controls

Under the Phase II program each permittee, or group, must develop a stormwater program that encompasses, at a minimum, six minimum stormwater program control areas. Because many permittees do not have ready access to the regulatory language, it is hard to read, and the guidance contained within the preamble is difficult to sort out — this section has been provided to give that information in a digestible format. Unlike most regulations, there are mandatory and guidance sections to each of the six minimum controls. The guidance section gives clues to how permit writers will generally look at a proposed program and its changes over the permit term.

General Permit Conditions

§ 122.34 As an operator of a regulated small MS4, what will my NPDES MS4 stormwater permit require?

I. Your NPDES MS4 permit will require at a minimum that you develop, implement, and enforce a stormwater management program designed to reduce the discharge of pollutants from your MS4 to the maximum extent practicable (MEP), to protect water quality, and to satisfy the appropriate water-quality requirements of the Clean Water Act. Your stormwater management program must include the minimum control measures described in paragraph (b) of this section unless you apply for a permit under § 122.26(d). For purposes of this section, narrative effluent limitations requiring implementation of best management practices (BMPs) are generally the most appropriate form of effluent limitations when designed to satisfy technology requirements (including reductions of pollutants to the maximum extent practicable) and to protect water quality. Implementation of best management practices consistent with the provisions of the stormwater management program required pursuant to this section and the provisions of the permit required pursuant to § 122.33 constitutes compliance with the standard of reducing pollutants to the maximum extent practicable. Your NPDES permitting authority will specify a time period of up to 5 years from the date of permit issuance for you to develop and implement your program.

The Six Minimum Controls

II. Minimum Control Measures

FIGURE 15-11
Public education program.

Control #1. Public education and outreach on stormwater impacts

1. You must implement a public education program to distribute educational materials to the community or conduct equivalent outreach activities about the impacts of stormwater discharges on water bodies and the steps that the public can take to reduce pollutants in stormwater runoff.

2. Guidance: You may use stormwater educational materials provided by your state, tribe, U.S. EPA, environmental, public interest, or trade organizations, or other MS4s. The public education program should inform individuals and households about the steps they can take to reduce stormwater pollution, such as ensuring proper septic system maintenance, ensuring the proper use and disposal of landscape and garden chemicals including fertilizers and pesticides, protecting and restoring riparian vegetation, and properly disposing of used motor oil or household hazardous wastes. The U.S. EPA recommends that the program inform individuals and groups how to become involved in local stream and beach restoration activities as well as activities that are coordinated by youth service and conservation corps or other citizen groups. EPA recommends that the public education program be tailored, using a mix of locally appropriate strategies, to target specific audiences and communities. Examples of strategies include distributing brochures or fact sheets, sponsoring speaking engagements before community groups, providing public service announcements, implementing educational programs targeted at school-age children, and conducting community-based projects such as storm drain stenciling, and watershed and beach cleanups. In addition, EPA recommends that some of the materials or outreach programs be directed toward targeted groups of commercial, industrial, and institutional entities likely to have significant stormwater impacts. For example, providing information to restaurants on the impact of grease clogging storm drains and to garages on the impact of oil discharges. You are encouraged to

tailor your outreach program to address the viewpoints and concerns of all communities, particularly minority and disadvantaged communities, as well as any special concerns relating to children.

Guidelines for Developing and Implementing this Measure

There are three main action areas that are important when implementing a successful public education and outreach program.

Forming Partnerships

Owners or operators of regulated small MS4s would be encouraged to enter into partnerships with other governmental entities to fulfill this minimum control measure's requirements. It is generally more cost-effective to use an existing program or to develop a new regional or state-wide education program, than to have numerous owners/operators developing their own local programs. Owners/operators would be encouraged to also look to nongovernmental organizations (e.g., environmental, civic, and industrial organizations) for assistance, because many already have educational materials and perform outreach activities.

Using Educational Materials and Strategies

Owners or operators of regulated small MS4s could use stormwater educational information provided by their state, tribe, U.S. EPA region, or environmental, public interest, or trade organizations instead of developing their own materials. Owners/operators should strive to make their materials and activities relevant to local situations and issues, and incorporate a variety of strategies to ensure maximum coverage.

Reaching Diverse Audiences

The public education program should use a mix of appropriate local strategies to inform a variety of audiences and communities, including minority and disadvantaged communities, as well as children. Printing posters and brochures in more than one language or posting large warning signs (e.g., cautioning against fishing or swimming) near storm sewer outfalls would help to reach audiences that are less likely to read standard materials. Some materials or outreach programs should also be directed toward specific groups of commercial, industrial, and institutional entities likely to have significant stormwater impacts. For example, information should be provided to restaurants on the effects of grease clogging storm drains and to auto garages on the effects of dumping used oil into storm drains.

> **Control #2. Public involvement/participation.**
>
> You must, at a minimum, comply with state, tribal, and local public notice requirements when implementing a public involvement/participation program.
>
> Guidance: EPA recommends that the public be included in developing, implementing, and reviewing your stormwater management program and that the public participation process should make efforts to reach out and engage all economic and ethnic groups. Opportunities for members of the public to participate in program development and implementation

FIGURE 15-12
Public involvement in stream cleanup.

include serving as citizen representatives on a local stormwater management panel, attending public hearings, working as citizen volunteers to educate other individuals about the program, assisting in program coordination with other preexisting programs, or participating in volunteer monitoring efforts. (Citizens should obtain approval where necessary for lawful access to monitoring sites.)

Guidelines for Developing and Implementing this Measure

Owners or operators of regulated small MS4s should include the public in developing, implementing, and reviewing their stormwater management programs. The public participation process should make every effort to reach out and engage all economic and ethnic groups. The U.S. EPA recognizes that there are challenges associated with public involvement. Nevertheless, the U.S. EPA strongly believes that these challenges can be addressed through an aggressive and inclusive program. Challenges and example practices that can help ensure successful participation are discussed below.

Implementation Challenges

The best way to handle common notification and recruitment challenges is to know the audience and think creatively about how to gain its attention and interest. Traditional methods of soliciting public input are not always successful in generating interest, and subsequent involvement, in all sectors of the community. For example, municipalities often use only advertising in local newspapers to announce public meetings and other opportunities for public involvement. Because there may be large sectors of the population who do not read the local press, the audience reached may be limited. Therefore, alternative advertising methods should be used whenever possible, including radio or television spots, postings at bus or subway stops, announcements in neighborhood newsletters, announcements at civic organization meetings, distribution of flyers, mass mailings, door-to-door visits, telephone notifications, and multilingual announcements. These efforts, of course, are tied closely to the efforts for the public education and outreach minimum control measure. In addition, advertising and soliciting for help could and should be targeted at specific population sectors, including ethnic, minority, and low-income communities; academia and educational institutions; neighborhood and community groups; outdoor recreation groups; and business and industry. The goal is to involve a diverse cross-section of people who could offer a multitude of concerns, ideas, and connections during the process.

Control #3. Illicit discharge detection and elimination.

1. You must develop, implement, and enforce a program to detect and eliminate illicit discharges [as defined at § 122.26(b)(2)] into your small MS4. You must develop, if not already completed, a storm sewer system map, showing the location of all outfalls and the names and location of all waters of the United States that receive discharges from those outfalls; to the extent allowable under state, tribal, or local law, effectively prohibit, through ordinance, or other regulatory mechanism, nonstormwater discharges into your storm sewer system and implement appropriate enforcement procedures and actions; develop and implement a plan to detect and address nonstormwater discharges, including illegal dumping, to your system; and inform public employees, businesses, and the general public of hazards associated with illegal discharges and improper disposal of waste.

   You need to address the following categories of nonstormwater discharges or flows (i.e., illicit discharges) only if you identify them as significant contributors of pollutants to your small MS4: water line flushing, landscape irrigation, diverted stream flows, rising groundwaters, uncontaminated ground water infiltration [as defined at 40 CFR 35.2005(20)], uncontaminated pumped groundwater, discharges from potable water sources, foundation drains, air conditioning condensation, irrigation water, springs, water from crawl space pumps, footing drains, lawn watering, individual residential car washing, flows from riparian habitats and wetlands, dechlorinated swimming pool discharges, and street wash water (discharges or flows from firefighting activities are excluded from the effective prohibition against

FIGURE 15-13
Illegal dumping of used motor oil.

nonstormwater and need only be addressed where they are identified as significant sources of pollutants to waters of the United States).

2. Guidance: EPA recommends that the plan to detect and address illicit discharges include the following four components: procedures for locating priority areas likely to have illicit discharges; procedures for tracing the source of an illicit discharge; procedures for removing the source of the discharge; and procedures for program evaluation and assessment. EPA recommends visually screening outfalls during dry weather and conducting field tests of selected pollutants as part of the procedures for locating priority areas. Illicit discharge education actions may include storm drain stenciling, a program to promote, publicize, and facilitate public reporting of illicit connections or discharges, and distribution of outreach materials.

Guidelines for Developing and Implementing this Measure

This is part of the program authorized by Congress wherein municipalities must be able to prohibit nonstormwater discharges to the municipal separate storm drain or drainage system. While some discharges of nonstormwater to the system are not illegal, per se, they must be covered by a separate NPDES permit and are subject to NPDES conditions (technology- or water-quality-based standards) similar to any other discharge covered under Sections 402 and 301 of the Clean Water Act.

Illicit connections and illegal dumping were not specifically sampled under the NURP program. However, it was concluded that illicit connections could result in high bacterial counts and the introduction of priority pollutants into the urban drainage system. Studies have shown that illicit connections to storm drains can cause severe and widespread contamination problems. The problem is seen as more severe in older neighborhoods and older industrial and commercial areas, where approval processes allowed for such connections. These areas should be targeted as a priority, because there is the possibility of dramatic improvement in water quality in these areas.

Development of a coherent illicit connections and illegal dumping investigation program includes the following steps (U.S. EPA, 1993a):

- Mapping and preliminary watershed evaluation
- Selection of tracer parameters
- Initial field screening activities
- Data analysis to identify problem outfalls and flow components
- Watershed surveys to locate inappropriate pollutant entries
- Corrective techniques

Low-cost techniques have been proven to be cost-effective in Ft. Worth, Texas (Falkenbury, 1987). Detailed manuals for field and laboratory procedures have been developed for a number of municipalities.

The types of pollutants to look for in an illicit connection program include sanitary wastewater sources, septic tank systems, automobile washing and maintenance centers, irrigation sources, laundry wastewaters, dewatering operations, improper disposal of household hazardous waste, spills, sump disposal, gas station materials and substances, industrial rinse water, cooling water, swimming pools, etc. Industrial sources can be related to raw materials, waste materials, and the final product (U.S. EPA, 1993a).

Stormwater Quality Management Programs

The objective of the illicit discharge detection and elimination minimum control measure is to have regulated small MS4 owners and operators gain a thorough awareness of their systems. This awareness allows them to determine the types and sources of illicit discharges entering their system, and establish the legal, technical, and educational means to attempt to eliminate these discharges. Permittees could meet these objectives in a variety of ways depending on their individual needs and abilities, but some general guidance for each requirement is provided below.

The Map

The storm sewer system map is meant to demonstrate a basic awareness of the intake and discharge areas of the system. It is needed to help determine the extent of discharged dry-weather flows, the possible sources of the dry-weather flows, and the particular waterbodies these flows may be affecting. Because the location of the major pipes and outfalls could be indicated on an existing topographical map, a new map would not need to be created specifically for this purpose, as long as the information is clearly presented on the existing map. The permittee would be allowed to choose the type and size of map that best fits its needs. The U.S. EPA recommends collecting all existing information on outfall locations (e.g., review city records, drainage maps, storm drain maps), and then conducting field surveys to verify locations. It probably will be necessary to walk (i.e., wade through small receiving waters or use a boat for larger waters) the streambanks and shorelines for visual observation. It may take more than one trip to locate all outfalls.

Legal Prohibition and Enforcement

EPA recognizes that some permittees may have limited authority under state or tribal law to establish and enforce an ordinance, or similar means, prohibiting illicit discharges. In such a case, the permittee would be encouraged to obtain the necessary authority, if at all possible. Otherwise, the NPDES permitting authority would assume the responsibility for implementation of this component of the minimum measure, yet the permittee would

FIGURE 15-14
A digital field screening system map.

remain ultimate responsible for the quality of its MS4 discharge. Model ordinances, including examples of amendments to local codes or existing ordinances, will be provided in the Phase II stormwater guidance for regulated small MS4s, which is part of EPA's planned implementation tool box for the final rule.

The Plan

The plan to detect and address illicit discharges is the central component of this minimum control measure. The plan would be shaped by several factors, including the permittee's available resources, size of staff, and degree and character of its illicit discharges. EPA envisions a plan similar to the one recommended for use in meeting Michigan's general stormwater NPDES permit for small MS4s. As guidance only, the four steps of a recommended plan are outlined below:

- *Locate problem areas* — The U.S. EPA recommends that priority areas be identified for detailed screening of the system based on the likelihood of illicit connections (e.g., areas with older sanitary sewer lines). Some methods that could be used to locate problem areas include public complaints and other input, visual screening, water sampling from manholes and outfalls during dry weather, and use of infrared and thermal photography.
- *Find the source* — Once a problem area or discharge is found, additional efforts usually would be necessary to determine the source of the problem. Some methods that could be used to find the source of the illicit discharge include dye-testing buildings in problem areas, dye- or smoke-testing buildings at the time of sale, tracing the discharge upstream in the storm sewer, employing a certification program that shows that buildings have been checked for illicit connections, implementing an inspection program of existing septic systems, and using video to inspect the storm sewers.
- *Remove/correct illicit connections* — Once the source is identified, the offending discharger would need to be notified and directed to correct the problem. Edu-

FIGURE 15-15
Dye test results showing illicit connection.

cation efforts and working with the discharger can be effective in resolving the problem before taking legal action.

- *Document actions taken* — As a final step, all actions taken under the plan should be documented. Doing so would illustrate that progress is being made to eliminate illicit connections and discharges. Documented actions should be included in the required annual reports and should include information such as the number of outfalls screened, any complaints received and corrected, the number of discharges and quantities of flow eliminated, and the number of dye or smoke tests conducted.

Educational Outreach

Educational outreach to public employees, businesses, property owners, the general community, and elected officials would be necessary to inform them of what they could do to detect and eliminate illicit discharges, but it would also help to gain support for the permittee's stormwater program. The educational outreach efforts should, at a minimum, include the following:

- Providing training programs for public employees
- Developing informative brochures, and guidance for specific audiences (e.g., carpet cleaning businesses) and school curricula
- Designing a program to publicize and facilitate public reporting of illicit discharges
- Coordinating volunteers for locating, and visually inspecting, outfalls or to stencil storm drains
- Initiating recycling programs for commonly dumped wastes, such as motor oil, antifreeze, and pesticides

Control #4. Construction site stormwater runoff control.

1. You must develop, implement, and enforce a program to reduce pollutants in any stormwater runoff to your small MS4 from construction activities that result in a land disturbance of greater than or equal to 1 acre. Reduction of stormwater discharges from construction activity disturbing less than 1 acre must be included in your program if that construction activity is part of a larger common plan of development or sale that would disturb 1 acre or more. If the NPDES permitting authority waives requirements for stormwater discharges associated with small construction activity in accordance with § 122.26(b)(15)(i), you are not required to develop, implement, or enforce a program to reduce pollutant discharges from such sites.

    Your program must include the development and implementation of, at a minimum: an ordinance or other regulatory mechanism to require erosion and sediment controls, as well as sanctions to ensure compliance, to the extent allowable under state, tribal, or local law; requirements for construction site operators to implement appropriate erosion and sediment control best management practices; requirements for construction site operators to control waste such as discarded building materials, concrete truck washout, chemicals, litter, and sanitary waste at the construction site that may cause adverse impacts to water quality; procedures for site plan review that incorporate consideration of potential water-quality impacts; procedures for receipt and consideration of in-

FIGURE 15-16
Construction site erosion problems.

formation submitted by the public; and procedures for site inspection and enforcement of control measures.

2. Guidance: Examples of sanctions to ensure compliance include nonmonetary penalties, fines, bonding requirements, and permit denials for noncompliance. The U.S. EPA recommends that procedures for site plan review include the review of individual preconstruction site plans to ensure consistency with local sediment- and erosion-control requirements. Procedures for site inspections and enforcement of control measures could include steps to identify priority sites for inspection and enforcement based on the nature of the construction activity, topography, and the characteristics of soils and receiving water quality. You are encouraged to provide appropriate educational and training measures for construction site operators. You may wish to require a stormwater pollution prevention plan for construction sites within your jurisdiction that discharge into your system. See § 122.44(s) (NPDES permitting authorities' option to incorporate qualifying state, tribal, and local erosion- and sediment-control programs into NPDES permits for stormwater discharges from construction sites). Also see § 122.35(b) (the NPDES permitting authority may recognize that another government entity, including the permitting authority, may be responsible for implementing one or more of the minimum measures on your behalf.)

Guidelines for Developing and Implementing this Measure

Runoff from construction sites can involve sediment runoff several orders of magnitude greater than that of the predeveloped site. Sediment surges can have a number of adverse impacts on the environment, including bonding of other pollutants to the sediment, clogging of fish gills, turbidity, burying of benthic organisms important in the food web, increasing the cost of drinking water treatment, filling and clogging stormwater conveyance and retention structures, aesthetic degradation, and changing downstream flow lines through bed and overbank deposition.

Construction sites can also generate high loadings of nitrogen and phosphorus from fertilizer, pesticide, and herbicide runoff, petroleum products, chemical solid wastes and other toxics.

The regulations include activities involving SIC Codes 15 and 16 (general building contractors and heavy construction contractors) and preconstruction activities with certain exceptions.

While a number of municipalities have some sort of erosion-control ordinance or program, many are not effective due to lack of inspection and enforcement capability. Also, there are often problems in the ordinances that allow, for example, residential subdivisions to sell off individual lots, which are often then regraded without erosion-control provisions.

Therefore, any erosion-control program must emphasize inspection and enforcement, meeting the requirements and limits of the regulations, and closing holes in existing ordinances. Under the regulations, a municipality must propose a plan for a program to implement and maintain structural and nonstructural BMPs for controlling stormwater runoff at construction sites meeting the requirements stated above. There are a large number of structural and nonstructural BMPs for use in erosion control.

Further explanation and guidance for each proposed component of a regulated small MS4's construction program are provided below.

Regulatory Mechanism

Through the development of an ordinance or other regulatory mechanism, the small MS4 owner/operator would need to establish a construction program that requires controls for polluted runoff from construction sites with a land disturbance of greater than or equal to 1 acre. In recognition of varying limitations on regulatory legal authority, the small MS4 owner/operator would be required to satisfy this minimum control measure only to the maximum extent practicable and allowable under state or tribal law. If an owner/operator is unable to establish an enforceable construction program due to a lack of legal authority and is unsuccessful in trying to obtain the necessary authority, the NPDES permitting authority would then assume responsibility. The U.S. EPA intends to develop a model ordinance that a small MS4 owner/operator could use as a basis for its construction program. Alternatively, amendments to existing erosion and sediment control programs, or other ordinances, could also provide the basis for the program.

Site Plan Review

The small MS4 owner/operator would be required to include in its construction program requirements for the implementation of appropriate BMPs on construction sites to control erosion and sediment, as well as various other wastes. To determine if a construction site is in compliance with such provisions, the small MS4 owner/operator would need to review the site plans submitted by the construction site owner/operator before ground is broken. Site plan review aids in compliance and enforcement efforts, because it alerts the small MS4 owner/operator early in the process to the planned use or nonuse of proper BMPs and provides a way to track new construction activities. The tracking of sites is useful not only for the small MS4 owner/operator's recordkeeping and reporting purposes, which would be required activities under their NPDES stormwater permit, but also for members of the public interested in ensuring that the sites are in compliance.

Inspections and Penalties

Once construction commences, the BMPs should be in place, and the small MS4 owner/operator's enforcement activities should begin. To ensure that the BMPs are properly installed, the small MS4 owner/operator would be required to perform regular inspections during construction and have penalties in place to deter infractions. Inspections would give the MS4 owner/operator an opportunity to provide additional guidance and education, issue warnings, or assess penalties. To conserve staff resources, one possible option for small MS4 owners/operators could be to have these inspections performed by the same inspector who visits the sites to check compliance with health and safety building codes.

Information Submitted by the Public

A final requirement of the proposed small MS4 program for construction activity would be the development of procedures for the receipt and consideration of public inquiries, concerns, and information submitted regarding local construction activities. This provision is intended to further reinforce the public participation component of the small MS4 stormwater program and to recognize the crucial role that the public can play in identifying instances of noncompliance. The small MS4 owner/operator would be required only to *consider* the information submitted, and may not need to follow-up and respond to every complaint or concern. Although some sort of enforcement action or reply would not be required, the small MS4 owner or operator would need to be able to demonstrate acknowledgment and consideration of the information submitted. A simple tracking process in which submitted public information, both written and verbal, is recorded and then given to the construction site inspector for possible follow-up would suffice.

Control #5. Postconstruction stormwater management in new development and redevelopment

1. You must develop, implement, and enforce a program to address stormwater runoff from new development and redevelopment projects that disturb greater than or equal to 1 acre, including projects less than 1 acre that are part of a larger common plan of development or sale, that discharge into your small MS4. Your program must ensure that controls are in place that would prevent or minimize water-quality impacts. You must develop and implement strategies that include a combination of structural and nonstructural best management practices (BMPs) appropriate for your community; use an ordinance or other regulatory mechanism to address postconstruction runoff from new development and redevelopment projects to the extent allowable under state, tribal, or local law; and ensure adequate long-term operation and maintenance of BMPs.

2. Guidance: If water-quality impacts are considered from the beginning stages of a project, new development and potentially redevelopment provide more opportunities for water-quality protection. The U.S. EPA recommends that the BMPs chosen be appropriate for the local community, minimize water-quality impacts, and attempt to maintain predevelopment runoff conditions. In choosing appropriate BMPs, the U.S. EPA encourages you to participate in locally based watershed planning efforts that attempt to involve a diverse group of stakeholders, including interested citizens. When developing a program that is consistent with this measure's intent, the U.S. EPA recommends that

FIGURE 15-17
Postconstruction bioretention area.

you adopt a planning process that identifies the municipality's program goals (e.g., minimize water-quality impacts resulting from postconstruction runoff from new development and redevelopment), implementation strategies (e.g., adopt a combination of structural and nonstructural BMPs), operation, and maintenance policies and procedures, and enforcement procedures. In developing your program, you should consider assessing existing ordinances, policies, programs and studies that address stormwater runoff quality. In addition to assessing these existing documents and programs, you should provide opportunities to the public to participate in the development of the program. Nonstructural BMPs are preventative actions that involve management and source controls such as policies and ordinances that provide requirements and standards to direct growth to identified areas, protect sensitive areas such as wetlands and riparian areas, maintain and increase open space (including a dedicated funding source for open space acquisition), provide buffers along sensitive water bodies, minimize impervious surfaces, and minimize disturbance of soils and vegetation; policies or ordinances that encourage infill development in higher density urban areas, and areas with existing infrastructure; education programs for developers and the public about project designs that minimize water-quality impacts; and measures such as minimization of percent impervious area after development and minimization of directly connected impervious areas. Structural BMPs include storage practices such as wet ponds and extended-detention outlet structures; filtration practices such as grassed swales, sand filters, and filter strips; and infiltration practices such as infiltration basins and infiltration trenches. The U.S. EPA recommends that you ensure the appropriate implementation of the structural BMPs by considering some or all of the following: preconstruction review of BMP designs; inspections during construction to verify BMPs are built as designed; postconstruction inspection and maintenance of BMPs; and penalty provisions for noncompliance with design, construction, or operation and maintenance. Stormwater technologies are constantly being improved, and the U.S. EPA recommends that your requirements be responsive to these changes, developments, or improvements in control technologies.

FIGURE 15-18
Housekeeping practices — street sweeping.

Guidelines for Developing and Implementing this Measure

This section includes some sample nonstructural and structural BMPs that could be used to satisfy the requirements of the postconstruction runoff control minimum measure. Because the proposed requirements of this measure are closely tied to the requirements of the construction site runoff control minimum measure, the U.S. EPA recommends that small MS4 owners or operators develop and implement these two measures in tandem.

Nonstructural BMPs

- *Planning and procedures* — Runoff problems can be addressed efficiently with sound planning procedures. Master plans, comprehensive plans, and zoning ordinances can promote improved water quality by guiding the growth of a community away from sensitive areas and by restricting certain types of growth (industrial, for example) to areas that can support it without compromising water quality.
- *Site-based local controls* — These controls can include buffer strip and riparian zone preservation, minimization of disturbance and imperviousness, and maximization of open space.

Structural BMPs

- *Storage practices* — Storage or detention BMPs control stormwater by gathering runoff in wet ponds, dry basins, or multichamber catch basins and slowly releasing it to receiving waters or drainage systems. These practices control stormwater volume and settle out particulates for pollutant removal.
- *Infiltration practices* — Infiltration BMPs are designed to facilitate the infiltration of runoff through the soil to groundwater, and, thereby, result in reduced stormwater quantity and reduced mobilization of pollutants. Examples include infiltration basins/trenches, dry wells, and porous pavement.
- *Vegetative practices* — Vegetative BMPs are landscaping features that, with optimal design and good soil conditions, enhance pollutant removal, maintain/improve natural site hydrology, promote healthier habitats, and increase aesthetic appeal. Examples include grassy swales, filter strips, artificial wetlands, and rain gardens.

Control #6. Pollution prevention/good housekeeping for municipal operations

1. You must develop and implement an operation and maintenance program that includes a training component and has the ultimate goal of preventing or reducing pollutant runoff from municipal operations. Using training materials that are available from the U.S. EPA, your state, tribe, or other organizations, your program must include employee training to prevent and reduce stormwater pollution from activities such as park and open space maintenance, fleet and building maintenance, new construction and land disturbances, and stormwater system maintenance.

2. Guidance: EPA recommends that, at a minimum, you consider the following in developing your program: maintenance activities, maintenance schedules, and long-term inspection procedures for structural and nonstructural stormwater controls to reduce floatables and other pollutants discharged from your separate storm sewers; controls for reducing or eliminating the discharge of pollutants from streets, roads, highways, municipal parking lots, maintenance and storage yards, fleet or maintenance shops with outdoor storage areas, salt/sand storage locations, and snow disposal areas operated by you, and waste transfer stations; procedures for properly disposing of waste removed from the separate storm sewers and areas listed above (such as dredge spoil, accumulated sediments, floatables, and other debris); and ways to ensure that new flood management projects assess the impacts on water quality and examine existing projects for incorporating additional water quality protection devices or practices. Operation and maintenance should be an integral component of all stormwater management programs. This measure is intended to improve the efficiency of these programs and require new programs where necessary. Properly developed and implemented operation and maintenance programs reduce the risk of water quality problems.

Guidelines for Developing and Implementing this Measure

The intent of this control measure is to ensure that existing municipal or facility operations are performed in the most appropriate way as to minimize contamination of stormwater discharges. EPA encourages the small MS4 owner/operator to consider the following components when developing their program for this measure:

- *Maintenance activities, maintenance schedules, and long-term inspection procedures* for structural and nonstructural controls to reduce floatables and other pollutants discharged from the separate storm sewers
- *Controls for reducing or eliminating the discharge of pollutants* from areas such as roads and parking lots, maintenance and storage yards (including salt/sand storage and snow disposal areas), and waste transfer stations; these controls should include programs that promote recycling (to reduce litter), minimize pesticide use, and ensure the proper disposal of animal waste
- *Procedures for the proper disposal of waste* removed from the separate storm/sewer systems and the areas listed above, including dredge spoil, accumulated sediments, floatables, and other debris
- *Ways to ensure that new flood management projects assess the impacts on water quality* and examine existing projects for incorporation of additional water-quality protection devices or practices (The U.S. EPA encourages coordination with flood-

control managers for the purpose of identifying and addressing environmental impacts from such projects. The effective performance of this control measure hinges on the proper maintenance of the BMPs used, particularly for the first two items listed above. For example, structural controls, such as grates on outfalls to capture floatables, necessitate that the outfalls be cleaned regularly, while nonstructural controls, such as training materials and recycling programs, need to be updated periodically.)

15.5 Costs of NPDES Phase II

There is naturally much speculation on the actual program elements and costs for a particular stormwater program developed under Phase II. There have been several attempts at estimating Phase II program costs based on current costs of similar programs.

Due to external pressures and directives from the current and past administrations, the U.S. EPA is conscious of attempting to make the current stormwater NPDES program cost-effective. For example:

> EPA believes this rule will cost significantly less than the existing 1995 rule that is currently in place, and will result in significant monetized financial, recreational and health benefits, as well as benefits that EPA has been unable to monetize, including reduced scouring and erosion of streambeds, improved aesthetic quality of waters, reduced eutrophication of aquatic systems, benefit to wildlife and endangered and threatened species, tourism benefits, biodiversity benefits and reduced siting costs of reservoirs (Federal Register, 1998, p. 1536).

> ...the Agency recognizes the continuing imperative to assure that environmental regulations accomplish statutory objectives in the least burdensome and most cost-effective fashion. As explained further in this preamble, the form and substance of NPDES permits to address the sources designated in today's proposal would provide greater flexibility for the newly covered sources than the existing "standard" NPDES permit (Federal Register, 1998, p. 1550).

While the benefit side of the proposed regulations exists in the realm of gross estimates, the cost side is also filled with unknowns. What will the mandated and negotiated stormwater program cost a local community? Are there ways to reduce costs?

But it is still not possible to say what the regulations will cost everyone in total. This is so because:

- There is great flexibility inherent in the regulations to create a stormwater quality program tailored to meet an individual community's needs and situations
- Each permit writer has preferences and "hot buttons" that will color what any particular program will look like
- Each community setting is different in terms of climate, topography, pollutants of concern, and current condition of local waters

In the draft regulations, EPA had provided estimates of the probable cost implications of the NPDES Phase II permit. These estimates were based on summary information from the permit applications from 21 Phase I cities. Very high and very low figures were thrown

Stormwater Quality Management Programs 1033

Measure	Percent of municipalities expected to incur costs (percent)	Low end of range of per capita costs	High end of range of per capita costs
First Permit Cycle:			
Public Education	39	$0.02	$0.34
Public Involvement	100	0.19	0.20
Illicit Discharge D&E	90	0.04	2.61
Const Site SW Runoff Control	83	0.04	1.59
Post Construction SW Mgt	4	1.09	1.09
PP/GH of Municipal Ops	71	0.01	2.00
2nd and 3rd Permit Cycles:			
Public Education	39	0.01	0.34
Public Involvement	100	0.12	0.12
Illicit Discharge D&E	73	0.04	2.17
Const Site SW Runoff Control	80	0.01	0.83
Post Construction SW Mgt	4	1.09	1.09
PP/GH of Municipal Ops	67	0.01	1.08

Exhibit 5.—Percentage of Municipalities Affected and Range of Per Capita Costs for Six Minimum Measures

FIGURE 15-19
EPA cost estimates of Phase II.

out by EPA in developing these estimates. Figure 15-19 shows the summary table developed by EPA.

The range depicted in Figure 15-19 is from $1.39 to $7.83 per person per year for the first permit 5-year period, and $1.28 to $5.63 for other permit cycles. For a city of 50,000, that is a wide range of $69,500 to $391,500 annually for the first permit cycle. This is clearly not helpful in attempting to estimate a specific community's costs.

There is question about the vagueness in the regulatory language and the high degree of potential flexibility inherent in briefly described program elements. For example, if you reread the first of the minimum control regulatory language you will note a great disparity between what is required and what is suggested. There is wide room for interpretation of the intensity and detail necessary to accomplish this minimum control. The devil is *always* in the details, and there will always be great variability in what two different programs intend to do to accomplish the same general goals.

NAFSMA (1999a, 1999b) published a survey on potential Phase II program costs responded to by 121 cities and counties nationally. Ten communities responded with programs that had three or more suggested elements in the first minimum control: public education and outreach. The annual per capita costs for these ten ranged from $0.04 to $1.17 — again, a wide range.

Of those responding, only one community stated that it had any program activity in each of the six minimum control measure areas, and it spent $15.11 per capita annually, well above the EPA estimate (the city has a population of about 25,000). Of the 121 respondents, only 26 had programs in at least three (most had only three) of the six mandatory minimum control areas, and these can be considered far from complete. The distribution of costs for these 26 programs is so wide as to be meaningless. The median was $1.44, and the average was $4.07. The low value was $0.04, and the high was $26.00. Figure 15-20 shows the results for these 26 communities.

We can speculate that, if many of these communities had a fully developed Phase II program, the average costs could more than double, because each community would be adding new program areas and upgrading its existing programs to make them comply with the details of the Phase II permit writers requirements.

In the final regulations, the U.S. EPA took a different approach in making estimates of the costs of compliance using the NAFSMA information and past experience with Phase I (USEPA, 1999). The U.S. EPA estimated annual costs for the municipal programs based on a fixed cost component and a variable cost component. The fixed cost component

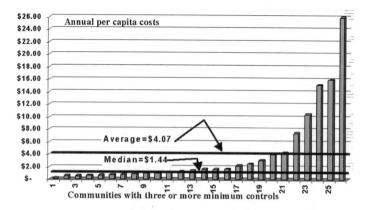

FIGURE 15-20
NAFSMA results summary for 26 communities.

included costs for the municipal application, recordkeeping, and reporting activities. On average, the U.S. EPA estimated annual costs of $1525 per municipality. Variable costs include the costs associated with annual operations for the six minimum measures and are calculated at a rate of $8.93 annually per household (assuming 2.62 persons per household). This equates to $3 to $4 per person per year in a local community Thus, the cost estimating equation (plotted in Figure 15-21) is as follows:

$$\text{Annual cost} = \$1525 + \text{population}/2.62 * \$8.93$$

Finally, rule-of-thumb estimates based on the author's experience working in over 100 communities indicate that comprehensive stormwater programs that include advanced stormwater quality programs cost between $7.00 and $20.00 per capita per year. The quality portion is normally between 20 and 30% of the total average program cost for a well-developed stormwater program.

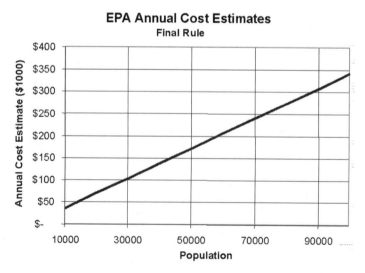

FIGURE 15-21
EPA final cost estimates.

The methods used above do not provide details of the components of the stormwater programs resulting in the costs, and thus are not very helpful in assisting other communities in their thinking about the regulations. An effort was made by Reese (2000) to develop cost estimate ranges based on a direct interpretation of the stormwater regulations as applied to two example communities at each end of the spectrum, in terms of size and intensity of water quality program. This has an advantage in that it deals directly with the stormwater regulatory requirements and illustrates specific program components so that we can control and define all details. This analysis was also completed for the same city but developing both a minimal and a more comprehensive stormwater program by the Denver Urban Drainage and Flood Control District staff.

The information from both of these analyses, verified by several cities, shows that there is indeed a wide range of potential programs for the same city, and that this reflects the inherent flexibility within the regulations. This indicates that, unless states constrain innovation, that there is lots of room for negotiating a permit that stays within available resources and meets the needs of the local community. There is also lots of room to have a program imposed on a local community that does not fit its needs. It pays to be proactive in Phase II SWQMP development.

15.6 The Ten Commandments in Developing a SWQMP

Overall Approach

Two priorities must be balanced in development of a SWQMP. It is desirable to identify and solve real problems, not simply spend money on paperwork or meaningless activities. If it is not known what those problems are (as is the case in most of the municipalities in the United States) there is a need to have a focus on identifying and characterizing the problems and developing strategies to solve them.

However, based on years of data and information, the U.S. EPA feels that there are certain stormwater runoff pollution problems that exist in almost any municipality. For example, it is expected that sediment from construction, litter, dumped motor oil, and other pollutants enter streams from virtually every urbanized area in the world: thus, there is a need for minimum controls.

There is a U.S. EPA mandated requirement for a minimal type of program even in areas where specific pollution problems are not known to exist but are generally suspected to be present in any urban area. Therefore, it will be necessary to compare any proposed SWQMP to the baseline minimums for these four areas to ensure overall program completeness. There is a two-pronged attack: (1) identify and solve real problems, and (2) address U.S. EPA program minimums.

The Ten Commandments

There has been much experience gained from implementation of Phase I stormwater programs as well as many communities transitioning to more ecologically based paradigms of stormwater management. Based on this experience, on the specific requirements of the Phase II program, and on the foundational principles on which the stormwater quality program is based, there are some basic understandings that can serve to guide any stormwater quality program in its inception. Reese (1993, 1994) developed the "Ten

Commandments" of stormwater quality programs in anticipation of Phase I programs. These have been updated for presentation here.

#1 — Think Paradigm Shift

Chapter 1 of the book spoke of a series of paradigms that urban stormwater has gone and is going through. For many Phase II communities, the shift from a sleepy stormwater quantity- and infrastructure-focused program to a full-blown Phase II program with strong ecological overtones and changes to development regulations will be a major paradigm shift. It should be managed as such. We cannot legislate and regulate ourselves into effective change. We must motivate and educate. We must vision, mentor, and train. And, we must fund.

Many past programs failed or were reduced to ineffective triviality, partially because the stormwater managers did not think deeply and long about how to bring about change. They simply responded to unfunded regulatory mandates with unfunded regulations of their own. If we begin the program thinking in terms of bringing about a long-term change in how we do the stormwater business, we will start in the right direction.

Changing from a traditional efficient drainage system approach to one that involves purposeful reductions in urban impacts, protection of urban resources, and creation of alternate forms of drainage infrastructure (and all the impacts on lifestyle and transportation that these may imply) will take time and must be done with wisdom and in steps. It is much easier to know what the next paradigm is than to know how to get there. Think ten years (or two permit terms).

#2 — Fix Real Things

One maddening aspect of several of the Phase I programs could be termed: ready, fire, aim. Some communities spent large sums of money on program aspects that seemed vague, ill focused, and without tangible results. The reason seemed to be that they were just "working on the minimum controls." While, as stated above, meeting U.S. EPA minimums is important, of much greater importance is focusing the stormwater program on compelling issues.

People are not interested in being regulated, and the fact that this program is an unfunded federal mandate does not insure cooperation (maybe the opposite). But in every community, there are some things having to do with the surface water system and its impacts (and impacts on it) that are important to citizens, or can be made to be so. The

FIGURE 15-22
Greenway trails rally citizen support.

Stormwater Quality Management Programs

focus of a stormwater quality program should be on those things that the community values. In one city, it might be greenways, in another a lake, in another drinking water supply, and so on.

The initial step in any Phase II program development should be to assess the local community and see what matters to it. It might not matter to all citizens, but even a vocal group can bring about effective change. Once these issues are identified, then an effective case must be made for any program to focus on them. Effective cases are made using statistics, facts, examples, and testimonials.

#3 — Bring Me in Early, I'm Your Partner; Bring Me in Late, I'm Your Judge

The eventual level of sophistication and success of the SWQMP will depend to some extent on the ability of the stormwater manager to convince local citizens that it is worth it. The smaller the town, the more this can be true. In larger cities, the local government can do more things without specific knowledge of the general population. But, under Phase II, the smaller towns and counties often need the support of leading citizens. And there is one major rule of public support: bring me in early, I'm your partner; bring me in late, I'm your judge.

It is important, especially under Phase II where there is so much emphasis on citizen input and environmental justice, to involve key representatives of the community in development of the Phase II program and its implementation. This is especially true if there will be a long-term reliance on citizen groups to carry out some aspects of the stormwater program. As stated in the preamble to the final regulations (CFR Vol. 65 No. 235, 1999, pp. 68740, 68755):

> Another concern for EPA and several stakeholders was that the program ensure citizen participation. The NPDES approach ensures opportunities for citizen participation throughout the permit issuance process, as well as in enforcement actions. The NPDES program provides for public participation in the development, enforcement, and revision of stormwater management programs. Citizen suit enforcement has assisted in

FIGURE 15-23
Early citizen involvement.

focusing attention on adverse water quality impacts on a localized, public priority basis. Citizens frequently rely on the NPDES permitting process and the availability of NOIs to track program implementation and help them enforce regulatory requirements.

Early and frequent public involvement can shorten implementation schedules and broaden public support for a program. Opportunities for members of the public to participate in program development and implementation could include serving as citizen representatives on a local stormwater management panel, attending public hearings, working as citizen volunteers to educate other individuals about the program, assisting in program coordination with other pre-existing programs, or participating in volunteer monitoring efforts. Moreover, members of the public may be less likely to raise legal challenges to a MS4's stormwater program if they have been involved in the decision making process and program development and, therefore, internalize personal responsibility for the program themselves.

Second, public participation is likely to ensure a more successful stormwater program by providing valuable expertise and a conduit to other programs and governments. This is particularly important if the MS4's stormwater program is to be implemented on a watershed basis. Interested stakeholders may offer to volunteer in the implementation of all aspects of the program, thus conserving limited municipal resources.

#4 — Do not Build New Problems

While figuring out how to fix existing problems, take the necessary steps to avoid building new problems — control and guide new development. Early in the permitting period, it is important to begin the process of changing the way new development looks or can look. Several Phase I communities have developed and implemented environmental overlay districts or Planned Unit Development guidance and then supplied carrots for such development to take place, including transferable development rights, density variances, tax credits, fee waivers, conservation easements, streamlined plans review, and public–private partnerships.

We already know that current development approaches cause adverse impacts. It is time for a change. Newer approaches to development that decrease the "environmental footprint" abound in the literature, including smart growth, conservation design, better site design, low-impact development, green neighborhoods, sustainable infrastructure, zero discharge, and from Europe, sustainable urban drainage systems (SUDS). Not all have been tried and found true, but the general concept is laudable, and even necessary, if urban impacts are to be held to an acceptable level. It has been found that many of these approaches are of equal or lesser cost due to reduced infrastructure costs (Frank, 1989; CWP, 2000; Clar and Rushton, 2001), and many of the practices are espoused by such diverse groups as Homebuilders' Associations and environmental groups (CWP, 1998). Figure 15-24 indicates economic consequences of changes in land use based on the use of smart growth practices (CWP, 2000).

Master planning is also key. No one approach ensures total success, but planning can integrate small- and large-scale approaches in a more seamless whole. Work extraterritorially if necessary in master planning and property acquisition. Staff and public education and team building is key, and often strange to public works engineers. Build teams with wide expertise across staff elements and various stakeholder groups. Biologists and ecologists should be on the planning team. Parks, planning, engineering, public utilities, and public works must work jointly in planning for and implementing development. Transportation planning, zoning modifications, and large area planning must consider environmentally sensitive areas and work around them and with them.

Watershed Protection Tools	Developer/Builder	Adjacent Property Owner	Community	Local Government
1. Watershed Planning and Zoning	(−) cost of land (−) locational constraints	(+) property value	(+) business attraction (+) protection from adverse uses	(−) staff and budget resources (+) reduced "clean up"
2. Protect Sensitive Areas	(+) natural area premium (−) permitting costs (−) locational constraints	(+) property value	(+) habitat (+) fisheries	(−) staff resources (+) reduced "clean up" costs (+) lower cost of services
3. Establish Buffer Network	(+) buffer premium (−) locational constraints	(+) property value	(+) flooding risk (+) wildlife (+) greenway (+) trails	(−) staff resources (+) fewer drainage complaints
4. Cluster and Open Space Development	(+) construction costs (+) marketability (+) no lost lots	(+) property value (−) HOA fees	(+) recreation (+) green space (+) natural area preservation	(−) staff resources (+) lower cost of services
5. Narrow Streets and Smaller Parking Lots	(+) reduced construction cost	(+) property value (−) parking	(+) better sense of place (+) pedestrian friendly	(−) staff resources
6. Erosion and Sediment Control	(−) higher cost (+) savings in cleaning/grading	(+) trees saved increase value (+) no off-site sediment	(+) water quality (+) tree conservation	(−) staff resources (+) reduced complaints from downstreamers
7. Stormwater Best Management Practices	(−) higher costs (+) pond/wetland premium	(−) maintenance (+) waterfront effect (if done right)	(+) protection of water supply (+) stream protection	(−) staff resources (+) reduced waterbody programs/problems
8. Treat Septic System Effluent	(−) higher design and engineering costs	(−) clean out costs	(+) protection of water supply	(−) staff resources
9. Ongoing Watershed Management	no impact	(−) annual fee for utility (+) continued healthy environment	(−) annual fee (+) involvement in watershed services	(−) staff resources
ECONOMIC TREND	MIXED	POSITIVE	POSITIVE	NEGATIVE

(−) negative economic consequence   (+) positive economic or environmental impact

FIGURE 15-24
Economic consequences of smart growth.

#5 — Get the Foundations Right

The old stormwater joke goes that, while nature follows the hydrologic cycle, municipalities follow a "hydro-illogical" cycle. They cycle repeatedly from flooding or environmental disaster to panic to planning to procrastination and back to disaster. The only way out of that cycle is through laying firm foundations. No SWQMP can flourish and grow without the key underpinnings of a sound organizational structure with sufficient legal authority and adequate, stable, and equitable financing. Actual physical urban pollution or flooding problems, when fully investigated, are usually found to be the result of misguided or missing institutional policies as further described in Chapter 1.

Financing is a key. Many municipalities (about 400 at last count) have gone to a stormwater utility form of financing, where the users of the drainage system pay in accordance with their use. Use of the system is defined according to how much water or pollution a property puts into the system. Other forms of financing may include use of state-revolving funds for stormwater quality purposes and proposed federal grants and loans for specific stormwater activity. Various confederations of environmental groups and state and local political leaders have in many cases been able to put together a patchwork of grants, federal or state matching funds, private sources, and tax-based financing for certain projects and studies. Concepts such as pollution trading and credits are being tried in some Eastern states to maximize resources through capitalism and natural working of market forces.

Another firm foundation is to get the organizational aspects of the program correct. Stormwater quality laws and regulations will (and should) bring together cities and surrounding urbanized areas. Interlocal agreements that provide consistent treatment of properties that are situated similarly are important from a physical environmental perspective and from a political standpoint. For example, a Southern city was unable to make such arrangements, eventually leading to newspaper headlines with local councilmen decrying the lack of consistent treatment and threatening to derail the whole stormwater quality and quantity program. In another situation, a state was unwilling to bring a set of watershed sharing co-applicants into the permit process, effectively crippling the efforts of a local municipality to effectively treat its whole stormwater pollution problem by treating the whole system.

Finally, it is important to have firm legal foundations. Many Phase I cities felt they did not have authority to require compliance, only to find that they were not taking advantage of the authority they had by charter or state legislation. Chapter 4 talks about legal foundations in more detail. Suffice to say that all aspects of the program must be legally enforceable, and the ordinances used to enforce them must be clear and thorough.

#6 — Integrate and Graduate

Do not forget that for the last 50 years or so, the purpose of the stormwater program was to avoid flooding and provide drainage. These two important needs do not go away just because stormwater quality has swum into focus. In fact, in our experience, most citizens in most municipalities are more concerned about flood protection than they are about clean water, though that is changing rapidly.

In addition, most stormwater staffs have long history in flood control and are organized to provide a suite of services for that control, including maintenance, capital construction, regulation and enforcement, engineering and planning, and administration. All of those skills are also needed for stormwater quality. Taking advantage of current assets and resources is the single largest cost savings a municipality can experience in instituting the stormwater quality program.

For a stormwater quality program to be successful, it cannot simply be an add-on or appendix to the stormwater quantity program. Under Phase I, many municipalities, not understanding this, created stormwater quality appendices to their design criteria manuals, created stormwater quality sections in their stormwater ordinances, and added a staff position to administer the permit. They all failed, because they forget that stormwater is one unified practice — quantity and quality.

To maximize staff acceptance, the stormwater administrator should use a realistic building-block approach and employ measures consistent with or extensions of other programs and regulations. Maximize what you have to get what you need. Most municipalities have inherent in their legal responsibility and staff resources the ability to transform to a stormwater quantity and quality organization, even if it is done in a small way. But, like everything else related to the new quality regulations, it will take some convincing, education, and hard teamwork. In many cases, there is a built-in resistance to a "tree-hugger" mentality among field personnel. One key is to remember that some of your most important stakeholders are staff. Bring key staff players to the table early in the process.

The stormwater permit will make strange bedfellows of departments that normally do not interface. Fire departments may perform stormwater inspections along with industrial and commercial HAZMAT inspections like in several Eastern cities. All field-related municipality employees could undergo basic environmental sensitivity training and could serve as the eyes and ears of the SWQMP reporting system as in one Southwestern city. Complaint lines can be turned into environmental hotlines if properly staffed and advertised. Utility companies will begin to interface much more with planners to guide development with utility availability and conditions of use.

#7 — Ride the Treatment Train

As explained in a previous section, studies and experience have shown that no single BMP is able to handle all the impacts of urban development on the receiving waters. Therefore, it is important to work in small incremental units with reliance on accumulated results. The five-fold treatment train described previously should be part of the overall planning process when implementing the SWQMP.

Initially, a basic set of programs and BMPs should be established in each of the five sectors of the train. These should maximize current resources, seem cost-effective, focus on the known problems, and fit as well as possible into the municipality's urban social culture and landscape. Then, on the basis of additional need for pollution control and monitoring results, additional programs and practices should be input into that part of the train where it makes most sense. Realize that typically most programs and practices are most effective at the source (at the top of the treatment train).

#8 — Get a Tool Set

There are at least two types of tools that are necessary to support a successful stormwater quality program: data and technology.

Livingston, long a leader in the Florida stormwater program, credits more recent success partially to the effective use of biocriteria and assessment and sediment criteria to help meet pollution-reduction goals (Livingston, 2000). In Ohio, Yoder has led in the development of state-specific biological criteria to help direct the stormwater and surface water program for that state (Yoder, 2000, 2001). Herricks has proposed a postexposure lethal exposure time (PE-LET50) toxicity test geared explicitly to assess the potential toxicity of runoff from wet-weather events (2001). Sweitlick proposes modifications to water-quality criteria on the basis of combinations of habitat and bioassessments (2000). Santa Clara Valley (2001) assessed the suite of 26 environmental indicators of the Center for Watershed Protection (Claytor and Brown, 1996), finding many of them useful as measuring devices.

In each of these cases, measurement and assessment tools are being developed and employed to provide initial information on stream condition and feedback on the progress of improvement. Currently in stormwater quality management it is difficult, if not impossible, to know when enough has been done. In this day of scarce resources, it is important to be able to target programs that are cost-effective and to make adjustments. Making judicious use of receiving waters monitoring can be very cost-effective. This is especially true if a receiving water is listed on the state's 303(d) list and may be scheduled for a TMDL. Many states intend to use the NPDES permit as the enforcement mechanism for the TMDL program.

In many cases, the data that placed that stream segment on the list, and the data that are being used to make allocations, are incomplete or even poor science. In those cases, the best defense is proactive monitoring of incoming and outgoing flows, and the establishment of better information. There may be a potential to change the terms of a permit, receive a delisting, or change TMDL allocations.

It is possible to spend significant amounts on monitoring and sample analysis without achieving the intended goals. See Chapter 6 for more information on monitoring programs. Great care should be taken to craft the program to cover all needs without undue duplication or inappropriate approaches.

Technology is neither good nor bad. Being a pioneering user of high technology can be costly. It may best be left to universities, government agencies, and those few municipalities that have the budget and technological wherewithal to pay for the product and process development.

But if high technology can be costly, continuing to use low technology will be costly. For example, you can collect dry-weather flow information on 3 × 5 cards and keep them in a file for much lower cost than some computer database management methods. But do you really want to manage the program that way, forever? You can continue to use Mylar overlays to base maps to keep track of your stormwater system (if you keep track of it at all). But can you stand all the erasures, changes, man-hours, and lost or misfiled maps? A prudent investment in automation and technological advancements will pay big dividends in the future. Computers should do the grunt work, while men do the thinking.

Even simple spreadsheet capabilities can be useful for stormwater management purposes. There are products on the market that allow a city with only simple CADD capabilities to keep track of its sampling and monitoring information or outfall locations and industrial activities. At the next layer of complexity, AM/FM systems tie mapping with complex database management capabilities and span the gap between simple CADD and more complex applications. Finally, geographic information systems (GIS) provide almost unlimited flexibility in using networks, shared databases, and cross-layer information computations and graphical displays. Many municipalities are entering the GIS world and using stormwater management as the lever to do it. Many off-the-shelf products are available for just this purpose. And the maps and coverages have many other uses throughout the municipality and, therefore, some other partners in paying for the system and data acquisition.

One word of caution. Many municipalities spend large amounts on hardware and software, larger amounts on database development, and a pittance on training, implementation, and real-world system applications. The result is great disillusionment with GIS and a few red faces. Be application- and long-term use-focused (Figure 15-25).

#9 — Not All that Glitters is Gold

As discussed above, the NPDES program can be costly. But all that cost does not have to be in terms of hard cash. There are many ways to help resource the NPDES program that

FIGURE 15-25
Monitoring can be cost-effective.

cost little. For example, a local community can look to the following resources prior to looking to the general fund:

- *Modify local programs* — Perhaps 25% of a typical Phase II program is already being done to some extent by current staff, or similar things are being done. With suitable adjustment and refocus, some responsibilities can be covered by current staff as part of, or a redefinition of, their current duties. For example, public relations specialists can handle mailings using slots available in bill stuffers. Standing citizens groups can take on a role with the program.
- *Share costs with neighbors or region-/state-wide* — Much of what can be done can be done more cheaply sharing the cost. Phase I saw large numbers of group permits issued, causing regional approaches to spring up. Some of these worked better than others, but most saved all participants money in the long run. Cooperation can be of several types, including sharing a permit, coordinating permit requirements, or simply identifying and sharing some mutual programs. One set of cities and counties defined a minimal program into which they would all contribute. Each entity was then free to embellish its permit as it wished over and above the minimal program. They developed the program through the use of a facilitator and shared cost on the basis of population. In other cases, a regional entity handled some aspects of the permit for all included entities, perhaps a public education program, monitoring, design criteria development, model ordinance, etc. In still other cases, a state program sufficed for much of a permit requirement, allowing the local government to focus on other aspects of the permit.
- *Free information on the Web* – The Internet has hundreds of sites giving examples of BMPs, manuals, ordinances, documents, guidance, pamphlets, etc. Literally almost every written document that might be necessary has been developed somewhere and is available free of charge. The experience of other Phase I cities is especially helpful for Phase II cities. Fort Worth (http://ci.fort-worth.tx.us/dem/sitemap.htm) has an especially helpful Web site with multiple links to other sites. The Center for Watershed Protection (http://www.cwp.org/) offers a multitude of helpful documents and links, and its stormwater center (http://www.stormwatercenter.net/) has hundreds of references and assistance tools. The U.S. EPA's Web site (best found from a search, as it changes quite often) offers significant Phase II guidance as well as information on many related programs.
- *Partner with nonprofits* — There are hundreds of nonprofit organizations created to accomplish various environmentally related functions. Often these groups will adopt a watershed, provide workers, perform monitoring, do public education and involvement campaigns (they are a public involvement campaign), and find sources of money not available to local governments [501(c)(3) grants to non-profits]. Some local communities actually assist them in finding and applying for grants. They also are less willing to file a lawsuit against a local government when they are partners with it.
- *Federal programs and consulting* — Various federal programs provide funding on a cost-share basis and also provide consulting either gratis or cost-share. For example, TVA supplies Stream Teams to any local community willing to pursue a watershed protection program. The National Park Service provides a Rivers, Trails and Conservation Assistance Program that provides meeting facilitators and planning assistance for river corridor development. Several Phase II com-

munities received significant assistance from the Corps of Engineers in their Phase II permit application and parts of their implementation. The USGS cooperative program will provide monitoring and data analysis.

- *State and regional grants* — Many states provide grant monies for local governments to pursue environmental projects. In addition, state-administered programs such as Section 319, 604(b), 104(b)(3), Coastal Zone, Well Head Protection, FEMA, Federal Highway Commission, etc., provide funds for various programs. Much of this information can be gleaned from federal Web sites, including http://www.epa.gov/efinpage/fundings.htm (the environmental finance program), http://www.epa.gov/OWOW/watershed/wacademy/fund.html (Watershed Academy funding site), and EPA regional sites.
- *Special fees for service* — Another source of funding is to charge special fees for added services, including inspection fees for BMPs, additional construction program-related fees, plans review fees, etc. These fees can be scaled to cover part of a or a whole program area.
- *Public/private partnerships* — "Clean environment" is a good business message. Whether it is sponsorships, advertising, or grants, businesses are often willing participants with local programs.

#10 — Check the Pool

It has been said that a lawyer can send his mistakes to jail, a doctor buries his, and a preacher's failures are sent to Sheol. But an engineer's mistakes stand as a monument to his ignorance forever. There is great potential in the planning and implementation of stormwater quality programs to build, or require to be built, many structural treatments for water quality which, in the end, will prove to have little worth or are unworkable over the long run. It is important to wade into stormwater quality, not dive in.

There is a great temptation to put something visible on the ground. Resist...reasonably. This is especially true in municipalities that have no history or experience with the "care and feeding" of structural BMPs. The landscape is littered with scum-covered ponds in dense residential neighborhoods, catch basin retrofits clogged with coarse sediments, bioretention areas abandoned clogged and weedy, short-circuited vegetative buffer strips, and linear stream corridor nature preserves full of trash. Yet there are many more success stories, many of which were learned the hard way.

There are two major ways to avoid violation of this commandment: use of nonstructural BMPs and the use of pilot studies and phased implementation of BMP projects. There have been lessons learned. One Eastern state had required infiltration practices for a number of years only to find that it experienced a high failure rate due primarily to poor construction and poor or nonexistent maintenance practices. A Midwestern city developed capital improvements in each one of its councilmanic districts but without a clear prioritization and consideration of the actual environmental needs to be met.

The most effective way to keep surface waters clean is to keep pollutants from those waters in the first place. That is, address the pollution at the source — the minds and hearts of the common man. Seek to address known pollution problems with cheaper and more flexible nonstructural and programmatic BMPs and institutional changes prior to or along with physical construction. Public education, awareness and reporting, and other nonstructural programs can be easily modified or abandoned if they prove to be ineffective or to address a nonproblem. Employ measures that have high public, political, and stakeholder acceptance such as the Stream Teams of a Northwestern city or the Thousand Points of Blight program for storm inlet stenciling. While adults may not willingly change their lifestyles to reduce pollution, their children will. Many school programs eventually influence parents to recycle, become citizen stream monitors, and volunteer in cleanup programs.

Define programs to be developed in well-defined phases and steps with checks of effectiveness at every major milestone. Proceed slowly and cautiously with in-course corrections. Look for methods with few environmental side-effects, and which do not introduce unacceptable risks to health or safety. The NPDES permit process allows for annual reporting. Use these reports to request and negotiate changes in the conditions of the permit based on information gleaned from the previous year(s). Define a permit application with several options based on results of pilot studies and phased programs.

15.7 The SWQMP Development Process

Each regulatory program touching on stormwater management has developed a planning process that can roughly be expressed as follows: (1) determine existing conditions, (2) quantify sources and assess impacts, (3) assess alternatives, (4) develop and implement the recommended plan, and (5) assess results (USEPA, 1993d). Inherent in any approach is the setting of goals and objectives that are quantitative, measurable and flexible on a site, watershed, or municipal-wide basis. There are several approaches to effectively establishing a SWQMP.

The Nine-Step Planning Approach

One approach (described in support of Phase I in the first edition of this text) to the development of a SWQMP used in several municipalities is a nine-step process, which proceeds rationally through narrowing down possible approaches to the few that have the most promise. It is important to have a systematic way to go about planning the stormwater quality management program. This will not only provide a logical, building-block program development but will provide ample support should litigation result from program implementation measures. The approach seeks to define goals, set objectives, take actions, and assess program effectiveness. Many factors working together can be thought of as a chain linked together. If one link is weak, the whole program suffers (Shaver, 1988; Ogden, 1992a). Four key characteristics of a successful SWQMP are that it is comprehensive, integrated, balanced, and dynamic. Many facets of municipal government work together, some of which may not have been thought of as part of a stormwater management team before. However, they can create synergy as they work together. The result is a program that is balanced between studying and doing, and effort is expended commensurate to the problems generated. The program also changes as new information is gleaned.

The nine-step process is illustrated in Figure 15-26. It is specific to the municipal NPDES permit requirements for stormwater in that it shows an initial two-pronged approach to the problem identification phase. But the process also serves those municipalities not involved in the NPDES process in two ways: (1) it positions them to more easily meet any new NPDES regulations for smaller municipalities or cooperate with adjacent municipalities already in the program, and (2) it takes advantage of limited prior knowledge of the pollution problems in the system through problem identification and doing those minimal things deemed necessary in almost every municipality.

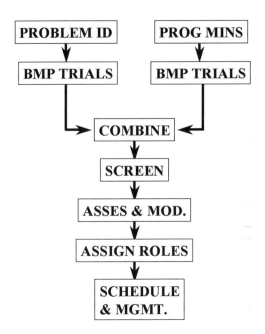

FIGURE 15-26
The nine-step process.

The process uses as a guiding philosophy the approach of solving real problems and meeting program minimum requirements. Several steps will go on simultaneously for various portions of the program development. There may be a need to cycle back through several steps one or more times to arrive at an acceptable MEP strategy.

Under Phase II, EPA has required a mandatory description for each proposed management program including BMPs, measurable goals schedules, and responsible parties. These items should be kept in mind and documented:

- The nature of the controls
- The sources addressed and the area served by the system that will be affected
- Intergovernmental coordination necessary
- The proposed nature of the implementation and the implementation schedule
- Staff and equipment necessary to implement the program
- Procedures for monitoring implementation and evaluating effectiveness of the management program component

Step 1

Identify and quantify (if possible) known problems. Problems can be defined as anything from aesthetic appearance to actual known environmental impairment. Often, the inability to attain a designated stream use is considered a problem. If urban runoff contributes to that nonattainment, then it will be addressed in the development of the SWQMP. The example below illustrates a worksheet approach to assist in defining known problems under Phase II.

Step 1 — Defining Known Problems
Purpose

To get a sense of the main stormwater quality related problems and needs your community now faces and your general philosophy in developing a stormwater NPDES permit. This is just to help you to think about your own community.

Instructions

The table below gives typical stormwater quality problems faced by many communities separated along the six minimum controls. Most of these problems will impact the things you will do under stormwater quality permitting and programs as well.

Place checkmarks where you know you have problems, and add additional items as appropriate.

Minimum Control Area	Some Potential Problems
Public education	☐ Citizens know nothing about how they cause pollution ☐ Citizens do not know how to report a problem ☐ Citizens do not know how to recognize a pollution problem ☐ Citizens dump pollution into streams or on the ground: pesticides, oil, fertilizer, household waste ☐ No materials available ☐ No public outreach ☐ Little respect or appreciation for streams or other riparian areas
Public involvement	☐ Citizens have little real input into stormwater policy ☐ Citizens are not interested ☐ Citizens have little way to get involved in stormwater pollution reduction ☐ ☐ ☐
Illicit discharges	☐ Leaky sanitary sewers ☐ Industrial pollution ☐ Nonstormwater plumbed or dumped into drains ☐ Failing septic systems ☐ Dirty or smelly outfall locations ☐ Unexplained fishkills ☐ Runoff from urban hot spots and dirty land uses ☐ Little knowledge of outfall locations or condition ☐ No system mapping ☐ Lack of ordinance controls ☐ Lack of local public entity industrial NPDES permits ☐ Poor enforcement of industrial NPDES permits ☐ Many dirty neighborhoods or industries ☐ Lots of used oil or other pollution dumping ☐ Sewage plumbed into stormpipe system ☐ Little employee or citizen knowledge of how to recognize and report problems ☐ Pollution from certain categories of dischargers (e.g., restaurants, gas stations, car washes, etc.) ☐ ☐ ☐

Minimum Control Area	Some Potential Problems
Erosion and sedimentation	☐ Poor erosion-control practices and regulations ☐ Channel erosion due to development ☐ Scour in key locations due to poor design ☐ Clogging of stormwater systems by construction site erosion ☐ Lack of ordinance control ☐ Lack of site inspections ☐ Poor or lax enforcement capabilities and practices ☐ Lack of developer knowledge of erosion-control practices ☐ Lack of technical guidance on erosion control ☐ Inability to receive and respond to complaints ☐ Poorly trained inspectors ☐ Poor site-inspection practices or holes in the inspection program ☐ Erosion from residential construction ☐ ☐ ☐
Postconstruction pollution	☐ Few or no permanent controls on stormwater quality ☐ General runoff is dirty ☐ Poor stream or river water quality ☐ Lack of ability to use innovative development concepts ☐ Cumbersome PUD process hinders innovative site design ☐ Lack of developer or staff education on alternatives ☐ Lack of ordinance controls ☐ Lack of technical guidance or training ☐ Mindset that other than ordinary developments are expensive and "weird" ☐ Anti-environmental mindset ☐ No program for stormwater quality (or quantity) controls ☐ Little data or information ☐ Little way to influence the rezoning and development or redevelopment process ☐ Riparian areas not protected or available ☐ Floodplain management not done well ☐ No incentives to reduce pollution ☐ No nonstructural practices common in community ☐ ☐ ☐
Municipal housekeeping	☐ Municipal maintenance operations cause pollution ☐ Municipal fertilizer and pesticide operation cause pollution ☐ Municipal NPDES permits not completed or implemented ☐ Vehicle maintenance operations cause pollution ☐ Materials storage and handling (i.e., salt) causes pollution ☐ Litter in streams ☐ Employees unaware of pollution issues ☐ Design criteria does not take into account pollution issues ☐ Flood-control projects do not take into account pollution-reduction practices and designs ☐ ☐ ☐
Some Institutional Problems	
☐ Lack of legal authority ☐ Lacking technology (mapping, data, CADD or GIS, etc.) ☐ Lack of stable and adequate funding ☐ Lack of interest from leadership ☐ Lack of having a plan of action ☐ Low priority of stormwater	☐ Lack of public awareness and/or involvement ☐ Scattered organization ☐ Little visible leadership ☐ Little articulated vision ☐ ☐ ☐

I would describe our basic philosophy and planned approach to stormwater quality as:

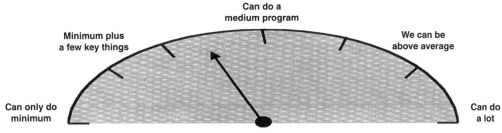

Enviromental Concern and Capacity Meter

Knowing what you know now, what would you say are the few key areas you would focus on in your stormwater quality management program?

Focus Area	Main Approach Toward Solution

If little information exists, the initial NPDES permit period may be used to develop more information and put off commitment to expensive BMP programs until a minimum of information is available to guide the development of these programs. Known problems can be gleaned from federal, state, and local reports and records; maintenance records or the knowledge of field personnel; field screening; citizen environmental groups; etc.

Seek, as possible, to determine cause and effect or at least possible sources. If possible, seek to quantify problems. One method of this is to develop a stream segment rating system for various types of pollution or impairment. One city (Ogden, 1992b) developed rating systems for key indicator pollutants that would place stream segments in "action," "watch," "no action," or "unknown" categories based on sample information from 47 receiving stream sites. The system was placed into a graphic information system (GIS) for display and long-term management. Stream segments could be switched on to display the category of each segment for each pollutant of concern. When overlaid with land use, industrial sites, outfall locations, and neighborhood age, a clear picture of cause and effect began to emerge. The U.S. EPA (1993d) gives more information.

At this stage, an initial development of goals and objectives is done. However, the details and final goals and objectives are reserved for later steps, when more information is available to realistically set limits and determine tradeoffs.

Step 2

Develop a preliminary brainstorm list of potential solutions. There are hundreds of BMPs and variations. A long laundry list helps stimulate discussion and imagination. The solution direction may, at first, appear obvious but, because of the integrated nature of such solutions, there may be other aspects (such as citizen monitoring or a new technology) that may add significantly to mitigation. A good source of information on education products is published by the U.S. EPA (1993b). There are many other sources of BMP information, including APWA (1992), City of Austin (1988, 1991), U.S. EPA (1990, 1992c, 1992d, 1992e, 1992f, 1993c), Stormwater Quality Task Force (1993), Ogden (1992a, 1992b), MWCOG (1987, 1992), Denver UDFCA (1992), Maine DEP (1990, 1992), State of Minnesota (1989), and many newer sites on the Web, including these few examples:

- EPA BMP Guidance — http://www.epa.gov/npdes/menuofbmps/menu.htm
- Georgia Stormwater Manual — http://www.georgiastormwater.com/
- Nashville BMP Manual — http://www.nashville.org/pw/bmp_manual.html
- Western Washington Manual — http://www.ecy.wa.gov/programs/wq/stormwater/manual.html
- California DOT Manual — http://www.dot.ca.gov/hq/oppd/stormwtr/index.htm
- Low Impact Development Manuals — http://www.epa.gov/owow/nps/urban.html
- Massachusetts Stormwater Manual — http://www.state.ma.us/dep/brp/stormwtr/stormpub.htm
- Maryland Stormwater Manual — http://www.mde.state.md.us/environment/wma/stormwatermanual/download_manual.htm
- New York State BMP Manual — http://www.dec.state.ny.us/website/dow/swmanual/
- St. Johns Watershed Management District Manual — http://sjr.state.fl.us/programs/outreach/pubs/index.html
- Texas Nonpoint Source Book — http://www.txnpsbook.org/
- Center for Watershed Protection — http://www.cwp.org/
- Stormwater Center — http://www.stormwatercenter.net/
- Fort Worth, Texas, Links Page — http://ci.fort-worth.tx.us/dem/stormcontacts.htm

The example below describes a worksheet methodology for this brainstorming step.

Step 2 — Brainstorm List Worksheets
Purpose

To develop a first-cut laundry list of the types of things we might do in our community under each of the 6 minimum controls.

Instructions

Fill out the table below, checking those areas you feel would be most useful and applicable to your community. Fill in the blank space under each minimum control with particulars about how to apply specific items.

Six Minimum Controls BMP List		
Public Education and Outreach BMP Ideas		
☐ Brochures ☐ Videos ☐ Slide shows ☐ Speakers bureau ☐ News articles ☐ TV shows or ads ☐ Billing inserts ☐ Door hangers ☐ Stencils on drains ☐ Billboards ☐ Matchbook covers	☐ In-school classes ☐ Special curriculum ☐ Adopt-a-stream ☐ Storm drain stencils ☐ Fact sheets ☐ Web sites ☐ Bumper stickers ☐ Recreational guides ☐ Refrigerator magnets ☐ Posters ☐ Restaurant placemats	☐ Special training programs ☐ Library of educational materials ☐ Hotline ☐ Advertised economic incentives ☐ Tributary signage ☐ Newspaper ads ☐ Displays ☐ Flyers
Particulars:		
Public Participation/Involvement BMP Ideas		
☐ Citizens groups ☐ Forums ☐ Special education days ☐ Festivals ☐ Booths ☐ Citizen instructors and speakers	☐ Citizen monitoring ☐ Adopt-a-stream ☐ Storm drain stenciling ☐ Clubs ☐ Scout troops ☐ Citizen watch groups ☐	
Particulars:		

Illicit Discharge Detection and Elimination BMP Ideas		
☐ System mapping ☐ GIS database of outfalls and watershed boundaries ☐ GIS databases of industries ☐ GIS database of actions taken ☐ Hotline ☐ Fire department inspections of industry ☐ Plan to detect and address illicit discharges ☐ Prepositioned spill equipment	☐ Priority neighborhood plan and maps ☐ Student dry-weather screening ☐ Complaint database ☐ Hot spot spill plans and inspections ☐ Ordinance controls ☐ Site inspections ☐ Dye or smoke testing ☐ Certification program ☐ Septic program ☐ Video sanitary and storm lines ☐ Storage and handling training and inspection	☐ Storm drain stenciling ☐ Brochures and other outreach and education materials (see first two minimum controls) ☐ Advertised hotline for public reporting ☐ Training for public employees ☐ Recycling programs ☐ Household hazardous waste collection ☐ Used oil programs ☐ Infrared testing ☐ Low-cost toxicity testing ☐ Pollutant prevention programs ☐ Product substitution

Particulars:

Construction Site Runoff Control BMP Ideas		
☐ Ordinance ☐ BMP manual ☐ BMP and site planning training ☐ Operator certification program ☐ Mandatory E&S plans ☐ Training videos with notes and exercises	☐ Deputize inspectors ☐ Environmental court ☐ Plan review checklists ☐ Example plans to give out ☐ Cross-train building inspectors ☐ Timed or milestone-based inspections ☐ Site prioritization map and scheduling ☐ Recordkeeping	☐ Fines ☐ Bonding requirements ☐ Building inspection holds for noncompliance ☐ Complaint tracking program and database ☐ Public complaint hotline ☐ See minimum controls 1 and 2 for other education and involvement ideas

Particulars:

Postconstruction Runoff Control BMP Ideas		
☐ Better site design guidance and ordinances ☐ Changed zoning or subdivision codes to encourage conservation site design practices ☐ Conservation easements ☐ Transfer of development rights ☐ Tax incentives ☐ Land trusts ☐ Incentives for PUD developments ☐ Open-space and vegetation ordinances ☐ Restrictive zoning	☐ Integrated site and stormwater infrastructure design standards ☐ Structural BMPs: infiltration practices, filtration practices, ponds and wetlands, buffers and vegetated channels, commercial BMPs, industrial BMPs ☐ Master planning ☐ Sensitive area overlay districts ☐ Extended floodplains ☐ Riparian buffer requirements	☐ Watershed groups ☐ Multiobjective regional BMPs ☐ Mandatory maintenance agreements ☐ Training in BMP design ☐ BMP postconstruction inspection program ☐ Low-impact landscaping ☐ Crediting mechanisms for good designs and impervious area reduction ☐ Preservation of key areas

Particulars:

Pollution Prevention/Good Housekeeping BMP Ideas		
☐ Maintenance program and procedures assessment ☐ Floatables controls at inlets and outlets ☐ Materials handling and storage assessment	☐ Waste disposal program for BMP sludge ☐ Retrofit of flood control structures ☐ Retrofit of design criteria ☐ Street and parking lot sweeping with high-efficiency sweepers	☐ Salt storage assessment ☐ Comprehensive review and modification of all municipal outdoor operations ☐ Employee training ☐ Inspection of public sites ☐ Recycling programs ☐ Adopt-a-highway programs ☐ Neighborhood or stream clean-up programs
Particulars:		

Step 3

Identify program minimums not adequately covered. After all known or suspected problems are dealt with, there may still be program areas thought to be important by the U.S. EPA and generally common to all municipalities that have not been adequately addressed.

Step 4

Develop a preliminary brainstorm list of potential solutions for these program minimums. Use the brainstorm list from above but apply it to gaps in the program not already filled to address known, real problems.

Step 5

Combine lists for problem solutions and program minimums, blend programs, and modify. This shortened list is organized and analyzed to determine how they will function singularly and in conjunction with other program elements, and how and by whom they will be implemented. Conflicts and overlaps are eliminated.

Another part of this analysis is to determine ranges of BMP application to allow for development of alternative programs and to get a feel for cost sensitivity. Normally, a minimal, moderate, and maximum program are assessed to some extent, keeping in mind realistic upper limitations to application (legal, financial, political, physical, social, etc.) and lower limits where the programs have no effectiveness. This is particularly applicable to the program minimums, where no specific known problems are targeted.

Step 6

Finalize goals and priorities and develop stormwater quality programs. Goals and objectives were set in a preliminary way during the problem and program minimums identification stages. In this step, they are analyzed and finalized based on the problems and program minimums identified in Steps 1 and 3 and the combined list of solution possibilities developed in Step 4. It is important to involve the public in each step of the goal and priority-setting and refinement process to ensure community support. See Chapter 3 for further information.

Goals, objectives, and priorities may be set on a municipal-wide basis, watershed basis, or to protect a specific resource or environmental amenity. For example, Austin, Texas, set a goal to maintain the quality of the water in Town Lake in the center of the city. Interim objectives and priorities were then set to attain this goal. Another city determined that illicit connections were a major problem for a local river and set goals to eliminate them by a certain date. Actions, objectives, and priorities were then set up around this goal. A Northwestern city set as a goal the maintenance of salmon runs within its jurisdiction. A Southern county wanted to maintain a particular wetland area for both passive recreation and habitat. A Midwestern city focused on reduction of erosion from its stream banks and the elimination of illicit connections and dumping from industrial hot spots.

Inevitably there will be fewer financial or manpower resources than are necessary to address each of the programs or needs fully. Because, at this point, there are so many options and subprograms being considered in a big-picture way, it is not advisable to get too detailed in cost estimation. There will be development of priorities and trade-offs based more on a feel for costs and other resources than a detailed estimate. Programs are adjusted and shifted with an emphasis on the overall goal of a sound and comprehensive program than to maximize cost-effectiveness, which is the next step.

There are several methods to attempt to assign priorities and goals to environmental programs (USEPA, 1987, 1993d), which include:

- Assessing the resources impacted by pollution in terms of types of resource (water, biological, aesthetic, etc.), values of resources, desired use of resource and extent of impairment
- Assessing institutional concerns related to the resource impairment, including legal authorities available, organizational capabilities, technological capabilities, financial resources available, legal penalties and enforcement issues of noncompliance, and public and political perceptions
- Assessing the pollutant source in terms of type, magnitude, impact, transport mechanisms, wet-/dry-weather trends
- Assessing how the problem or program fits overall goals and objectives, including land-use goals, water resource goals, site- or watershed-specific goals, and planning and development goals

Screening of the combined BMP list to meet the goals can be done in a number of ways. Table 15-2 provides a list of screening criteria used in some form by several municipalities to provide a consistent rating methodology. The next section gives more information on Table 15-2 and screening methods. Some programs are scaled back, delayed, or planned to be accomplished in phases to make way for higher-priority programs and activities.

The end result is a preliminary comprehensive SWQMP, which attempts to meet the goals and objectives set by the municipality in light of EPA requirements using the best combination of BMP strategies in a cost-effective and responsible manner. The next step refines this process using cost determinations.

Evaluation of Best Management Practices

Best management practices (BMPs) are used to control and reduce the discharge of pollutants. There are many ways to categorize BMPs, including structural or nonstructural; source control or treatment control; structural or programmatic; etc. However BMPs are categorized, there must be a consistent way to assess their effectiveness. The evaluation of best management practices is more an art than a science. EPA (1993d) and others describe several methods for the evaluation of BMPs. These methods are most often used together to assess various parts of a SWQMP. Significant guidance on the selection and

TABLE 15-2

Best Management Practice Screening Criteria

BMP Criteria Description	+	0	−	Comments
1. Human risk, public safety, and potential liability				
2. Physical and regulatory suitability				
3. Ability to control key targeted pollutants				
4. Costs to implement and continuing costs				
5. Acceptability to public, stakeholders, staff, political leadership				
6. Equitability to impacted persons				
7. Reliability and consistency over time				
8. Sustainability of maintenance or program management				
9. Ability to be applied universally throughout the municipality, or on a specific watershed basis				
10. Fit with other operations and programs				
11. Relationship to federal, state, or local requirements				
12. Environmental risk and implications				
13. Amenity or multiuse value				
Totals				

use of BMPs is provided by U.S. EPA and will not be repeated here (http://www.epa.gov/npdes/menuofbmps/menu.htm).

There are many structural BMP screening approaches, usually contained in the BMP manual describing each of them. However, physical applicability is only one facet of a more comprehensive screening. The holistic approach is a combination of intuition, engineering judgment, basic analysis, and cost assessment, which can lead to a set of BMPs and a strategy. This method can be effective but requires the involvement of experienced personnel from a number of technical disciplines and from senior staff or political leadership. Often, an ad hoc committee is selected that develops a set of objective criteria and then goes through a process, like the nine-step process above, to arrive at a final program.

Cost–benefit analysis is also an approach that can be used to develop SWQMPs. Such measures as cost per pound of pollutant removed or cost per day of activity can be used. This method is good for making decisions on which approach to use first and for how long. For example, it may be possible to achieve a reduction in a certain pollutant (such as petroleum hydrocarbons) using public education and regulatory processes. At some point, though, structural controls may be necessary to complete the program. Cost–benefit analysis is typically used for more site- or project-specific activities rather than municipal-wide or more subjective nonstructural type applications.

Optimization has been used in other sciences for decision-making. Optimization consists of finding the maxima or minima of an objective function subject to a series of constraints, each expressed in terms of decision variables. The objective function can be expressed in terms of cost or some complex combination of cost, benefit, and judgment. Nonlinear relationships among the variables and the fact that the numerical complexity can hide the subjectivity of many of the inputs often limits this method to narrow applications targeted toward a single site or pollution concern.

Matrix comparison is a more subjective method (unless numeric ratings are used), whereby a table is constructed with various BMPs or BMP programs along the horizontal axis and evaluation criteria along the vertical axis. Each alternative program is then rated in terms of each of the criteria. A numeric rating (say, 1 through 5) or a simple + or - rating can be used. Table 15-2 provides a list that has been used in various forms in several locations for BMP program development. Some localities total an actual score and use the form for a numeric ranking of BMPs. Others simply use the pluses and minuses as flags to alert planners to particular BMP strengths or potential problems. Not all criteria apply to each BMP. Another list may be developed specifically for subcategories of BMPs. Following is a discussion of the items included in Table 15-2.

1. *Human risk, public safety, and potential liability* — Some BMPs carry with them some inherent risk (such as the risk of drowning in a wet pond) as well as value. In many cases, this risk can be minimized through sound design. Risk must be assessed even informally to limit potential human risk and liability.

2. *Physical and regulatory suitability* — This is the ability of the BMP (primarily structural) to fit the site, given the set of constraints for the site (e.g., bedrock, slope, soil type, etc.), and the ability of the BMP to be constructed, given any regulatory controls (e.g., in-stream blockage, groundwater contamination).

3. *Ability to control key targeted pollutants* — Each BMP should be targeted toward certain pollutants and certain problems. The BMP should be able to reduce to an acceptable extent the targeted pollutant. This may require consideration of chemical forms the pollutant may take, long-term ability, extreme environmental conditions, etc.

4. *Costs to implement and continuing costs* — Every BMP must be scrutinized for its cost-effectiveness. First costs may be low, but, if life-cycle costs are high, the BMP may not prove effective or economically sustainable.

5. *Acceptability to the public, stakeholders, staff, and political leadership* — Some BMPs may not have a high acceptance in certain areas or for certain uses. For example, some types of zoning control may not be acceptable if seen as a taking of land. Use of some BMPs can be made more palatable through the use of incentives such as user fee credits for the use of detention ponds.

6. *Equitability to impacted persons* — All BMPs inconvenience someone. The level of this inconvenience must be weighed against the public good and ways found to reduce the impact on those most affected by the BMP. The application of the BMP must be fair and reasonable and have considered ways that would reduce the impact.

7. *Reliability and consistency over time* — Reliability refers to the ability of the BMP to continue over a relatively long period of time without failing. Consistency refers to its ability to reproduce its mitigation ability without changing or varying appreciably in its ability to mitigate the pollution problem.

8. *Sustainability in terms of maintenance or program management* — All BMPs require program or structural upkeep. If the program cannot be sustained, modifications in its application should be made. If maintenance cannot be assured, reasons for nonassurance should be eliminated or other BMP alternatives should be sought.

9. *Ability to be applied universally throughout the municipality or on a specific watershed basis* — BMPs have an advantage if they can be applied widely. This supports fairness, flexibility, and possible economies-of-scale. While the use of certain BMPs may be appropriate in only some areas, it is desirable to be able to use them wherever necessary.

10. *Fit with other operations and programs* — Stormwater quality management must be integrated with, not simply added onto, the existing stormwater management program. BMPs have a better prospect for success if existing programs can be simply modified to incorporate stormwater quality aspects.

11. *Relationship to other federal, state, or local regulatory requirements* — There are many overlapping program requirements. Watershed protection, zoning, wetlands permitting, and other programs may have bearing on how and where the BMP is used. It is advantageous if a BMP fits within and does not violate the requirements of these other programs.
12. *Environmental risk and implications* — Some BMPs carry little human risk but may carry a high environmental risk. For example, infiltration basins in a groundwater recharge zone may be unacceptable if the pollutants infiltrated would enter the drinking water supply. If the risk of BMP failure is high, the use of the BMP may need to be avoided or modified.
13. *Amenity or multiuse value* — Structural BMPs should be functional and beautiful. Program BMPs should mutually support each other and other community objectives. Parks and greenways can be easily integrated with structural BMPs. Education programs and public schools or citizen volunteer organizations often gain synergy from each other and can assist in gaining acceptance of multiuse projects.

Step 7

Estimate overall program costs and pollution-reduction effectiveness, modify program as necessary. Some measure of effectiveness is better than none. From a municipality's point of view, the best measure of permit condition attainment is one that simply looks at accomplishment of prescribed tasks rather than actual pollution reduction or expenditure of funds. Actually, measuring pollution reduction should be accomplished but not used as a standard for permitsm at least until the level of cause and effect is improved in stormwater quality management.

Cost estimates have been made for each of the municipal permits already applied for. They are, at best, first-order accuracy. Experience has shown that the addition of a medium level of stormwater quality to an advanced stormwater quantity program will increase costs by from 30 to 50%. APWA (1992) provides a comprehensive table of cost estimates for various types of BMPs. Experience with municipalities having similar subprograms indicates that the costs can also be quite variable.

It should be remembered that the MEP standard is not normally numerical. Specific objectives, performance specifications, or other means of measurement of performance for pollution reduction have not been specified by regulatory agencies. In many cases, particularly for nonstructural BMPs, it is difficult to assign pollution reduction numbers without better data and information. One is often compelled to solve a problem, which has not been fully defined, and use poorly documented means, which have not been well measured in terms of effectiveness. Great care and engineering judgment must then be exercised.

There is a need in any comprehensive program development to go back and look at the whole assembled puzzle after suitable examination of each of the pieces. There will almost always be a need for adjustment, compromise, and combining of programs at this stage.

Step 8

Describe roles and responsibilities to implement the program. All key tasks must be identified and clear direction given as to responsibility and actions expected. The goal is to get all parties to attain a sense of ownership of their portion of the program and a sense of teamwork in execution of the overall program.

Step 9

Develop a schedule for implementing the program including management and feedback loops. Feedback information may allow for changes in direction or emphasis in any program element. These changes can be negotiated during preparation of the annual report. Decision points can be built right in as permit conditions. The annual reporting requirement is a good time to make adjustments. Such freedom to modify proposed programs should be built into the schedule, if possible.

The end result of this step is a schedule and budget for implementation of the program. It is important to assess the success of the programs at every step and build into each program ways to measure that success. This may be through specially designed feedback from the program implementors, through data collection and monitoring, public awareness polls, or other means.

EPA's Guidance Approach

EPA, based partially on initial work by Reese (2000) and others, developed guidance for a self-assessment and plan for implementation. This same approach, modified, can be used for repermitting at 5-year intervals.

Self-Analysis

You should conduct a comprehensive self-analysis to help you gain a better understanding of your current situation with respect to complying with the Stormwater Phase II Rule. The self-analysis should consist, at a minimum, of the following components:

1. *Understand the stormwater regulations and your stormwater responsibilities.* First, you should obtain a copy of the Phase II rule and your state's Phase II permits. Before you undertake the process to develop a Phase II program, you should have a clear understanding of what you are required to do. Begin by asking yourself the following questions:
    a. Am I in an urbanized area as designated by the 1990 Census?
    b. Could I be included in an urbanized area as designated by the 2000 Census?
    c. If I am not in an urbanized area, is my population greater than 10,000 people (potential designation by the permitting authority)?
    d. Does my city government own or operate a facility with industrial activity as defined by EPA's stormwater regulations (e.g., wastewater treatment plants, vehicle maintenance facilities, etc.)?
    e. Does my city government own or operate construction activity that disturbs greater than 1 acre?
    f. Do I understand what the stormwater regulations require (the development of a stormwater management program that includes the six minimum measures and measurable goals)?
    g. Do I understand the deadlines, and when I am required to submit a permit application?
2. *Understand how your city currently manages its stormwater runoff.* Make an assessment of your city's stormwater management and conveyance system. Get copies of maps, inventories, or other assessments of the physical infrastructure in place. Begin by asking yourself the following questions:

a. Do you have an inventory of stormwater inlets, pipes, ditches, and open channels?

b. Do you know how many outfalls your city discharges to and where they are located?

c. Do you know if someone else is discharging stormwater into your system?

d. Do you know the major pollutant sources in your city (industrial, commercial, residential)?

e. What types of flood-control or water-quality practices are currently in place in your city?

3. *Know the condition of your receiving waters.* Stormwater programs should be designed to address the specific needs of the community and water resources they are intended to protect. If you have not done so already, collect information on your city's receiving waters and what pollutants and sources are impacting those waters. You should also know the various uses of your receiving waters so you can design a program to protect those uses. Begin by asking yourself the following questions:

   a. Do you know the names and locations of the waters that receive a discharge from your MS4?

   b. Do you know the character and quality of these waters?

   c. Are any of these waters listed as impaired on your state's 303(d) list?

   d. What are the pollutants impacting these waters?

   e. Do you know the designated uses of these waters?

4. *Assess your current programs and practices to determine what needs to be changed.* The Phase II program provides an opportunity to identify and change programs and practices that are or could be impacting water quality. Begin by asking yourself the following questions:

   a. What are your current practices that contribute to water-quality problems?

   b. What are your current practices that will help meet NPDES stormwater requirements?

   c. Do you have an existing educational program on water-quality?

   d. Do you have an erosion- and sediment-control program established?

   e. Do you have procedures to address illegal dumping and spills?

   f. What legal authority do you already have, and what legal authority will you need to develop?

5. *Identify stakeholders who can help you develop and implement your stormwater program.* These can include people who are impacted by city ordinances, concerned citizens, and groups who would be expected to pay for stormwater management (as part of a stormwater utility, for example). Begin by asking yourself the following questions:

   a. Are there other Phase II communities in your area willing to cooperate with you?

   b. Is there a Phase I city in your area with which you can work?

   c. Are there groups or associations, such as environmental, industry, or community associations, that can help you?

6. *Determine the overall objectives for your stormwater program.* These objectives could include improving water quality, decreasing flooding, increasing citizen aware-

ness and cooperation, and increasing funding. You should develop an objective for each of the six minimum control measures to help guide you in selecting and targeting BMPs and measurable goals. Your stormwater management plan should be designed with these goals in mind.

Action Plan

Your next step will be to develop an action plan to help you determine what to do and when. An action plan is a tool to help guide you as you develop your stormwater management program and is not required under the U.S. EPA regulations. The first step in developing an action plan is to complete the self-analysis previously described.

1. *Assemble your team.* This will include stakeholders and city departments that may have a role in stormwater management.
2. *Develop a time schedule.* This would ideally identify the date your permit application is due, March 10, 2003, for all initial permittees, and work backwards from there. You should set interim milestones to assess your progress. Key dates could be included for public comment and review, local authority approval, stakeholder meetings, and acquiring funding. Your time schedule should also accommodate a stormwater management program plan approval process. Your stormwater management plan will probably need to be approved by local authorities, regulatory authorities, and stakeholders.
3. *Determine your strategy for compliance.* What does a good program look like? Try to determine what type of program your city managers want and what type of program you can realistically develop. Begin by asking yourself the following questions:
    a. What benefits do you want to achieve?
    b. What is your tolerance for risk? The Phase II program includes a lot of flexibility, but inherent in that flexibility is uncertainty. You will need to balance your tolerance for risk in developing a stormwater management plan.
    c. What is the best program approach for you? For example, you can develop a minimal program that meets legal requirements, an aggressive proactive program, a "the best we can afford program," or a "the best that the city council will approve" program.
    d. What is realistically achievable? You should determine your financial resources and limitations by asking the following questions:
    e. What is realistic given your current program and legal constraints?
    f. What is realistic in terms of your receiving water quality?
    g. What goals should I set? Setting clear goals for your stormwater program will help you set clear measurable goals and document your program's success to regulators and the public.
4. *Network with other local governments.* Talk to other cities in your area to find out what they are doing. Consider establishing regular meetings with these cities to share information, and, if your goals are compatible, consider partnering with some of these cities to share resources or join as co-permittees. If there is a Phase I community nearby, investigate what it has been doing and consider working with some of its ideas and using materials it has already developed.

5. *Determine the main elements of your program.* Using the information from your self-analysis and the items above, start to formulate the major elements of your program. Identify how you will address each of the six minimum control measures. First, identify the BMPs and measurable goals that will be used to implement the six minimum measures. Second, identify practices that will require ongoing operation and maintenance. Finally, plan for developing and maintaining public support through education and outreach.

6. *Establish an implementation plan.* This plan will describe how will you develop your Phase II stormwater management program, including public participation components. The following are factors you should consider when implementing your stormwater management program:

    a. Determine program funding and staff requirements. Assess whether you will do the work in-house or contract it out.

    b. Develop your institutional framework. Identify a lead city department or agency. Develop MOUs, if necessary, and consider designating or establishing a regional group, such as a council of governments, to help coordinate activities.

    c. Identify your permitting approach. Will you choose a general permit or an individual permit? Will you join as a co-permittee with another city?

    d. Assign an individual or group to be responsible for submitting the permit application, developing annual reports, etc.

References

American Public Works Association, Study of Nationwide Costs to Implement Municipal Stormwater Best Management Practices, May 1992.

American Public Works Association, Water Quality: Urban Runoff Solutions, APWA Spec. Report #61, May 1991.

Atlanta Regional Commission (ARC), Georgia Stormwater Manual. 2002.

Barbour, M.T., Gerritsen, J., Snyder, B.D., and Stribling, J.B., Rapid Bioassessment Protocols for Use in Streams and Wadeable Rivers: Periphyton, Benthic Macroinvertebrates, and Fish, 2nd ed., USEPA, Office of Water, Washington, DC, EPA 841-B-99-002, 1999.

Bay Area Stormwater Management Agencies Association (BASMAA), Start at the Source, 1999.

Bledsoe, B.P., Relationships of Stream Response to Hydrologic Changes, Proc. ASCE Eng. Found. Conf.: Linking Stormwater BMP Designs and Performance to Receiving Water Impact Mitigation, Urbonas, B., Ed., August 19–24, Snowmass Village, CO, 2001, pp. 127–144.

Center for Watershed Protection, The Economics of Watershed Protection, in *The Practice of Watershed Protection*, Schueler, T.R. and Holland, H.K., Center for Watershed Protection, Ellicott City, MD, 2000.

Center for Watershed Protection, Better Site Design, Site Planning Roundtable, 1998, www.cwp.org.

City of Austin, Inventory of Urban Nonpoint Source Pollution Practices, June 1988.

Clar, M. and Rushton, B., Low Impact Development Case Studies, Proc. ASCE Eng. Found. Conf.: Linking Stormwater BMP Designs and Performance to Receiving Water Impact Mitigation, Urbonas, B., Ed., August 19–24, Snowmass Village, CO, 2001, pp. 484–488.

Clark, S., Field, R., and Pitt, R., Wet-Weather Pollution Prevention through Product Substitution , Proc. ASCE Eng. Found. Conf.: Linking Stormwater BMP Designs and Performance to Receiving Water Impact Mitigation, Urbonas, B., Ed., August 19–24, Snowmass Village, CO, 2001 pp. 266–283.

Claytor, R.A. and Brown, W.E., Environmental Indicators to Assess Stormwater Control Programs and Practices, Final Report, CWP, Silver Spring, MD, 1996.

Cloak, D. and Bicknell, J.C., Use of Environmental Indicators to Assess Stormwater Program, Proc. ASCE Eng. Found. Conf.: Linking Stormwater BMP Designs and Performance to Receiving Water Impact Mitigation, Urbonas, B., Ed., August 19–24, Snowmass Village, CO, 2001, pp. 305–315.

Deletic, C., Maksimovic, C., Loughriet, F., and Butler, D., Modeling the Management of Street Surface Sediments in Urban Runoff, Proc. of the 3rd Int. Conf. on Innovative Tech. in Urban Storm Drainage, NOVATech, Lyon, France, 1998, pp. 415–422.

Denver Urban Drainage and Flood Control District, Urban Storm Drainage Criteria Manual, Vol. 3 Best Management Practices, Denver, CO, September 1992.

Falkenbury, J., Water Quality Standard Operating Procedures, City of Fort Worth Public Health Department, Ft. Worth, TX, 1987.

Federal Register, January 9, 1998, pp. 1536, 1550.

Federal Register, December 7, 1988, pp. 49416–49487.

Frank, J.E., The Costs of Alternative Development Patterns: A Review of the Literature, Urban Land Institute, Washington, DC, 1989.

Harrison, D., Zero Impact Policy Issues and Concerns — Municipal Perspective, Proc. ASCE Eng. Found. Conf.: Linking Stormwater BMP Designs and Performance to Receiving Water Impact Mitigation, Urbonas, B., ed., August 19–24, Snowmass Village, CO, 2001, pp. 181–195.

Herricks, E.E., Observed Stream Responses to Changes in Runoff Quality, Proc. ASCE Eng. Found. Conf.: Linking Stormwater BMP Designs and Performance to Receiving Water Impact Mitigation, Urbonas, B., Ed., August 19–24, Snowmass Village, CO, 2001, pp. 145–157.

Karr, J.R., Fausch, K.D., Angermeier, P.L., Yant, P.R., and Schlosser, I.J., Assessing Biological Integrity in Running Waters, Illinois Natural History Survey, Special Pub. No. 5, 1986.

Lee, G.F., and Jones-Lee, A., Comments on Consolidated Hotspot Cleanup Plan, California Water Quality Resource Board, Sacramento, 1999.

Livingston, E.H., Protecting and Enhancing Urban Waters: Using all the Tools Successfully, in Proc. of Natl. Conf. on Tools for Urban Water Resour. Manage. and Protection, Chicago, IL, February 7–10, USEPA EPA-625-R-00-001, 2000, pp. 94–123.

MacRae, C.R., Experience for Morphological Research in Canadian Streams: Is Control of the Two-Year Frequency the Best Basis for Stream Channel Protection, Effects of Watershed Dev. and Manage. on Aquatic Ecosystems, Proc. of an Eng. Conf., ASCE, 1997.

MacRae, C.R., An Alternate Design Approach for the Control of Instream Erosion Potential in Urban Watersheds, Urban Storm Drainage: Proc. of the 6th Int. Conf., Niagara Falls, Ontario, Canada, IAHR/IAWQ Joint Committee on Stormwater Drainage, 1993.

Maine DEP, Environmental Management: A Guide for Town Officials, Maine Department of Environmental Protection, April 1992.

Maine DEP, Watershed: An Action Guide to Improving Maine Waters, Maine Department of Environmental Protection, April 1990.

Malcom, H.R., A Study of Detention in Urban Stormwater Management, Water Resources Research Institute of the University of North Carolina, Report No. 156, July 1980.

Metropolitan Washington Council of Governments, A Current Assessment of Urban Best Management Practices, March 1992.

Metropolitan Washington Council of Governments, Controlling Urban Runoff, Schueler, T.R., July 1987.

The National Association of Flood and Stormwater Management Agencies (NAFSMA), Municipal Separate Storm Sewer (MS4) Permit Application Costs, June 1992.

The National Association of Flood and Stormwater Management Agencies (NAFSMA), Phase II Survey Raw Data Report, Washington, DC, 1999a.

The National Association of Flood and Stormwater Management Agencies (NAFSMA), Survey of Stormwater Phase II Communities, Washington, DC, 1999b.

Ogden Environmental, Stormwater Quality Management Plan Development, Nashville, TN, Memorandum, 1992a.

Ogden Environmental, Stormwater Quality Permit Application for the City of Charlotte, NC, May 1992b.

Pitt and Field, Hazardous and Toxic Wastes Associated with Urban Stormwater Runoff, 16th Annual Hazardous Waste Research Symposium, U.S. EPA-ORD, Cincinnati, OH, 1990.

Practice of Watershed Protection (PWP), Hydrocarbon Hotspots in the Urban Landscape, Schueler, T.R. and Holland, H.K., Eds., Center for Watershed Protection, Ellicott City, MD, 2000.

Prince Georges County (PGC), Low Impact Development Design Strategies, Department of Environmental Resources, 1999

Reese, A.J., The 10 commandments of municipal stormwater quality management programs, *APWA Reporter*, March 1994.

Reese, A.J., Developing Municipal Stormwater Quality Management Programs, Proc. 6th Int. Conf. on Urban Storm Drainage, September 12–17, Ontario, Canada, 1993.

Reese, A.J., NPDES Phase II Cost Estimates, in Proc. of Natl. Conf. on Tools for Urban Water Resour. Manage. and Protection, Chicago, IL, February 7–10, *USEPA EPA-625-R-00-001*, 2000, pp. 532–546.

Riley, A.L., Restoring Streams in Cities, Island Press, Washington, DC, 1997.

Rosgen, D., A Classification of Natural Rivers, CATENA 22, 1994, pp. 169–199.

Schueler, T.R. and Caraco, D.S., Prospects for Low Impact Development and Watershed Level, Proc. ASCE Eng. Found. Conf.: Linking Stormwater BMP Designs and Performance to Receiving Water Impact Mitigation, Urbonas, B., Ed., August 19–24, Snowmass Village, CO, 2001, pp. 196–209.

Schueler, T. and Shepp, D., The Quality of Trapped Sediments and Pool Water within Oil Grit Separators in Suburban MD, Metropolitan Washington Council of Governments, 1993

Shaver, E. and Hatton, C., Auckland Experience with BMP's Mitigating Adverse Impacts, Proc. ASCE Eng. Found. Conf.: Linking Stormwater BMP Designs and Performance to Receiving Water Impact Mitigation, Urbonas, B., Ed., August 19–24, Snowmass Village, CO, 2001, pp. 387–402.

Shaver, H.E., Stormwater management issues, in *Design of Urban Runoff Quality Controls*, ASCE, Roesner, L.A., Urbonas, B., and Sonnen, M.B., Eds., 1988.

Simons, D.B. and Senturk, F., *Sediment Transport Technology*, Water Resources Publications, Littleton, CO, 1991.

Sovern, D.T. and MacDonald, A., Can Instream Integrity be Obtained through On-Site Controls?, Proc. ASCE Eng. Found. Conf.: Linking Stormwater BMP Designs and Performance to Receiving Water Impact Mitigation, Urbonas, B., Ed., August 19–24, Snowmass Village, CO, 2001, pp. 47–59.

State of Minnesota, Protecting Water Quality in Urban Areas, Minnesota Pollution Control Agency, October, 1989.

Stormwater Quality Task Force, California Stormwater Best Management Practice Handbooks — Vol. 1, Municipal, State of California, Water Control Board, March 1993.

Strecker, E.W., Low Impact Development (LID) — Is It Low or Just Lower?, Proc. ASCE Eng. Found. Conf.: Linking Stormwater BMP Designs and Performance to Receiving Water Impact Mitigation, Urbonas, B., Ed., August 19–24, Snowmass Village, CO, 2001, pp. 201–222.

Sutherland, R.C. and Jelen, S.L., Contrary to Conventional Wisdom, Street Sweeping can be an Effective BMP, Advances in Modeling the Management of Stormwater Impacts, James, W., Ed., Vol. 5, CHI, Guelph, Canada, 1997, pp. 179–190.

Swietlick, W.F., Urban Aquatic Life Uses — A Regulatory Perspective, Proc. ASCE Eng. Found. Conf.: Linking Stormwater BMP Designs and Performance to Receiving Water Impact Mitigation, Urbonas, B., Ed., August 19–24, Snowmass Village, CO, 2001, pp. 163–181.

U.S. Environmental Protection Agency, Measurable Goals Guidance for Phase II Small Ms4s, .2002, http://www.epa.gov/npdes/stormwater/measurablegoals/part2.htm.

U.S. Environmental Protection Agency, Report to Congress on the Phase II Stormwater Regulations, Office of Water, EPA 833-R-99-001, October 1999.

U.S. Environmental Protection Agency, Economic Benefits of Runoff Controls, EPA 841-S-95-002, September 1995.

U.S. Environmental Protection Agency, Investigation of Inappropriate Entries into Storm Drainage Systems, EPA/600/R-92/238, January 1993a.

U.S. Environmental Protection Agency, Urban Runoff Management Information/Education Products, Office of Wastewater Enforcement and Compliance, Permits Division, February 1993b.

U.S. Environmental Protection Agency, Guidance Specifying Management Measures for Sources of Nonpoint Pollution in Coastal Waters, EPA 840-B-92-002, January 1993c.

U.S. Environmental Protection Agency, Handbook — Urban Runoff Pollution Prevention and Control Planning, EPA/625/R-93/004, September 1993d.

U.S. Environmental Protection Agency, Environmental Impacts of Stormwater Discharges, EPA 841-R-92-001, June 1992a.

U.S. Environmental Protection Agency, NPDES Permit Application Workshop for Stormwater Discharges from Municipal Separate Storm Sewer Systems and Industrial Sources — Speaker's Notebook, USEPA Region IV, July 1992b.

U.S. Environmental Protection Agency, Guidance Manual for the Preparation of Part 2 of the NPDES Permit Applications for Discharges from Municipal Separate Storm Sewer Systems, The Cadmus Group, August 1992c.

U.S. Environmental Protection Agency, Stormwater Management for Construction Activities — Development of Pollution Prevention Plans and Best Management Practices, EPA 833-R4-92-005, September 1992d.

U.S. Environmental Protection Agency, Stormwater Management for Industrial Activities, Stormwater Permit Manual Reprint, Thompson Publishing Group, September 1992e.

U.S. Environmental Protection Agency, Stormwater Management for Construction Activities — Development of Pollution Prevention Plans and Best Management Practices, Summary Guidance EPA 833-R4-92-001, October 1992f.

U.S. Environmental Protection Agency, Urban Targeting and BMP Selection, Region V Water Division, November 1990.

U.S. Environmental Protection Agency, Setting Priorities: The Key to Nonpoint Source Control, Office of Water Regulations and Standards, 1987.

U.S. Environmental Protection Agency, Results of the Nationwide Urban Runoff Program, Vol. 1 — Final Report, PB84-185552, December 1983.

Virginia DCR, Stormwater Management Regulations and Act, State of Virginia Department of Conservation. and Recreation, Richmond, VA, 1990.

Yoder, C.O. and Miltner, R,.J., Using Biological Criteria to Assess and Classify Urban Streams and Develop Improved Landscape Indicators, in Proc. of Natl. Conf. on Tools for Urban Water Resource Manage. and Protection, Chicago, IL, February 7–10, USEPA EPA-625-R-00-001, February, 2000, pp. 56–82.

Yoder, C.R., The Biological Condition Axis, State of Nevada Bioassessment Workshop, May 21–22, Desert Research Inst., Sparks, NV, 2001.

16
Site Design and Construction

16.1 Introduction

Stormwater management increasingly includes a consideration of site design and layout, and the constructability and maintainability of the structures. The comprehensive treatment-train concept implies that the first line of defense against flooding and water pollution is a site design concept that incorporates features that intrinsically reduce pollution and flood generation. But, these features must be able to be built in the real world. Complex lot-by-lot grading, microfeatures, etc., require significantly more oversight by the local government than standard practice. And the planning and design of sites to incorporate these features require a higher level of skill and staff involvement than has heretofore been available in many cities.

16.2 Site Design Concepts — Overview

Site planning is the intentional thought given to the location and siting of human uses onto the landscape before they are built. Site planning deals with how to alter, use, and preserve various land features and resources based on the need of the development program, regulatory controls, and the desires and vision of the developer and owner. Among the most important natural resources to take into account and protect during site development is surface water — the network of lakes, ponds, streams, channels, swales, wetlands, etc., that, together, comprise a complex system vital to living resources within the watershed and valuable to eventual dwellers and users of the land — postdevelopment.

In the past, this surface water system has often been considered a nuisance to be dealt with rather than an amenity to be valued and used carefully. Recently in the United States, there has been increasing emphasis on states' water resources. Such diverse factors as court cases, private citizen groups, regulatory changes, water shortages, environmental damage, "smart growth" movements, political recognition and initiatives, and other factors have led to a greater focus on changing the way development occurs to better preserve and use of surface water resources. These forces have led to a balanced focus on three important aspects of site development: financial viability, environmental sensitivity, and aesthetic attractiveness.

This chapter serves as a framework for including effective and environmentally sensitive stormwater management into the site development process.

The first step in addressing stormwater management begins with the site planning and design process. Development projects reduce their impact on watersheds through careful efforts to conserve natural areas, reduce impervious cover, and better integrate stormwater treatment. By promoting a combination of environmentally sensitive design approaches, a community can help developers reduce the amount of runoff and pollutants that are generated from a development or redevelopment site, and provide for some nonstructural on-site treatment and control of runoff. The goals of better site design include (Atlanta Regional Commission, 2002):

- Managing stormwater (quantity and quality) as close to the point of origin as possible and minimizing collection and conveyance.
- Preventing stormwater impacts rather than mitigating them.
- Utilizing simple, nonstructural methods for stormwater management that are lower cost and lower maintenance than structural controls.
- Creating a multifunctional landscape.
- Using hydrology as a framework for site design.

There are a number of names given to a variety of related site design concepts including Better Site Design, Conservation Design, Smart Design, Sustainable Development, Green Infrastructure, Zero Discharge Design, Low-Impact Development, etc. Each family of concepts stresses slightly different approaches. Conservation or environmental site design techniques include such things as preserving natural features and resources, effectively laying out the site elements to reduce impact, reducing the amount of impervious surfaces, and utilizing natural features on the site for stormwater management. The aim is to reduce the environmental impact "footprint" of the site, while retaining and enhancing the owner's and developer's purpose and vision for the site. Many of the better site design concepts can reduce the cost of infrastructure, while maintaining or even increasing the value of the property.

These concepts can be viewed as water-quantity and water-quality management tools and can reduce the size and cost of required structural stormwater controls — sometimes eliminating the need for them entirely. The site design approach can result in a more natural and cost-effective stormwater management system that better mimics the natural hydrologic conditions of the site, has a lower maintenance burden, and provides for more sustainability.

Conservation of Natural Features and Resources

Identifying and preserving the natural features and resources that can be used in the protection of water resources is a key initial step in site design. These features contribute to stormwater management goals by reducing stormwater runoff, providing runoff storage, reducing flooding, preventing soil erosion, promoting infiltration, and removing stormwater pollutants. Some of the natural features that should be taken into account include areas of undisturbed vegetation, floodplains and riparian areas, ridge tops and steep slopes, natural drainage pathways, intermittent and perennial streams, aquifers and recharge areas, wetlands, soils, and other natural features or critical areas.

Delineation of natural features is typically done through a comprehensive site analysis and inventory before any site layout design is performed. Approaches that should be followed in conserving natural features and resources include preserving undisturbed natural areas, preserving riparian buffers, avoiding floodplains, avoiding steep slopes, and minimizing siting on porous or erodible soils.

Lower-Impact Site Design Techniques

After conservation areas have been delineated, there are additional opportunities in the preliminary stages of a site design for avoiding downstream impacts from the development. These primarily deal with the location and configuration of lots or structures on the site and include the following recommendations and options (USEPA, 2000): fitting the design to the terrain, reducing the limits of clearing and grading, locating development in less sensitive areas, utilizing open-space development and nontraditional lot designs for residential areas, and considering creative development design.

Reduction of Impervious Cover

Reducing the area of total impervious surface on a site directly reduces the volume of stormwater runoff and associated pollutants that are generated. It can also reduce the size and cost of necessary infrastructure. Impervious reduction can also have great impacts on the lifestyles of residents, as they may switch from larger lot subdivisions to more clustered housing and smart growth or sustainable development land use and development concepts (Smart Growth Network, 2001). Some of the ways that impervious cover can be reduced in a development include reducing roadway lengths, reducing roadway widths, reducing the footprint of buildings through clustering and other methods, reducing the number of mandated parking spaces, reducing the parking footprint, reducing setbacks and frontages, and having fewer or alternative cul-de-sacs.

Using Natural Features for Stormwater Management

Traditional stormwater drainage design tends to ignore and replace natural drainage patterns and often results in overly efficient hydraulic conveyance systems. Structural stormwater controls are costly and often can require high levels of maintenance for optimal operation. Through use of natural site features and drainage systems, careful site design can reduce the need and size of structural conveyance systems and controls (DNREC, 1997). Some of the methods of incorporating natural features into an overall stormwater management site plan include the following: using buffers and undisturbed areas, using natural drainage ways instead of storm drain systems, use vegetated swales instead of curb and gutter, and draining runoff to pervious areas.

Figures 16-1 and 16-2 illustrate the application of some of these concepts to two typical sites (ARC, 2001).

Site-Planning Goals

It is important to consider the goals of stormwater integrated site development. Generally, the stormwater portion of the site development should help meet the long-term water quality goals for the watershed and receiving waters, mitigate the anticipated impacts of the development (both quality and quantity), and integrate well with other aspects of the site development in a cost-effective and sustainable manner.

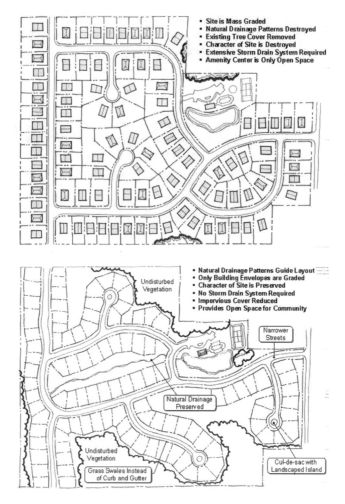

FIGURE 16-1
Comparison of a traditional residential subdivision design (above) and an innovative site plan developed using better site design practices (below).

Performance Goal #1: Site designs should preserve and utilize natural drainage systems and reduce the generation of additional stormwater runoff to the maximum extent practicable.

Performance Goal #2: Stormwater runoff generated from new development should be adequately treated and controlled prior to discharge into a jurisdictional wetland or waters of the state. Note that adequately treated means that the designated water-quality volume has been treated through one or more of the approved stormwater controls and site design practices that are detailed in this book.

Performance Goal #3: Local communities should require, as necessary, additional or site-specific management measures to control and treat stormwater runoff from certain types of development and areas draining to sensitive receiving waters.

Performance Goal #4: Stream channel protection should be provided by adopting three general approaches: (1) upland source control and detention; (2) bank protection measures such as energy dissipation and velocity control; and (3) riparian corridor preservation and conservation.

Site Design and Construction 1069

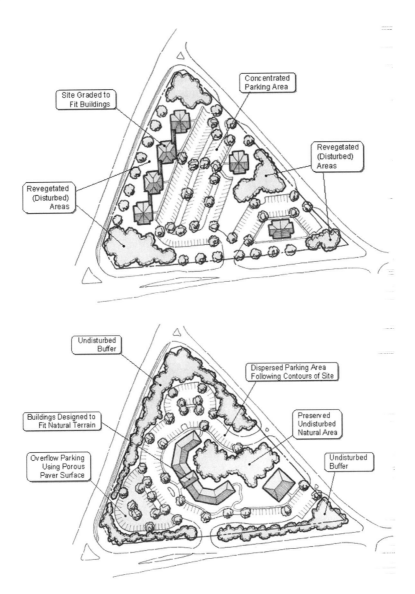

FIGURE 16-2
Comparison of a traditional office park design (above) and an innovative site plan developed using better site design practices (below).

Performance Goal #5: Overbank flood protection should be provided for by all sites discharging water to a stream or river.

Performance Goal #6: All habitable structures and major transportation arteries (roads, railroads, etc.) should be protected from flooding to at least the 100-year flood level for the expected life of the structure from all flooding sources, major and minor.

Performance Goal #7: Annual groundwater recharge rates should be maintained to the extent practicable by promoting infiltration through the use of structural and nonstructural methods.

Performance Goal #8: Local communities should require effective short- and long-term maintenance of all of the drainage system and structural stormwater controls.

Performance Goal #9: Regional stormwater management facilities should be evaluated as an alternative for on-site water quantity and water-quality controls. Regional stormwater management refers to facilities designed to manage runoff from multiple projects and properties through a local jurisdiction-sponsored program, where the individual properties assist in the financing of the facility, and the requirement for on-site controls is eliminated or reduced.

Performance Goal #10: To the maximum extent practicable, development projects should strive to implement nonstructural pollutant prevention practices such as material use, exposure, disposal, and recycling controls, spill prevention and cleanup, illegal dumping controls, illicit connection controls, and conservation and preservation measures.

Many of these goals can be achieved through the practice of mimicking predevelopment hydrology. That is, instead of creating an efficient drainage system that rapidly conveys runoff to receiving waters through direct connection of all paved areas to a piped or channel system, the designer attempts to preserve the natural hydrologic functions of the site. This is best done through a combination of better site designs, nonstructural controls, and the use of structural controls disbursed throughout the site.

Most of the adverse impacts of urban development on stormwater runoff are directly related to the amount of effective impervious area that is on the site. By effective impervious area, we mean the impervious area that is directly connected to the conveyance system without intervening mitigation measures. Reduction of effective impervious area through the measures discussed later in this chapter is then a primary goal for integrated stormwater site design techniques. If stormwater can be controlled near the source through the use of simple, integrated, multifunctional methods, then the chances for successfully mitigating adverse impacts on a long-term basis is good.

Watershed Basis

The most effective means of managing stormwater is on a watershed basis, as illustrated in Figure 16-3. A somewhat different stormwater management strategy is often used for each part of the watershed subbasin:

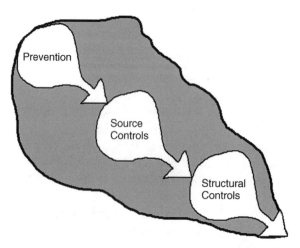

FIGURE 16-3
Watershed basis for application of controls.

- In the upper end of the drainage system, prior to the point where rainfall is traditionally collected and concentrated, pollution and flow *prevention* methods are most effective. These include the better site design techniques outlined in this chapter. Prevention methods involve the avoidance of development of flow or pollution-generating land uses, or the modification of standard construction practices to reduce the impacts of development, rather than the construction of structural devices.

- In the parts of the watershed where flow is beginning to be concentrated and conveyed, the focus changes from prevention to *source controls* through retention, detention, infiltration, and treatment of stormwater on a disbursed microscale basis. That is, there may be many opportunities to develop filter strips, grassed channels, small storage areas, bioretention areas, natural holding areas, infiltration trenches, filtration devices, etc. Some communities even recommend rain barrels, under-sidewalk retention, dry wells (in sandy areas), rooftop meadows, and underground tanks. In some cases, commercial devices work well for removal of solids, trash, and some hot spot pollutants, or as pretreatment devices for general application stormwater controls further down in the drainage system.

- Finally, after the flow has been concentrated, there is the opportunity to employ *structural control measures* to capture and further reduce the pollutant load in the runoff and to mitigate peak flow and volume increases due to urban development. There may be an overall master plan for the watershed into which this development nestles. In that case, there may be an opportunity to take advantage of regional type controls, of coordinated land uses, and of cost sharing.

16.3 Site Design Concepts — Implementation

Communities should actively promote the use of environmental site design as a way to protect watersheds and water resources, and implement cost-effective and lower maintenance stormwater management. However, in order to make better site design a reality, local jurisdictions will often need to review their regulations, development rules, community plans, and review procedures to ensure that they support the better site design concepts outlined above.

Often, communities have in place development rules that work against better site design and create needless impervious cover and unnecessary environmental impact. Examples include the minimum parking ratios that many communities require for retail or commercial development, and zoning restrictions that limit cluster development designs. Some of the policy instruments that need to be reviewed for compatibility with the better site design principles include zoning ordinances and procedures, subdivision codes, stormwater management or drainage criteria, tree protection or landscaping ordinances, buffer and floodplain regulations, erosion- and sediment-control ordinances, grading ordinances, street standards or road design manual, parking requirements, building and fire regulations and standards, septic and sanitary sewer regulations, and the local comprehensive plan.

Table 16-1 presents a set of questions that can be used to review a community's local development codes and ordinances with the goal of making it easier to implement stormwater better site design. The questions are organized by better site design categories (ARC, 2001).

TABLE 16-1

Questionnaire for Reviewing Local Development Regulations

Conservation of natural features and resources	**Land Conservation Incentives**
	Does the community have a viable greenspace program?
	Are there any incentives to developers or landowners to preserve nonregulated land in a natural state (density bonuses, conservation easements, stormwater credits or lower property tax rates)?
	Natural Area Conservation
	Is there an ordinance or requirements for the preservation of natural vegetation on development sites?
	Tree Conservation
	Does the community have a tree protection ordinance?
	Stream Buffers
	Is there a stream buffer ordinance in the community that provides for greater buffer requirements than the state minimums?
	Do the stream buffer requirements include lakes, freshwater wetlands, or steep slopes?
	Do the stream buffer requirements specify that at least part of the buffer be maintained with undisturbed vegetation?
	Floodplains
	Does the community restrict or discourage development in the full buildout 100-year floodplain?
	Steep Slopes
	Does the community restrict or discourage building on steep slopes?
Lower-impact site designs	**Fitting Site Designs to the Terrain**
	Does the community provide preconsultation meetings, joint site visits, or technical assistance with site plans to help developers best fit their design concepts to the topography of the site and protect key site resources?
	Clearing and Grading
	Are there development requirements that limit the amount of land that can be cleared in a multiphase project?
	Locating Development in Less Sensitive Areas
	Does the community actively try to plan and zone to keep development out of environmental sensitive areas?
	Open Space Development
	Are open-space or cluster development designs allowed?
	Are the submittal or review requirements for open-space designs greater than those for conventional development?
	Are flexible site design criteria (e.g., setbacks, road widths, lot sizes) available for developers who utilize open-space or cluster design approaches?
	Does a minimum percentage of the open space have to be managed in an undisturbed natural condition?
	Does the community have enforceable requirements to establish associations that can effectively manage open space?
	Nontraditional Lot Designs
	Are nontraditional lot designs and shapes allowed?
	Creative Development Design
	Does the community allow and promote planned unit developments (PUDs) that give the developer or site designer additional flexibility in site design?
Reduction of impervious cover	**Roadway Length**
	Do road and street standards promote the most efficient site and street layouts that reduce overall street length?
	Roadway Width
	What is the minimum pavement width allowed for streets in low-density residential developments that have less than 500 average daily trips (ADT)?

TABLE 16-1 *(Continued)*

Questionnaire for Reviewing Local Development Regulations

	Building Footprint
	Does the community provide options for taller buildings and structures that can reduce the overall impervious footprint of a development?
	Parking Footprint
	What is the minimum parking ratio for a professional office building (per 1000 ft^2 of gross floor area)?
	What is the minimum parking ratio for shopping centers (per 1000 ft^2 of gross floor area)?
	What is the minimum required parking ratio for single-family homes?
	If mass transit is provided nearby, are parking ratios reduced?
	What is the minimum stall width for a standard parking space?
	What is the minimum stall length for a standard parking space?
	Are at least 30% of the spaces at larger commercial parking lots required to have smaller dimensions for compact cars?
	Is the use of shared parking arrangements promoted?
	Are there any incentives to developers to provide parking within structured decks or ramps rather than surface parking lots?
	Can porous surfaces be used for overflow parking areas?
	Is a minimum percentage of a parking lot required to be landscaped?
	Is the use of bioretention islands and other structural control practices within landscaped areas or setbacks allowed?
	Setbacks and Frontages
	What is the minimum requirement for front, rear, and side setbacks for a 1/2-acre residential lot?
	What is the minimum frontage distance for a 1/2-acre residential lot?
	Alternative Cul-de-Sacs
	What is the minimum radius allowed for cul-de-sacs?
	Can a landscaped island be created within a cul-de-sac?
	Are alternative turnarounds such as "hammerheads" allowed on short streets in low-density residential neighborhoods?
Utilization of natural features for stormwater management	**Using Buffers and Undisturbed Areas**
	Are requirements in place and guidance provided in using level spreaders to promote sheet flow of runoff across buffers and natural areas?
	Using Natural Drainageways
	Are storm drain systems required for all new developments? Are natural systems allowed?
	Using Vegetated Swales
	Are curb and gutters required for residential street sections?
	Are there design standards for the use of vegetated swales instead of curb and gutter?
	Rooftop Runoff
	Can rooftop runoff be permanently designed to discharge to pervious yard areas?
	Do current grading or drainage requirements allow for temporary ponding of runoff on lawns or rooftops?

16.4 Site Design Concepts — Design Steps

The flowchart shown in Figure 16-4 depicts a typical site development process from the perspective of the developer. A developer has an idea about a type of project he or she would like to build. A real estate broker helps the developer find a piece of land. After an initial site visit, the developer begins the process of establishing the preliminary feasi-

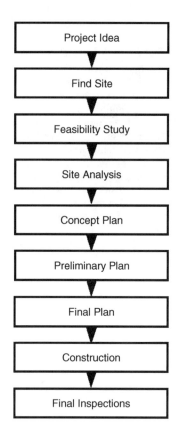

FIGURE 16-4
Site design flowchart.

bility of the project. If the project appears initially feasible, the developer proceeds to get the property under contract, often through a purchase option pending further analysis. This additional analysis includes a detailed financial and legal analysis and, normally, a detailed engineering site analysis or site assessment. The goal is to get the project underway quickly if it appears feasible.

If the project is deemed to be feasible, a survey is completed, and a title transfer takes place. The design team prepares a concept plan (often called a sketch plan) and, if necessary, completes the rezoning or other permitting processes. Then, a preliminary plan is prepared and submitted for necessary reviews and approvals. Federal, state, and local permits are applied for at various stages in the process. The developer's goal is to expedite the process and insure that it proceeds without unnecessary delay or cost, and without unexpected expense.

After negotiations with the reviewer, necessary changes, public hearings, etc., the final construction plan is prepared. There may be several iterations between plan submittal and plan approval. Bonds are set and placed, contractors are hired, and construction of the project takes place. During and after construction, numerous types of inspections take place. Erosion control or grading permits are usually obtained first. Depending on the complexity of the project, there might be several local, state, and federal requirements to be met, some with individual permits and inspections. Normally, there is a final inspection and a use and occupancy permit issued for the structure. If it is a residential development, lots are sold, sometimes in phases. The developer's goal is to complete construction and sell or lease the property as quickly as possible.

The actual process is never so orderly, and it always balances the need of the developer to make decisions at certain times, the need of the local government to have adequate review documents and time, and the vagaries of nature, the economy, and the physical construction process.

The process of stormwater planning and design is a subset of overall site development and must fit into the overall process if it is to be successful. Each development-related element (such as water and wastewater, transportation and traffic circulation, building layout and functional program, pedestrian traffic, communication, etc.) has an impact on stormwater management and vice versa. Each local government is different in terms of the submittal requirements, checklists, inspections, and other procedures. But, there are certain basic principles and practices of stormwater site design that should be common to all localities.

Each local government should strive to have an integrated and comprehensive stormwater and site design process and to include the four key elements of the stormwater site design approach discussed in this chapter: better site design practices, design criteria, stormwater credits, and proper selection and sizing of stormwater controls.

Stormwater planning and design is a consideration throughout the development process, with the key elements fitting primarily into the overall site development process in the five areas of feasibility study, site analysis, concept plan, the combined areas of preliminary and final plan, and construction (Dewberry et al., 1995). Table 16-2 shows the highlights of stormwater site design and how it fits into each of these areas. Details are given in the next section.

Feasibility Study

For each of the following sections, stormwater objectives and key actions will be listed at the beginning of the section, as shown in Table 16-2.

- Stormwater Objectives:
  - Create a vision for better site design and integrated stormwater management
  - Understand major site constraints and opportunities
  - Initiate relationship with local government reviewers and inspectors
- Key Actions:

Step 1: Preconsultation with developer and plan reviewer

Overview

At this stage of the project, the developer is interested in whether there are any constraints to development of the planned program that would cause the project to be infeasible for that particular site. This requires a design team to understand the intent of the development and the characteristics of the site pertinent to that intent. Three main avenues of investigation are pursued:

- Confirmation of the program components and development intent
- Confirmation of the site characteristics (micro and macro)
- Identification of applicable planning and regulatory controls

The most important action the local government can take at this time is to hold a preconsultation with the developer and his team to assist them in assessing constraints,

TABLE 16-2

Stormwater Site Planning and Design

Feasibility Study

Summary
Develop an overview of all factors that would influence the decision to proceed with the site development, including the basic site characteristics, local and other governmental requirements, area information, surrounding developments, etc. Includes preconsultation with local government. Build a vision for the site.

Stormwater Objectives
- Create a vision for better site design and integrated stormwater management.
- Understand major site constraints and opportunities.
- Initiate relationship with local government reviewers and inspectors.

Key Actions
- Step 1: Preconsultation with developer and plan reviewer

Major Activities
- Base map development
- Review of local and regional requirements
- Joint site visit with local government
- Secondary source information
- Water-based regulatory controls and limitations
- Master plans or comprehensive plans
- Other factors or constraints impacting feasibility

Site Analysis

Summary
Gain an understanding of the constraints and opportunities associated with the site through identification, mapping, and assessment of all types of resources, and through discovery of regulatory and design requirements. Identify primary building locations and primary and secondary conservation areas.

Stormwater Objectives
- Identify key site physical, environmental, historical, and cultural resources
- Determine all stormwater requirements
- Build on preliminary vision for stormwater concept

Key Actions
- Step 2: Delineation of resource protection areas and site evaluation
- Step 3: Determination of stormwater management requirements

Major Activities
- Overlays of key physical data
- Mapping of natural resources: soils, vegetation, streams, topography, slope, wetlands, floodplains, aquifers
- Identification of other key cultural, historic, archeological, or scenic features, orientation and exposure
- Habitat, plant, or animal species constraints
- Adjacent land uses
- Mapping of legal constraints such as easements and utilities
- Identification of conservation areas
- Adjacent transportation and utility access
- Location within watershed
- Integration of all layers — "sieve mapping"
- Other constraints and opportunities

Concept Plan

Summary
Develop a concept of the site that honors the original idea of development while making provision for constraints and taking advantage of resources available on the site. Attempt various alternatives in a "what-if" design mode. The most promising is taken to a more detailed stage and public review.

Stormwater Objectives
- Develop concept of better site design and water-quality credit locations
- Develop concept of structural and nonstructural controls
- Perform preliminary drainage layout and design calculations
- Gain approval from developer and local government of concept

TABLE 16-2 (CONTINUED)
Stormwater Site Planning and Design

Key Actions
- Step 4: Runoff characterization and stormwater control selection
- Step 5: Development of the preliminary concept plan

Major Activities
- "Blob" sketches of functional land use
- Major transportation routes and circulation
- "What-if" analysis on land conservation and innovative smart growth concepts
- Identification of primary and secondary conservation areas
- Layout of drainage and controls
- Design volume preliminary calculations
- Utilization of better site design concepts and crediting mechanisms in layout concept
- Other public facilities
- Basic parcel layout or basic building layout for single nonresidential site
- Public review process

Preliminary and Final Plan

Summary
Perform basic layout and calculations to ensure that the concept plan can be approved and constructed in a cost-effective manner that retains the goals of the developer. Leads to development of an approved set of construction plans.

Stormwater Objectives
- Integrate all design concepts
- Perform design volume final calculations
- Perform preliminary and final stormwater control and conveyance designs

Key Actions
- Step 6: Securing local and nonlocal permits
- Step 7: Final design and permit acquisition

Major Activities
- Final land-use configuration
- Final drainage and controls
- Final plans and specifications
- Final site design credits
- Drainage calculations
- Permits

Construction

Summary
Ensure that all elements are being built according to plan, and that all resource or conservation areas are suitably protected during construction. Ensure erosion control is in place and functional throughout the process.

Stormwater Objectives
- Ensure stormwater controls and site design practices are built as designed
- Pass all inspections
- Leave a fully functional and clean drainage system

Key Actions
- Step 8: Preconstruction meeting
- Step 9: Inspection during construction

Major Activities
- Execute bonds
- Inspection during key phases or key installations
- Protection of stormwater controls
- Protection of primary conservation areas
- Erosion control
- Proper sequencing

Final Inspection

Summary
Ensure that all elements are built according to plan. Structural controls should be built according to design, as-built certification completed, and final clean-out of controls and establishment of filter or infiltration media completed. Maintenance agreements should be executed and the controls handed off to owners.

TABLE 16-2 (CONTINUED)
Stormwater Site Planning and Design

Stormwater Objectives
- Ensure long-term maintenance of stormwater controls
- Ensure long-term protection of conservation areas
- Submit drawings of stormwater controls for record

Key Actions
- Step 10: Final inspection and submission of record drawings
- Step 11: Maintenance inspections

Major Activities
- Final stabilization
- As-built survey
- Final cleaning of controls
- Execute maintenance agreements
- Final inspection and use permit

opportunities, and potential for integrated stormwater design concepts. This meeting can be the most important action taken in directing a developer along a path that will preserve the environment, while addressing the needs of the developer for an economically viable site development.

This recommended step helps to establish a constructive partnership through the development process. A joint site visit, if possible, can yield a conceptual outline of the stormwater management plan and strategies. By walking the site, the two parties can identify and anticipate problems, define general expectations, and define general boundaries of site resource protection and conservation areas. A major incentive for preconsultation is that permitting and plan approval requirements will become clear at an early stage, thus increasing the chance that the approval process will be smoother and more rapid.

Thus, it is important for the stormwater manager to be conversant in the various aspects of site design that could be employed to bring about a less-impacting site development. It is also important for the local government to make this type of development easy to implement through clear and efficient regulatory controls, design criteria, and guidance literature.

Program Components

The design team should not only understand the intent of the developer but also the specific needs of the development program. For residential sites, it might include the number of lots necessary (yield) to make the development viable. For small-lot residential and attached residential development, the architectural footprint for the structures often dictates the need for other site elements such as parking and open space. For nonresidential sites, it would include the assumptions on building(s) size, bulk, parking needs, open space, etc., to meet the needs of the developer.

This is the point in the process where the vision for the development is formulated. The stormwater planner can have an important part in this visioning process by supplying information about better site design practices, the need to preserve and conserve key environmental features, the cost advantages of better site design, the use of green spaces, integrated stormwater management, etc.

Site Characteristics

A base map is the vehicle that conveys pertinent physical information to assist in confirmation of the site characteristics. The scale of the base map should be sufficient to convey the necessary attributes of the site at an accuracy adequate for decision making. USGS mapping can suffice for larger developments. Local governments may have mapping at a scale of 1:200 or 1:100 with 2- or 5-ft contour intervals that may be better. Many plans are developed at 1:50 scale. This base map serves as the foundation for later site analysis. The use of geographic information systems (GIS) or computer-assisted design and drafting (CADD) greatly facilitates the process.

Detailed analysis is rarely done at this point. Therefore, the primary elements usually considered necessary to be shown on the map are topography and property boundaries. This information should be augmented with road rights-of-way and pavement, utility rights-of-way and easements, regulatory floodplains and wetlands, existing structures, vegetation, bodies of water and drainage features, and adjacent property boundaries. Any factor that would contribute materially to a go or no-go decision or that represents a limit or boundary should be identified.

The stormwater planner can ensure that pertinent natural, historic, archeological, or cultural features are considered, inventoried, and mapped as necessary in preparation for the site analysis step that follows. Pertinent features to consider might include the following:

- Soils
- Vegetation
- Slope
- Woodlands
- Wetlands
- Floodplains
- Aquifers and recharge areas
- Significant wildlife habitat
- Ruins
- Old buildings
- Endangered species
- Earthworks
- Stone walls
- Burial grounds
- Views into and out of the site

A more detailed discussion of some of these variables is included in the next section. Figure 16-5 shows an example of a base map for a mixed-use site.

Planning and Regulatory Controls

Current land-use plans, comprehensive plans, zoning ordinances, road and utility plans, state laws or overlay areas, and public facility plans should all be consulted to determine compliance requirements and imposed constraints. It is important to understand not only local regulations, but also state and even federal or regional government or special district overlays of regulations.

Not only do regulations impose constraints, but they also can provide opportunities. Opportunities for special types of development (e.g., clustering) or special land-use opportunities (e.g., conservation easements or tax incentives) should be investigated. There may also be an ability to partner with a local community for the development of greenways or other corridor-type developments. The stormwater manager plays a key role in inves-

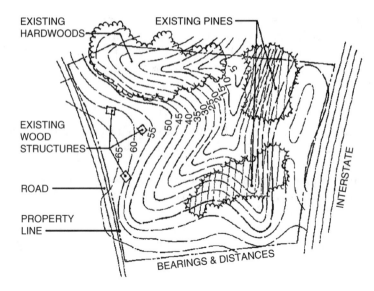

FIGURE 16-5
Example base map. (From North Carolina Department of Environment and Natural Resources, Division of Water Quality, Site Planning, 1998.)

tigating the ability to use environmental site design concepts and nonstructural land-use controls in a local community.

The plan developer should be familiar with the most current stormwater management requirements that apply to the site. Representative requirements might include control of water-quality and water-quantity design storms, conveyance design criteria, floodplain criteria, buffer and setback criteria, wetland permits, construction site NPDES permits, maintenance plans, infiltration tests, geotechnical evaluations, etc. Much of this guidance can be fleshed out at the preconsultation meeting with the plan reviewer and should be detailed in various local ordinances (e.g., subdivision codes, stormwater and drainage codes, etc.). This information could be contained in a checklist that would be provided to the developer.

In addition, the developer and designer need to be sure that the proposed site layout, drainage, and stormwater management practices are consistent with any existing and applicable watershed management plans. It is possible that the specific water-quality and water-quantity goals and guidance may exist at the receiving stream level that will require additional protection measures to be implemented. For example, a downstream reservoir may have a eutrophication problem due to nonpoint source phosphorus loading, and there may be an associated phosphorus limit that has been set as a goal that stormwater runoff from development in the watershed should strive to meet. It should also be a responsibility of the plan reviewer to be familiar with any broader watershed goals that should be complied with.

Site Analysis

- Stormwater Objectives:
    - Identify key site physical, environmental, historical, and cultural resources
    - Determine all stormwater requirements
    - Build on preliminary vision for stormwater concept

- Key Actions:
  - Step 2: Delineation of resource protection areas and site evaluation
  - Step 3: Determination of stormwater management requirements

Overview

The purpose of the site analysis is to identify all constraints and opportunities on the site through development of an overlay system of all physical, environmental, historical, and cultural information and resources on the site. The collection of these data was begun in the previous step and is completed and analyzed here.

Using approved field and mapping techniques, the developer's engineer would identify and map all previously unmapped site resource areas such as wetlands, critical habitat areas, intermittent and perennial streams, forest boundaries, floodplain boundaries, steep slopes, required buffers, proposed stream crossing locations, proposed stormwater treatment and conveyance areas, and other required protection areas (e.g., drinking water well setbacks). Some of this information may be available from previously performed studies or from the previous feasibility study. For example, if a development site requires a permit under the state's erosion and sedimentation regulations, most of the resource protection features will likely have been mapped as part of the land disturbance activity plan.

Other recommended site information to map or obtain includes soils data (i.e., NRCS hydrologic soil group designation, infiltration rates, erosion potential, etc.), seasonal groundwater levels, geologic mapping, and physical characteristics (e.g., geometry, capacity, composition) of existing natural or man-made drainage features on or adjacent to the site. This information will be useful and necessary in the selection and design of structural stormwater controls and conveyances.

Sometimes, individual layers can be developed in such a way as to facilitate analysis. For example, soils can be plotted according to hydrologic soils grouping, or slopes can be plotted in terms of steepness instead of simple contours. Figure 16-6 illustrates this concept.

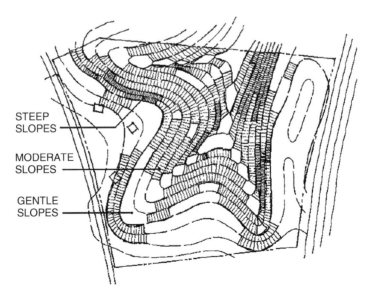

FIGURE 16-6
Slope map. (From North Carolina Department of Environment and Natural Resources, Division of Water Quality, Site Planning, 1998.)

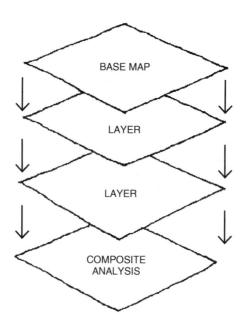

FIGURE 16-7
Sieve analysis. (From North Carolina Department of Environment and Natural Resources, Division of Water Quality, Site Planning, 1998.)

The engineer should be aware of and look for innovative and creative ways to incorporate and integrate some of the above features into the site design. For example, stream buffers, wetlands, and forest areas can be established as conservation areas that are a focal point of a site and provide passive recreational opportunities. Cluster development is an effective technique to accomplish this type of site design.

Typically, this type of analysis is done through the development of map overlays of key features (sometimes called sieve analysis). Figure 16-7 illustrates sieve analysis. These overlays can best be developed through the use of GIS or CADD. Each layer (or group of related information layers) is placed on the map in such a way as to facilitate comparison and contrast with other layers. Sometimes, a composite layer is developed to show all the layers at the same time. With this analysis, buildable sites as well as opportunities for synergy using several layers quickly become apparent. For example, an historic item can be combined with a wooded area and integrated into a regional control site.

This stage will help the stormwater manager to see the limitations and opportunities of the site. Some portions of the site will have relatively few development constraints, while others will have many.

The point of this stage of the site analysis is to identify primary and secondary areas for conservation or preservation. Primary areas for conservation are not buildable due to legal, physical, or practical considerations. Layers that will present physical limitations to development include steep slopes, soils constraints, and vegetation patterns. Other layers will dictate where development can or cannot occur regardless of physical limitations; these areas include property lines, setbacks, rights-of-way, easements, buffers, or other restricted zones (NCDENR, 1998).

Secondary conservation areas have more variability, and their analysis often depends on the feeling of significance of a particular feature. For example, a certain stand of trees may provide striking beauty. But if it stands between a site being feasible or not, it may become relatively insignificant as a conservation feature. As stated by Arendt (1996):

Priorities for conserving or developing certain kinds of resources should be based upon an understanding of what is more special, unique, irreplaceable, environmentally valuable, historic, scenic, etc., compared with other similar features, or compared to different kinds of resources altogether.

The composite analysis should note site features that offer advantages as well as disadvantages to development. Attractive views, access points for circulation, access to utilities, site amenities, or climatic features, as well as those portions of the site that have existing drainage features and significant vegetation should be noted, so these elements can be incorporated into the site plan. Figure 16-8 demonstrates the results of a sieve analysis and a composite drawing.

At this point, these layers are simply information. So even if there does not seem to be room for the development concept, this does not mean the site is not feasible, but that some secondary preservation areas may have to give way. This is the point in the analysis that the stormwater manager investigates, in a preliminary way, the use of stormwater site design credits or special arrangements, if available.

The stormwater designer also considers the land needs of the regulatory stormwater management requirements. Any kind of BMP and stormwater control screening analysis also begins at this point to establish feasibility of key sites. For example, there will normally be a limited number of locations that are amenable to the use of some general application structural controls. It is important at this point to identify potential sites and to set them aside as primary conservation areas — at least on a temporary basis until the concept plan (the next step) is formulated.

Dewberry and Davis (1995) provide a checklist for a typical site assessment.

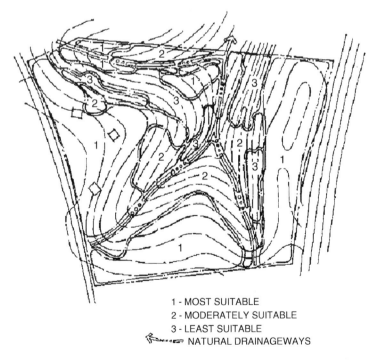

FIGURE 16-8
Composite site analysis. (From North Carolina Department of Environment and Natural Resources, Division of Water Quality, Site Planning, 1998.)

Topography, Slope, and Soils

- Is the topography fairly uniform throughout the site?
- Are there areas where the existing topography is not conducive to the proposed program?
- Are there problem soils or soil and slope relationships that reflect potential instability, unusual construction practices, or excessive costs or development restrictions imposed by local government?
- Do soils indicate the potential for wetlands or other unique surface characteristics?
- What implications does the topography have for utility and drainage considerations?
- Confirm the location of major and minor ridge lines or drainage divides.

Property Configuration

- Are there constrictions or dimensions that may inhibit the utility or usability of the site for the intended program?

Existing Vegetation

- Confirm the location and character of existing vegetation.
- Does the vegetation type vary within the site?
- Do the quality and distribution of existing vegetation afford opportunities to enhance the intended use(s) by incorporating the existing vegetation in the design?
- Do the prevalent species indicate the probability of poor or wet soils?
- Does the existing vegetation afford natural opportunities for screening or buffering of internal or external views within the site?

Hydrology, Drainage, Water, Wetland, and Floodplains

- Does the size, location, distribution, or quality of water-related site elements provide opportunities or constraints in regard to the intended use program?
- Confirm general condition of site runoff. Are there indications of erosion or intermittent ponding?
- Is stormwater detention or retention required on site? If so, what type and size facility will be required, and will its location serve as a site constraint or opportunity?

Views and Visual Characteristics

- Document and qualify the existence of internal and external viewsheds.
- Evaluate the potential for openness or enclosure associated with development opportunities.
- Will the removal of existing vegetation extend viewsheds or detract from the site's visual quality?
- Will it be possible to screen undesirable views?

Climate, Site Orientation, and Exposure

- What are the prevailing wind directions and patterns?
- Are there on-site conditions that prompt microclimate nuances particular to the subject property?
- Are there on-site opportunities to promote passive solar techniques?

Adjacent Land Uses

- Are the adjacent land uses compatible with the program considerations for the subject site?
- Is there a need or opportunity to extend existing or planned community systems into the site, such as parks or pedestrian trails?
- Are there visual or acoustic conflicts associated with surrounding uses or activities? Traffic noise? Airport flight patterns? Conflicts that will require mitigation strategies focusing on the provision of additional setback and buffers, or unique building design or site layout concessions?

Access and Circulation Patterns

- Does the property have existing public street frontage?
- Where are opportunities to access the community vehicular (and pedestrian) circulation systems?
- Will potential connections meet applicable standards?
- Does the site terrain suggest or mandate desirable alignments for internal circulation systems?

Utility Locations and Existing Easements

- Are there existing utilities available in the vicinity to serve the development program? How and where would these systems access the site? Is the topography conducive to these routings?
- Would some utilities require individual systems on-site (i.e., well and septic), and are areas of the site more suitable for on-site utilities?
- Will additional easements be required to route utilities to the site?
- What are the cost implications in getting the utilities to the site?

Existing Development Encumbrances

- Are there additional design or development criteria that will govern the site design response, such as deed restrictions, covenants, or design guidelines?

Other Regulatory Requirements

- The Fair Housing Accessibility Guidelines and the Americans with Disabilities Act Accessibility Guidelines have significant consequences on layouts. These and similar issues should be reviewed as part of the site analysis step.

The Concept Plan

- Stormwater Objectives:
  - Develop concept of better site design and water-quality credit locations
  - Develop concept of structural and nonstructural controls
  - Perform preliminary drainage layout and design calculations
  - Gain approval from developer and local government of concept
- Key Actions:
  - Step 4: Runoff characterization and stormwater control selection
  - Step 5: Development of the preliminary concept plan

Overview

The concept plan represents the initial efforts at satisfying the vision of the development program while staying within the constraints identified during the site analysis. It also considers synergies created and identified during the site analysis. The concept plan represents the distribution of land uses and major circulation requirements using approximate shapes to perform rapid "what-if" analyses.

Elements of the site are listed in terms of size, numbers, type, and requirements. They are then placed within the site and within the building envelopes developed during the site analysis step. The relationships among the various elements are determined, and the need for access and circulation among the elements is developed. Figure 16-9 shows a typical spatial relationships diagram developed during the concept plan formulation. In this figure, it is shown that residential land uses are simply called out in outline form, while nonresidential uses are represented by squares and "blobs" representing the specific

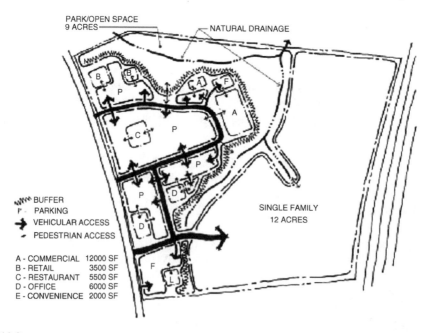

FIGURE 16-9
Spatial relationship diagram. (From North Carolina Department of Environment and Natural Resources, Division of Water Quality, Site Planning, 1998.)

needs of buildings, parking, open space, etc. The most promising of the concept plans is taken to the preliminary design stage.

During the concept plan stage, the stormwater designer will perform most of the conceptualization of the stormwater design and site layout. This will include the following:

- Development of preliminary estimates of the volumes for water quality, channel protection, overbank flooding protection, infiltration requirements, and extreme flood protection as applicable
- Environmental site design concepts and features to be integrated into the site layout
- Site design credits or other arrangements to be accounted for in design of any structural controls handling the water-quality volume
- Screening and preliminary selection of general application stormwater controls

Development of the Stormwater Concept

It is extremely important at this stage that stormwater design be integrated into the overall site design concept. If stormwater considerations are tacked on at the end, it will probably not be cost-effective or effective in reducing impacts of development.

The designer begins with consideration of how to arrange the various elements to allow for environmental design features and for an ability to gain any water-quality credits. The designer needs to work at the macro- and the microscale, and consider the locations within the watershed. The designer considers, in order:

- Ways to arrange site elements to reduce impacts through better site design practices
- Use of primary conservation areas for site design credits and other multiuse
- Calculation of the design criteria volumes — water quality, channel protection, infiltration/recharge, overbank flooding protection, and extreme flood protection
- Use and placement of microscale controls and pretreatment among the various site elements

At the macroscale, the designer considers:

- The major conservation areas have been set aside and are now considered for their best use and for stormwater credits, natural buffer areas, recreational use, etc.
- The major locations for general application structural controls at the lower end of the site are identified and reserved.

At the microscale, flow is beginning to be concentrated and conveyed. The focus is source controls through retention, detention, infiltration, and treatment of stormwater on a disbursed microscale basis. That is, there may be many opportunities to develop filter strips, grassed channels, small storage areas, bioretention areas, natural holding areas, infiltration trenches, filtration devices, etc. See Figure 16-10 for an example of a residential layout. Commercial devices are now located in strategic areas as applicable. Pretreatment of stormwater runoff is considered, as is the use of the treatment train concept.

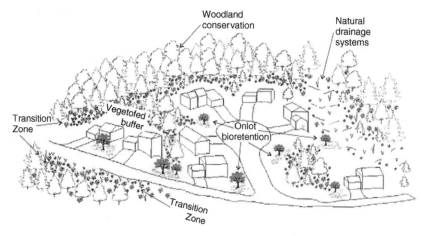

FIGURE 16-10
Microscale considerations. (From Prince George's County, 2000.)

Recall that the overall goal is to minimize the impacts of development through reduction of effective impervious areas. Using hydrology calculations and the goal of mimicking predevelopment conditions, serves a useful purpose in setting sizes and locations. For example, the designer can:

- Store the water-quality volume in a small grass channel between two buildings
- Filter runoff from a parking lot into an infiltration trench or perimeter sand filter
- Cause runoff from all rooftops to run through a grass filter prior to entering the drainage system
- Store runoff in an underground vault prior to discharge to the drainage system
- Create parking lot detention
- Use bioretention areas for all residences

The designer should consider the following general guidelines for each site element (NCDENR, 1998; Prince George's County, 2000):

- Minimize disturbance to natural vegetation by reducing the limits of clearing and grading
- Keep natural drainage patterns intact
- Minimize use of impervious surfaces to promote infiltration
- Route flow over longer distances
- Keep runoff velocities low
- Do not discharge runoff directly into surface waters
- Use nonstructural controls where possible
- Use overland sheet flow
- Maximize on-site runoff storage
- Incorporate best management practices
- Reduce impacts by clustering development
- Use existing contours as much as possible (Figure 16-11)

Site Design and Construction

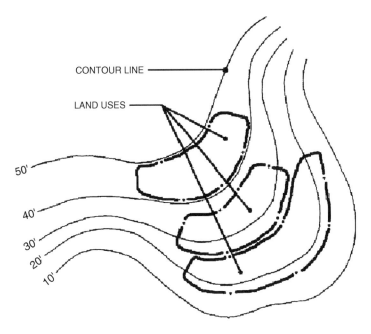

FIGURE 16-11
Use of existing contours. (From North Carolina Department of Environment and Natural Resources, Division of Water Quality, Site Planning, 1998.)

Preliminary/Final Plan

- Stormwater Objectives:
  - Integrate all design concepts
  - Perform design volume final calculations
  - Perform preliminary and final stormwater control and conveyance designs
- Key Actions:
  - Step 6: Securing local and nonlocal permits
  - Step 7: Final design and permit acquisition

Overview

All the parts of the design come together in the preliminary plan (sometimes called the master plan) and final plan. It is here that final design decisions are made, permits are obtained, and the final construction set of drawings is developed.

Preliminary Plan

The preliminary plan step is another means of communication throughout the planning and design process between the developer and engineer and plan reviewer. It helps to ensure that primary requirements and criteria are being complied with and that opportunities are being taken to minimize adverse impacts from the development.

The preliminary plan should consist of mapping and a narrative that illustrates and provides supporting design calculations (hydrologic and hydraulic) for the proposed stormwater drainage and treatment system. The information should be consistent with

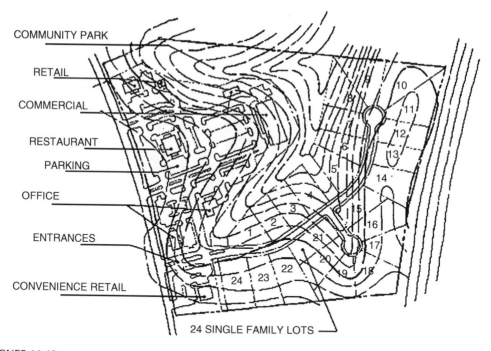

FIGURE 16-12
Preliminary plan. (From North Carolina Department of Environment and Natural Resources, Division of Water Quality, Site Planning, 1998.)

the local ordinance requirements and should follow recommended guidance provided in an approved stormwater design criteria manual. It should be demonstrated that appropriate and effective stormwater controls have been selected and adequately designed. Figure 16-12 illustrates a typical preliminary plan.

The preliminary plan should also include, among other things, street and site layout, delineation of site resource areas as described above, soils data, existing and proposed topography, relation of site to upstream drainage, limits of clearing and grading, proposed methods to manage and maintain resource protection areas (e.g., easements, maintenance agreements and responsibilities, etc.), landscaping plan for stormwater and drainage facilities, and any requests for waivers or variances. Depending on the land use of the site, the plan reviewer may also request information on proposed location of hazardous waste areas, underground and aboveground storage tanks, and on-site wastewater treatment facilities.

It is recommended that a licensed design professional develop or oversee the development of the preliminary plan. The plan, along with the requisite fee, should be submitted to the appropriate review agency to receive preliminary approval.

Calculation of Final Stormwater Control Volumes

Based on the site conditions, proposed land use, and the delineated resource protection areas, it is necessary to quantify the stormwater control volumes using the local design criteria and rates of discharge for the specified water-quality and water-quantity design events and to assess the potential treatment options available. Approved hydrologic and hydraulic models and procedures should be used in the determination of pre- and post-development design volumes and flows. Frequently used public domain hydrologic models include HEC-1 HEC-HMS, TR-20, TR-55, and the rational formula. Public domain hydraulic models include HEC-RAS and HY-8. There are also many privately developed models. Care should be taken to ensure the designer has an understanding of the techniques and assumptions that are part of the model.

Once the necessary stormwater control volumes have been calculated, any applicable stormwater credits for better site design practices can be applied. In calculating runoff volumes and discharge rates, consideration may also need to be given to upstream activities currently impacting the site as well as any planned future upstream land-use changes. Depending on the site characteristics and given design criteria, upstream lands may need to be modeled as "existing conditions" or "projected buildout/future conditions" when sizing and designing on-site conveyances and stormwater controls. Additional analyses (such as the 10% downstream analysis) may also be required to ensure safe passage of design flows on downstream properties, including in-channel conditions.

Many decisions can be made at the preliminary plan level, which will have an effect on the efficiency and cost-effectiveness of the stormwater runoff management system. Because the detail of the master plan allows the more exact placement of elements onto the site, the stormwater management components of the plan should be shown in a more detailed manner on this drawing than on the conceptual plan. As details of various pavements and surfaces are shown, impervious areas can be measured more accurately. Measurements can also be made to establish the sizes of various constructed components of the system. Sizes of individual components are based on the quantities of runoff that will be generated by each of the drainage basins of the site. Calculations and formulas are used to determine the quantities and velocities of runoff that must be accommodated by each component of the stormwater system. These formulas are explained in great detail in other chapters of this book and are readily available for those who have a need or interest to use them (NCDENR, 1998).

Using the site data obtained in the site assessment and the runoff characterization performed in this step, a screening process should be implemented that assists in the selection of appropriate structural stormwater controls that will meet water-quality and water-quantity goals and considers constraints such as terrain, community acceptance, and permitting.

The developer should obtain any applicable nonlocal environmental permit such as 404 wetland permits, 401 water-quality certification, or construction NPDES permits prior to or in conjunction with final plan submittal. In some cases, a nonlocal permitting authority may impose conditions that require the original concept plan to be changed. Developers and engineers should be aware that permit acquisition can be a long, time-consuming process.

Final Design

This step adds further detail to the preliminary plan and reflects changes that are requested or required by the reviewer. This process may be iterative. The reviewer should ensure that all submittal requirements have been satisfactorily addressed, and permits, easements, and pertinent legal agreements (e.g., maintenance agreements, performance bond, etc.) have been obtained and executed. The final design step culminates in a set of construction documents as illustrated in Table 16-3 (NCDENR, 1998).

Approval of the final plan is the last major action in the stormwater planning process. The remaining steps are to ensure that the plan is installed, implemented, and maintained properly. Figure 16-13 illustrates the increased detail that occurs at the final design stage.

Construction

- Stormwater Objectives:
    - Ensure stormwater controls and site design practices are built as designed

TABLE 16-3
Final Design Documents

Site Survey (Boundary and/or Topographic)
May include boundary and topographic
May be a record survey for platting purposes

Demolition and Clearing Plan
Indicates portions of the site that must be removed before construction can occur
Should indicate areas to be protected and the method of protection — field staking and protection should occur prior to grading initiation
Natural areas should be so noted so equipment will not enter these areas
Use the site assessment as the basis for the plan

Layout Plan
Measured and dimensional drawing showing how the various elements are to be surveyed and staked
Point of beginning is established to serve as the control point
Road center lines and building corners are located relative to the point of beginning

Grading and Drainage Plan
Shows existing and proposed contours — fill and excavation sites
Gives instructions on how to perform clearing
Shows drainage system layout, slopes, dimensions, sizes, etc.
Lays out the structural controls
Lays out the engineering function of open spaces and microcontrol areas such as bioretention, commercial devices, pretreatment, filters and buffers, etc.
May have a separate stormwater management plan

Sediment and Erosion Control Plan
Must contain all the elements specified in the state erosion and sediment control regulations (if applicable) and local ordinances and regulations
Should specify phasing of construction and temporary stabilization measures
Should note those temporary structures that will be converted into permanent controls
Should specify a final cleaning of the system prior to final inspection.

Architectural Plan
Design and layout of the buildings and other architectural elements of the site plan
Should illustrate methods both specific and typical for the reduction of runoff and control of stormwater
May contain references to the stormwater plan when controls are integrated with buildings and other architectural features
Should integrate aesthetic features with stormwater controls

Planting/Landscape Plan
Landscape plans illustrate the arrangement of planted areas, natural areas, exterior pavements, exterior landscape structures, lighting, irrigation and other landscape features onto the site plan
The landscape construction details explain and illustrate information necessary to construct the elements shown on the plan drawings
Landscape specifications provide descriptions and standards for the methods and materials to be used in the construction
Should integrate aesthetic features of stormwater controls based on the site assessment (e.g., use of flow spreader or berm integrated with landscaping)
Selection of plants and methods should be coordinated with the design of the stormwater controls to ensure engineering function (e.g., for filter strips) and tolerance to inundation
Requires close coordination with stormwater designer

Construction Details
Include details for the stormwater controls
Show portions of the master plan at a larger scale to allow for more details — especially for the use of smaller disbursed microcontrols

Specifications
Written description of the materials and methods to be used
Establish standards for quality of workmanship
Important to have specifications and workmanship for stormwater controls that may be new to the contractor
This is especially true for devices that rely on filtration and infiltration, where substitution of an inferior product may render the control inadequate

Site Design and Construction

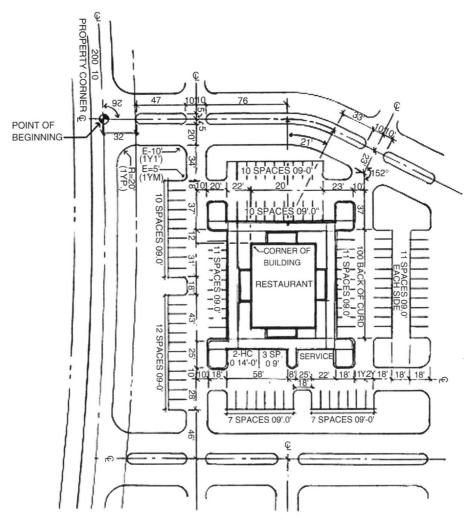

FIGURE 16-13
Final plan. (From North Carolina Department of Environment and Natural Resources, Division of Water Quality, Site Planning, 1998.)

- Pass all inspections
- Leave a fully functional and clean drainage system
- Key Actions:
  - Step 8: Preconstruction meeting
  - Step 9: Inspection during construction

Preconstruction Meeting

This step again assists in the communication process, where the contractor, engineer, inspector, and plan reviewer can be sure that each party understands how the plan will be implemented on the site. A preconstruction meeting should occur before any clearing or grading is initiated on the site. This is the appropriate time to ensure that site resource and natural feature protection areas and limits of disturbance have been adequately staked, and adequate erosion- and sediment-control measures are in place. The next section will discuss stormwater issues with construction in more detail.

Inspection/Observation during Construction

Sites should periodically be inspected by local agencies to ensure that resource protection areas have been adequately protected and that stormwater control and conveyance facilities are being constructed as designed. Inspection frequency may vary with regard to site size and location; however, monthly inspections are a good target. In addition, it is recommended that some inspections occur after larger storm events (e.g., 0.5 in and greater). The inspection process can prevent later problems that result in penalties and added cost. An added benefit of a formalized and regular inspection process is that it should help to motivate contractors to internalize regular maintenance of sediment controls as part of the daily construction operations. If necessary, local agencies can consider implementing a penalty system, whereby fines can be assessed or stop-work orders can even be issued. Local governments need to establish appropriate legal authority prior to implementing a penalty system.

Final Inspections

- Stormwater Objectives:
  - Ensure long-term maintenance of stormwater controls
  - Ensure long-term protection of conservation areas
  - Submit drawings of stormwater controls for record
- Key Actions:
  - Step 10: Final inspection and submission of record drawings
  - Step 11: Maintenance inspections

Final Inspection and Submission of Record Drawings

A final inspection is needed to ensure that the construction conforms to the intent of the approved original design. Prior to issuing an occupancy permit and releasing any applicable bonds, the reviewing agency should ensure that temporary erosion-control measures have been removed, stormwater controls are unobstructed and in good working order, permanent vegetation cover has been established in exposed areas, any damage to resource protection areas has been restored, conservation areas and buffers have been adequately marked or signed, and any other applicable conditions have been attended to. Record drawings of the structural stormwater controls and drainage facilities should also be acquired by the review agency, as they are important in the long-term maintenance of the facilities. Local review agencies should keep copies of the drawings and associated documents and develop a local stormwater control inventory and data storage system. With geographic information systems (GIS) becoming more widely used, much of these data can be stored electronically in a streamlined and efficient manner.

Maintenance Inspections

This is often the weakest component of stormwater plans. It needs to be clearly detailed in the stormwater plan what entity is responsible for operation and maintenance of all structural stormwater controls and drainage facilities. Often, the responsibility for maintenance is transferred from the developer and contractor to the owner. Communication about this important responsibility is usually inadequate, so local governments may need to consider ways to notify property owners of their responsibilities. For example, notification can be made through a legal disclosure upon sale or transfer of property, or public outreach programs may be instituted to describe the purpose and value of maintenance.

Ideally, preparation of maintenance plans should be a requirement of the stormwater plan preparation and review process. A maintenance plan should outline the scope of activities, schedule, and responsible parties. Vegetation, sediment management, access, and safety issues should also be addressed. It is important that the maintenance plan contain the necessary provisions to ensure that vegetation establishment occurs in the first few years after construction. In addition, the plan should address testing and disposal of sediments that will likely be necessary.

Annual inspections of stormwater management facilities should be conducted by an appropriate local entity. Where chronic or severe problems exist, the local government should have the authority to remedy the situation and charge the responsible party for the cost of the work. This authority should be well established in an ordinance.

16.5 Stormwater Aspects of Site Construction

Lack of constructability and lack of construction supervision are two of the major reasons stormwater systems do not function properly. In most municipalities, different personnel perform the design and construction functions. Adequate communication between the two is essential. Design personnel should be aware that there are construction-related design considerations, and construction personnel should be aware that there are design-related construction considerations. When problems arise, they should be discussed and a workable solution decided upon and documented. A forum for design- and construction-related discussion sessions should be provided on a regular basis. Redundancy and adherence to archaic methodology could be overcome by using this forum to develop solution(s) and utilize expertise from both functions. Primarily, these sessions would be at the pre-bid stage. Open communication between design and construction personnel should lead to good working relationships. These relationships will help improve designs and ensure that projects are constructed as envisioned without causing major problems or field revisions.

Construction-related hydraulic considerations are a necessary part of the planning and design phases. Factors that will affect construction timing and methods need to be kept in mind as project development proceeds. Those responsible for contract administration and actual construction may need to coordinate their scheduling and construction procedures with the designer in order to achieve results intended. Any special or unique construction requirements should be communicated to the designer prior to the final design phase of the project.

The designer should be present at the preconstruction conference to explain special features and planned construction phasing, where these considerations are necessary to the proper functioning of the design. It may be advisable and necessary to specify certain time limits and special instructions as to how the work will be accomplished. Phased construction to accommodate seasonal variations, floods, fish passage, irrigation, etc., may be needed. In addition, the need for special considerations related to needed temporary work, detours, and public safety issues can be outlined and discussed. It should be emphasized by the designer that any and all revisions of stormwater management designs as contained on construction plans should be discussed with the designer prior to execution.

Construction Costs

Cost is an important consideration in any design. The primary components of first costs are related to materials, land, and labor. Future maintenance costs are an important related design consideration. Life-cycle costing and value engineering are used to make design and materials decisions. The designer must achieve the proper balance of material and construction costs. Ordinarily, material costs are optimized by using available materials in a consistent manner; recycling materials; researching programs to identify potential construction materials, and how they may be utilized efficiently; using reasonable safety factors in design; and encouraging and allowing alternatives where possible. In some cases, the least expensive material may not be the proper choice, because construction costs are greater than those for a more expensive material or future maintenance/replacement costs override the material cost advantage (ASCE, 1992).

Construction costs are affected by the following:

- Relative difficulty of construction
- Laws, rules, and regulations governing construction procedures
- Degree of competition among contractors
- Construction latitude allowed by the specifications
- Quality of the construction plans
- Degree of supervision and inspection provided by the municipality

The choice of a more complicated and expensive construction procedure may be proper if it allows the use of more economical materials, decreases maintenance costs, and eliminates or reduces the need for replacements.

Cost estimating and benefit–cost analysis should be performed for a reasonable set of project alternatives after constructability and suitability screening is accomplished. Initial cost estimates are normally done on a unit or rule-of-thumb basis. These are refined as the project moves through a sketch plan phase, through preliminary design, and to final design and specifications. Actual costs based on bidding still vary considerably (often more than 20%) based on individual construction company circumstances and equipment. Construction cost estimates must look beyond first costs and should include realistic consideration of such factors as (AASHTO, 1992; ASCE, 1992):

- Engineering, surveying, mapping, and design
- Land and easements
- Site preparation
- Materials
- Installation requirements such as trenching or tunneling
- Sizes of structures of all types
- Excavation requirements and special excavation needs such as blasting
- Water-quality enhancement and permitting costs

- Dewatering or flow rerouting costs
- Replacement costs over the life of the project for different material options
- Annual costs such as operation and maintenance
- Annual flood or other losses for various options
- Public involvement and input costs
- Costs related to liability and legal costs
- Construction-related costs such as traffic delays, service disruption, etc.

Life-cycle costing has some advantages in this type of consideration in that it allows a comparison basis for all the various cost factors mentioned above, each reduced to some base year, taking into account the impacts of estimated interest rates and inflation. The actual interest rate and inflation factors are not often as important as the fact that they are consistent across all options.

Construction Plans

The designer must be aware of the above relationships and how they affect costs, and consider them in design. Additionally, construction plans should reflect these considerations by containing:

- Suggested construction sequences that consider construction costs and environmental considerations as well as public convenience
- Subsurface soil borings
- Complete descriptions of utilities
- Consistent plan format that enhances the contractor's ability to assimilate and understand the municipalities' plans

Despite the best of efforts, construction changes occasionally occur. The designer should be consulted when these changes may affect the proper functioning of the stormwater management facilities.

The postconstruction inspection following completion of the project should document any deviations from the original plans as well as an initial assessment of the hydraulic performance. Construction personnel should be encouraged to inform the designer of any design-related difficulties encountered and suggestions to improve future designs. Changes should be incorporated into as-built plans for future reference.

Plans should be checked to verify that site conditions have not changed between location surveys and construction, and also between location survey and preliminary plan completion. Meander migration, bank caving, aggradation, headcutting, or other natural or man-induced changes in the channel may have occurred that would require that the designer reconsider decisions made on the basis of conditions that were different from those that existed at the beginning of construction. Design, construction, and maintenance personnel best accomplish this with a joint field inspection. Additional objectives of the inspection are to assure location survey accuracy and to ascertain if the designer has properly visualized existing situations and designed accordingly.

The changed conditions may require river control works, revisions to pier locations and orientation, rearrangement of spans, or other modifications of the design to accommodate the changes that have occurred. Plan changes required because of differences between location surveys and construction field inspections should be made in consultation with the designer. Some changes could significantly affect the hydrology or the hydraulic performance of the stormwater management facility designed for the site.

Land-use changes in the watershed can modify the hydrology and debris considerations used in the design. New development located near the project could change damage risk considerations for the design. Dependent upon the time that has elapsed between completion of the design plans and the beginning of construction, changes in land use could significantly affect the validity of design considerations. Commercial mining of materials for construction is a rather common practice that can change flow velocities, volume and character of bedload, and flow direction and distribution at the crossing site. Land clearing for agricultural purposes may create a need to reconsider the location and size of waterway openings and the need for spur dikes. Land development near the site could change damage risk considerations for the crossing. The designer should be consulted regarding the need to modify the design at any stormwater management facility site, if the site conditions have changed significantly from the conditions that existed during design.

Changes in stream alignment and profile can result in different flow conditions than those used in the outfall or cross-drain design. Drainage area changes due to diversions or site grading can affect inlet and outlet locations and type, as well as storm drain or roadside ditch designs.

Utilities added after the survey may require extensive redesign of storm drain systems to avoid conflicts; this reinforces the need for good utility surveys prior to design in order to forestall costly redesigns and delays. See Table 16-4 for additional discussion of some typical changes that may affect stormwater management facility design.

In some cases, a considerable amount of time may elapse between design and construction. In other cases, designs may change before construction is begun. Any changes in the plans, specifications, and estimates should be reflected in the final plans. If questions arise, the construction personnel should check with the designers to determine if changes have been made and how construction should proceed.

TABLE 16-4

Changes that May Affect Stormwater Management Facility Designs and are Detectable during Field Inspection

---

Possible Errors or Omissions
Incorrect existing structure and/or invert elevations
Incorrect drainage area size
Channel alignment, profile
Unreported utility
Existing structure condition as related to service life, outlet scour, siltation, etc.
Local flooding not documented
Existing slope erosion not reported
Sensitive receiving waters not reported
Unreported debris problems
Attractive nuisance problems

Possible Changes
Increased development
Channel improvements
Diversion or site grading changes
New utility
Loss of outfall due to development

Miscellaneous
Incorrect typical section choice, and/or incorrect grade

Construction Considerations

Problems may be avoided during construction when important drainage or other water-related factors are considered during the location and planning phases of the project. If at all possible, problem locations should be avoided. A site may be considered a problem location because of geological aspects, environmental concerns, other existing facilities, or other reasons that might conflict with the proposed project.

The concerns of erosion and sediment, where they might occur, and how to control them, must be considered, at least in broad terms, during the early phases of location. As an example, the designer may be involved in the geological investigations because of underground water, so that proper measures can be taken to prevent problems before they occur.

The time of the year and total construction time should be taken into consideration. Certain elements, such as embankments along a stream, should be completed before the anticipated flood season. In some areas, work cannot be performed in the streams during fish spawning runs. In other areas, the stream may be an irrigation supply, and flows cannot be interrupted and the pumping and distribution system cannot be contaminated with sediment.

The use of temporary structures must also be planned. A temporary crossing can be smaller than normal if it is only going to be utilized during the dry months. If it will be used for more than 1 year, perhaps it needs to be sized for a lower-frequency flood. This consideration may change the concept of the project or at least the type of structure designed.

Many construction-related hydraulic problems are related to scheduling. Although these problems will be studied in more detail during the design phase, they should be initially considered, at least in a preliminary manner, as early as possible. Commitments regarding water resource-related items made in the environmental impact statement (EIS) must be made known to the personnel who will be involved in the actual construction. Some commitments that sound nice may not be feasible to build. In other cases, construction occurs so long after the EIS has been prepared that those commitments are forgotten or not included in the plans or contract documents. A commitment list that follows the project through the various stages of development should be prepared to ensure these items are, in fact, incorporated into the project.

The designer can work with other disciplines to devise and construct mitigation measures that reduce adverse effects. The designer can recommend locations and sizes for stormwater management facilities (such as for culverts, bridges, and channels) and spoil disposal areas and geometry; and various construction alternatives can be identified. The designer may also assist in developing programs for protecting surface waters during construction. These programs include such things as levees and ponds to collect various types and quantities of pollutants, including those from construction equipment or which are accidentally spilled, methods to reduce erosion and sedimentation, and replacement of surface waters that assimilate the hydrologic and hydraulic regime of those that are affected.

Hydrology Considerations

Construction and maintenance of stormwater management facilities may require knowledge of low-flow discharge properties such as discharges, flow stages, flow durations, and related flow variables. For example, the construction of a culvert or a bridge may require knowledge of the time frame at which flows are below certain levels or below certain magnitudes. This knowledge might be useful in scheduling construction or designing temporary construction facilities. With some facilities, it is often necessary to avoid long periods in which the facilities are unavailable to the user due to prolonged occupation of a portion of the facility by frequent low flows.

Annually, the USGS publishes water resources data for gauged streams, listing mean daily discharge. Based on these daily records, a low-flow analysis may determine an acceptable discharge for the hydraulic design of temporary construction facilities. A rigorous flood frequency analysis is not generally required for these low-flow studies. Flow discharges may be cursorily determined based on a visual examination of monthly mean discharge data as determined from the mean daily discharge values for all years of record and with consideration given to construction timing and degree of risk. Data for the monthly mean discharge may be obtained from the local USGS field office.

If hydrographs or other hydrologic data or temporary structure sizes are furnished to the contractor or included in the plans for the contractor's use in planning and scheduling operations, the contract documents should indicate that the plots or data are for information only and that the municipality assumes no responsibility for conclusions or interpretations made from the records.

Water quality of streams and lakes has become a sensitive issue. Stormwater management facility construction may deliver such things as sediment and chemicals to streams, rivers, and lakes unless precautions are taken. Annual runoff hydrographs may indicate that very low stream flows will occur during the late fall and winter months. During these periods, even small amounts of additional sediment or chemicals entering the stream from construction areas could be detrimental, because of the low dilution effect provided by the receiving waters. The effects of sediment or chemicals due to construction during the low-flow periods should be investigated for those sensitive areas such as where stream flow is used for municipal water supply. This investigation may include periods of water-quality monitoring and testing. If the investigation concludes that the amount of sediment or chemicals will exceed an acceptable threshold value, the construction periods may have to be rescheduled or mitigation measures taken.

Erosion- and Sediment-Control Considerations

By the time construction begins, all erosion and sediment control considerations made during the planning, location, and plan development phases should be contained in the plans, specifications, and special provisions provided to the contractor and municipal personnel for accomplishment of the project construction. The resident engineer and the municipal inspection staff should be thoroughly familiar with the erosion- and sediment-sensitive areas of the project and the control measures contained in the plans. This information should be shared with the contractor for formulation of a work plan.

The contractor should utilize an erosion- and sediment-control schedule that sets forth the proposed construction sequences and the erosion-control measures that will be employed. This schedule allows the contractor and engineering personnel to plan ahead and control erosion and sediment before it becomes a problem rather than adding measures after damage has occurred.

Adequate inspection during construction is essential for erosion and sediment control. If deficiencies in the design or performance of control measures are discovered, the supervising engineer should take immediate steps for correction, including notification of the designer to avoid a reoccurrence of the problem. Periodic field reviews and inspections by the design and construction personnel to correct deficiencies and improve control procedures are highly recommended.

An important consideration in the decision to utilize any erosion- or sediment-control measure is its effectiveness in the particular circumstances of planned use. There is no better way to answer this question than through experience. For this reason, it is critical to the development of a good erosion- and sediment-control program that communication

Site Design and Construction

exist between design and construction personnel. One method of establishing communication is to have regularly scheduled project field reviews or meetings involving those responsible for design and construction. During these meetings, problems and successes with particular items can be evaluated. Different ideas and procedures that have been successfully employed by a contractor can be studied to determine if they warrant consideration for widespread use. Also of importance for discussion is possible modification to standard design items that would facilitate their construction or perhaps reduce their cost.

This feedback procedure extends beyond construction into the long-term maintenance of erosion-related items. Maintenance personnel must check and correct any deficiencies in the permanent erosion-control measures. Design personnel should be apprised of any persistent problems so that an analysis can be made to determine if any alteration of design or construction practices is warranted to reduce maintenance problems.

Culvert Considerations

The plans, specifications, and other construction documents should be reviewed to ensure that the design fits current site conditions. Design personnel should be informed and involved in all changes to the plans and specifications.

As soon as final locations are determined, the contractor should be furnished with a revised culvert list, including those culverts that have been added or altered by change order. Assembly or construction, bedding and backfill are as important to culvert service as the hydraulic and structural design.

Culverts should be protected from damage during construction operations and should be periodically inspected. A particularly critical time for inspection is upon completion of grading operations and prior to the start of surfacing operations. It is as important to inspect culverts that are not under the roadway as it is for those structures that are under the roadway. Prior to the acceptance of the installation, all culverts should be inspected and cleaned as necessary.

Records should be kept of the construction of each culvert installation. The final location and slope of the culvert should be recorded on the as-built plans. This information is useful for evaluating overall performance of the installation. The following records should be kept for each installation:

- Inspection tags
- Location and layout including:
  - Station
  - Skew(s)
  - Location of inlets, outlets, and junctions
  - Flowline elevations in inlets, outlets, and junctions
  - Camber
  - Alignment
  - Grade
- Daily reports
- Structure summary sheet containing:
  - Measurements
  - Calculations
  - Pay quantities

Bridge Considerations

The responsibility for construction-related hydraulic considerations of bridge construction ordinarily rests with the contractor, but, in some cases, the municipality may include construction-related details in the plans and specifications in order to mitigate potential environmental effects or to assume or reduce the risk of failure during construction. In addition, other special provisions related to the construction phase of the bridge construction may be specified in the plans. Whether the municipality or the contractor assumes the risk and responsibility, hydraulic considerations during construction usually differ from the design considerations for the completed facility.

A hydrograph of superimposed mean daily flows and a plot of the rating table for a stream gauging station near the crossing site are useful for the design of cofferdams, falsework, and temporary crossings, in the municipality scheduling of the work, and in selecting the location of work and material storage areas. If the hydrograph is furnished to the contractor or included in the plans for the contractor's use in planning and scheduling operations, the contract documents should include that the plot is for information only and that the municipality assumes no responsibility for conclusions or interpretations made from the records. In the event a gauging station is not located near the stream-crossing site, records from upstream or downstream gauges may be useful as an indication of the usual magnitude, duration, and time of flood events.

Any site conditions that might impact the foundation design or create unusual scour problems should be discussed with design personnel to determine if design changes are indicated. If specific elements of the bridge design are dependent on special foundation or scour considerations, construction personnel should be informed so they can identify possible problems.

Cofferdams, falsework, and occasionally contractor's equipment, such as barges, constrict the stream channel more than the completed substructure, and, consequently, have greater potential for causing scour and bank caving and for collecting debris. Scheduling of work to avoid flood seasons is especially important if these types of operations will be involved.

Temporary stream crossings necessary for the construction of bridges are usually the responsibility of the contractor. It may be desirable in some instances, however, for the municipality to design such crossings in order to minimize or mitigate the adverse effects on the stream environment, to facilitate securing permits, or to reduce the risk assumed by the contractor and thereby reduce construction costs. In addition, stream crossings for detours are built to much lesser standards than permanent crossings. The criteria used for the hydraulic design of detour stream crossings should be based on risk factors that should be evaluated considering the probability of flood exceedance during the anticipated service life of the detour, the construction period for the crossing, the risk to life and property, and traffic service requirements. Figure 16-14 can be used to assist in designing temporary stream crossings and detours.

As in the case of the design of highway stream crossings, detour designs should accommodate floods larger than the event for which they are designed in order to avoid undue liability for damages from excessive backwater and to reduce the probability of losing the detour stream-crossing structure during a larger flood. In most instances, the conveyance of floods larger than the detour design flood is provided for by a low roadway profile that allows overflow without creating excessive velocities or backwater.

Minimum disturbance of the banks and bed of a stream during the construction period will reduce erosion damage to the banks, sedimentation, and harm to fish and wildlife. Embankments in or along streams should be constructed of erosion-resistant material and protected against erosion to avoid adverse sediment concentrations that contribute to the turbidity of the stream.

Site Design and Construction

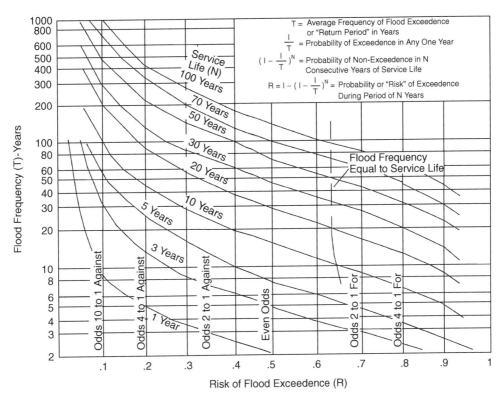

FIGURE 16-14
Risk of exceedance — flood event versus service life of a highway encroachment. (From American Association of State Highway and Transportation Officials, Model Drainage Manual, 1991.)

Consideration should be given to precluding in-stream operations that would cause turbidity during the spawning season of certain types of fish. Detours and construction roads are other sources of turbidity and should be constructed at a time that fishery activities will not be disturbed, or provisions should be made to control any harmful effects of erosion. Silts and clays will generally flush out of the substrate over a period of time, but sands tend to become embedded. Gravel and rock similar to the gradations found in the existing substrate will do the least damage to the aquatic habitat. Municipalities may want to identify local agencies that can be contacted to obtain information about the presence of fish and the seasons during which protection is necessary.

Pumping of cofferdams and other dewatering operations may have a discharge of unacceptable quality to the receiving stream. Mitigation measures such as settling basins may be necessary if the ecosystem of the stream would be upset by the temporary degradation of water quality.

Most designers do not have an opportunity to participate in the construction of the works that they have designed. For this reason, designs that could be improved upon for construction purposes tend to be perpetuated simply because the designer is not informed of the deficiencies. Designers are encouraged to visit construction sites to discuss problems with designs and possible improvements in future designs. This is especially important for major projects like bridge construction. Upon completion of a project, a design critique conducted jointly by designers and field personnel can be a useful learning experience for both. This critique should include difficulties encountered in the construction and possible design changes to prevent such difficulties in the future. This will also give the designer an opportunity to present why some difficulties in construction are necessary because of specific design considerations.

Open-Channel Considerations

Many of the construction considerations for open channels are the same as for culverts and bridges (i.e., plans, specifications, and special provisions; hydrologic information, timing and scheduling; environmental and ecological aspects; feedback); thus, Sections 16.8 and 16.9 should be reviewed, as they relate to open-channel construction. The following discussions will concentrate on those construction considerations that are unique to open channels.

Bank stabilization is an important aspect of open-channel construction. Because, in some cases, a considerable length of a stream or channel system may be disturbed by construction, great care should be exercised in scheduling and implementing stabilization measures. Immediately prior to the commencement of construction of bank stabilization measures, the designer should inspect the site to ensure that measures proposed are not inappropriate because of bank movement subsequent to completion of design surveys. Recognizing that an entire reach of stream may require stabilization, municipal responsibility may well be much more limited in scope, and a total solution may not be possible.

Channel excavation work on some projects may be completed several months before total project completion. The time between completion of channel excavation and total project completion is usually longer when grading and structure projects are separated from the contract for paving or stabilization. During this period, vegetative erosion and control measures are not well established, and maintenance to correct erosion and sediment deposition in the newly constructed channels is important to achieving the results intended. The municipality should provide for maintenance by the contractor during the term of his contract, require interim protective measures, and advance its own maintenance schedule to assure that minor damage will not develop into major damage that will require costly repairs or replacement when it assumes the permanent maintenance responsibility.

Damaged channels can be expensive to repair and hazardous to the public. To facilitate repair and maintenance, channels should be designed, recognizing that periodic maintenance, inspection, and repair will be required. Where possible, access should be incorporated for personnel and equipment during the construction period and afterward. Consideration should be given to the size and type of equipment that will ordinarily be required in assessing the need for access easements, entrance ramps, and gates through right-of-way fences, and purchase of right-of-way.

Storm Drain Considerations

Many storm drain construction considerations are similar to those encountered in culvert and open-channel construction. Thus, these sections should be reviewed, as they relate to storm drainage construction considerations. The following discussions will concentrate on those considerations unique to storm drainage systems.

Plans for subsurface drains are seldom as complete as those for culverts. The discovery of damaging amounts of groundwater during preliminary materials investigation is difficult. During dry seasons, or following a long dry cycle, indications of groundwater problems may be missing entirely. However, with the return of a wet season, serious problems may occur if needed subsurface drains are not installed.

Site Design and Construction

Installations should be carefully reviewed and plans revised as necessary to fit field conditions, in consultation with the designer. It is seldom necessary to decrease the number of planned subsurface drains; the contrary is usually the case. Also, the location of subsurface and other drains may need to be changed to locate these facilities in stable areas and at low points or other locations where the drainage of surface water can be intercepted and allowed to efficiently enter the storm drain system.

During the clearing and grading operations, groundwater problems may become evident. Swamps, bogs, springs, and areas of lush growth are possible indicators of excess groundwater. Fill foundation areas should be inspected minutely before starting embankments. Ravines and draws are especially suspect. As excavation progresses, perched water or various aquifers may be encountered in the area of slopes or at grade.

Temporary Stormwater Management Facilities

Temporary stormwater management facilities include all channels, culverts, or bridges required for haul roads, channel relocations, culvert installations, bridge construction, temporary roads, or detours. They are to be designed with the same care that is used for the primary facility. These designs are to be included in the plans for the project. Approval is required from the municipality having jurisdiction for those designs that they regulate.

It is recommended that drainage systems for these facilities be designed for a 2-year frequency if the roadway is required for a year or less and a 5-year frequency if required for longer than a year. All other temporary stormwater management facilities connected with these roads are to be designed for frequencies as determined by using municipal guidance.

Stormwater management facilities for haul roads, which cross or encroach into a watercourse, are to be designed for a frequency as determined by using a design risk of 50%. As a general rule, to avoid excess upstream flooding, the profile of the road should connect the tops of the channel embankments and the road designed to be overtopped by those events that exceed the design discharge. Sufficient cover must be provided over the temporary conduit to ensure structural integrity. The structural analysis of the conduit is to be included with the design. The plan is to include a warning to the contractor that this road is expected to be under water during certain rainfall events for undetermined lengths of time.

Selection Factors

The selection of a design flood frequency for the remaining temporary stormwater management facilities involves consideration of several factors. Following is one procedure that can be used to quantify these factors. In this procedure, the factors are rated considering their severity as 1, 2, or 3 for low, medium, or high conditions. Following are the selection factors:

Potential loss of life — If inhabited structures, permanent or temporary, can be inundated or are in the path of a flood wave caused by an embankment failure, then this item will have a multiple of 15 applied. If no possibility of the above exists, then loss of life will be the same as the severity used for the average daily traffic.

Property damages — Private and public structures (houses, commercial, or manufacturing); appurtenances such as sewerage treatment and water supply; utility structures above or below ground are to have a multiple of ten applied. Active cropland, parking lots, and recreational areas are to have a multiple of five applied. All other areas should use the severity determined by site conditions.

Traffic interruption — This includes consideration for emergency supplies and rescue, delays, alternate routes, busses, etc. Short duration flooding of a low-volume roadway might be acceptable. If the duration of flooding is long (more than a day), and there is a nearby good quality alternate route, then the flooding of a higher-volume highway might also be acceptable. The detour length multiplied by the average daily determines the severity of this component traffic projected for bidirectional travel.

Detour length — This is the length in miles of an emergency detour by other roads should the temporary facility fail.

Height above streambed — This is the difference in elevation in feet between the traveled roadway and the bed of the waterway.

Drainage area — This is the total area contributing runoff to the temporary facility, in square miles.

Average daily traffic (ADT) — This is the average amount of vehicles traveling through the area both ways in a 24-hour period.

Example Format

Table 16-5 gives an example format that illustrates a method of determining the design discharge. The severity and rating of each component is determined and entered in the impact rating table. The total impact rating determines the percent design risk, and the construction time is then considered to find the design frequency. A ratio corresponding to the frequency is used with the 50-year and 100-year storms to determine the design discharge. The municipality may wish to change the rating selection criteria to fit local conditions.

Note that, if sufficient discharges have been developed by the designer or a flood insurance study, then a frequency curve should be plotted to determine the design discharge instead of the final formula using the ratio.

As-Built Plans

As-built plans serve many functions related to the design and construction process, including documentation of the following:

- The final location of all elements of the drainage system and stormwater management facilities
- Any changes that were made in the design during the construction process (i.e., size of facilities, materials used, addition or elimination of facilities)
- Any variation between the original plans and specifications and the final installed facilities

The completion of accurate and complete as-built plans can be invaluable in documenting changes that can be incorporated in future designs and to future investigations of the project, if problems are encountered or there is some need to analyze the facilities' performance. Possible future legal action makes the documentation of as-built plans important.

TABLE 16-5

Impact Rating Table

Factor	Rating 1	Rating 2	Rating 3
Rating Selection			
Loss of life	See Instructions		
Property damage	See Instructions		
Traffic interruptions	0–2000	2001–4000	4001–6000
Detour length (mi)	<5	5–10	>10
Height above streambed	<10	10–20	>20
Drainage area (sq mi)	<1	1–10	>10
Rural ADT	0–400	401–1500	>1500
Suburban ADT	0–750	751–1500	>1500
Urban ADT	0–1500	1501–3000	>3000

Impact Rating
Loss of life x 15 = _____
Property damage x 10 or x 5 = _____
Traffic interruption = _____
Detour length = _____
Height above streambed = _____
Drainage area = _____
Average daily traffic = _____
Total impact rating = (sum of the above) = _____

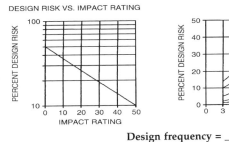
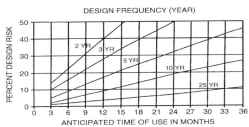

Design frequency = _____ years

Year	Ratio	Year	Ratio
2.0	8.0	10.0	1.9
2.33	1.0	25.0	2.7
3.0	1.2	50.0	3.7
4.0	1.3	100.0	5.0
5.0	1.4		

Ratio = _____ x 0.27 (Q_{50} _____) = _____ cfs

Ratio = _____ x 0.20 (Q_{100} _____) = _____ cfs

References

American Association of State Highway and Transportation Officials, Model Drainage Manual, 1991.

American Society of Civil Engineers, Design and Construction of Urban Stormwater Management Systems, Manual and Report of Eng. Practice No. 77, 1992.

Arendt, R., *Conservation Design for Subdivisions: A Practical Guide to Creating Open Space Networks,* Island Press, Washington, DC, 1996.

Atlanta Regional Commission, Georgia Stormwater Manual, 2002.

Center for Watershed Protection, The Do-it-Yourself Local Better Site Planning Roundtable Kit, 2001.

Center for Watershed Protection, Frederick County Recommendations: A Consensus Agreement, 2001.

Dewberry, D., Dewberry, S.O., (Ed.), Matusik, J.S., and Dewberry, S.D., Land Development Handbook, Dewberry & Davis, Inc., 1995.

DNREC, Delaware Conservation Design for Stormwater Management Guidance Manual, Division of Soil and Water Conservation, Sediment and Stormwater Program, 1997.

North Carolina Department of Environment and Natural Resources (NCDENR), Division of Water Quality, Site Planning, 1998.

Prince George's County, Stormwater Management Design Manual, 2001.

Smart Growth Network, Getting to Smart Growth — 101 Policies for Implementation, 2001, www.smartgrowth.org.

U.S. Environmental Protection Agency, Low-Impact Development Design Strategies: An Integrated Design Approach, EPA 841-B-00-003, 2000.

U.S. Environmental Protection Agency, Low-Impact Development Hydrologic Analysis, EPA 841-B-00-002, 2000.

17
Maintenance

17.1 Introduction

Stormwater management facilities perform the function of removal of water from street, highway sections, parking areas, and other drainage areas and the protection of the facilities from the effects of the water. These stormwater management facilities include drop inlets, storm drains, culverts, underdrains, ditches, slope protection, detention facilities, and erosion-control devices. In order for these facilities to function as designed and constructed, they must be properly maintained. In a recent survey of North Carolina cities, 20% of the cities stated that local flooding problems within their jurisdiction were attributed to maintenance problems (Roenigk et al., 1992). Detention pond surveys conducted in Maryland found a 70% failure rate for one set of detention ponds and about 50% failure rate for ponds from all three surveys conducted (Maryland, 1986). Another Maryland survey found high failure rates for infiltration basins (Galli, 1992). A major reason was poor or nonexistent maintenance.

Full consideration must therefore be given to maintenance during the design process. Designing stormwater management facilities that are as maintenance-free as practical will often result in cost savings that, over the service life of the facility, equal or exceed initial construction cost. Good stormwater management facility design practices recognize that all structures require periodic maintenance inspections and repairs. Reasonable access for maintenance personnel and equipment must be provided for this necessary function. For example, proper access spacing in a storm drain system provides a relatively easy way to clear sediment and debris blockages or to isolate a portion of the system for repairs. Provision of a sediment trap and machinery access area for detention design greatly reduces annual maintenance cost for larger detention and retention ponds.

Communications between designers and maintenance personnel are essential. Design personnel are encouraged to contact maintenance personnel for their input on difficulties they identify in maintaining stormwater management facilities. Conditions that appear to require extensive repair or that incur frequent recurring maintenance should be investigated. Investigation may reveal that a complete redesign is more cost-effective than repetitive repair. Reports by the maintenance forces of effective and noneffective installations aid designers in future work.

Improper maintenance of a stormwater system does not affect just the system but has spillover effects on surrounding property and other infrastructure. For example, typical stormwater management facility maintenance problems that affect streets are ponding of water that softens the subgrade, secondary ditches along the permanent edge that erode the material that supports the pavement edge, and breaks in stormwater management facilities that lead to erosion of pavement-supporting material.

Maintenance Categories

Maintenance generally falls into three categories — routine, remedial, and capital improvements:

- Routine maintenance includes those activities that happen on a periodic basis, which may be driven by the passage of time, not the specific deterioration of the system.
- Remedial maintenance corrects specific deficiencies in the existing system without upgrading its capacity. Remedial maintenance makes the best use of a system even if it is deficient.
- Capital improvements replace deficient systems with larger or improved designs. They become, in effect, new systems.

Routine maintenance of stormwater management structures and facilities include the following:

- Keeping water courses free from accumulations of debris and vegetation, and storm drains free of silt, sand, and debris
- Correcting malfunctioning parts of the systems (Settlements and breaks are the most common types of failure.)
- Anticipating problems and making minor modifications
- Inspecting detention facilities periodically to ensure that siltation and debris have not decreased the storage volume and that the outlet structure is not clogged (They should also be checked after major storms.)

The routine cleaning and minor repairs of drainage features often require that labor-intensive hand methods be used. Adequate access for maintenance personnel and equipment to get to the site and do work on stormwater management facilities should be provided for in the designs. In addition, most costly maintenance work might easily be avoided, or more efficiently accomplished, if designers were to give more attention and thought to the shape and location of drainage features. For example, a V-shaped roadside gutter that is contiguous to the shoulder can be efficiently reshaped and cleaned with a motor grader. Small trapezoidal and other shaped roadside ditches may require hand cleaning or special equipment.

Inspection

Inspection of stormwater facilities serves as a method for scheduling and controlling work crews and for responding or anticipating specific problems within the system. It is often cheaper to send a single inspector to a complaint site or suspected trouble spot than a whole crew. Also, in some municipalities, inspectors serve as scouts, determining when, for example, inlets need cleaning or ditches need mowing or erosion control.

Maintenance of stormwater management facilities is important during the development and construction of a project and afterwards. In some areas, maintenance of natural drainage systems presents minimal problems, while in other areas, major resources will need to be allocated for maintenance-related tasks. Stormwater system inspections should be made quarterly in other than arid climates and during and after each major storm to confirm that satisfactory conditions exist, or to evaluate the need for cleanup and repair. Inspection schedules should include mandatory inspection of known trouble areas and inspection of other areas as time and resources permit.

The best time to look at stormwater management facilities is often during a storm. It is easy then to see where water ponds and where stormwater management facilities are overflowing. There is no gainful work to be performed at this time, so personnel are available for this inspection. It is felt in some municipalities that the same individual should always inspect the same drainage area. In this way, the inspector can spot any changes that might have occurred. The inspector should be alert to any pavement cracks or ground settlements that appear after a severe storm, even if these defects are small, as they may be evidence of an erosion caused by a break in the pipes. Areas that generate large amounts of sediment and debris and locations within the drainage system where debris and sediment accumulate should be identified and included in any preventive maintenance schedule. A record of the inspection should be kept with any deficiencies referenced by street name and house number.

Specific inspection items are contained under a description of each type of stormwater management component in later sections and in Chapter 13. From these sections, inspector checklists can be developed according to the preferences and needs of a specific municipality. For example, the Corps of Engineers has developed an inspection guide for local flood-control projects that has specific questions and a form for inspectors to fill out (U.S. COE, 1973). It has also published a four-volume inspector's guide, which includes all types of stormwater components (U.S. COE, 1960).

17.2 Municipal Maintenance Concepts

Introduction

The development of an effective maintenance program begins with the development of clear goals and objectives, along with a number of supporting policies. While there are a number of ways to consider grouping of these policy issues, the grouping used by several of the leading stormwater management programs in the United States has been combined and serves as a convenient and intuitive structure.

The remainder of the program is the myriad details required to carry out the objectives and policies and to make appropriate adjustments as the program continues to grow. The resources and policies to carry out the policies can be organized into the categories of organization and staffing, technical approach and activities, and financial resources and procedures.

Goals

Typical goals related to the maintenance of a stormwater system can be developed along the lines of the major duties of local governments and can be stated as follows (Louisville MSD, 1988):

- Provide, operate, and maintain a system of stormwater management facilities, controls, criteria, and standards that will mitigate the damaging effects of uncontrolled and unplanned stormwater runoff, to improve public health, safety, and welfare, protect property and lives, and maintain and enhance the environment.
- Establish consistent levels of protection for local and regional collection and transmission systems against flooding for existing and future land development conditions, and to correct existing physical drainage problems.

Objectives

Comprehensive objectives developed to meet these goals can include such typical considerations as the 14 objectives listed below:

- Adopt, review, and update regulatory and other legal measures that control stormwater rates, volumes, and pollution loadings into streams and groundwater.
- Review all plans and permit applications to ensure compliance with design criteria, master plans, and sound engineering judgment in design.
- Inspect effectively all construction of stormwater management facilities to ensure compliance with all design criteria, conditions, and plans.
- Develop and implement specific levels of service for design and operation of all stormwater systems on public and private property to provide satisfactory protection of individuals, property, and the environment.
- Enforce all standards, ordinances, and policies in an effective manner to deter, reduce, or eliminate activities that would harm life, property, or the environment.
- Conduct public education and awareness programs that emphasize protection of the natural and structural stormwater management facilities.
- Train maintenance crews to be able to respond effectively to the full range of drainage and flooding maintenance complaints and activities.
- Develop and implement an operation and maintenance program designed to achieve the lowest life-cycle costs for stormwater management facilities consistent with adopted standards and criteria without violation of environmental standards and regulations.
- Develop and maintain an up-to-date inventory of the complete stormwater system, including identified deficiencies.
- Seek to automate and integrate system maintenance operations with planning, engineering, and environmental management through the use of joint databases and geographical information systems.
- Develop and implement a prioritized annual capital construction and remedial maintenance program based on documented needs.
- Develop and implement design and maintenance standards and practices that are effective, reduce the need for maintenance, and take into account environmental considerations and opportunities for multiobjective and integrated uses of stormwater systems and facilities.
- Provide adequate resources to ensure the stormwater system is operated and maintained according to standards, criteria, and policies.

- Develop and implement operations and maintenance financing mechanisms that maximize equitability, stability of funding, and the targeting of special charges and fees to those requiring or causing the need for special services.

Policies

To carry out the maintenance objectives, a series of basic and more specific policies need to be developed. These policies cover a large number of generic and municipal-specific topics under the general categories of organization and staffing, technical approach and activities, and financial resources and procedures. The more generic policies include the following:

- Organizational responsibility — Which personnel will be responsible for which activities, and how will they be organized, staffed, and controlled?
- Privatization of functions — Which activities will be conducted with in-house resources, and which will be contracted out?
- Intergovernmental coordination – How will the municipality coordinate with surrounding entities or regional authorities, state government, and federal agencies?
- Physical extent of service — What are the geographic limits and categories of drainage facilities for which the municipality will have some responsibility?
- Levels of service — Which services will be provided where, or which standards will apply where?
- Setting maintenance priorities — How will priorities for use of scarce resources be set and implemented?
- Financing operations, maintenance, and capital projects — Which mix of funding methods (including cost sharing) will be used for maintenance and capital improvements for which types of projects?
- Use of geographic information systems (GIS) and infrastructure management techniques — How will modern automation be utilized to manage the program, and how will it integrate with other programs?

There are also a large number of procedural policy decisions on inspections, enforcement procedures, maintenance procedures, activity types and frequencies, etc.

Policy Resolution

Resolving some policy issues can be a complicated exercise, especially when it must be done in the midst of major changes like those lying ahead for many municipalities facing growing water-quality demands. However, adequate maintenance is essential in order to successfully upgrade the program, and can only be established on the basis of clear and consistent policy decisions.

Although policy-making in the highest sense is reserved to the chief executive and council, day-to-day policy decisions are in fact made at several levels in municipalities. The municipal Council formally adopts many of the major policy decisions that guide the municipality. The chief executive makes policy decisions based on council positions. However, the municipal manager and staff administrators pursuant to the general directives spelled out by the chief executive and council make other decisions. Recognizing this dispersed policy-making environment, a simple hierarchy is recommended for the level of review of the issues listed in this section.

An initial screening of possible issues must consider:

- Impacts of policy decision alternatives on costs and manpower
- Appropriate level(s) of municipal government at which the issue should be addressed and resolved
- Relationship of each specific issue to other policy issues
- Priority and timing associated with the issue given the municipality's objectives

Extent and Level of Service

Extent of service refers to how much of the system the municipality will take responsibility for, in some way. Level of service refers to what type of service or service standards will be incorporated in its extent of service. Service levels can be design-based (e.g., design, construct and maintain for the 10-year storm), condition-based (e.g., culverts will be cleaned if they become more than 1/3 blocked), or performance-based (e.g., clean culverts every 5 years).

Extent of Service

Extent of service includes a definition of how much of the system in some way belongs to the municipality and what the municipality will provide services for. It is important to recognize that the stormwater network and flow conveyance system is made up of many components that can be characterized in different ways, including the following:

- Some of the system is located on private property, while some is in the public right-of-way or on other public lands.
- Some of the system carries significant runoff, which comes from public property such as streets, while some carries runoff only from private property.
- Some of the system is an important part of the roadway system (such as curb, gutter, storm drains, side ditches, cross ditches, etc.), while some is remote from the road system.
- Some of the system can be considered major drainageways that carry large amounts of water, while some is relatively minor, draining small areas.
- Some of the system is underground and out of sight, while some is on the surface, such as ditches, swales, and streams.
- Some of the system is considered by the U.S. EPA to be waters-of-the-state, while some is considered to be part of the public or private system.
- Some of the system may be considered by the courts to be the legal responsibility of the municipality, while some is not.

Each of these ways of characterizing the stormwater infrastructure system may have implications for the municipality in deciding how much of the system should be maintained, regulated, or planned for and at what level.

Level of Service

The basic tenet of level of service is that similarly situated properties are treated in a similar manner (Reese and Suggs, 1992). Level of service can be used for all sorts of stormwater management-related purposes including such things as regulatory functions or development plans review, engineering standards and design criteria (termed performance levels of service), capital project programs, overall condition descriptors, and others. One practical use of level of service is in assisting day-to-day routine maintenance management operations.

Conditions Standards and Performance Standards

When considering routine maintenance operations, two types of level-of-service standards are used: condition standards and performance standards:

- Condition standards refer to the physical condition and function goals or objectives of the various parts of the drainage system.
- Performance standards refer to rate of work or production goals or objectives of maintenance crews.

Condition standards have the advantages in that they are physically based, are directly related to the ability of the system to fulfill its design purpose, and tend to reflect the way customers would look at or lodge complaints about a system. Performance standards have the advantage in that they are easy to manage and monitor, although there is not always a good correspondence between meeting some performance objective (such as cleaning 50 catch basin inlets per crew per day) and the actual condition (of the municipality's catch basin inlets).

Condition measures that can be addressed through the routine maintenance process (as distinguished from the remedial maintenance and capital improvement processes) can be identified for each type of stormwater management facility. For example, conditions that can be assessed and addressed for channels and ditches might include erosion, channel obstruction, and vegetation growth factors. The single standard for storm drains may be obstruction, or the municipality can also include minor structural damage as a standard (if this can be addressed through the routine maintenance process).

Each of these condition factors are given a standard or objective to be achieved and maintained individually or for a certain percent of the structures in an area. For example, a conditional level-of-service standard might state that "95% of all grate inlets must be less than 20% obstructed at any one time." Cities such as Tampa, Florida, and Bellevue, Washington, have used such condition-based levels of service with some success.

To be useful, condition levels of service must have a strong basis in the actual adverse impacts of exceeding the condition. Engineering, economic, and policy judgments must be made on, for example, how much obstruction can be allowed in a culvert opening before it must be cleaned out. Furthermore, the adverse impacts of even minor obstruction of some culverts may be deemed to be intolerable due to location near flood-prone structures and lack of adequate size. In other cases, a system could be almost totally obstructed without significant consequences. Thus, there may be differing conditional levels of service for the same type of structural elements but with differing situations and locations. The basic philosophy that similar situations are treated in a similar manner requires judgment that must be carefully applied. Initially, a uniform level of service is applied to the entire service area. In practice, as experience is gained in system capacity and function, some portions of the system may be cleaned more often or to a more exacting standard.

When a program is developing these conditional levels of service, initial estimates must be made on such things as numbers of structures, crew production rates, and activity interval to maintain desired standard. From these estimates, program costs can be developed. Developing these estimates relies on experience in other municipalities, local experience, pilot studies, and other factors.

In practice, after sufficient experience and knowledge of the system is gained, leading municipalities usually shift from operating on a conditional basis to operating on a more easily managed performance basis supplemented with inspections (Reese and Suggs, 1992). These performance standards are derived empirically to meet the conditional levels of service. Thus, for example, it might be found that in a certain neighborhood or watershed, annual cleaning of catch basins or biweekly mowing of ditch banks is adequate to maintain desired conditional standards. In a recent survey of North Carolina cities, less than 1/2 of the 86 cities stated they did any routine maintenance except mowing and catch basin cleaning (Roenigk et al., 1992).

The operation of such a program must be integrated with an effective inspection, complaint reception, and work order processing program. Obviously, some culverts would clog faster than others. Inspections trigger work orders for some parts of the system (such as obstruction of larger culverts), while some parts are serviced on a routine basis without the need for independent precursor inspection (such as changing filtration media in first-flush devices). Some maintenance activities are triggered by specific need (such as clearing an obstruction from a detention basin), while some are routine (mowing).

17.3 Developing Municipal Maintenance Programs

Maintenance Approaches

The ideas and concepts above can be combined into the establishment or improvement of a stormwater infrastructure maintenance program. But first, a cautionary note. It takes time to establish an effective maintenance program — 5 years minimum — and it takes dedicated resources. To increase the effectiveness of stormwater maintenance will also require a commitment to technology, such as infrastructure management software, GIS mapping, efficient communications, and dedicated specialized equipment capable of going off the right-of-way and into backyards.

There are four basic approaches to stormwater maintenance. Every stormwater program contains a mixture of these four. They are reactive, predictive, periodic, and proactive:

> *Reactive maintenance* is essentially complaint-driven. It is episodic in nature, wherein the maintenance staff attempts to meet normally-overwhelming demands for stormwater services in some sort of a priority basis — often politically or relationally driven. It is cheap maintenance in the short term, as it involves minimal expenditure of planning resources. But, in the long term, it maximizes community costs. In order for a problem to be addressed, it must be severe enough to make it to the top of the list. Thus, relatively more extensive damage must be suffered by a property owner or public infrastructure prior to a fix being authorized. In essence, the community waits until the system has caused the most damage it can cause through deterioration or blockage, and then fixes the problem.
>
> *Predictive maintenance* is driven by inspections and inspection standards. The level-of-service discussion above applies here, as inspectors move through the system ahead of crews, performing inspections and generating work orders. It, in theory, is the most cost-effective way to perform some system maintenance. Inspectors catch a system prior to damage and failure. For example, if an inspector notices

FIGURE 17-1
Reactive maintenance is costly.

a headwall that is cracked and sagging, or a partially clogged culvert, and gets it repaired, the inspector has saved possible damage from a blocked system and potentially replacing the whole headwall or culvert.

Periodic maintenance is calendar-driven. It is efficient when maintenance needs are known without inspection. For example, it might be known through experience that channels need to be mowed three times in the growing season. In that case, contracts can be let without resorting to inspection costs. Predictive maintenance works best when there is long-term knowledge of the system and its need for regular service. It can be inefficient when crews are sent out to areas not needing maintenance.

Proactive maintenance gets at the root cause of problems. For example, it might be discovered through a study that a large reason for clogged systems is the ineffectiveness of the erosion-control program. In that case, making changes to that program would bring about large cost savings throughout the municipality. Another example might be the use of certain kinds of materials. In some situations, perhaps due to the acidity of the flow or type of sediment flux, certain types of conveyance structures may not last long. Substitution of different materials for that type of structure can save long-term replacement costs (Figure 17-2).

As shown in Figure 17-3, an immature program must typically depend on reactive maintenance approaches. But a more mature program shifts more effort toward the other three approaches, though reactive maintenance capabilities must be retained, because stormwater is an unpredictable natural system prone to episodic failure.

Different maintenance approaches are used over time as a system ages. Figure 17-4 illustrates this concept over the life of a pipe system. Periodic and predictive maintenance approaches are used when the system is in good condition and simply needs cleaning. The system is maintained in a routine way until its condition cannot be maintained using these methods. At that time, it is scheduled for remedial maintenance to restore function. At some point, the whole system has decayed to the point, or has been overwhelmed with increasing demand, that a whole new and improved system is put in place. Remedial maintenance will not handle it.

FIGURE 17-2
Need for proactive maintenance.

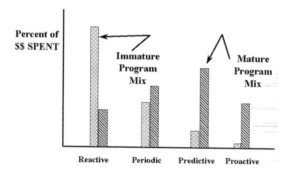

FIGURE 17-3
Maintenance approach mix.

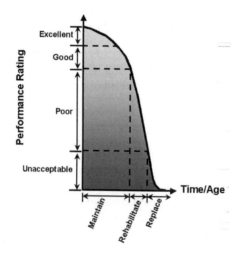

FIGURE 17-4
Life cycle of stormwater infrastructure.

LEVEL OF SERVICE	EXTENT OF SERVICE			
	IN RIGHT-OF-WAY	OUTSIDE RIGHT-OF-WAY		PRIVATE SYTEM
		PUBLIC WATER		
		EASEMENT	NO EASEMENT	
LOS 1				
LOS 2				
LOS 3				

FIGURE 17-5
Drainage segment maintenance matrix.

Organizing the Drainage System

The paired concepts of extent of service and level of service can be used to organize a stormwater system for maintenance operations. Figure 17-5 depicts a table used in organizing the drainage maintenance program.

The extent of service is indicated in a series of columns across the top. Every drainage segment is within or outside the right-of-way. For those segments outside the right-of-way, segments carry public water or do not. By public water, we mean water that drains from a public street or other public property. Courts have shown that, when public water significantly flows within a drainageway, there is a public responsibility for maintenance and protection of downstream property owners (Whalen, 1998). Based on this, many municipalities are determining that they must define their extent of service to include all public water systems.

Finally, each segment in which public water flows is within an easement or not. During an infrastructure inventory and related legal research, each segment could be placed into such a set of categories.

Segments that may be similar as far as extent of service is concerned may have different considerations in terms of the type of service they need. These considerations depend not on legal location, but on location with respect to potential damage should they become clogged or otherwise inoperative. Thus, the vertical axis of the table indicates three levels of service. Level of service one may indicate those segments that, if inoperative, would cause damage to habitable property or are a high safety risk. Level of service two might indicate segments that would cause only minor damage if conveyance capacity is deteriorated. And level of service three might indicate those segments that would cause little, or no, damage if conveyance were degraded.

Written policies are then developed for each of the blocks within the table. The maintenance services within each block may be different from column to column and level of service to level of service. The eventual key is that similarly situated properties are treated in a similar and consistent fashion.

Over time, a community could begin to develop such a maintenance approach, slowly extending its recognized official public system into priority areas. Care must be taken that the policies do not outstrip the ability of the municipal's field crews and other resources to meet demands.

Infrastructure and Asset Management

There are a number of drivers that, together, are forcing municipalities to begin to manage infrastructure in a more automated and proactive manner. For many years, local municipalities have been investing in maintenance and replacement of deteriorating stormwater infrastructure at a rate that does not match the rate of decay of those systems (Doyle and Rose, 2001). New mandated accounting rules, NPDES permitting, and citizen complaints are conspiring to force a change in approach — to modern asset management.

Asset management attempts to maximize system performance at least cost, staying within a set of predefined levels of service for the system, defined for activities (cleaning frequency, for example) and performance (design conveyance, for example).

The use of computerized databases and GIS or facilities management (FM) systems is becoming more commonplace in stormwater management. These computerized systems have grown out of water, wastewater, and pavement management practices. Various firms have developed commercial applications of complex work order and tracking systems, with some tied to sophisticated GIS software. Others, for smaller towns or simpler applications, can be operated as a shell over simple and popular CADD software. All these systems will become even more popular as probabilistic and reliability concepts in maintenance management are brought into general practice (Lapay, 1990; Mays, 1991; Wonderlich, 1990, 1991; Quimpo and Shamsi, 1991) and data become available for more reliable estimates of replacement and maintenance scheduling (Lund, 1990).

Maintenance management is a subset of general infrastructure or facilities management. It is a systematic approach, complex or simple in function, to develop work programs, budget and allocate resources, schedule work, and report and evaluate performance and cost (NACE, 1992; Karaa, 1989):

- Work programs specify the kinds and amounts of maintenance work required to provide a desired level of service. Work programs can be used to measure results by expectations and to make suitable adjustments in activities, schedules, equipment, or other aspects of the program.

- Budgeting is a result of a well-developed work program. Labor, equipment, and material needs are derived from the work plan. The budget allocates these resources among field-operating units. Unit costs are derived from the work program and experience, and used to feed the next year's budget process. In this way, the efficiency of the program can be reviewed, and realistic estimates of new or lost labor and equipment can be made in the face of annexation or budget reductions.

- Work program objectives must be communicated effectively to field supervisors who develop schedules. Through good communication, the well-developed work program will be used as a tool by the field supervisors for scheduling. They translate the plan into weekly and daily work schedules or work calendars that consider program objectives and goals balanced with unexpected needs and cost demands.

- Timely reporting of pertinent and accurate data is of vital importance to any maintenance program. This allows management decisions to be made on the basis of facts and in-course corrections to be made prior to problems getting out of hand. It also allows for more accurate input into the budgeting process. Accounting for the cost of work by activity can lead to significant cost savings. This type of reporting greatly assists in convincing public officials of the need for infrastructure rehabilitation (Martin, 1984).

Maintenance

TABLE 17-1

Relationship between Mapping GIS and FM Systems

Topic	Automated Mapping	Facilities Management
Technology	Picture processing	Data processing
Strengths	Graphics management	Text management
Weaknesses	Text management	Graphics management
Transaction volume	Initial drafting and design	Daily work orders
User focus	Building facilities	Managing facilities
Database	Hierarchical and flat file	Relational and extended
Product focus	Work print	Work order
Management focus	"What if"	"What now"

Source: From Network Magazine, Hansen Software, Sacramento, CA, Fall, 1989.

Experience has shown that the benefits of a well-designed and implemented maintenance management system include the following (NACE, 1992):

- Improved resource utilization
- Equitable resource allocation
- Timely and accurate budget evaluation
- Increased employee morale

Common problems in the implementation of maintenance management systems include:

- Insufficient management support to maintain the system or use it as a management tool
- Unrealistic refinement and paperwork to meet too-detailed expectations
- Lack of hands-on orientation and training of supervisory and worker staff
- Labor union opposition to new and unproven work habit changes

Components of a typical facilities management system may include a database of the system's elements, database management capabilities, a complaint processing and management system, a work order processing system, mapping capability, report generating, financial and cost accounting and reporting, customer service systems, remote field input capabilities, GIS, internal graphics or CADD linkages, spreadsheet capabilities, multiuser access using LAN, and system security. Future systems will increasingly involve remote data collection and real-time operations. They will continue to merge all these overlapping operations into one seamless package with a user-friendly interface for the day-to-day user. Table 17-1 shows the relationships between mapping and facilities management systems.

There must also be a detailed and maintained system inventory and database. Collection of this information is faster and more error free with the use of automated data collection systems. Consisting of programmed hand-held data collectors, these systems allow for plain-English inputs, logical question-and-answer systems, and complex database schemes. They can then dump data directly into other databases accessible by GIS and database management software. More information on inventories of systems is contained in Chapter 2.

Privatization of Services

Because stormwater maintenance for most municipalities is on an as-needed basis, when the system fails, contracting for stormwater maintenance services has been mainly equated

TABLE 17-2

Summary of Stormwater Maintenance Contracts

Topic	Percent of Respondents	Topic	Percent of Respondents
No work contracted	33	Mainline replacement	28
Pump repairs	27	Emergency repairs	22
Channel cleaning	19	Dredging	18
Vegetation control	18	Concrete repair	15
Catch basin repair	12	Litter/debris removal	12
Pipe cleaning	12	Catch basin cleaning	12
Grass maintenance	10	Sewer repairs	10
Erosion control	10	Inlet repairs	10
Leak detection/sealing	9	Culvert cleaning	9
Headwall repair	7		

Source: From American Public Works Association, Contracting Maintenance Services, Special Report 58, Chicago, IL, 1990.

to contracting for remedial and capital construction (APWA, 1990). The privatization of stormwater routine maintenance services has been limited in the past to activities such as mowing, channel cleaning, litter removal, and other seasonal unskilled labor jobs. However, all aspects of privatization are growing in importance in municipalities across the country. In some municipalities, there has been serious consideration given to a totally privatized stormwater management function similar to a wastewater treatment function. Some municipalities now contract for activities such as pressurized pipe cleaning and vacuum trucks, infiltration studies and repairs, channel and ditch cleaning using advanced and high-cost equipment, etc. Of over 1000 survey responses, APWA (1990) categorized the types of services contracted out for stormwater management. Table 17-2 gives a summary of its findings.

Statements of work are normally based on standard specifications, and unit pricing is the norm. Charlotte, North Carolina, developed a unit pricing and contracting scheme that has proven effective for all types of system remedial maintenance (Charlotte, 1993). Denver, Colorado, has used a privatization scheme for routine and remedial maintenance on major flood-control channels for years (Denver UDFCD, 1983). It solicits bids and selects contractors based on rates, experience, and past performance. A weighting scheme has been used with unit costs counting about 43% and experience and past performance splitting the other percentage points evenly.

17.4 Detention Facilities

Detention facilities are often used in urban drainage systems to temporarily store or detain excess stormwater runoff and then release it at a regulated rate to downstream areas. Using this approach, runoff is stored in constructed or natural basins from which it is released continually until the water elevation in the facility reaches its design dry-weather stage. To function properly, the storage volume must be maintained at its design level, and outlet facilities must be kept open and free from obstructions or clogging. A North Carolina survey found that detention problems were the most pronounced of all problems for stormwater systems (Roenigk et al., 1992).

Detention facilities are susceptible to many maintenance problems that can affect how the facility will function, including:

- Weed growth
- Grass maintenance
- Sedimentation buildup
- Detention basin and streambed deterioration
- Mosquito, rodent, and snake control
- Outlet stoppages
- Soggy surfaces
- Inflow water pollution
- Algal growth
- Fence maintenance
- Inadequate emergency spillway capacity
- Dam leakage and failures

These facilities will require more maintenance than other elements within the drainage system. Maryland found that detention basins fail most often for five major reasons: vegetation overgrowth, debris clogging, erosion, sedimentation, and outlet or riprap failure (Maryland, 1986).

Inspection of major detention facilities should be made as frequently as experience shows necessary, perhaps quarterly as a minimum and more often in wet seasons. Where debris is a problem, inspections must be spaced according to debris generation. In any event, it is important to conduct inspections and cleanup work following major individual runoff events. It is sometimes necessary to make inspections during rainstorms, when intense rainfall occurs.

During inspections, besides identifying facilities where debris blockages should be removed, mechanical equipment such as generators, float valves, pumps, discharge controls, and other electrical and mechanical equipment should be checked and adjusted as necessary.

Maintenance Tasks

Maintenance tasks related to detention facilities can be grouped in three general categories:

1. Aesthetic maintenance
2. Nuisance maintenance
3. Operation maintenance

Of these, the most important, from the standpoint of health and safety, is operation maintenance. Operation maintenance can be characterized as that level of maintenance required ensuring against failure of major structural components and flowing controls, and to ensure that the facility continues to function as designed. Neglecting this level of maintenance could cause dam failure and subsequent property damage as well as possible loss of life. In addition, neglect often causes a facility to cease functioning as it was originally designed to do. A program of scheduled, periodic inspections of the facility is essential to recognize potential structural maintenance needs. The following is a partial list of items that should be checked periodically and corrective action taken as required:

- Settling of dam
- Woody growth on the dam (roots can create channels for dam leakage and eventual failure)
- Signs of piping (leakage)
- Signs of seepage or wet spots on the downstream face of a dam (may require toe drains or chimney drains to solve problems)
- Riprap failures
- Deterioration of principal and emergency spillways
- Various stage/outlet controls
- Effectiveness of debris racks
- Outlet channel conditions
- Safety features (access controls to hazardous areas)
- Mechanical and electrical equipment (pumps, generators, automatic controls, etc.)
- Access for maintenance equipment
- Availability of manufacturer's mechanical and electrical information manuals
- Availability of design information such as rating curves and tables for spillway flow, bypass flow, total flow, and storage and pumpout calculations

In addition, the following actions should be taken on each dam as required to ensure that the dam will not present operation or safety problems:

- Replace soil removed by rodent burrows.
- Inspect drainage systems and relief wells annually for proper functioning, and clean or replace as necessary.
- Maintain riprap or other wave-protective measures, and replace as needed.
- Remove and stabilize slide material as soon as practical. It may be necessary to construct a berm or flatten the slope.
- Replace eroded material and establish vegetation in eroded areas in emergency spillways, swales, and other areas.
- Repair any unusual seepages, boils, or settlements in fill areas, or sinkholes in pool areas.

Observations should be made of any changes in topography, downstream drainage systems, or upstream land uses that may have a bearing on the operational effectiveness and safety of the detention facility.

Storage Volume Control

One of the most important variables in the design of a detention facility is the volume available for storage of runoff. If a detention facility is allowed to accumulate sediment and debris, decreasing the storage volume, the ability of the facility to function as designed will be reduced. To facilitate the inspection of the facilities for volume control, it is recommended that a sediment accumulation gauge marker be installed in the detention

Maintenance

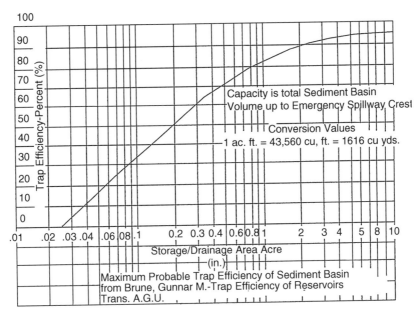

FIGURE 17-6
Efficiency of sediment basins. (From AASHTO, 1992.)

facility to indicate the maximum level for silt buildup before the facility must be dredged or cleaned. This marker could be a small pipe with a stripe or suitable indicator at the cleanout level. A suitable indicator could also be placed on the inlet or outlet device or in some location that can be easily identified during the inspection process.

Detention facilities are often used as temporary sediment basins. To control the maintenance of these facilities, some criteria must be established to determine when these facilities should be cleaned and how much of the available storage can be used for sediment storage. Following is an example of possible criteria and procedures that can be used for sediment basin maintenance.

Figure 17-6 can be used to estimate sediment trap efficiency for sediment basins with different volumes. Following is the procedure for using this figure:

1. Establish sediment generation criteria (e.g., 0.5 acre-in per disturbed area draining to the sediment basin).
2. Estimate total volume available for sediment storage from the geometric shape of the basin (e.g., 0.446 acre-ft).
3. Calculate minimum silt storage needed given the silt generation criteria (e.g., 0.5 acre-in per disturbed area x 10 acres of disturbed area x 12 in/ft = 0.417 acre-ft).
4. Trap efficiency can be estimated from Figure 17-6 as follows:

    0.446 available storage x 12 in/ft = 5.35 acre-in.

    5.35 acre-in/10 acres = 0.535 in - storage/drainage area.

    From Figure 17-6 at 0.535 in - Trap Efficiency = 72%.

5. If you were required to clean out the facility when the efficiency reached 50%, the cleanout elevation could be determined as follows:

    From Figure 17-6 at 50% trap efficiency, the storage/drainage area acre = 0.19 in.

    Storage = 0.19 in x 10 acres x 12 in/ft = 0.158 acre-ft.

Given the basin geometry, try different depths until you have 0.158 acre-ft of storage still available for sediment storage. This is then the depth where the basin should be cleaned to ensure that the trap efficiency does not go below 50%.

This gives a procedure that can be used to estimate trap efficiency and also establish cleanout levels for a given efficiency. Different efficiencies might be established according to the damage potential downstream. Another procedure is found in Schueler (1987).

17.5 Storm Drains

Storm drains move the water collected from catch basins, drop inlets, and sometimes detention facilities to the natural watercourses. The maintenance involved in storm drain systems is the removal of any sand, silt, or debris and the maintenance of a tight seal at each pipe joint. There are occasions where abrasive material is present in the water (or some chemical that has a deleterious effect on the pipe) that causes the pipe material to be worn away. This necessitates relining the pipe to preserve its integrity. It is recommended that the entire storm drain system be inspected every 10 years and the catch basins twice yearly. In a North Carolina survey of 86 cities, 50% of the cities indicated that pipe failure accounted for a significant portion of system failures and complaints.

When storm drains do not adequately accommodate the stormwater, they may be of inadequate size; but they may be partially clogged. Water flushing and heavy-duty vacuum equipment can remove some clogs. Storm drains can be cleaned by inserting a rodding machine (heavy-duty sewer snake) in one manhole and running it through to the next manhole. A line is attached, and the tool is pulled back through to the first manhole.

A cable machine can also be used by placing it at each manhole. The cable from the drum of the first machine is attached to a line, and then it is pulled through to the opposite end. A torpedo-shaped cleaning bucket is hung from the cable on the second machine. The cable is then attached to the bottom of the cleaning bucket, making it the connecting link between the cables of the two machines. A sheave is suspended from each machine and braced directly over the center of the pipe to facilitate changing the direction of the cable. As the first machine pulls the bucket through the storm drain, the jaws of the bucket automatically open as it meets resistance. The first machine continues pulling the bucket through the material until it is full. When the direction of the cable pull is reversed to retrieve the bucket, the jaws close to retain the material and form the bottom of the bucket. A laborer in the manhole empties the bucket when it reaches the manhole. The traditional cleaning machines can be used with various bucket sizes depending upon the diameter of the pipe and the amount of material that must be removed.

17.6 Culverts

Culverts are openings under a roadway or embankment that permit the natural flow of water from one side of the roadway to the other. They may be constructed of corrugated metal, reinforced concrete, or other materials. Culverts must be kept free of obstructions.

Sand or sediment deposits should be removed as soon as possible. During storms, critical areas should be patrolled and the inlets kept free of debris. Inlet and outlet channels should be kept in alignment, and vegetation should be controlled in order to prevent any significant restriction of flow. In some areas, the culverts are normally self-cleansing when large storms occur. Inspection in this case identifies permanence of the clogging material indicated by vegetation or a compacted material.

Culverts may become clogged if the flow-line grade prevents self-cleaning. A permanent correction is to relay the pipe on a steeper grade, but this is not always possible and is often expensive. The alternative is to clean the pipe frequently.

Reinforced concrete box culverts require little maintenance, but they should be inspected annually for cracks, bottom erosion, and undermining. Undermining is the result of high outlet velocities or poor control of seepage. Correction of tailwater undermining usually requires adding an energy dissipator.

Flushing away debris with water pressure may clean small culverts. A water truck or other vehicle equipped with a pump and hose attachment is used to direct the stream from the hose into the outlet end of the culvert. Thus, the water dislodges and washes away debris and sand.

Culverts over 3 ft in diameter must usually be cleaned by hand. A small sled or wagon is useful for transporting material from inside the barrel to the culvert ends. A small rubber-tired tractor, equipped with a push blade, may be used to remove sand and silt deposits from the larger reinforced concrete box culverts.

If the invert of a metal or concrete culvert becomes worn or eroded, relining with concrete grout, Gunite, or asphalt cement can repair it. If the hydraulic capacity of the culvert is not critical, it can be lined or a smaller pipe can be placed inside, and the space between the pipes filled with pressure-pumped Portland cement grout, or Gunite.

High-velocity flows containing large quantities of stone and rock scour the culvert bottom. Securing steel plates longitudinally along the bottom may reduce scour. Scour around footings, cutoff walls, and headwalls is repaired by replacing the eroded material in kind or by filling the void with riprap or sacked concrete. In an emergency, a bituminous mix may be used.

When concrete pipe culverts settle, joints pull apart. Joints are repaired by tamping or rodding grout into the cracks. Grout consists of 1 part Portland cement, 2 parts sand, and 1/5 part hydrated lime, with sufficient water to produce a plastic mix. Sand must be well graded and of such size that all will pass a No. 8 sieve.

In order to prevent erosion, energy dissipators are sometimes placed at outlets of culverts and drains. It is important that these be inspected periodically, particularly after major flows, to ensure that they are in place and functional.

17.7 Bridges

Bridges must be kept free of obstructions. Debris and vegetative growth under a bridge may contribute to scour, create a potential fire hazard, and reduce freeboard for ice and debris during high-water flows, resulting in a serious threat to the bridge. A reduced effective flow area under the bridge may also result in excessive bridge backwater damage, more frequent roadway overtopping, and a hazard to the traveling public.

Maintaining a channel profile record and revising it as significant changes occur provides an invaluable record of the tendency toward scour, channel shifting, and degradation or aggradation. A study of these characteristics can help predict when protection of pier and abutment footings may be required. Being able to anticipate problems and taking adequate protective steps will avoid or minimize the possibility of future serious difficulties.

Maintenance inspection must be commensurate with the risk involved. Where probing and diving are necessary, the inspection should be scheduled at the season of lowest water elevation. High-water, high-ice, and debris marks with the date of occurrence should be recorded for future reference.

Following are some of the maintenance problems that can be encountered related to bridges:

- Clogging of bridge deck drains and scuppers, which may create a hazard to traffic and contribute to deck deterioration
- Discharges of bridge deck drains that are detrimental to other members of the bridge, and those spilling onto a traveled way below, plus, discharges that may cause fill and bank erosion
- Clogging of air vents in the superstructure or deck of bridges subject to overtopping, which may increase buoyancy forces and the possibility of bridge washouts
- Accumulation of debris in the open space between the handrails of bridges subject to overtopping, which may induce additional lateral forces on the bridges and increase the risk of washouts
- Channel aggradation or degradation
- Scour at piers and abutments caused by accumulation of debris and excessive velocities
- Damage to bridge approach embankment caused by channel encroachment
- Loss of riprap due to erosion, scour, and wave action
- Damage to bridge elements due to debris, ice jams, and excessive velocities
- Missing navigational signs and lights over navigable channels

Maintenance measures that should be undertaken to ensure bridge integrity and design standards include the following:

- Repair of damaged bridge elements
- Scheduling for removal of debris after major floods
- Removal of sand and gravel bars in the channel that may direct stream flow in such a manner as to cause harmful scour at piers and abutments
- Cleaning of bridge deck drains and keeping their outlets away from traffic underneath; also, providing riprap or other means of protection at outlets to avoid fill and bank erosion
- Removal of debris caught between bridge handrails and opening vent holes . designed to reduce buoyancy
- Changing a channel when necessary to redirect the flow away from bridge approaches and keep in line with the bridge skew
- Dredging of channels that are subjected to a high degree of aggradation in order to maintain waterway adequacy
- Constructing cutoff walls to reduce or stop progressive channel degradation

- Replacing lost dirt in scour holes and constructing riprap mats or other means of protection for undermined piers and abutments
- Replacing missing riprap on embankment slopes, channel banks, spur dikes, etc.
- Replacing missing or damaged navigational signs and lights
- Constructing additional openings to accommodate increased urbanization in the drainage area upstream from the bridge
- Modifying or increasing existing protective measures when needed

17.8 Ditches

Ditches convey water away from roadways and other areas to locations where the water can flow without causing erosion or ponding. Ditches may be unlined or lined with Portland cement concrete, Gunite, masonry, riprap, or bituminous concrete. Ditches must be kept free of silt, debris, large amounts of vegetation, or any other material that restricts the flow of water.

The flow lines of unlined roadside ditches can be maintained by motorized equipment supplemented with handwork. A pass is made with a motor grader having the blade positioned about 120 degrees to the direction of travel and with the blade set approximately to the slope between the outside edge of the shoulder and the ditch flow line. This removes unwanted material from the ditch and deposits it in a windrow near the edge of the shoulder. Then, this material is loaded into a dump truck with a rubber-tired, front-end loader, or by hand shovels, and it is hauled to a disposal site. Handwork will be required to remove unwanted material at locations inaccessible to the motor grader, e.g., near pipe culverts.

Large roadside ditches are sometimes located at an elevation well below the roadway and not accessible to a motor grader. These may be reached with a truck-mounted hydraulic excavator operated from the shoulder. In this situation, unwanted material is placed directly into a dump truck and hauled away. The equipment operator should exercise care to prevent undercutting the flow line grade. Such undercutting would result in undesirable ponding.

Interceptor ditches on slopes, and along excavation or embankment benches, and outlet ditches from culverts may require hand-cleaning by using shovels and wheelbarrows.

Joint separation is a common problem associated with concrete-lined ditches. Once water gets under the concrete or asphalt, the underlying soil is removed, and the deterioration may be rapid, so frequent inspection is vital and fast repair a necessity if the investment is to be protected. Joints are sealed with hot rubber asphalt. Heating kettles, for hot rubber asphalt compound, should be the double-jacketed, oil bath-type to avoid damage to the compound by overheating. Before any compound is used, joints and cracks should be cleaned. Enough sealing material should be placed to fill the crack. When filling deep cracks, the cooled sealer may shrink, and additional sealer must be added to fill the joint so the sealer is flush with the surface.

Ditch erosion is the loss of soil caused by rapid flow of water. It is controlled by paving the ditch with bituminous asphalt aggregate mix, placement of masonry, grouting rock, establishing erosion-resistant vegetation, or by constructing wash checks. Ditches lined with bituminous material oxidize or weather rapidly and should be sprayed with asphalt emulsion.

The growth of vegetation is often encouraged. The vegetation may be maintained by adjoining property owners in residential areas, but, in other areas, the municipality often must maintain it. One of the major problems when vegetation is used to control erosion in ditches is the control of weeds. Weeds become a major problem in turf when the grass loses its vigor and density and cannot compete with them. Clover and knotweed may take possession in areas where nitrogen levels are low. Crabgrass is a serious pest in many areas where high summer temperatures check the growth of grass. Weed encroachment is often the result and not the primary cause of poor turf. Weed eradication often will not result in permanent improvement unless conditions that weakened the turf are corrected.

The best weed-control chemicals available are nitrogen, phosphorus, and potassium (i.e., fertilizers) applied in the correct amounts, at the proper times, and in the correct ratio. A healthy turf will compete with and drive out most weeds. Turf specialists agree that the best weed control is a dense vigorous growth of grass. Herbicide chemicals, however, have a place in any turf maintenance program. Weeds in the right-of-way are unsightly, and a good program of spraying and mowing can eliminate most of them. A good program of spraying the entire right-of-way for three consecutive years will eliminate most of the weeds, except possibly for small areas of weeds that may require spot spraying. A good mowing program goes hand-in-hand with a good spraying program in elimination of weeds before they go to seed.

Where weeds have been destroyed and short grasses cover the unsurfaced areas of the roadway, the mowing expense can be reduced and the municipality will still have a neat, well-kept right-of-way. Certain steep slopes are not to be mowed, so these must be sprayed to control weed growth. Weed spraying should not be done on new seeding, except to kill noxious weeds and then only by spot spraying. The spraying of new seeding will kill out the desirable legumes and young grass. New seeding should be at least three years old before an overall spray is given, so, if the area is weedy, the area should be mowed instead of sprayed.

There are many different formulations of herbicides on the market under different trade names. Many of these are special-purpose herbicides that may or may not have application in ditch maintenance. Also, concern for stream protection is producing a rapidly changing picture of the use of pesticides and fertilizers. The municipality should keep up with new products, equipment, and the findings of studies, and should at various times recommend chemicals and equipment for test or general usage.

17.9 Stormwater Inlets

Stormwater inlet structures are designed to intercept water in gutters and drainage courses. They also act as settling basins to collect heavy solids, and they prevent debris from entering culvert systems. Mobile heavy-duty industrial vacuum equipment is used to clean sediments from catch basins.

Grates on catch basins are used to prevent large objects and debris from entering the system. Frequent inspections during run-off periods are required, because debris such as pieces of cardboard, newspapers, or flat metal can prevent water from entering the catch basin. Many times, the time of rainfall is too short to allow for self-cleaning of the system, and debris tends to build up over time. Grates are usually designed to be placed with bars parallel to the flow. However, they can be turned perpendicular to the curb and sized

Maintenance

so that they do not allow bicycle tires to drop in the opening, although this will affect the efficiency of the grate.

Large catch basins constructed without a grate may collect large quantities of rock. This rock may be removed by lowering a clam or backhoe bucket into the catch basin. Handwork will be required to load the bucket. The loaded bucket is lifted from the catch basin, and the rock is dumped into a truck and hauled to a disposal site. An orange-peel bucket may remove muck.

17.10 Slope Drains

Slope drains are paved, or metal troughs or pipes used to carry water from a collector drain, gutter, or ditch, into a roadside channel or natural watercourse. They should have firm contact with the supporting surface. If connected to pavement or dike, a tight seal shall be maintained. Cracks should be sealed with hot rubber asphalt compound. The outlet end of slope drains should be inspected regularly for erosion. Eroded areas should be repaired with broken concrete, sacked concrete, riprap, etc. Sometimes the repair may include extending the drain.

Metal pipes used for slope drains on high embankments or benched excavations should be rigidly attached to the surface with pipe anchors. Anchors are designed to prevent the pipe from separating at joints, but, in spite of this, separation occasionally will occur. Repairs consist of removing, reinstalling and anchoring all pipe below the separation.

Once water gets under a paved drain, it quickly erodes the bedding material and dislodges the gutter. This erosion can be reduced or eliminated by installing a cutoff wall at the beginning of the gutter and at intervals throughout the length. Those cutoffs should be 12 to 18 in deep and extend the full width at the gutter, and should be made of Portland cement or asphalt concrete.

17.11 Wash Checks and Energy Dissipators

Wash checks are used to collect water, redirect its flow, provide sediment basins for siltation control, and control rate of flow. They are constructed of reinforced concrete or grouted stone. Wash checks are inspected for undermining after each period of heavy water runoff. Filling the void with sacked concrete should repair undermining.

Energy dissipators are used at the outlets to culverts and pipes flowing with inlet control and, therefore, supercritical velocities. All energy dissipators lessen outlet velocities so that downstream channels can more nearly sustain their integrity. The energy dissipator may be a level spreader, discharge apron, or hydraulic jump. They should discharge into an area stabilized by vegetation as a minimum. They should be inspected for undermining after periods of heavy runoff, and any undermining should be repaired.

Undermining is the major cause of failure of this type of bank protection. Undermining leaves a void under the grouted riprap that ruptures and may cause the riprap to collapse into the void. Repair consists of filling the voids and collapsed surfaces with rock and grouting.

17.12 Bank Protection

Bank protection is required when stream flow or wave action endangers highway embankments or structures. Erosion may be controlled by rock slope protection, grouted riprap, sacked concrete riprap, concrete slope paving, Gunite slope paving, retards, permeable jetties, retaining walls, and cribs. The protection should make optimum use of local materials. Velocity of flow and direction of currents are critical factors in selecting materials. With any bank protection, care should be exercised to prevent the protection from being undercut by erosion. The leading edge of the protection should be protected against erosion.

Undermining causes failure of many bank protection materials. Undermining leaves a void under the bank protection that may lead to the failure of the protection to stabilize the banks. Repair consists of filling the voids and repairing the bank protection. Cracks and other damage to the bank protection should be repaired to prevent seepage and further damage.

17.13 Underdrains

Underdrains consist of a trench lined with filter cloth and backfilled with filter material. Perforated galvanized corrugated metal pipe, semicircular pipe, and tile pipe may be placed near the bottom of the trench. Underdrains are used to remove groundwater and to lower the water table. Local conditions determine whether the installation should be along the shoulder line, or the toe of slope, or herringbone system under the traveled way.

Underdrains should be checked in the early part of the wet season to ensure that they have not become clogged with sand or roots, and those outlets are free to drain. Presence of silt or dirty water coming out of an underdrain indicates a possible break in the pipe. This should be reported to the municipality at once so that materials investigation and remedial measures can be initiated.

Suction pumps, high-pressure pumps, or powered rotary sewer cleaners are required to clean long sections of clogged pipe. When the pipe becomes clogged, the filter material probably has become silted, and its effectiveness has been reduced to a level that make it necessary to consider replacement.

Filter cloth is available that reduces silting when used in underdrain construction. Filter cloth made of polypropylene, polyester, or nylon is highly permeable, lightweight, strong, and flexible. It can be used with or without pipe. The cloth is filled with crushed stone or open-graded gravel. There are many types of filter cloths; choose a type suitable for the purpose. Some types cannot be used in sunlight because they will disintegrate.

17.14 Pavement Edge Drains

Water enters the pavement structure in several ways: cracks in the pavement, infiltration through the shoulders, melting of ice layers, water through the pavement base, and a high water table. This free water under the pavement is of particular concern, because it can decrease the strength of the pavement by reducing the cohesion, reducing the friction between particles, and by increasing more pressure. As a result, the bearing capacity is reduced, and the pavement then must support loads without the proper support from below, resulting in cracking and deflection.

New pavements are designed with subdrainage systems for infiltrated water. The infiltration of free water into the pavement structure, its effect on material strength, and its normal removal by vertical flow or a lateral subsurface drainage system should be an integral part of the pavement structural design process.

Where infiltration of water has become a problem on an existing pavement, some action should be taken to prevent this infiltration or remove it. The prevention includes sealing of cracks in the pavement, positive slope on shoulders, and deeper ditches. Where this prevention does not result in the elimination of the subsurface water, longitudinal drains installed as edge drains may be successful. A 9- to 12-in slotted plastic tubing is laid in the trench, and the trench is then backfilled with permeable material. The tubing must have a positive outlet, and rodent screens should be provided. The edge support for the pavement must not be damaged when the drain is installed, and the material adjacent to the drain, which needs to be drained, must be sufficiently permeable to allow the free water to reach the longitudinal drain.

References

American Association of State Highway and Transportation Officials, Model Drainage Manual, 1991.

American Association of State Highway and Transportation Officials, Manual for Maintenance Inspection of Bridges, 1983.

American Public Works Association, Contracting Maintenance Services, Special Report 58, Chicago, IL, 1990.

American Public Works Association, Managing Public Equipment, Special Report 55, Chicago, IL, 1989.

American Public Works Association, Urban Stormwater Management, Special Report No. 49, American Public Works Association, Chicago, IL, 1981.

Charlotte, Stormwater Maintenance Contracts, personnel communication memorandum, 1993.

Denver Urban Drainage and Flood Control District (UDFCD), Contracting for restoration maintenance, *Flood Hazard News*, 1983.

Doyle, M.J. and Rose, D., Protecting your assets, *Water Env. and Tech.*, ,pp. 43–47 July 2001.

Galli, J., Analysis of BMP Performance and Longevity, Metropolitan Washington Council of Governments, 1992.

Karaa, F.A., Infrastructure maintenance management system development, *ASCE J. Prof. Issues in Eng.*, 115, 4, October 1989.

Lapay, W.S., Probabilistic Basis for Managing Maintenance, Proc. ASCE Symp. on Water Resour. Infrastructure, Ft. Worth, TX, April 1990.

Louisville MSD, Stormwater Drainage Master Plan, 1988.

Lund, J.R., Cost-Effective Maintenance and Replacement Scheduling, Proc. ASCE Symp. on Water Resour. Infrastructure, Ft. Worth, TX, April 1990.

Martin, J.L., Selling elected officials on infrastructure needs, *ASCE J. Prof. Issues in Eng.*, 110, 2, April 1984.

Maryland, Sediment and Stormwater Division, Maintenance of Stormwater Management Structures, a Departmental Summary, July 1986.

Mays, L.W., Models for Optimal Maintenance of Hydraulic Structures, Proc. ASCE Symp. on Water Resour. Infrastructure, Ft. Worth, TX, April 1990.

National Association of County Engineers, Maintenance Management, Action Guide Series, 1992.

Network Magazine, Hansen Software, Sacramento, CA, Fall 1989.

Quimpo, R.G. and Shamsi, U.M., Reliability-based distribution system maintenance, ASCE J. Irrig. and Drainage, 117, 3, 1991.

Reese, A.J. and Suggs, J.B., Discussion of levels of service applied to urban streams, *ASCE J. Water Resour. Plann. and Manage.*, September–October, 118, 5, 582–584, 1992.

Roenigk, D.J., Paterson, R.G., Heraty, M.A., Kaiser, E.J., and Burby, R.J., Evaluation of Urban Storm Water Maintenance in North Carolina, Report No. UNC-WRRI-92-267, Department of City and Regional Planning, University of North Carolina, Chapel Hill, June 1992.

Schueler, T.R., Controlling Urban Runoff, Metropolitan Washington Council of Governments, July 1987.

U.S. Army Corps of Engineers, Inspection of Local Flood Protection Projects, ER 1130-2-339, October 1973.

U.S. Army Corps of Engineers, Construction Inspector's Guide, 4 Volumes, 1960.

U.S. Department of Transportation, Federal Highway Administration, Culvert Inspection Manual, FHWA-IP-86, July 1986.

U.S. Department of Transportation, Federal Highway Administration, Integration of Maintenance Needs into Preconstruction Procedures, FHWA-TS-78-216, 1978.

Whalen, A., 1998, Legal Implications in Creating a Stormwater Management Utility, unpublished memorandum.

Wonderlich, W.O., Probabilistic Concepts in Hydroproject Maintenance, Proc. ASCE Spec. Conf. on Hydraul., Nashville, TN, 1991.

Wonderlich, W.O., Probabilistic Analysis of Maintenance Operations, Proc. ASCE Symp. on Water Resour. Infrastructure, Ft. Worth, TX, April 1990.

Index

A

Aerial photography, 159
Antecedent moisture conditions, 208
Areal reduction equation, 172
As built plans, 1106
Austin first flush basin, 833

B

Baffled outlets, 707
Balanced storm approach, 225
Best management practices (BMPs), 717
    Design criteria, 735
    Design guidance, 769
    Pollutant removal effectiveness, 746
    Structural and nonstructural, 723
    Usage, 718
Better site design, 740
Bioretention areas, 819
Bridge decks, 304

C

Calibration, 221
Channel design, 515
    Bank protection, 554
    Best hydraulic section, 527
    Bioengineering, 563
    Criteria, 517
    Critical flow, 520,534
    Direct step method, 545
    Environmental features, 585
    Equations, 519
    Erosion control, 568
    Flexible linings, 516
    Floodplain management, 53
    Floodplain setback limits, 541
    Froude number, 520,535
    Gabions, 558
    Grade control structures, 583
    Gradually varied flow, 520,544
    Hydraulic jump, 521,552
    Hydraulic terms and definitions, 519
    Irregular channels, 533
    Manning's equation, 521
    Manning's handbook, 524
    Manning's values, 522
    Maximum permissible velocity, 569
    Modification, 565
    Neill Method, 576
    Preservation and restoration, 594
    Regime equations, 574, 583
    Rigid linings, 516
    Riprap design, 554
    Simons and Albertson equations, 575
    Soil cement, 561
    Stability, 564,566,568,581
    Standard step method, 546
    Stream health assessment, 598
    Tractive stress method, 572,581
    Transitions, 518
    Types, 515
    Uniform flow calculations, 528, 541
    Vegetative design, 536
    Vegetative linings, 516
Channel protection volume estimation, 282
Citizen group or stakeholder, 66
Civil law doctrine, 84
Clean Water Act, 81, 995
Combination inlet, 299
Common enemy doctrine, 83
Composite gutter sections, 310
Computer models
    GIS, 22, 29
    Paired software, 22
Construction
    Bridges, 1102
    Costs, 1096
    Culverts, 1101
    Erosion and sediment control, 1100
    Hydrology, 1099
    Open channels, 1104
    Plans, 1097, 1106

Preconstruction meeting, 1093
　　Storm drains, 1104
　　Temporary facilities, 1105
Continuity equations, 522
Continuous simulation models, 220
Convective systems, 169
Critical depth, 362
Cross drainage, 214, 362
Culvert design, 361
　　Beveled-edge, 388,390
　　Buoyancy protection, 367
　　Debris control, 367
　　Design criteria, 365
　　Design equations, 376
　　Design procedures, 378
　　Design steps, 363
　　Environmental considerations, 375
　　Erosion and sediment control, 375
　　Flood frequency, 366
　　Flood routing, 391
　　Headwalls, 371
　　Headwater limitations, 367
　　Ice buildup, 367
　　Improved inlets, 369, 371, 387
　　Inlet control, 379
　　Inlets, 369
　　HY-8 culvert model, 393
　　Length and slope, 367
　　Long span, 386
　　Material selection, 371
　　Nomographs, 379
　　Outlet control, 379
　　Outlet protection, 375
　　Performance curve, 384
　　Procedures, 378
　　Safety considerations, 375
　　Sizes and shapes, 372
　　Skews, 372
　　Slug flow, 378
　　Storage, 369, 384
　　Tailwater conditions, 368, 379
　　Velocity limitations, 366
　　Weep holes, 374
　　Wingwalls and aprons, 371
Curb and gutter sections, 303
Curb-opening inlet, 299

D

Damage-frequency curves, 972
Data, 151
　　Accuracy, 163
　　Acquisition, 33, 159
　　Bases, 23
　　Categories, 153
　　Collection effort, 27, 151, 168
　　Environmental, 182
　　Equipment and techniques, 174

　　Evaluation, 163
　　Extreme event analysis, 166
　　Field investigation form, 204
　　Field reviews, 162
　　Flow collection, 174
　　Flow measurement devices, 175
　　Geostatistical analysis, 168
　　Global positioning systems, 33
　　Inventory, 24
　　Impacts, 191
　　Key constituents, 185
　　Land use data, 192
　　LIDAR, 35
　　Major topics, 153
　　Management, 33
　　Pollution sources, 189
　　Precipitation, 167
　　Remote sensing, 34
　　Satellite, 160
　　Sediment routing, 391
　　Sensitivity studies, 164
　　Sources and types, 152,201
　　Statistical analysis, 165
　　Stream gage, 217
　　Telemetry, 161
　　Time series analysis, 165
　　Topics, 153
　　Urban runoff quality, 194
Debris and ice, 158
Depression storage frequency, 208
Design frequency, 213
Design manual, 97
Detention, 88, 215, 305, 605
　　Chambers and pipes, 644
　　Extended, 649
　　Parking lot, 644
　　Preliminary calculations, 653
　　Rooftop, 641
　　Routing, 655
Developer extension/late-comer fees, 127
Diffused surface water law, 83
Documentation, 105
Downstream analysis limits, 94, 291
Drainage services charges, 125
Drainage systems, 299
Drop inlet, 299, 639
Drop structures, chutes and flumes, 584

E

Energy dissipators, 687
　　Baffled outlets, 707
　　Computer model, 714
　　Design criteria, 688
　　Design procedure, 691
　　Downstream channel transitions, 711
　　HY-8 – energy dissipators, 714
　　Material selection, 690

Procedure, 691
Riprap aprons, 695
Riprap basins, 699
Scourhole estimation, 693
Type selection, 688
Energy equation, 522
Energy grade line, 363
Energy losses in pipe system, 328
Entrance losses, 328
Friction losses, 328
Junction losses, 329
Velocity head losses, 328
Environmental design considerations, 341
Erosion and sediment control, 81
Event mean concentration, 192, 726, 1001
Extended dry detention basin, 772

F

Federal flood insurance program, 12
FEMA, 76
Filter strips and flow spreaders, 898
Financial aspects, 38, 117
Financing needs, 117
Financing questionnaire, 146
Financing study, 130
First flush, 1004
Five whys methodology, 11
Flanking inlet, 299
Flood insurance, 55
Floodplain environmental activities, 592
Floodplain management, 53
Flumes, 174
Free outlets, 363
Frequency, 211
Froude number, 535

G

Gabions and rock mattresses, 558
Gaging-station selection, 178
General facilities charges, 126
Geographical information systems, 160
Grade control structures, 583
Gradually varied flow, 544
Grassed swales, 888
Grate inlet, 300
Gravel bed equations, 575
Gravimetric and volumetric, 175
Gravity flow equations, 333
Greenway planning, 592
Gutter, 300

Gutter flow calculations, 308

H

Huff distributions, 227
Human pathogens, 187
HY-8 culvert program, 393
Hydraulic grade line, 300, 337, 363
Hydraulic jump, 521
Hydraulic roughness, 208
Hydrograph, 208
Hydrologic
Calibration, 221
Design policies, 209
Procedure selection, 216
Roughness, 208
Soil groups, 240
Hydrology, 207
Hyetograph, 208
Huff distribution, 227

Illegal dumping, 1021
Illicit connections, 1021
Impact fees, 126
Impervious areas, 247
Improved inlets, 363
In lieu of construction fee, 126
Infiltration, 209
Infiltration trenches, 863
Informed consent, 70
Inlets, 306
Intensity-duration frequency, 223
Inlets, 306
Combination inlets, 327
Composite gutter sections, 310
Continuous grade, 308
Curb inlets in sag, 318
Curb inlets in sump, 324
Curb inlets on grade, 315, 320
Design criteria, 307
Design frequency and spread, 307
Grate inlets in sag, 318
Slope, 303
Sump, 324
Uniform cross section, 308
Interception, 209
Isohyetal method, 173

K

Kerby formula, 237
Kringing technique, 173

L

Lag time, 209
Legal basis and documentation, 113
Liability and damages, 84
Log Pearson type II distribution, 217

M

Maintenance, 1109
    Bank protection, 1132
    Bridges, 1127
    Concepts, 1111
    Culverts, 1126
    Detention, 1122
    Developing program, 1116
    Ditches, 1129
    Extent and level of service, 1114
    Infrastructure and asset management, 1120
    Inspection, 1110
    Objectives, 1112
    Pavement edge drains, 1132
    Policies, 1113
    Privatization and services, 1121
    Programs, 1116
    Slope drains, 1131
    Standards, 1115
    Storage volume control, 1124
    Storm drains, 1126
    Stormwater inlets, 1131
    Underdrains, 1132
    Wash checks and energy dissipators, 1131
Manning's equation, 256
Manning's n, 522
Master planning, 945
    Benefit and cost analysis, 972
    Computer model choice, 952
    Constraints, 950
    Damage frequency curves, 972
    Ecological study plan, 980
    Flood study, 953
    Holistic master plan, 983
    Hydraulic model development, 966
    Hydrologic model development, 962
    Major creek flooding, 957
    Minor system flooding, 970
    Needs analysis, 948
    Rapid watershed planning, 985
    Role, 945
    Scoping study, 947
    Stormwater quality plan, 976
    Successful master planning, 746
    Technical approach, 951
Maximum extent practicable, 1013
Media, 68, 74
Median barriers, 305
Modular paving blocks, 883

N

National Flood Insurance Act, 81, 214
National pollution discharge elimination system (NPDES), 7, 995
National urban runoff program (NURP), 186, 191
Nationwide regression equations, 728
Neill method, 576
Normal flow, 363

O

Oil/grit separators, 908
One hundred (100) year elevations, 540
Open channel design, 515
Open flow nozzles, 634
Ordinances and regulations, 81, 86
    Developing policies, 92
    Drafting the ordinance, 95
    Field inspection, 103
    Flexibility, 98
    Formulation of objectives, 90
    Identify problems and issues, 89
    Legal basis, 82
    Liability and damages, 84
    Municipal role, 102
    Technical requirements, 103
    When to adopt, 100
Orifice meters and nozzles, 634

P

Pavement drainage, 300
Peak discharge, 209
Perforated riser, 637
Phase I WQA, 1010
Phase II WQA, 1011
Plan review and inspection fees, 126
Policy papers, 71
Pollutant loadings, 726
Pollution accumulation, 719
Pollution availability and fate, 720
Pollution loads, 726
Pollution mitigation measures, 718
Pollution sources, 189, 719
    Impacts on receiving waters, 191
    Land use baseline data, 192
Porous pavement, 644, 872
Precipitation and losses, 222
Precipitation data collection, 167

Pressure flow formulas, 333
Prior appropriation doctrine, 82
Probably maximum flood, 213
Proprietary structural controls, 921
Pubic awareness, 63
    Dealing with the media, 74
    Participation, 64
    Plan, 61
    Policy papers, 71
    Public program techniques, 63
    Technical communications, 79
    Volunteer programs, 73
Public education and awareness, 50, 59, 1019
Public program techniques, 63

Q

Quality (runoff) , 995

R

Radar precipitation data, 170
Rain gages, 170
Rainfall
    Durations and time distributions, 224
    Distribution, 225
    Estimation, 172
    Excess, 209
    Limits, 233
Intensity, 237
Rational method, 219, 232
    Runoff coefficient, 238
    Formula, 232
    Time of concentration, 235
Reciprocal-distance method, 173
Reasonable use doctrine, 84
Regression equations, 218
Remote data acquisition, 159
Retention pond, 781
Return period, 302
Revenue bonding, 125
Riparian doctrine, 82
Riprap aprons, 695
Riprap basin, 699
Riprap design, 554
Risk-based analysis, 211
Risk communications, 76
Roadside and median channels, 303
Roadway pollution, 341
Routing with water quality removed, 768
Runoff quality, 995

S

Safe dams act, 612
Sand filters, 829
Santa Barbara unit hydrograph method, 275
Satellite data, 159
Scour potential, 157
Scourhole estimation, 693
Scupper, 300, 304
Shoulder gutters, 304
Sills, 584
Site Design, 1065
    Concepts, 1065, 1071
    Concept plan, 1086
    Construction, 1091
    Design steps, 1073
    Feasibility study, 1075
    Final design, 1091
    Final inspection, 1094
    Goals, 1067
    Plan, 1089
    Site analysis, 1080
    Stormwater aspects, 1095
    Stormwater control volumes, 1090
Simons and Albertson equations, 575
Simulation models, 735
Simple method, 733
Slotted drain inlet, 300
Soffit, 363
Soil cement, 561
Soil Conservation Service method (SCS) , 220, 242
    Antecedent moisture conditions, 247
    Connected impervious areas, 247
    Curve number, 245, 249
    Hydrograph generation, 265
    Hydrologic methods, 265
    Hydrologic soil groups, 240
    Lag time, 254
    Limitations, 260
    Peak discharges, 259, 270
    Peak rate factor, 272
    Rainfall distributions, 249
    Rainfall excess, 265
    Rainfall-runoff equations, 243
    Sheet flow, 257
    Storm, 225
    Time of concentration, 249, 256
    Twenty-four hour storm, 226
    Unconnected impervious areas, 247
    Unit hydrograph, 268
Special assessments, 126
Spread, 300, 302
Stage, 209
Stage-discharge curve, 615
Stage measurement, 178
State-storage curve, 614
State revolving fund, 123
Storage facilities, 605
    Chainsaw routing, 662

Combination outlets, 645
Construction and maintenance, 679
Design criteria, 609
Design procedure, 615
Drywells, 643
Hand outing method, 668
Land-locked, 676
Location criteria, 611
Modified Rational Method, 642, 663
Outlet works, 611, 617, 646
Perforated risers, 617
Pipes and culverts, 645
Preliminary detention calculations, 653
Preliminary volume calculations, 657
Protective treatment, 680
Retention facilities, 677
Rooftop detention, 641
Routing calculations, 613, 655, 657
Safety grates, 680
Section 208 area-wide studies, 5
Side-channel weirs, 630
Stage-discharge, 615
Stage-storage, 614, 661
Trash racks, 680
Two-way drop inlets, 619
Uses and types, 605
Weirs, 622
Storm drains, 299, 305, 330
Storm movement and decay, 169
Storm types, 169
Stormwater
  CADD, 32
  Computer models, 22
  Concept plans, 36
  Costs/funding, 37, 118, 123
  Databases, 23
  Day-to-day aspects, 19
  Feasibility study, 52
  Financial aspects, 38
  GIS, 29
  GPS, 33
  Inventory, 24
  Legal, 20
  Long-range aspects, 18
  Management, 19, 37
  Master plan, 945
  Organizational aspects, 36
  Overview, 36
  Paradigms, 1
  Problem areas, 40
  Programs, 17
  Remote sensing, 34
  Successful programs, 50
  Technical aspects, 21
  Treatment trains, 764
  User fees, 120
  Utilities, 119, 128
Stormwater management manual, 21, 97
Stormwater management program, 995
Stormwater pollution minimum control measures, 1017
Stormwater quality programs, 995
  Basic findings, 997
  Construction sites, 1025
  Costs of NPDES, 1032
  Categories, 997
  Development process, 1045
  EPA's guidance approach, 1058
  Evaluation of BMPs, 1054
  Illicit connections and dumping, 1021
  Impacts of urban development, 998
  Minimum controls, 1017
  Permit conditions, 1017
  Pollution prevention, 1031
  Post construction, 1028
  Ten commandments of developing program, 1038
  Water Quality Act, 1009
Stormwater utility, 119
  Costs, 120
  Credits, 129
  Equivalent residential unit, 122
  Financing questionnaire, 146
  Financing study, 130
  Funding methods, 118, 123
  Overview, 119
  Policy issues, 127
  Scope of services, 138
  User fee, 120
Stratiform systems, 168
Stream cross-sections, 158
Stream gage data, 217
Stream profile, 157
Structural BMPs, 723
  Alum treatment, 795, 917
  Austin first flush basin, 833
  Bioretention areas, 819
  Design criteria, 735
  Design guidance, 769
  Detention systems, 741
  Effectiveness, 746
  Extended dry detention, 772
  Filtration systems, 742
  Filter strips and flow spreaders, 898
  Grass channels, 896
  Grass swales, 888
  Infiltration trenches, 863
  In series, 767
  Mitigation measures, 718
  Modular paving blocks, 883
  Multi-chambered treatment train, 932
  Oil/grit separators, 908
  Online versus offline, 759
  Overview, 741
  Pollution concentration and loads, 726
  Pollution dynamics, 719
  Porous pavement, 872
  Proprietary structural controls, 921
  Regional versus on-site, 761
  Retention pond, 781
  Sand filters, 829, 839, 854, 856, 860
  Screening, 749
  Selection, 742

Specific measures, 769
Types, 741
Vegetative filtration, 742
Unified sizing criteria, 739
Water quality inlet, 908
Wetlands, 801
Structural roadway BMP, 344
Submerged inlets and outlets, 363
Surface sand filter, 839
Survey information, 162
Synder's unit hydrograph, 219
System development charges, 125

T

Technical communications, 79
Technical manual, 21, 97
Telemetry, 161
Ten-percent rule, 293
Thiessen polygon method, 172
Time of concentration, 209
Tractive stress method, 572
Treatment train concept, 1005

U

Unified sizing criteria, 740
Uniform flow, 363
Unit hydrograph, 209, 219
Urban hot spots, 719, 1002
Urban hydrology, 207
Urban runoff quality data, 194
Urban runoff stressors, 722

V

Velocity measurement, 179

Salt or color-dilution method, 182
Salt-velocity method, 182
Tracer methods, 182
Velocity meters, 175
Venturi meter tubes, 175

W

Water balance calculations, 286
Water quality and water law, 86
Water Quality Control Act, 6, 81, 1009
Water quality inlet, 908
Water quality standards, 1007
Water quality volume and peak flows, 280
Watercourse law, 82
Watershed characteristics, 153
Wetlands, 801
Weirs, 622
Broad-crested, 627
Inverted triangle, 628
Ogee shapes, 628
Proportional, 629
Sharp-crested, 622
Sharp-crested rectangular, 623
Side-channel weirs, 630
Submerged, 622, 628
Trapezoidal, 626
Triangular channel broad-crested, 628
V-notch, 626
Volunteer programs, 73

Y

Yen and Chow's method, 230